AF518630

Hitech Horticulture

S. Prasad
Dharm Singh
R. L. Bhardwaj

AGROBIOS

Published by:
AGROBIOS (INDIA)
Agro House, Behind Nasrani Cinema
Chopasani Road, Jodhpur 342 003
Phone: 91-0291-2643993
Website: *www.agrobiosonline.com*
E. mail: *agrobiosindia@gmail.com*

Authors:
S. Prasad
Ex-Associate Professor (Horticulture)
Rajasthan Agricultural University, Bikaner
Presently: Registrar, Gandhi Vidhyaminder, Sardarshar (Rajasthan)

Dharm Singh
M.Sc. Ag., Ph.D., FASA (USA), M.H.Sc. (London)
Managing Director
Indian Medicinal Plants Marketing Federation, New Delhi

R. L. Bhardwaj
Subject Matter Specialist (Horticulture)
Krishi Vigyan Kendra, Sirohi (Rajasthan)

ISBN No.: 978-81-8854-434-3 (HB)

Published by: Dr. Updesh Purohit for Agrobios (India), Jodhpur
Laser Typeset at: Yashee Computers, Jodhpur
Cover Design by: Shyam
Printed at: Agrobios Digitals, Jodhpur

Preface

The term horticulture is derived from two Latin words, *hortus*, meaning a garden, and *cultura*, meaning cultivation. It may be described as the practice of growing and manipulating plants in a relatively intensive manner. Horticulture involves eight areas of study, which can be grouped into two broad sections - ornamentals and edibles:

- **Arboriculture** is the study of, and the selection, planting, care, and removal of, individual trees, shrubs, vines, and other perennial woody plants.
- **Floriculture** includes the production and marketing of floral crops.
- **Landscape horticulture** includes the production, marketing and maintenance of landscape plants.
- **Olericulture** includes the production and marketing of vegetables.
- **Pomology** iincludes the production and marketing of fruits.
- **Viticulture** includes the production and marketing of grapes.
- **Postharvest physiology** involves maintaining the quality of and preventing the spoilage of horticultural crops.

Horticulture needs to redefine and/or focus itself because horticulture is an industry and its aim is economic success. It is an emerging sector of economic importance in India. Horticulture is defined as the cultivation of an orchard, garden, or nursery on a small or large scale, or the science and art of growing fruits, vegetables, flowers or ornamental plants. In the past, horticulture was practiced in order to produce food to survive. Ornamentals and flowers were added to improve well-being. People were proud to produce their own fruits, vegetables, medicinal and aromatic plants, and flowers. Gardening was a most welcome job. Nowadays, gardening (i.e. Horticulture) is thought of as a hobby, although it's also a profession. However, horticultural industry has reached unprecedented growth in the size, diversity, and value worldwide, but still horticultural science within academia is experiencing a crisis. In modern age, Technology has become a pervasive element in modern horticultural cropping systems. In this context it is define technology as "*advanced systems that manage, control, communicate, sense, and report information on crops at all stages of development and utilization*." The technology includes many faces like: (1) Improved plant genotypes, some developed by recombinant DNA technology, (2) Value-added plant propagation, (3) Advanced hybrids, coated and primed seed, artificial seeds, micropropagation, advanced plug technology, (4) Computer-based environmental sensing and control specially of green houses, (5) High speed data interchange between critical production points, (6) Environmental sensors and equipment for monitoring plant status, (7) Advanced production techniques (hydroponics), (8) Integrated Pest Management, (9) Use of cultural, biological, chemical, host plant resistance/tolerance, physical-mechanical control and regulatory control methods, and (10) Integrated nutrient management (INM).

There needs to be an expansion and/or introduction of new high-income careers in horticulture. For example, floriculture has great potential to generate high income jobs in the perfume industry; new careers in medicinal use of horticultural crops (traditional as well as allopathic and homeopathic uses), organic fruit and vegetable production, and other less traditional careers need to be emphasized. Hi-tech horticulture is the need of all the developing countries. Controlled-environment horticulture certainly has a promising future within multi-disciplinary plant-science research as well as many exciting applications within controlled-environment agriculture. There are various ways to make horticulture hi-tech.

The present book has been designed to provide comprehensive text to the readers. Various aspects of hi-tech horticulture have been included like, Hi-tech Horticulture: an introduction, Nature, Scope and Progress of Horticulture in India, Hi-tech Horticulture for Agriculture Reform, Perspectives of Horticulture, The Power of Biotechnology in Hi-tech Horticulture, Techniques of Biotechnology for Horticulture, Applications of Genetic Engineering Hi-tech Horticulture, Techniques for Genetic Engineering, Genetic Engineering Techniques for Producing Transgenic Crop Varieties, Plant Tissue Culture: Principles and Methodology, Protoplast Technology, Micropropagation in Plants, Somatic Embryogenesis, Mutation and Plant Breeding Techniques for Hi-tech Horticulture, Marker Assisted Selection (MAS) Technique for Hi-tech Horticulture, Gynogenesis in Hi-tech Horticulture, Synseeds or Synthetic Seeds for Hi-tech Horticulture, Plant Tissue Culture some Related Aspects, Greenhouse Technology for Hi-tech Horticulture, Hydroponics for Hi-tech Horticulture, Hi-tech Horticulture for Postharvest Technology, Hi-tech Horticulture and Organic Production, nutrient Management in Horticultural Crops, Plasticulture for Hi-tech Horticulture, Value Added Products of Horticultural Crops, Hi-tech Horticulture: Harvest Technology and Value Addition of Medicinal Plants, Marketing of Horticultural Produce, e-marketing of Horticultural Produce, export of Horticultural Crops. In the end Appendix and Selected Bibliography have been appended.

We hope that the text of the present book will be helpful to horticulturists/ growers of India to increase export potential. The authors are thankful to various sources from where the literature is cited. We express our sincere thanks to Dr. S. S. Purohit for his help in preparing the book.

Authors

Contents

1. Hi-Tech Horticulture: An Introduction **1**
Markets for Horticultural Biotech 2
How to Make Horticulture as Hi-Tech Horticulture 3
Role of Biotechnology in High-tech Horticulture *3*
Genetic Resources *4*
Tissue Culture *5*
Postharvest Handling, Storage and Processing *10*
Integrated Production Systems (IPSs) *10*
Economic and Market Aspects *10*
Role of Public and Private Agencies *11*
Intellectual Property Rights (IPRs) *11*
Extension Plan *11*
Quality Control Mechanism *12*
Need for Education Reform *12*
Future Research and Development *13*

2. Nature, Scope and Progress of Horticulture in India **17**
Prevalence of Horticulture in India 18
Geography *18*
Climate *19*
Scope, Production and Research Development of Horticultural Crops in India 19
Thrust Areas *20*
Achievements *20*
Research Achievements *21*
Needs and Expectations *22*
Constraints to Horticultural Production *24*
Fruits *26*
Vegetables *29*
Flowers and Floriculture *30*
The Customer's Need for Flowers and Ornamental Plants *31*
Flowers in Industry *31*
Per Capita Consumption of Flowers *33*
Floriculture: Production, Marketing Businesses and Trade Organizations *33*
Floriculture Business Competition *33*
Sources of Supply for Commercial Flower *34*
Education and Research in Floriculture *34*
Spices and Plantation Crops *35*
Coconut *37*

Areca Nut 37
Oil Palm 38
Cashew Nut 38
Mushroom 38
Horticulture Research Institutes in India 41
Indian Council of Agricultural Research 41
Indian Institute of Horticultural Research 41
Research Institutes of Horticulture Research 43
Organizations / Institutes/ University 44
All India Coordinated Research Projects 44
National Horticulture Board 45
Investment Subsidy Schemes 45
National Horticulture Mission 46
Mission Objectives 46
Strategy 47
National Mission On Bamboo Technology And Trade Development 47
Central Institute Of Horticulture, Medziphema, Dimapur, Nagaland 48

3. Hi-tech Horticulture for Agriculture Reform 49
Markets for Horticultural Biotech Products 50
Ways to Make Horticulture as Hi-Tech Horticulture 51
Role of Biotechnology in High-tech Horticulture 51
Genetic Resources 53
Tissue Culture 54
Seed Technology and Management for Horticultural Crops 56
Improvement in Postharvest Handling, Storage and Processing of Fruits and Vegetables 58
Promotion of Cut Flower Industry 61
Integrated Production Systems (IPSs) 61
Indigenous Knowledge for Horticultural Crops 62
Popularizing Irrigation Practices for Water Conservation 62
Use of Plasticulture 63
Improvement in Mechanics of Horticultural Products Marketing 63
Role of Public and Private Agencies 64
Intellectual Property Rights (IPRs) 64
Extension Plan 65
Quality Control Mechanism 65
Need for Education Reform 66
Future Research and Development 67

4. Perspectives of Horticulture 71
Fruit Culture 71
Vegetable Culture 72
Ornamental Horticulture 72
Floriculture 72
Landscape Architecture 73
Allied Horticultural Industries 73
The Nursery Industry 73
The Seed Industry 74

Horticulture and Society 74
Source of Food 74
Ornamentals 75
Medicinal Plants and Horticulture 75
Jobs 75
Indirect Jobs 76

5. The Power of Biotechnology in Hi-Tech Horticulture **78**
Biotechnology: Strategies for Horticultural Development 80
Biotechnology Interventions in Control of Ornamental Crop Diseases 81

6. Techniques of Biotechnology for Horticulture **84**
Genetic Engineering Techniques 84
Gene Transfer to Enhance Horticulture 85
Isolation of Important Genes 86
Gene Transfer Technology 86
Molecular Markers 87
Marker-Assisted Selection: Fast Track to New Crop Varieties 88
Molecular Shortcut 88
Marked Benefits 89
Marker-Assisted Selection 89
DNA Fingerprinting 90
Genetic Engineering of Microorganisms Associated with Plants 90
Arabidopsis Genomics and Horticulture 90
Genetically Modified (GM) Crops 90
Genomics 91
Protein (or Enzyme) Engineering 92
Tissue Culture Techniques 92
Tissue Culture: Production of True to Type Plant 97
Tissue Culture Techniques to Increase Genetic Variability 98
Protoplast Culture and Micropropagation 98
Somatic Hybrids and Cybrids 98
Haploid Production 99
Production of Virus-Free Plants 100
Shoot Tip Micrografting 100
Micropropagation Techniques 100
Propagation by Axillary Shooting 101
Propagation by Direct or Indirect Organogenesis 101
Improvement of Axillary Branching 101
Plant Tolerance to Drought and Salinity Stress 101
Somatic Embryogenesis Propagation 102
Synseeds 102
Somaclonal Variations 103
Haplodiploidisation 103
Embryo Rescue 103
Germplasm Conservation (Gene Bank) 104
Cryopreservation of Plant Germplasm 104
Microbial Biopesticides, Biofungicide, Herbicides and Agricultural Antibiotics 105

Use of Biotechnology in Biocide Production *106*
Biosensors, Biochips, Biofilms and Biosurfactents 106
Biosensors *106*
Biochips *106*
Biofilm *107*
Biosurfactents *107*
Monoclonal Antibody Technologies 107
New Approaches to Crop Production 108
The Genetic Engineering of Plants *108*
The Genetic Engineering of Microorganisms Associated with Plants *109*
Genetic Engineering for Crop Protection *110*
Bioprocessing Opportunities 111
Other Products *111*

7. Applications of Genetic Engineering Hi-Tech Horticulture 113
Biological Molecular Markers Assisted Breeding 113
Transgenic Plants *113*
I. Herbaceous Dicotyledons *114*
II. Woody Dicotyledons *114*
Engineered Resistance against Herbicide 115
Modification of the Target *115*
Resistance against Glyphosate *115*
Resistance against Sulfonylurea and Imidazolinone Herbicides *115*
Resistance against Phosphinothricin *116*
Resistance against Atrazine *116*
Detoxification or Degradation of Herbicide *116*
Engineered Resistance against Pest 117
Bacillus thuringiensis *Endotoxins* *117*
Pseudomonas fluorescens *that Secretes Bt Toxin* *118*
Protease Inhibitors *118*
Gene for other Insecticidal Secondary Metabolites *118*
Engineered Resistance against Viral Infection 119
Cross Protection *119*
Coat Protein Mediated Protection *120*
Antisense RNA Mediated Protection *120*
Satellite RNA Mediated Protection *120*
Defective Interfering Mediated Protection *120*
Replicase Mediated Protection *120*
Engineered Resistance against Fungal Pathogen 121
Genetic Engineering and Plant Lipids 121
Genetic Engineering and Storage Proteins 122
Genetic Engineering for Nitrogen Fixation 123
*Transfer of **nif**-genes from **Klebsiella pneumoniae** to **E. coli*** ***123***
*Transfer of **nif**-genes of Eukaryotic Organisms* *123*
*Transfer of **nod**-genes to Increase Host Range* *123*
Plants that Produce Plastic *123*
Miscellaneous Applications 123

8. Techniques for Genetic Engineering **125**
Isolation of Total Cellular DNA 125
Nucleic Acid Hybridization 125
Methods for Labelling Nucleic Acids 127
Methods of Labelling Nucleic Acid and Probes 127
Nick Translation *127*
Primer Extension Method *128*
Methods Based on RNA Polymerases *130*
End Labelling of Nucleic Acids 130
Choice of Label 131
Radioactive Labels *131*
Non Radioactive Labels *131*
Mapping Genes on Specific Chromosomes 132
Somatic Cell Hybrids *132*
In–Situ Hybridization *133*
Transposon Tagging *135*
Genetic Linkage Mapping *135*
Genomic Library 135
Colony Hybridization *136*
Plaque Hybridization *136*
Chromosome Walking *136*
Chromosome Jumping *137*
Cleaving DNA with Restriction Enzymes 138
Gel Electrophoresis *138*
Blotting Techniques 140
Analysis of DNA by Southern Blotting *140*
Analysis of RNAs by Northern Blot Hybridizations *141*
Analysis of Proteins by Western Blot Techniques *141*
Detection of RFLP 143
DNA Sequencing 143
Methods for Analyzing DNA Sequence *144*
Sequencing by Chemical Degradation *144*
Sequencing by Chain Termination *145*
Automated DNA Sequence Analysis *148*
Mechanical Shearing of DNA 148
Transcript Mapping Techniques 148
Primer Extension *149*
S1 Mapping *150*
Synthesis of cDNA from mRNA 151
Preparation of Double–Stranded cDNA *151*
cDNA Library *153*
Searching for Gene using Computers 153

9. Genetic Engineering Techniques for Producing Transgenic Crop Varieties **154**
The Goal for Improved Crop Varieties 154
Transgenic Techniques 155
Steps for Developing New Crop Varieties 155

Bacteria: Nature's Genetic Engineers 157
Bacteria and Gene Transfers in Plants *158*
How Bacteria are used to Modify Plants *158*
How do they do it? *158*
Agrobacterium Mediated Gene Transfer 161
Plant–Transformation Vectors *161*
Gene Cloning 163
Creating Recombinant DNA *163*
DNA Ligase *163*
T4 DNA Ligase *163*
Linkers *164*
Stop Linker *164*
Adaptors *164*
Double–Linkers *164*
Terminal Transferase *166*
Cloning in Bacteria and Eukaryotes 166
Cloning in Bacteria *166*
Cloning in Eukaryotes *166*
Amplification of DNAs by Polymerase Chain Reaction (PCR) 168
Procedure *168*
Applications of PCR *171*
Gene Transfer Technology 172
Sexual Method *172*
Asexual Methods *172*
Biological Delivery System 175
***Agrobacterium tumefaciens**: An Important Vector for Genetic Engineering* *175*
Disarmed Ti Vectors *178*
Vectors *178*
Gene Transfer Using Particle Bombardment Technique 178
History *179*
Applications of Biolistic *179*
Particle Gun Design *179*
Preparation of DNA–Coated Microprojectiles *180*
Gene Gun *180*
Modified Bombardment Devices *180*
Components of the Introduced Plasmid DNA *182*
Target Tissue *185*
Tissue Treatment *186*
The Fate of the Introduced DNA(s) *187*
Gene Transfer by Microinjection Technique 188
Applications of Microinjection Technique *189*
Technique *190*
Silicon Carbide Fiber–Vortex (Silicon Whiskers) 191
Sonication 191
Electroporation 192
Poly–Cation Mediated DNA Uptake 193
Ca–DNA Co–Precipitation Method 193

Ultrasonication 194
UV Laser Microbeam 194
Genetic Transformation in Monocotyledonous Plants 195
*Factors Involved in **Agrobacterium**–Mediated Transfer of Genes to Monocots* *195*
Direct Gene Transfer Monocotyledonous Plants *195*
Problems Involved in Gene Transfer 196
Expression System *196*
Cellular Localization *196*
Proteolysis, Protein Folding and Prosthetic Group Acquisition *196*
Precursor Availability *197*
Inhibitory Environments *197*
Side Reaction of New Compounds *197*

10. Plant Tissue Culture: Principles and Methodology **198**
Importance of Tissue Culture Technique 198
Global Picture of Plant Tissue Culture *199*
History of Plant Tissue Culture Researches 199
Tissue Culture Researches in India 201
Plant Tissue Culture: Principles 204
Callus Culture *204*
Meristem Culture *205*
Organ Culture *205*
The Concept of Totipotency of Cells 205
Contrasts between Plant and Animal Cells *206*
The Autonomous Organelles: Their Behaviour during Growth Induction and Morphogenesis *208*
Sites and Modes of Action of Growth Regulating Substances *208*
The Behaviour of Carrot Cells in vitro *208*
Clonal Development from Animal and Plant Cells *208*
Plant Tissue Culture: Methods 209
Media Preparation: Area/Room *209*
Cabinets or Shelves *210*
Culture Media, Washing Powder/ Liquid Detergent, Disinfectants *210*
Aseptic Transfer Chamber Area *210*
Environmentally Controlled Culture Room *210*
Analytical Room *211*
Acclimatization Room *211*
Miscellaneous Items *211*
Specifications of Laboratory Equipments *212*
Aseptic Technique 212
Sterilization of Plant Tissues 213
Cleaning (Preparation of Glasswares/Plastic Wares (Autoclavable) *214*
Sterilization *214*
Surfactants *214*
Control of Bacterial and Fungal Contaminants by Antibiotics in Plant Tissue Culture *215*
In Vitro Environment 216
Pretreatment to Explant Tissues Prior to Culture 218

Procedure 218
Micropropagation through Organogenesis 220
Culture Media and Preparation 221
Media Components 222
Inorganic Salts 222
Plant Growth Regulators 222
Carbon Source 223
Gelling Agent 223
Amino Acids and Amides 224
Antibiotics 224
Natural Complexes 224
Additional Requirements 224
Sterilization of Media 229
Use and Storage of Coconut Water 229
Related Procedures 229
Ultraviolet Light 229
Preparation of Phenol 230
Working with ^{32}p Labeled Compounds 230
Silanization of Plastic and Glassware 230
Preparation of Dialysis Tubing 231
Initiation of Embryogenesis in Suspension Culture 233
Establishment of Embryogenic Suspension Culture 234
Maturation of Embryos 236
Important Parameters for Consideration 236
Simplification of the Technique 236
Different Causes of Failures (and Remedies) 237

11. Protoplast Technology 239
Importance of Protoplast Isolation and Culture 239
Isolation of Protoplast from Various Plant Parts 239
Enzymatic Method 240
Mechanical Method 240
Enzymic Isolation of Protoplasts: Method 240
Sequential Method 243
Methods of Protoplasts Culture 245
Laboratory Facilities 245
Enzyme Mixture and Osmotic Stabilizer (Osmotica) 245
Purification of Isolated Protoplasts 247
Sedimentation and Washing 247
Flotation 247
Other Purification Methods 247
Protoplast Viability Testing 248
General Steps of Protoplast Culture 248
Culture Media for Protoplast Culture 248
Protoplast Culture Media for PC I Group 248
Protoplast Culture Media PC II Group 250
Protoplast Culture Media PC III Group 250
Agar Embedded Culture 251
Microchambers 251

Hanging Drop Cultures (HDC) Techniques 251
Multidrop Array (MDA) Techniques 251
Growth and Division of Protoplast 252
Protoplast Culture: Regeneration of Cell Wall 253
Development of Callus/Whole Plant 253
Determination of Protoplast Plating Efficiency 254
Technique for Isolation of Sub–Protoplasts 254
Handling of Regenerated Plantlets 255
Protoplast Fusion 255
General Steps of Protoplast Fusion 256
Protoplast Culture: Regeneration of Plants 258
Organogenesis 259
Embryogenesis 259
Handling of Regenerated Plantlets 260
Protoplast Culture: Uptake of Foreign Materials 260
Incorporation of Foreign DNA 260
Incorporation of Nuclei 260
Incorporation of Chloroplasts 261
Incorporation of Cyanobacteria 261
Incorporation of Bacteria 261
Incorporation of Virus 262
Incorporation of Non–Biological Materials 262
Selection of Fusion Hybrids 262
Visual Selection 262
Fluorescent Labels 263
Fluorescence Activated Cells Sorting 263
Nutritional Selection 263
Drug Sensitivity and Resistance 263
Somatic Cell Hybridization or Cybrid or Cytoplast 263
Protoplasts for Isolation of Cell Components 265
Plasmalemma 265
Chloroplasts 265
Mitochondria 265
Vacuoles 266

12. Micropropagation in Plants **267**
Advantages of Micropropagation 267
Other Advantages of Micropropagation Methods 267
Advantages of Micropropagation over Traditional Methods 268
Uses of Micropropagation in Horticultural Industry 270
Commercial Uses of Micropropagation 271
Materials for Micropropagation 271
Plant Health 274
Germplasm Storage 274
Types of Micropropagation 274
Micropropagation through Shoot Bud Proliferation 274
Micropropagation through Adventitious Buds 278
Meristem Tip Culture 279
In Vitro Tuberization 279

Somatic Embryogenesis *279*
Methods of Micropropagation (General) 282
Which Material should be used? *282*
Stages of Micropropagation *282*
Materials *282*
Methods *283*
Important Considerations and Precautions 286
Problems and Contamination in Culture *286*
Micropropagation in Forest Trees 288
Conifers *289*
Commercial Aspects of Micropropagation 289
Large Potential Market *290*
Futuristic Outlook of Micropropagation *291*
13. Somatic Embryogenesis **292**
Carrot as a Embryogenic System 293
Somatic and Zygotic Embryos 293
Totipotency and Role of Auxin 293
Proembryogenic Masses and Role of Auxin *293*
Acquisition of Totipotency *293*
Expression of the Programme in the Absence of Auxin *295*
Methylated Cytosine Level and Development *295*
The Genetic Approach of Somatic Embryogenesis 295
The Need *295*
The Process of Somatic Embryogenesis 296
Induction of the Embryogenic State *296*
Recurrent Embryogenesis 297
Problems *297*
Maintenance *297*
Auxin vs Cytokinin for Induction of Embryogenesis *298*
How to Obtain Embryos from Embryogenic Cultures? 299
Embryo Maturation and the Development of Germinability 299
Auxin Levels *299*
Sucrose Level *299*
Heat Shock Treatment *300*
ABA Levels *300*
Water Saturation *300*
Genetic Control of Embryo Cloning 300
Genetic Variability *301*
Cytoplasmic Effects *301*
Applications of Somatic Embryogenesis 301
Mass Propagation *301*
Scale–Up Potential *302*
Use of Bioreactors *302*
Protoplast Culture *303*
Embryo Cloning and Gene Transfer 305
Somatic Embryogenesis in Trees 306

14. Mutation and Plant Breeding Techniques for Hi-Tech Horticulture **309**
Principles of Breeding for Improved Plant Cultivars 309
Techniques of Mutation Breeding 311
Genetic and Biological Basis of Mutation *312*
Induced Mutation and Plant Breeding 314
Techniques of Induced Mutation 314
Plant Materials Used in Radiation Breeding *315*
Separation and Selection of Mutants *316*
In Vitro Technique *317*
Raising of M_1 Generation *319*
Utilization of Induced Mutations *320*
Mutagens *321*
Doubled Haploid Production *321*
Ion Beam Mutation *322*
Characters to be improved *324*
Radiation Breeding of Fruit Trees and Ornamental Plants *325*
Practical Aspects of Induced Mutation *325*
Factors affecting Induced Mutation *326*
Induced Mutagens and their Doses 328
Screening *329*
Other Parameters *329*
Crop Improvement 330
Somaclonal and Gametoclonal Variant Selection 330
Genetic Variations and Crop Improvement *331*
Why Variations Occur? *332*
The Mechanism of Somaclonal Variation *332*
Source Material and Culture Conditions *333*
Forms of Somaclonal Variation *333*
Detection and Isolation of Somaclonal Variants *333*
Mutagenesis and Somaclonal Variation *334*
Somatic Genetics of Nitrogen Metabolism *335*
Method for Isolation of Desired Variant Cells for NaCl -Tolerant from Callus/ Suspension Cultures *335*
New Cultivars Developed by Mutation Breeding 336
Applications in Plant Breeding *337*
Future Role 341
Mutant Stocks as Gene Repositories *342*
Outlook for the Future *342*

15. Marker Assisted Selection (MAS) Technique for Hi-Tech Horticulture **344**
Introduction 344
Fundamentals of MAS *346*
Benefits of Molecular Markers *347*
Demerits of Morphological Markers *347*
Steps for MAS *349*
Use of Mas for Backcross Breeding *350*

Marker Assisted Gene Pyramiding 350
Applications of MAS in Agriculture 351

16. Gynogenesis in Hi-Tech Horticulture **354**
Gynogenesis Induced by Irregular Pollination In Situ 354
Development of Useful Crop Varieties 357
Gynogenesis Induced by In Vitro Culture of Unpollinated Female Reproductive Organs 357
Origin of Gynogenic Regenerants and Methods of Chromosome Doubling 359
Conclusions 360

17. Synseeds or Synthetic Seeds for Hi-Tech Horticulture **361**
The Natural Seed 361
Development of the Concepts of Tissue Culture and Artificial Seeds 361
Discovery of Synthetic Seeds 363
Uses and Limitations of Artificial Seeds 366
Potential Uses of Artificial Seeds 366
Production of Synthetic Seeds 369
Encapsulation or Coating of Synthetic Seed 370
Gel Complexation via a Dropping Procedure 370
Automate Encapsulation Process 370
Mass Balance Concept 371
Steps of Commercial Artificial Seed Production 371
Hydrogel Encapsulation of Artificial Seeds 372
Synthetic Seed and Forest Trees 373
Somatic Embryogenesis and Synseeds 373

18. Plant Tissue Culture some Related Aspects **374**
Bio Village Concept 374
Qualifications Required to Join the Training Course 375
Employment for Rural Youth 375
Efforts of Public Research Institutes 375
Production Criteria and Economics 376
Selection of Crops for Micropropagation 376
Selection of Location 376
OPTIMIZING Size of Production 377
Available Facilities and Production Potential 377
Economic Analysis of the Project 377
Production Management 378
Order Processing 378
Planning for Production 378
Multirate 379
Passage 379
Operator Efficiency 380
Production Monitoring 380
Contamination: Economic Assessment 380
Calculation of Cost Price 380
Quality Control in Commercial Plant Tissue Culture 380

19. Hi-Tech Conservation of Horticultural Crops **382**
Why Conservation of Genetic Resource? 382
Objectives of Conservation *383*
Genetic Resources 383
Types of Conservation 383
Cryopreservation Techniques *396*
Application of Cryopreservation Techniques *396*
Conservation and Recovery of Germplasm *399*
*Limitations of **In Vitro** Conservation* *399*
Other Methods *401*
Documentation *402*

20. Bioinformatics **404**
Sequencing DNA, RNA and Proteins 404
Techniques for Determination of Proteins 405
Pattern Databases for Protein Families *405*
Principles of Genome Annotation 406
Bioinformatics: Internet and WWW 407
Database Searching and available Database 408
Annotated Sequence Database *409*
Genome and Organism-specific Database *410*
Miscellaneous Databases *410*
Data Retrieval with ENTREZ and DBGET/ LinkDB *411*
Data Retrieval with Sequence similarity Searches (SRS) *411*
Transcription Regulatory Regions Database (TRRD) *412*
Applications of Bioinformatics 422
Bioinformatics in Study of Phylogenetics *423*
Bioinformatics Helps in Building Phylogenetic Trees *423*
Bioinformatics and Molecular Phylogeny *424*
Bioinformatics in Discovery of Drugs *424*
Pharmainformatics *425*
Cheminformatics *425*
Protein Interaction Informatics *426*
Protein Interaction Databases *427*

21. Plant Growth Regulators for Hi-Tech Horticulture **428**
Auxins 429
Gibberellins 431
Cytokinins 432
Abscisic Acid 433
Ethylene 434
Non-Traditional Plant Growth Regulators 436
Jasmonic Acid *436*
Brassinosteroids *437*
Oligosaccharins *438*
Glucose as a Hormone *439*
Polyamines *440*
Salicylate *441*
Nitric Oxide (NO) *442*

Florigen, Anthesin and Vernalin 443
Morphactins 443
Use of Plant Growth Regulator's in Hi-Tech Horticulture 444
Special Use of Plant Growth Regulators in Hi-Tech Horticulture 453
Use in Tissue Culture Technique 453
(i) Auxins 453
(ii) Cytokinines 453
(iii) Gibberellins 455
(iv) Ethylene 455
(v) Abscisic Acid 455
Role of Plant Growth Regulators on Sex Expression of Cucurbits in Hi-Tech Cultivation 455

22. Modern Nursery Management **457**
Introduction 457
Necessity of a Nursery 458
Type of Nurseries 458
Nursery Management 459
Selection of Suitable Site 460
Layout of Model Nursery 461
Miscellaneous Propagating Structures 466
Selection of Suitable Crops 468
Selection of Suitable Plant Propogation Technique 469
Improvised Nursery Management 471
Nursery Facilities 471
Soil and Soil Preparation 473
Seed Treatment 475
Media for Raising Seedlings 476
Nursery Bed Preparation 480
Advantage of Raised Nursery Bed 481
Sowing of Seeds in the Nursery 483
Seed Covering Material and its Treatments 483
Seedling Management 484
Effect of Hardening 486
Potting Procedure 490
Container and Tools for Propagation and Growing Young Plants 493
Use of Hormones 494
Disposal and Packing of Produce 496
Cost-Efficiency of Nursery 496

23. Greenhouse Technology for Hi-Tech Horticulture **498**
Introduction 498
Advantage of Greenhouse 498
Types of Greenhouses 499
Classification of Greenhouses 499
Greenhouse Type based on Utility 502
Greenhouse Type based on Construction 502
Greenhouse Type based on covering Materials 503
Rigid Panel Greenhouses 504

Glass Greenhouses 504
Film Plastic Greenhouses 505
Role *505*
Types of Film Plastic *506*
Quonset Style Greenhouse *509*
Gutter Connected Greenhouse *509*
Greenhouse with Double Layer Covering *510*
Rigid Panel Greenhouses 511
Polyvinyl Chloride *511*
Fiberglass Reinforced Plastic *511*
Acrylic and Polycarbonate *513*
Greenhouse Construction: Related Aspects 513
Location *514*
Framework for Various Types of Greenhouse 518
Types of Greenhouse Covering Materials *526*
Covering Materials for Greenhouse 527
Fiberglass *527*
Poly Film *529*
Heating and Cooling of Greenhouse 529
Sensing Temperatures *530*
Sources of Heat *530*
Methods of Distributing Heat in Greenhouse *531*
Conservation of Heat Energy in Greenhouses *531*
Insulation of Greenhouse Roofs *531*
Ventilation and Air Movement Equipment for Greenhouses 532
The Principle of Ventilation and Air Movement *532*
Air in the Greenhouse *533*
Carbon Dioxide Generation Equipment in Greenhouse 533
Importance of Carbon Dioxide in Greenhouse *534*
Greenhouse Benches 534
Materials used for Greenhouse Benches *534*
Greenhouse Benches for Cut-Flowers *534*
Greenhouse Benches Pot Plant *535*
Arrangement of the Benches *536*
Greenhouse Walks *538*
Working Room 538
Conveyors 539
Cloth Houses 539
Lath Houses 539
Bulb Storage 540
General Storage 540
Boilers for the Greenhouse Heating System 541
Operation of the Boiler *541*
Draft System *541*
Operation of the Boiler *541*
Emergency Heating Equipment for Greenhouse 542
Emergency, Standby Power *542*
Figuring the Heat Requirement of the Greenhouse *542*

Location of Heating Lines 543
Arrangement and Placement of Heating Lines 544
Steam Traps 544
Control of Heating Line 544
Special Heat Requirements for Greenhouse 544
Providing Steam for Treatment of Soil and Equipment 545
Lower Down Temperatures in the Greenhouse 546
Design of the Fan and Pad Cooling System 546
Cool Storage 546
Other Essential Equipments for Greenhouse 547
Equipment for the Control of Light 547
Kinds of Screening Materials 547
Photoperiod for Flowering 547
Shading for the Control of Photoperiod 548
Electric Light Equipment 548
Fluorescent Light 548
Electric Lights for Photoperiod Control 548
Intermittent Light 549
Problems with Fluorescent Lights 550
Equipment for Moisture Control in Greenhouse 550
How to Reduce Moisture in the Greenhouse 550
Water Distribution Systems for Pot Plants in Greenhouse 551
Capillary Irrigation System 551
Water Distribution Systems for Cut-Flower Crops 551
Equipment for Regulation of Irrigation 551
Equipment for Handling Fertilizer 552
Application of Fertilizer in Liquid Forms 552
Proportions for Uniform Supply of Fertilizers 552
Selection of Fertilizer Proportioner 553
Pest and Pathogen Control Equipment 553
Hydraulic Spraying 553
Stationary and Portable Sprayer 553
Dusting 554
For Fumigators 554
Equipment for Application of other Agro Chemicals 554
Equipment for Preparation of Soil for Planting 555
Soil Shredders 555
Materials Handling Equipment 556
Trolley Type Conveyors 556
Repair and Construction Equipment 556
24. Hydroponics for Hi-Tech Horticulture **557**
Introduction 557
Advantages of Hydroponics 557
Modern Hydroponics 559
Future Strategy of Hydroponics 559
Limitations of Hydroponics 559
Hydroponic Techniques and their Classification 560
Classification 560

Hydroponic Techniques 561
Liquid Hydroponics *561*
Aggregate Systems *566*
Aeroponics *569*
Basic Requirements of Hydroponics 570
Growing Media *570*
Nutrient Solutions *570*
Nutrient Solution 571
Temperature *571*
Fresh Air *572*
Shelter and Support *572*
Water *572*
Light *572*
Growing Media *573*
Equipments and Appliances 577
Management Requirements of Hydroponics 577

25. Hi-Tech Horticulture for Postharvest Technology 579
Introduction 579
Postharvest Technology 581
Reasons of Postharvest Losses *581*
Constraints Faced by The Postharvest Processing Industry 583
Maturity Indices and Harvesting of Fruits 584
Maturity Indices *584*
Physical Methods *586*
Chemical Methods *588*
Physiological Methods *589*
Harvesting 590
Bruising in Fruit *590*
Harvesting Methods *592*
Harvesting Containers *592*
Harvesting Tools *593*
Pre-Harvest Factors affecting Postharvest Life 596
Pre-Harvest Production Practices *596*
Time of Fresh Produce Harvest 597
Perishability and Produce Losses 598
Postharvest Management of Mushrooms 598
Postharvest Technologies *598*
Future Thrust and Export *600*

26. Organic Horticulture 601
Why Organic Farming? 602
Organic Farming ***Verses*** *Conventional Farming* *602*
Organic Farming Tactics 603
How to Convert to Organic Farming? 603
Principles of Organic Farming 605
Terminology used in Organic Farming 605
Organic Fruit Production 607
Orchard Management *608*

Land Management 608
Manure Management 612
Sustainable Water Management (SWM) 616
Organic Farming and Water Use Efficiency 620
Organic Farming and Soil Fertility 622
Organic Farming Practices for Soil Fertility 623
Required Elements for Plant Nutrition 625
Organic Farming: Nutrient Recycling 626
Organic Manures 633
Concentrated Organic Manures 636
Economic Value of Organic Waste 636
Composting 636
Use of Biofertilizer 637
Biological Weed Control 640
Biological Control of Diseases and Insects 643
Organic Control of Diseases 645
Biological Control of Pests 650
Integrated Pest Management (IPM) 651
Organic Vegetable Crop Production 653
Carrots 653
Pumpkins and Squash 653
Potatoes 653
Sweet Corn 653
Onions 654
Cole Crops (Cabbage Family) 654
Other Vegetables 654
Herb Production 655
Culinary Herbs 655
Medicinal Herbs 655

27. Nutrient Management in Horticultural Crops **656**
Essential Elements in Plant Nutrition 657
Classification of Essential Nutrients 658
Criteria of Essentiality 658
Essential Nutrient Elements and their Sources 659
Essential Elements from the Soil 659
Antagonistic Effects of Micronutrients 660
Chelate 660
Synthetic Chelating Agents 660
Functions of Essential Nutrients 660
Nitrogen (N) 660
Phosphorus (P) 660
Potassium (K) 661
Calcium (Ca) 661
Magnesium (Mg) 661
Sulphur (S) 661
Boron (B) 662
Manganese (Mn) 662
Iron (Fe) 662

Zinc (Zn) 662
Molybdenum (Mo) 662
Copper (Cu) 663
Chlorine (Cl) 663
Cobalt (Co) 663
Sodium (Na) 663
Vanadeum (V) 663
Silicon (Si) 663
Selenium (Se) 664
Aluminium (Al) 664
Role of Essential Plant Nutrient 664
Hidden Hunger 664
Deficiency Symptoms 664
Deficiency Symptoms on Old Leaves 664
Symptoms on New Leaves 665
Deficiency Symptoms on both Old and New Leaves 666
Toxicity Symptoms 666
Mobility of Nutrients 666
Mobility in Soil 666
Mobility in Plants 666
Classification of Diagnostic Methods 666
Nutrient Uptake by Plants 667
Nutrient Absorption by Plants 668
Sources of Plant Nutrients 669
Manures, Fertilizers and Biofertilizers 669
Green Leaf Manure Crops 674
Management of Green Manuring 674
Vermicompost 675
Concentrated Organic Manures 675
Biofertilizers 676
Fertilizers 681
Precautions in Mixing Fertilizers 682
Complex or Mixed Fertilizers 690
Micronutrients 693
Deficiencies of Micronutrients 693
Conditions Conducive to Micronutrient Deficiency 694
Fertilizer Application 694
Principles of Fertilizer Application 694
Frequency of Application 695
Balanced Fertilizer Application 697
Integrated Nutrient Management (INM) 698
Soil Sampling for Nutrient Management 700
When to Collect Soil Samples 701
Sampling Frequency 701
Sampling Strategy 701
Sampling Pattern 702
Sampling Equipment 702
Sampling Depth 702

Making a Composite Sample 703
Sample Handling and Containers 703
Labeling 703
What Analyses are required? 703
Fertilizer Analyses and Calculating Fertilizer Rates 703
Soil pH Modification 704
Interpreting Lab Results for Nutrient Management Planning 704
Liming 704

28. Hi-Tech Horticulture: Micro-Irrigation (Drip and Sprinkler Irrigation) **706**
Introduction 706
Economic Characteristics of Water 706
The Dominance of Agricultural in Water Use 707
Micro Irrigation System (MIS) 708
Types of Micro-Irrigation System (MIS) 709
Components of Drip Irrigation System (Listed in Order from Water Source) 710
Plastic Mulching with Drip Irrigation System 714
Where to use Sprinkler Irrigation 717
Operating of Sprinkler Systems 719
Sprinkler Drop Sizes 720
Control of Sprinklers Cycle 721
Advantages of Sprinkler System 721
Disadvantages of Sprinkler System 721
Importance of Micro-Irrigation System 722
Irrigation Criteria and Irrigation Scheduling 723
Irrigation Scheduling Methods 725
Fertigation 726
Preparation and Uses of Fertilizers 729
Interaction between the Fertilizers and Irrigation Water 730

29. Plasticulture for Hi-Tech Horticulture **733**
Use of Plasticulture 734
Plastic Mulch 734
Weed Control 736
Low and High Tunnels 737
Runner Control 740
Pest Management 740
Strip Fumigation and Soil Solarization 741
Windbreaks 741
Management of Ripening Times and Season Extension 741
Clean Picking Surface 742
Plastic Flower Pots and Field Harvesting Containers 742
Disposal of the Used Plastics from Plasticulture 743
Current and Future Status of Biodegradable Plastics 743

30. Value Added Products of Horticultural Crops **745**
What is Value Added? 746
What is Customer Value? 746

What Creates Added Value? 746
i) Creating a Value Added
Product 747
Post-Harvest and Value Addition 747
Capturing Value and Creating Value 750
Learning New Production and Marketing Skills 750
Launching of Food Business 751

31. Hi-Tech Horticulture: Harvest Technology and Value Addition of Medicinal Plants **752**
Harvesting 752
Harvesting from the Wild 752
Time of Harvest 752
Harvesting Correct Useful Parts 752
Processing 753
Decoction 753
Filtering Operation 753
Extraction 753
Infusion 754
Maceration 754
Extraction 754
The Problem 755
Forms of Preparations 755
The Solvents 755
Syrups Preparation 758
Distillation Process 759
Distillation 760
Fractional Distillation 764
Parts of Medicinal Plant to be Used 765
Other Method of Processing 766
Preservation 767
Standardization of Medicinal Plants 767
Active Constituents of Medicinal Plants 771
Tests for Standardization 773
Preparation of Herbal Remedies 773
Other Preparations 776
Standardization of Medicinal Plants 776
Decoction 778
Filtering Operation 779
Extraction 779
Infusion 779
Maceration 779
Extraction 780
The Problem 780
Forms of Preparations 780
The Solvants 781
Syrups Preparation 783

32. Marketing of Horticultural Produce **785**
Market 785
Marketing 786
Selling *787*
Functions of Marketing *787*
Agricultural Marketing 788
Marketing Environment *788*
Emerging Alternative Agricultural Marketing Institutions in India 790
Various Reform Measures *790*
Reforms through Model APMC Act *790*
Strategy Proposed during the State's 10th Plan in Agriculture Marketing and Agri-business *792*
Food Chain Revolution *792*
Role of Department of Agricultural Marketing and Agri-Business 793
Development of Market Committees and Regulated Markets *793*
Creation of Infrastructure for Marketing *794*
Post-harvest Management Formarketing of Horticultural Crops 794
Need for Drying Yards in the Villages *794*
Development of Product Specific Markets *794*
Agricultural Production and Marketing Information Centre *795*
Agro Food Processing Units *795*
Contract Farming *795*
Development of Food Parks *797*
Development of Agri Export Zones *797*
Role of Seed Certification Department *799*
Crop Insurance Scheme in India 802
India/State Government Organizations for Agricultural Marketing 805
Marketing of Horticultural Crops 807
Characteristics of Horticultural *808*
Other Aspects *809*
Marketing Channels 810
Direct Marketing *812*
Indirect Marketing *813*
Fixing of the Price Range of Horticultural Products *814*
Evaluation of Customer Demand *814*
Quality of Farm Produce 815
Quality Factors / Refractions *815*
Standardisation 815
Inspection and Sampling 817
Inspection *817*
Sample *817*
Grade and Grading of Samples *817*
Analysis *820*
Certification *820*
Pricing Horticulture Products 822
The Cost to Grow Product *822*
Profit Goals *823*
Important Products of Horticulture Industry 823

33. e-Marketing of Horticultural Produce **824**
e-Learning 824
e-Education in Agriculture 825
e-Content Development through Web Pages 826
Web Site Optimization *827*
e-Commerce (Marketing) 829
Types of Electronic trading or e-Commerce *830*
Opportunities and Challenges 831

34. Export of Horticultural Crops **833**
World Exports from India 834
Fruits *834*
Vegetables *834*
Government of India's Efforts 836
Major Development Programmes in Central Sector *836*
Vegetables *836*
Root and Tuber Crops *837*
Mushroom *837*

35. Hi-tech Horticulture and Intellectual Property Rights **838**
HISTORY OF IP Rights Related to Food and Agriculture 839
Plant Variety Rights Protection *839*
Definitions *840*
Salient Features of Protection of Plant Varieties and Farmers' Right Act 841
Protection of Plant Varieties and Farmers' Rights Authority *841*
Implications of PPV and FR Act 843
Criteria for Deciding Protection of Plant Varieties *845*
Advantages of Intellectual Property Rights 849
Intellectual Property Rights and Agriculture *849*
Intellectual Property Rights in India *849*
The relationship Between Intellectual Property Rights (TRIPS) Agreement *850*
Intellectual Property, Plant Breeders and Farmers *850*
Farmers as Innovators *851*
Features of UPOV Systems of Protection *852*
Is UPOV Appropriate for Developing Countries? *854*
An Alternative to UPOV *854*
Indian Laws and protection of plant varieties 855
IPRs and Biotechnology *856*

CHAPTER 1

Hi-Tech Horticulture: An Introduction

Globalization is becoming a world phenomena and it is especially true in horticulture as compared to other agricultural sectors. Horticulture combines art and skill with science and technology for growing and distributing ornamental plants, fruits and vegetables. Our landscapes, homes and public buildings are adorned with horticulture. It enriches our lives through varied diets and physical activities from gardening improves our environment and enhances our mental well-being by providing shade, air filters and noise barriers.

Agriculture has long been a staple of India's diverse economy, with the region offering farmers exceptional environmental conditions to grow crops and raise livestock. Now, with advances in technology and research, the region is poised to play a major role in developing the nascent agri-technology industry. With an abundance of farm land, nurseries and greenhouse facilities, India is the cornerstone of the agri-tech business.

High-tech horticulture soon will empower other fields of agriculture as well. Horticulture shall have a growing commercial importance in the near future as world population is aging, people devote more attention to health and most communities better emphasize the quality of life and environment.

Horticulture needs to redefine and/or focus itself because horticulture is an industry and its aim is economic success. It is an emerging sector of economic importance in India. Horticulture is defined as the cultivation of an orchard, garden, or nursery on a small or large scale, or the science and art of growing fruits, vegetables, flowers or ornamental plants. In the past, horticulture was practiced in order to produce food to survive. Ornamentals and flowers were added to improve well-being. People were proud to produce their own fruits, vegetables, medicinal and aromatic plants, and flowers. Gardening was a most welcome job. Nowadays, gardening (i.e. Horticulture) is thought of as a hobby, although it's also a profession. Though horticultural industry has reached unprecedented growth in the size, diversity and value worldwide, but still horticultural science within academia is experiencing a crisis. In modern age, Technology has become a pervasive element in modern horticultural cropping systems. In this context it is define technology as "a*dvanced systems that manage, control, communicate, sense and report information on crops at all stages of development and utilization*." The technology includes many faces like:

1. Improved plant genotypes, some developed by recombinant DNA technology.
2. Value-added plant propagation.
3. Advanced hybrids, coated and primed seed, artificial seeds, micropropagation, advanced plug technology.
4. Computer-based environmental sensing and control specially of green houses.
5. High speed data interchange between critical production points.

6. Environmental sensors and equipment for monitoring plant status.
7. Advanced production techniques (hydroponics, CEA, ebb and flood, etc.)

Horticulture is becoming more and more a low-tech, low-income activity. For high technology production systems needs highly trained managers and workers. They may be expensive but needs periodic updating. The aims of high-tech horticulture can be achieve to make mission successful. It is essential to ensure an end-to-end holistic approach covering production, post harvest management, processing and marketing (to assure appropriate returns to growers/producers). The grower should promote R&D technologies for production, post-harvest management and processing. It needs to enhance acreage, coverage, and productivity through diversification, from traditional crops to plantations, orchards, vineyards, flower and vegetable gardens and extension of appropriate technology to the farmers for high-tech horticulture cultivation and precision farming. High-tech horticulture also needs assist setting up post harvest facilities such as pack house, ripening chamber, cold storages, Controlled Atmosphere (CA) storages etc, processing units for value addition and marketing infrastructure. For this one needs to adopt a coordinated approach and promotion of partnership, convergence and synergy among R&D, processing and marketing agencies in public as well as private sectors, at the National, Regional, State and sub-State levels. It can be possible if we promote capacity-building and Human Resource Development at all levels.

MARKETS FOR HORTICULTURAL BIOTECH

On the supply side, "horticulture" includes an enormous diversity of fruit and vegetable crops, but it also includes many nonfood species, such as ornamentals, flowers and recreational turf grass. Collectively these horticultural crops compare well with major agronomic crops in terms of total value. However, they use much less acreage, and the market size for some biotech products depends on both acreage and production value.

Horticulture includes a small number of larger-scale crops (such as potatoes and onions, apples and wine grapes) as well as a large number of smaller-scale crops (such as brussels sprouts and persimmons). At current costs for R&D and regulatory approval, it is unlikely that biotechnology products will be developed and achieve market acceptance for many of these smaller-scale crops in the near future.

Further, experimentation with perennials such as grapes, nuts and fruit trees is comparatively expensive (because the experimental unit is larger and takes more time), and it is costly to bring new acreage into production or replace an existing vineyard or orchard with a new variety. On the demand side, the market for horticultural products, especially fresh fruits and vegetables, is undergoing important changes associated with the changing structure of the global food industry. Increasingly fewer and larger supermarket chains have been taking over the global market for fruits and vegetables, especially fresh produce and changing the way these products are marketed. Because fresh produce is perishable and subject to fluctuations in availability, quality and price, it presents special problems for supermarket managers compared with packaged goods. Supply chain management and the increasing use of contracts that specify production parameters as well as characteristics and price, is replacing spot markets for many fresh products. A desire for standardized products, regardless of where they are sourced around the world, could limit the development and adoption of products targeting smaller market segments, unless retailers perceive benefits and provide shelf space for diversified products such as biotech and non-biotech varieties of particular fruits and vegetables. On the other hand, an increasingly wealthy and discriminating consuming public can be expected to continue to demand increasingly differentiated products with an ever evolving list of characteristics such as organic, low fat, low carbohydrate and farm fresh.

Hence retailers will have to balance the cost savings and convenience associated with global standardization against the benefits from providing a greater range of products, which will include GM products when retailers begin to perceive benefits from stocking them. Unlike other types of foods, fruits and vegetables are often consumed fresh and in clearly identifiable and recognizable form. This has implications for perceptions of quality and food safety that may influence consumer acceptance perhaps favorably, for instance, if a genetically modified sweet-corn could be marketed as reduced pesticide. Other elements of GM horticulture such as nonfood products, ornamentals or turf grass have advantages in terms of potential market acceptance. GM trap crops, which provide pesticide protection for other crops and GM sentinel crops, which signal the presence of pests or provide other agronomic indicators may be used in food production without overcoming barriers of acceptability to market middlemen or consumers.

Biotechnology products designed for home gardeners may be more readily accepted because the grower is the final consumer. Market acceptance in India may also be linked to continued access to export markets. The relative importance of the domestic market could help to account for the success of the GM food-grain technologies in India, and it may also help to account for the success of these and other GM technologies. China is comparatively important in horticultural biotechnology its investment in agricultural biotechnology is second only to the United States.

HOW TO MAKE HORTICULTURE AS HI-TECH HORTICULTURE

There needs to be an expansion and/or introduction of new high income careers in horticulture. For example, floriculture has great potential to generate high income jobs in the perfume industry new careers in medicinal use of horticultural crops (traditional as well as allopathic and homeopathic uses), organic fruit and vegetable production, and other less traditional careers need to be emphasized. Hi-tech horticulture is the need of all the developing countries. Controlled environment horticulture certainly has a promising future within multi-disciplinary plant science research as well as many exciting applications within controlled-environment agriculture. There are various ways to make horticulture hi-tech. Following the means for it.

Role of Biotechnology in High-tech Horticulture

Biotechnology tools to plant science, regularly working on the synthesis, isolation and cloning of new genes and the genetic transformations of plants. Worldwide, consumption and cultivation of fruits and vegetables is increasing. Between 1990 and 2002, global fruit and vegetable production grew from 0.89 billion tons to 1.3 billion ton and per capita availability expanded from 342 pounds to 426 pounds (FAO 2003). Much of this growth has occurred in China, which is aggressively pursuing agricultural biotechnology. The global fresh fruit and vegetable marketing system is increasingly focused on adding value and decreasing costs by streamlining distribution and understanding customer demands. The year round availability of fresh produce is now seen as a necessity by both food service and retail buyers. Product form and packaging are also changing as more firms introduce value added products, such as fresh cut produce, salad greens and related products in consumer-ready packages. Fresh cut sales are even higher in Europe and beginning to develop in Latin America and Asia as well. The implications of this trend may become as important to the biotechnology industry as the changes in market structure, since fresh cut processors are increasingly demanding specific varieties bred with attributes beneficial to processing quality.

Technological change has driven economic progress in agriculture and will continue to play a crucial role in the 21st century. The latest wave of technological change in agriculture is based in molecular biology. Genetically engineered crop varieties have been adopted on a wide scale in some agronomic crops, but horticultural crops face larger hurdles. High costs for research, development and regulatory approval combined with the small acreages planted and the diversity of varieties, will limit the potential for profitable applications of biotechnology to many fruits, vegetables, nuts and nursery crops. In addition, there are so many market barriers for GM products. Like most important changes in agriculture, modern biotechnology has met with spirited political opposition from some quarters. Threats of political action may discourage food manufacturers and retailers from adopting biotech products that are wanted by some consumers and may be profitable for growers.

Agriculture has been an important engine of economic development and the mainspring of economic progress in agriculture has been productivity improvements driven by technological change that is fueled by research and development (R&D). These gains can be attributed to new biological, mechanical and chemical technologies, including improved genetic material, machines, fertilizers, pesticides and knowledge. The current wave of technological progress continues this pattern, while emphasizing information technologies and biotechnology in particular genetically modified (GM) crops. For many, GM crops represent the hope for a future with less hunger and malnutrition and for a more sustainable agriculture with more varied cheaper and safer food.

Biotech crops have been a commercial reality only for a few years but they have made very rapid inroads in some parts of the market.

In the case of horticulture the cultivation of fruits and vegetables, tree fruits and nuts, turf grass, flowers and ornamental crops these issues are sharply drawn because the private sector has not found it profitable to develop or commercialize many GM crops in the current political, legal and market environment. What will happen in biotechnology applied to horticultural crops is up to the government, for a variety of economic reasons. Some of these aspects may be unique to GM horticultural crops but many are common to GM crops generally and similar issues arise with some new non GM technologies.

Genetic Resources

The identification, preservation, use and control of plant genetic resources play a vitally strategic role in horticulture for the future develop of produce diversification. It is no accident that multinationals are searching the world for unknown or forgotten species for the potential value and added value they may represent. Aromatic and medicinal herbs and plants, dietetic uses of fruits and vegetables, and industrial non-food for chemicals, cosmetics, pharmaceuticals and biodegradable manufactures all fall into this scheme of things. Though still relatively abundant, these resources are in evident diminution and research must play a major role in all countries in the field, as has been and is being done by FAO, IPGRI and other international agencies and NGOs.

Conserving and protecting germplasm, as experts call it, means first of all safeguarding biodiversity, i.e. the natural genetic variability that is the savings bank for the future needs of mankind. Here too political controversy is no stranger, despite the general consensus on the basic objectives. As is well known, most of these resources are found in developing countries, whereas the technologically advanced countries have the ways and means to exploit the potential of these resources. Whence the inevitable claims of the former to a greater share of the benefits derived therefrom without having to comply with the intellectual property rights (IPRs) the latter have acquired through the development of resistant selections, transgenic material and other derivatives subject to patenting and the payment of royalties.

Research can play a leading role in the field of new crop development, of enhancing the value of forgotten or unknown species of producing new industrial spin-offs, of discovering new natural molecules for the food, pharmaceutical, health-care and other industries and can make invaluable contributions in developing ways to preserve *in & ex situ* germplasm both *in vivo* and *in vitro* through cryoscopic and molecular techniques as well as gene banks. Indeed, it is for the future that we must now lay the groundwork by avoiding the pitfalls which could reduce horticulture to the level of a few crops or highly specialized species and with them spell the end of its biodiversity.

Molecular Genetic Characterization of New Floricultural Germplasm

Understanding the genetics of floricultural germplasm represents a value-added component of managing collections. Molecular approaches collectively represent a potential goldmine of important information that can be applied to programs of genetic improvement. In fact, they can be applied to almost every component of a unified germplasm program. The use of molecular markers for genetic diversity analysis and as a selection tool is a high priority for new efforts in the development of ornamental cultivars and the exploitation of species diversity. Molecular marker and DNA sequence analysis of extant and new floricultural germplasm collections should allow a more complete characterization and understanding of the genetic relationships between species and cultivars. With the advent of polymerase chain reaction (PCR) technology, numerous genetic approaches such as random amplified polymorphic DNA (RAPD), amplified fragment length polymorphisms (AFLP), microsatellite DNA (SSR), expressed sequence tags (EST) and single nucleotide polymorphisms (SNP) in candidate genes are available for germplasm characterization. In addition to estimating genetic diversity, these tools can be utilized to establish molecular markers for phenotypic traits of interest. Once obtained, such markers can be further applied in marker assisted breeding, developing various sorts of genetic maps and as quantitative trait loci (QTLs). Molecular markers can also be used to confirm the parentage of individuals in a breeding population, for genetic "finger-printing" with application towards intellectual property rights, and also for understanding systematic relationships of cultivated species.

Tissue Culture

For all horticultural crops, propagation is one of the main business activities and a basic element to a grower's success. It involves seed and plant production, i.e. nursery activities, selection work, bioassays for disease and pest control/certification, and nursery techniques on a large and specialized scale. Some research stations specialize only in these fields, widely using *in vitro* culture, micropropagation and biotechnological approaches.

Tissue culture, an important area of biotechnology can be use to improve the productivity of planting material through enhanced availability of identified planting stock with desired traits. Tissue culture has helped to develop new strain of food crops, cereals, vegetables flowers, oilseeds and plantation crops such as spices, coffee, tea and rubber. Tissue culture in ornamental plants constitute a major part of the total lancing material produced in state out of 31.25 lakh ornamental and other plantlets, gerbera a alone is a major flower crops produced in the state. Although, this technology has been adopted by the farmers, considering area covered, adoption is at a very low level. Plant tissue culture has been promoted by government of India since last 2 decades to enhance the production land availability of disease free, true to type, quality planting material.

Micropropagation

Micropropagation is one of the important contributions of plant tissue culture to commercial plant propagation and has vast significance. The technique provides a rapid reliable system for a production of large number of genetically uniform disease free plantlets. Micropropagation is the technique of developing plants from very small portion of plants such as shoots tip root tip, embryo, stem, pollen grain, callus or single cell. Plant tissue culture owes its origin to the revolutionary concept of totipotency of plant cell propounded by the famous German plant physiologist, Haberlandt in 1902. This technique has opened a vast scope for improve of fruits and plantation crops though micropropagation, creating genetic diversity, germplasm conservation, virus elimination, development of somatic hybrids and gene transfer.

Micropropagation holds a great promise in fast multiplication of fruit and nut crops, which are invariably propagated asexually to meet the ever increasing demand for adequate and timely supply of clean planting material. It is possible through micropropagation to produce millions of identical plants under controlled and aseptic conditions, economy of time and space, affording greater output and augmentation of elite, disease free propagules, safer and quarantined movements of germplasm across nations. It also induces precocity in flowering, precision timing uniformity and often increases yield.

In fruit and plantation crops, comparatively difficult to micropropagate, protocol have been developed for citrus, apple, banana, papaya, pineapple, grape, peach, plum, almond, walnut, strawberry, oilpalm and date palm. In India, commercial exploitation of micro-propagation is limited to oilpalm, strawberry and banana. It is primarily because of the highly heterozygous nature of the material which requires independent protocol for the different genotypes, problems of clonal fidelity, involvement of rootstock and overall costs. However, attempts are on to standardize protocol for crops like mango, cashew, walnut, oil palm, coconut, litchi, sapota and cocoa.

Virus Free Planting Material

Viruses have been the major constraint in improving productivity of several horticultural crops. Thermotherapy, meristem culture and micrografting are commonly used technique to obtain virus free planting material of both scion and stock. Micrografting is more suitable than thermotherapy because it is less stressful to plant material. Virus free planting material can be achieved through meristem culture with the assumption that extreme tips of the meristem are devoid of viruses and the material thus generated are virus free. It is has been achieved for apple, apricot, banana, citrus, peach and fig. Micrografting, which involves the excision of the extreme tip of the shoot and its eventual grafting onto the rootstock *in vitro* and growing the graft in liquid medium is another technique of elimination of viruses.

Germplasm Conservation

In vitro plantlets of oilpalm, banana and pineapple are now routinely exchanged between institutions of different countries. Embryo culture technique in coconut saves bulk transportation of nuts from distant areas. The material stored may be protoplast, isolated cell grown in suspension or on semi-solid medium meristem cultures or organized plantlets. Conservation is through repeated sub culturing. It is expensive and vulnerable to loss due to the invasion of pathogens and the danger of genetic instability. However, the technique of cryopreservation of propagules capable of regeneration *in vitro* in liquid nitrogen is the major breakthrough towards developing a scientifically viable system of long term conservation of the germplasm.

Somaclonal Variation

Genetic instability of the apical meristem, adventitious bud formation from axillary buds, chimeric nature of explant and tissue culture induced disorganization results in somaclones which show considerable variations from their mother plants. These can be used as useful economically genetic material. Somaclonal variations are being exploited for carotenoids, chlorophyll, sugar terpenes, phenolic compounds and anthocyanins. Somaclonal variation is now widely used as a means of crop improvement e.g., salt tolerant citrus.

In Vitro Screening

Tissue culture technique can be used for screening larger batches of plant material for its reaction to biotic (insect, disease) and abiotic (salinity and stress) from a large germplasm pool for speeding up evaluation of progenies in resistance breeding programmes. In horticulture, it has been successful in indexing for citrus viruses, screening of citrus line for resistance to *Phytophthora* and peach for bacterial disease. This has a major application in screening of transgenic plants carrying disease or insect resistance genes before being transferred to soil and for international plant exchange.

Embryo Rescue

Embryo rescue refers to *in vitro* culturing of embryos at different stages of development. It is normally used in hybridization of distantly related species to overcome post zygotic incompatibility. Hybrids have been developed through embryo culture in pineapple. Seedless varieties of grapes have been developed through embryo rescue method from a cross between two seedless parents. Micro cloning has been achieved in embryo culture of banana involving a cross between *Musa acuminata* and *Musa balbisiana*. The technique is immense significance in triploids or sterile crops like banana. It is also useful in plants where dormancy is prolonged or growth is retarded.

Haploid Breeding

Anther culture is extremely used for haploid production and has shown great promise in custard apple, sapota, papaya and guava. Started in early seventies, it is used for rootstock breeding, efficient parental line for hybridization, homozygous diploid selections can be used pure line for heterosis and directly as scion cultivars.

Protoplast Fusion

Protoplast fusion and somatic hybridization technique provide the opportunity to facilitate mixing of both nuclear and cytoplasmic genetic traits between species through the by passing of reproductive isolation barriers. The long juvenile cycle and existing *hybridization* barriers make the traditional breeding approach lengthy and inappropriate. The technique for regeneration of plants from both sexually compatible and incompatible combinations in citrus is established. This technique is standardized for *Carica papaya, Olea europea*, *Persea americana,* grape and apple. The first cybrid plants using donor recipient was reported in *Citrus* and *Microcitrus* genotype. Somatic *hybridization* is an alternative to sexual reproduction in order to combine the entire genome from otherwise incompatible parents. Distant somatic hybrids reported include *Citrus sinensis* + *Poncrus trifoliate*, *Citrus sinensis* + *Citrus unshiu* and *Citrus sinensis* + *Citrus paradisi.*

Molecular Markers

Markers are available to develop genetic map, identification of genotype and distinguish between closely related species and cultivars. They are less influenced by the environment and are free from epistasis and various gene actions. RFLP, RAPD and AFLP are important techniques to construct genetic maps for rapid crop improvement.

Present Scenario of Plant Tissue Culture in India

Plant tissue culture has been promoted by government of India since last 2 decades to enhance the production land availability of disease free, true to type, quality planting material. At the end of 1996 about 76 commercial micro propagation units have registered for the production with government of India of which 30 are functional within the country. Total 40 million plants annually are produced against the installed capacity of 110 million plants.

Department of bio-technology, government of India has promoted the activity with the help of which two pilot plans were set up of the capacity or 1 million plants per annum of economically important plant species. National Chemical Laboratory, Pune (Maharashtra) is one of the units making extensive efforts for the development of cost effective indigenous technologies with felid evaluation. This unit is now upgraded to Micropropagation Technology Park to serve the entrepreneurs / industries institutes / universities so king for mass scale production of plants.

The potential for the domestic market is enormous and by conservative estimates it is around Rs. 150 cores with the annual growth rate of 10 per cent. In order to fulfill the demand of domestic market even at the meager lever of 10 to 15 per cent efforts needs to concentrate not only to increase the units at private sector but also the back up at national and state level from the government is necessary.

In Maharashtra there are 17 active public and private sector dealing with production of tissue cultured planets producing banana, sugarcane, arid ornamentals and orchids etc. Currently, the focus of the companies is mainly in the floriculture sector however, micro propagation in banana, sugarcane is also gaining popularity. Recently Bhabha Atomic Research Centre, Mumbai have also done initial work on micropropagation cardamom, mulberry and sandalwood. Tissue culture in ornamental plants constitute a major part of the total lancing material produced in the state out of 31.25 lakh ornamental and other plantlets, gerbera alone is a major flower crops produced in the state.

Potential of Tissue Culture in India

Flower industry is an emerging sector of economic importance in India. Floriculture has already established a place in economic development of farmers and private sectors of India. Some awareness has been created in use of micropropagation plants in the cultivation of other crops of horticultural importance. Existing laboratories are also producing flowering species. Since number of farmers for poly house horticulture is increasing annually there is a scope for various use of tissue culture planting material of flowering plants and vegetable crops. Tissue culture units involved in production of tissue culture seedlings of flowering species like gerbera, carnation, lily, arcadias etc. The number of green house units in India should be increased for production of floriculture plants. The other potential crops may also be propagated through tissue culture technique to enhance the production.

Suggestions for Improvement

1. Strategy should be framed to establishment, encourage and strengthen of tissue culture laboratories. For strengthening the existing units a grant may be given to the units so as to produce additional 25 per cent plants over the existing production capacity at a rate lower by at least Rs. 2.00 per plant. This assistance could be in the form of interest subsidy on the present interest burden. In order to make these units viable certain policy decision need to be taken up by the state government.
2. Collection and consolidation of information related to present scenario of micropropagation technology and to study the utility of this technology.
3. To evaluate the potential for increasing agriculture production with the help of tissue culture and to decide the strategy for next 5 years for use of this technology.
4. Establishment of tissue culture laboratories along with criteria regarding financial assistance.
5. To suggest extension plan for use of micropropagated plants by the farmers.
6. To suggest quality control measures for micropropagated plants.
7. To decide future the direction for research and development in the field of tissue culture considering needs of the state.
8. **Standardized package of practices:** Though the private sector has good tissue culture plants, but they are not ready with relevant standardized package of practices to meet field requirement raising field crop with tissue cultured plants. Lack of an adequate R&D infrastructure at unit level makes such type of recommendation impossible to private units arm an turn, grower has accepted to use tissue cultured plaids with traditional package of practices. This situation not only leads to wastage market purchased inputs, but also reduces per rupee net return. It needs improvement.
9. Tissue culture plants are charged Sales Tax. Although the producer has to pay the tax, the ultimate burden of this tax is shifted on the farmer. Sales tax on tissue culture plantlets should be waved at least for initial 10 years.
10. The electricity tariff should be brought at par with tariff for agriculture purpose.
11. The tissue culture units extending hardening facilities in captive areas should be given an incentive for potential paining. Sheds at the rate or 50 percent of erection cost.
12. For establishment of new unit in private sector, grant of Rs. 10 lakhs may be given for the production capacity of minimum 5 lakhs plant per annum.
13. The job it venture may be encouraged with the help of agriculture universities, on the government's lands.

Transgenic Produce to Evolve Global Marketplace

If and when genetically engineered (GE) horticultural products become more widely available and adopted, they will enter an expanding marketplace that is becoming globally integrated and more consolidated. GE produce must meet the challenges of the quickly evolving market for fruits and vegetables.

To date, the successful GM crop varieties have emphasized "input traits," related to reducing the use of chemical pesticides or making them more effective, rather than "output traits," related to product quality.

Postharvest Handling, Storage and Processing

Throughout the world postharvest crop losses are high. Many research programs are active in this field. Of course, postharvest is linked not only to growers but also to others involved in the distribution chain, although storage (physiology and maintaining crop quality) and packaging are paramount.

There are also alternative techniques to improve the healthiness of edible produce, or to reduce costs of storage and handling, that are part of this research. They involve the very large field of processing (with new techniques that make processed fruits competitive for quality and price with the fresh). For some developing countries processed produce is more important than fresh, also because it is better suited to long distance transport (market globalization).

Integrated Production Systems (IPSs)

Integrated production systems are a leading concept for developing counties and should be encouraged throughout the world for the realm of sustainable agriculture. The organization of the production process and related research can be included are (1) Crop management (training, pruning, thinning, etc.), (2) Crop protection, (3) Soil management, nutrient, etc. (4) Crop protection, (5) Soil management, nutrition, (5) Water supply and irrigation, (6) Ripening and picking and (7) Crop quality assessment.

Economic and Market Aspects

The countries in Asia, South America and part of Africa that have development trends of 3-5% for an array of produce commodities, a rate spread superior to the others and induced by increased domestic demand spurred by a growing economy. These countries are now ready to enter the export arena to compete for share in the global marketplace, especially countries like Chile, South Africa, Morocco, Egypt, Israel, India, and Thailand and so on.

The development of new technologies through R&D is only one element of the picture. The technologies must also be approved for commercial application and economically attractive enough to be adopted by farmers. The experience with other biotech crops has lessons for horticultural biotechnology. Biotech crops have been a commercial reality only for a few years but they have made very rapid inroads in some parts of the market. In particular, pest resistant and herbicide tolerant corn, soybeans, canola and cotton were rapidly developed and adopted in the United States and to a lesser extent in other countries. To date, the successful GM crop varieties have emphasized "input traits," related to reducing the use of chemical pesticides or making them more effective, rather than "output traits," related to product quality. Why has there been rapid development and adoption of GM cropping technologies for these crops and not other important crops, such as for many, (transgenic) crops represent the hope for a future with less hunger and malnutrition, and for a more sustainable agriculture with more varied, cheaper and safer food. For others they are cause for serious concern about the environment and food safety.

It is common knowledge that promotion and marketing are essential in the commercializing horticultural products. They are expensive and require specific organization of growers and supermarket/retail groups. They address three basic needs:

1. Establishing an 'image' (brand names and trademarks, labels, etc.) and building confidence in the consumer through education.
2. Providing a guarantee of produce to the consumer through education.

3. Providing a guarantee of the production process as related to food safety.

Role of Public and Private Agencies

The balance in agricultural R&D between the private and public sectors varies among types of research. Without government intervention, the rate of innovation will be too slow, reflecting both underinvestments in research and under adoption of some research results. Both problems are related to the nature of property rights for research results. "Free-rider problems" occur when property rights are incomplete, and private investors can capture only part of the returns to their investments in certain types of research (such as developing new crop varieties) as a result, their incentives to invest are reduced.

Intellectual Property Rights (IPRs)

Intellectual property rights (IPRs) are a double edged sword to the extent that they provide a greater incentive for investing in research they are also likely to result in lower adoption rates. Governments have addressed the incentive problems in agricultural research in several ways. Federal and state governments (as well as industry) have funded agricultural research at public institutions such as the ICAR and state agricultural experiment stations associated with land grant colleges. This approach allows an increase in total research without the problems associated with monopoly pricing of inventions. However, even though the investment has paid handsome dividends, it is increasingly difficult to sustain the past levels of funding for public agricultural R&D, in the face of general budget problems and declining political support for public science funding, including agricultural science. Governments have also acted to strengthen IPRs applied to plants and animals.

Extension Plan

The extension system today is partially successful in creating general awareness among farmers about benefits of hi-tech horticulture. For example for tissue culture the farmers are not clear about the entire package of technology. The presumption that tissue cultured seedlings itself will increase the yield needs to be addressed in effective manner. Entire package of practices required for getting optimum yield needs to be demonstrated effectively on the farmers fields in order to spread the proper message of benefits of micro propagation, extension machinery itself needs to be equipped and trained in this area. The entire extension program for bio technology has to be organized in such a manner that extension workers and the farmers get simultaneous training and could upgrade their skills for quality production from tissue culture plantlets. Agronomic practices for tissue culture plantlets have to be finalized through research and then have to propagate through extension machinery of the state.

Training

Intensive training program for extension staff may be organized first to train the master trainers and generate in louse resource persons. This training may be organized for identified persons in the institutes of repute. The trained manpower or resource persons should train the field functionaries so that the field level workers will also be able to develop technology. During the first year, there has to be a major thrust this training of staff. The farmers also need training in these aspects of biotechnologies. Private sector also needs to be involved in the process of farmers training. Training farmers should be mandatory on the part of entrepreneur for which 10 percent financial assistance on the amount of effective sale may be given as assistance from the government. There needs to be a regular interaction arid continuous dialogue between the developed agencies, extension workers and regular work shops, seminars, symposium needs to be organized for making farmers aware and also getting feed back from them.

Technology Demonstrations

Through there is awareness about use of tissue culture in agriculture there is hardly a demonstration on tissue culture technology exhibiting total package of practices. Extension machinery and the private sector will have to be aggressive in organizing demonstration on the farmer's fields. Listing out the technology generated in various crops will be the first stage in this process. Identification of user farmers will be the second stage.

Quality Control Mechanism

Although tissue culture laboratories have established in the state during last decade producing tissue culture plants for distribution to the farmers, there is no mechanism to check the quality of planting material produced by them. Since the quantum of tissue culture plant would be increasing at higher rate. The quality of plants will have to be ensured for getting desired type of quality planting material. It should be mandatory for all the tissue culture units to register themselves with the competent authority.

Need for Education Reform

Recently, the horticultural industry in India has grown to a great extent however, the academic side has to be improved. Academics are mostly limited to Indian agriculture. India is a country with diverse climatic conditions. Horticulture students have an opportunity to study tropical, subtropical, temperate and arid horticulture. This is the time to revise the horticultural curricula to include a global perspective. More emphasis needs to be given to horticultural marketing and economics. Farmers are able to produce crops using appropriate technologies if they are assured the market is there. Therefore, the objectives of horticultural education in India should be redefined to broaden the scope of the subject and should not be locale specific. This will help the students/growers to become globally competitive and will meet the requirements of employers from any part of the world. The horticulture industry should be encouraged to institute fellowships for bright students. An international student and faculty exchange program can be put into place to promote collaborative research projects.

The Indian Council of Agricultural Research (ICAR), New Delhi, has a strong network of research institutes and agricultural universities in India. Most of the agricultural universities offer courses in horticulture at the undergraduate and graduate level. The number of horticultural science departments in Indian universities is declining over the last 15 years. Students often fail to perceive horticulture as a science and certainly not as a career option. Horticultural science professionals around the world have to share their thoughts on this issue. The science of horticulture should now be linked with biochemistry, biotechnology and molecular biology. There is a need for initiation of a campaign for making students aware of biotechnology in relation to high-tech horticulture. The revision of course curricula for undergraduate and graduate programs is required to meet the changing needs of the industry. For developing countries, horticultural science or technology still lags behind, therefore we need to improve the higher education program at our universities.

If students fail to perceive horticulture as a science, they are correct. The fundamental task of scientists in horticulture is to contribute to this success. This can be achieved by improving the product or the methods of production by all available means a fascinating and never ending challenge. For this purpose, scientists must have a good overall knowledge of the crops they are dealing with, of their needs, their weaknesses and of the methods to better serve these needs and to overcome these weaknesses. The general approach of a scientist working in and for horticulture should be a holistic one, trying to understand the plant as a system, the function of this system and the specific manner of its interaction with the larger system of environment.

We Need

1. To promote an awareness in the general community and in particular primary and secondary school students, of the importance of horticulture and horticultural science to the economy of India and to the health, well-being and quality of life of Indians.
2. To review the quality of science teaching and the qualifications of science teachers, in graduate and post graduate level and to upgrade these, providing rewards, status and incentives for better qualifications, so that students will be instilled with a love of science and inspired to pursue a career in science, including horticultural science.
3. To coordinate the different levels and types of horticultural education to ensure that their individual objectives are defined to more effectively meet the training and knowledge needs of students by the efficient delivery of complete and appropriate courses, avoiding wasteful duplication.
4. To revise the horticultural curriculum to ensure that content is appropriate to current needs and that areas such as processing, marketing and value adding are adequately catered for.
5. To review the accessibility of education and training options in horticulture to ensure that industry participants have the opportunity to acquire and upgrade their skills and knowledge through appropriate on going educational opportunities.

Future Research and Development

Equally important as the research initiatives themselves are the role they play in determining the balance of disciplinary expertise in modern horticulture departments.

At present the cost of tissue culture planted are quite high due to how multiplication rates in many protocol efforts need to take place in this field. Agronomic practices commensurable with tissue culture technology should be finalized. A new protocol should be established to propagate horticultural importance crops which have tremendous potential for India, e.g., medicinal and aromatic plants, spices, mango and cashewnut for each state and agroclimatic conditions.

Horticulture shall have a growing commercial importance in the near future as world population is aging, people devote more attention to health and most communities better emphasize the quality of life and environment. Development and adequate share of new discoveries across the scientific horticulture community through adequate training of horticulture specialists at all educational levels shall insure bright specialists at all educational levels shall insure a bright future to horticulture and the humanity. Networking of world horticulture scientists working on the same commodities, disciplines, geographical areas, theme of research, etc shall help to support such development.

Following are the major need to make horticulture – hi-tech horticulture:

1. Basic versus applied research.
2. The needs for research on biodiversity and molecular genetics.
3. The growing importance of biotechnologies.
4. Organic or agrochemical growing systems.
5. Promoting biological control of horticultural crop predators.
6. Quality of horticultural products.
7. Human nutrition and nutraceuticals development.
8. Globalization and distribution concentration.
9. Evolution of consumer needs.
10. Urban horticulture and environmental quality.
11. Internationalization of training and availability of knowledge.
12. Characterization of the impact of fruit and vegetable consumption on human health and well being, with correlative attention focused on plant genetic modification, specialized cropping systems, postharvest physiology and technology, food safety and public education to maximize the beneficial impact of fruit and vegetable consumption.
13. More attention to the field of "environmental horticulture" in its broadest dimensions, including study of the psycho-social value of human plant interactions. Investigations should include landscape ecology on the micro and macro scale, restoration biology and the environmentally sensitive management of home and urban landscapes. Identification and development of plants resistant to common pests to decrease use of chemicals by consumers will be important.
14. Ornamental horticulture, including nursery and greenhouse crop production, floristry, landscape design, construction and maintenance, has been overlooked entirely by national competitive programs, except for minor elements that focus on integrated pest management or invasive species. Ornamental horticulture is the most profitable and rapidly expanding sector of plant agriculture.
15. Sustainable cropping systems encompass many horticultural crops and small to medium sized operations look to fruits and vegetables for economic success because of their high value per unit of land and resource inputs. These systems are economically feasible options for re-directing the "tobacco economy" into more socially acceptable production alternatives. The search for alternative crops and new uses for current crops should be accelerated. Increased attention should be focused on the needs of small full and part time growers, as the sustainable systems of the future likely will involve farming as supplementary rather than primary income.
16. The emerging field of Controlled Environment Agriculture (CEA) increasingly will produce our salad, vegetable and specialty crops as the twenty first century progresses. Coupling this high-tech, "clean" horticulture with bio-derived, renewable energy sources will be the key to developing a profitable CEA growth industry that is sustainable and environment friendly.
17. Need for high density plantations, which include adoption of appropriate plant density/canopy management, quality planting material, support and management system with appropriate inputs.
18. Micro propagation (Tissue Culture), for mass production of true-to type, in possible shortest time and disease-free plantlets.
19. Hi-tech cultivation under controlled climatic conditions i.e. in poly houses, green houses, net-houses, etc. Rainfed production through efficient water management techniques, mulching for soil moisture conservation, use of barriers in soil to reduce percolation, irrigation by drip, sprinklers, fertigation and water harvesting

structures etc.

20. Nursery management for quality seed/plant production of vegetables, flowers, ornamentals, fruit etc.
21. Hybrid Seed production.
22. Organic farming.
23. Hydroponics for year round quality production.
24. Use of plastics in horticulture.
25. Bio-technology.
26. Genetically modified organisms (GMOs).
27. Development of infrastructure for production, post-harvest handling, processing and marketing.
28. Development of markets and introduction of new primary processing of products.
29. Development of horticulture ancillary industry for improved packaging, equipments, plastics, corrugated boxes, horticulture machinery/tools, etc.

Future Needs for Production and Post Harvest Management

1. To develop high quality horticultural farms in identified belts and make such areas vibrant with horticultural activity which in turn will act as hubs for developing commercial horticulture by adopting high-tech horticulture techniques.
2. To develop post-harvest management infrastructure.
3. To improve linkages between horticulture producers and marketers.
4. To create integrated network for marketing of horticulture produce.
5. To increase producer's share in consumer price.
6. To encourage networking of schemes for resource mobilization with all other related agencies/organizations both of government, financial institutions and private agencies engaged in the field of horticulture promotion in the country.

Production Related

- High quality commercial horticulture crops
- Indigenous crops/ produce, herbs
- Aromatic plants
- Seed and Nursery
- Bio-technology, Tissue culture
- Bio-pesticides
- Organic foods
- Establishment of Horti-health clinics/laboratory
- Consultancy services
- Beekeeping

Primary Processing Related

- Grading/packing/washing/waxing/sorting/drying centres
- Pre-cooling unit/cool stores
- Reefer van/containers (with multi-chamber, multi-product facility)
- Specialized Transport Vehicle
- Retail outlets
- Auction platform
- Ripening curing chamber
- Market yards/rope ways
- Radiation unit/dehydration unit/vapour heat treatment unit

- Primary processing of products fermentation, extraction, distillation, juice vending pulping, dressing cutting, chopping etc.
- Horticulture ancillary industry e.g. tools, equipment, plastics, packaging, etc.
- Plastic crates, cartons, baskets, aseptic packaging and nets (50% subsidy).

Priority Areas

- Export oriented units/projects
- Project from cooperative sector
- Projects in North Eastern Region, Hilly and Tribal areas
- Introduction of new processes, products or markets including new technology / equipment
- Project involving women entrepreneurs and ex-serviceman

CHAPTER 2

Nature, Scope and Progress of Horticulture in India

The term horticulture is derived from two Latin words, *hortus*, meaning a garden, and *cultura*, meaning cultivation. It may be described as the practice of growing and manipulating plants in a relatively intensive manner. In ancient times, the gardens of large estates were surrounded by high walls or similar structures and the crops usually cultivated within these areas were tree fruits such as apple, pear, peach, date, pomegranate, fig, and/or many vegetables, flower and ornamental plants. As a result, the term horticulture in the original sense referred to the cultivation of crops within a protected enclosure. Quite frequently this area was called a garden. Thus, horticulture referred to the culture of crops grown in gardens.

At present, fruits, vegetables, flowers and ornamentals are grown not only within the home grounds, but also in large quantities on a commercial scale. In other words, many horticulture crops throughout the world are grown primarily as business enterprises. Thus, horticulture has at least two aspects: the amateur and the professional. In addition, with the establishment of agriculture experiment stations as an integral part of colleges and universities, combined with the use of the scientific method in the solutions of problems, a comparatively large amount of scientific information on the behaviour of horticultural plants and products has become available. This information or literature, constitutes the science of horticulture and the application of this scientific information constitutes the technology of horticulture. Thus, present day horticulture may be defined as the science and technology involved in the production, processing and merchandising of fruits, vegetables, flowers and ornamentals.

Horticulture is a wide field and includes a great variety and diversity of crops. In general, these crops are classified as follows:

1. Crops grown for the food and/or vitamin and mineral content, for example, apple and sweet potato,
2. Crops grown for their beverage properties, for example tea and coffee.
3. Crops grown to beautify the environment for example the grasses of a well kept lawn or to express human sentiments, for example cut roses and miscellaneous group, including crops grown for perfume or spices.
4. Each crop has its own science and technology. Consequently, horticulture may mean one thing to one person and another thing to another. To the teacher, research worker, extension specialist or landscape gardener, it is an applied science and profession, to the commercial producer, the wholesaler or the retailer of horticulture products or to the manufactures or seller of horticulture supplies, it is a business and to the home owner, it is a fascinating hobby.

PREVALENCE OF HORTICULTURE IN INDIA

The horticulture sector has assumed paramount importance in the economic development of the Indian Farmer and is contributing over 29.5 percent to agricultural GDP with just only 8.5 percent area under horticulture crops.

To remain self sufficient in food grains, we need another green revolution or rather, a greener, revolution. Innovative agro practices need be adopted towards transformation of Indian agriculture to precision farming practices which will result in stretching our agro input resources main-fold to increase agricultural productivity both in quality and quantity.

All aspects of horticulture are prevalent in India, i.e. the production of fruits, vegetables, potato, tubers, ornamentals, medicinal and aromatic plants, spices, plantation crops and mushroom. Both in fruits and vegetables, India is the second largest producing country in the world.

Temperate fruits are grown in the north Himalayan region while subtropical and tropical fruits, vegetables, ornamentals, mushroom, spices are cultivated in the rest of India. Spices and plantation crops are found in peninsular region. Arid zone crops are grown in western India.

TABLE 2.1: Area, production and productivity of different horticultural crops in India

(Area in 000 ha., Production in 000 MT, and productivity =MT/ha.)

Crops	2006-07			2007-08			2008-09		
	Area	Production	Pdy.	Area	Production	Pdy.	Area	Production	Pdy.
Vegetable	7581	114993	15.2	7848	128449	16.4	7981	129077	16.2
Fruits	5554	59563	10.7	5857	65587	11.2	6101	68466	11.2
Plantation Crops	3207	12007	3.7	3190	11300	3.5	3217	11336	3.5
Spices	2448	3953	1.6	2617	4357	1.7	2629	4145	1.6
Flowers*	144	880	6.1	166	868	5.2	167	987	5.9
Aromatic & Med. Plants	324	178	0.5	397	369	1.0	430	430	1.0
Almonds & Walnut	132	150	1.1	132	177	1.3	136	173	13
Mushroom	-	37	-	-	37	-	-	37	-
Honey	-	51	-	-	65	-	-	65	-
Total	**19389**	**191813**	**9.9**	**20207**	**211234**	**10.5**	**20661**	**214716**	**10.4**

Source: National Horticulture Board, (Website: www.nhb.gov.in), Database-2009, page-4
* Flower production figures for loose flowers only
Note: Total may slightly differ due to rounding of figures.

Geography

The geological regions may be grouped into three regions, i.e., the Himalayas and their associated group of mountains, the Indo-gangatic plain and the peninsular shield. Himalayan region has temperate climate, Indo-gangatic plain, sub-tropical and peninsular region tropical climate. The soils are predominantly sandy loam, alluvial, black cotton soil and red laterite.

Climate

Almost all types of climatic conditions exist in India. In the North-West and North-East Himalayan region, temperate climate prevails, thereby, occurrence of snow fall and minimum temperature of -20°C. In the West, desert exists denoted by sand dunes, very low rainfall and arid vegetation. In the rest of India, the climate is tropical monsoon type with four seasons: (i) winter (January-February), (ii) hot weather summer (March-May), (iii) rain (June-September) and (iv) Autumn or Post monsoon (October-December). The climate of the country is affected by two seasonal winds, the North-East monsoon and the South-West monsoon. Thus, all types of horticultural crops are grown in the country.

SCOPE, PRODUCTION AND RESEARCH DEVELOPMENT OF HORTICULTURAL CROPS IN INDIA

Horticultural crops play a unique role in India's economy by improving the income of the rural people. Cultivation of these crops is labour intensive and as such they generate lot of employment opportunities for the rural population. Fruits and vegetables are also rich source of vitamins, minerals, proteins, carbohydrates, etc. which are essential in human nutrition. Hence, these are referred to as protective foods and assumed great importance as nutritional security of the people. Thus, cultivation of horticultural crops plays a vital role in the prosperity of a nation and is directly linked with the health and happiness of the people.

India has a good natural resource base, an adequate R&D infrastructure and excellence in several areas. As a result, the horticultural scenario of the country has been changing fast. Both production and productivity of several crops has increased manifold and India can boast itself as a leading horticultural country of the world.

The horticulture scenario of the country is rapidly changing. The production and productivity of horticultural crops have increased manifold. According to S. P. Ghosh, Former Deputy Director General (Horticulture), Indian Council of Agriculture Research, New Delhi, 110001 (*J. Applied Hort., 1(1):64-69, January, 1999*), a major shift in consumption pattern of fresh and processed fruits and vegetables is expected in the coming century. There will be greater technology adoption both in traditional horticultural enterprise as well as in commercial horticulture sectors. Diversification and value addition will be the key words in the Indian horticulture in the 21st century.

The recent emphasis on horticulture in our country consequent to the recognition of the need for attaining nutrition security and for more profitable land use, has brought about a significant change in the outlook of the growers. The need for great utilization of available wastelands against the background of dwindling water and energy resources has focused attention to dry land, to arid and semi-arid tracts and to horticultural crops which have lesser demands on water and other inputs besides being 3 to 4 times more remunerative than field crops.

It is estimated that India has 240 million acres of cultivable wasteland, which is lying idle, which can be brought under orchard crops without curtailing the area under food crops. The country has abundant sunshine throughout year, surplus labour and widely varied agro-climatic conditions, offer high potential for successful and profitable commercial horticulture.

Under technology generation programme various steps have been promoted like varietal development, seed/planting material propagation, agro-techniques, protection technologies, post harvest management technologies and assessment of production constraints.

He further indicated that horticulture research in India is about four decades old.

Systematic research on fruit, vegetable and ornamental crops began in 1954 with the initiation of independent institutions and programmes. The research agenda is designed relevant to national plans and priorities for the horticulture development. Today, eight ICAR institutes are also involved in horticultural research. All the State Agricultural Universities (35) have departments and faculties in horticulture which offer undergraduate and postgraduate specialization in horticulture and carry out the research on local horticultural crops. There is one university exclusively on horticulture and forestry at Solan. A few private seed companies also have their own research activities. Information on horticulture generated through research is published in various journals issued by societies of various horticultural commodities and research reports of various institutes.

In addition, a few R&D establishments of crop/commodity boards and private sectors are providing research support to Indian horticulture. Research system in horticulture is now geared to provide necessary technological support to the expanding horticultural industry.

Under research strategies and programmes the important fields are rationalization of research, safeguard for intellectual property rights (IPR), genetic resource management, horticultural biotechnology, integrated production system, export promotional research, quality planting material production, post harvest management and human resource development.

Thrust Areas

1. Development of improved varieties/hybrids of fruits, vegetables, plantation crops, medicinal and aromatic crops, flowers and ornamental crops, spices, cashew, oilpalm with high production potential, biotic and abiotic resistance and export value.
2. Standardization of techniques for rapid propagation of planting materials, agro-techniques, water management, integrated plant nutrient management system and integrated disease and pest management for horticultural crops.
3. Evolution of appropriate horticultural based cropping systems for different agroclimatic areas.
4. Development of post harvest handling, storage and processing system, product diversification and value addition.
5. Utilization of the waste, infertile, eroded land, unfit for the production of other crops asks for little improvement in the existing commercial cultivars of fruits.
6. Regulation of the market to avoid glut through staggered production through raising varieties of different maturity periods.
7. To improve productivity of fruit crops. Most of the fruit crops lack ideal cultivars for intended use and are faced with one or the other production or post harvest constraints.
8. Extension of the fruit crops to the non-traditional area e.g. temperate fruit in sub-tropical area and sub-tropical to tropical and *vice versa*

Achievements

1. India has emerged as the second largest producer of fresh fruits and vegetables in the world.
2. India is presently the largest exporter of spices and cashew.
3. A total of 460 high yielding varieties and hybrids of horticultural crops developed. As a result, productivity of banana and potato has gone up three times each and cassava two times.
4. Regular bearing mango hybrid, export quality grapes, multi-disease resistant

vegetable hybrids, high value spices and tuber crops of industrial use developed.
5. Substantial increase in production of banana plants through tissue culture, use of drip irrigation, chemical regulation of mango flowering and high density orcharding in fruit crops.
6. True potato seed (TPS) technology standardized.
7. Low cost environment friendly cool chamber for on-farm storage of fruits and vegetables developed.

Research Achievements

According to Dr. K. L. Chadha (2006) following are the research achievements in various fields of horticulture:

Introduction and Cultivation of New Crops

1. Kiwi fruit in sub-mountain areas of North India
2. Olive in mid hills of North Western Himalayas
3. Low chilling stone fruits in the North Western plains
4. Oil palm in coastal states of Karnataka, Andhra Pradesh, *etc.*
5. Gherkin in south and west India
6. Baby corn and sweet corn in certain specific pockets, and
7. Broccoli, brussels' sprouts, asparagus, celery, parsley near the cities.

Crop Improvement

1. A large number of high yielding varieties developed in several horticultural crops *e.g.* fruits (76), vegetables (160), potato (29), other tuber crops (24), ornamental crops (300), palms (20), spices (51), cashew (33) and betel vine (1).
2. First seedless variety of mango developed.
3. 40 F_1 hybrids developed in brinjal, tomato, chillies, cauliflower, carrot, capsicum and muskmelon.
4. Self incompatible lines in cauliflower, gynodioecious lines in cucumber and muskmelon, genetic male sterile lines in tomato and temperature tolerant strains of button mushroom developed.

Propagation of Quality Planting Material

1. Standardized propagation technique for many fruits hitherto propagated by seed. *e.g., aonla, bael, ber,* black pepper, cardamom, cashew, cassia, cinnamon, clove, custard apple, jack fruit, *jamun,* nutmeg, sapota and walnut.
2. Standardized Seed Plot Technique resulting in successful disease free potato seed production in the tropics and sub tropics of the country. Standardized method of micro-propagation and *in vitro* micro-tuber production in potato.
3. Identification of suitable parental lines for production of True Potato Seed (TPS) and standardized technology for raising commercial crops.
4. Micropropagation protocols developed in banana, black pepper, betel vine, cardamom, ginger and turmeric.
5. Production of coconut hybrids through establishment of Seed gardens of Tall (T) x Dwarf (D) and D x T hybrids.
6. Standardized rootstocks in citrus, grape and apple.

Agro-Techniques

1. Standardized high density plantations in banana, citrus, mango and pineapple

and high production technology in several crops *e.g.*, pineapple, black pepper and cardamom.
2. Year round production technology in tomato and "off season" cultivation of onion and cauliflower developed.
3. Areca nut, coconut and potato based cropping systems developed to maximize productivity under high management conditions.
4. Standardized use of several plant growth regulators and chemicals now commercially employed in production and quality improvement of horticultural crops *e.g.*, paclobutrazol for induction of flowering in mango; gibberellic acid for improving berry size and quality in grape; Maleic hydrazide for preventing sprouting in onion and potato; Dormex for hastening bud burst in grapes and Boron and Calcium for changing flower cycle in some cucurbits.

Crop Protection

1. Developed improved disease detection techniques such as ELISA and ISEM for improving seed quality and tissue culture technique for rapid multiplication of potato.
2. Developed IPM for fruit borer in brinjal, diamond back moth in cabbage, thrips in chillies, phytophthora foot rot in black pepper, "Katte", rhizome rot in cardamom, rhizome rot in ginger, late blight and bacterial wilt in potato.
3. Developed biological control measures for mealy bug in grape, fruit borer in tomato and okra.

Post Harvest Management

1. Preharvest treatments to control post harvest losses in citrus, mango and grape standardized.
2. Maturity standards for mango, guava, grape, litchi and ber standardized.
3. Chemical treatment for regulation of ripening in mango, sapota and banana standardized.

Optimum storage temperatures worked out for several fruits, vegetables and tuber crops.

1. A mango harvester, fruit peeler, hand and pedal operator cassava chipping machine, harvesting tools (5-14 times efficient). Implements for mechanization of potato cultivation *e.g.*, oscillating tray type potato grader, fertilizer application cum line marker, potato culti-ridger, soil crust breakers, potato digger and automatic potato planter/diggers developed.
2. Low cost environment friendly storage system for fruits, vegetables, potato and onion developed.

Needs and Expectations

Though India has made significant achievements in the field of horticulture production, however, still there are various challenges to be faced:

1. Inadequate supply of quality planting material,
2. Heavy losses caused by several biotic and abiotic stresses, and
3. Several unresolved chronic disorders.

In India, the productivity per unit area is low. This results in high cost of production. The farmers are not getting high quality of produce. There is a high level of post harvest losses. The advantages of various techniques like biotechnology, protected cultivation, computer aided management of inputs, integrated nutrient management, leaf nutrient standards, biofertilizers, integrated pest management and mycorrhiza, etc. have not reached to laboratory to land. The involvement and partnership with private sector in various fields like production of hybrids, green house production of flowers, biotechnology, value addition and export is still needed. The horticulturist and farmers should largely depend on new and globally competitive technologies. As such, ambitious research programme is called for in horticultural crops in the following thrust areas.

Genetic Resources

1. Introduce fruits like mangosteen, durian, rambutan, longan, macadamia and berries not yet commercially exploited in India.
2. Widen genetic base in mango (*Mangifera* species of South East Asia), Citrus (newly developed rootstocks), apple (scab resistant cultivars), guava (coloured varieties), papaya (species and varieties) and litchi (varieties from South East Asia and China). Also, widen genetic base for improvement of medicinal and aromatic plants required by pharmaceutical industries.
3. Enrich germplasm collections with species/ cultivars which are high yielding and resistant/tolerant to different biotic and abiotic stresses.
4. Exploit full potential for cultivation of avocado, kiwi and olive.
5. Standardize long term techniques for cryopreservation of propagating materials and pollen grains.
6. *In situ* conservation of endangered genetic wealth.

Crop Improvement

1. Develop dwarf rootstocks/ scion varieties for high density planting and export in mango, litchi, sapota, citrus, *ber,* coconut, areca nut, oil palm and cashewnut.
2. Induce and exploit useful genetic changes through mutations in commercial cultivars.
3. Develop rootstocks and scion varieties in fruits resistant/tolerant to major biotic and abiotic stresses, *e.g.* malformation in mango, guava rootstocks for wilt, citrus rootstocks against phytophthora, root wilt resistance in coconut, frost and PRV resistance in papaya, weevil in sweet potato, cassava mosaic and stress tolerance in areca nut and spices.
4. Develop F_1 hybrids in vegetable crops for yield, nutritional quality and export.
5. Development of virus resistant potato varieties having high tuber dry matter and low accumulation of sugars during low temperature storage and varieties with high temperature tolerance.
6. Evolving varieties with bold nuts and higher shelling percentage in cashew.
7. Domestication of indigenous medicinal plants with bulk demands both in the modern and traditional systems of medicine.
8. Studies on the quality control, seed health, packing and storage of seeds of annual crops.

Crop Production

1. Standardize rootstocks for all important fruit crops *e.g.* mango, guava, litchi, sapota, *ber* and walnut.
2. Develop horticultural crop based cropping systems for different agro-ecological regions.
3. Develop integrated nutrient management system *i.e.* efficient utilization of

chemical fertilizers, use of bio-fertilizers and organic material using leaf nutrient standards.

4. Standardize water management practices in major crops including microirrigation and fertigation.
5. Develop techniques of organic farming for export oriented horticultural crops.
6. Standardize practices for production of horticulture crops particularly flowers and vegetables under protected cultivation involving computer aided supply of inputs.
7. Standardize production technology for quality crops for export *e.g.* mango, grape, litchi, cashew, potato and spices.

Crop Protection

1. Develop IPM strategies for important pests of commercial crops.
2. Develop biological control for important diseases and insects affecting commercial crop production.

Post Harvest Management

1. Reduce losses occurring during harvesting, storage and transport and improve shelf life of perishable horticultural commodities.
2. Conduct basic and applied research in CA/ modified atmospheric storage of high value perishable crops.
3. Improve indigenous low cost storage systems developed for onion, potato, fruits and vegetable crops.
4. Develop techniques for bulk preservation of fruit pulps, improve drying systems for raisins, mushroom and different vegetables, and standardize technique for frozen vegetables.
5. Initiate post harvest research on ornamentals with special emphasis on export market.
6. Utilize wastes for development of economically viable products.
7. Develop/fabricate low energy requiring machinery for horticultural crop production.

Biotechnology

1. To standardize *in vitro* culture techniques for mass multiplication of rootstock/ scion of difficult to micro propagate plants like mango, guava, sapota, litchi, walnut, date palm, coconut and oil palm.
2. Development of protocols for rapid propagation of seedless water melon; vegetable crops with male sterility and self incompatibility barriers *e.g.*, tomato and capsicum and selected ornamental crops.
3. Use of anther culture system for production of di-haploids for integration with breeding programmes.
4. To identify molecular markers based on RAPD and RFLP in important varieties of horticultural crops.
5. Develop transgenic plants with endogenous resistance to insect pests *e.g.* bacterial canker in kagzi lime, salt and drought tolerance in tomato and capsicum.

Constraints to Horticultural Production

Horticulture commodities are predominantly produced by small to medium scale farmers. For example, smallholder farmers in India generate less income. The major constraints facing horticulture smallholder farmers include:

1. High inputs costs.
2. Insecurity.
3. Poor building, farm and road infrastructure.
4. Inadequate extension support services
5. Limited awareness on market access or market standards.
6. Poor and inadequate rural infrastructure especially feeder access roads, power/ electricity, telecommunication and market facilities in major urban centers.
7. Lack of access to affordable credit.
8. High cost of agricultural inputs that include seeds, planning materials, fertilizers and other agrochemicals.
9. Reduced government support to extension.
10. High air freight rates to our traditional markets in Europe and other countries.
11. Strict food safety, traceability and good agricultural practice requirements as stipulated in the certification requirement.
12. Lack or poor dissemination of local and external market information.
13. Increasing competition from imports in to the local market.
14. Poor quality agricultural inputs like seeds and appropriate technologies for smallholders
15. Inadequate access to new varieties of crops, information on market demands and other market information.
16. Several chronic production problems due to major disorders such as mango malformation, alternate bearing, spongy tissue in mango, bunchy top of banana, guava wilt, citrus decline, root wilt in coconut, yellow leaf disease of areca nut, etc.
17. Lake of linkages between production, marketing and processing.
18. Inadequate research support particularly location specific field oriented problems.
19. Inadequate or lack of assessment of successful stories in the field of production, post harvest handling, marketing and processing.

Those smallholder farmers producing for export face additional challenges such as:

1. Inaccurate trade data
2. Limited compliance to regulatory standards i.e. Good agricultural practices or trade standards
3. Food miles
4. Increasing freight charges

Production

- Poor dissemination and adaptation of the available technologies
- Lower tree density, gapes and wastage of space
- Poor tree canopy management
- Inefficient water utilization in irrigation areas
- Inadequate nutrition
- Lake of integrated pest management
- Inadequate credit and data base
- Poor mechanization of fruit production

Post Harvest Handling and Marketing

- Acute shortage of infrastructure and manpower for post harvest handling,

processing and marketing of produce
- Wide gap in technologies available and in vogue
- Inadequate post harvest quality control
- Unorganized marketing. The produce handling and marketing is unscientific and more middlemen than producer and consumer oriented.
- Absence of pre-cooling and cold storage
- Inadequate market facilities, market intelligence and market information service (MIS)

Processing

- Absence of linkages between the processor and grower
- Poor consumption base and sale of the products
- High duties and taxes
- Non availability of raw material specially suited to the industry.
- Lack of incentives for the use of horticulture produce in processing as against synthetic products, which are cheap.
- Inadequate sales and publicity campaigns.

Fruits

India is the second largest producer of fruits (68.47 million tones) in the world. However with its growing population it is in the position to meet the recommended per capita consumption norms. There is vast scope of growing fruit crops in India, because the total area under fruit crops is very small, about 61.01 lac ha is being cultivated.

The fruit industry in India has made remarkable progress during last 4 decades (1961-2008). The area under fruits has increased from 1.22 million ha in 1961 to 6.10 million ha in 2008-09. According to Indian Horticulture Database-2009, India ranks second after China by producing 68.47 million tones, the India's share in the total world production of fruits being 8 per cent. India has been exporting fresh fruits since several decades. However, in the more recent past export promotional activities for fresh fruits received greater attention. The fresh fruits and vegetables exports which were only 27,028 tones valued at Rs. 17.4 million in 1983-84 rose to 2646267.78 tonnes valued at Rs. 36591.55 million during 2008-09. (*Source:* National Horticulture Board, Database - 2009, page-212).

Area under fruits increased from 1.22 million hectares in 1961 to 6.10 million hectares in 2008-09. India with a production of 68.47 million tonnes (2008-09) is the second largest fruit producer next to China (110.42 million tonnes) with a share of 8% in world fruit production. India produces 65% and 11% of world's mango and banana, respectively, ranking first in the production of both the crops. It has the highest productivity in grape in the world. Significant expansion has taken up in apple, *aonla, ber,* and pomegranate and sapota cultivation.

The majority of Indian population is vegetarian and the production is far less. There is a great demand from all the classes of people for fresh fruits and also for cained fruit products because fruits are the rich source of minerals and vitamins. The climatic condition of India is quite suitable for growing various types of fruits in various zones, throughout the year. Certain fruit crops are well adapted to varied edaphic and climatic conditions, e.g., datepalm, ber, pomegranate, guava, lasora, phalsa, etc. The productive life of fruits is quite long, however, banana and papaya have short productive life. Comprehensive economic survey reveals that numbers of fruit based agro-industries have been established in past years because they are highly remunerative.

TABLE 2.3: Some major horticultural products

Tree fruit	Orange, lemon, lime, mandarin, grapefruit, apple, mango, banana, guava, soursop, lichee, peach, apricot, pear, plum, rambutan, fig, quince, persimmon, durian, chichu, pawpaw, pomegranate, mangosteen, loquat, carambola, cherimoya, cherry, date, mulberry.
Vine fruit	Grape, passion fruit, kiwi fruit (or Chinese gooseberry).
Other fruit	Strawberry, pineapple, Cape gooseberry, watermelon, sweet melon, raspberry, blackberry, blackcurrant, gooseberry, cranberry, blueberry, rhubarb, loganberry.
Tree nuts	Cashew, walnut, hazelnut, macadamia, pistachio, pecan, coconut, almond, Brazil.
Fruit that are normally considered as vegetables	Breadfruit, avocado, tomato, egg plant (brinjal or aubergines) hot pepper, sweet pepper, karella (or bitter gourd), squash, marrow, gourd, cucumber, luffa, pumpkin, plantain, christophine or choyote.
Vegetables derived from seeds and flowers	Broccoli, cauliflower, artichoke, pea, bean, lentil, chickpea, broad bean, okra, mangetout pea, asparagus pea, yardlong bean, sweetcorn.
Leaf and stem vegetables	Lettuce, cabbage, spinach, chard, brussels sprout, endive, watercress, celery, asparagus, celeriac, green onion, leek, amaranthus, bean sprout, bok choy, Chinese cabbage, Chinese celery, spinach, chicory, kohlrabi, fennel.
Root vegetables	Arrowhead, onion, potato, sweet potato, cassava, yam, taro, garlic, radish, carrot, turnip, parsnip, beetroot, Jerusalem artichoke, dasheen, eddoe.
Herbs	Parsley, mint, coriander, dill, basil, rosemary, thyme, sage.
Spices	Black pepper, chili pepper, cardamon, ginger, clove, cinnamon, bay leaf, turmeric.
Cut flowers	Rose, chrysanthemum, carnation, gladiolus, tulip, narcissus, orchid.
Cut foliage	Asparagus tern, leather feat fern, soft ruscus.
Tropical plants for house and garden	Dieffenbachia, coleus, yucca, cordyline, dracaena, monstera, fatshedera, ficus, maranta.
Temperate plants for garden	Roses, ornamental shrubs, herbaceous flowers, bedding plants, conifers, flowering bulbs.

Apart from above advantages, fruit trees provide plantation cover for ecological balance improvement. Many fruits are also used for medicinal purposes.

Fruits such as mango, banana, citrus, guava and papaya are cultivated throughout the country and others in certain pockets. For instance apple is mainly confined to the temperate and certain extent sub-tropical zones and pineapple to heavy rainfall areas.

However, India's share in global exports of horticulture production is negligible being only one per cent. But given the volume of world trade in fresh and processed horticulture products and the country's resources, India has the potential of more than doubling the present level of exports within the next five years. Currently, potential technologies have been largely directed towards higher production with little consideration for quality which has to meet export standards. Thus emphasis is required on export directed production technology.

TABLE 2.4: Area, production and productivity of major fruits crops in India

(Area in 000 ha., Production in 000 MT, and productivity =MT/ha.)

Crops	2006-07			2007-08			2008-09		
	Area	Production	Pdy.	Area	Production	Pdy.	Area	Production	Pdy.
Mango	2154	13734	6.4	2201	13997	6.4	2303	12750	5.5
Apple	252	1624	6.5	264	2001	7.6	274	1966	7.2
Banana	604	20998	34.8	658	23823	36.2	709	26217	37.0
Citrus	798	7145	9.0	867	8015	9.2	923	8608	9.3
Guava	176	1831	10.4	179	1981	11.1	204	2270	11.1
Grapes	65	1685	26.0	68	1735	25.4	80	1878	23.6
Litchi	65	403	6.2	69	418	6.1	72	423	5.9
Papaya	72	2482	34.4	83	2909	35.1	98	3629	37.1
Pineapple	87	1362	15.7	80	1245	15.5	84	1341	16.0
Pomegranate	117	840	7.2	124	884	7.2	109	807	7.4
Sapota	149	1216	8.2	152	1258	8.3	156	1368	8.4
Other	1016	6244	6.1	1112	7321	6.6	1083	7249	7.1
Total	**5554**	**59563**	**10.7**	**5857**	**65587**	**11.2**	**6101**	**68466**	**11.2**

Source: National Horticulture Board, (Website: www.nhb.gov.in), Database-2009, page-5

The Government of India is too keen to give much greater attention to the development of horticulture. This is indicated from the fact that plan allocation for horticulture sector has received a quantum jump from little over Rs 1000 crore in the Eighth Five-year Plan to almost its double in the Ninth Plan.

In the Ninth Five Year Plan, a modest beginning will be made with an ambitious programme aimed at covering over two lac hectares under drip irrigation. 35,000 hectares under mulching and setting up of over 16,000 green houses. The programme would also have a major component of transfer of technology through intensive training and extension activities and provision of efficiency after sales service network.

So far low productivity, lack of infrastructural facilities, inadequate post harvest management arrangements and over dependence on conventional items were among the important reasons for poor export performance. An Export Enhancement Programme for horticulture produce has now been chalked out. It envisages investment in infrastructural facilities with special reference to post harvest management techniques, encouraging export oriented production and export promotion market development measures.

TABLE 2.5: Yield potential of fruit crops (t/ha)

Crop	Present yield	Potential
Mango	5.5	15
Banana	37.0	58
Citrus	9.3	30
Apple	7.2	30
Guava	11.1	20

Crop	Present yield	Potential
Pineapple	16.0	85
Sapota	8.4	80
Papaya	37.1	80
Grapes	23.6	25
Ber	5.00	20
Passion fruit	4.50	15
Fruits	11.20	42

Source: National Horticulture Board, (Website: www.nhb.gov.in), Database-2009, page-5.

TABLE 2.6: Major fruit producing countries in the world (2008-09)

Country	Area in ha.	Production in mt.	Productivity mt/ha
China	11042621	107837629	9.8
India	6100900	68465530	11.2
Brazil	2455867	38988300	15.9
USA	1160595	28202514	24.3
Italy	1243225	17653457	14.2
Mexico	1201897	16122211	13.4
Indonesia	720260	15917953	22.1
Spain	1895782	15835000	8.4
Philippines	1118530	15420664	13.8
Iran	1354381	13604303	10.0
Turkey	1094357	12824806	11.7
Uganda	1822590	10038600	5.5
Others	24247046	217285895	9.0
World	**55458051**	**578196862**	

Source: National Horticulture Board, (Website: www.nhb.gov.in), Database-2009, page-233.

Vegetables

Vegetable growing is an important branch of horticulture. India has shown tremendous progress in the increase of vegetable production after independence and is the second largest vegetable producer of the world, only next to China. Vegetable production has increased three times during the last 50 years. A large area is now covered with F_1 hybrids in vegetable crops resulting in increased yield and better socio-economic status of farmers. Vegetables like tomato, cabbage, cauliflower, radish and onion are now produced almost round the year. India has attained self sufficiency in seed production of temperate vegetables.

In India, out of 187.4 million ha of gross cultivated area about 7.98 million ha are under vegetable cultivation. The net irrigated area is about 33.3 million ha. India produces about 129 million tones of vegetables compared with 15 million tones at the time of independence. It is sufficient to provide about 140 g vegetable/day/capita.

Vegetables are essential to human diet particularly in India, as a large section of people are vegetarian. Various nutritive elements like calcium, phosphorous and iron are generally present in cabbage, cauliflower, beans, peas, tomato, carrot, onion, etc. Green leafy vegetables have high quantities of cellulose necessary for digestion and prevent constipation. vegetables also provide vitamins.

TABLE 2.7: Area, production and productivity of major vegetable crops in India.

(Area in 000 ha., Production in 000 MT, and productivity =MT/ha.)

Crops	2006-07			2007-08			2008-09		
	Area	**Production**	**Pdy.**	**Area**	**Production**	**Pdy.**	**Area**	**Production**	**Pdy.**
Potato	1743	28600	16.4	1795	34658	19.3	1828	34391	18.8
Onion	768	10847	14.1	821	13900	16.9	834	13565	16.3
Tomato	596	10055	16.9	566	10303	18.2	599	11149	18.6
Brinjal	568	9453	16.6	561	9678	17.2	600	10378	17.3
Cabbage	249	5584	22.4	266	5910	22.2	310	6870	22.1
Cauliflower	302	5538	18.3	312	5777	18.5	349	6532	18.7
Okra	396	4070	10.3	407	4179	10.3	432	4528	10.5
Peas	297	2402	8.1	313	2491	8.0	348	2916	8.4
Tapioca	255	8232	32.2	270	9056	33.6	280	9623	34.3
Sweet Potato	123	1067	8.7	123	1094	8.9	124	1120	9.0
Other	2282	29146	12.8	2414	31402	13.0	2275	28006	12.3
Total	**7581**	**114993**	**15.2**	**7848**	**128449**	**16.4**	**7981**	**129077**	**16.2**

Source: National Horticulture Board, (Website: www.nhb.gov.in), Database-2009, page-6.

Flowers and Floriculture

Protected cultivation of cut flowers started a decade back and India has already entered the world cut flower market. Micro propagated ornamental foliage plants are being exported in millions internationally. Export of dried flowers from India is increasing.

Floriculture was practised only on small farms but now few large ones have been set up. The estimated area under flowers is 167,000 ha and the major states being Karnataka, Tamil Nadu, West Bengal, Andhra Pradesh and Maharashtra. The production is estimated to be nearly 987,000 tones of loose flowers and over 4794 million cut flowers with stem.

The per capita consumption of flowers is going up around the world. The business of floriculture is surging forward, the latest estimate of the annual global flower trade is worth 2,500 million US dollars and the Indian contribution at present is about $0.36 million. Indian participation will in due course show a marked difference for the better in the next two years.

The international floriculture market trade is estimated to be 40 billion US dollars (Rs. 125,000 crores) of which cut-flowers account for nearly 60 per cent i.e., 25 billion US dollars (Rs. 80,000 crores). Countries like Columbia, Israel, Kenya and Italy have made an entree and have created a modicum of competition in a field not even recognized as a major industry. But the world floriculture trade is still controlled by Holland which has the lion's share of 67 per cent to its credit. Amongst the many changes which were ushered in after and due to the Indian economic reforms is the budding of the floriculture industry. The *bloom boom* has caught on the imagination of the entrepreneurs.

Most of the floriculture units are situated in south part of India with a few scattered in the north India., Bangalore and Hyderabad have become the hub of activity and will be cultivating roses of every conceivable variety on a massive scale. There are also some projects in the country which are venturing into cultivation of more exotic and expensive flowers, companies like Natural Synergies grow Orchids, while Oriental Floritech is exporting Carnations.

Floriculture is a wide field and includes a great variety and diversity of job opportunities. In general, ornamental flowers are classified variedly. To the commercial producer or wholesaler or the retailer of flower products, or to the manufacturer or seller of floricultural supplies, it is a business and to the homeowner it is a fascinating hobby.

TABLE 2.8: Area and production flowers in India.

Years	Area (In 000 ha)	Production	
		Loose (in 000 mt)	Cut (Million flowers)
2006-07	144	880	3718
2007-08	166	868	4365
2008-09	167	987	4794

Source: National Horticulture Board, (Website: www.nhb.gov.in), Database-2009, page-24

The Customer's Need for Flowers and Ornamental Plants

Regardless of the product, purchases are made to satisfy needs and wants. These needs and wants may be classed as cultural, religious, emotional or physical, and the decision to make the purchase may be made on emotional or economic bases.

Indian Cultural and Religious Practices

The use of flowers has a central place in person of the Indian cultural and religious practices. Religious worship does not take place without offering of flowers. Occasions such as marriage even welcoming someone calls for an offer of flowers. In addition, women in some parts of the country adorn themselves with flowers daily. Arrangements of expensive cut flowers are an inherent part of decoration in luxury hotels and affluent homes Flowers are increasingly given on happy occasions and on sad occasions to cheer those effected.

Emotional Needs

Flowers and ornamental plants are bought to satisfy emotional needs. To some individuals emotions, often described as feelings, is a condition in another person caused by imagination. This view, however, is inaccurate. The emotional needs of human beings are just as real as their physical needs, and each is affected by the other. Flower and plant producers should understand the effect that the purchase of their products can have on customers and evaluate their products on that basis. Customers want to improve their own and/or someone else's emotional well-being, and in this instance accomplish it by the purchase of flowers or ornamental plants. The flower and plant grower who can sense the real emotional needs of the customer will do a much better job of supplying the right product.

Flowers in Industry

These flowers are traded round the year all-over the country and a bulk of them are used in making garland, bouquets, bracelets, veins and other adornments. Flowers are also used as decorations in marriages and parties and for offerings to gods in temples, *poojas* etc. Some essential flowers are also used in extraction of essence by the perfumery industry.

Perfumery Industry

Extracts from fragrant flowers such as rose (*Rosa* spp.) and lavender champak (*Cananga odorata*) are generally used in the perfume industry. The Mughals were known to be ardent admirers of the perfumes and empress Noor Jehan was credited with the discovery of *attar* of roses. Jasmine oil extract from the flowers of *Jasminium grandiflorum* forms the base for many of the internationally renowned perfumes. The ultimate refined product which is called *concrete* from jasmine oil may cost about Rs. 10,000 or more per kg in the international market, while tuberose oil from *Polyanthes tuberosa* blossoms is used in high-grade perfumery and is valued more than that of jasmine. The commercial viability of raising these flowers in a large scale through modern technology and the production of high-grade concrete through industry is envisaged.

Flowers as Blooming Industry

The world trade in cut-flowers is estimated to be at Rs. 15,750 crores a year. Alsmeer, close to Amsterdam, the Netherlands, a small village is considered to be the world's largest wholesale flower market. An average of 12 million cut-flowers from 4,000 nurseries are sold here each day at the market building with a total investment of about 1000 crores which includes building, staff, conveyance, electronic computers and the like. More than 100 people work at the auction site and florists come from far-off places like Japan and the US to buy flowers. Here, the total sale of flowers is about one-third of the world's trade per year. The major international centres of flower production are the Netherlands, UK, US and Israel. The developing countries such as Brazil, Colombia, Kenya, Thailand and Singapore export flowers which assumed an important place in their economy.

Some industry people have recognized the need, and better foliage plants for use indoors have been developed. There have been many improvements in bedding plants that better satisfy customer needs in outdoor landscaping. New rose, poinsettia, and begonia cultivars are more acceptable to customers. They are more satisfied-they feel better from using these products.

There was a time when the manufacturers of plastic flowers and plants were able to satisfy customer needs with their products. Eventually, however, the customers found that these lifeless products had very little value in their emotional betterment, and they returned to purchasing live flowers and plants.

Flower and plant products do have much competition in the marketplace from inanimate products. These competing products may be clothing, jewelry, cosmetics, candy, food appliances, and various other items, and this does need to be recognized. Although some of these products have direct physical effects, they also have emotional values. The customer may put more value on the combined physical and emotional effect of a steak dinner than he would on the emotional effect of a vase of flowers or a potted plant. The economic factor enters here, too. A $10. 00 piece of jewelry may be equated as having similar emotional value for a Valentine gift as a $25.00 vase of roses.

The floriculture industry should be affected less by generally depressed markets than some other industries because the emotional value of flowers or plants is greater than more expensive appliances or clothing.

Per Capita Consumption of Flowers

The per capita consumption of flowers is going up around the world. The business of floriculture is surging forward, the latest estimate of the annual global flower trade is worth 2,500 million US dollars and the Indian contribution at present is about $0.36 million. Indian participation will in due course show a marked difference for the better in the next two years.

The international floriculture market trade is estimated to be 40 billion US dollars (Rs. 125,000 crores) of which cut-flowers account for nearly 60 per cent i.e., 25 billion US dollars (Rs. 80,000 crores). Countries like Columbia, Israel, Kenya and Italy have made an entree and have created a modicum of competition in a field not even recognized as a major industry. But the world floriculture trade is still controlled by Holland which has the lions share of 67 per cent to its credit. Amongst the many changes which were ushered in after and due to the Indian economic reforms is the budding of the floriculture industry. The *bloom boom* has caught on the imagination of the entrepreneurs.

Most of the floriculture units are situated in south part of India with a few scattered in the north India., Bangalore and Hyderabad have become the hub of activity and will be cultivating roses of every conceivable variety on a massive scale. There are also some projects in the country which are venturing into cultivation of more exotic and expensive flowers, companies like Natural Synergies grow Orchids, while Oriental Floritech is exporting Carnations.

Floriculture: Production, Marketing Businesses and Trade Organizations

The national trade organization is the Society of American Florists (SAF). It represents the floriculture industry and has divisions for producers, retail marketers, and wholesale marketers. The SAF is involved in an advisory capacity with the federal legislative bodies. Other national trade organizations are Roses, Inc., which represents the greenhouse rose producers, and Wholesale Florist Suppliers of America (WF&FSA), which represents the wholesale marketers. In most states, members of the industry have banded together in organization that may include the producers or the retailers or both groups. Some of these state organizations are very active. In addition to sponsoring meetings, they may represent their members in legislative or other matters. Some of the state organizations publish bulletins regularly. Frequently the state extension service works closely with the organization in organizing and conducting meetings. Most metropolitan areas have organizations called allied that represent producers, retailers, and wholesalers in the promotion of industry business in the area. In many localities they have effectively increased business.

The agricultural industry generally is understood to be involved in the production of various plant crops and livestock, and sometimes the term is restricted to farming or crops produced on farms-food and fiber. But agriculture does include the production of fruit crops (pomology), vegetable crops (olericulture), and landscape ornamental plants.

Floriculture Business Competition

The flower and plant manufacturer does have competition from many other types of businesses that make products which can provide emotional betterment in human beings. These competing fields may be entertainment, literature, welfare, cosmetics, memorials, appliances, jewelry, foods, and several others. This is a big field, but the grower should look on it not as a business threat but as an opportunity to participate in the obviously large market for emotional improvement. There are a large number of potential customers that need the help that flowers or plants could give them.

The flower and plant producer can have competition from business peers within and outside of the United States. These competitors may be able to produce a better product or sell the same product more reasonably because of climatic or business advantages at their locations. Sometimes the local grower can meet the competition successfully, but in other situations must change crops or methods in order to keep the business viable. Distant producers always do have the disadvantage of shipping costs and shipping time, but in spite of this they may be difficult to compete with.

The highly competitive areas within the United States have been California, Colorado, and Florida, primarily because of some climatic advantages over some other areas. Most of the competition is in some of the cut-flower crops, but there are some highly competitive situations in pot-plant crops also.

More recently there has been increased competition from outside the country. With continued improvements in transportation, competition from flower and plant producers in any part of the world could be expected. The effect of this competition varies considerably with location and air transport facilities. The competitive advantage that these growers have is favorable climatic conditions or some economic advantages. Some of the countries and regions that ship flowers or plants into the United States are Canada, Mexico, Central America, South America, Holland, Belgium, Israel, and Australia. In some instances the imports into the United States have had such great effects on the market that some domestic producers have either discontinued business or changed their crop production.

Sources of Supply for Commercial Flower

Some supplies can be obtained locally: general industrial or agricultural kinds of things such as plumbing and pipe-fitting, electrical, automotive, heating, and general construction materials, farm fertilizers and some pesticides.

Many of the items needed are specific to the business and cannot be obtained locally. Greenhouse structures and some of the related equipment are available from a few (less than six) firms in the United States that specialize in that equipment.

Other specific items that are required, such as nozzles, timers, fertilizers, growth regulators, pesticides, and conveyors, can be obtained from specialty supply firms (possibly fewer than 24) in various parts of the country. Some of these suppliers (approximately 12) also handle plants that are needed by the growers. The plants are handled as a brokerage item they are shipped directly to the customer by the plant producer. These plant sellers, however, do provide scheduling and advisory services for their customers.

Some plant suppliers sell their own plants, and they may have one or more sales representatives traveling in the United States who call on their greenhouse customers once or more times a year.

Most of these firms that handle plants or supplies for commercial greenhouse operations advertise regularly in the trade papers.

Education and Research in Floriculture

Because of federal encouragement and monetary grants, vocational high schools have been established in most cities in the United States. Some of these schools offer courses in horticulture. The results of this instruction have been variable because of several factors. Some of the schools have graduated individuals who are interested in and well-qualified for employment in the greenhouse. If the local school is not making good progress in development of vocational students, possibly the administrators and educators would welcome suggestions from industry members.

Some states have post-high schools that offer 2-year courses in commercial floriculture instruction. A good source of information on these schools and the courses of study that they offer should be available in most high school principals' offices. College registrars' offices also should have this information.

Many states have colleges that offer degree work in floriculture. This is academic rather than vocational education, but in spite of this some educators do an excellent job of acquainting the students with the realities in commercial floriculture production. The strictly academic courses of study are more suitable for the instruction of teachers and research workers. The individuals who are most interested in thorough understanding of commercial procedures will get part-time employment in the industry before or while they attend college.

Usually the state-supported university that offers a course of study in floriculture also will have a research program in floriculture. This work may be done right at the university or it may be conducted at another site in the state. Some of the work will have federal as well as state funding.

The United States Department of Agriculture (USDA) conducts some research in floriculture at Beltsville, Maryland, and at other locations in the country. The results of this work are published in the trade papers. They have 1-day meetings at Beltsville periodically in which the current research work is displayed and discussed.

State universities have an extension staff that is responsible for disseminating pertinent information and advice on technical matters to agriculturists in the state. Each county also has extension personnel. The extension program has federal funding. The size of the extension staff is determined by the amount of agriculture industry in the state, and the specialties of the staff members correspond to the kind of agriculture business that is predominant in the state. If there will be one or more floriculture specialists on the state extension staff. The extension specialists organize and conduct meetings for trade members, and they may make visits to a place of business if assistance is requested.

Spices and Plantation Crops

Spices are considered essential in the culinary art all over the world. They play a definite role in enhancing the taste and flavour of foods. Many of the spices are carminative, digestive, stimulant, stomachic and resoresolvent. Some of them act as antioxidants and preservatives also and thus enhance the keeping quality of food. Some of them have medicinal value also, therefore, they are used in ayurvedic medicines and tonics.

India is the largest producer, consumer and exporter of spices. It has over 2.6 million hectares of land under cultivation of spices. On an average, the country produces about 4.14 million tones of spices of which about 93 per cent are consumed in the domestic market and only 7 per cent are exported.

India produces over 60 different spices valued at more than Rs. 5,000 crores. Till a few years ago, India dominated the global spices market, now estimated at 5 lake tones and valued at $ 1.5 billion (Rs. 4,5000 crore), with a 95 per cent share.

At present the production of many important spices is confined almost completely to Kerala. There is very good scope for their cultivation in Tamil Nadu, Karnataka and Andhra Pradesh also due to similar agro-climatic conditions. In the past, India was exporting spices mostly in raw and bulk form. Now it has started exporting part of the spices in branded consumer packs also.

Among the major spices produced in India, Kerala, Karnataka and Tamil Nadu are the main producers of pepper and cardamom. Ginger and chillies are grown in almost all states of India except Jammu & Kashmir and Himachal Pradesh. Turmeric is grown in Uttar Pradesh, Bihar, West Bengal, Orissa, Maharashtra and Andhra Pradesh. In 1993-94, the area under spices in Madhya Pradesh rose to 19,000 hectares from 18,000 hectares in 1992.93.

The production of spices in India is largely in the hands of small and marginal farmers with limited land holdings. The level of productivity in India, in the case of most of the spices, is below the levels prevailing in other countries. The lower productivity is attributed to lack of scientific support for production and quality improvement, adoption of primitive and crude agro-techniques and poor management.

India has been the largest producer, exporter and consumer of species. Besides export, domestic consumption is also steadily increasing. During Eighth Plan, the Spices Board has fixed a target for export of about Rs. 1,000 million worth of spices by the end of the Eighth Plan mainly by value addition through exports in consumer packs. During 1996-97, 2.19 lac tones of species worth Rs. 1.180 million were exported (Source: E. Edison, High Productivity Vital, page 77, The Hindu Survey of Indian Agriculture, 1997).

TABLE 2.9: Area, production and productivity of major spices crops in India.

(Area in 000 ha., Production in 000 MT, and productivity =MT/ha.)

Crop	2007-08			2008-09		
	Area	Production	Productivity	Area	Production	Productivity
Chillies	808.17	1294.15	1.60	779.05	1269.85	1.63
Garlic	206.12	1068.50	5.18	166.21	831.10	5.00
Turmeric	174.51	794.19	4.55	181.09	821.16	4.53
Ginger	104.36	390.08	3.74	108.64	380.10	3.50
Coriander	384.21	229.95	0.60	396.85	242.13	0.61
Tamarind	55.04	182.08	3.31	54.63	177.68	3.25
Cumin	429.38	172.47	0.40	429.38	172.47	0.40
Fenugreek	55.20	55.48	1.01	68.29	76.58	1.12
Fennel	54.29	67.78	1.25	47.16	84.29	1.79
Pepper	197.33	47.01	0.24	238.71	47.40	0.20
Ajwan	19.29	11.12	0.58	19.59	6.41	0.33
Cardamom	81.93	13.65	0.17	91.99	5.45	0.06
Nutmeg	15.26	11.37	0.75	15.27	11.37	0.74
Other seed spices	17.71	10.71	0.60	17.71	0.71	0.04
Tejpata	2.45	5.29	2.16	2.44	4.98	2.04
Cinnamon	0.87	1.67	1.92	0.88	1.67	1.90
Clove	2.25	1.01	0.45	2.57	1.33	0.52
Other	9.01	0.23	0.03	9.01	0.23	0.03
Total	**2617.36**	**4356.71**	**1.66**	**2629.44**	**4144.91**	**1.58**

Source: National Horticulture Board, (Website: www.nhb.gov.in), Database-2009, page-8

TABLE 2.10: Area, production and productivity of major plantation crops in India.

(Area in 000 ha., Production in 000 MT, and productivity =MT/ha.)

Crops	2006-07			2007-08			2008-09		
	Area	Production	Pdy.	Area	Production	Pdy.	Area	Production	Pdy.
Coconut	1939.9	10893.8	5.6	1903.2	10148.0	5.3	1903.2	10148.3	5.3
Cashewnut	854.0	620.0	0.7	868.0	665.0	0.8	893.0	695.0	0.8
Areca nut	382.7	483.3	1.3	386.6	476.0	1.2	387.1	481.3	12.0
Cocoa	30.3	10.2	0.3	31.8	10.6	0.3	34.0	11.8	0.3
Total	**3206.9**	**12007.3**	**3.7**	**3189.6**	**11299.6**	**3.5**	**3217.3**	**11336.4**	**3.5**

Source: National Horticulture Board, (Website: www.nhb.gov.in), Database-2009, page-7

The Government of India is taking a number of steps to improve processing and packaging of spices to boost their export and meet the international requirements. The measures include organizing training for growers and exporters in different states in the field of modern post harvest techniques of processing to enable them to manufacture spices of international quality and specifications. Training is also being imparted to small exporters on modern concepts of packaging, retail consumer packaging for international markets and conducting studies to streamline packaging standards and requirements for Indian spices under EEC assistance.

The United Nations Development Programme (UNDP) is providing aid worth $ 1 million (Rs. 3 crore) to the Indian government over the a period of three years (1994-95 to 1996-97) to enable India to upgrade quality and export value added spices. Consequent to the recommendations of the Spices Enquiry Committee in 1953, research work on spices in India was initiated in the 1950s' on cardamom, pepper, ginger, turmeric, clove, nutmeg and coriander at Panniyur, Chethalli, Dergaon, Sirsi, Kandaghat, Thodupuzha, Ambalavayal, Burliyar and Fulia. The formation of an `All-India Coordinated Research Project on Spices' in 1971, however, was the first major step taken by the government towards a concerted programme on spices research. Under this project, 15 centres for research have been established in almost all the major spice growing states in the country. The main crops covered under the project are black pepper; cardamom (small and large), ginger, turmeric, cumin, coriander, fennel and fennel and fenugreek. In the Eighth Plan, it is proposed to include tree spices like clove, nutmeg and cinnamon also under this project.

Coconut

In coconut, area has increased from 1 million hectares in 1980 to 1.90 million hectares in 2008-09. India is the largest coconut producing country of the world. Its production has gone up from 5677 million nuts to 13968 million nuts and the productivity has increased from 5249 nuts/ha to 7808 nuts/ha. Coconut contributes about 700 billion rupees to the GDP of the country. The contribution of the crop to the total edible oil pool in India is around 6 per cent. India also exports coir and coir products derived from coconut husk to the tune of 2260 million Rupees.

Areca Nut

India continues to dominate the world in area (387,000 ha), production (481,000 tonnes) and productivity (1.2 mt/ha) of areca nut and has achieved self sufficiency in areca nut production. Most of the production is domestically consumed.

Oil Palm

Area under oil palm has gone up from 200 ha in 1965 to 40,700 ha in 1996-97. Average productivity in oil palm plantations is now 4-5 tonnes/ha which compares favorably with other countries.

Cashew Nut

In cashewnut, area has increased from 0.176 million hectares in 1961 to 0.893 million hectares in 2008-09. The production in cashew has gone up from 0.079 million tonnes to 0.695 million tonnes in 2008-09. India exported cashew kernels worth Rs. 13000 million (362 million $US during 1996-97). Export of cashew rising @ 27% per annum. These export earnings are exceeded only by coffee and rice among agri-exports.

Mushroom

Mushroom cultivation has spread to almost all parts of the country. Its production has increased from 100 tonnes in 1970 to 30,000 tonnes in 1996-97. The productivity has increased from 10-12 kg m^{-2} in 1985 to 18-22 kg m^{-2} in 1995.

Mushrooms have a huge export potential. There is a world market for 14 lac tones mushroom per annum against which India is exporting only about 2,400 tones. The world demand is likely to go up to more than 20 lac tones by 1998. Before 1990, only dried morels (*Morchella* spp.) and some amount of black ear mushroom (*Auricularia*) were exported to the tune of about 50 MT to France and Switzerland. In 1993-94, there had been about 50% increase in the export of preserved mushrooms compared to the exports in 1992-93. Northern States of Himachal Pradesh, Haryana and Punjab are the largest producers of white button mushroom. Mushroom production in Himachal Pradesh has already crossed 1500 MT. Mushroom production in other States is catching up fast. One 100% EOU in Dehradun is producing more than 2500 MT of mushroom per annum.

Mushrooms are reproductive structures of edible fungi that belong to Ascomycotina and Basidiomycotina. They occur under various ecological conditions from desert to forest. They comprise a large heterogeneous group with different shapes, sizes, colour and edibility. Of the 2000 known edible species, only 20 are commercially cultivated.

Mushrooms can produce highest quantity of protein per unit area and time from the agro-wastes. They are good source of high quality proteins and are rich in vitamins and minerals. They contain 20-35% protein (dry weight basis) (higher than in vegetables and fruits), lysine and tryptophan (the two essential amino acids that are deficient in cereals). They contain good amount of vitamin C and B complex (thiamine, riboflavin and niacin), potassium, phosphorus and sodium, also low but available form of iron. In mushrooms, potassium: sodium ratio is very high which is ideal for patients of hypertension. They are a low calorie food with very little fat and sugars and without starch and cholesterol. Mushrooms have medicinal properties. Pharmaceuticals worth $ 700 million are produced annually in Japan from *Lentinus*, *Coriolus*, *Schizophyllum* and *Ganoderma*. Further, compounds extracted from mushrooms have antifungal and antibacterial properties.

Mushrooms are capable of agro-waste degradation. Mushrooms are grown on organic substrates, either raw or composted. These substrates are mostly waste materials from farms, plantations, or factories. Further, used compost from mushroom growing may also be recycled for use as animal feeds, soil conditioning and fertilizer. Mushrooms grow independent of sunlight without fertile land. They do not compete with field or fruit crops and provide an additional avenue for increasing food supply.

TABLE 2.11: Various horticultural products and their production States

Products	State	District/Area
Apples	Jammu & Kashmir	Districts of Srinagar, Baramula, Anantnag, Kupwara, Badgaum and Pulwama
Apples	Himachal Pradesh	Shimla, Sirmour, Kullu, Mandi, Chamba and Kinnaur
Banana	Maharashtra	Jalgaon, Dhule, Nandurbar, Buldhana, Parbhani, Hindoli, Nanded and Wardha
Cashewnut	Tamil Nadu	Cuddalore, Thanjavur, Pudukottai and Sivaganga
Chilli	Andhra Pradesh	Guntur
Coriander	Rajasthan	Kota, Bundi, Baran, Jhalawar & Chittoor
Cumin	Rajasthan	Nagaur, Barmer, Jalore, Pali and Jodhpur
Darjeeling Tea	West Bengal	Darjeeling
Dehydrated Onion	Gujarat	Districts of Bhavnagar, Surendranagar, Amreli, Rajkot, Junagadh and Jamnagar.
Flowers	Tamil Nadu	Dharmapuri
Flowers	Uttaranchal	Districts of Dehradun, Pantnagar
Flowers	Maharashtra	Pune, Nasik, Kolhapur and Sangli
Flowers	Karnataka	Bangalore (Urban), Bangalore (Rural), Kolar, Tumkur, Kodagu and Belgaum
Flowers	Tamilnadu	Nilgiri District
Flowers (Orchids) & Cherry Pepper	Sikkim	East Sikkim
Fresh & Processed Ginger	Assam	Kamrup, Nalbari, Barpeta, Darrang, Nagaon, Morigaon, Karbi Anglong and North Cachar districts
Gherkins	Karnataka	Tumkur, Bangalore Urban, Bangalore Rural, Hassan, Kolar, Chtradurga, Dharwad and Bagalkot
Gherkins	Andhra Pradesh	Districts of Mahboobnagar, Rangareddy, Medak, Karinagar, Warangal, Ananthpur and Nalgonda
Ginger	Sikkim	North, East, South & West Sikkim
Ginger and Turmeric	Orissa	Kandhamal District
Grapes and Grape Wine	Maharashtra	Nasik, Sangli, Pune, Satara, Ahmednagar and Sholapur
Horticulture Products	Kerala	Districts of Thrissur, Ernakulam, Kottayaam, Alappuzha, Pathanumthitta, Kollam, Thiruvanthapuram, Idukki and Palakkod
Kesar Mango	Maharashtra	Districts of Aurangabad, Jalna, Beed, Latur, Ahmednagar and Nasik
Lentil and Grams	Madhya Pradesh	Shivpuri, Guna, Vidisha, Raisen, Narsinghpura, Chhindwara
Lychee	Uttaranchal	Udhamsingh Nagar, Nainital and Dehradun
Lychee	West Bengal	Districts of Murshidabad, Malda, 24 Parganas (N) and 24 Parganas (S)

Products	State	District/Area
Lychee	Bihar	Muzaffarpur, Samastipur, Hajipur, Vaishali, East and West Champaran, Bhagalpur, Begulsarai, Khagaria, Sitamarhi, Sarannd Gopalganj
Mango & Grapes	Andhra Pradesh	Districts of Ranga Reddy, Medak & Parts of Mahaboobnagar Districts
Mango Pulp & Fresh Vegetables	Andhra Pradesh	Chittor District
Mangoes	Uttar Pradesh	Sahranpur, Muzaffarnagar, Bijnaur, Meerut, Baghpat, Bulandshar and Jyotifulenagar
Mangoes	Maharashtra	Districts of Ratnagiri, Sindhudurg, Raigarh and Thane
Mangoes	Andhra Pradesh	Krishna District
Mangoes	West Bengal	Malda and Murshidabad
Mangoes	Tamil Nadu	Districts of Madurai, Theni, Dindigul, Virudhunagar and Tirunelveli
Mangoes & Vegetables	Gujarat	Districts of Ahmedabad, Khaida, Anand, Vadodra, Surat, Navsari, Valsad, Bharuch and Narmada
Mangoes and Vegetables	Uttar Pradesh	Lucknow, Unnao, Hardoi, Sitapur & Barabanki
Medicinal & Aromatic Plants	Uttaranchal	Districts of Uttarkashi, Chamoli, Pithoragarh, Dehradun and Nainital
Medicinal Plant	Kerala	Wayanad, Mallapuram, Palakkad, Thrissur, Ernakulam, Idukki, Kollam, Pathanamittha, Thiruvananthapuram
Onion	Maharashtra	Districts of Nasik, Ahmednagar, Pune, Satara and Solapur
Oranges	Maharashtra	Nagpur and Amraoti
Oranges	Madhya Pradesh	Chhindwara, Hoshangabad, Betul
Pineapple	West Bengal	Darjeeling, Uttar Dinajpur, Cooch Behar and Jalpaiguri
Pineapple	Tripura	Kumarghat, Manu, Melaghar, Matabari and Kakraban Blocks
Pomegranate	Maharashtra	Districts of Solapur, Sangli, Ahedabagar, Pune, Nasik, Osmanabad, Latur
Potatoes	Uttar Pradesh	Agra, Hathras, Farrukhabad, Kannoj, Meerut, Aligarh and Bagpat
Potatoes	Punjab	Singhpura, Zirakpur (Patiala), Rampura Phul, Muktsar, Ludhiana, Jallandhar
Potatoes	West Bengal	Districts of Hoogly, Burdwan, Midnapore (W), Uday Narayanpur and Howrah
Potatoes, Onion and Garlic	Madhya Pradesh	Malwa, Ujjain, Indore, Dewas, Dhar, Shajapur, Ratlam, Neemuch and Mandsaur
Rose Onion	Karnataka	Bangalore (Urban), Bangalore (Rural), Kolar
Seed Spices	Madhya Pradesh	Guna, Mandsaur, Ujjain, Rajgarh, Ratlam, Shajapur and Neemuch
Sesame Seeds	Gujarat	Amerali, Bhavnagar, Surendranagar, Rajkot, Jamnagar
Vanilla	Karnataka	Districts of Dakshin Kannada, Uttara Kannada, Udupi, Shimoga, Kodagu, Chickamagalur
Vegetables	Punjab	Fatehgarh Sahib, Patiala, Sangrur, Ropar & Ludhiana

Products	State	District/Area
Vegetables	Jharkhand	Ranchi, Hazaribagh and Lohardaga
Vegetables	West Bengal	Nadia, Murshidabad and North 24 Parganas
Walnut	Jammu & Kashmir	Kashmir Region – Baramulla, Anantnag, Pulwama, Budgam, Kupwara and Srinagar / Jammu Region - Doda, Poonch, Udhampur, Rajouri and Kathua

HORTICULTURE RESEARCH INSTITUTES IN INDIA

Indian Council of Agricultural Research

The ICAR is an apex body at the national level mainly responsible for adding, promoting and coordinating agricultural research in the country. It has the concurrent responsibility of research, education and extension education. The objectives of the ICAR during the IX Five Year Plan are as follows (source: 50 Years of Crop Science in India, ICAR, 1996, page 13-14):

1. Conservation, planned enhancement and utilization of agro-biodiversity
2. Enhancing productivity through evolution of high yielding hybrids and varieties
3. Research on diversification, quality improvement, post-harvest technology, value addition and export-oriented commodities
4. Sustaining enhanced productivity of irrigated agriculture and judicious development and use of energy, especially renewable sources of energy
5. Characterization and development of sustainable land use models for rainfed agriculture in high rainfall areas
6. Development of the integrated pest management (IPM) and integrated nutrient management system (INMS) approaches and systems for sustainable agriculture
7. Fostering excellence in relevant basic and strategic research
8. Generating research and technologies geared to promote equity among regions, sectors of society and gender
9. Strengthening social science, policy planning, agri-business, research monitoring mechanisms, administration and personnel reforms, publications and information dissemination system
10. Strengthening Agricultural Research Information System (ARIS)
11. Promoting Agricultural Human Resource Development (AHRD)
12. Linking scientists with the farmers through Institute Village Linkage Programme (IVLP) an innovative technology transfer model
13. Institutionalization and strengthening linkages/partnership with the CGIAR and other national and international agencies and research and development establishments, non-governmental organizations (NGOs), farmer organizations, private sector etc.
14. Optimization of resources through planning, prioritization and coordination

The translation of these objectives into programmes and activities would certainly develop horticultural research in India.

Indian Institute of Horticultural Research

The Indian Institute of Horticultural Research was established by the Indian Council of Agricultural Research (ICAR) to conduct research on various aspects of horticultural crops viz. fruits, vegetables, ornamentals and medicinal plants in order to improve their productivity, quality and utility.

The Institute was established at ICAR Headquarters at New Delhi on 5th September, 1967 and later shifted to Bangalore during February 1968 with late Dr. G. S. Randhawa as the Founder Director. Initially, the Institute had 24.7 ha land under the National Hortorium, which was transferred to Indian Institute of Horticultural Research. Later, an additional land of 238 ha was transferred to the Institute by Government of Karnataka. The Institute had Central Mango Research Station at Lucknow as its Regional Station, during 1972 which was further upgraded and separated as Central Institute for Sub-Tropical Horticulture.

Similarly, the Institute had another Regional Station on Citrus at Nagpur under the administrative control of the Institute which was separated and upgraded as National Research Centre for Citrus during 1988. During 1972 two Central Horticultural Experiment Stations at Chethalli and Gonikoppal were transferred to the Institute. Later the Citrus Experiment Station, Gonikoppal was converted into a full fledged Krishi Vigyan Kendra during 1991. During 1979 another Regional Station was sanctioned at Godhra which was later merged with National Research Centre for Semi-Arid Horticulture, Bikaner (Rajasthan). Similarly, the Central Horticultural Experiment Station, Ranchi was sanctioned during the year 1979 which was later merged with ICAR Research Complex for Eastern Region, Patna. The Central Horticultural Experiment Station, Bhubaneshwar was sanctioned during the year 1992 to cater to the needs of tribal / coastal belts of Orissa, West Bengal and coastal Andhra Pradesh. During the year 2002, Hirehalli (Karnataka) which was the Regional Station of Central Plantation Crops Research Institute, Kasaragod has been shifted to IIHR, Bangalore and renamed as Central Horticultural Experiment Station, Hirehalli. Apart from this, the Institute has been instrumental in establishment of various other research institutes like National Research Centre for Banana, Trichy (Tamil Nadu); National Research Centre for Grapes, Pune (Maharashtra) and National Research Centre for Litchi, Muzaffarpur (Bihar) which were for sometime under the administrative control of IIHR in the beginning years. At present, the Institute has got only three regional stations and one Krishi Vigyan Kendra under its administrative control.

Main Institute- Experiment Station, Hessaraghatta

Though the Institute was established on 5th September, 1967 at the ICAR Head quarters, New Delhi, the main institute was established in February, 1968 at Bangalore with the experimental station at Hessaraghatta, located about 25 kms. from the city. The entire laboratory complex, the administrative unit, farm and few essential residential quarters are housed in the Hessaraghatta campus.

Central Horticultural Experiment Station (CHES), Chettalli

This station situated in Kodagu district of Karnataka was transferred to IIHR from the Government of Karnataka with effect from February 1, 1972. Occupying an area of 92 ha, the mandate crop of the station has been Coorg Mandarin with major emphasis on citrus die-back. The station also works on fruit crops like sapota, guava, papaya, pineapple, passion fruit, minor fruits like pummelo, avocado, mangosteen, and of late has diversified into floricultural crops. The station has a strong nursery unit for production and supply of disease free citrus planting material, other plant materials and *trichoderma* cultures. All the research work along with research laboratories of the erstwhile Citrus Experiment Station, Gonikoppal were shifted to CHES, Chettalli with effect from 1.1.1992.

Krishi Vigyan Kendra (KVK, Gonikoppal)

This center is situated in the Kodagu district of Karnataka. Established in 1954 by the Karnataka State Government, and was transferred to IIHR on February 1, 1972. The center has an area of 17.5 ha. It was started with the objectives of investigating the nature and causes of citrus dieback disease in Kodagu and nearby areas. Till 1991, this was functioning as the Citrus Experiment Station and was converted into a full-fledged KVK in 1992. Research work on Citrus has been shifted to Chettalli.

Central Horticultural Experiment Station, (CHES), Bhubaneswhar

This station was established on 6th November, 1992. The objective of the Station is to cater to the research and development needs in horticulture of the costal belts of Orissa and adjoining eastern region. The station has an area of about 40 ha. A field laboratory building has been constructed and developmental activities are in progress.

Central Horticultural Experiment Station (CHES), Hirehalli

The Regional Station of Central Plantation Crops Research Institute, Kasaragod at Hirehalli (Tumkur), Karnataka has been transferred to Indian Institute of Horticultural Research, Hessaraghatta, Bangalore and is renamed as Central Horticultural Experiment Station, Hirehalli with effect from 1.4.2002.

The objective of the station is to undertake applied research for developing strategies to enhance productivity for utilization in vegetable and floriculture crops and to take up breeder seed and foundation seed production programme for vegetable crops and few flower crops. The station would also work on per urban horticulture.

Research Institutes of Horticulture Research

1. Indian Institute of Horticultural Research, Karnataka
2. Indian Institute of Vegetable Research, Uttar Pradesh
3. Central Institute for Sub-Tropical Horticulture, Uttar Pradesh
4. Central Institute Of Temperate Horticulture, J&K
5. Central Institute of Arid Horticulture, Rajasthan
6. Central Potato Research Institute, Himachal Pradesh
7. Central Tuber Crops Research Institute, Kerala
8. Central Plantation Crops Research Institute, Kerala
9. Indian Institute of Spices Research, Kerala
10. Central Agricultural Research Institute, A&N Gr of Islands
11. NRC for Cashew, Puttur, Karnataka
12. NRC for Citrus, Nagpur, Maharashtra
13. NRC for Mushroom, Solan, Himachal Pradesh
14. NRC for Oilpalm, Pedavegi, Andhra Pradesh
15. NRC for Orchids, Pakyang, Sikkim
16. NRC for Grapes, Pune, Maharashtra
17. NRC for Medicinal & Aromatic Plants, Boriavi, Anand, Gujarat
18. NRC for Onion & Garlic, Rajgurunagar, Pune, Maharashtra
19. NRC for Banana, Trichi, Tamilnadu
20. NRC on Seed Spices, Tabiji, Ajmer, Rajasthan

Organizations / Institutes/ University

1. Indian Council of Agricultural Research, New Delhi.
2. C.S.S. Haryana Agricultural University, Hisar, Haryana
3. H.P. Krishi Vishvavidyalaya University
4. Y.S. Parmar University of Horticulture and Forestry, Solan, Nauni, Himachal Pradesh
5. G.B. Pant University of Agriculture and Technology, Pantnagar, Distt. Udham Singh, Nagar, Uttaranchal
6. Assam Agricultural University, Jorhat, Assam
7. Orissa University of Agriculture and Technology, Bhubabeshwar, Orissa
8. Mahatma Phule Krishi Vidyapeeth, Rahuri, Maharashtra
9. Marathwada Agricultural University, Parbhani, Maharashtra
10. Acharya N.G. Ranga Agricultural University, Hyderabad, Andhra Pradesh
11. University of Agricultural Sciences, Bangalore
12. University of Agricultural Sciences, Dharwad
13. Kerala Agricultural University, Vellanikara, Trichur, Kerala
14. Dr. Panjabrao Deshmukh Krishi Vidyapeeth, Akola, Maharashtra
15. Rajendra Agricultural University, Pusa Samastipur, Bihar
16. Bidhan Chandra Krishi Viswa Vidyalaya, Mohanpur, Nadia, West Bengal
17. Maharana Pratap Agricultural University, Udaipur
18. Indira Gandhi Krishi Vishwa Vidyalaya, Krishak Nagar, Raipur, Chhattisgarh
19. Punjab Agricultural University, Ludhiana, Punjab
20. Tamil Nadu Agricultural University, Coimbatore, Tamil Nadu
21. S K Rajasthan Agricultural University, Bikaner
22. Anand Agriculture University, Anand, Gujarat
23. Birsa Agriculture University, Kanke Ranchi-834006, Jharkhand
24. Central Agricultural University, P.O. Box 23, Imphal-795004, Manipur
25. Chandra Shekar Azad University, Kanpur-208002, Uttar Pradesh
26. CSK Himachal Pradesh Krishi Vishvavidyala, Palampur-176062, Himachal Pradesh
27. Dr Balasaheb Sawant Konkan Krishi Vidyapeeth, Dapoli Distt, Ratnagiri 415712, Maharashtra
28. Jawaharlal Nehru Krishi Viswavidyalaya, Jabalpur-482004, Madhya Pradesh
29. Junagadh Agriculture University, Junagadg-362001, Gujarat
30. Narendra Deva University of Agriculture & Technology, Faizabad, Uttar Pradesh
31. Navsari Agriculture University, Navsari, Gujarat
32. Sardarkrushinagar-Dantiwada Agricultural University, Sardar Krushinagar Distt. Banaskanth, Gujarat
33. Sardar Vallabh Bhai Patel University of Agriculture & Technology, Modipuram, Meerut-250110 Uttar Pradesh
34. Sher-E-Kashmir University of Agricultural Sciences & Technology, Railway Road, Jammu
35. Sher-E-Kashmir University of Agricultural Sciences & Technology of Kashmir Shalimar Campus Shrinagar, Jammu Kashmir
36. Uttar Banga Krishi Viswavidyalaya, P.O. Pundibari, Distt. Coach Bihar-736165, West Bengal

All India Coordinated Research Projects

1. All India Coordinated Research Project on Betelvine, NRC for Medicinal and Aromatic Plants, Boriavi, Gujarat
2. All India Coordinated Research Project on Cashew, NRC Cashew, Puttur,

Karnataka

3. All India Coordinated Research Project on Floriculture, Division of Floriculture & Landscaping, IARI, New Delhi
4. All India Coordinated Research Project on Mushrooms, National Centre for Mushroom Research & Training, Chambaghat, Solan, Himachal Pradesh
5. All India Coordinated Research Project on Palms, CPCRI, Kasaragod, Kerala
6. All India Coordinated Research Project on Post Harvest Technology, Division of Fruits & Horticulture Technology, IARI, New Delhi
7. All India Coordinated Research Project on Potato, Central Potato Research Institute, Shimla, Himachal Pradesh
8. All India Coordinated Research Project on Subtropical Fruits, Central Institute for Subtropical Horticulture, Lucknow, Uttar Pradesh
9. All India Coordinated Research Project on Spices, Indian Institute of Spices Research, Marikunnu, Calicut, Kerala
10. All India Coordinated Research Project on Fruits, CISTH, Lucknow
11. All India Coordinated Research Project on Tropical Fruits, IIHR, Hessarghatta Lake Post, Bangalore, Karnataka
12. All India Coordinated Research Project on Tuber Crops, Regional Station of CTCRI, Thiruvananthapuram, Kerala
13. All India Coordinated Research Project on Vegetable, IIVR, Gandhinagar, Sunderpur, Varanasi, Uttar Pradesh.

NATIONAL HORTICULTURE BOARD

National Horticulture Board has the mandate to take up programmes right from the stage of production to development of post-harvest infrastructure including marketing, processing and storage facilities for ensuring integrated development of horticulture. Major programmes of the Board include establishment of cold storage for horticulture produce and strengthening of post harvest inputs where back ended capital subsidy is provided. The board has also been given the mandate of creating a database for horticultural crops. Other areas entrusted to the Board are providing market information service on fruits and vegetables, development of captive plantations for export and processing, market support operations for horticultural crops, development of export of horticultural products through equity participation, etc.

1. Popularization of new technologies / tools / techniques for commercialization / adoption.
2. Introduction of new concepts to improve farming systems.
3. Upgradation of skills by exchange of technical know-how.
4. Consolidating research efforts for specific problems.
5. Identification/collection, rapid multiplication & popularization of indigenous crops and other flora of horticultural importance with emphasis on domestic and export promotion.
6. Familiarization and exposure towards the newer scientific concepts/temper and research & development from hitherto unknown, unexplored and traditional status of farming and post harvest management on to the modern lines.

Investment Subsidy Schemes

The main objectives set forth for the National Horticulture Board are:

1. To encourage, promote, stimulate and develop the growth of the diverse

2. Horticulture Industry.
3. To coordinate the activities of different departments and organizations at the central and the State level engaged in activities pertaining to Horticulture Industry.
4. To assist in the establishment and growth of infrastructure for the development of post harvest technology and development of market intelligence and information system;
5. To provide technological, financial and other assistance in organization of consultancy service, preparation, monitoring and evaluation of projects related to the Horticulture Industry, including but limited to the transfer of improved technology for production, processing, quality control and marketing and matters allied or incidental thereto;
6. To promote integrated development of the horticulture industry with particular reference to potato, onion, tomato, cauliflower, cabbage, ginger, turmeric, apple, pineapple, mango, grapes and citrus fruits and other horticultural crops on priority basis;
7. To take appropriate measures for assisting farmers and growers to get incentive prices, having due regard to the interest of consumers in this behalf.

NATIONAL HORTICULTURE MISSION

The Government of India launched National Horticulture Mission during the X Plan with effect from 2005-06, for holistic development of horticulture duly ensuring horizontal and vertical linkages, with the active participation of all the stake-holders. National Horticulture Mission (NHM) will be implemented in all the States and Union Territories of India except the North Eastern States, Himachal Pradesh, Jammu & Kashmir and Uttaranchal (for which a separate Technology Mission for integrated development of horticulture exists) to promote holistic growth of the horticulture sector covering fruits, vegetables, root and tuber crops, mushroom, spices, flowers, aromatic plants, cashew and cocoa. Programmes for the development of coconut will be implemented by the Coconut Development Board (CDB), independent of the Mission. This will be a Centrally sponsored scheme in which Government of India shall provide 100% assistance to the State Missions during Tenth Plan. During the XI Plan, the Government of India assistance will be 85% with 15% contribution by the State Governments.

Mission Objectives

The main objectives of the mission are:

1. To provide holistic growth of the horticulture sector through an area based regionally differentiated strategies which include research, technology promotion, extension, post harvest management, processing and marketing, in consonance with comparative advantage of each State/region and its diverse agro-climatic feature.
2. To enhance horticulture production, improve nutritional security and income support to farm households.
3. To establish convergence and synergy among multiple on-going and planned programmes for horticulture development.
4. To promote, develop and disseminate technologies, through a seamless blend of traditional wisdom and modern scientific knowledge.
5. To create opportunities for employment generation for skilled and unskilled persons, especially unemployed youth.

Strategy

To achieve the above objectives, the mission would adopt the following strategies:

1. Ensure an end-to-end holistic approach covering production, post harvest management, processing and marketing to assure appropriate returns to growers/producers;
2. Promote R&D technologies for production, post-harvest management and processing;
3. Enhance acreage, coverage, and productivity through:
   a) Diversification, from traditional crops to plantations, orchards, vineyards, flower and vegetable gardens;
   b) Extension of appropriate technology to the farmers for high-tech horticulture cultivation and precision farming.
4. Assist setting up post harvest facilities such as pack house, ripening chamber, cold storages, Controlled Atmosphere (CA) storages etc, processing units for value addition and marketing infrastructure.
5. Adopt a coordinated approach and promotion of partnership, convergence and synergy among R&D, processing and marketing agencies in public as well as private sectors, at the National, Regional, State and sub-State levels.
6. Where appropriate and feasible, promote National Dairy Development Board (NDDB) model of cooperatives to ensure support and adequate returns to farmers.
7. Promote capacity building and Human Resource Development at all levels.

The thrust of the Mission will be on area based regionally differentiated cluster approach for development of horticulture crops, having comparative advantage. A number of activities are proposed to be taken up for improving production and productivity besides ensuring proper post harvest management of produce so that the farmer is able to harvest and sell his products at competitive rates. National Horticulture Mission will also involve the active participation of various agencies at the national level and those of the State Government, Research Institutes and Organizations, farmer associations, Self Help Groups, Cooperative Societies in Horticulture, Marketing Boards, and many others.

NATIONAL MISSION ON BAMBOO TECHNOLOGY AND TRADE DEVELOPMENT

Bamboo, a fast growing, versatile woody grass is found across the country. It is an economic resource having immense potential for improving the quality of life of rural and urban communities with environment regeneration qualities like carbon sequestering. Bamboo provides raw material for large industries like paper and pulp as well as for cottage and handicrafts industry. The bamboo shoot is highly nutritious and is considered as a delicacy world over. It holds the potential to benefit the poor by means of providing sustainable livelihood in farming and in industrial products. However, the potential has remained largely untapped in the country. Recognizing its potential, the Planning Commission had prepared a report on National Mission on Bamboo Technology and Trade Development (NMBTTD), which will be administered through the Ministry of Agriculture with the association of other concerned Ministries/ Departments. The Mission will address the issues relating to the holistic development of bamboo with linkages for its cultivation, post harvest management, processing, establishment of entrepreneurial ventures through Private Sector participation and marketing.

CENTRAL INSTITUTE OF HORTICULTURE, MEDZIPHEMA, DIMAPUR, NAGALAND

During February 2006, Government of India has approved the setting up of a Central Institute of Horticulture, Dimapur, Nagaland with the following objectives for development of horticulture especially in NE States:

1. Capacity building by training of trainers, extension officers, farmers, entrepreneurs, processors and exporters.
2. Demonstration of improved technologies such as use of improved varieties/ hybrids, adoption of INM/IPM practices, Hi-tech farming, precision farming, protected cultivation, post harvest technologies, etc.
3. Follow on extension support in the field of horticulture.
4. Promotion of organic cultivation of horticultural crops.
5. Establishing convergence and synergy among programmes in the field of horticultural research and development.
6. Monitoring of Centrally sponsored programmes in the area of horticulture.

The Institute shall have a Board of Management (BOM) with representative from North Eastern Council (NEC), NE States, industry/entrepreneurs, two horticultural experts as members, Horticulture Commissioner as the Chairman and the Director of the CIH as the Member-Secretary for approval of programmes and regular monitoring of the activities of the Institute. The Institute will be networked with other Institutes and Departments involved in development of horticulture in the country and abroad.

CHAPTER 3

Hi-tech Horticulture for Agriculture Reform

The agricultural produce sector has been one of the most important components of the Indian economy. The increasing trend of agricultural production has brought, in its wake, new challenges in terms of finding market for the marketed surplus. There is also a need to respond to the challenges and opportunities, that the global markets offer in the liberalized trade regime. To benefit the farming community from the new global market access opportunities, the internal agricultural marketing system in the country needs to be integrated and strengthened. Government of India is striving to prepare the Indian agricultural markets and marketing environment so as to provide maximum benefit to the producers and in turn, compete with the global markets. Agriculture and agricultural marketing need to be re-oriented to respond to the market needs and consumer preferences. Agricultural marketing reforms and creation of marketing infrastructure has been initiated to achieve the above purpose in India.

Globalization is becoming a world phenomena and it is especially true in horticulture as compared to other agricultural sectors. Horticulture combines art and skill with science and technology for growing and distributing ornamental plants, fruits, and vegetables. Our landscapes, homes, and public buildings are adorned with horticulture. It enriches our lives through varied diets and physical activities from gardening; improves our environment and enhances our mental well-being by providing shade, air filters and noise barriers.

Agriculture has long been a staple of India's diverse economy, with the region offering farmers exceptional environmental conditions to grow crops and raise livestock. Now, with advances in technology and research, the region is poised to play a major role in developing the nascent agri-technology industry. With an abundance of farm land, nurseries and greenhouse facilities, India is the cornerstone of the agri-tech business.

High-tech horticulture soon will empower other fields of agriculture as well. Horticulture shall have a growing commercial importance in the near future as world population is aging, people devote more attention to health and most communities better emphasize the quality of life and environment.

Horticulture needs to redefine and/or focus itself because horticulture is an industry and its aim is economic success. It is an emerging sector of economic importance in India. Horticulture is defined as the cultivation of an orchard, garden, or nursery on a small or large scale, or the science and art of growing fruits, vegetables, flowers or ornamental plants. In the past, horticulture was practiced in order to produce food to survive. Ornamentals and flowers were added to improve well-being. People were proud to produce their own fruits, vegetables, medicinal and aromatic plants, and flowers. Gardening was a most welcome job. Nowadays, gardening (i.e. Horticulture) is thought of as a hobby, although it's also a profession. Though horticultural industry has reached unprecedented growth in the size, diversity, and value worldwide, but still horticultural science within academia is experiencing a crisis. In modern age, Technology has become a pervasive element in modern horticultural cropping systems. In this context it is define technology as "*advanced systems that manage, control, communicate, sense, and report information on crops at all stages of development and utilization.*" The technology includes many faces like:

1. Improved plant genotypes, some developed by recombinant DNA technology.
2. Value-added plant propagation.
3. Advanced hybrids, coated and primed seed, artificial seeds, micropropagation, advanced plug technology.
4. Computer-based environmental sensing and control, specially of green houses.
5. High speed data interchange between critical production points.
6. Environmental sensors and equipment for monitoring plant status.
7. Advanced production techniques (hydroponics, CEA, flood, etc.)

Horticulture is becoming more and more a low-tech, low-income activity. For high technology production systems needs highly trained managers and workers. They may be expensive but needs periodic updating. The aims of high-tech horticulture can be achieve to make mission successful. It is essential to ensure an end-to-end holistic approach covering production, post harvest management, processing and marketing (to assure appropriate returns to growers/producers). The grower should promote R & D technologies for production, post-harvest management and processing. It needs to enhance acreage, coverage, and productivity through diversification, from traditional crops to plantations, orchards, vineyards, flower and vegetable gardens and extension of appropriate technology to the farmers for high-tech horticulture cultivation and precision farming. High-tech horticulture also needs assist setting up post harvest facilities such as pack house, ripening chamber, cold storages, Controlled Atmosphere (CA) storages etc, processing units for value addition and marketing infrastructure. For this one needs to adopt a coordinated approach and promotion of partnership, convergence and synergy among R & D, processing and marketing agencies in public as well as private sectors, at the National, Regional, State and sub-State levels. It can be possible if we promote capacity-building and Human Resource Development at all levels.

MARKETS FOR HORTICULTURAL BIOTECH PRODUCTS

On the supply side, "horticulture" includes an enormous diversity of fruit and vegetable crops, but it also includes many nonfood species, such as ornamentals, flowers and recreational turf grass. Collectively these horticultural crops compare well with major agronomic crops in terms of total value. However, they use much less acreage, and the market size for some biotech products depends on both acreage and production value.

Horticulture includes a small number of larger-scale crops (such as potatoes and onions, apples and wine grapes) as well as a large number of smaller-scale crops (such as brussels sprouts and persimmons). At current costs for R & D and regulatory approval, it is unlikely that biotechnology products will be developed and achieve market acceptance for many of these smaller-scale crops in the near future.

Further, experimentation with perennials such as grapes, nuts and fruit trees is comparatively expensive (because the experimental unit is larger and takes more time), and it is costly to bring new acreage into production or replace an existing vineyard or orchard with a new variety. On the demand side, the market for horticultural products, especially fresh fruits and vegetables, is undergoing important changes associated with the changing structure of the global food industry. Increasingly fewer and larger supermarket chains have been taking over the global market for fruits and vegetables, especially fresh produce, and changing the way these products are marketed. Because fresh produce is perishable and subject to fluctuations in availability, quality and price, it presents special problems for supermarket managers compared with packaged goods. Supply-chain management, and the increasing use of contracts that specify production parameters as well as characteristics and price, is replacing spot markets for many fresh products. A desire for standardized products, regardless of where they are sourced around the world, could limit the development and adoption of products targeting smaller market segments, unless retailers perceive benefits and provide shelf space for diversified products, such as biotech and non-biotech varieties of particular fruits and vegetables. On the other hand, an increasingly wealthy and discriminating consuming public can be expected to continue to demand increasingly differentiated products - with an ever-evolving list of characteristics such as organic, low fat, low-carbohydrate and farm-fresh.

Hence retailers will have to balance the cost savings and convenience associated with global standardization against the benefits from providing a greater range of products, which will include GM products when retailers begin to perceive benefits from stocking them. Unlike other types of foods, fruits and vegetables are often consumed fresh and in clearly identifiable and recognizable form. This has implications for perceptions of quality and food safety that may influence consumer acceptance perhaps favorably, for instance, if a genetically modified sweet-corn could be marketed as reduced-pesticide. Other elements of GM horticulture such as nonfood products, ornamentals or turf grass have advantages in terms of potential market acceptance. GM trap crops, which provide pesticide protection for other crops, and GM sentinel crops, which signal the presence of pests or provide other agronomic indicators may be used in food production without overcoming barriers of acceptability to market middlemen or consumers.

Biotechnology products designed for home gardeners may be more readily accepted because the grower is the final consumer. Market acceptance in India may also be linked to continued access to export markets. The relative importance of the domestic market could help to account for the success of the GM feed-grain technologies in India, and it may also help to account for the success of these and other GM technologies. China is comparatively important in horticultural biotechnology its investment in agricultural biotechnology is second only to the United States.

WAYS TO MAKE HORTICULTURE AS HI-TECH HORTICULTURE

There needs to be an expansion and/or introduction of new high-income careers in horticulture. For example, floriculture has great potential to generate high income jobs in the perfume industry; new careers in medicinal use of horticultural crops (both traditional as well as allopathic and homeopathic uses), organic fruit and vegetable production, and other less traditional careers need to be emphasized. Hi-tech horticulture is the need of all the developing countries. Controlled-environment horticulture certainly has a promising future within multi-disciplinary plant-science research as well as many exciting applications within controlled-environment agriculture. There are various ways to make horticulture hi-tech. Following are the means for it.

Role of Biotechnology in High-tech Horticulture

Biotechnology tools to plant science, regularly working on the synthesis, isolation and cloning of new genes, and the genetic transformations of plants. Worldwide, consumption and cultivation of fruits and vegetables is increasing. Between 1990 and 2002, global fruit and vegetable production grew from 0.89 billion tons to 1.3 billion tons and per capita availability expanded from 342 pounds to 426 pounds (FAO 2003). Much of this growth has occurred in China, which is aggressively pursuing agricultural biotechnology. The global fresh fruit-and-vegetable marketing system is increasingly focused on adding value and decreasing costs by streamlining distribution and understanding customer demands. The year-round availability of fresh produce is now seen as a necessity by both food service and retail buyers. Product form and packaging are also changing as more firms introduce value-added products, such as fresh cut produce, salad greens and related products in consumer-ready packages. Fresh-cut sales are even higher in Europe and beginning to develop in Latin America and Asia as well. The implications of this trend may become as important to the biotechnology industry as the changes in market structure, since fresh-cut processors are increasingly demanding specific varieties bred with attributes beneficial to processing quality.

The Minister of Agriculture, Food, Public Distribution & Consumer Affairs. Shri Sharad Pawar has underlined the urgent need to stimulate the direction of institutional reforms in Indian agriculture. Inaugurating the National Convention on "*Knowledge-driven agricultural development: Management of change* (March 29, 2006), he said that biotechnology has tremendous potential of dealing with the intricate problems of falling natural resource health and rising expectations on quantity and quality of food. In view of its importance, we have recently launched a major Indo-US Knowledge Initiative, in which there is considerable focus on biotechnology.

Technological change has driven economic progress in agriculture and will continue to play a crucial role in the 21st century. The latest wave of technological change in agriculture is based in molecular biology. Will horticulture participate? Genetically engineered crop varieties have been adopted on a wide scale in some agronomic crops, but horticultural crops face larger hurdles. High costs for research, development and regulatory approval combined with the small acreages planted and the diversity of varieties, will limit the potential for profitable applications of biotechnology to many fruits and vegetables, tree fruits and nuts, and nursery crops. In addition, there are market barriers. Like most important changes in agriculture, modern biotechnology has met with spirited political opposition from some quarters. Threats of political action may discourage food manufacturers and retailers from adopting biotech products that are wanted by some consumers and may be profitable for growers.

Agriculture has been an important engine of economic development, and the mainspring of economic progress in agriculture has been productivity improvements driven by technological change that is fueled by research and development (R & D). These gains can be attributed to new biological, mechanical and chemical technologies, including improved genetic material, machines, fertilizers and pesticides, and knowledge. The current wave of technological progress continues this pattern, while emphasizing information technologies and biotechnology - in particular genetically modified (GM) crops. For many, GM crops represent the hope for a future with less hunger and malnutrition, and for a more sustainable agriculture with more varied cheaper and safer food.

Biotech crops have been a commercial reality only for a few years but they have made very rapid inroads in some parts of the market.

In the case of horticulture - the cultivation of fruits and vegetables, tree fruits and nuts, turf grass, flowers and ornamental crops - these issues are sharply drawn because the private sector has not found it profitable to develop or commercialize many GM crops in the current political, legal and market environment. What will happen in biotechnology applied to horticultural crops is up to the government, for a variety of economic reasons. Some of these aspects may be unique to GM horticultural crops but many are common to GM crops generally, and similar issues arise with some new non-GM technologies.

Genetic Resources

The identification, preservation, use and control of plant genetic resources play a vitally strategic role in horticulture for the future develop of produce diversification. It is no accident that multinationals are searching the world for unknown or forgotten species for the potential value and added value they may represent. Aromatic and medicinal herbs and plants, dietetic uses of fruits and vegetables, and industrial non-food for chemicals, cosmetics, pharmaceuticals and biodegradable manufactures all fall into this scheme of things. Though still relatively abundant, these resources are in evident diminution, and research must play a major role in all countries in the field, as has been and is being done by FAO, IPGRI and other international agencies and NGOs.

Conserving and protecting germplasm, as experts call it, means first of all safeguarding biodiversity, i.e. the natural genetic variability that is the savings bank for the future needs of mankind. Here too political controversy is no stranger, despite the general consensus on the basic objectives. As is well known, most of these resources are found in developing countries, whereas the technologically advanced countries have the ways and means to exploit the potential of these resources. Whence the inevitable claims of the former to a greater share of the benefits derived there from without having to comply with the intellectual property rights (IPRs) the latter have acquired through the development of resistant selections, transgenic material and other derivatives subject to patenting and the payment of royalties.

Research can play a leading role in the field of new crop development, of enhancing the value of forgotten or unknown species, of producing new industrial spin-offs, of discovering new natural molecules for the food, pharmaceutical, health-care and other industries and can make invaluable contributions in developing ways to preserve *in* and *ex situ* germplasm both *in vivo* and *in vitro* through cryoscopic and molecular techniques as well as gene banks. Indeed, it is for the future that we must now lay the groundwork by avoiding the pitfalls which could reduce horticulture to the level of a few crops or highly specialized species and with them spell the end of its biodiversity.

Molecular Genetic Characterization of New Floricultural Germplasm

Understanding the genetics of floricultural germplasm represents a value-added component of managing collections. Molecular approaches collectively represent a potential goldmine of important information that can be applied to programs of genetic improvement. In fact, they can be applied to almost every component of a unified germplasm program. The use of molecular markers for genetic diversity analysis and as a selection tool is a high priority for new efforts in the development of ornamental cultivars and the exploitation of species diversity. Molecular marker and DNA sequence analysis of extant and new floricultural germplasm collections should allow a more complete characterization and understanding of the genetic relationships between species and cultivars. With the advent of polymerase chain reaction (PCR) technology, numerous genetic approaches such as random amplified polymorphic DNA (RAPD), amplified fragment length polymorphisms (AFLP), microsatellite DNA (SSR), expressed sequence tags (EST) and single nucleotide polymorphisms (SNP) in candidate genes are available for germplasm characterization. In addition to estimating genetic diversity, these tools can be utilized to establish molecular markers for phenotypic traits of interest. Once obtained, such markers can be further applied in marker assisted breeding, developing various sorts of genetic maps, and as quantitative trait loci (QTLs). Molecular markers can also be used to confirm the parentage of individuals in a breeding population, for genetic "finger-printing" with application towards intellectual property rights, and also for understanding systematic relationships of cultivated species.

Tissue Culture

For all horticultural crops, propagation is one of the main business activities and a basic element to a grower's success. It involves seed and plant production, i.e. nursery activities, selection work, bioassays for disease and pest control/certification, and nursery techniques on a large and specialized scale. Some research stations specialize only in these fields, widely using *in vitro* culture, micropropagation and biotechnological approaches.

Tissue culture, an important area of biotechnology can be use to improve the productivity of planting material through enhanced availability of identified planting stock with desired traits. Tissue culture has helped to develop new strain of food crops, cereals, vegetables flowers, oilseeds and plantation crops such as spices, coffee, tea and rubber. Tissue Culture in ornamental Plants constitute a major part of the total lancing material produced in state out of 31.25 lakh ornamental and other plantlets, gerbera a alone is a major flower crops produced in the State. Although, this technology has been adopted by the farmers, considering area covered, adoption is at a very low level. Plant tissue culture has been promoted by Government of India since last 2 decades to enhance the production Land availability of disease free, true to type, quality planting material.

Micropropagation is one of the important contributions of plant tissue culture to commercial plant propagation and has vast significance. The technique provides a rapid reliable system for a production of large number of genetically uniform disease free plantlets.

Present Scenario of Plant Tissue Culture in India

Plant tissue culture has been promoted by Government of India since last 2 decades to enhance the production Land availability of disease free, true to type, quality planting material. At the end of 1996 about 76 commercial micro propagation units have registered for the production with Government of India of which 30 are functional within the Country. Total 40 million plants annually are produced against the installed capacity of 110 million plants.

Department of Bio-technology, Government of India has promoted the activity with the help of which two pilot plans were set up of the capacity or 1 million plants per annum of economically important plant species. National Chemical Laboratory, Pune (Maharashtra) is one of the units making extensive efforts for the development of cost effective indigenous technologies with felid evaluation. This unit is now upgraded to Micropropagation Technology Park to serve the entrepreneurs / industries institutes / universities so king for mass scale production of plants.

The potential for the domestic market is enormous and by conservative estimates it is around Rs. 150 cores with the annual growth rate of 10%. In order to fulfill the demand of domestic market even at the meager lever of 10 to 15% efforts needs to focus not only to increase the units at private sector but also the back up at National and slate level from the government is necessary.

In Maharashtra there are 17 active public and private sector dealing with production of tissue cultured planets producing Banana, Sugarcane arid Ornamentals and Orchids etc. Currently, the focus of the companies is mainly in the floriculture sector however, micro propagation in banana arid sugarcane is also gaining popularity. Recently Bhabha Atomic Research Centre, Mumbai have also done initial work on micropropagation Cardamom Mulberry and Sandalwood. Tissue Culture in ornamental Plants constitute a major part of the total lancing material produced in the state out of 31.25 lakh ornamental and other plantlets, gerbera a alone is a major flower crops produced in the State.

Potential of Tissue Culture in India

Flower industry is an emerging sector of economic importance in India. Floriculture has already established a place in economic development of farmers and private sectors of India. Some awareness has been created in use of micropropagation plants in the cultivation of other crops of horticultural importance. Existing laboratories are also producing flowering species. Since number of farmers for ploy house horticulture is increasing annually there is a scope for various use of tissue culture planting material of flowering plants and vegetable crops. Tissue culture units involved in production of tissue culture seedlings of flowering species like Gerbera, Carnation, Lilly, Arcadias etc. The number of Green house units in India should be increased for production of floriculture plants. The other potential crops may also be propagated through tissue culture technique to enhance the production.

Suggestions for Improvement

1. Strategy should be framed to establishment, encourage and strengthen of tissue culture laboratories. For strengthening the existing units a grant may be given to the units so as to produce additional 25 per cent plants over the existing production capacity at a rate lower by at least Rs. 2.00 per plant. This assistance could be in the form of interest subsidy on the present interest burden. In order to make these units viable certain policy decision need to be taken up by the State Government.
2. Collection and consolidation of information related to present scenario of micropropagation technology and to study the utility of this technology.
3. To evaluate the potential for increasing agriculture production with the help of tissue culture and to decide the strategy for next 5 years for use of this technology.
4. Establishment of tissue culture laboratories along with criteria regarding financial assistance.
5. To suggest extension plan for use of micropropagated plants by the farmers.
6. To suggest quality control measures for micropropagated plants.
7. To decide future the direction for research and development in the field of tissue culture considering needs of the state.
8. **Standardized package of practices:** Though the private sector has good tissue culture plants, but they are not ready with relevant standardized package of practices to meet field requirement raising field crop with tissue cultured plants. Lack of an adequate R & D infrastructure at unit level makes such type of recommendation impossible to private units arm an turn, grower has accepted to use tissue cultured plaids with traditional package of practices. This situation not only leads to wastage market purchased inputs, but also reduces per rupee net return. It needs improvement.
9. Tissue culture plants are charged Sales Tax. Although the producer has to pay the tax, the ultimate burden of this tax is shifted on the farmer. Sales tax on tissue culture plantlets should be waved at least for initial 10 years.

10. The electricity tariff should be brought at par with tariff for agriculture purpose.
11. The tissue culture units extending hardening facilities in captive areas, should be given an incentive for potential paining. Sheds at the rate or 50 percent of erection cost.
12. For establishment of new unit in private sector, grant of Rs. 10 lakhs may be given for the production capacity o~ minimum 5 lakhs plant per annum.
13. The job it venture may be encouraged with the help of agriculture universities, on the government's lands.

Transgenic Produce to Evolve Global Marketplace

If and when genetically engineered (GE) horticultural products become more widely available and adopted, they will enter an expanding marketplace that is becoming globally integrated and more consolidated. GE produce must meet the challenges of the quickly evolving market for fruits and vegetables.

To date, the successful GM crop varieties have emphasized "input traits," related to reducing the use of chemical pesticides or making them more effective, rather than "output traits," related to product quality.

Challenges for Horticultural Biotechnology

Applications of biotechnology in crop improvement have attracted the most controversy.

The application of biotechnology to crops has transformed the landscape of agriculture for soybeans, corn, cotton and canola by providing genetic resistance to herbicides and insects. However, biotechnology has had limited commercial success to date in horticultural crops, including fruits, vegetables, flowers and landscape plants.

A number of technical, economic, regulatory and market factors have combined to create hurdles for the utilization of biotechnology in horticultural crops, which are more diverse than field crops. Horticulture includes hundreds of distinct plants, the majority of which are grown on small acreages and which individually represent relatively small market values. Even the vegetable crops with the largest gross revenues, such as lettuce and tomatoes, are minor crops compared to major field crops like corn or soybeans.

Their limited acreage makes it more difficult to recover the research and development costs of any new technology specific to these crops. At the same time, consumer concerns and the related reluctance of food processors and marketers to accept new biotech commodities are delaying the introduction of horticultural products already developed. These barriers are exacerbated by the globalization of fresh produce markets.

Seed Technology and Management for Horticultural Crops

The seed is the carrier and catalyst of agro technologies. Seed is the cheapest and basic input for sustained agricultural production. Seed technology is an interdisciplinary science, encompassing a broad range of subjects, and includes seed production, handing, processing storage, testing, certification, quality control/ marketing and distribution. It also includes in research in seed development and maturation, seed physiology, seed dormancy and germination, seed viability and longevity, seed pathology and entomology, seed enhancement based on modern botanical and agricultural sciences and somatic embryo genesis and synseed production.

The global domestic markets have grown slowly, international seed exchanges (including seeds of horticultural crops) have quadrupled in the last 25 years from a modest US$ 880 million to over US$ 30 billion at present. Thus the trade between the countries now accounts for over 11 per cent of the total seed trade. Due to liberalization in seed import and export and signing of the GATT agreement in 1994 by the GOI, the Indian seed trade has also shown a positive and rising response. The total seed trade has reached Rs. 1600 crore during 1995-96.

It provides carrier of new technology, basic tool for secured food supply, principle means to secure crop yields in less favorable production areas and a medium for rapid rehabilitation of agriculture during times of natural disaster. Therefore the goals for seed technology should be rapid multiplication, timely supply, assured high quality seeds and reasonable price.

TABLE 3.1: Horticultural Seed exchange

Crops	Seed Exports (Million US$)
Herbage Crops	427
Potato	400
Beet	308
Other Agricultural Crops	750
Horticultural Crops	1150
Total	**3038**

TABLE 3.2: Seed exports of main exporting countries (million US$)

Country	Horticultural crops
USA	200
Netherlands	200
France	100
Brazil	1
Denmark	40
Germany	35
Belgium	NA
Chile	44
Italy	30
Canada	NA
Argentina	1
Austria	NA
Japan	40
Hungary	3
New Zealand	6
Australia	10
India	NA

Good quality seed in agriculture pay more but do not cost more. The important characteristics of quality seed are physical purity, germinability and vigour, genetic purity and seed health should be taken in to account.

At the time of release of a variety, small quantity of seed normally known as nucleus seed is available with the plant breeder. Commercial quantity of seed is produced after a series of multiplication steps. Starting from maintenance programme in which nucleus seed is multiplied in a generation system of multiplication as breeder, foundation and certified seed. The seeds may be breeder seed, foundation seed, certified seed.

Indian seed export figures are showing healthy trend. Our share in International seed trade is very meager but there has been a phenomenal increase in exports from 13.9 to 83.9 crores in a span of only five years. It is interesting to note that our exports are shifting to high value quality seeds. We started with export of open pollinated wheat seeds and now the major component of export is F_1 hybrid seeds. Many of the multinational seed companies have now identified the Indian potential in seed trade.

Seed Village Concept – A New Approach for Quality Seed Production

The seed village concept will go long way in quality seed production. The main objective of this concept is to involve farmers in seed production, and thereby make quality seed available at a reasonable price within the village. Isolation and synchronization of flowering can be solved to some extent by adoption of this concept. Seed crop harvesting, threshing and processing can be done in collaborative manner.

Improvement in Postharvest Handling, Storage and Processing of Fruits and Vegetables

India is the second largest producer of fruits after Brazil and the second largest producer of vegetables after China. The total estimated annual production is 70 million tonnes of which fruits constitute 25 million tonnes and vegetables 45 million tonnes. Due to inadequate facility of processing the produce nearly 35 percent of the total production is wasted. Recent development in technology has been limited to production level only. At rural level very little attention has been given on the handling, processing and preservation aspects of fruits and vegetables. The fruits and vegetables having some other specific characteristics differ from cereals. Fruits and vegetables are highly perishable in nature because of high moisture content (70 to 95 per cent), soft texture, unit size etc. A substantial return can be obtained by processing and marketing these products at small scale level in India. These are valuable crops having great potential of generating income and employment in rural areas and also high export values provided proper and scientific means of handling transportation and processing technologies are used.

In India different types of fruits and vegetables are produced because of diverse agroclimatic conditions, Considering major fruits and vegetables production in the country it has been found that mango production stands first in the country amounting 9.2 Mt followed by Banana and orange respectively at 4.5 and 1.35 Mt-annually. Similarly in major vegetables potato is the maximum produced crop amounting 12.7 Mt annually followed by okra (3 Mt), onion (2.7 Mt) and tomato (2 Mt). There is a great scope of primary handling and processing of these produces at rural level by applying appropriate Post Harvest technology.

The country's share of export of processed food product is not more than one per cent mainly because of inadequate capacity, technology, scale of production, and the packaging used does not match the major exporting countries. Besides productivity of the basic raw materials like horticultural produce being low, adequate processable quality of raw materials are not available to the industry over a longer period of time at reasonable prices for producing quality processed food products at competitive prices for larger export.

India has registered Rs. 676915.44 lakhs worth of export of horticultural products during the year 2008-09 as against Rs. 482182.25 lakhs in the corresponding period 2007-08. Production of processed fruits and vegetables increased by 40.00 per cent in the year 2008-09 and – 6.03 per cent in the year 2007-08. Growth rate of soft drinks was around 8 percent that of bear averaged at 10-15 per cent. Fruit exports comprise fresh fruits as well as fruit pulp, jams, jellies, squashes and syrups. Though the overall mango pulp production has been static its export from Bangalore was good. It has bee reported by APEDA (Agricultural and Processed Food Products Export Development Authority) the export of fruits has been growing by 37 percent annually. During the last financial year fruits and vegetables work more than Rs. 365915.53 lakhs were exported.

Export Potential

Though India is the second largest producer of fruits and vegetables in the world after China, it is still unable to tap vast export market due to lack of proper quality and reliability of processes. Huge post harvest losses and processing of very small fraction of produce are other major constraints to the extraction of full potential.

TABLE 3.3: Export of horticulture produce in India

(QTY IN: MTS, VALUE IN: Rs. LAKHS)

PRODUCTS	2006-07		2007-08		2008-09	
	Qty.	Value	Qty.	Value	Qty.	Value
FLORICULTURE & SEED						
Floriculture	42545.28	65269.73	36240.71	34014.42	30798.34	36881.41
Fruit & vegetable seed	8104.09	12158.95	10157.13	14212.29	8535.53	11999.09
Total for Floriculture & vegetable seed	2597031.11	77428.68	46397.84	48226.71	39333.87	48880.5
FRESH FRUITS & VEGETABLES						
Fresh Onion	1378373.17	116330.57	1008606.48	103577.89	1670186.29	182752.21
Other fresh vegetable	276824.60	43314.38	350235.47	48949.01	505285.47	68020.32
Dried nuts (walnuts)	5062.86	11803.79	6716.48	16207.80	5696.34	14123.63
Fresh Mango	79060.88	14193.95	54350.80	12741.76	83703.18	17071.25
Fresh Grapes	85897.79	30192.45	96963.57	31782.51	124627.97	40861.28
Other fresh fruits	177638.30	30997.39	207700.78	30452.60	256768.53	43086.84
Total for fresh fruits and vegetables	2002857.6	246832.53	1724573.58	243711.57	2646267.78	365915.53
PROCESSED FRUITS AND VEGETABLES						
Dried and preserved vegetables	119270.43	42754.17	125726.28	42993.81	147861.21	49641.51
Mango pulp	156835.51	50582.79	166752.17	50968.51	173013.60	75298.90

PRODUCTS	2006-07		2007-08		2008-09	
	Qty.	Value	Qty.	Value	Qty.	Value
Other processed fruits and vegetables	318067.57	95550.82	311756.29	96281.65	387126.42	137179.00
Total for processed fruits and vegetables	594173.51	188887.78	604234.74	190243.97	708001.23	262119.41
GRAND TOTAL	**5194062.22**	**513148.99**	**2375206.16**	**482182.25**	**3393602.88**	**676915.44**

Source: National Horticulture Board, (Website: www.nhb.gov.in), Database-2009, page-212

TABLE 3.4: Item-wise export of spices from India

Spices	2006-07		2007-08	
	Quantity (tonnes)	Value (Rs in lakhs)	Quantity (tonnes)	Value (Rs in lakhs)
Pepper	28,750	30,620	35,000	51,950
Cardamom (S)	650	2,236	500	2,475
Cardamom (L)	1,500	1,695	1,325	1,500
Chilli	148,500	80,775	209,000	109,750
Ginger	7,500	3,975	6,700	2,800
Turmeric	51,500	16,480	49,250	15,700
Coriander	20,500	7,462	26,000	11,025
Cumin	26,000	20,150	28,000	29,150
Celery	3,550	1,321	2,900	1,325
Fennel	3,575	2,380	5,250	2,850
Fenugreek	8,500	2,699	11,100	3,300
Other seeds (1)	8,000	2,240	8,850	3,125
Garlic	11,500	2,128	675	400
Nutmeg & Mace	2,100	4,274	1,300	2,875
Vanilla	125	1,996	200	1,775
Other spices (2)	19,500	7,280	19,000	8,100
Curry powder/Paste	9,500	8,693	11,500	11,100
Mint Products (3)	16,250	110,095	21,100	128,050
Oils and Oleoresins	6,250	51,079	6,600	56,300
Total	**373,750**	**357,575**	**444,250**	**443,550**
Value in million US $		**792.95**		**1,101.80**

(1) Include Ajwan seed, Dill seed, Poppy seed, Aniseed, Mustard etc.
(2) Include Asafoetida, Cinnamon, Cassia, Cambodege, Saffron, Spices (NES) etc.
(3) Include menthol, menthol crystals and mint oils.
Source: SPICES BOARD, Cochin

There is a great demand for fresh as well as processed fruit and vegetable products. The requirement can be met by production of good quality produce, proper grading, storage, packaging, transportation and marketing network. So far very little work has been done for these perishables from an engineering point of view. To exploit the crops the produce processing is the only solution. There is an urgent need to develop specific technologies to reduce losses from harvesting till consumption in order to meet domestic demands and to exploit huge export markets.

Throughout the world postharvest crop losses are high. Many research programs are active in this field. Of course, postharvest is linked not only to growers but also to others involved in the distribution chain, although storage (physiology and maintaining crop quality) and packaging are paramount.

There are also alternative techniques to improve the healthiness of edible produce, or to reduce costs of storage and handling, that are part of this research. They involve the very large field of processing (with new techniques that make processed fruits competitive-for quality and price-with the fresh). For some developing countries processed produce is more important than fresh, also because it is better suited to long-distance transport (market globalization).

Promotion of Cut Flower Industry

The flower industry suffers from logistical bottlenecks and inadequate post-harvest infrastructure, but government support has focused mostly on the growers. Delhi is the silent hub of the nation's enormous flower traffic. The flowers moving in and out of Delhi in all directions of the globe are as diverse as the various stages of handling they pass through. However, while roses from Pune decorate coffee tables in a plush German café, the perennial struggle of flower growers, suppliers and distributors never make it to the desktops.

Problems surfaced five years ago with the establishment of the market, and their magnitude has not changed much since then. Today, of the total of about 500 flower sellers, a measly 37 hold authorized licenses.

Delhi may have the maximum 'flower traffic' in Asia. Flowers from Bangalore arrive in Delhi and are then transported to Bhopal. Orchids are imported from Thailand and then exported to Singapore. Despite the growth and sale of flowers having spread over most of the country, India's annual flower production stands at around 1,000 tonnes and the country's floriculture industry has a miniscule 0.01 per cent share in the international market.

Looking into the International perspective (a report of the International Labour Organization - China and India dominate in terms of area under cultivation but their yield per hectare is low. Germany is the typical example of a second group of countries with sizeable markets where imports satisfy most demand (up to 70 per cent). Roses are the main traded product and Germany is the largest import market, followed by the United States. A third group, which includes Columbia and Kenya, have a small home market but a large volume of exports. Lastly, The Netherlands is a member of the group with a large home market combined with a large export share. It is the world's leading exporter.

A majority of Indian flowers blooming in the fields are certainly at par in quality to the ones grown in Europe, the problem begins the moment they are cut and commence their journey to the market. Improper handling after harvest, unavailability of optimum temperature and moisture conditions during storage and transportation, under or over-packaging leading to damage and wiltering are a few factors which have an adverse impact on the quality of Indian flowers reaching the consumer - both domestic and international. Therefore, there is an urgent need to promote cut flower industry for hi-tech horticulture.

Integrated Production Systems (IPSs)

Integrated production systems are a leading concept for developing counties and should be encouraged throughout the world for the realm of sustainable agriculture. The organization of the production process and related research can be included are (1) Crop management (training, pruning, thinning, etc.), (2) Crop protection, (3) Soil management, nutrient management, etc.), (4) Crop protection, (5) Soil management, nutrition, (5) Water supply and irrigation, (6) Ripening and picking and (7) Crop quality assessment.

Indigenous Knowledge for Horticultural Crops

Indigenous Knowledge (IK) is defined as the knowledge that, the people in a given community have developed overtime, and continue to develop. It is based on experience, often tested over centuries of use, adapted to local culture and environment and it is dynamic and changing. Indigenous knowledge is not confined to tribal groups or the original inhabitants of an area. It is not even confined to rural people. Rather any community possesses IK - rural and urban, settled and nomadic, original inhabitants and migrants. As defined by Grenier (1998) indigenous knowledge is the unique, traditional, local knowledge existing within and developed around the specific conditions of women and men indigenous to a particular geographic area. The term 'indigenous knowledge' is used synonymously with 'traditional' and 'local knowledge' to differentiate the knowledge developed by a given community from the international knowledge system, sometimes also called as 'Western' system generated through Universities, Government research centers and Private industries. Further IK is dynamic and not static, as the word 'traditional' commonly 'implies'. This is supported by the inference of Berkes (1999) that 'traditional' does not mean an inflexible adherence to the past; it simply means time-tested and wise. IK of horticultural crops should be taken into consideration while making horticulture hi-tech.

Popularizing Irrigation Practices for Water Conservation

Irrigation, in agriculture, is artificial watering of the land. Although used chiefly in regions with annual rainfall of less than 20 in. (51 cm).

An adequate water supply is important for plant growth. When rainfall is not sufficient, the plants must receive additional water from irrigation. Various methods can be used to supply irrigation water to the plants. Each method has its advantages and disadvantages. These should be taken into account when choosing the method which is best suited to the local circumstances. There are three commonly used methods: surface irrigation (basin irrigation, furrow irrigation and border irrigation), sprinkler irrigation and drip irrigation. Following methods of irrigation are commonly adopted for horticultural crops for water conservation:

Surface irrigation is the application of water by gravity flow to the surface of the field. Either the entire field is flooded (basin irrigation) or the water is fed into small channels (furrows) or strips of land (borders). This kind of irrigation is not helpful for water conservation.

Sprinkler irrigation is a method of applying irrigation water which is similar to natural rainfall. Water is distributed through a system of pipes usually by pumping. It is then sprayed into the air through sprinklers so that it breaks up into small water drops which fall to the ground. The pump supply system, sprinklers and operating conditions must be designed to enable a uniform application of water. It is suited for most row, field and tree crops and water can be sprayed over or under the crop canopy. However, large sprinklers are not recommended for irrigation of delicate crops such as lettuce because the large water drops produced by the sprinklers may damage the crop.

Drip irrigation was used in ancient times by filling buried clay pots with water and allowing the water to gradually seep into the soil. The concept of drip irrigation systems is simple. Tiny holes inserted at various points along a hose, allow small quantities of water to trickle slowly into the soil over long periods. The disadvantage of sprinkler systems is that a lot of water is wasted, owing to evaporation of the spray droplets as they fly through the air. Drip irrigation systems use less water, because it is applied directly where the plants need it most. Drip irrigation also known as trickle irrigation or microirrigation is an irrigation method that applies water slowly to the roots of plants, by depositing the water either on the soil surface or directly to the root zone, through a network of valves, pipes, tubing, and emitters. The goal of drip irrigation is to minimize water usage. It is used by farms, commercial greenhouses, and residential gardeners.

There is a need for educating farmers for proper irrigation practices. They should sprinkler and drip irrigation systems. These will help to conserve water which can be used for new crop production.

For micro-irrigation The Government of India has been providing focused attention for the development of Micro Irrigation in the country for which initiative was taken during the 8th Plan under the scheme "Use of Plastics in Agriculture". Development of Sprinkler Irrigation was also promoted through various programmes of the Crops Division. Due to this efforts, it has been possible to cover 0.5 million hectares under Drip Irrigation and 0.07 million hectares under the Sprinkler Irrigation till 2003-04.

With a view to cover more areas in the country under this Technology, the Ministry of Finance constituted a task force with the mandate to give the recommendation on the strategy. The Task force on Micro Irrigation (TFM) submitted its report on 4th February 2004 to the Hon'ble Prime Minister. The TFM has recommended the coverage of 3 million hectares under Micro Irrigation during the Xth Plan involving an investment of Rs. 10,500 crores. The Department of Agriculture & Cooperation is contemplating to launch a new scheme on Micro-Irrigation during Xth Plan. The scheme was launched in January 2006 and funds to the tune of Rs. 280.00 crores were approved for release to the Implementing Agencies by 31st March 2006.

Plasticulture applications are one of the most useful indirect agricultural inputs which hold the promise to transform Indian agriculture and bring in the "Second Green Revolution"

Use of Plasticulture

Plasticulture is a technique for the use of plastics in agriculture, horticulture, water-management, food grains storage and related areas. The plastics materials and end products are used in Plasticulture applications like water conservation, irrigation efficiency, crop and environment protection, as well as end product storage and transportation. Plastics play a major role in energy conservation because they require minimum energy in production and conversion to finished products. Plastics have definite advantages over conventional materials. Plastics have higher strength / weight ratio, superior electrical properties, superior thermal insulation properties, excellent corrosion resistance, superior flexibility, impermeability to water, gas, etc., resistance to chemicals and less friction due to smoother surface.

Improvement in Mechanics of Horticultural Products Marketing

Marketing involves finding out what your customers want and supplying it to them at a profit. The importance of good marketing can be conveniently considered from the four different perspectives of the national economy, the farmer, the product and the consumer. The whole marketing process has to be customer oriented. Production must supply customers with what they want or need. In fact, marketing is a commercial process and is only sustainable if it provides all the participants with a profit. Agricultural marketing is achieved by a series of processes which include harvesting techniques, the grading and sorting of crops and the packing, transport, storage, processing, distribution and selling of products. These are the mechanics of marketing. For this purpose proper market executive should be appointed or can be organized at own level. This role includes:

- Collection, evaluation and dissemination of market information;
- assistance in the planning and scheduling of production;
- securing the market for producers, e.g. through contracts with buyers;
- advice on the best practical post-harvest practices;
- coordination of inputs, transport, storage, credit and post-harvest facilities.

Therefore, for high-tech horticulture proper marketing strategies should be prepared. Since horticultural products are mainly sold fresh; some are eaten raw while others are cooked. Some horticultural products have traditionally been processed when no other form of storage was available, e.g. dried fruit and jams. Marketing may be done through society. As society develops and becomes more affluent the market for processed and prepared horticultural products develops. A market also develops for horticultural products such as flowers and house and garden plants which are sold for purely aesthetic reasons. Increased wealth also brings with it an increased demand for product diversity in the form of new crops, off-season supplies and different flavours.

It is common knowledge that promotion and marketing are essential in the commercializing horticultural products. They are expensive and require specific organization of growers and supermarket/retail groups. They address to following basic needs:

1. Deliver produce regularly and reliably to the market;
2. Provide a feed-back of information between the market and the growers so that production can be constantly oriented towards consumer demand
3. Establishing an 'image' (brand names and trademarks, labels, etc.) and building confidence in the consumer through education.
4. Providing a guarantee of produce to the consumer through education.
5. Providing a guarantee of the production process as related to food safety.

Role of Public and Private Agencies

The balance in agricultural R & D between the private and public sectors varies among types of research. Without government intervention, the rate of innovation will be too slow, reflecting both under investment in research and under adoption of some research results. Both problems are related to the nature of property rights for research results. "Free-rider problems" occur when property rights are incomplete, and private investors can capture only part of the returns to their investments in certain types of research (such as developing new crop varieties); as a result, their incentives to invest are reduced.

Intellectual Property Rights (IPRs)

Intellectual property rights (IPRs) are a double-edged sword: to the extent that they provide a greater incentive for investing in research they are also likely to result in lower adoption rates. Governments have addressed the incentive problems in agricultural research in several ways. Federal and state governments (as well as industry) have funded agricultural research at public institutions such as the ICAR and state agricultural experiment stations associated with land-grant colleges. This approach allows an increase in total research without the problems associated with monopoly pricing of inventions. However, even though the investment has paid handsome dividends, it is increasingly difficult to sustain the past levels of funding for public agricultural R & D, in the face of general budget problems and declining political support for public science funding, including agricultural science. Governments have also acted to strengthen IPRs applied to plants and animals.

Extension Plan

The extension system today is partially successful in creating general awareness among farmers about benefits of hi-tech horticulture. For example for tissue culture the farmers are not clear about the entire package of technology. The presumption that tissue cultured seedlings itself will increase the yield needs to be addressed in effective manner. Entire package of practices required for getting optimum yield needs to be demonstrated effectively on the farmers fields In order to spread the proper message of benefits of micro propagation, extension machinery itself needs to be equipped and trained in this area. The entire extension program for bio technology has to be organized in such a manner that extension workers and the farmers get simultaneous training and could upgrade their skills for quality production from tissue culture plantlets. Agronomic practices for tissue culture plantlets have to be finalized through research and then have to propagated through extension machinery of the state.

Training

Intensive training program for extension staff may be organized first to train the master trainers and generate in louse resource persons. This training may be organized for identified persons in the institutes of repute. The trained manpower or resource persons should train the field functionaries so that the field level workers will also be able to develop technology. During the first year, there has to be a major thrust this training of staff. The farmers also need training in these aspects of biotechnologies. Private sector also needs to be involved in the process of farmers training. Training farmers should be mandatory on the part of entrepreneur for which 10 percent financial assistance on the amount of effective sale may be given as assistance from the Government. There needs to be a regular interaction arid continuous dialogue between the developed agencies, extension workers and Regular work shops, seminars, symposium needs to be organized for making farmers aware and also getting feed back from them.

Technology Demonstrations

Through there is awareness about use of tissue culture in agriculture there is hardly a demonstration on tissue culture technology exhibiting total package of practices. Extension machinery and the private sector will have to be aggressive in organizing demonstration on the farmer's fields. Listing out the technology generated in various crops will be the first stage in this process. Identification of user farmers will be the second stage.

Quality Control Mechanism

Although tissue culture laboratories have established in the state during last decade producing tissue culture plants for distribution to the farmers, there is no mechanism to check the quality of planting material produced by them. Since the quantum of tissue culture plant would be increasing at higher rate. The quality of plants will have to be ensured for getting desired type of quality planting material. It should be mandatory for all the tissue culture units to register themselves with the competent authority.

Need for Education Reform

Recently, the horticultural industry in India has grown to a great extent; however, the academic side has to be improved. Academics are mostly limited to Indian agriculture. India is a country with diverse climatic conditions. Horticulture students have an opportunity to study tropical, subtropical, temperate, and desert horticulture. This is the time to revise the horticultural curricula to include a global perspective. More emphasis needs to be given to horticultural marketing and economics. Farmers are able to produce crops using appropriate technologies if they are assured the market is there. Therefore, the objectives of horticultural education in India should be redefined to broaden the scope of the subject and should not be locale specific. This will help the students/growers to become globally competitive and will meet the requirements of employers from any part of the world. The horticulture industry should be encouraged to institute fellowships for bright students. An international student and faculty exchange program can be put into place to promote collaborative research projects.

The Indian Council of Agricultural Research (ICAR), New Delhi, has a strong network of research institutes and agricultural universities in India. Most of the agricultural universities offer courses in horticulture at the undergraduate and graduate level. The number of horticultural science departments in Indian universities is declining over the last 15 years. Students often fail to perceive horticulture as a science and certainly not as a career option. Horticultural science professionals around the world have to share their thoughts on this issue. The science of horticulture should now be linked with biochemistry, biotechnology and molecular biology. There is a need for initiation of a campaign for making students aware of biotechnology in relation to high-tech horticulture. The revision of course curricula for undergraduate and graduate programs is required to meet the changing needs of the industry. For developing countries, horticultural science or technology still lags behind, therefore we need to improve the higher education program at our Universities.

If students fail to perceive horticulture as a science, they are correct. The fundamental task of scientists in horticulture is to contribute to this success. This can be achieved by improving the product or the methods of production by all available means; a fascinating and never-ending challenge. For this purpose, scientists must have a good overall knowledge of the crops they are dealing with, of their needs, their weaknesses, and of the methods to better serve these needs and to overcome these weaknesses. The general approach of a scientist working in and for horticulture should be a holistic one, trying to understand the plant as a system, the function of this system and the specific manner of its interaction with the larger system of environment.

We Need

1. To promote an awareness in the general community, and in particular primary and secondary school students, of the importance of horticulture and horticultural science to the economy of India and to the health, well-being and quality of life of Indians.
2. To review the quality of science teaching, and the qualifications of science

teachers, in graduate and post graduate level and to upgrade these, providing rewards, status and incentives for better qualifications, so that students will be instilled with a love of science and inspired to pursue a career in science, including horticultural science.

3. To coordinate the different levels and types of horticultural education to ensure that their individual objectives are defined to more effectively meet the training and knowledge needs of students by the efficient delivery of complete and appropriate courses, avoiding wasteful duplication.
4. To revise the horticultural curriculum to ensure that content is appropriate to current needs and that areas such as processing, marketing and value-adding are adequately catered for.
5. To review the accessibility of education and training options in horticulture to ensure that industry participants have the opportunity to acquire and upgrade their skills and knowledge through appropriate on-going educational opportunities.

Future Research and Development

Equally important as the research initiatives themselves are the role they play in determining the balance of disciplinary expertise in modern horticulture departments.

At present the cost of tissue culture planted are quite high due to how multiplication rates in many protocol efforts need to take place in this field. Agronomic practices commensurable with tissue culture technology should be finalized. A new protocol should be established to propagate horticultural importance crops which have tremendous potential for India, e.g., medicinal and aromatic plants, spices, mango and cashewnut for each state and agroclimatic conditions.

Horticulture shall have a growing commercial importance in the near future as world population is aging, people devote more attention to health and most communities better emphasize the quality of life and environment. Development and adequate share of new discoveries across the scientific horticulture community through adequate training of horticulture specialists at all educational levels shall insure a bright specialist at all educational levels shall insure a bright future to horticulture and the humanity. Networking of world horticulture scientists working on the same commodities, disciplines, geographical areas, theme of research, etc shall help to support such development.

Following are the major need to make horticulture – hi-tech horticulture:

1. Basic versus applied research.
2. The needs for research on biodiversity and molecular genetics.
3. The growing importance of biotechnologies.
4. Organic or agrochemical growing systems.
5. Promoting biological control of horticultural crop predators.
6. Quality of horticultural products.
7. Human nutrition and nutraceuticals development.
8. Globalization and distribution concentration.
9. Evolution of consumer needs.
10. Urban horticulture and environmental quality.
11. Internationalization of training and availability of knowledge.
12. Development of commercial horticulture through production and post-harvest management
    a) To develop high quality horticultural farms in identified belts and make such areas vibrant with horticultural activity which in turn will act as hubs for developing commercial horticulture by adopting high-tech horticulture

techniques.

b) To develop post-harvest management infrastructure.
c) To improve linkages between horticulture producers and marketers.
d) To create integrated network for marketing of horticulture produce.
e) To increase producer's share in consumer price.
f) To encourage networking of schemes for resource mobilization with all other related agencies/organizations; both of Government, Financial Institutions and private agencies engaged in the field of horticulture promotion in the country.

13. Characterization of the impact of fruit and vegetable consumption on human health and well-being, with correlative attention focused on plant genetic modification, specialized cropping systems, postharvest physiology and technology, food safety, and public education to maximize the beneficial impact of fruit and vegetable consumption.
14. More attention to the field of "environmental horticulture" in its broadest dimensions, including study of the psycho-social value of human-plant interactions. Investigations should include landscape ecology on the micro- and macro- scale, restoration biology, and the environmentally-sensitive management of home and urban landscapes. Identification and development of plants resistant to common pests to decrease use of chemicals by consumers will be important.
15. Ornamental horticulture, including nursery and greenhouse crop production, floristry, landscape design, construction and maintenance, has been overlooked entirely by national competitive programs, except for minor elements that focus on integrated pest management or invasive species. Ornamental horticulture is the most profitable and rapidly expanding sector of plant agriculture.
16. Sustainable cropping systems encompass many horticultural crops, and small to medium-sized operations look to fruits and vegetables for economic success because of their high value per unit of land and resource inputs. These systems are economically feasible options for re-directing the "tobacco economy" into more socially acceptable production alternatives. The search for alternative crops and new uses for current crops should be accelerated. Increased attention should be focused on the needs of small full- and part-time growers, as the sustainable systems of the future likely will involve farming as supplementary rather than primary income.
17. The emerging field of Controlled-Environment Agriculture (CEA) increasingly will produce our salad, vegetable and specialty crops as the twenty-first century progresses. Coupling this high-tech, "clean" horticulture with bio-derived, renewable energy sources will be the key to developing a profitable CEA growth industry that is sustainable and environment-friendly.
18. Need for high density plantations, which include adoption of appropriate plant density/canopy management, quality planting material, support and management system with appropriate inputs.
19. Micro-propagation (Tissue Culture), for mass production of true-to type, in possible shortest time, disease-free.
20. Hi-tech cultivation under controlled climatic conditions i.e. in Poly houses, Green houses, net-houses, etc. Rainfed Production through efficient water management techniques, mulching for soil moisture conservation, use of barriers in soil to reduce percolation, irrigation by drip, sprinklers, fertigation, and water harvesting structures etc.
21. Nursery management for quality seed/plant production of vegetables, flowers, ornamentals, fruit etc.
22. Hybrid Seed production.

23. Organic farming.
24. Hydroponics for year -round quality production.
25. Use of plastics in horticulture.
26. Bio-technology.
27. Genetically modified Organisms (GMOs).
28. Development of horticulture ancillary industry for improved packaging, equipments, plastics, corrugated boxes, horticulture machinery/tools, etc.

Future Needs for Production and Post Harvest Management

1. To develop high quality horticultural farms in identified belts and make such areas vibrant with horticultural activity which in turn will act as hubs for developing commercial horticulture by adopting high-tech horticulture techniques.
2. To develop post-harvest management infrastructure.
3. To improve linkages between horticulture producers and marketers.
4. To create integrated network for marketing of horticulture produce.
5. To increase producer's share in consumer price.
6. To encourage networking of schemes for resource mobilization with all other related agencies/organizations; both of Government, Financial Institutions and private agencies engaged in the field of horticulture promotion in the country.

Production Related

- High quality commercial horticulture crops
- Indigenous crops/ produce, herbs
- Aromatic Plants
- Seed and Nursery
- Bio-technology, Tissue culture
- Bio-pesticides
- Organic Foods
- Establishment of Horti-health clinics/laboratory
- Consultancy services
- Beekeeping

Primary Processing Related

- Grading/Packing/Washing/waxing/sorting/drying centres
- Pre-cooling unit/Cool Stores
- Reefer Van/Containers (with multi-chamber, multi-product facility)
- Specialized Transport Vehicle
- Retail outlets
- Auction platform
- Ripening curing chamber
- Market yards/rope ways
- Radiation unit/Dehydration Unit/Vapour Heat Treatment Unit
- Primary processing of products fermentation, extraction, distillation, Juice vending pulping, dressing cutting, chopping etc.
- Horticulture ancillary industry e.g. tools, equipment, plastics, packaging, etc.
- Plastic Crates, Cartons, Baskets, Aseptic Packaging and Nets (50% Subsidy).

Priority Areas

- Export Oriented Units/Projects
- Project from cooperative sector

- Projects in North Eastern Region, Hilly and Tribal areas
- Introduction of new processes, products or markets including new technology/ equipment
- Project involving women entrepreneurs and ex-serviceman.

CHAPTER 4

Perspectives of Horticulture

Horticulture (garden cultivation) is the branch of agricultural plant sciences that deals with the production of fruits, vegetables, nuts, and ornamentals. It is major source of food and employment in society. Operationally, there are several divisions of horticulture: olericulture (vegetable production), floriculture (flower production), turf culture (turf production), pomology (tree fruit production), arboriculture (tree production), and landscape architecture (design and use of plants in the landscape). Horticultural foods are rich in minerals and vitamins. Horticulture can be undertaken on a small scale by homeowners on their property, producing flowers and food plants. As an industry, horticulture is supported by a large number of service providers that supply equipment, chemicals, and implements. Nurseries provide plant materials for growers. Numbers jobs are available to persons with formal training in horticulture, but an equally large number of jobs in this branch of agriculture require little or no formal training.

Horticultural activities may be divided into several broad categories based mainly on the kinds of plants involved. These divisions form the basis of certain academic programs in horticulture:

FRUIT CULTURE

Fruits vary in numerous ways. Some are borne on trees, others on bushes. Some fruits are succulent and juicy, whereas others are dry. Growing from is a long-term operation. Fruit trees take a long time (several years) to come into bearing. They also require more growth space per plant than vegetables. An area of land on which fruit trees are grown in a significant concentration is called an orchard. The branch of horticulture involved with the production (including growing, harvesting, processing, and marketing) of fruit trees (including nuts) is called *pomology.*

Pomology (from *pome*, meaning fruit, and *ology*, meaning branch of learning) is the science of fruit growing and is divided into three following classes:

- **Practical pomology:** It is referred to as fruit production. Modern scientists refer to pomology as the actual growing of fruits.
- **Commercial pomology:** It is primarily the marketing and disposition of fruits, including their storage, preparation and market and many other areas, transportation and outlets, etc.
- **Systematic pomology:** It is the knowledge of plants as they grow, their distinctions and their habitat in growth.

Fruit trees such as apple, orange, and per are operationally distinguished from small fruits such as grape, blueberry, and strawberry.

VEGETABLE CULTURE

Vegetable production is one of the most popular horticultural activities indulged in by homeowners, often in the backyard or private section of the property. The branch of horticulture involved with the production of vegetables is called olericulture.

Olericulture is science of vegetable production the ancient art of vegetable gardening involves judicious timing and many skills. The knowledge and understanding of planning, planting and producing vegetables reward a gardener with an abundant supply of quality vegetables. They can be grown with use of man-made products or organically.

Some vegetable plants are grown for their fruits (e.g., tomato), leaves (e.g., spinach), roots (e.g., carrot), or pods (e.g., bean). Unlike fruits, vegetables are generally short-duration plants that need to be restarted each growing season.

Vegetables may be harvested and used fresh. However, they are also processed in a variety of ways.

ORNAMENTAL HORTICULTURE

The production and use of ornamentals is the branch of horticulture generally called *ornamental horticulture.* The term *use* is included definition because it is an integral part of this branch of horticulture. Ornamentals may be cultivated in open space (or landscape) or in indoor containers. They may also be grown, arranged, and displayed in a variety of ways. Subdivisions of this branch of horticulture involve distinct activities. *Floriculture* is the production and use of flowering plants and one of the areas most readily identified with when horticulture is mentioned. It is primarily the art of growing flowers and shrubs used to beautify the landscape.

Floriculture

The operations and practices relating to the care of plants cultivated or grown for their flowers or foliage. Floriculture deals with cultivation, marketing and arranging of flowers and foliage plants. Flowers and ornamental plants have gained an important position in present day society. In fact, man has utilized flower to express his sentiments on occasions of social functions, wedding, birth-day, funerals, etc. In past, flowers were grown outdoors only and their quality and quantity was very much limited by the season.

This specialized food and ornamental-plant production businesses are grouped together in the section of agriculture that is known as horticulture. Ornamental horticulture more appropriately covers the complete industry of flower and ornamental plant production and landscaping.

Flowers and ornamental products are produced commercially and their markets exist throughout the country in which flower and other products are sold.

This contrasts with agriculture, which, in most Western European countries, relies on a high level of machinery use over an extensive area of land, consequently involving few people in the production process. However, the boundary between the two is far from clear, especially when considering large-scale *vegetable production.* There is also a fundamental difference between productive horticulture, whether producing plants themselves or plant product, and upkeep of gardens and landscape for their amenity, cultural and recreational values. Where the tending of plants for leisure moves from being horticulture to countryside management is another moot point.

In contrast, the change associated with replacing plants with alternative materials, as in the creation of artificial playing surfaces, tests what is meant by horticulture in a quite different way. This book concerns itself with the principles underlying the growing of plants in the following sectors of horticulture:

An important aspect of the landscape is the ground covering, which is usually grass. *Turf grass* science has developed into a full-fledged program at many colleges. A lawn is the basic landscape element in most cases. Other plants are then playgrounds, and home grounds. Flowers in the landscape may be herbaceous or woody. The branch of horticulture involved with the production of tress is called *arboriculture.*

The other branches are as follows:

- **Turf culture:** It includes decorative lawns and sports surfaces for football, cricket, golf etc.
- **Garden construction and maintenance:** It involves the skills of landscaping together with the development of planted areas. Closely associated with this aspect the maintenance of trees and woodlands (arboriculture), specialist features within the garden such as walls and patios (hard landscaping) and the use of water (aquatics). Interior landscaping is the provision of semi-permanent plant arrangements inside conservatories, offices and many public buildings and involves the skills of careful plant selection and maintenance.
- **Greenhouse Production:** It enables plant material to be supplied outside its normal availability, e.g. chrysanthemums all the year round, tomatoes to a high specification over an extended season, and cucumbers from areas where the climate is not otherwise suitable. Plant propagation, providing seedlings and cuttings, serves outdoor growing as well as the greenhouse industry. Protected culture, mainly using low or walk-in polythene tunnels, is increasingly important in the production of vegetables, salads and flowers.
- **Nursery Stock:** Nursery stock is concerned with the production of soil-grown or container-grown shrubs and trees. Young stock of fruit may also be established by this sector for sale to the fruit growers: soft fruit (strawberries, etc.), cane fruit (raspberries etc.) and top fruit (apples, pears etc.).

Trees are perennial elements in a landscape design. They usually are large in size and hence require more space than annual plants.

Landscape Architecture

Landscaping is the use of ornamental plants in conjunction with other elements to beautify a given area. The professionals who design such plants are called *landscape architects.* Since landscaping can enhance a property, it has become an integral part of home construction. Commercial facilities and other public areas are also appropriately landscape. Malls, playgrounds, boulevards, and parks are example of public places where ornamental plants are used to enhance the environment aesthetically and make it more functional. The use of plants indoors is called interioscaping (as opposed to landscaping).

ALLIED HORTICULTURAL INDUSTRIES

As previously stated, horticulture as an industry, supported by other allied industries. The major ones are the nursery industry and seed industry.

The Nursery Industry

The growth in the horticultural industry today is attributable in part to the growth in the nursery industry. Nurseries provide seedlings for growers who do not want to raise plants from scratch and prefer to take advantage of their convenience. In fact, certain plants are difficult to propagate without special conditions that the homeowner ordinarily cannot provide. Nurseries also grow and sell mature plants in containers for use indoors and outdoor. Nurseries facilitate the work of landscape architects and contractors by providing materials that are ready to be installed on-site enabling a bare ground to be instantly transformed into a lawn with tress and other ornamental plants. Commercial nurseries are equipped to provide ideal conditions for plant growth. By growing plants under a controlled environment, nurseries provide growers a head start on plant production for the season. They start the plants in the greenhouse in winter when growing them outside is impossible. These plants are timed to be ready for transplanting into the field when spring conditions arrive. Nurseries produce a variety of plants-bushes, trees, tubers, roots, and cool-season) plants because they are equipped to control the plant growth environment. The small-scale home grower can purchase portable plant growth chambers for use at home.

The Seed Industry

Researchers (geneticists and breeders) are continually developing new plant cultivars. These new types may be higher yielding, more resistant to environmental stresses (such as moisture, temperature, and light) and diseases, higher in nutritional value, or aesthetically more pleasing, among other qualities. Seeds from the research domain reach the consumer after going through several steps I the seed release process. Once certified and released as a cultivar, seed growers in the seed industry become responsible for multiplying the seed of the new cultivar, processing it, and packaging it for sale. The role of the seed industry is crucial to the success that the horticultural industry currently enjoys. Seed packets come with instructions about how the plant should be raised to maturity. These instructions are of tremendous help, especially to novice growers. The seed industry has eliminated the need for growers to produce their own seed for planting, unless they so desire. The price of commercial seed is reasonable, and mail-order purchases are possible in many cases. Seed production is usually concentrated in areas where the growing season is most favorable for cropping.

HORTICULTURE AND SOCIETY

Horticulture is important to society in a variety of ways, including as a source of food, ornamentals, and jobs.

Source of Food

Society depends on horticulture for a substantial portion of its food needs in the form of vegetables, fruits and nuts. These types of food sources are high in complex carbohydrates and rich in vitamins and minerals. Leguminous vegetable plants are high in both carbohydrates and protein. Horticultural products are hence part of a balanced diet for humans.

Commercial producers account for most of the horticultural products in the nation However, numerous homeowners are engaged in gardening on their property as a source of fresh produce for the table and as a hobby for recreation and exercise. Horticultural products may be purchased fresh or processed. For example, certain vegetables (such as carrot, lettuce, tomato, and pepper) may be eaten fresh and raw in salads, and fruits (such as apple, orange, and grape) may be eaten fresh or processed into beverages.

Horticultural crops are an important source of revenue through exports for a country. The United States exports both fresh and processed horticultural crops worldwide.

Ornamentals

As previously indicated, landscaping has become an integral part of home construction. Ornamentals are found both indoors and outdoors. Plants in the landscape include trees, shrubs, bedding plants, and grasses. Ornamentals are formally displayed for public enjoyment in places such as arboretums, parks and botanical gardens. Botanical gardens are designed to exhibit a large variety of plant types for the pleasure and education of visitors. Public areas such as malls, playgrounds, and cemeteries are places where plants are displayed for specific purposes. In many societies, certain flowers are associated with specific social events. For example, roses are associated with Valentine' Day, carnations with graduation and Mother's Day, poinsettias with Christmas, and lilies with Easter, which is not to say that these flowers cannot be used for other purposes. Flowers feature very prominently say that these flowers cannot be used for other purposes. Flowers feature very prominently at funerals and weddings. The golf course industry is a major horticultural enterprise that involves not only turf grasses but also a wide variety of trees, shrubs, bedding plants, and other ornamentals.

Medicinal Plants and Horticulture

Horticulture is known to have therapeutic value that can be derived by participating in it or simply enjoying what has been created by others. Walking through a botanical garden can be very relaxing and healthy.

People with emotional and mental problems have been helped when they were deliberately exposed to ornamental plants. For the visually impaired, horticulture can be enjoyed by touching the plant parts and enjoying the sweet scents.

Gardening can be undertaken by people to keep fit or to relieve boredom and other negative emotional feelings. People who are incarcerated or severely limited in their movement are prone to frustration. For such individuals, horticultural activities can be helpful in better managing their emotions.

JOBS

The ornamental industry provides a wide variety of jobs for many categories of people, directly or indirectly. Conducting a computer search via the Internet reveals a wide variety of advertised jobs.

Direct Jobs

A large number of jobs require knowledge and training in horticulture. The level of training could be vocational or at the college level. The work may be indoors or outdoors. Intense manual labor or paperwork in the office may be involved. Many jobs in horticulture require a high school diploma and a short course in horticulture or agriculture. A college education provides more in-depth knowledge of the field and offers job opportunities at supervisory or managerial levels and to conduct research. The following are selected categories of jobs that require varying degrees of familiarity with horticulture:

- Greenhouse manager or worker

- Nursery manager or worker
- Florist
- Golf course manager or worker
- Landscape designer or architect
- Tree surgeon
- Groundskeeper
- Garden center manager or technician
- Vegetable
- Fruit grower
- Flower grower
- Researcher
- Extension officer
- Sales or marketing officer
- Teacher
- Farm manager

As mentioned earlier, certain jobs do not require any familiarity with horticulture by way of formal training. For example, one can find numerous fobs in the greenhouse that require only an ability to follow directions and instructions directions and instructions and a sense of responsibility. Many workers in the greenhouse perform jobs such as watering, transplanting, filling pots with media, harvesting produce, and so on. job prospects for those who pursue formal training in agriculture or horticulture are very bright.

Indirect Jobs

The ornamental industry has spawned a number of supporting or service industries, including the following:

Research

Many scientists are engaged in developing new and improved types of vegetables, fruits, and ornamentals. These new cultivars may have wider and better adaptation, be higher yielding and of higher nutritional quality, and have other qualities depending on breeding objectives. Research is conducted in both private and public sectors (at universities, research institutes, and research companies) to find solutions to problems in the horticultural industry. College-level training (often graduate level) is required to adequately prepare for a career in research. Research institutes invest a great amount of human and financial resources in developing new cultivars, which is why commercial seed companies sell their improved seed (such as *hybrid seed*) at premium prices. Apart from improving the agronomic and nutritional qualities of plants, horticultural scientists also devote considerable time to improving the aesthetics of ornamentals and the quality of products.

Chemical Industry

The horticulture industry depends on large amounts and varieties of chemicals, including fertilizers, pesticides and growth hormones. Many companies involved in producing chemicals that are used to enhance plant production and the quality of produce. Chemicals are an integral part of modern high-input production practices. The increasing trend toward ensuring a safer environment has been the impetus for the creation and enforcement of laws and guidelines for the judicious and safe use of chemicals. Crop production using little or no chemicals, called organic farming, is gradually gaining popularity.

Machinery

Engineers design and produce tools and machinery for use in the production of horticultural plants. Machinery and implements are available for preparing land, planting cultivating, spraying, harvesting, storing, and packaging. These aids enable large-scale production of horticultural plants to be undertaken. Home garden versions of some of this machinery and equipment are available.

Distribution

Horticultural products are transported from the areas of production to marketing outlets. Because of their largely perishable nature, horticultural produce and products require special handling in transportation to retain their quality for a long time. Certain items require refrigeration during storage. Horticultural has spawned an elaborate transportation and distribution network. Because most horticultural products are harvested and used fresh, the ability to preserve quality in transit is critical to the industry. In certain cases, the produce is harvested before it ripens in order to increase its shelf life. Home gardeners have the advantage of ready access to vine-ripened and fresh produce.

Numerous jobs are available in these four general areas at various levels. These jobs can be obtained by persons trained in fields other than horticulture, such as basic science, engineering, economics, marketing, agribusiness, genetics, and post-harvest physiology.

CHAPTER 5

The Power of Biotechnology in Hi-Tech Horticulture

Biotechnology is a tool, a powerful tool with remarkable potential for change. But with biotechnology also come the dilemmas that are always invoked by change. Will it be appropriately regulated? Will it be safe? Is it ethical? What will be its economic impact? Should we take the risks associated with new technologies or eschew their benefits to avoid the unknown?

W. R. Gomes

The human population of the world is still growing and at a total of around 6 billion there are 800 million people that have a chronical shortage of food. Yearly around 10-15 million people die from malnutrition, the majority of which are children. The latest FAO report on the state of the world food situation indicates that, as in the two years before, there are no improvements in this situation for the year 2001. World food production is sufficient to feed all people. Much is lost in transport, storage and distribution. There are many political issues that hamper an efficient food trade and distribution. The loss of fertile land due to mismanagement and especially in coastal regions due to salination is increasingly endangering food production. Predictions are that even in the most optimistic scenarios, the human population will need more food then can be produced with the presently available set up around the year 2040. It took 10.000 years to expand food production to the current level; within 25-30 years we will have to double that' (Norman Borlaug). So, among other things we have to aim at developing agriculture further by improving methods, equipment, storage/ transport/distribution, but also by selecting improved varieties by classical breeding methods and indeed, where appropriate, by biotechnological means. A recent example of a very promising development in this area is the creation of rice varieties that produce trehalose in their plant body. Trehalose provides protection against salt stress, against cold stress and against drought. As such it will enable cultivation in precisely the increasingly endangered agricultural areas.

Horticultural scientists study crops that are used for food, drugs, or aesthetics. Many of these crops are unique or so highly specialized that they are no longer propagated sexually by seeds but rather asexually by such methods as cuttings, grafting, layering, and tissue culture. Seedless grapes, potatoes, maple trees, and roses are some examples of crops that are propagated by some or all of these methods. Although some cultivars are very well known and widely grown, problems related to disease susceptibility, fruit or flower quality, or growth habits arise occasionally.

Traditional sexual-breeding programs can improve some of these crops. The process can be slow, however, especially in tree-breeding programs.

Many of our most popular flower, fruit, and vegetable cultivars are very old. The public often rejects a new cultivar in favor of a better-known cultivar that has endured. The asexually propagated 'Bartlett' pear, for example, introduced in 1770, remains the world's standard for pear quality. Because most new cultivars fail this test, they remain as local cultivars or are discarded.

Another method to improve cultivars is to screen "sports" or mutations that occur naturally, but rarely, on cultivars grown in a field, greenhouse, or laboratory. For example, the green-fruited 'Bartlett' pear sported to yield the red-fruited 'Red Bartlett'. Although this natural variability is useful for plant improvement, the process is random and many plants must be inspected to find a desirable mutant type.

Variability is a universal feature of plant tissue-culture, a method whereby plant parts are grown aseptically on artificial medium. Variation that develops under tissue-culture conditions is now called somaclonal variation. In tissue culture, whole plants can develop from single cells under the proper stimuli of growth regulators, growing medium, and light. Although the precise cause of somaclonal variation remains unknown, its induction and use have become important research topics.

The exploitation of natural and induced variation seems especially applicable for improving older asexually propagated cultivars. During the past 200 years, the clone has been exposed to natural radiation - ultraviolet and cosmic rays - and various sprays and pollutants. These exposures have probably resulted in mutations that stabilized as islands or regions within the body of these plants. When a cultivar possesses mutant cells (natural or artificially induced), whole plants derived from these cells may express the mutation and be classified as somaclones.

Researchers in the Department of Horticulture have used somaclonal variation to obtain thornless blackberries and red pear trees. 'Lincoln Logan', one of our thornless blackberries derived from tissue culture already has been released as a cultivar. Another thornless blackberry is at an advanced stage of testing. We also hope to find variations in fruit color in our small orchard of apple trees that were derived from tissue culture.

Because there may be an upper limit to natural variation within a cultivar, biotechnology protocols are important to supplement and direct the type of variation that we encounter. Researchers at the Department of Horticulture have worked with tomato plants that possess a gene from bacteria (Bt) that gives resistance to certain caterpillars. Other scientists in the department have produced Solanaceous plants with good resistance to certain herbicides and are trying to introduce virus resistance to apple trees.

How much progress can be expected using biotechnology will vary with the clone, the age of the culture- older cell lines often show more somaclonal variation than newer lines--use of mutagenic agents such as radiation or chemical mutagens, and selection pressure applied to single-cell clones for stress conditions such as salt level, herbicides, and microorganisms or their by-products, and specific metabolites. For instance, by placing a variable population of cells in a medium with herbicide, the cells that survive should have some resistance to the herbicide. Progress will also depend on the availability of useful genes for transfer to horticultural crops.

The power of biotechnology is no longer fantasy. Biotechnology – the use of technologies based on living systems to develop commercial processes and products – now includes the techniques of recombinant DNA, gene transfer, embryo manipulation and transfer, plant regeneration, cell culture, monoclonal antibodies, and bioprocess engineering. Using these techniques, we have begun to transform ideas into practical applications. For instance, scientists have learned to genetically alter certain crops to increase their tolerance to certain herbicides. Biotechnology has also been used to develop safer vaccines against viral and bacterial diseases such as pseudorabies, enteric colibacillosis (scours), and foot-and-mouth disease. Yet we have barely scratched the surface of the many potential benefits the tools of biotechnology will bring. Biotechnology offers new ideas and techniques applicable to agriculture. It offers tools to develop a better understanding of living systems, of our environment, and of ourselves. Yet, continued advances will take a serious commitment of talent and funds. Biotechnology offers tremendous potential for improving crop production, animal agriculture, and bioprocessing. It can provide scientists with new approaches to develop higher yielding and more nutritious crop varieties, improve resistance to diseases and adverse conditions, or reduce the need for fertilizers and other expensive agricultural chemicals. In animal agriculture, its greatest immediate potential lies in therapeutics and vaccines for disease control. Bioprocessing – the use of living systems or their components to create useful products – offers opportunities to manufacture new products and foods, treat and use wastes, and use renewable resources for fuel. Biotechnology could also improve forestry and its products, fiber crops, and chemical feed stocks.

There are numerous examples of biotech horticultural crops that have performed well in the laboratory and in field tests, but have never been brought to market or were removed from the market after commercialization, These include fungus-resistant strawberries, virus-resistant pumpkins and potatoes, and "trap crops" for tree fruits and nuts. Trap crops divert insect pests from the main crops. Horticultural crops consist of numerous, diverse varieties (such as dozens of kinds of lettuce), which increases research and development costs. Agronomic crops, by contrast, often have fewer varieties that are planted over larger acreages. Too few acres of horticultural crops are planted to make the business model profitable for large life-sciences companies, as opposed to millions of acres planted in field crops. Each gene-insertion "event" - even in different varieties of the same crop - must receive separate regulatory approval from three U.S. government agencies, an expensive and time-consuming process. Commodity groups have been hesitant to pursue genetically engineered varieties, which they often believe may jeopardize sales of non-biotech varieties. Intellectual property rights for genetically engineered crops are owned by many different people and firms, and are difficult to acquire.

BIOTECHNOLOGY: STRATEGIES FOR HORTICULTURAL DEVELOPMENT

India needs to frame important strategies for agricultural development through biotechnology, which will offer new opportunities for increased sustainability, profitability, and international competitiveness in agriculture. Such a strategy should address improving the full spectrum of activities, from the quality and direction of research to the realization of the benefits of this research in agricultural production.

The potential benefits of biotechnology will not be realized without a continued commitment to basic research. Six research areas merit emphasis are:

Gene Transfer to Enhance Agriculture	
Gene identification	Locating and identifying agriculturally important genes and creating chromosome maps
Gene regulation	Understanding the mechanisms of regulation and expression of these genes and refining the methods by which they may be genetically engineered
Structure and function of gene products	Understanding the structure and function of gene products in metabolisms and the development of agriculturally important traits, specially genetically modified organisms (GMO).
Tissue Culture Techniques	
Cellular techniques	Developing and refining techniques for cell culture, cell fusion, regeneration of plants, and other manipulations of plant and animal cells and embryos.

Environmental Biotechnology	
Development in organisms and communities	Understanding the complex physiological and genetic interactions and associations that occur within an organisms and between organisms.
Environmental considerations	Understanding the behavior and effect of genetically engineered organisms in the environment.

Apart from it microbial biotechnology also plays important role in horticulture. The important aspects are:

- Utilization of arbuscular mycorrhizal fungi in Horticulture.
- Nitrogen (N2) Fixation in Vegetable Legumes: Biotechnological Perspectives
- Microbial control of plant pathogens of horticultural crops.
- Biological control of plant diseases by natural and genetically engineered fluorescent Pseudomonas spp.
- Microbial control of postharvest diseases of fruits and vegetables.
- Use of Bacillus thuringiensis in horticulture.
- Microbiology of fresh and minimally processed vegetables.
- Microbial technology of fermented vegetables.
- Post harvest microbial biotechnology of tropical root and tuber crops.
- Cassava fermentation in fermented starch and flour.
- Microbial starch converting enzymes of the a-amylase family.
- Mushroom: production, disease, management and processing.
- Production of renewable energy from agricultural and horticultural substrates and wastes.

BIOTECHNOLOGY INTERVENTIONS IN CONTROL OF ORNAMENTAL CROP DISEASES

According to Sunita Chandel and Chhaya Sharma (*Agrobios Newsletter, June, 2010*)

In modern terms biotechnology is defined as the manipulation and genetic modification of living organism through novel technologies such as tissue culture and genetic engineering resulting in the production of improved or new organisms and products that can be used in variety of ways. Plant biotechnology is based on a through understanding of plant molecular biology on the use of a variety of plant tissue culture techniques and ability to identify isolate and transfer specific genes from one kind of organism (plant) to another plant or plant organism. Floriculture and ornamental crops are grown for aesthetic purpose likely to be less concern in bio safety issues compared to other food crops. Advances in transgenic technology provide new opportunities for manipulation of the genome in floral crops. This will have a significant impact on expanding and diversifying the gene pool of crop plants, introducing specific genes and shortening the time required for the production of new varieties of hybrids.

More than any other sector of agriculture and horticulture, the visual quality of the ornamental product at the retail level is critical. This is especially true for cut flowers and potted plants. Visible symptoms of disease caused by pathogen therefore have a major impact on quality. Direct and indirect effect of pathogens includes reduction in growth, vigor, quality and market value, as well as overall crop health which is of prime concern to the ornamental industry. Many ornamental crops are susceptible to multiple diseases each of which may cause serious economic losses. In addition infected plant material may not be acceptable for export. Control of diseases of floral crops usually is based on use of clean propagative material that have been indexed and shown to be free of known pathogens. Other methods include chemical control, biological control, resistant varieties and cultural methods. The use of resistant varieties is most recommended approach for control of diseases in many crop species

Conventional breeding strategies require the identification of sources of disease resistance gene. This is difficult task in ornamental crops due to the diversity of floral and nursery crop species that are susceptible to such a large diverse group of pathogens. The large numbers of cultivars which are produced per crop and the rapid turn over in cultivars have made breeding for disease resistance extremely difficult in floral crops. Even where effective disease resistance can be identified in related germplasm. Introgression of a single gene into horticulturally desirable plant lines requires multiple backcrossing which may reduce expression of the resistance compared to the source material. New tools and genes have been developed for use in the genetic engineering of plants to introduce effective resistance to plant diseases and to understand the mechanisms of resistance. This approach should allow increase in both productivity and quality of ornamental plants in an environmentally friendly manner.

As technologies for regeneration and gene transfer in plants improve and broaden across diverse genera, transgenic approaches to develop disease tolerant ornamental crops become more promising. Transgenic approaches for breeding woody ornamentals are attractive. Less space is required to grow out and evaluate germplasm in identifying disease resistance traits. Also, long generation time of backcrossing of woody plants can be eliminated. The costs, potential benefits, and risks of using genetic engineering to create disease resistant ornamental plants must be carefully evaluated. Unlike many agronomic crops, woody and perennial ornamental crops remain in the field for years or decades. Thus the expression of transgene must be durable in all season that plant could under attack and must also be stable enough with stand multiple cycle of plant dormancy. The lived natures of these perennial ornamental also poses unique problems for environmental risk assessment. The plant could serve as reservoir either for an escaping transgene or plant pest that has overcome engineer resistance. On the other hand few ornamental plants are cultivated in typical monoculture. The selective pressure of pathogen to the overcome resistant not as strong as could be monoculture agronomic crops. Because the fruits and seed of many ornamental plants serve as food of wild life, the aspects of ingestion and seed dispersal by wildlife must also be considered. Issues regarding gene recombination and seed production are minimized in the many ornamental crops that are vegetative propagated.

Combining resistance against different diseases is also possible by transgenic approaches, as multiple genes can be introduced into a horticultural desirable genotype on a single construct. Introduction of multiple transgene against different viruses has been demonstrated. The use of multiple transgene constructs appear to be much more efficient means of introducing multiple resistance genes into well adapted genotypes than combination by conventional breeding and multiple cycles of selection. There are various examples of gene transfers against viruses in ornamental crops such as carnation, chrysanthemum, lily, gladiolus and various orchids etc using coat protein gene, CP antisense, replicase viral genes, defective interfering RNA and DNA, ribozyme, antiviral peptides, ribonucleases, ribosome inactivating proteins. A recent scientific breakthrough that has revealed the possibility of controlling the plant diseases is the use of transgenic plants that produce antibodies to a specific plant virus. Complete antibodies or single chain variable fragment (scFv) antibodies have been expressed in plants by transient expression by using viral vector, agro-infiltration or biolistic or after stable integration of transgene directly into the plant genome.

Thus advances in biotechnology have provided new opportunities to solve practical horticultural problems pertaining to ornamentals. The development of technologies for gene identification and gene transfer in plants has provided the opportunity for genetically disease resistance into horticulturally desirable cultivars without altering critical quality traits.

CHAPTER 6

Techniques of Biotechnology for Horticulture

GENETIC ENGINEERING TECHNIQUES

Genetic engineering is the most fundamental mechanics of biotechnologies and is a recent off-shoot of biotechnological research. It involves gene splicing, recombinant DNA cloning and tissue culture technology. The technique overall involves into two steps:

1. The *in vitro* incorporation of the gene or segment of DNA of interest into a small, self-replicating chromosomes and,
2. The introduction of the recombinant *minichromosome* into a host cell where it will replicate. Step one involves synthesis of **recombinant DNA** and step two is the **Gene cloning**.

These two, recombinant DNA and gene cloning technologies are the most powerful tools developed in field of biology. Genetic Engineering involves manipulation of the genetic material of an organism to give an altered expression of our choice. It deals with identification and isolation of desired gene and then joining this gene of interest into another organism. The desired gene expresses itself in that organism by the gene product. Various steps of genetic engineering are:

Gene isolation	Desired gene is identified, isolated and purified. This DNA of interest is also called donor DNA or target DNA.
Selection of vector	A vector is a self-replicating molecule of DNA or replicon to which desired gene is linked. Vector molecule with foreign DNA inserted is known as chimeric DNA. Vector acts as carrier and transports the gene into the host cell. Thus, it is also known as a cloning vehicle or a carrier molecule. Suitable vector is identified for a system; commonly used vectors are plasmids and viral DNA molecule. However, recent techniques involve physical delivery of DNA by various methods.
Cloning of desired gene	Multiple copies of desired gene can be obtained by placing them in host cell with the help of vectors. Here, the desired gene along with the vector is amplified. Large number of identical copies of gene of interest are produced for subsequent gene transfer into target cell. Now, gene cloning is a fast and mechanized process, using polymerase chain reaction (PCR) machines.
Specific gene transfer	The gene of interest is finally transferred to host cells. Transformed cells are selected, multiplied and produce transgenic plants by tissue culture technique.
Expression of desired gene	The desired gene produce, the product in new environment of host, thus the desired traits.

Genetic Engineering includes the propagation of chimeric DNA in a different host organism. The ability to cross natural species barriers and place genes from any organism in an unrelated host organism is one important feature of gene manipulation. A second important feature is the fact that a defined and relatively small piece of DNA is propagated in host organism. Thus genetic engineering opened the door to a range of molecular biological opportunities including nucleotide sequence determination, site-directed mutagenesis, and manipulation of gene sequences to ensure very high-level expression of an encoded polypeptide in a host organism. If used wisely genetic engineering promises to enhance the quality of human life. However, only future will determine the scope and final outcome of this technology.

Genetic engineering techniques have so far had limited impact in the field of ornamental horticulture. The transformation systems and potential genes of interest are available. As the development of new, novel varieties is an important driving force in the industry, there are, therefore, good prospects for the development of genetically modified ornamental varieties. The few products in the market to date may simply be a reflection of the relatively small scale of the industry compared to the major food crops, and the wide diversity of species within it. Commercial issues attendant to the use of gene technology in ornamental plants need careful consideration. These include careful choice of crop and background variety, the international regulatory process and freedom to operate

Gene Transfer to Enhance Horticulture

Throughout the history of agriculture, humans have taken advantage of the natural process of genetic exchange through breeding that creates variation in biological traits. This fact underlies all attempts to improve agricultural species, whether through traditional breeding or through techniques of molecular biology. In both cases, people manipulate and natural process to produce varieties of organisms that display desired characteristics or traits, such as disease-resistant crops or food animals with a higher proportion of muscle to fat.

The major differences between traditional breeding and molecular biological methods of gene transfer lie neither in goals nor processes, but rather in speed, precision, reliability, and scope. When traditional breeders cross two sexually reproducing plants or animals, tens of thousands of genes are mixed. Each parent, through the fusion of sperm and egg, contributes half of its genome (an organism's entire repertoire of genes) to the offspring, but the composition of that half varies in each parental sex cell and hence in each cross. Many crosses are necessary before the *right* chance recombination of genes results in offspring with the desired combination of traits.

Molecular biological methods alleviate some of these problems by allowing the process to be manipulated one gene at a time. Instead of depending on the recombination of large numbers of genes, scientists can insert individual genes for specific traits directly into an established genome. They can also control the way these genes express themselves in the new varieties of plant or animal. In short, by focusing specifically on a desired trait, molecular gene transfer can shorten the time required to develop new varieties and give greater precision. It also can be used to exchange genes between organisms that cannot be crossed sexually.

Gene transfer techniques are keys to many applications of biotechnology. The essence of genetic engineering is the ability to identify a particular gene – one that encodes a desired trait in an organism – isolate the gene, study its function and regulation, modify the gene, and reintroduce it into its natural host or another organism. These techniques are tools, not ends in themselves. They can be used to understand the nature and function of genes, unlock secrets of disease resistance, regulate growth and development, or manipulate communication among cells and among organisms.

Isolation of Important Genes

The first step in an effort to genetically engineer an organism is to locate the relevant gene(s) among the tens of thousand that make up the genome. Perhaps the researcher is searching for genes to improve tolerance to some environmental stress or to increase disease resistant. This can be a difficult task – similar to trying to find a citation in a book without an index.

This task is made easier with restriction enzymes that can cut complex, double-stranded macromolecules of DNA into manageable pieces. A restriction enzyme recognizes a unique sequence in the DNA, where it snips the strands. By using a series of different restriction enzymes, an organism's genomic DNA can be reduced to lengths equivalent to one or several genes. These smaller segments can be sorted and then cloned to produce a quantity of genetic material for further analysis. The collection of DNA segments from one genotype – a gene library – can be searched to locate a desired gene. Patterns can also be analyzed to link a particular a desired gene. Patterns can also be analyzed to link a particular sequence – a marker – to a particular traitor disease, even though the specific gene responsible is still unknown.

Restriction enzymes are also used in cloning genes. To clone a gene, a small circle of DNA that exists separate from an organism's main chromosomal complement – a plasmid is cut open using the same restriction enzyme that was used to isolate a desired gene. When the cut plasmid and the isolated gene are mixed together with an enzyme that rejoins the cut ends of DNA molecules, the isolated gene fragment is incorporated into the plasmid ring. As the repaired plasmid replicates, the clones gene is also replicated. In this way, numerous reproductions of the clones gene are produced within the host cell, usually a bacterium. After replication, the same restriction enzyme is used to snip out the cloned gene, allowing numerous copies of that gene to be isolated.

The ability to isolate and clone individual genes has played a critical role in the development of biotechnology. Cloned genes are necessary research tools for studies of the structure, function, and expression of genes. Further, specific gene traits could not be transferred into new organisms unless numerous gene copies were available. Cloned genes also are used as diagnostic test probes in medicine and agriculture to detect specific diseases.

Gene Transfer Technology

Genes transfer from one organism to another, molecular biologists use vectors. Vectors are the carriers used to pass genes to a new host, and they can mediate the entry, maintenance, and expression of foreign genes in cells. Vectors used to transfer genes include viruses, plasmids, and mobile segments of DNA called transposable elements. Genes can also be introduced by laboratory means, such as chemical treatments including injection with microneedles. The basic principles behind these technologies are the same for animals, plants, and microbes, although specific modifications may be necessary.

Vectors based on viruses, plasmids, and transposable elements have been adapted from naturally occurring systems and engineered to transfer desired genes into animals, plants, and microbes. For plants, the classic example is the Ti plasmid from the soil bacterium *Agrobacterium tumefaciens,* which in nature transfers a segment of DNA into plant cells, causing the recipient cells to grow into a tumor. Scientists have adapted this plasmid by eliminating its tumor-causing properties to create a versatile vector than can transfer foreign genes into many types of plants.

Similarly, the transposable P-element of the fruit fly *Drosophila melanogaster* is an effective vector for gene transfer into *Drosophila.* This or similar transposable elements should prove to be adaptable to insects of agricultural importance. Animal viruses such as simian virus 40 (SV40), adeno, papilloma, herpes, vaccinia, and the retroviruses, all originally studied because of their role in disease, are now being engineered as vectors for gene transfer into animal cells and embryos. Plant viruses such as cauliflower mosaic virus, brome mosaic virus, and geminiviruses are similarly being exploited for their abilities to transfer genes.

Molecular Markers

Markers are protein based (such as isozymes, allozymes) or DNA based (such as RFLP, AFLP, SSR). They are commonly used for many purposes in pre-breeding such as the identification and control of material obtained (e.g. somatic hybrids, haploids, vitro-variants) or maintained (e.g. genetic resources of vegetatively propagated crops) by *in vitro* methods. They are increasingly used for the management of genetic resources collections, because they are complementary tools to conventional descriptors for characterizing and structuring the genetic diversity for pea, banana, apricot, and *Hydrangea*.

Marker assisted selection (MAS) involves preferably simple products of the polymerase chain reaction (PCR) as markers, is increasingly associated to breeding schemes, in particular for vegetables and fruit trees. A number of technical pre-requisites are necessary to use MAS. Most important is the availability of markers linked to the traits of interest for a given target species. Synteny features between phylogenetically related crop species (such as the similarity of the molecular structure of genetic maps, and of gene positions and functions) can contribute to enrich markers availability. For *Prunus* species, the synteny between the genomes and the set up of a consensus map, offer the potentiality to use markers linked to interesting traits in one species for screening for the same traits in another species.

For instance, plums SSR are transferable to peach and almond. Second, the markers need to be closely linked (distance of 1 cm or less) or even to be located within the major genes or major QTL controlling traits of interest, such as gametophytic self incompatibility, and fruit quality traits (flesh colour, flesh adherence to stone, fruit skin pubescence) in *Prunus* species.

MAS methods have been developed for many horticultural crops including potato, and have become routine procedures for many monogenic traits of tomato, pepper, melon, apple, and *Prunus* species. Markers are increasingly integrated in horticultural practices. Cultivars can be identified by fingerprint for peach, cherry and strawberry. These techniques are particularly useful in nurseries where numerous cultivars, difficult to differentiate on a morphological basis at the vegetatively propagated stage, are simultaneously multiplied at a large scale.

Markers become increasingly part of the procedure for cultivar certification. They are routine tools for seed quality control (genetic identity and purity) operated by seed companies, in particular for the costly and high performing vegetable hybrids. The use of DNA markers for plant registration testing (distinction, uniformity, stability), essential derivation determination, as well as for plant breeder's rights is still experimental, but it provides an additional tool over conventional morphological identification. The updated state of the art is available on the International Seed Federation web site (http://www.worldseed.org/).

Molecular markers and genome mapping are used in detecting new desirable genes in plant improvement specially for tomato, potato, pea, melon, cabbage, etc. The markers help in identifying important quantitative trait loci (QTL). The QTLs affecting salt tolerance in tomato during germination have been identified. As many as 700 tomato markers were known in 1990 (De Verna and Paterson, 1990). DNA fingerprinting helps in identification, characterization and evaluation of genetic uniformity of regenerated new plants of vegetatively propagated fruit crops at cultivar level.

Marker-Assisted Selection: Fast Track to New Crop Varieties

Back in 1930, when a Manitoba farmer came across a few stems of rust-free wheat in a rust-infested field, it took plant scientists several years of work before they could develop those rust-resistant kernels into a new wheat variety. Almost 25 years later, the new variety called Selkirk was released. It happened at a good time, though. Selkirk arrived just as another rust outbreak was occurring on the Prairies in 1954.

The process of developing new crop varieties still requires many steps. But now, applications of biotechnology have considerably shortened the time it takes to bring them to market. Where it once took almost 25 years, it currently takes about seven to 10 years.

One of the tools which is making it easier for scientists to select plant traits, further reducing the time needed to develop a new variety, is called marker-assisted selection (MAS).

Molecular Shortcut

The differences which distinguish one plant from another are encoded in the plant's genetic material, the DNA. The DNA occurs in pairs of chromosomes (strands of genetic material), one coming from each parent. The genes, which control the plant's characteristics, are specific segments of each chromosome. All of the plant's genes together make up its genome. Some traits like, flower color, may be controlled by only one gene. But other more complex characteristics, like crop yield or starch content, may be influenced by many genes.

Traditionally, plant breeders have selected plants based on their visible or measurable traits, called the phenotype. But, as indicated by the development of the Selkirk wheat variety, this process can be difficult, slow and costly‹not only in the development itself, but also for the economy, as farmers suffer crop losses.

As a shortcut, plant breeders now use marker-assisted selection. To help identify specific genes, scientists use what are called molecular markers. The markers are a string or sequence of nucleic acid which makes up a segment of DNA. The markers are near the DNA sequence of the desired gene. Since the markers and the genes are close together on the same chromosome, they tend to stay together as each generation of plants is produced. This is called genetic linkage. This linkage helps scientists to predict whether a plant will have a desired gene. If researchers can find the marker for the gene, it means the gene itself is present.

As scientists learn where each of the markers occurs on a chromosome, and how close it is to a specific gene, they can create a map of the markers and genes on specific chromosomes. These genetic linkage maps show the location of markers and genes, and show their distance from other known genes. Scientists can produce detailed maps in only one generation of plant breeding.

Previously, scientists produced very simple genetic maps using conventional techniques. It was long ago observed that as generations of plants were crossed, some traits consistently appeared together in the new generations (genetic linkage). However, since researchers could concentrate on only a few traits in each attempt at cross-breeding, it took many crosses to obtain even a very simple genetic map. Using very detailed genetic maps and better knowledge of the molecular structure of a plant's DNA, researchers can analyze a tiny bit of tissue from a newly germinated seedling. They don't have to wait for the seedling to grow into a mature plant so that they can test for a specific characteristic. Once the tissue is analyzed, scientists know whether that seedling contains the appropriate gene. If it doesn't, they can quickly move on and concentrate analysis on another seedling, eventually working only with the plants which contain a specific trait.

Marked Benefits

Some crops, like canola, have relatively simple genomes. This means it is easier for scientists to develop markers for specific traits and produce genetic maps. Some crops, like wheat, have complicated genomes. Each wheat cell contains approximately 10 times more DNA than a human cell. As well, there are very few differences in the DNA sequences that distinguish varieties. In these cases, it may be much more difficult to develop markers.

However, scientists still consider mapping the wheat genome to be an important effort. A detailed understanding of wheat genetics will help to identify desirable qualities like disease and drought resistance. Wheat varieties with these traits will make it easier to grow more wheat in drier climates, such as Africa, and will help to increase world food production.

Marker-assisted selection can also be used to pyramid genes in a new plant variety. Pyramiding genes means putting two or more desirable genetic traits into one crop variety. The goal is to produce varieties with not just one, but several, desirable characteristics.

In Saskatchewan, scientists with major crop research organizations like the University of Saskatchewan, Agriculture and Agri-Food Canada and Plant Biotechnology Institute use molecular markers and marker-assisted selection for their plant research. There are many types of molecular markers. For example, scientists at the University of Saskatchewan and Agriculture and Agri-Food Canada in Saskatoon are beginning to concentrate their research on a type of marker called microsatellites. These are very short sequences of DNA that are repeated over and over throughout the chromosome. By using microsatellites, scientists are working to create more detailed genetic maps of crops and better gene markers.

More and more pathologists are using molecular probes for early detection of diseases. However, the main contributions of modern biotechnology remain probably in the hands of the plant breeders. Indeed, molecular markers are of great use in detecting desirable new genes, or to identify important QTL. This marker-assisted selection is very useful for many vegetable plants. Other applications are DNA - fingerprinting and genetic engineering. This recent technology aims to insert and to express new specific genes into selected plants.

Marker-Assisted Selection

Molecular markers and genome mapping will reach a large extension in the next breeding programmes of tomatoes, peas, cabbages, melons, potatoes,

It will be possible to improve the speed and the incorporation efficiency of desirable new genes by utilizing of closely linked selectable molecular markers. The number of backcross will be reduced. Tanksley et al. (1981) show that only with 12 markers, one on each tomato chromosome, the composition of the recurrent parent genome is similar to that observed in the absence of selection after the third backcross two years later. Therefore, it's easy to understand the high improvement that we can get with the 700 tomato markers already known in 1990 for this crop (De Verna and Paterson, 1990).

The utilization of molecular markers could also identify important quantitative trait loci (QTL). Recently Foolad and Chen (1998) have identified 13 RAPD markers at eight genomic regions, that were associated with QTLs affecting salt tolerance during germination in tomato.

DNA Fingerprinting

The DNA polymorphism observed by RFLP, AFLP, or RAPD, allows cultivar identification of fruit trees, apple, and citrus. or other vegetatively propagated plants. Although some mutants, as coloured fruits, escape these identification techniques, some people would evaluate the genetic uniformity of regenerated plants in this way.

Genetic Engineering of Microorganisms Associated with Plants

Microorganisms in the environment affect the growth of plants in a variety of ways, many of which are still poorly understood. Their effects can be either beneficial or harmful. For instance, some microorganisms protect plants from bacterial or fungal infections. Others protect plants from environmental stresses such as acidity, salinity, or high concentrations of toxic metals. Still others attack weeds that compete with crops. The best known association between microorganisms and plants is the symbiotic relationship between nitrogen-fixing bacteria of the genus *Rhizobium* and members of the legume family, such as soybeans.

As our understanding of the relationships between microorganisms and crops improves, the genes controlling these relationships – whether in the microorganism or in the plant – can be engineered to enhance the abilities of beneficial microorganisms or inhibit the effects of harmful microorganisms.

Arabidopsis Genomics and Horticulture

Arabidopsis is considered as the major model plant for molecular and genetic studies. Especially the ease to do genetics and transformation experiments together with the resources available (including the complete genome sequence) make that research in Arabidopsis leads to faster results than similar research using crop plants. The cloning of genes controlling complex quantitative traits has been illustrated already on flowering time genes. An important question is, if plants are so similar that what we can learn from Arabidopsis can be applied to crop plants. The comparisons of the DNA sequence of genomes (especially the genes that are expressed = EST's) and the study of individual genes controlling specific traits show strong similarity between plants. This implies that Arabidopsis knowledge often can be used to identify functional similar genes in crop plants. However, differences between plants also exist, probably especially in the regulating factors. Examples that show similarity and differences between *Arabidopsis* and crop plants will be demonstrated for flowering time, plant length, pigmentation and glucosinolate biosynthesis.

Genetically Modified (GM) Crops

The development in the direction of plant cell and tissue culture research occurred after the discovery of restriction endonuclease enzymes in the early 1970s. These enzymes are responsible to cleave DNA molecules at predictable sites. They help in removing specific genes of desire, which can be modified, and inserted into other DNA strands. The development of watershed of technology has provided the basic tools needed for genetic engineering. The availability of totipotent plant cells that, conceivable, could be altered by insertion of specific genes caused a revolution in plant research. The potential commercial value of such genetically modified (or GM) plants attracted an unprecedented amount of industrial and investor interest in the late 1970s and 1980s which have developed many new plant biotechnology-based companies around the world.

The establishment and commercialization of GM crops has been very rapid in past years. In 1995, however, there were no commercial plantings of GM crops. In USA, out of 80 million acres of corn planted yearly in 400,000 acres of GM corn were planted in 1996, increasing to 3 million acres in 1997 and 17 million acres in 1998. IN 1997, 71 million acres of genetically modified soybean were planted in 20 million acres and it is hoped that nearly 100% of the soybean crop will be composed of GM varieties within the next few years. Similarly in 1998 (in USA), about 45 per cent of the cotton crop was genetically modified. Corn and soybean account for about 75 per cent of all GM crops planted, followed by cotton (12%), Canola (11%), and potato (<1%). The above 75 per cent plant genetically modified varieties are now grown in Argentina, Canada, Mexico, Spain, France and South Africa. China also grows GM crops, data as to the extent are not available.

About 99 per cent of the two important transgenic traits - herbicide and insect resistance (Bt) are of common use. Less than 1% of GM acreage was devoted to "quality" traits, such as controlled ripening and enhanced oil profiles. Globally, herbicide-resistant soybean dominates GM crop acreage (52%), followed by Bt corn (24%). The success of these crops is evident in that it is estimated that $200 million worth of pesticide was eliminated by implementation of insect-resistant GM crops alone. Herbicide usage has dropped because applications can be delayed until weeds are commingled with the crop plant and are most susceptible.

Genetically modified crops have also been developed for food industries, such as enzymes from microorganisms in processed products like soft drinks, cakes, cheese, bread, fish, and meats, etc. About 90 per cent of common foodstuffs contain GM-components.

Genomics

Genomics' basic goal is to unravel the organization (structural genomics) and functioning (functional genomics) of entire genomes. Based on the study of DNA and RNA, it uses molecular tools (e.g. various types of markers, bacterial artificial chromosome (BAC) libraries, expressed sequence tag (EST) libraries, micro-arrays) and techniques (e.g. mapping, cloning, sequencing) that yield masses of data, analyzed thanks to the progresses of bio-informatics.

Applied to various segregating progenies, mutants, plant organs at precise developmental stages, genomics is bringing a revolution to the understanding of crop genomes and development:

- by yielding genetic and physical maps where genes and quantitative trait loci (QTL) controlling traits of interest are identified and located,
- by characterizing gene functions and gene interactions, and
- by giving access to the simultaneous analysis of the expression of thousands of genes.

Because of high cost, scientists attempt to use the knowledge obtained on model species for applications to other species. Comparative mapping in plant families such as Rosaceae and Solanaceae, which have been studied extensively, indicates that genome evolution within a family concerns mostly chromosome restructuring (inversions or translocations) and that large chromosomal fragments are conserved between related species and genera. Molecular maps of *Prunus* species (almond, peach, apricot, cherry, wild material) constructed with different markers [restriction fragment length polymorphisms (RFLP), simple sequence repeats (SSR), amplified fragment length polymorphisms (AFLP) and isozymes] present a very strong colinearity without any major chromosomal rearrangement.

The study of genome evolution as well as the use of molecular tools in research and breeding is also advanced in the Solanaceae (potato, tomato, and eggplant. For tomato, many genes have been cloned, and molecular dissection and candidate gene strategy are commonly used for identifying and characterizing major quantitative trait loci (QTL) involved in fruit quality components. For pepper, QTL of resistance to several pests and diseases are currently mapped and their function investigated at the molecular level.

PROTEIN (OR ENZYME) ENGINEERING

Protein engineering is an important area of biotechnology, through which it is possible to produce superior enzymes and storage proteins. In this area, a protein engineer first a computer aided protein model for a specific function and then prepares a synthetic gene which produces the desired protein in a predictable manner. Use of immobilized enzyme has made possible the production of a variety of substances e.g., high-fructose corn syrup (as sweetening agent for soft drink industry), glucose isomerase, etc.

Enzyme technology is an area of considerable current interest and development. Enzymes are biological catalysts and have been used for many years as isolated agents particularly in food industry, e.g., rennin, papain, and invertase. Microorganisms are very convenient enzyme sources, because the enzymes concentration may be markedly increased by acting on environmental conditions and by genetic manipulation. These enzymes have increasingly replaced plant and animal enzymes: thus, amylases from *Bacillus* and *Aspergillus* have substituted those of malted wheat and barley in brewing, backing and biscuit-making, and in the textile industry, etc. Most of the enzymes are lost in product or the effluent after process, therefore, re-usability is essential to reduce the cost of enzyme production and to improve the economics of enzyme catalyzed reactions in industry. In 1980, around 530 tonnes of protease were produced, 350 tonnes of glucoamylase, 320 tonnes of α-amylase and 70 tonnes of glucose isomerase. This production was obtained by most of the European companies, the important one are, Novo Industry (Denmark) and Gist Brocades (The Netherlands) who had 60 per cent of the world market.

TISSUE CULTURE TECHNIQUES

The ability to regenerate plants from single cells is important for progress with gene transfer into plans. Animals cannot be regenerated asexually, so the only way to introduce a foreign gene into all cells of an animal is to insert it into the sperm, egg, or zygote. Cell culture techniques are important for the regeneration of plants. They are also critical for fundamental studies on both plant and animal cells, and for the manipulation of microorganisms.

- Tissue-culture could be a useful way of circumventing or eliminating disease which can accrue in stock plants.
- of Tissue Cultured Plants (TCPs) may have increased branching and flowering, greater vigour and higher yield, mainly due to the possibility of elimination of diseases.
- The method may succeed to propagate plants where seeds or vegetative propagation is not possible or difficult or undesirable.
- The method saves space and energy. For example, 2500 m2 of heated green house can be replaced by a climatised room of 10 m2

- The flexibility of nurseries can be improved. As the capital investment on mother plants is reduced to almost zero, it may be easier to adapt to changing conditions. Additionally, a better programming of the production is possible, because of the greater plant uniformity and the availability in the mass at any time.
- Tissue culture can be utilised for breeding new varieties, preservation of germplasm and in vitro synthesis of metabolites

The vegetative propagation of stem cuttings or other growing plant parts to produce genetic clones is common for some agricultural crops. Potatoes, sugarcane, bananas, and some horticultural species, for example, are cultivated by vegetative propagation. Techniques exist to propagate and regenerate whole plants from tissues, isolated plant cells, or even protoplast (plant cells from which the cell all has been enzymatically removed) in culture. This set of techniques is complete for some agricultural species, such as alfalfa, carrots, oilseed rape, soybeans, tobacco, tomatoes, and turnips. Progress on other crops, including major food species such as many cereals and legumes, has been slower.

Propagation by tissue-culture offers good commercial prospect in ornamental plants, vegetables and also fruit plants, where value of the products is high. The technique has reportedly been successful in more than 100 species of plants. It has been estimated that more than 350 million TCPs are being produced annually through tissue culture.

In India there are at least ten commercial organizations, which have developed technical competence for tissue culture with or without foreign tie-ups. The present installed capacity is about 50 million TCPs and the export is of the order of Rs. 5 crores. The working group appointed by the Ministry of Commerce has proposed an export target of Rs. 30 crores i.e. about 60 million TCPs over a five year period as against the present production of 8-10 million TCPs. Tissue culture method of propagation is highly labour intensive, 55-60% of the cost is on account of labour. India's potential for export of tissue culture plants is rated very high because of abundant and cheap labour.

The domestic market for TCPs, though nascent at present, is likely to develop, since tissue culture method of propagation can multiply an elite plant very rapidly. It is well known that one of the major constraints of horticultural development in our country is inadequate availability of quality planting materials. The demand of planting materials for major fruit crops has been estimated to be of the order of Rs. 175.4 crores during the Eighth Plan period.

Cell culture techniques have taken on added importance as biotechnology has progressed. Genetic engineering requires an ability to manipulate individual cells as recipients of isolated genes. Cell culture techniques allow scientists to maintain and grow cells outside the organism and thus expand their ability to perform gene transfer and study the results. In addition, cell culture allows scientists to regenerate numerous copies (clones) of the manipulated varieties, which is easier, more efficient, and more convenient, especially for producing significant quantities of stock plants. A third use of cell culture is to regenerate "somaclonal variants," plants with altered genetic traits that can prove useful as new or improved crops. Thus, cell culture techniques are important to increasing the productivity and versatility of agriculture.

However, there are some important limitations. Chromosomal abnormalities appear as cultures age. These changes are related to the phenomenon of somaclonal variation, which may prove useful to agriculture, but in many instances the changes are undesirable. Therefore, scientists must learn how to prevent chromosomal changes in cell cultures. Second, long-term cultures lose regenerative potential. As biotechnology expands, it will be critical to understand why different species have differing abilities to regenerate from cell cultures into plants and how factors such as the genetic or physiological origin of the cells and the culture conditions affect growth. Most plant cells appear to be totipotent, that is, they are in a reversible differentiated state that will permit them to regenerate into a whole plant under appropriate conditions. Understanding what these appropriate conditions are remains a fundamental question in the study of plant development and its genetic control.

Tissue culture can serve a number of purposes, and growers have started their own commercial production laboratories for a variety of reasons. Growing plants from seeds or cuttings can be unacceptable or impractical due to some of the following (and other) factors:

1. Seed-grown products lack uniformity.
2. Seed-grown products are not true to type.
3. Seeds take too long to grow to mature plants.
4. Seeds are difficult to handle.
5. Seeds are not available.
6. Cuttings grow too slowly.
7. Cuttings have poor survival rate.
8. Cuttings require too much care.
9. Cuttings are too vulnerable to disease.
10. Tissue culture is often the only practical way to produce the large numbers required.
11. There is a shortage of stock plants from which to take cuttings because there is: (a) only one hybrid, (b) only one virus-free plant, (c) only one desirable mutant.
12. There is insufficient room for stock plants and it is not cost effective to maintain the stock plants.

Plant Species Reported to Respond to Propagation by Tissue Culture

1. **Ornamental plants:** *Philodendron, Gladiolus, Diffenbachia, Lily, Monstera, Kalanchoe, Maranta, Pentunia, Alocasia, Narcissus, Aloe vera, Rose, Fern, Mimosa, Ficus benjamina, Gypsophila, F. elastica, Aster, F. lyrata, Aglaonema, Ficus robusta, Hydrangea, Ficus mini, Amarylis, F. compacta, Nerine, F. foliole, Tulip, Iris, Nephrolepis, Freesia, Calathea, Hyacinth, Heliconia, Anemone, Syngonium, Begonia, Dracaena, Eucharis, Peperomia, Caladium, Euphorbia, Carnation, Pelargonium, Chrysanthemum, Platycerium, Gerbera, Pteris, Saintpaulia, Davallia, Streptocarpus, Osmunda, Anthurium, Mammillaria, Aechmea, Christmas Cacti, Cyclamen, Easter Cacti, Kalanchoe, Agapanthus, Episcia, Asparagus, Spathiphyllum, Gloxinia, Alstroemeria, Guzmania, Hamamelis, Cordyline, Hemerocallis, Schefflera, Dendrobium, Liriope, Cymbidium, Strelitzia, Cattleya, Nandina, Odontoglossum, Rhododendron, Phalaenopsis, Bougainvillea, Vanda, Buddelia, Weigela, Magnolia, Ribes, Deutzia, Epidendrum, Crocsmia, Forsythia*
2. **Vegetables:** Onion, *Asparagus*, *Brassica*, Tomato, Chicory, *Capsicum* etc.
3. **Fruit plants:** Banana, Strawberry, Apple, Pear, Cherry, Citrus, Rubus, Gooseberry, Grapes, Papaya, Pineapple, Ananas etc.
4. **Forest species:** *Poplar*, *Eucalyptus*, *Bamboo*, *Pinus*, *Cupressus*, *Thuja*, *Sequoia*, *Ulmus*, *Spiraea*, *Betula*, *Salix*, *Ilex*, *Fagus*, *Picea* etc.
5. **Others:** Coffee, Cardamom, Ginger, Turmeric, Vanilla, Hops, Oil Palm etc.

TABLE 6.1: In vitro responses of important fruit crops

Common name	Botanical name	Type of explant	Responses elicited
Apple	*Malus pumila*	Shoot tip, lateral bud, apical meristem, seedling explants	Multiple shoots, callus and adventitious shoot
Banana	*Musa spp.*	Rhizome tip, floral tissue (excised embryo)	Axillary shoot proliferation, Hybrid plantlets and multiple shoot proliferation
Ber	*Zizyphus mauritiana*		
Cape gooseberry	*Physalis peruviana*	Anther	Triploid plant regeneration
Grapefruit	*Citrus paradisi*	Stem and leaf segments	Callus, embryoids and adventitious shoots
Lemon	*C. limon*	Nucellus	Direct adventive embryo formation
Lime	*C. aurantifolia*	Nucellus, anther	Callus, deformed shoots and roots plantlets
Mandarin orange	*C. reticulata*	Shoot segment	Callus and shoot differentiation
Pumello	*C. grandis*	Shoot segment	Callus and shoot differentiation
Rough lemon	*C. limon var. Jambhiri*	Nucellus	Callus and embryogenesis
Sweet orange	*C. sinensis*	Unpollinated ovary, shoot segment	Differentiation
Custard apple	*Anona squamosa*	Leaf, anther	Organogenesis, haploid plants
Date palm	*Phoenix dactylifera*	Shoot tip, seedling	Multiple shoots, somatic embryogenesis
Fig	*Ficus carica*	Shoot tip	Elongation and rooting of shoot tips
Grape	*Vitis spp.*	Shoot tip	Multiple shoots, callus formation and somatic embryogenesis
Guava	*Psidium guajava*	Shoot tip	Multiple shoots
Jackfruit	*Atrocarpus heterophyllus*	Stem tip	Multiple shoots
Jamun	*Syzygium cumunii*	Leaf axillary bud	Regeneration of plantlets
Mango	*Mangifera indica*	Nucellus, embryo	Somatic embryogenesis, hybrid plantlets and somatic embryogenesis
Papaya	*Carica papaya*	Shoot tip, ovules embryo, anther	Multiple shoots, callus and somatic embryogenesis, hybrid plantlet production, haploid plantlets
Passion-fruit	*Passiflora edulis*	Stem and leaf	Adventitious buds
Peach	*Prunus persica*	Shoot tip	Axillary shoots
Pear	*Pyrus communis*	Seedling tissues	Callus and adventitious shoots and roots
Pineapple	*Ananas comosus*	Lateral meristem, shoot tip	Multiple shoots, callus
Plum	*Prunus domestica*	Shoot tip	Adventitious shoot proliferation

Common name	Botanical name	Type of explant	Responses elicited
Pomegranate	*Punica granatum*	Mature stem tip	Multiple shoots
Sapota	*Achrus sapota*	Endosperm	Callus formation
Sweet cherry	*Prunus spp.*	Meristem tip	Axillary shoot proliferation
Strawberry	*Fragaria spp.*	Shoot tip	Axillary and adventitious shoots

If an ample supply of seeds is available, or if plants from cuttings are acceptable and cutting stock is available and not a problem to maintain, then tissue culture may not be the most practical option because it can be an expensive, labor-intensive process, especially if less than a thousand plants are needed. If plants from seeds are acceptable but there are not enough seeds, then seeds or excised embryos can be used as starting material for tissue culture.

Every year excessive amounts of growers' time, labor, and space are spent on unproductive seeds, cuttings, and grafts. Significant numbers of young plants are lost to pests, diseases, or other environmental factors. Tissue cultured plantlets are less subject to such attacks and disasters because in the sterile environment of the laboratory they are not exposed to the pathogens or extreme conditions that afflict many plants grown in the field or greenhouse. Material usually comes out of culture as well-started plantlets or microcuttings with a stockpile of nutrients and vigor often superior to that of conventional cuttings. It is no secret that healthy plants are the first line of defense against disease.

Tissue culture avoids and enormous amount of the daily care that is required with cuttings and seedlings. Cultures usually need to be divided and transferred to a fresh medium every 2 to 6 weeks, but between transfers there is no need to water or tend to the cultures, other than casual surveillance. How different this is from the daily watering and weeding requirements accompanying greenhouse growing!

The simplest tissue culture hobby is the multiplication of easy, fast-growing plant material, such as *Kalanchoe*, Boston ferns (*Nephrolepis*), African violets (*Saintpaulia*), or *Begonia*; next in order of complexity are carnations (*Dianthus*), strawberries (*Fragaria*), or *Syngonium*. The chemist looking for a tissue culture hobby may be challenged to explore the field of plant by products; dyes, flavorings, medicinals, and oils are just some of the by-products of certain plants.

The earlier researchers explored problems of academic interest, then the emphasis shifted to applied aspect such as haploid from pollen, triploids from endosperms, somatic hybridization, tissue culture of cereals, legumes and oil crops, and clonal multiplication of elite species of plants. Our knowledge of cell and tissue culture is developing with a fast speed specially in many areas, like totipotency, differentiation, cell division, cell nutrition, metabolism, radiobiology, cell preservation, etc. We are now in position to cultivate cells in quantity, or as clones from single cell, to grow whole plant from isolated meristems, to induce callus or even single cell to develop into complete plant either by organogenesis or directly by embryogenesis *in vitro*. The production of haploid through tissue culture from anthers or isolated microspores and of protoplasts from higher plant cells has served as the basic tools for genetic engineering and somatic hybridization. Protoplasts can also be used as genetic material present in nuclei and chloroplasts as well as isolated DNA molecules. This technique provides the opportunity to combine by fusion the genotype of species which are sexually incompatible and to introduce foreign genetic material such as organelles or DNA into the genome.

Tissue Culture: Production of True to Type Plant

Plant tissue culture is a field of many facets. The term *tissue culture* is actually a misnomer borrowed from the field of animal tissue culture. It is a misnomer because plant micropropagation is concerned with the whole plantlet and not just isolated tissues, though the explant may be a particular tissue. The terms *plantlet culture* or *micropropagation*, therefore, are more accurate. Plant tissue culture is the technique of growing plant cells, tissues and organs in an artificial prepared nutrient medium static or liquid, under aseptic conditions. Better control of environmental conditions like light, temperature, gas mixtures and nutrients can be achieved for plant tissues growing *in vitro*. But whether we call it cloning, tissue culture, micropropagation, or growing *in vitro*, the process remains the same; it is a vegetative method for multiplying plants. Tissue culture is the ever-ready tool for specialists who hybridize plants by either sexual or asexual means. It is a clean and rapid way for genetic engineers to grow material for identifying and manipulating genes or to transfer individual characteristics from one plant to another. It plays a role in a wide array of fields, such as botany, chemistry, physics, genetic engineering, molecular biology, hybrid development, pesticide testing, and food science.

Regeneration protocols for haploids and homozygous diploids, and woody perennials mango, litchi, apple, walnut, cashew, coconut and oil palm are required to be developed. Meristem culture, callus culture, anther culture and organ cultures are usually employed.

The use of meristem culture technique successfully been used in banana, citrus, cardamom, papaya, carnation, chrysanthemum, etc. to produce virus-free plants in a large number. Cloning technique is being used in a number of horticultural crops and shoot-tip grafting for virus cleaning in citrus.

Propagation of horticultural crops through somatic embryogenesis important for perennial tree species where a tap root is desirable. Micropropagation of parental lines for production of F_1 hybrid seed and exploitation of somaclones in ornamentals are important aspects for development of new horticultural crops. For stable and profitable variants of horticultural crops the callus regeneration, cell suspensions and protoplast cultures are being used. Ovule culture (for gerbera), and anther culture (for eggplant, petunia and cabbage) which help in getting haploids through different t *in-vitro* techniques are also successful for horticultural crops. The haploid plantlets after using colchicine can produce fertile homozygous lines which give elite genotype. They are known as *double haploid lines*. This technique develops hybrid parents after *in-vitro* propagation, as super males in asparagus.

A piece of a plant, which can be anything from a piece of stem, root, leaf, or bud to a single cell, is placed in that tiniest of greenhouses, a test tube. In an environment free from microorganisms and in the presence of a balanced diet of chemicals, that bit of plant, called an *explant*, can produce plantlets that, in turn, will multiply indefinitely, if given proper care. The *medium* (plural, *media*) is the substrate for plant growth, and in the context of plant tissue culture it refers to the mixture of certain chemical compounds to form a nutrient-rich gel or liquid for growing cultures, whether cells, organs, or plantlets. The process of tissue culturing plants from the explant stage to the final stage of transferring a mature plant to field or greenhouse conditions involves 4 basic stages. The four stages of culture growth are: Stage I, explant establishment or initiation; Stage II, multiplication; Stage III, rooting; and Stage IV, acclimatization or hardening off. These stages can overlap in certain cases, and the requirements of each stage vary widely from plant to plant.

Tissue Culture Techniques to Increase Genetic Variability

The haploid plantlets (n chromosomes) are treated with colchicine to produce fertile homozygous lines, called "double haploid lines". This haplodiploidization technique could give immediately new elite genotypes, hybrid parents after in vitro propagation, as asparagus super males (MM), or useful genetic material to establish gene mapping. A good synthesis book concerning all these possibilities was published by Bajaj (1990).

Protoplast Culture and Micropropagation

Protoplasts are the smallest units able to regenerate a whole plant. Therefore protoplasts cultures can serve to enlarge genetic variability by introducing somaclonal variation. However, the main interest of protoplasts is their capacity to fuse and to produce hybrids or cybrids, new organisms most often unknown in nature. Indeed, protoplasts allow transgressing the barrier of the botanical species or genus. Naked protoplasts can accept without rejection external elements: nuclei, cytoplasmic organelles, liposomes, containing genetic information.

The hybridization programme is hampered in some cases because of sexual incompatibility. Synthetic production of hybrids has, however becomes possible through the novel technique of protoplast culture and their fusion. The isolation, culture and fusion of protoplast are one of the most fascinating fields of research, though still in developing stage. The protoplast culture technique can be suitably used for microinjection and other genetic engineering experiments. The techniques are important, specially, because of their far-reaching effects on crop improvement by somatic hybridization and cell modification. The protoplast culture can be regenerated into an entire plant. It is imperative that the technique in future will be one of the most frequently used research tools for tissue culturists, physiologists, pathologists, molecular biologists, cytogeneticists, and biotechnologists. The discovery of enzymes which could separate the cells and remove the cell wall revolutionized the concept of plant tissue culture. It has provided a tool for isolating protoplasts and exploring the possibilities of genetic engineering.

Somatic Hybrids and Cybrids

The first protoplast fusion application was a cytoplasmic transfer from one genotype to another to induce male sterility (CMS) from mitochondrial origin. These male sterile hybrids are interesting to produce F_1 hybrids (*Brassica, Cichorium*).

A plant cell, when rid of its pecto-cellulosic wall after an enzymatic treatment (cellulase and pectinase) becomes a protoplast and is able to fuse with another protoplast under appropriate conditions, such as electric pulse or presence of polyethylene glycol combined with high pH.

Mechanical or enzymatic removal of cell wall from a plant cell yields, a protoplast, which is bound by plasma membrane. An isolated protoplast, under suitable culture condition can regenerate entire plant. Plasmolyzing cells prior to enzymatic treatment help facilitate protoplast isolation. Isolated protoplast, having lost the protective cell wall, has to be protected against the osmotic shock by keeping them in the isotonic state. Thirteen- per cent mannitol is a good agent in this context. A variety of enzymes are available for protoplast isolation. More common are cellulase, pectolyase and macerozyme. The viable protoplast tend to synthesize new cell wall within few hours to few days of culture divide and re-divide, produce clumps of cells which finally produce plantlets.

When suitable conditions are then applied to the fusion product, cell divisions, callogenesis, and organogenesis can take place giving rise to a totally new plant, which has theoretically the nuclear genomes and the cell organelles of both parents. In practice, after nuclear fusion some chromosomes are often eliminated and the ploidy status of the somatic hybrid can be highly variable; generally only one of the parent chloroplast populations is retained (the other one being eliminated) and the mitochondria of both parents most often recombine. When nuclear fusion occurs, regenerated plants are called somatic hybrids, but when one of the parental nuclei is completely eliminated, the regenerated plants are called cybrids because only their cytoplasm is hybrid.

The genetic interest of cybrids lies in the transfer into the cultivated species of traits that are coded by the cytoplasmic genome of a *donor* plant. There are various techniques used for obtaining cybrids, for instance by inactivating temporarily the nucleus of the protoplasts of the recipient parent with iodo-acetamide, and by killing by X- or gamma rays the nucleus of the protoplasts of the donor parent. The distinction between the allo-fusion products and the auto fusion products can be achieved by markers, such as a nuclear resistance to kanamycin of one parent, and defective chloroplasts of the same parent: in this case, the allo-fusion products are resistant to kanamycin and green. Somatic hybridization has been used in research programmes on potato, eggplant, lettuce and *Pelargonium* for introducing disease resistance available in distantly related species.

Chrysanthemum intergeneric somatic hybrids have been obtained with rust resistance. Somatic hybrids have also been obtained with *Allium, Asparagus, Hibiscus, Lilium, Mentha, Passiflora*, and *Primula* species. For *Citrus* species, this method is successfully used in breeding programmes for creating tetraploid rootstocks that combine desirable traits from both diploid parents; it is also useful for creating tetraploid material that is either used directly as cultivars or in interploid crosses for generating seedless triploid cultivars.

Cybridization also has horticultural applications, for simplifying the production of hybrid seed by creating cytoplasmic-male-sterile (CMS) seed parents, which are then pollinated by male-fertile plants. Cytoplasmic male sterility discovered by Ogura (1968) in radish (*Raphanus sativus*), was first transferred into cabbage (*Brassica oleracea*) and rapeseed (*B. napus*) by sexual hybridization, but the "ogura" sterility was linked to other cytoplasmic traits, such as chlorophyll deficiency, atrophied nectaries, and ovary abnormalities. Cybridization proved to be an outstanding tool for correcting this dysfunctional nucleo-cytoplasmic interaction. The fusion of cabbage (or rapeseed) protoplasts carrying the "ogura" cytoplasm with normal protoplasts of cabbage (or rapeseed), produced a normal chlorophyll efficiency (after exchange of the chloroplasts) and a normal flower morphology (nectaries and ovaries).

Haploid Production

Haploid plants can be obtained naturally or by numerous artificial in situ or *ex situ in vitro* methods applied to male gametophyte (immature pollen grains in or out of the anthers) or to the female gametophyte (embryo sac). In this latter case, direct in vitro culture of ovules is possible, as well as the in situ induction of embryo development by pollination with a distant species or with irradiated pollen having lost its fertility, followed later on by in vitro embryo rescue from the endosperm-free seeds.

Haploids are very often sterile, and the doubling of their chromosome number (spontaneous or induced, if needed) is necessary for the restoration of their fertility. Haplo-methods present many advantages for geneticists and breeders, namely:

1. The rapid obtaining of homozygous material;
2. the obtaining of direct products from meiosis (which increases the probability to

obtain multi-recessive genotypes);
3. the simplification of genetic analysis; and
4. the possibility of realizing interspecific hybridizations between species of different ploidy levels (very useful for tuberiferous *Solanum* species).

Androgenesis in the broad sense (anther or microspore culture) has been applied very successfully in breeding programmes of vegetable crops such as asparagus, diverse Brassicaceae such as cabbage, cauliflower, broccoli and pakchoi, eggplant, pepper, potato; apple and pear.

Gynogenesis is successful for melon, *Allium* species, apple, pear, and *Gerbera*. Haploids have been used for genetic analysis of pepper and apple

Production of Virus-Free Plants

In 1952, Morel and Martin were successful in regenerating a virus-free dahlia plant by the excision of some meristematic domes from virus infected shoots. Three years later, the same authors were able to eliminate virus A and Y from virus infected potato. Semal and Lepoivre (1992) reported that a virus-free sweet potato was producing 40 T/ha in China by comparison of the 20 T/ha produced before meristem culture.

Virus eradication is dependent on several parameters. But to take advantage of the non uniform and imperfect virus distribution in the host plant body, the size of the excised meristem should be as small as possible. For Stone (1963), only tips between 0.2 and 0.5 mm most frequently produce virus free carnation plants. The explants smaller than 0.2 mm can't survive, and those larger than 0.7 produce plants that still contain mottle virus.

Today, no definitive explanation can be given to understand this virus eradication (Wang and Charles, 1991). Various explanations have been given: absence of plasmodesm in the meristematic domes, competition between synthesis of nucleoproteins for cellular division and viral replication, inhibitor substances, absence of enzymes present only in the cells of the meristematic zones, and suppression by excision of small meristematic domes. This last proposal could explain why some potato plant showing virus particles in the meristematic domes, could regenerate a virus free plant.

Shoot Tip Micrografting

When meristem tip culture fails, it is possible to graft small meristematic domes on young seedlings growing in vitro. In this way, Navarro et al. (1975) eradicated all the virus diseases from Spanish Citrus orchards. This technique was also very successful in eliminating virus diseases from peach trees (Mosella et al 1980).

Micropropagation Techniques

Micropropagation can be exploited for long term conservation of valuable genetic stock such as *Allium* spp. male sterile material, diverse heterozygous or haploid plants, as well as interspecific hybrids. It is currently used for storing genetic resources of vegetatively propagated plants such as coffee and can interfere with biodiversity protection, by rescuing endangered species. Micropropagation can be combined with cryo-conservation, which means the storage of plant material (meristems or embryogenic cell suspensions) in liquid nitrogen at very low temperatures (-196°C). It is a promising option for long term storage of vegetatively propagated crops and for rescuing endangered crops. This method requires only limited space, protects the material from contamination, and involves little maintenance. This technique is used for crops such as potato, garlic, *Pelargonium*, *Dianthus*, *Azalea*, *Malus* sp., and oil palm.

During the micropropagation process, the genetic stability of the new shoots is dependent upon their origin. Axillary shoots issue from pre-existing buds and are normally true to type, indeed the meristematic cells are genetically very stable. Adventitious shoots, such as somatic embryos, are neo-formed buds developed directly on some organs, or indirectly through a callus phase formed on this organ. So, if the mother plant presents a cell mosaic or chimaeric tissues, risks of genetic variation exists. It is similar in the case of an indirect regeneration, when the callus phase is too long.

Propagation by Axillary Shooting

This technique has proved to be the most applicable and reliable method of in vitro propagation. Axillary shoot growth is stimulated by overcoming apical meristem dominance. Commercial tissue culture laboratories are now able to propagate a large number of herbaceous ornamental species and several woody plants in this way. However, the propagation of *Pelargonium*, Howea, and a few other horticultural plants are always difficult to propagate by axillary branching.

Propagation by Direct or Indirect Organogenesis

Adventitious shoots could arise directly from the tissue of explants without callus formation. Several plants of the family gesneriaceae (*Saintpaulia, Streptocarpus*) regenerate directly buds on leaf explants, likewise *Lilium* regenerates on scales. However more often, like for *Ficus lyrata*, adventitious buds appear on callus. While coffee, cocoa trees, and many conifers are produced by somatic embryogenesis developed on callus or cell-suspensions.

Improvement of Axillary Branching

The cost of micropropagated plantlets is also an important limitation of the techniques. In New Zealand, where they produce 2-3 million micropropagated radiata pine per annum, the relative cost of micropropagated planting stock had dropped from 13,8 times the cost of seedlings in 1988 to 6,9 by 1993 (Smith, 1997). To reduce manpower costs, several improvements have been proposed. The more simple method was in vitro layering developed by Wang (1977) to clone PVX-free potato plants. The first plantlets placed on the medium in a horizontal position developed axillary shoots. They are harvested by cutting one centimeter above the medium surface, at 3 weeks intervals. A similar technique called 'hedging system ' by Aitken Christie and Jones (1987) was later used to produce *Pinus radiata*. Ziv (1990) proposed for corm plants, Gladiolus and Nerine, a very rapid propagation system. She reduces the internodes and leaves by introduction of an anti-gibberellin agent in the medium. Finally, only aggregates of buds are formed, then they are divided and introduced in bioreactors for mass production. Similar systems were developed in Gembloux, to propagate carub trees and asparagus. Since 1988, Duhem was producing very large quantities of *Eucalyptus plantlets* in Petri dishes without anti-gibberellin but in complete darkness. Transfers from one Petri dish to another is made by a simple squashing (Boxus et al., 1991).

Plant Tolerance to Drought and Salinity Stress

Many countries face periodical serious drought periods and a continuous deficit of good-quality fresh water resources. Most countries will face it within 20-25 years.

Therefore, agriculture must use alternative water resources such as saline water and recycled sewage effluents. It is clear that salt accumulation in irrigated soils can increase salinity to levels that will reduce growth and yield of even the more tolerant crops. In non-irrigated areas, salinization is even more devastating, causing famine and death. Therefore, breeding for drought and salinity tolerance in crop plants (for food supply) and in forest trees (being a central component of the global ecosystem) should be given the highest research priority in all future plant biotechnology programs. The molecular basis of drought and salinity tolerance is rather limited. Discovery of new genes for abiotic stress tolerance, combined with controlled molecular breeding will have an important role in shaping agricultural plants in the post-genomic era.

Examples of several molecular control mechanisms of drought and salinity tolerance will be presented, including our recent findings on the molecular characterization of a novel class of plant chaperone-like stress-associated proteins (SP1 and related proteins), which are able to protect the structure and activity of other proteins and to sustain growth. Biotechnological-oriented breeding for drought and salinity tolerance includes:

1. Functional genomics analysis,
2. Use of molecular probes and markers for selection among natural and bred populations,
3. Transformation with specific genes and generation of transgenic plants. Biotechnology cannot solve all problems, however it is a most powerful technique that can and will be integrated in all classical plant breeding programs. At the end of the day, the only criterion to evaluate the efficiency of genetic engineering will be its cost effectiveness and benefits to society.

Somatic Embryogenesis Propagation

For genetically stable species, somatic embryogenesis offers a very fast scaling-up system, especially when it's possible to produce embryos in bioreactors. Unfortunately this production via fermenters was not as simple as first envisaged. And today, only a few model plants are successfully produced by such technology: carrot, celery. Other applications remain at the experimental stage: coffee, oil, palms, conifers, *Euphorbia pulcherrima*,. and several other horticultural species.

Several bottlenecks limit the use of this interesting technology. One of the main problems is genetic stability. So, despite the clonal nature of nucellar embryos, different morphological anomalies can occur among mango somatic embryos, as it was also observed in Citrus plants derived from nucellar cultures (Litz et al., 1993). Another difficulty is the loss of embryogenic capacity over time, a phenomenon observed with different species. It is also important to note that somatic embryogenic lines of conifers are always originated from immature embryos.

Synseeds

Another very interesting possibility of the somatic embryogenesis technology has been developed during the past fifteen years by Redenbaugh and his team (1991, 1993). They were able to encapsulate somatic embryos by hydrogel coatings (sodium alginate), producing single embryo artificial seeds. To date, some improvements offer the possibility to directly plant the artificial seeds in the greenhouse on special substrates (vermiculite, sans,). This methodology will provide in the future a good technique to reduce the cost of transplants.

Somaclonal Variations

The production of plantlets by callus regeneration, cell suspensions, protoplast cultures, could present some deviations with regard to the mother plant. This is a way to increase the genetic variability. Associated with a selective pressure (stress to toxins, pH, salinity, cold) it's used to obtain resistant lines. Indeed, after regeneration, plants can express new potentialities rarely obtained another way.

Creation of Genetic Variability

Artificial mutagenesis with physical or chemical methods can be applied to a great diversity of isolated cells such as microspores (haploid cells), protoplasts, or plant organs (buds, stem or Industrial scale micropropagation: (A) growth room; (B) media preparation; and (C) transfer room. The results are variable depending on the species. The potential of mutagenesis has been explored with *Weigela*, and applied to lettuce protoplasts. New fruit flesh colours and seedlessness have been produced this way in *Citrus*, and improved organoleptic quality has been obtained in banana.

Plant cells that pass through the callus stage can undergo somaclonal variations due to genetic modifications such as chromosome number changes, chromosomal deletions or translocations, activation of transposons, gene mutations, or epigenetic variations. When diverse pressures such as cold, toxins or salinity are applied during the regeneration process, it is possible to direct the selection towards plants with new interesting traits, which could not have been otherwise obtained.

However, commercial use of somaclonal variants is possible only when the mutations are stable. Several ornamental new cultivars having original traits have been obtained with this technique, such as the *Pelargonium* 'Velvet Rose', a callus clone of *P. graveolens* 'Robert's Lemon Rose', with thick grey-green and densely pubescent leaves.

A new cultivar of *Lavatera thuringiaca*, obtained from somaclonal variation of petiole fragments of the cultivar 'Bredon Springs'. 'Inovera' is tetraploid, and displays a compact growth habit, generous flower production, special leaf shape and larger flowers than those of the mother cultivar.

Other valuable somaclonal changes include seed weight variants for pea, and disease resistance, increased food quality and productivity for banana. Somaclonal variation is extensively exploited for foliage plants improvement, as it is the case of *Syngonium* (22 somaclonal cultivars originated from 'White Butterfly'), *Anthurium*, *Aglaonema* and *Dieffenbachia*.

Haplodiploidisation

Many species are able to produce haploids through different in vitro techniques. The oldest was anther culture of androgenesis. To date, the results vary considerably from one species to another. Solanae (datura, tobacco, red pepper, eggplant, petunia), cereals (wheat, barley, rice, triticale, maize) or crucifers (soybean, cabbage) are species easy to regenerate by anther culture. To the contrary, tomatoes, leguminous, or Compositae are recalcitrant. Ovule culture, or gynogenesis, was a successful technique for Gerbera, beet, courgette. However, the most common technique is pollination with irradiated pollen, in order to induce in vivo parthenogenesis. The oospheres developed in embryos without fertilization are saved by embryo rescue.

Embryo Rescue

In some cases zygotic embryos are unable to reach their full development *in situ* for various reasons. *In vitro* culture can save such embryos when isolated at an early developmental stage, and allows them to become whole plants. This technique is widely applied to embryos issuing from interspecific hybridizations, for vegetable species such as eggplant, lettuce and tomato, ornamentals such as *Pelargonium* and many other horticultural crops. For *Citrus* species, polyploid embryos from various origins were rescued (Ollitrault and Luro, 2001). In banana, the fragile embryos that derive from manual pollinations have been rescued for breeding programmes since the 1960s.

Germplasm Conservation (Gene Bank)

Tissue culture methods offer the opportunity for in vitro collecting, rapid multiplication and distribution of important, elite, or rare plants that are threatened with extinction. The two major in vitro storage strategies are slow growth and cryopreservation. Since the first results of Seibert (1976), who was able to initiate shoots from carnation shoot apices frozen to -196°C, this technique is now successful for many of horticultural species. Dereuddre et al. (1991) have provided a very simple technology to freeze encapsulated meristems in dried alginate beads. It works for pear, strawberry, eucalyptus, potato. Some international institutions have very large collections of old and current varieties, available for exchange and introduction in crop improvement programmes. The International Potato Center (CIP) in Lima, Peru, has a large world potato collection. Germplasm of sweet potato and cassava is at the International Institute of Tropical Agriculture (IITA), Ibadan, Nigeria.

The movement of germplasm involves the risks of accidentally introducing plant quarantine pests along with the host plant material. To limit these risks, the plant material should be transferred from one country to another as in vitro culture through a transit Centre, where it should be indexed. For bananas, in the framework of INIBAP, the transit Centre is the Catholic University of Leuven in Belgium where a very large in vitro germplasm exists. The germplasm conservation (*vitro* preservation) of asexually propagated horticultural crops for storage under conditions closer to ambient temperature to suit the local conditions are important. An *in-vitro* storage slow growth and cryopreservation have been standardized for many horticultural crops. Freezing of encapsulated meristems is possible in pears, strawberry and potato. For the movement of germplasm, *in-vitro* technique is very handy. At world level exchange of germplasm in tissue culture forms is in routine practice for banana, potato, cassava and sweet potato. The molecular taxonomy (like RAPD and RFLP) in genomic analysis and classification and development of molecular tools for disease indexing in *in-vitro* propagated materials, in checking clonal fidelity, germplasm classification and identification of markers linked are important priority areas in this field. Molecular characterization of indigenous germplasm, application of DNA markers for identification of cultivars and molecular linkage maps for developing new varieties are important aspects of biotechnological activities. Work in Indian laboratories is still is progress to develop molecular biology of horticultural crops, particularly in DNA markers.

Cryopreservation of Plant Germplasm

Cryopreservation refers to the storage of biological material at ultra-low temperatures i.e. at –196°C, the temperature of liquid nitrogen. At this temperature, all chemical and metabolical and most physical processes are arrested. For this reason, cryopreservation is the method of choice for the long term conservation of plant germplasm since both the problem of genetic instability and the risk of loosing accessions due to contamination or human error during subculturing are overcome. Most cryopreservation endeavours deal with recalcitrant seeds, *in vitro* tissues from vegetatively propagated crops, species with a particular gene combination (elite genotypes) and dedifferentiated plant cell cultures. Other applications of cryopreservation in plant science are also discussed, i.e. cryo-therapy and its potential for storing the causal agents of pest and diseases. So far, cryopreservation procedures are developed for *in vitro* tissues and recalcitrant seeds of about 100 and 40 species, respectively. There is still a limited number of cases where cryopreservation is used routinely for plant germplasm conservation, mainly because the techniques need to be adapted for each species in function of its natural freezing resistance, explant size and type, and its water content. Care must be taken to avoid ice crystallisation during the freezing process, which otherwise would cause physical damage to the tissues. The existing cryogenic strategies rely on air-drying, freeze dehydration, osmotic dehydration, addition of penetrating cryo-protective substances and adaptive metabolism (hardening) or combinations of these processes.

MICROBIAL BIOPESTICIDES, BIOFUNGICIDE, HERBICIDES AND AGRICULTURAL ANTIBIOTICS

The contribution of biotechnology to the practice of medicine through the discovery and development of a number of potent antibacterial antibiotics is widely recognized. A major effort concerned with the discovery of new structures with antibacterial activities and the chemical modification of existing natural products is still ongoing. Although these antibiotics have found widespread use as growth promotors in the animal health area, only more recently were microbial products produced primarily for animal growth or agricultural purposes, for instance monensin as a coccidiostat for poultry and a growth permittant for ruminant animals. Bacteria, fungi and viruses are the most commonly researched as potential microbial biocides. The development of low cost production methods and successful formulation of products such as microbial insecticides represents the commercial cutting edge of biotechnology.

The term insecticide usually brings thoughts of poisonous, non-selective, energy-consuming chemicals that are used to kill pest insects. However, not all insecticides fit this generally accepted stereotype. Those microorganisms which often suppress and naturally control populations of pests which injure man and causes extensive damage to food and fibre crops. Formulation of living viruses, bacteria, fungi and protozoa are called microbial insecticides.

Revolutionary techniques in biotechnology offer tantilizing new prospects for the future. Biological control has a long history, reaching into the undated past when the Chinese used Pharoah's ants to control pests of stored grain. Others date the start of biological control to the domestication of the cat. The introduction of an exotic species to control a pest, the classical form of biological control, is claimed to date from 1762 when the Indian Mynah was brought to Mauritius to control the Red Locust. Subsequent attempts at introduction occasionally had unwanted results; the introduction of cats to control rats on Ascension Island in 1815 resulted in the extermination of virtually all the sea birds on the island. Successful control of cotton cushion scale by Australian ladybirds in California in 1888 encouraged practitioners to continue to develop biological pest control by the introduction of exotic enemies.

Introduction are not theoretically limited to insects and vertebrate predators; the use of the myxomatosis virus to control rabbits in Australia suggests that microbes may be able to play similar roles. However, introductions have two fundamental limitations: (1) They have an extra dimension of unpredictability because an organism, hitherto unknown, is being brought to a new area which it might find futile in unanticipated ways and/or may not acclimatized in changed conditions; (2) There is no clear way to make a profit from one of introductions.

Use of Biotechnology in Biocide Production

Substantial contributions can also come from fundamental biotechnological research.

1. Increases in potency by modification of the plasmid complement of the bacterium, which controls the synthesis of the protein both reduce the cost of production and provide the starting place for new, more easily used formulations.
2. Changes in the spectrum of activity of the toxic protein by genetic engineering can open wholly new markets.
3. Production of asporogenic strains may reduce production costs by avoiding the waste of metabolic energy on spore production and may make the product more acceptable in certain countries.

Biopesticides are living organisms or products derived from microorganisms and plants. Technologies for developing biopesticides – *Heliothis* NPV, *Spodoptera* NPV, *Chilo* GV, *Trichograma* and *Trichoderma* – have already been standardized. The most successful biopesticide *Bt*-based insecticides is expected to grow further. Some of the biopesticides like *Trichoderma* spp. against root-infecting fungi in citrus and black pepper, viral biopesticide like Ha HPV for tomato fruit-borer and bacterial antagonist to *Pseudomonas solanacearum* to control bacterial wilt in potato have been successfully used. Indian still need development of bioinsecticides derived from insecticidal toxins of *Bacillus thuringiensis.*

BIOSENSORS, BIOCHIPS, BIOFILMS AND BIOSURFACTENTS

Biosensors

The natural biosensors are the *sense organs* primarily the chemical sensors of smell and taste. Biosensors technically in their various forms share a reliance of biological materials as sensing elements. Technical biosensors have been under intense development since the middle of 1960 with the prospects of commercial potential offered by biotechnology. Biosensors are the combinations of biologically active material displaying characteristic specifically with chemical or electronic sensor to convert the response into electrical signals. Biosensors have their roots in military research, as means of detecting nerve gases and other chemical warfare toxins. Their applications have branched out to include simple-t-use *alternate site* diagnostic devices for home, doctor's office, or drug use screening; medical and surgical monitors (small enough to fit inside a blood vessel) and environmental quality monitors.

Biochips

Biochips are made from different biological materials and can control the computer by replacing silicon chips. Biomolecular computers thus made, promise to be ten to thousand times smaller than the best super computers with much faster switching times and extremely low power dissipation. They are made up of **semi-conducing molecules** inserted into the protein framework and fix the whole on to a protein support. They can be used in implanting of several sorts in human body, like regulation of heart beats, responses to nerve impulses by artificial limbs (bearing such computer device), overcoming of blindness and deafness, etc. Biochips may be infected by microbes since they are made up of proteins.

Biofilm

A biofilm is an accumulation of microbial cells and inorganic components held together in a polymeric matrix and firmly attached to a substratum. Accumulation of biofilms is encountered in many natural and modulated environments. It may be fundamental to process performance i.e., for fixed-film biological waste water treatment, sudden deterioration of water quality and deterioration of substrata.

Biosurfactents

Biosurfactents are surface-active compounds produced by microorganisms. Biological surfactants possess a number of potential advantages over their chemically manufactured counterparts, including lower toxicity, biodegradability, a wide variety of possible structures, and ease of synthesis from inexpensive, renewable feed stocks. Consequently, biosurfactents may have applications in numerous areas, particularly for enhance oil recovery and in foods, beverages, cosmetics, pharmaceutical preparations, etc.

MONOCLONAL ANTIBODY TECHNOLOGIES

The development of monoclonal antibody technology is based on advances in our ability to culture cells. Antibodies are the protein components of the immune system found in the blood of mammals. They have a unique ability to identify particular molecules and select them out. When a foreign substance (an antigen) enters the body, specialized cells called B lymphocytes produce a protein (an antibody) to combat it. To envision how antibodies work, think of a lock and key: The antibody key *fits* only the specific antigen lock. This marks the antigen for destruction. Each of the specialized B lymphocyte cells produces only a single type of antibody and thus recognizes only one antigen.

Apart from their natural role in protecting organisms via the immune response, antibodies are important scientific tools. They are used to detect the presence and level of drugs, bacterial and viral products, hormones, and even other antibodies in the blood. The conventional method of producing antibodies is to inject an antigen into a laboratory animal to evoke an immune response. Antiserum (blood serum containing antibodies) is then collected from the animal. However, antiserum collected in this way contains many types of antibodies, and the amount that can be collected is limited.

Modern biotechnology has opened a door to amore efficient, more specific, and more productive way of producing antibodies. By fusing two types of cells, antibody-producing B lymphocytes and quasi-immortal cancer cells from mice, scientists found that the resulting hybrid cells, called hybridomas, secreted large amounts of homogeneous antibodies. Each hybridoma has the ability to grow indefinitely in cell culture and thus can produce an almost unlimited supply of a specific "monoclonal" antibody. By immunizing mice with specific antigens, researchers can create and select hybridomas that produce a culture of specific, desired monoclonal antibodies. Thus, biotechnology has produced a way of creating pure lines of antibodies that can be used to identify complex proteins and macromolecules. Monoclonal antibodies are powerful tools in molecular analysis, and their uses in detecting low levels of disease agents such as bacteria and viruses are rapidly expanding.

Beyond many diagnostic uses, hybridoma technology shows promise for immunopurification of substances, imaging, and therapy. Immunopurification is a powerful technique to separate large, complex molecules from a mixture of either unrelated or closely related molecules. For imaging, easily visualized tags can be attached to monoclonal antibodies to provide images of organs and to locate tumors to which the antibody will specifically bind. Finally, new therapeutic methods have been developed that use monoclonal antibodies to inactivate certain kinds of immunological cells and tumor cells or to prevent infection by certain microorganisms.

Although many applications of this technology are still in the experimental stages, the commercial agricultural use of monoclonal antibodies has begun. For example, monoclonal antibodies are now on the market as therapeutics against calf and pig enteric colibacillosis, which causes neonatal diarrhea (scours). This approach is often more effective than conventional vaccines, and it supplements genetically engineered vaccines. Monoclonal antibody-based diagnostic kits can detect whether scouring animals are infected with a particular strain of an *Escherichia coli* bacterium that causes scours, and thus help veterinarians determine the appropriate therapeutic monoclonal antibody to use on an infected herd.

NEW APPROACHES TO CROP PRODUCTION

In the past 50 years, agricultural production in the United States has more than doubled while the amount of land under cultivation has actually declined slightly. This impressive agricultural success is the result of many factors: an abundance of fertile land and water, a favorable climate, a history of innovative farmers, and a series of advances in the science and technology of agriculture that have made possible more intensive use of yield-enhancing inputs such as fertilizer and pesticides. Yet the productivity successes brought about the farm mechanization, improved plant varieties, and the development of agricultural chemicals may be harder to repeat in the future unless new approaches are pursued.

Biotechnology offers vast potential for improving the efficiency of crop production, thereby lowering the cost and increasing the quality of food. The tools of biotechnology can provide scientists with new approaches to develop higher yielding and more nutritious crop varieties, to improve resistance to disease and adverse conditions, or to reduce the need for fertilizers and other expensive agricultural chemicals. The following paragraphs highlight some examples of how genetic engineering can be used to enhance crop production.

The Genetic Engineering of Plants

Perhaps the most direct way to use biotechnology to improve crop agriculture is to genetically engineer plants – that is, alter their basic genetic structure – so they have new characteristics that improve the efficiency of crop production. The traditional goal of crop production remains unchanged: to produce more and better crops at lower cost. However, the tools of biotechnology can speed up the process by helping researchers screen generations of plants for a specific trait or work more quickly and precisely to transfer a trait. These tools give breeders and genetic engineers access to a wider universe of traits from which to select.

Although powerful, the process is not simple. Typically, researchers must be able to isolate the gene of interest, insert it into a plant cell, induce the transformed cell to grow into an entire plant, and then make sure the gene is appropriately expressed. If scientists were introducing a gene coding for a plant storage protein containing a better balance of essential amino acids for human or animal nutrition, for example, it would need to be expressed in the seeds of corn or soybeans, in eh tubers of potatoes, and in the leaves and stems of alfalfa. In other words, the expression of such a gene would need to be directed to different organs in different crops.

There are already successes that demonstrate how plans can be genetically engineered to benefit agriculture. Herbicide resistance traits are being transferred to increase options for controlling weeds. Soon, the composition of storage proteins oils, and starches in plans may be altered to increase their value.

One plant gene that has been isolated, cloned, and transferred is for the sulfur-rich protein found in the Brazil nut, *Berthalletia excelsa.* This protein contains large amounts of two nutritionally important sulfur-containing amino acids: methionine and cysteine. These are the very nutrients in which legumes, such as soybeans, are deficient. If the sulfur-rich protein gene were transferred into soybeans, it might enhance this legume's role as a protein source throughout the world.

By purifying the Brazil nut protein and determining the order and kind of amino acids in the protein, scientists were able to synthesize an artificial segment of DNA coding for a section of this protein. This DNA "probe" was used to find and pull out the natural gene from the Brazil nut. Researchers then transferred the gene into tomato and tobacco plants, which were chosen because they are easier to manipulate than soybeans. Researchers have also transferred the gene into yeast cells. Early results show that the genetically engineered yeast do produce the sulfur-rich protein.

Scientists have taken another important step in using genetic engineering to improve crop production: They have for the first time engineered plants to be resistant to powerful herbicides. One example is glyphosate (trade name: *Roundup*), a common, effective, and environmentally safe herbicide. However, glyphosate indiscriminately kills crops as well as weeds. Thus, it must usually be used before crop plants germinate. Yet by engineering crops to be resistant to glyphosate, scientists hope to expand the range of the herbicide's applications.

Scientists have isolated a glyphosate-resistance gene and successfully transferred it into cotton, poplar trees, soybeans, tobacco, and tomatoes. The gene was derived from the bacterium *Salmonella typhimurium.* Similarly to other accomplishments in biotechnology, this success depended on extensive prior basic research on biochemical pathways in bacteria and plants, and sophisticated gene cloning and transfer techniques. Field testing and commercialization of glyphosate-resistant crops should follow soon. Analysis of tomato growers' costs in California predicts that farmers could save up to $100 per acre in weed control costs if they used glyphosate in place of current herbicides, with concomitant reductions in labor, equipment, and environmental damage. This advance would also give farmers improved flexibility, yield, quality, and spectrum of weed control.

The Genetic Engineering of Microorganisms Associated with Plants

Microorganisms in the environment affect the growth of plants in a variety of ways, many of which are still poorly understood. Their effects can be either beneficial or harmful. For instance, some microorganisms protect plants from bacterial or fungal infections. Others protect plants from environmental stresses such as acidity, salinity, or high concentrations of toxic metals. Still others attack weeds that compete with crops. The best known association between microorganisms and plants is the symbiotic relationship between nitrogen-fixing bacteria of the genus *Rhizobium* and members of the legume family, such as soybeans.

However, some microorganisms, particularly certain bacteria and fungi, are pathogens that attack crops and cause disease, sometimes in epidemic proportions. The Irish Potato Famine of the mid-1800s, the Dutch Elm disease of the twentieth century, and the southern corn leaf blight of 1970 are dramatic examples of losses caused by pathogens.

As our understanding of the relationships between microorganisms and crops improves, the genes controlling these relationships – whether in the microorganism or in the plant – can be engineered to enhance the abilities of beneficial microorganisms or inhibit the effects of harmful microorganisms. Yet to successfully engineer microorganisms, scientists must understand the molecular mechanisms by which they interact with their plant hosts. Much remains to be learned about both the plant and microbial genes involved, their regulation, and the intricate relationships between microorganisms and their hosts.

Initial discoveries in genetic engineering technologies were made with microorganisms because they are simpler life-forms than higher plants and animals, and thus are easier to manipulate in the laboratory. Methods developed in medical research with bacteria and viruses are now being adapted to agriculturally significant microorganisms. One example that has progressed to the point of field testing involves genetically altered bacteria designed to prevent frost damage. *Pseudomonas syringae* is a bacterial species with many members that are normally harmless and commonly inhabit the outer surface of plant cells. However, some of these bacteria contain a protein that initiates the formation of ice crystals at temperatures below freezing. The growing ice crystals can rupture and damage plant cells. If the bacteria are not present, plants can withstand colder temperatures without damage. Researchers have now created an *ice-minus* strain of *P. syringae* by removing the gene that makes the protein. In laboratory tests the ice-minus strain has been sprayed on plants to displace the wild strain and thereby provide the crop with some measure of frost protection. Although the genetically engineered, ice-minus *Pseudomonas* is already several years old, field tests necessary to test its commercial application have been blocked by public apprehension that has led to court actions and confusion over the types of precautions needed to regulate such environmental testing.

Another practical application involves the use of DNA probes to detect plant viruses and viroids. Detection permits rapid screening to eliminate infected stock and thus halt the spread of disease.

Nearly 60 years ago scientists found that a mild strain of tobacco mosaic virus (TMV) could protect tobacco plants against the adverse effects of a subsequently inoculated, severe strain of the virus. This phenomenon, termed cross-protection, has been applied on a limited scale to protect greenhouse tomatoes and a few orchard crops. There are potential problems with the conventional cross-protection approach, however, because the mild, protecting virus might spread to other crops or mutate to a more virulent form. Recently, scientists installed fragments of the TMV genome in tobacco and tomato plants. Because these transgenic plants have only a portion of the genetic information that is needed for TMV replication, the problems of conventional cross-protection are avoided. Some transgenic plants appeared to be completely resistant to the TMV virus. Tests show that virus resistance introduced by recombinant DNA technology can be transmitted through seed as a simple Mendelian trait and can thus be transmitted by conventional breeding techniques.

Genetic Engineering for Crop Protection

Another strategy to improve crop production through genetic engineering involves protecting crops from pests. Insects, viruses, bacteria, fungi, nematodes, and weeds can all impair agricultural productivity. Yet in a natural ecosystem, organisms typically serve many functions. Insects, for example, can be pests – destroying crops and stored products and transmitting disease. They can also be benefactors – pollinating plants, eating other pests, and recycling organic wastes.

Most chemical insecticides, herbicides, and other pesticides that have been the primary methods of controlling pests are not selective enough to affect only harmful organisms. As biotechnology becomes more refined, methods for handling bothersome pests and beneficial organisms will be created.

BIOPROCESSING OPPORTUNITIES

Several familiar age-old procedures are forms of bioprocessing – fermenting grape juice or leavening bread dough, for example. Yet bioprocessing also includes a range of technologies in which living cells or their components, such as enzymes, are used to cause the desired physical and chemical changes.

Bioprocessing to produce industrial chemicals began during World War I when researchers developed alternative ways to produce acetone and butanol using microorganisms. However, the growth of petrochemical industry during World War II replaced the microbial production of industrial solvents, and industrial bioprocessing for bulk chemicals practically disappeared. The climate changed again, however, when it was discovered how well biological processes could synthesize complex molecules such as antibiotics, vitamins, and enzymes. The industry was transformed from one that produced high-volume, low-value industrial chemicals to one that produced lower-volume, high-value products.

Advances in biotechnology have renewed interest in industrial uses of agricultural and forestry commodities. Bioprocessing offers innovative opportunities to create new products and foods, treat and use wastes, and use renewable resources (biomass) for fuel. Once developed, such processes could prove more economical as well as less environmentally damaging than current industrial processes.

Other Products

Bioprocessing already contributes to our ability to produce vitamins, amino acids, enzymes, and more recently hormones and this role should increase in the future. For instance, much of the supply of vitamins B_2 (riboflavin) and B_{12} (cobalbumin) comes from microbes. Researches have adapted wild strains of a mold, *Ashbya gossypii,* to produce 20,00 times its original output of vitamin B_2. Research has also intensified microbial production of vitamin B_{12} over 50,000 times.

Most cereal grains are deficient in two essential amino acids, lysine and methionine. These are usually added to animal feed to ensure an adequate diet. Methionine is made by chemical processes, but 80 per cent of all lysine is produced by fermentation using bacteria. The amino acid derivative monosodium glutamate, which is used as a flavor enhancer in cooking, is produced by two bacteria through a bioprocess.

In the past, the use of enzymes has been limited by the expense of isolating them from natural sources and by their instability. Recent advances have provided ways to immobilize enzymes and use whole microorganisms as catalytic systems, thus yielding more stable and reusable enzymes and increasing the opportunities for their use. The biotechnological production of sugar substitutes is one example of a growing industry that has been made possible largely because of our increased ability to manufacture enzymes through microbial processes.

Another area with potential for bioprocessing is waste treatment. As mentioned in previous examples, some bioprocessing systems can transform plant debris and other wastes into useful products, in effect creating an inexpensive and abundant renewable resource. Another current example is a new strain of yeast genetically engineered with an enzyme that converts the lactose in whey, a dairy industry waste product, into ethanol, which has fuel energy value. On another front, bioprocesses are being developed to more efficiently treat municipal, industrial, and agricultural wastes. However, some problems remain in improving the dependability and design of these systems.

To develop new approaches toward bioprocessing and to bring them into widespread use will require a great deal of research. First, systems must e designed to accomplish each goal. Successful systems will require (1) a solid understanding of the organism involved, (2) an effort to develop the most productive strain of the organism and isolate the appropriate enzymes, and (3) intensive, specific research on the dynamics of each individual bioprocess. Next, the bioprocessing systems must be improved and perfected to offer economically competitive products. Research is needed to develop new industrial-scale methods to isolate products at the degree of purity appropriate for commercial use. Concentration of the final products is also important because separating them out after microbial conversion is often a major cost.

CHAPTER 7

Applications of Genetic Engineering Hi-Tech Horticulture

Genetic engineering has unlimited scope for application, which include technologies that cater to human health and prosperity. The goal of plant, animal and microbial genetic engineer is to improve the quality and yield of products.

BIOLOGICAL MOLECULAR MARKERS ASSISTED BREEDING

Molecular markers are the most advanced tools of genetic research. Breeders have used phenotypic markers to follow segregation of genes but most of the agronomic characters are not associated with scorable phenotypic characters. Discovery of restriction enzymes that cleave DNA at sequence-specific sites has evolved the molecular marker technology by splicing any specific segment of DNA and the traits associated with it (Thro *et al.*, 1994). Genetic analysis for morphological markers is few (Hahn *et al.*, 1989). Application of molecular and biochemical markers may help breeders to tag agronomic traits and to screen those at early stages of growth.

The DNA molecular markers exist in every genotype; their number may vary from hundreds to thousands. They are stable makers and are independent of plants growth stage and external environment. As a result, molecular markers can be scored using DNA from seedlings and plants at any stage of growth in any environment. For example if a set of markers established for high starch/protein, a breeding population could be screened for that character even at the seedling stage. Interdisciplinary research is required to establish molecular markers, associated with the traits of agronomic or market interest.

In this chapter, various applications have been fragmented into three major subdivisions viz. transgenic plants, transgenic animals and miscellaneous applications.

The term 'transgenic' is used to describe plants and or animals which have had DNA introduced into them by means other than transfer of DNA from a sperm cell to an egg. The DNA transferred by these non-sexual processes may be DNA from same species, from other plant or animal species or even from a prokaryote. Today many transgenic plants as well as animals have been engineered by the scientists with special beneficial properties; few of them are discussed in this chapter.

Transgenic Plants

Up till 1980s, production of transgenic plants was restricted mainly to tobacco, petunia or tomato of the family solanaceae. However, now in more than 40 dicot species transgenic plants have been successfully produced. The production of transgenic plants has now been extended to several monocotyledons like wheat, maize, rice and oats.

Table 7.1: A list of higher plants were transgenic plants have been produced, using different methods

I. Herbaceous Dicotyledons

1. *Actinidia* sp. (kiwi)
2. *Apium graveolens* (celery)
3. *Arabidopsis thaliana*
4. *Armoracia* sp. (horse radish)
5. *Beta vulgaris* (sugarbeet)
6. *Brassica napus* (oilseed rape: canola)
7. *Brassica oleracea* (cauliflower)
8. *Brassica oleracea* var. capitata (cabbage)
9. *Brassica rapa* (syn. *B. campestris*)
10. *Carica papaya* (papaya)
11. *Chrysanthemum* sp. (chrysanthemum)
12. *Cichorium intybus* (chicory)
13. *Cucumis melo* (muskmelon)
14. *Cucumis sativus* (cucumber)
15. *Daucus carota* (carrot)
16. *Dianthus caryophyllus* (carnation)
17. *Digitalis purpurea* (foxglove)
18. *Fragaria* sp. (strawberry)
19. *Glycine max* (soybean)
20. *Glycyrrhiza glabra* (licorice)
21. *Gossypium hirsutum* (cotton)
22. *Helianthus annuus* (sunflower)
23. *Ipomoea batatas* (sweet potato)
24. *Ipomoea purpurea* (morning glory)
25. *Lactuca sativa* (lettuce)
26. *Linum usitatissimum* (flax)
27. *Lotus corniculatum* (lotus)
28. *Lycopersicon esculentum* (tomato)
29. *Medicago varia*
30. *Medicago sativa* (alfalfa)
31. *Nicotiana plumbaginifolia* (wild tobacco)
32. *Nicotiana tabacum* (tobacco)
33. *Petunia hybrida* (petunia)
34. *Pisum satium* (pea)
35. *Rosa sp*. (rose)
36. *Solanum melongena* (eggplant)
37. *Solanum tuberosum* (potato)
38. *Vaccinium macrocarpon* (crtanberry)
39. *Vigna aconitifolia*
40. *Vitis vinifera* (grape)

II. Woody Dicotyledons

1. *Populus sp*. (poplar)
2. *Malus sylvestris* (apple)
3. *Pyrus communis* (pear)
4. *Azadirachta indica* (neem)
5. *Juglans regia* (walnut)

ENGINEERED RESISTANCE AGAINST HERBICIDE

The development of crop plants that are tolerant to herbicide is an important approach to control weeds. Most of the modern herbicides interfere with amino acid biosynthesis in plants. Once the mode of action of a herbicide is established, the transfer of resistance to the herbicide can be done by two different approaches:

1. The first approach is modification of the target side.
2. Detoxification of herbicide

Transgenic plants carrying the genes for herbicide tolerance has been produced in tobacco, petunia, tomato, potato, *Populus* etc.

Modification of the Target

This approach has been made for developing resistance against atleast three herbicides viz. glyphosate, sulphonyl urea and imidazolinones.

Resistance against Glyphosate

Glyphosate, an active component of round up herbicides, is used as non-selective post-emergence herbicides. It is suggested that glyphosate inhibits aromatic amino acid biosynthesis by competitive inhibitor of enzyme 5-enol-pyruvyl shikimate-3 phosphate synthase [EPSPS]. EPSPS catalyzes the synthesis of phenylalanine, tyrosine and tryptophan. The aro A gene which codes EPSPS has been isolated and sequenced from *E. coli*, *Salmonella typhimurium* and *Arabidopsis thaliana*.

There are two possible methods for producing glyphosate resistant plants by genetic engineering:

1. transfer of a glyphosate sensitive EPSPS under the control of a powerful promoter causing over expression of the protein and/or
2. transfer of gene which codes a mutated, glyphosate resistant EPSPS. cDNA of the EPSPS gene with 35 S cauliflower mosaic virus promoter was constructed and the transformed petunia cell lines with this construct exhibited 40-fold increase in EPSPS activity.

An alternative method is insertion and expression of mutated bacterial aro A gene coding for glyophosate resistant EPSPS. The source of the aro A gene was a mutagenized strain of *S. typhimurium* or *E. coli* and was transferred to tomato and/or tobacco. Transgenic tobacco plants, showing tolerance to commercial levels of glyphosate are field tested in USA.

Resistance against Sulfonylurea and Imidazolinone Herbicides

The herbicidal action of sulfonylureas and imidazolinones is based on their ability to inhibit non-competitively an enzyme of branched-chain amino acid biosynthesis, acetolactate synthase (ALS). ALS is the first enzyme in the biosynthetic chain resulting in the synthesis of the branched chain amino acids valine, leucine and isoleucine.

Mutated strains resistant to sulfonylureas have been obtained from bacteria, fungi and plant cell cultures. The resistance of all these mutants can be attributed to possession of mutated forms of the ALS. Transgenic tobacco plants expressing a mutant ALS gene from tobacco or *Arabidopsis*, were produced that were tolerant to sulfonylurea herbicides.

TABLE 7.2: Mechanisms of action of different herbicides and basis of achieving resistance against them in transgenic plants

Active principle of herbicide	Inhibited pathway	Target product	Use	Basis of resistance
Amino acid biosynthesis inhibitors				
Glyphosate (Roundup)	Aromatic amino acid biosynthesis	EPSPS	Broad spectrum	Over-expression of EPSPS gene; bacterial aroA gene
Sulphonylurea and Imidazolinones	Branched chain amino acids	ALS	Selected crops	Mutant ALS gene
Phosphinothricin (Basta)	Glutamine	GS	Broad spectrum	Gene amplification; bar gene; detoxification
Photosynthesis inhibitors				
Atrazine (Lasso)	Photosystem II	Qb (32 kDa protein)	Selected crops	Mutant PsbA gene; GST gene; detoxification
Bromoxynil (Buctril)	Photosynthesis	-	Selected crops	bxn gene; detoxification

Resistance against Phosphinothricin

L-Phosphinothricin (PPT) is a naturally occurring amino acid with herbicidal activity and is active ingredient of herbicide 'Basta'. PPT is a potent inhibitor of glutamine synthase and thus causes rapid increase in ammonia concentration in plants. This leads to death of plant cell. 'bar' gene from *Streptomyces hygroscopicus* encodes phosphinothricin acetyl transferase enzyme that acetylates the free-NH_2 groups of phophinothricin and renders it inactivates the herbicide Basta. This 'bar' gene with CaMV 35S promoter has been transferred to tobacco, tomato and potato plants using *Agrobacterium* Ti plasmid vector system. The transgenic plants have been tested for 'Basta' resistance.

Resistance against Atrazine

Atrazine is another herbicide, which inhibits photosynthesis. A 32kDa protein, Q_B is responsible for herbicide binding and is encoded by psb A gene. The DNA sequences of psb A are highly conserved in different plants. More than 20 weed species have developed resistance to atrazine herbicides, for example, *Amaranthus hybridus*.

Detoxification or Degradation of Herbicide

This approach is based on selective use of herbicide, which will kill the weeds and not the crop. A number of detoxifying enzymes have been identified in plants as well as microbes. Few examples are given here: *(i) glutathione-S transferase* or GST (in maize and other plants): It detoxifies the herbicide atrazine; *(ii) nitrilase*: It is coded by gene bxn in *Klebsiella penumoniae*, and it detoxfies the herbicide bromoxynil, and *(iii) phosphinothricin acetyl transferase* or PAT (coded by *bar* gene in *Streptomyces* and transgenic plants using the *bxn* gene from *Streptomyces* spp).: It detoxifies the herbicide PPT (L-phosphinothricin).

Transgenic tomato plants using the *bxn* gene from *Klebsiella ozaenae* and *bar gene* from *Streptomyces,* as well as, transgenic potato, oilseed rape (*Brassica napus*) and sugarbeet using *bar* gene from *Streptomyces* have been obtained and were found to be herbicide resistant. Other crops for engineered herbicide tolerance include soybean, cotton and corn.

ENGINEERED RESISTANCE AGAINST PEST

Pests and insects cause severe damage to our crops and thus use of insecticides and pesticides is a common measure. With genetic engineering, crops can now be protected with insects without applying any synthetic chemical. More than 30 crop species have been successfully transformed including corn, cotton, soybean, and rice with insecticidal properties.

For transgenic plant production with pest resistance properties, two major approaches have been developed: a) the introduction of protein delta endotoxins from Bt and, b) introduction of protease inhibitors in plants.

Bacillus thuringiensis Endotoxins

The entomocidal bacterium *Bacillus thuringiensis*, upon sporulation normally produces a parasporal crystalline toxin. When ingested by a susceptible insect, a combination of the high pH and proteinases of the insects midgut solubilizes the crystal and yields active toxin. The effects of the toxin occur within minutes of indigestion, leading to disruption of insect's midgut cells. Bt toxin activity has been found against many species of insects within the orders of Lepidoptera, Diptera and Coleoptera.

TABLE 7.3: Insecticidal efficacy of CpTi in artificial diets and in transgenic plants

Insects killed by CpTi in artificial diets	Insects, against whom resistance noticed in transgenic tobacco plants (with CpTi)
Lepidoptera	
Heliothis virescens	*Helothis virescens*
H. zea	*H. zea*
Spodoptera littoralis	*Spodoptera littoralis*
Chilo partellus	*Manducta sexta*
	Autographa gamma
Coleoptera	
Callosobrichus maculatus	
Anthonomus grandis	
Diabrotica undecimpunctata	these insects do not attack control tobacco plants
Tribolium confusum	
Costelytra zealandica	
Orthoptera	
Locusta migratoria	

These delta endotoxins are very specific in their action and are safe insecticides, but their use is limited due to high production cost and due to instability of crystal proteins when exposed in the field.

The above toxin gene (*ht2*) from *B. thuringiensis* has been isolated and used for *Agrobacterium* plasmid mediated transformation of tobacco, cotton potato and tomato plants. These transgenic plants were resistant to the feeding damage of *Manducta sexta* larvae (tobacco horn worm), a pest of tobacco. Experiments of feeding the larvae of *M. sexta* on leaves of these transgenic plants showed 75%-100% mortality of the larvae, while the control plants carrying no transgenes were severely damaged.

Pseudomonas fluorescens that Secretes Bt Toxin

The crystalline protein toxin produced by *Bacillus thuringiensis* var. *israelensis* is toxic to the larval stage of several mosquitoes and black fly species. Normally this toxin is produced by fermentations of *B. thuringiensis* and the toxin is sprayed onto crops. Scientists have produced genetically engineered *Pseudomonas fluorescens* living in the roots of corn plants. This bacterium can secrete Bt-toxin to kill black cut worm.

Field tests using transgenic insect resistant plants were also conducted with tobacco and tomato and may be used for commercial cultivation in the near future.

Protease Inhibitors

Protease inhibitors are proteins with anti-metabolic activity against wide range of insects. Such inhibitors are widely distributed within the plant kingdom and can accumulate to high levels in seeds and storage organs. Other major advantages of protease inhibitors is their inactivation with cooking.

In cowpea (*Vigna unguiculata*), trypsin inhibitor (CpTi) level was shown to be responsible for its resistance to attack by the major storage pest of its seeds (i.e. bruchid beetle = *Callosobruchus maculatus*) A variety of insects have shown to be toxic to CpTi. This CpTi gene was joined with a CaMV 35S promoter, and one marker gene with the transgene and was used to infect tobacco leaf discs. *Agrobacterium* was used for transformation and the transgenic tobacco plants express a high level of CpTi. The CpTi gene in transgenic plants is stably inherited and there is no serious 'yield penalty'. Thus like Bt toxin, CpTi can also be used as a protectant against insect attack in transgenic plants. However, extensive field trials are necessary before releasing these transgenic plants to the farmers.

Gene for other Insecticidal Secondary Metabolites

The production of secondary metabolites by plants can also implicate the resistance to insect attack (Table 7.4). However, biosynthesis of each of these metabolites involves many steps, each controlled by a separate gene. Furthermore, these genes are tissue specific in expression. These features make the production of transgenic plants difficult in this case. Utilizing these facilities, transgenic plants with insect resistance due to secondary metabolites will be available in the near future.

TABLE 7.4: Examples of secondary compound from legume seeds with demonstrated insecticidal properties

Compounds	Bruchid (insect)	Legume
Non-protein anti-metabolites		
Alkaloids		
(i) 2,5-dihydroxymethyl-3,4-dihydroxypyrrolidine (DMDP)	*Callosobruchus maculatus*	*Lonchocarpus sp.*
(ii) Castanospermine	*Callosobruchus maculatus*	*Castamospermum austrole*

Compounds	Bruchid (insect)	Legume
Non-protein amino acids		
p-Aminophenylalanine	*Collosobruchus maculantus*	*Vigna sp.*
	Zabrotes subfasciatus	
Rotenoids	*Callosobruchus maculatus*	*Lonchocarpus salvadoreensis*
Saponins	*Callosobruchus chinensis*	*Phaseolus vulgaris*
	Callosobruchus chinensis	*Glycine max*
Polysaccharides		
Pectosans	*Callosobruchus chinensis*	*Phaseolus vulgaris*
Heteropolysaccharide	*Callosobruchus chinensis*	*Phaseolus vulgaris*
Polysaccharide (?)	*Acanthosclides obtectus*	*Phaseolus vulgaris*
	Callosobruches chinensis	*Vigna spp.*
Protein anti-metabolites		
Lectins (phytohaemagglutinins)	*Callosobruchus maculatus*	*Phaseolus vulgaris*
α-Amylase inhibitors	*Zabrotes subfascitus*	*Phaseolus vulgaris*
	Callosobruchus chinensis	*Phaseolus vulgaris*
Protease inhibitors	*Callosobruchus maculatus*	*Vigna unguiculata*
Arcelin / LLP	*Zabrotes subfascitus*	*Phaseolus vulgaris*
	Zabrotes subfasciatus	*Phaseolus vulgaris*

ENGINEERED RESISTANCE AGAINST VIRAL INFECTION

Over seven hundred plant viruses are recognised causing various diseases and significant crop losses. Due to the lack of 'viricides', the practical control depends on methods that prevent or restrict virus infection. Various strategies have been applied for protecting viral infection; few are described here like:

Cross Protection

If susceptible strain of a crop is inoculated with a mild strain of a virus, the susceptible strain, as a result, develops resistance against more virulent strains. The phenomenon is known as **cross protection.** This method has been used to reduce yield losses in crops like tomatoes against **tomato mosaic virus** (TMV), in potato against **potato spindle tuber viroid** and in citrus against **citrus tristeza virus**. In most cases of cross protection, the symptoms of infection are delayed, and even the replication of virus is suppressed, although eventually the severe strain may be able to overcome the protection.

The disadvantages of practicising cross protection includes

1. Possibility of mutation in inducing mild virus strain.
2. Possibility of synergism between inducing virus and another unrelated virus.
3. Possibility of unnecessary spread of mild virus causing threat for future yield losses.
4. Possibility of some yield losses due to mild strain. Transgenic plants have been produced in tobacco, tomato, and potato with single gene using a broad spectrum of plant viruses.

Coat Protein Mediated Protection

Coat protein - mediated resistance to viruses has been one of the successes of plant genetic engineering. Coat protein gene from tobacco mosaic virus (TMV), a positive strand RNA virus, has been transferred to genome of tobacco plants. In the transgenic plant, expression of coat protein (CP), as well as, low and delayed infection was observed when inoculated with TMV. However, in such transgenic plants, coat protein gene should be constitutively expressed and may thus have effects on the nutritional value of plants. Subsequently Cp genes isolated from different plus strand RNA plant viruses have been cloned into plant expression vectors and transferred to various crop plants like tobacco, tomato and potato, using *Agrobacterium* mediated plant transformation system. Potatoes have been produced which have coat protein genes and are tolerant in field tests to both potato virus X (PVX) and potato virus Y (PVY). A target DNA coding for a component of TMV replicase enzyme was also transferred to tobacco plants, conferring resistance to TMV.

Antisense RNA Mediated Protection

Antisense constructs also generate protection against RNA and DNA viruses. Transgenic tobacco plants expressing antisense transcript (antisense - coat protein gene) were protected from CMV infection.

Transgenic tobacco plants with Nucleocapsid (N) gene, CaMV 35S promoter and a leader sequence (to enhance expression) showed expression of N-gene. Such transgenic plants were resistant to tomato spotted wilt virus. (TSWV) TSWV is a negative strand RNA virus with RNA joined firmly to the nucleocapside protein.

Satellite RNA Mediated Protection

Satellite RNAs are associated with specific strains of some plant RNA viruses, though they are unrelated to viral genome. These extragenomic RNA species depend on a helper virus for replication. Some satellite RNAs are known to attenuate the symptoms of their helper virus. Thus satellite RNA might be used to protect crop plants against the effect of CMV infection. The nature of protection in transgenic plants by satellite RNA is different from coat-protein protection.

Defective Interfering Mediated Protection

Defective interfering (DI) particles are deletion mutants of genomic viral sequences, which depend on their parent virus for replication. Like satellite RNAs, DI particles can attenuate the disease symptoms of their helper virus by interfering with its replication. Artificial DI molecules of TMVY, TMV and BMV have been constructed and inserted in plant protoplast e.g., deletion mutants of BMV RNA 2 exhibited reduced replication of MBV RNA1 and RNA2 in barley protoplasts.

Replicase Mediated Protection

It has been discovered that transgenic tobacco expressing part of replicase gene of TMV strain was highly resistant to this and closely related strains.

Currently, atleast three approaches are available to broad spectrum resistance. The first is based on the replication and cell to cell movement of virus. Second approach relies on the transgenic expression of antivirus protein of non-plant origin including antibodies. The third approach is based on genetic manipulation of natural viral defenses in plants.

ENGINEERED RESISTANCE AGAINST FUNGAL PATHOGEN

Genetic engineering has opened vistas with isolation of resistance genes and gene transformation methods to develop transgenic plants. Resistance to fungal pathogen is achieved by isolating and transferring fungal resistance genes to the desired susceptible crop. Compounds toxic to fungus can also be used to produce transgenic plants as done in case of tobacco. Antifungal proteins like **endochitinase** and **ß-1,3-endoglucanases** are thought to inhibit fungal growth. Transgenic tobacco plants with endochitinase CH5B gene with Ca MV 35S promoter exhibits reduced seedling mortality when grown in presence of *Rhizoctonia solani*.

Another method is using ribosome inactivating proteins (RIPs). These RIPs do not interfere with the functioning of self-ribosomes but exert inhibitory activity towards fungal ribosomes. When RIPs are mixed with endochitinase or ß-1,3-glucanases, the antifungal activity exhibits synergistic increase in *in vitro* assay. Transgenic tobacco with barley type 1 RIP gene and potato Wun1 promoter showed an increased resistance to *R. solani*. Table 7.5 gives a brief account of transgenic crops showing resistance against bacterial and fungal pathogens.

TABLE 7.5: Some pathogens for which resistance has been transferred in some crop plants.

Pathogen	Disease	Resistance gene	Source of gene	Transgenic crop
Pseudomonas syringae	Wild fire	Acetyl transferase gene	-	Tobacco
Alternaria longipes	Brown spot	Citinase gene	Serratia marcescens (soil bacterium)	Tobacco
Rhizoctonia solani		Chitinase gene	Bean	Tobacco
Phytophthora infestans	Late blight	Osmotin gene	Potato	Potato

GENETIC ENGINEERING AND PLANT LIPIDS

Plant lipids are major components of the human diet and are present in plants as storage oil. Engineered vegetable oils have been tested as field trials. A high stereate phenotype has been engineered in rapeseed by antisense approach. Stearoyl-acyl carrier protein (Stearoyl - ACP) desaturase is a chloroplast enzyme which catalyzes the first desaturation step in seed oil biosynthesis, converting stearoyl-ACP to oleayl-ACP.

This coding region of the gene for rapeseed desaturase was fused in an inverted position to the napin storage protein promoter and to the promoter of seed-specific acyl carrier protein gene, and then introduced into rapeseed. This reduced the activity of desaturase enzyme in engineered plant resulting in increased levels of stearate in seeds (upto 40% of the total fatty acids), with corresponding decrease in oleate. Due to tissue specific expression of antisense RNA, integrity of membrane lipids in leaf remained unaffected.

Length of fatty acids can be manipulated with thioesterase enzyme, which is responsible for the accumulation of short chain fatty acids in triglycerides. The gene has been cloned and expressed in developing seed of *Arabidopsis*, where it caused the accumulation of C12 storage lipids. It is useful for application in confectionery production.

The field of storage lipid modification via the engineering of the fatty acid biosynthesis enzymes (Table 7.6) promises to be a very fertile one that will contribute to production of healthier food and of chemical feed stocks.

TABLE 7.6: Modifications of lipid metabolism in transgenic plants

Modification achieved	Enzyme engineered	Source of gene
Lauric acid production	Acyl-ACP thioesterase	California bay
Increased Stearic Acid	Antisense of stearoyl ACP desaturase	Brassica napus
	Stearyl-CoA desaturase	Rat
Reduced saturated fatty acids	Stearyl-CoA desaturase	Yeast
	3-Ketoacyl-ACP synthase II	Castor
	Acyl-ACP thioesterase	Soybean
Reduced saturated fatty acids in phosphatidyl glycerol	Acyl-ACP glycerol-3-phosphate acyltransferase	Squash, Arabidopsis
Increased saturated fatty acid in phosphatidyl glycerol	Acyl-ACP glycerol-3-phosphate acyltransferase	Escherichia coli
Petroselinic acid production	Acyl-ACP desaturase	Coriander
Increased-linolenic acid	ω3 desaturase	Arabidopsis
Cylopropane fatty acid production	Cylopropane synthase	Escherichia coli
γ-Linolenic acid production	Linolenic Δ6 desaturase	Synechocytis
Increased oleic acid	ACP: protein a fusion	Spinach

GENETIC ENGINEERING AND STORAGE PROTEINS

Human nutrition requires a balanced source of amino acids and the amino acid balance of many plant products is unsatisfactory. Genetic engineering may be carried out in crops making following three approaches: (a) expressing a desirable and heterologous storage protein, (b) increasing the level of a desirable, but little expressed, endogenons protein and, (c) suppressing the expression of anti-nutritional proteins.

A heterologous species of sulphur-rich storage protein in Brazil nut containing 18% methionine and 8% cysteine has been expressed in tobacco and rapeseed. In transgenic tobacco, phaseolin gene has been expressed, leading to accumulation of methionine rich phaseolin.

Another approach for increasing amino acid synthesis is by engineering of a rate-limiting enzyme. For example, in a tobacco mutant lysine overproduction was caused by a lysine-insensitive dehydro-picolinate synthase. Forty fold increase in free lysine has been obtained by expression of a bacterial dehydropicolinate synthase in plant chloroplasts.

Genomic clones encoding two subunit of soybean ß-conglycin (7S protein) genes were transferred into petunia plants. The temporally regulated expression of these two genes in transgenic plants mimic their temporal regulation during soybean embryo development. The gene encoding for 42 kDa potato storage protein, patatin was introduced into tobacco. Stable synthesis of patatin protein was confirmed in the heterologons environment. Furthermore, transgenic tobacco possessing the 5' and 3' flanking sequences of the bean ß-phaseolin gene fused with 15 kDa *zein* gene could synthesize up to 1.6% of total seed protein as zein. Thus, zein is deposited and accumulates in vacuolar protein bodies of the tobacco embryo and endosperm.

GENETIC ENGINEERING FOR NITROGEN FIXATION

Transfer of *nif*-genes from *Klebsiella pneumoniae* to *E. coli*

Klebsiella pneumoniae is a non symbiotic nitrogen fixing bacterium. R. A. Dixon and J. R. Postgate (1972) transferred nif-genes of *K. pneumoniae* into *E. coli* cells by conjugation. The transgene in the new host produced functional nitrogenase system. The nif-genes of *K. pneumoniae* have been introduced into several other organisms, e.g. *K. aerogenes*, *Salmonella typhimurium*, *Erwinia herbicola*, *Serratia marcescens*. These organisms are genetically modified and are now able to fix atmospheric nitrogen.

Transfer of *nif*-genes of Eukaryotic Organisms

Nif-genes from Rhizobium have been transferred to *Agrobacterium tumefaciens*. The expression of transferred gene has been successful in cereal crops like wheat, paddy etc. through the Ti-plasmids of *A. tumefaciens*.

Transfer of *nod*-genes to Increase Host Range

Cloned nod-genes of *Rhizobium leguminosarum* that nodulate pea have been transferred to *R. phaseoli* (this normally nodulate bean). The transformed *R. phaseoli* can now nodulate pea plants also.

Plants that Produce Plastic

Chris Somerville and coworkers have evolved a transgenic *Arabidopsis thaliana* to polyhydroxy butyrate (PHB). It is a natural and totally biodegradable plastic material. In future, plastic producing sugar beet and potatoes could also be produced. These scientists have isolated two genes from a bacterium *Alcligenes eutrophus* which produced PHB in nature and accumulates it as a coating. These two genes were inserted separately into two *A. thaliana* plants with the help of *Agrobacterium tumefaciens*.

MISCELLANEOUS APPLICATIONS

The knowledge of genetic engineering has been exploited in industries to deliver products and processes of economic and social value. Products of biotechnology include organisms and substances made by it that are consumed directly as such or serve as components for other industries. Processes of social value include technologies aimed at social hygiene and well-being, such as purification of drinking water, pollution control and waste disposal. The actinomycetes or soil microorganisms that produce the plethora of miracle drugs the **antibiotics** have also been added to the list. The new techniques permit harnessing of single cells to produce non-native gene products from exogenously supplied genes. Industrial use of single cells and enzymes for the production of stated products may require special tailoring of the cells for adaptation to mass-scale production processes.

The pharmaceutical and chemical industries were the first to exploit the gene engineering technology. A futuristic application is envisaged in the field named *bioelectronics* or *bionics*. The potential of biological material to serve as extra sensitive conductors, sensors, and support matrices for semiconductor molecules is already being explored. The drugs and pharmaceuticals may be synthesized in heterologous cells or in animal cell lines that produce them. For instance, human chorionic gonadotropin (HCG), erythropoietin and calcitonin synthesizing cell lines have been reported. Of these, only HCG is successfully used for commercial production. The drugs and pharmaceuticals that are candidates for the genetic engineering are: (i) hormones, (ii) growth factors, (iii) blood-derived factors, (iv) enzymes, (v) antibiotics, (vi) immunoproteins, (vii) immunodiagnostics, and (viii) DNA probes for molecular diagnosis.

CHAPTER 8

Techniques for Genetic Engineering

Recombinant DNA technology deals with isolation and manipulation of genes. Before nucleic acids can be cut or otherwise manipulated, they must be either isolated and purified or artificially synthesized. The isolated genes are also sometimes sequenced for a better understanding of its structure that needs to be manipulated.

ISOLATION OF TOTAL CELLULAR DNA

DNA is very easily damaged by shear forces; even rapid stirring of solution can break even with high DNA molecular weight into much shorter fragments. Consequently, DNA is recovered from cells by gentlest possible method of rupture, in the presence of EDTA to chalate the Mg^{2+} ions needed for DNase activity. Cell walls, if present, are digested enzymatically (e.g. lysozyme treatment of bacteria) and the cell membrane is solubilized using detergent. Cell disruption is performed at 4 °C using autoclaved glassware and solutions to destroy DNase activity.

After release of nucleic acids from the cells, RNA can be removed by treatment with ribonuclease (RNase). The other major contaminant, protein, is removed by shaking the solution gently with water saturated phenol, or with phenol/chloroform mixture, either of which will denature protein but not nucleic acids. Centrifugation of the emulsion formed by this mixture produces a lower, organic phase, separated from the upper aqueous phase by an interface of denatured proteins. The aqueous solution is recovered and deproteinised repeatedly, until no matter is seen at interface. Finally, the deproteinised DNA preparation is mixed with ethanol and allowed to precipitate out in freezer. After centrifugation, the DNA pallet is redissolved in a buffer containing EDTA for protection against DNases and this solution can be stored at 4°C for at least a month. DNA solutions can be stored frozen, but repeated freezing and thawing tends to damage long molecules by shearing; so preparation in frequent use are normally stored at 4 °C.

This procedure is suitable for total cellular DNA. If the DNA from specific organelle or viral particle is needed, it is best to isolate the organelle or virus before extracting its DNA. Now we will discuss several basic techniques related with gene isolation and identification.

NUCLEIC ACID HYBRIDIZATION

If a double–stranded DNA molecule is exposed to high temperature, or to very alkaline conditions, then the two strands will come apart. The molecule is said to have become denatured. The temperature at which denaturation occurs is termed as melting temperature or T_m. If the denatured DNA is returned to a temperature below its T_m, or to neutral pH when alkali was used to denature it, each strand will, after a time, find its complementary strand. The two strands will 'zipper' back together to re–form a double–stranded DNA molecule. This ability of complementary sequences to anneal, or hybridize, to one another is called **nucleic acid hybridization**. This technique helps in determining the gene structure and in identifying molecules which contain same sequences of nucleotides. In a complex mixture of nucleic acid molecules, nucleic acid hybridization technique helps in separation of complementary sequence.

When employed analytically, hybridization is normally performed using one labeled sequence, termed the probe, and an unlabelled sequence called the target. Probe is a short synthetic oligo deoxyribonucleotide which is complementary to target DNA sequences. Target DNA may be obtained from total cellular DNA preparations. The probe is labeled by incorporation of either radioactively labeled nucleotides or with some chemicals (described later). The probe is the known, pure species in the hybridization and the target is the unknown species to be identified.

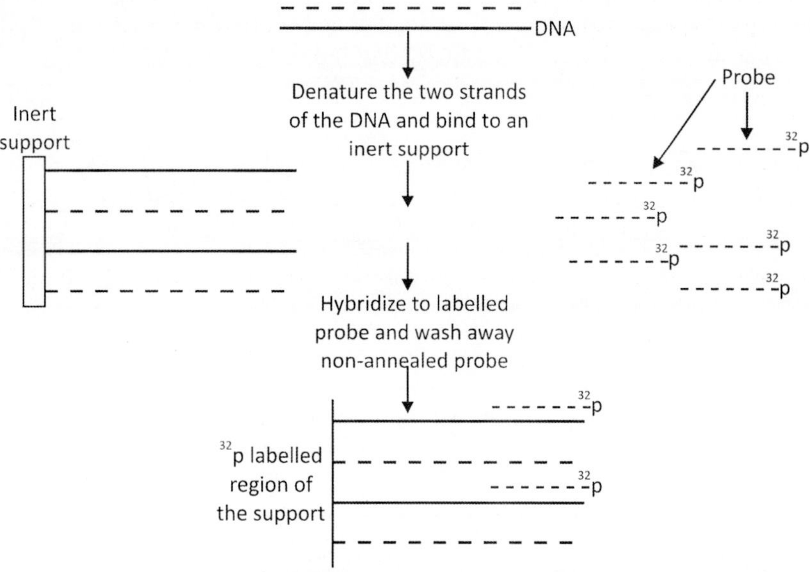

FIG. 8.1: The principle of hybridization using an inert support. The target DNA is most often localized to one part of the support and, in most cases, the aim of the hybridization experiment is to identify this region. The probe is prelabeled by some method, most often by the incorporation of radioactivity. After hybridization, the annealed probe is detected by autoradiography.

The target will most often form part of a mixture of unrelated nucleic acid sequences. The probe is usually added in considerable excess over the target, and its concentration determines the rate of reaction. Hybridization reaction time is extended to ensure complete reaction. The target and probe can be hybridized to each other in solution, or alternatively the target can be immobilized on an inert support such as nitrocellulose. When the target is immobilized the rate of annealing is greatly reduced.

The rate of hybridization depends upon the DNA concentration, the concentration of salt in the solution and the temperature. Hybridization reactions are normally performed at 10–20 °C below the T_m of the duplex. Under any given conditions of ionic strength and temperature, the T_m of a duplex also depends on the base composition of the annealed partners.

Hybridization involving RNA is performed in the presence of formamide as RNA cannot withstand higher temperatures. Formamide disrupts hydrogen bonding and thus reduces the effective T_m of the duplex, allowing annealing to occur at a lower temperature (e.g. 37 °C).

METHODS FOR LABELLING NUCLEIC ACIDS

For identification of hybridized nucleic acid duplexes, labelling of probe or target is necessary. As probe is short synthetic oligonucleotide it can be easily labeled. A ^{32}P isotopic label can be added to the 5' hydroxyl group using polynucleotide kinase to a probe (Table 8.1). Alternatively, labeled nucleotides can be incorporated at the 3' end using terminal transferase. Because label is added at only one place within the oligonucleotide, such end–labeled probes contain relatively little radioactivity per unit weight of DNA. This low specific activity limits their sensitivity as probes. If a high specific activity probe is required, then some method of uniform labelling must be used. Here the probe sequence is copied along part or all of its length using DNA polymerase and labeled nucleoside triphosphate.

TABLE 8.1: Some methods of in vitro labelling of nucleic acids

Labelling method	Enzyme used	Template	Product	Structure at end of reaction*	Specific activity
5' end labelling	Polynucleotide kinase	DNA or RNA	DNA or RNA	d.s. or s.s.	Low
3'end labelling	Large (Klenow) subfragment of E. coli DNA polymerase–1	DNA	DNA	d.s.	Low
3'end labelling	Terminal transferase	DNA or RNA	DNA or RNA	s.s.	Low
Nick translation	E. coli DNA polymerase–1	DNA	DNA	d.s.	High
Random priming	Klenow subfragment of DNA polymerase–1	DNA or RNA	DNA	d.s.	High
In–vitro transcription	Bacteriophage (e.g. T7) RNA polymerase	DNA	RNA	s.s.	High

*d.s. = double stranded; s.s. = single stranded.

METHODS OF LABELLING NUCLEIC ACID AND PROBES

There are five basic methods for labelling nucleic acids. These are: nick translation, primer extension, methods based on RNA polymerase, end labelling methods and direct labelling methods.

Nick Translation

This is done by making single–strand cuts ("*nicks*") in the double–stranded DNA molecule by brief exposure to a dilute solution of an endonuclease (usually deoxy ribonuclease I of *E. coli*). DNA polymerase I is then used in the presence of at least one radioactive precursor (e.g., Â–^{32}P–dATP) to "*translate*" the nick along the molecule in the 5' to 3' direction. This occurs by the concerted action of the 5'→3' polymerase and the 5'→3' exonuclease activities built into DNA polymerase I. The

5'→3' polymerase adds nucleotides to the 3'–OH terminus at each nick, while the 5'–phosphate terminus at the nick. The net result is that a nonradioactive strand of DNA is replaced by a radioactive strand. The DNA is then denatured and used as a radioactive probe in hybridization experiments (Southern blots, northern blots, etc.).

Nick translation can be used with a variety of labels to generate probes suitable for most hybridization applications. It is also appropriate for the generation of biotinylated probes (described later).

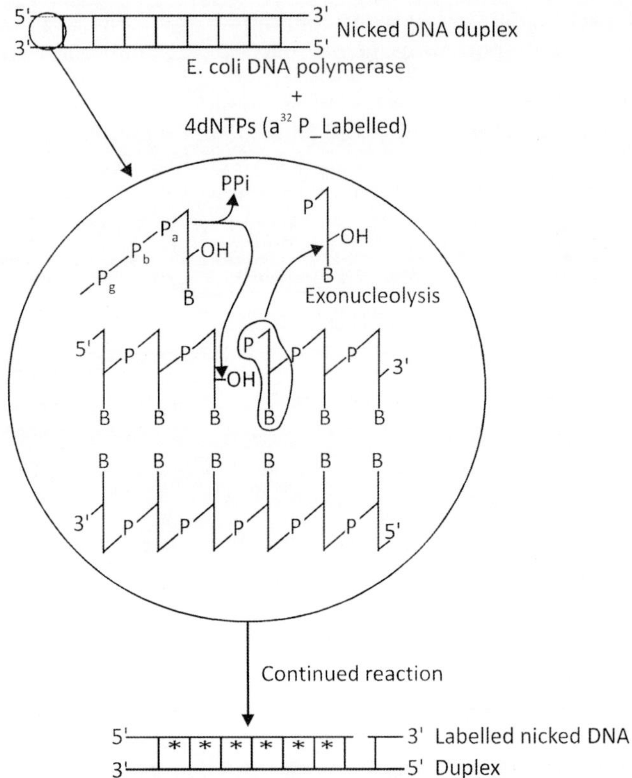

FIG. 8.2: ^{32}P–labelling of duplex DNA by nick translation. Asterisks indicate radiolabeled phosphate groups.

A major advantage of nick translation over other uniform labelling methods is its flexibility with respect to probe size, specific activity and concentration. It is particularly suitable for the production of large quantities of probes for use in multiple hybridization and/or where a high probe concentration is required.

Primer Extension Method

Primers are synthetic oligodeoxyribonucleotides which are complementary to specific regions of known vector DNA. Some commonly used primers are M13 sequencing primer, pBR_{322} primer etc. The 3' termini of these primers serve as initiation site for template dependent DNA synthesis by enzymes like DNA polymerase I.

DNA polymerase works by extending a short double–stranded region made by annealing an oligonucleotide primer to the single–stranded template. Thus this method of uniform labelling requires a primer which matches the probe sequence. Radiolabeling of primers can be done with two methods.

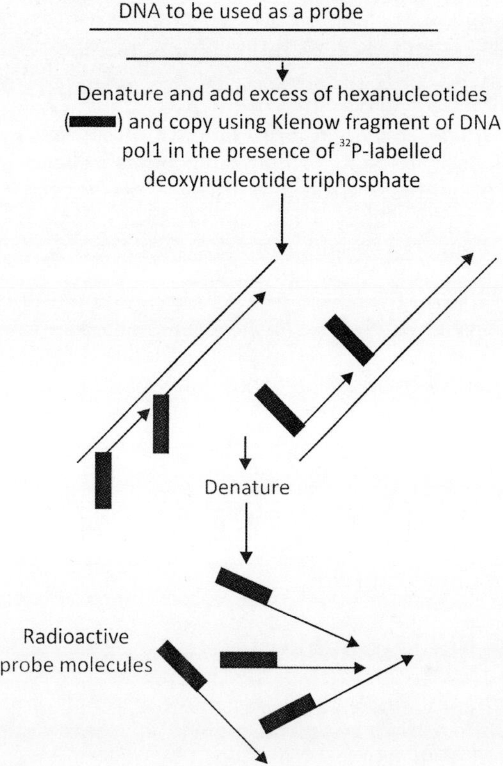

FIG. 8.3: The principle of random primed (oligo–) labelling. The DNA to be used as a probe is denatured by heating and mixed with hexanucleotides of random sequence which then act as primers for DNA synthesis.

If the probe sequence is not known then random oligonucleotide labelling can be used. It is often in the case when natural cellular DNA is used. These primers are made by adding a mixture of all four bases at each step in the chemical synthesis reaction. The DNA is denatured and the two complementary strands are copied in the presence of labeled primers as well as, nucleoside triphosphates. The polymerase used is Klenow fragment derived from DNA polymerase–I of *E. coli*. Chance homology ensures that these primers anneal to the separated DNA strands at many points along their length, thus providing a base for polymerase to initiate DNA synthesis. This is only one of several uniform labelling methods.

The second method uses a unique primer to restrict labelling to a particular sequence of interest. In the primer extension method it is essential to use a polymerase lacking a 5'→3' exonuclease activity otherwise degradation of the primer will occur. Both the Klenow fragment of *E. coli* DNA polymerase I, which lacks the 5'→3' exonuclease activity and reverse transcriptase have been used successfully.

An advantage of random priming method is that is may be used to label relatively impure DNA. In primer extension method, there is not synthesis of DNA and labeled nucleotides are incorporated in concentrations. It is an ideal method for situations where high specific activity and low probe concentration are frequently employed.

Methods Based on RNA Polymerases

RNA polymerase catalyzes the synthesis of RNA from nucleoside triphosphates using a DNA template. Thus they can incorporate labeled ribonucleotides into RNA during transcription if such labeled nucleotides are provided to it. If a specific site of a vector or DNA is transcribed in such way, RNA probes (or transcripts) of defined length and sequence can be obtained.

Initially RNA probes were obtained from RNA polymerase from *E. coli*. This enzyme when used *in vitro* exhibited very little template and promoter specificity and therefore produces transcripts which have been initiated and terminated at random. Today, RNA polymerases from a number of phages have been used, which possess a high degree of specificity. Thus transcription can be limited to sequences cloned downstream of an appropriate promoter.

RNA probe's specific activity is determined solely by the specific activity of labeled nucleotide and moreover such probes are need not to be denatured before hybridization. These probes are more sensitive than nick translated probes because of avoidance of probe reannealing during hybridization (as RNA–RNA hybrids are more stable than RNA–DNA hybrids)). A further advantage of RNA probes is that non-specifically bound probe can be removed with RNase A. RNase A is very specific for single stranded RNA.

The only difficulty with RNA polymerase–based–labelling is that a specific promoter is read by RNA polymerases before transcripting the downstream. This promoter has to be subcloned and inserted into the vector. However, once an insert has been cloned into such a vector, DNA probes may still be prepared using nick translation or random primer labelling.

END LABELLING OF NUCLEIC ACIDS

A wide variety of techniques is available for introducing label at either the 3' or 5' ends of linear DNA or RNA. Usually only a single label is introduced at the terminus. Nucleic acid can be 5' end labeled using T4 polynucleotide kinase. The enzyme catalyzes the transfer of the γ–phosphate of ribonucleoside 5'–triphosphate donor to the 5' hydroxyl group of a poly– or oligo–nucleotide or even to nucleoside–3–triphosphate. Radiolabeled phosphate group is donated by [γ^{32}–P] ATP to DNA or RNA containing a 5'–hydroxyl terminus. This is termed as forward reaction.

If 5'–phosphate group is present in DNA or RNA, then it is removed with alkaline phosphatase. This reaction is driven by excess ADP which causes the enzyme to transfer the terminal 5'–phosphate from DNA to ADP. This is known as exchange reaction. The DNA is rephosphorylated by transfer of labeled γ–phosphate from [γ^{32}–P] ATP.

The major advantages of 5'–end labelling are:

- Both DNA and RNA can be labeled.
- Oligonucleotide can be conveniently labeled e.g. for detection of mutants following site–directed mutagenesis.
- Location of labeled group is known.
- Very small fragments can be labeled.
- Restriction digest fragment can be labeled.

DNA fragments also can be labeled at their 3'end by using terminal deoxynucleotidyl transferase, however, this method is not of wide use.

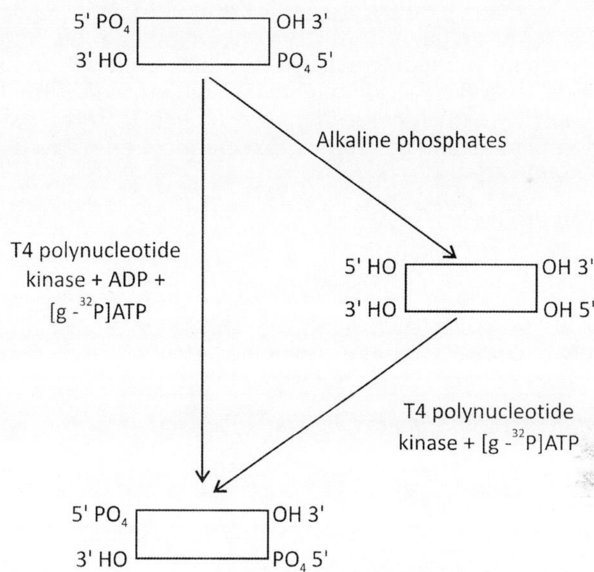

FIG. 8.4: The 5'–end labelling reaction using T4 polynucleotide kinase.

CHOICE OF LABEL

There are two technical parameters, resolution and sensitivity, which determines the success of probe application. High degree of resolution is required to known the relative position of a nucleic acid fragment. High sensitivity is necessary because sequence of interest may be present at low abundance.

Other factors are probe stability, safety and ease of use. Broadly labels can be categorized into radioactive and non radioactive types.

Radioactive Labels

These labels have wider applications as they can be easily detected with autoradiography. Their detection gives two important information, firstly about occurrence of hybridization between probes and target DNA and secondly about their position. Radioactive methods using ^{32}P are easily detectable. They are used often.

Non Radioactive Labels

A number of non radioactive labels for probes are available but biotin is widely used.

Biotin Labeled Probes

Biotinylated probes are prepared through a nick–translation reaction by replacing nucleotides with biotinylated derivatives. After hybridization and washing, detection of hybrids is done by adding avidine and going through a series of cytochemical reactions which finally give a blue colour whose intensity is proportional to the amount of biotin in the hybrid. There are several advantages of using biotinylated probes. For major advantages of using biotinylated probes are: (a) assays employ non– toxic materials, with longer half–life (b) can be prepared in advance in bulk and stored at –20 °C for repeated uses (c) detection of hybrids is much faster than by radioactive probes. A limitation of this technology is that biotinylated sites are limited, thus decreasing the intensity of signal. It has been solved by adding long 'tails' of biotinylated nucleotides to the probes through enzymatic methods. Another disadvantage of biotin labeled probes is that the cytochemical visualization reactions lead to precipitation of insoluble material which can not be removed and therefore, the filter can not be reused.

Digoxigenin Labeled Probes

Another method is to label DNA with digoxigenin–labeled deoxyuridine triphosphate (dig–ÂUTP). After hybridization to the target DNA the hybrids are detected by enzyme–linked immunoassay using an antibody conjugate e.g. anti digoxigenin alkaline phosphatase. Enzyme reporter molecules with enzymes can also be used as label.

Alternatives to Biotin and Digoxigenin Labeling

Other ligands (e.g. hapten determinants, 2, 4–dinitrophenol, arsenate derivatives, etc.) are attached to specific proteins that can be tagged with enzymes or fluorescent molecules. It is possible to monitor many probes simultaneously by using several different ligands since each ligand would yield a different signal. A chemiluminescent (light emitting) probe system has also been developed in which two different probes complementary to a continuous segment of DNA hybridize to adjacent of a gene. The first label is a chemiluminescent complex which emits light at a specific wavelength. This emission excites the label molecule on the second probe to emit light at a different wave length which can be detected using a photomultiplier device. The processes called **non–radioactive energy transfer**, and can occur only if the two probes hybridize correctly and the two labels are close to each other.

The major disadvantage of biotin, and of other nonradioactive labels, is low sensitivity in comparison to radioactive labels.

MAPPING GENES ON SPECIFIC CHROMOSOMES

One of the most important applications of nucleic acid hybridization is construction of genetic maps. This helps in determining the chromosomal location of individual genes. There are three commonly used methods that differ in the degree of resolution they afford viz somatic cell hybrids, *In situ*, hybridization and genetic linkage mapping.

Somatic Cell Hybrids

Somatic cell hybrids are prepared by fusing together human and rodent cells using reagents, such as polyethylene glycol or Sendai virus. The reagents help in membrane fusion. The resulting hybrid cells are then grown in culture medium containing selective agents which allow only the hybrid cells to grow. In hybrids produced using human and rodent cells the rodent chromosomes are retained, while the human chromosomes are lost. This random chromosome loss occurs during continued culture and shows no apparent selectivity for individual human chromosomes. Thus it is possible to produce a panel of hybrid cells containing different combinations of human chromosomes. Genomic DNA is prepared from panels of such hybrids and used in hybridization analysis. The human probe to be mapped is hybridized with such hybrids. By determining which of the panel of hybrids shows a hybridization signal, and correlating this information with the karyotype of the cells, it is possible to deduce which chromosome carries the test gene.

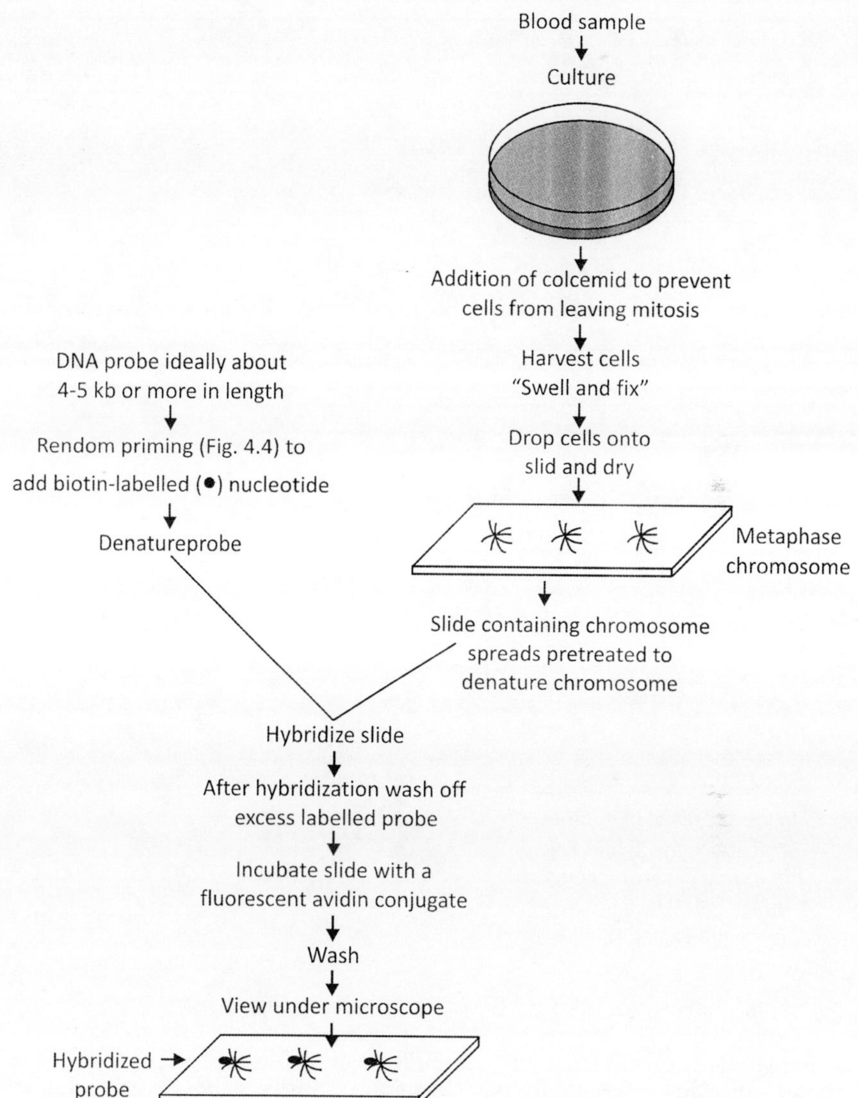

FIG. 8.5: FISH is used where gene mapping is done with in–situ hybridization. It uses a non–radioactive labelling system involving biotin attached. When viewed under special filters the specific sites of DNA probe can be seen on a chromosome.

In–Situ Hybridization

In this technique a labeled DNA probe is hybridized to metaphase chromosome. The position of hybridization is visualized at the light microscope level. Metaphase chromosomes are liberated from cells and bound to the surface of a cover slip. The DNA within the chromosomes is denatured in such a way that the individual chromosomes remain visually identifiable.

The probe is labeled, either by incorporation of a ^{3}H–nucleotide or non–

isotopically. The chromosomes are incubated with the labeled probe for many hours and, after hybridization, the excess unbound probe is washed away and the annealed probe detected.

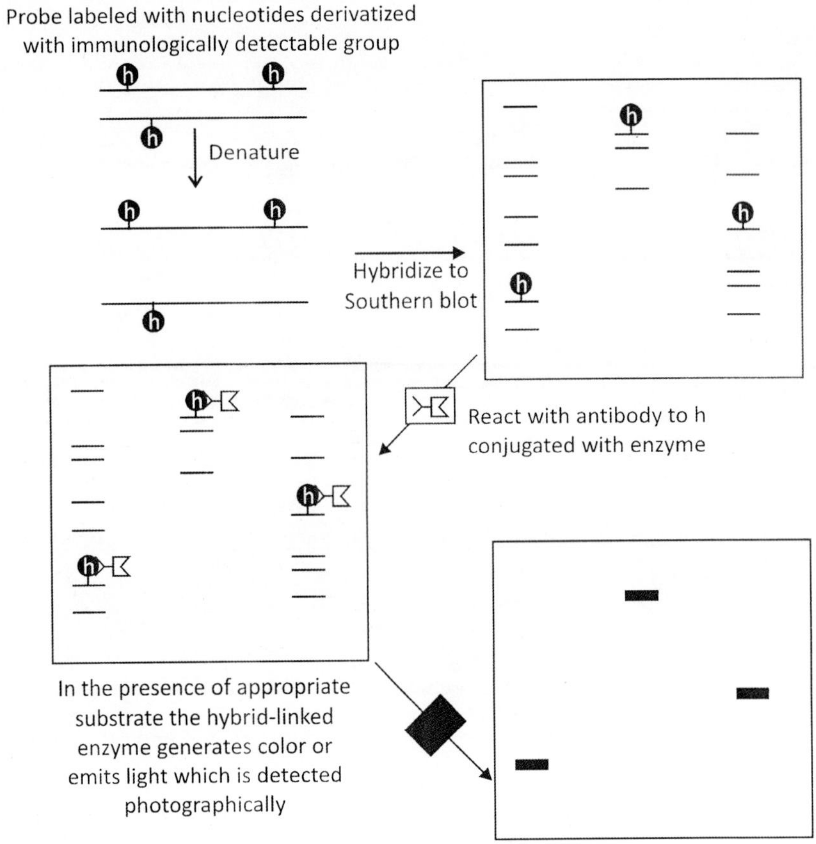

Probe labeled with nucleotides derivatized with immunologically detectable group

Denature

Hybridize to Southern blot

React with antibody to h conjugated with enzyme

In the presence of appropriate substrate the hybrid-linked enzyme generates color or emits light which is detected photographically

FIG. 8.6: The principle of non–radioactive detection methods for in–situ hybridization.

If radiolabeled, probe used then the slides are autoradiographed. Deposited silver grain after autoradiography will lie in the approximate area at which the probe bounds to the chromosome. The accuracy of mapping by this technique is dependent on the size of the grains which, in the case of small chromosomes, can cover a significant portion of the total chromosome length.

If the probe is labeled non–isotopically then it is possible to determine the precise position of hybridization with much greater accuracy. The signal is detected using a fluorescent molecule thus is called fluorescent *in–situ* hybridization (FISH). It is more sensitive than radiolabeling and can be used to give a fairly precise position of hybridization.

Thus FISH is particularly useful for determining the respective locations of two markers on the same chromosome. The two probes are labeled with different immunologically labeled adducts and a single hybridization reaction is performed. The two probes are then detected using fluorochromes which emit at different wave lengths when excited by UV light, such as rhodamine (red fluorescence) and fluorescein (green fluorescence). This technique has been used for probes which are separated by few hundred kilobases of DNA on same chromosome. The FISH technique can also be applied to interphase nuclei, which allows accurate ordering of probes along a chromosome when they are separated by no more than 20–30 kilobases

Transposon Tagging

Isolating a eukaryotic gene by transposon tagging involves inserting into target DNA, fractionating fragments of the latter on southern blots and hybridizing the transposon with a probe. Cleaving out the transposon, together with flanking DNA, reveals the adjoining sequences of a gene. The probes are made for these adjoining sequences to find the fragment having gene of interest.

Presence of transposon is confirmed by altered phenotype. This method was first employed to isolate *white* locus in *Drosophila*. The mutants with *white– apricot* eyes harboured a *copia* retrotransposon as well as accompanying *white* locus.

Genetic Linkage Mapping

This form of mapping is used to construct genetic linkage maps of individual chromosomes and also to localize genes involved in genetic disease. Genetic linkage analysis relies on an ability to estimate the frequency of crossing over (recombination) that occurs between homologous chromosomes during meiosis. The closer two genetic markers are to each other then the lower is the chance that they will be separated during meiosis. The statistical probability that two markers are linked is measured as the logarithm of the odds (Lod). A Lod score of 3 is usually taken as the threshold of significance, while a Lod score below –2 suggests that the two loci are not linked. Lods between –2 and +3 are not significant.

GENOMIC LIBRARY

The complete set of cloned fragments of DNA of an organism is known as **genomic library**. This is also called **gene library**. Such a library contains all or most of the genes present in the cell or the organism. It is difficult to identify a particular gene from an organism and then clone it. Hence the organism's entire genome as DNA fragments is cloned and the clone containing the sequence of interest is identified.

Cloning experiments can be divided into two types (1) those where the cloned DNA segment is of unknown composition and function and, (2) those where the cloned DNA segment is of known composition and function. In the first approach the entire cellular DNA has to be fragmented into clonable size by either digestion with restriction endonucleases or, mechanical sheering.

These experiments with randomly cloned fragments are known as 'shot gun cloning experiments'.

The chromosomal DNA of the organism of interest is isolated. It is then treated with a known restriction enzyme to obtain DNA fragments of clonable size. The fragments are cloned using appropriate cloning vectors in any organisms like bacteria, virus or yeast. For identification, a marker gene is inserted while multiplying in vector. Hybridization of such genes can be carried out by two major techniques viz. colony hybridization and plaque hybridization depending on the type of vector used.

Colony Hybridization

Once a genomic library or cDNA library is available, gene sequence can be isolated by colony hybridization techniques. It involves replica plating the colonies formed by transformed cell into nitrocellulose filter and their *in situ* hybridization with a radioactive probe. This method was first developed by Hogness and Grunestein (1975). Procedure is described in following steps:

1. The cloned yeast colonies, bacterial colonies or phage plaques to be tested are transferred from the culture plate on to a nitrocellulose filter paper by replica plating.
2. The filter with the colony replicas is treated with NaOH to lyses the cells/phages and to denature DNA.
3. The filter is then baked to fix the DNA.
4. It is then treated with a radioactive ^{32}P labeled single stranded DNA probe that is complementary to the ssDNA of interest.
5. The filter is washed to remove the unbound excess probe, dried and then autoradiographed.
6. The DNA fragments hybridized with the radioactive probe can be detected on the autoradiograph film.
7. This indicates that the cells in the corresponding colonies contain the desired gene. The colony is taken out from the master culture plate for mass culture.

Plaque Hybridization

This technique is applicable where a chimeric phage particle is carrying the gene of interest. In such a situation, a bacterial lawn is infected with a mixture of chimeric phage particles (i.e. the library) and a large number of plaques develop overnight. These plaques can be treated just like the colonies in colony hybridization to identify and isolate the chimeric phage particle carrying the gene of interest. Two new techniques have shown their potential in gene isolation and characterization. They are chromosome walking and chromosome jumping. Let us have a small account on both of these techniques.

Chromosome Walking

Identification of fragments with overlapping sequences may be a key to the reconstruction or characterization of large chromosome regions. This is achieved by the technique popularly described as chromosome walking.

The technique of chromosome walking involves the following steps:

1. From the genomic library a alone of interest (identified by a probe is selected.
2. One end of the clone is subcloned.
3. The subcloned fragment of the selected clone may be hybridized with other clones in the library and a second clone hybridizing with the subclone of the first clone is identified due to presence of overlapping region.
4. The end of the second is then subcloned and used for hybridization with other clones. This helps in identifying a third clone having overlapping region with the subcloned end of the second clone.
5. Identified third clone is also subcloned and hybridized with clones in the same manner and the procedure may be continued.
6. Restriction map of each selected clone is prepared and compared to know the region of overlapping, so that identification of few overlapping restriction sites resembles the walking along the chromosome or along a long chromosome

segment.

Regions of chromosomes approaching 1000 kb have been mapped following the above technique. Restriction maps of entire chromosomes can be prepared in this manner following the technique of chromosome walking.

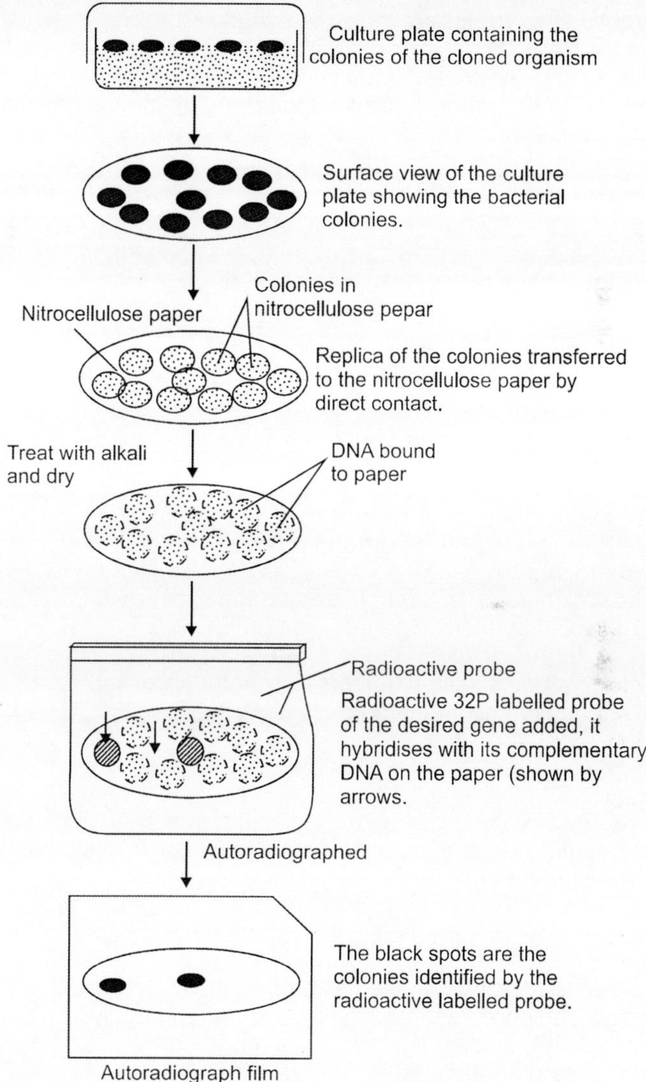

FIG. 8.7: Colony hybridization to identify the clones containing the DNA of interest.

Chromosome Jumping

In order to bring the molecular marker close to the gene of interest, *chromosome jumping* approach has been utilized. This technique can be described in the following steps:

1. The distance of 'jumps' or 'hopsize' (e.g. 100 kb or 200 kb) is determined.
2. Genomic DNA molecules in the range of selected size are selected through pulsed–field gel electrophoresis.
3. For circularization of DNA segments, ligation between two ends of each long linear DNA molecule is allowed using T4 ligase).
4. These circular DNA circles are digested with *Eco*RI.
5. Phage vector Ch3A lac, is also cut with *Eco*RI and used for cloning small DNA fragments.
6. Such cloned DNA fragments represent the jumping library, which can be plated on a bacterial host and screened through the technique of plaque hybridization.

The technique of chromosome jumping helps in narrowing the gap between the gene and available molecular markers. After several cycles of chromosome jumping followed by cloning the regions proximal to the gene, it will be possible to approach very close to the desired gene capable approached and cloned. The gene for cystic fibrosis has recently been isolated using this technique.

CLEAVING DNA WITH RESTRICTION ENZYMES

After locating gene of interest in colony or on chromosome, the DNA is extracted and cleaved with enzymes. Most restriction enzymes cleave DNA molecules in a site specific manner. The genomic DNA from an individual organism is digested with either a single or more restriction enzymes. This results in fragmentation of DNA molecules with variation in length. These fragments are known as **restriction fragment length polymorphisms** or RFLPs. Since restriction enzymes cut DNA in a sequence specific manner, every homologous DNA molecule will be cut at exactly the same sites. This allows isolating large quantities of specific DNA fragments for subcloning and isolating the DNA sequence of interest.

The sizes of the restriction fragments can be determined by polyacrylamide or agarose gel electrophoresis. It also helps in separation of various RFLPs. Here few electrophoretic and blotting techniques are described.

Gel Electrophoresis

Nucleic acid molecules can be distinguished by differences in their chain length, and study of gene organization and expression includes nucleic acid hybridization with a chain length measurement.

DNA and RNA have a net negative charge. At neutral pH therefore, these molecules migrate towards the positive terminal when placed in an electric field. This process of electrophoresis is performed in an agarose or polyacrylamide gel. Nucleic acids are loaded into slots in the gel and allowed to migrate towards the positive terminal. The pores in the gel act to sieve the molecules. So that, the mobility of a particular nucleic acid species depends on its length. All the molecules of a particular size move at approximately the same rate through the gel, forming a band which gradually increases in width during electrophoresis because of diffusion. The size of the molecules can be estimated by running DNA molecules of known size in a parallel track on the gel, and determining their migration position relative to the species of unknown size.

Polyacrylamide has a smaller average pore size than agarose and so is effective at separating molecules of 10–1000 nucleotides in length (1000 nucleotides is termed 1 kilobase or 1 kb), with very high resolution. The larger pores in agarose allow it to resolve much bigger molecules, up to 100 kb in length. However, the separated molecules migrate as a wider band so that the resolution (i.e. the ability to separate molecules which differ only slightly in size) is lower with agarose than with polyacrylamide. In pulsed–field gel electrophoresis (PFGE) the electrical field across the gel is repeatedly switched back and forth between two or more directions. This allows very long DNA molecules, several million base pairs in length, to be separated, since very big molecules change direction less rapidly as they move through the agarose gel.

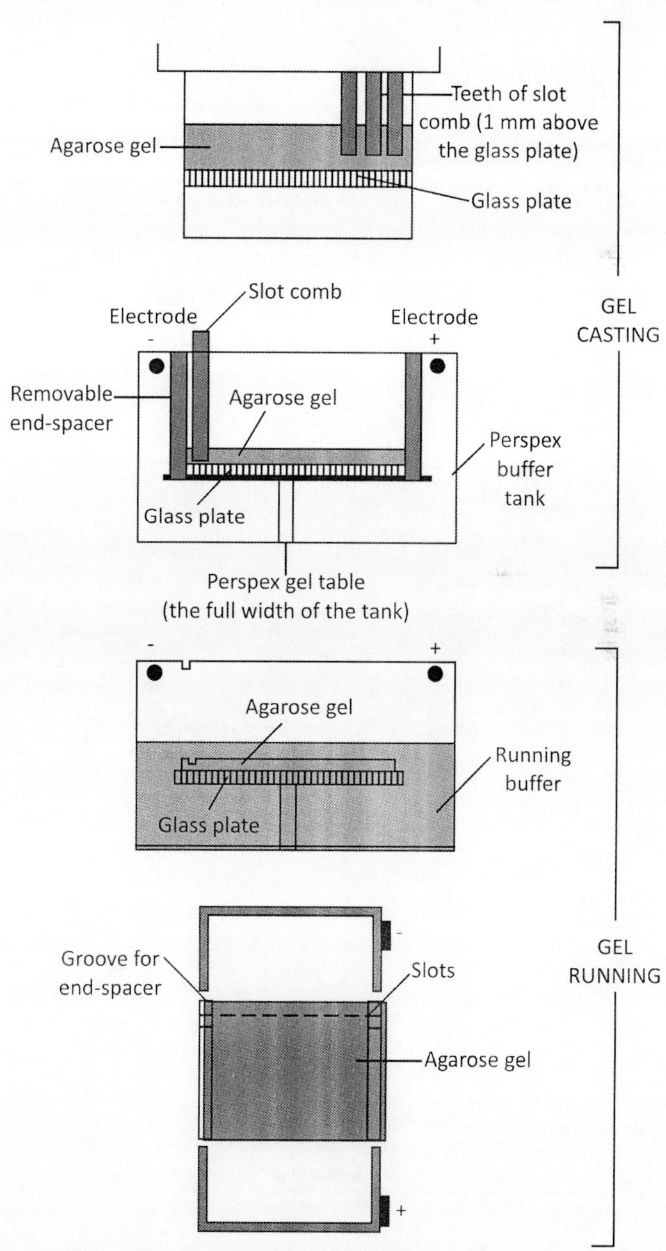

FIG. 8.8: Agarose gel electrophoresis.

After gel electrophoresis, the nucleic acid molecules must be detected. They can be visualized directly by including ethidium bromide in the polyacrylamide or agarose gel. This dye binds to double–stranded nucleic acids by intercalating between adjacent base pairs, and emits a red fluorescence when exposed to UV radiation. As little as 5–10 ng of DNA can be detected. If the nucleic acid molecules are radioactively labeled, then individual resolved species can be detected by autoradiography.

BLOTTING TECHNIQUES

Analysis of DNA by Southern Blotting

In 1975, E. M. Southern invented a procedure to identify the locations of genes and other DNA sequences on restriction fragments separated by gel electrophoresis. The essential feature of this technique is the transfer of the DNA molecules separated by gel electrophoresis to a nitrocellulose or nylon membrane.

The DNA is denatured either prior to or during transfer by placing the gel in an alkaline solution. After transfer is complete, the DNA is immobilized on the membrane by drying or UV–induced cross–linking to the filter. A radioactive DNA ("*probe*") containing the sequence of interest is then hybridized or annealed with the immobilized DNA on the membrane. The probe will anneal to form a double helix only with complementary DNA molecules on the membrane. Nonannealed probe is then washed off the membrane, and the washed membrane is exposed to X–ray film that detects the presence of the radioactivity in the bound probe. After the autoradiogram is developed, the dark bands show the position(s) of DNA sequences that have hybridized with the probe.

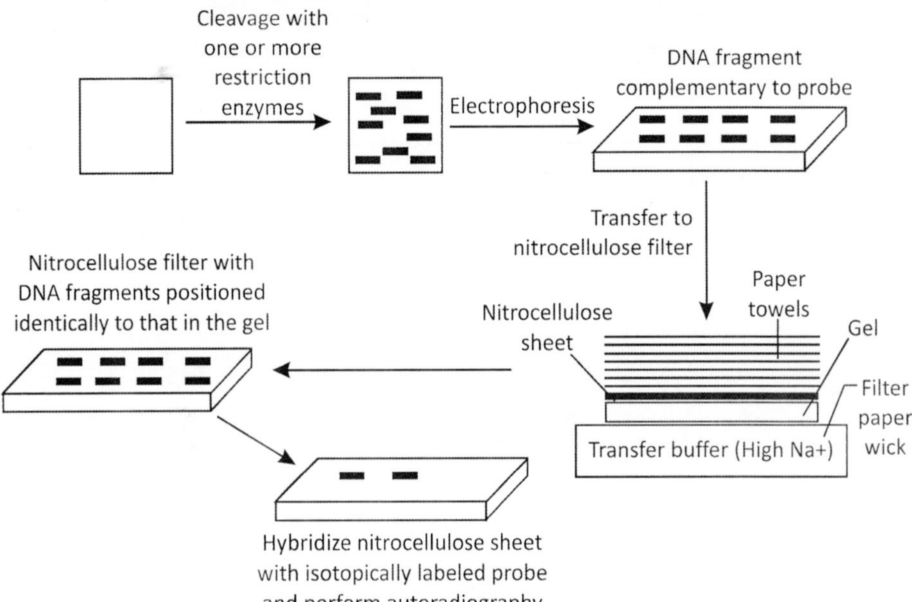

FIG. 8.9: Southern blotting method of analyzing DNA segment that share homology with a nucleic acid probe.

The ability to transfer DNA restriction fragments or other DNA molecules that have been separated by gel electrophoresis to nitrocellulose or nylon membranes for hybridization studies and other types of analyses has proven to be extremely useful. Such transfers of DNA to membranes are called Southern blots after E. M. Southern, who developed the technique.

Analysis of RNAs by Northern Blot Hybridizations

As DNA molecules can be transferred from agarose gels to nitrocellulose or nylon membranes for hybridization studies, RNA molecules can also be separated by agarose gel electrophoresis similarly and transferred for analyses analyzed. Such RNA transfers are used routinely in molecular genetics laboratories and are called ***northern blots*** as it is the mirror image of the Southern blotting technique.

The northern blot procedure is essentially identical to that used for Southern blot transfers however, degradation by RNA molecules are more sensitive to degradation by RNases. Thus, gloves must be worn at all times to prevent contamination of solutions with RNases on one's fingers. In addition, all glassware must be baked overnight to inactivate RNases, and all chemical reagents must be baked or be highly purified to ensure that they are RNase–free. RNase molecules are extremely stable enzymes; they are not inactivated by boiling or other treatments that would totally destroy most enzymes. Most RNA molecules contain considerable secondary structure. Therefore, they must be kept denatured during electrophoresis to separate them on the basis of size. Denaturation is accomplished by adding formaldehyde or some other chemical denaturant to the loading buffer and the electrophoresis running buffer. After transfer to an appropriate membrane, the RNA blot is hybridized to either RNA or DNA probes just as with Southern blots. Now an alternative to nitrocellulose, DBM or diazobenzyloxymethyl paper is used. It may be reused several times for different probes, without appreciable loss of nucleic acid signal.

Northern blot hybridizations are extremely useful in studies of gene expression. They can be used to determine whether a particular gene is transcribed in all tissues of an organism or only certain tissues. They can also be used to study the temporal pattern of expression of individual genes during growth and development. Major disadvantage with northern blot hybridizations is that it only measures the accumulation of RNA transcripts. but explains nothing about why the observed accumulation occurs.

Analysis of Proteins by Western Blot Techniques

Polyacrylamide gel electrophoresis has been used as an important tool for the separation and characterization of proteins.

Since many functional proteins are composed of two or more subunits, individual polypeptides are separated by electrophoresis in the presence of the detergent sodium dodecyl sulfate (SDS), which denatures the proteins.

After electrophoresis, the proteins are commonly detected by treating with Coomassie blue or silver stain. However, the separated polypeptides in the gel can also be transferred to a nitrocellulose membrane, and individual proteins can be detected by using specific antibodies. This transfer of proteins from acrylamide gels to nitrocellulose membranes is called western blotting and is performed by using an electric current–so–called ***electroblotting***. After transfer, a specific protein of interest is identified by placing the membrane with the immobilized proteins in a solution containing an antibody to the protein. Non-bound antibodies are then washed off the membrane, and the presence of the initial ("*primary*") antibody is detected by placing the membrane in a solution containing a "*secondary*" antibody.

This "secondary" antibody reacts with immunoglobulins (the group of protein containing all antibodies) in general. The secondary antibody is conjugated to either a radioactive isotope (permitting autoradiography) or an enzyme that produces a visible product when the proper substrate is added. The western blot procedure is a very powerful tool for identifying and characterizing specific gene–products.

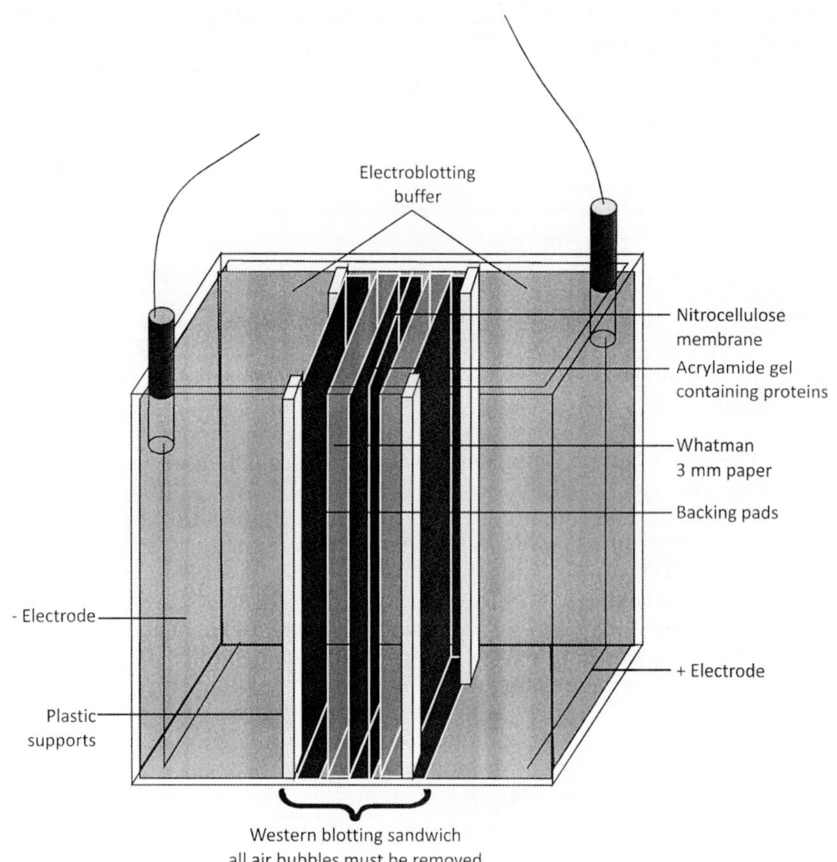

FIG. 8.10: Diagram of a typical Western blotting or electroblotting apparatus. An electric current is used to transfer the proteins from the gel to the nitrocellulose membrane placed next to it in the blotting sandwich. All other components of the sandwich function to provide gentle but firm support; tight contact between the gel and the membrane is essential for good transfer. All bubbles must be carefully removed from the sandwich to assure uniform flow of current and complete transfer of all proteins.

Dot Blots and Slot Blots

Another variety of Southern or Northern blots is dot blots or slot blots. In dot blots, cloned or purified extract of DNAs is tested by spotting them adjacent to each other on a nitrocellulose membrane. DNA blots thus produced are immobilized and denatured so that they are bound on the membrane as single stranded DNA blots. The membrane is then hybridized with radioactively labeled probe (DNA or RNA). The dot representing sequence related to probe will light up by autoradiography. The intensity of the dot will indicate the relative concentration of a sequence related to the probe in the DNA sample used for a particular dot. No steps involving digestion with enzyme, gel electrophoresis or transfer from gel to membrane are needed in this technique. Dot blot are useful in screening viral genomic RNA, RNA transcripts and especially in samples where restriction sites are not considered.

Sometimes, an apparatus is used for placing spots on the membrane, through slots made in this equipment. The spots made thus are in the form of oblong slots rather than round blots. These slots are used just like dot blots and are described as slot blots.

DETECTION OF RFLP

A RFLP can be demonstrated using the following steps:

1. Extract and purify DNA from several individuals which may differ in some respects among themselves.
2. Digest DNA samples with a restriction endonuclease.
3. Subject each DNA digest to gel electrophoresis separately, but on the same gel slab.
4. The DNA fragments of different lengths resulting from digestion can be separated by gel electrophoresis, but the differences in the distribution patterns of fragments in several DNA samples due to digestion by several endonucleases can not be detected directly. This is because the number of fragments is large and the range in size is rather continuous, such that they form a continuous smear on the gel. Electrophoresis is, therefore, followed by the following steps.
5. Fragments are transferred from gel to nitrocellulose filters using the technique of **Southern blotting**.
6. The filters with small fragments, fully or partially homologous to the probe, can be detected by autoradiography after hybridization.

With a sufficient number of digests, it is possible to deduce the origin of each band on the gel, that is, which enzyme or combinations of enzymes have generated it. This yields a restriction enzyme cleavage map or restriction map. Obtaining such a map is very often essential first step in analyzing a piece of DNA of unknown structure. Another method is by labelling with probe followed by autoradiography. Since DNA clones having repetitive DNA may hybridize with many fragments, they will give a smear on autoradiography. Therefore, unique DNA sequences are generally used as probes for detecting RFLPs. The DNA fragments hybridizing with the probe, can be visualized by autoradiography, and will be called RFLPs at the phenotypic level. RFLPs can be studied in a set of related species, using a random or a specific DNA probe. The similarities and differences be used to infer phylogenetic relationships. This has been done in a number of cases, both in plants and in animals. RFLPs have also been used to prepare chromosome mops in humans mice, fruitfly, maize, tomato, lettuce and rice.

DNA SEQUENCING

Restriction enzyme mapping and Southern blotting help establish the gross organization of a piece of DNA, and allow the approximate position of sequences of interest to be determined. The next step in analysis is normally to determine the nucleotide sequence.

Methods for Analyzing DNA Sequence

Our present ability to sequence essentially any DNA molecule is the result of four major developments. (1) The most important breakthrough was the discovery of restriction enzymes and their use in preparing homogeneous samples of specific segments of chromosomes. (2) A second major advance was the improvement of gel electrophoresis procedures to the point where DNA fragments that different in length by a single nucleotide could be resolved. (3) The development of gene cloning techniques greatly facilitated the preparation of large quantities of a particular gene or DNA sequence of interest for use in sequencing studies. (4) Finally, two different approaches to the rapid sequencing of DNA molecules (i.e. by chemical degradation and by chain termination) were developed in the period from 1974 to 1977.

Sequencing by Chemical Degradation

Determination of DNA sequence depends upon the resolution obtained from polyacrylamide gels which are run in the presence of denaturant e.g. urea. If two single–stranded DNA molecules, identical in sequence over most of their length but differing in just one base on one of the two molecules, are subjected to electrophoresis they will be separated.

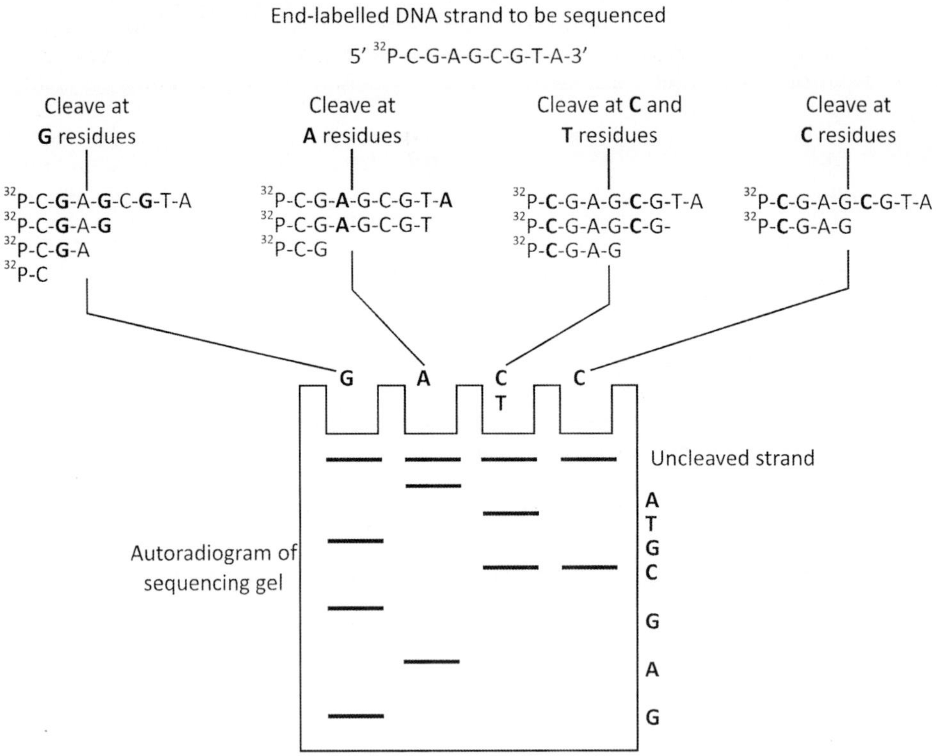

End-labelled DNA strand to be sequenced

5′ ^{32}P-C-G-A-G-C-G-T-A-3′

Cleave at **G** residues

^{32}P-C-**G**-A-**G**-C-**G**-T-A
^{32}P-C-**G**-A-**G**
^{32}P-C-**G**-A
^{32}P-C

Cleave at **A** residues

^{32}P-C-G-**A**-G-C-G-T-**A**
^{32}P-C-G-**A**-G-C-G-T
^{32}P-C-G

Cleave at **C** and **T** residues

^{32}P-**C**-G-A-G-**C**-G-T-A
^{32}P-**C**-G-A-G-**C**-G-
^{32}P-**C**-G-A-G

Cleave at **C** residues

^{32}P-**C**-G-A-G-**C**-G-T-A
^{32}P-**C**-G-A-G

G A C T C

Uncleaved strand

A T G C G A G

Autoradiogram of sequencing gel

FIG. 8.11: DNA sequence analysis by the chemical degradation (Maxam and Gilbert) technique.

In the chemical degradation procedure a DNA restriction fragment is isotopically labeled with P at either its 5′ or its 3′ termini and each of the two strands is separated and isolated. The DNA is then partially modified with chemical reagents specific for the different bases, and cleaved at the modified nucleotides. This generates a set of molecules differing in length but with the same isotopically labeled terminus. These fragments are subjected to denaturing gel electrophoresis in parallel slots of high resolution gel, and the sequence of the DNA can be deduced from the resulting autoradiogram.

Sequencing by Chain Termination

In practice the chemical degradation procedure has been replaced by chain termination method. In this procedure DNA is synthesized rather than degraded, but the principle is similar to the chemical method. DNA polymerases function by adding a primer, a short synthetic oligonucleotide, to pre–existent DNA strand, to provide a fixed 5' terminus for the growing chain.

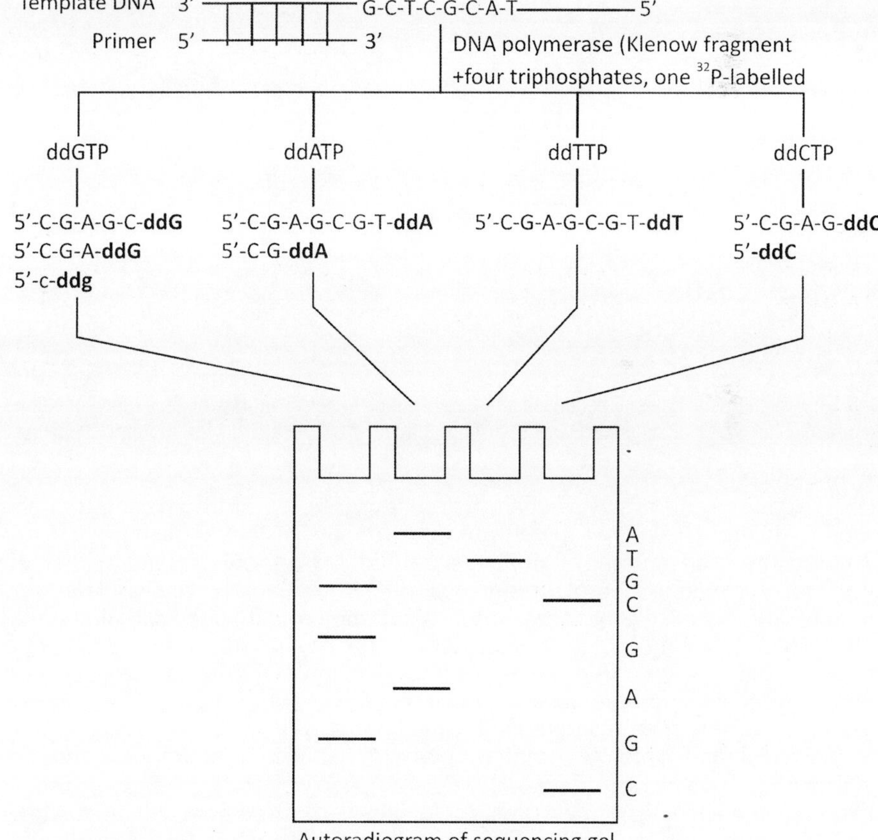

FIG. 8.12: DNA sequence analysis by the chain termination (Sanger) method.

The reaction is performed in the presence of chain terminators. These nucleotide derivatives act as blocking group at their 3' hydroxyl terming and once there are incorporated into the growing chain, they prevent further addition of nucleotides. For example, it dideoxy GTP, is mixed with unmodifies GTP and the other three nucleoside triphosphates, the terminates at a subset of G residues in the newly synthesized strand. Four parallel reactions are run, with dideoxy radiolabeled nucleoside triphosphates of all four kinds are included. The nucleotide sequence of the template strand can be deduced by comparison of the pattern of autoradiographic bands in the four lanes of the gel.

A suitable template for single–strand DNA sequence analysis can be created by inserting the double–stranded template DNA into the genome of a bacterial virus which naturally packages only one of the two strands into viral particles. Another method is denaturing of double–stranded DNA template by treatment with alkali. An excess of primer is added before re–annealing occurs, so that primer will compete with the complementary strand for annealing to the template. This is a preferable method over insertion of DNA into viral vector.

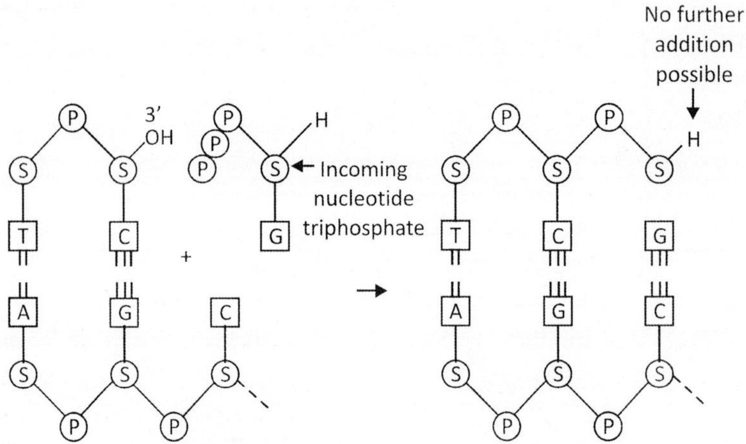

FIG. 8.13: The chemical basis of sequence analysis using chain terminators. This is a representation of a reaction in which a dideoxy G residue incorporated into the growing chain blocks further incorporation.

DNA sequence analysis using chain terminators is best performed using a single–stranded template. This allows the primer to be free accessible to the template.

The major disadvantage of double–stranded DNA sequence analysis is the presence of competing complementary strand. This leads to a higher level of sequencing artefacts like presence of more than one band. In such situation it may be impossible to decide which the correct residue at that position is. Artifacts of this kind are not restricted to double–strand DNA sequence analysis; some level of artifacts is a universal problem in sequence analysis.

The DNA sample then resequenced using another primer, or by using the other strand (i.e. the strand which is complementary to that used in the original analysis) as template.

Even when a single–stranded DNA, molecules with more than 500 nucleotides, clearly it is not possible on a polyacrylamide gel to resolve when the molecules are larger than about the nucleotides. This sets the upper limit to the length of sequence to be determined. One way to extend from a known into an unknown region of sequence is to synthesize an oligonucleotide primer which is complementary to a tract near the end of the region whose sequence has already been established. In this way it is possible to 'walk' along a sequence, at each step synthesizing a new primer which is used to extend the region of known sequence. Alternatively, it is possible to delete sequences enzymatically from the template at a position downstream of the annealing site of the primer, to generate a deletion series. The nucleotide sequence of a set of such molecules can be used to deduce the sequence of the intact template.

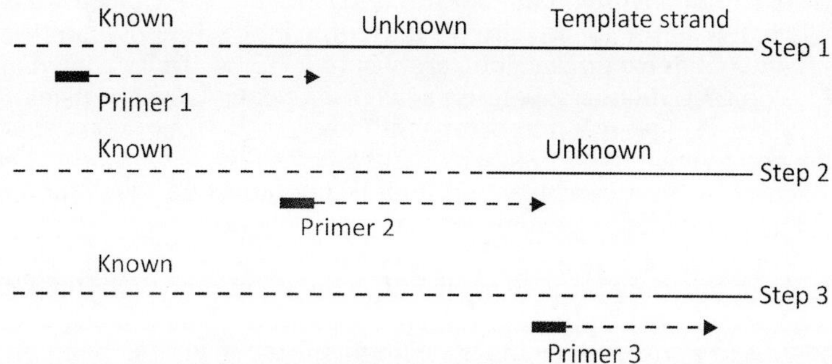

FIG. 8.14: DNA sequence analysis by primer walking.

By determining the sequence of both strands of a DNA molecule, and confirming that the two deduced sequences are perfectly complementary, it is possible to detect and eliminate artefacts generated in the reactions or during electrophoresis. When this check is performed, DNA sequence can be established with a very high degree of accuracy.

Direct DNA Sequencing Using PCR (Also Called Ligation Mediated PCR=LMPCR)

Polymerase chain reaction (PCR) has also been used for sequencing the amplified DNA product. This method of DNA sequencing is faster and more reliable and can utilize either the whole genomic DNA or cloned fragments for sequencing a particular DNA segment. The DNA sequencing using PCR involves two steps: (i) generation of sequencing templates (double stranded or single stranded) using PCR and (ii) sequencing of PCR products eight with the thermolabile DNA polymerase or with the thermostable Taq DNA polymerase. Thus the DNA sequencing method using PCR eliminates the need of cloning the DNA in single stranded DNA phage vector i.e. M13. Although double stranded DNA product of PCR can be utilized for sequencing, this may cause difficulty due to reassociation after denaturation thus preventing the sequencing primer from annealing to its complementary sequence to allow extension. To reduce this problem, either a variant of the standard method for sequencing double stranded DNA is employed or single stranded DNA templates are produced following **asymmetric PCR**.

A number of thermolabile DNA polymerases have been used for sequencing of *in vitro* amplified DNA. Alternatively, Taq polymerase (thermostable enzyme) used for PCR, can also be used for sequencing reaction. In either case, Sanger's synthetic method involving incorporation of dideoxynucleotide (ddNTP) for chain termination is used for sequencing. However, Maxam and Gilbert's chemical degradation method can also be used.

In Sanger's method, as usual, four mixtures ar prepared, each using one of the four ddNTP. The sequencing primer is labeled with ^{32}P and the mixtures with amplified DNA, Taq polymerase and appropriate buffer are incubated at 70 °C for 5 min. The reaction is stopped by addition of formamide stop solution in each tube and mixtures are run on polyacrylamide sequencing gel to obtain the ladders, which can be read by computer or manually. This method using PCR helped in automation of DNA sequencing.

Automated DNA Sequence Analysis

Using the chain termination method, DNA sequencing can be mechanized to make the process rapid. In a conventional sequence reaction described above four separate single–strands extension reactions are performed, with each containing a different chain–terminating dideoxynucleoside triphosphate. The radiolabeled reaction products are separated on four separate tracks of a polyacrylamide gel. In machine–based sequencing a single reaction is run, with each of the four dideoxynucleoside triphosphates containing a different fluorescent label.

When excited by an argon laser, the fluorescent labels emit light at a different wavelength. The same chain–termination reactions are carried out; however, it is no longer necessary to run the products in separate lanes of a gel. The reaction products are mixed and run in a single lane in a gel and as the different length fragments pass the laser each fluorescent label gives a unique signal which is stored in the computer.

Machine–based DNA sequencing is beginning to change the speed and scope of sequencing projects dramatically. The first complete genomic sequences were those of simple viruses such as the SV40 tumor virus, containing around 500 based of DNA. Recently, the DNA sequence of over 106 000 bases of DNA from human chromosome 19 was determined in a few months using DNA sequencing machines. The sequence of a significant fraction of the *E. coli* genome has been established, the entire sequence of yeast chromosome 3 is known, and there are plans eventually to determine the sequence of the entire human genome.

MECHANICAL SHEARING OF DNA

In addition to digestion DNA with restriction endonucleases to produce discrete fragments, there are a variety of treatments which result in non specific breakage. Non specific endonucleases and chemical degradation can be used but the only method that has been much applied to gene manipulation involves mechanical shearing.

The long, thin threads which constitute duplex DNA can not be easily broken by shear forces in solution. Intense sonication with ultrasound can reduce the length to about 300 nucleotide pairs. More controlled shearing can be achieved by high–speed stirring in a blender. Breakage occurs essentially at random with respect to DNA sequence. The termini consist of short, single–stranded regions which may have to be taken into account in subsequent joining procedures.

TRANSCRIPT MAPPING TECHNIQUES

Analyzing the nucleotide sequence of a piece of DNA is a necessary preliminary to understanding its precise organization and function, but is not normally sufficient for these purposes. Because eukaryotic gene transcripts are subjected to a complex series of post–transcriptional processing events, the structure of a gene can be understood only by correlating the organization of the genomic DNA with the organization of its mRNA transcript. This has required the establishment of sophisticated cloning and mapping techniques which can, when used in conjunction, give a complete description of the structure of a gene.

It is possible to isolate, by gene cloning, the double–stranded DNA copies of individual mRNA sequences. These cDNA clones will normally contain a DNA copy of the poly(A) sequence at the 3' terminus. Hence, by determining which strand of the cDNA clone contains a poly(A) tract at one end, it is possible to identify the coding strand of the cDNA clone. Provided a sufficient length of sequence has been established to make their presence statistically likely (in practice 300–500 nucleotides), termination codons will occur in two of the three potential coding frames, ruling them out as translational open reading frames (ORFs).

Sometimes this level of analysis, that is, identification of the reading frame and deduction of the encoded protein sequence, will be all that is required. However, it is often desirable to determine the structure of the gene itself. It will have been isolated as part of a genomic clone. This will contain the gene and a variable amount of flanking DNA depending upon the method of cloning used.

By correlating the sequence of the gene with the sequence of the mRNA deduced from the cDNA clone, it is possible to identify the site of poly(A) addition and the positions of introns. However, technical limitations in the cDNA cloning procedure ensure that the cDNA clone will not normally contain sequences derived from the extreme 5' terminus of the mRNA. There are two potential methods to determine the position of the 5' end, primer extension and S1 mapping.

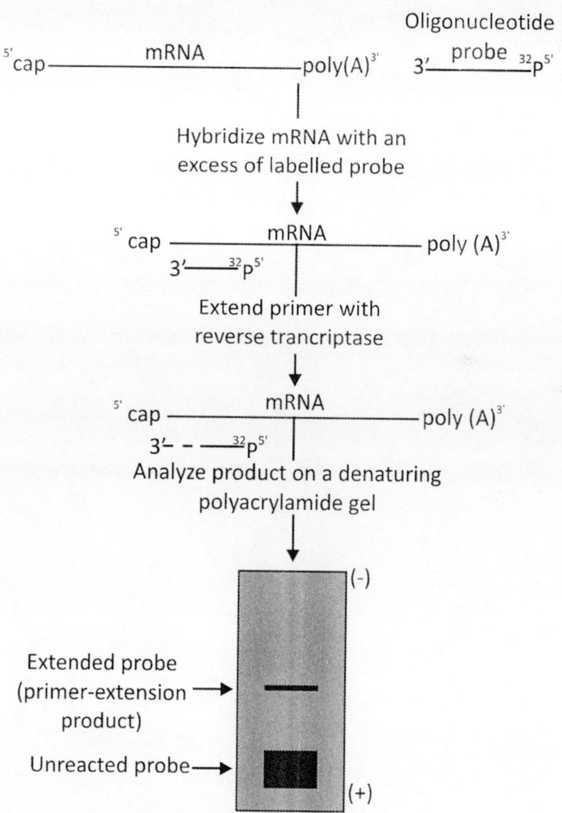

FIG. 8.15: Identification of the 5' end of an mRNA by primer extension.

Primer Extension

In primer extension, a short oligonucleotide complementary to a sequence tract near to the 5' end of the mRNA is annealed to the mRNA. This is then used to prime synthesis with the enzyme reverse transcriptase. This enzyme produces a DNA copy of the mRNA (a cDNA) which extends to the 5' terminus of the mRNA. By correlating the length of the cDNA primer extension product with the sequence of the genomic DNA, it is possible to deduce the start site of transcription.

S1 Mapping

The alternative technique is called S1 mapping. An isotopically labeled probe is generated from the genomic clone. This probe is complementary to the mRNA over a region encompassing part or all of the first exon, and extending upstream of the region where the cap site is expected to be located. The cap site may have already been putatively localized using primer extension. The probe is annealed to the mRNA and digested with an enzyme which destroys single–stranded nucleic acids but leaves double–stranded molecules intact.

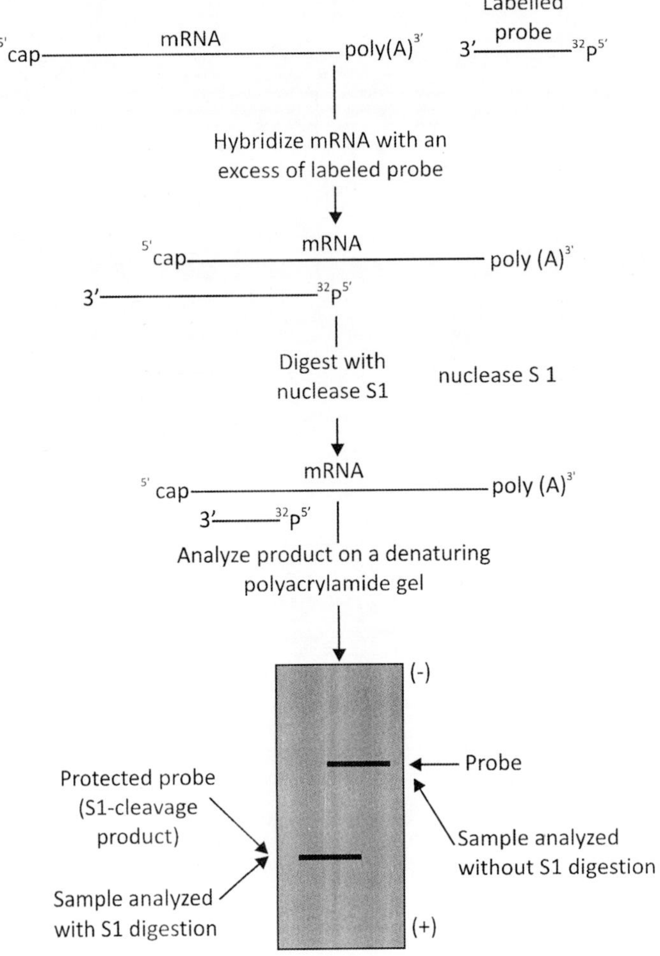

FIG. 8.16: Identification of the 5' end an mRNA by S1 mapping.

In early experiments an enzyme called nuclease S1 was generally used, and the name S1 mapping has been retained, even though other nucleases are now sometimes used. During digestion, the 3' proximal region of the probe, which has no complement in the mRNA molecule and which therefore remains single–stranded, is degraded. A molecule is generated which is the complement of the region between the cap site and the position at which the 5' terminus of the probe annealed to the mRNA. Again, by determining the length of this molecule it is possible to infer the position of the cap site.

These two techniques are often used in parallel to determine the cap site. They are both subject to artefacts, but of different kinds. Therefore if both methods agree, it greatly increases confidence in the assignment of the cap site. If the hybridization is performed with an excess of probe, the amount of primer extension or nuclease–resistant product is proportional to the amount of primer extension or nuclease–resistant product is proportional to the amount of complementary mRNA. Both techniques can therefore be used an alternative to Northern transfer to quantitates mRNA levels.

SYNTHESIS OF cDNA FROM mRNA

Complementary DNA can be prepared by isolating mRNAs from tissues which are actively synthesizing proteins, like roots and leaves in plants, ovaries or reticulocytes in mammals etc. Isolated and purified mRNA can then be directly used to synthesize complementary DNA strand using the enzyme reverse transcriptase. This results in a DNA/RNA hybrid. The RNA part of the hybrid is then removed with alkali. Thus a single–stranded DNA is generated which can act as the template for the synthesis of its complementary strand in the presence of DNA polymerase. The double stranded DNA molecule thus formed is called **complementary DNA** (cDNA), as it is synthesised complementary to the mRNA. In the prokaryotes the double stranded cDNA will be identical to the gene (DNA) from which the mRNA is formed. But, most of the eukaryotic genes contain intervening sequences (introns) which interrupt the coding sequences. Hence, the cDNA of the eukaryotic cells and their native DNA are not collinear.

As bacterial genes do not have the intron – exon organization found in the eukaryotic genes they find the expression of eukaryotic genes in them (bacterial) difficult. Therefore one has to depend upon cDNA of eukaryotes, which does not have introns, for cloning.

Preparation of Double–Stranded cDNA

The mRNA that is specific for a particular kind of protein is isolated and purified. All naturally occuring mRNAs of eukaryotes have about 200 adenylate residues at their 3'–termini. This is called **polyadenylate tail**. The enzyme reverse transcriptase can mediate the synthesis of a new DNA strand only when it has a growing point in the form of a pre–existing primer, which is base paired with the mRNA template and has a free 3' OH group. Such a primer is provided by hybridizing a short complementary oligo (dT) molecule with the poly adenylate (poly–A) sequence.

The 3' OH end of the poly–T primer is made double standard by reverse transcriptase forming RNA–DNA hybrid. The RNA strand of this RNA – cDNA hybrid is digested by alkaline hydrolysis (DNA is resistant to alkali). At the 3'–OH end of the newly formed single stranded cDNA, a hairpin is formed which is stabilised by base – pairing. This hairpin which has a 3'–OH free end acts as a self–priming structure. Forming second strand of the cDNA in the presence of the fragment of the *E. coli* DNA polymerase–1 (Kornberg enzyme) and the four deoxynucleotide triphosphates.

The enzyme reverse transcriptase or T4DNA polymerase can also be used for the synthesis of the second strand of cDNA. The synthesised double stranded DNA is treated with the single–strand specific nuclease S1. This enzyme digests the curved region of the hair pin and any polynucleotide tails. Thus, finally a double stranded cDNA molecule is formed. This DNA can now be inserted into a suitable vector and can be further cloned in the desired host.

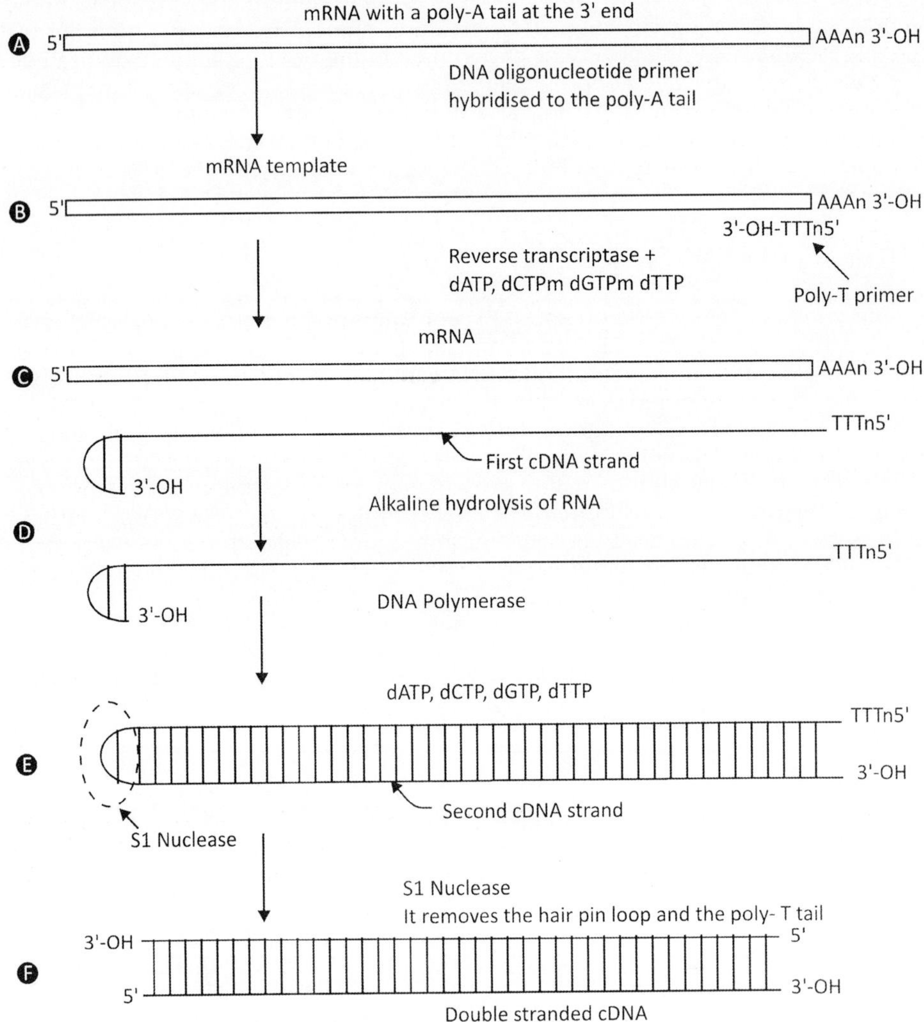

FIG. 8.17: Synthesis of double stranded cDNA from an mRNA.

Complementary DNA of a given mRNA is inserted into a suitable vector can then be cloned in a host. The cDNA is blunt ended. It has no cohesive ends that are complementary to the cohesive ends of any of the vectors. This is overcome by homopolymer tailing. Once inside the host cell, the cDNA in the hybrid will express the gene(s) it carries.

cDNA Library

Complementary DNAs (cDNAs) of the genome of an organism can be synthesised and cloned to generate a cDNA library. Thus, cDNA library is a collection of cDNA clones of an organism. The starting material for the synthesis of cDNA is the mRNA. Hence the resulting cDNAs will represent the mRNAs present in the cells from where these mRNAs were isolated. In the case of multicellular organisms there is specialization of individual cells. For example, a mammal is made up of a large number of cell types: brain cells, mammary cells, blood cells, liver cells, cells of the pancreas, nerve cells etc. Each type of cell will contain the same complement of genes. However, only a particular set of genes will be expressed in a given type of cell. Other genes will remain silent. Only those genes that are expressed are transcribed into mRNA. Hence the cDNA clones of this type of cell will be only a selection of the total number of genes in a cell. Hence for such organisms cDNA clones of different cell are prepared.

SEARCHING FOR GENE USING COMPUTERS

Although the series of analyses described above are necessary ultimately to establish the organization and mode of expression of a particular gene, genes can be sought by less laborious means using computer programs. These searches for regions of a DNA sequence which look like coding regions flanked by the signals that direct gene transcription and processing of their RNA products. This procedure is of greatest value when large regions of genomic DNA for which cDNA clones are not available are being analyzed. It is thus an essential part of the Human Genome Project.

One of the principles, that of searching for long ORFs (= open reading frames), has already been described. Although the sequence (the Kozak sequence) that precedes the ATG initiation codon is only weakly conserved between different genes it can provide a helpful guides as to the start site of translation. The start site of transcription often lies downstream of a TATA box, and it is also possible to search for the consensus binding sites of other transcription factors.

RNA processing signals can also be used to help establish the likely gene organization. The sequence that signals polyadenylation is well conserved, and splice junctions have a consensus sequence comprising an almost perfectly conserved core (5' GU——AG 3'), with moderately well conserved flanking sequences.

An identification gene by analysis of DNA sequences is an evolving science. The structures of new genes are constantly adding and the sophistication of the search programs is increasing. The databanks exists which contain many thousands of gene sequences e.g. Genbank and the EMBL. Although some scientists initially submit their gene sequence data to one rather than other, the two organizations that compile the databases then exchange data.

CHAPTER 9

Genetic Engineering Techniques for Producing Transgenic Crop Varieties

THE GOAL FOR IMPROVED CROP VARIETIES

By harnessing nature's methods of gene transfer, scientists have been able to genetically alter a number of common crops, creating new varieties that are better suited to farmer's needs. Certain crops respond more readily to biotechnology than others. One of these crops is canola. Already, several new varieties of canola have been developed through genetic engineering. For example, we now have canolas that are resistant to certain kinds of herbicides that used to kill them. Other canolas have been modified to produce specific kinds of oil suitable for special uses, including enhanced nutrition, medicines, and the production of lubricants. Grasses, like wheat, or legumes, such as peas, are more difficult to modify. Recently, scientists at the National Research Council's Plant Biotechnology Institute in Saskatoon have used *A. tumefaciens* biotechnology to produce their first genetically modified peas. Although there are many new biotechnologies available to solve basic problems in plant transformation, scientists continue to use nature's method of genetic modification to further their research.

Special techniques in genetic engineering allow scientists to produce new varieties of plants with selected characteristics. These new varieties are known as transgenic varieties and they have features that improve production and give farmers an opportunity to grow stronger, more productive crops.

Inside the cells of all living things are tiny strands called chromosomes which contain bits of protein called genes that hold a unique set of codes. These codes act as a recipe for growth and development. Genetic engineering transfers genes from one plant to another in order to create new characteristics.

When two plants are crossbreed, they each contribute all of their genetic information to the new plant. The new plants that result from this breeding may receive almost 50,000 pieces of genetic information from their "*parents*" and could have thousands of new characteristics. Sometimes the desired characteristic is lost amongst the thousands of extra genes and the trait plant breeders are looking for may not be able to express itself in the plant.

Traditional plant breeding uses the pollination of plants to create new crop varieties and takes many years to accomplish. Pollination can only be done with closely related plant species because nature has systems, which stop unrelated plants from pollinating with one another. Because nature was so cautious, breeders used to be limited to working with a small selection of traits that were available only closely related species.

TRANSGENIC TECHNIQUES

Genetic engineering overcomes the limitations of traditional breeding and allows scientists to use new traits from many kinds of plants and other living things such as fish, insects, bacterium and even humans. Because all organisms use the same genetic language, different kinds of organisms can understand each other's genetic code. If a new gene is introduced to another organism's cells, that cell will begin to produce that gene's particular protein and will display the new trait directed by the gene.

There are four basic steps to creating a transgenic plant.

1. First, biotechnologists isolate a gene which has the recipe for the characteristic that the new plant must have.
2. Second, a vector is chosen which can carry the new gene into the plant's cells.
3. Third, the gene is clipped from its chromosome and loaded onto the vector (usually bacteria) which acts like a genetic gun that shoots the gene into the cell.
4. Fourth, once the gene has been delivered into the cell it travels into the chromosome strand.
5. After it is shot into the cell, the new gene becomes part of the plant's recipe book. All of the new plants which are produced from the altered plant will pass on the gene to their offspring.

This genetic engineering process has made plant breeding more precise. Using transgenic methods, biotechnologists can be certain that if they have selected the correct gene, the new characteristic will be permanently established in the new crop variety.

Transgenics has made it possible to create a whole new range of genetic diversity. Using this tool scientists help a farmer's crop become more productive.

STEPS FOR DEVELOPING NEW CROP VARIETIES

Genetic engineering to enhance a crop variety involves most if not all of the following steps. The process can take several years, but often moves much more quickly, especially as the various techniques become perfected. In any case, the process is still faster than transferring genes through conventional plant breeding, if that is even possible for the particular desired trait.

Step 1

Finding a plant or other organism showing the desired characteristic.

Step 2

Identifying the gene controlling that trait and its location on a chromosome. In recent years, scientists have been working to produce genetic maps of the chromosomes of many species, including humans, livestock, crops and potentially useful microorganisms. (These are often referred to as genome projects, genome being the term used to refer to an organism's entire genetic complement.) Through this process of genetic mapping, scientists have been able to determine the location of many specific genes on the chromosomes of different species and to determine their respective functions.

Step 3

Marking the gene so its presence can be detected quickly and simply. Identifying where a gene is located on a chromosome and confirming that it has been isolated or transferred is done through a process called **marker–assisted selection**. Specific genes or segments of DNA, called markers, are used to mark the location of the genes to be transferred. Markers that can be detected easily and that appear to always be associated with the desired trait are chosen for this process.

To some extent, marker genes have been used in conventional plant breeding for many years. Plant breeders observed that certain traits almost always appeared together as plants were cross–bred and their offspring tracked through the generations of selection required developing a crop variety. Such sets of genes are said to be linked. By choosing easily detectable marker genes that were linked to desirable traits that were more difficult to measure, plant breeders were able to speed the process of incorporating desirable traits into new varieties. As they identify where each marker occurs on a chromosome, and how close it is to a specific gene, scientists can create maps of the markers and genes on specific chromosomes.

In the past, plant breeders usually had to grow plants through to maturity in order to observe their marker genes. Today, marker–assisted selection is applied at the cellular level. For example, cellular resistance to a specific chemical or antibiotic is a commonly used marking system. Thus, plant breeders do not need to wait for a seedling to grow into a mature plant to be able to test for a marker and its linked trait. By analyzing small amounts of tissue, perhaps even single cells, researchers can determine which contain the desired gene and concentrate on only those cells or tissues that contain the desired trait.

Step 4

Isolating the gene from the rest of the DNA so that only the desired information will be transferred. Scientists are able to isolate individual genes using enzymes that cut DNA at specific sites along the DNA strand. Several hundred of these enzymes have been discovered and catalogued according to the point at which they cut a DNA molecule. By using selected enzymes, scientists are able to cut the DNA strand at particular points and isolate specific genes.

Step 5

Producing large numbers of the gene and introducing the desired gene into cells of the organism to be enhanced. One common means of transferring genes into plants is by employing the natural ability of some bacteria to alter the DNA of organisms which they infect. These bacteria have the ability to inject a segment of their DNA into that of an infected cell, thereby affecting its behaviour. For example, the common soil bacterium, *Agrobacterium tumefaciens*, causes a disease called **crown gall** in susceptible plants by inserting its genes into a plant's DNA. The bacterial genes disrupt the genetic code of the plant's cells in the infected area causing them to grow into an abnormally shaped mass called a gall. Scientists have discovered ways to use this ability of *A. tumefaciens* to transfer genes between species in the laboratory. *A. tumefaciens* and some other bacterial species produce small, circular pieces of DNA called plasmids. Plasmids function independently of other bacterial DNA and can enter cells of other organisms, carrying genetic information with them that then merges with that of the host cell.

To transfer genes from one cell to another, researchers use an enzyme to cut a gap in a bacterial plasmid. They then insert a gene from the donor DNA strand into the plasmid. Because the cut ends of both the plasmid and the donor gene segment are chemically 'sticky', they attach to each other or recombine to form a plasmid containing the new gene. (Hence the name, recombinant DNA [rDNA] technology). The recombined plasmid now carries the gene that can result in the expression of the new trait when transferred into a plant cell by the plasmid.

Plasmids are also important as the basis of gene cloning in biotechnology. Gene transfer is a 'numbers game.' Not every attempted transfer will result in the successful combining of the desired gene with the DNA of the new host, and in only a few combinations will the new gene be located so that it is expressed properly without disrupting the function of the rest of the host's genes. Thus, a great many cells have to be transformed in order to get a few that function properly.

Once a gene is inserted into a plasmid, it can be introduced into a bacterium that is easily reproduced in the lab (e.g., *E. coli*). When the altered bacteria multiply, the plasmid containing the desired gene is also copied or cloned in every cell. Because bacteria grow quickly, this method makes it possible to make millions of copies of the desired gene in a short period.

Using the appropriate enzymes, the plant genes are then extracted again from the bacteria and inserted into the plasmids of *A. tumefaciens*. Finally, the altered *A. tumefaciens* is mixed with the target plant cells, where the engineered plasmid enters the target plant cells. The 'foreign' gene is then incorporated into DNA of some of the target plant cells.

For some crops, researchers use a particle gun instead of *Agrobacterium* to transfer desirable traits from one plant to another. Researchers coat microscopic pellets of gold or tungsten with DNA that has the desired trait. The coated pellets are fired through plant cells with a special gun. As the pellets go through the cells, they leave some of the DNA coating behind to combine with the plant cell's DNA and a new beneficial trait.

Step 7

Identifying which plant cells now contain the desired gene, through the use of markers as described above.

Step 8

Growing the altered cells into full plants, using a process called **tissue culture**, to confirm that the desired trait is expressed properly.

Step 9

Using traditional plant breeding techniques to transfer the trait into a usable variety.

BACTERIA: NATURE'S GENETIC ENGINEERS

As far back as 1869, the Swiss scientist Fredrick Meischer discovered and isolated DNA, the substance inside the cells of any organism which contain the genetic "*blueprint*" controlling all of its characteristics. Like his contemporary Gregor Mendel, who first established the principles of heredity, Meischer's discovery was largely ignored by the scientific community of his time. It was in the mid–1920s that the British pathologist Frederick Griffith learned that the pneumonia–causing bacterium he was studying was capable of transforming its basic characteristics. His experiments showed that a harmless strain of bacteria could transform into a lethal one. Several decades later, researchers discovered that viruses living in the bacteria are the agent of its genetic transformation from one type to another. The scientists observed that bacterial and viral DNA can integrate, and then *ferry* genes between different kinds of bacteria.

Bacteria and Gene Transfers in Plants

A similar discovery about natural gene transfers in plants was made by agricultural scientists in 1907. Erwin Smith and Charles Townsend discovered that plant traits can be changed in nature in much the same way bacteria are altered by the activity of viruses. The scientists discovered this when they observed the formation of a plant tumor called a crown gall. Crown galls form on a plant at the site of a fresh wound when the wound is infected by a common soil bacterium. Although at the time they didn't understand the genetic science involved, the researchers were observing the activity of *Agrobacterium tumefaciens*, which is capable of inserting its genes into plants. The foreign genes can then *hijack* the genetic code of the plant, making it a better host for the bacteria. What Smith and Townsend did come to realize was that the cell mechanisms of all living things, whether plant or animal, are fundamentally the same. This discovery later made it possible to apply the process of gene transfer, first observed in the plants changed by *A. tumefaciens*, to the modification and improvement of many plants useful in agriculture. In time, this was accomplished by finding a way to delete the genes that cause the crown gall disease and substitute other genes that produce desirable traits. Scientists found that they could harness this unusual ability of *A. tumefaciens* to carry out gene transfers in the lab.

How Bacteria are used to Modify Plants

Once scientists have discovered the genes responsible for the trait they want to introduce into another plant, they face several difficult problems in transforming the target plant.

- First they must accumulate enough pure DNA needed to genetically alter a number of target plant cells.
- A second problem is finding a way to deliver genes through the walls of the cell and into the nucleus. After all, the cell is designed by nature to keep foreign material out.

Several ways have been found to solve these problems, but one of the most useful methods harnesses bacteria to both increases the amount of DNA as well as to *deliver* genes through cell wall and into the nucleus.

How do they do it?

Scientists have discovered that a number of enzymes (as described previously), complex organic substances originating in living cells, can be used as tools to *cut* genes from plants. Some 300–400 enzymes are now available for use in biotechnology. Different enzymes are used to cut and isolate specific genes. Once scientists have discovered and isolated the genes that control a desired trait in a plant (such as resistance to a disease), they have to find a way to replicate or clone enough exact copies of the genes to transform a large number of cells from the target plants.

Their first step is to take the genes responsible for a desired trait from a plant and insert them, using enzymes, into a plasmid taken from a common type of bacterium called *E. coli*. Plasmids (as discussed previously) are very important in biotechnology because they are the basis of gene cloning. Once a gene is inserted into a plasmid, the plasmid can be reintroduced into the bacterial cell. When the altered bacteria is grown in a culture (a medium that promotes rapid growth), the plasmid containing the desired gene is copied or cloned in every cell. Because bacteria grow quickly, this method makes it possible to make millions of copies of the desired gene in a short period. Using enzymes, the plant genes are now removed from the E. coli plasmids and inserted into the plasmid of *A. tumefaciens*, the soil microorganism that can move foreign genes into plant cells.

Finally, the altered *Agrobacterium tumefaciens* is mixed with the target plant cells. The bacteria ferry the engineered plasmid into the target plant cells. The "*foreign*" plant gene is then incorporated into the plant cell's DNA. New plants can be grown directly from these modified cells in a special growth chamber. Some of the plants grown from these cells will have the trait specified by the new gene. By growing the plants, scientists can now test to see which of the plants have been genetically transformed.

Three gene–transfer techniques have been developed to such a stage of reproducibility and practical applicability that they can be recommended for future practical work. These are *Agrobacterium–mediated* gene transfer, *ballistic* gene transfer, and *direct* gene transfer to *protoplasts.* These techniques are described in some detail in Table 9.1 which summarizes which crop plants have been transformed with which of these techniques. These are additional technique that have yielded confirmed transgenic plants and that show some promise for future development, such as *microtargeting, microinjection, electroporation* into tissues, and *whiskers.* These techniques are described to a smaller extent and the literature citations will support further reading. There is a further group of techniques for which unequivocal evidence for the production of transgenic plants is still missing such as the pollen tube–pathway, and for which the reader is referred to a more detailed assessment (Potrykus, I.).

TABLE 9.1: Transgenic Crop Plants Obtained by Agrobacterium–Mediated Transformation.

English name	Latin name	Method	References
Apple	*Malus pumila*	A	(James et al., 1989), (Lambert and Tepfer 1991)
Apricot	*Prunus armeniaca*	A	(Laimer et al., 1992)
Asparagus	*Asparagus officinalis*	A	(Delbreil et al., 1993)
Aspen	*Populus tremula*	A	(Block 1990), (Tsai et al., 1994)
Barley	*Hordeum vulgare*	B	(Wan and Lemaux 1994)–(Jähne et al., 1994)
Birdsfot trefoil	*Lotus corniculatus*	A	(Webb et al., 1990)
Black currant	*Ribes nigrum*	A	(Graham and McNicol 1991)
Carnation	*Dianthus sp. caryophyllus*	A	(Nugent et al., 1991)
Carrizo Citrange	*Citrus*	A	(Moore et al., 1992)
Carrot	*Daucus carota*	A	(Scott and Draper 1987)
Celery	*Apium graveoleus*	A	(Catlin et al., 1988)
Chicory	*Cichorium endivia*	A	(Vermeulen et al., 1992)
Chrysanthemum	*Chrysanthemum sp.*	A	(Ledger et al., 1991)
Cranberry	*Vaccinium macrocarpon*	B	(Serres et al., 1992)
Cualiflower	*Brassica oleracea*	A	(Block et al., 1989)

English name	Latin name	Method	References
Cucumber	*Cucumis sativus*	A	(Chee 1990)
Fescue, red	*Festuca rubra*	P	(Spangenberg et al., 1994)
		B	(Spangenberg et al., 1995)
Fescue, tall	*F. arundinacea*	P	(Wang et al., 1992)
		B	(Spangenberg et al., 1995)
Gerbera	*Gerbera hybrida*	A	(Elomaa et al., 1993)
Grapevine	*Vitis spp.*	A	(Baribault et al., 1990), (Torregrosa et al., 1994)
Kiwi fruit	*Actinidia chinensis*	A	(Uematsu et al., 1991), (Rugini et al., 1991)
Lettuce	*Lactuca sativa*	A	(Michelmore et al., 1987)
Muskmelon	*Cucumis melo*	A	(Gang and Grumet 1990), Dong et al., 1991)
Orchardgrass	*Dactylis glomerata*	P	(Horn et al., 1998)
Orchid	*Dendrobium sp.*	B	(Chia et al., 1994)
Papaya	*Carica papaya*	A	(Fitch et al., 1993)
		B	(Manshardt et al., 1992)
Passionfruit	*Passiflora edulis*	A	(Manders et al., 1994)
Pea	*Pisum sativum*	A	(Puonti–Kaerlas et al., 1990)–((Schroeder et al., 1993)
Pecan	*Carya ovata*	A	(McGranahan et al., 1993)
Penennial ryegrass	*Lolium perenne*		(Spangenberg et al., 1995)
Pepino	*Solanum muricatum*	A	(Atkinson and Gardner 1991)
Petunia	*Petunia hybrida*	P	(Krüger–Lebus and Potrykus 1987)
Plum	*Prunus domestica*	A	(Mante et al., 1991), (Scorza et al., 1994)
Poplar	*Populus sp.*	A	(Filatti et al., 1987)
		B	(McCrown et al., 1991), (Wilde et al., 1992)
Potato	*Solanum tuberosum*	A	(Sheerman and Bevan 1998), (Stiekema et al., 1998)
		P	(Masson et al., 1989)
Rye	*Secale cereale*	B	(Castillo et al., 1994)
Strawberry	*Fragaria x Anannassa*	A	(James et al., 1990), (Nehra et al., 1990)
Stylo	*Stylosanthes humilis*	A	(Mammers 1988)
Subterranean clover	*Trifolium subterraneum*	A	(Khan et al., 1994)
Sugar beet	*Beta vulgaris*	A	(Lindsey and Gallois 1990)
Sunflower	*Helianthus annuus*	A	(Everett et al., 1987)–(Knittel et al., 1994)
		B	(Bidney et al., 1992)
Sweetgum	*Liquidambar styraciflua*	A	(Sullivan and Lagrimini 1993)
Tamarillo	*Cyphomandra betacea*	A	(Atkinson and Gardner 1993)
Tobacco	*Nicotiana tabacum*	A	(Horsch et al., 1984)
		P	(Paszkowski et al., 1984)
		B	(Tomes et al., 1990)
Tomatillo	*Physalis ixocarpa*	A	(Assad–Gracia et al., 1992)

English name	Latin name	Method	References
Tomato	*Lycopersicon esculentum*	A	(Fillatti et al., 1987), (Shahin et al., 1986)
Trifoliate orange	*Poncirus trifoliata*	A	(Kaneyoshi et al., 1994)
Turf grasses: creeping bentgrass; Redtop	*Agrostis alba, Agrostis palustris*	P B	(Asano and Ugaki 1994) (Hartman et al., 1994)
Walnut	*Juglans regia*	A	(McGranahan et al., 1988), (Leslie et al., 1990)
Watermelon	*Citrullus vulgaris*	A	(Choi et al., 1994)
White clover	*Trifolium repens*	A	(White and Greenwood 1987), (Voisey et al., 1994)
White mustard	*Sinapis alba*	A	(Hadfi and Batschauer 1994)
White spruce	*Picea glauca*	B	(Ellis et al., 1993)

(A) = Agrobacterium mediated transformation; (B) = Ballistic Methods; (P) = or Direct Gene Transfer to Protoplasts.

Several approaches have been used to produce transgenic plants. They areas follow:

AGROBACTERIUM MEDIATED GENE TRANSFER

For genetic transformation in plants *Agrobacterium tumefaciens* (a soil bacterium) is used. It transfers T–DNA into plant genome. Foreign genes are inserted in T–DNA through *Ti*–plasmid, co–transferred and integrated in the host genome. Only some sequences of T–DNA in *cis*–position are required for its transfer. By deleting T–DNA encoded genes and adding selectable markers a *disarmed plasmid vector* is prepared without disturbing endogenous balance. Disarmed strains with *Ti*–plasmid can be used to produce transgenic plants.

Plant–Transformation Vectors

These vectors are designed on the basis of naturally occurring gene–transfer mechanisms. They are widely employed in plant molecular biology and in the genetic engineering of plants (Fraley *et al.*, 1985, 1986; Hiel *et al.*, 1997; Tyagi *et al.*, 1999). Chilton (1993) described these vectors as *poor man's vector.* They are used for gene delivery to higher plants.

Advantages of Vector

1. They are convenient and efficiency.
2. They allow the transfer of large segments of DNA with minimal rearrangements.
3. They are the integration of small numbers of copies of genes into plant chromosomes (Hamilton *et al.*, 1996; Klee *et al.*, 1987; Zambryski, 1992).
4. Foreign genes delivered by this method are usually transmitted to progeny plants in a Mendelian manner.
5. In the *Ti*–plasmid, the genes responsible for the transfer of T–DNA are clustered together in a region. The transfer process is active even when the virulence genes and the T–DNA are located on separate replicon in an *A. tumefaciens* cell.
6. In the widely used binary system, a *Ti*–plasmid services as helper; providing the virulence functions, and artificial T–DNA, which contains a selectable marker

gene and other genes of interest, is placed on a second, small plasmid, which is often referred to as binary vector (An *et al.*, 1998).

7. In dicotyledonous plants the systems of *Agrobacterium* mediated transformation have been well established for many. However, monocotyledonous plants (cereal/grasses) were originally outside, the host range of *A. tumefaciens.*

TABLE 9.2: References on Selectable Marker Genes Used in Transformation of Crop Plants

Species	npt2	hpt	Bar
Wheat	(Nehra et al., 1994)	–	(Vasil et al., 1992).
Barley	(Ritals et al., 1994)	–	(Wan and Lemaux 1994)
Maize	(D'Halluin et al., 1992)	(Walters et al., 1992)	(Gordon–Komm et al., 1990)
Rice (japonica)	(Battraw and Hall 1992)	(Shimamoto et al., 1989)	(Christou et al., 1991)
Rice (indica)	(Peng et al., 1992)	(Datta et al., 1990)	(Datta et al., 1992)
Oat	–	–	(Somers et al., 1992)
Sorghum	–	–	(Casas et al., 1993)
Sugarcane	(Bower and Birch 1992)	–	
Soybean	(McCabe et al., 1988) (Hinchee et al., 1988)	(Finer and McMullen 1991)	
Pea	(Davies et al., 1993)	(Puonti–Kaerlas et al., 1990)	(Schroeder et al., 1993)
Peanut	(Mansur et al., 1993)	(Ozias–Akins et al., 1993)	(Brar et al., 1993)
Alfalfa	(Shahin et al., 1986)	(D'Halluin et al., 1990)	(D'Halluin et al., 1990)
Rapeseed	(Fry et al., 1987)	–	(Block et al., 1987)
Arabidopsis	(Valvekens et al., 1988)	(Llyod et al., 1986)	(Sawasaki et al., 1994)
Cotton	(Firoozabady et al., 1987)	(Finer and McMullen 1990)	
Sugar beet	(Lindsey and Gallois 1990)	–	(Denecke et al., 1992)
Sunflower	(Schrammeijer et al., 1990)	–	
Potato	(Shahin and Simpson 1986)	–	(De Block et al., 1987)
Tobacco	(Horsch et al., 1984)	(Waldron et al., 1985)	(De Block et al., 1987)
Tomato	(Shahin et al., 1986)	–	(De Block et al., 1987)
Lettuce	(Michelmore et al., 1987)	–	
Carrot	(Pawlicki et al., 1992)	–	
Cucumber	(Trulson et al., 1986)	–	
Melon	(Fang and Grumet et al., 1990)	–	
Apple	(James et al., 1989)	–	
Citrus	(Moore et al., 1992)	–	
Poplar	(Klopfenstein et al., 1991)	–	(De Block et al., 1990)

GENE CLONING

Gene cloning is a technique, where an isolated single copy of a gene is cloned to obtain its indefinite identical copies. In gene cloning, we introduce a vector containing the desired gene into the host cell. Cloning can be done in bacterial host (e.g. *E. coli*) or in eukaryotes (e.g. yeast). Basically, cloning involves four steps:

The vector DNA is cleaved with one or more restriction enzymes;

- The DNA to be cloned, the target or insert is joined to the vector, generating a recombinant molecule;
- The recombinant DNA molecules are introduced into the host bacterial cell. The host cell is said to be transformed by the introduction of such vector molecule;
- Transformed colony is selected and amplified.

Another method of amplifying the desired gene is PCR, which offers this possibility in test tube. The two methods PCR and gene cloning are complementary techniques. Molecular cloning of a PCR product is now commonly used to isolate and manipulate genes.

Creating Recombinant DNA

Recombinant DNA molecules are produced by inserting a foreign DNA segment into the DNA molecule of a vector. This involves joining of the broken ends of vector DNA with the two ends of sequence to be cloned. These recombinant DNA molecules (also called as rDNA molecule) are subsequently cloned. There are three methods of joining fragments *in vitro*. The first uses ability of DNA ligase to join the cohesive ends produced by certain restriction enzymes. The second depends on the ability of T4 ligase to form phosphodiester bonds between blunt–ended fragments. The third utilizes the enzyme terminal deoxynucleotidyl transferase to synthesize homopolymeric tails at the ends of fragments. The three enzymes and their mechanism of joining DNA are separately dealt below.

DNA Ligase

The sticky end created by restriction enzyme has the nick after a few base pairs in opposite strands. DNA ligase repairs these nicks and forms intact duplex. The optimum temperature for ligation of nicked DNA is 37 °C, but at this temperature the hydrogen–bonded joint between the sticky ends is unstable. Thus temperature is kept in range of 4 to 15 °C.

This technique has the advantage of regenerating two restriction sites in recombinant DNA. Thus, the foreign DNA segment can be retrieved from cloned DNA copies by cleaving again with the same enzyme. The disadvantages within this technique include:

- The cohesive ends of foreign DNA may join end to end and self–circularize before getting inserted in vector.
- The recognition site, particularly in sequence to be cloned, may not lie at a convenient position so that sometimes only a part of gene is inserted.

T4 DNA Ligase

The *E. coli* DNA ligase can not catalyse ligation of blunt ends except under special reaction conditions of macromolecular crowding. Such blunt end ligation can be done with T4 DNA ligase. T4 DNA ligase can join the broken blunt ends of two DNA molecules irrespective of the sequences. The disadvantage of this technique is that any two broken ends may join including those belonging to same DNA molecule. However, blunt–ended ligation is a useful way of joining together DNA fragments, which have not been produced by the same restriction enzyme and therefore probably have incompatible sticky ends. These ends are removed prior to ligation using the enzyme S1 nuclease, which digests single stranded DNA.

Blunt ligation is most usefully applied where blunt–ended fragments are joined via *linker* molecules (described later). The linker molecule can be ligated to both ends of foreign DNA to be cloned, and then treated with restriction enzyme to produce sticky ended fragments. The vector molecule is also cut with same restriction enzyme. Finally, the foreign DNA is ligated to vector via linker created sticky ends. These linkers also help in excising and recovering the foreign DNA (gene of interest) after cloning and amplification in host bacterium. Here a brief account on various types of linkers is given.

Linkers

They are self complementary synthetic oligonucleotides which contain restriction endonuclease recognition sites. They may be used to clone DNA fragment into plasmids or other vectors.

These blunt ended linkers are first phosphorylated using polynucleotide kinase and labeled or unlabelled ATP molecules. These are then ligated to the blunt ended DNA fragment using T4 DNA ligase. The DNA fragment is finally treated with restriction enzyme to generate sticky ends and ligated into vector which has been cut by the same enzyme.

Stop Linker

They are 5'–phosphorylated double–stranded synthetic oligo–deoxynucleotides which can be cloned into an expression vector to ensure transcription termination. Such linkers contain stop codons in each reading frame from both directions. Each stop linker contains an SspI restriction site, which can be used to screen clones for linker insertion.

Adaptors

They are linkers which contain more than one restriction enzyme recognition sites. They are used to insert new restriction sites into cloning vectors or to join restriction fragments to vectors with incompatible cohesive ends.

Double–Linkers

In some plasmid vectors like pUC8, duplex cDNA molecules can be cloned. This is done by double–linker approach, in which different linker molecules are added to the opposite ends of the cDNA. This is an advantageous method as:

- Orientation of inserted DNA is fixed.
- Vector re-closure without insertion of foreign DNA can be done.
- Efficient for cloning of cDNA obtained from rare mRNA molecules.
- Better than homopolymers where expression from vector–borne promoter is desirable.

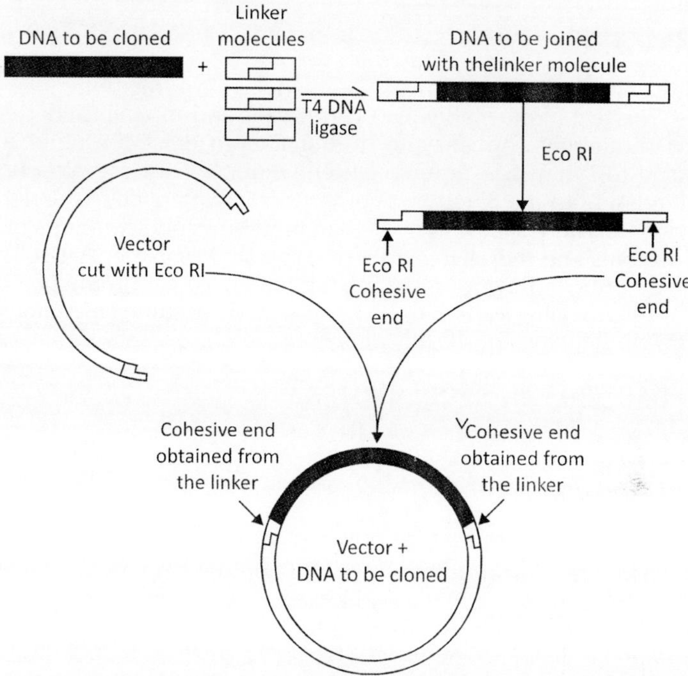

FIG. 9.1: Linker molecules containing Eco RI cleavage site are joined by T4 DNA ligase to both the ends of blunt–ended DNA–to–be–cloned. Cohesive ends are then generated by cutting with Eco RI. This DNA is then inserted into a vector molecule cut by the same restriction enzyme.

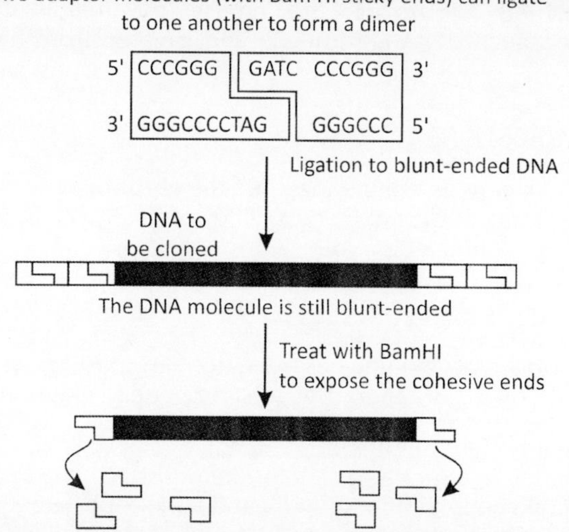

FIG. 9.2: Diagram to explain the problems associated with the use of adoptor molecules. A. Two adaptor molecules with Bam HI cohesive ends have ligated to each other to form a dimer. B. Such dimer of the adaptor molecules can ligate to a blunt–ended DNA molecule. The DNA molecule thus obtained is still blunt–ended. C. The cohesive ends are recreated by treatment with Bam HI.

Terminal Transferase

Besides adding cohesive ends or linkers, blunt ended DNA fragment and vector can also be ligated. This technique makes use of producing and annealing complementary homopolymer sequences. An enzyme, terminal transferase (purified from calf–thymus), is used to synthesize homopolymeric extensions. Using precursor dGTP, poly dG is added at both the 3' cut ends of vectors while poly dC is added to 5' cut end of foreign DNA. The vector and clone can then be joined by annealing the poly dG with poly dC tails and then ligating them using DNA ligase enzyme. The cloned sequence can be retrieved by using Pst I restriction enzyme. The method works out the chances of re-annealing between the two cut ends of same DNA molecule or of two similar DNA molecules, thus is advantageous.

CLONING IN BACTERIA AND EUKARYOTES

Cloning in Bacteria

After ligation of target into vector, the recombinant DNA is multiplied in a host. Various bacterial strains are available as host. Most common prokaryotic host is *E. coli*, which is also termed as **workhouse of genetic engineering**. In this technique the recombinant DNA is subjected to *in vitro* packaging and used to transform bacteria for its amplification. The host bacteria are modified and selected on the basis of high transformation efficiency. The type of bacterial host is selected on the basis of plasmid vector. Various bacterial hosts commonly employed are:

- Gram positive bacteria e.g. E. coli: host for plasmids only.
- Gram negative bacteria e.g. Bacillus subtilis: host for plasmid, phages and cosmid vectors.

Chimeric DNA is inserted into bacterial cells to create transformed bacterial cells. Colonies of transformed cells are selected and used for multiplication in suspension cultures. The recombinant DNA is retrieved and isolated from the host cell when required.

Cloning in Eukaryotes

E. coli is workhouse of genetic engineering but, the chromosome structure and gene expression in prokaryotes and eukaryotes differ radically. Therefore, *E. coli* offers limitation to its use in cloning eukaryotic genes. The split genes can not be correctly expressed in prokaryotic host as latter do not have introns and thus splicing of mRNA can not take place. Therefore, eukaryotic cells are used to clone and express eukaryotic genes.

Among eukaryotes, DNA cloning has been done in yeast, mouse and some plant species. Yeast has following attractions as host for genetic engineering:

- Short doubling time (90 minutes).
- Can be maintained in either a haploid or diploid state.
- Can be grown on simple defined media and auxotrophs can be selected.
- The wall can be removed and foreign DNA entry can be facilitated with PEG and CaCl2 mixture.
- Confers extra–chromosomal replication of chimeric DNA just as plasmids in E. coli.

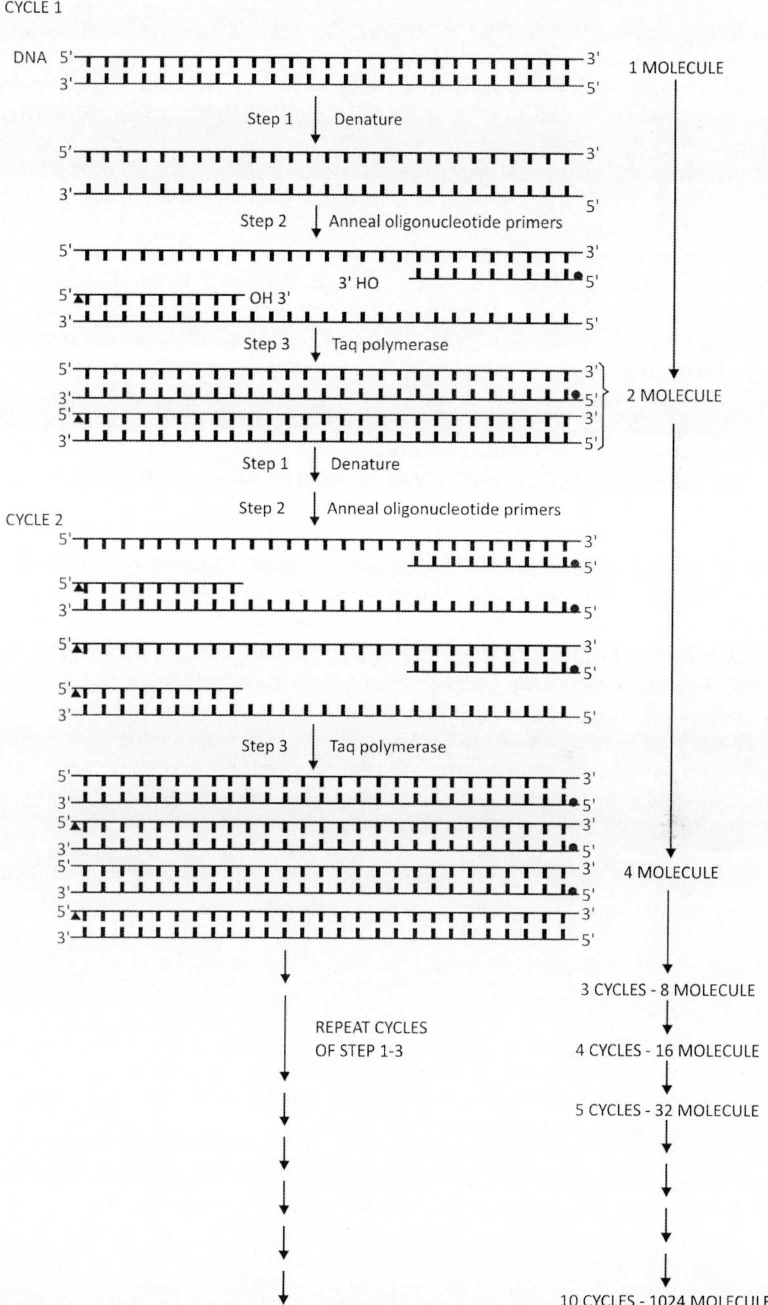

FIG. 9.3: Diagram illustrating the use of PCR to amplify DNA molecules in vitro. Each cycle of amplification involves three steps: (1) denaturation of the genomic DNA being analyzed, (2) annealing of the denatured DNA to chemically synthesized oligonucleotide primers with sequences complementary to sites on opposite sides of the DNA region of interest, and (3) enzymatic replication of the region of interest by Taq polymerase.

The simplest kind of yeast vector contains a selectable marker such as TRP1, joined to plasmid vector. TRP1 encodes an enzyme in the tryptophan biosynthetic pathway and when introduced into TRP^- cell by DNA transformation, it renders the cell capable of growth in absence of added tryptophan. Vector used is a hybrid or shuttle type.

When transformed into yeast, such a vector integrates into the yeast genome by recombination and makes the transformed yeast selectable on the basis of growth in absence of added tryptophan.

Yeast is becoming an important organism for the commercial production of medically important proteins such as viral vaccines. It is of great value as an organism for basic research, for producing heterologous proteins and as a vehicle for cloning large segments of DNA. For some purpose, however, it is essential to be able to transform mammalian cells. While yeast and mammals share many basic properties, there are important different cells in the precise mechanics of gene expression. Mammalian RNA transcripts are not spliced in the yeast cell, nor the mammalian proteins are glycosylated correctly.

In animal cells like mouse cells, special animal viruses are used as cloning vehicle. Simian virus 40 (SV40) with foreign DNA can be transcribed and translated in mouse kidney. Retroviral vectors are used for transformation of mammalian cells. These animal viruses comprise single stranded RNA genome, gene of interest and G418R gene for resistance against protein synthesis inhibitor G418.

AMPLIFICATION OF DNAS BY POLYMERASE CHAIN REACTION (PCR)

The major scientific development of year 1989, as named by prestigious journal 'Science' was **Polymerase Chain Reaction** (PCR). Kary Mullin is the inventor of PCR, who works in Cetus Corporation, a biotech company, of United States. Since its introduction, the polymerase chain reaction has revolutionized the method of DNA analysis in both research and clinical laboratories.

The PCR, is an extremely powerful new procedure that allows one to make million fold copies of a selected DNA sequence in a genome. PCR can be used to clone a given DNA sequence *in vitro* without using living cells during the cloning process. However, the procedure can only be applied when the nucleotide sequence of at least one short DNA segment on each side of the 'region of interest' is known. The PCR procedure makes use of synthetic oligonucleotides complementary to the known sequences spanning the region of interest to fetch *enzymatic amplification of this segment of DNA in the test tube*.

PCR may eventually replace gene cloning technique. This technology is much easier and requires much smaller quantity of DNA (in ng) whereas for cloning, DNA needed in microgram (μg) quantities.

Procedure

The PCR procedure involves three steps, each repeated many times to produce cycles of amplification.

- **Step 1:** The genomic DNA containing the sequence to be amplified is denatured by heating.
- **Step 2:** The denatured DNA is annealed to an excess of the synthetic oligonucleotide primers.
- **Step 3:** DNA polymerase is used to replicate the DNA segment between the sites complementary to the oligonucleotide primers. The primer provides the free 3'–OH required for covalent extension; the denatured genomic DNA provides

the required template function. These three steps (denaturation, primer binding and DNA synthesis) represents a single PCR cycle.

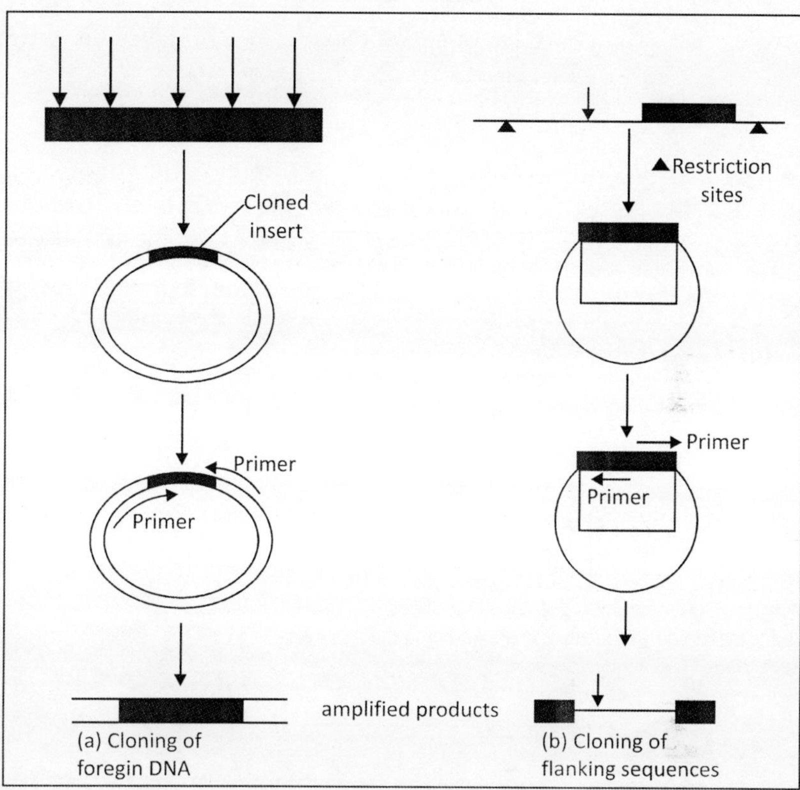

FIG. 9.4: Inverse PCR (a) amplification of genes using flanking vector sequences as primers; (b) amplification of sequences flanking the gene, using sequences at the two ends of the gene as primers.

The products of the first cycle are replicated for further amplification. This reaction can be performed many times to supply unlimited copies of amplified DNA. Repetition is made possible by regular denaturation of freshly synthesized double stranded DNA molecules by heating it to 90–98 °C.

At this temperature, the two strands separate. Once the double stranded DNA is made single stranded, it is mixed with two primers and is cooled to 40–60 °C.

This allows the primers (which are in excess) to bind to their complementary strands through renaturation followed by replication with DNA polymerase. The cycle is repeated many times until the desired level of amplification is achieved.

PCR for Confirming the Presence of Transferred Gene

When a chimeric gene is transferred to cultured cells or organisms, primers can be designed to conduct PCR for amplification of the gene sequence so transferred.

This technique allows the confirmation of transfer and maintenance of the gene of interest.

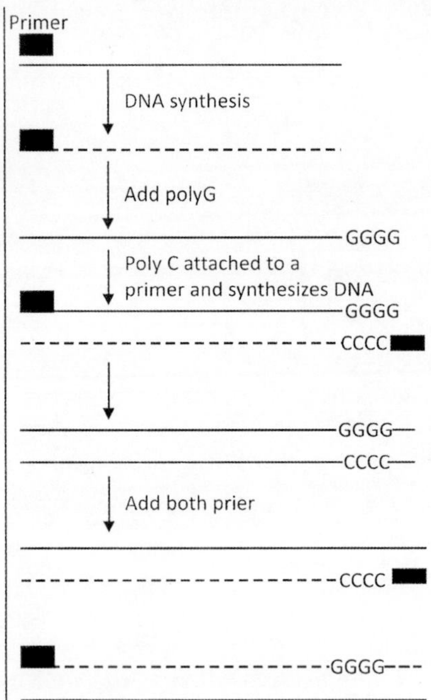

FIG. 9.5: Anchored PCR, using a single primer in the first cycle.

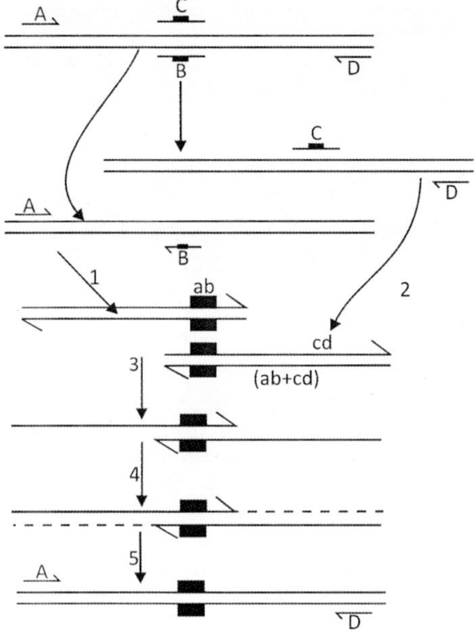

FIG. 9.6: Introduction of mutation at an internal site using overlap extension (two pairs of primers A & B and C & D are used. B and C represent internal sites and overlap each other).

Applications of PCR

During the last few years, with the improvement of PCR protocols and due to the availability of automatic thermal cyclers commercially, the applications of PCR have increased manifold. Some of the applications are listed here.

Study of DNA Polymorphism Using PCR

DNA polymorphism is amplified using PCR and can be used for study. Loci either with known DNA sequence or at random sites both can be amplified. PCR can be used to study RFLPs as well as RAPDs.

Loci with poly (A), poly (G), poly (TC) and poly (TG) etc., known as short tandem repeats or simple sequence repeats (SSR) can also be amplified with PCR. Some of the these SSR loci such (CA), or (GT)n occur in the human genome with many copies. SSRs can be utilized in preparing saturated molecular maps.

Gene Tagging Using PCR

PCR has also been utilized for developing molecular markers closely linked to specific genes of economic importance. For instance in tomato, 144 random primers were used to produce 625 PCR products from a set of near isogenic lines. Near isogenic lines (= NIL's) differ only for the presence and absence of a specific gene for *Pseudomonas* resistance. Atleast three of these PCR products were present in one and not in the other of a pair of NIL's showing linkage with *Pseudomonas* resistance gene (Pto). These PCR products could be mapped close to Pto.

DNA Fingerprinting Using PCR

DNA fingerprinting is use in forensic medicine for identification of criminals and disputed parentage. Some well characterized sequences of microsatellites are being used for designing primers. PCR allows amplification of DNA from individual hairs, stains of blood or seminal fluid having partially degraded DNA. Characterization of individuals with such small amount of DNA could not be done earlier.

PCR in Gene Therapy

In gene therapy experiments, the transfer and presence of a marker gene for neomycin resistance (NeoR) could be detected in the blood of patients, even after 60 days, although the frequency of marked lymphocytes gradually decreased necessitating a second transfusion of such marked cells. Similarly, ADA (adenosine deaminase deficiency) gene was transferred for therapeutic purposes and PCR was used to detect the presence of this gene. Thus PCR proves to be of immense help in monitoring a gene in gene therapy experiments.

PCR technologies thus provide shortcuts for many cloning and sequencing applications. These procedures permit scientists to obtain definitive structural data on genes and DNA sequences when very small amounts of DNA are available. One important application involves the diagnosis of inherited human disease, especially in cases of prenatal diagnosis, where very limited amounts of fetal DNA are available. A second major application occurs in forensic cases involving individual identities. No criteria can provide more definitive evidence of identity than DNA sequences; after all, other phenotypic traits are controlled by the expression of these DNA isolated from small blood samples, sperm, or even individual human hairs. Thus, PCR DNA printing experiments are likely to play a major role in legal cases involving disputed identity.

GENE TRANSFER TECHNOLOGY

Gene transfer is of basic as well as applied interest. It enables to study the structure and function of isolated genes via modification, reintroduction and subsequent analysis of their function in transgenic plants, cell lines or even cell populations. Besides these, transfer of specifically designed hybrid genes to plant genomes creates novel varieties with designed characters.

Gene transfer in plants can be accomplished by sexual and asexual methods. Asexual methods include:

- Non–specific gene transfer by protoplast techniques.
- Specific gene transfer methods.

In animals, gene transfer is done using egg and cultured stem cells or by specific gene transfer.

Sexual Method

Gene Transfer Using Pollen or Pollen Tube

Pollen grains are treated with heavy dose of X–ray which pulverizes chromosomes. The pollen, if able to pollinate, assures incorporation of changed chromosomes into daughter progeny. It is assumed that DNA can be taken up by the nuclei or reach the zygote through the pollen tube pathway. Both these approaches have been tried, though several problems exist in this method like presence of cell wall, nucleases, callose plugs in pollen tube etc.

Incubation of Dry Seeds, Embryos, Tissues or Cells in DNA

This method has been tried in many cases and expression of defined genes has been witnessed. In all the cases, plant cell walls work as efficient barriers, thus DNA uptake can be enhanced by PEG or electroporation or by any other device.

Asexual Methods

Non Specific Gene Transfer

Fusion of protoplasts to form somatic hybrid and partial hybrids are the known methods. The most direct method of gene transfer to plant cells is the uptake of plasmid DNA but at present its application relies on the availability of isolated genes. Agronomic traits are not coded for by single genes but rather by families of interacting genes. Thus, there is an interest in techniques which allow the transfer of large blocks of intact DNA encoding polygenic traits.

Protoplast membranes may be induced to fuse with nucleated or mitotically inactivated protoplasts to form hybrids; they may be induced to take up isolated nuclei or chromosomes; or they may be fused with bacterial spheroplasts, or liposomes.

Protoplast fusion may be carried out with three different methods; by using chemicals, by an electric field or by using a UV laser microbeam. Electrofusion is reported to be free of cytotoxicity, associated with chemical usage such as polyethylene glycol (PEG) or polyvinyl alcohol (PVA). Electrofusion appears superior to laser fusion as former is commercially available and affordable for a typical biological laboratory while this is not presently true for the laser microbeam. Furthermore, the number of hybrids that can be produced in a given time is much higher using electrofusion.

Foreign DNA can be directly introduced into mammalian cells. Firstly, DNA is precipitated with $CaCl_2$ and then this precipitate is mixed with the cells to be transformed. The DNA will be taken up by the cells. Inside the cell, the new DNA fragments get ligated to give a concatemer, which is then integrated into the nuclear DNA as one single piece. The integration occurs at random sites. In order to select the transformed cells several marker genes are also introduced along with the genes to be cloned. The marker genes will be linked with the genes of interest prior to integration into the host genome. One such marker is the neomycin resistance gene which would normally inhibit protein synthesis.

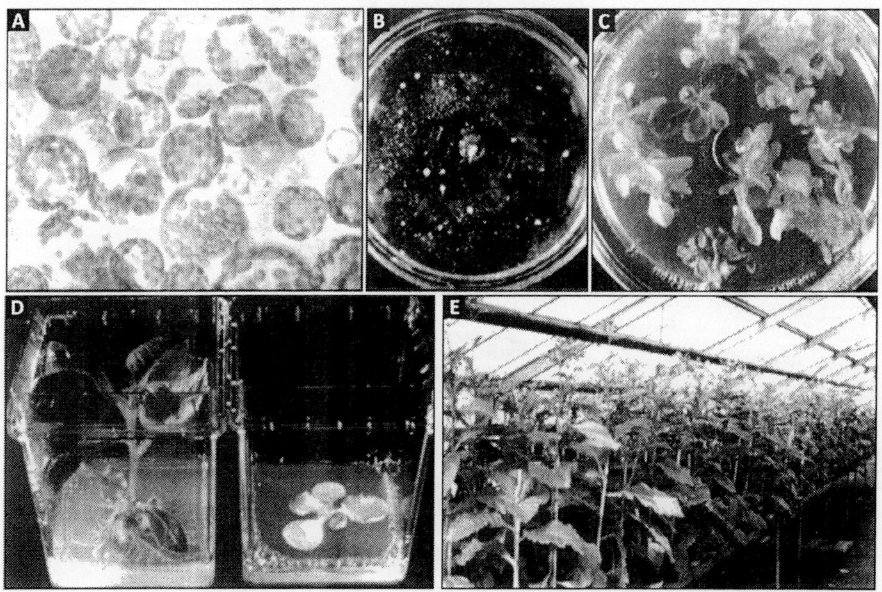

FIG. 9.7: Direct transfer to leaf protoplasts and regeneration of transgenic plants in tobacco (Nicotiana tabacum). (A) Freshly isolated leaf protoplasts of tobacco treated with polyethylene glycol and plasmid DNA bearing a chimeric npt2 gene as selectable marker; (B) selection of kanamycin–resistant transformed colonies from protoplasts in bead–type culture; (C) shoot differentiation from tobacco calli obtained from transformed protoplasts; (D) rooting of transformed (left) and untransformed (right) shoot from protoplasts on selective medium; (E) fertile transgenic tobacco plants obtained from protoplasts.

Specific Gene Transfer

Specific gene transfer in plants is mostly done with *Agrobacterium* species having natural biological delivery system. There are several other techniques developed recently for introducing cloned genes into plant cells. These are categorized as artificial delivery system transferring gene directly. Some examples of artificial delivery systems are direct DNA uptake, electroporation, fusion with liposomes etc. Unlike with *Agrobacterium* species mediated gene transfer (where cells with intact cell walls and tissue explants can be used), artificial delivery systems can be applied to protoplasts only.

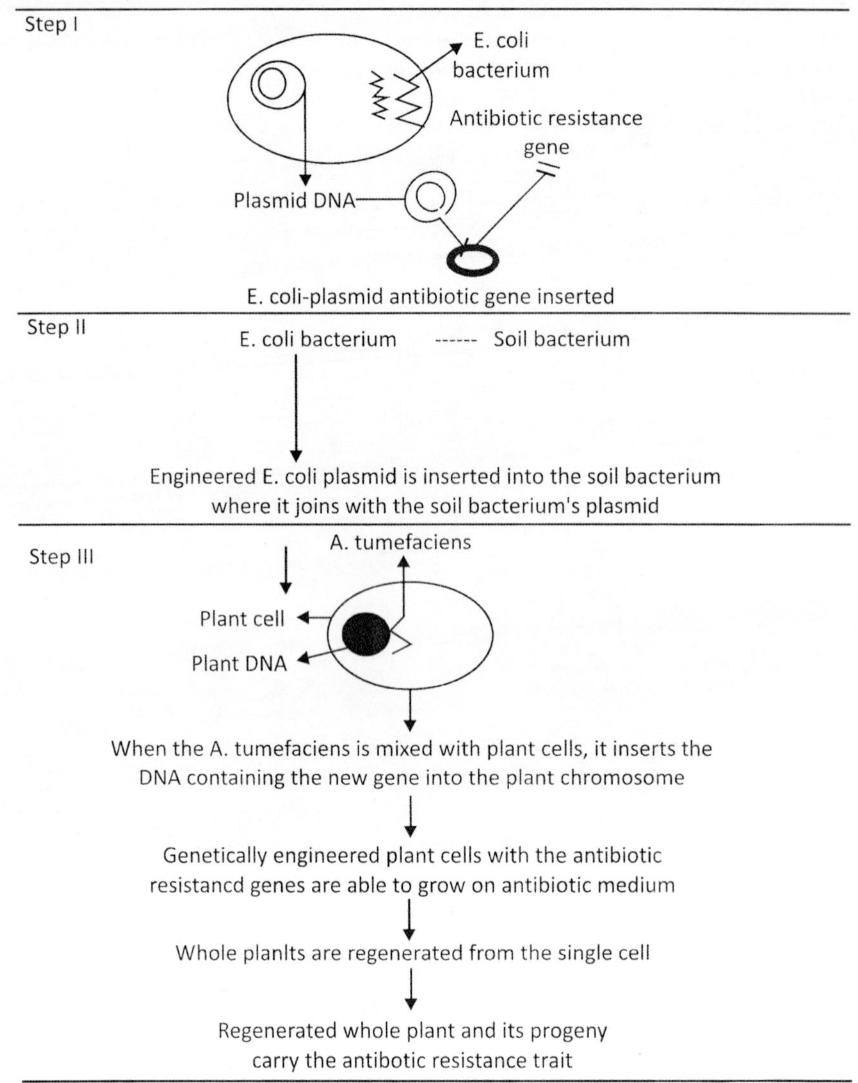

FIG. 9.8: Various steps (I–III) used in plant genetic engineering system.

The basic pre–requisites for successful gene transfer are:

1. selectable marker gene, which could be expected to yield a clear phenotype in transformed cell clones;
2. expression signals, which are independent of physiological state of host cell;
3. Precise linkage of the protein–coding region of the selectable gene to the expression signals. This will guarantee the expression of hybrid gene in the host genome;
4. DNA stretches on both sides of the hybrid gene for its integration into the host genome without disrupting the gene construct;
5. an absolutely tight selection protocol for presumptive transformants; and

6. Where of artificial delivery systems are utilized, the protoplast system taken should have routine high plating efficiency and plant regeneration capacity.

BIOLOGICAL DELIVERY SYSTEM

To date, *Agrobacterium tumefaciens* mediated transformations have been in wider practice and a much broader range of transgenic plants have been produced with this technique. It is a method of choice for routine production of transgenic species since it can transform regenerable tissues with relative ease.

Agrobacterium tumefaciens: An Important Vector for Genetic Engineering

Agrobacterium tumefaciens is the most important tool in the genetic engineering of plants (the Ti plasmid of the soil bacterium *A. tumefaciens*). This bacterium is the causative agent of crown gall disease of dicotyledonous plants only (unable to develop gall in monocotyledonous species).

The name refers to the galls or tumours that often form at the crown (junction between the root and the stem) of infected plants. Since the crown of the plant is usually located at the soil surface, this is where a plant is most likely to be wounded (e.g., due to a soil abrasion from the plant blowing in a strong wind) and infected by a soil blowing in a strong wind) and infected by a soil bacterium such as *A. tumefaciens*. However, *A. tumefaciens* can infect a plant and induce a tumour at any wound site.

In the normal disease process the bacterium enters the plant at the site of a wound. The infection often occurs at the crown, but it may involve an upper part of the stem, the petiole, or the leaf. The bacterium attaches to the wall of a cell and transforms the cell. Infected cells proliferate at the site of infection, forming a tumor or gall. The tumor can be fatal to young plants if it grows big enough to crush the vascular system, although often the tumor is merely debilitating and unsightly. Transformed cells also have the ability to grow in culture on a medium that lacks the hormones auxin and cytokinin. Finally, transformed cells produce opines, such as octopine and nopaline. These are chemicals that are formed from two amino acids. Argenine and alanine form octopine, and argenine and glutamine produce nopaline. Octopine and nopaline are not found in normal plant cells.

Infection of Agrobacterium tumefaciens

During the infection of a wound site by *A. tumefaciens*, two key following events occur:

- The plant cells begin to proliferate and form tumours.
- They begin to synthesize an arginine derivative called an opine.

The synthesized opine is usually either nopaline or octopine depending on the *A. tumefaciens* strain involved.

These opine are catabolized and used as energy sources by the infecting bacteria. Interestingly, *A. tumefaciens* strains that induce the synthesis of nopaline can grow on nopaline, but not on octopine, and vice versa. Clearly, an interesting interrelationship has evolved between *A. tumefaciens* strains and their plant hosts. *Agrobacterium tumefaciens* is able to divert the metabolic resources of the host plant to the synthesis of opines, which are of no apparent benefit to the plant, but which provide sustenance to the bacterium.

Genetic Information Needed for Crown Gall Development (Ti Plasmid)

The ability of *A. tumefaciens* to induce crown gall disease in plants is controlled by genetic information carried on a large (about 200,000 nucleotide–pairs) plasmid called the Ti plasmid for its tumor–inducing capacity.

Tumour–Inducing Principle

Armin Braun at Rockefeller University explained that the bacteria had provided a tumour–inducing principle. The nature of this principle remained obscure until the 1970s, when Jeff Schell and his co–workers at the State University of Ghent in Belgium found that virulent bacteria all contain a large circular DNA plasmid. When strains of bacteria lose the plasmid, they lose the ability to infect plants. When bacteria regain a plasmid, they regain virulence. Further experiments by Rob Schilperoort's group in Leiden, Holland, demonstrated that many of the genes that control

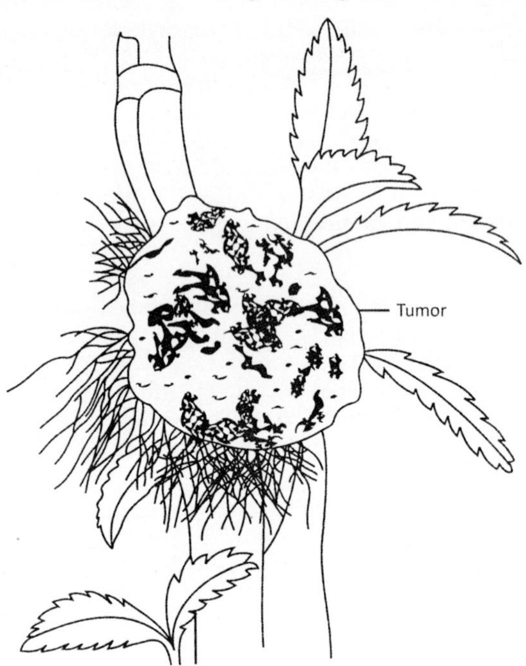

FIG. 9.9: A plant tumor induced by Agrobacterium tumefaciens.

Tumour characteristics (morphology, opine synthesis) lie on the plasmid. Finally, Mary–Dell Chilton *et al.* at the University of Washington in Seattle showed that a segment of the plasmid, the T (transferred) DNA, integrates into the genome of transformed cells. The plasmid (Ti plasmid) is thus identified as the tumour–inducing principle.

Two components, the T–DNA and the vir region, of the Ti plasmid are essential for the transformation of plant cells.

T–DNA (Transferred DNA)

During the transformation process, the T–DNA (for Transferred DNA) is excised from the Ti plasmid, transferred to a plant cell, and integrated (covalently inserted) into the DNA of the plant cell. The available data indicate that integration of the T–DNA occurs at random chromosomal sites; moreover, in some cases, multiple T–DNA integration events occur in the same cell. In nopaline–type Ti plasmids, the T–DNA is a 23,000–nucleotide–pair segment that carries 13 known genes. In octopine–type Ti plasmids, there are two separate T–DNA segments.

Some of the genes on the T–DNA segment of the Ti plasmid encode enzymes that catalyze the synthesis of phytohormones (the auxin–IAA and the cytokinin isopentenyl adenosine). These phytohormones are responsible for the tumorous growth of cells in crown galls. The T–DNA region is bordered by 25–nucleotide pair imperfect repeats, which are required in cis for T–DNA excision and transfer. The deletion of either border sequence completely blocks the transfer of T–DNA to plant cells.

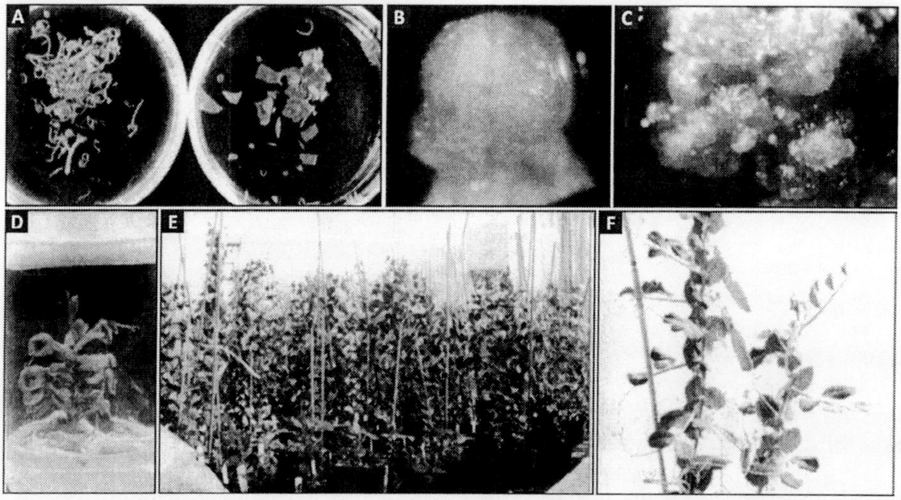

FIG. 9.10: Agrobacterium–mediated gene transfer and regeneration of transgenic plants in pea (Pisum sativum). (A) cocultivation of leaf and shoot explants of pea with A. tumefaciens strain bearing Ti–plasmid with chimeric hpt gene as stable marker; (B) Selection of transformed pea callus on hygromycin–containing proliferation medium; (C) induction of shoot formation from transformed pea callus; (D) in vitro growing transformed pea plant; (E) adult transgenic pea plants growing in soil under greenhouse conditions; (F) fertile transgenic pea plants obtained from Agrobacterium–mediated gene transfer.

vir Region (for Virulence)

The *vir* (for virulence) region of the Ti plasmid contains the genes required for the T–DNA transfer process. These genes encode the DNA processing enzymes required for excision, transfer, and integration of the T–DNA segment. *vir* genes can supply the functions needed for T–DNA transfer when located either *cis* or *trans* to the T–DNA. The *vir* genes are normally expressed at very low levels in *A. tumefaciens* cells growing in soil. However, exposure of the bacteria to plant cells or exudates from plant cells induces increased levels of expression of the operons containing the vir genes. Surprisingly, this induction process is very slow for bacteria, taking 10 to 15 hours to reach maximum levels of expression. The phenolic compounds like acetosyringone have been purified from plant exudates and shown to act as inducers of the *vir* operons. Thus, the induction process can now be studied *in vitro* by using these purified inducer molecules.

Disarmed Ti Vectors

When the T–DNA region of the Ti plasmid of *Agrobacterium tumefaciens* is transferred to plant cells and becomes integrated in plant chromosomes, the potential use of *A. tumefaciens* in plant genetic engineering seems to be obvious. The foreign genes can be introduced into the T–DNA region and, hopefully, these genes can be transferred to the plant with the rest of the T–DNA segment. In fact, this works very well. All the available data indicate that any DNA inserted any where between the T–DNA border sequences is transferred to plant cells and integrates in plant chromosomes with the rest of the T–DNA. The problem is that the transformed plant cells that have received the T–DNA from a wild–type Ti plasmid lose their normal control of cell division and form tumours. This oncogenicity of T–DNA renders wild type Ti plasmids incompatible with the goals of most gene–transfer experiments.

Vectors

A shuttle vector is commonly used, which can replicate in *E. coli* (where recombinant manipulations can be easily handled), and then transferred into *A. tumefaciens* in preparation for transfer into plants. There are two basic modes of maintenance of the shuttle vectors in *Agrobacterium*:

- either by integration into Ti plasmid by recombination at a region of DNA homology
- or, by autonomous replication in trans to the Ti plasmid.

The former type of vector is referred to as cis or integrating vector while the latter is called a trans or binary vector.

A selectable markers is required to recognize the transformed cells and on other hand, suppress the growth of wild type cells. This helps in identification of transformed cells, which are in small proportion.

Limitations: *A. tumefaciens* based gene transfer is hampered by host range limitations, which are especially disturbing when aiming at gene transfer into cereals or other graminaceous monocots.

Artificial Delivery Systems

Escape from host range limitations can be made by transferring the genes *directly* without using any biological vectors, into plant cells. Cell wall constitutes an effective barrier for DNA molecules to pass; naked plant cells, the 'protoplasts', have been found ideal experimental system to study vectorless gene transfer. Various artificial delivery systems are given here.

GENE TRANSFER USING PARTICLE BOMBARDMENT TECHNIQUE

The first genetically engineered plants were produced in middle 1980s (Horsch *et al.* 1985) and developed interest of genetic engineers in the production of transgenic plants for both basic and applied work. The major efforts were in the development and refinement of methods for transgenic plant production and also in the isolation and characterization of useful gene for introduction into plants. Through gene transfer, plants have already been produced that contain genes for disease and insect resistance, and modified fruit and grain quality. Particle bombardment genes transfer (or biolistics) is a method of transferring genes. Biolistics has been used to transform a large number of different plant species (reviewed by Christou 1994), as well as some animals, fungi, and bacteria.

The basic principle of particle bombardment for gene transfer is to use high velocity microprojectile to incorporate the introduced genes. The outer cell layers are penetrated in order to introduce genetic material into living cells.

History

Biolistics (a combination of *biological* and *ballistics*), is a method in which high–velocity microprojectiles are used to deliver exogenous genetic material into plant tissue. It was developed, and first described by John Sanford in collaboration with Ted Klein, Ed Wolf, and Nelson Allen (Sanford *et al.* 1987). The first used penetration particles was onion tissue, where a monolayer of large epidermal cells was used as the target tissue. By observing cytoplasmic streaming in these cells, this tissue could be easily evaluated for viability following bombardment. Onion epidermal cells, transformed with the chloramphenicol acetyl transferase (CAT) gene were the first transiently transformed cells produced using this method.

The first stably transformed plants obtained using biolistics were reported by Christou *et al.* (1980) and Klein *et al.* (1988) using soybean and tobacco respectively. Since these initial reports of successful transformation, a broad range of species has been transformed including bacteria (Smith *et al.* 1992), fungi (Toffaletti *et al.* 1993) and even animals *et al.* 1988). In addition, organelles such as chloroplasts (Boynton *et al.* 1988) and mitochondria (Johnston *et al.* 1988) have also been transformed using Biolistics.

This method of particle acceleration, has been described as the *particle gun, the gene gun, the bioblaster*, and the *microprojectile bombardment method.* It initially christened the biolistic method by its inventor (Sanford, 1988). Biolistics is a technique which utilizes instrumentation to accelerate DNA coated microprojectiles into cells, past the cell wall and cell membrane. The microprojectile is small enough (0.5–5μm) to enter the plant cell without too much damage, yet large enough to have the mass to penetrate the cell wall and carry an appropriate amount of DNA on its surface into the interior of the plant cell. The important aspects of biolistics are:

- It is a direct gene transfer method.
- It does not require biological vector.
- It has been used successfully to transform cells that were once impossible to transform by other means.
- DNA delivery appears to be genotype and tissue independent.
- With this technique, the microprojectile is shot through the physical barriers of the cell.
- There are no biological limitations to DNA delivery.
- Transformation via the biolistics method is simple and safe.

Applications of Biolistic

- The utility of the Biolistic method is broad and the technology can be used for basic studies on transformation.
- Modifications of gene expression.
- Understanding essential components of genes.
- The production of transgenic plants of economic value.

Particle Gun Design

Several particle gun have been designed in various laboratories. The basis of all of these designs is to coat DNA onto small dense particles and accelerate the particles towards a target tissue. The particles usually consist of either gold or tungsten.

Gold particles are chemically inert. They do not produce cytotoxic effects. They are more uniform in size than tungsten particles. However, the cost of gold particles may limit their accessibility and the more affordable tungsten particles are sufficient for most studies.

Tungsten particles are somewhat phytotoxic (RUSSELL *et al.* 1992a) and tend to be more variable in size than gold particles. Ideally the particles used for bombardment should have good initial affinity for the DNA, yet freely release it once in the cytoplasm or nucleus of the target cell.

Preparation of DNA–Coated Microprojectiles

- Wash gold or tungsten particles and mixed them with plasmid DNA.
- The DNA is bound on the particles using either ethanol or $CaCl_2$ precipitation methods.
- Spermidine may be added to the mixture, possibly protecting the DNA from degradation and/or altering its conformation.
- After precipitation, the particles may be washed, resuspended and either dried or stored on ice as an aqueous suspension until use.

Use of Macrocarrier

Many of the particle guns utilize a macrocarrier. They support or carry the particles and gun is accelerated along with the particles towards the target. The macrocarrier is usually retained by a stopping plate or screen before it collides with the target, whereas the particles continue along their course. In most cases, the particles are accelerated under partial vacuum in a vacuum chamber to reduce air drag.

The penetration of particle is controlled by modifying the intensity of the explosive burst. It is done by changing the distance that the particles must travel to reach the target tissue or by using different sized particles.

Gene Gun

The original gene gun described in the literature used DNA–coated tungsten particles in an aqueous slurry loaded on the end of a small plastic bullet, which served as the macrocarrier. The plastic macrocarrier was placed into a 0.22 caliber barrel in front of a gun powder cartridge. Once the cartridge was fired, the plastic bullet was propelled towards a solid stopping plate containing a small opening in the center. The microprojectiles continued through the hole towards the target tissue while the macrocarrier was retained. Both the stopping plate and the macrocarrier were discarded afterwards. The initial blast from the gunpowder cartridge necessitated frequent cleaning of the chamber and the target variations in results were common with this device, this first unit was used successfully by a number of different laboratories.

Modified Bombardment Devices

Helium–Modified Bombardment Device

The first important modification of the original device included the substitution of a helium blast for the gunpowder discharge (Sanford *et al.*, 1991). Helium pressure continually builds in a reservoir and is released using *rupture discs.* The release of helium produces a shockwave which travels to a second disc or macrocarrier. It holds the DNA–coated particles. The macrocarrier carries the DNA coated particles a short distance into a stopping screen. It retains the macrocarrier. The microprojectiles continue traveling, ultimately penetrating the target tissue which is held in a partial vacuum. The PDS–1000/He is the most commonly used device.

Accel Particle Gun

This gun uses a high voltage electrical discharge to vaporize a water droplet which produces a controlled shock wave (Mccabe and Christou, 1993). The device is constructed so that the initial shock wave is reflected to produce secondary shock waves. This in turn accelerates a Mylar sheet coated with particles. The Mylar sheet is accelerated towards a retaining screen, stops the carrier and allows the particles to continue onto the target tissue. Particle speed, and the resulting depth of particle penetration are accurately modified by adjusting the intensity of the voltage passing through the water droplet. It is the most efficient particle bombardment device in current. However, it is not commercially available.

Particle Inflow Gun

It is a costly device due to manufacturing expenses. This created a need to develop more affordable, simple, yet safe and effective devices such as the **Particle Inflow Gun** or PIG (Finer, *et al.*, 1992). With this device, a solenoid controlled by a timer relay is used to generate a burst of low–pressure helium. The particles, supported by a screen in a reusable syringe filter unit, are accelerated directly in a stream of helium without the need for a macrocarrier. The target tissue is held in a chamber under a partial vacuum. To minimize the localized impact of the particles and to aid in dispersal, a nylon mesh baffle may be placed over the tissue. For this device there is no need to purchase macrocarriers or rupture membranes. It is less cleanup and minimal down time between shots.

Microtargeting Device

This device accelerates small amounts of a DNA/particle mixture in a focused stream of high–pressure nitrogen. It was originally developed to allow precise delivery of particles to the shoot meristem. DNA is not precipitated on the gold particles but is used as a DNA/particle mixture. If the shoot meristem could rapidly give transgenic progeny.

Helios Gene Gun

In these guns targeted tissue held in a vacuum, and apparatuses are not very portable. The Helios Gene Gun is a hand–held device which uses low–pressure helium to accelerate DNA coated particles from the interior of a small plastic cartridge towards the target tissue. A spacer at the tip of the gun maintains optimal target distance and minimizes cell damage by venting the helium gas away from the tissue. They could not be easily used to transform target tissue that was sensitive to those conditions created by an evacuated chamber. The Helios Gene Gun can accommodate up to 12 loaded cartridges at once thus allowing for multiple firing of the device before reloading is necessary. The greatest utility of this device seems to be *in situ* transformation of animal cell and tissues (Sundaram *et al.*, 1996).

FIG. 9.11: PDS 1000 particle gun.

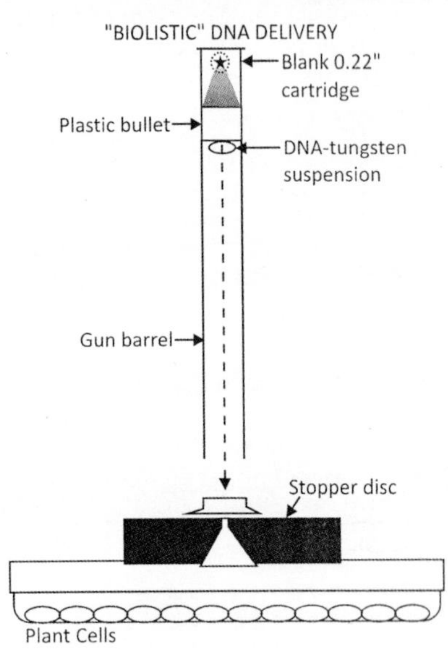

FIG. 9.12: The gene gun.

TABLE 9.3: Comparison to Agrobacterium–Mediated Transformation

Particle Bombardment Technique	Agrobacterium–Mediated Transformation
No need of the biological incompatibilities found when using this biological vector	Agrobacterium medicated transformation needs biological incompatibilities
A good utility for transformation of conifers, dicots and monocots	It is host specific ii relies on a host–parasite interaction in order to be successful. Agrobacterium appears to have a limited host range with a preference for some dicots while other dicots and most monocots and conifers are, in general more recalcitrant to Agrobacterium–mediated transformation. Agrobacterium tumefaciens is a gram–negative, soil–borne pathogenic bacterium that has the unique and natural ability to transfer part of its DNA to the cells of a host plant
It may not be the method of choice for all gene transfer work	It may be the method of choice for all gene transfer work

Components of the Introduced Plasmid DNA

The introduced plasmid DNA(s) typically contains the necessary plasmid backbone (allows growth and selection in the bacterial host), plant marker genes and the gene(s) of interest.

The *plant marker genes* can be selectable markers (which allow growth of transformed plant cells in the presence of a selective agent) and/or scorable markers (reporter genes). The genes themselves are composed of the promoter, coding region and terminator. Introns, enhancers and attachment regions are defined regions of additional DNA that can be used to enhance, modulate or stabilize gene expression.

FIG. 9.13: Microprojectile bombardment of cultured immature zygotic embryo and regeneration of transgenic plants in wheat (Triticum aestivum): (A) Cultured immature zygotic embryo of wheat before microprojectile bombardment; (B) microprojectile bombardment precultured wheat embryos after proliferation and transfer to regeneration medium in absence of selective agent; (C) phosphinothricin–resistant regenerating callus obtained from precultured immature zygotic embryos bombarded with microprojectiles coated with plasmid DNA bearing a chimeric bar gene as selectable marker; (D) methotrexate–resistant plantlet regenerated from precultured immature zygotic embryos bombarded with microprojectiles coated with plasmid DNA bearing a chimeric dhfr gene as selectable marker; (E) Transformed (center) wheat plantlet obtained from microprojectile bombardment using a chimeric bar gene as selectable marker and untransformed controls (left and right) after transfer to phosphinotricin–containing rooting (left); (F) in vitro growing methotrexate–resistant wheat plantlets from microprojectile bombardment using a chimeric dhfr gene as selectable marker; (G) fertile transgenic wheat plants from microprojectile bombardment to precultured immature zygotic embryos.

Selectable Marker Genes

The selection of proper marker genes of choice is critical for the successful recovery of stably transformed tissues. The selectable marker allows the transformed cells to proliferate in the presence of a selective agent while nontransformed cells either do not grow or multiply at a much reduced rate. Selectable marker genes may be either antibiotic or herbicide resistance genes. The resistance genes can act either by modifying the selective agent itself to inactivate it or by producing an insensitive from of the normal target of the selective agent.

Selectable Marker Gene: NPTII

- It is most commonly used selectable marker gene.
- It imparts resistance to the antibiotic kanamycin.
- The NPTII gene product is neomycin phosphorylation, which inactivates the antibiotic by phosphorylation.
- Other antibiotics that are commonly used with selectable marker genes in plants include hygromycin, phleomycin and gentamycin.

Selectable Marker Gene: Bar Gene

- It is most commonly used herbicide resistance marker gene.
- It is typically used with species of the Gramineae.
- The bar gene encodes phosphinothricin acetyl transferase, which inactivates phosphinothricin herbicides by acetylation.
- The bar gene can be used with bialaphos of glufosinate; again, selection of the proper herbicide must be empirically determined as with antibiotics.
- Other commonly used herbicide resistance genes impart resistance to glyphosate, sulphonyl ureas and dalapon.

Reporter Genes

1. β–glucuronidase (GUS) is the most widely used reporter gene for plant transformation.
2. The GUS gene has proven to be very useful in monitoring transient expression in a number of species.
3. Plant tissues could be evaluated at various times following bombardment by reporter genes whose expression can be monitored without causing tissue damage. Such a system could greatly reduce the effort and time involved in selection of stable transformants.
4. The firefly **luciferase gene** is a nontoxic reporter gene that has been successfully used to monitor gene expression in plants. After activity bombarded tissues are treated with luciferin (substrate) solution, luciferase activity is documented using either X–ray film.
5. Luciferase activity can also be assayed using extracts of plant tissue that are incubated with the luciferin substrate.
6. The detection techniques for luciferase severely limit the flexibility of using this reporter gene.

Introns

1. The introns are present between the promoter and coding region of a gene.
2. They significantly affect gene expression. Monocotyledonous species, in particular the Gramineae, show significant enhancement of gene expression when using introns.
3. Introns may act by stabilizing mRNAs, thus leading to greater amounts of translation products.
4. Introduction of different introns between the cauliflower mosaic virus 35S promoter and GUS gene caused significant enhancement of transient gene expression in maize. The degree of expression varied depending on the intron.

Promoter Analysis

The use of reporter genes such as GUS or GFP under control of strong promoters (e.g., CaMV 35S) allows for optimization of transformation efficiency of crop plants. Particle bombardment also provides a rapid and convenient means for evaluation of promoter strength, analysis of the function of different regions of promoters, and for examination of tissue specificity of gene expression. The promoter, either in its entirety or with various deletions/mutations, is fused to a reporter gene and introduced into the target tissue. This approach has successfully been used in a number of different species.

Codon Usage

The ability to introduce genes into plants from nonplant sources had led to the realization that genes from different organisms can be processed very differently with respect to their expression when compared with have native plant genes (Diehn, *et al.*, 1996). Unaltered DNA from nonplant sources may contain DNAs that are recognized by plants as signals for splicing or polyadenylation. Changes for codon usage resulted in a 300–fold increase in expression levels of a synthetic Bt gene.

Target Tissue

A variety of different plant tissues have been used as targets for particle bombardment–mediated transformation. Selection of the appropriate target tissue is dependent on the nature of the research. For rapid gene expression studies, various plasmid constructions can be introduced into different tissues and transient expression can be quickly analyzed to assess promoter activity without the production of stably transformed plants (IIDA *et al.* 1995). Almost any tissue can be used for transient expression studies as long the cell wall is penetrable by the DNA–coated particles. The choice of the appropriate target tissue, physiological state of the plant material, and pre– and post-bombardment treatments are critical to the successful regeneration of stable transformants.

Embryogenic Cultures

Embryogenic cell cultures have been very useful for the production of stably transformed plants. Particle bombardment of embryogenic maize and wheat cultures was used to generate the first transgenic plants of these species. Embryogenic cultures provide a reliable source of tissue for optimization of bombardment conditions and can be used for year–round production of transgenic plants.

Shoot Apical Meristems

Shoot meristem cells are most effective target for production of stable transgenics. Theoretically, introduction of DNA into would result in the production of chimeric plants, where the transformed cells would directly give rise to germ–line tissue and the introduced DNA would be passed on to the progeny. This approach could potentially avoid or reduce tissue culture manipulations but there are many problems with shoot meristem bombardment.

In sunflower, bombarded shoot apices of particles alone (no DNA) was done and then inoculated the tissue with *Agrobacterium*. In this case, particle bombardment was used as a method to wound the target tissue for transformation via *Agrobacterium* rather than for direct DNA introduction itself. Although the resultant plants were chimeric for the introduced gene(s), it is important to note that many of the transformation events were germ–line and transgenic progeny were recovered.

Other Target Tissues

Particle bombardment can be used to introduce DNA into any target tissue that is penetrable by the DNA–coated particles. The other tissues that have been subjected to particle bombardment include leaves, pollen, styles, cereal aleurone cells and tassel primordial.

FIG. 9.14: Direct gene transfer to embryogenic cell suspension–derived protoplasts and regeneration of transgenic plants in rice (Oryza sativa): (A) Embryogenic rice callus used for initiation of of suspension cultures: (B) established embryogenic cell suspension culture; (C) protoplasts isolated from embryogenic cell suspensions and treated with polyethylene glycol and plasmid DNA bearing a chimeric hpt gene as selectable marker; (D) selection in bead–type culture of rice protoplasts treated with (upper) and without (lower) plasmid DNA and polyethylene glycol; (E) In vitro growing plant regenerated from hygromycin–resistant callus obtained from transformed protoplasts; (F) fertile transgenic rice plants obtained from protoplasts.

Tissue Treatment

One treatment that greatly influences gene expression is osmotic conditioning. Osmotic conditioning can be both pre and post bombardment. Tissues are subjected to either partial drying in a laminar flow hood or cultured in a medium containing an osmotic agent such as mannitol and sorbitol at relatively high concentration (~0.4 M). Osmotic treatment can enhances both transient expressions as well as significantly increase stable transformation. The osmotica may act by causing plasmolysis of the target cells that lessens or eliminates extrusion of the protoplasm in cells that are penetrated by particles.

Other media additives are silver thiosulphate along with mannitol. They increase GUS expression in bombarded scutellare tissue of wheat, presumably by reducing the negative effects of wound ethylene. Inclusion of abscisic acid with an osmoticum (myo–inositol) increased transient GUS expression 1 day after bombardment of *Picea* embryogenic cultures.

The Fate of the Introduced DNA(s)

Although particle bombardment has been used for DNA delivery to plant cell for some time, it is still nuclear what happens to the majority of DNA that is introduced. Transient expression studies indicate that large numbers of cells can receive the particles resulting in approximately 10,000 GUS–positive cells per bombardment. In cells that transiently express the introduced DNA, particles are most often seen either in or directly adjacent to the nucleus.

Recombination

Plasmid DNA which is introduced into plant cells via particle bombardment undergoes recombination with other plasmid molecules as well as with the host chromosomal DNA. Recombination occurs whenever new gene arrangements are formed through exchange, elimination or insertion of DNA. Plasmid DNA can either insert cleanly into plant chromosomal DNA as a single entity or large linkages of plasmids can form and subsequently integrate (the full range of intermediate combinations is possible). Integration of a large chain of plasmids can result in a single insertion event with a high copy number of introduced plasmid DNA. With transformed event tissue obtained via particle bombardment, a single integration event rarely results in the introduction of a single plasmid. When large numbers of plasmids integrate into the same site, the plasmids may recombine with each other prior to integration into genomic DNA.

Although the exact fate of DNA introduced via particle bombardment is not clear, it appears that plasmid DNA molecules rearrange or recombine with each other to yield linkages of both whole plasmids and plasmid fragments.

Position Effect

The expression of the introduced gene(s) is dependent on the nature of the DNA as well as the final physical arrangement of the foreign DNA in the plant genome. The influence that adjacent, native plant DNA has on expression of the introduced DNA is called *position effect.*

If the foreign DNA integrates in a region of DNA that is active, the introduced gene may be expressed at relatively high levels. If the introduction occurs in a region that is not transcriptionally active, gene expression may be reduced.

Although the T–DNA from *Agrobacterium* is generally inserted into transcriptionally active regions, DNA that is introduced via particle bombardment show no known preference for insertion sites. Position effects control the variation in gene expression in different transformed clones. Methylation of introduced DNA has been shown to occur in cases where the introduced gene shows minimal activity. However, the precise association of methylation with position effects is still not clear.

Co–Suppression

A reduction in expression of a gene resulting from the introduction of multiple copies of the gene is called *co–suppression*. Co-suppression is useful in some situations when a reduction in expression of a native gene is desirable. Unfortunately, co–suppression is also observed as an undesirable product of particle bombardment–mediated transformation when multiple copies of the same gene are introduced.

Scaffold Attachment Regions

To stabilize gene expression in transgene plant tissue, scaffold attachment regions

(SARs; also known as **matrix attachment**/associated regions, MARs) have shown some utility. SARs are regions of DNA which were isolated based on their ability to bind to the nuclear scaffold. For transformation, the SARs are ligated to the flanking regions of the gene of interest. They apparently act by stabilizing or normalizing gene expression rather than enhancing overall activity (Breyne, *et al.* 1992). SARs from yeast (Allen, *et al.* 1993) and tobacco (Allen, *et al.* 1996) have helped to stabilize gene expression in plants while a human and soybean SAR were nonfunctional. Particle bombardment–mediated transformation efficiency could potentially be increased using SARs as SAR activity was found in Transformation Booster Sequence isolated from petunia. SARs appear to be especially useful in cases of high copy number DNA introductions, which would normally result in highly variable gene expression profiles.

Targeted Recombination Systems in Plants

In order to have better control on the expression of the introduced DNA(s), it would be beneficial to predict or *direct the site integration*. This can be accomplished through either *agrolistic transformation* or homologous recombination

Agrolistic Transformation or Homologous Recombination: For agrolistic transformation, plant tissues are bombarded with plasmids that contain some of the DNA excision components from the *Agrobacterium* gene transfer system. Approximately 20% of agrolistically transformed plant cells contained precisely defined DNA inserts that would be expected for *Agrobacterium*–mediated transformation.

For targeting using homologous recombination, genes could potentially be placed in certain areas of the genome based on sequence homology between the introduced and the genomic DNA. Although targeted or homologous recombination is fairly well–characterized in some organisms, it has not been well–studied in higher plants. Very little is known about native plant recombinases with most recombination studies in plants relying on information and genes that are available from nonplant systems.

Homologous Recombination Systems: Two homologous recombination systems have been evaluated in plants.

1. *Cre–lox system* from bacteriophage P1
2. *FLP/FRT system* from yeast

In these systems, a target site containing specific DNA sequences must be introduced specific recombinase, a second piece of DNA using transformation. Following activation with a specific recombinase, a second piece of DNA replaces the target DNA through homologous recombination.

GENE TRANSFER BY MICROINJECTION TECHNIQUE

Intranuclear microinjection is a direct physical approach that overcomes many of the biological and other obstacles. Originally developed for animal cells, it is now applied to plant cells also. The microinjection technique uses a fine capillary needle to deliver DNA precisely into cells (even nuclei of cells) in such a way that the cells are not killed but survive the treatment and can develop further into clones.

For transformation, microinjection systems are designed for speed and operational simplicity as these factors govern the total number of cells that can manipulated. To facilitate rapid injection rates, protoplasts with partially reformed cell walls can be attached to solid support (glass cover slips or slides etc.) with poly–L–lysine or artificially bound substrate without damaging the cells. After one day in culture, the cell walls of protoplasts are firm enough to provide support and protection yet they can be easily penetrated with micropipettes worth 0.5–10 μm diameter tip). Each protoplast is microinjected, segregated and transferred to optimal condition to obtain transformed cell line. An alternative method employs a blunt holding pipette to hold each protoplast under suction and segregating it directly into microdrop culture after microinjection.

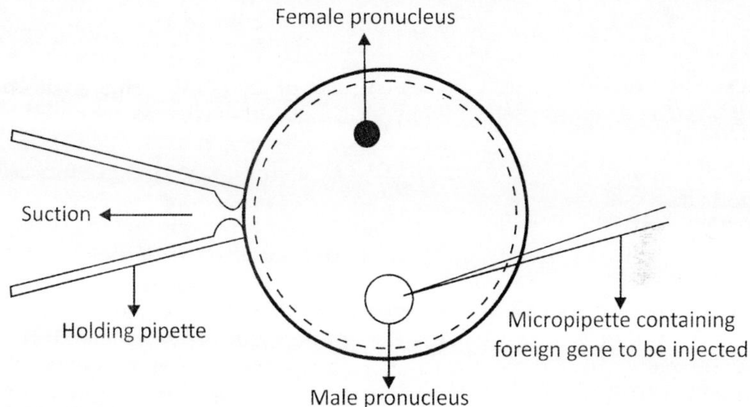

FIG. 9.15: Microinjection system.

Several advantages are associated with the microinjection technique. First of all, the amount of DNA delivered per cell is not limited by the technique and can be optimized. This improves the chance for integrative transformation. Secondly, the delivery is precise, even into the nuclei of the target cell, again increasing the chance for integrative transformation. Thirdly, the small structures can be injected containing only a few cells and with high regeneration potential, for example, microspore derived and zygotic proembryos. Lastly, it is a direct physical approach and therefore host–range independent.

However, there are some major drawbacks like, only one cell is targeted per injection and the handling requires specialized skill and instrumentation. Is hundreds of injections can be performed per working day, low transformation rate is yet not a serious disadvantage.

Applications of Microinjection Technique

Different applications of microinjection technique in plant science have been reported. Some experiments dealt with the introduction of fluorescent dyes into differentiated cells to study intracellular transport. Efforts have been made to microinject microorganisms into suspension cells. Attempts to transfer chromosomes or even cell organelles have been made. A common feature to these experiments is low numbers of manipulated cells.

DNA has also been microinjected in mice and fish. In mice, foreign DNA is injected into the mole pronucleus, residing in fertilized ovule. The injected zygotes are cultured *in vitro* to blastocysts and then placed inside a foster mother. This technique has been used to introduce rat growth hormone gene. Transgenic fishes have been obtained in many species like carp, cat fish, gold fish, salmon, Tilapia, rainbow trout and Zebrafish. As the male pronucleus is difficult to locate in fish zygote, the DNA is injected into cytoplasm of fertilized ovule within first few hours of after fertilization. As the fishes undergo external fertilization, transfer of embryos into foster mothers is not needed. Transformed embryos are instead hatched in trays. Various gene transferred into fish are human or rat growth hormone, chicken delta–crystalline protein, *E. coli* β–galactosidase gene, *E. coli* neomycin resistance gene, *E. coli* hygromycin resistance gene, winter flounder anti freeze protein (AFT) gene, and, cDNA of rainbow trout growth hormone. Cytoplasmic microinjection is a tedious and time consuming procedure and is unsuitable for mass gene transfer.

Technique

The macroinjection technique involves the use of injection needles with diameters far wider than cell diameters for delivering DNA into wound sites within tissues. However, at the same time it destroys those cells which are in direct contact with the injected DNA. To integrate, the DNA must gain entry to intact cells adjacent to the wound site.

There are a variety of problems associated with experiments that require DNA passage through single, double (in the case of pollen) or even many cell walls. The uptake of DNA across cell walls is a very rare event. Majority of DNA molecules applied get attached to cell walls before they have a chance to enter the cell. Penetration is further reduced by subsequent presence of cell wall. Moreover, DNA does not normally move from cell to cell. Thus, recombination occurs at relatively low frequency by this method. Microinjection has been mostly applied to ovule, embryos and other tissues

Lipofection

The use of liposomes for transformation of transfection system is called **lipofection**. Liposomes, as defined by Bongham, in 1965, are special type of lamellar phase in which water is self contained and the lipid molecules are disposed in bimolecular layers attached by their non–polar interfaces.

Liposomes possess many properties akin to those of biological membranes. They are easy to manipulate, their lipid contents can be varied at will and many substances may be trapped inside the interlamellar spaces. They are found to be ideal carrier systems and are now used in genetic engineering to deliver foreign DNA molecules into a desired cell.

DNA or RNA, encapsulated in liposomes, can be transferred into plant protoplasts by either direct fusion with the naked plasma membrane or through endocytosis of the liposome. Although it is efficient in terms of DNA encapsulation and transfer, the method once again requires protoplasts and its application is limited to species for which protocols are available for protoplast regeneration. Foreign macromolecules, for example impermeant fluorescent dyes, RNA and DNA, can be introduced into walled plant cells by an alternative strategy which involves encapsulation of the desired probe into liposome that can be pressure–injected into the vacuole. There are two advantages of this method. First of all, the liposome protects the DNA from being damaged by the acidic pH and proteases present in the vacuole. Secondly, when liposomes are constructed from material that is compatible with the lipids of the tonoplast, liposome–tonoplast fusion occurs (Madore *et al.*, 1986) and so the encapsulated DNA should be successfully delivered into the cytoplasm.

Liposomes would appear to offer the following advantages as vectors for gene delivery into plant cells:

- Enhanced delivery of encapsulated DNA by membrane fusion.
- Protection of nucleic acids from nuclease activity.

- Delivery into a variety of cell types besides protoplasts entry through plasmodesmata.
- Delivery of intact small organelles.
- Protect the genetic material from immunogenic reaction (liposomes are non–immunogenic).
- Composition of lipid can be manipulated to confer specific properties to liposomes.

The use of liposome for delivery of gene was first described for delivering polio virus RNA. Later liposomes were used in combination with polyethylene glycol for delivery of plasmid DNA into tobacco protoplast.

SILICON CARBIDE FIBER–VORTEX (SILICON WHISKERS)

Silicon carbide forms microscopically small needles (silicon whiskers) that can be used to perforate plant cell walls and plasmalemma of suspension cells or protoplasts. Advantages of the silicon carbide fiber–mediated transformation method over other procedures includes the ability to transform walled cells (thus avoiding protoplast isolation), relative ease of the procedure, and very low equipment costs.

Technically, the method is very simple: a mix of whiskers and plant cells in a small test tube is shaken. This method involves a mixture of plasmid DNA encoding a selectable or screenable marker gene, silicon carbide fibers, and tissue of the species to be transformed in an Eppendorf tube. Mixture is vortexed to facilitate cellular penetration by enhancing frequent and forceful intercellular collisions. DNA deliver in this system is presumably due to cell wall penetration by DNA coated silicon carbide fibers. The silicon carbide fibers apparently act as microinjection needles facilitating DNA delivery.

Only Eppendorf tubes and a vortex is sufficient equipment. This opens the way for DNA diffuse into the cells (Kaeppler *et al.*, 1992). Although, details of the mechanisms by which silicon carbide wiskers work is unclear, stable transformation with maize has been recently published (Frame *et al.*, 1995). The overall efficiency of the wiskers approach is between one–fifth and one–tenth that of conventional microprojectile shooting, but is advantage would be its *low tech* character, dispensing with any sophisticated or expensive instrumentation. However, currently, only maize has been transformed and a highly regenerating system was used. It remains an open question whether the whiskers approach will be useful in organized tissue.

A procedure has been recently developed for the injection of DNA into *Zea mays* and *Nicotiana tabacum* tissue cultures using silicon carbide fibres.

SONICATION

The latest method developed uses ultrasound to facilitate the uptake of nucleic acids into plant cells. By this technique, plant protoplasts and cells in suspension and intact pieces of plant tissue can readily be transformed. The protoplasts, suspension cells or small pieces of tissue are suspended in a few ml of sonication medium in a microcentrifuge tube. Plasmid DNA (and possibly carrier DNA) is then added and after rapid mixing the samples are ready for sonication.

The pulses of ultrasound are delivered by ordinary machines. The microtip (normally tapered) is immersed 2–3 mm into the suspension and pulses of selected intensity and duration are delivered. The cells are finally transferred to fresh growth medium.

There are two possibilities for permeabilization of cell and uptake of DNA.

Firstly the violent collapse of activation bubbles, generating high-pressure and high-temperature shock waves could possibly cause localized rupture of the plasmalemma and lead to uptake of exogenous solute. This may be followed by re-establishment of membrane integrity. The second hypothetical possibility may be the existence of a critical hydrostatic pressure at which the intrinsic membrane potential is sufficiently high to induce mechanical breakdown of the membrane (60–100 MPa) in a hyperbaric chamber. Consequently, it is possible that either the high oscillating pressure generated by the ultrasonic field and/or the high-pressure shock waves originating from collapse cavitation could produce such higher hydrostatic pressures that reversible membrane breakdown would occur.

This method has been applied for transferring plasmid DNA into protoplasts, suspension cells and tissues of sugarbeet and potato.

ELECTROPORATION

This method uses an electrical pulse to introduce DNA into cells. By exposure to brief electric shock, the cells become temporarily permeable to DNA. If the required DNA fragment is kept in direct contact with the protoplast membrane, the entry of DNA molecules into the cells is facilitated. Successful DNA transfer through this technique has been shown in bacterial, yeast, fungal, plant and mammalian cells. Among plants, successful gene transfer using electroporation method has been reported in protoplasts of tobacco, petunia, maize, rice, wheat and sorghum. In carrot, maize and tobacco protoplasts, bacterial CAT gene with different plant promoters has been introduced and efficiently expressed. Electroporation is applicable to both monocot and dicot protoplasts.

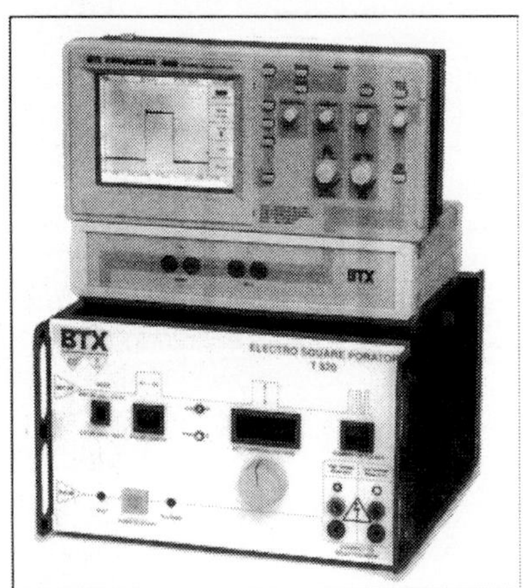

FIG. 9.16: Electroporation apparatus

It has been recently hypothesized that the mechanism of DNA entry into the cell during electroporation may be either due to unspecific membrane process, or it may be specifically mediated by permease. The activation of permease may be due to high electric field, and thus, may lead to thinning of membrane. Hole may be created as a result and stabilized by a favourable dipole interaction with the electric field. Various advantages of electroporation have been listed. It is a convenient, simple, fast and efficient method applicable to a wide range of plant protoplasts and involves low cell toxicity.

The main disadvantage of electroporation is its requirement for protoplasts and therefore a system where regeneration of these protoplasts to calli or whole plants have been established. Consequently, it may not be applicable to some plant systems. Recently an electroporation system using organized and easily regenerable tissues such as immature maize embryos has been developed.

POLY–CATION MEDIATED DNA UPTAKE

Direct DNA uptake by protoplasts can be stimulated by using hydrophilic, long chain polycations such as polyethylene glycol (PEG), poly L–lysine, poly L–ornithine or DEAE–dextran with or without metal ions such as Zn^{2+}, Li^{+}, Cs^{+}, Rb^{+}, K^{+}, Na^{+}, Ca^{2+} or Mg^{2+}. The technique is efficient and can be applied virtually to every protoplast system and for transfer of even small quantities of DNA.

Polyethylene–glycol is an extremely hydrophilic molecule and it removes 'free water' when placed in a solution. Free water indicates such molecules, which interact with charged (usually ionic) molecules soluble in water.

Thus, under high PEG concentration (15–25%), ionic macromolecules such as DNA can no longer stay in solution and get precipitated. The protoplast membrane as well as DNA is normally negatively charged, and therefore, charge repulsion is obvious. This repulsion limits the interaction between DNA and protoplasts. At very high concentration, PEG seems to minimize charged repulsion effects thus stimulating DNA uptake by endocytosis without any gross damage to protoplasts. Studies have been conducted with *Petunia, Nicotiana*, rice, maize etc. In these experiments, PEG was used in combination with pure Ti plasmid or calcium phosphate precipitated Ti plasmid mix with a carrier DNA. Mechanism of cross membrane transport in presence of PEG is still to confirm.

DNA–uptake may be enhanced as a result of the interaction of the polycations at one or more of the following levels:

- Protection of the exogenous DNA against nuclease degradation.
- Increase in the permeability of the plasma membrane.

DNA transformation by using the polycation "*polybrene*" (hexamidethrine bromide) is new, although similar techniques have been used for several years in the transformation of animal cells. The major advantages of the polybrene technique are the following:

- It is less toxic than other polycations
- The high transformation efficiency allows very small quantities of plasmid DNA to be used, thus it is relatively easy to introduce only one sequence of a genome. Even with quite high plasmid concentrations only one or a few copies of the plasmid are integrated and concatemers of integrated plasmid are rarely observed.
- Since no carrier DNA is used to integrate plasmid DNA into the host genome thus, the sequences surrounding the site of integration can be analyzed.
    - The transformation frequency appears to be independent of nature of plasmid (supercoiled or linear).
    - The polybrene method can be used for both stable and transient expression.

CA–DNA CO–PRECIPITATION METHOD

The basis of the Ca–DNA technique is the formation of a co–precipitate of plasmid

DNA with calcium phosphate, formed by adding $CaCl_2$ to a solution of DNA in phosphate buffer. On contact with protoplasts, the co–precipitate is transferred across the plasma membrane in a calcium requiring process; and in order to encourage the endocytotic uptake of the precipitate, the protoplast/Ca–DNA complex is treated with polyvinyl alcohol and high pH calcium. This method was first used for plant transformation by Hain *et al.*, (1985) using a tobacco leaf protoplast system.

ULTRASONICATION

Gene transfer by ultrasonication is the latest and simple method, where, plant protoplasts, cells in suspension and intact pieces of plant tissues can be transformed. The target cells are suspended in sonication medium in a microcentrifuge tube. Carrier DNA is added and microtip (generally tapered) is immersed in the sonication medium. Pulse of selected intensity and duration are delivered by the ordinary homogenizing machines. Transformed cells are finally transferred for selection.

There are two hypotheses for permeabilization of membrane by sonication method. First assumes that reversible localized rupture of plasmalemma results in is due to violent collapse of activation bubbles which results in generating high–pressure and temperature shock waves to plasmalemma. The second hypothesis predicts the existence of critical hydrostatic pressure responsible for reversible mechanical breakdown of membrane.

This method has been used in tobacco, wheat and sugarbeet. In tobacco, transgenic plants were obtained at 22% frequency with ultrasonication. Ultrasound transformation is unique among existing methods because of its versatility as plant protoplasts, cell in suspension and intact pieces of plant tissue can be readily transformed. This feature, alongwith the fact that the equipment for ultrasound transformation is simple, cheap and multifunctional should allow for substantial future applications.

UV LASER MICROBEAM

A new method of introducing exogenous gene materials into cells is laser microbeam irradiation. DNA can be incorporated into whole plant cells and has been successfully applied to tobacco. DNA enters the cell through a self–healing hole made with a laser microbeam.

There are several advantages to this method. A laser beam can be used inside of tissues, cells and organelles, thus increasing the rate of transformation and the speed of operation. Wide controllability of the laser beam intensity, flexibility in optical manipulation wide choice for laser wave length and pulse width are other advantages. It is also applicable to cells with hard walls. Its wide application in the field of cell biology is the ability to transfer cytoplasmic material. Finally, future versions of this system are expected to be more automated which should greatly increase output as well as the success rate.

This technique can be used for genetic manipulation of plant tissue, individual cells, protoplasts and organelles. Furthermore, it permits selective fusion in a of part of a subcellular organelle present inside a single living cell.

A number of techniques exist today for transferring genes into plant protoplasts and cells. Most of these techniques use protoplasts because the cell wall is not sufficiently permeabilized for DNA uptake. As protoplasts of many species are often difficult to regenerate efficiently into fertile plants it can not be applied to any system. These techniques can be preferably used as an analytical tool, for example, in testing the function of gene constructs.

Moreover, most of the systems require sophisticated and costly equipment. In addition, it does not allow the combination of both accurate quantitative estimates of gene expression and efficient generation of transgenic plants.

However, successful gene transfer has been reported in few systems using above techniques, and much research in the context is still in progress.

GENETIC TRANSFORMATION IN MONOCOTYLEDONOUS PLANTS

The initial development of system for genetic transformation of monocotyledonous was slow. Formation of tumors and production of opines in certain monocotyledons suggested that at least some monocotyledonous species might be susceptible to infection by *A. tumefaciens. Agrobacterium tumefaciens* (*Ti*–plasmid).

Factors Involved in *Agrobacterium*–Mediated Transfer of Genes to Monocots

1. This work is mainly done on cereals (particularly rice).
2. Induction of *vir* genes, one of the most commonly used techniques in the transformation is the addition of phenolic compounds such as acetosyringene to co-culture or bacterial cultures.
3. The cell division should be active.
4. Bacterial strains and vectors
5. Composition and quality of the culture medium.
6. High osmotic pressure.
7. Host genotypes, and
8. Acidic pH.
9. Selectable marker
10. Temperature of the room below 28°C.

TABLE 9.4: Genetic Transformation in monocotyledonous plants

Work	Reference
Transformation of monocots and cereals	Heii et al. (1997)
few transgenic rice plants by inoculating immature embryos with a strain of A. tumefaciens	Chua et al. (1993)
Reported a method for efficient production of transgenic rice plants from calli of japonica cultivars that had been co–cultivated with A. tumefaciens	Hiei et al. (1994)
The successful application of calli of japonica cultivars method to Basmati cultivars of indica rice	Rashid et al. (1996)
Achieved efficient transformation of maize by A. tumefaciens	Ishida et al. (1996)

Direct Gene Transfer Monocotyledonous Plants

These methods are used for direct DNA transfer in monocots (cereals). Various methods are:

Polyethylene Glycol (PEG) Medicated Gene Transfer

PEG acts as an important fusion agent in somatic cell hybridization. It is being used in obtaining transformed cells by delivering DNA into protoplasts (Davey *et al.*, 1980; Krens *et al.*, 1982). The foreign DNA in the form of plasmid DNA is inserted in plant protoplast resulting in the recovery of fertile regenerated tobacco (Paszkowski *et al.*, 1984).

Electroporation

Electroporation means the use of high voltage electrical pulses to a solution containing a mixture of protoplast and foreign DNA. The pulses reverse pores created in membrane by the short electrical pulses. Transgenic plants of maize have been obtained by this method (Rhodes *et al.*, 1988).

Microinjection

It is the injection of DNA into plants cells including cultured cells and intact plant organs to produce transgenic plants (Neuhaus *et al.*, 1987; Dela Pena *et al.*, 1987). For example, selected embryoids derived from microscope culture of *Brassica* are transferred for microinjection to a minute droplet of culture medium (Neuhaus *et al.*, 1987). Embryos were injected with **NPT– II** gene constructs. Injected DNA is considered to be transported by plants vascular system to the germ line.

Micro Projectile Bombardment

This method helps in delivery of DNA into cells by projective bombardment. Micro projectiles (small high density particles) are accelerated to high velocity by a particle gun apparatus. These particles acquire sufficient kinetic energy to penetrate cell membrane and carry foreign DNA into bombarded cells. Klein *et al.* (1987) used this method for *Allium cepa* epidermal tissues. Treated tissues showed varied levels of chloramphenicol acetyl transferase (CAT).

PROBLEMS INVOLVED IN GENE TRANSFER

Expression System

Adequate expression of gene is essential with high expression at right time and in a particular tissue. For high expression strong promoters like 35–S promoter of Cauliflower Mosaic Virus (CaMV) is used which has a wide host range. More specific promoters are required for tissue specific attachment regions (SAR's) are used to increase expression. SAR's act by forming loops in chromatin structure thereby reducing gene silencing in cells having limited number of transgenes. But high amount of m–RNA may not have corresponding amount of product.

Cellular Localization

Abiotic stress causes injury to membrane (by formation of reactive oxygen intermediates) affecting membrane bound process finally influencing metabolism of plant. It is desirable to locate membrane modifying, osmoprotectant producing or radical scavenging enzymes to defined organelles. Transit peptides are required to ensure correct localization of enzymes expressed. Typical transit sequence used to direct polypeptides to chloroplast is from small subunit of Rubisco (Schreier *et al.*, 1985; Allen *et al.*, 1996).

Proteolysis, Protein Folding and Prosthetic Group Acquisition

Short polypeptides after expression may undergo proteolytic degradation. This can be avoided by processing and folding of proteins by co–expression of chaperon proteins in transgenic plants especially in case of large enzymes. Oxygen radical metals like, Fe, Zn, Cu, Mn, Therefore, the metal ions should be present in adequate amount.

Precursor Availability

Amount of precursor has a great effect on product formation. Precursor can be extracted from environment or synthesized in cell from existing metabolic pathways.

Inhibitory Environments

Intracellular condition in transgenic plants (pH, temperature, ionic strength) affect foreign proteins. It is usually operated during varied environmental conditions. Therefore, the introduced gene should be taken from closely related plant.

Side Reaction of New Compounds

Endogenous enzyme activities may deplete the product pool or form toxic compounds. Therefore, analysis of transgenic plants using NMR is important for detection of any metabolic changes.

CHAPTER 10

Plant Tissue Culture: Principles and Methodology

Plant tissue culture is the technique of growing plant cells, tissues and organs in an artificial prepared nutrient medium static or liquid, under aseptic conditions. It has advanced the knowledge of fundamental botany specially in the field of agriculture, horticulture, plant breeding, forestry, somatic cell hybridization, phytopathology and industrial production of plant metabolites, etc.

IMPORTANCE OF TISSUE CULTURE TECHNIQUE

All the cells in an organism carry the same genetic information, yet show variations in expression. Our knowledge of cell and tissue cultures has been developing with full swing, specially in biotransformation, forestry, genetic engineering, morphogenesis, somatic hybridization, secondary metabolite production, hybridization, variety development and their conservation, maintaining pathogen free plants and rapid clonal propagation, totipotency, differentiation, cell division, cell nutrition, metabolism, radio biology, cell preservation, etc. It is now possible to cultivate cells in quantity, or as clones from single cells; to grow whole plant from isolated meristems and to induce callus or even single cell to develop into complete plant either by organogenesis or directly by embryogenesis *in vitro*.

The production of haploid through tissue culture from anthers or isolated microspores and of protoplasts from higher plant cells has served as the basic tools for genetic engineering and somatic hybridization. Tissue culture technique helps to propagate plants of economic importance such as orchids and other ornamental plants in large numbers by their meristem culture or by other *in vitro* methods. This provides them virus–free plantlets. Propagation of valuable economic plants through tissue culture based on the principle of totipotency (*every cell within the plant has potential to regenerate into a whole plant*).

In plant breeding, embryo, ovary and ovule culture as well as *in vitro* pollination have been employed to overcome morphological and physiological sterility and incompatibility. In recent years, plant tissue culture technique is in increasing use for producing haploids from anthers or isolated microspores, and of protoplasts from higher plant cells and the recognition of the potential of these materials in genetics and plant breeding. One of the most significant developments in the field of plant tissues culture during recent years are the isolation, culture and fusion techniques which have their special importance in studies of plant improvement by cell modification and somatic hybridization.

Plant tissue culture technique is a boon in the studies of the biosynthesis of secondary metabolites and provides an efficient means of producing economically important plant products (fine chemicals). Plant tissue culture also provides raw material to pharmaceutical, cosmetic and confectionary industries. Examples are berberine, ginseng, shikonin and vanilla.

Global Picture of Plant Tissue Culture

The potential value of plant tissue culture technology is being commercially exploited by various organizations all over the world. The Horticulture Industry responded very quickly to the micropropagation research. At present more than 400 million plants are produced through tissue culture in different parts of the world. There is a potential market for billions of US dollars per year worldwide for tissue culture products. There are more than 65 laboratories all over the world producing more than a million plants per year. The total production capacity of North American laboratories probably exceeds 150 million plants per year. The production in Asian countries is nearing 100 million plants which includes orchids, temperate and tropical flowering crops, foliage plants and plantation crops.

HISTORY OF PLANT TISSUE CULTURE RESEARCHES

The first works in the field of excised plant tissue cultivation date back to the end of the last and the beginning of this century and are linked with the names of the three outstanding scientists: Vochting, Rechinger, and Haberlandt. Vochting (1878) studied plant polarity, and attempted to grow small lumps of tissue *in vitro*. He demonstrated that polarity was present even in the plant cell itself. Rechinger (1893) placed segments of poplar stem and pieces of beet and candeloin roots on a wet filter paper, and observed callus formation. Since the start of 19th century people started thinking about the culture of plant tissues. Haberlandt (1902), a German botanist tried but invein to obtained plant tissues from the isolated mesophyll cells of *Lanium* on artificial Knop's medium *in vitro*. Though he did not succeed in his attempts, he pointed out towards a new technique– *Plant tissue culture.* In 1904, Hanning succeeded in culturing nearly mature embryos of *Raphanus* and *Cochlearia* to maturity. Realizing the complexity of nutritional requirements of younger embryos, van Overbeek and coworkers, for the first time, used **coconut milk** for *Datura* embryos and demonstrated its stimulatory effect. Subsequently, coconut milk and other substances were added to the nutrient medium (containing mineral salts, vitamins, amino acids and sugars) but it failed to promote the growth of cultured tissues. Lampercht (1918), Kundson (1919) and Nemec (1924) have made numerous attempts to find suitable media and optimal conditions for growing organs, tissues, and cells excised from whole plants.

Zoologists have also tried to cultivate mammalian tissues. The workers like Skvortsov (1886), Garrison (1907), Carrel and Burrows (1911), and Krontovskii (1917), developed the nutrient media for growing mammalian tissue. Zoologists like Czech (1927), Prat (1927) and other workers attempted to grow excised plant tissues on plant extracts.

Knop and Prdifer developed synthetic media. The approach of Mlliard (1921) is considered to be an original. He used segments of root and hypocotyl of young radish shoots. These tissues, possessing embryonic activity, grew in the culture, but the cell did not divide, and new tissues were not formed. The utility of embryo culture technique soon became apparent when Laibach (1925, 1929) reared hybrid embryos (*Linum perenne* x *L. austriacum*) from non–viable seeds to maturity. Impetus was thus provided for further work on other tissues as well, and prompted Robbins (1922) and Kotte (1922) to try root culture. By using a synthetic medium containing inorganic salts, pyridoxine, thiamin, nicotinic acid, an iron source, yeast extract and sucrose, White (1937) could maintain tomato root culture for almost 30 years (1934–1968).

TABLE 10.1: Land marks in plant tissue culture research*

Worker	Year	Contribution
G. Haberlandt	1902	Was first to culture isolated plant cells in vitro on artificial medium.
W. J. Robbins, W. Kotte	1922	Culture of isolated roots–organ culture.
P. R. White	1934	Demonstrated capability of indefinite culture of tomato roots (long periods).
R. J. Gauhtheret, P. Nobecourt	1939	First culture cambial tissues isolated from carrot.
P. R. White	1939	Callus culture of tobacco tumour tissue from interspecific hybrid of Nicotiana glauca x N. langsdorfii.
J. Van Overbeck	1941	Discovered nutritional value of liquid endosperm of coconut (coconut milk) for culture.
P. R. White, A. C. Braum	1942	Experiments on crown–gall and tumour formation in plant growth of bacteria–free crown–gall tissue
A. Caplan, F. C. Steward	1948	Use of coconut milk and (2, 4–Dichlorophenolacetic acid) for proliferation of cultured carrot and potato tissues.
G. Morel	1950	Culture of monocot tissues using coconut milk.
W. H. Muir	1953	Developed technique for culture of single isolated cells (nurse culture method).
W. Tubecke	1953	Haploid cultures from pollen of gymnosperm (Ginkgo sps.)
C. O. Miller, F. Skoog	1955	Discovery of cytokinins, e.g. kinetin, as potent cell division factor
E. Ball	1955	Culture of gymnosperm tissue (Sequioa)
F. Skoog, C. O. Miller	1957	Predicted hypothesis that shoot and root initiation in cultured callus is regulated by the proportion of auxin and cytokinin in the culture medium
E. C. Cocking	1960	Enzymatic isolation and culture of protoplasts
G. Morel	1960	Development of shoot apex culture technique
G. Morel	1964	Use of modified shoot apex technique for orchid propagation
S. G. Guha, S. C. Maheshwari	1966	Cultured pollen and anthers can produce haploid embryos
J. P. Nitsch	1974	Culture of microspores of Datura and Nicotiana, to double the chromosome number, and to harvest seeds, from the homozygous diploid plants just within 5 months
R. R. Hendre and coworkers	1975	Established technique for obtaining virus free sugarcane, citrus, potato and cassava.
G. Melchers	1978	Production of somatic hybrid from fusion of protoplasts isolated from potato and tomato.
S. K. Mukherjee, S. Bhaskaran	1980	Isolated auxotrophic mutants for vitamin B components in Nicotiana.
A. Mascarenhas, V. Jagannathan	1982	Isolated spontaneous varients from wheat callus cultures.
K. A. Barton, W. J. Brill,	1983	Insertion of foreign genes attached to a plasmid vector into naked plant protoplasts
M. D. Chilton	1983	Production of transformed tobacco plants following single cell transformation or gene insertion

Worker	Year	Contribution
H. Kohn and coworkers	1985	Somatic hybrids in tobacco mediated by electrofusion.
E. Sundberg and K. Glemelius	1986	Somatic hybrids in Brassicaceae
R. Nadgauda, A. Mascarenhas	1986	Isolated and planted variants of sugarcane resistance to mosaic virus and turmeric plants of the variety 'Tekurpeta' in the field.

*for details of more developments please refer to the work done by various researchers in the text.

Philip White (USA) and R. Gautheret (France) devoted many years to conduct tissue cultivation experiments. They could get success because of their fortunate choice of material for investigation, careful selection of nutrient media suitable for growing plant tissues. They are regarded as the founders of methods for cultivating excised tomato roots, root tips, root meristem and callus of cambial origin. The workers from France, USA, Czechoslovakia, Hungry, Germany, Italy, Switzerland, China, India and Russia, etc. have done sincere efforts in developing tissue culture techniques in various fields of normal and pathological plant physiology, biochemistry, cytology, and genetics which have now been used in biotechnological studies. R. Gautheret (1937) cultivated undifferentiated carrot tissues.

Isolation and culture of single cell, remains elusive until Muir (1953) devised the technique of agitating callus tissue in liquid medium (on a shaker). Isolated plant cells were grown in culture in 1954 by Muir and coworkers. The technique developed by them was perfected in France by Lutz. used bacterial culture techniques to grow plant cell suspensions. H. E. Street (UK.) developed bacterial culture techniques for tissue culture. Skoog and Miller (1957) obtained roots and stems from callus treated with auxin and kinetin. Morel used gibberellin for induction of proliferation of meristems and their differentiation into whole plants. The regeneration process has also been developed by use of tissue culture techniques.

The single cell thus obtained could be mechanically picked up and grown by using the nurse culture method. Later, a micro culture method was designed by Jones *et al.* (1960). F. C. Steward of Cornell University has shown that it is possible to produce a complete plant from a single cell. Steward isolated cells of carrot root, which grew in a medium containing coconut milk, the fluid that nourishes the coconut embryo. He found that these isolated cells, which under ordinary circumstances would not divide again, began to grow. Some of them showed highly abnormal growth patterns, but others, as they divide, organized themselves into perfect duplicates of normal carrot embryos. Such embryos can develop into normal mature carrot plants, with normal roots, stalks, flowers, and seeds. Vasil and Hilderbrandt (1975) used this technique to raise complete plant from single cell culture of *Nicotiana*. Besides, the academic interest, the significance of tissue culture technique in rapid and clonal propagation of plants was soon realized.

Observations over the past several years have led to a general belief that cells grown in culture for an extensive period of time are genetically unstable due to increased polyploidy and spontaneous mutation. This often results into decrease or loss of morphogenic potential, decline in biosynthetic capability of secondary metabolites and/or reversion of valuable mutant in wild types.

TISSUE CULTURE RESEARCHES IN INDIA

In India, the significance of *in vitro* culture technique was realized quite early by

Professor P. Maheshwari. He initiated investigation with complex tissues such as ovary, ovule, and control of fertilization. The initial exploratory work on these organs laid the foundation of several achievements in plant cell, tissue and organ culture in India. During 1956 to 1966, attempts were initiated to culture the pollinated ovaries of *Aerva tomentosa*, *Allium cepa*, *Althaea rosea*, *Anethum graveolens*, *Foeniculum vulgare*, *Hyoscyamus niger*, *Iberis amara*, *Linaria maroccana* and *Tropaeolum majus*. Success with the culture of pollinated ovaries made it possible to culture ovaries and ovules at even earlier stages. Cultured ovaries from pollinated flowers of *Zephyranthes* developed seeds with fully differentiated embryos. Excised ovules containing zygote and primary endosperm nucleus also matured into normal–looking seeds but were without fully–differentiated embryos. The seeds were however always smaller than the natural seeds (M. Kapoor, 1959). The *in vitro* formed fruits of *Anethum* were larger than the *in vivo* fruits. Another important feature was the occurrence of polyembryony due to cleavage and/or budding of zygotic embryos, on a culture medium. Polyembryony is useful in the multiplication of plants producing a limited number of seeds. This is of a special significance in umbellifers because of low seed–set and germination.

The method of ovary and ovule culture is also appropriate for rearing hybrid seeds. In *Gossypium hirsutum*, Joshi (1962) was able to grow ovules with 12–celled pro-embryo and 500 endosperm nuclei to maturity. The hybrid ovules of a cross between *G. arboreum* and *G. hirsutum* could be grown to maturity and viable seeds were obtained.

Of considerable importance are the adventive embryos from nucellus which genetically uniform, and reproduce the characters of the maternal parent. Excised nucellus from micropylar–half of pollinated ovules of *Citrus microcarpa*, a polyembryonic species, proliferated on a medium and the callus differentiated *pseudobulbils* which, eventually, developed into plantlets (Rangaswamy, 1961). Embryogenesis from nucelli of unfertilized ovules of *Citrus sinensis* and *C. aurantifolia* is reported by Mitra and Chaturvedi (1972). The embryos arose from nucellar cells or the callus. It was demonstrated that the pollination/fertilization stimulus is not essential.

The organogenic potentiality of endosperm has been demonstrated by Johri and Bhojwani (1965) in *Exocarpus cupressiformis*. In *Scurrula* and *Taxillus*, shoot buds differentiated on a cytokinin medium, whereas auxin is also essential for *Dendrophthoe* and *Tolypanthus*. Continuously–growing tissue and organogenesis have also been observed in the mature endosperm of autotrophic taxa such as *Croton honplandianum*, *Jatropha panduraefolia*, *Putranjiva roxburghii*, *Codiaeum variegatum*, *Santalum album*, *Oryza sativa*, *Nigella damascena*, *Euphorbia geniculata* and *Emblica officinalis*.

Khuspe *et al.* (1980) described methods for *in vitro* growth of immature hybrid embryos of *Carica papaya* x *C. cauliflora* and development of F_1 plants. The resultant seedlings could be transferred to pots and field. The culture of young floral primordia has provided a potential tool for understanding the morphogenic sequences leading to flowering. Studies in floral morphogenesis have been conducted in *Ranunculus sceleratus*, *Phlox drumondii*, *Kalanchoe pinnata* and *Browallia demissa*. In *Ranunculus* spp., cultured floral buds proliferated rapidly, and differentiated roots, shoot buds and embryoids. The buds developed into shoots terminating in a flower bud. Flowers were also born directly on the callus. Rarely, mature seeds were also formed. The embryoids passed through the stages comparable to *in vivo* embryogenesis. These plants also flowered *in vitro*.

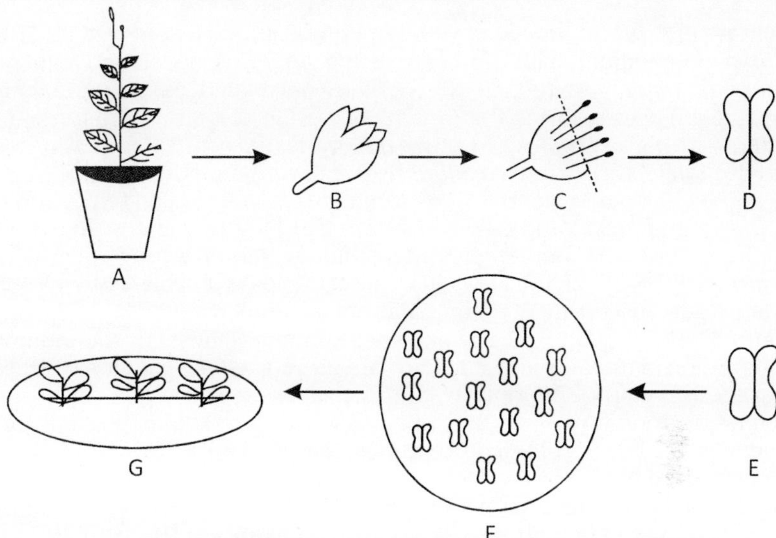

FIG. 10.1: Diagrammatic representation of the anther culture protocol for tobacco. (A) Flowering diploid plant. (B) Floral bud excised from plant. Calyx and corolla re of approximately equal lengths. (C) Floral bud with calyx and corolla removed. Buds normally contain five anthers, which are shown still attached to filaments. (D) Isolated stamen with anther and filament. (E) Anther removed from filament. (F) Petri dish with cultured anthers. (G) Plantlets emerging from anthers.

In the past few years many commercial ventures in plant tissue culture have come up in Pune. They are listed in Table 10.2.

TABLE 10.2: Some commercial achievements in plant tissue culture at Pune (India)

Name of the Firm	Plant Species Produced	Capacity (Millions)
Florida Plant Cultures	Foliage	2.0
Green Laboratories	Foliage	1.0
Indrayani Biotech	Strawberry, Roses	5.0
Kalyani Agrotech	Microtubers	–
Khushboo	Flowering plants	–
Kumar Gene Tech	Banana, Flowering plants	2.0
Microplantae	Banana, Flowering plants, Strawberry	10.0
Padamji Paper Mills	Foliage	–
Pocha Seeds	Ferns, Foliage	–
Raj Agrobase Pvt. Ltd.	Orchids, Trees	1.0
Western Indian Tissuetech Ltd.	Sugarcane	1.0

The work on haploids provided a novel experimental system for speeding up of homozygosity, analyses of genetic combinations, and detection of mutations. In cultured anthers of *D. innoxia* (with uni– or bi–nucleate pollen), numerous embryoids emerged on Nitsch–medium. The embryoids germinated to form plantlets (Guha and Maheshwari, 1967). George and Rao (1979) induced triploids through anther cultures of *Physalis* spp. Direct microspore–derived embryos in anther culture of two cultivars of *Brassica juncea* have been demonstrated by Sharma and Bhojwani (1985). On B5–medium with 2% sucrose the embryos developed into plantlets. Anthers of *B. nigra* cultured on B5–medium developed callus and/or embryos. The most significant result is the *in vitro* flowering of these plants in 2–months–old cultures. Other examples of anther culture leading to the formation of pollen embryoids/plantlets are *Petunia axillaris* by Doreswamy and Chacko (1973), *Solanum melongena* by Raina and Iyer (1973), *Petunia obconica* by Bajaj (1981), and *Physalis ixocarpa, Datura innoxia* and *Petunia hybrida* by Chandra and coworkers (Chandra and Kothari, 1987). By anther culture technique, homozygous recombinants of *Hyoscyamus niger* with higher alkaloid content could be obtained. Such a technique has proved very useful in the improvement of crop plants. Haploids of cereals are used for one–step transfer of genotypes of inbred lines. For the induction and characterization of mutants also, haploids are used. The pollen grains are extensively used for the isolation of protoplasts. Shrivastava and Johri (1988) reviewed Indian contribution to morphogenesis and plant tissue culture studies (Indian Review of Life Sciences, 8:249). A. F. Mascarenhas has compiled a Hand Book of Plant Tissue Culture published by ICAR, New Delhi (Feb., 1993).

Majority of the private firms are involved in production of foliage, flowering plants (roses, etc.), and ornamentals. Few deals in plants like *Musa* (Banana), *Fragaria* spp. (Strawberry), Orchids and Tree species. They mostly cater to the foreign market. Some of these private laboratories are also involved in R & D work. But majority of the private tissue culture companies get the know how from a R & D institution either from India or abroad and put it in operation in mass produce the plant species.

PLANT TISSUE CULTURE: PRINCIPLES

The technique has developed around the concept that a cell is **totipotent** that is has the capacity and ability to develop into whole organism. The principles involved in plant tissue culture are very simple and primarily an attempt, whereby an explant can be to some extent freed from inter–organ, inter–tissue and inter–cellular interactions and subjected to direct experimental control.

The most common culture in plant tissue is callus, which is wound tissue composed of undifferentiated, highly vacuolated and unorganized cells.

Callus Culture

For raising the callus tissues, a tissue culturist must have clear understanding of some basic principles. A cell from any part of the plant like shoot apex, bud, leaf, mesophyll cells, epidermis, cambium, anthers, pollen, fruit etc., when inoculated in a suitable medium under aseptic laboratory conditions can able to differentiate and multiply. This results into the formation of an amorphous mass of cells known as **callus**, which can be induced to re–differentiate on appropriate medium to develop **embryoids** which directly develop into the plantlets, eventually giving rise to a whole viable plant.

The term clone (from the Greek *klon,* meaning: a slip or twig suitable for plant propagation) was suggested by Webber (USA) in 1903 to explain those plants which were obtained by a sexual reproduction; it is even applied to DNA multiplication (cloning of genes in bacteria). In strict scientific sense, cloning means *an organism obtained from a single cell through mitotic divisions.*

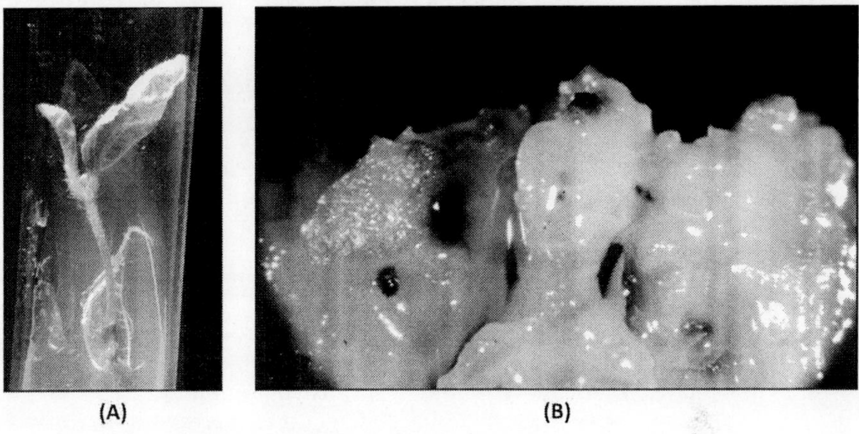

FIG. 10.2: (A) In vitro cultured plant material, (B) A callus.

Meristem Culture

When a meristem is cultured *in vitro*, then it produces a small plant bearing 5 or 6 leaves. This could be obtained within a few weeks. Then the stem is cut into 5–6 small micro cuttings, which under favourable conditions, become fully grown plants.

Organ Culture

A body of higher plants has complex inter–relationships between different organs like root, shoot, apical meristem, leaf primordia, floral buds, ovary, ovule, anther lobs, pollen grains, fruit, seed, etc. In this method a particular organ is isolated and cultured under laboratory conditions in a chemically defined medium where they retain their characteristic structures and other features and continue to grow as usual. In organ culture, organs are not induced to form callus, therefore, it differs from the callus culture where the organization of the intact tissues is lost.

This technique provides an experimental system to define the nutrients and growth factors that are usually received by the organ from other organs of the plant body and from surrounding environment. It also helps us in understanding the inter–dependence of organs with respect to various physical and chemical growth factors including growth hormones. Organ culture technique also provides the knowledge about the various problems of morphogenesis and the sites of biosynthesis of specific metabolites and growth compounds. It may be used as a tool for improvement of various economically important crops.

Organ culture may be grouped into two major categories: vegetative organs (root culture, leaf culture, and shoot tip culture) and reproductive organs (complete flower culture, isolated ovary culture, isolated ovule and embryo culture, pollen mother cell culture, seed and fruit culture).

THE CONCEPT OF TOTIPOTENCY OF CELLS

Some concepts in science become inherently acceptable long before their practicality is demonstrable. This was so in the concept of the totipotency of cells of higher plants. Even in the mid–twenties one encountered the tacit view that, apart from inherent practical difficulties, there was no theoretical reason why one should not rear a begonia plant from a single leaf hair cell. This view was traceable **first** to the then well recognized principle that as cells divide mitotically, they do so equationally to produce daughter cells in facsimile.

Secondly, the concept was due to G. Haberlandt's historic but then unsupported claim, expressed in 1902, that one day it should be possible to rear plants from isolated surviving cells of flowering plants. Haberlandt's insight was, in fact, the more remarkable because he even stated that out of surviving somatic cells artificial embryos would be reared asexually. In higher plants, embryos do develop *in situ* from appropriately stimulated somatic cells but without any prior sexual act. At the outset, the special significance of the fertilized egg in plants should be seen as restricted to its role as the genetically unique product of a fusion of male and female gametes. Having established the genetic constitution of a given individual, the zygote (or fertilized egg) really behaves developmentally as a very general kind of living cell– a cell with a built–in capacity to grow in an organ, the ovule, that fosters that growth. Moreover, as will be shown, and in a satisfying number of cases, isolated somatic cells may return to a simulated zygotic state and grow into embryos and to plants under the appropriately applied conditions which furnish the requisite nutrients and stimuli.

Contrasts between Plant and Animal Cells

Some inherent differences between plants and animals are relevant here, for these go far to explain why the clonal development of individuals from somatic cells is intrinsically more difficult in animals than it is in plants. The much heralded new biology rightly emphasizes that the molecular basis of metabolism, whether of plants, animals, or bacteria, is essentially the same. The classical events are essentially the same from microorganisms to man. However, the differences in plants and animals are organized and provide their cells and organs *in situ* with the requirements for what they do?, where they are?; moreover, they profoundly affect the ability of animal cells to develop in isolation.

The green flowering plants are autotrophic and can live in an inorganic world. Even so, the non–green cells of plants, which are necessarily heterotrophic for carbohydrate, have such simple other requirements that they may often be grown (even though slowly) in tissue explants or organ cultures (for example, of roots) in a relatively simple, largely inorganic, medium. Plant cells remain capable of receiving their requirements from the outside world (CO_2 and H_2O, NO_3, SO_4, and PO_4 and mineral elements), where they are very dilute. They can survive in varied composition of their ambient media. Plants are bounded by a cellulose wall and so may utilize their turgor to maintain a thin protoplasmic lining around large enclosed vacuoles; this thin, living layer is, however, in easy contact with the outer air or medium with which the cells of plants are in constant interchange.

In contrast, higher animals are heterotrophic for a wide range of metabolites, and they are adapted to receive, digest, and assimilate their nutrients from complex sources and to distribute these via the blood in a complex way, without direct parallel in plants, to the cells and organs of the body. Moreover, the animal body creates and maintains within narrow limits an internal environment to which its living cells are exposed and which furnishes their requirements in relatively concentrated form. For example, a mammalian embryo develops from a fertilized egg, it rapidly establishes a complex nutrient supply with the parent via the placenta; or, in other forms, the embryo develops first a free living larval form at the expense of the nutrients prestored in the egg.

In higher plants the provision of nutrients to the embryos of higher plants in the embryo sac and ovule, via the endosperm for their attachment in the ovule via a suspensor serves a mere mechanical purpose, while the pro–embryo and its later formed cotyledons absorb nutrients and stimuli directly over their outer surfaces. Thus, if somatic plant cells can be induced to form immature embryos, it is very much simpler to provide for their nourishment externally than for embryos of higher animals, which needs blood supply.

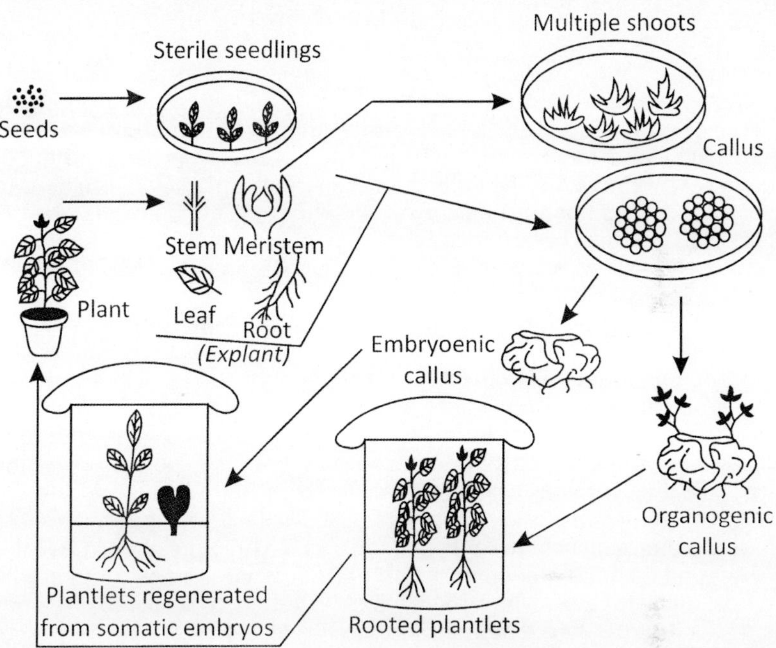

FIG. 10.3: General presentation of steps used for plant tissue culture.

This may well account for the fact that testicular teratomas of the mouse, lacking a blood supply, fail to develop. However, these structures do embark upon a simulated embryogeny and indeed similar asexual forms have long been known to occur in other animals. By contrast, and as will be shown below, proembryos of several plants which originate from free somatic cells can now be caused to form embryos, and plantlets by the thousands; this occurs even in isolation and in a limited volume of an appropriate aqueous solution which behaves, in this respect, like an artificial ovule or embryo sac.

There is, however, another obvious reason why totipotent development from somatic cells is more feasible in plants than in higher animals. Such animals develop their organs early, they assign to them highly specialized functions and then integrate them by means of the circulating tissue of the blood, by a nervous system spinal cord, and brain, and by highly specialized hormones which are the product of discrete glands. Moreover, these developmental events in animals essentially occur but once. By contrast, higher plants create their organs repetitively and, often, virtually indefinitely. Plants do all this by arranging, in their growing regions, living cells which retain the essential characteristics of the plant and which, during development, depart from, or return to, the situation to be seen in the primary growing apices of shoot and root by relatively small and often easily reversible steps.

The Autonomous Organelles: Their Behaviour during Growth Induction and Morphogenesis

The stimuli that cause quiescent cells of carrot to proliferate and grow also affect all their organelles (mitochondria, plastids, dictyosomes, reticulum and polysomal aggregates of ribosomes in the ground cytoplasm) in various ways. The same stimuli affect the metabolism of the tissue and its ability to form residual or secondary biochemical products.

For example, deep orange–red and carotene–rich plastids occur in the carrot root, but, even so, they become bright green in explants cultured on coconut milk in the light. Combinations of light intensity and of temperature and of diurnally fluctuating light and temperature all have their effects on the pigmentation, but none have yet produced in free cells or proliferating explants the bright orange–red, carotene–rich cells of the normal carrot root. Thus, it is not only what the potentially totipotent cells are that determines what they will do, nor in fact how readily they may be caused to grow, but where, morphologically, they develop.

Sites and Modes of Action of Growth Regulating Substances

How the totipotent cells respond to the exogenous messages that they receive? It would be gratifying if, in response to the appropriate stimuli, the nuclear genes were to pour out messenger RNA's into cytoplasm and there stimulate the appropriate proteins and enzymes to cause the cells to do what they should, when and where they are. This mechanism, essentially bacterial in nature, may in due course be observed to operate in flowering plants, but, so far, attempts to demonstrate it have had meager success. In fact, it seems as if carrot cells on a basal medium may possess all the idling mechanisms of protein synthesis and require only certain external chemical triggers, which have been described, to set it into directed motion. All this raises the question *whether plant cells really need to ask the permission of their nuclei*, in terms of activated or de–repressed genes, for every task they need to perform.

The Behaviour of Carrot Cells in vitro

The genetic inheritance of the zygote, transmitted at cell division to all the daughter cells, conveys to every organelle the inherent capacity to carry out its various functions. However, thereafter the autonomous organelles, each with their own genetic information, may be called into appropriate activity by the great array of simple substances which, singly or in combination, act as growth regulators; these release, wholly or in part, the inherent **totipotency** of angiosperm cells. Thus, there is a vast area open to exogenous chemical control over the responses of otherwise totipotent cells.

Clonal Development from Animal and Plant Cells

There is now good evidence that the nuclei of certain animal cells (e.g. of the intestinal wall of a tadpole) retain intact all the genetic information, but they only release this, as yet, in contact with the cytoplasm of an enucleated egg. Perhaps eventually the equivalent of the plant techniques here described may induce somatic cells of animals to embark directly upon an embryological development, and, perhaps, if this is done, they may even be furnished in some way with the equivalent of a parental blood supply.

Such steps present, however, obviously greater orders of difficulty, even if they do not constitute insuperable barriers to the free clonal development of higher animals from somatic cells. Thus, until these problems show signs of being resolved, it seems best to regard the free clonal growth of higher animals from somatic cells as a distant, even if it is a desirable, goal. Meanwhile, however, the techniques of clonal development of plants from free cells open up various technically rewarding possibilities, which are being pursued in this and many other laboratories.

PLANT TISSUE CULTURE: METHODS

The setting up of a tissue culture laboratory needs proper planning. It depends upon space availability, volume of work to be carried out and funds. Usually, the available laboratory space, is be divided into five distinct laboratory areas. These are: *Media preparation area/room, Aseptic transfer chamber area, Environmentally controlled culture room, Analytical room and Acclimatization room.*

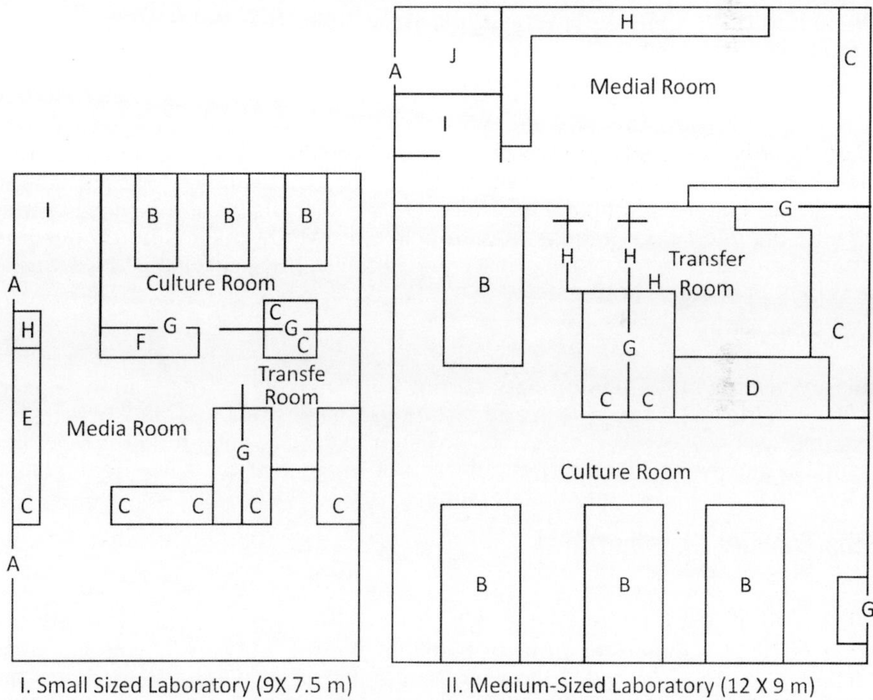

FIG. 10.4: Various designs of tissue culture laboratory (A) entrance (B) shelves (C) counter (D) laminar airflow cabinet (E) sink (F) gas outlet/burner (G) window (H) refrigerator and deep freezer (I) store (J) conference room (K) library

Media Preparation: Area/Room

Requirements

Laboratory tables or benches and revolving stools with adjustable height, Hot plate and magnetic stirrers, analytical loading single pan balance with precision of ± 0.001 g, weighing range 0.1 mg – 180 g, digital read out, Top pan loading balance for quick weighing, range 100 mg – 500 g capacity, sensitivity 0.1 g, Refrigerator and freezer,

Water purification and storage system, Glassware washing facility with proper drainage, Gas outlet, Electric hot–air oven range upto 250 ± 2°C and microwave oven, Digital pH meter, Range 0–14 pH, accuracy 0.1 pH temperature compensation 0–100°C, Autoclave preferably horizontal., Continuous supply of single and double distilled water

Cabinets or Shelves

For storing glasswares, plastic wares, chemicals, plugs and appliances required for media preparation are as follows:

Requirements: Culture tubes/conical flasks/petri dishes of various capacities, Measuring cylinders 25 ml, 100 ml, 500 ml, 1000 ml, Cotton for plugs/plastic caps (autoclavable), General glassware's/plastic wares of various capacities such as volumetric flasks, beakers, reagent bottles, pipettes, vacuum filtration system and glass rods.

Culture Media, Washing Powder/ Liquid Detergent, Disinfectants

Requirements

Powder or liquid detergents or wetting agents such as Tween–20, 70% alcohol and absolute alcohol, Glass distilled water, Stock solutions of nutrients of tissue culture media or ready–made/pre–mixed powdered media, Sucrose, Agar (tissue culture grade), Sterile culture vessels with distilled water, Chemicals of analytical grades (Inorganic, organic salts, vitamins, amino acids, growth regulators/hormones and activated charcoal), Coconut milk, yeast extract, malt extract, casein hydrolysate and extracts of potato, carrot and tomato.

Other Requirements

Spatulas, weighing butter paper/boats, stirring bars (magnetic) for magnetic stirrers and stirring retrievers, Brush (flask and test tubes), gloves (disposable 23 cm), mop (household), scoop, towels (household), wastebasket (large and medium), Service lines such as gas, water, electricity, vacuum pump and generator, A low bench/table/ desk for culture evaluation and data recording.

Aseptic Transfer Chamber Area

Requirements

Laminar air–flow cabinets to provide constant flow of air across working area, Dissecting and stereo zoom microscopes and lens magnifier, Micro–dissecting scissors, scalpel handles with blades, forceps (various sizes), catheter trays, needles and inoculating loops, Gas outlet, vacuum facility, Tissue paper/filter paper, Sterilizer (dry heat with glass beads) and pipette dispenser (automatic).

Environmentally Controlled Culture Room

Requirements

Racks with light arrangements on timers and controlled temperature (25 ± 2°C) maintained with air cooler or by window air conditioners fitted with temperature indicators, Incubators having dark and light/dark photoperiod with controlled temperature, Rotary shakers of variable speed from 80–220 rpm to take 100 ml or

250 ml Erlenmeyr flasks with arrangement of lighting to provide an intensity of 2000 to 4000 lux, Lux meter to measure intensity of light in culture room, incubator and shaker.

The cultures are usually incubated at 25 ± 2°C under 16:8 light: dark photoperiod. The source of light for the cultures in racks should make available with cool–day–light (fluorescent tube lights of 40 watts, 2000 lux). It is advisable to connect the controlled culture room with power generator for emergency power supply. Some space should also be kept for incubating cultures in continuous darkness.

Analytical Room

Requirements

Inverted microscope for bright field, dark field with compensating wide angle eye pieces, preferably with photomicrographic attachment, Colorimeter for chemical estimation such as chlorophyll, starch, nucleic acid, phenols, oxidizing enzymes etc. Low speed centrifuge with continuous variable electronic speed control, Chemical reagent racks for qualitative and quantitative chemical analysis. Viscosity meter, Gas outlet.

Acclimatization Room

The hardening chamber needs high illumination (4,000–10,000 lux) and high humidity (90–100%, through mist and fog systems). Humidity is required for conditioning tissue culture plants after taken out from rooting media and transfer to pots under greenhouse.

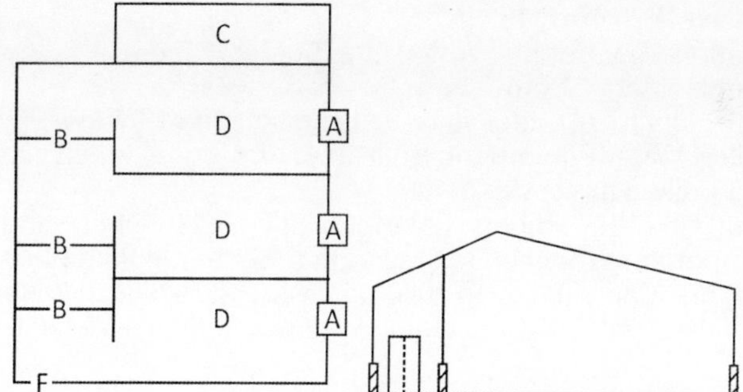

FIG. 10.5: Design and elevation of greenhouse with three compartments having desert coolers. (A) Desert–cooler (B) Air–vents (C) Water tank (D) Greenhouse chamber (E) Entrance.

Miscellaneous Items

Requirements

Air conditioners, uninterrupted power supply (UPS) and emergency light, Bunsen burners, Permanent markers, tapes (autoclave indicator), tape label (self–adhesive), aluminium foil and parafilm, Fluorescent lamps/tubes, Trays and baskets for cultures, Plastic carboys to store water and other solvents, UV germicidal lamp, Metal racks to keep test–tubes in culture room, Gas lighter/match box, Fire saving, equipment and first–aid box etc., Filter paper, culture trays, culture boxes and culture tube racks.

To ensure the growth and development of an explant, it has to be provided with a suitable nutrient medium and proper laboratory conditions for culturing. These operations have to be carried out under aseptic conditions.

Specifications of Laboratory Equipments

The following laboratory equipments are required for tissue culture.

- pH Meter (digital): 230V 50Hz, single phase supply with combined pH electrode, range 0–14 pH/0–1400 mv, temperature compensation 0–100°C.
- Balances: Manual – single pan, capacity 100–200 g, sensitivity 0.1 mg. Electronic – top loading, precision 0.005 g, range – 0 to 1200 g.
- Electric Hot Air Oven: Thermostatically controlled, range 5°C to 250 ± 2°

Microscopes

- Dissecting: two lenses (eyepieces) 10X, 20X.
- Laboratory: with facilities for bright field, dark field, phase contrast, wide angle eyepiece with photomicrographic attachment.
- Inverted: Wide angle eyepieces, quintuple objective turret adjustable specimen 55 mm petridish holder and photomicrographic attachment.
- Centrifuge (low speed): Variable electronic speed control, speed indicator, Amp. meter, timer, dynamic break, 0 starting switch, 230 V 50 Hz.

Electric Autoclaves

- Vertical: with safety valves, pressure gauge, steam–release cock.
- Horizontal: mounted to tubular stand, heavy hings
- Steamers: With immersion heaters of ejection safety device, size: 31 cm ´ 1 3 cm x 10 cm to 61 cm ´ 20 cm ´ 15 cm.
- Filter Sterilization Equipment: Syringe filter holder: 2.5 cm
- Pressure Filter Holder: 4.7 cm
- Manestry Stills: Electrical water stills, 2 lit capacity/hr of 1.5 kw.
- Double Distilled Water Equipment: Built–in energy regulator all glass water stills, heat resistant boiling flask with heater.
- Environmental Growth Cabinets: Cabinet with controlled temperature, light and humidity, temperature range 4° to 45°C with timer to regulate photoperiods.
- Gyratory Shakers: Capacity: 50 to 1000 ml flasks, speed: 80 to 200 rpm.
- Laminar Air–flow Cabinets: Constant flow of purified air, sizes: 0.6 m, 1.2 m and 1.8 m.

The other miscellaneous equipments which are required for tissue culture are: air conditioners, arrow heads, bunsen burners, deep–freeze, dissecting needles, glasswares, forceps, florescent lamps/tubes, heaters, hot plates with magnetic stirrer, inoculation cabinets, metal trays and bowls (for transport of cultures), tubes, refrigerators, UV germicidal lamps, wooden or metal racks, etc.

ASEPTIC TECHNIQUE

In *in vitro* condition plant cells and microbes have basically same requirements. When the culture medium contains sugar (as carbon source) it attracts a variety of microorganisms which grow fast than that of the cultured tissue in medium and they ultimately kill the plant cells. It is, therefore, necessary to have complete aseptic condition around the culture equipments which prevents contamination of the culture medium. Following are the three main sources of contamination of the medium and the subsequent methods to check them:

1. The microorganisms may be present in the nutrient medium at the time of its preparation. These microorganisms can be destroyed by proper plugging and autoclaving the culture tubes/flask. The medium can be completely sterilized by maintaining it at 120°C for about 20 minutes at 15 lb pressure in the autoclave.
2. The explant (plant part to be cultured) may carry microorganisms with it, therefore, the plant part should be surface–sterilized by mercuric chloride (1 to 2%) or by sodium hypochloride solution for 30 minutes.
3. Precautions must be observed to prevent the entry of microorganisms when the plug of a culture is removed during transfer of the plant material to the medium or from one medium to another. The inoculation chamber may be sterilized by UV–radiations.
4. Correct pH of the medium is important. Highly alkaline or acidic pH affects the nutrient uptake in culture tissues. Therefore, the tissue culture medium is adjusted to a pH of 5.6 to 6.0 before autoclaving.
5. Semi–solid and liquid media are most commonly used for growing plant cells. A high concentration of gelling agent (agar–agar, gelatin, silica gel) makes the medium very hard and decreases the nutrient uptake by the tissues. Agar at 0.8% to 1.0% concentration is widely used.

STERILIZATION OF PLANT TISSUES

It is essential to remove dirt and debris from the plant tissue and should be washed in a weak detergent solution and rinsed several times with distilled water prior to sterilization. Some woody tissues, such as buds and twigs, are cleaned by immersing them briefly in a 70% ethanol solution, which wets and spreads over the tissue surfaces more effectively than a higher concentration alcohol solution.

- Sodium Hypochlorite (NaOCl): It is the most common chemical agent used to sterilize plant tissues (0.025%–0.25% NaOCl). Diluted household bleach can also be used for this purpose, which normally contains 5.25% NaOCl. It is equally effective and considerably less expensive.

FIG. 10.6: Portable autoclave.

- Calcium hypochlorite (CaOCl): It can be used as a substitute for NaOCl. CaOCl causes slightly less damage to plant tissues but tends to precipitate out of solution. To avoid the accumulation of CaOCl on the plant tissue surfaces, sterilization solutions should be filtered or decanted prior to use.
- Hydrogen Peroxide (H2O2) Solution: Plant tissues can also be surface sterilized using a H2O2 (3%–10%). It is much easier to remove from tissues than NaOCl and CaOCl.
- Other Substances: Plant tissue can also be surface–sterilized by bromine water (1%–2%), silver nitrate (AgNO3 1%) and mercuric chloride (MgCl2 0.1%–1%).

Cleaning (Preparation of Glasswares/Plastic Wares (Autoclavable)

1. Clean the glassware's/plastic wares in 10% commercial detergent liquid or powder for 1 hr and then in HCl for 2 hr.
2. Remove the traces of detergent and acid by thorough washing with tap water
3. Rinse vessels with double distilled water and allow them to dry over night at room temperature.

Sterilization

1. Plug glasswares such as conical flasks or test tubes with non–absorbent cotton or cover by the plastic caps. Wrap the petri dishes with aluminium foil. Place forceps and scalpels in test tubes, plug the tubes with cotton or cover with aluminium foil. Plug the mouth end of the pipettes with cotton. Wrap them individually in aluminium foil.
2. Autoclave glasswares and instruments at 121°C for 1 hr.
3. For dry–heat sterilization, metal instruments should be sterilized in an oven at 140–160°C for 2 hr.
4. Filter sterilization for heat labile amino acids, vitamins, growth regulators, antibiotics, natural complexes should be through millipore filtration assembly using filter membranes of 0.45 or 0.22 μM porosity. Plug the receiver flask with cotton. Assemble the filtration assembly and wrap the filtration unit with paper or foil. Autoclave the receiver and filtration unit at 121°C for 1 hr. Attach the filtration unit with a receiver flask with vacuum pump in a laminar flow bench pour solution to be sterilized into the filtration unit. Apply slight air pressure to start filtration. Transfer the desired volume to sterile flasks under laminar air–flow bench. Use a sterile pipette for drawing filter sterilized solution to autoclaved medium.

Surfactants

- Tween 20 or Triton X–100: These are scientific reagent–grade surfactants and are often added in low concentrations (0.05%) to chemical sterilization solutions. Their use ensure that the sterilizing agent come in contact with the entire plant tissue surface.
- Stirring of the Tissues: Good surface contact is also facilitated by stirring the tissues during sterilization.
- Ultrasonic Bath: It is an effective method to ensure good surface contact during sterilization treatment in an ultrasonic bath like those used to clean dentures.

This technique is particularly useful for sterilizing buds and woody tissues that have many small surface crevices and cracks.

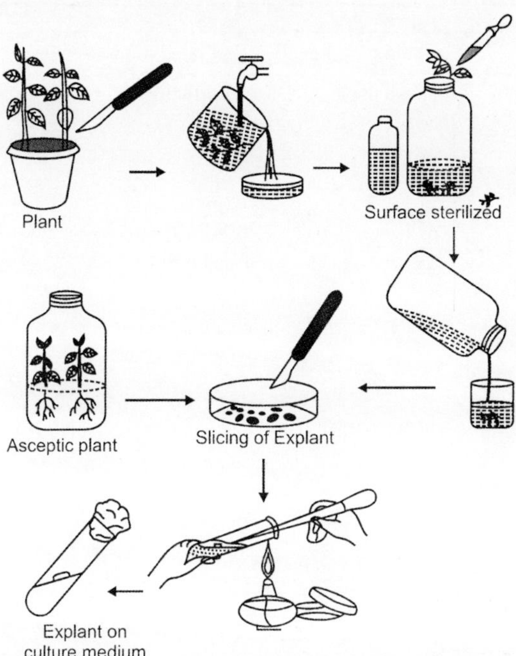

FIG. 10.7: Diagrammatic presentation of the procedure for surface sterilization of plant material and inoculation of explant for culture.

After surface sterilization, a minimum of three sequential rinses with sterile distilled water are recommended to remove any remaining chemical sterilizing agent.

Control of Bacterial and Fungal Contaminants by Antibiotics in Plant Tissue Culture

Antibiotics are used to reduce, control, or eliminate contaminants in plant tissue cultures.

It is suggested that antibiotics should not replace careful aseptic technique. Most antibiotics are heat–sensitive and cannot be autoclaved. They are usually dissolved in water or another suitable solvent, filter–sterilized, and added to autoclaved culture media that has cooled to 45 °– 50 °C. Falkiner (1990) suggested the list of antibiotics which are suitable for plant tissue culture.

TABLE 10.3: Antibiotic agents and their mode of action

Antibiotic	**Mode of Action**
Amino glycosides	Inhibit protein synthesis by interaction with 30S or 50S ribosomes.
Manamycin	
Neomycin	
Streptomycin	

Antibiotic	Mode of Action
ß–Lactams	Inhibit bacterial cell wall synthesis.
Ampicillin	
Carbenicillin	
Penicillin	
Chloramphenicol	Inhibits protein synthesis by acting on 50S ribosome.
Glycopeptides	Interfere with bacterial cell wall synthesis.
Vancomycin	
Macrolides and lincosamides	Inhibit protein synthesis by acting on 50S ribosome.
Erythromycin	
Lincomycin	
Polymixins	Attach to cell membrane and modify ion flux, resulting in cell lysis.
Polymixin B	
Polymixin E	
Quinolones	Interfere with DNA replication by inhibition of DNA gyrase.
Nalidixic acid	
Norfloxacin	
Ofloxacin	
Rifampicin	Interferes with mRNA formation by binding to RNA polymerase.
Tetracyclines	Inhibit protein synthesis by acting on 30S ribosome.
Trimethoprim and Sulphonamides	Inhibit synthesis of tetrahydrofolate.

TABLE 10.4: Goal and type of explant tissue.

Goal	Explant Tissue
Bud culture	Apical and axillary bud
Meristem culture	Apical meristem
Micropropagation	Shoot apex or lateral bud or embryo
Root culture	Lateral roots from adult or seedling
Callus culture/ Somatic embryogenesis	Cotyledons, hypocotyl, stem, leaf, root, any part of seedling
Haploid culture	Anther and pollen

IN VITRO ENVIRONMENT

A piece of plant tissue taken out from original site of plant and transferred to an artificial tissue culture media for the growth or maintenance, is called as **explant material**. The choice of tissue depends upon on ultimate goal of the tissue culture project. Any piece of the plant tissue can be used as an explant material.

Various factors of an explant tissue source influence the culture on tissue culture media. These are:

1. Physiological and ontogenetic age of organ or tissue
2. Quality of source plant
3. Season in which explant tissue is obtained

4. Size of the explant
5. Aim of the culture

FIG. 10.8: Tissue culture / growth room with culture vessels.

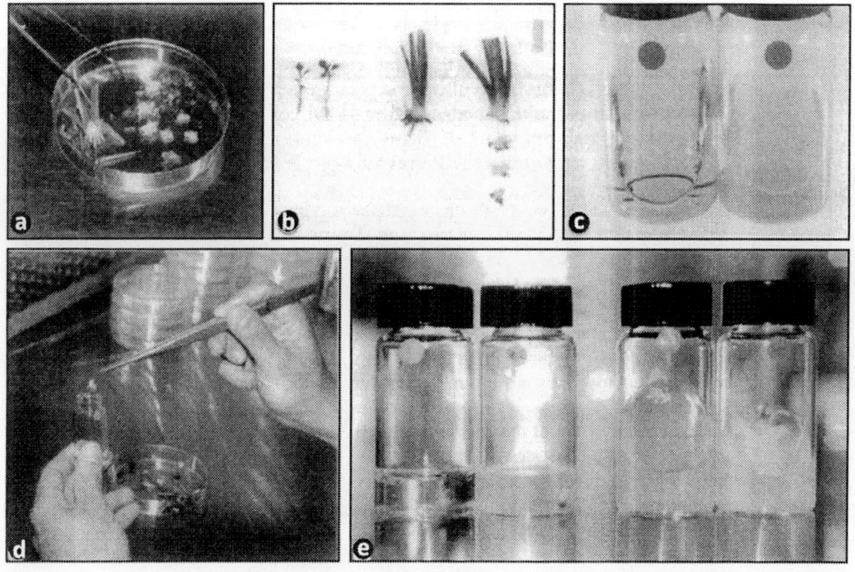

FIG. 10.9: Culture indexing for cultivable contamination. (a) Aseptic technique is use to obtain tissue sections from the Stage I shoot culture. (b) Shoots with upright stems (left) are sectioned into nodal segments. The shoot tip is either cultured in the indexing medium or transferred to fresh Stage I medium. Cross–sectional stem disks are made from shoots with basal meristems (right). Scale base = 1 cm. (c) Indexing medium is prepared as a liquid and agar–solidified slants. (d) Several tissue samples are directly placed in the liquid indexing medium. (e) Clouding of inoculated liquid indexing medium is a positive indication of the presence of cultivable contaminants in the tissue sample. Development of colonies on the surface of the medium and/or halo in the medium where the tissue sample was stabbed indicated culture contamination.

PRETREATMENT TO EXPLANT TISSUES PRIOR TO CULTURE

Many explant tissues specially woody tissues produce phenolics into tissues culture medium, which cause browning of explant material. Antioxidants such as L–ascorbic acid, free acid 0.1 g/l; citric acid, free acid anhydrous 0.15 g/l; L–cysteine HCl 0.05 g/l are commonly used for checking browning problem. The browning pigments are toxic to the plant growth in cultures. Best results are obtained when explants are treated with freshly prepared antioxidant solutions.

Procedure

1. Prepare the mixture of L–ascorbic acid (100 mg/l) and citric acid (150 mg/l) in double distilled water
2. Filter sterilize the solution through a 0.22 μm filter unit
3. Store the explant material in a cold antioxidant mixture and incubate explants in refrigerator at 0°C for 5–30 min. to allow the tissue to soak in antioxidant solution
4. Commercial bleach contains about 5% sodium hypochlorite and thus may be used at a concentration at 10–20%, which is equivalent to 0.5–1.0% sodium hypochlorite.

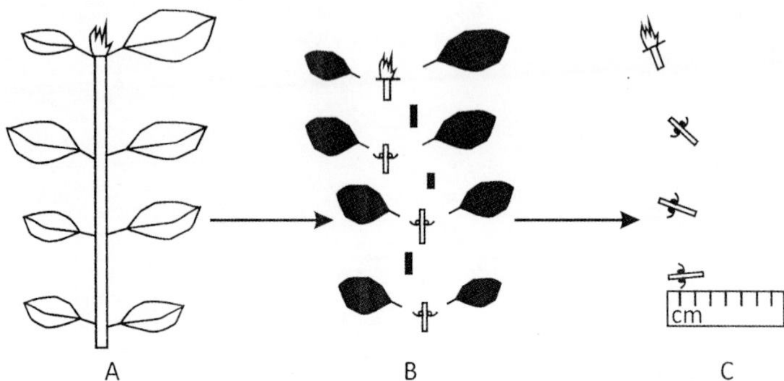

FIG. 10.10: Diagram illustrating how to cut shoots (A) into single–node segments 2 to 3 cm long, removing the leaves (B), and keeping a small part of the petiole subtending the bud (C). Shaded sections are discarded.

TABLE 10.5: Commonly used disinfectants for plant tissue culture

Disinfectant	Concentration (%)	Exposure (min)
Calcium hypochlorite	9–10	5–30
Sodium hypochlorite	0.5–5	5–30
Hydrogen peroxide	3–12	5–15
Ethyl alcohol	70–95	0.1–5.0
Silver nitrate	1	5–30
Mercuric chloride	0.1–1.0	2–10
Benzalkonium chloride	0.01–0.1	5–20

Sterilization Procedures may be enhanced by:

1. Place the material in a 70% ethyl alcohol solution before treating with another

disinfectant solution. The use of a two step (two sources) sterilization procedure is beneficial.

2. Use of a wetting agent (Tween 20 or 80 respectively) is beneficial. It reduces surface tension and allows better surface contact.
3. Conduct the sterilization process under vacuum. This helps in the removing air bubbles and provides efficient sterilization process.

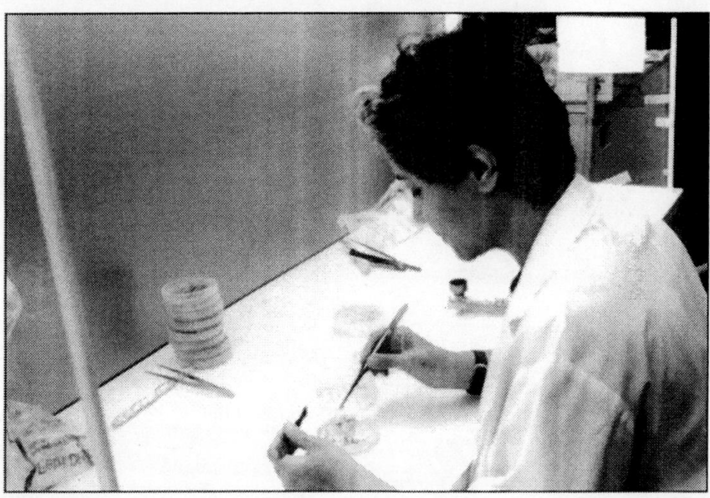

FIG. 10.11: Inoculation of explants on a laminar air flow bench.

Procedure

1. Wash explants in a mild detergent before treatment with disinfecting solution (exclude herbaceous material which may not need this treatment).
2. Within a few days, cells proliferate and a callus culture is obtained. In this, dividing cells form a layer of meristem and build a globular mass of non–dividing parenchyma. Alternatively, they may form small meristematic and build a globular mass of non–dividing parenchyma. Alternatively, they may form small meristematic zones interspersed in non–meristematic regions, yielding a sort of nodulated callus.

Organogenesis is the development of adventitious organs or primordia (embryoid) from undifferentiated cell mass (callus) in tissue culture. It is controlled mostly by a balance between cytokinin and auxin. A relatively high ratio of auxin: cytokinin induces root formation in callus tissues whereas, a low ratio induces shoot formation. Caulogenesis is a type of organogenesis by which only adventitious shoot bud initiation takes place in the callus tissue. When it is applicable for root, it is known as rhizogenesis. Anomalous structures when develop during organogenesis is called organoids. The localized meristematic cells on a callus which give rise to shoots and/or roots is termed as meristemoids.

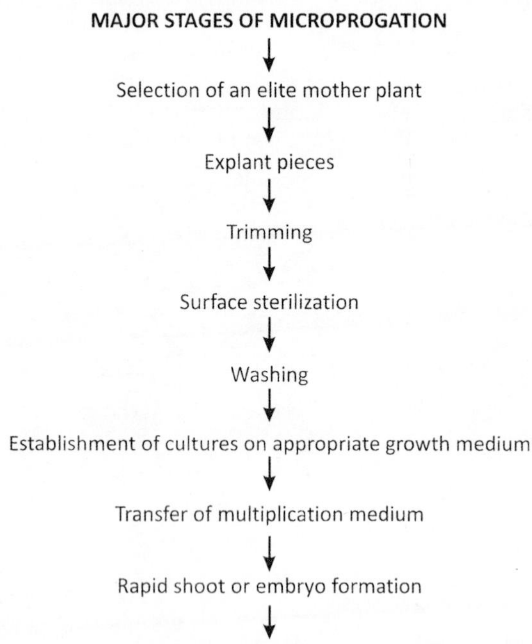

FIG. 10.12: Procedure of removing plant parts.

MICROPROPAGATION THROUGH ORGANOGENESIS

The principle of micropropagation is based on the phenomenon that shoot tip when cultured on tissue culture medium, can develop large number of shoots identical to the parent plant.

The technology provides many advantages over conventional propagation methods such as product development in highly speedy manner, uniforming in product development, large population can be produced in small growing space, quick multiplication of elite and genetically engineered products, and *in vitro* storage of germplasm. The technique also helps in enhancing tissue culture products via specific product format, rooted or unrooted micro cutting, bulblets, and embryos, certified plant material free of pathogens and production can be obtained throughout the year, as most of the multiplication phase is completed under artificially controlled conditions.

Through the use of auxin and cytokinin, hormonal control of organogenesis became feasible. Skoog and Miller (1957) demonstrated that the differentiation of root, shoot, and both root and shoot in tobacco pith tissue is the result of auxin–cytokinin ratio (i.e., the differential regulatory role is concentration–dependent). A high level of cytokinin promoted shoot, and auxin root formation. An equal concentration of cytokinin: auxin resulted in callusing. However, the requirement of exogenous supply of growth hormones for differentiation is dependent upon the endogenous level of these substances. To prove the efficacy of somatic cells to behave as zygote, Reinert (1958) and Steward (1959) reported the first somatic embryogenesis from the tissues of carrot.

By culturing shoot tips of *Lupinus* and *Tropaeolum*, Ball (1946) raised complete plants which could be transplanted. Shoot tip culture helped Morel and Martin (1952) to recover virus-free plants of *Dahlia* and, subsequently, in rapid multiplication of orchids.

CULTURE MEDIA AND PREPARATION

The success in cell, tissue and organ culture technology is related to the selection or development of the culture medium. As no single medium will support the growth of all tissue

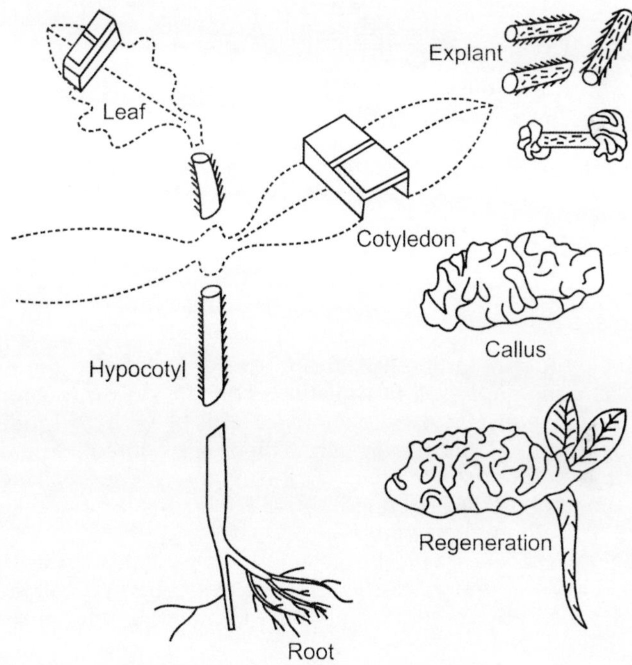

FIG. 10.13: Schematic representation of static callus culture of plant tissue cultures therefore modifications in the nutritional component including growth regulators are often necessary for different types of growth responses in a single explant material.

Various media compositions which are frequently used for tissue culture. A literature search is useful for selecting the appropriate culture medium as a starting points in developing a medium for specific purpose such as callus induction, axillary bud proliferation, organogenesis, somatic embryogenesis, anther culture etc. A nutrient medium generally contains inorganic salts, vitamins, growth regulators, a carbon source and gelling agent. Other components added for specific purposes include organic nitrogen compounds, hexitols, amino acids, antibiotics and plant extracts. The Murashige-Skoog medium (MS) (1962), Revised Murashige-Skoog medium, White's medium (1963), Linsmaier and Skoog (LS) (1965), B5 (Gamborg et al, 1968), Nitsch and Nitsch (1969), Woody plant medium (Llyod and McCown, (1981), Somatic embryogenesis medium and derivatives of these media have wide applications for different plant species and for different culture objectives. The decision on using type of media for the metabolic needs of the cultured cells and tissues, is a major factor of success in plant regeneration process.

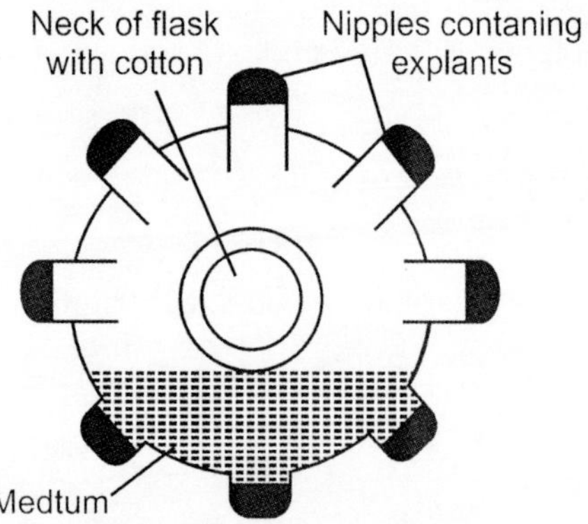

FIG. 10.14: Tissue culture techniques are now practiced on a grand scale. Special nippled flasks have proved very useful for the liquid culture of single cells and cell clumps.

MEDIA COMPONENTS

Inorganic Salts

A relatively small number of mineral salts are used as component of media for plant tissue culture. The inorganic salt formulations can vary in various reported media, however MS formulation is most widely used with or without modifications. The distinguishing feature of MS inorganic salts is their high content of nitrate, potassium and ammonium in comparison to other salt formulations. The stocks are prepared at 100 X (times) the final medium concentration. Each stock is added at the rate of 10 ml per 1000 ml of medium prepared.

The Na–FeEDTA stock should be protected from light stored in bottle that is amber coloured or wrapped in aluminium foil. Concentrated salt stocks enhance the accuracy and speed of media preparation. Guidelines for maintaining stock solutions.:

1. All salt stocks should be stored in the refrigerator and are stable for several months
2. Always prepare stocks with glass distilled or demineralized water
3. Label the stock solutions clearly with date
4. Reagent grade chemicals should be used to ensure maximum purity
5. The nitrate stock usually precipitate out and must be heated until crystals are completely dissolved before using
6. Any stock showing cloudy or has bacterial or fungal growth should be discarded
7. Do not combine the stock to other stocks unless they are stable and compatible.

Plant Growth Regulators

The four classes of growth regulators are commonly used in tissue culture media i.e., auxins, cytokinins, gibberellins and abscisic acid. The type of growth regulators and concentration used will vary according to the cell culture purpose.

An auxin (IAA, NAA, 2, 4–D or IBA) is required for the induction of cell division and root initiation in cultured tissues. The auxins are mostly used in combination with cytokinins. The 2, 4–D is used for callus induction where as IAA, IBA, and NAA are used for root induction. Auxin stocks are usually prepared by dissolving in ethanol, I N NaOH or 1 N KOH until crystals dissolved (not more than 0.3 ml/10 mg of auxin), rapidly adding 90 ml of distilled water and increasing the volume to 100 ml in a volumetric flasks. Though the auxins are thermostable, however IAA is destroyed by low pH, light, oxygen and peroxidases. The NAA and 2, 4–D are most stable form of auxin.

The cytokinins (KN, BAP, 2iP and Zeatin) are adenine derivatives, promote cell division, shoot proliferation, organogenesis and somatic embryogenesis. They have essential role in differentiation and micropropagation of most plant species.

The cytokinin stocks are prepared in a few drops of In HCl and water to dissolve crystals. Gentle heating is usually required for complete dissolving of crystals. Bring the stock up to the desired volume by adding double distilled water in a volumetric flasks. Cytokinin stocks can be stored for several months in the refrigerator. Cytokinins are thermostable during autoclaving in media.

The gibberellins are infrequently used in plant tissue cultures as it can inhibit callus growth but for meristem culture after shoot primordia formation are used in plant regeneration and elongation.

The stock solution of GA_3 can be prepared by dissolving in water and adjusting the pH 5.7. They are not thermostable therefore should be filter sterilized.

The abscisic acid is useful in embryo culture and somatic embryogenesis. Abscisic acid is heat stable but light sensitive. Stock solutions can be prepared in double distilled water and stored in coloured bottle in refrigerator. Dilution of stock solutions may be as per the requirement.

Vitamins: Vitamins have catalytic functions in enzyme reactions. The most commonly used vitamins in tissue culture media are nicotinic acid (B_3), thiamine (B_1) and pyridoxine (B_6). They are added in medium before autoclaving. The stocks are usually prepared in water at 100X or 1000X (10 ml per 1000 ml medium or 1 ml per 1000 ml medium) and stored in a freezer.

Carbon Source

The carbohydrates in form of sucrose or glucose (2–5% W/V), as a carbon source are essentially required in tissue culture as cells or tissues are generally not photosynthetically active. Lower levels of a carbohydrate may be used in protoplast culture but higher levels are required for embryo or anther culture.

Sugars undergo caramelization prolonged on autoclaving (too long period) and will react with amino acid compounds. Sugars are degraded and form melanoidin, which are brown, high molecular weight compounds that can inhibit cell growth.

Hexitols: Among hexitols, myo–inositol has been found very important ingredient in tissue cultures, it is considered as growth promoter in tissue cultures. This has an action like carbon source as well as vitamin like. Mannitol or sorbitol are good osmotica for protoplast isolation. It is water soluble and stock can be made up at the strength of 100 X (10 ml aliquots are used for 1000 ml medium).

Gelling Agent

In tissue cultures, washed or purified agar of TC grade or Difco–bacto agar grade is used. The agar must be kept in motion while dissolving otherwise it will burn on the bottom of the flask. The agar must be completely dissolved before it is dispensed into the culture vessels.

The agar can also be melted in a autoclave or in a foil capped Erlenmeyer flask for 15 min at 121 °C and dispensed aseptically into sterile containers by using laminar air–low bench before solidification of agar.

Amino Acids and Amides

The amino acids and amides are very important in tissue cultures specially for the morphogenesis. All L forms of amino acids are commonly used, as L–tyrosine can contribute to shoot initiation, L–arginine can facilitate rooting, and L–serine can be used in haploid embryos induction in microspore cultures. L–cysteine is used for controlling phenol leaching from explant tissues. Amides such as L–glutamine and L–asparagine can induce somatic embryogenesis.

Antibiotics

The various fungicides and bactericides are used in case plant explants on cultures excessively contaminated. These chemicals are toxic not only to contaminants but also to cultures or explant materials so restricted use should be made for additions into the culture medium. The antibiotics are soluble in water should be made fresh and be added to the medium. The antibiotics are soluble in water should be made fresh and be added to the medium after autoclaving by filter sterilization.

Natural Complexes

The natural complexes such as coconut (endosperm) milk (CM), yeast extract (YE), malt extract (ME), tomato juice, potato extract, casein hydrolysate (use enzyme digest) and fish emulsion are used in tissue cultures for various purposes. Addition of these complexes in the medium make the medium undefined, since variation in growth promoting or inhibiting compounds in these complexes.

Antoxidants: The antioxidants such as citric acid, ascorbic acid, pyrogallol, phloroglucinol and L–cysteine are used in tissue culture to reduce excessive browning of the explants. Adsorbents like PVP and activated charcoal are also used for checking excessive browning.

ADDITIONAL REQUIREMENTS

Quality of water, chemicals and natural complexes: Demineralized or double distilled water of high purity are used in making stocks and medium. Glass distilled water is most desirable and stored in clean containers.

Callus–induction medium

Murashige and Skoog medium 1.01
2, 4–D .... 1.0 mg
Agar .... 8.0 g

Prepare 1 litre of standard MS medium. Add the 2, 4–D and adjust the medium pH to 5.5 using 0.1 M NaOH. Dissolve the agar in the medium in a steam bath. Dispense the medium into culture tubes or vessels and autoclave for 20 min at 121 °C.

Chlorate Selection Medium

$MClO_3$ .... 600.0 mg
$Ca(NO_3)_2\ 4H_2O$ .... 118.0 mg
$MgSO_4\ 7H_2O$ .... 19.5 mg

K_2HPO_4 .... 19.7 mg
P–IV trace metal solution .... 3.0 ml

Dissolve the salts in 900 ml of distilled water. Add the P–IV trace metal solution and adjust the medium pH to 7.0. Add additional distilled water to make 1 litre of medium.

Chlorate Selection Medium Overlay

Chlorate selection medium .... 500.0 ml
Gel–rite ™ gelling agent .... 4.0 g

Slowly add the Gel–rite a little at a time to the chlorate selection medium while stirring the mixture with a magnetic stirrer. Set the mixture in a steam bath to dissolve the Gel–rite. Dispense 4 ml of the overlay medium into each culture tube; 4 ml should spread out as a very thin layer over the surface of the media in plates.

Embryo Culture Medium

Murashige and Skoog medium 1.01
Agar .... 8.0 g

Prepare 1 litre of standard MS medium. Adjust the medium pH to 5.5 using 0.1 M NaOH. Dissolve the agar in the medium in a steam bath. Autoclave for 25 min at 121 °C. Allow the medium to cool to 50 °C in a temperature controlled water bath. Pour the medium into sterile 100 mm petri plates.

L/10 Green Algae Medium

$Ca(NO_3)_2\ 4H_2O$ .... 0.118/ g
$MgSO_4\ 7H_2O$ .... 0.0195 g
K_2HPO_4 .... 0.0197 g
P–IV trace metal stock .... 3.0 ml

Dissolve all of the salts in 1 litre of distilled water. Adjust the medium to pH 7.0 by adding 1 M HCl or 1 M NaOH.

Micropropagation Medium

Murashige and Skoog medium 1.01
Indolebutyric acid (IBA) .... 1.0 mg
Benzylaminopurine (BAP) .... 3.0 mg

Prepare 1 litre of standard MS medium. Add the IBA and BAP and adjust the medium pH to 5.5 using 0.1 M NaOH. Dissolve the agar in the medium in a steam bath. Dispense the medium into culture tubes or vessels and autoclave for 20 min at 121 °C.

Murashige and Skoog (MS) Medium

Macro Salts

NH_4NO_3 .... 1.65 g
KNO_3 .... 1.90 g
$CaCl_2\ 2H_2O$ .... 0.44 g

$MgSO_4$ $7H_2O$ .... 0.37 g
KH_2PO_4 .... 0.17 g

Micro Salts

$FeSO_4$ $7H_2O$ .... 27.80 mg
Na_2EDTA $2H_2O$ .... 33.60 mg
KI .... 0.83 mg
H_3BO_4 .... 6.20 mg
$MnSO_4$ $4H_2O$ .... 22.30 mg
$ZnSO_4$ $7H_2O$ .... 8.60 mg
Na_2MoO_4 $2H_2O$ .... 0.25 mg
$CuSO_4$ $5H_2O$ .... 0.025 mg
$CoCl_2$ $6H_2O$ .... 0.025 mg

Organic Supplements

Myoinositol .... 100.00 mg
Nicotinic acid .... 0.05 mg
Pyrodoxine HCl .... 0.05 mg
Thiamine HCl .... 0.05 mg
Glycine .... 0.20 mg
Sucrose .... 20.00 g

Dissolve the salts and organics in 800 ml of distilled water. Adjust the medium pH to 5.7 by adding 1 M NaOH. Add additional distilled water to adjust the final volume to 1 litre.

MS/C Medium

MS Salts and Organic Supplements

IAA solution .... 8.0 ml
Kinetin solution .... 2.5 ml
Agar .... 8.0 g

Dissolve the salts and organics in 800 ml of distilled water. Add the IAA and kinetin solutions. Adjust the medium pH to 5.7 by adding 1 M NaOH or 1M HCl. Add additional distilled water to adjust the final volume to 1 litre. Add the agar and heat the medium on a hot plate or in a steam bath until the agar melts. Stir the medium occasionally until all the agar is dissolved and the solution is clear. Do not let the medium boil. Dispense 8 ml aliquots in 20 X 150 mm culture tubes (approximately 120 tubes of medium).

Myriophyllum aquaticum Shoot–Induction Medium

Murashige and Skoog medium .... 1.0 litre
[2–isopentenyl] adenine (2iP) .... 2.0 mg
Agar .... 8.0 g

Prepare 1 litre of standard MS medium. Add the 2iP and adjust the medium pH to 5.7 using 0.1 M NaOH. Dissolve the agar in the medium in a steam bath. Dispense the medium into culture tubes or vessels and autoclave for 20 min at 121 °C.

M. aquaticum Stock Plant Medium

Murashige and Skoog medium .... 1.0 l
Agar .... 8.0 g

Prepare 1 litre of standard MS medium. Adjust the medium pH to 5.7 using 0.1 M NaOH. Dissolve the agar in the medium in a steam bath. Dispense the medium into culture tubes or vessels and autoclave for 20 min at 121 °C.

P–IV Trace Metal Solution

Na_2EDTA .... 0.750 g
$FeCl_3$ $6H_2O$ .... 97.0 mg
$MnCl_2$ $4H_2O$ .... 41.0 mg
$ZnCl_2$ .... 5.0 mg
$CoCl_2$ $6H_2O$ .... 2.0 mg
Na_2MoO_4 .... 4.0 mg

First dissolve the Na_2EDTA in 500 ml of distilled water, then dissolve the remaining metal salts. For greater accuracy, it may be easier to prepare 10 X concentration stock solutions of the Zn, Co, and Mo salts and add 1/10 of these stocks to the P–IV stock solution.

Pandorina Ammonium Medium

NH_4Cl .... 27.0 mg
$CaCl_2$ $2H_2O$ .... 100.0 mg
$MgSO_4$ $7H_2O$ .... 19.5 mg
K_2HPO_4 .... 19.7 mg
P–IV trace metal solution .... 3.0 ml

Dissolve the salts in 900 ml of distilled water. Add the P–IV trace metal solution and adjust the medium pH to 7.0. Add additional distilled water to make 1 litre of medium.

Pandorina Nitrate Medium

$NaNO_3$ .... 35.0 mg
$CaCl_2$ $2H_2O$ .... 100.0 mg
$MgSO_4$ $7H_2O$ .... 19.5 mg
K_2HPO_4 .... 19.7 mg
P–IV trace metal solution .... 3.0 ml

Dissolve the salts in 900 ml of distilled water. Add the P–IV trace metal solution and adjust the medium pH to 7.0. Add additional distilled water to make 1 L of medium.

Pandorina Hypoxanthine Medium

Hypoxanthine .... 68.0 mg
$CaCl_2$ $2H_2O$ .... 100.0 mg
$MgSO_4$ $7H_2O$ .... 19.5 mg
K_2HPO_4 .... 19.7 mg
P–IV trace metal solution .... 3.0 ml

Dissolve the salts in 900 ml of distilled water. Add the P–IV trace metal solution and adjust the media pH to 7.0. Add additional distilled water to make 1 litre of medium.

Pandorina Uric Acid Medium

Uric acid .... 84.0 mg
$CaCl_2$ $2H_2O$ .... 100.0 mg
$MgSO_4$ $7H_2O$ .... 19.5 mg
K_2HPO_4 .... 19.7 mg
P–IV trace metal solution .... 3.0 ml

Dissolve the salts in 900 ml of distilled water. Add the P–IV trace metal solution and adjust the medium pH to 7.0. Add additional distilled water to make 1 litre of medium.

Potato Dextrose Agar

White potatoes, sliced .... 250 g
Dextrose .... 20 g
Agar .... 15 g

Boil the potatoes in 500 ml of distilled water for 15 min until soft. Filter this mixture through cotton to remove most of the particulate matter. Dissolve the dextrose in 200 ml of the potato infusion. Add 800 ml of distilled water. Adjust the final pH to 3.5–4.0. Dissolve the agar in a steam bath or on a hot plate. Autoclave at 121 °C for 25 min.

Trypticase–Soy Broth Medium

Trypticase .... 17.0 g
Phytone .... 3.0 g
NaCl .... 5.0 g
K_2PO_4 .... 2.5 g
Glucose .... 2.5 g

Dissolve the ingredients in 1 litre of distilled water. Adjust the medium pH to 7.3 by adding 1 M NaOH. Dispense 10 ml aliquots in 20 X 150 mm culture tubes. (approximately 100 tubes of medium).

Yeast Extract Broth (YEB)

Yeast extracts .... 1.0 g
Beef extract .... 5.0 g
Peptone .... 5.0 g
Sucrose .... 5.0 g
$MgSO_4$ $7H_2O$ .... 0.5 g

Dissolve all the ingredients in 1 litre of distilled water. Adjust the medium pH to 7.0 by adding 1 M NaOH. Dispense and autoclave for 25 min at 121 °C.

Yeast Extract Indicator Medium (YI)

Yeast extracts .... 1.0 g
Lactose .... 10.0 g
Agar .... 20.0 g

Dissolve the yeast extract and lactose in 1 litre of distilled water. Adjust the medium pH to 7.0 by adding 1 M NaOH. Add the agar and dissolve it by heating the mixture in a steam bath or on a hot plate. Autoclave for 25 min at 121 °C.

Sterilization of Media

Tissue culture media are generally sterilized by autoclaving at 121 °C and 1.05 kg/ cm^2 (15–20 psi). The time required for sterilization depends upon the volume of medium in the vessel. Dispense medium in small aliquots whenever possible as many media components are broken down on prolonged exposure to heat. Medium exposed to temperatures in excess of 121 °C may not properly get or may result in poor cell growth.

Minimum autoclaving time includes the time required for the liquid volume to reach the sterilizing temperature (121 °C) and 16 min at 121 °C. Times may vary due to differences in autoclaves. Validation with system is recommended. Several medium components are considered thermolabile and should not be autoclaved. Stock solutions of the heat labile components are prepared and filter sterilized through a 0.22 mm filter to sterile container. The filtered solution is aseptically added to the culture medium which has been autoclaved and allowed to cool to approximately 35–45 °C. The medium is then dispensed under sterile conditions.

TABLE 10.6: Minimum Autoclaving Time

Volume of Medium per Vessel (ml)	Minimum Autoclaving (min)	Volume of Medium per Vessel (ml)	Minimum Autoclaving (min)
25	20	500	35
50	25	1000	40
100	28	2000	48
250	31	4000	63

Use and Storage of Coconut Water

Coconut water has been shown to stimulate shoot proliferation in many species of plants. Coconut water is prepared from selected coconuts by filter sterilized and frozen prior to use. This precipitation should not effect the growth of the plant tissue. Coconut water can be divided into smaller aliquots, corresponding to standard medium batch size, and refrozen until needed. Coconut water should be used at a concentration of 5–20% (v/v).

RELATED PROCEDURES

Ultraviolet Light

UV light may be divided into three wave length groupings near UV (315–400 nm), mid range UV (280–315 nm) and far UV (200–280 nm). Maximal sensitivity in humans is at about 280 nm. Exposure to direct or indirect mid–range or for UV can cause acute eye irritation after a latent period of 2–24 hrs. Because retina is not sensitive to UV eye damage may result without the subject being aware of the exposure. Skin is also sensitive to UV which may cause for skin cancer. Hence protect your eyes and skin from the effects of UV irradiation by wearing goggles with side shields by clothing, and by limiting exposure.

Preparation of Phenol

All crystalline phenol must be redistilled at 160 °C to remove contaminants that cause or cross linking of DNA or RNA. Soon after distillation add 0.1% hydroxyquinoline. The melted phenol is extracted several times with an equal volume of 1.0 M Tris pH 8.0 followed by 0.1 M Tris pH 8.0 and 0.2% ß–mercaptoethanol, until pH of the aqueous phase is 7.6. Phenol is stored in aliquots at 4 °C under equilibration buffer for periods upto 1 month.

Phenol is widely used as a disinfectant and germicide. It is a dangerously toxic material that can produce poisoning when ingested, inhaled or absorbed trough the skin. The toxic effect include headache, dizzines, nausea, weakness, difficulty in breathing, unconciousness and death. Phenol is corrosive to skin, initially producing a softened area followed by severe burns.

1. If phenol is spilled on the skin, flush immediately with large amounts of water. Do not use ethanol.
2. If eyes are contaminated, wash them with running water for about 15 min, call for medical help.

Working with ^{32}p Labeled Compounds

ß–particles with energy of 1.71 MeV (6.1 meter range in air) is emitted by ^{32}P. Hence ^{32}P labeled compounds must be handled carefully with much caution using shields. When ß particles hit targets, electromagnetic radiation known as **Bremsstrahlung** is produced, the yield of which is directly proportional to the density of material used for shielding. Therefore, a low density material may be added to absorb the Bremsstrahlung emitted. Always ear gloves (two pairs if necessary), protect eyes and use dosimeters. Always cover the work area with absorbant papers and use survey meter to check spillage. Eating, smoking or drinking while handling radioactive compounds should be banned. Use special tape to label containers and tubes in which radioactive materials are kept. The maximum permissible burden of ^{32}P is 30 uCi but the maximum permissible burden for bone is only 6 μCi.

Silanization of Plastic and Glassware

Place the plastic and glasswares to be silanized in a desiccator. Add about 1 ml of dichlorodimethyl silane in a small container in the desiccator. Pull vacuum on the system until dichlorodimethyl silane boils. Close the system and allow to sit for about 2 hrs. Open the desiccator in a fume hood and allow to air out several hours. Rinse the plastic and glassware with water and autoclave.

10 X Restriction Endonuclease Buffers (Refrigerate)

Low............100 mM Tris–HCl (pH 7.5)
............100 mM $MgCl_2$
............10 mM Dithiothreitol (DTT)
Medium............500 mM NaCl
............100 mM Tris–Cl (7.5); 10 mM DTT
High............1 M NaCl
............500 mM Tris–Cl pH 7.5
............100 mM $MgCl_2$
............10 mM DTT

Preparation of Dialysis Tubing

Cut dialysis tubing into convenient length. Boil them for 10 min in a large volume of 2% sodium bicarbonate and 1 μM EDTA. cool and rinse thoroughly in distilled water and again boil for 10 min in distilled water. Cool and store inrefrigerator submerged in water. Just before use rinse with water. Wear gloves while handling the dialysis tubing.

Lengths and Molecular Weights of Common Nucleic Acids

Nucleic Acid	Number of Nucleotides	Molecular weight
LAMBDA DNA	48, 502 (Circular, dsDNA)	3.0×10^7
pBR322 DNA	4, 363 (ds DNA)	2.8×10^6
28SrRNA	4, 800	1.6×10^6
23S rRNA	3, 700	1.2×10^6
18S rRNA	1, 900	6.1×10^5
16S rRNA	1, 700	5.5×10^5
5S rRNA	120	3.6×10^4
tRNA (E. coli)	75	2.5×10^4

Standards

1 kb of ds DNA (sodium salt) .... 6.6×10^5 Daltons
1 kb of ss DNA (sodium salt) .... 3.3×10^5 Daltons
1 kb of ss RNA (sodium salt) .... 3.4×10^5 Daltons
The average MW of a deoxynucleotide base ....324.5 Daltons

Common Conversions of Nucleic Acids and Proteins

I. Spectrophotometric Conversions
IA_{260} unit of ds DNA ....50 mg/ml
IA_{260} unit of ss DNA ....33 mg/ml
IA_{260} unit of ss DNA....................40 mg/ml
II. Protein Molar Conversions
100 pmoles of 100,000 MW protein ....10 mg
100 pmoles of 50,000 MW protein ....5 mg
100 pmoles of 10,000 MW protein ....1 mg
III. Protein/DNA Conversions
1 kb of DNA = 333 amino acids of coding capacity .... 3.7×10^4 MW
10,000 MW protein.................... 270 bp DNA
30,000 MW protein.................... 810 bp DNA
50,000 MW protein....................1.35 kb DNA
100,000 MW protein....................2.7 kb DNA

Half life of Important Radioisotopes Used

Radionucleotide .... Half life
Tritium....................12.43 years
Carbon–14 ....5.730 years
Sulphur–35....................87.4 years

Phosphorous–32 ....14.3 years

Problems and Possible Solutions in Plant Tissue Culture Work

Symptoms	Possible Causes	Possible Solution
Culture contamination	Source heavily infested/ infected with micro organisms Poor sterilization	Improve sterilization method (70% Ethanol, drop of detergent, antimicrobial agents) clean plant material/ select unexposed tissue.
Explant dies	Strong disinfectants.	Use weaker disinfectant/ change sterilizing agents
	Media too strong	Use 1/2 or 1/4 strength
	Wrong stage of growth	Obtain explants at different stage of growth
Culture blackens and dies	Contaminated by micro–organisms	Discard with care. Review sterile technique and sanitation
	Bleeding	Subculture immediately. Transfer more frequently Try different agar concentrations
	Agar problem	Check water purity Try different formula
	Water problem	
	Wrong formula media constituents	
Explant live but no growth	Dormant	Chill for a month
	Media too harsh	Use explant at different stage of growth
	Wrong formula	Lower salts and hormones try different formula
Culture live but no growth	Too cold	Increase temperature
	Wrong medium	Try different medium
Shoots tool long (leggy), and poor multiplication leaves are yellow, watery	Too little cytokinin	Increase cytokinin Change cytokinin/ auxin ratio
Leafy shoots too short	Hormones too strong	Decrease or omit hormones
No multiplication	Too little cytokinin	Increase cytokinin
	Needs chilling	Cold store 4–8 weeks
	Too cold	Run cytokinin/ auxin grid
	Requires dormancy period	Increase temperature Cold treat 3–8 weeks
Fat stems, small and pale leaves	Too much cytokinin	Decrease cytokinin/ increase auxin
Unwanted callus	Wrong hormones	Decrease or omit hormones Run cytokinin/ auxin grid
Leaves chlorotic	Contaminant	Index for contaminants
	Too hot	Decrease temperature
	Wrong formula	Try different medium
Leaves succulent (watery), virtrification, abnormal stem, embryos	Osmotic potential upset	Decrease temperature
	Too high cytokinin	Increase agar strength

Symptoms	Possible Causes	Possible Solution
	Wrong agar Culture too old	Decrease hormones Try different agar Transfer more often
Premature rooting	Wrong hormone balance	Transfer more often Run cytokinin/ auxin grid Increase cytokinin/ decrease auxin
Red stems/ embryos/ cells	Stress	Change light and/ or temperature Lower sucrose content Increase nitrate (NO_3) Transfer more frequently
Non–friability of callus	Too much sugar Not enough NO_3 Culture too old Media composition Plant source	Using number of media having range of hormones. Select most rapidly growing callus. Change to liquid culture and after friable to agar medium.

Source: Bhansali, R. R. (1995)

INITIATION OF EMBRYOGENESIS IN SUSPENSION CULTURE

Steward and coworkers (1958) observed the initiation and development of embryos from differentiated tissue (somatic tissue). However, earlier researchers had already observed these formations in culture, without recognizing them as embryos.

These embryos are named as somatic embryos, and their origin is different from zygotic embryos, which result from the fusion of gametes. This phenomenon is also called adventive or asexual embryogenesis.

Somatic embryos closely resemble zygotic embryos in their structure, as well as in their physiological properties, especially with wild carrot. Nevertheless, the organization of these somatic embryos is generally more variable than that of zygotic embryos. Abnormalities in shape and size are frequent, and because of this variation they were, at first, called embryoids.

In contrast to buds, somatic embryos have a bipolar axis with an apical meristem and a root, and as they have no vascular connection with the mother callus, they are easily detached by the swirling action of the agitated medium.

Three types of medium are required to induce somatic embryogenesis in suspension.

1. The first medium is solidified by agar and contains auxin; this dedifferentiates thesomatic tissue and provokes initiation of embryogenic cells (primary culture).
2. The second medium, which is liquid and also contains auxin, ensures the multiplication of these cells.
3. The third medium, which is also liquid but without auxin, allows the cells to express their embryogenic potential.

It is not very easy to carry out culture in a liquid medium. When the tissue is inoculated, it sinks and rapidly dies because of lack of oxygen. To prevent asphyxia, the liquid must be agitated to dissolve air in the medium. The culture of somatic embryos in a liquid medium, however, has numerous advantages. The swirling medium naturally separates the embryos, which are then easily observed. Thus, the embryos can be fractionated according to their stages. They can be obtained in great quantity and used as a basis for a large–scale micropropagation.

Materials

1. Growth chamber for plant culture. Temperature (25 °C) and lighting are regulated. Lighting must be bright (5 W/m^2), discontinuous (16 h of lighting/d), and provided by Grolux fluorescent tubes (F36W/GRO). This growth chamber is optional, since plant material can also be collected from outdoors.
2. Culture room for tissue culture. Temperature has to be regulated at approximately 25 °C. Continuous light (1 W/m^2) can be supplied by ordinary fluorescent tubes placed above the rotary shaker.
3. Actively growing callus in tubes.
4. Culture medium: The mineral solution of Murashige and Skoog can be used for culture of somatic embryos. The solution described in chapter was evolved for the culture of zygotic embryos because of its low toxicity. It must have a pH of 5.5. In addition, it contains NH_4^+ and K^+ ions, which are considered to be promoters of somatic embryo growth.
5. For mature embryo–growth, use half–strength MS medium containing sucrose at 5 g/l.
6. Sieves (1–mm grid) made of stainless screens (tea sieves from hardware store) to fraction suspension culture and transfer only the smaller embryos into the new flask. Sieves can also be made with nylon mesh filtration cloth (industrial nylon mesh filter).
7. Subsequent plantlet development requires filter–paper bridge apparatus and small pots containing soil mixture.

Method

Generally, three steps are needed to obtain embryogenic suspension culture. First, embryogenic tissue is initiated by culture on an agar medium with auxin; it is then transferred to liquid medium with auxin, and eventually transferred into a liquid medium without auxin.

Initiation of Embryogenic Tissue

Growth initiation is better on solidified medium containing 2, 4–D (1 mg/l). When the callus has developed the potentiality for embryogenesis, that is to say 5 wk at least after explant isolation, it is transferred to a liquid medium.

ESTABLISHMENT OF EMBRYOGENIC SUSPENSION CULTURE

1. The callus should be removed from the culture tube with sterile forceps and transferred to a Petri dish containing filter paper.
2. Discard brownish parts of the callus, and inoculate a fragment of about 2 g in a 250–ml, Erlenmeyer flask containing 50 ml of medium.
3. Smaller Erlenmeyer flasks can be used, but the volume of liquid in relation to the size of the flask must be, for adequate aeration, about 20% of the volume of the flask. These flasks are placed on a gyratory or orbital shaker and are agitated at 100–150 rpm.
4. When the plant material is first placed in the medium, there is an initial lag period prior to cell division. This is followed by an exponential rise in cell number. Finally, the cells enter a stationary stage.
5. In order to maintain the viability of the culture, the cells should be subcultured at the beginning of this stationary phase. It is reached in 2–3 wk, and the suspension

has to be transferred to fresh medium at regular intervals within this period. In addition, a minimum density must be achieved when cells are transferred to fresh medium, to maintain embryogenic potential.

6. To subculture embryogenic suspension, wait several minutes until the cells are settled at the bottom of the flask and decant almost all the medium. Resuspend the suspension by gently rotating the flask, and transfer one–fourth of the entire population to fresh medium.
7. All these operations can also be carried out, under safer conditions, using a micropipet. In that case, when the cells have settled, aspirate almost all the supernatant above the cells. Then, aspirate one–fourth of the cell population and transfer it into a new medium.
8. A population of cells typically shows a wide range of proembryonic stages visible under the dissecting microscope. It is a mixture of single cells, proembryonic cell clusters, and new embryogenic centers and older embryos.
9. To obtain a certain degree of uniformity, it is necessary to sieve the inoculum when transferring the suspension.
10. The proembryo suspension is passed through a 200 mm sieve and then a 100 mm sieve. In the second sieve, differentiated embryos from the late globular stage are retained and may be examined, while proembryonic cells pass through.
11. When the suspension, which has passed the sieves, is settled decant most of the medium to obtain a suspension with a high density of cells.

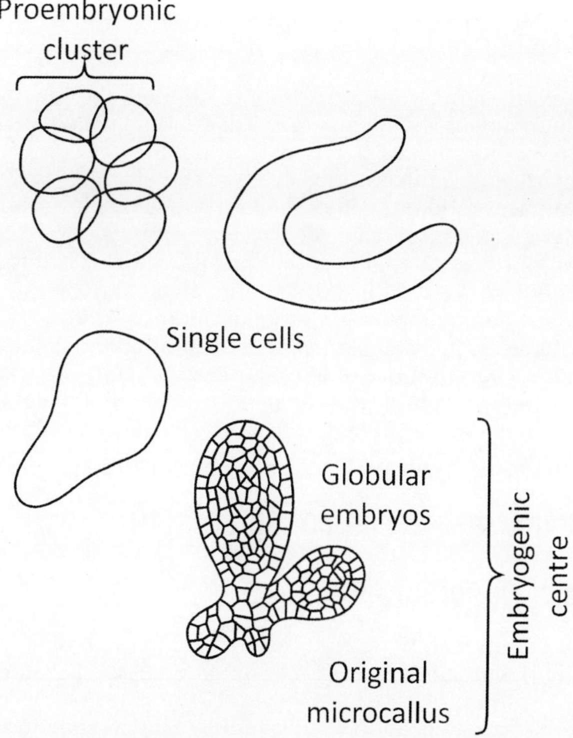

FIG. 10.15: The embryogenic cells divide in the liquid medium and constitute proembryonic clusters. A certain number of new embryogenic centers will give rise to globular embryos of different sizes (wild carrot).

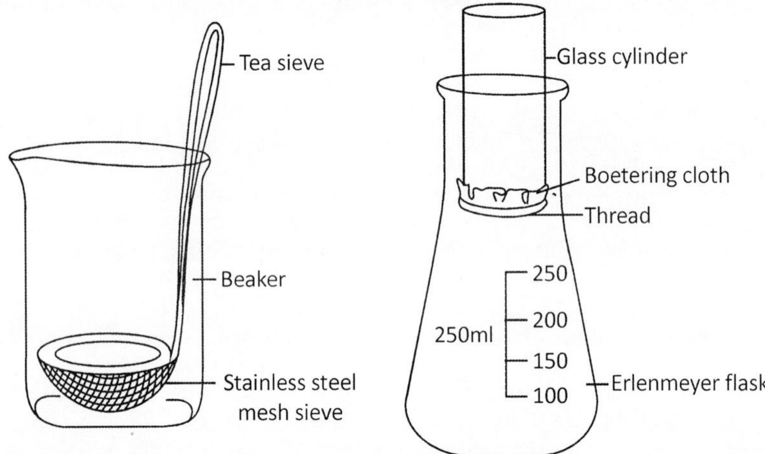

FIG. 10.16: The metal sieve, on the left, is used principally to remove the larger embryos before transferring the embryonic population into a fresh medium. The nylon sieve, on the right, is used to obtain a homogeneous population of embryogenic cells and synchronize the different stages of embryo development.

Maturation of Embryos

When embryo expression and maturation are needed, the inoculum is transferred to a liquid medium having the same composition, but lacking auxin and glutamine. Cells express their embryogenic potentiality, and if the cells have not been carefully homogenized by sieving, a mixture size in 2 wk. These embryos can be dispersed, in a Petri dish, on an agar medium without auxin, but with small amounts of cytokinins (0.1 mg/l of BAP or zeatin).

When the embryos are mature, they are transferred to tubes on filter paper bridges. The medium is composed of Murashige and Skoog mineral solution diluted by half and containing a low concentration of sucrose (5 g/l). To stimulate the formation of leaves, the tubes must be well illuminated. When the plantlets are tall enough, with a large number of leaves and roots, they are transferred into pots with a mixture of peat, soil, and vermiculite for subsequent development. After their transfer, plantlets have to be protected from desiccation by covering the pots with plastic film for a few days.

IMPORTANT PARAMETERS FOR CONSIDERATION

Simplification of the Technique

1. Suspension culture can be started by inoculating the liquid medium containing 2, 4–D (1 mg/l) with an explant of differentiated tissue isolated directly from the plant.
2. The initiation of multiplication of cells requires the inoculation of a large amount of tissue– about 3 g for 100 ml of medium.
3. The dividing cells will gradually free themselves from the inoculum because of the swirling action of the liquid.

Different Causes of Failures (and Remedies)

Darkening of the Medium

In a number of cases, especially with explants isolated from woody plants, the liquid medium becomes dark after a few days. The radicle ends of the embryos turn brown, and the embryos rapidly stop growing and show necrosis.

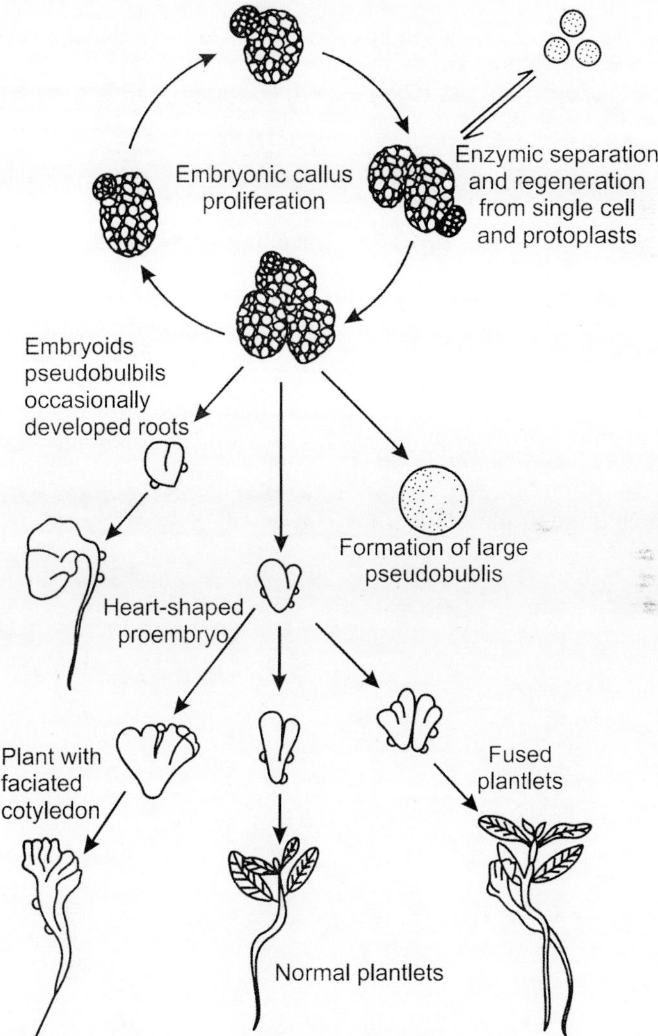

FIG. 10.17: Presentation of embryonic callus proliferation, enzymic separation and regeneration from single cell and protoplast.

Several procedures can help to reduce this process. More frequent transfer into fresh medium can prevent browning. After transfer to liquid, for the first days, when the proembryos are particularly sensitive, the culture must be performed in complete darkness. After this period, the culture can again be exposed to light. This lethal browning is the result of formation of oxidized polyphenols, so antioxidants, such as ascorbic acid and cysteine, may be used, but the results are variable.

No Appearance of Embryos

Not all plant species respond to culture and produce somatic embryos. Some plants give only adventitious buds, and others, no organogenesis at all. if a good experimental model is sought, it is preferable to select some species in families like Umbelliferae an Ranunculaceae, since they offer an easy way to obtain somatic embryogenesis. The explant has to be carefully chosen, since different regions of the plant body do not have the same embryogenic potential. Young material like zygotic embryos and reproductive tissue is an excellent source to initiate embryogenesis. For example, the diploid tissue of an immature anther of grapevine develops into somatic embryos.

Insufficient Number of Embryos

If a culture produces few embryos, suspending the cells in a plasmolyzing solution containing a high concentration of sucrose (1 M) for a short time (45 min) is recommended. When these are replaced in normal conditions, somatic embryogenesis is efficiently promoted. The effect may be because of the rupture of intercellular connections and isolation of cells that are reinstated into the condition of the zygote.

Alterations of the composition of the medium can be beneficial. Some authors demonstrated that NAA (naphthaleneacetic acid) is sometimes better to initiate somatic embryogenesis than 2, 4–D. Cytokinins like zeatin can help the embryos to mature. The addition of some amino acids like serine, proline, and glycine is generally considered to stimulate embryo growth.

CHAPTER 11

Protoplast Technology

The field of tissue culture has expanded at such a rate that the time has now arrived for individual techniques such as protoplast culture to break off as their own individual areas.

Protoplasts are plant cells with a plasma membrane but without the cell wall. The absence of cellulosic cell wall permits advantages to fuse protoplast of similar or on different species and the fused product can generate into whole plants. This is called somatic fertilization or fusion of two vegetative cells to form hybrids.

IMPORTANCE OF PROTOPLAST ISOLATION AND CULTURE

The isolation, culture and fusion of protoplast are one of the most fascinating fields of research. The techniques are important for the following reasons:

1. To develop novel hybrid plant through protoplast fusion, genetic engineering would continued to be an exciting area of research in modern plant biotechnology This technology holds great promises to synthesize a plant of desired characteristics.
2. This helps in crop improvement by somatic hybridization and cell modification.
3. The protoplast in culture can be regenerated into an entire plant.
4. The technique in future will be one of the most frequently used research tools for tissue culturists, physiologists, pathologists, molecular biologists, cytogeneticists, and biotechnologists.
5. It provides a tool for isolating protoplasts and exploring the possibilities of genetic engineering.

Isolation of Protoplast from Various Plant Parts

Plant tissues	Young leaves, roots, stems
Flower parts	Petal, reproductive organs
Callus/cells	Friable callus tissues/cells
Suspension cultures	Fast growing cells
Aseptically grown plantlets	Leaves, stems, roots

Protoplast is the living material of the cell where as an isolated protoplast is the cell from which the cell wall is removed. Protoplasts can be isolated from almost all plant parts viz., roots, leaves, fruits, tubers, root nodules, endosperms, crown gall tissues, pollen mother cells, pollens, pollen tetrads and cells of callus tissues grown *in vitro*. Protoplasts are isolated from cells by two methods:

1. Mechanical isolation, and
2. Enzymatic isolation

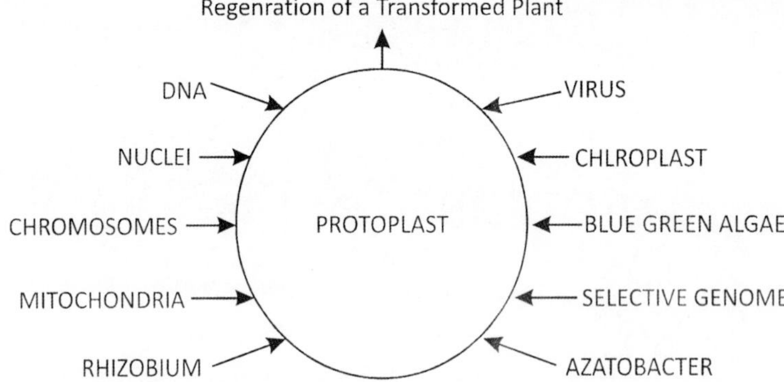

FIG. 11.1: Hypothetical illustration for the uptake of different cell organelles and introduction of genetic material into plant protoplasts. This technique results into the regeneration of new plant bearing desired characters.

TABLE 11.1: Application of plant protoplast in fundamental research of plant sciences

1.	Structure	Cell wall, Plasma membrane, Organelle– structures and biosynthesis
2.	Function	Biochemistry, Photosynthesis, Absorption
3.	Genetics	Fusion, Organelle and DNA uptake
4.	Plant Pathology	Virus, RNA uptake, Pathotoxin
5.	Differentiation	Morphogenesis, Embryogenesis

Enzymatic Method

Enzymatic method is most often used in isolation of protoplasts from various plant parts. Enzymatic methods generate very large numbers of protoplasts in comparison to mechanical methods, but in some instances, the enzymes have deleterious effects on plant metabolism. However, the effects of enzymes used on protoplasts (which is unknown) are overlooked in this method. Enzymatic isolation of protoplast has some advantages over mechanical method such as: (i) cells is not damaged, (ii) Osmotic shrinkage of protoplast is much less.

Mechanical Method

Mechanical method, though not most often is in use, still has its merit. It is rarely used because it is an extremely tedious process that results in the yield of only very small numbers of protoplasts. The cells are plasmolysed, causing the protoplast to shrink away from the cell wall. The protoplasts obtained from this method are then cultured on suitable culture medium.

ENZYMIC ISOLATION OF PROTOPLASTS: METHOD

The starting material is sterile *in vitro* shoot cultures or any plant part. In these cases protoplasts are enzymically isolated in a two–step or a one–step method.

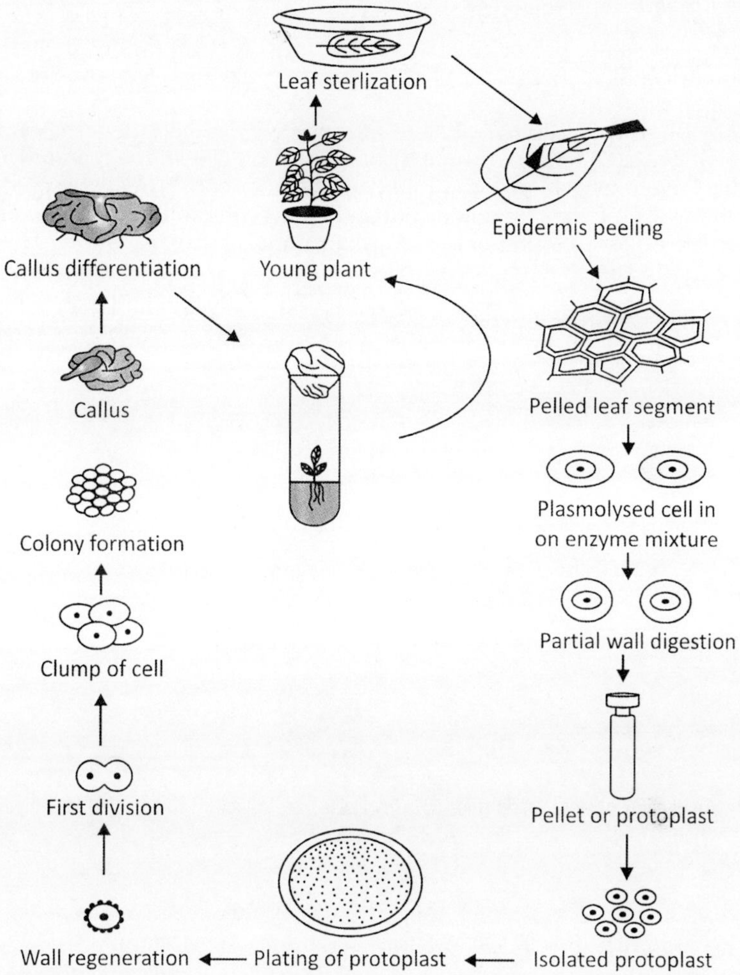

FIG. 11.2: Systematic illustration of the technique used for isolation, culture and regeneration of plants from leaf protoplasts.

First

Isolated cells are prepared (maceration) by treating the leaf or other tissue segment with macroenzyme (a pectinase) in 13% mannitol. The isolated cells obtained from this incubation are purified by filtration through a nylon mesh.

Second

The protoplasts are prepared by incubating these isolated cells in 2 per cent cellulase for about 90 minutes. In the *one–step* method both enzymes (cellulase + pectinase) are used simultaneously. In *one–step* method 2 to 3 hours at 20 to 22°C is adequate to release significant numbers of protoplasts.

Isolation of Protoplast from Leaves

The process of isolating protoplasts from leaves involves a number of steps; these include surface sterilization, removal of epidermis, enzymic digestion and cleaning of the protoplasts.

For isolation, the physiological state of the leaves is extremely important. The leaves normally chosen for experimental purposes should be fully expanded and from young plants (vegetative state, e.g., 10–14 weeks in tobacco). The lower epidermis is removed and is used to isolate epidermal and guard cell protoplasts. The remaining peeled leaf is used for isolation of mesophyll protoplasts. The isolation of protoplast involves four main steps which are as follows:

1. Surface sterilization of leaves.
2. Peeling off the epidermis.
3. Enzymatic treatment.

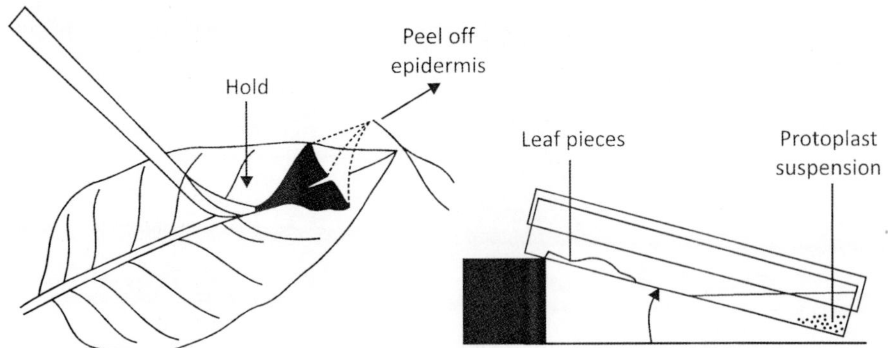

FIG. 11.3. (A) Peeling off leaf epidermis. (B) Release of protoplasts from peeled leaf pieces after enzyme treatment.

Method

1. Fully expanded young leaves are surface sterilized by dipping them into 70% ethanol for one minute and then in 2% solution of sodium hypochlorite for 20–30 minutes.
2. The leaves than washed three times with sterile glass distilled water to remove traces of sodium hypochlorite.
3. The subsequent steps to isolate sterile protoplast are taken up under aseptic conditions.
4. Lower epidermis of sterilized leaves in carefully peeled off and the stripped leaves are cut into small pieces.
5. Peeling off process is easier with flaccid leaves than those with turgid leaves.
6. The protoplast is generally isolated by two methods: (i) Direct Method and Sequential Method.

Direct Method

1. Peeled leaf pieces are placed with lower surface down, into 20 ml solution of 13% mannitol and inorganic salts contained in Petri dishes (90 mm) for one hour.
2. The mannitol–inorganic salt solution is pipette out with the help of Pasteur pipette and an enzyme mixture (20 ml) in mannitol (0.5% macerozyme + 2%

Onozuka cellulase in 13% sorbitol or mannitol at pH 5.4) is added and incubated overnight (15 to 18 hours) at 20-22 °C.

3. The petri plates are then gently agitate to facilitate the release of the protoplasts, pushing the larger pieces of leaf–material to one side with a sterile forceps and keeping the Petri dish at an angle of 15°C.
4. After sometimes protoplasts will settle down in the Petri dish and then protoplast–enzyme mixture is transferred to 15 ml screw-capped centrifuge tubes using Pasteur pipette.
5. Centrifuge the material at 100 g for one minute. This process is repeated for 2 to 3 times and protoplasts are washed with 13% sorbitol solution.
6. For the final washing, 20% sucrose solution is used in place of sorbitol and centrifuged at 200 g for one minute.
7. The cleaned protoplasts float and debris settles down.
8. The floating protoplasts are carefully pipette out with a Pasteur pipette leaving behind the remains of mesophyll cells.
9. This method is useful for the isolation of protoplast both from spongy and pallisade parenchyma.

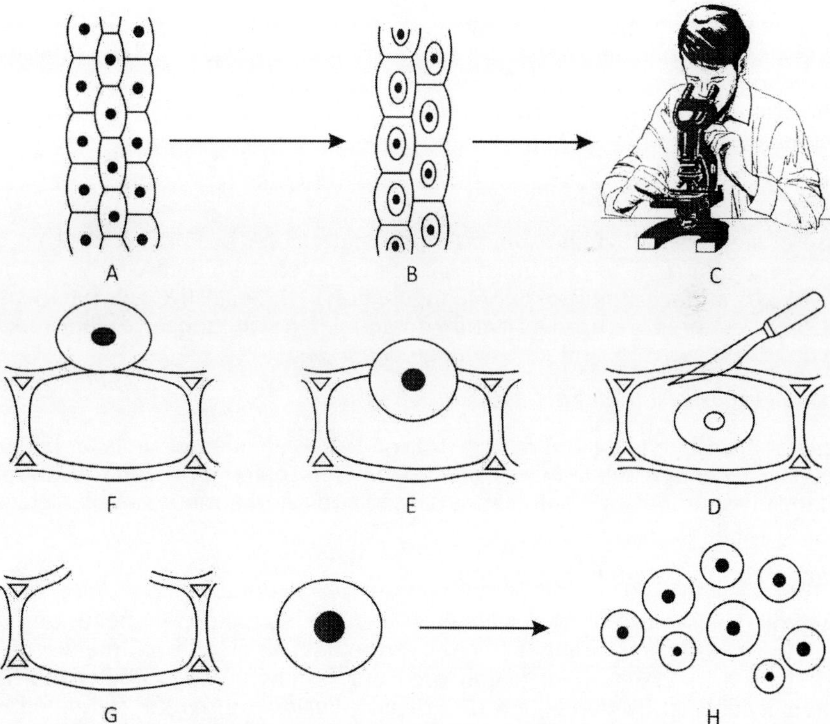

FIG. 11.4: Method of mechanical isolation of protoplasts. (A) A small piece of peeled epidermis, (B) Plasmolysis of the same cells, (C–F) Steps of removal of protoplasts from the cell after incesion, (G&H) Free isolated protoplasts.

Sequential Method

In this method of protoplast isolation cells are first separated and then transformed into protoplasts. This method is completed into two steps:

Step I. Isolation of Cells

1. The peeled leaf segments are placed into a Petri dish (140 mm) containing an enzyme mixture (mecerozyme 0.5%, potassium dextran sulphate 0.3% in 13% mannitol at pH 5.8).
2. The petridishes are put into a desiccator for vaccum infiltration of enzyme mixture for about 4 to 5 minutes then transferred to a water bath at 25°C.
3. Agitate them slowly.
4. After 15 to 20 minutes the enzyme mixture is replaced by fresh one and is again incubated for one hour. By this process cells are completely separated from one another.
5. The isolated cells are filtered through a nylon mesh, then centrifuge at 100 g for one minute.

Step II. Isolation of Protoplasts

1. Isolated cells are now treated with an another enzyme mixture containing 2% cellulase in a 13% solution of mannitol at pH 5.4 for 90 minutes at 30°C and then centrifuged at 100 g for one minute.
2. The supernatant is decant off and the protoplasts are washed three times with mannitol.
3. The protoplasts are finally cleaned with 20 to 25% sucrose solution. This method can be used for isolation of protoplasts from pallisad cells.

Isolation of Protoplast from Shoot Apex

The use of sterile shoot cultures as a starting material for leaf and stem protoplast isolation is popular. One key advantage is that the material is already sterile and does not have to be surface sterilized prior to incubation in the digestive enzyme mixture. Sterile shoot cultures are used as starting material in a wide range of plant species, from potato and tobacco, and a wide range of tree species.

Isolation of Protoplast from Root and Storage Organs

Protoplasts have been isolated directly from a wide selection of roots and storage organs. These include potato tubers, artichoke tubers, onion roots and root nodules. The enzyme treatments are sometimes slower and often a more complex enzyme mixture is required.

Isolation of Protoplast from Pollen Grains

Pollen protoplasts have some advantages over the somatic cells being found in uniform in chromosome number (ploidy). Protoplast obtained from mature pollen could be used in the study of pollen ontogeny, mutation, somatic hybridization, cell modification and regeneration of haploids. Fusion causes the formation of a normal diploid instead of doubling the chromosome number (to tetraploid) of the new hybrid. It is also possible to regenerate haploid plantlets from isolated pollen protoplasts.

Pollen grain has highly sculptured and chemically complex cell wall. It is covered in a highly resistant coat of sporopollenin, as polymer of carotenoid esters and can be removed by a pretreatment with strong oxidizing solutions.

Pollen grains taken out of the anther lobes are sterilized with 2% sodium hypochlorite solution for 10 to 12 minute and subsequently washed 2 to 3 times with sterile distilled water. The concentration and combinations of the enzyme incubation to pollen grains depend upon their stage of development. Protoplasts are released in many ways described as follows:

Protoplast ooze out through germ pores due to dissolution of pectocellulosic membrane by enzyme action. The enzyme mixture used normally contains high concentrations of β–1, 3–gluconase, which is capable of digesting both cellulose and callose.

1. Due to mechanical rupture of the exine.
2. By partial dissolution of the exine.

Isolation of Protoplast from Cultured Cells

Actively dividing cells in suspension cultures are proved as most suitable material for isolation of non–germ protoplasts in large amount. Moreover, protoplasts isolated from various somaclones can also be used by plant breeders as raw material for somatic hybridization and cell manipulation. Protoplast isolation through enzymatic activity from older plant parts is very difficult. Protoplasts have been isolated from suspension culture of many plant species like, carrot, rose, *Atropa* sp., sugarcane, etc. Protoplasts can be easily isolated from such cultures by treating the filtered suspension with 2 to 4% Onozuka cellulase in 0.6 M mannitol, for 4 to 6 hours at 32°C in a gently shaking water bath.

Other Sources: Protoplast has also been isolated from coleoptiles, flower petals, aleurone layers and plant cell tumours or galls. The basic isolation technology is the same in all cases.

METHODS OF PROTOPLASTS CULTURE

Laboratory Facilities

Protoplasts can be cultured in liquid as well as solid medium. Concentration of enzyme, intensity of light (300–3000 lux) and temperature (28–45°C) play vital role in protoplasts culture.

Enzyme Mixture and Osmotic Stabilizer (Osmotica)

The composition of the cell wall indicates that the digestion mixture should be able to degrade cellulose, hemicellulose and pectin, and in some cases callose. The number of commercial enzyme preparations available is somewhat limited and very often they are contaminated with a wide range of biologically active impurities of other enzymes– i.e., proteases, nucleases and lipases. They have a degrading effect on the plasma membrane of the isolated protoplasts.

Following are the commercial enzyme preparations used for isolation of protoplasts.

1. Onozuka R10 cellulase (Kinki Yakult Biochemical, Nishinomiya, Japan)
2. Cellulase R10 (Kinki Yakult Biochemical, Nishinomiya, Japan)
3. Macerozyme R10, pectinase (Calbiochem–Behring, San Diego, California, USA)
4. Hemicellulase (Rohm and Hass Co. Philadelphia, PA., USA)
5. Driselase, cellulase, (Kyowa Hakko Kogyo Co., Tokyo, Japan)
6. Cellulysin, cellulase (Calbiochem–Behring, San Diego, California, USA)
7. Pectinase (Sigma, St. Louis, MO, USA)
8. Macerase (Calbiochem–Behring, San Diego, California, USA)

9. Rhozyme HP 150, Hemicellulase (Rohm and Hass Co. Philadelphia, PA., USA)

TABLE 11.2: Plant species in which protoplast have been successfully cultured

Source	Enzyme mixture used	Culture medium	Growth response
MESOPHYLL CELLS			
Brassica napus L.	Cellulase Onozuka P 1500 (0.5%) + Rhozyme HP 1500 (0.5%) + Driselase (0.5%) + Hemicellulase (0.5%)+ Driselase (0.5%)+ Hemicellulase (0.5%) in sorbitol (4.5%)+ mannitol (4.5%)	B + NAA (10–6M) + BA (10^{-6} M)	Plants
Cucumis sativus	Preplasmolysis in a salt medium and cellulase (0.3%) +pectinase (0.4%)+ PDS (0.5%) in 11% mannitol	Harad's medium (1973)	Cell clusters
Datura innoxia	Macerozyme (1%)+ cellulase Onozuka (3%) in 0.5M mannitol BAP (0.4 mg/l) + NAA (0.4 mg/l)	Durand et al. (1973)	Plants
Hyscyamus niger	Pectinol (0.02%) + cellulase Onozuka R 10 in KCl (2.5%) + MgSO (1%) + 0.6 M mannitol	Durand et al. (1973)	Callus
Nicotiana tabacum cv. xanthi (2x)	Macerozyme (0.5%) + cellulase Onozuka (2%) in NAA (3 mg/l)+ 6–BAP (1 mg/l) + 0.7M mannitol	Nagata & Takebe (1971)	Plants
CALLUS			
Saccharum officinarum	Cellulase Onozuka 4S (5%) + glusulase (1.5%)+ sorbitol + 0.4M + glucose 5.5 mM	Mod. White's medium + yeast extract	Callus
STEM CALLUS			
Atropa belladonna	Cellulase R10 (1.5%) + in 0.5M sorbitol + NAA (2 mg/l) + kinetin (0.5 mg/l)	Mod. MS +	Embryoids & plants
OVULAR CALLUS			
Citrus sinensis	Cellulase (1%) + pectinase (1%) + PDS (0.3%) in 0.14 M sucrose + 0.28M mannitol + 0.28M sorbitol	Murashige & Tucker's (1969) basal medium	Embryoids
PERICARP CALLUS			
Vitis vinifera	Cellulase Onozuka SS (2%) + meserozyme (1%) in 0.1M $CaCl_2$ + 0.14M KCl + (2000 mg/l) kinetin + 0.2 mg/l) +NAA (0.1 mg/l)	B5 + Casein hydrolysate	Callus
CROWN GALL			
Parthenocissus tricuspidata	Mecerozyme R 10 (0.01%)+ cellulase + NAA (0.1 mg/l) + Onozuka R 10 (2%) in 13% mannitol	Nagata & Tabeke (1971)	Callus

The enzyme mixture is normally dissolved in culture media together with an osmotic stabilizer. Calcium (2 to 10 mM) is a necessary component and phosphate (0.5 to 2.0 mM) appears to stabilize the isolated protoplasts.

After the digestion of cell wall the isolated protoplast is subject to osmotic stress. If an osmotic stabilizing agent is not included in the medium the isolated protoplasts would take in water by the process of osmosis and would eventually burst as there is no cell wall to constrain the cells. The osmotic agents may be sugar, alcohols, sorbitol or mannitol (13% w/v). Sucrose can be used for this purpose. The concentration of osmoticum should not be high because it causes protoplast shrinkage.

Protoplast is cultured in liquid medium in Erlenmeyer flasks or in microchambers. Microchamber is prepared by placing two 22 mm^2 coverslips about 18–22 mm apart on a drop of heavy mineral oil on a slide. A drop of medium containing protoplasts is placed in between the two coverslips. A third coverslip is then gently placed on the medium and finally sealed with mineral oil or parafilm. Through this method protoplast can be cultured in a small amount and for a limited time.

PURIFICATION OF ISOLATED PROTOPLASTS

Protoplasts contain cell debris and broken cell organelles in medium. A number of methods have been used to separate the intact protoplasts from this surrounding cellular debris.

Sedimentation and Washing

1. The crude protoplast suspension in osmotically adjusted medium is decanted into a conical tip centrifuge tube.
2. It is centrifuged at low speed (50–100 x g/5 min). Under these conditions the intact protoplasts accumulate in the form a soft pellets in the tip of the tube.
3. The supernatant, containing the broken cells debris, cell wall, and cell organells etc., are then carefully pipetted off.
4. The pellets are then gently resuspended in fresh culture medium containing mannitol, and rewashed. This process is repeated two or three times.

Flotation

As intact protoplasts have a relatively low density as compared to other cell organelles, many types of gradients have been used that allow the protoplasts to float and for the sedimentation of cell debris.

A concentrated solution of mannitol, sorbitol or sucrose may be combined with the enzyme–protoplast mixture in a Babcock bottle and then centrifuged at low speed.

Protoplasts can be pipetted off from the top of the tube. The concentration of sucrose used for flotation varies from 0.3 to 0.6 M. The flotation method causes less loss due to damage than the pelleting and washing method. However, the high osmotic concentrations of solution may cause osmotic stress to the protoplasts which may delay the wall formation or leads to eventual loss of the viability.

Other Purification Methods

Other more complex pelleting and flotation systems are also used for protoplast purification. Some workers have layered crude protoplast preparation of *Hyoscyamus muticus* onto 20% Percol (polymer of solicos). After centrifugation debris was in the lower half of the gradient and protoplasts floated to the top.

After obtaining a reasonably pure isolated protoplast preparation, it is important to determine what proportion of the protoplasts are alive and to be able to culture.

PROTOPLAST VIABILITY TESTING

In early days the viability of protoplast is determined by observing its cytoplasmic streaming, exclusion of Evans Blue dye, changes in protoplast size induced by changes in the level of osmoticum and measure of photosynthetic and respiratory activity. Most of these methods have proved to be rather unreliable for determining viability.

The most frequently used methods for estimation of viability is the use of fluorescein diacetate (FDA). As the FDA accumulates within the plasma membrane, viable protoplasts fluoresce green or white. Protoplast preparation treated with FDA (normally 0.01%) should be observed within 5 to 15 minutes, because after this period FDA dissociates from the membrane.

Another stain commonly used is phenosafranine (0.01%). This stain detects dead protoplasts, which turn red in the presence of the dye. Viable protoplasts remain unstained even after 2 hours of staining.

Isolated protoplast can be cultured in several ways of which agar embedding technique in small petridish is commonly followed. In petridishes, embedding of protoplasts in solid agar medium is known as plating of protoplasts.

General Steps of Protoplast Culture

Dilute the protoplast culture medium (1 □ 10^5 protoplasts/ml) and transfer 2 ml of culture medium containing protoplasts in an air–tight petri dish. Add 1 ml of protoplast culture medium containing double agar concentration (1.6%) to each petri dish containing protoplasts in liquid. For this, melt the medium having agar and store at 30–35°C until required. Higher temperature is injurious to protoplasts. Mix gently but quickly by rotating petri dish. Allow the medium to set and seal petri dishes with parafilm. Incubate petri dishes down in an incubator at 25 ± 2°C with continuous illumination at 2000 lux. Protoplasts can also be cultured by suspension culture (without agar) and droplet culture method by placing of protoplast suspension containing protoplasts plate (1 x 10^4 or 1 x 10^5 ml). Incubation procedure is same as described above.

The plated protoplast can be handled very easily and the agar medium gives good support to the protoplast. Though various techniques have been developed for the culture of plant protoplasts however, the style of the culture often depends upon the objective of the experiment.

CULTURE MEDIA FOR PROTOPLAST CULTURE

Protoplast Culture Media for PC I Group

F.5 Medium (Frearson et al., 1973)

Constituents	Concentrations (mg/l)	Constituents	Concentrations (mg/l)
$CaCl_2.2H_2O$	850.0	$ZnSO_4.7H_2O$	4.3
$CuSO_4.5H_2O$	0.025	BAP	1.0
$CoSO_4.7H_2O$	0.15	Biotin	0.05
$FeSO_4.7H_2O$	13.9	Folic acid	0.5
H_3BO_3	3.1	Glycine	1.0

Constituents	Concentrations (mg/l)	Constituents	Concentrations (mg/l)
KH_2PO_4	353.6	Mannitol	130,000.0
KI	0.498	Meso–inositol	100.0
KNO_3	525.0	NAA	2.0
$MgSO_4.7H_2O$	739.0	Nicotinic acid	5.0
$MnSO_4.4H_2O$	11.15	Pyridoxine HCl	0.5
Na_2EDTA	18.75	Sucrose	10,000.0
$Na_2MoO_4.2H_2O$	0.125	Thiamine HCl	1.0
NH_4NO_3	412.25	pH	5.8

Nagata and Takeba Medium (1971)

Constituents	Concentrations (mg/l)	Constituents	Concentrations (mg/l)
$CaCl_2.2H_2O$	220.0	Na_2EDTA	37.3
$CuSO_4.5H_2O$	0.025	$Na_2MoO_4.2H_2O$	0.25
$CoSO_4.7H_2O$	0.030	NH_4NO_3	825.0
$FeSO_4.7H_2O$	27.8	$ZnSO_4.4H_2O$	8.6
H_3BO_3	6.2	BAP	1.0
KH_2PO_4	680.0	Mannitol	130,000.0
KI	0.83	Meso–inositol	100.0
KNO_3	950.0	NAA	3.0
$MgSO_4.7H_2O$	1233.0	Sucrose	10,000.0
$MnSO_4.4H_2O$	22.3	Thiamine HCl	1.0
		pH	5.8

V 47 (Binding, 1974)

Constituents	Concentrations (mg/l)	Constituents	Concentrations (mg/l)
$CaCl_2.2H_2O$	735.0	BAP	0.4
$CuSO_4.5H_2O$	0.015	Biotin	0.04
$CoCl_2.6H_2O$	0.015	Folic acid	0.4
H_3BO_3	2.0	Glycine	1.4
KH_2PO_4	68.0	Glucose	99,000.0
KI	0.25	Mannitol	8, 19,000.0
KNO_3	1480.0	Meso–inositol	10.0
$MgSO_4.7H_2O$	984.0	NAA	1.52
$MnSO_4.H_2O$	5.0	Nicotinic acid	2.0
Na_2EDTA	37.0	Pyridoxine HCl	0.7
$Na_2MoO_4.2H_2O$	0.1	Sucrose	1, 70,000.0
NH_4NO_3	1444.0	Thiamine HCl	4.0
$ZnSO_4.7H_2O$	1.5	pH	5.8

Murashige and Skoog Medium

For protoplast culture, use MS culture medium as described for tissue culture supplemented with 13% mannitol (w/v).

Modified B5 Medium (Kartha et al., 1974)

Constituents	Concentrations	Constituents	Concentrations
+ Sorbitol	4.55% w/v	$CaCl_2.2H_2O$	0.0875%
Mannitol	4.55% w/v	2, 4–D	2.3×10^{-6} M
d–glucose	0.25% w/v	BAP	10^{-6} M
D–ribose	0.125% w/v	NAA	10^{-6} M
N–Z anime	0.015% w/v	Calculated molarity	0.52

Protoplast Culture Media PC II Group

Media calculated under PC I Group with minor changes. The changes are: the concentration of mannitol/sorbitol present in media individually or in combination reduced by half. Concentration of agar in the medium is 0.8%.

Protoplast Culture Media PC III Group

Media described PC I Group with following minor changes:

1. Mannitol/sorbitol/D–glucose/D–ribose individually or in combination is omitted in the media.
2. The sucrose concentration in all media is adjusted to 2% w/v.
3. PC III Group media may be prepared with 1.6% or 0.8% agar depending on the mode of protoplast culture.

Composition of the digestion mixture, wash medium and regeneration medium used for isolation and culturing of protoplasts.

Digestion Mixture

Salts and Vitamins of Murashige & Skoog (MS)

Cellulysin 1.0%
Macerase 0.5%
Rhozyme 1.0%
$CaCl_2$ .... 4.5 mM (0.5 g/l)
Mannitol .... 0.3 M (54.7 g/l)
Sorbitol .... 0.3 M (54.7 g/l)
MES .... 0.3 mM (640 mg/l)
pH 5.7

Filter Sterilize Washing Medium (MS)

$CaCl_2$ .... 4.5 mM
Sorbitol .... 0.3 M
Mannitol .... 0.3 M
pH 5.7

Autoclave Regeneration Medium (MS)		Vitamins, Sugar and sugar alcohols (pH 5.7)		Sugar and Sugar Alcohols (g/l)	
Sucrose	20 g/l	Nicotinamide	1.0	Fructose	0.125
2, 4–D	0.5 mg/l	Pyridoxine HCl	1.0	Fibose	0.125
Kinetin	1.0 mg/l	Thiamine HCl	1.0	Xylose	0.125
Coconut water	2.0% (v/v)	D–calcium pantothenate	0.5	Mannose	0.125
Inositol	100 mg/l	Folic acid	0.2	Cellobiose	0.125
Casein hydrolysate	200 mg/l	p–aminobenzoic acid	0.01	Sorbitol	54.7
Sodium pyruvate	5.0 mg/l	Choline chloride	0.5	Mannitol	54.7
Citric acid	10.0 mg/l	Riboflavin	0.1		
Malic acid	10.0 mg/l	Ascorbic acid	1.0		
Fumaric acid	10.0 mg/l	Vitamin B12	0.01		

Agar Embedded Culture

In this method protoplasts may be allowed to regenerate a new cell wall in liquid culture before embedding or may be embedded directly after isolation. The original method of Nagata and Takebe for the isolated protoplasts is mixed with 1.0% agar–culture medium and maintained at 40 to 45°C. Small amount of the agar (liquid) protoplasts mixture is then poured into sterile petriplates.

Microchambers

This method for the observation of plant cells is quite common. The number of possible designs of chamber is enormous. A chamber is constructed of a small plastic ring with a coverslip on top. The joints of the system are sealed with mineral oil. The simplest form is to hold a droplet between two cover slips with a ring of mineral oil around to form a seal. The mineral oil prevents evaporation of media; this can be a major problem with protoplast cultures as the osmotic concentration of the media rises. Another good sealing substance which is easy to handle is clear nail polish.

Hanging Drop Cultures (HDC) Techniques

To avoid evaporation of the media while observing protoplast, this method is used. Drop can be placed in the depression of a specially prepared hanging drop–slide which is then undersealed with a coverslip and oil, or the drops are placed on the lid of a Petri dish. The Petri dish contains mannitol solution to maintain the humidity. The dish is sealed with parafilm.

Multidrop Array (MDA) Techniques

This is a refinement and amplification of the hanging drop technique. Through this technique it becomes easier to screen a wide range of nutritional and hormonal factors rapidly (using a small amount of plant material, less than 1 g of plant tissue). The system can be described as follows:

1. Prepare a dilution series of two variables A and B, so as to have tubes A_1 to A_{10}, B_1 to B_{10}, likewise 100 possible combinations can be prepared, i.e., A_1B_1, A_2B_2 ..... $A_{10}B_{10}$, etc.
2. To these 100 different media (1.0 cm^3 each) add 0.1 cm^3 of protoplast suspension prepared in basal media at 20 times the minimum planting density.

3. For each combination, i.e. A_1B_1, A_1B_2, etc. make, 49 drops + 4 control drops in order on the lid of a Petri dish and label the dish, i.e. A_1B_1.
4. To each drop add 20 μl of another variable media component previously prepared C–C7, 11–17. Thus each dish contains 49 different media combinations.
5. A total of 4900 different media compositions each inoculated with cells have thus been prepared on only 100 Petri plates.
6. 20 cm^3 of mannitol is added to the base of the dish. The lid with attached droplets is inverted into position and the plates are sealed and incubated.
7. After a suitable incubation period the plates can be easily evaluated microscopically to determine which media combination, has given the most successful growth.
8. The best method for the culture of protoplast is by plating them in a medium containing 1.2% agar. In this condition, protoplasts grow best by good number of divisions and can be handled conveniently. Moreover, it could be observed under the microscope.

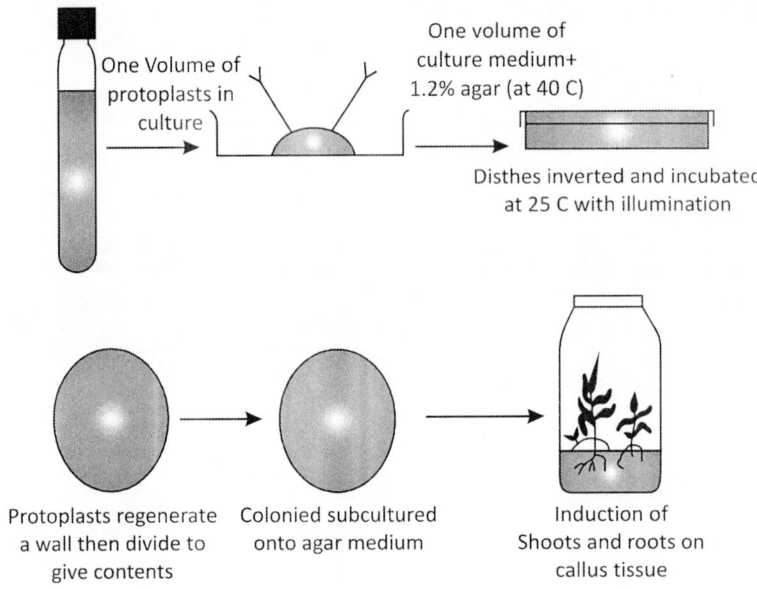

FIG. 11.5: Diagrammatic presentation of culture of protoplast by agar embedding technique in small Petri dishes.

GROWTH AND DIVISION OF PROTOPLAST

For the growth and nuclear division, the regeneration of cell wall is not seems to be a prerequisite in protoplast culture. In *Convolvulus* sp. protoplasts undergo one or two nuclear division prior to cytokinesis. The same was observed in *Haplopappus* spp.

It has been observed that for the rapid growth of protoplasts frequent sub-culturing in mannitol free medium is necessary. Protoplasts don't grow in large colonies on high osmotic medium. Its growth gradually shows a downward trend which ultimately inhibited altogether. The colonies, therefore, should be picked up along with small pieces of agar and then transferred on the top of another medium to avoid any damage to protoplasts by rough handling with the forceps.

Protoplast Culture: Regeneration of Cell Wall

In culture, protoplasts start developing a wall around itself within few hours and it takes only few days to complete the process. Wall materials are progressively deposited at the surface of the plasmalemma. The cellulose is deposited either between the plasmalemma and the multilamellar wall material or directly on the plasmalemma. The nature of biosynthesis of the cell wall depends on the plant material and the system of protoplast culture.

The newly built cell wall can be observed either by plasmolyzing the protoplast by transferring it in a hypertonic solution, or by staining the cell wall with calcofluor white fluorescent stain (0.1%). However, electron microscopic studies and freeze etching studies have revealed much about the structure and progressive development of cell wall around the protoplast in culture medium.

1. Observe regularly the regeneration of cell wall, cell division and small callus formation under inverted microscope
2. Examine cell wall formation in protoplasts with a droplet of 0.1% calcofluor white R, American Cyanamid, Bound Brook, N. J. USA, in 0.4 M sorbitol solution on a slide. The cell wall regenerated protoplasts fluoresce
3. Small cluster of calli are observed after 2–3 weeks of culturing protoplasts
4. Subculture the cell clusters on a freshly prepared protoplast culture medium with or without 1/2 the intial mannitol and 0.8–1.6% agar.

Development of Callus/Whole Plant

Soon after the formation of wall around the protoplasts, the reconstituted cells show considerable increase in size and first divisions usually occur within 7 days. Subsequent divisions give rise to small cell colonies. After 2–3 weeks macroscopic colonies are formed which can be transferred to an osmotic free medium to develop a callus. The callus may be induced to undergo organogenic differentiation, or whole plant regeneration.

TABLE 11.3: Name of some of the families and species in which shoot differentiation of plant regeneration has been achieved from cultured protoplasts

Family	Species
Compositae	*Cichorium intybus, Lactuca sativa cultivars (L. serriola, L. saligna), Petasites japonicus, Senecio vulgaris*
Cruciferae	*Arabidopsis thaliana, Brassica campestris (B. carinata, B. juncea, B. napus, B. nigra, B. oleraceae var. capitata), Sinapis alba*
Cucurbitaceae	*Cucumis sativus*
Euphorbiaceae	*Manihot esculenta*
Gramineae	*Bromus inermis, Oryza sativa, Pennisetum americanum, Saccharum spp., Triticum aestivum*
Leguminosae	*Glycine argyrea, G. canescens, G. clandestina, G. max, Medicago arborea, M. coerulea, M. difalcata, M. falcata, M. glutiniana, M. hemicyla, M. sativa cultivars (M. varia), Psophocarpus tetragonolobus, Trifolium hybridum, T. repens, T. rubens*
Liliaceae	*Asparagus officinalis, Hemerocallis sp.*
Linaceae	*Linum usitatissimum, L. strictum, L. lewissii*
Magnoliaceae	*Liriodendron tulipfera*

Family	Species
Ranunculaceae	*Ranunculus sceleratus*
Rutaceae	*Citrus aurantifolia, C. grandis, C. limon, Citrus medica, C. paradisi, C. reticulata, C. sinensis*
Salicaceae	*Populus tremula, P. alba x P. grandidentata, P. nigra var. Butulifolia x P. trichocarpa*
Santalaceae	*Santalum album*
Solanaceae	*Atropa belladonna, Capsicum annuum, Datura metel, D. meteloides, D. innoxia, Hyoscyamus muticus, Lycopersicon esculentum, Nicotiana acuminata, N. alata, N. debneyi, N. glauca, N. langsdorffii, N. longiflora, N. otophora, N. paniculata, N. plumbaginifolia, N. suaveolens, N. sylvestris, N. tabacum, Petunia hybrida, P. inflata, P. parodii, P. parviflora, P. violacea, Salpiglosssis sinuata, Solanum aculeatissimum, S. aviculare, S. brevidens, S. chacoense, S. dulcamara, S. etuberosum, S. fernandezianum, S. gilo, S. khasianum, S. luteum, S. lycopersicoides, S. melongena, S. nigrum, S. pennellii, S. phureja, S. phureja x S. chacoense, S. pinnatisectum, S. torvum, S. tuberosum cultivars (S. tuberosum tetraploid clones, diploid clones), S. uporo, S. viarum, S. xanthocarpum*
Ulmaceae	*Ulmus species*
Umbelliferae	*Daucus carota, Foeniculum vulgare*

DETERMINATION OF PROTOPLAST PLATING EFFICIENCY

Calculate the average number of cell colonies per cm^2 of the petri dish by using following formula:

$$\text{Plating efficiency} = \frac{\text{X} \times \text{Area of Petri Dish} \times 100}{\text{Number of Protoplast Plated / Perti Dish}}$$

$$X = \frac{S_1 + S_2 + S_3 + S_5}{5}$$

X = Average number of cell colonies per cm^2 and S_1, S_2, S_3, S_4, S_5 = the number of cell colonies in 1 cm^2 of Petri dish.

TECHNIQUE FOR ISOLATION OF SUB–PROTOPLASTS

Subprotoplasts are fragments derived from protoplasts and do not contain the entire content of plant cells. **Cytoplasts** are subprotoplasts containing most or fractions only of the original cytoplasmic material but are lacking the nucleus. Subprotoplasts containing the nucleus are called **karyoplasts** or **miniprotoplasts**. In this case the nucleus is surrounded by only a small, amount of cytoplasm and the plasma membrane. The term **microprotoplast** was suggested for subprotoplasts containing not all but a few chromosomes. The term **microplast** was used to describe protoplast fragments containing only minor fractions of the cytoplasmic material surrounded by an inner membrane of the cell. In contrast to microplasts, subprotoplasts normally surrounded by the original outer membrane, the plasma membrane.

After obtaining satisfactory quantities of isolated intact and viable protoplasts, that are able to regenerate cell wall, undergo mitotic and cellular divisions and form a callus. The next important step is to regenerate a whole plant by suitable manipulation in culture medium.

HANDLING OF REGENERATED PLANTLETS

Plantlets regenerated by the process of embryogenesis or organogenesis are then used to transfer to non–sterile conditions. This can be very critical period when the plantlet is removed from sterile controlled environment and transplanted to the real world of agriculture. As the plantlet is moved to non–sterile conditions care must be taken to allows gradual release of these careful controls.

Anatomical analysis of these plantlets have shown that under these conditions often the cuticle (covering the leaf surfaces), is extremely thin and the root hairs are poorly developed. If these plantlets are suddenly move to a situation where they are subjected to water stress they would be unable to control their water balance adequately and will die. Both the plantlets should first allow to develop a good root system before transferring them *in vivo* conditions and care should be taken for least possible damage of the roots. If the plants have been grown on agar solidified medium, this can be removed by gentle washing in a basin of warm water.

These plantlets are then carefully taken out from the test tubes or flasks and adhered agar medium is removed from them. The *in vitro* plants with roots are then carefully planted in small pots. The care is to be taken that roots should make quick and good contact with the potting mix because the plant may suffer water stress. The planting mix should be well balanced for water retention and sterile. Small, sterile peat pots (Jiffy pots) are excellent for this purpose.

PROTOPLAST FUSION

Plant protoplasts represent the finest single cell system that could offer exciting possibilities in the fields of somatic cell genetics and crop improvement. The isolated protoplasts are surrounded by plasma membrane only. The lake of the cell wall allows the plasma membrane of two or more protoplasts to come into intimate contact. The important aspect has been that incompatibility barriers do not exist during the cell fusion process at interspecific, intergeneric, or even interkingdom levels. Under certain conditions they will stick together rather like two soap bubbles. Later, if they are given an appropriate stimulus they will fuse together forming a single mass surrounded by a common membrane. Kuster was able to see occasional spontaneous fusion of protoplasts. However, with the development of enzymic methods for producing large numbers of isolated protoplasts, great interest has now been developed in the use of protoplast fusion (**somatic hybridization**) as a possible tool of plant breeding. Many people believe that protoplast fusion offers a useful tool to plant breeders to make crosses between sexually incompatible species for transfer of nuclear or cytoplasmic characters. Protoplast fusion can be used to make crosses within species (intraspecific), between species (interspecific), within genera (intrageneric) and between genera (intergeneric).

Protoplasmic fusion permits the formation of cytoplasm containing organelle of 2 different parents like chloroplast of one parent and mitochondria of the other, but coexistence of a particular organelle like chloroplast of both the parents in a same cytoplasm usually does not persist due to sorting during mitotic divisions. Potoplasmic fusion is helpful in several graminae, solanaceae and cruciferae members to overcome sexual barriers, to produce plants with resistance to diseases or tolerance to selective herbicides, to produce somatic hybrids having efficient organelle system and widen to genetic base (Glimelius *et al.*, 1991). Protoplast system can be transformed by incorporation of pure DNA or calcium phosphate–DNA coprecipitate or DNA encapsulated in liposomes or by direct micro injection of DNA (Schlperoort and Wullems, 1983). The DNA uptake can be further enhanced under high pH or by heat shock treatment and / or by addition of polyethylene glycol, poly –1 ornithine or polyvinyl alcohol.

By inactivation of nucleus of one parent and subsequent fusion with normal protoplasts or by fusion of normal and subprotoplasts or microprotoplasts or by fusion of normal and subprotoplass or microprotoplasts or by using mutants and anti metabolites it is possible to produce 'cybrids' having nuclear genes of one parent and cytoplasmic genes of other parent. Cytoplasmic organelle like chloroplasts and mitochondria possess DNA which control the expression of male sterility. (Levings *et al.*, 1980). Thus the production of cybrids will be helpful in seed propagated annual ornamentals to exploit hybrid vigour by heterosis breeding. It may also be possible to achieve postponement of pollen induced senescence in flowers.

General Steps of Protoplast Fusion

An efficient method employing polyethylene glycol (PEG) has found wide application in experiments on protoplast fusion. Many fusion inducing agents such as sodium nitrate, high Ca^{++}, high pH 10.5, high temperature (37°C) and polyvinyl alcohol have been used for fusion of protoplasts.

Procedure

1. Place a drop (1 drop = 50 μl) of each protoplast suspension in sterile petri dish, gently shake the plate to ensure proper mixing of suspensions from callus or suspension culture with leaf protoplast. The concentration of each species should be about 1 x 10^5/ml before mixing.
2. Wait for 5 min to allow the protoplasts to settle to the bottom of petri dish
3. Slowly add 300–450 μl PEG solution to the edge of the protoplast suspension, then in the centre of the protoplast, and wait for 15 min
4. Add drop by drop 1 ml of 0.7 M mannitol solution to dilute the PEG solution
5. Raise one side of the plate and wash the protoplast clumps adhering to the plastic surface of plate upto with 9 ml of 0.7 M mannitol
6. Remove the PEG and mannitol residues from petri dish. Add a few drop of mannitol solution to the fused cells
7. After 5 min observe the fused product on the inverted microscope. The process of adhesion or agglutination of protoplasts in various combinations can be seen
8. Study the fused and unfused protoplasts for the synthesis of cell wall, first cell division, formation of homokaryons/heterokaryons, frequency of multinucleate homokaryons from heterokaryons and mitotic index, cytologically examination under microscope.

The technique of protoplasts fusion has developed new vistas for the plant breeders where conventional methods failed or the plant species are incompatible. Protoplasts fusion results in somatic hybrids, the phenomenon is known as somatic hybridization. Protoplast fusion may be spontaneous or induced.

The protoplasts fusion may be of three kinds:

1. Spontaneous fusion
2. Mechanical fusion
3. Induced fusion

Spontaneous Fusion

Only interspecific protoplasts have been found very often fusing through their plasmodesmata during enzyme treatment. Nearby protoplastsin a Perti dish have been observed to form large multinucleate (coenocytic) mass. Protoplasts of young leaves are prone to fuse spontaneously with one another.

Mechanical Fusion

The giant protoplasts of *Acetabularia* have been fused mechanically) This kind of fusion is not dependent upon the presence of fusion–inducing agent. However, in this procedure protoplasts are likely to get injury.

Induced Fusion

This kind of fusion can be done with the help of (a) $NaNO_3$ treatment, (b) High pH/Ca^{++} treatment, (c) Polyethylene glycol (PEG) treatment, (d) Electrofusion. This method can bring together both intra– and interspecific protoplasts. For this, some inducing agent is necessary. The inducing agent is called **fusogens**. In animals **sendai virus** plays this role but in plants both physical as well as chemical methods are used.

PROTOPLAST FUSION

Protoplasts (Leaf)	⇓	**Protoplasts**
	⇓	**(Cell suspension culture)**
?	Combine	
	⇓	PEG High Ca^{2+} and
	⇓	pH (10.5) PEG Dilution
,	Fusion	
	ß	Culture Medium
	Mitosis	
	ß	
	Hybrids	

FIG. 11.6: Presentation of stages in protoplast fusion and production of hybrids. 1 = Parental protoplasts, 2 = Aggulation adhesion.

Several chemicals like sodium nitrate, PEG, poly–L–ornithine, poly–D–lysine, cytocholasin B, protomine sulphate, lysozyme, glycerol, dimethyl sulphoxide, proteins and calcium ions at high pH have been used as fusogens. Kuster (1909) for the first time used sodium nitrate to plasmolyse onion protoplasts which underwent fusion on deplasmolysis. Thereafter, Power *et al.* (1970) employed the same technique and successfully fused root tip protoplasts of maize and oat.

Keller and Melchers (1973) employed calcium ions at high pH (0.05 M $CaCl\ 2H_2O$ in 0.4 M mannitol at pH 10.5) for fusion of tobacco protoplasts. They spinned the isolated protoplast in the fusion inducing solutions for 3 minutes at 50 g and then kept the tubes in water bath at 37°C for 40–50 minutes nearly 20–50% protoplasts undergone fusion process.

The one most widely used fusogen is polyethylene glycol. The effects of PEG are not specific and it promotes the aggregation and fusion of protoplasts from the same or different species or intergeneric protoplasts.

Zimmermann has shown that if protoplasts are placed into a small culture cell containing electrodes and a potential difference is applied, then the protoplasts will lie between the electrodes. Protoplasts fusion can be induced after applying extremely short, square wave electrical stocks. This method has become very popular for its highly controllable fusions property. It is possible to fuse two single protoplasts.

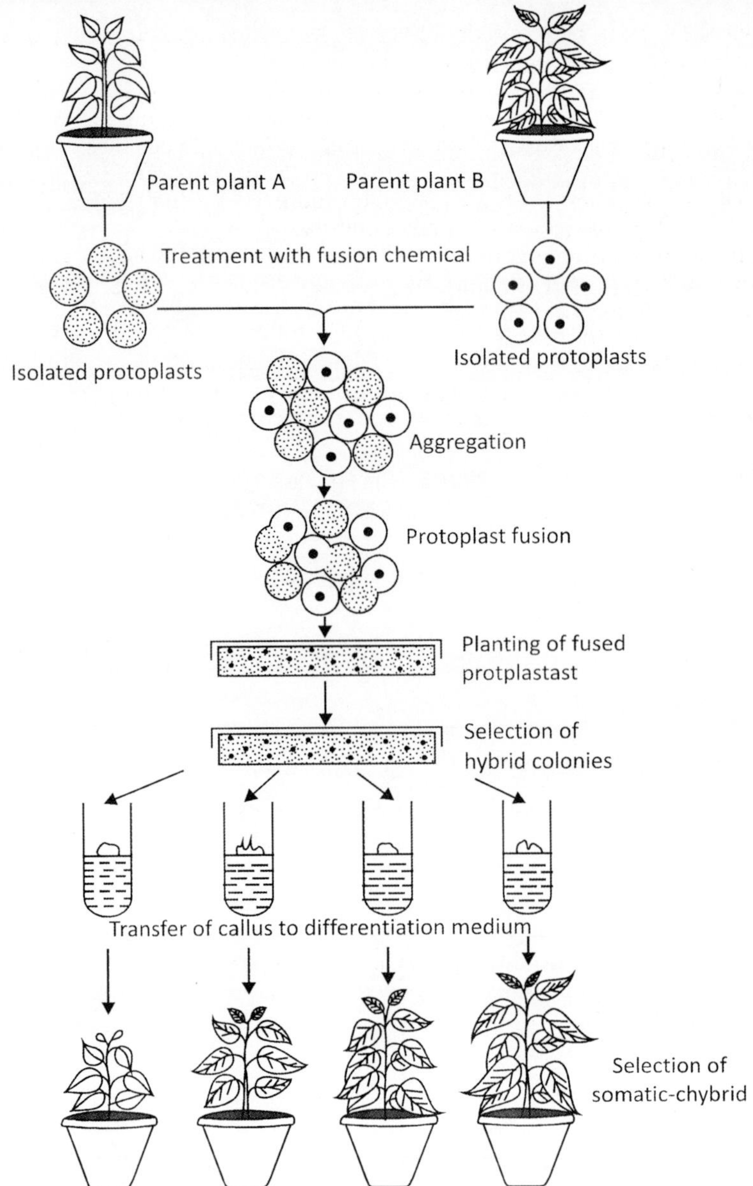

FIG. 11.7: Diagrammatic presentation of fusion of protoplast from two different plant species and the regeneration of somatic hybrid plants.

PROTOPLAST CULTURE: REGENERATION OF PLANTS

The idea that cell contains all the information necessary for the regeneration of a whole organism was given in the cell theory, proposed by Schwann. At that time, however, the technology did not exist to study regeneration of plants from any single cell other than a zygote, and even this developmental process takes place within plant body and not as an isolated single cell *in vitro*.

It has been a long practice to regenerate vegetative plants by taking stem cuttings. Haberlandt successfully regenerated whole plant from small meristematic pieces, however, he could not obtain any growth of a single cell system because of its inability to regenerate plants from single cell. This was not due to genetic reason, but due to lack of adequate nutrients in the medium. He further suggested the use of embryo sac fluids in the medium for such culture.

Protoplasts in culture regenerate cell wall around themselves to reconstitute cells. These cells in proper medium by repeated cell divisions form callus. Callus tissue by manipulation of the nutritional and physiological conditions can be induced to regenerate into the morphogenic form of plant.

About 100 plant species have successfully been regenerated from single cell or protoplast. Regeneration of plants from isolated cells or protoplasts is by one of two following methods:

1. By Organogenesis
2. By Embryogenesis

Organogenesis

Organogenesis takes place from a callus and not directly from a single cell. When isolated protoplasts are put into culture under appropriate conditions of plant inoculum, medium and environmental factors they go through a set series of events.

1. Wall regeneration
2. Early mitotic division and callus formation
3. Organogenesis

Embryogenesis

Plant tissue *in vitro* can induce to form somatic haploid embryos. Steward and coworkers first observed the phenomenon of somatic embryogenesis in carrot. Somatic embryos can be induced in cultural conditions from three different sources:

1. Vegetative cells of mature plants;
2. Reproductive tissues other than the zygote; and
3. Hypocotyls and cotyledons of embryos

Embryoids are initiated in callus from small superficial clumps of cells associated with dense cytoplasm and large starch grains. The developing embryoids *in vitro* then pass through a sequence of growth stages that is exactly the same as seen in the development of a seed embryo.

Like cells, protoplasts are genetically unstable *in vitro*. Somaclonal variations have also been detected in the plants regenerated from protoplasts. Chromosomal variation has also been observed in plants regenerated from protoplasts. Shepard and his associates conducted detailed studies of plant regeneration from protoplasts of Russet Burbank potatoes. Significant variations were detected for growth habit, tuber colour, fruit production and tuber uniformity. Several of these traits are considered to be improved over the widely grown commercial variety. They further reported somaclones with resistance to early blight (*Alternaria solani*) and late blight (*Phytophthora infestans*). Our knowledge is insufficient to understand the mechanism of such instability in culture.

In *Nicotiana tabacum*, protoplast cultures were regenerated into cells and callus tissues. These calli were then transferred to fresh medium supplemented with IAA

(4 mg/l) and cytokinin (2 mg/l). Calli differentiated into shoots and occasionally in roots within 3 to 4 weeks. The plantlets are then subsequently transplanted to pots.

Handling of Regenerated Plantlets

Plantlets regenerated by the process of embryogenesis or organogenesis are then used to transfer to non–sterile conditions. This can be very critical period when the plantlet is removed from sterile controlled environment and transplanted to the real world of agriculture. As the plantlet is moved to non–sterile conditions care must be taken to allows gradual release of these careful controls.

Anatomical analysis of these plantlets have shown that under these conditions often the cuticle (covering the leaf surfaces), is extremely thin and the root hairs are poorly developed. If these plantlets are suddenly moved to a situation where they are subjected to water stress they would be unable to control their water balance adequately and will die. Both the plantlets should first allow to develop a good root system before transferring them *in vivo* conditions and care should be taken for least possible damage of the roots. If the plants have been grown on agar solidified medium, this can be removed by gentle washing in a basin of warm water.

These plantlets are then carefully taken out from the test tubes or flasks and adhered agar medium is removed from them. The *in vitro* plants with roots are then carefully planted in small pots. The care is to be taken that roots should make quick and good contact with the potting mix because the plant may suffer water stress. The planting mix should be well balanced for water retention and sterile. Small, sterile peat pots (Jiffy pots) are excellent for this purpose.

PROTOPLAST CULTURE: UPTAKE OF FOREIGN MATERIALS

Isolated protoplast of plant cells have the remarkable property of taking up micro– as well as macromolecules like ferritin, polystrene, latex, protein, DNA, chromosomes, whole nuclei, chloroplasts and even virus and bacteria. This unique property of protoplast can be utilized for the incorporation of desirable traits into plants to get what we call *transferred* or *novel varieties* from wild ones.

Incorporation of Foreign DNA

Unlike transformation of bacteria by exogenous DNA, transgenosis in cells or protoplasts of higher plants are now possible and have been tried successfully in some plant species. Doy and his associates transformed haploid tomato callus culture by lambda phage. The transformed higher plant cells can grow on a medium containing galactose and lactose as the only carbon source that is considered to be due to phage DNA incorporation. This kind of transformation phenomenon is termed as **transgenosis**. These studies have given a major break through in agricultural biotechnology.

Incorporation of Nuclei

Intra– and inter-specific nuclear incorporation into protoplasts of *Nicotiana tabacum*, *Petunia hybrida* and *Zea mays* have been observed. These exogenous nuclei enter the protoplasts without any damage to plasmalemma. However, the role of these incorporated nuclei in the metabolism of protoplasts in culture is yet to be confirmed.

Incorporation of Chloroplasts

In addition to nuclear material and nuclei as a whole protoplast can take up certain cell organelles like chloroplasts, mitochondria etc. Transplantation of chloroplasts into albino plant species could solve a great physiological deficiency. Potrykus (1973), Bonner and Eriksson (1974) have reported the uptake of chloroplasts by albino *Petunia* and carrot protoplasts. Nass (1969) and Gills and Sarafis (1971) have incorporated spinach chloroplast into animal cell cultures. They have also reported that this chloroplast survive and remain metabolically active in animal cells. Transplantation of chloroplast in plant having insufficient or deficient photosynthetic system will have far– reaching implications in plant improvement programme.

Several methods for the induction of chloroplast into plant protoplasts have been reported, all of them depending either on some form of osmotic effect or on hypothesized changes in the charge pattern on the plasmalemma.

Bonnett and Eriksson used a method involving plasmolysis by polyethylene glycol (PEG) to induce uptake of *Vaucheria* chloroplasts by carrot protoplasts. A dense mixture of protoplasts and chloroplasts was made up to a volume of 0.5 ml with protoplast culture medium. A volume of 1.5 ml of the solution containing 3 ml of protoplast culture medium and 7 ml of a 56% aqueous solution of PEG was added to the protoplast–chloroplast mixture. The final concentration of PEG was about 30%, a concentration known to promote cell aggregation and cell fusion without reducing cell viability. After 10 min the mixture was diluted to 10 ml with a solution of 0.1 M $CaCl_2$ in 0.3 M sorbitol, and centrifuged for 3 min at 150 x g. The pellet was resuspended in 10 ml of the same solution, recentrifuged and resuspended in 2 ml of the protoplast culture medium.

Incorporation of Cyanobacteria

Experiments have been performed to incorporate certain cyanobacteria or blue–green algae (*Gleocapsa*, *Anacystis*, etc.) in protoplast of higher plants. The method is used to co–incubate the algal preparation with the isolated protoplast preparation in the presence of 25% polyethylene glycol, when high Ca concentration is added the protoplasts begin to engulf the algal cells. The method of uptake has been shown to be by invagination of the plasma membrane.

Incorporation of Bacteria

Incorporation of nitrogen fixing bacteria (*Rhizobium, Azotobacter*) blue green algae and *nif*–genes into non–leguminous plants specially cereals is a important task in the hands of geneticists and agronomists. Davey and Cocking (1972) introduced a symbiotic nitrogen fixing bacterium (*Rhizobium*) in legume chloroplasts. These chloroplasts containing nitrogen fixing bacteria could be fused with non–legume chloroplast to impart the capability of nitrogen fixation in non–leguminous cash crops. *Agrobacterium tumefaciens* have been adapted to become efficient and reliable gene vectors for genetic engineering of dicotyledonous plants. *Chimeric*–genes have been constructed which upon transfer into differentiated plant cells allow the identification of DNA sequences involved in the regulation of gene expression. Recently, direct DNA transfer and expression of a bacterial gene in protoplasts of *Triticum monococcum* and *Nicotiana tabacum* have been reported; and in the case of tobacco transformed protoplasts were regenerated into plants which obeyed Mendelian law of inheritance.

Incorporation of Virus

Virus, an obligate parasite in nature enters the plant cells, leaving behind its protein coat only in those plant cells whose cells wall is damaged. The virus genome (RNA) enters the protoplasts of the cell and then multiply in cytoplasm. Takebe and Otsukic (1974) reported that protoplasts can be infected by one or more than one virus. Cocking (1966) observed the uptake of TMV by tomato fruit protoplast.

The inclusion of these micro and megamolecules in plant protoplasts have opened a new path to solve the genetic deficiency in economically important plants through biotechnology.

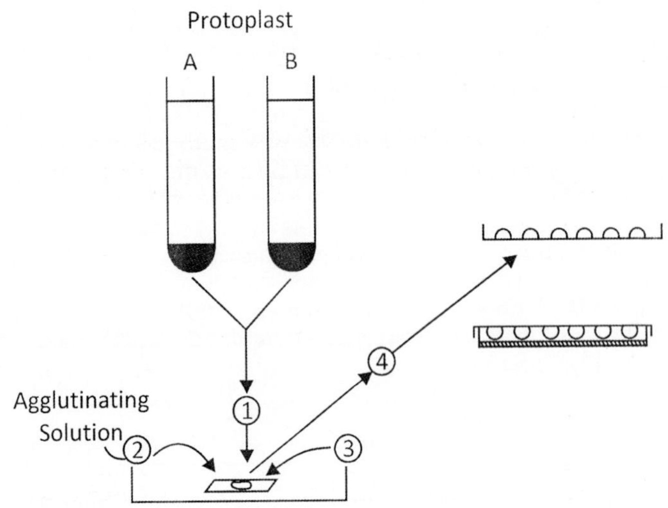

FIG. 11.8: Typical protocol for protoplast fusion of potato using PEG. (1) Dilute protoplast mixture to 10^4 cells per milliliter in 0.5 M glucose, 3 mM $CaCl_2$,0.7 mM KH_2PO_4, pH 5.7; (2) add 3 to 200 μl volumes of agglutinating solution (0.09 mM PEG 6000, 10 mM $CaCl_2$,0.7 mM KH_2PO_4, pH 5.5); (3) add five 100–μl volumes of diluting solution (50 mM $CaCl_2$,0.5 M glucose, 50 mM glycine buffer, pH 10.0); (4) wash cells with culture medium and culture in hanging drop suspensions.

INCORPORATION OF NON–BIOLOGICAL MATERIALS

Incorporation of non biological materials have been tried such as polystylene in to plant protoplasts. These types of studies have proved useful as a method of determining the maximum size of an object that can be engulfed by endocytosis. However, no direct biological value can be seen to these studies at the present time.

SELECTION OF FUSION HYBRIDS

Visual Selection

This method can be used to select fusion products of protoplasts that have distinct physical characters. For example, fusion between leaf mesophyll protoplasts of green colour and a cultured cell that has an anthocyanin– containing red protoplast. A single fusion product between the two will have both chloroplasts and an anthocyanin vacuole together with two nuclei. The same method can be applied to fusions between leaf mesophyll and hyaline cultured cell protoplasts.

Obviously, the drawback with this method is that it is very laborious and only a very limited number of fusion products can be selected.

Fluorescent Labels

1. In this method fluorescent labeled dyes are used to detect fusion products.
2. If the two original protoplast cultures are pre–incubated for 12–15 hours, one in octadeconyl aminofluorescein and the other in octadecyl palamine B each group of protoplasts takes on a specific fluorescence colour. The dyes are non–toxic and do not affect viability, wall regeneration or growth.
3. After fusion of the protoplasts fusion products may be identified by their fluorescence characteristics under a fluorescence microscope.

Fluorescence Activated Cells Sorting

Protoplasts are fluorescently labeled as already described. After fusion three types of fluorescence can be found; two original fluorescence of the two parental populations and the fluorescence of the fused hybrids. A machine now exists to separate these populations. A droplet containing a single cell is held between two fluorescence detectors. If it is one of the two parent populations the machine charges two electrical plates, as the droplet falls it is deflected by the electrical charge. If the fluorescence is of the hybrid, no electrical charge is applied to the plates and the droplet fall vertically. Thus the protoplasts are automatically separated into the parent types and the fusion products. The fusion products can then be plated into an appropriate culture system.

Nutritional Selection

In some cases, Carlson *et al.*, were able to separate fusion products by their ability to grow on a media that would not permit growth of the parent protoplast lines. For example hybrid of *Nicotinia glauca* and *N. langsdorfii* could be cultured in auxin free culture medium in contrast to its parents protoplast.

Drug Sensitivity and Resistance

Power and co–workers used sensitivity to actinomycin–D to select fusion products between *Petunia hybrida* and *Petunia parodii*. The two parental species are sensitive to actinomycin–D; however, the hybrid is resistant. Again it would be of great benefit to the field of somatic cell hybridization if a wider range of resistance mutants could be isolated. Some research groups are now attempting to produce large numbers of resistant mutants for use in protoplast fusion studies.

SOMATIC CELL HYBRIDIZATION OR CYBRID OR CYTOPLAST

Success in protoplast fusion between diversified plant species gave great impetus to geneticists and plant breeders though the technique has some shortcomings. Nevertheless, it has stimulated research activities in many laboratories around the world.

When two protoplasts fuse along with their nuclei true hybrid is formed. However, often the two nuclei will survive independently in the mixed cytoplasms, forming a heterokaryon. Chromosome loss can occur from one or both of the nuclei; if all the nucleus of one parent is lost, what is left is one nucleus in a mixture of both cytoplasms. This product is called a cytoplast or cybrid.

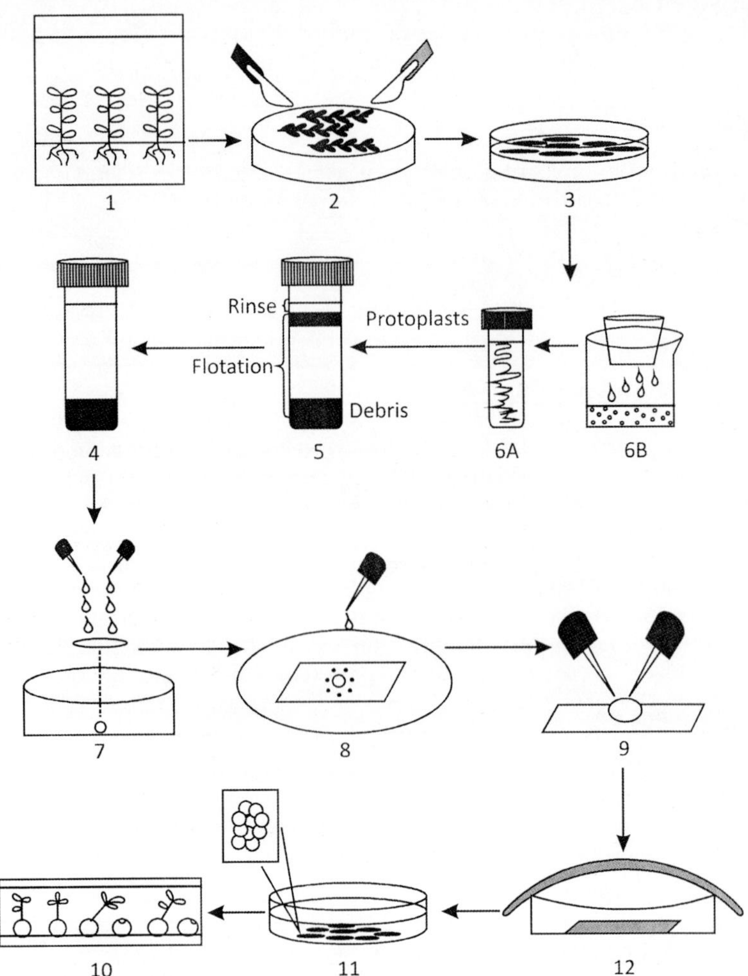

FIG. 11.9: Experimental procedures for fusing potato protoplasts. (A) Propagated potato plantlets in vitro until 15 copies of each genotype are obtained. (B) cut shoots and leaves into fine pieces using two scalpels. (3) Incubate cut pieces in plasmolysis medium for 1 to 8 h followed by 4 to 16 h in enzyme medium. (D_1) Pour protoplast suspension through a sterile 63–μm filter into a 250–ml beaker. (D_2) Transfer protoplast suspension into a sterile centrifuge tube and centrifuge. (E) Resuspend pellet in 10 mL flotation medium overlaid with 1 ml rinse medium and centrifuge (protoplasts collect at the interface of the two medium during centrifugation). (F) Rinse protoplasts and resuspend in CPW 13 M at a density of 1×10^6 per milliliter. (G) (1) In the center of a sterile petri dish add one small drop of sterile mineral oil; (2) position a sterile cover slip over the droplet; (3) add three drops of each of the two protoplast suspensions. (H) Fuse protoplasts by gradually adding six drops of Fusogen PEG 22.5 medium. (I) Wash protoplasts with Ca^{2+} washing medium. Replace washing medium with culture medium. (J) Incubate protoplasts in dim light at 25°C. (K) After 2 to 3 weeks embed multicellular colonies in medium with agarose. (L) Regenerate shoots and transfer to MS basal medium for rooting.

Cybrids can also be prepared by fusing a normal protoplast with an enucleated protoplast. Presently, researchers are studying the use of cybrids to transfer resistance to some herbicides. This resistance character is carried on the chloroplast genome. Transfer of cytoplasmic male sterility can be controlled by transfer of the mitochondrial genome. The study of cybrids is also important for investigation of possible recombination within these extra chromosomal genomes.

The scenario presented in this chapter gives an overall view of protoplast culture. The technique has unraveled the complexities of protoplast culture and its application to agricultural biotechnology. The protoplast technique is used in studies of plant cell genetics specially in somatic embryogenesis and transfer of genetic information by DNA uptake and organelle implantation. Though tissue culturists have developed many plantlets and new variants through this technique, care must be taken to achieve cost–effective results so that the *in vitro* technique of protoplast culture can be adapted for commercial exploitation of agricultural crops in general and plantation crops in particular.

PROTOPLASTS FOR ISOLATION OF CELL COMPONENTS

The danger of erroneous interpretation of protoplast metabolism as truly representing intact cell processes is hardly encountered when the aim is merely the isolation of cell components. The objective of isolating such components as plasmalemma, nuclei, and chloroplasts from protoplasts, rather than from cells or intact tissues, is in most cases to obtain them by convenient methods which will cause the least destruction. Therefore, if the functionality of these components is of primary importance, the evaluation of the rate and quality of damage (e.g. in nuclei) as well as their unquestionable identification and purity (e.g. of plasmalemma) are important considerations. Most of the problems involved in the isolation of cell components from protoplasts are specific to plant cells because of the latter's rigid cell wall and their specific organelles, e.g. chloroplasts and vacuoles. These problems are very rare in animal cells but are common to plant cells.

Plasmalemma

Isolation of plasmalemma from cells of higher plants is a difficult task. It is hampered by two major factors: (a) high shear force used for cell homogenization will damage the plasmalemma (b) good markers are essential to identify the plasmalemma fraction. Therefore, protoplasts obtained from leaf mesophyll or from cell suspensions were isolated by enzyme treatment. Plant protoplasts bind concanavalin a (con A) is used as a plasmalemma marker.

Chloroplasts

Chloroplasts were first isolated from protoplasts by Wagner and Siegelman as a by–product of their isolated vacuoles. The protoplasts were ruptured osmotically by suspending them into a phosphate buffer containing Mg^{2+} and dithiothreitol. Chloroplasts were then separated on discontinuous sucrose gradients and were found to be photochemically active. Isolation of cereal chloroplasts is especially problematic, chloroplasts of these plants are obtained by suspending the protoplasts in an appropriately buffered osmoticum (containing bovine serum albumin and dithiothreitol) and passed through a 20μ nylon net. The chloroplasts are then collected by centrifugation without fractionation. Nishimura developed a new technique for spinach leaf protoplasts, using a syringe needle to rupture the protoplasts.

Mitochondria

Large number of mitochondria can be obtained from plant tissues and cultured cells. Some of the methods mentioned above for chloroplast isolation from protoplasts may also be useful for mitochondria and the two types of organelles can be separated from the same sample of plant materials. When 0.05–0.1% bovine serum albumin is included and a sucrose gradient is employed for fractionation of the disrupted protoplasts, mitochondria will band quite sharply at a density of 1.18 g. cm^{-3}, and they can be identified by their fumerase activity, while chloroplasts will peak at an appreciably higher sucrose density (1.22 g.cm^{-3}).

Vacuoles

Vacuoles are most fragile cell organelles. Wagner and Siegelman reported that when protoplasts from various plant tissues are ruptured gently by transferring it into phosphate buffer, vacuoles could then be separated by a low–speed centrifugation and be further purified by layering over 5% Ficol containing 0.55 M sorbitol and 1 mM tris–MES buffer. The homogeneity of the vacuole preparation can be nicely demonstrated by the use of anthocyanin–containing tissues, in Tulipa petals.

CHAPTER 12

Micropropagation in Plants

Men has been vegetatively propagating superior plant material because of its beauty, growth form or yield for centuries. It is only during the past 20 years that micropropagation has been available to increase the levels of efficiency and control over these systems.

It is important, if one is to produce many plants from a single individual that the starting material be of the right clonal selection. For effective micropropagation plant must be in an active, vigorous, and healthy state. Micropropagation cannot be regarded as a rehabilitation phase for poor, sickly material; if it is poor and sick on its own roots, it is going to be equally substandard in culture.

ADVANTAGES OF MICROPROPAGATION

Micropropagation is cash fetching biotechnological technique being practiced in crop improvement programmes in several commercial labs and nurseries all over the world. The exigency for this multi–billion dollar technique of tissue culture has risen from the increased demand for floriculture plants such as flowers and foliage plants.

Micropropagation offers many advantages over conventional methods of plant propagation such as rapid large scale multiplication of true–to–type plants, through the year available technique, disease free plants, and conservation of space and time. Main methods of micropropagation are:

1. Micropropagation through shoot bud proliferation
2. Micropropagation through adventitious shoots
3. Meristem tip culture
4. *In vitro* tuberization
5. Somatic embryogenesis

This technique is an alternative method of vegetative propagation and is applied with the objective of enhancing the rate of multiplication. The numbers of plant propagules that can be developed in a short time are extremely large.

A major advantage of micropropagation happens to be the minimum growing space is required in commercial nurseries. Several thousand million plants can be maintained inside vials on a shelf space built into a room of about 3 m × 3 m × 5 m. This makes possible the propagation of clones on a commercial scale for large number of horticultural species (African violet, banana, eucalyptus, ferns, orchids, gloxinia, gerbera and rhododendrons).

Other Advantages of Micropropagation Methods

1. Millions of plants of the elite, productive types(s) can be produced.
2. All the plants are of the same physiological age and hence of uniform growth, otherwise, it is extremely difficult to obtain uniformity in vegetatively propagated

material, unlike in the seed raised crops.

3. Planting material can be made free from viral and bacterial disease which is specially important in vegetatively propagated species.
4. A large number of plants can be produced in a short time and space.
5. Cells grown *in vitro* may be used for biotransformation to a new chemical (drug).
6. Continuous and reliable sources of plants and products round the year irrespective of seasonal variation.
7. Genetic and phenotypic uniformity can be maintained.
8. Induction of genetic variability and high yield of desired products (Bajaj *et al.*, 1988).
9. Micropropagation is the only quick way to produce products developed by genetic engineering in the pharmaceutical industry. Again the advantage of using *in vitro* regeneration systems is the possibility to multiply the initial transformation event by cell proliferation and to select the transformed cell production.
10. Plant cells and tissues can be manipulated under cultural conditions so that chemical potentiality of specific phytochemicals can be induced (Staba, 1990).
11. Quality and desirability of the product can be maintained as the plantlets produced are free from bacteria, fungi and viruses.
12. This method can solve many fundamental problems associated with the biosynthetic pathway of phytochemicals.

The most popular use of micropropagation is as an alternative to the conventional methods of vegetative propagation with the objective of enhancing the rate of multiplication. Through this method, millions of plants can be produced from a small, even microscopic piece of plant tissue within a limited time. Such a prolific rate of multiplication cannot be expected by any of the *in vivo* methods of clonal propagation. Micropropagation is applied on a commercial scale on a wide range of ornamentals, bulbs and corms, flowering and pot plants, foliage plants, fruit and nut crops, vegetable crops, plantation and forest trees like:

Aglaonema	**Alpinia**	**Alstroemeria**	**Ananus**
Anthurium	Asparagus	Bambusa	Begonia
Carica	Coffea	Cordyline	Curcuma
Dalbergia	Dieffenbachia	Elettaria & Amomum	Eryngium
Eucalyptus	Ficus	Gerbera	Gladiolus
Hydrangea	Impatiens	Iris	Lilies
Maranta	Mimosa	Musa	Narcissus
Narcissus	Orchids	Philodendron	Rosa
Saccharum	Solanum	Spathiphyllum	Syngonium
Tamarindus	Vanilla	Vitis	Zingiber

Advantages of Micropropagation over Traditional Methods

In vivo techniques have the following advantages over:

1. Certified virus free plants can be produced in large number.
2. It is possible to produce clones of some kinds of plants that are otherwise slow and difficult to propagate vegetatively.
3. Only limited space is required to maintain and produce large number of plants.
4. Plant material requires less attention between subcultures and there is no labour or material requirement for watering, weeding, spraying, etc. Micropropagation

is most advantageous when the cost is less than traditional methods of multiplication.

5. Plants may acquire a temporary vigour through micropropagation which makes them more desirable to the grower than conventionally raised stock.
6. Predictability of yield and uniformity the product.
7. Production can be continued all the year round and is independent of seasonal changes.
8. Propagation is ideally carried out under aseptic conditions, free from pathogens. Once cultures have been started there should be no loss through disease and the plantlets finally produced should be free from bacteria, fungi and other microorganisms.
9. The rate of propagation is much greater than in micropropagation and many more plants can be produced in given time. This will enable newly selected varieties to be made available quickly and widely.

TABLE 12.1: Comparison of conventional and in vitro breeding system

Factors	Conventional breeding	in vitro breeding
Growth cycle	Several decades	25–26 hr
Size		One to several dozen metres, 50–100 mm in diameter
Space (106 offspring per year)	Several hectares	101 of suspended cells
Quantities required for mutation rate investigation	107 trees	107 cells
Quantities required for character investigation	103 trees	103 cells
Time required for seed production	5–15 years	One month produces 106 somatic embryos for use as artificial seeds
Predictability of seed production	Seed–setting has on– and off–years and is affected by natural factors	Seed production is completely under artificial control
Genetic uniformity	Much variation occurs during sexual propagation and pollination, and selection should be controlled	Less variation in somatic embryos
Multiplication	Cutting or grafting at slow speed	Propagation coefficient of in vitro culture can increase by millions
Ploidy	Difficult to obtain haploid and homozygous lines; diploid materials	Easy to get haploid and pure lines
Breeding methods	Selection is most common; hybridization and mutagenization are available in some species; self–cross and back–cross are difficult or impossible	Combination of hybridization and anther culture can accelerate breeding process and increase selection efficiency; mutation and mutant screening are conducted at cellular level with high efficiency; pure lines can be achieved through haploid culture and chromosome doubling; genetic engineering methods can be used

Uses of Micropropagation in Horticultural Industry

Micropropagation provides new varieties and novelty in ornamental plants. Floriculture has become popular among small farmers and multinationals. Micropropagation technique id gaining importance in crop improvement programmes in several commercial labs and nurseries all over the world. This a multi–billion dollar technique of tissue culture which has risen from the increased demand for floriculture plants such as flowers and foliage plants. Currently, worldwide use of micropropagated plants is about 50 million per annum. It is likely that the market itself could absorb more than 250 million per annum, if sufficient facilities are available. The significance of the system lies in the fact that if one sprout or hybrid plant becomes available from a grower or breeder, it is possible to use hundreds of thousands of plants within a relatively short time possibly 18 months to 24 months through micropropagation. This allows the rapid introduction of new lines and varieties into the marketplace. It is estimated that this system shortens the average length of time taken to introduce new lily varieties by about 50% from 15 to 16 years down to 7 to 8 years. The same is true of many other plant species.

It is an estimate that floriculture private industries of USA, UK and Netherlands produce about 5 million plants per year. In India, about 51 companies including Indo American Hybrid seeds, SPIC, Khoday's, AV Thomas and Co, Hindustan Levers Ltd. are also making good efforts in this directionusing tissue culture techniques for different horticultural crops. Cut flowers and foliage ornamentals which are in great demand such as Rose, Carnation, *Anthurium, Ficus, Syngonium,* Lilies, *Diffenbachia*, are routinely grown on commercial basis using micropropagation. National Botanical Institute, Lucknow; Indian Institute of Horticulture Research, Bangalore (Government of India organizations) are also working on ornamental plants. In India, AV Thomas and Co. Ltd. is pioneer in the field with tissue culture of Banana upto 1984. Now this company is involved in tissue culture technique of ornamental plants like Gerbera, lilies, orchids.

Indo–American hybrid seeds, Bangalore was set up in consultation with Sunki Pvt. Ltd. (Australia) which supplies 6 million ornamentals like lilies, Alpine and Mimosa raised through micropropagation. They are in process to develop diagnostic kits for disease indexing and going to develop new varieties using genetic engineering technique. Protocols have been developed for fruit crops such as strawberry, cherry, apple through successful micropropagation technique

Through micropropogation, protocols for *in vitro* mass production of some important forest trees such as *Tectona, Tamarindus, Poplars*, Teak and Bamboo have developed.

Micropropagation is also being used for development of new profitable species varieties of cinnamon, clove, and nutmug at Kerala Agricultural University and Calicut University.

Cinnamon is a cross pollinated crop and referred to as queen of spices. It has considerable export earnings. It is propagated vegetatively as well as through seeds. Its seed progenies show variations and hence vegetative propagation through tiller is carried out. However, the rate of multiplication is slow and only 3 tillers are produced per plant per year. Through micropropagation, rate of multiplication has been shown to be more than 30 per cent over open pollinated seedlings.

Biotechnology division of Bhabha Atomic Research Centre (India) has standardized protocols for Banana. Field performance of these plants have been encouraging with vigorous growth, early maturity and increase in yield.

Palms are usually propagated through seeds except date palm which is propagated through seeds as well as vegetatively by off shoots. Seedling derived plants often show a lot of variation which affects the total yield. Hence the mass multiplication of high yielding clones from elite palms through tissue culture is advantageous.

Sustainable development and improvement Pulses, grain legumes and oil seeds has also been done in these crops. Application of tissue culture in legume is difficult proposition. However, recent successful regeneration of whole plants from variety of legumes including pigeonpea and chickpea (Eapen and Rao, 1993) has paved the way for successful application of these techniques for genetic improvement and mass multiplication of these important crops. The successful protocols developed for regeneration and transformation can be used for productivity (Mansur *et al.*, 1993; Eapen and George, 1994) in legumes.

Disadvantages of Micropropagation

1. The changes of producing genetically aberrant plants may be increased.
2. The major disadvantage is the need for advanced skills, expensive production facility availability of specific systems for each species and variety, labour intensive and high cost of the propagules.
3. The plants derived from the cultures are not initially able to produce their own requirement of organic matter by photosynthesis and have to undergo a transitional period before they are capable of independent growth.
4. The young plants produced through micropropagation are more susceptible to water loss in an external environment. They therefore have to be hardened in an atmosphere by slowly decreasing the humidity and increasing the light.

COMMERCIAL USES OF MICROPROPAGATION

Micropropagation is used for production of various commercial of planting material. Micropropagation helps in rapid multiplication of plants and a large numbers of plants can be produced quickly. Most of the systems involve subculture every 30–45 days. The plant material at least doubles at each subculture. There is a gestation period (range from 6–12 months) between the start of the culture procedure and the production of plantlets in quantity. Since cultures are grown under controlled conditions (in an incubation room at 20–25°C and specific photoperiod, say 16 hr/8 hr light/dark photoperiod), the plant production can go on without any influence of the variations in the climate outside. It has rapid propagation potential, therefore, the stock area can be considerably reduced. Micropropagation also helps in producing disease–free planting material very rapidly. Micropropagation can also be used for introducing new species and varieties in large numbers for commercial plantations. Moreover, the plant production can be programmed in such a way that seasonal peaks in the demand are correctly met or a steady year round demand is adequately taken care of. By selecting the stock plants the quality of the planting material can be enhanced and maintained, in micropropagation. It is the answer for the propagation of difficult to root species. Micropropagated plants are grown under aseptic conditions that are without soil, fertilizer (biological and inorganic) and climatic influences can be freely moved between countries without fear of spreading diseases.

MATERIALS FOR MICROPROPAGATION

The starting material for a micropropagation programme is a shoot tip, not necessarily as small as the meristematic apical dome, but something a little larger with leaf primordia already developing around it. This material must initially be surface sterilized. Any tissue affected by exposure to sterilize material is trimmed and the remaining material placed on one of a variety of different media. Since it is initially unknown which medium will be suitable for a specific variety or species involved. Concentration of growth regulators should be adjusted in order to allow the plant tissue to initiate into culture. This is the *establishment phase* of the tissue *in vitro* conditions and can take, in some species, as long as 3 or 4 months, usually it can be completed within a few weeks. Under the optimal conditions the small shoot tip will initiate to grow in normal way, giving rise to a small, unrooted shoot in culture.

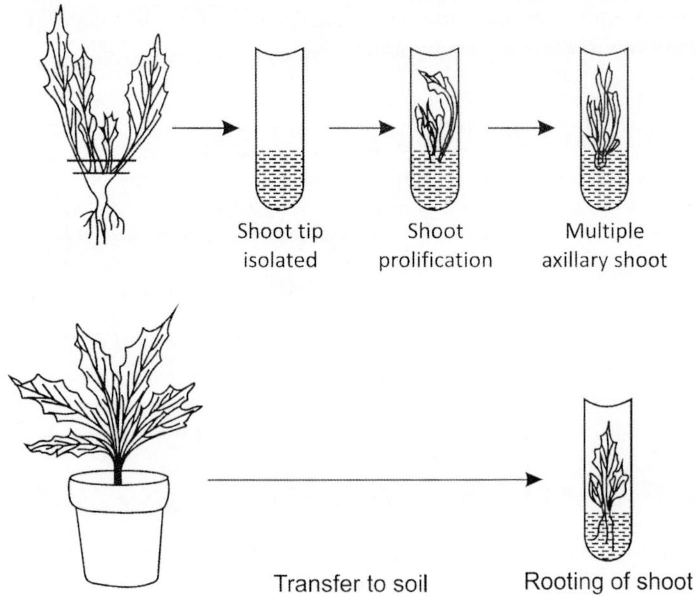

FIG. 12.1 (A): Micropropagation by axillary bud method.

The initiated shoot, after a suitable period in the initiation medium can be subcultured such that nodal sections of the shoot are placed onto a multiplication medium. This medium differs from the initiation medium in that, it contains higher levels of cytokinins which are likely to give rise to precocious shooting of the axillary buds. The normal period between subcultures is 3 weeks and during this time the number of shoot tips available for subculture multiply as a result of breaking of nodal axillary buds anywhere from 2 to 12 times. Thus, after the 3 weeks on multiplication medium, somewhere on an average of about 5 shoot tips can be placed onto fresh medium, and these in turn will multiply 5 times in the subsequent 3 weeks, and thus micropropagation has led to a 25–fold increase in the plant material available over a total of 6 weeks.

For rooting in micro–shoots, transfer the shoot to media which is rich in auxins and will allow the induction of *in vitro* roots. The roots formed *in vitro* are usually different morphologically than that of normal. They are be slightly swollen and brittle, and upon transfer to the soil new roots develop by the plant; before any extensive growth of the shoot can occur and the initial *in vitro* roots die back. This lag in the establishment of the plants can be critical and certainly costly to a grower. It can be avoided by the direct rooting of the micropropagated shoots into a natural or artificial potting mix. If the micropropagated shoots are treated as micro–cuttings, and placed in artificial potting mix maintained at high humidity, roots will develop within 2 to 3 weeks. These are normal physiological roots and not like the *in vitro* roots mentioned above. There is no lag phase when the material is potted on. It represents important part in the overall strategy of plant establishment.

Once the micropropagated plants are satisfactorily rooted in soil or soil substitutes, they must be weaned to a normal humidity level. During the rooting process they are maintained at close to 100% humidity, and this does not allow the normal development of cuticle and stomatal activity associated with drier conditions. The rooting normally takes place under reduced light. After rooting plants are exposed to full sunlight. This is done slowly for a period of 7 to 10 days. Such small plants are now autonomous, photosynthesizing and can fixed carbon. During the rooting phase under the reduced lighting conditions, they are living on carbohydrate reserves built up during micropropagation period *in vitro*. Once this weaning process is completed, the hardened can be incorporated into normal growing systems for their further development.

The investment in commercial tissue culture business will depend to a large extent on cost of the laboratory set–up, type of plant to be propagated and the skill involved. It is advisable that a commercial nursery man start with those crop species for which published methods are available.

TABLE 12.2: List of Orchids which have been clonally propagated in vitro

Plant	Explant
Anacamptis pyramidalis	Shoot tip
Aranda	Shoot tip, axillary bud
(Arachnis hookeriana x Vanda lamellata)	Shoot tip
Aranthera	Shoot tip
Arundina bambusifolia	Shoot tip (from young seedling)
Ascofinetia	Inflorescence segment (with flower primordia)
Brassocattleya	Axillary bud
Calanthe	Shoot tip
Cattleya	Shoot tip, axillary bud, lateral bud, leaf base, leaf tip
Cymbidium	Shoot tip
Dendrobium	Shoot tip, Nodal segment, Flower stalk segment (with vegetative buds)
Epidendrum	Leaf tip
Laelia	Axillary bud
Laeliocattleya	Axillary bud
Lycaste	Shoot tip
Miltonia	Shoot tip
Neostylis	Inflorescence segment (with flower primordia)
Neottia nidus–avis	Root
Odontioda	Shoot tip
Odontoglossum	Shoot tip
Odontonia	Shoot tip
Oncidium	Shoot tip
Oncidium papilio	Flower–stalk segment (with dormant apical buds)
Phajus	Shoot tip

Plant	Explant
Phalaenopsis	Flower–stalk segment, Shoot tip, Leaf segment, stem segment, root segment
Pleione	Shoot tip
Rhynchostylis gigantea	Shoot tip, lateral buds
Schomburgkia superbiens	Lateral bud
Vanda (Terete–leaf)	Shoot tip, Stem section
Vanda (Strap–leaf)	Shoot tip
Vanda hybrid	Shoot tip
(V. teres x V. hookeriana)	Axillary bud, root segment
Vascostylis	Inflorescence segment (with flower primordia)
Vuylstekeara	Shoot tip

Plant Health

Good plant health is the major advantage in the development of elite mother stock material for both the herbaceous ornamental producer and the woody plant grower. Before micropropagation, plant material can be put through stringent tests to bacterial and virus index. The presence of fungal contaminants or pathogens can also be detected and eliminated prior to micropropagation. Thus, the plants produced by the system can be certified as being disease indexed, and they represent clean, elite mother stock material for future conventional propagation methods.

Germplasm Storage

The concept of a tissue bank (a collection of valuable breeding or growth material) in a relatively confined space under controlled environment conditions is particularly attractive for plant breeders. Many plant breeders wish to maintain specific breeding or hetrogenous individuals in the field or greenhouse for subsequent years for testing or seed production. Micropropagation techniques applied to these materials allow them to over winter the material in the safety of a controlled environment and also give rise to the potential for multiplying up individuals for specific crossing or testing.

This elimination of the seasonal nature of much breeding and growing work has been another advantage often suggesting with respect to micropropagation. However, a word of caution is necessary here – although it is possible to micropropagate plant species throughout the year, in the case of biennial species requiring vernalization, it is necessary to do the micropropagation at a suitable time of the year to allow the tissue to become vernalized for flowering for next year.

TYPES OF MICROPROPAGATION

Micropropagation through Shoot Bud Proliferation

Mass propagation through axillary and apical shoot bud proliferation is the common and reliable technique. Explants are obtained from actively growing or dormant plants. The buds are collected from plants growing in the field or greenhouse. Such buds are less contaminated.

Method

1. The leaves are removed carefully to minimize damage to the stem.
2. An initial thorough wash under tap water, then the buds are washed in a very dilute detergent solution.
3. It is followed by sterilization in dilute mercuric chloride or sodium or calcium hypochlorite solution containing a drop of wetting agent, viz., Tween–20 or teepol etc.
4. Rinse them for several times in sterile water before transferring on to the nutrient medium.
5. For actively growing shoot tips quick dips in ethanol is practiced (Lane, 1978).
6. The aseptic cultures are then transferred to proliferation medium with a hormonal combination.
7. The shoots obtained by this method are then transferred to rooting medium under *in vitro* conditions.
8. The well rooted plants are transferred to green house for acclimatization.
9. This method helps to increase multiplication rate of a plant (approximately 10,000 times) by two methods:
   a) The axillary buds present in the explant and in the newly developing branches are stimulated to develop into branches (an axillary bud is present in the axil of each leaf), this is achieved by a relatively higher concentration of a cytokinin in the medium (enhanced axillary branching). After suitable each axillary branch is excised and transferred to a fresh appropriate medium for enhanced axillary branching. The branches may be transferred to a rooting medium and subsequently transferred to soil.
   b) This method is applicable to those plant species where enhanced axillary branching cannot be achieved (each axillary bud/shoot tip develops into a single shoot). Nodal cuttings (each containing a single node) are excised from these stems and cultured for further multiplication.

The other company which entered ornamental tissue culture in a big way is Indo–American hybrid seeds, Bangalore. The lab was set up in consultation with Sunki Pvt. Ltd. (Australia). Now, they are at their own and have orders for the supply of 6 million ornamentals like lilies, Alpine and Mimosa. The company also proposes to come up with diagnostic kits, for disease indexing and go in a big way in genetic engineering. Successful micropropagation protocols have been developed for fruit crops such as strawberry, cherry, apple.

Forest resources in India are getting depleted day by day due to deforestation. One sided approach to agriculturalization, urbanization, industrialization has resulted in miserable disbalance to forest land. This calls for immediate and urgent efforts to start tree improvement programmes. In this regard tissue culture has great potential. Today, protocols for *in vitro* mass production of some important forest trees such as *Tectona, Tamarindus, Poplars*, Teak and Bamboo are available.

Micropropagation work on spices is being done at Kerala Agricultural University and Calicut University. The important spices are cinnamon, clove, and nutmug. Cinnamon referred to as queen of spices has considerable export earnings, is propagated vegetatively as well as through seeds. Since it is a cross pollinated crop, seed progenies show variations and hence vegetative propagation through tiller is carried out. However, the rate of multiplication is slow and only 3 tillers are produced per plant per year. Through micropropagation, rate of multiplication has been shown to be more than 30 per cent over open pollinated seedlings.

Biotechnology division of Bhabha Atomic Research Centre (India) has standardized protocols for Banana. Field performance of these plants has been encouraging with vigorous growth, early maturity and increase in yield.

Micropropagation work on spices is being done at Kerala Agricultural University and Calicut University. The important spices are cinnamon, clove, and nutmug. Cinnamon referred to as queen of spices has considerable export earnings, is propagated vegetatively as well as through seeds. Since it is a cross pollinated crop, seed progenies show variations and hence vegetative propagation through tiller is carried out. However, the rate of multiplication is slow and only 3 tillers are produced per plant per year. Through micropropagation, rate of multiplication has been shown to be more than 30 per cent over open pollinated seedlings.

Biotechnology division of Bhabha Atomic Research Centre (India) has standardized protocols for Banana. Field performances of these plants have been encouraging with vigorous growth, early maturity and increase in yield.

Palms are usually propagated through seeds except date palm which is propagated through seeds as well as vegetatively by off shoots. Seedling derived plants often show a lot of variation which affects the total yield. Hence the mass multiplication of high yielding clones from elite palms through tissue culture is advantageous.

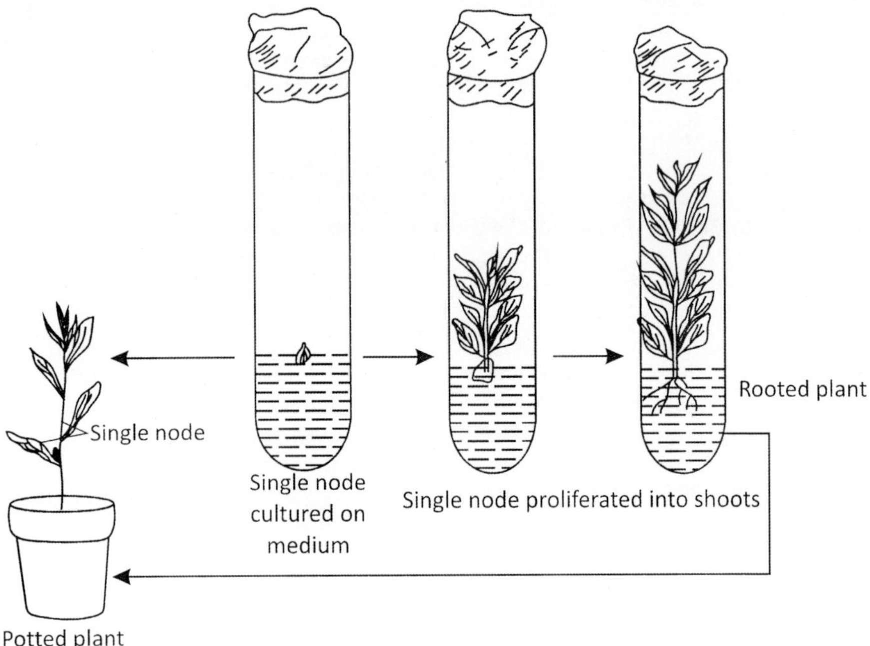

FIG. 12.1(B): Micropropagation by single node culture method.

Pulses, grain legumes and oil seeds form an important constituent for food, fiber and oil for human and animal population. Sustainable development and improvement in these crops is vital for feeding the millions. Application of tissue culture in legume is difficult proposition. However, recent successful regeneration of whole plants from variety of legumes including pigeonpea and chickpea (Eapen and Rao, 1993) has paved the way for successful application of these techniques for genetic improvement and mass multiplication of these important crops. The successful protocols developed for regeneration and transformation can be used for productivity (Mansur *et al.*, 1993; Eapen and George, 1994) in legumes.

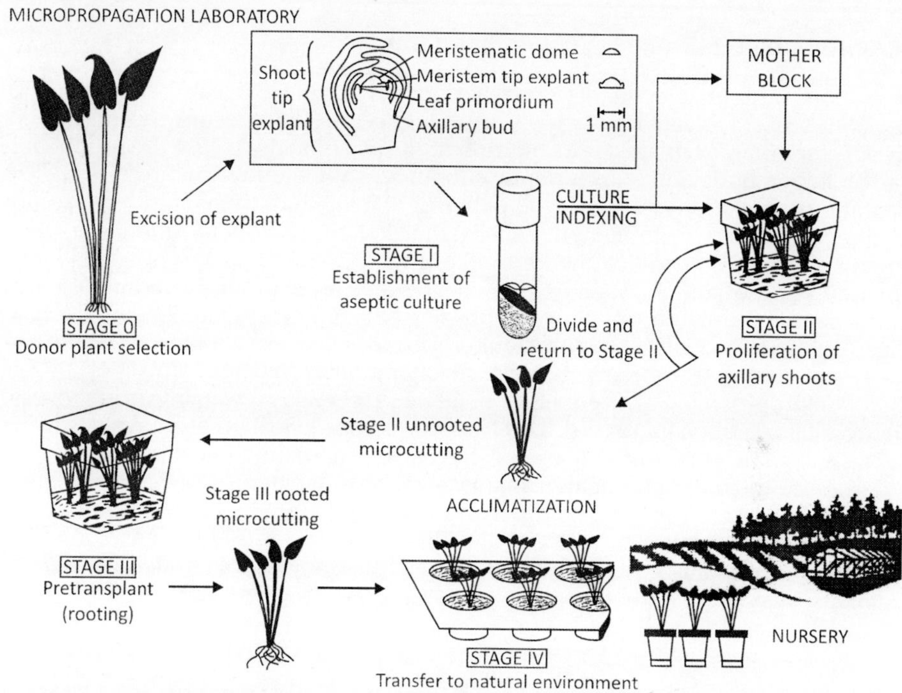

FIG. 12.1 (C): Micropropagation stages for production by shoot culture.

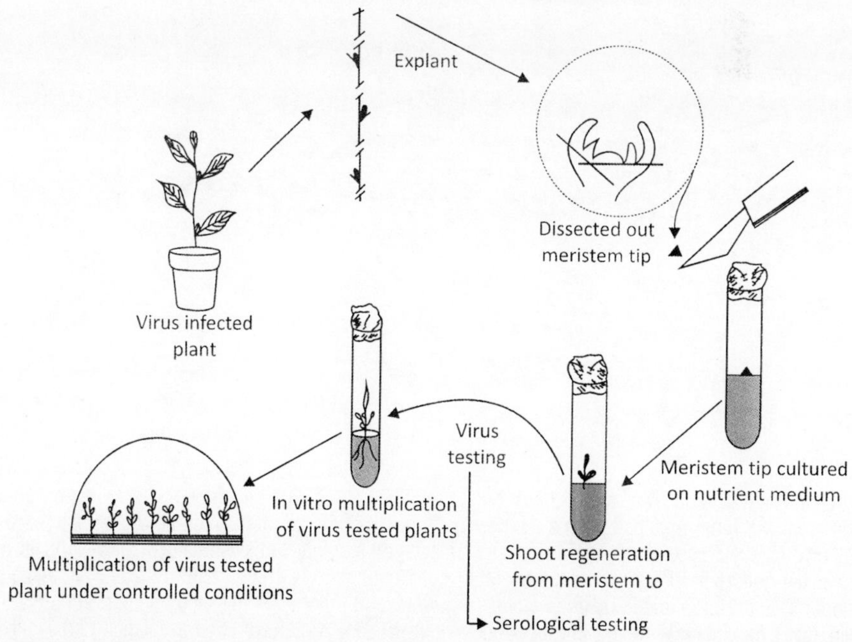

FIG. 12.2: Diagrammatic representation of meristem tip culture.

Micropropagation through Adventitious Buds

Adventitious buds arise on plant parts other than in normal leaf axils. Propagation by direct induction of adventitious shoots on stems, leaves, bulbs, scales, corms, tubers and rhizomes of intact plants, is a common method. Similarly, *in vitro* the production of adventitious buds and subsequently shoots be simulated to that in *in vivo*. The adventitious shoot buds may develop either a) directly from the cultured explant or b) indirectly from the callus generated from the explant. Thus factors namely explant and growth hormones in the culture medium control the induction of adventitious buds and subsequent shoot formation. Deliberate induction of adventitious buds may also be done by an organogenic phase in an undifferentiated tissue such as callus. Callus tissue is the form of culture in which adventitious meristems are usually induced and from which plant regeneration is initiated. However, in this type of micropropagation, frequency of genetic changes, resulting from mitotic abnormalities, is often increased. This type of micropropagation also involves usual stages i) explant culturing and establishment ii) shoot multiplication iii) rooting iv) hardening of micropropagated plants and transplantation. Adventitious shoot prediction has been used successfully for ornamental and herbaceous horticultural crops. Bulbs and corms grown adventitiously from meristems at the base of leaves and can be continuously propagated by adventitious shoots proliferation.

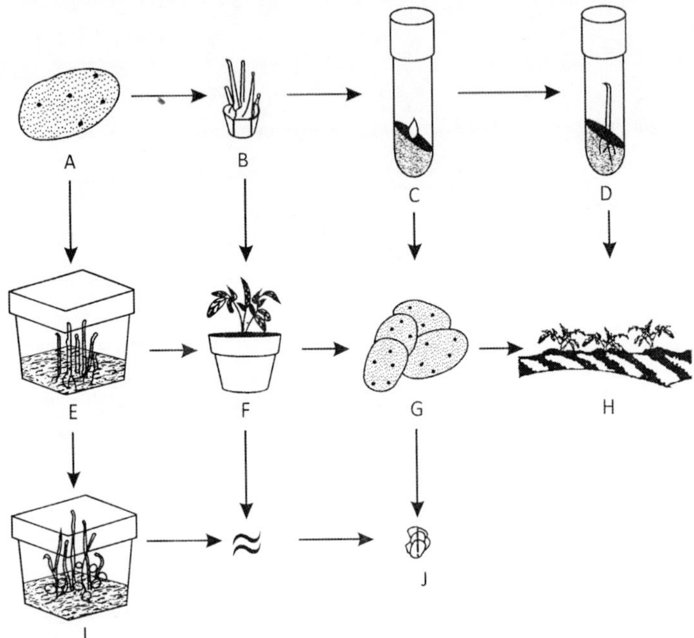

FIG. 12.3: Diagrammatic representation of potato micropropagation by node culture and micro–tuber production. (A) Tuber obtained from a multigeneration pathogen–infected field potato population; (B) meristem tip explants excised from sprouted tuber sections; (C) establishment of meristem tip in culture; (D) culture is SPT and indexed for other contaminants; (E) shoots determined to be specific pathogen free are propagated by node culture to increase numbers; (F) microcuttings are rooted and acclimatized ex vitro; (G) tubers produced ex vitro from STP plants are used as planting stock; (H) field planting of STP seed tuber; (I) shoots may be transferred to a microtuber induction medium and promotive conditions; (J) microtubers produced are stored or used as STP seed tubers for field planting.

Meristem Tip Culture

Meristem tip refers to the region of shoot apex lying distal to the youngest leaf primordia, whereas shoot apex includes meristem tips plus a few sub-adjacent leaf primordia. Meristem tip culture is a method in which shoot meristem (meristem tip) is cultured *in vitro* with a few leaf primordia. The meristem when cultured *in vitro* on a suitable medium, differentiates into whole plants identical to the parent plant. Ball (1946) was the first to obtain rooted plants of *Tropaealum majus* and *Lupinus albus* from excised shoot apical dome.

Meristem tips are carefully excised from shoot tip. Shoot tips taken from *in vitro* growing plant material need not to be sterilized. The shoot buds or tips are excised from field grown plant material. The bud is taken under stereozoom microscope. Meristematic region at the apex is gradually uncovered by gradually removing the scales, immature leaves and pubescence with the help of very fine needles and blades. Then the apical meristem was gently removed from parental tissue by giving a horizontal cut at the base and cultured on solid medium supplemented with growth regulators which induce plantlet regeneration.

Though, meristem culturing has been extensively used for vegetative propagation of many crops and fruit trees, it is mainly employed to obtain virus free plants since in infected plants the apical meristems are generally either free or carry a low concentration of viruses (Quak, 1977). The question need to be answered is why the meristems escape the viruses and how meristem culture leads to the production of virus free plants. Various reasons, explaining the use of meristem culture to obtain virus free plants, have been proposed. The most plausible explanation appears to be that there is a competition in a meristem between cell production and virus multiplication. Another explanation is the virus deactivating system in the plant body if any has higher activity in the meristems than in any other region. Absence of vascular system and plasmodesmata in the meristem greatly hinders the transport of virus particles.

In Vitro Tuberization

Another method of mass propagation is the *in vitro* tuberization. The tuberization *in vitro* can be induced by culturing axillary shoots of potato to a culture medium supplemented with cytokinins and growing the cultures under an 8 hour photoperiod, at 22°C. *In vitro* produced tubers or minitubers have the following advantages i) A continuous year round production in large quantities ii) this enables an easy shipment and convenient supply of material iii) they can be stored for several months iv) *in vitro* tuberization provides an opportunities to carryout an year round operation with year round scheduling according to the market demand. It has been estimated that higher yields of the order of 5–6 times was obtained by *in vitro* tuberization over conventional methods. Similar, *in vitro* tuberization can also be induced in tropical tuber crops like *Dioscorea alata* (Yam) by culturing them under controlled conditions.

Somatic Embryogenesis

Somatic embryogenesis is the development of embryos from immature embryos and other somatic tissues cultured *in vitro* on defined media or from cells cultured in suspension. It is a process by which embryo like structures, appear to recapitulate the typical stage in the embryogenesis of a zygote, which ultimately give rise to whole plants (Raghvan, 1983). Such type of embryos are referred to as somatic embryos since they are formed from sporophytic or somatic cells, in contrast to gametophytic or gametic cells. Somatic embryogenesis was first described in carrot

(*Daucus carota*) by Reinert, 1958 and Steward *et al.* (1958). This phenomenon has been observed in over 200 species (Ammirato, 1983).

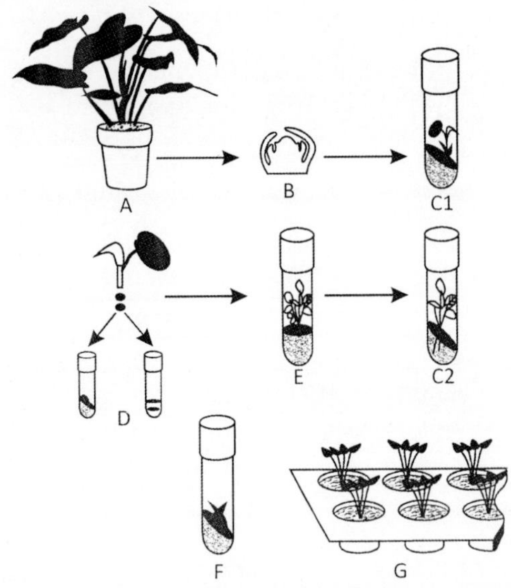

FIG. 12.4: Diagrammatic representation of the Syngonium shoot culture laboratory exercise protocol. (A) Nodal and terminal vine tip explants are obtained from potted plants and then surface disinfested. (B) Successive inner leaf layers are removed to isolate the meristem and shoot tip explants. (C) Primary explants are inoculated onto SEM (C1) and SMM (C2). Growth responses are compared after 6 weeks of culture. (D) Thin basal stem cross sections, excised from 6–week–old shoots generated on SEM, are cultured in both liquid and agar–solidified AC Broth for 7 days to screen for the presence of cultivable contaminants. (E) Shoots are transferred to SEM. (F) Cultures screening positive for contaminants are discarded. Sterile cultures are transferred to SMM and cultured for 4 weeks. (G) Microcuttings are rooted and acclimatized ex vitro (Stage IV).

Somatic embryogenesis can be induced from a variety of explants such as leaf, floral parts, young inflorescences, anther, immature embryos, nucellus, anthers etc. Both zygotic and somatic embryos share similar ontogeny passing through globular, heart shaped, torpedo and cotyledonary stages in the case of dicotyledons; globular, scutellar and coleoptilar stages for monocotyledons; embryonal suspensor mass, globular, torpedo and cotyledonary for conifers. There are two different routes for somatic embryogenesis.

Direct Somatic Embryogenesis

Somatic embryos arise directly from a cell or tissue without an intermediary phase of callus.

Indirect Somatic Embryogenesis

Somatic embryos arise from callus. The process of somatic embryogenesis normally takes place in two stages: first, the induction of cells with embryogenic competence referred to as proembryonic cell masses or clumps, proembryos etc. in the presence of high auxin and second, the development of embryogenic masses or proembryos into embryos in the absence or lower concentrations of auxins (Thorpe, 1994). Several factors such as cultural medium, growth regulators, carbon source, development stage of the explant and cultural conditions influence somatic embryogenesis.

Mass propagation through somatic embryogenesis, can best be achieved using suspension cultures. For mass propagation of some of the ornamental species and vegetables like celery and carrot, which quickly form somatic embryos in suspension culture, has been adopted on commercial scale. Date palm has successfully been multiplied through somatic embryogenesis and some of the commercial tissue culture labs have already exploited this technique. Tissue culture of oil palm is commercially important and is cloned at present on large scale through embryogenic cultures (Bromton and Black, 1985). Protocol for mass propagation of oil palm through somatic embryogenesis is now available in many labs. Plant produced through somatic embryogenesis had more uniform growth than seedling derived plants. It is estimated that higher yields of the order of 30 per cent and improved oil quality would be obtained by cloning palms possessing the desired characters.

In addition to their use in mass propagation, somatic embryos have been recently used to prepare synthetic seeds by their encapsulation in an appropriate matrix. Early reports in this case have been in carrot, celery and alfalfa (Reddenbaugh and Paash, 1986). The main thrust idea is to prepare a simple delivery system by which the embryos survive sowing process, develop further and finally germinate. Although the technique is quite recent, it offers many advantages on a commercial scale. The resultant plant population, which originates from the synthetic seed will be uniform. The direct delivery of somatic embryos will save many subcultures to obtain plantlet from regenerated embryos. The encapsulated material could be packed with pesticides, fertilizers, nitrogen fixing bacterial and even microscopic parasite destroying worms. Thus, an artificial or synthetic seed may be defined as an analoguous to a true seed in which a somatic embryo is encapsulated with a coating material which can be a thin or thick covering. There is another term 'conversion' which is generally used in literature alongwith artificial seeds. Conversion is defined as the germination (elongation of radicle) of a somatic embryo followed by the development of a vigorous root system. The root is directly connected to a shoot.

However, to start programmes on synthetic seeds, several basic studies are to be conducted. The plant should have the capacity to produce large number of somatic embryos over a long time period. The next step is to synchronize the development of somatic embryos since somatic embryos usually develop asynchronously from a cell mass comprising embryogenic tissue.

Two types of artificial seeds have been reported in literature: desiccated and non–desiccated. It varies from species to species whether a somatic embryos is to be desiccated or not since in soybean conversion has been improved from 30 per cent to 60 per cent (Obendort and Slaminska, 1986) in desiccated somatic embryos whereas conversion was reported 59 per cent in *Triticum* (Carman *et al.*, 1987) which was less than achieved with non desiccated embryos.

Somatic embryos have successfully encapsulated in hydrated gels. The most practical encapsulation system has been found to be dripping of two per cent sodium alginate from a separatory funnel into a 100 mM calcium nitrate solution. The moment sodium alginate drops emerge at the tip of the funnel, a somatic embryo is inserted. These encapsulated embryos are complexed in calcium salt for about 20 minutes after which these are rinsed in water and stored in a closed container. The capsules containing somatic embryos generally dry out, within 24 hours if these are not placed in air tight containers. The coating material is desirably packed with nutrients, hormones and biofertilizers.

Artificial seed research has successfully been done in Banana, sandal wood, in some cereal crops viz., barley, rice and finger millet. The production of synthetic seeds in rice might be useful in the propagation and delivery of high yielding, individual hybrids.

So far the most advanced synthetic seed system has been developed for alfalfa where direct field planting has been accomplished. Despite the fact that somatic embryogenesis has been achieved for many plant species the efforts are called for research on somatic embryogenesis in valuable crop plants. Synthetic seed technology in years to come would certainly find its application in Agriculture, Forestry and Horticulture.

METHODS OF MICROPROPAGATION (GENERAL)

Shoot tip is the starting material for micropropagation which is surface sterilized then trimmed and the remaining material placed on one of a variety of different media. The selection of specific medium is necessary with special reference to concentrations of particular growth regulator. This is the phase for acclimatization of the tissue to the *in vitro* conditions which requires about 3 or 4 months and varies from species to species. Under the suitable laboratory conditions the small shoot tip grows well in a normal fashion, giving rise to a small, unrooted shoot in culture.

The initiated shoot can, after a suitable period on the initiation medium, be subcultured so that the nodal sections of the shoot are placed onto a multiplication medium. This medium differs from the initiation medium in that it contains higher levels of cytokinins which are likely to give rise to precocious shooting of the axillary buds. The normal period between subcultures is 3 weeks and during this time the number of shoot tips available for subculture will multiply as a result of breaking of nodal axillary buds anywhere from 2 to 12 times. After 3 weeks on multiplication medium, somewhere on an average of about 5 shoot tips can be placed onto fresh medium, and these in turn will multiply 5 times in the subsequent 3 weeks, and thus micropropagation has led to a 25–fold increase in the plant material available over a total of 6 weeks. Rooting of above cultured micro–shoots can occur in two ways. One can transfer the shoot to media rich in auxins which will allow for the induction of *in vitro* roots. In some species it is necessary since they need to be actively induced to root.

Which Material should be used?

1. The starting material for a micropropagation programme is a shoot tip, not necessarily as small as the meristematic apical dome, but something a little larger with leaf primordia already developing around it.
2. This material must initially be surface sterilized.
3. Any tissue affected by exposure to sterilize material is trimmed and the remaining material placed on one of a variety of different media.

Stages of Micropropagation

1. Selection and sterilization of elite plants
2. Establishment of axillary buds in culture
3. Multiplication in culture
4. Rooting of *in vitro* plants and transfer to compost

Materials

Stage I: Selection and Sterilization of Elite Plants

1. The selected elite stock plants should be visibly free from any sign of disease,

stress, or surface blemishes. This material is grown in an environmentally controlled growth cabinet or clean glasshouse. It should be tested for the presence of specific viruses by using ELISA techniques. It is used for identification of viruses.

2. Sterilization is done by using 70% ethanol plus four drops of surfactant (manoxol)/100 ml solution at room temperature or 2% chloros plus four drops of surfactant (manoxol)/100 ml solution at room temperature should be used.

Stage II: Establishment of Axillary Buds in Culture

1 mm diameter sterile Petri dishes containing approximately 10 ml of medium 1 or 85 mm diameter sterile petri dishes containing approximately 25 ml of medium should be used. Use incubator or environmentally controlled growth chamber allowing environmental conditions of 20–24 °C (depending on variety or species), a 16–h photoperiod, and a light intensity of 4000–6000 lux.

Stage III: Multiplication in Culture

1 ml sterile jars containing approximately 30 ml of medium 2 or medium 3. Environmental conditions as described for Stage II.

Stage IV: Rooting of In Vitro Plants and Transfer to Compost

1 ml sterile jars containing approximately 50 ml of medium 4, Propagation trays, plastic bags, or seed trays covered with glass sheets or a misting or fogging system, 50 mm diameter pots containing compost (3:2, peat : sand) supplemented with nutrients.

For potato (*Solanum tuberosum*) these are:

Osmocote	57 g
Sulphated potash	21 g
Epsom salts	28 g
Frit 252 A	14 g
Lime	170 g
Phosphate per 300 l peat sand mix	85 g

Methods

Stage I: Selection and Sterilization of Elite Plants

1. Use a clean sharp blade, carefully excise axillary buds of the desired variety and store in distilled water until enough buds have been obtained.
2. Commercial preparations of M&S and B5 salt (minus sucrose glutamine and hormones) are available in packets that make up 1 l of media.
   a) pH adjusted using. 1 M HCl and 0.1 M KOH before agar added and prior to autoclaving.
   b) The amount of agar needed varies depending on the brand used.
   c) Varying the hormones allows the above media to be used for axillary bud culture of a wide range of species.
3. The practical should be conducted in a laminar flow bench.
4. Put a maximum of 20 buds into a sterile test tube.
5. Fill to the brim with ethanol solution and leave for 1–1.5 min.
6. Decant the ethanol.
7. Fill to the brim with chloros solution, replace the top, and agitate at 120 strokes/

min for 12 min, either manually or with a shaker.
8. Decant the chloros solution and refill with sterile distilled water.
9. Rinse in sterile distilled water 3x.
10. Store in sterile distilled water until reedy to continue (not longer than 2h).

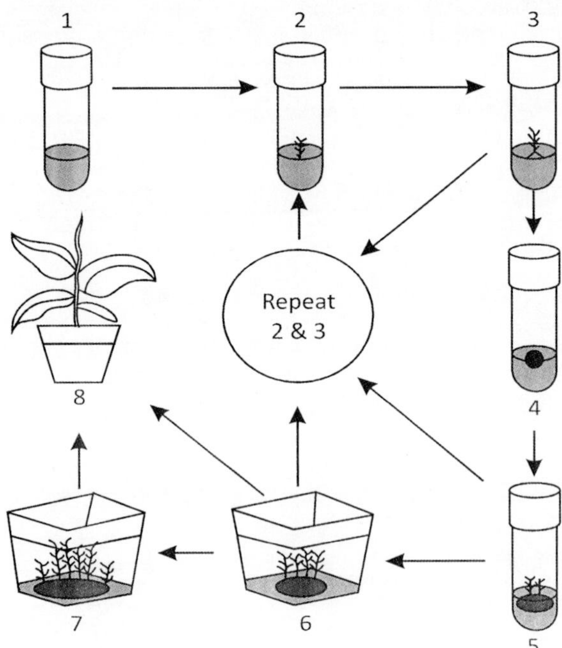

FIG. 12.5: Diagrammatic representation of the micropropagation protocol for Dieffenbachia. Lateral bud or terminal growing point after surface disinfestations and excision on shoot initiation medium (1). Depending on cultivar differences, after 12 to 18 weeks (two or three transfer cycles) the developing plant is multiplication medium for 6 weeks (2) when the enlarged callus base is separated from the apex (3). The shoot tip is used to repeat steps 2 and 3 or it can be planted out to test for genetic integrity. The callus base is left on multiplication medium to develop adventitious shoots (4). Increasing size of callus base and newly developing adventitious shoots (5) requires transfer to larger culture vessel (i.e., babyfood jar or GA–7 magenta vessel). Several shoots developing from callus base (6) may be harvested and planted out or returned to steps 2 and 3 to renew the cycle. Callus base is maintained in culture to produce additional adventitious buds and cuttings on a 6–week transfer cycle (7). Unrooted cuttings are transferred directly to soil; rooting occurs within 6 to 8 weeks (8). Transfer of Dieffenbachia cultures is normally on a 4– or 6–week cycle.

Stage II: Establishment of Axillary Buds in Culture

1. Empty the water plus axillary buds into an empty sterile Petri dish, for ease of handling.
2. Place up to 4 buds in a 50 mm Petri dish or 10 buds in an 85 mm Petri dish containing medium 1.
3. Make sure that the base of the bud is stuck firmly in the medium.
4. Take care that the bud is not buried.
5. Seal the Petri dish with laboratory sealing film, ensuring adequate gaseous exchange by puncturing the film 2–4x with a fine sterile needle.
6. Transfer to growth room or incubator, and leave for between 1 week and 2

months depending on the variety.

Stage III: Multiplication in Culture

1. When shoot extends, cut internodes, and transfer apical cutting and internodal cuttings to sterile jars containing medium 2 or 3.
2. Make sure that the basal portion of the stem is firmly pressed into the medium without burying the explant.

Stage IV: Rooting of In Vitro Plants and Transfer to Compost

1. Cut off the top 5 mm of *in vitro* plant (apex plus 2 or 3 nodes).
2. Transfer to medium 4. (12 plants/250 ml jar).
3. Leave for 3–5 d until 3 or 4 small, study roots approximately 5 mm long are visible.
4. Remove each plantlet carefully from the jar, removing as much agar as is possible without harming the root structure.
5. Transfer to damp compost in 50 mm pots.
6. Keep in high humidity conditions 12–24 h (propagator, plastic bag, misting, or the like).
7. Transfer to the glasshouse, preferably placing on top of capillary matting. Shade from direct sunlight.
8. When the plants are approximately 70 mm high and have begun to lose their juvenile characteristics, they should be hardened off and transferred to larger pots or to the field.

TABLE 12.3: Amounts of Chemicals (gL^{-1} or mg L^{-1}) in Media used in axillary bud culture of Solanum species based on either Murashige and Skoog medium or Gambourg B5 medium.

Chemicals	Medium 1 (M&S)	Medium 2 (M&S)	Medium 3 (B5)	Medium 4 (M&S)
NH_4NO_3	1.65 g	1.65 g	–	1.65 g
KNO_3	1.9 g	1.9 g	2.5 g	1.9 g
$CaCl_2 2H_2O$	0.44 g	0.44 g	0.15 g	0.44 g
$MgSO_4 7H_2O$	0.37 g	0.37 g	0.25 g	0.37 g
KH_2PO_4	0.17 g	0.17 g	–	0.17 g
$(NH_4)_2SO_4$	–	–	13.4 g	–
$NaH_2PO_4.H_2O$	–	–	0.15 g	–
FeNa EDTA	36.7 mg	36.7 mg	40.0 mg	36.7 mg
H_3BO_3	6.2 mg	6.2 mg	3.0 mg	6.2 mg
$MnSO_4.4H_2O$	22.3 mg	22.3 mg	10.0 mg	22.3 mg
$ZnSO_4.7H_2O$	8.6 mg	8.6 mg	2.0 mg	8.6 mg
KI	0.83 mg	0.83 mg	0.75 mg	0.83 mg
$Na_2MoO_4.2H_2O$	0.25 mg	0.245 mg	0.25 mg	0.25 mg
$CuSO_4.5H_2O$	0.025 mg	0.025 mg	0.025 mg	00.025 mg
$CoCl_2.6H_2O$	0.025 mg	0.025 mg	0.025 mg	0.025 mg
Nicotinic Acid	0.50 mg	0.50 mg	1.0 mg	0.50 mg
Thiamine HCl	0.10 mg	0.10 mg	10.0 mg	0.10 mg

Chemicals	Medium 1 (M&S)	Medium 2 (M&S)	Medium 3 (B5)	Medium 4 (M&S)
Pyridoxine HCl	0.50 mg	0.50 mg	1.0 mg	0.50 mg
Glycine	2.0 mg	2.0 mg	–	2.0 mg
Sucrose	20.0 g	20.0 g	20.0 g	20.0 g
Inositol	0.1 g	0.1 g	0.1 g	0.1 g
Glutamine	0.1 g	0.1 g	0.1 g	–
BAP (6–benzyl–amino purine)[4]	0.25 mg	0.25 mg	0.25 mg	0.25 mg
GA_3 (gibberellic acid)[4]	0.1 mg	–	–	–
NAA (1–napthalene–acetic acid)	–	–	–	0.05 mg l^{-1}
pH	5.64	5.64	5.64	5.64
Agar (Difco Bacto)	8.0 g	8.0 g	8.0 g	8.0 g

IMPORTANT CONSIDERATIONS AND PRECAUTIONS

Problems and Contamination in Culture

Contamination by Bacteria and Fungi

Bacteria (*Pseudomonas* sp., *Erwinia* sp. and *Bacillus* sp.) are the major causative factor of contamination during large–scale micropropagation of some plants. These slow growing bacteria contaminants and persist even after initial surface sterilization of explants. Such contaminants may persist for many generations without being noticed and cause reduction in vigour or chlorosis in propagated plantlets. Addition or antibiotics or fungicides to the culture medium may control the infection by microbial contaminants but for propagation of plants normally infected with latent or symptomless types of viruses or mycoplasmas, it is necessary to maintain stock of plants of them which are disease free.

Surface sterilization should remove most contamination. Frequent scrutiny of cultures should allow healthy buds to be transferred to fresh medium if occasional contaminated buds are seen. Alternatively, placing fewer explants in smaller Petri dishes reduces the problem and infected plants can then be discarded. The use of antibiotics to suppress bacterial or fungal infections is generally ineffective.

Contamination by Viral Pathogens

The danger can be minimized by testing for infections by known viruses before initiating the culture. Once in culture, viral pathogens generally do not produce any symptoms, but are rapidly spread by axillary bud cultures. The smaller the piece of tissue used to initiate cultures, the less risk there is of transferring viral infection. To reduce this risk to a minimum, meristems can be used to initiate cultures.

Browning of the Medium

This kind of problem is commonly found during micropropagation of woody perennials. This occurs due to the accumulation of inhibiting substances in the growth medium specially during initiation of cultures. Explants from adult tissues of these species often produce excessive amounts of phenolic substances which turn the medium dark brown. Such a medium is toxic to the tissues and inhibits their growth. Quick transfer of explants to a fresh medium at short intervals could alleviate this problem in orchid bud cultures. Browning of the medium may also be prevented by dissecting explant tissues under the surface of liquids or by incorporating ascorbic acid or citric acid in culture media. For teak tissue cultures, polyvinyl pyrrolidone has proved an effective amendment to the culture medium. Certain phenolic compounds e.g., phloroglucinol (PG) have been found essential for shoot multiplication and rooting in a number of rosaceous cultivars. PG (found in the xylem sap of apple) is known to promote shoot multiplication and rooting in Pissardi plum and *Cinchona*.

Precautions for Sterile Culture

Use a laminar flow bench, and swab down the bench, roof, and sides with disinfectant. Always wear gloves (thin latex gloves are best), and swab gloves and bench surface frequently during any experimental work. Sterilize and flame all instruments before use. Finally, avoid passing anything over the top of opened containers.

1. **Stage I:** Choose as juvenile a tissue as possible to initiate cultures. Do not try to sterilize too many different genotypes at once. Start with 1 or 2, and build up to about 6. Problems of contamination and toxic effects of the sterilant increase with increased numbers of genotypes being treated simultaneously.
2. **Stage II:** Condensation can be a problem in Petri dishes. This can be minimized by stacking the Petri dishes and rotating the position of each dish in the stack daily. The intensity of light reaching each culture does not appear to be critical at this stage, although light is necessary.
   Gibberellins tend to produce elongated, abnormal looking shoots. These revert to normal growth when transferred to medium that does not contain gibberellins. Callusing will occur around the base of the bud; only shoots that can be clearly seen to have derived from extension of the preexisting axillary bud should be transferred. Adventitious shoot formation in callus tissue generally produces high levels of variation and is the basis for producing somatic variants.
3. **Stage III:** Murashige and Skoog medium and Gambourg B5 medium are those most frequently used in the culture of axillary buds. They often need to be modified, mainly by varying the vitamins and growth regulators, depending on the species or variety being used.
   If the intent is to use the axillary bud cultures as a source of protoplasts, then all media need to be made up using distilled deionized water or water of equal purity other additions may also have to be made to the culture medium.
   Vitrification can be a serious problem in *in vitro* cultures. In this condition, the shoots tend to elongate and look glassy, the leaves become reduced or malformed, and if not treated the culture will die. The causes of vitrification are not fully understood. However, the risk of vitrification can greatly reduced if there is adequate gaseous exchange between the culture and the external environment which is achieved by piercing laboratory film seals or by loosening jar lids slightly.
4. **Stage IV:** Rooting and transfer of plants to compost can be difficult. If the auxin level is too high, subsequent root development after transfer to compost is reduced. Some species can be transferred to compost without a rooting step. Some fruit tree species need a dark treatment as well as additions of auxin for successful rooting to take place (5). The transition from culture to compost is critical. High humidity must be maintained, but care must be taken not to waterlog the compost. If plants are transferred to the field, they need protection from wind damage, drought, and birds.

TABLE 12.4: Relative cost components associated with micropropagation of a typical crop

Cost component	Per cent
Laboratory, direct labour	15
Utilities	9
Depreciation	7
Supervision (laboratory and greenhouse)	23
Planting, direct labour	9
Other production costs	37
Total	**100**

Once the micropropagated plants are satisfactorily rooted in soil or soil substitutes, they must be grown to a normal humidity level. During the rooting process, they require about 100% humidity, which inhibits the normal development of cuticle and stomatal activity associated with drier conditions. The rooting normally takes place under conditions of reduced lighting, therefore, the plants must not be exposed for the first time to full sunlight. This is done slowly for a period of couple of weeks. This procedure will develop small, autonomous and photosynthesizing plants. During the rooting phase under the reduced lighting conditions, they live on carbohydrate reserves built up during micropropagation period *in vitro*. Once this weaning process is completed, the hardened can be incorporated into normal growing systems for their further development.

MICROPROPAGATION IN FOREST TREES

Micropropagation is also feasible in forest; however, with some exceptions, traditional *in vitro* methods are not as yet practical or economically viable for most forest trees. Therefore, improvement of current procedures and their scaling-up are required. A major constraint on the use of *in vitro* procedures in forestry is the common belief that, from the environmental perspective, the genetic diversity of forests should be conserved; hence, the traditional use of mixed populations of seedlings for forestation. Three major exceptions, for which applications of micropropagation becomes most relevant, do exist:

1. *In vitro* propagation methods are of crucial importance in breeding work, both to accelerate the process of provenance selection, and to establish selected mother trees for further breeding.
2. Uniformity of the trees (i.e., clonal propagation) is desired where forestation is practiced for industrial applications, such as for pulp or specialized timber production (e.g., certain poplar and eucalyptus genotypes). In these cases, micropropagation can complete with traditional methods of vegetative propagation.
3. Whenever a need exists for rapid, large–scale production of selected trees for forestation.

Forest trees are no exception as far as the basic techniques are concerned, but proper adaptation of existing procedures should be worked out for each plant genotype. In the following, the application of micropropagation to some conifers will be reviewed.

Conifers

Micropropagation has been applied over the last 25 years to an increasing number of conifer species including Norway spruce, white and black spruces, radiata pine, and *Pinus pinaster*, Southern pines, longleaf and Virginia pines. Through micropropagation numerous transplantable plants could be generated. Studies on other pines, including Mediterranean species, such as Aleppo and brutia pines, are very limited, and practical success has not yet been reported. Four major types of micropropagation have been employed for conifers.

Direct Organogenesis of Adventitious Buds from Organized Explants

In general, the more juvenile the tissue, the better it will respond to treatment s leading to organogenesis. Thus, mature isolated embryos are frequently used as the explants of choice, as well as cotyledons and epicotyls. Shoot tips, immature needles, needle fascicles, and short shots (brachyblasts) were used in other cases. Adventitious bud formation offers a greater potential for shoot production than axillary bud breading, because it involves the induction of localized meristematic tissues, and results in many adventitious buds at sites other than axillary bud meristems. Another potential approach (*meristematic nodule multiplication*), has been successful with radiata pine. In this example, over 5000 pieces of meristematic tissues were obtained from one embryo of radiata pine in 13.5 months, resulting in numerous shoots. By using this method, shoots are multiplied so that the potential for large–scale clonal propagation can be realized.

Fascicular and Axillary Bud Breaking

Micropropagation of pines by this process usually begins with initially stimulating axillary or fascicular meristems on field grown or potted trees by hedging and/ or repeated cytokinin applications. The resulting shoots are then excised, and restimulated in culture to produce additional axillary or fascicular shoots. Sequential elongation and restimulation can produce many shoots. Although not as frequently used as induction of adventitious buds, the process of axillary bud breaking has become useful for multiplying dormant and nondormant buds of some mature conifers.

Organogenesis by Callus

Indirect adventitious bud formation, from callus, has shown only limited success in conifers. A few instances of shoot formation in conifer callus have been reported, and regeneration from long–term subcultured callus of *Pinus eldarica* has only more recently been established. (Also refer the Chapter on Synseeds).

COMMERCIAL ASPECTS OF MICROPROPAGATION

About thirty companies in the country who are setting up plant tissue culture facilities for micropropagation. More than a dozen are already exporting tissue cultured plants and also selling in the domestic market. Recently one of the State Seed Corporations– the Maharashtra State Seed Corporation (MSSC) has set up a tissue culture facility for micropropagation of banana at Akola in Vidarbha. The SPIC facility near Coimbatore is going to be the largest such facility in Asia. Tissue cultured planting material is available for *Elettaria cardamomum, Amomum subalatum* (Cardamom), *Musa* spp. (Banana), *Fragaria* spp. (Strawberry), *Carica papaya* (Papaya), *Pyrus malus* (Apple), *Prunus persica* (Peach), Lilies, *Rosa* spp. (Roses), *Dianthus* spp. (Carnations),

Gerbera, Eucalyptus, Tectona grandis L (Teak), *Populus deltoids* (Poplars) and *Bambusa* spp. (Bamboo). Tissue culture Cardamom plants have given over 29.5% increased yield were conducted in three States Karnataka, Tamil Nadu and Kerala. Tissue cultured banana plants matured uniformly and a month earlier in comparison to the plants raised by suckers in the experimental plantings of BARC and MSSC.

The Department of Biotechnology established two large facilities for tissue culture propagation of forest tree species, one at the National Chemical Laboratory (NCL), Pune and other at the Tata Energy Research Institute (TERI), New Delhi. Both these units have transferred over 3.5 lakh plants to the Forest Department in the State of Haryana, Rajasthan, Uttar Pradesh, Maharashtra, Karnataka, Andhra Pradesh and Tamil Nadu. They cover an area of 200 ha. The initial growth parameters show a distinct advantage of the tissue cultured plants. Quantitative data on cost:benefit ratio would be available in few years.

Cost of tissue cultured plants would always be higher. Ultimately the success will depend on the cost–benefit ratio for each species. This is most crucial for any commercial activity. What can be the profitably sold in the market will attract investment from the business community? Scientists, in general, have no or very little knowledge of the market and hence identification of the right 'product' which is most important should be done in consultation with the industry. So far, the export market is largely for the ornamentals, especially the foliage plants and also floriculture. Globally the floriculture market is supposed to be Rs 750 Billion. In considering the cost–benefit ratio, the social and environment benefits for which it is difficult to assign a monetary value are equally important. This is specially true in forestry and agro–forestry.

Large Potential Market

There is a large potential for horticultural plants in domestic market, spices, and vegetatively propagated crops like sugarcane, potato, etc. Good planting material enhances the productivity but only to a limited extent. Good management, good agronomy, pest and disease control are equally important. All this would lead to good production. Essentially a package of production technology is required which may also include drip irrigation system for efficient utilization of water resources. Horticultural products are highly perishable commodities. Therefore, it is important to have the market for the produce. Adequate transport facilities and processing facilities to utilize what cannot be sold free in the market are required. The whole package of production, processing and marketing is essential for success. In a commercial venture the success is measured by the bottom line of the profits. The in turn depends on productivity. *How efficiently the material and human resources can be utilized to produce saleable planting material*? Research goals in the tissue culture area should aim at cost reduction, i.e., cost per plant.

At present, India has the advantage of low cost manpower which is attracting the foreign companies to India for buy–back arrangements with local companies. Abroad, tissue culture research is aiming at automation and robotics. This may, in future, take away the advantage we have in India, at present.

The plant tissue culture industry seeks a complete technology package on a turn key basis from a single source. None of our laboratories are able to provide that. For example, the technology for micropropagation of banana is available at BARC, and NCL. Two other companies have it. Yet a company recently imported the technology package which included the tissue culture facility, hardening facility, mist chamber, glass house, dematerialized water plant and all the other components.

Futuristic Outlook of Micropropagation

The transformation of self–sufficient agriculture to commercial and industrial agriculture and on to agribusiness will be the future task for Indian economy. Agri–exports are going to contribute a large share in countries foreign trade. In the eastern region you already have the large tea and jute industries. Genetic wealth of the fruit crop in the region can be realized for economic growth. The potential of the region for growing high value orchids and other ornamental plants in yet to be exploited. The role of tissue culture laboratories in the University and Colleges is in the budding stage. It can be enlarged into production oriented training units. They can accept contractual work for the propagation of specific plants from industry and nurseryman. This would provide opportunity to earn while gaining practical experience for the Botany students. All the same time they will also be exposed to the working of the industry. In the area of plant tissue culture and micropropagation, though we have the competence, knowledge and essential infrastructure, but still we have to go far to develop standard low cost technology to gain a source of immense wealth for India.

CHAPTER 13

Somatic Embryogenesis

Somatic embryogenesis is the process by which the somatic cells or tissues develop into differentiated embryos, and each fully developed embryo is capable of developing into a plantlet (young or miniature plant). Embryos can be obtained either directly from cultured explants (the organized structures, for example, leaf, hypocotyl, stem and other plant parts) and anthers (or pollen) or indirectly from callus (an unorganized mass of parenchymatous tissue derived from explant culture as a result of wound response) and isolated single cells in culture. The process of embryogenesis involves various stages of differentiation and development such as proembryo, globular, heart–shape and torpedo embryo.

In fact, somatic embryogenesis is essentially similar in structure, function and biochemistry to that of zygotic embryogenesis (embryo derived from fertilized egg cell the zygote), and somatic embryos share many characteristic features with zygote embryos; for instance, both are bipolar structures and originate from a single cell and do not show vascular differentiation. Nevertheless, somatic embryos are large in size and sometimes show more than two cotyledons (pluricotyledony) as compared to zygote embryos which are smaller in size and exhibit two cotyledons (in case of dicots). Significantly, a large number of embryos potential use in volume and rapid micropropagation of high value crops, including crops propagated by seeds. Despite its potential, this method has not yet been commercialized, except in case of oil palm *(Elaeis guineensis)*.

More recently, the embryos derived from somatic cells (mostly) or pollens are being utilized in the preparation of artificial seeds which appear to be of great value in plant propagation programme.

There are several differences between the two pathways of regeneration: a bipolar somatic embryo arises from a single cell and shows no vascular connection with the maternal tissue or explant; in fact in develops a closed radicular end. In organogenesis, on the other hand, the origin is multicellular and the monopolar structure develops procambial strands which establish a connection with the pre–existing vascular tissue dispersed within the callus or the cultured explant.

The main difference between embryogenesis and organogenesis can be appreciated at the theoretical level. Normally, organogenesis is a property of the somatic tissues, whereas embryogenesis pertains to the reproductive tissues.

The properties of a reproductive tissue which are acquired, *in vivo*, during the phase transition that occurs when a vegetative meristem becomes a floral bud, are acquired by the somatic cells *in vitro* through a process which is normally called acquisition of totipotency, but whose molecular mechanism is far from being understood.

Somatic embryogenesis has been described in more than a hundred species and it is likely that with the passage of time more, hitherto recalcitrant species, will be made to exhibit this potentiality. Of the several factors that affect the induction of somatic embryogenesis, foremost are the genotypes of the starting material, origin of the explant and media composition, particularly with respect to hormones.

CARROT AS A EMBRYOGENIC SYSTEM

There are now a number of embryogenic systems that work satisfactorily but carrot, the first to be described, is still the best. A number of embryogenic lines of carrot have been maintained in suspension cultures, in some cases for more than ten years. When hormones are removed from the medium and the cell concentration is lowered, these lines start the embryogenic programme that leads to a plantlet through the morphologically defined stages of *globular, heart* and *torpedo–shaped* embryos.

SOMATIC AND ZYGOTIC EMBRYOS

Somatic embryos show a few differences from a zygotic one. In the initial stages, the suspensor is less developed in somatic embryos (perhaps a reflection of its uselessness under *in vitro* conditions).

However, an important difference is that whereas the zygotic embryos after forming embryonal axis, with the apical meristems, prepare for dormancy, the somatic embryos do not undergo desiccation and dormancy and embryogenesis goes on from initial cell to plantlet without interruption. *In vivo*, development is similar to somatic embryogenesis up to the torpedo stage. Later, development of somatic embryos, particularly the developmental programme of the meristems, follows a path which, in a sense, shows more analogies with organogenesis.

TOTIPOTENCY AND ROLE OF AUXIN

A tissue explanted *in vitro* consists of several cell types. The explant is treated with growth regulators so that its cells (or, rather, a fraction of them capable of responding to hormonal stimuli) de–differentiate and start to proliferate. In the presence of an auxin, after several days (sometimes weeks) a population of small, compact cells emerges in this tissue which can be hand picked or concentrated by a process of differential centrifugation through a Ficoll gradient.

Proembryogenic Masses and Role of Auxin

Those small, compact cells divide in an asymmetrical way and their daughter cells, sticking together, give rise to typical cell clumps which have been called **proembryogenic masses** (Halperin, 1966) or **embryogenic clusters** (McWilliam *et al.*, 1974). From these proembryogenic masses develop embryos upon dilution of cells and removal of hormones. Totipotency, once acquired, is a long–lasting capacity in the sense that a cell population that has acquired this property can be subcultivated in the presence of auxin for months and years and yet retain the capacity to generate embryos upon removal of the hormone.

Acquisition of Totipotency

It is clearly one of the most critical steps in somatic embryogenesis. However, the details of the process are not known. Treatment with auxin at high concentration is needed to cause de–differentiation and to elicit totipotency; carrot plantlets regenerated via somatic embryogenesis give rise to adventive embryos directly from the epidermal cells of the stem, but this has never been observed in the plants originated from zygotic embryos.

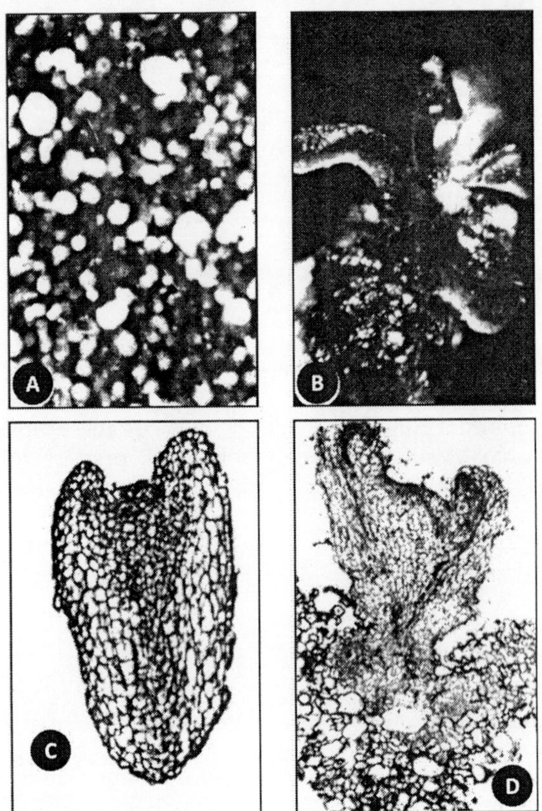

FIG. 13.1: (A) Suspension culture of carrot somatic embryos. (B) An organogenic callus of tomato. (C) Section of a carrot somatic embryo (torpedo stage). (D) Section of a primordium arising from an organogenic tobacco callus.

Various types of auxins have been used for this purpose, viz., the natural auxin IAA and the synthetic ones NAA and 2, 4–D. The latter is the most efficient and the most commonly used auxin for the promotion of somatic embryogenesis in carrot tissue cultures.

The natural auxin (IAA) at a low concentration (10^{-6} M) causes its polar and induces the adjacent cells to become specific for its transport; at a higher concentration (10^{-5} M), however, polar transport is switched off and IAA slowly diffuses in all directions (Goldsmith, 1982). In nature, a shift from low to high concentration of auxin (as it may occur on cutting a root and, thus, disrupting the communications and increasing the concentration of IAA in the cells of the wound) causes a block in polar transport and, thus, favours the onset of morphogenetic processes (for example, from callus to a new root). Other examples of re–programming that require the presence of auxin are gall formation and induction of tumours.

LoSchiavo *et al.* (1989) found a positive correlation between the level of added auxin and the level of methylation in DNA. Among the auxins, 2, 4–D can reach the highest intracellular concentration as free auxin, and is also the most efficient in promoting hypermethylation and in morphogenetic activity.

If, instead, auxin is retained in the medium, the embryogenic progression stops somewhere before globular stage. The point of arrest is more or less typical of each cell line. Therefore, there will be carrot lines which proliferate showing various stages from undifferentiated cells to globular embryos and others showing only undifferentiated cells and PEM (pro–embryogenic masses). It should be noted that auxin affects differentiated cells of an explant by inducing de–differentiation on one hand and formation of embryo primordia (PEM) on the other. This apparent contradiction can be understood if one considers that once PEM are formed they become insensitive to auxin. Auxin sensitivity is then regained at a later stage (post–globular) when the embryos, in the presence of auxin, will stop their differentiative programme and revert back to unorganized tissue.

Experiments on embryogenesis in the presence of TIBA (2, 3, 5 tri–iodobenzoic acid), a drug that blocks the polar transport of auxin from the cells with consequent increase in the intracellular hormonal levels revealed that during the PEM–globular stages the embryos do not produce auxin, because embryogenesis proceeds normally up to globular stage embryos. Thereafter, the embryos start callusing, which can be explained by assuming the production of IAA, block of its transport, internal accumulation, and callus formation. Hence, after the globular stage the embryos start producing auxin and become sensitive to it and, thus, behave like the somatic tissue (LoSchiavo, unpublished results).

Expression of the Programme in the Absence of Auxin

The onset of embryogenesis leads to a profound rearrangement of the cell metabolism. Such studies are usually made by comparing unorganized callus tissue with embryos. The rate of protein synthesis as well as the rate of poly (A)$^+$RNA synthesis increases in embryos. Also, DNA synthesis and polyamine levels increase. Tubulin genes were also found to be expressed at higher levels during embryogenesis. Lipids also shows a certain amount of developmental variation.

Methylated Cytosine Level and Development

At the DNA level, changes in the level of methylated cytosine were recorded as a function of the developmental stage. As soon as the hormones are removed (the act that operationally marks the onset of embryogenesis) there is a sudden demethylation that may interest up to 30% of the sites. Then, with progress in the embryogenic process, the level of methylated cytosine goes up again and reconstitutes the level typical of the starting cell population. This methylation is, in part at least, site–specific so that DNA fragments from a genomic DNA library can be used as probes to analyze development–associated variations in cytosine methylation. The development–associated variations were found and, moreover, the pattern of methylation may, in turn, be associated with RNA transcription as measured in Northern blots.

THE GENETIC APPROACH OF SOMATIC EMBRYOGENESIS

The Need

- Somatic embryogenesis is studied because it is an interesting process, liable to produce results of economic interest.
- The field of plant embryogenesis is not well developed and somatic embryogenesis looks a promising model system.
- Somatic embryos provide an interesting possibility as it is possible to mutagenize a cell population, make it to embryogenize and look for mutants of conditional type, e.g. temperature–sensitive mutants.
- The seed embryos, being wrapped in various coats, are not amenable to

microsurgery as in case of animal embryos. Moreover, for biochemistry a large number of synchronized embryos is needed and this is not so easy to obtain, particularly if one is interested in early embryonic stages.

This has been done in three laboratories, with or without mutagenic treatment, starting from haploid or diploid cell lines, using 24–25 °C as permissive temperature and 31–32 °C as non–permissive one.

Now that genetic manipulation of cells has reached such a high level of refinement plant regeneration from cultured cells is extremely important. In this respect, somatic embryogenesis has many advantages over organogenesis because embryos, unlike shoots, originate from single cells, and the embryogenic cultures can be synchronized and purified so that one can deal with practically pure cultures of homogeneous material. Moreover, once embryogenesis has started, it goes by itself; no further intervention is needed to adjust the auxin/cytokinin ratio, to remove the embryos from undifferentiated callus and so on. Haploid embryos can be obtained by cultivating anthers and the possibility of raising triploids from endosperm has been suggested and, to a very limited extent, exploited.

The formation of transgenic plants and the use of artificial seeds, which are the main applications of this technique, are so well established.

THE PROCESS OF SOMATIC EMBRYOGENESIS

Induction of the Embryogenic State

Induction of the embryogenic state in differentiated explants often requires extensive proliferation through unorganized callus cycles, death or disruption of surrounding explant cells, and/or high levels or a synthetic auxin such as 2, 4–dichlorophenoxyacetic acid (2, 4–D) or picloram. Plasmolysis of explant cells was also shown by Wetherell (1984) to enhance somatic embryogenesis in wild carrot (*Daucus carota*). These factors are presumed to alter the epigenetic state of cells, and may be related in their ability to disrupt the cell–cell interactions required to maintain coordinated patterns of development. The role of cell isolation in induction of somatic embryogenesis was reviewed by Williams and Maheswaran (1986). Smith and Krikorian (1989) showed that breakage or wounding of zygotic embryos at explanting led to formation of somatic embryos on a hormone–free medium. Constituent cells of the damaged tissues were presumably released from positional or chemical restraints and were thus able to express their innate embryogenic potential on a medium permissive for growth. A similar observation was made for *Dysosma pleiantha* by Chuang and Chang (1987). Embryogenic callus was induced from wounded zygotic embryos, while intact embryos failed to respond. Cell isolation is manifest early in somatic embryogenesis by the formation of a cuticle.

In embryogenic cultures, each proembryonic group of cells becomes separated from surrounding cells by thickened, cutinized walls on the outer surface (Williams and Maheswaran, 1986).

Direct somatic embryogenesis, especially in the absence of exogenous auxin, is normally associated with a relatively brief developmental period between the time of cotyledon initiation and the beginning of seed maturation (Maheswaran and William, 1986). During this time, embryonic cell division appears to directly clone the existing early embryonic epigenetic state. Even within this developmental window, some differentiation away from the embryogenic state may be inferred from the nature of structures produced by cellular proliferation. In flax (*Linum usitatissimum*), late cotyledonary stage embryos produced numerous accessory cotyledons in addition to well formed somatic embryos (Pretova and Williams, 1986).

A similar observation was made by Young *et al.* (1987) for *Lycopersicon*. Older immature zygotic embryos gave rise to accessory cotyledons and shoot apices rather than recognizable somatic embryos.

Developmental gradients within a zygotic embryo are also evident by their response to different auxins. For example, in soybean (*Glycine max*) the tissue that preferentially produces somatic embryos in response to the auxin. When the stronger 2, 4–D is used as the inducing auxin, somatic embryos are initiated from a more extensive subset of cotyledon cells.

Starting with a culture of embryogenic cells in the form of proliferating proembryonic masses, the process of regeneration requires, the initiation of bipolar differentiation to produce cotyledons, shoot apex and root pole, and then, in sequence, maturation of embryos, germination or conversion to plantlets, and transfer of plants out of culture into soil.

RECURRENT EMBRYOGENESIS

It is also known as repetitive, accessory, proliferative, or **secondary embryogenesis**. The power of embryo cloning techniques and their exploitation for mass propagation, metabolite production, or genetic transformation have recurrent embryogenesis as their basis. It also occurs when primary somatic embryos fail to mature normally into plantlets and instead give rise to successive cycles of embryos, most commonly from superficial cells of the cotyledons or hypocotyl. The process is probably homologous with the proliferation of globular proembryos in standard embryogenic cultures, differing only with respect to the stage at which integrated control of development is lost.

Expressions of recurrent embryogenesis are best viewed as a continuum, with proliferation of globular PEMs or early globular stages at one extreme, and the development of early embryogenic stages on bipolar embryos or germinating plantlets at the other extreme.

Problems

Recurrent embryogenesis may become a problem if it cannot be controlled when germination and normal growth are required. Where it can be stimulated or prevented, it offers the advantage of greatly facilitating mass propagation.

Maintenance

The maintenance of recurrent cycles of somatic embryogenesis can be spontaneous as is the case with alfalfa (*Medicago sativa* L, Lupotto, 1983, 1986). The cycles are maintained in the absence of growth regulators.

More frequently, however, the initiation of recurrent cultures requires that the developing embryos be locked into a developmental stage beyond which they cannot proceed, thereby repeating a cycle. This can be achieved by initial exposure to a very high auxin concentration such as 40 mg/l of 2, 4–D, followed by maintenance of the recurrent system using a lower level of auxin, such as 5 mg/l of 2, 4–D (Finer and Nagasawa, 1988), which prevents the transition from proembryonic to embryonic development.

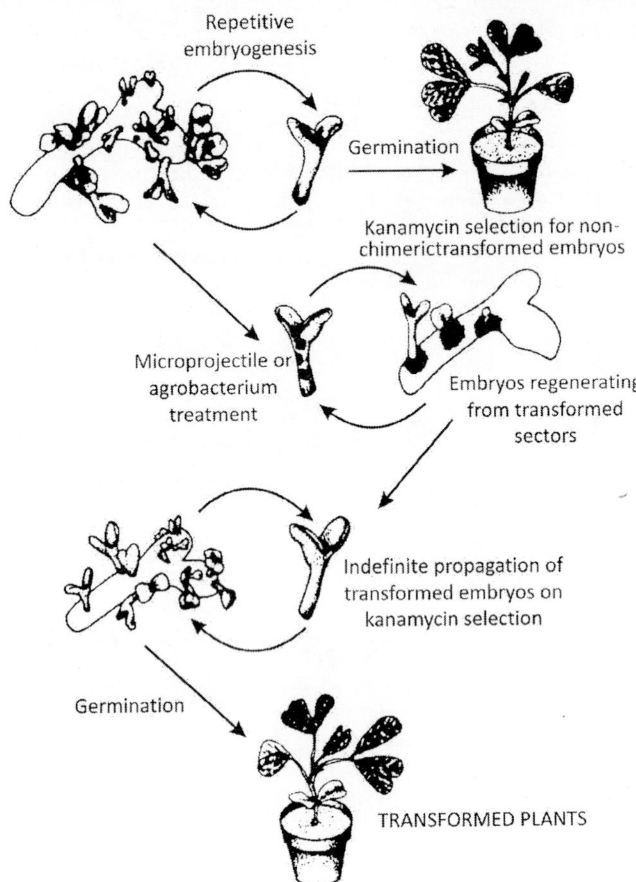

FIG. 13.2: Recurrent embryogenesis and its use in genetic transformation and mass propagation. Although new embryos can form from older embryos at any stage of development, this example depicts recurrent embryogenesis occurring from cotyledonary stage embryos. Developing embryos can be exposed to Agrobacterium or bombarded with microprojectiles, transforming individual epidermal cells. As the embryo continues to grow and develop, transformed cells give rise to patches of transformed tissue from which transformed embryos develop on selection medium. In this example, kanamycin is used as the selection agent, but several other agents could be used, depending on the vector used in transformation. As long as the cycles of recurrent embryogenesis are maintained, transformed or non-transformed embryos can be propagated indefinitely.

Auxin vs Cytokinin for Induction of Embryogenesis

Although a majority of studies have employed auxins, particularly 2, 4–D and NAA, for the induction of somatic embryogenesis from immature embryonic explants, cytokinins have also been used as inducing agents in some instances. When young cotyledonary embryos of white clover (*Trifolium repens*) are explanted and exposed to the cytokinin BAP (6–benzylamino purine), the hypocotyl responds with the formation of somatic embryos (Maheswaran and Williams, 1984). When 2, 4–D is used instead of BA, it is the cotyledons that form somatic embryos. Likewise, pea (*Pisum sativum*) shows a positive response from hypocotyl cells in the presence of cytokinin, but it is the cotyledon cells that respond in the presence of auxin (Kysely *et al.*, 1987). Genotype, tissue type, and developmental stage may all be determining factors in the comparative ability to respond to auxin or cytokinin.

HOW TO OBTAIN EMBRYOS FROM EMBRYOGENIC CULTURES?

As stated earlier, a removal or decrease of the auxin concentration in the growth medium can break the cycle of **continuous proliferation** of PEMs, and permit embryos to develop to maturity.

Brawley *et al.*, (1984) showed that in globular somatic proembryos of carrot (*Daucus carota*), ionic currents flowed inward at the site of the future shoot and out at the site of the future root pole. These currents were identified as being largely a K^+ influx and H^+ efflux (Rathore *et al.*, 1988). Similar ionic currents have been detected around haploid embryos forming from immature pollen grains of tobacco.

In carrot somatic embryogenesis, electrical polarity is accompanied by asymmetry in the distribution of activated calmodulin. Timmers *et al.* (1989), after using fluphenazine fluorescence, showed that polarity in the distribution of activated calmodulin already exists in the globular proembryogenic masses before morphological polarity is visible. Activated calmodulin concentration remains higher in the region of the root pole. During the later stages of bipolar development, fluphenazine fluorescence is also strong in the region of the forming shoot apex. Polarity in development of somatic embryos has also been detected as asymmetry in DNA synthesis by Nomura and Komamine (1986).

In direct somatic embryogenesis from immature zygotic embryos, suppression of normal polarity may play a role in embryo induction. Zygotic embryos in which growth of the main axis is weak or suppressed are more likely to give rise to somatic embryos from superficial cells.

EMBRYO MATURATION AND THE DEVELOPMENT OF GERMINABILITY

Auxin Levels

It is essential that for germination embryos must have functional shoot and root apices with the activity of meristematic growth. **High auxin** levels can inhibit development and growth of the shoot meristem if young proembryos are not transferred to a low auxin or zero auxin medium after induction. Activated charcoal may be added to the medium to remove excess auxin from the somatic embryos (Buchheim *et al.*, 1989). At **low auxin** levels, shoot meristem formation is usually achieved early after the initiation of cotyledons, so that under inappropriate culture conditions, germination can occur prematurely to give weak or inviable plantlets.

Sucrose Level

For production of vigorous plantlets, a period of embryonic growth and maturation is needed before germination. This can be achieved by culturing at sucrose levels of 3 to 6%, although progressively increasing levels up to 40% have been used for some species. High concentration of sucrose in the medium may cause osmotic desiccation in the embryos, however, for some species, efficient conversion to plantlets also requires the imposition of temporary desiccation before germination. This procedure, which mimics seed maturation *in vivo*, may be necessary to trigger metabolic processes needed for germination and seedling growth. The gradual reduction in osmotic potential through desiccation of mature somatic embryos of wheat (*Triticum aestivum*) improved germination percentages has also been reported.

Heat Shock Treatment

The vigor of the plantlets from dried somatic embryos remains greater than that of plantlets derived from embryos. Heat shock treatment induced a degree of desiccation tolerance in alfalfa somatic embryos that was equivalent to that conferred by ABA treatment but did not have detrimental effects on subsequent growth of plantlets.

ABA Levels

Ammirato (1974) showed that ABA at 10^{-7} M prevented precocious germination of somatic embryos of caraway (*Carum carvi*) in suspension. Ammirato (1983) later reported that the same level of ABA had a similar effect on suspension–cultured carrot somatic embryos, producing embryos more similar to their zygotic counterparts than those grown without ABA. Based on these results, he proposed that regulation of embryo maturation by ABA might be used to facilitate large–scale batch cultures, mechanized planting, artificial induction of dormancy and incorporation into artificial seeds.

The role of ABA in initiating the accumulation of storage reserves has not been ruled out, especially as the initiation of reserve accumulation coincides with the highest levels of endogenous ABA. The role of ABA in the initiation of reserve accumulation has also been observed (Roberts *et al.*, 1990). They found the presence of ABA essential for the stimulation of storage protein accumulation in somatic embryos of interior spruce (mixtures of *Picea glacua*, and *P. engelmannii* and their hybrids).

Water Saturation

Plantlets grown *in vitro* in a water saturated atmosphere show reduced development of cuticular waxes and abnormal stomatal function. On removal from culture, losses of such plantlets can be high if they are not protected from transpirational water loss while roots and normal leaves are developing. Acclimatization of culture–grown plantlets remains a problem in commercial micropropagation, since plantlets must usually be subjected to progressively reduced humidity over a period of week. Somatic embryogenesis offers some hope of avoiding or minimizing acclimatization problems if embryos can be removed from culture at physiological maturity and germinated under normal growing condition.

GENETIC CONTROL OF EMBRYO CLONING

The capacity to undergo somatic embryogenesis from immature zygotic embryos is under some degree of genetic control. This control appears to be the result of a low number of genes, and is, therefore, highly heritable and amenable to selection. Consistent with the observation that low numbers of genes are responsible for regeneration capacity. Regeneration capacity appears to consist of two major parameters which are commonly measured in the literature.

1. The first is the frequency of explants which regenerate.
2. The second is the average number of somatic embryos formed per regenerating explant.

The mass propagation and genetic transformation of crops are most likely dependent on the breeding and development of germplasm with a high capacity to undergo recurrent somatic embryogenesis. Once capacity for regeneration has been backcrossed into elite lines and agronomically superior cultivars, embryo cloning techniques will finally be sufficiently efficient to play an important role.

Genotype is one of the factors that influences regeneration from cell culture, however, very little is known about the genetic components of somatic embryogenesis from immature zygotic embryos. The genetics of regeneration of alfalfa is especially well documented and individual genes have been identified and named.

Genetic Variability

Genetic variability for regeneration via somatic embryogenesis has been documented for a wide variety of species, including soybean, maize, rice, barley, wheat, etc. Genetic control of regeneration capacity is largely additive and highly heritable in maize, rice, and wheat.

Cytoplasmic Effects

Cytoplasmic effects have also been important in maize, rice, wheat. In these crops cytoplasmic effects are sufficient to necessitate careful selection of maternal parents to ensure regeneration success. Non–embryogenic callus can be derived from embryogenic callus initiated from immature zygotic embryos of the cultivar Chinese Spring. The use of defined cytogenetic stocks has made it possible to further elucidate the nature of the genetic control of regeneration in wheat.

APPLICATIONS OF SOMATIC EMBRYOGENESIS

The high–volume multiplication of embryonic propagules is the most commercially attractive application of *in vitro* somatic embryogenesis. As commercially conceived, the system involves harvesting maturing embryos from a continuously proliferating embryogenic culture of elite genotype, and converting the harvested **clonal embryos** to **seedling** transplants or synthetic seeds for delivery to the grower. Although the induction of normal embryo physiology, scale–up of culture volume, and design of field delivery systems have so far prevented industrial applications, recent intensive work on synthetic seed systems shows commercial promise.

Mass Propagation

Somatic embryos have powerful advantages for mass propagation in comparison to both conventional clonal propagation methods (e.g., rooted cuttings, grafting) and other *in vitro* regeneration systems (e.g., micropropagation).

One advantage of propagation via somatic embryogenesis is the very high multiplication rates possible with many embryogenic systems. Unlimited numbers of embryos can be generated from a single explant. It depends on type of plant species.

In comparison, multiplication by rooted cuttings is limited to the amount of material available from the mother plant, and for most species, micropropagation also is characterized by relatively low multiplication rates. A second advantage of somatic embryogenesis is that, for many species, both growth of the embryogenic tissue and development of the somatic embryos can be carried out in liquid medium, making possible the handling of enormous numbers of embryos at one time.

Drew (1980) estimated that one liter of a carrot suspension culture contained 1.35 million somatic embryos. Thus, in comparison to rooted cuttings and micropropagation, somatic embryos offer the potential for high volume, large–

scale propagation systems that can be translated into significant labor savings. Even greater economics of scale may be possible if bioreactor and continuous culture technologies can be applied to embryogenic systems.

Plants derived from somatic embryos are less variable than those derived via organogenesis. This may reflect an intolerance of somatic embryos to mutations in any of the numerous genes that must be necessary for ontogeny to be successfully completed. In contrast, vegetative meristems may be more tolerant to mutations and epigenetic changes.

Probably the most obvious advantage of somatic embryogenesis in comparison to other clonal propagation methods is the fact that the product is an embryo. The morphological and physiological similarity of somatic embryos to zygotic embryos means that they are almost complete propagules in themselves, with embryonic roots, shoots and leaves (or at least cotyledons) and, most importantly, the programme to make a complete plant. Thus, unlike other clonal propagation systems, no separate shoot growth or rooting steps are required for plantlet production, again providing savings in labor.

Unlike organogenic or axillary branching systems, many embryogenic systems produce discrete embryos, and thus require no physical separation from mother tissue or other embryos in order to be handled, which once again means savings in labor.

Over the past few years, some of these special characteristics of somatic embryogenesis have been examined for possible commercialization purposes. The potential for somatic embryos to be grown in large volumes in continuous culture and employed as direct–delivered propagules has received much attention.

Scale–Up Potential

The fact that both the growth of embryogenic cells and subsequent development of somatic embryos can be carried out in **liquid medium** gives somatic embryogenesis the potential to be combined with engineering technology to create large–scale mechanized or automated culture systems. Such systems are capable of producing huge numbers or propagules with low labor inputs. With the application of this technology, costs per propagule have the potential to be reduced to the point where they may be competitive with seed–derived plants, depending on the crop.

Use of Bioreactors

The first report of large scale embryogenic cultures described an attempt to grow carrot cells in 20 liter carboys, which resulted in the formation of few embryos (Backs–Husemann and Reinert, 1970). The biological/mechanical system most often described for application to embryogenic systems is the stirred tank bioreactor, a mass culture system originally developed for microbial fermentations, but more recently adapted for growing plant cells on a large scale (Wilson *et al.*, 1971; Martin, 1980; Kurz and Constabel, 1981).

A major problem with adapting these bioreactor designs for use with plant cells is the high shear that stirring generates in these systems. Air driven bioreactors, with lower shear levels, have been tested as possible alternatives to the stirred–tank design, and have supported successful growth of a number of plant cell types. Air–lift bioreactors gave slightly higher yields of alfalfa somatic embryos compared to propeller–stirred bioreactors or cultures grown in flasks on a shaker.

Plant Biotech Industries, Ltd. has developed an automated system for large scale commercial propagation of plants, which makes use of somatic embryos as well as other propagules such as microtubers and bulblets. The system integrates a bioreactor with a bioprocessor in a closed system for separation, sizing and distribution of propagules into a culture vessel, and even employs an automated transplanting machine which transfers plantlets to soil mix in greenhouse trays at the rate of 8000 per hour. The so developed bioreactor–based system could cut production costs of plantlets by as much as 60% compared with conventional tissue culture propagation methods. Other benefits of bioreactor technology are lower contamination rates, savings in space, time and labor, accurate monitoring and control of temperature, pH, and gasses.

To date, the application of bioreactor technology has apparently not met its potential to produce hundreds of thousands of clonal embryos capable of growing into plants. To improve the capability of bioreactors to produce competent embryos, a group of researchers has recently developed a kinetic model of carrot somatic embryo development in suspension culture by monitoring substrate utilization, culture growth and embryo development over the time course of an embryogenic culture.

Protoplast Culture

Embryogenic callus and suspension cultures, as well as somatic embryos themselves have been employed as a source of protoplasts for a range of species. The logic of this approach is that isolation of protoplasts from cells or tissues that are themselves regenerable will likely yield protoplast cultures capable of forming whole plants (Shillito *et al.*, 1989). Earliest application of the regenerative potential of protoplasts isolated from embryogenic material was made with embryogenic carrot suspension cultures.

Since then in three groups of plant species, viz. graminaceous species, citrus species, and forest trees (especially conifers), embryogenic cultures have proven to be especially valuable in providing a source of regenerable protoplasts.

In the Gramineae, regeneration of callus or even sustained cell divisions in mesophyll derived protoplasts could not be achieved following methods that had previously proven successful with mesophyll protoplasts of solanaceous species. Although there were many reports of sustained cell divisions in protoplasts isolated from nonmorphogenic cell suspension cultures of the Gramineae, the protoplast derived calli failed to undergo morphogenesis. Therefore, Vasil and Vasil (1980) turned to embryogenic cultures derived from immature embryos of pearl millet (*Pennisetum glaucum*) as a source of protoplasts. These protoplasts could be cultured to give rise to cell masses, from which embryoids and eventually plantlets could be regenerated. Similar success was subsequently reported using embryogenic suspensions of several other graminaceous species.

Embryogenic citrus suspension cultures have not only provided a source of regenerable protoplasts, but also made possible the production of interspecific and even intergeneric somatic hybrid plants. Interspecific somatic hybridization in citrus was first achieved by Kobayashi *et al.* (1988). They fused protoplasts isolated from an embryogenic suspension culture of navel orange (*C. sinensis*) cultivar Washington with leaf protoplasts of satsuma mandarin (*C. unshiu*) cultivar Hayashi. Interspecific somatic hybrids have since been produced between a number of citrus species using similar techniques.

The ability to isolate protoplasts from embryogenic cultures of forest trees has had a large impact on regeneration studies for this group of plants, in particular coniferous species. Although a few researchers reported the growth of protoplasts isolated from conifer cotyledons, leaves, or suspension cultures to the colony or even callus/suspension stage.

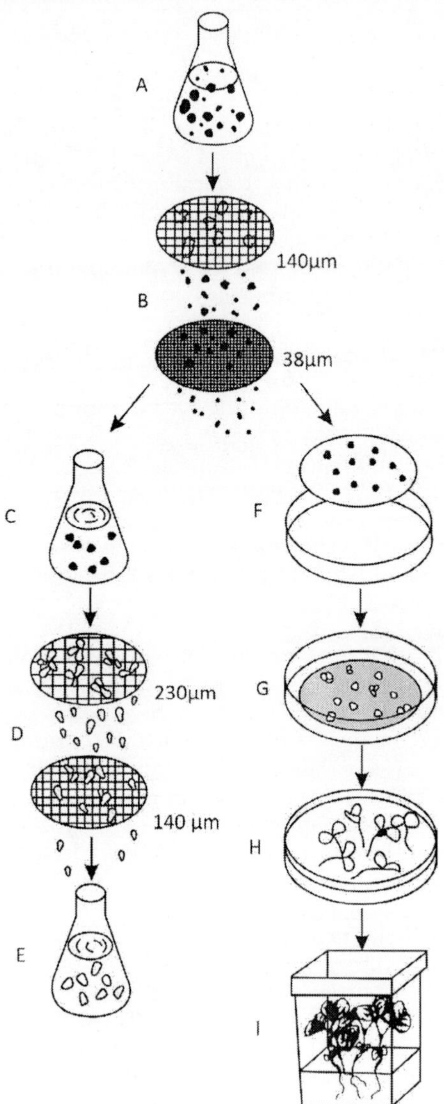

FIG. 13.3: Fractionation/synchronization of embryogenic yellow–poplar suspension cultures for mass propagation. A Embryogenic suspension are grown in shaken flasks of liquid induction medium. B. PEMs are fractionated on stainless steel sieves, saving the fraction that passes through 140 µm, but not 38 µm mesh. C. Saved fraction is cultured for one week in liquid basal medium. D. Globular stage embryos are fractionated again to eliminate clusters and free cells, saving the fraction that passes through 230 um, but not 140 µm mesh. E. Globular stage embryos are cultured an additional 7–10 days in basal medium to obtain synchronous heart–torpedo stage embryos. Alternatively, following the first fractionation, F. PEMs are immediately placed on filter paper, which is then plated on semisolid basal medium. G. PEMs are cultured on filter paper/basal medium for 12–14 days to obtain synchronous, mature embryos. H. mature embryos are transferred to basal medium without filter paper to promote germination. I. Germinants are transferred to plantlet development medium, from which plantlets are ready for transfer to soil mix in 6–8 weeks.

The development of embryogenic callus and suspension cultures proved to be the key to the production of morphogenic protoplasts in coniferous.

Embryogenic cultures have also been shown to be a valuable source of regenerable protoplasts in some hardwood forest tree species. Rao and Ozias–Akins (1985) isolated Protoplasts from embryogenic cell suspension cultures derived from proliferating shoot segments of a 20–year–old sandalwood tree (*Santalum album*). The protoplasts could be cultured to form embryogenic cell aggregates, somatic embryos and eventually plantlets.

Similarly, embryogenic suspension cultures of yellow poplar provided protoplasts capable of regenerating whole plants via embryogenesis (Merkle and Sommer, 1987).

EMBRYO CLONING AND GENE TRANSFER

Regeneration in several species, especially tress and large seeded legumes, is limited to regeneration via direct somatic embryogenesis from immature zygotic embryos. The embryos form directly on the original explant tissue. If callus is present, it grows concomitantly with the somatic embryos. The embryos do not originate from it, thereby bypassing any opportunity that a callus phase provides to sort transformed cells from non–transformed cells. Confounding the problem is the fact that these somatic embryos may have originated from groups of cells rather than from single cells within the explant.

As current gene transfer techniques transform single cells, not clumps of cells, the recovery of chimeric embryos consisting of transformed and non transformed tissues is virtually assured. Whereas the absence of a callus phase is the main factor that has limited genetic transformation in these species, this barrier has been overcome by the use of recurrent embryogenesis.

In recurrent embryogenesis, a cycle is initiated whereby somatic embryos continuously proliferate from previously existing somatic embryos. The cycles of embryo proliferation effectively substitute for a callus phase.

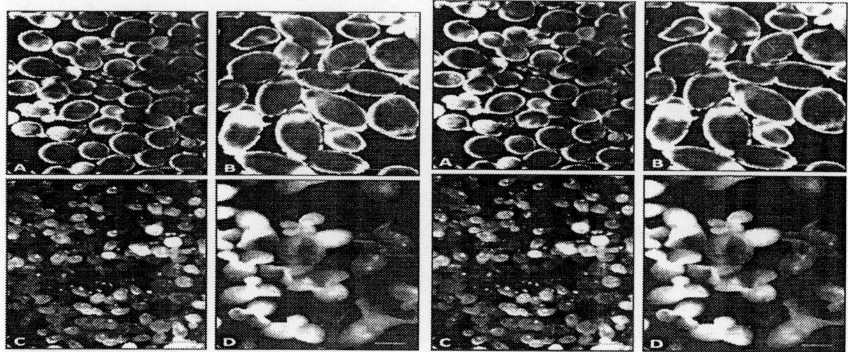

FIG. 13.4: Synchronous yellow–poplar somatic embryo populations obtained by fractionation of PEMs on stainless steel sieves. A Globular–stage yellow poplar somatic embryos derived from fractionated PEMs 3 days after second sieving (bar=500 μm). B. Early torpedo–stage yellow–poplar somatic embryos derived from fractionated PEMs 10 days after second sieving (bar = 500 μm). C. Roughly synchronous population of yellow–poplar somatic embryos 6 days after fractionation of PEMs and plating on filter paper placed on semisolid medium (bar = 500 μm). D. Mature yellow–poplar somatic embryos 14 days after fractionation and plating on filter paper placed on semisolid medium (bar = 500 μm).

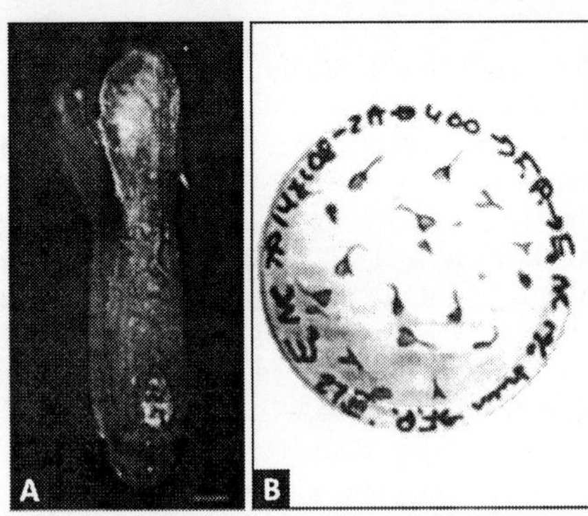

FIG. 13.5: Yellow–poplar and magnolia somatic embryo maturation, conversion, and plantlet acclimatization. A. Mature yellow–poplar somatic embryo, at the onset of germination, obtained from fractionation and plating (bar = 200 μm). B. Germinating yellow–poplar somatic embryos obtained from fractionation and plating of PEMs, following transfer from filter paper to fresh medium.

Even if a chimeric embryo is obtained in the first cycle of regeneration, it becomes possible to obtain a non–chimeric embryo from the patch of transformed tissue on the original embryo. The recurrent embryos appear to have an epidermal or subepidermal origin. The recurrent embryos have single cell origins. Consequently, if a transformation technique is applied to a primary somatic embryo instead of a zygotic embryo, it should become possible to obtain totally transgenic somatic embryos, and this has, in fact, been observed for walnut.

The nature of recurrent embryogenesis also makes it ideally suited to particle gun mediated transformation (Klein *et al.*, 1987). Instead of relying on *Agrobacterium* to mediate the transfer of genes into plant cells, the particle gun literally shoots into plant cells DNA that has been precipitated onto particles of a heavy metal.

SOMATIC EMBRYOGENESIS IN TREES

The technique and pattern of the development of somatic embryogenesis in tree species differs, depending on whether the species is angiospermous or coniferous (Gymnosperms). Conifer somatic embryogenesis is analogous to zygotic cleavage polyembryony and for most examples, depends on an *in vitro* proliferation of non–callus, suspensor like cells derived from mature or immature zygotic embryos. In *Picea abies*, the cells, known as the **embryonal suspensor mass** (EMS), proliferate on BM–1 medium containing high concentrations of kinetin. Differentiation of ESM to cotyledonary stage, somatic embryos occurred on BM–1 medium containing 10 per cent of the original kinetin. However, few somatic embryos have been converted to whole plants.

Somatic embryogenesis in angiosperm tree species is either analogous to adventive in crassinucellate angiosperms, nuclear polyembryony as in *Citrus*, or adventive zygotic polyembryony as in immature zygotic embryos of *Juglans*. The key tissue culture manipulation in a number of species was the use of immature zygotic embryos or nucellar tissue as the original explant source. The cell proliferation stage in tissue culture is in the form of an undifferentiated callus.

Somatic embryogenesis has, also been obtained from non–zygotic explant tissue (coffee leaf), suggesting that somatic embryogenesis may be obtained from sexually immature tree, zygotic embryo or ovule tissue is not available, thus it accelerate the process of tree genetic engineering.

Difficulties in developing somatic embryogenesis systems for tree species are similar to those for herbaceous species. However, tree species may also exhibit a unique set of tissue culture–dependent variabilities. Genetic variation may be higher for tree species than cultivated herbaceous species. The size of the genome is quite large for many trees, which is particularly important for genetic engineering. A further limitation for tree improvement using genetic engineering is that mature material must be collected from the field rather than from a controlled environment. In addition, the problem of ecotypic or physiological variation from individual to individual is much greater due to the overall heterozygosity of most tree species. These additional variables further increase the potential difficulty of developing repeatable somatic embryogenesis in tree species.

For the production of artificial seeds in tree species, the recovery of plants (conversion) is crucial. Possibly, mango is the first tree species for which the viable artificial seeds have been produced. *Citrus* and coffee, which already have reasonable conversion frequencies.

For the production of artificial seeds in tree species, the recovery of plants (conversion) is crucial. Possibly, mango is the first tree species for which the viable artificial seeds have been produced. *Citrus* and coffee, which already have reasonable conversion frequencies.

TABLE 13.1: Somatic embryogenesis in trees of economic importance.

Name	Common name	Germination	Conversion
PULP, PAPER AND TIMBER			
Gymnosperms			
Abies balsamea	Balsam fir	no	no
Larix deciduas	European larch	yes	1% of calli
Picea abies	Norway spruce	5–10%	no
		24% of calli	no
		15% of calli	no
P. glauca	White spruce	yes	no
Pinus labertiana	Sugar pine	1–2%	no
P. taeda	Loblolly pine	yes	1–5/g callus
Pseudoesuga menziesii	Douglas–fir	no	no
Angiosperms			
Albizia lebbeck	Indian walnut	yes	yes
Eucalyptus citriodora	Gum	50%	yes
Fraxinus Americana	White ash	yes	yes
Juglans regia	English walnut	yes	yes
J. hindsii	Walnut	yes	yes
Leucaona diversifolia	Popinac	yes	no
Liquidambar styraciflua	Sweet gum	yes	yes
Populus ciliata	Poplar	yes	yes
Pterocarya spp.	–	yes	yes
Quercus rubra	Red oak	no	no

Name	Common name	Germination	Conversion
FRUIT TREES			
Citrus aurantifolia	Lime	yes	no
C. aurantium	Sour orange	no	no
C. limonia	Lemon	yes	yes
C. medica	Citron	no	no
C. paradisi	Grapefruit	yes	32%
C. reticulata	Mandarin	yes	yes
C. sinensis	Sweet orange	yes	20%
Cocos nucifera	Coconut	no	no
Eriobotrya japonica	Loquat	no	no
Eugenia malaccensis	Malay apple	yes	no
Malus spp.	Apple	yes	yes
Mangifera indica	Mango	yes	yes
Musa ABB	Plantain	no	no
Myrciaria cauliflora	Jaboticaba	no	no
Phoenix dactylifera	Date palm	yes	no
Prunus cerasus	Sour cherry	no	no
Pyrus spp.	Pear	no	no
Syzygium jambos	Rose apple	yes	no
OILS, PHARMACEUTICALS, AND ORNAMENTALS			
Chamaedorea costaricana	Palm	no	no
Coffea arabica	Arabian coffee	yes	yes
C. arabica var. catimor	Coffee	yes	20–27%
C. canephora	Robusta coffee	yes	no
Elaeis guineensis	Oil palm	yes	no
Hevea brasiliensis	Para rubber	yes	no
Howea forsteriana	Paradise palm	no	no
Liriodendron tulipifera	Yellow poplar	yes	yes
Paulownia tomentosa	Paulownia	50%	no
Santalum album	Sandalwood	yes	yes
Theobroma cacao	Cacao	yes	no
Thuja orientalis	Oriental arborvitae	no	no
Veitchia merrilli	Christmas palm	yes	no

CHAPTER 14

Mutation and Plant Breeding Techniques for Hi-Tech Horticulture

PRINCIPLES OF BREEDING FOR IMPROVED PLANT CULTIVARS

Breeding for improved plant cultivars is based on two principles genetic variation and selection. The process is extremely laborious and time consuming with high inputs of intellectual and manual work. However, the development of plant cell and tissue culture over the last 20 years has made it possible to transfer part of the breeding work from field to laboratory conditions.

TABLE 14.1: Modification made in various crops through plant biotechnology.

Crop	Modifications
Apple	Disease resistance, slower ripening
Field vegetables	Pest resistance, hybridity*
Maize	Insect resistance, herbicide tolerance
Oilseed rape	Improved oil quality, herbicide resistance, hybridity*
Potato	Improved starch quality, insect/disease resistance, processing benefits
Soft fruit	Slower ripening
Sugar beet	Modified starch content, disease resistance, herbicide tolerance
Tomato	Slower ripening, processing benefits
Wheat	Modified starch, disease resistance

* improves crop uniformity and field performance

Extensive research has resulted in new areas of plant breeding namely plant biotechnology and genetic engineering. They are based on cellular totipotency, or the ability to regenerate whole, flowering plants from isolated organs (meristems), pieces of tissue, individual cells and protoplasts. The isolated plant parts are aseptically grown in test tubes on artificial media of known chemical composition (*in vitro* culture). Under strictly controlled conditions, they form plantlets that subsequently can be transferred to soil where they grow to maturity.

Tissue culture has been exploited commercially for micropropagation of disease free stocks of horticultural plants (strawberry, potato, and ornamentals, for example). *In vitro* techniques also are useful in various steps of the breeding process, such as germplasm preservation, clonal propagation, and distant hybridization.

Radiation mutation breeding and isotope techniques, combined with tissue culture, have made a significant contribution to plant breeding. They have introduced new techniques for inducing genetic variation, by improving selection technology, and by accelerating breeding time.

Other methods, known as anther, or pollen culture, make it possible to regenerate plants from male gametes with half the number of chromosomes haploids. Compared to plants with full chromosomal content (diploids), the use of haploids in mutation breeding is advantageous since it allows detection of mutations immediately after their induction. Haploid method have proven to significantly speed up the breeding of new varieties of rice, barley, and vegetables for instance.

Genetic engineering procedures allow the transfer of genetic material (DNA) from the cell of one species to that of another genetically unrelated organism. For example, a piece of DNA from a bacterial cell may be integrated into the genome of a plant cell to form a transgenic plant. The new DNA (gene) expresses itself in the plant phenotype regenerated from the transgenic cell. Nuclear techniques, based on nucleic acid bases labeled by isotopes, are employed in genetic engineering, to identify and isolate suitable genes for transfer as delivery systems to introduce genes into recipient cells and to detect new genetic material in recipient organisms.

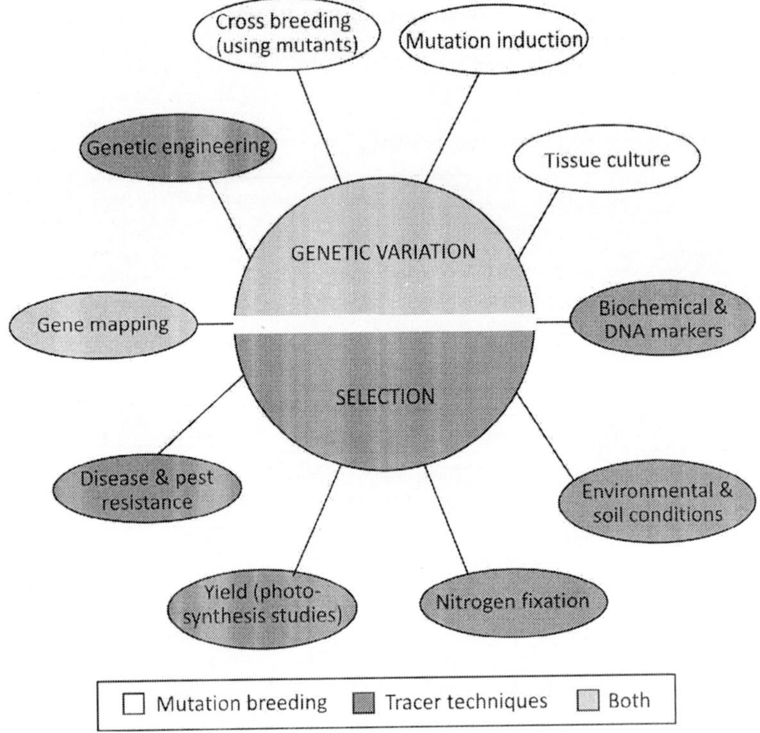

FIG. 14.1: Crop improvement is based on two basic principles: genetic variation and selection. Serving as invaluable tools are mutagenic irradiation and isotope tracer techniques, which are incorporated in to the various breeding methods.

Genetic engineering already has resulted in the production of plants with new desirable traits, such as insect resistance, virus disease resistance, and better ripening properties. However, early enthusiasm is being tempered by the growing discussion of the potential hazards of releasing transgenic plants to the environment.

Another issue that has emerged concerns the commercialization of this technology, and the access of developing countries to it. Recent developments in plant biotechnology have resulted in enormous capital investments and in a concentration of highly qualified human resources into the commercial sector of many industrial countries. In the process, scientific knowledge and its technological applications increasingly becoming the subject of commercial legislation, involving patents, industrial secrecy, and licensing policy. As a result, developing countries have experienced difficulties in gaining access to biotechnological results and their applications in national programmes.

In this context, specialized agencies of the United Nations including the Food and Agriculture Organization (FAO), United Nations Educational, Scientific and Cultural Organization (UNESCO), United Nations Industrial Development Organization (UNIDO), and IAEA are playing an important role. Programmes have been set in place whose main concerns are to identify and transfer appropriate biotechnologies, and to train personnel, in developing countries. In this way, national capabilities in research and development are being strengthened in this field.

TECHNIQUES OF MUTATION BREEDING

Crop improvement by deliberate attempt to cross two parents of desirable characters to incorporate such characters in next generations is called breeding. The entire process of pollination, fertilization and regeneration is kept under control and selected parents are allowed to reproduce only with pollination from desired parent. This method is popularly called as crossing. Crossing has been extensively used in most pollinated crops. Selection and breeding are normally combined to evolve new varieties and hybrids of food crops. The green revolution in our country was possible mainly because of release of high yielding varieties and hybrids. There are three methods of breeding hybridization, ploidy breeding and mutation breeding.

The technique of mutation is a tool for geneticists and plant breeders which they use it for study of nature and function of genes. Genes are the building blocks for the development of new crop cultivars through breeding. Induced mutant and selection methodology provide viable variability in crop plants. Induced mutation is a tool which in combination with other tools (cross-breeding) gives us new cultivars. The technique of mutation breeding has given many genetic variants, at least in 50 species. The induced mutants have been directly or indirectly utilized via crosses with other gene sources for creation of new crop cultivars. The use of radiation in plant breeding has allowed the obtention of 1300 new varieties worldwide, corresponding 559 to cereals, 136 to legumes, 67 to industrial crops, 397 to ornamentals and 80 to fruits and vegetables.

Notwithstanding that the application of radiation in plant breeding goes back to 1922 (Pirovano, 1924), and that several studies have been done on grapes, pears, apples, and other fruit trees by Granhall, *et al.*, (1949), and Donini *et al.*, (1982), which have conducted to the development of more than 40 new varieties in fruit trees.

Mutation induction continues to contribute to crop improvement, using physical mutagens such as gamma ray, X-ray, fast neutron, and chemical mutagens such as EMS (ethyl-methane-sulphonate) and sodium azides. Recently, new physical mutagens, such as ion beam radiation and cosmic rays, have been proven to be effective for inducing mutations. Below is an overview of recent developments related to plant mutation breeding; induced mutations for harnessing genomics resources; bio-/molecular technologies for enhancing the efficiency of mutation induction and utilization; and ion beam radiation as a novel mutation induction technique.

Use of mutagenic chemicals and radiation for induced mutation is very traditional, now it has been supplemented by the use of *in vitro* culture and transposable elements. Though, mutation technique does not provide an omnipotent solution for crop improvement because it has limitations in classical plant breeding, however, *in vitro* culture and transposable elements may be helpful in bringing out new cultivars.

Under the Joint FAO/IAEA programme, radiation-induced mutations are used for genetic improvement of both seed and vegetatively propagated plants. The FAO/IAEA programme maintains a database of officially released mutant varieties worldwide (http:www-mvd.iaea.org/). Currently, over 2300 mutant varieties are registered in our database. Coordinated Research Projects (CRPs) and Technical Co-operation Projects (TCP) are two major activities at IAEA that serve Member States at the national, regional and interregional levels. This article highlights CRPs on banana, underutilized and neglected crops, and tropical and subtropical fruits. CRPs on banana and underutilized and neglected crops have already been concluded. TCPs in South East Asia (Thailand, and Malaysia), Africa (Algeria, Morocco, Tunisia, Ghana), and the Middle East (Yemen) are discussed.

Genetic and Biological Basis of Mutation

Mutations are induced in single cells, and mutants are observed in the progeny of the mutated cells. Induced mutation is mainly done in haploid pollen grains followed by **self-fertilization**. The dominant mutations (if occurs) are expressed in the first M1 generation. The mutagenic plant will be heterozygous and will segregate into mutant and non-mutant phenotypes in M2 generation. Recessive mutations will occur in the M1 generation as heterozygotes and will not normally be expressed. Self-fertilization of the M1 plants will permit expression of the homozygous recessive mutants in the M2 generation.

In cross-fertilizing species, homozygous mutants will only be produced by enforced selfing of heterozygotes. Where self-fertilization is not possible as in self-incompatible species, homozygous mutants cannot be expressed before the M3 generation, and then only following sib-mating of sister M2 plants.

Induced mutation is also applied in multicellular organs e.g., seeds, buds, or growing plants. It also follows the same genetic principles. They show complexity of cellular heterogeneity (chimera) in the M1 plants.

Seed is the most convenient organ for induced mutation because it affects embryo (multicellular), the resultant M1 plant will be chimeric. Throughout the growth of the M1 plant, the opportunity for competition between the mutant and wild-type cells exists, and the results of this diplontic selection influences the frequency of mutants observed in the M2 generation.

Dominant mutations is expressed in the M1 plant, the characters are visible only in the chimerical sectors as chlorophyll-deficient stripes on M1 plants, where the effect of diplontic selection can sometimes be observed as a progressive change in the size of the mutated sector as the plant develops. In M2 generation, Mendelian segregation ratios is followed in the M2 generation that is: dominant mutant 3:1, recessive mutant 1:3. The M2 generation is influenced by the mutation frequency and by the number of cells in the treated seed that contribute to the M2 generation (genetically effective cells).

TABLE 14.2: Mutant expression after mutagenic treatment of different tissues (after S. K. Sharma and T. R. Sharma, 2000)

Tissue	Parent genotype	Mutation	M1 generation		M2 generation			
					Self-fertilized		Cross-fertilized with a non-mutant genome	
			Genotype	Phenotype	Genotype	Phenotype	Genotype	Phenotype
Pollen					AA	Parental	AA	Parental
	AA	A- a	Aa	Parental	Aa	Parental		
					aa	Mutant	aA	Parental
					aa	Parental	aa	Parental
	aa	a - A	aA	Mutant	aA	Mutant		
Seed					AA	Mutant	Aa	Mutant
					AA/	Parental		
			AA/	Parental	AA	parental	AA	Parental
	AA	A- a	Aa		Aa	Parental		
					aa	Mutant	aA	Parental
					aa/	Parental		
			aa/	Mutant	aa	Parental	aa	parental
	aa	a-A	aA	Chimera	aA	Mutant		
					AA	Mutant	aA	Mutant
Vegetative tissue	AA	A-a	AA/	Parental	Sexually reproduction same as for seed vegetative propagation of mutant sector			
	Aa	A- a	Aa/aa	mutant Chimera				
	aa	a-A	aa/aA	Mutant Chimera				

A variety of procedures and materials are used for this purpose e.g., pollen is mutagenized and used in pollination for crop improvement. The dominant mutations will be expressed in the next generation, and further generations of selfing reveal recessives. Similarly, seeds are also mutagenized. A cell in the enclosed embryo of a seed may become mutant, and then it may become part of germinal tissue or somatic tissue. If the mutation is in somatic tissue, any dominant mutations will show up in the plant derived from that seed, but this generation will be the end of the road for such mutations.

INDUCED MUTATION AND PLANT BREEDING

Induced mutation is induced by the use of ionizing radiations (X-rays, gamma rays and neutrons) and chemical mutagens for inducing variation in plants. Induced mutations have been used to improve major crops such as wheat, rice, barley, cotton, peanuts, and beans, which are seed propagated. Since the establishment of the Joint FAO/IAEA Division of the Nuclear Techniques in Agriculture, more than 1800 cultivars obtained either as direct mutants or derived from their crosses have been released worldwide in 50 countries. In vegetatively propagated plants, many of mutants were derived from irradiating rooted stem cuttings, detached leaves, and dormant plants. According to the FAO/IAEA database, of the 465 mutants released among the vegetatively propagated plants, most are in the floricultural plants and a few in fruit trees. These include chrysanthemum, *Alstroemeria*, dahlia, bougainvillea, rose, *Achimenes*, begonia, carnation, *Streptocarpus*, and azalea. The irradiation of *in vitro* cultured date palm, apple, potato, sweet potato and pineapple now provides a means to treat large populations which would not have been possible before. Irradiation of micropropagated plants, axillary and adventitious buds, apical meristems, regenerative callus cultures, anthers and microspores, and somatic embryos provides a miniaturized version of trees and seeds in the Petri dish instead of the field. During the last decade, the use of radio-actively labeled probes in recombinant DNA research for cloning and mapping plant genes and transgenesis, particularly for RFLP, micro satellite based DNA fingerprinting, has become a routine procedure. Many homeotic mutants that change floral development have been isolated in *Arabidopsis, Petunia, Antirrhinum* and *Lycopersicon*. Mutants of *Arabidopsis* are being used to analyze genes, which determine response to auxins, cytokinins, gibberellin, abscisic acid and ethylene in plant growth, floral development and senescence, fruit formation and ripening. These mutants are facilitating the isolation, identification and cloning of the genes, which would ultimately help in designing crops with improved yield, increased stress tolerance, longer shelf-life and reduced agronomic inputs. The identification and analysis of mutants by using molecular techniques of DNA fingerprinting and mapping with PCR based markers, such as RAPDs, AFLP and STMS, and mutant tagging shall bring a new dimension in gene technology. Already, mutations can be linked to changes in DNA sequences for some plant traits and to establish molecular maps in structural and functional genomics of crop plants. These in turn would lead to a rapid enhancement of crop yields and quality. (Ahloowalia, B. S. and Maluszynski, M., 2001) Induced mutations A new paradigm in plant breeding, *Euphytica*, 118: (2) 167-173).

TECHNIQUES OF INDUCED MUTATION

Several approaches have been efficiently used to increase the economic feasibility of artificial mutagenesis in plant breeding, these include (a) the use of pollen, zygotes, and embryos in mutation induction; (b) diploidization of loci in polyploids by mutagen treatment; (c) alteration of reproductive mechanisms, especially self-incompatibility and apomixis, by mutagen treatment; and (d) development of efficient recurrent mutagen treatment techniques.

Most of these are concerned with the appearance, detection, and selection of mutations rather than with the induction of mutations *per se*. However, they will be considered briefly here since progress along these lines can extend the usefulness of mutagen techniques.

Most mutation breeding experiments have been initiated either with seeds of sexually propagated crops or with buds of cuttings in vegetatively propagated plants. Seeds and buds have multi-cellular, complex tissues, therefore, there are certain difficulties for induced mutation detection and selection. These are as follows:

- The chimeral nature of mutated sectors.
- A rapid loss of induced cytogenetic changes during subsequent mitoses.
- In sexual crops, during meiosis of the M1 plant.
- Overgrowth of mutated cells by nonmutated cells

In general, induced mutations offer less prospect for the improvement of cross-fertilizing species than autogamous ones. This is partly because of the difficulty of selecting, in operating and maintaining recessive mutations in such populations, and partly because the plant breeding problems in cross-fertilizing species are more often problems of handling the existing variability than lack of variability. Where the lack of variability exists for specific and simply inherited traits, the basis of choosing between induced mutations and hybridization is essentially the same in self- and cross-fertilizing species. However, the genetic consequences of the failure of recessive mutations to express in cross-fertilizing systems without forced selfing or sibmating must be taken into account in assessing the cost of such ventures.

Plant Materials Used in Radiation Breeding

According to breeding targets, three principles were followed to choose the plant materials in radiation breeding:

- Chose materials with comprehensive good characters and only one or two defects.
- Chose materials with high genetic heterozygosity.
- Chose materials having genotypes which were easy to induce gene mutations.

Only by doing so, mutant characters could be seen more easily and separated more efficiently in later generations.

The type of plant organs used in irradiation depends on the propagation characteristics of plants. In vegetatively propagated plants bud sticks, pollens, calli and seeds are often used in fruit trees. Organs used in chrysanthemum radiation breeding are rooted plants, shoots, calli and *in vitro* plants. Stem sections, 1 or 2 year plants and F_1 seeds are used in rose. Organs used in other ornamental species are dormant root-tubers, dormant rhizomes, bulbs, cuttings, potted plants and seeds.

In woody fruit trees and ornamental plants bud sticks for grafting and cuttings are used. On the other hand, growing plants and other vegetative organs are used more often in herbaceous plants.

In vegetatively propagated plants seeds are used in radiation breeding. The great advantage of using seeds is that it is very convenient to treat a large amount of seeds at the same time, hence can provide a large population for selection of useful mutants. The disadvantages are that mutations from irradiated seeds are mostly chimeras and the time from sowing to fruiting is long, especially for woody plants. Seeds have proved good materials in the combined project of long-term and short-term breeding or in herbaceous plant breeding.

In recent years, tissue-cultured materials, such as *in vitro* protoplasm, suspended-cultured cells, *in vitro* calli and explants are also used in studies of radiation breeding. The advantage of using tissue-cultured materials for radiation is to avoid or restrict the formation of chimera and therefore obtain homogeneous mutants directly. But it had higher demand in *in vitro* culture techniques, especially the techniques of organ generation, morphogenesis and plant regeneration under in vitro conditions.

In the asexually propagated species chimera formation occurs after irradiation of a multicellular meristem. It creates problem, therefore, to eliminate this problem, adventitious bud technique is used. The apex of the adventitious buds originates from only one epidermal cell. Consequently, adventitious plantlets either are completely normal or complete solid mutants.

Many plant species can be propagated vegetatively by adventitious shoot formation on detached leaves *in vivo*. In some ornamentals adventitious bud technique has demonstrated to be of great significance for mutation breeding, since irradiation of detached leaves resulted in the production of solid (non-chimeral) mutants. In most plant species, however, an *in vivo* adventitious bud technique is not available. In view of mutation breeding of certain crops it has been successfully done.

In chrysanthemum a fast propagation has been realized by adventitious shoot formation on pedicel explants and proved to be employable for the long-term storage of plant material. After irradiation of pedicels, this *in vitro* adventitious bud technique results in the production of solid mutants, which differed from the original genotype with respect to visible characteristics like the colour, size and/or form of the flower heads. The in vitro procedure further enabled the rapid clonal multiplication of mutant genotypes.

On rachis explants of potato, adventitious shoots can be regenerated through callus and thousands of plantlets can be produced *in vitro*.

In begonia and carnation *in vitro* procedures of vegetative propagation has been developed. The significance of these procedures is still under evaluation for mutation breeding.

In other cases, where adventitious buds cannot be forced to develop the breeder has no choice but to give mutagen treatment to multicellular propagules like tubers, bulbs, corms, suckers, stolens, dormant cuttings, scion wood, rooted cuttings, bud wood or even growing plants. Anatomical considerations may indicate the most suitable stage for irradiation so that large sectors of mutated tissue are obtained from which a maximum number of mutants can be recovered. The primordia to be used for treatment should consist of as few cells as possible. If growing plants are irradiated, decapitation of the main shoots will force new buds to develop through regeneration and thereby increase the chance for recovering a greater number of mutants.

Separation and Selection of Mutants

Mutations induced by radiation in fruit trees and ornamental plants are mostly chimeras. Making the mutations appear, separating and selecting useful mutations are the most important measures in radiation breeding. Effective methods to make mutants express on individual plant level are adventitious bud techniques, continuous pruning, grafting and cutting techniques and *in vitro* culture techniques. Stable mutants are usually not able to acquire until after several generations. Using *in vitro* culture techniques in the procedure of mutation breeding to help separate somatic mutation could effectively avoid the formation of chimera and diploid selection. It opened up a vast range of prospects for speeding up the process of mutation breeding.

In addition, some morphological and biochemical indexes could help identify and select mutations in early stage. Tang (1993) found that in the process of developing new lines of seedless 'Hongjiangcheng' orange, the deformity of the young plants could help separate and select mutants. Chen (1992) found that compared with the control, all the seedless mutant strains of citrus lacked one band on peroxidase zymogram in leaves. According to this, the number of the bands on peroxidase zymogram was suggested as a biochemical index to early identification and selection of mutants. With the development of molecular biology, people will be able to conduct early identification of mutants on molecular level.

In Vitro Technique

Induced mutations technique is a valuable tool not yet fully exploited in fruit breeding. Tissue culture makes it more efficient by allowing the handling of large populations and by increasing mutation induction efficiency, possibility of mutant recovery and speediness of cloning selected variants. Some vegetatively-propagated species are recalcitrant to plant regeneration, which can be a limit for the application of gene transfer biotechnology, but not for mutation induction breeding. Mutagenesis offers the possibility of altering only one or a few characters of an already first-rate cultivar, while preserving the overall characteristics. Traits induced by mutagenesis include plant size, blooming time and fruit ripening, fruit color, self-compatibility, self-thinning, and resistance to pathogens. The combination of *in vitro* culture and mutagenesis is relatively inexpensive, simple and efficient. The availability of suitable selection methods could improve its effectiveness and potential applications. The molecular marker technology available today already provides tools to assist in mutation induction protocols by investigating both genetic variation within populations and early detection of mutants with desired traits. However, cost still represents a major limitation to their application.

Among the techniques and sources of genetic variation available for tissue culture mutation induction, physical mutagens have already shown potential for application in fruit breeding. The types of radiation suitable for mutagenesis are ultraviolet radiation (UV) and ionizing radiation (X-rays, gamma-rays, alpha and beta particles, protons, and neutrons). X-rays and gamma-rays are the most convenient and easiest types of radiation to use with regards to application methods and handling and have been both the most widely used ionizing radiation types and the most effective for fruit breeding purposes.

Technique

Manually the seeds are surface sterilized by immersion in 50% v/v commercial bleach containing 5.25% sodium hypochlorite (NaOCl) for 30 minutes followed by three washings with sterilized distilled water. The seeds are placed on CI medium (modified MS medium containing MS salts and vitamins as described by Murashige and Skoog, 1962) supplemented with 2, 4- dichlorophenoxyacetic acid (2, 4-D) 9.05 μM, sucrose 30 g/l and bacto agar (Difco) 10 g/l. The pH of the medium was adjusted to 5.8. The cultures are placed in dark at a temperature of 25 ± 1 °C for callus induction for four weeks.

Observations regarding the morphology of the callus are noted. Non-embryogenic callus are discarded. Embryogenic callus is excised aseptically and divided into two portions. Half portion is kept as non-irradiated (control) while remaining half is irradiated at 50 Gy of gamma rays of 60 Co. Both non-irradiated and irradiated calluses are divided into about 1 mm. in diameter and proliferated onto non-saline and saline CI medium for three consecutive phases with interval of 45 days. Four electrical conductivity (EC) levels of NaCl (4.0, 6.0, 8.0 and 10.0 d/Sm) were established in CI medium with the same amount of 2, 4-D, sucrose, bacto agar and with similar culture conditions. CI medium is replaced with fresh medium at each proliferation phase. Growth of tissue is expressed by value determined by the ratio of final weight to the initial weight. Calluses with maximum growth value are used as source of inocula in the next phase. Thus adapted callus lines at different EC of NaCl are obtained. Finally all adapted calli are transferred to MSR medium (modified MS medium containing MS salts and vitamins supplemented with kinetin, 9.2 μM; naphthalene acetic acid, 5.37 μM; sucrose 30 g/l; agar 10 g/l; pH 5.8) for regeneration. All the cultures are placed in a growth room under 16/8 h. light / dark period at 25±1 °C. Cultures are transferred to fresh MSR medium after three weeks to regenerate plants (M_1 generation).

Emergence of green shoots above 3.0 cm is considered to calculate regeneration frequency (RF) as under:

$$RF = \frac{\text{Number of embryo derived calli showing green shoots}}{\text{Total number of calli transferred x 100}}$$

The regenerated mutants are transferred to hormone free MS medium for vigorous root development and then to hydroponic solution (Yoshida, *et al.*, 1976) for root hardening. Each regenerated plant is assigned a plant line number and grown to maturity in NaCl-free field conditions. Agronomic data of interest are recorded and seed is collected at maturity.

If the screening for salt tolerance in M_2 generation is to be conducted, it is done as follows:

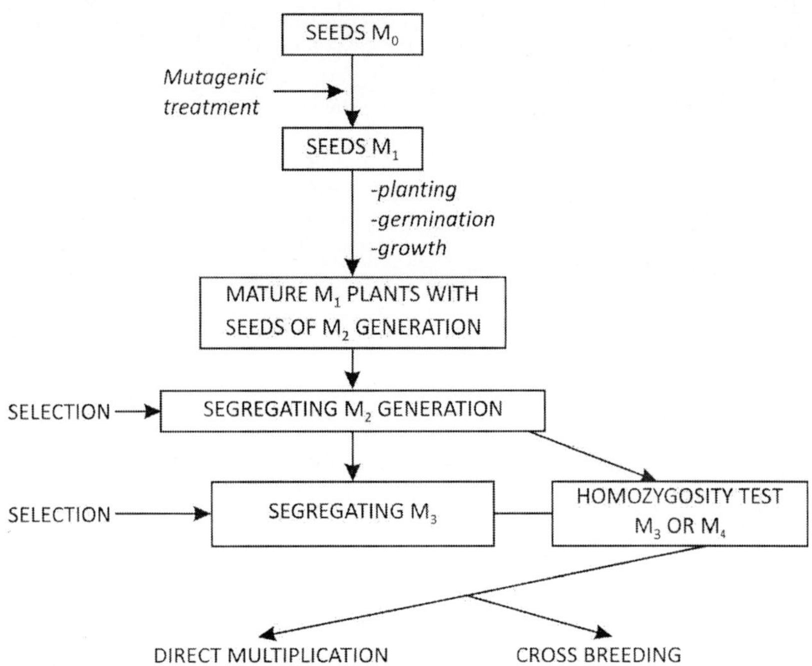

FIG. 14.2: Classical scheme of induced mutation techniques in self-pollinated plants.

Gregorio, *et al.* (1997) conducted rapid screening of ten seed from each M_2 plant line and parent cultivar Basmati 370. The seeds were surface sterilized with fungicide vitavax. These were placed in Petri dishes with moistened filter paper and incubated at 30°C- 48°C to germinate. Two germinated seeds were sown per hole of a Styrofoam float placed on rectangular tray filled with distilled water. Total number of rows on float was 10 in fashion of 10 holes per row. Thus 8 M_2 lines were compared with 2 lines of control. When seedlings were established within 3-4 days, the distilled water was replaced by nutrient solution (Yoshida *et al.*, 1976). Nutrient solution was initially salinized at 2 d/ Sm EC of NaCl. Later on salinity was increased to 4 and 6 d/ Sm with gradual interval of three days by adding NaCl while stirring up to desired level of EC. The pH of the solution was adjusted to 5.0 and checked twice a day i.e., at 0900 and 1500 hour. Adding distilled water compensated the loss of solution due to evaporation and transpiration. Old nutrient solution was replaced with fresh solution after every 8-day and pH was maintained at 5.0 daily. Each M_2 line was rated at 10 and 16 day after initial salinizaiton. Shoot and root length of seedlings of M_2 line and Basmati 370 was recorded to note tolerance level. Modified standard evaluation score (Gregorio, *et al.*, 1997) of visual salt injury at seedling stage were employed for rating M_2 line along with original variety using score of 1 (highly tolerant), 3 (tolerant), 5 (moderately tolerant), 7 (susceptible) and 9 (highly susceptible).

Raising of M_1 Generation

For raising of M_1 generation from seeds, they are sown. The mutated seeds require careful handling. The seeds treated with physical mutagens are stored before sowing, but the seeds treated with chemical mutagens are first washed thoroughly and be planted as early as possible. In case the seeds are not planted immediately due to weather conditions or transportation facilities, then they should be dried in shade and the seed moisture should remain about 13%. The seeds should not be damaged.

Soil conditions, nitrogen fertility and other nutrients of the soil have considerable influence on the survival and growth of the M_1 generation. Soil condition should remain optimal. Vegetative growth should remain normal, excessive vegetative growth should be avoided.

Isolation of M_1 Material

The isolation of the M_1 is essential to avoid the introduction of genetic variability in the population, other than that induced with the mutagen treatment. The mutagen treatments induce pollen sterility, hence, amount of out crossing is increased the. Isolation in M_1 can be achieved by the following methods.

1. **Space M1:** The plants (population) are planted 75-100 m apart from the parental or other genotypes of the same crop species. In insect pollinated crop (legumes) should be planted in isolation.
2. **Time:** In treated M_1 materials flowering takes long time than that of untreated control materials. A slightly later sowing of M_1 material than the parental genotype will permit separation of flowering times.
3. **Genetic:** When mutation breeding is practiced with a limited number of M_1 types of a crop, it may be possible to grow M_1 treatments side by side. This will help to recognize F_1 hybrids by means of marker traits (flower, plant or spike colour).
4. **Mechanical protection:** In mutation experiments, it is important to avoid cross-pollination, and bird damage therefore, bagging of spikes in cereals with plastic or paper bags is essential to prevent cross-pollination and. M_1 generation can also be grown in a green house or in the screen enclosures to achieve control

over pollination by insects.

5. **Harvest of M1:** The time and methods of harvesting of M_1 populations depends on the growth pattern of a particular species. The screening of mutants should be done. The seed yield from each primary branch should be done separately. In cereals, individual plant or spike can be harvested.
6. **Management of M2 populations:** The obtained seeds are sown for M_2 generation as follows:
   a) Firstly, M_1 plant to row, where all seeds produced from a single plant are grown in row. The success of its use will depend, to a large extent, on how well the branching has been controlled because it tends to dilute the yield of M_2 mutants.
   b) Second method is of M_1 spike or branch to row, which offers the greatest precision with regard to the origin of a mutant when the material treated is genetically homogeneous as regards the non-mutant allele and when out-crossing is controlled.

Screening Techniques

The following selection techniques can be employed for the selection of mutants in M_2 and subsequent generations.

- Visual observations: For identification of mutant phenotypes visual selection in a population is done observing disease resistance, earliness, plant height, colour changes, non-shattering, adaptation to soil, climate, growing period etc.
- Mechanical or physical: This kind of selection is to know seed size, shape, weight, density etc. For this purpose appropriate sieving machinery is used.
- Other selection methods: The mutants plant are also screened for chemical, biochemical, physiological, physio-chemical parameters. For biochemical analysis (e.g., alkaloid content or protein variants of mutants) calorimetric, chromatographic or electrophoresis techniques may be used.
- Management of M3 generation: M3 generation is raised from all M2 plants as plant progeny rows. In M3 generation, the inheritance of most induced mutations varies from dominant to recessive; M2 mutants are ordinarily considered homozygous for the trait. Progeny tests are done for the identification of all mutant lines useful for plant improvement and re-selection from M3 where further progeny tests may be necessary for the detection of mutants.
- Propagation and evaluation of mutants: When all above parameters meet the requirement of a breeder and a mutant appears promising, seed are sown for multiplication in field for extensive field testing. The mutant, the mother strain and other varieties with which it is intended to compare, should produce comparable seed properties for the basic trial seed. Mutants of vegetatively propagated plants are multiplied using cuttings, grafting, budding, layering, bulbs, tubers etc.

For a newly developed strain it is essential to make comparative trials in the field because it is necessary to evaluate the success of mutant and their range of adaptability to varied range of environmental conditions like, locations, climate, soil, water and nutrient conditions, seed rates, planting, distances, sowing dates etc.

Utilization of Induced Mutations

The induced mutant can be used with following possibilities:

- Multiplication of a desired mutant without any genetic manipulation leading to the direct release of selected mutant.

- The use of such a mutant in cross breeding.

It is assumed that a portion of the selected mutants will carry no other relevant genetic alternations than the induced genetic trait for which the mutant was selected. After evaluating the performance of the mutant varieties they are certified and released for multiplication to farmers. Further, record should be maintained of various morphological, physiological and biochemical parameters of mutant and the parental genotype.

It is always presumed that mutants will show positive features and responses, therefore, more promising use of mutations can be obtained through cross-breeding. This approach is receiving increased attention, as the genetic analysis and breeding potential of useful mutants have become known. Many cultivars now carry useful induced mutant traits.

At present the plant characters are being improved by mutation breeding in seed propagated crops. These characters are flowering and ripening time, adaptability, plant type and growth habit, resistance to lodging and stem breakage, shattering and shedding resistance, disease and pest resistance, quality, yielding ability, etc.

Mutagens

Radiations are most commonly used with vegetatively propagated crops. Chemical mutagens are not favoured because of poor uptake and penetration of the chemical. Besides, bulky material like bulbs, stolons etc. are difficult to treat in a reproducible way with chemicals. Unless a particular technique is developed such as in vitro culture to treat material with chemical mutagens, it is better to irradiate the material. The procedures are rather simple with good repeatability and high mutation frequency. All types of ionizing radiations are effective.

Doubled Haploid Production

Another bottleneck to induced crop mutations relates to quality and the inherent recessive nature of mutations. This leads to the masking of the mutation events in the appearance of the mutants by the dominant allele at the same gene locus. In a heterozygous background therefore, phenotypic manifestations of mutations are practically impossible to detect in the early progenies necessitating several cycles of crossing the plant with itself in order to produce homozygous recessives that express the recessive phenotype. Again, totipotency is exploited in the regeneration of doubled haploids (DHs), when the chromosome number of gametic cells, i.e. pollens or anthers and egg cells, is doubled prior to regeneration of a plant to mitigate this problem. This process could be incorporated into induced mutagenesis by the treatment of these gametic cells prior to regeneration of the doubled haploids. With spontaneous and/or induced doubling of the haploid chromosomes, homozygous individuals are produced, availing the researcher of the most rapid route to attaining homozygosity without having to cross the plant with itself.

By facilitating the possibility of targeting either the haploid or doubled haploid cells for mutation treatment, a mutation is captured in a homozygous, pure line. These mutants are homozygous for all loci including the mutated segments of the genome being targeted for modification and subsequent detection. For seed propagated crops, doubled haploid strategies provide the fastest method for achieving homozygosity, as compared to self-pollination. The savings in time and cost are significant as recessive mutations usually are not detectable till the first self-pollination generation or later generations. Rapid advances in cellular and tissue biology techniques have resulted in the availability of reproducible DH protocols for over 250 plant species covering most plant genera. The Agency has, through its coordinated research activities, supported the development of easily applicable DH protocols for many crop species. The DH methodology has been successfully used to expedite the pace for generating true breeding mutants in crops such as barley, wheat, rice. Salt tolerant wheat was produced in China by combining mutagenesis with anther culture and at the Agency's laboratories, DH was also used to generate a semi-dwarf (and hence lodging resistant) rice mutant from a salt tolerant but uncultivated wild relative of rice.

Ion Beam Mutation

An ion beam is a type of particle beam consisting of ions. Ion beams have many uses in electronics manufacturing (principally ion implantation) and other industries. Today's ion beam sources are typically derived from the mercury vapor thrusters developed by NASA in the 1960s.

Heavy-ion beam mutagenesis is generally accepted as an effective method for producing mutations, as well as for functional studies of genes. Heavy-ion beams have high linear energy transfer (LET). In contrast to electron irradiation, which has a low LET, heavy ion beams are thought to produce more DNA damage and double strand breaks (DSBs).

Ion beams have been widely used in the research on material surface modification since the 1970s. Ion beams as a mutagen are different from other physical mutagens such as gamma or X rays in that they not only involve energy transfer (as gamma or X rays), but also mass deposition and charge exchange; hence could result in complex DNA damage and changes that are not found when gamma or X rays are used (high percentage of double strand breaks and subsequent chromosome aberrations). Ion beams are produced by particle accelerators, i.e. cyclotrons.

DNA double strand breaks are believed to be the most important consequence of ion beam radiation. Very complex repair mechanisms have been unveiled but are prone to errors due to double-strand breaks and lead to deletions, insertions, inversions and translocations. Studies on the mutant gene alleles induced by ion beam radiation showed that most mutations are deletions and that the size of DNA deletion is LET dependant. Most complex DNA damage caused by the intricate set of effects of heavy ion beams (HIB) escapes the repair efforts and thus is described as more biologically effective and mutagenic than X rays and gamma rays. A wider mutation spectrum and less collateral physiological damage (i.e. effect on plant survival and growth) is commonly reported for ion beam radiation as compared to other mutagens, which is considered an important advantage. In China, 23 new rice and wheat mutant varieties have been bred using ion beam technology and released for large scale commercial production (more than one million ha per annum). The wheat variety 'Wanmai 54' displayed excellent resistance to head scab disease and rust disease and recorded the highest yield in the national new wheat variety yield trial, with yield increases over control variety of 7~10.6%.

In Japan, ion beam technology has been used for generating mutants for a vast number of plant species by various researchers; for example, a consortium of more than 90 user groups was established to utilize the ion beam technology available in RARF (Japan). Six new flower varieties have been developed using this technology and marketed in Canada.

Handling the Material after Treatment

The irradiated bud will give rise to a primary shoot (M_1V_1), first generation, at the nodes of which, in the lower and medium zone, the leaf and axillary buds originate from primordia already pre-formed in the treated bud, whilst the leaf and bud of the distal zone arise from new-formed primordia during shoot growth. The distinction and relative length of pre-formed and 'new-formation' M_1V_1 regions depend on the number of primordia and axillary buds already present in the irradiated bud and ion shoot ontogeny, which is species and variety characteristics. Knowledge of the number and size of the primordia already present in the irradiated bud can be useful for proper handling the M_1V_1 generation and further vegetative propagation in M_1V_2 generation.

The question arises as to which buds to force or which, and how many, to propagate. By taking into consideration the general principle that a mutational event occurs randomly and with the same probability in any cell of the various primordia and meristem of the bud and as consequence the number of mutated cells will be higher in the larger primordia (more cells) but the size of the mutated sector will be smaller and, of course, the contrary in small primordia (few cells). As a result it could be expected that for each species or variety a given 'portion' of the M_1V_1 shoot will give higher chance of recovery in somatic mutations. This portion in several fruit plants has been localized. Data are available from literature indicating that in apple the greatest frequency of mutations in the M_1M_2 was obtained in the propagation of the buds from 8 to 12 of M_1V_1 shopt in cherry the mutation frequency was the highest in the propagation of 5 to 8 buds of the M_1V_1 shoot; which were associated with the youngest leaf primordia in the treated bud. It is reasonable in conclusion to say that those buds coming from the pre-formed primordia in the treated bud which are located somewhere along the M_1V_1 shoot are worthwhile for propagation in the M_1V_2. shoots coming from those buds will have an acceptable mutation frequency and larger enough mutated sector.

Because of the apical dominance prevailing in nearly all fruit plants, most of the buds on the primary shoot, particularly in the lower region, will remain dormant if the shoot is left intact (without pruning). In order to recover the potential mutations in the buds of the lower region, they must be forced into growth by cutting back the primary shoot or by using the axillary buds in propagation. Preferential selection for the induced somatic mutation has to be reserved in the M_1V_2 generation. However, also in this generation we may expect the occurrence of different types of chimeric situation periclinal as well as sectorial and mericlinal or luckily uniform solid mutant.

Consequently the isolation of somatic mutations is the most difficult phase of mutation breeding. Mutations affecting gross morphological changes of the plants such as size and type of growth, change in leaf color, fruit size, ripening period etc., can be recognized without great difficulty. While other mutations based on quantitative genetically controlled factors and finally deviations affecting characteristics such as fruit flavour, productivity, season of fruit maturity, regularity in fruit bearing, are more difficult to isolate, requiring many years of observations and repetitive observations and measurements. Therefore, the M_1V_2 material has to be observed throughout the vegetative cycle, visual selection and measurement will allow to identify mutants affecting the growth habit, the shoot length, internode length, compact type, branching type, change in leaf size and morphology, flowering period, flower fertility, pollen abortion, fruit characteristics, fruit ripening period etc. A number of buds (up to 12) for each individual M_1V_2 selected mutant have to be further propagated to give wider basis for observation, to confirm their characteristics and to assess their uniformity and stability in M_1V_3 generation. Those M_1V_2 normal looking shoots, which may possess mericlinal or sectorial undetected chimera, have to be backpruned to force the growth of axillary buds or a number of consecutive buds along the shoot (up to 5) had to be propagated for the further selection and purification of the mutants in the M_1V_3 generation. In the M_1V_3 generation, one should assess the uniformity among the individual plants originating from the M_1V_2 selected mutants, mutants of those clone which are not uniform have to be again propagated in the next generation to reach the uniformity. Only clones that are uniform have to be selected for planting in the first test orchard.

Characters to be improved

Mutation breeding has been used successfully for the improvement of following characters in the vegetatively propagated crops.

Plant Height and Growth Habit

Growers are looking for fruit tree variety with reduced and compact growth suitable for density planting and intensive production. In addition, reduced size of tree facilitates the fruit harvest by hand or by machine and reduce the pruning cost making economically profitable the cultivation. Mutation breeding has been very successful in inducing compact mutants in several fruit species. In sweet cherry, several compact mutants have been induced. In apple a large number of semi-dwarf and dwarf mutants have been reported. A compact growth mutant was selected in the apricot without altering the fruit characteristics.

Time of Flowering

Induced changes of the flowering time (early or late) are frequently reported in several fruit and ornamental plants. The modification of this character, which apparently is easily changed, may be desired to synchronize a variety and its pollinator, to escape the spring frost damage and, in general, to diversify the time of flowering in ornamental and flower plants. Several mutations affecting time of flowering were induced in sweet cherry, almond and blackberry.

Flower Fertility and Seedlessness

Mutations affecting genes controlling gametic fertility are very frequently reported. Partial gametic sterility due to chromosomal disturbances at meiosis is persisting in mutants, which normally are vegetatively propagated to equilibrate fruit setting, to achieve a more regular annual fruit bearing and increase of fruit size. In Citrus desired seedless mutants have been reported.

Time of Fruit Ripening

Early or late ripening mutants are induced in peach, cherry, apple etc.

Fruit characteristics: Mutations with an increase amount of red fruit skin were among the first and most widely reported in apple, also changes in fruit size, skin or flesh colour, texture and flavour are frequently observed in a number of fruit plants. Several fruit colour mutants of Delicious apple have been induced. Likewise, mutants in grape fruit are induced.

Self-incompatibility: The first self-compatible mutation was induced in sweet cherry and the mutants have been used in crosses. Mutagenic treatment have induced self compatibility at the locus S in the pine apple and have resulted in increased self-fertility in apple and grape.

Disease resistance: In general, resistance for a specific disease is controlled by more than one gene and frequently is dominant. Under such circumstances the change of inducing useful mutations may be limited. However, a limited number of successful results have been reported. Disease resistant mutants in sugarcane (smut), pepper mint (Verticilium wilt), pear (black spot), apple (powdery mildew) etc., have been obtained.

Other characters: The hazelnut cultivars produce a large number of suckers, which vegetatively affect growth, compelling their removal. By irradiation, no sucker mutants have been obtained. Besides, mutants with increased rooting ability have been isolated in the root stock of sweet cherry.

Radiation Breeding of Fruit Trees and Ornamental Plants

Fruit Trees

The important fruit trees are few-seeded or seedless orange cultivars, apple and pear cultivars. Good fruit quality is the most important target of fruit production and breeding. Other desired variations in the obtained fine mutants are short-shoot form (or compact form) variation, early maturing variation, resistance variation and seedless variation. The character of short-shoot form (or compact form) is very important in fruit dwarfing and close planting.

Radiation breeding has proved to be an effective method to acquire cultivars of short shoot form (or compact form). In breeding practice some short shoot (or compact) mutants are obtained in apple, chestnut and hawthorn through this technique.

Early maturing variation not only can enrich the fruit types but also can extent the supplying time of fruits in the market. By use of the radiation-induced mutation technique, early maturing lines of Hawthorn, chestnut, pear and apple have been obtained. Their mature periods remain half to one and a half months earlier than their parent materials.

Resistance variation can help enhance the resistance of fruit trees and improve the quality of fruits. Apple cultivar 'Dongyuanhong' is the mutant of 'Golden Delicious'. 'Dongyuanhong' is stronger in overwintering ability than 'Golden Delicious' and is resistant to early defoliating and powdery mildew. Using Fast neutron to irradiate dormant bud sticks of chestnut, three fine mutant lines resistant to fruit mildew and rot can be obtained.

Induction and selection of seedless variation has been concentrated on orange. So far some few-seeded or seedless orange cultivars or lines have been obtained by radiation-induced mutation technique. In order to reveal the causes of few seeds or no seed, meiosis of chromosome was studied. It showed that chromosome translocation and inversion caused by radiation were the basic reasons of few seeds or no seed in fruits.

In addition, skin color variation and store-tolerance variation of apple, and large-fruit variation of hawthorn were also found in the radiation-induced mutants.

Ornamental Plants

Radiation breeding of ornamental plants has been focused on rose and chrysanthemum, canna, lotus, bougainvillea and dahlia. The purpose of radiation breeding of ornamental plants is mainly to enhance the ornamental characters, such as flower color, flower type, flower size, petal type, petal number and plant form, etc. Mutant cultivar Ziyuntuoyue of chrysanthemum is a good example of flower color change by mutation induction technique. It was developed by Institute of Isotopes, Henan Academy of Sciences in 1991. Besides the ornamental characters, the useful mutations in flowering time and stress-resistance were paid attention to at the same time.

Practical Aspects of Induced Mutation

Any proposal to use induced mutations in plant improvement must consider the likelihood of success when compared with conventional techniques and the efforts required to obtain the desired genotypes. The induction of recessive mutation is a much more likely event than the induction of dominant mutation.

Factors affecting Induced Mutation

Some progress toward both objectives has been achieved through the control and manipulation of secondary factors that influence the response of plant tissues to ionizing radiation. Progress has been achieved also through the use of certain chemical mutagens, such as diethyl sulfate and ethyl methane Sulfonate, which induce a spectrum of effects different from radiation.

Plant radiobiological studies have shown that there are many factors that alter the response of tissues to the sparsely ionizing radiations such as X and gamma rays. These factors are genotype, age, stage of cellular development, chromosome number and size, nuclear volume, moisture, temperature, atmosphere (oxygen, nitrogen, carbon dioxide, hydrogen sulfide, etc.), infra red radiation, and chemicals (colchicine, cysteamine, etc.).

Most modifying factors, however, do not seem to change the response of plant tissues to neutrons and other densely ionizing types of radiation. Therefore, in mutation breeding, neutrons give repeated and predictable results regardless of the physiological condition and environment of the plant tissue. In spite of this advantage, neutron treatments cannot be experimentally modified for increasing total mutation yield and altering the induced mutation process.

Seed has been used as chief experimental material for induced mutation. This is because the seed has several unique properties not possessed by the more actively metabolizing tissues, such as the vegetative bud or fresh pollen. One important characteristic of the seed in these radiobiological studies is its adaptability. It can be irradiated under a range of conditions that greatly alter the cellular environment. For instance, the seed can be desiccated, soaked, or frozen. It can be maintained under a vacuum, almost free of oxygen, or under high pressures of oxygen or other gases for extended periods. When dry it is resting, almost biologically inert, and the severe environmental treatments apparently cause little or no biological damage. After such treatments and controlled rehydration, the irradiated seed can be measured for damage using several biological criteria.

Level of Tolerance

Under normal conditions, plant tissues have distinct levels of tolerance to doses of ionizing radiation. Doses of radiation above these levels lead to low cell survival that most mutations are lost within the plant. These levels vary from species to species. The tolerance is governed to a large extent by numerous secondary factors, only a few of which are understood. Nevertheless, through proper manipulation of a few of them in irradiated seeds, it has been possible to increase the radiation dose tolerance, and, hence, mutation yield.

After-effects of mutation are related to time, oxygen, moisture, and temperature and degree of damage by X- and gamma-rayed seeds. By proper control and manipulation of oxygen, moisture, and temperature before, during, and after irradiation, the dose tolerance of the seed can be greatly altered.

Oxygen Effect

For example, the oxygen-effects and after-effects in irradiated barley seeds increase radiation-induced damage, which in turn causes low cell and plant survival. A typical dose response is exhibited by the seeds planted immediately after radiation, but no dose response occurred in the irradiated, stored seeds. These oxygen- and after-effects can be controlled slightly in seeds with over 12 per cent moisture. Controlling both the oxygen- and after-effects also depends on the rehydration of the irradiated seeds in oxygen-free water. The length of time necessary for rehydration depends upon temperature of the water. Longer rehydration may be required for very dry seeds.

Alteration in Dose

The dose tolerance of plant tissues is the application of certain compounds that affect plant response to radiation. For example, soaking pine seeds in extracts of mustard seeds increased their LD50, where soaking mustard seeds in extracts of pine seeds reduced their LD50. Protective effects can also be observed after treatment with both acidic and basic fractions from mustard seeds. Furthermore, the phenolic and neutral fractions of the mustard seed extracts contained germination inhibitors. The alteration may be due to production of peroxides by irradiation of the unsaturated fatty acids may be the primary cause of the sensitization.

Causes of Cell Death

Cell death after mutation occurs due to high dosages of irradiation and chromosome aberration. For example, chromosome aberrations in barley increase exponentially, whereas the mutations increase linearly with dose. This means that at high doses, chromosome aberrations concurrently induced limit the mutation yield. Therefore, for radiation to become more efficient for inducing mutations in plant breeding, the frequency of chromosome aberrations in irradiated cells must be decreased. Such a decrease will lead to a more favorable ratio of mutations to chromosome aberrations.

Various factors of seed environment may decrease X-ray-induced chromosome aberration frequencies and maintain or increase the M_1 plant survival and M_2 mutation frequency.

Heat-shock post-treatments may control the radiation-induced mutations to aberrations. However, the mechanism conditioning this more favorable ratio of mutations to chromosome aberrations is yet unknown.

Others

The effectiveness and efficiency of the chemical agents are greatly influenced by modifying factors. For example, the uniformity in the growth of seedlings from seeds treated with diethyl sulfate seems to be measurably increased by the addition of certain divalent cations to the treatment solution. The hydrolysis product of ethyl sulfuric acid, in the treatment solution may chelate divalent cations in the cell membranes and facilitate the penetration of the active agent into the cells of the seed. It is also interesting to speculate that the hydrolysis products from both diethyl sulfate and ethyl methane sulfonate may have important roles in the mutagenic and cytogenetic activity of these agents. The ethyl sulfuric and methane sulfonic acids, weak chelating agents, may affect cell membrane permeability.

Selection of Varieties

Selection is a process of selective improvement of a crop. In selection, a plant is selected based on its genotypic and phenotypic characters in order to improve some of its characters in its next generation.

Natural selections have occurred in many crops, where some characters were retained and some others rejected by nature. In the process of agricultural development over centuries, many crops or their varieties were selected in order to improve specific character in future generations. Farmers, early scientists, naturalists have preformed the process of selection in many crops. Actual selection process involves selecting a desirable variant among genetically divergent crop and allowing it to reproduce - in order to consolidate and intensify a specific desirable character. Selection may have to be done over many generations rejecting the undesirable variants at every stage.

In selecting the parent variety for mutations are selected ideally based on the best base germplasm sources. The plant selected which has greatest potential in the area or region such as advanced breeding lines or recently released cultivars. It is essential to ensure about the purity of the base germplasm stock.

Choice of a Mutagen

Ionizing radiations have the advantages of good penetration, precise dosimetry and a wide random spectrum of mutations. A higher ionizing density leads to more chromosomal aberrations. Various chemical mutagens given higher mutation rates, some produce a higher ratio of gene mutations *vs* deletions or other chromosome mutations. But, there are several practical problems with chemical mutagens such as soaking of seeds, penetration to the relevant target cell, safety of handling and disposal, poor reproducibility, persistence of mutagens or its metabolites, which may counter balance the advantages. In general, it would seem advisable to use several mutagens for a spectrum of mutations as wide as possible.

INDUCED MUTAGENS AND THEIR DOSES

Mutation induction continues to contribute to crop improvement, using physical mutagens such as gamma ray, X-ray, fast neutron, and chemical mutagens such as EMS (ethyl-methane-sulphonate) and sodium azides. Recently, new physical mutagens, such as ion beam radiation and cosmic rays, have been proven to be effective for inducing mutations. Below is an overview of recent developments related to plant mutation breeding; induced mutations for harnessing genomics resources; bio-/molecular technologies for enhancing the efficiency of mutation induction and utilization; and ion beam radiation as a novel mutation induction technique.

The selection of an effective and efficient doze of a mutagen for the starting material is important. The effectiveness of a mutagen is measured in terms of biological effect that it produces. A relationship between observed biological effect to a well defined and easily measurable physical quantity characterizing the amount of radiation or chemical mutagen should be established. The mutagenic effect in biological targets is commonly and conveniently described in terms of dose-effect relationships.

The simple dose is called D and the special unit of D is rad (1 rad = 100 erg/g = 102 Joule/kg) and is expressed in terms of time as rad/h, rad/min. and rad/sec. That is the duration of exposure in hour or minute or for seconds. The biological material exposed to the irradiation is important parameters of physical mutations.

The chemical dose of treatment for mutagenesis is determined by the concentration, duration of treatment and temperature during treatment.

Concentration: The concentration of chemical mutagen is determined by the per cent strength of the chemical in the solvent (distilled water).

Volume: The volume of treatment solution is also equally important. How much of number of moles of mutagen is absorbed by the seed or organ. It is generally accepted that a treatment giving 30-40% growth reduction is likely to give an optimal mutation yield in cereal crops.

Duration: The duration provide an opportunity for hydration and full penetration through the treated tissue of the mutagen. Long treatment is advisable, but it can be shortened by using pre-soaked seeds. The treatment of duration depends on the hydrolytic rate of the mutagen. Tf the duration is short, high concentration of mutagen may be used after pre-soaking at high temperatures.

Temperature: It affects the mutagenic process. Usually, LD50 dose results in 50% mortality of treated seeds. With ionizing radiations, a dose which restricts survival to 50 per cent (LD50) or growth to 50 per cent (GR50) is a good treatment. The dose varies from plant to plant. Gamma-ray doses needed for 50% survival (LD50) varies from families to families as shown in following Table.

Family	LD50
Arachis hypogaea	35-45
Cicer arietinum	18-26
Cruciferae (*Brassica oleracea*)	120-140
Cucurbitaceae	35-60
Gramineae	20-40
Liliaceae (*Allium cepa*)	20-30
Phaseolus aureus	65-100
Phaseolus vulgaris	15-30
Pisum sativum	10-27
Solanaceae	30-60
Vicea faba	4-6

Similarly, the range of concentration of different chemical mutagens for cereal seeds also varies as follows:

Mutagen	Concentration Range
EMS	0.05 to 0.3 M (0.03 to 1.5%)
dES	0.015 to 0.02 M (0.1 to 0.6%)
EI	0.85 to 9.0 mM (0.05 to 0.15)
SaN3	0.001 to 0.004 M

Screening

Mutation breeding is dependent on the effectiveness and variations in mutant plant which can be recognized in M_2 or M_3 generation. It is essential to reduce the population of potential variants so that it is easy to sufficiently evaluate the effectiveness and variations.

Other Parameters

The production of useful genotype by mutation breeding or by conventional breeding is labour and time consuming. Therefore, to start with the mutation breeding programme the assessment of plant populations should be made to obtain the desired mutation.

Probability: The results of mutation are random and in a mutagenesis-derived population one can not have the clue of kind or magnitude of genetic change. A particular gene can be expected to mutate once in about 10,000 mutagen treated cells, provided they have been treated effectively. On an average, it appears that only about five cells of the embryonic shoot apex may become part of the germline and thus relate to the next generation.

In seed propagated species the M_2 generation of 2,000 mutagen treated (M_1) plants need examination in order to have reasonable change of finding mutation in a particular gene. For example, if a plant genome has 10-100 × 10^3 relevant genes. Based upon the single locus mutation rate estimate of 1 × 10^{-4} this would mean that in an experiment using 2,000 M_1 plants, there would be 10-100 × 103 mutations or 1-10 per treated cell. Of these, some may be easily recognizable, others not; few usable, most of them not. Admittedly such estimates cannot be very accurate but the order of magnitude should be acceptable and, therefore, be taken into account.

The other parameters are population size and M_2 family size.

CROP IMPROVEMENT

For crop improvement in 1850s cross pollination between varieties of the same species (wheat) was done, followed by hybridization between different species and genus e.g. wheat + rye = triticale in 1930s. In 1975 cell fusion was conducted to overcome species barriers by combining individual plant cells. In 1985, DNA technology - modification of crop characteristics at the level of individual genes. This was the time to initiate the role of plant biotechnology in crop improvement.

These efforts resulted in an increase in cereal yield up 250% in the past 40 years and quality improvements (e.g. wheat for bread making, barley for brewing and potatoes for crisping) and development of new crops, such as oilseed rape and maize, etc.

Biotechnology involves managing biological processes to make useful products. Everyday goods, such as bread, cheese and beer, are all made using biotechnology. Recent progress in explaining the science behind biological processes has significantly enhanced our ability to exploit them for the benefit of mankind and the environment.

The evolution of plant breeding is a classic example of how improved biological understanding has been adapted to provide more effective methods of meeting the demands of a changing world.

All forms of plant breeding involve the improvement of crops by selection. Since man began farming he has selected seeds from the best plants for the next generation. Over the centuries, this selection process has become more scientific, bringing major improvements in the yield, quality and diversity of agricultural and horticultural crops.

Biotechnology allows crop improvement to take place at the level of individual genes. Techniques such as genetic modification enable plant breeders to identify the single gene responsible for a particular trait, and insert, delete or modify it in a plant variety. This extends the range of characters at the breeder's disposal, and enables specific genes to be expressed in a crop plant without the uptake of unwanted characteristics which must then be selected out.

SOMACLONAL AND GAMETOCLONAL VARIANT SELECTION

In plant cell, genetic variability develops spontaneously during tissue cultures. The term *somaclonal variation* was introduced by Larkin and Scowcroft (1981). Plant variant obtained from tissue culture is called **somaclones**. Evans et al (1984) coined the term **gametoclonal** variation for variant clones specifically raised from gametic or gametophytic cells. Occurrence of somaclonal variation in regenerated plants is considered a rule rather than exception. Somaclonal variation can be of two types:

- Genetic (i.e. heritable) variability is caused by mutations or other changes in

DNA. During genetic variability various molecular mechanisms are responsible for genetic variability which are associated with somaclonal variation.

- Epigenetic (i.e. non-heritable) variability is caused by temporary phenotypic changes. Epigenetic changes somaclonal variation can be temporary and over time are reversible. However, sometimes they can persist through the life of the regenerated plant. One common phenotypic change seen in plants produced through tissue culture is rejuvenation. Rejuvenation causes changes in morphology such as earlier flowering and enhanced adventitious root formation. Epigenetic changes may be caused by DNA methylation and thus may be one of the important causes of somaclonal variation.

Studies concerning different aspects of somaclonal variation are important for several reasons.

1. **First**, it is hailed as a novel source of genetic variation. However, successful utilization of somaclonal variation heavily depends upon its systematic evaluation and judicious utilization in breeding programmes. This necessitates appropriate experimentation.
2. **Second**, somaclonal variation is of interest as a basic genetic process, since it contradicts the concept of **clonal uniformity**. The cells and tissues which are expected to produce true to type plants through the processes of de-differentiation, division and re-differentiation, possibly perceive the whole process as stress, as a result of which the genome, known for its plasticity, restructures itself to modulate the expression of genes as demanded by the *in vitro* conditions.
3. **Third**, somaclonal variation is unwanted when the objective is micropropagation of elite genotypes or genetic transformation, which partly involves tissue culture. Under such circumstances, prevention or at least minimization of variation is of utmost importance. To achieve this, the frequency, nature and magnitude of somaclonal variation in relation to manipulation of media components, explant source, culture conditions etc. should essentially be understood.

Majority of studies undertaken on somaclonal variation are confined to early generations of somaclones. Therefore, information on the nature, inheritance pattern and stability of morphological and molecular changes expressed in the advanced generations of somaclones is lacking.

The different aspects of somaclonal variation investigated so far are as follows:

1. Generation of variation
2. Characterization of variants for morphological traits.
3. Analysis of biochemical and chromosomal basis of variation.
4. Relating the variations to alterations in DNA.

Genetic Variations and Crop Improvement

Genetic variation appears during or after culture *in vitro*. It may occur in undifferentiated cells, isolated protoplasts, calli, tissues and morphological traits of regenerated plants. Variant selected in tissue cultures have been referred as calliclones (from callus culture), or protoclones (from protoplast culture). The changes occur because of variation in chromosome number and structure. Cytological heterogeneity in cultures develops due to the following reasons:

1. The expression of chromosomal mosaicism or genetic disorders in cells of the initial explant. The chromosomal mosaicism may be pre-existing mosaicism (polysomaty - the phenomenon of polyploidisation of body cells) in explants

used for culture initiation, or by nuclear fragmentation associated with first cell division of callus initiation or endoreduplication or endomitosis occurring during culture initiation and/or abnormalities of the mitotic process leading to aneuploidy.

2. New irregularities brought about by culture conditions.

Somaclonal variation causes problem for plant propagationists, since their objective is to reserve a specific plant genotype. While for plant Geneticists somaclonal variation is a valuable source of new genetic information. It leads to crop plant improvement and a deeper understanding of the biochemical and molecular basis of inheritance.

Somaclonal variation is one of the aspects of tissue culture technology and is widely recommended for crop improvement specially of desired traits for the salt, drought, temperature and disease tolerance. The method refers to heritable changes, which accumulate in the callus from a somatic explant and express in the progeny of *in vitro* regenerants obtained from callus. Even though relative data from deliberatly designed experiments is not available, indications of possible advantages of somaclonal *vis-a-vis* induced mutagenesis are:

- The frequency of variation seems to be far greater than the yield of induced mutations.
- The changes are very subtle and may not involve drastic altration in the genetic background.
- Somaclonal variation occurs for trait of both nuclear and cytoplasmic origin. The variation of cytoplasmic genes obtained by this method is a distinct advantage.

In wide crosses somaclonal variations provide a mechanism of gene introgression. Immature embryos of the wide cross can be callused and plants with the introgressed desired genes (or gene complex) are selected among the regenerants of their progenies.

Why Variations Occur?

Variants may also arise as a result of more suitable changes due to single gene mutations in cultures which have cells apparently showing no karyological changes. Recessive mutations are not detected in plants regenerated *in vitro* from any cell or tissue, but expressed in progeny. This shows that variants are the mutants. Single gene mutation responsible for somaclonal variation relates to transposable elements. Transposon-induced changes have been observed in maize, tobacco and wheat. Somaclonal variation may also be due to changes caused by mitotic crossing over in regenerated plants. Such changes may also occur due to changes in organelle DNA, isoenzyme and protein profiles *e.g.*, in wheat, potato, maize, barley and flax, etc. Changes in cytoplasmic genome have also been observed in somaclones.

The Mechanism of Somaclonal Variation

According to Bhaskaran (1985) variations in somaclons occur due to the following reasons:

- The pre-existing genetic variation in the explant tissue,
- The spontaneous mutation which can accumulate during the many division cycles that cells of the explant go through before differentiating into an in vitro plant. The recessive mutations will naturally require a method by which they can express even diploid cells. Somatic crossing over followed by segregation is a likely mechanism, for the homozygosity and thus phenotypic expression of

the recessive.

- Numerical and structural changes in chromosomes during in vitro growth.
- Intracellular mutagenic agents produced during in vitro growth.
- Activation of transposable elements or jumping genes, are genetic entities which have the locus at which they get integrated is matured.

Agriculturists are very hopeful about practical advantages of somaclonal variations and they are waiting when this technique is fully integrated with the conventional plant breeding procedures.

Source Material and Culture Conditions

Plant cell, tissues and somatic embryos developed from various explant sources for generating somaclonal variation. Explants are generally taken from any tissue, namely leaves, internodes, ovaries, roots and inflorescence. The source of explant has often been considered a critical variable for somaclonal variation.

Determination of Cell Number

Take an aliquot of suspension and filter off the cultures through a wire mesh (300 mm). Note the volume of the filtrate (F) containing single cells and small clumps and place the drop of this suspension to haemocytometer to determine the number of cells by the equation

$$N = \frac{P \times 100 \, X \, F}{.01 \text{ mm}}$$

Where, N = total number of cells and clumps, P = number of cells in the squares of the haemocytometer, F = volume of the filtrate.

Forms of Somaclonal Variation

Many different forms of somaclonal variation arise. The most common forms include point mutations, chromosomal aberrations, and increase or decrease in the number of nuclear chromosomes. It is important to realize that not all forms of variability that arise *in vitro* are heritable. Some morphological and biochemical variants are due to physiological effects and are not exhibited in subsequent generations.

Detection and Isolation of Somaclonal Variants

There are several different approaches to detecting and isolating somaclonal variants from cultured plant cell populations.

- Morphologically distinct cells such as non-photosynthetic (non-green) cells or cells that accumulate anthocyanins and other plant pigments are detected visually.
- To isolate herbicide- and antibiotic-resistant variants, plant cells are simply grown on media containing of the wild-type cells in a culture.
- The surviving cells are then sub-cultured and retested for growth on herbicide or antibiotic supplemented medium.
- Through this method one can eliminate any remaining wild-type cells that may have inadvertently survived the first round of selection.

TABLE 14.3: Somaclonal variation induced in morphological traits in some crop plants

Crop	Characters
Sugarcane	Cane diameter, stalk length and weight Cane yield, sugar yield, stalk number, length diameter, volume, density the weight
Potato	Growth habit, maturity period, tuber uniformity and skin color. Photoperiod requirement, fruit production.
Tobacco	Yield, Days to flowering, plant height, stem diameter, leaf number, leaf length, leaf width and yield Leaf Shape, leaf number, plant height, type of inflorescence and yield.
Rice	Number of tillers per plant, number of fertile fillers per plant, average panicle length, frequency of fertile seed, plant stature and flag leaf length. Flowering period, plant height, seed fertility and heading date.
Oats	Plant height, heading date, twin culms, yellow leaf stripe, awn morphology and fertility.
Maize	Twin stalks from a single node, reduced pollen fertility and male sterility.
Brassica	Altered leaf way, multiple branching of the stem, precocious flowering from the apex, stem or leaf, abnormal leaves, reduced lamina in leaves, spontaneously absorbing vegetative buds, slow growth, failure to flower, large pollen grains. Delayed flowering altered growth habit and gross morphology.
Pelargonium	Leaf shape, size and form, flower morphology, plant height, fasciation, pubescence and anthocyanin pigmentation.
Pineapple	Leaf colour, foliage density, leaf width and leaf spine formation.
Tomato	Male sterility, jointless pedicel, tangerine virescent leaf, flower and fruit colour, lethal chlorophyll deficiency mottled leaf appearance, fruit ripening and growth habit.

We can also use this direct selection technique for isolation of temperature-resistant variants because those cells which survive in an extended incubation periods at abnormally high or low temperatures are temperature-resistant.

Those somaclonal variants that cannot be detected visually or selected directly are isolated by indirect means.

Those auxotrophic plant cells which unable to survive in absence of specific nutrient supplements not required by sensitive cells, which will not survive at temperatures above or below a certain threshold. This threshold will not affect normal wild-type cells. In absence of the necessary nutrient supplement or when grown at excessive temperatures, these cells become very weak and at this weakened state, these cells assimilate exogenously supplied cytotoxic compounds (arsenate, bromodeoxyuracil, or fluorodeoxyuridine) at lower rate than that of wild-type cells. Thus, treating a cell culture under restrictive growth conditions with one of these substances will tend to favor short term survival of the weaker auxotrophic or temperature-sensitive cells.

Mutagenesis and Somaclonal Variation

Mutagenic agents (chemical and physical) are used to produce particular heritable forms of somaclonal variation *e.g.*, ultraviolet (UV) radiation, ethyl-methane sulfonate (EMS), nitrosuguanidine, and sodium azide, etc.

Nitrate
⇓
Nitrite
⇓
Ammonia

FIG. 14.3: The biochemical reduction of nitrate to ammonia.

UV radiation induces dimer formation between adjacent thymine residues in DNA. This produces lesions that cause frameshift mutations and base pair substitutions during DNA replication. EMS and nitrosoguanidine are alkylating agents that cross-link and sever DNA molecules, often resulting in gross DNA alterations. In contrast, sodium azide induces single base pair substitutions or deletions, resulting mostly in small point mutations.

Somatic Genetics of Nitrogen Metabolism

Plant cells in culture medium can metabolize ammonium, nitrate, and nitrite sources of inorganic nitrogen. The cells use ammonium directly while the nitrate is first reduced to nitrite (enzyme nitrate reductase, NR) and then nitrite is reduced to ammonium (enzyme nitrite reductase NiR). NR needs a molybdenum-containing cofactor (MoCo) for proper enzyme activity as follows:

Hypoxanthine
⇓
Uric acid
⇓
Urea

FIG. 14.4: The biochemical reduction of hypoxanthine to urea.

For isolation of variants, which unable to utilize nitrate, plant cells are grown on chlorate-supplemented media. Chlorate is a close analog of nitrate and will be reduced to chlorite, a patent cytotoxin that accumulates internally and eventually kills the cell. Certain metabolic mutants can survive chlorate treatment. Some of these mutant cells carry mutations affecting NR expression or assimilation. Other mutations may affect MoCo expression or chlorate assimilation.

Method for Isolation of Desired Variant Cells for NaCl -Tolerant from Callus/ Suspension Cultures

1. Semi-solid liquid media is prepared and NaCl (500 mg/l) is added to the medium, pH is adjusted at 5.6. Distribute the medium in 30 ml aliquots into 150 ml flasks or 50 ml aliquots into 250 ml flasks and grow the callus on NaCl containing medium. Select the granular friable callus for further subculture. 2 g of callus is inoculated to each 250 ml flask containing 50 ml of liquid medium then incubate the cultures on a gyratory shaker (100 rpm) for 21 days at 25 ± 2 °C.
2. Pass the cell suspension through a stainless steel wire mesh (300 μm) and collect filtrate in 15 ml centrifuge tubes. Examine the filtrate for single cells and small aggregates (5-20 cells). Centrifuge the filtrate at 100 g for 5 min and discard the supernatant, add 5 ml of the sterilized medium to the pellet to obtain a cell suspension, and maintain the cell density of the suspension about 0.5-2.5 × 10^5 cell/ml. Add to the flasks containing with or without NaCl.
3. Incubate the cultures at 100 rpm at 25 ± 2 °C and centrifuge the cultures grown

in control and NaCl (500 mg/l) at 100 rpm for 5 min.

4. Transfer the pellets to fresh medium of the same composition. Incubate at 25 ± 2 °C for 21 days and repeat these steps for 2-3 times to get a good growth rate of the NaCl tolerant cell line
5. Grow cultures on 1000 mg/l NaCl and follow the same steps, as carried out at 500 mg/l NaCl till a salt concentration beyond which the cells cease to grow. It is suggested that the salt concentration should be increased gradually by repeating steps
6. Regenerate variant cells into whole plants. Transfer 0.5 ml of cell suspension onto the surface of the regeneration medium gelled with agar by a wide mouth graduated sterile pipette.
7. In case of callus, inoculate 500 mg piece onto the surface of the organ/shoot forming medium having 500 mg/l NaCl
8. In most of the plant species, the differentiation of shoots and roots occurs on different media. Hence once the shoots are obtained, they are transferred to the rooting medium. Transfer the selected NaCl tolerant plantlets to soil.
9. Use NaCl solution to water the salt tolerant plants. The concentration of NaCl should be the same as used in the regeneration medium. This procedure can also be used to develop tolerant plants of alkali, high temperature, drought, disease, herbicide can be developed by using *in vitro* selection system.

NEW CULTIVARS DEVELOPED BY MUTATION BREEDING

Sustained efforts to use induced mutations for crop improvement were initiated during the later half of the 1950s. These early results had a high visible impact in the form of fully awned mutants in awnless wheat cultivars, and in early flowering and plant height variants. This raised high hopes and generated a great deal of interest in mutation research with crop plants. Most of the experiments carried out during the 1960s were aimed at estimating the frequency and spectrum of mutations obtained following treatment with different physical and chemical mutagens, and combination of these mutagens. This exploratory phase continued in the 1960s and early 1970s was replaced by mutation breeding programmes with specific and well defined objectives. A large number of new cultivars have been developed in a variety of crop plants over the past three decades, and mutation breeding is now an accepted method for crop improvement in the country.

TABLE 14.4: Examples of plants that were produced via mutation breeding

Crop	Cultivar Name	Method Used to Induce Mutation
Rice	Calrose 76	Gamma rays
Wheat	Above	Sodium azide
	Lewis	Thermal neutrons
Oats	Alamo-X	X-rays
Grapefruit	Rio Red	Thermal neutrons
	Star Ruby	Thermal neutrons
Burmuda grass	Tifeagle	Gamma rays
	Tifgreen II	Gamma rays
	Tift 94	Gamma rays
	Tifway II	Gamma rays
Lettuce	Ice Cube	Ethyl methanesulphonate

Crop	Cultivar Name	Method Used to Induce Mutation
	Mini-Green	Ethyl methanesulphonate
Common bean	Seafarer	X-rays
	Seaway	X-rays
Lilac	Prairie Petite	Thermal neutrons
St. Augustine grass	TXSA 8202	Gamma rays
	TXSA 8212	Gamma rays

Quite a few flower cultivars have been developed via mutation breeding, among them some of the cultivars of *Alstroemeria*, begonia, carnation, chrysanthemum, dahlia, and snapdragon.

Mutation breeding was particularly popular in the United States during the 1970s. Although interest has waned somewhat in recent years, occasional varieties continue to be produced using these methods. Mutation breeding efforts continue around the world today. Of the 2,252 officially released mutation breeding varieties, 1,019, or almost half, have been released during the last 15 years.

Cultivars obtained as direct mutants as well as those developed by using induced mutants in recombination breeding in almost all crop plants.

TABLE 14.5: Crop cultivars developed by mutation breeding in India (1980-1989)

Crops	No. of cultivars developed		
	1960-69	1970-79	1980-89
Cereals	4	18	19
Pulses		4	27
Oilseeds	2	6	10
Fibre crops		5	5
Vegetables		4	7
Sugarcane		2	3
Fodder crops			1
Medicinal plants		1	
Sericulture		1	
Fruits		2	2
Ornamentals		81	20
Total	**6**	**104**	**95**

Release of 205 cultivars over the past three decades in the country shows that the number of cultivars accepted for commercial cultivation has increased gradually. The initial scepticism of plant breeders towards use of mutation techniques for genetic improvement no longer exists. Mutation breeding is used in a number of National Institutes and Department of Genetics or Plant Breeding at Agricultural Universities with a strong commitment to plant breeding programmes. Most of the improved cultivars from mutation breeding have originated from such places.

Applications in Plant Breeding

Somaclonal variation and gametoclonal variation are the important sources of introducing genetic variations that could be of value to plant breeders. Single gene mutation in the nuclear or organelle genome usually provides the best available variety *in vitro* which has a specific improved character. Somaclonal variation are used to uncover new variants retaining all the favourable characters along with an additional useful trait, *e.g.*, resistance to diseases or an herbicide. These variants can then be field-tested to ascertain their genetic stability. Gametoclonal variation is induced by meiotic recombination during the sexual cycle of the F_1 hybrid, results in transgressive segregation to uncover unique gene combinations.

Various cell lines selected *in vitro* and plant regenerated through it prove potentially applicable to agriculture and industry specially resistance to herbicide, pathotoxin, salt or aluminium, useful in the synthesis of secondary metabolites on a commercial scale, etc.

The techniques used for development of somaclonal and gametoclonal variation are relatively easier than recombinant DNA technology and is the appropriate technology for genetic manipulation of above cited crops.

TABLE 14.6: Examples of gametoclonal variation among some plants raised from gametophytic tissue cultures

Species	Altered characteristics observed	Genetic basis	Androgenetic (A) or Callus Mediated (CM)
Brassica napus	Time to flower, glucinolate content, leaf shape and colour, flower type, pod size and shape	Mutation	A
Hordeum vulgare	Plant height, yield, fertility, neck length, days to maturity	Not examined	CM
	Grain yield, culm length, heading date	Transgressive segregation, gametophytic selection	CM
Nicotiana sylvestris	Leaf colour, leaf shape, growth rate, DNA content, repeated sequences of DNA	Crumpled leaf is nuclear, germ amplification	A
N. sylvestris (several cycles)	Flower length, plant height, leaf shape, leaf colour, exerted stigma, capsule weight, foliar outgrowths	Nucleus genes	A
N. tabacum	Yield, grade, time to flower, plant height, number of leaves, leaf shape, total alkaloids, % reducing sugars	Not examined	A
	Yield, plant height, number of leaves, % nornicotine	Not residual heterozygosity	A
	TMV resistance, root knot resistance	Variation in expected segregation ratios	A
	Days to flower, total alkaloids, yield per plant, leaves per plant	Residual heterozygosity; no cytoplasmic effects	A
	Yield reduction, plant height, days to flower, total alkaloids	Nuclear mutations except plant height is maternal; not residual heterozygosity	A

Species	Altered characteristics observed	Genetic basis	Androgenetic (A) or Callus Mediated (CM)
	Yield reduction, alkaloid content, leaf size, time of flowering, leaf quality	Induced changes, not residual heterozygosity	A
	Heterochromatin content, nuclear DNA content	Gene amplification	A
	Time of flowering, plant size, leaf shape, alkaloid content, number of leaves	Residual heterozygosity	A
	Altered temperature sensitivity	Cytoplasmic mutant	A
Oryza sativa	Chloroplast content, time to flower, plant height, fertility	Some single genes induced in culture	CM
	Plant height, seed and leaf size, seed weight, tillering, flowering, % open hulls	Transmitted to progeny	CM
	Seed size, seed protein level, plant height, level of tillering	Transmitted to progeny	CM
	Waxy mutant	Single gene change	CM
Saintpaulia ionantha	RuBCase activity	Not examined	A
Secale cereale x S. vavilotrii	Fertility	Not analyzed	CM

TABLE 14.7: Aspects of improvement desired in some crop plants through somaclonal variation

Crop plants		Improvement desired
Cereals	Rice	Tungro virus and leaf hopper resistance
	Wheat	Quality, high temperature tolerance
	Sorghum	Borer and shoot fly resistance
	Maize	Delinking of phytohormone effect from reproductive development
Legumes	Pigeon-pea	Resistance to pod borer, fusarium wilt, high protein content
	Bengal gram	
Oilseeds	Groundnut	Resistance to Cercospora, Fusarium, Aspergillus flavus
	Rapeseed	Male sterility, low pungency combined with insect resistance
	Sunflower	Self compatibility, resistance to Alternaria leaf blight, Rhizoctonia, Fusarium complex
	Cotton	Insect resistance to gossypol free lines
Vegetables	Potato	Virus and Phytophthora resistance
	Tomato	Resistance to disease
Fruits	Grapes	Resistance to disease, short duration
	Citrus	Higher solid content, increased shelf life
Other (s)	Sugarcane	Resistance to disease, short duration
	Castor	Mutants with high fatty acids; oleic acid, ricinoleic acid

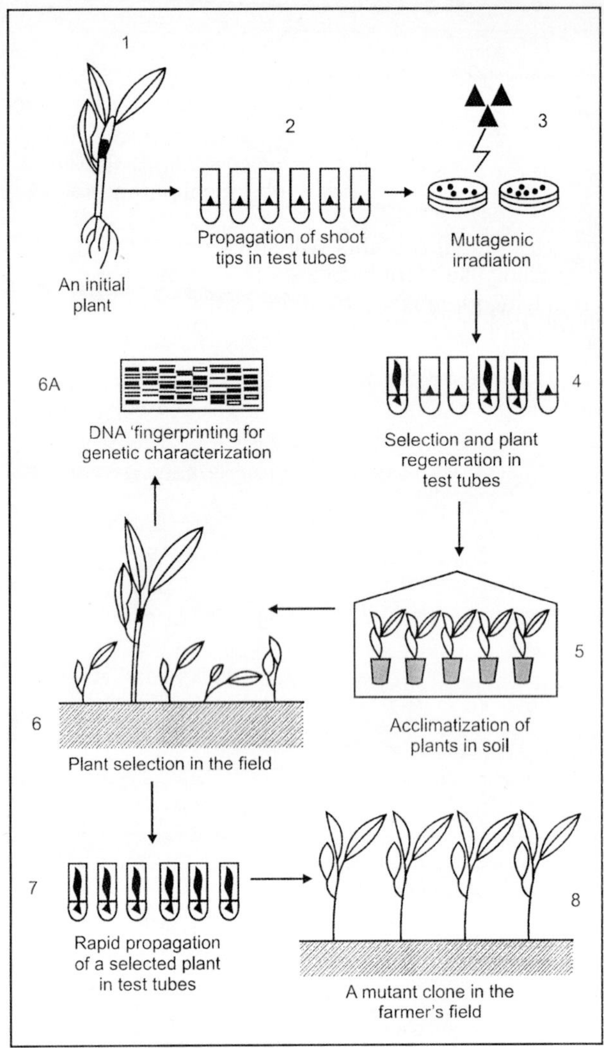

FIG. 14.5: Method for crop improvement through mutation and tissue culture.

Mutation Induction and Breeding Technology for Banana and Plantain

Low genetic variation and sterility handicap genetic improvement of banana and plantains (*Musa* spp.) by conventional breeding techniques. Shoot-tip culture and *in vitro* plant regeneration are being investigated for use in mutation induction and mutant selection. Somatic embryogenesis and plant regeneration from cell suspensions of *Musa* are used to develop somatic cell manipulation procedures for banana and plantain breeding. Methods of screening such plants for resistance to Panama disease are studied in tissue culture, and biochemical markers (peroxidase) are applied for the identification of tolerant genotypes. DNA markers are used for identifying mutants and characterizing cultivars and species of *Musa*.

Mutation Breeding to Improve the Tolerance to Environmental Stress of Azolla

Azolla is a small aquatic fern that lives in symbiotic relationship with the nitrogen-fixing cyanobacterium *Anabaena*. Under suitable field conditions *Azolla* can double its weight every 3-5 days. The *Azolla-Anabaena* symbiotic system provides green manure for flooded crops, particularly rice. Induced mutagenesis has produced *Azolla* variants tolerant to high salinity, toxic aluminium levels, and/or to herbicides. Tolerant plants are being investigated under field conditions to confirm that heritable changes cause the increased tolerance to environmental stress.

Mutation Induction and Breeding of Tropical Root and Tuber Crops (Cassava and Yam)

Cassava and yam are among the most important staple food crops of the lowland tropics. Mutation breeding technology is being developed to increase variation in plant stature, cyanide contentx, disease, and pest resistance. *In vitro* techniques are used for the propagation of healthy plants and improved clones. Somatic embryogenesis is being developed for cassava and yam improvement through *in vitro* mutagenesis and later on by somatic cell manipulation. Mutant and polyploid clones are prepared for field testing in Member States.

Tissue Culture in Cocoa as a System for more Efficient Mutation Breeding

Attempts to breed cocoa for disease resistance have yielded very limited success. A major constraint is that little variation exists in currently available cultivars. Somatic embryogenesis is being developed for propagation of desirable genotypes and, through *in vitro* mutagenesis and pollen mutagenesis, is being applied for induction of virus-resistant cocoa trees in Ghana.

FUTURE ROLE

Contemporary knowledge, tools and techniques have been used in the past, and will always be used in the future for crop improvement programmes. Since funds, time and manpower are limited, the most economic approach towards reaching the desired objective will be used among the alternatives available. The classical method of mutagenising seeds is still, and will remain, relevant as a relatively simple tool in plant breeding programmes. However, emphasis would shift from visible morphological mutants to those governing the response to the biotic and abiotic stresses, and physiological traits.

With the imminent threats posed by global climate change to crop production and the ever increasing and more sophisticated demands of agricultural products, crop improvement efforts have to be more powerful and precise in developing new crop varieties. Breeders therefore require tools that permit achieving subtle changes to the genetic make-up of otherwise superior crop varieties e.g. high yielding but lacking in specific quality traits, and yet leaving the genome largely intact in order not to disturb already stacked alleles of genes. The availability of genomics information in the public domain coupled with recent advances in molecular and cellular biology techniques have paved the way for transforming old mutation techniques into state of the art technology for both crop improvement and basic genomics research.

Development of novel and more efficient genomic tools have become routine and the pursuit of new physical mutagens continues. Some technologies are already in place and when integrated into mutation research, they will greatly increase the efficiency and application of mutation techniques in plant research.

Mutant Stocks as Gene Repositories

Generations of induced mutants have long been driven by plant breeders, who would then use them directly or indirectly (through further cross breeding) in the process of developing new varieties. The plant breeder's ability to improve a crop is as good as the available genetic variation, which can be exploited in cross breeding and other forms of selection. Mutation induction offers the possibility to for a trait, known as the gene, can be identified by the deductive process of inducing a series of mutants, identifying which traits have been modified and relating such observed modifications to changes in genomic regions of such mutants compared to the "normal" types is important. The mutated genomic region that exerts control over the trait in question is therefore identified as the gene responsible for the expression of the trait.

The mutant genetic resources as one of the outputs of mutation induction is an indispensable tool not only in plant breeding research but also in fundamental genetic research activities related to gene discovery and gene function analyses.

Significant efforts are currently being made not only in the generation of genetic variations through induced mutations but also in the systematic characterization of the induced mutants for observed phenotypic and genomic changes and the cataloguing of the variations in searchable databases. Many of these databases are available to interested parties. Climate change may well lead to significant losses of genetic diversity within species critical for food and agriculture.

Roughly 20 - 30% of species assessed are likely to be at high risk of extinction, if global mean temperature exceeds 2 - 3°C above pre-industrial levels. These include crops' wild relatives, which are essential to increasing agricultural productivity, but also serve as the foundation for adaptation strategies required for adjusting to abiotic changes such as heat, drought and salinity as well as pests and diseases. Loss of genetic diversity would have important negative consequences locally as well as internationally, because key traits for climate change and variability adaptation may be lost forever. Mutation induction can help to meet this challenge.

Most National Governments have also established germplasm collections that contain sizable amounts of characterized mutant stocks that are distributed to plant breeders and geneticists. The U.S. Department of Agriculture, Agricultural Research Service (USDA/ARS) runs several germplasm repositories (GRIN) that also include mutant genetic stocks under the National Plant Germplasm System and is considered to be one of the most comprehensive and best organized collections.

In addition to these genetic resources with mostly single point mutations, another type of mutant genetic resource, radiation hybrids (RH), has been developed for genomic research. RHs are produced by exposing somatic cells to lethal doses of gamma radiation or X-ray, in order to fragment the chromosomes. They are then rescued by introduction into host cells, which are subsequently fused with suitable recipient cells for the assessment of their expressions providing unique materials for the establishment of physical maps, a process known as radiation hybrid mapping. RH maps have been developed in a number of crops, such as barley, maize, wheat and cotton for gene discovery and detailed linkage analysis. This may soon lead to the identification and transfer of genes affecting useful agronomic, quality and stress tolerance traits.

Outlook for the Future

Scientific interest in mutation research on plants has definitely passed its peak and is now on the decline. A revival, with new supporters, came with the discovery of somaclonal variation in the progeny of plants raised from in vitro cultured cells.

Somaclonal variants for characters that can be screened in vitro, and in vegetatively propagated species, may have an advantage, there are not many agronomically desired traits that can be reliably selected in cultures.

A number of new cellular and molecular techniques now form part of the rapidly changing scene of genetic manipulation. They have evoked immense interest and generated high expectations. The new cellular methods are essentially random chance events and have limited and distinct advantages over the classical methods of mutagenizing seeds or other meristem. Mutagen treatment of single cells, use of haploids, in vitro screening etc: can certainly enhance the efficiency of mutation breeding experiments, but these techniques have not been used extensively in breeding programmes.

Molecular approaches are more precise and specific gene transfers across the species can be made, provided the DNA sequences governing the trait can be isolated. Technically, it is possible to isolate, clone and modify the promoter or structural gene sequences, and re-introduce them back into the genome. This would still fit into the present definition of mutation breeding. Experts in each technique will advocate the applications of their specialty. However, successful breeding programmes can only be goal oriented and cannot rely on a single technique.

CHAPTER 15

Marker Assisted Selection (MAS) Technique for Hi-Tech Horticulture

INTRODUCTION

Molecular markers are specific fragments of DNA that can be identified within the whole genome. They are found at specific locations of the genome. They are used to 'flag' the position of a particular gene or the inheritance of a particular characteristic. In a genetic cross, the characteristics of interest will usually stay linked with the molecular markers. Thus, individuals can be selected in which the molecular marker is present, since the marker indicates the presence of the desired characteristic.

It is an identifiable physical location on a chromosome whose inheritance can be monitored (e.g. gene, restriction enzyme site or RFLP marker).

Molecular marker is a term with a number of uses. It is any kind of molecule indicating the existence of a chemical or physical process. In particular, in the fields of geology and astrobiology, biomarkers (also known as biosignatures) are sometimes understood as molecules indicating the presence (past or present) of living organisms. It can also be a molecule used as fiscal marker to distinguish lower-taxed items from higher-taxed ones when they otherwise can be substituted.

In medicine, a molecular marker can be a substance that is introduced in an organism as a means to examine something. For example, rubidium chloride is used as a radioactive isotope to evaluate perfusion of heart muscle. In biology and medicine, a molecular marker (biomarker means an indicator of a particular disease state or a particular state of an organism) can be a substance native to the organism whose detection indicates a particular disease state (for example, the presence of an antibody may indicate an infection). In genetics, a molecular marker (identified as genetic marker) is a fragment of DNA sequence that is associated to a part of the genome. A genetic marker is can be described as a variation, which may arise due to mutation or alteration in the genomic loci, that can be observed. A genetic marker may be a short DNA sequence, such as a sequence surrounding a single base-pair change (single nucleotide polymorphism, SNP), or a long one, like minisatellites.

A single nucleotide polymorphism (SNP, pronounced *snip*), is a DNA sequence variation occurring when a single nucleotide - A, T, C, or G - in the genome (or other shared sequence) differs between members of a species (or between paired chromosomes in an individual). For example, two sequenced DNA fragments from different individuals, AAGCCTA to AAGCTTA, contain a difference in a single nucleotide. In this case we say that there are two alleles: C and T. Almost all common SNPs have only two alleles. For a variation to be considered a SNP, it must occur in at least 1% of the population.

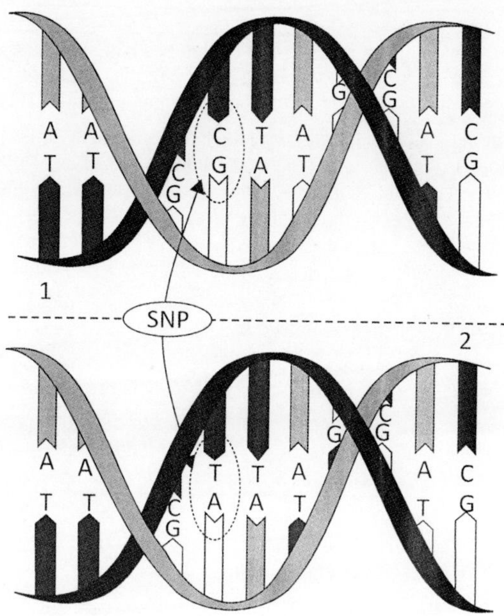

FIG. 15.1: DNA strand 1 differs from DNA strand 2 at a single base-pair location (a C/T polymorphism).

Molecular markers are used in molecular biology and biotechnology experiments where they use to identify a particular sequence of DNA. As the DNA sequences are very highly specific, they can be identified with the help of the known molecular markers which can find out a particular sequence of DNA from a group of unknown.

DNA-based markers are used in cultivar identification, germplasm management, characterization of genetic diversity, population structure of wild relatives, phylogenetic analyses, and marker assisted selection.

Molecular markers and genome mapping are used in detecting new desirable genes in plant improvement specially for tomato, potato, pea, melon, cabbage, etc. The markers help in identifying important quantitative trait loci (QTL). The QTLs affecting salt tolerance in tomato during germination have been identified. As many as 700 tomato markers were known in 1990 (De Verna and Paterson, 1990). DNA fingerprinting helps in identification, characterization and evaluation of genetic uniformity of regenerated new plants of vegetatively propagated fruit crops at cultivar level.

In past few years there is a considerable developments in biotechnology have led plant breeders to develop more efficient selection systems to replace traditional phenotypic-pedigree-based selection systems.

Marker assisted selection (MAS) is the indirect selection process where a trait of interest is selected not based on the trait itself but on a marker linked to it. For example if MAS is being used to select individuals with a disease, the level of disease is not quantified but rather a marker allele which is linked with disease is used to determine disease presence. The assumption is that linked allele associates with the gene and/or quantitative trait locus (QTL) of interest. MAS can be useful for traits that are difficult to measure, exhibit low heritability, and/or are expressed late in development. Following are the types of markers:

Biological	Different pathogen races or insect biotypes based on host pathogen or host parasite interaction can be used as a marker since the genetic constitution of an organism can affect its susceptibility to pathogens or parasites.
Morphological	First markers loci available that have obvious impact on morphology of plant. Genes that affect form, coloration, male sterility or resistance among others have been analyzed in many plant species. Examples of this type of marker may include the presence or absence of awn, leaf sheath coloration, height, grain color, aroma of rice etc. In well-characterized crops like maize, tomato, pea, barley or wheat, tens or even hundreds of such genes have been assigned to different chromosomes.
Biochemical	A gene that encodes a protein that can be extracted and observed; for example, isozymes and storage proteins.
Cytological	The chromosomal banding produced by different stains; for example, G banding.
DNA-based and/or molecular	A unique (DNA sequence), occurring in proximity to the gene or locus of interest, can be identified by a range of molecular techniques such as RFLPs, RAPDs, AFLP, DAF, SCARs, microsatellites etc.

The gene of interest is directly related with production of protein(s) that produce certain phenotypes whereas markers should not influence the trait of interest. They are genetically linked and so go together during segregation of gametes due to the concomitant reduction in homologous recombination between the marker and gene of interest. In many traits genes are discovered and can be directly assayed for their presence with a high level of confidence. However, if a gene is not isolated marker's help is taken to tag a gene of interest. In such case there may be some false positive results due to recombination between marker of interest and gene (or QTL). A perfect marker would elicit no false positive results.

Fundamentals of MAS

All living cells are programmed by genetic material called DNA. This molecule is made up of a long chain of nitrogen-containing bases (there are four different bases – adenine [A], cytosine [C], guanine [G] and thymine [T]). Only a small fraction of the DNA sequence typically makes up genes, i.e. that code for proteins, while the remaining and major share of the DNA represents non-coding sequences, the role of which is not yet clearly understood. The genetic material is organized into sets of chromosomes (e.g. five pairs in *Arabidopsis thaliana*; 30 pairs in *Bos taurus* [cow]), and the entire set is called the genome. In a diploid individual (i.e. where chromosomes are organized in pairs), there are two alleles of every gene – one from each parent.

Molecular markers are not normal genes and do not have any biological effect, however, they can be thought of as constant landmarks in the genome. They are identifiable DNA sequences, found at specific locations of the genome. They are transmitted by the standard laws of inheritance from one generation to the next. They rely on a DNA assay, in contrast to morphological markers that are based on visible traits, and biochemical markers that are based on proteins produced by genes. Following are the types of molecular markers:

1. Restriction fragment length polymorphisms (RFLPs)
2. Random amplified polymorphic DNA (RAPDs) markers
3. Amplified fragment length polymorphisms (AFLPs)
4. Microsatellites and single nucleotide polymorphisms (SNPs)

They differ with their technical requirements each (e.g. whether they can be automated or require use of radioactivity); the amount of time, money and labour needed; the number of genetic markers that can be detected throughout the genome; and the amount of genetic variation found at each marker in a given population.

An ideal marker has following important characteristics:

1. Demonstrates measurable differences in expression between trait types and/or gene of interest alleles, early in the development of the organism.
2. Easy recognition of all possible phenotypes (homo- and heterozygotes) from all different alleles.
3. It has low or null interaction among the markers and allows to use of many at the same time in a segregating population.
4. It has no effect on the trait of interest that varies depending on the allele at the marker loci.
5. They are abundant in number.
6. They are polymorphic.

MAS can be used for manipulating both qualitative and quantitative traits. A highly saturated marker linkage map is necessary for effective marker based selection. Basically, MAS consists of identifying association between molecular markers and genes controlling agronomic traits, and using these to improve lines or populations. Otherwise, based on the heritability of molecular markers, which is essentially 100 %, when the selection is performed for a low heritable trait, it will be more effective and potentially less expensive than phenotypic selection. It is necessary to consider that effectiveness and cost of MAS is greatly influenced by the marker system used, therefore it must be chosen carefully.

Benefits of Molecular Markers

The potential benefits of using markers linked to genes of interest in breeding programmes, thus moving from phenotype based towards genotype-based selection, have been obvious for many decades. However, realization of this potential has been limited by the lack of markers. With the advent of DNA-based genetic markers in the late 1970s, the situation changed and researchers could, for the first time, begin to identify large numbers of markers dispersed throughout the genetic material of any species of interest and use the markers to detect associations with traits of interest, thus allowing MAS finally to become a reality. This led to a whole new field of academic research, including the milestone paper by Paterson *et al.* (1988). This showed that with the availability of large numbers of genetic markers for their species of interest (tomato), the effects and location of marker-linked genes having an impact on a number of quantitative traits (fruit traits in their case) could be estimated using an approach that could be applied to dissect the genetic make-up of any physiological, morphological and behavioural trait in plants and animals.

Demerits of Morphological Markers

Morphological markers have certain demerits for example, the delay of marker expression until late into the development of the organism, dominance, deleterious effects, pleiotropy, confounding effects of genes unrelated to the gene or trait of interest but which also affect the morphological marker (epistasis), rare polymorphism and frequent confounding effects of environmental factors which affect the morphological characteristics of the organism.

These demerits of morphological markers have been rectified by developing the DNA-based markers which are highly polymorphic, simple inheritance, abundantly occur throughout the genome, easy and fast to detect, minimum pleiotropic effect and detection is not dependent on the developmental stage of the organism. Such markers have been mapped to different chromosomes in several crops including rice, wheat, maize, soybean and several others. Those markers have been used in diversity analysis, parentage detection, DNA fingerprinting, and prediction of hybrid performance. Molecular markers are useful in indirect selection processes, enabling manual selection of individuals for further propagation.

The major genes responsible for economically important characteristics are common in various plants which include disease resistance, male sterility, self-incompatibility, others related to shape, color, and architecture of whole plants and are often of mono- or oligogenic in nature. The marker loci which are tightly linked to major genes can be used for selection and are sometimes more efficient than direct selection for the target gene. Such vantages in efficiency may be due for example, to higher expression of the marker mRNA in such cases that the marker is actually a gene. Alternatively, in such cases that the target gene of interest differs between two alleles by a difficult-to-detect single nucleotide polymorphism, an external marker (be it another gene or a polymorphism that is easier to detect, such as a short tandem repeat) may present as the most realistic option.

There are several indications for the use of molecular markers in the selection of a genetic trait. In such situations that:

1. The selected character is expressed late in plant development, like fruit and flower features or adult characters with a juvenile period.
2. The expression of the target gene is recessive.
3. There is requirement for the presence of special conditions in order to invoke expression of the target gene(s), as in the case of breeding for disease and pest resistance (where inoculation with the disease or subjection to pests would otherwise be required). This advantage derives from the errors due to unreliable inoculation methods and the fact that field inoculation with the pathogen is not allowed in many areas for safety reasons. Moreover, problems in the recognition of the environmentally unstable genes can be eluded.
4. The phenotype is affected by two or more unlinked genes (epistatis). For example, selection for multiple genes which provide resistance against diseases or insect pests for gene pyramiding.

TABLE 15.1: Comparison of Molecular Markers for their advantages and disadvantages

Type of markers	Advantages	Disadvantages
Restriction Fragment Length Polymorphism (RFLP)	• High genomic abundance • Co-dominant markers • Highly reproducible • Can use filters many times • Good genome coverage • Can be used across species • No sequence information • Can be used in plants reliably (well-tested) • Needed for map based cloning	• Need large amount of good quality DNA • Laborious (compared to RAPD) • Difficult to automate • Need radioactive labeling • Cloning and characterization of probe are required

Type of markers	Advantages	Disadvantages
Randomly Amplified Polymorphic DNA (RAPD)	• High genomic abundance • Good genome coverage • No sequence information • Ideal for automation • Less amount of DNA (poor DNA acceptable) • No radioactive labeling • Relatively faster	• No probe or primer information • Dominant markers • Not reproducible • Can not be used across species • Not very well-tested
Simple Sequence Repeat (SSR)	• High genomic abundance • Highly reproducible • Fairly good genome coverage • High polymorphism • No radioactive labeling • Easy to automate • Multiple alleles	• Can not be used across species • Need sequence information • Not well-tested
Amplified Fragment Length Polymorphism (AFLP)	• High genomic abundance • High polymorphism • No need for sequence information • Can be used across species • Work with smaller RFLP fragments • Useful in preparing contig maps	• Very tricky due to changes in patterns with respect to materials used • Cannot get consistent map (not reproducible) • Need to have very good primers
Sequence-Tagged Site (STS)	• Useful in preparing contig maps • No radioactive labeling • Fairly good genome coverage • Highly reproducible • Can use filters many times	• Laborious • Cannot detect mutations out of the target sites • Need sequence information • Cloning and characterization of probe are required
ISOZYMES	• Useful for evolutionary studies • Isolation lot easier than that of DNA • Can be used across species • No radioactive labeling • No need for sequence information	• Laborious • Limited in polymorphism • Expensive (each system is unique) • Have to know the location of the tissue • Not easily automated

Budak et al., KSÜ Fen ve Mühendislik Dergisi, KSU Journal of Science and Engineering 7(1)-2004.

Steps for MAS

At the initial stage map the gene or quantitative trait locus (QTL) of interest by using different techniques and then use this information for marker assisted selection. It is essential that markers to be used should be close to gene of interest (<5 recombination unit or cM) in order to ensure that only minor fraction of the selected individuals will be recombinants. Two markers are used in order to reduce the chances of an error due to homologous recombination.

QTL Mapping Techniques

Bi-parental cross populations method is used for QTL mapping in plants. A cross between two parents which have a contrasting phenotype for the trait of interest are developed. The used populations are recombinant inbred lines (RILs), doubled haploids (DH), back cross and F_2. Linkage between the phenotype and markers which have already been mapped is tested in these populations in order to determine the position of the QTL. Such techniques are based on linkage therefore called **linkage mapping**.

Single Step MAS and QTL Mapping

In contrast to two-step QTL mapping and MAS, a single-step method for breeding typical plant populations has been developed. In the first step few breeding cycles, markers linked to the trait of interest are identified by QTL mapping and later the same information in used in the same population. In this approach, pedigree structure are created from families that are created by crossing number of parents (in three-way or four way crosses). Both phenotyping and genotyping is done using molecular markers mapped the possible location of QTL of interest. This will identify markers and their favorable alleles. Once these favorable marker alleles are identified, the frequency of such alleles will be increased and response to marker assisted selection is estimated. Marker allele(s) with desirable effect will be further used in next selection cycle or other experiments.

High-Throughput Genotyping Techniques

Recently high-throughput genotyping techniques are developed which allows marker aided screening of many genotypes. This will help breeders in shifting traditional breeding to marker aided selection. One of example of such automation is using DNA isolation robots, capillary electrophoresis and pipetting robots. One of recent example of capillary system is Applied Biosystems 3130 Genetic Analyzer. This is the latest generation of 4-capillary electrophoresis instruments for the low to medium throughput laboratories.

Use of Mas for Backcross Breeding

A minimum of five or six-backcross generations are required to transfer a gene of interest from a donor (may not be adapted) to a recipient (recurrent – adapted cultivar). The recovery of the recurrent genotype can be accelerated with the use of molecular markers. If the F_1 is heterozygous for the marker locus, individuals with the recurrent parent allele(s) at the marker locus in first or subsequent backcross generations will also carry a chromosome tagged by the marker.

Marker Assisted Gene Pyramiding

Gene pyramiding has been proposed and applied to enhance resistance to disease and insects by selecting for two or more than two genes at a time. For example in rice such pyramids have been developed against bacterial blight and blast. The advantage of use of markers in this case allows to select for QTL-allele-linked markers that have same phenotypic effect.

MAS has also been proved useful for livestock improvement.

APPLICATIONS OF MAS IN AGRICULTURE

Successful release of new and better crop varieties increasingly requires genomics and molecular biology. The main reasons supporting the utilization of molecular markers in breeding programs are the heritability of the markers (100 %), and their cost, which is potentially lower than the conventional selection. Additionally, the perspective proposed uses for the molecular markers into plant improvement are their application onto selection for traits with low heritability, identification of resistance or tolerance for biotic or abiotic stresses in plants, and the gene introgression coming from native or exotic germplasm. Other uses include estimation of genetic relatedness among accessions, cultivar description and the identification of quantitative trait loci (QTL) linked to molecular markers, which control important agronomic traits.

Previous efforts to develop large numbers of molecular markers, high density genetic maps, and appropriately structured mapping populations have now made routine for many crop species the ability to simultaneously define gene action and breeding value at hundreds and often thousands of loci distributed relatively uniformly across entire genomes. The results from such mapping studies provide greatly improved estimates for the number of loci, allelic effects, and gene action controlling traits of interest. More importantly, genomic segments can be readily identified that show statistically significant associations with quantitative traits (quantitative trait loci [QTLs]). In addition to genetic mapping in families derived from biparental crosses, new advances in association genetics with candidate genes and approaches that combine linkage disequilibrium analysis in families and populations further enhance power for QTL discovery.

Information about QTLs can be used in a number of ways to increase heritability and favorable gene action. For traits exhibiting low to moderate heritability, such as grain yield, QTLs, and their associated molecular markers often account for a greater proportion of the additive genetic effects than the phenotype alone. Furthermore, knowledge of genetic architecture can be exploited to add or delete specific alleles that contribute to breeding value. When either genetic linkage or epistasis among loci with antagonistic effects on a trait limits genetic gain, QTL information can be used to break these undesirable allelic relationships (Moose, S. P. and Mann, Rita H., Plant Physiology 147: 969-977, 2008).

Success in using information about QTLs to increase genetic gain depends greatly on the magnitude of QTL effects, precise estimation of QTL positions, stability of QTL effects across multiple environments, and whether QTLs are robust across relevant breeding germplasm. Prediction of QTL positions is enhanced by further fine mapping, which facilitates testing QTL effects and breeding values in additional populations. When the density of observed recombinations approaches the resolution of single genes, the causal genetic change for a QTL can be determined. Molecular isolation of QTLs permits the development of perfect or functional molecular markers at the potential resolution of the fundamental unit of inheritance, the nucleotide, and dramatically increases the specificity and precision by which genetic effects are estimated and manipulated in breeding programs (Moose, S. P. and Mann, Rita H., 2008).

The use of transgenes can further simplify the genetic architecture for desirable traits, in ways that may be superior to or not possible even when perfect markers are available for robust QTLs of large effect. Transgenes typically condition strong genetic effects at operationally single loci, which also exhibit dominant gene action where only one copy of the event is needed for maximal trait expression in a hybrid cultivar. These features of transgenes can reduce complex quantitative improvement to a straightforward, often dramatic, solution. Excellent examples are provided by the expression in transgenic corn hybrids of insecticidal toxin proteins from *Bacillus thuringiensis* (*Bt*) to reduce feeding damage by larvae of the European corn borer (*Ostrinia nubilalis*) or the corn rootworm beetle (*Diabrotica* spp.). Partial resistance in maize germplasm to these insect pests had been previously characterized as quantitatively inherited traits with low heritability, but the *Bt* transgenic events offer a simply inherited alternative that is efficiently manipulated in breeding programs (Moose, S. P. and Mann, Rita H., 2008).

Molecular markers and more recently, high-throughput genome sequencing efforts, have dramatically increased knowledge of and ability to characterize genetic diversity in the germplasm pool for essentially any crop species. Using maize as one example, surveys of molecular marker alleles and nucleotide sequence variation have provided basic information about genetic diversity before and after domestication from its wild ancestor teosinte, among geographically distributed landraces, and within historically elite germplasm. This information enriches investigations of plant evolution and comparative genomics, contributes to our understanding of population structure, provides empirical measures of genetic responses to selection, and also serves to identify and maintain reservoirs of genetic variability for future mining of beneficial alleles. In addition, knowledge of genetic relationships among germplasm sources may guide choice of parents for production of hybrids or improved populations (Moose, S. P. and Mann, Rita H., 2008).

While molecular markers and other genomic applications have been highly successful in characterizing existing genetic variation within species, plant biotechnology generates new genetic diversity that often extends beyond species boundaries. Biotechnology enables access to genes heretofore not available through crossing and creates an essentially infinite pool of novel genetic variation. Genes may be acquired from existing genomes spanning all kingdoms of life, or designed and assembled de novo in the laboratory. Both subtle and extreme examples of the power of transgenes to introduce novel phenotypic variation can be found in the three different transgenes developed for resistance to glyphosate herbicides in maize and other crops. The first glyphosate-tolerant maize hybrids used a modified version of the endogenous maize gene encoding 5-enol-pyruvylshikimate-3-P synthase, which was followed later by events produced with a 5-enol-pyruvylshikimate-3-P synthase gene isolated from *Agrobacterium*. More recently, a synthetic gene with enhanced glyphosate acetyltransferase activity was created via gene shuffling and selection in a microbial system. Each of these glyphosate-tolerant maize events also illustrates another benefit of biotechnology, where new combinations of regulatory sequences (e.g. the cauliflower mosaic virus 35S and rice *actin1* promoters) may be used to achieve optimal trait expression with respect to overall activity and tissue distribution relative to what might be possible with endogenous genes.

Marker-assisted selection has enhanced genetic gain for traits where the phenotype is difficult to evaluate because of its expense or its dependence on specific environmental conditions. Molecular markers may be used to increase the probability of identifying truly superior genotypes, by focusing testing resources on genotypes with the greatest potential (i.e. early elimination of inferior genotypes), by decreasing the number of progeny that must be screened to recover a given level of gain, and by enabling simultaneous improvement for traits that are negatively correlated. Successful examples include resistance to soybean cyst nematode, resistance to cereal diseases, and drought tolerance in maize.

The efficiency of phenotypic selection for some complex traits can be enhanced by including physiological or biochemical phenotypes as secondary traits, if these exhibit strong genetic correlations with the target trait and possess high heritability. Recent advances in functional genomics permit the population-scale profiling of RNA abundance, protein levels and activities, and metabolites that are associated with important traits. In addition to molecular markers that tag DNA sequence variation, such genetical genomics approaches may provide additional secondary phenotypes as selection targets, particularly for traits defined by responses to environmental, developmental, or physiological cues.

Marker-assisted selection also accelerates the deployment of transgenes in commercial cultivars. Typically, this has been achieved through marker-assisted backcrossing. However, for future biotechnology improvements such as tolerance to drought or nutrient limitation, forward breeding may be required to cooptimize transgene expression and genetic background because endogenous genes and environmental factors may have the potential to influence the phenotypes resulting from transgenic modifications. Of course, use of molecular markers could aid forward breeding efforts as well. Alternatively, discovery efforts for additional genes or QTLs that are necessary for dependable trait performance may suggest design of new transgene constructs that stack primary transgenes with known genetic modifiers into second-generation transgenic events (Moose, S. P. and Mann, Rita H., 2008).

CHAPTER 16

Gynogenesis in Hi-Tech Horticulture

Gynogenesis or **haploid parthenogenesis** is frequently used terms which describe the possibility of obtaining haploid plants derived from female gametophyte. Spontaneous induction of haploid plants *in situ* have been known since 1922, when Blakeslee described this phenomenon in *Datura*. Spontaneous haploid induction is today regarded as a rare event of limited practical use. The main progress in induced haploid production has been made since 1964, when Guha and Maheshwari described the first successful *in vitro* haploid induction by means of anther culture.

As reviewed by Yang and Zhou (1982) as a result of this finding, studies of haploid induction from female gametophyte were neglected for about a decade. Later, gynogenesis was again frequently studied in several agricultural plants, and successful induction of gynogenic haploids has currently been reported in several species or genera. In most cases, gynogenesis is used in plants where androgenesis has failed. In some other studies gynogenesis is favoured over androgenesis on the basis of its higher genetic stability. Such, for instance, has been well studied in tobacco but little evidence exists to support the conclusion that this phenomenon is expressed also in other species.

Methods of inducing haploid plants via gynogenesis can be divided into two categories. The first is *in situ* induction by irregular pollination, which can be assisted by *in vitro* rescue of haploid embryos, and the second is *in vitro* cultivation of unpollinated female reproductive organs. Irregular pollination methods can be divided into pollination with pollen of the same species with or without selected haploid inducer genes, to pollination with pollen from wild relatives or other genera, and to pollination with irradiated pollen of the same species. *In vitro* culture methods include the culture of unpollinated ovules, placenta attached ovules, ovaries or whole flower buds.

The number of reports on gynogenesis differs among species. Only on a few do a relatively large number of published papers exist, while in others, only one or very few references could be found. In this paper different techniques are analysed on a limited number of better studied plant species.

GYNOGENESIS INDUCED BY IRREGULAR POLLINATION IN SITU

Induction of parthenogenesis *in situ* is mainly limited to the use of irregular pollination. There is only limited evidence that a chemical toluidine–blue can be used as a haploid inducer. A positive effect of toluidine–blue has been described for Populus but was not successful in *Vitis* and apple. All other efficient approaches use pollen as an inducer of haploidy.

The first possibility is to use the pollen of the same species, so as to provide an actually legitimate pollination, but the genotype of the pollen parent includes specific genes which may cause an increased percentage of haploid embryo development. The mechanism of induction is not yet completely analyzed but it has been suggested that maternal haploids result after the failure of fertilization, and the consequent development of a reduced egg cell into the embryo. In some species, parthenogenic seeds can be recognized as twin seeds, where one is haploid and another normal diploid. There are also reports of haploids being detected on the basis of the morphology of the seedlings, in pear for instance described haploid genotypes as smaller, having shorter internodes, slender stem and narrow thin leaves.

TABLE 16.1: Induction of haploid plants induced by legitimate of illegitimate pollination, including irradiated pollen

Plant species	Pollinator and some experimental characteristics
Zea mays	maize Stock 6 pollen, on plant seed maturation
Triticum aestivum	maize pollen, chrom. eliminat. on plant seed maturation
	maize pollen, spikelet culture 2 days after pollination
	Maize pollen, 2, 4–D injection, embryo rescue after weeks 2
Hordeum vulgare	Hordeum bulbosum crosses
	H. bulbosum, GA_3 spraying, embryo rescue after 14 days
Gossypium hirsutum	Selected cotton lines, on plant seed maturation
Nicotiana sp.	Irrad. pollen (>200–5— Gy) on plant seed maturation
	irrad. pollen 180 Gy on plant seed maturation
	N. africana, on plant seed maturation
Malus x domestica	embryo rescue after 8 days
	irrad. pollen (500 Gy), embryo rescue
	after 3 months, cold treatment same as above
Pyrus communis	legitimate pollin and visual selec. or use of irrad. pollen (500 Gy)
Cucumis sativus	irrad. pollen (400–600 Gy), embryo rescue after 2–6 weeks
Cucumis melo	irrad. pollen (300, 1000 Gy), embryo rescue after 5–9 days
	seasonal effects, summer opt.
	irrad. pollen (1000 Gy, X–ray) embryo rescue after 3 weeks
Allium cepa	irrad. pollen (150 Gy) in situ seed maturation
Petunia hybrida	irrad pollen (>600 Gy) embryo rescue after 2 weeks
	irrad. pollen (500–700 Gy) embryo rescue after 2 weeks
Brassica oleracea	irrad. pollen (800–1200 Gy) on plant maturation or in vitro rescue
Solanum tubersum	Solanum phureja crosses

Maize is a typical example of haploid induction as result of legitimate crossings. Several approaches have been tested to induce a higher percentage of haploid seeds in maize. The discovery of a genetic strain 'Stock 6' by Coe (1959) was the most important, since it was found that this line produced 3.2% haploid seeds upon selling. This was even improved by hybridization and further selection. Lashermes and Beckert (1988) reported that 2–5% maternal haploids were obtained when crossed by line WS14 (hybrid between W23 x Stock 6). One of the main problems of using this approach is distinguishing haploid seedlings from true hybrids. This can be done by the selection of female parents, which should possess homozygous recessive gene(s) which are not present in the pollinator line, so converting to dominant type can be recognized easily (ligule less or glossy characters lg_1, lg_2, gl_1). Phenotypes with the traits of the pollen parent are true hybrids, others are haploids. Evaluations of agronomic characteristics of doubled haploid lines produced by this method in comparison with pedigree inbred lines have shown (Lashermes *et al.* 1988) that there was no significant selection of haploid lines according to agronomic characters during this process, so it was suggested that such lines can be successfully included in breeding programmes. It should be noted that no *in vitro* culture is needed for this approach, so the only chemical used is colchicine or other genome doubling agent.

The second possible approach is pollination of the studied species with wild relatives or with other genera of cultivated crops. This technique mainly requires *in vitro* embryo rescue to ensure the survival of induced haploid embryos. We shall analyse the specificity of this technique on the examples of wheat, barley and potato.

Potato is another plant species in which haploids are induced by means of wide crosses, only the mechanism of induction is different. It was recognized that crosses among commercial tetraploid varieties and the wild diploid relative *Solanum phureja* used as male parent can result in reducing the ploidy of tetraploid *Solanum tuberosum* to diploid level. In some genotypes of *Solanum phureja*, irregular meiosis occurred, resulting in only one sperm nucleus, which can therefore fuse with the egg nucleus or embryo sac nucleus, resulting in up to 40% diploids. Such diploid potatoes can be used for breding purposes, including alien germplasm introgression or selection on diploid level, but such plants are not homozygous. There is some recent evidence that mechanism of elimination is more complicated and that several *Solanum phureja* genes can be included in regenerated dihaploids. In potato, further reduction to monohaploid level is mainly achieved by androgenesis, in which highly genotype dependant responses have been reported (Meyer *et al.*, 1993). Culture of unfertilized ovules have also been reported (Table 16.1) but it seems that this method has not attracted further attention.

A the third possible approach is the use of irradiated pollen of the same species for provoking haploid embryo development, which can be further grown on plant until maturity, or more often be recovered by *in vitro* culture. Hertwig (1911) first noticed this phenomenon studying irradiated frog spermatozoa. He found that lower doses caused high amount of embryonic death or abnormalities while fertilisation with spermatozoa irradiated with higher doses resulted in apparently normal offspring. It was suggested that the penetration of the egg by highly damaged spermatozoa stimulated haploid egg development, to form a haploid organism. The same Hertwig Effect is also efficient in plants, as has already been found in many species examples in apple and melon.

Haploid induction in apple by the use of irradiated pollen was studied by Huang and Lespinasse. Their method, as published in several articles uses the following procedure. Pollen should be collected from a donor plant (clone RR) that carries a homozygous dominant marker gene coding for anthocyanin production in whole plant organs.

Gynogenic production has been studied in melon by several groups. Sauton and Dumas de Vaulx (1987) used genetic markers present in female parents as homozygous recessive for *gl* (smooth) and *nsv* (resistance to muskmelon necrotic virus) genes and homozygous dominant in male parents. Pollen was irradiated at doses from 100 to 1500 Gy, while later, only a 1000 Gy dose was used. The ovaries were extracted 5–9 days after pollination and cultured *in vitro*. Only a 100 Gy dose resulted in hybrid plants, others were haploid. Increasing the amount of pollen applied improved haploid induction. Sauton (1988) found that a season of inoculation has an effect on gynogenic haploid production– donor plants grown during the summer months being superior to spring or autumn grown. Katoh *et al.* (1993) used the same dose (only they used X rays instead of gamma rays) and was successful in all 12 cultivars studied, with 2.5% being the highest induction percentage. Chromosome analysis of root tips in comparison to flow cytometry analysis showed that spontaneous chromosome doubling occurs on root tips but not in other plant parts.

DEVELOPMENT OF USEFUL CROP VARIETIES

Even when successful, interspecific and intergeneric transfers made by sexual methods are laborious and time consuming moreover, other problems must be solved before a genotype useful for crop production is obtained. Even after 6 backcross generations in an interspecific gene transfer, the natural process of recombination frequently will not separate tightly linked genes. This means that undesired traits that affect crop quality, yield, or adaptation may be carried along with the desired gene. Even if the desired gene is successfully separated from linked deleterious genes, its inheritance and expression may be unpredictably altered in the new genetic background. Le Deneuf and Sauton (1994) claimed that irradiated pollen induces stenospermocarpy (fruit development after embryo abortion) in melon, but they cannot clearly determine whether early embryogenesis is gynogenic or abnormally zygotic. In any case, the efficiency of haploid induction using this method is high enough to be routinely used in plant breeding programmes (Dore, 1994).

GYNOGENESIS INDUCED BY IN VITRO CULTURE OF UNPOLLINATED FEMALE REPRODUCTIVE ORGANS

Cultures of unpollinated ovules, ovaries or flower buds have been successfully used to induce haploid plants in several crops. The first successful induction was performed by San Noeum 1976 in barley. This technique was subsequently efficient in several other species In comparison to gynogenesis induced by the use of irregular pollinations, *in vitro* culture of unpollinated female reproductive organs has the advantage that no hybrid plants can develop from legitimate pollination but regenerants can be induced from other sporophytic tissues, so confirming the haploid origin is still required. For the induction of unpollinated female reproductive organs, the most important elements are the composition of media and the genotype used.

Allium cepa and *Beta vulgaris* are the most studied plant species, in which gynogenesis is induced by *in vitro* culture of unpollinated female reproductive organs.

Beet cultivars are mainly triploid, resulting from crosses of tetraploid and diploid lines. Since beets are allogamous with expressed high inbreeding depression, the resulting hybrids are relatively heterogeneous. Approaches have therefore been tested to produce gynogenic haploid plants induced from ovule cultures. Hosemans and Bossoutrot (1983) succeeded in regenerating the first 17 plants from cultured unpollinated ovules. The basal medium was MS, several growth hormones were tested, and sucrose ranged from 2 to 8%. In a subsequent study, the same authors (Bossoutrot and Hosemans 1985) tested three other basic media supplemented with 1 mg l^{-1} BA and 3% or 8% sucrose. As a result, they achieved 0.2% regeneration of gynogenic plants.

Van Geyt *et al.* (1987) succeeded in producing up to 6.1% gynogenic plants. They used MS medium supplemented with various concentrations of BA, NAA and 2, 4–D and 3% sucrose. Substantial improvement was achieved when the induced calli were removed from the developing ovule and transferred to a medium without 2, 4–D. Ovary culture resulted in a much higher frequency of regeneration then ovule culture. Doctrinal *et al.* (1989) tested several media and culture conditions for ovule cultures and found the best to the N6 medium supplemented with IAA and BA, or

KIN. They also used a two step induction process, in which sucrose was higher (6 or 8%) in the first induction medium and lower in the second (3 or 6%). They noticed a seasonal effect and the influence of genotype, which resulted in 6 to 10% of induction frequency.

TABLE 16.2: Haploid induction from unpollinated ovules, ovaries or flower buds

Plant species	Organ cultured and some experimental characteristics
Zea mays	ovary culture, MS and N6 media
Triticum aestivum	ovary culture, N6 medium
	ovary culture, N6 medium
Hordeum vulgare	ovary culture, Miller's medium
	ovary culture, Miller's medium
	ovary culture, MCPA with cytokinin
Oryza sativa	ovary culture, Miller's medium
	unhusked flowers, N6 medium, MCPA
Fagopyrum esculentum	ovule culture, B5 medium, BAP or 2iP
Nicotiana tabacum	ovary culture, H medium
Cucurbita pepo	ovule culture, 1 day before anthesis
Solanum tuberosum	ovary culture, uninucleate stage of pollen
Beta vulgaris	ovule culture
	ovule culture, preselection of fertile. ovules
	ovule and ovary culture, two step proc.
	ovule culture, two step cult. proc.
	ovule culture, dark – light induction
	ovule culture, 5 days cold treatment
	ovule culture, genotype dependent response
	ovule culture, embryogenesis
Helianthus annuus	ovule culture, N6 MCPA
	ovule culture, MS, NAA
	ovule and ovary culture, MS medium, NAA
Allium cepa	ovule culture, flower precultue
	ovary culture MS, 2, 4–D, 3–5 days before anthesis
	ovule, ovary, flower c. BDS medium
	ovule culture, MS medium
	ovule, ovary, flower c. MS, B5, BDS
	ovule and ovary culture, effect of TDZ
Allium porrum	ovule culture, CM2 medium
Allium tuberosum	ovule culture
Petunia hybrida	placenta att. ovules Nitsch medium no effect of in vitro pollination
Gerbera jamesonii	ovule culture
	ovule culture, MS, BA, IAA, Ga_3
	ovule culture, two step cult. proc.
Morus indica	ovary culture, no acclimatized plants

Generalization of successful approaches for efficient *in vitro* induction from unfertilized female reproductive organs is difficult, since results differ even within studies of the same species. Choice of macro and micro elements is very different, practically all media, which are frequently used in tissue culture procedution have been found to be successful for induction of gynogenesis. Similarly, hormonal requirements differ even among genotypes of the same species. Like in androgenesis gynogenic plants can be induced in very low frequencies even on no hormone media, but generally hormones are needed for induction step and sometimes (in lower doses) also in regeneration step. In most cases auxins (especially 2, 4–D) are efficient in the first induction period, but also other hormones can have the same inducing effect. Requirements for carbohydrates are very important what is again similar to androgenesis, but beneficial effect is mainly limited to higher dose of sucrose, while other carbohydrates like maltose were not so efficient. The same is true for cold pretreatments, where attempts to use it for improved haploid induction have almost always been unsuccessful. Embryo sac should in most cases be in mature developmental stage, very few exceptions are reported. This situation is therefore also different from androgenesis, where immature pollen grains are used.

ORIGIN OF GYNOGENIC REGENERANTS AND METHODS OF CHROMOSOME DOUBLING

The origin of gynogenic regenerants has been studied in several species. It seems that the main source of haploid embryogenesis s the egg cell. Such development has been confirmed in sunflower with the additional observation that also other reduced cells such as the antipodals, can divide but no organized structure occurred in this case. The egg cell, synergids and antipodal cells of sunflower could be induced to undergo sporophytic development. Similarly in *Beta* viable gynogenic embryos originated only from the egg cell. It seems that the main problem in studies of gynogenic origin is that a high frequency of induction should first be achieved to enable the detection of dividing cells and their developmental pathway.

General to all gynogenic approaches is a need to double the genome from a haploid to doubled haploid state. Plant species differ in endogenous potential for chromosome doubling. While in some plants like apple and pear, spontaneous chromosome doubling occurs very frequently, in others like onion, beet, barley, wheat, maize, melon and tobacco, spontaneous chromosome doubling is mainly limited to 10% or less. Several approaches have been used for induced chromosome doubling. Colchicine has been far the most widely used chemical. In addition to previously known applications of colchicine on plants *in situ*, several techniques exist for applying colchicine *in vitro*. Colchicine is for the most part included in the media, where concentration and duration of treatments differ. Xu *et al.* (1991), for instance, found a 0.05% to 0.5% colchicine supplemented medium with additional vibration effect to be optimal for doubling haploid sugarbeet shoots, while Hansen *et al.* (1994) found the exposure of beet ovules to 0.4% colchicine for 5 hours in liquid medium to be optimal. Since colchicine is known to have a potential mutagenic effect, other chemical agents have been tested for efficiency for genome doubling and in relation to genetic stability. While Oryzaline has been recognized as a stronger inhibitor of spindle formation than colchicine (Verhoeven *et al.*, 1990) and has been efficiently used in doubling chromosomes of several species, the use of another herbicide trifluralin has only been recently proposed (Eikenberry, 1994) as being highly efficient in canola.

CONCLUSIONS

It has been shown that gynogenesis is already an efficient tool for induction of haploid plants. Technically, approaches differ more than those applied for androgenesis. In some plants, such as sugarbeet, onion, sunflower, melon and others, gynogenesis is already used in plant breeding programs, while for some others induction frequencies should first be improved. Gynogenesis should always be used to produce (di) haploid lines in dioecious plants like asparagus or hop, where hybrids are crosses among male and female plants. In asparagus, such system has already been applied for breeding new varieties while success in induction of gynogenesis in hop is at the moment still limited

In future, experiments should, therefore, be focused on finding the most appropriate low labour intensive method that would yield the highest number of (di) haploid lines regardless of genotype. The genetic stability of gynogenic lines should also be evaluated including comparisons among androgenic and gynogenic origin of haploids.

CHAPTER 17

Synseeds or Synthetic Seeds for Hi-Tech Horticulture

THE NATURAL SEED

The seed stage of seed plants represents a unique developmental phase of the spermatophyte life–cycle, and as such involves structures, not characteristic of other stages of development. The essential structure of seed is defined as a *ripened ovule consisting of an embryo and its coats*. Anatomically a seed consists of some old or parental sporophyte tissue viz. the seed coats, which are derived from the integument's and nucleus: some endosperm, which may be either gametophytic tissue or fertilized tissue: and the embryo, the new young sporophyte. The normal seed contains materials, which it utilizes during the process of its germination. These substances are frequently found in the endosperm. Thus endosperm may contain variety of stored materials such as starch, oils, proteins etc. In some plants, however, the reserve food material is present in cotyledons.

DEVELOPMENT OF THE CONCEPTS OF TISSUE CULTURE AND ARTIFICIAL SEEDS

P. R. White is acknowledged as the father of tissue culture in the United States. He was the first to grow excised root tips of tomatoes (*Lycopersicon*) in continuous culture. When new material is started in culture, grown *in vitro* (literally, in glass), it develops very small juvenile shoots, which are reminiscent of seedlings. A plantlet continues to produce and maintain small stems and leaves throughout its duration in culture. This is fortunate because most mature material would be too unwieldy for micropropagation to succeed in a test tube. After multiplication in culture and when transferred to soil outside the laboratory, the plantlets will produce leaves of normal size and assume the mature features of the plants from which they originated.

When plants are multiplied vegetatively (as distinguished from those grown from seeds) whether by tissue culture or by cuttings all the offspring from a single plant can be classified as a clone. This means that the genetic make–up of each offspring is identical to that of all the other offspring and to that of the single parent. On the other hand, plants propagated by seed, resulting from sexual reproduction, are not clones because each seed (and the resultant plant) has a unique genetic make–up–a mixture from 2 parents, different from either parent and different from one seed to another. The term *cloning*, with respect to tissue culture, refers to the process of propagating in culture large numbers of selected plants with the same genotype (the same genes or hereditary factors) as their respective parent plant.

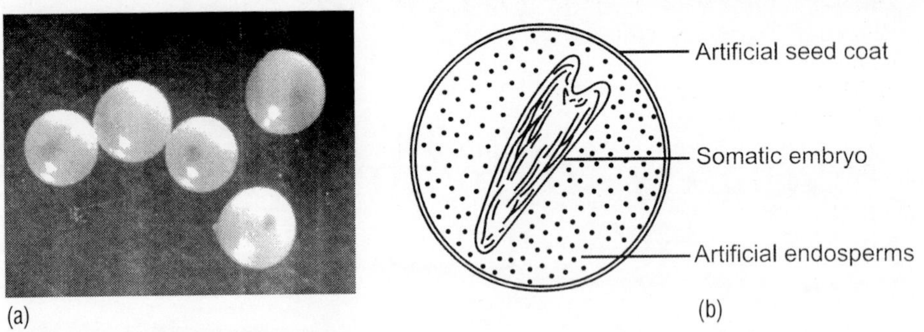

FIG. 17.1: (a) photograph of synthetic seed. (b) Capsule gel with hydrophobic membrane.

The liquid cell suspension cultures have particular significance for the mass production of cells. One common source of cells for cell suspension is from friable callus, although specific cells, such as from leaf mesophyll (the thin, soft tissue between the upper and lower epidermis of the leaf), are also grown in suspension. Cells in suspension can form embryoids (somatic embryos) in the process of embryogenesis. Embryos may multiply and/or be induced to form plantlets in the process of morphogenesis. Many hybrid plants produce embryos that do not mature to viable seeds. These embryos can be rescued, removed from the seed at immature stage, and then grown in culture.

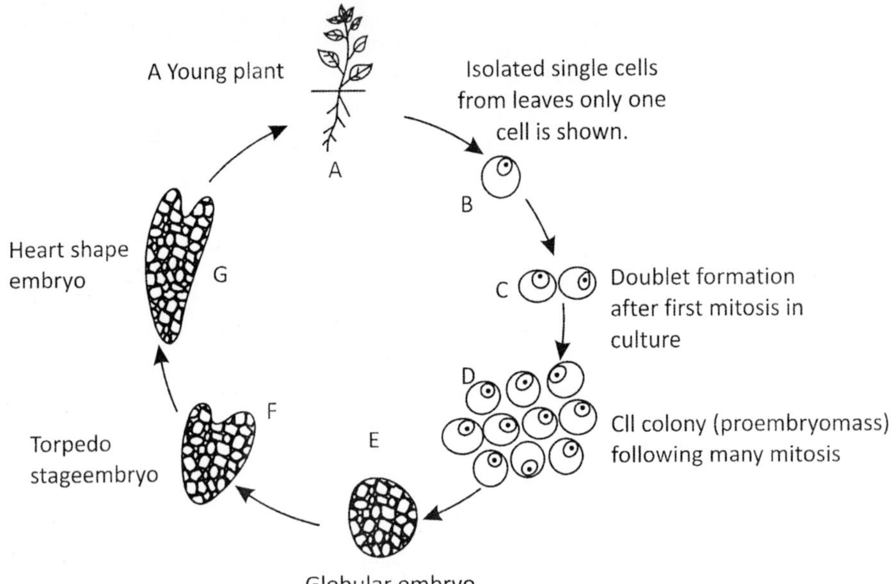

FIG. 17.2: Different stages of somatic embryogenesis.

Suspension cultures have been enhanced by new methods, which can continuously introduce fresh medium into the suspension culture, thereby, enabling the production of thousands of cells or embryos in a single container with a minimum of manual transfer. This is one way that tissue culture can compete with the plentiful seed production in nature.

Interest in anther and pollen culture– the tissue culturing of anthers or pollen to obtain haploid (cells with half the normal number of chromosomes of vegetative cells) clones– is spurred by the practical applications of such haploid cultures. Haploid (n) plants are sterile, but if the chromosomes duplicate, either spontaneously or by induction, the plants will be diploid (2n, which is normal for the vegetative state), and their progeny will be true to form. Considering the fact that it takes several generations of inbreeding to obtain a pure line by conventional means, it is little wonder that plant breeders are interested in anther culture.

DISCOVERY OF SYNTHETIC SEEDS

The origin of the idea of an artificial seed is difficult to determine. Certainly, those who first produced somatic embryos may have considered such an application (Steward, *et al.*, 1958 and Reinert, 1958). The discovery of somatic embryogenesis in carrot in the year 1958 almost simultaneously by F. C. Steward (USA) and J. Reinert (Germany). F. C. Steward, a renowned plant physiologist, at Cornell University in New York, was so impressed by the dramatic effects of coconut milk in carrot culture media that he set beside his other objectives in order to dedicate himself to the study of growth factors in this and other liquid endosperms. Among the active materials he extracted from the coconut milk were several ingredients that are now commonly included in purified form in many tissue culture media. Coconut milk is still used in some orchid culture media. They provided a new way of propagating the plant species.

Later, S. Guha and S. C. Maheshwari (University of Delhi, Delhi, India) in 1964 discovered the formation of pollen embryos from cultured anthers of wild *Datura innoxia*.

However, it was not until the early 1970's that the concept of using somatic embryos began to be presented as a potential propagation system for seed–sown crops. Toshio Murashige gave a number of seminars on tissue culture propagation where he concluded with this concept. For a period of time he conducted research in his laboratory that was focused on the developmental physiology of somatic embryos which he felt to be the limiting factor for large scale propagation. He formally presented his ideas on artificial seeds at the Symposium on Tissue Culture for Horticultural Purposes in Ghent, Belgium, September 6–9, 1977. His terse comments in the proceedings, however, were to be applicable, *the cloning method must be extremely rapid, capable of generating several million plants daily, and competitive economically with the seed method* (Murashige, 1977).

During the mid 1970's, two separate research groups began work on somatic embryogenesis for crop propagation. Keith Walker, then at Monsanto Company, directed a group of scientists that identified basic concepts of delivery of cloned, agricultural crops. Since the focus was to develop thrifty somatic embryo systems that would recapitulate zygotic embryogenesis, their choice was the advanced system developed for *Medicago sativa* L. (alfalfa) using a line (Regen S) identified by Bingham, *et al.* (1975). Soybean and vegetable crops were also of interest to them. Walker cited two reports that had a strong impact on their thinking about the use of somatic embryos for crop propagation. Early in 1980, Walker moved to Plant Genetics, Inc. where Redenbaugh, *et al.* (1984 and 1986) discovered that hydrogels such as sodium alginate which could be used to produce single embryo artificial seeds. In a few experiments, the artificial seeds were planted in the greenhouse with plant production (7% for alfalfa and 10% for celery).

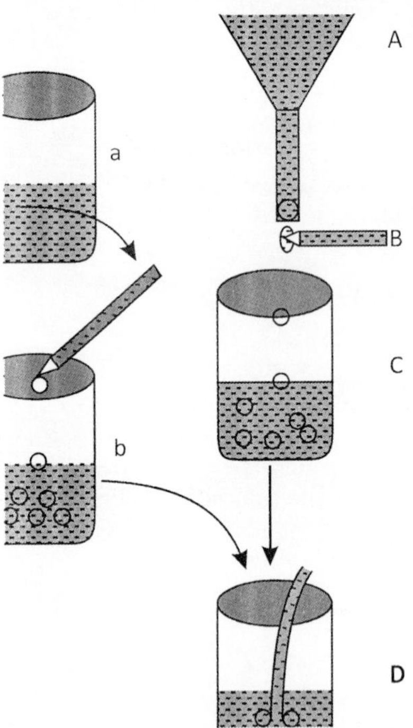

FIG. 17.3: Preparation of artificial seeds via encapsulation with calcium alginate beads using a dropping method. Alginate (2%) is either mixed with somatic embryos (a) or poured in a separatory funnel (A), dropping of alginate beads along with embryo into a bath of calcium nitrate (100 mM) solution (b) or a single embryo is inserted into alignate drop using a plastic pipet having 4 mm inside diameter (B), falling of alginate beads into a bath of calcium nitrate (C) and washing of the capsules in water (D). Diagram modified from Redembaugh et al., 1991.

FIG. 17.4: Encapsulation of shoot tips of cardamom. A – multiple shoot cultures. B – Shoot tips encapsulated in 3% sodium alginate matrix, C – emerging shoot roots from the encapsulated shoot tips and D– plantlets derived from encapsulated shoot tips paper cups. (After Rao et. al, 1997).

Street (1977) advocated the problem of reliability in embryogenesis. According to him morphogenic competence is determined from the time of culture initiation, such that there is the need to have an initiation medium that will ensure that the competent cells are involved in callus formation. Sunderland (1977) demonstrated that the production of hundreds of morphologically uniform embryos from *Datura* and *Nicotiana* pollen.

Robert Lawrence (of Union Carbide) started to develop various methods for cloning forest trees. It was difficult for him to produce hybrids for crops such as celery and lettuce. This group focused on delivery of somatic embryos using fluid drilling technology (Lawrence, 1981) and using polyoxyethylene to form seed tapes or sheets. Lawrence and Walker's groups came together at a symposium workshop, Advances in Methods of *In Vitro* Cloning for Large Scale Propagation of Plants, held September 21–22, 1981 at the W. Alton Jones Cell Science Center in Lake Placid, New York. They discuss the various concepts like how low cost, high–volume propagation system can be developed for vegetable and agronomic crops using somatic embryos and delivered by fluid drilling (in a seed tape, or as an artificial seed). The Lawrence group (then at Agrigenetics) subsequently began to use alginate for encapsulating carrot and celery somatic embryos (Lutz, *et al.*, 1985). Because their encapsulation results with alginate were similar to those at Plant Genetics, they focused on the problems of somatic embryo physiology. In this way, they were successful in obtaining germination of carrot somatic embryos in vermiculite in a growth chamber.

Drew (1979) was active in developing methods to commercially propagate crops using somatic embryos. He suggested delivering carrot somatic embryos in a fluid drilling system, but was able to produce only three plants from carrot embryos on a carbohydrate free medium. He could not get success in producing many plants through this system. He faced a crucial problem and found the very slow rate of development of plantlets derived from culture.

Kitto and Janick (1982) coated clumps of carrot embryos, roots, and callus with polyoxyethylene. Some embryos survived the coating process as well as a desiccation step (Kitto and Janick, 1985a and 1985b).

The early assessments of Murashige and Street (1977) on the difficulty of somatic embryogeny are still valid today. The quality and fidelity of somatic embryos are the limiting factors for development and scale–up of artificial seeds.

TABLE 17.1: Important crop plants in which artificial seed production and plant conversion has been demonstrated

In vitro propagules for encapsulation	Crop
Somatic embryos	Alfalfa, Celery, Brinjal, Carrot, Brassica, Lettuce, Sandalwood, Rice, Horseradish
Axillary buds/ Adventitious buds	Mulberry, Eucalyptus, Vitis
Shoot tips	Banana, Cardamom, Carum carvi

Interestingly, artificial seeds prepared from shoot buds can also be used for plant propagation, and this was reported by P. S. Rao's group from BARC, Bombay. Rice is the world's most important food crop and a primary food source for more than one third of the World's population. This crop has received considerable attention in biotechnological research programmes. Research on artificial seeds in rice is still in infancy and this technology through somatic embryogenesis would offer a great scope for large scale propagation of superior, elite hybrids (Brar and Khush, 1994). P.

S. Rao and his associates have reported high frequency somatic embryogenesis from indica rice cultivars (Suprasanna *et al.*, 1995) and utilized this embryogenic system for the production of artificial seeds.

Since then the induction of somatic and/or pollen embryogenesis has been reported in a wide array of plants, including several crop plants such as rice, wheat, triticale, maize, pearl–millet, sorghum, sugarcane, potato, sweet potato, eggplant, lettuce, carrot, alfalfa, soybean, cucumber, *Brassica* species, asparagus, coffee, tobacco and cotton. However, the production of high quality and uniform embryos (which is very important for the preparation of artificial seeds) has been limited to only certain crops like carrot and alfalfa.

Somatic embryogenesis has, also been obtained from non–zygotic explant tissue (coffee leaf), suggesting that somatic embryogenesis may be obtained from sexually immature tree, zygotic embryo or ovule tissue.

Difficulties in developing somatic embryogenesis systems for tree species are similar to those for herbaceous species. However, tree species may also exhibit a unique set of tissue culture–dependent variabilities.

USES AND LIMITATIONS OF ARTIFICIAL SEEDS

Plants are traditionally propagated either by seeds or the vegetative propagules like stem cutting. Now most plants can be multiplied through tissue culture techniques, particularly shoot tip culture (clonal micropropagation).

Micropropagation through artificial seeds may be commercially exploited on a large scale, generating millions of plants in a few days, and this may become a profitable multibillion rupees industry in near future. This technology would be feasible and even competitive economically with the seed method.

Artificial seeds offer the possibility of a low–cost, high–volume propagation system that will compete with true seeds and transplants. Most likely, the technology will first be used with hybrid vegetable crops such as celery or with high–value flower and ornamental species. Because of the relative ease of producing large numbers of somatic embryos, artificial seeds will be applicable for monoculture as well as mixed genotype methods of agriculture. The artificial seed coating also has the potential to hold and deliver beneficial adjuvants such as growth–promoting rhizobacteria, plant nutrients and growth control agents, and pesticides.

Potential Uses of Artificial Seeds

1. Delivery Systems.
2. Reduced costs of transplants.
3. Direct greenhouse and field delivery of elite, select genotypes, hand–pollinated hybrids, genetically engineered plants, sterile and unstable genotypes, large–scale monocultures, mixed–genotype plantations.
4. Carrier for adjuvants such as microorganisms, plant growth regulators, and pesticides Protection of meiotically–unstable, elite genotypes.
5. Analytical Tools.
6. Comparative aid for zygotic embryogeny.
7. Production of large numbers of identical embryos.
8. Determine role of endosperm in embryo development and germination.
9. Study of seed coat formation.
10. The synthetic seeds so developed are breed true.
11. There are potential advantages of artificial seed technology specially for tree genetic engineering.

12. The artificial production of seeds has already been obtained successfully in *Zea mays*, *Apium graveolens*, *Daucus carota*, *Lactuca sativa*, *Madicago sativa*, *Brassica* spp., *Gossypium hirsutum*, *Velerina* sp., *Santalum* spp., etc.
13. The encapsulation of somatic embryos (hydrated or desiccated) provides a potential method to combine the advantages of clonal.
14. Propagation with the low–cost, high–volume capabilities of seed propagation.
15. These seeds can be produced within a short time (one month) whereas natural seeds are the end product of complex reproductive process and breeders have to wait for a long time for development of new varieties.
16. Artificial seeds can be produced at any time and in any season of a year.
17. Dormancy is the common feature of natural seeds, but by means of artificial seeds the dormancy period can be reduced to a great extent, thereby shortning the life cycle of a plant.
18. They are useful in preserving germplasm.
19. Synthetic seeds are applicable for large scale monocultures as well as mixed genotype plantations.
20. Such seeds give the protection of meiotically unstable, elite genotypes.
21. The synthetic seeds provide us knowledge to understand the development, anatomical characteristics of endosperm and seed coat formation.

Both these propagation methods have certain limitations such as the need of intensive labour, rooting of regenerated shoots and transplantation, slow and small scale multiplication. Micropropagation has some additional problems like the need of acclimatization of tissue culture derived plants before they are transferred into field conditions (hardening) because of their tenderness due to the absence of lignification and low cuticle formation. By contrast, plant propagation via artificial seeds has several advantages over classical methods as well as micropropagation (with shoot tip culture). The advantages of this technology include the rapid and large scale multiplication, minimal labour and low cost propagation. In addition, artificial seeds can be directly delivered to the field, thus eliminating transplantation and tissue hardening steps. They can also be provided with various kinds of adjuvants like plant growth regulators, useful microorganisms and pesticides to tailor a field specific, plantable unit for a desired crop. However, the genetic uniformity is maintained in all these propagation methods.

Artificial seed technology can be very useful for the propagation of a variety of crop plants, especially crops for which true seeds are not used or readily available for multiplication (e.g. potato) or the true seeds are expensive (e.g. cucumber and geraniums), hybrid plants (e.g. hybrid rice) and many vegetatively propagated plants which are more prone to infections (e.g. day lily, garlic, potato, sugarcane, sweet potato, grape and mango).

This newly emerging technology would also be useful for multiplying genetically engineered plants (transgenic plants), somatic and cytoplasmic hybrids (obtained through protoplast fusion techniques), sterile and unstable genotypes. Besides, artificial seeds would be useful material for preservation of desirable elite genotypes (cryopreservation). They would also be valuable tools in experimental research to study the process of zygotic embryogenesis and understanding the role of endosperm in normal embryo development and germination.

Synthetic seed technology offers many useful advantages on a commercial scale. The resultant plant population from the synthetic seed will be uniform and the direct delivery of somatic embryos will save many subcultures to obtain plantlets from regenerated embryos. The encapsulated embryos could also be packed with pesticides, fertilizers, nitrogen fixing bacteria and even microscopic destroying worms.

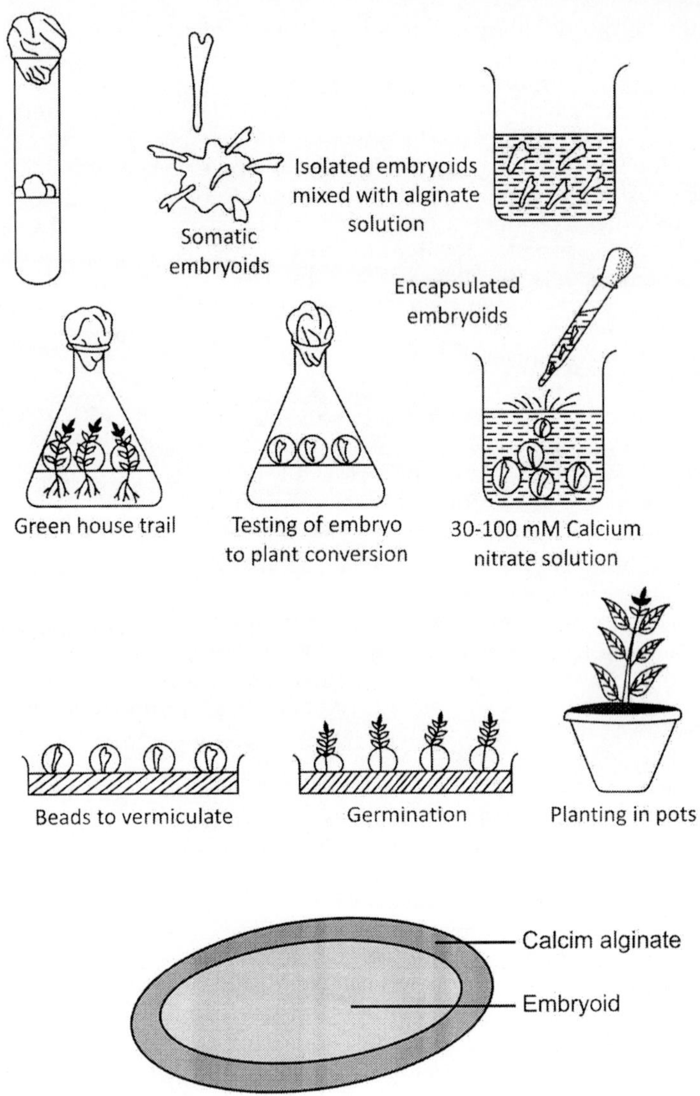

FIG. 17.5: (A) Flow diagram presenting the procedure of synthetic seed production. (B) Cross section of synthetic seed.

At the Biotechnology Division of BARC, research on the development of protocols for synthetic seeds using somatic embryos, axillary buds and shoot tips is in progress in five economically important plants, sandalwood, rice, mulberry, banana and cardamom. The following pages describe the results obtained in this direction.

The artificial seed systems coupled with artificial intelligence and microcomputer systems like the most advanced robots which can mimic the motions and functions of a living being (i.e., automated encapsulation) would tremendously increase the efficiency of encapsulation and production of artificial seeds, and revolutionize the plant propagation method in the years ahead.

ESTABLISHMENT OF CALLUS CULTURE

↓

Induction of Somatic Embryogenesis

↓

Maturation of Somatic Embryos

↓

Encapsulation of Somatic Embryos

↓

Evaluation of Embryoid and Plant Conversion

↓

Planting in Fields/green House

FIG. 17.6: Schematic presentation of steps of synthetic seed production.

This technology is gradually moving towards the commercial propagation of high value crops. However, there is a great need for refinement of this technology by the tackling of certain technical problems such as the need to produce high–quality and high–fidelity somatic embryos, and to avoid the genetic instability and variability of tissue culture derived plants (somaclonal variation, which is not preferred for crops where true–to type plants are important). These problems can be overcome if the process and regulation of somatic embryogenesis and origin of somaclonal variation are well understood. Intensive research is being carried out in several laboratories to address these vital issues. Further, the understanding about the storage, transport, handling, growth habit and harvest index of artificial seeds is essential. Similarly, the efforts to increase the output of embryos/plants per gram callus tissue of input (mass balance) and conversion frequency of artificial seeds are also needed. If all these problems are rectified with technical progress, no doubt this novel method can become valuable tool in agriculture to propagate crop species.

Although synthetic seed research is being carried out in many laboratories, the principle limitation for commercialization has been the somatic embryogenesis. Despite the fact that somatic embryogenesis has been achieved for many species, during the last decade for many of the important, high value genotypes, somatic embryogenesis has not been reported. Therefore there is a need to shift the focus on somatic embryogenesis in valuable crop plants. Research on encapsulation of axillary buds, shoot tips or any other propagules should be taken up. Equally important is the need to develop new synthetic seed coatings for encapsulating embryos and other vegetative propagules. Synthetic seed technology in years to come would certainly find its application in plant propagation and delivery of tissue cultured plants.

PRODUCTION OF SYNTHETIC SEEDS

Synthetic or artificial seeds are the living seed–like structure derived from somatic embryoids *in vitro* culture after encapsulation by a hydrogel. *The preserved embryoids are termed as synthetic seeds.* Somatic embryoids are identical with zygotic embryos and give rise to plants only under controlled laboratory conditions. Somatic embryoids are without seed coats. *In vitro* embryoid develops from callus tissue and their induction is initiated by somatic embryogenesis supplementing the medium with auxin and cytokinins in proper ratio.

Such seeds are contaminated with microbes and desiccate quickly when they are subjected to field conditions. Therefore, to get rid from this problem, they are encapsulated by a protective gel like calcium alginate. These encapsulated embryoids can resist unfavourable field conditions without desiccation. These seeds so developed behave like a true seed and are used as a substitute of natural seeds. They can also be sown directly in the greenhouse or in fields.

ENCAPSULATION OR COATING OF SYNTHETIC SEED

Encapsulation is necessary to produce and to protect synthetic seeds. The encapsulation is done by various types of hydrogels which are water soluble. The gel has a complexing agent which is used in varied concentrations.

TABLE 17.2: Various type of hydrogels

Gel (Concentration) (% w/v)	Complexing Agent Concentration (μM)
Sodium alginate (0.5–5.0)	Calcium salts (30–100)
Sodium alginate (2.0) with Gelatin (5.0)	Calcium chloride (30–100)
Carragenan (0.2–0.8)	Potassium chloride
Locust Beam Gum (0.4–1.0)	Ammonium chloride (500)

The encapsulation or coating of carrot and celery somatic embryos is done by polyoxyethylene (Polyox) and dried the embryo in polyox mixture.

Two standard methods have been used for encapsulation of somatic embryos.

Gel Complexation via a Dropping Procedure

The most useful encapsulation system is to drip two per cent LF60 sodium alginate from a separatory funnel into a 100 mM calcium nitrate solution. As the sodium alginate drops from at the tip of the funnel, the somatic embryos are inserted. The encapsulated embryos complex in calcium salt for 20 min, after which they are rinsed in water and then stored in a air tight container otherwise the capsule will dry out within 24 hour. It is a slow method of seeds production.

Automate Encapsulation Process

Automate encapsulation process is the recent and quick method of artificial seed production. An encapsulation machine can be used successfully to encapsulate somatic embryos, e.g., for alfalfa. Blank alginate capsules were planted in Speeding□ trays using a vaccum seeder. The blank capsules are planted in the field using a Stanhay planter. However, because for the rapid drying and the thickness of the alginate capsules, a hydrophobic coating is required for mechanical handling. For coating, an Elvax 4260 Copolymer™ (Dupont), is suitable for producing a slow–drying, non–tacky coating which allows embryo conversion Based on this system of production of artificial seeds the cost factor is not high for most of the cash crops.

Alginate or carrageenan is also used for artificial seed production, e.g., of carrot, asparagus, Norway spruce, etc. Mascarenhas (India) reported the encapsulation of *Eucalyptus* somatic embryos. He obtained 50 per cent germination from such seeds. Alginate artificial seeds are spherical and transparent. The alginate capsule is generally non–inhibitory.

MASS BALANCE CONCEPT

An additional concept that has greatly aided the improvement of artificial seed performance is *mass balance*. Mass balance considers the amount of tissue at the beginning of the experiment (or production run) and the number of high quality plants produced at the end of it. Simply emphasizing the number of embryos per gram of fresh callus or the number of embryo–producing calli is not adequate. In fact, treatments that lead to higher numbers of embryos may actually produce fewer superior quality of embryos than another protocol. At this time, the artificial seed package, consisting of a calcium alginate bead coated with a hydrophobic Elvax polymer, appears to be sufficient.

STEPS OF COMMERCIAL ARTIFICIAL SEED PRODUCTION

The following steps are needed for commercial synthetic seed production:

1. Production of embryogenic tissue from transformed cells or tissue.
2. Large–scale production of synchronous somatic embryos.
3. Maturation of somatic embryos.
4. Non–toxic encapsulation/coating process.
5. Artificial endosperm/megagametophyte, depending on species.
6. Storage capability of artificial seeds.
7. High frequency, direct green house/nursery field conversion, depending on production requirements.
8. Low genetic and epigenetic variation.
9. Appropriate expression of engineered trait.
10. Artificial Seed Propagation
11. Artificial seed propagation could potentially reduce the time needed to insert a desirable gene into a productive forest, as compared to using seed as the propagation method. A considerable advantage would be to eliminate or minimize the requirements for seed production using the following process:
12. Production of large–scale embryogenic tissue from genetically engineered cells;
13. Concurrent plant regeneration, confirmation of transformation, and *progeny* testing;
14. Cryogenic storage of potential superior lines;
15. Scale–up production and maturation of somatic embryos;
16. Encapsulation of somatic embryos as artificial seeds;
17. Either greenhouse/nursery establishment, growth, and transplanting into the field or direct seedling.
18. The final stage will be the evaluation of production plantations for increased yield/ performance due to engineered trait.

Unlike herbaceous species, the growth and maturity of trees far exceed the time required for tissue culture manipulations. If the engineered trait is expressed only in the mature tree and is not stable during meiosis, then mass clonal propagation would be accomplished only through tissue culture methods (or, by traditional relatively low–volume ramet production).

Cryopreservation of desirable genotypes is one of the key components for artificial seed propagation of tree species. It retains the genetic gains from genetically engineered tree species without having to establish clonal orchards. Once the progeny tests are completed, then embryogenic tissue corresponding to the superior clones can be thawed for rapid scale–up production via somatic embryogenesis.

Although the use of artificial seed technology should be extremely valuable for the rapid introduction of genetically engineered material into production of forests.

HYDROGEL ENCAPSULATION OF ARTIFICIAL SEEDS

Water soluble hydrogels have been found suitable for making artificial seeds. Two methods have been used to coat somatic embryos: gel complexation via a **dropping procedure** and **molding**.

Redenbaugh *et al.* (1986) mixed alfalfa somatic embryos with sodium alginate (2% w/v) and dropped them into a calcium nitrate solution (100 mM). Surface complexation began immediately and the drops were gelled completely in 30 minutes. Alternatively, the embryos could be mixed in a temperature–dependent gel such as Gel–rite™, placed in the well of a microtiter plate, and gelled as the temperature was lowered. Somatic embryos from several crops have been encapsulated in alginate with plants recovered *in vitro*.

Initially, the effect of encapsulation was difficult to assess because of the overall poor quality of the somatic embryos. Although visually normal embryos were produced (i.e. bipolar, root and shoot axis, cotyledons), the germination and continued development of the embryos was very inconsistent. In fact, the use of germination (root elongation and emergence) as an efficacy assay was found to be unsuitable when shoot production and further growth was not observed concomitantly. Consequently, the approach for developing artificial seeds was shifted away from one focused on somatic embryogenesis (initiation of embryo formation) to one of somatic embryogeny (initiation, development, and maturation of embryos).

TABLE 17.3: Useful gels for encapsulation of somatic embryos

Gel	Conc. % w/v	Complexing Agent	Concentration (mM)
Sodium Alginate	0.5–5.0	Calcium Salts	30–100
Sodium Alginate with Gelatin	2.0	Calcium Chloride	30–100
Carrageenan with	0.2–0.8	Potassium	
Locust Bean Gum	0.4–1.0	or Ammonium Chloride	500
Gelrite	0.25	Temperature lowered	

The concept of somatic embryogeny and the production of high quality embryos is not one that is widely followed by many researchers who have either focused on the production of a somatic embryogenesis system that results in some plant recovery or who have interest only in the study of the early stages of embryogenesis. This focus separated from an emphasis on producing mature, true–to–type, high quality embryos can possibly lead to conclusions based on abnormal somatic embryogenesis. To overcome this problem and to achieve high quality embryo production for scale–up of artificial seeds, measure the embryo response in terms of embryo–to–plant development or *conversion*. Essentially, embryo conversion frequency is the per cent of the somatic embryos that produce green plants having a normal phenotype. This assay has been critical for developing conditions and media that select for uniform plant production. Following are the events which are associated with the process of embryo–to–plant conversion.

1. Germination (radicle elongation)
2. Development of a vigorous root system
3. Growth and development of the shoot meristem
4. Production of true leaves
5. A direct shoot–to–root connection

6. Absence of hypertrophy of the hypocotyl
7. Minimization of callus growth in the hypocotyl
8. A green plant with a normal phenotype

SYNTHETIC SEED AND FOREST TREES

The use of biotechnological approaches in forestry may be greatly enhanced and considerable time could be saved by using artificial seed technology. Genetic engineering in forestry will be similar to that for field and horticultural crops. Desirable genes should be identified, cloned and inserted into the tissue (protoplasts, cells pollen, zygotic embryos, needle tissue, etc.). The putatively transformed tissue will be regenerated to plants and tested for expression of the genes. With annuals, the transformed individual plants can then be backcrossed with the original population for large scale production of transformed seed within one to few years. However, for most tree species, after adequate gene expression is confirmed, scale up production of transformed seeds deviates significantly at this point from that of annual and biennial crops because of the very long generation time for trees, particularly conifers.

Plain steps in genetic engineering of tree species focus on transformation, followed by traditional forest seed production and tree evaluation are as under:

1. Regeneration of genetically engineered tissue;
2. Confirming presence and expression of engineered gene;
3. Bulking up propagules through vegetative propagation or seed production;
4. Greenhouse/nursery establishment and growth;
5. Seed orchard establishment with concurrent progeny testing, including evaluation of the engineered trait;
6. Seed orchard roguing;
7. Seed production in seed orchards and;
8. Either nursery production with transplanting into the field or direct seedling.
9. The final stage will be the evaluation of plantations for increased performance due to the engineered trait.

SOMATIC EMBRYOGENESIS AND SYNSEEDS

Somatic embryogenesis in many woody species is less successful than other methods of micropropagation. The earliest report of somatic embryogenesis in a coniferous tree is that for spruce *Picea abies*. Since then, somatic embryogenesis has been reported in several other *Picea* species. Mainly immature, and occasionally mature embryos, were used successfully in conifers, yielding somatic embryos and plantlets. Although the number of species that can be propagated through this method has been increasing, the percentage of recovery for plantlets remains relatively low. Studies of somatic embryo formation in pines show that they can arise either in the conventional manner, as for spruce, or by a process termed somatic polyembryogenesis, as found with sugar and loblolly pines. The resulting embryos elongated and formed numerous cotyledons in a repetitive and true–to–type polyembryonic process. In the loblolly pine, the somatic embryos thus produced developed transplantable plants. The rate of somatic embryo formation has been improved recently, making it commercially viable for some species. Procedures for induction of somatic embryogenesis and subsequent conversion of the embryos to artificial seeds have been developed in a limited number of spruce and pine genotypes, such as Norway spruce and loblolly pine, with treated embryos being stored for further use as artificial seeds.

CHAPTER 18

Plant Tissue Culture some Related Aspects

BIO VILLAGE CONCEPT

The tissue culture could become an essential component of the **bio village** concept being mooted by noted agricultural scientists Dr. M. S. Swaminathan. His foundation is already into development of such villages in Pondicherry.

Dr. C. R. Raju of Kerala has created some sort of history by transforming the science of tissue culture into an art that is being practiced by school going children. Raju is turning matriculates into tissue culturists in a matter of 12 days. Thanks to Raju's ingenious the practice which was limited to botanists, is now within reach of the common man. The training can help villagers to start cottage industry of test tube grown flowers and herbs.

Tissue culture technology as practiced in the scientific laboratories is very expensive. Contrary to this the **Raju technology** is low cost, appropriate one. Dr. Raju is associated with Centre of Science and Technology for Rural Development (COST–FORD) in Kochi, Kerala.

This technology can be applied in any place where the year round ambient temperature does not fluctuate beyond 18 and 34°C. This is specifically to avoid the need to have air–conditioning which involves high capital and recurring expenditure. Normal electricity supply is not required. A germicidal ultraviolet lamp can be run alternate lyon a 12 volt battery connected to an inverter. All the expensive equipments used usually in a regular laboratory are done away with.

Currently, Raju is trying to grow virus free healthy coconut seedlings from selected tissue in a test tube for the Central Plantation Crops Research Institute. He is translating his experience of working with coconut to other crops and is using coconut–water and coconut–milk to fortify the medium to grow his plants. Coconut–pith has been used to grow orchids.

Once perfected, the system is conveyed to the trainees in simple terms. The trainees practice the procedure leading to mass propagation step by step. Once trained, all that remains is the mechanical process of mixing chemicals, preparing culture media, collecting, disinfecting and introducing plant parts into the culture medium. From the test tube the healthy seedlings are taken out into pots with subsequent transfer to the fields.

He has used all kind of household glass–utensils including bottles to replace the conventional test tubes. These are properly disinfected with hot water. After evaluating different kinds of plants, Raju has found tropical orchids and medicinal plants as the best candidates to start with. He has valid reasons to choose the orchids and medicinal plants. Says he, growing orchids and medicinal herbs is profitable. It costs Rs. 5,000 to set up a tissue culture facility for 100,000 plants.

The technique has proved successful in multiplying several orchids which include 22 varieties of Dendrobium, 3 varieties of Phalaenopsis, 3 varieties of Vanda and one variety of *Asconcenda*. He has trained young boys and girls in at least four types of medicinal herbs.

Qualifications Required to Join the Training Course

Any matriculate who can pick up the names of chemicals is fit for the course. Besides, one must have a love for plants. Two basic aspects of the technique of tissue culture, however, must be learned thoroughly. These are the composition of the medium required at different stages of multiplication and the method of handling tissues and culture in sterile– germ–free conditions.

Employment for Rural Youth

Raju is confident that his short term course can open wide opportunities for unemployed rural youth to seek self employment. Following are advantages of popularizing tissue–culture in Kerala.

- Climate is suitable to grow test tube plants.
- There is plenty of coconut–water or coconut–milk used to nourish the growing plantlets in the test tube.

It was Kerala which provided an opportunity for the first Plant Tissue Culture Company of India. It was set up as a small scale laboratory at Manalaroo in Kerala. It has used tissue culture technology to grow cardamoms plants in test tubes. The technology was developed by National Chemical Laboratory (NCL), Pune.

The company at one time was making just about Rs. 2.5 crores in profit. It now hopes to net a profit of Rs. 12 crores. The phenomenal growth of the first biotech company has encouraged many others to join the fray. If companies can make profit, why can't poor villagers.

Dr. Raju's technique is worth emulating by research institutions and extension agencies in the country.

EFFORTS OF PUBLIC RESEARCH INSTITUTES

In India, research facilities in public research institutes on biotechnology are being improved with the assistance of UNIDO and other agencies. Private sector companies are also showing keen interest in the technology as is evident from their increased investments in research and development as well as the collaborative arrangements made with foreign companies. The research effort so far is concentrated more on plantation crops like tea, coffee, rubber, spices besides vegetables, flowers and fruits.

A great deal of biotechnology research in the West is looking into the replacement of reduced use of flavour and fragrance plants with laboratory processes. A produce called **Nocordia** is being developed by two companies which would eliminate the need for castor oil. Biotechnology offers the potential to displace sugar as an industrial sweetener through the development of new natural sweeteners from plants. Experiments in cocoa, rubber, vanilla are aiming to reduce the dependence on the developing countries. All this would have a profound impact on the livelihood of the Third World farmers. Such development should be constantly monitored so that timely and appropriate policy actions can be taken before they swamp the Indian farmers.

PRODUCTION CRITERIA AND ECONOMICS

Plant tissue culture is the only way of modern plant propagation where a large volume of plants needs to be produced regularly, as in the case of cut flowers. Plants produced by tissue culture technique are much healthy with desirable characteristics. *In vitro* culture systems have the potential for long distance shipment of propagules and allow long–term storage of clonal material.

In the conventional propagation system, about 75% of the greenhouse area is committed to mother plant maintenance. Micropropagation would obviate such necessity and the entire space of the greenhouse could be effectively planted with plants of high value received from tissue laboratories on a contractual basis. Micropropagation offers an environment–friendly industry to flourish and is a source of potential employment.

Selection of Crops for Micropropagation

Tissue culture production is market oriented. Therefore, it is essential that one should have a sound market–place reason for contemplating the use of plant tissue culture for a particular crop or clone. The use of plant tissue culture should offer economic advantages to the commercial laboratory as well to the grower. The tissue culture method of production can be employed only if it assures as increase in the net economic return, compared to that of the conventional propagation method, for a certain unit of land, water, time and capital.

For economic viability producer should know the costs of not only propagating a crop through tissue culture, but also the costs of propagation through conventional means. Under congenial conditions certain plants can be produced more economically, using conventional techniques. The micropropagation system should offer propagules at a cheaper rate as against conventional methods of propagation. Propagation through plant tissue culture is the only method wherein conventional propagation is not possible, as in the case of runner less strawberries. The greater impact of tissue culture can be felt where the cost of field production of micropropagation progeny can be lowered, rather than in those crops whose productivity can be increased with yield increases.

Growing varieties resistant to pests, diseases, salinity or drought would save corresponding input costs or decrease operating costs associated with energy, irrigation and labour. Therefore, clones or varieties with such traits should be selected for micropropagation.

For example ten flowering plants in the floriculture trade can be arranged in the descending order of their turnover are: rose, *Chrysanthemum*, carnation, tulip, *Freesia*, *Gerbera*, lily, orchid, *Gypsophila* and *Iris*. Similarly, ten foliage plants are *Ficus*, *Dracaena*, *Begonia*, *Saintpaulia*, *Yucca*, *Azalea*, *Poinsettia*, *Kalanchoe*, *Dieffenbachia* and *Cyclamen*.

Selection of Location

It is essential to consider following factors before selecting the location of a laboratory:

1. **Availability of water and electricity:** These are the major requirements of a tissue culture laboratory.
2. **Transport of sapling:** As cost incurred on transport is a limiting factor. The laboratory should be established near the place of a potential market. If the unit is export–oriented it should be near an international airport, which will reduce the time lag between packing and shipment. This will help in quick delivery of

quality plants.

3. **Climate:** Maintenance of stock plants and the hardening and weaning of tissue–cultured plants require a conducive microclimate, either natural or controlled. The laboratory should be established in a place where conducive natural environment prevails. This will save considerable cost on cooling or heating, obviating the need for sophisticated greenhouses.
4. **Infrastructure:** Availability of inputs, skilled manpower and infrastructural facilities such as availability of engineering skill to repair the unit without losing time, should also form one of the major factors in selection of locality.

OPTIMIZING SIZE OF PRODUCTION

Maximum benefits from investments can be obtained by producing the maximum number of plants per unit capital and by proper management and labour availability. Whether, the facility is small or large (which are only relative terms), our aim should be to produce plants at competitive price levels.

The investments made in most of the capital equipment is not dependent on the volume of production. For example if we produce 1 million or 2 million plants, the investment made on a pure water system or autoclave remains constant. Large scale production necessitates investment in certain facilities or equipment (such as change room and horizontal double door autoclave) either to reduce the contamination or to increase the efficiency of operation. Such investments are beneficial to the industry in the long run. With increased volume of production the capital burden per plant will be minimal, reiterating that micropropagation is a production emphasized industry with the ability to produce plants well beyond the demand.

The advantage of large scale production is that the range of crops micropropagated can be larger compared to the small scale production unit with restricted range of crops. A wide range of products hold the price of produce stable, assuring higher profits for the large scale units. Mass production of restricted crop range would result in heavy competition and slashing of prices well below the break even point.

AVAILABLE FACILITIES AND PRODUCTION POTENTIAL

Financial capability and market potential always decide the size of the production facility. Our next step is to arrive at a combination of various facilities or equipment that would go hand in hand. The important items of investment that need careful study are growth room, laminar airflow cabinets and autoclave. These will decide the peak accumulated culture capacity to get the maximum growth room space requirement. As demand for tissue–cultured plants is greater in certain months, therefore, one should take into account the peak growth room space requirement of these months. The number of cultures handled to achieve such accumulated growth room capacity will give the idea to install the number of laminar airflow cabinets or autoclaves in the unit; and these equipments should be subjected to use on shifts or 3–4 cycles in a day to reduce capital burden per plant. As this industry is fast growing and changing, enough thought should be given to further expansion or to adopt changes to produce a different kind of end product.

ECONOMIC ANALYSIS OF THE PROJECT

The economy of any micropropagation project is market driven and production emphasized and is affected by (Cumulative) volume of dispatch, number of dispatches and frequency of dispatches. These factors are governed by market demand and beyond our control, however, the right combination of these factors will increase the profitability of the unit. An increase in the volume of dispatch will increase the profitability throughout the year. The investment made in capital items can be minimized and manpower utilization can also be uniform.

Though the tissue culture method of propagation is possible throughout the year, the market demand is not uniform and is characterised by peaks and falls. Many fruit and vegetable species which are being targeted by micropropagation businesses as crops for the future, have product delivery windows of only 2–3 weeks of the year. Even many flowering crops, for example *geraniums*, have production windows of less than two months for the deliver of 15–20 million units. Because of this erratic market trend it is necessary to have capacious growth rooms, extra autoclaves to prepare media in the peak work period and increased manpower during such peaks. In subsequent lean months, the facilities and manpower are underutilized. This calls for a constant search for a continuous market and efficient planning to produce the maximum number of plants, utilizing the resources available.

PRODUCTION MANAGEMENT

The implementation of any micropropagation project starts with the layout of physical and financial programmes in tune with the presumed production volume. The maximum possible production, per unit investment, can be possible only through efficient production management. The success of production management lies in understanding the art and science of propagation. Order processing, production planning and production monitoring are the main aspects of production management.

ORDER PROCESSING

This needs macrolevel planning because all orders on hand are to be put into a master plan. This will to assess the requirement of growth–room space, cabinets, media and other consumables and manpower at a particular time. This also helps to know the potential production capacity and limitations of the laboratory. Orders should be accepted at least 6 months in advance (if the laboratory has enough starter cultures) and preferably with about 25% advance payment. The down payment indicates a commitment by the buyer to the laboratory in buying plants. It also enables the laboratory to meet a part of the requirement of recurring expenditure.

PLANNING FOR PRODUCTION

Production planning forms the base for producing various crops, taking into consideration the market demand, limited facilities of the laboratory and limitation of the nature of the plants themselves (that they multiply exponentially and ought to be sub-cultured at regular intervals). Tissue culture production forms the base for the entire plant production industry and hence all laboratory operations are time bound. Production planning tends to tap the maximum possible benefit from the existing market by the best use of available scarce resources of the laboratory. The production plan should be prepared for every individual crop, comprising all the crops of the laboratory. It is essential that producer should be aware of the nature of the end product which is to be delivered, e.g., microcuttings or ex–agar rooted plants. Therefore, it is essential to consider certain parameters like (i) multiplication rate, (ii) passage or transfer cycle, and (iii) operator efficiency. The plan should be finalized before considering these parameters.

Multirate

Currently, the *in–vitro* production methods are the extension of traditional vegetative multiplication methods. Therefore the expectation is to have a mutation frequency comparable to that of classical *in vivo* methods. Chances that variations will occur are much greater if the system is based on adventitious shoot formation and still greater with embryogenesis, callus and cell systems. Hence the multiplication ratio adopted should be the one that would give product plants having minimum variation and are repeatable in nature, i.e., the multirate is constant on each sub-culturing for a particular stage of plant. The system development of any crop should aim at arriving at an optimum multirate. Too low a multirate would not be economical and a high multirate, though it would help the laboratory to build up the stock in the initial period, would not be advisable as it would render the manpower insufficient to handle the huge volume of culture build up later.

It is important that on each culturing the cumulative multirate be adopted for further planning. Production planning prepared based on the multirate of the latest transfer alone would not be practical, as it might under– or overestimate the subsequent volume of production.

The multiplication rate of four stages are to be borne in mind in working out the plan:

1. Multiplication stage
2. Multiplication to shooting stage
3. Shooting to rooting or multiplication to rooting, and
4. Rooting to dispatch.

Passage

The other important parameter in production planning is the transfer cycle or passage, which varies from crop to crop as well from stage to stage within a particular crop. The transfer interval should be repeatable and reliable. The optimum number of cycles required to produce plants with minimum variation is to be inferred from field experiments. For most of the crops the number of subcultures would be 15–20. Hence the initiation programme is to be meticulously planned to meet the periodic starter culture requirement of the production division.

Whenever the passage is shorter, more stress falls on manpower as frequent subculturing is required. On the other hand, whenever the passage is longer, more stress falls on growth–room space.

During the peak season of production of shorter passage crops (say *gerbera*). we can manipulate the production by increasing the number of shifts or number of operating hours of the laminar airflow bench.

But if crops of longer passage are chosen (for instance lily), the requirement of growth–room space would be a limiting factor, which cannot be manipulated as in the previous case. We are left with no other option but to have capacious growth rooms. Hence a combination of crops having various transfer cycles is to be chosen for effective utilization of cabinet space and growth–room space throughout the year.

The transfer cycle of the following stages is to be borne in mind in working out the plan.

1. During multiplication stage
2. During shooting stage
3. During rooting stage

Once we know the stage wise multiplication rate and transfer cycle of a particular crop, we have to work out a plan backwards, from the orders received, to check whether we have enough starter cultures or not. If we do not have enough starter cultures. we should plan forward and inform the customer about the quantity we can deliver in the periods of his requirement.

Operator Efficiency

After finalizing the individual crop it is essential to calculate the requirements of manpower in handling, multiplying, shooting and rooting cultures.

PRODUCTION MONITORING

Subsequent to the formulation of the annual plan, weekly work schedules should be prepared for implementation. After the completion of work, a statement of completed work is prepared and compared with the original plan. The deviation from the original plan in terms of cumulative culture position (growth–room stock), culture handled, efficiency of operators, multiplication rate. contamination etc., are constantly monitored by the date–processing section.

Frequent small experiments are to be conducted along with the mass production. to fine–tune the production system. If proper monitoring is not done or efforts are not made to smoothen the production process, deviations may occur, coming as a great shock. As micropropagation forms the base for the entire crop production, the laboratory should inform the buyer frequently of possible changes in the delivery volume or the time of delivery, according to the availability of stock.

CONTAMINATION: ECONOMIC ASSESSMENT

Contamination is a serious problem plaguing the micropropagation industry. Losses can be ascribed to carry–over of inoculum on the explant surface, failure of contamination control system in the laboratory, infestation with mites and thrips or poor sterile technique. More serious losses could occur in stage 2 as compared to stage 1 where the value of inputs is very high. Large–scale losses of this kind are most likely to be a consequence of poor monitoring of production combined with the clonal multiplication of sublimally infected stock. Production should only be based on multiplication of axenid stocks and should involve continuous monitoring. Whether an axenic status can be achieved for all the crops at an economic cost is still in question (Cassells, 1991).

CALCULATION OF COST PRICE

With a variety of plants on the production line. each exacting various production parameters, it is necessary that the producer should understand per plant production cost and various strategies to lower the (per plant) production cost. Wary eyes should be kept on the costs of inputs whose change will greatly reflect on the cost of the end product, such as the price of agar.

QUALITY CONTROL IN COMMERCIAL PLANT TISSUE CULTURE

The term quality is defined as *fitness for purpose*. In the commercial manufacturing environment it is important that the quality of the product match the expectations of the eventual customer. The object of quality control, therefore, is to produce a product of quality that will reflect to the following aspects:

1. satisfy the customer's need,
2. be deliverable according to scheduled requirements.
3. be deliverable according to scheduled requirements.

Thus it is clear that the object of quality control in the manufacturing process is not to produce an item of the highest possible quality, but, rather to satisfy the customers' needs.

The concept of quality within the environment of a plant tissue culture laboratory has been largely overlooked for a number of years. This is probably due to the fact that commercial plant tissue culture has evolved from fundamental research into plant propagation systems. Examples of this are the salt mixtures which were originally developed to study the growth and development of tobacco (Murashige and Skoog, 1962; Linsmaier and Skoog, 1965). Such salts still form the basis of many media formulations and only a limited number of media have been formulated specifically for certain plants, e.g. rhododendron.

The research and development of plant propagation systems is often a complicated process wherein each stage presents a number of problems to be overcome:

1. Efficient sterilization of explants,
2. Initiation of multiplying cultures,
3. Formation of rooted plantlets,
4. Establishment of weaned plantlets.

The results from micropropagation research projects are often published but their direct transference to a commercial plant propagation system is seldom possible without further development and refinement.

In a commercial environment a micropropagation system must be capable of producing a product of the desired quality. A number of factors come into defining the plant quality but basically the plant must satisfy the customers' requirements. It is not sufficient for the laboratory to define its own quality standards; it must work closely with its customers to ascertain exactly what they require of the plant; otherwise their own standards may be too low or too high.

If the laboratory sets too low a standard of quality, the customer will go elsewhere to find a more suitable, better quality product.

If the laboratory sets too high a quality standard, the customer may be very pleased with the product but may not be willing to pay the increased price required to cover the increased production costs associated with very high–quality standards.

Obviously, there is a very wide difference between the aims of the research micropropagation laboratory and the commercial production laboratory. The production laboratory has to be able to produce a plantlet of sufficient quality within the constraints of the commercial operating environment.

CHAPTER 19

Hi-Tech Conservation of Horticultural Crops

Indian region is a treasure house of wild genetic resource. Wild species and relatives of crop plants contain valuable genes that are of immense genetic value in crop improvement programmes. The important wild related species and types in various crop groups, prevailing under different phytogeographic zones of the country needs particular attention in the agro-biodiversity management system for a sustainable use to help maintain food, nutritional and agricultural economic security. In horticultural crops, genetic diversity is well represented in India due to the prevalence of varied ecological condition, there are 190 species of economic importance in fruits (109), vegetable (54) and spices and condiments (27) available. Fruit species existing in the country of which aonla, bael, banana, ber, few *Citrus spp*., jackfruit, jamun, mango and phalsa are indigenous and many others have become commercially important. The genetic variability of these fruits needs to be properly conserved for use in the crop improvement programme to cater to the national and regional needs. The growing demand for superior cultivars can also be met through introduction from other countries either for direct use or in breeding. The rich genepools incorporating extensive variability from basic ingredients of improvement programmes.

WHY CONSERVATION OF GENETIC RESOURCE?

It is assumed that these genetic resources shall play a positive and unique role in the development of new cultivars, including restructuring the existing ones that are handicapped due to poor expression or lack of one or the other attribute. The most important characters derived from the wild relatives have been the resistance to biotic and abiotic stresses. Such resistance has been mostly observed to be invariably simply inherited, easily transferable and clearly expressed. The need to conserve the genetic resources is far greater in present time than ever before because the number of cultivated types has narrowed to only the best available ones with regard to productivity and quality. Even these types face pest and disease hazards and on introduction in regions away from their centres of origin, they are often subjected to abiotic stresses. Thus, the breeders need to depend on the conserved genepools to look for the desired sources of resistance. On the other hand, due to the advent of modern agriculture, population explosion, rapid urbanization and implementation of developmental projects, the genetic diversity, evolved as a result of thousands of years of natural selection and domestication is getting eroded rapidly. The related wild and weedy species of fruits are being ruthlessly cut for fodder, fuel and timber because these have no other profitable immediate use and do not receive the patronage of forest officials. In India, the fruit genetic resource have been collected and maintained by feudal landlord, nurserymen and fruit research stations.

Objectives of Conservation

- According to this, the main objectives of the conservation are:
- The conservation of biological diversity.
- The sustainable use of components of biodiversity.
- The fair and equitable sharing of benefits arising out of the utilisation of genetic resources.

The objective of equitable sharing of benefits has to be further seen in the reaffirmation of sovereign rights of states over their own biodiversity, which entitles a country to equity and fairness in sharing of benefits arising out of the utilisation of their genetic resources.

GENETIC RESOURCES

The genetic resources of a fruit consist of primary, secondary and tertiary genepools. The cultivated forms along with some wild species which can be intercross freely for transferring genes are included in a primary genepool whereas gene flows is possible only through interspecific hybrids in the secondary genepool. Hybridization generally results in a sterile progeny in the tertiary genepool. But, unlike in other cross the secondary and tertiary genepools in fruits can be utilized as rootstocks. The following material is included in the fruit genetic resources.

Commercial and modern bred cultivars: Fruit research stations and progeny orchards maintain collection of commercial cultivars and recently evolved hybrids (modern bred cultivars) which are the basis the fruit industry.

Seedlings, landraces and primitive cultivars: A great deal of fruit germplasm is existing in the form of seedling, landraces and primitive cultivars because many fruits are endemic in nature and have been propagated by seeds generating a lot of allelic variability, e.g., in almond, apricot, aonla, bael, ber, citrus, jackfruit, jamun, guava, mango and walnut. This, however, is factor for the greatest genetic erosion either due to natural mortality or replacement by commercial cultivars.

Special genetic stocks: The genetic materials are being continuously identified as donor's specific characters, e.g., Neelum and Bangalora as donors of regular bearing character in mango. Such donors are available in almond and all fruits in which breeding have been undertaken.

Related wild and weedy species: A large number of wild and weedy species of fruits have originated through natural hybridization and introgression in the form of subspecies, botanical varieties and ecotype and these exist in almost all fruits and specifically in citrus, temperate fruits, ber and banana.

TYPES OF CONSERVATION

The conservation of fruit germplasm is done for current use as well as in the future and comprises entire diversity of primary, secondary and tertiary genepool. The germplasm is mainly used in breeding of superior cultivars and rootstocks besides research and education. The related wild and weedy species are not of immediate use in breeding but the great strides taking place in the field of genetic engineering may open up new vistas in their utilization. The two major elements involved are conservation and utilization of biodiversity. Both are most important and have to be pursued with equal emphasis.

TABLE 19.1: Most common methods used for germplasm conservation and the corresponding PGR categories

S.No	Methods of conservation	Predominantly conserved PGR categories by corresponding method
1	Biosphere reserve	Ecosystem/ biodiversity by and large
2	Nature reserve	Specific habitat/wild and/or weedy species gene pool
3	Gene sanctuary	Ecosystem (specific)/ wild species gene pool
4	On farm conservation (mass reservoirs, bulk hybrid populations)	Agro-ecosystems / land races
5	Botanical garden/ arboretum	Wild species, obsolete cultivars, tree crop germplasm
6	Field gene bank	Wild species, vegetatively propagated crops, tree crop germplasm
7	Plant organ storage	Vegetatively propagated crops, mainly in the form of roots, tubers and bulbs
8	Seed storage	All plant species which produce fertile and orthodox seeds
9	Pollen storage	In principle all species which produce long living pollen
10	In vitro storage	Wild and cultivated species which produce recalcitrant or no seeds, vegetatively propagated crops, disease free germplasm as well as orthodox seeds
11	Cryopreservation	Germplasm mentioned above which permits cryopreservation
12	DNA and gene libraries	Special genetic stocks; in principle applicable for all germplasm

Most horticultural species needs to be conserved at three broad levels. This would include conservation at:

i) Ecosystem level (in situ)
ii) Genotype level (ex situ)
iii) Gene level (molecular)

In situ conservation of plant genetic resources of horticultural crops- This is an approach, which is applied mainly to wild species related to crop plant, forest and pasture species. It is often recommended that these species should be preserved, maintaining the genetic integrity of their natural state, as communities in stable environments. Most nature conservation programmes are aimed at the level of ecosystems or multi species communities. Conservation of horticultural crops has been possible primarily through *in situ* conservation, i.e. *in situ* conservation in the natural habitats. A natural genepool has buffered *in situ* by a surrounding zone of minimum human activity. Such conservation is relatively inexpensive and takes care of all the species (plants, animals and microorganisms as an interacting system) in a particular habitat. *In situ* conservation also warrantees the continuation of the ecological processes and organic evolution, thus helping the concerned biota to continuously evolve in response to newer environmental changes. This is also known as dynamic conservation because natural evolutionary process continues increasing the genetic variability.

Ex situ conservation of plant genetic resources of horticultural crops - This type of conservation includes the use of botanic gardens and arboreta on the one hand and gene bank on the other. Botanic gardens and arboreta represent the oldest forms of conservation in Europe data back to the 15th and 16th centuries. The conservation of horticultural crops has been primarily possible through *ex situ* conservation ever since agriculture evolved. Today there are many ways to achieve this by seed, pollen, sperm, ovule/ova, embryo, tissue and even DNA banks. These banks are working for conservation. In response to this concern, the International Agricultural Research Centres (IARCs) of the Consultative Group on International Agricultural Research (CGIAR) started to assemble germplasm collections of the major crop species within their respective mandates. It is in this contest that the International Board for Plant Genetic Resources (IBPGR) was established in 1974 to coordinate the global effort to systematically collect and conserve the world's threatened plant genetic diversity. Today, as a result of this global effort, there are over 1300 genebanks and germplasm collections around the world, maintaining estimated 5,554,505 accessions entries of plant materials of over 140 crop plant species (FAO, 1996). No country, howsoever rich in biodiversity, can be self-sufficient in genetic resources. There are instances where crops introduced form other countries have become major crops in the countries of adoption, e.g. wheat, maize, red pepper, tobacco, cashew, rubber, tomato, potato, etc. in India. Therefore, countries are dependent on genetic resources from other countries. It is also important that each country has not only to ensure conservation of its own biodiversity but also seek to enrich genetic diversity of species, which have, over a period of time, become naturalized crops and have entered their trade. Thus it is an interdependent effort rather than an independent one where each country each country is concerned with itself. Following are the four alternative objectives for *ex situ* conservation:

1) To establish and maintain conservation stands characterised as far as possible by the same genotype frequencies as the original populations (provenance)-static conservation of genotypes).
2) To establish and maintain conservation stands characterised as far as possible by the same gene frequencies as the original populations, thus avoiding total loss of any allele static conservation of gene pools.
3) To establish conservation stands in which gene frequencies are allowed to change freely according to natural selective forces evolutionary conservation.

To establish conservation stands in which gene frequencies are deliberately changed by man in order to conserve characteristics important for plantation economy in a region and at the same time eliminate undesirable characteristic-selective conservation.

TABLE 19.2: Ex situ conserved PGRFA by storage type

Storage type	Number of accession	Distribution per cent
Seed banks	3,610,428	65
Field banks	526,300	9
In vitro	37,600	1
Unknown storage facilities	1,380,177	25
Total	**5,554,505**	

Source: Calculated according to FAO, 1996 a WIEWS, 1996

The following methods have been adopted for *ex situ* conservation of plants:

Orthodox seed conservation: Storage of orthodox seeds is the most widely practiced method of *ex situ* conservation of plant genetic resources, as indicated in the Report on the State of the World's Plant Genetic Resources for Food and Agriculture (FAO, 1996), since 90% of the 6.1 million accessions stored in genebanks are maintained as seed. Over the years, techniques have been devised which allow seeds of many species to be conserved in this way for several decades. These techniques involve drying seeds to low moisture content (3–7% fresh weight basis, depending on the species) and storing them in hermetically sealed containers, at low temperature, preferably at -18°C or cooler (FAO/ IPGRI, 1994). There are species (most of them horticultural species) for which conservation as seed is problematic. Firstly, there are those that do not produce seeds at all and, consequently, are propagated vegetatively, for example banana and plantain (*Musa* spp.). Secondly, there are crops such as potato (*Solanum tuberosum*), other root and tuber crops such as yams (*Dioscorea* spp.), cassava (*Manihot esculenta*) and sweet potato (*Ipomoea batatas*), and sugarcane (*Saccharum* spp.) that have either some sterile genotypes and/or some that produce orthodox seed. However, these seeds are highly heterozygous and, therefore, of limited utility for the conservation of particular genotypes.

Field genebanks: The entire fruit germplasm is preserved in field genebanks, which were earlier known as varietal collections, clonal repositories or living collections. Field genebank is a facility where clonal materials are conserved as living collection in field, orchard or plantation. The primary objective of a field genebank is to make the authentic genetic resources of a conserved fruit available to the breeders besides pollen, fruit samples and plant parts. Germplasm of many horticultural crops, particularly tree fruits, is usually conserved in field collections, due to the perennial nature of the plants. However, because of disease and insect problems, weather factors and improper management, many collections may be at risk and germplasm may be lost. There is a need, above all, for systematic duplication of material, accompanied by information systems that effectively tie together collections in different sites for safety reasons. These are used for the conservation of clonal crops, where seed is either recalcitrant or seed is rarely produced. The 'rule of thumb' is to use the same propagation techniques as the farmers, for example, not disrupting adapted clones through genetic segregation in seed cycle. Management may be the same as used during routine farming, so the method can be adapted to local circumstances. Major constraints to field gene banks include cost and all the natural hazards of farming. Among the *ex situ* methods employed for germplasm conservation, high priority has been given to the maintenance of collections in fields, orchards, plantations, herbal gardens, and more recently in gene banks. Field gene banks have both advantages and limitations. In many cases, conservation methods also involve storage as seeds backed by a limited number of clones in field or *in vitro* gene banks.

Location and functions: As far as possible, the field genebanks have been located in agro-climatic region congenial for the cultivation and in the proximity of research institutions so that the breeders can take advantage of the assembled materials. The sites should normally be free from the occurrence of natural hazards, *viz.*, cyclones, floods, drought, hailstorms, landslides and earthquakes. Assured irrigation facilities are desirable. The accessions are maintained in duplicate sets. These are given slightly more spacing than usual and a minimum of training and pruning in order to allow them to assume normal shape and size.

The functions of a genebank are as under:

i) To assemble the genetic resources of the conserved fruit from all over the world
ii) To propagate and conserve the collected genetic resources under proper cultural and phytosanitary conditions.
iii) To carry out preliminary evaluation of the genetic resources by recording

horticultural characters on standard descriptors lists.

iv) To maintain liaison with the National Bureau of Plant Genetic Resources (NBPGR) and concerned horticultural research institute for technical for technical expertise and documentation.

v) To distribute the genetic resources in the form of plants, seeds, cuttings, scionwood and pollen to other research stations in small quantities.

Genetic Resources in Field Genebanks

(a) Tropical and subtropical fruits: The total number of accessions of tropical and subtropical fruit genetic resources along with their locations have been given in Table 19.3. The maximum number of accessions is available in grape, which is an exotic genetic diversity, the germplasm collection of mango banana and citrus needs to be strengthened by exploring the various endemic regions in the country and introduction from abroad.

TABLE 19.3: Conservation of genetic resources of tropical and subtropical fruits (Lyer, 1987)

Fruit (including species)	No. of accessions	Location of field genebanks
Banana	296	Bhubaneswar, Chinsurah, Coimbatore, Hajiur, Hassaraghatta, Kahikuchi, Kalyani, Kannara, Navsari
Citrus spp.	606	Hisar, Ambalayayal, Chethalli, Hessaraghatta, Katol, Ludhiana, New Delhi, Pantnagar, Periyakulam, Sritampur, Tinsukhia, Tirupati,
Grape	1209	Abohar, Hessaraghatta, Hisar, Ludhiana, New Delhi, Pune
Guava	123	Anantrajupet, Basti, Hessaraghatta, Pune, Saharanpur
Jackfruit	66	Trichur, Krishnagar
Litchi	36	Dehra Dun, Krishnagar, Muzaffarpur, Pusa, Saharanpur
Mango	1106	Bhubaneswar, Hessaraghatta, Lucknow, Navsari, Delhi, Panjim, Sabour, Sangareddy, Saharanpur
Papaya	150	Coimbatore, Hessaraghatta, Pantnagar, Pusa
Pineapple	33	Kahikuchi, Pantnagar, Periyakulam, Vellanikkara
Sapota	26	Coimbatore, Dharwar, Navsari, Rahuri

(b) Temperate fruits: Most of the pome and stone fruit germplasm has been introduced from abroad while valuable germplasm of almond and walnut existing in the country need to be exploited. Most of germplasm is conserved at the NBPGR, regional station, Phagli and at the regional fruit research stations of Mashobra, Shalimar and Chaubatia.

TABLE 19.4: Conservation of genetic resources of temperate fruits

Fruit (including species)	No. of accessions	Location of field genebanks
Apple	750	Chaubattia, Mashobra, Phagli, Shalimar, Shimla
Pear	145	Chaubattia, Mashobra, Phagli, Shalimar, Shimla
Apricot	81	Chaubattia, Mashobra, Phagli, Shalimar, Solan
Cherry	77	Chaubattia, Mashobra, Phagli, Shalimar

Fruit (including species)	No. of accessions	Location of field genebanks
Peach	162	Chaubattia, Mashobra, Phagli, Shalimar, Solan, Ludhiana
Plum	211	Chaubattia, Mashobra, Phagli, Shalimar, Solan
Almond	313	Chaubattia, Phagli, Ludhiana, Shalimar, Solan
Walnut	436	Chaubattia, Phagli, Shalimar,
Pecan	16	Phagli, Palampur
Hazelnut	13	Chaubattia, Phagli,
Kiwi-fruit	7	Phagli, Solan, Bajaura
Olive	18	Phagli, Solan
Persimmon	6	Solan
Cotoneaster spp.	11	Phagli, Shimla
Crataegus spp.	11	Phagli, Shimla

(c) **Arid zone fruits:** These fruits were earlier in a state of neglected condition but have now got good attention. Apparently, the germplasm holding are very small and should be increased by exploration and exchange. Pareek (1988) has also listed a number of other unexploited fruits.

Table 19.5: Conservation of genetic resources of arid zone fruits (Pareek, 1981)

Fruit (including species)	No. of accessions	Location of field genebanks
Annona spp.	36	Bangalore, Rahuri, Sangareddy, Udaipur
Aonla	16	Allahabad, Anand, Faizabad, Rajendranagar, Saharanpur, Udaipur
Bael	12	Basti, Faizabad
Ber	177	Bahadurgarh, Bawal, Faizabad, Hisar, Jobner, Jodhpur, Junagadh, Saharanpur, Udaipur
Date palm	45	Abohar, Bikaner, Junagadh
Fig	23	Bangalore, Rahuri, Rajendranagar, Udaipur, Yercaud
Phalsa	9	Allahabad, Bangalore, Faizabad, Hisar, Junagadh, Udaipur
Pomegranate	18	Abohar, Allahabad, Bangalore, Faizabad, Hisar, Jodhpur, Junagadh, Ludhiana, Rahuri, Saharanpur, Solan, Udaipur, Yercaud

Pollen storage: Since the collection as well as preservation of the pollen is relatively easier and cheaper and does not require much infrastructural establishment, it has been suggested as a potential method of genetic conservation in fruit crops. Long term conserved pollen in the form of nuclear genetic diversity (NGD) of citrus, papaya, grape, mango, tomato, eggplant, onion, capsicum, rose, gladiolus, gerbera, carnation and Rare Endangered and Threatened species of medicinal plants are continued to be cryopreserved in liquid nitrogen (-196°C). The pollen cryobank is maintained and managed by periodic replenishment of the cryogen, for maintaining a constant cryogenic temperature throughout the storage duration. Successful storage of papaya and grape pollen for a period of four years under cryogenic conditions. Citrus pollen could be cryopreserved for only two years. The longer period of storage has been reported to be nine years for apple pollen. Like seeds, pollen is stored at low temperature, low relative humidity

In vitro Conservation

Since conservation of fruit germplasm in field genebanks is very expensive, requires a lot of land and is vulnerable to natural disasters, pests and diseases, *in vitro* technique has great potential. While sixteen hectares area is required to maintan apple germplasm in field genebank, only two square metres space is sufficient for maintaining 800 grape cultivars under *in vitro* conservation. Therefore, *in vitro* conservation of fruit germplasm has been suggested as an alternative or to complement field genebanks. The *in-vitro* conservation methods are considered to be particularly interesting for those horticultural species that are normally vegetatively propagated as well as those with recalcitrant seeds, and continued efforts are needed in this field to develop protocols for a wider range of species. When one method of conservation is subject to unavoidable hazards, as with FGBs, an alternative method should also be used. The role of in vitro conservation techniques in the overall conservation strategies should be indicative of the fact that it should complement other conservation strategies within the total program of a given species or population. The methods chosen should be carefully considered taking into account the feasibility, practicality, economy and security. *In vitro* collections of species could be maintained at the same or separate site, but should have clear linkages with the field gene banks. In particular in vitro conservation is used for vegetatively propagated material species with recalcitrant seed and material biology. This later material may have been costly and limited in quantity and have characteristics whose stability during repeated mitotic divisions and following meiosis are unknown. It is, therefore prudent to conserve stocks of this material to ensure consistency in subsequent experimentation of production processes, and as definitive source of reference material. A recent FAO survey (FAO, 1994) indicated that 37 600 accessions are conserved *in vitro* worldwide. Slow growth storage is routinely used for the conservation of a number of species-notably Musa, potato, yam and cassava- in regional and international germplasm conservation centers such as INIBAP, CIP, FAL, IITA, CIAT and NBPGR. FAO has recently received reports from most countries in the world on the current status of plant genetic resources utilization and conservation. In relation to *in vitro* conservation in general, 64 out of 107 countries listed have indicated that they have facilities for *in vitro* conservation. However, very little routine cryopreservation work is carried out. In India the NBPGR has established the National Facility for Plant Culture Repository in 1986. *In vitro* conservation has been initiated for banana and citrus. By using meristem and shoot tip culture technique, 43 cultivars of banana were established and preserved at 15°C for over twenty months and were then transferred to the field with 100 per cent survival. Among citrus species, lime (*Citrus aurantifolia*) plants were established from root culture and maintained for two years without losing morphogenetic potential.

Objectives of In Vitro Conservation

1. To ensure the maximum level of genetic stability of the conserved material that is consistent with practical storage periods and with the available technology and resources.
2. To maintain accurate and detailed records concerning the biology of the conserved material, and its know history prior to collection. It is important to known the disease status of the material, and its full culture history including media details, the use of growth regulators, frequency of transfer etc. Cultures and derived plants should be characterized at some level of morphology

performance and any known genetic characteristics, including analysis of gene products and DNA fingerprinting where this is possible and appropriate. There is a responsibility to disseminate this information as widely as possible.

3. To develop and improve technology at all levels, and to increase the efficiency and utility of in vitro techniques in the conservation of plant germplasm.
4. To construct and maintain a policy for the in vitro conservation of those plant genetic resources appropriate to the collection with due regards to Internationals. National and Regional/Local priorities.
5. Conservation of somaclonal and gemetoclonal variations in cultures.
6. Maintenance of recalcitrant seeds.
7. Conservation of cell lines producing medicines.
8. Storage of pollen for enhancing longevity.
9. Conservation of rare germplasm arising through somatic hybridization or other methods of genetic manipulations.
10. Delaying the process of ageing.
11. Storage of meristem culture for micro propagation, micro grafting and production of disease free plants.

Applications of In Vitro Conservation

Techniques for in vitro conservation of plant germplasm may be applied in any situation where the basics of tissue culture and cell culture are available, and technologies discussed above can be applied to wild species, the crops of agriculture and horticulture and tress. Methods for minimal growth storage are available to most laboratories, and it becomes a matter for local decisions as to whether attempts at conservation should be made and in some way are formalized. This would mean, a commitment to adequate documentation for conservation purpose and to disseminating information about the collection, however limited, to suitable parties.

Cryogenic storage requires relatively simple equipment and facilities, but these are often costly which limits its application. Liquid nitrogen is a relatively common material that can be obtained in most parts of the world and its availability only occasionally restricts the use of cryopreservation in conservation. As this technique provides dramatically extended storage periods, it is perhaps even more important that care is paid to the quality of documentation and record keeping associated with the conservation processes. Major organisations such as IBPGR and ATCC (American Type Culture collection) are in a position to apply, develop and promote in vitro technology in the service of genetic conservation. Importantly, they are sufficiently large and international to be a vehicle for collection, collation and dissemination of information. They have a responsibility to improve in vitro techniques and to be involved in edition whilst such organisations can be seen as an effective worldwide model with regard to in vitro conservation. Other national and regional organisations should be prepared to take on a similar rate within their sphere of influence. There are also obvious applications of this technology in all aspects of research and development, including the work of commercial organisations.

Progress in the development of plant cell and tissue culture techniques for long-term germplasm conservation has been quite significant especially in view of the level of activity and small number of workers involved. The techniques of cell culture in particular have been so refined that only a few species are likely to the recalcitrant to known procedures. During the past decade studies on the freeze preservation of cell culture have trended from basic aspect.

The evidences available so far indicate that freeze preservation is the most reliable approach to the long term preservation of cell culture which possesses the biosynthetic capacity for synthesis and accumulation of secondary metabolites.

Accumulation of secondary metabolites requires a certain degree of differentiation, often accompanied by increase in cellular and volume as well as lengthening of generation period. Therefore the identification of appropriate growth stages most conductive to freezing becomes critical in devising an appropriate cryoperservation strategy of cell producing secondary metabolite.

Germplasm conservation using cell, tissue, and organ culture has been reported in 60 species including various crops. From the information available, conservation is possible in both monocotyledonous and dicotyledonous families. Temperate species figure more prominently then tropical species.

In biotechnology laboratories, freeze preservation and storage of plant material can be of enormous value in maintenance of stock cultures and samples that await screening for future used. Almost an indefinite number of replicates from germplasm of large number of plant species can be stored in vitro using little space in contrast to the thousand of hectares of land that would be required for the same number of plants if maintained in-situ. Thus, the material that would otherwise be discarded due to resource can certainly be stored without reservations.

TABLE 19.6: In vitro regeneration of threatened/rare/ vulnerable /endangered plant species

Name of Plants	Explant Used	Medium	Mode of Regeneration	Reference
Aconitum carmichaeli	Shoot tips & axillary buds	MS	Clonal multiplication	Hatano et. al. (1987)
Aconitum nepellus	Nodal shoot	MS	Axillary shoot proliferation	Robert cervelli (1981)
Aconitum novebora-acense	Nodal shoot	MS	Axillary shoot proliferation	Robert Cervelli (1981)
Anogeissus sericea	Nodal segment (seedling ex-plant)	MS	Axillary shoot proliferation	Kaur et. al. (1992)
Caralluma edulis	Shoot segments	MS	Clonal multiplication (shoot tip)	Hatanco et. al. (1987)
Commiphora wightii	Shoot segments	MS	Axillary shoot proliferation	Barve and Metha (1993)
Coptis teeta	Hypocotyl segment	MS	Callus culture	Tandon and Rathore (1991)
Gerbera aurantiaca	Axillary bud	MS	Axillary shoot proliferation	Mayer Vanstaden, (1998)
Nepenthes khasiana	Mature nodal segment	MS	Axillary shoot proliferation	Rathore and Tandon (1989)
Nepenthes khasiana	Axillary bud	MS or Wood Pt. Medium	Axillary shoot proliferation	Latha and Seeni (1994)
Ocotea catharinensis	Zygotic embryos	MS	Somatic enbryogenesis	Casta et. al. (1993)
Oncidium varicosum	Seeding	Knudson medium	Root tip culture	Kerbeay (1983)
Picrorhiza Kurroa	Axillary bud	MS	Axillary shoot proliferation	Upadhyay et. al. (1989)
Podophyllum hexandrum	Zygotic embryos	MS	Somatic embryogenesis	Arumugan and Bhojwani, (1989)

Name of Plants	Explant Used	Medium	Mode of Regeneration	Reference
Rauvolfia serpentina	Low temperature	MS	Storage	Sharma and Chandel (1992)
Rheum embodi	Axillary bud	Liquid	Axillary shoot proliferation	Lal and Ahuja, (1989-93)
Rumex acetosella	Axillary bud	MS	Somatic embryogenesis	Calafic et. al. (1987)
Trichopus zeyanicus	Axillary bud	MS	Axillary shoot proliferation	Krishan et. al. (1995)
Valeriana wallichii	Axillary bud	MS	Callus culture	Mathur and Ahuja (1991)
Wrightia tomentosa	Axillary bud	MS	Axillary shoot proliferation	Purohit et. al. (1994)

Recent Development of In Vitro Conservation

The role of the gaseous environment within the culture vessel in growth and development has been considered with view of extending the relatively low-technology approaches available for in vitro conservation. Volatile compounds in the gaseous culture environment, together with carbon dioxide, will affect growth rates in vitro. Other physical factors will be also moderate to the biological effects of the gaseous environment. The most significant of these to date are the total volume of the gas space above the culture medium and the diffusion distance from the culture tissue to the point of gas exchange with the external atmosphere i.e. the seal of the vessel.

In a study using regenerated carrot plantlets range of culture, vessels were constructed with identical gaseous volumes but differing diffusion distances. Biomass measurements at various stages during a 30 day culture cycle showed that growth increase was independent of the gaseous volume above the cultures at least within the range studies, but was significantly affected by changes in the diffusion distance. Increasing the diffusion distance from 50 to 150 nm, which effectively doubled the period between culture transfers, halved biomass increase. This provides an inexpensive technique that provides some degree of extended storage for in vitro conservation systems.

It is also possible that the advent of gas permeable culture vessels might assist conservation *in vitro* at the level of unfrozen storage. These containers of various designs are made of relatively inexpensive plastic. Their benefits are for organised plantlet cultures in the main, as well as they allow autotrophic metabolism to develop to a significant extent, reducing the dependence of the culture upon high levels of exogenous carbohydrates, and similarly reducing the need of complex growth regulators regime. This simplification of the culture environment allows for significantly prolonged growth in culture vessel, with basic nutrients being added using, for example, a hypodermic syringe for sterile transfer of basic medium.

Recent progress in the techniques of induction and manipulation of somatic embryogenesis may also have considerable implication of prolonged maintenance of tissue-derived in vitro, with the possibility of a high degree of genetic stability. The technique may be particularly applicable when the material of interest is an isolated cell or protoplast population that can be manipulated into somatic embryogenesis relatively simple and at high frequency. In particular, techniques have been developed that allow for the desiccation of somatic embryos in a way that reflects the physiology and development of the zygotic embryo of the orthodox seed. The desiccation system relies upon pre-treatment of the somatic embryos with compound such as ABA, IBA and sucrose, added to the culture medium and has been successfully demonstrated for species as diverse as grape, soybean and wheat, whilst not applicable to species whose zygotic embryos are in any case recalcitrant. These developments offer the potential for extended storage of germplasm derived from in vitro system by storing dried, somatic embryos in a manner similar to that used for conventional seed storage. This might provide for long-term storage without an absolute requirement for the relatively advanced technology associated with liquid nitrogen storage systems. If the degree of drying and its associated cellular adaptations do accurately reflect the characteristics of the orthodox seed it might be possible that the material could be kept at room temperature, or the more conventional at –29°C of the mechanical freezer and yet achieve the level of genetic stability required for effective in vitro conservation (Roberts et. al. 1981). Further investigation will be required to see if dried somatic embryos can indeed produce the storage characteristics, without genetic alterations of the zygotic orthodox counterparts, coupled with a coating technique used to produce artificial seeds, this system has even greater potential.

The relatively recent development of a field collection system where material in the field is taken directly into a protective by possible temporary *in vitro* environment is a significant enabling technology to be used in conservation. The technique employs various levels of sterilization for explants in the field and incorporates antibiotics into the growth medium. Explants could be collected in situ even inmost remote habitats and transported back to the laboratory with the minimum of loss, and without the need for transported back to the laboratory with the minimum of loss, and without the need for transporting whole plants or large pieces of them. To date the successful cryopreservation options available for in vitro plants tissues cultures have largely been based upon slow cooling techniques although faster rates have been beneficial at times. At slow cooling rates there is degree of cellular dehydration following extracellular freezing and then the plunge into liquid nitrogen that typically concludes such slow looking protocols.

To take pure water through the glass transitions temperature to form a glass, as opposed to crystalline ice requires or ultra rapid cooking rates under practical circumstances complete vitrification of any significant volume of water is rarely, it ever obtained. A working definition of the vivification does allow for some crystalline ice embedded in a glassy matrix where ice crystal size does not exceed 10 nm diameter. Aqueous solutions such, as those within and surrounding cells in suspension, will vitrify more readily than water alone, and at lower cooking rates, as the solute concentration increases solutions undergo vitrification more readily and so cells incubated with very high levels of cryoprotectant solution can e taken through vitrification as an alternative to the more conventional techniques of cryopreservation. The benefit is that following the appropriate pre-treatments required to achieve vitrification, the need for a controlled rate of cooling into liquid nitrogen is removed and simple quenching is likely to be sufficient.

Recently, the techniques of vitrification have been successfully applied to *Asparagus* and *Brassica* cells cultured *in vitro* adding another potential valuable option to the methodology for cryogenic storage.

In the case of *Brassica campestris* suspension cells were first equilibrated in 1.5 M ethylene glycol solution and then dehydrated in a complex medium estimated as being approximately 19 osmolal. The high concentration of extra-cellular solute well result in water loss from the cell, leaving a high concentration of intra-cellular solute also. In this instance vitrification was achieved by plunging a 0.5 ml plastic straw containing cells and medium directly into liquid nitrogen. To avoid crystallization during thawing rapid warming rates has been employed, which was close to 600°C per min. in this instance. Significant levels of survival were recorded using tripheny tetrazolium chloride (TTC) assay as indicator, and these results confirmed by callus regrowth. *In vitro* genetic conservation is accomplished in two ways. First the active *in vitro* culture bank the material is maintained as tissue culture by limiting the growth. The use of minimal or modified growth media, growth inhibitors in media, mineral oil sprays and low temperature is done for this purpose. The second is the base *in vitro* genebank or cryopreservation, i.e., storage of seed or plant parts in liquid nitrogen at -196°C.

***In vitro* culture bank (Tissue Culture):** Generally, in vitro maintained plant cells, tissues and organs are preserved under appropriate conditions for long terms storage as *gene banks*. *In vitro* methods are the most useful for conservation vegetatively propagated plant, species with recalcitrant seeds and genetically engineered materials. The potential advantages of in vitro conservation include:

1. Requirement for little space for preservation of a large number of clonally multiplied plants.
2. Maintenance of the material in an environment, free of pest or pathogens.
3. Protection against dangers of natural environmental hazards.
4. Availability of nucleus stock to propagate a large number of plants rapidly whenever necessary and
5. Minimizing the obstacles generally imposed by quarantine system on the movement of live plants across national boundaries, since they are raised and maintained in aseptic environment. The present chapter deals only with the application of plant cell and tissue culture in germplasm conservation. Plant tissue culture based biotechnology has opened new avenue for the propagation and improvement of plant species. A wide range of plants has now been regenerated through techniques of tissue culture. Many palm species including *Cocus nucifera*, Oil palm and *Phoenix dactylifera* (Data palm) multiplies mainly through seeds, and there is no efficient method of vegetative propagation.

The development and application of plant tissue culture system of propagation and conservation would therefore have obvious benefits.

1. Many long lived forest trees including both angiosperm and gymnosperm do not produce seed until a certain age and must be propagated vegetatively when it is desired to reproduce similar to the parental genotype.
2. One area of research currently being investigated is the possibility of very rapid clonal propagation of fruit trees by means of tissue culture techniques. These techniques have special relevance to certain fruit tree cultivars, which are either difficult and/or expensive to propagate by conventional means.
3. Somatic embryogenesis for higher frequency plant regeneration and synthetic seed production.
4. Generating genetic variability for better genetic traits (Larkin and Scowcraft, 1981) somatic hybrid production for increased fields.
5. Cryopreservation of valuable gene pool for future use. (Withers, 1985).
6. Production of cell line and recovery of plants for disease resistance, drought resistance, salinity resistance.
7. *In vitro* propagation, improvement and conservation of rare/vulnerable/ endangered plant species (Withers, 1986).

Modes of Propagation

There is a large number of threatened/rare/endangered plant species on which in vitro propagation of threatened plants could be achieved by the following methods:

Clonal Propagation: A variety of plants species can be propagated through forced axillary shoot bud proliferation. Thus plants produced through these methods are morphologically and genetically similar to the parental plant. Main benefits of these methods are:

1. Rapid multiplication of superior clones and maintenance of uniformity.
2. Multiplication of disease free plants.
3. Multiplication of sexually derived sterile hybrid.

Clonal propagation is achieved by placing sterilized shoot tip or axillary bud on culture medium that it's sufficient to induce formation of multiple buds.

In vitro propagation of plant using apical shoot or nodal segments was obtained in *Anogeissus sericea*. Propagation of *Nepenthes khasiana*, a threatened plant species endemic to Meghalaya was obtained using nodal segment from the field grown plants and from seedling raising explants. *Gerbera auranitca*, threatened ornamental species recovered by the culturing axillary shoot buds on nutrient medium (M.S. medium) containing 5 MMBAP, which differentiated into shoots and differentiated shoots include root on IBA (5 to 10 MM) containing medium. A wide range of plants has now been regenerated through this technique.

Somatic Embryogenesis: The process of development of embryogenesis. This was first demonstrated in Carrots where bipolar embryos developed from single cell. Further, the tissue derived from these somatic embryos spontaneously developed into embryos again. It is based on the concept of plant extracts, growth regulator and by the physiological state of calli.

In vitro propagation of threatened medicinal plants through somatic embryogenesis has been obtained in *Podophyllum hexandrum*, and *Santalum album*. Research work is also being done to produce artificial seeds via this pathway, but a lot of work has to be done in the area of embryo encapsulation, storage and conversion into plants.

Organogenesis: The process of regeneration of plants from the callus called as organogenesis. In *Meconopsis panicutlata*, regeneration was observed through organogenesis (Sulaman, 1994) and same type of mode of regeneration has been observed in *Meconopsis simplicifolia*, an endangered plant species of Himalayan region. *Rheum embodi* - a threatened plant was regenerated from shoot tip and leaf explant through organogenesis. A number of such examples can be sited with have developed the complete plant through regeneration by organogenesis thus helping in conservation of germplasm.

Callus Differentiation: *In vitro* regeneration has been reported in *Coptis teeta*, an endemic and endangered plant of Arunachal Pradesh through callus differentiation into shoots and subsequent induction. In *Valeriana wallichii*, a threatened plant was propagated through callus differentiation and recovered complete plantlets.

Cryopreservation

Cryopreservation offers a viable and economical method for the long-term conservation of genetic resources. Various techniques have been developed to minimize desiccation and freezing damage, thus ensuring high propagule recovery. In most cases, shoot apices obtained from *in vitro*-grown shoots are the plant material used for cryopreservation of vegetatively propagated plants. Cryopreservation techniques are based on either freeze-induced dehydration (classical methods), or vitrification of internal solutions (new methods). Various considerations should be taken into account when cryopreservation techniques are actually used for germplasm conservation.

The strategy of conservation depends on the nature of the material and on the objective and scope of the activity. The length of the life cycle, the mode of reproduction, the size of the individuals and its ecological status define the nature of the material, whether wild, weedy or domesticated. The purpose to which the material is put will determine the degree of integrity, which is essential or desirable to maintain. Conservation strategy must take into consideration the time dimension whether it is for short-medium or long-term storage and where the storage will be located. The genetic make-up of the material and the type of sample collected will reflect the breeding system of the material and how it will be regenerated.

Cryopreservation is, so far, the only viable procedure for long-term germplasm conservation of vegetatively propagated species and therefore, can be used for base collections. Cryopreservation implies the conservation of plant propagules at very low temperatures (below –150°C), thus ensuring that all metabolic processes stop. In this way, subcultures are not required and the threat of somaclonal variation is reduced.

Liquid nitrogen (–196°C) is usually used as refrigerant, although freezers with working temperatures of –150°C are now available. Cryopreserved material requires a limited space. However, power or liquid nitrogen supply must be guaranteed. Different types of plant cell, tissues and organs can be cryopreserved, including cell suspensions, pollen, embryogenic cultures, somatic and zygotic embryos, shoot apices or meristems. For vegetatively propagated species, the most widely used organs are shoot apices excised from *in vitro* plants. In some woody species, cryopreservation of dormant buds has also been developed, and recovery was achieved by grafting. Cryopreservation of *in vitro* cultured apices has certain advantages, among them that cryoprotective treatments can be applied to the shoots or apices while cultured *in vitro*. Besides, *in vitro* culture can be used for clonal propagation of the starting plant material and for virus-free plant production. Plant germplasm distribution in the form of *in vitro* cultures is usually less voluminous and improves health status. Furthermore, somaclonal variation is less probable to occur when recovery of plants is carried out directly from apices compared to other methods, for example by direct or indirect organogenesis. In several cases, embryogenic cell lines have shown to be highly stable, although that stability may differ among species and could decrease with time in culture. Besides, propagation procedures through somatic embryogenesis are not fully developed in many species.

Cryopreservation Techniques

Most plant living cells have high quantities of water and they are extremely sensitive to temperatures below 0°C. Therefore, cells should be dehydrated to avoid ice crystal formation. However, extreme desiccation also produces damages (on cell membrane, due to high concentration of internal solutes, protein denaturation). Cryopreservation techniques have been developed to minimize both types of damages.

Application of Cryopreservation Techniques

The number of plant genetic resources centers where cryopreservation is used is still low and usually concerns few accessions only. The new cryopreservation techniques do not require expensive equipment (programmable freezer). In all cases, personnel should be properly trained in, *in vitro* and cryopreservation techniques, otherwise results may vary greatly. Some preliminary testing should be carried out with a few cultivars.

Vitrification usually requires less handling than encapsulation-dehydration, but solutions may be toxic. Encapsulation-dehydration protocols are usually simple, but more handling of alginate beads (although easier due to their size) is required and some species do not tolerate the high sucrose concentrations employed. The advantages and disadvantages of each technique should be considered. However, other factors such as personnel, available facilities; and type of plant could influence the selection of the technique. Those first tests with few cultivars could give indications on how the protocol should be improved to obtain high percentage recovery. For genebank management, recovery as normal apex regrowth giving healthy *in vitro* plants should be the goal.

As genotypes usually respond differently to the same protocol, curators should make the following decisions:

1. Either modify the protocol for each genotype or group of genotypes to obtain high recovery in most of them, or
2. Store more propagules (to compensate for low recovery) with a protocol that may be not optimal for all genotypes.

More recently, probabilistic method has been developed to help curators to take certain decisions, for example, on the number of cryopreserved propagules to be tested in order to get a precise estimation of the expected recovery percentage, or on the minimum number of propagules which should be rewarmed per accession to obtain a fixed number of viable plants.

A major concern of curators is the genetic stability of conserved material. *In vitro* culture has been reported to induce genetic changes (somaclonal variation) in some cases. Care should be taken to test that the *in vitro* culture phase does not produce genetic instability. Although not many studies have been performed yet, cryopreservation protocols seem to ensure genetic stability of the plant material. Genetic stability of cryopreserved potato shoot tips has been studied through ploidy status or ribosomal genes stability. Study of the DNA variation through amplified fragment length polymorphism assays have been carried out in strawberry, apple, grape, and kiwi. To date, there is no substantial evidence to suggest that plants regenerated from cryopreserved shoot apices are genetically changed. In potato, DNA polymorphism observed was not induced by cryopreservation but by the whole process employed, including the tissue culture phase. In other works where variants were regenerated from cryopreserved apices, those were due to their chimeric structure.

Classical Methods

Plant cryopreservation procedures were firstly developed following the success of animal cell cryopreservation. They were based on chemical cryoprotection and slow cooling, followed by rapid immersion in liquid nitrogen. It is named controlled freezing, slow freezing or two stepfreezing method. By decreasing temperature at a relatively slow rate, ice crystals are formed in the extracellular solution and water is removed from the intracellular one, leading to cellular dehydration and therefore avoiding intracellular ice formation. Such cryopreservation procedures have been highly successful with callus and cell cultures, consisting of rather uniform and small units. For example, sugarcane and banana embryonic calli were successfully cryopreserved with this method and the field performance of the recovered plants studied. In both cases, no differences between cryopreserved and control plants, on agronomic traits and/or growth descriptors, were found after a period (two culture cycles or 12 months) of growth in the field. This technique is usually not very effective to cryopreserve larger structures, which comprise different cell types, such as shoot apices. However, there are some successful examples of two-step shoot-tip cryopreservation, some of them with vegetatively propagated species such as potato or chrysanthemum. In this method, shoot apices are treated with cryoprotective substances such as dimethyl sulfoxide (DMSO), ethylene glycol, polyethylene glycol, mannitol, sorbitol or sucrose, either alone or in mixtures. Subsequently, they are slowly cooled, usually at rates of 0.5-2°C min-1, to –40°C and then rapidly immersed in liquid nitrogen. Rapid rewarming is usually required to avoid recrystallisation phenomena. Transfer to fresh medium after one day in culture is recommended due to the toxic effect of cryoprotective mixtures. Controlled cooling rate can be achieved with the use of (expensive) programmable freezers or by more simple devices. A cooling rate of 0.4-0.6°C min-1 was obtained with an ethanol bath and a –40°C freezer for cryopreservation of sugarcane cultures.

New Methods

For the last ten years, new plant cryopreservation procedures have been developed, which are based on vitrification. Vitrification can be defined as «the solidification of a liquid brought about not by crystallization but by an extreme elevation in viscosity during cooling. During vitrification, the solution is said to become an amorphous glassy solid, or glass. Vitrification of solutions is achieved by reducing intra- and extracellular freezable water, either by exposure of plant tissues to highly concentrated cryoprotective mixtures or by physical desiccation, and subsequent very rapid cooling, generally by direct immersion in liquid nitrogen. Cryopreservation techniques based on vitrification are usually simple to carry out and applicable to complex structures such as embryos and shoot apices.

There are two main types of new cryopreservation techniques, although combinations of them have also been used. The first one is actually termed vitrification (*sensu stricto*) and the second one encapsulationdehydration.

In the vitrification technique, the plant material is exposed to highly concentrate cryoprotectant solutions for short periods. Previously, to induce desiccation tolerance, tissues are cultured on medium with high sucrose (e.g. 0.3 M) or sorbitol (e.g. 1.4 M) concentration and subsequently transferred to a glycerol-sucrose solution, called loading solution (e.g. 2 M glycerol + 0.4 M sucrose). The droplet technique can be considered a modification of the previous one. It was developed for potato germplasm cryopreservation at FAL (Germany) and consists on treating shoot apices in drops of a 10% DMSO solution placed on aluminium foil strips, which are rapidly immersed in liquid nitrogen.

The encapsulation-dehydration technique is based on the artificial seed technology. This technique was developed by Fabre and Dereuddre (1990), and consists on the inclusion of apices in alginate beads and their subsequent culture in a highly concentrated (0.7-1.5 M) sucrose solution followed by physical dehydration, and direct immersion in liquid nitrogen.

Culture of apices on sucrose enriched medium (0.3- 0.7 M), prior to encapsulation, usually improves survival after desiccation and freezing. Physical desiccation is carried out either with silica gel or in the airflow of the laminar flow cabinet. Water contents of around 20% (fresh weight basis) have resulted appropriate for high survival after freezing of vegetative explants in several species.

More recently, protocols combining the above techniques have been developed and named encapsulation vitrification techniques. Apices are firstly encapsulated and then submitted to vitrification, with no requirement for physical desiccation. In many temperate species, incubation of the *in vitro*-shoots, nodal segments or apices at low temperature (generally 4° to 10°C), for periods ranging from days to weeks, increases survival after freezing for both types of techniques, classical and new. During cold acclimation cellular changes (numerous smaller vacuoles, more abundant mitochondria and rough endoplasmic reticulum) and accumulation of certain proteins occur, as it has been shown in cryopreservation studies of peach cell suspensions. During shoot tip cryopreservation of non cold-hardy tropical species, preculture on sucrose enriched medium (0.3-0.8 M) improved survival after cooling either by encapsulation dehydration or by vitrification.

In some cases, an apex pregrowth on sucrose rich medium has proven sufficient for survival after cooling in liquid nitrogen. That is the case of banana meristem clumps cultured on 0.4 M sucrose for 2 weeks.

Conservation and Recovery of Germplasm

Long-term conservation can take place in liquid nitrogen (–196°C) or in the vapour phase. However, storage temperature should not be higher than –130°C to prevent solution devitrification and large ice crystal formation. Those temperatures can be also achieved by deep freezers running at –150°C. For the recovery of apices after cryostorage, rapid rewarming is usually required to avoid recrystallisation. Vials containing the apices are usually immersed in a water bath at 35-40°C. When apices are not included in vials (e.g. the droplet technique) rewarming usually takes place in liquid medium at room temperature. In many species, recovery of apices cryopreserved with the new techniques is direct, without callus formation. By contrast with the classic techniques, the structural integrity of most cells is well preserved

Limitations of *In Vitro* Conservation

There are practical limitation to any conservation system that does not provide a maximum genetic resource whilst, at the same time, conserving the potential of the system to produce whole plants. The level of limitation that is acceptable depends upon individual circumstances. However, where resources are not themselves limiting and the pressure of the time allow for significant research and development, the technique of cryopreservation must be the current in vitro conservation method of choice.

Yet, application of drying to somatic embryos, and the construction of artificial seeds, presents and exciting prospect for the future. Once successfully processed, the dried somatic embryos might be effectively stored at -20°C using a simple domestic refrigerator, which is globally available technology. However, the pre-treatments necessary to achieve desiccation without loss of viability are often complex and currently it is only the tissue of somatic embryos that can be treated in this way. It is likely that the somatic embryo shares much of its genetic expression and developmental pathways with the zygotic embryo. Both have a potential to withstand extremes of desiccation when the zygotic embryo, and that both have a potential to withstand extremes of desiccation when the appropriate sequence of genes are in operation. In a developing orthodox seed (Roberts, 1973) the hydrated, zygotic embryo will respond to chemical stimuli from the parent plant to develop desiccation resistance and thus dry to the low water content characteristic of the see. It appears that a somatic embryo can be triggered artificially in much the same way. This technique has obvious potential in circumstances where embryogenesis is both possible and desirable, and where the zygotic embryo does not show recalcitrant characteristics. However, it is not clear whether the techniques for drying could be feasible for other somatic cell and tissue types, that have patterns of development and differentiation far removed from that of the embryo whilst it is true that leaf callus cell for instance, contain the genes used in the developing embryo to tolerate desiccation during seed formation. It is another matter to be able to exploit this aspect of the genome independently of the rest of the embryo development.

As cryopreservation has developed as a technology, and surviving material examined more critically, the concepts of optimized procedures and modified culture conditions by support recovery growth following thawing have been considered in more depth.

It is probably an exceptional occurrence when a cryopreserved cell is thawed to physiological temperatures with structure and functions completely unaltered, ready to resume typical activity. More commonly, surviving cells will have varying cells will have varying degree of damage that may will be able to repair, depending upon available reserves and metabolic integrity, and may require specific "recovery" media or past thaw conditions to ensure a reliable return to a known and predictable performance. Those that cannot effect satisfactory repair will die, sometime after thawing. Different circumstances of cooling and warming may well generate different lesions and consequently differing requirements for repair, and it is not improbable that, even after repair, recover ells might show a range of alteration in both structure and function.

Future Strategies of *in vitro* conservation: With regard to horticultural crop germplasm, *in-situ* conservation strategies involving farmers more directly need to be considered, particularly for populations of wild crop relatives and wild food plants in the areas where they are found naturally. The term *in-situ* conservation also takes in the maintenance, by farmers, of domesticated landraces or traditional crop varieties. In the case of horticultural crops, home gardens in many parts of the world are examples of on-farm germplasm conservation efforts, which deserve to be given greater support. In case of horticultural species, the tasks to be undertaken to achieve effective *in-situ* conservation of crop relatives and wild food plants are really too vast to be contemplated. Hence there is a need for setting priorities and identifying those taxa and habitats for which protection is most urgently needed, and determining the most important, effective and efficient interventions to be undertaken. *Ex-situ* conservation strategies, need to be based on rationalizing criteria in order to i) maintain collections of genotypes that include most of the traits needed for genetic improvement programmes with the necessity, ii) to reduce gene bank maintenance costs and iii) minimize storage space requirements. In this regard, the concept of the core collection, i.e. the selected part of an entire collection, which effectively represents the genetic diversity of a crop species and its wild relatives, fulfils a useful purpose for breeders and other users. Needless to say, the selection work entailed in establishing a core collection requires an intricate knowledge of the genetic variability of the taxa in question, and this can only be derived from exhaustive studies on characterization and evaluation of the contents of collections. In future IIHR will be concentrating on

- Consolidation and inventorization of PGRs related to horticulture at IIHR-to be further geared up in the post CBD regime
- Emphasis on exploration and collection of indigenous PGRs related to horticulture at IIHR and other national institutes
- Ex situ conservation in the form of FGBs, seed, pollen, in vitro and cryopreservation for maintenance of maximum genetic diversity.
- Participation in regional and national activities related to in situ conservation
- Establish links with NBPGR for management of PGRs related to horticulture crops.

DNA storage: This is rapidly increasing in terms of importance. DNA can also be used for conservation of plant species. Preservation in the form of extracted DNA or as genomic DNA libraries can be a major thrust area. However, this techniques needs to have further studies to establish as a practical conservation strategy.

Other Methods

Cold Storage: Germplasm conservation by storing material in cultures at low and no-freezing temperature the ageing of plant material is slow down but not completely stopped as in freeze prevention.

Low pressure and low oxygen Storage: Attempts have been made to develop low pressure Storage (LPS) and Low Oxygen Storage (LOS) as feasible techniques for future conservation of cultured plant materials. In LPS, the atmospheric pressure surrounding the tissue cultures is reduced resulting in partial decrease of the pressure created by gases in contact with the plant material. On the other hand, in LOS, the atmospheric pressure (760 mm Hg) is not reduced but the inert gases (Particularly nitrogen) are combined with oxygen to create low oxygen pressure.

These are found useful in both short term and long term storage. There is also belated recognition that nature is infinitely more inventive than chemists and that natural molecules have with stood many tough tests of survival for millions of years. An effort is now onto revive culturing of plant in vitro. A biotechnology firm in USA. *Phytera*, has for example set up and stored plant cell cultures of 5,000 species in glass vials.

Genotype frequency gene frequency conservation: It is truly achievable in vegetatively propagated material. The aim is to prevent the loss of genetic information.

Static conservation: It takes the form of the storage of seeds or vegetative material in gene banks. It reduces the life processes to a low level. It is safe and the cheapest method.

Seed Banking: Similarly, the efficient and economical way by seed conservation has been developed as seed banks. Seeds of most crops can be stored for long periods under low relative humidity and low temperature conditions. However, seed storage is not feasible in some cases. Seeds of some species are recalcitrant or intermediate, i.e. they cannot stand desiccation below a relatively high critical water content value (10-12% or 20% fresh weight basis, respectively, and cold storage without losing viability. Species with recalcitrant or intermediate seeds are usually native from tropical or aquatic habitats, some of them with agricultural importance such as *Hevea brasiliensis* (rubber tree), and *Elaeis guineensis* (oil palm). Some forest tree species from temperate climates have also recalcitrant seeds (*Juglans* spp., *Quercus* spp.). Intermediate seeds can be stored by partial drying, although for shorter periods than those of orthodox seeds.

Conservation of recalcitrant seeds under humid conditions can be carried out only for short periods, due to germination onset, fungal attack or viability loss. A second group of species for which seed banking is not feasible are those vegetatively propagated. They are usually highly heterozygous and, in some cases, do not produce seeds. Some important crops belong to this group, including *Dioscorea* spp. (yam), *Solanum* spp. (potato), *Musa* spp. (banana), *Manihot* spp. (cassava), *Colocasia esculentum* (taro), and *Ipomoea batatas* (sweet potato). The genetic conservation of these species must be carried out in field collections.

Conservation of Plant Genetic Resources in India

India's area is 3,287, 263 km^2, 2.4 per cent of worlds area of which 60% used for agricultural production, 20000 vascular plants, out of which 7000 are endemic, 160 species are cultivated, 35 species are domesticated in India. Distribution of 144000 *ex situ* conserved accessions as (3% medicinal/ spices, 54% cereals, 18% pulses, 14% oil seeds, 1% improved varieties, 5% vegetables, 5% duplicates, 2% fibre crops). Of these *in vitro* conservation accounts for 0.6% (849 accessions)

1994 –95 total budget US $ 2 million the overall total of Indian domestic expenditure for the national conservation program amounts US $ 6.8 million, which is comparatively high for a developing country. NBPGR distribution rate is only 2.3% in 1988 (basic, if its is more than 10% the standard of distribution is judged as good). The investments in the conservation of PGRFA are undertaken without clear optimality criteria. This again relates to the uncertainty of expected benefits. Only a well-balanced combination of ex situ and in situ conservation seem to meet all relevant objectives. Conservation activities seldom managed accurately and effectively. And information on number and what kind of varieties conserved *ex situ* is lacking. Without a global national and local survey and inventory of existing crop species and varieties, there is no scientific means to decide how many and which crop species must be conserved on a global level.

Documentation

For proper utilization of the conserved genetic resource, meaningful documentation is essential. Detailed information on each accession is made available in an easily accessible form with regards to various horticultural characteristics, such as plant vigour yield, ripening period, fruit quality, resistance against pests, diseases and nematode and tolerance to heat, cold, drought and water logging. The documentation is usual done in the form of standard crop descriptor lists, inventories, directories, catalogues and international check lists which help in finding out and filling the gap in the collection, identification and discarding the duplicates. Gene banks or centres of plant genetic resources are sources of plant diversity. The plants maintained include economically important food crops, horticultural plants, forages, medicinal plants and trees. Gene banks do not maintain all types of plants, as this would be a Himalayan task. Hence, only selected plants are maintained.

A national gene bank may have all or a few of the following activities:

- Acquisition of new germplasm samples
- Multiplication/regeneration of germplasm
- Characterization and preliminary evaluation of germplasm
- Documentation and exchange of information on germplasm
- Conservation of germplasm
- Supply of germplasm
- Collaboration with other plant genetic resource centres
- Organization of technical meetings and training workshops
- Research

The data are normally recorded under varietal evaluation at various researchstations but on uniform pattern. Such data are neither comparable nor can be computerized, hence give inadequate information. In order to facilitate the data management through computers, it is necessary to use the standard descriptors which have been prepared by the International Board for Genetic Resource (IBPGR) for all important crops. The processes and procedures involved in the documentation of a given gene bank depends on the activities and also the type of collections it deals with. The gene bank may be involved in different types of collections. They are,

a) **Base collection** – a collection of germplasm, which is kept for long-term secure conservation and not to be used as a routine distribution source where seed is usually stored at sub-zero temperatures with low seed moisture content.

b) **Active collection** - a collection of germplasm used for regeneration, multiplication, distribution, characterization and evaluation. Ideally, germplasm in the active collection should be maintained in sufficient quantity to be available on request. Active collection germplasm is commonly duplicated in a base collection and is often stored under medium to long-term storage.

c) **Field collection or Field gene bank** - a collection of living plants, germplasm which would otherwise be difficult to maintain and often the only type of collection maintained by the gene bank as a part of active collections.

d) **In vitro collection** - a collection of germplasm kept as plant tissue grown in active culture on solid or in liquid medium as active collections and in some cases at very low temperatures such as in liquid nitrogen as a part of base collections.

The need for a documentation system:

Gene banks will have problems managing their collections if they do not have up to date, accurate and reliable information stored in a systematic and easily accessible way. Such a documentation system allows gene banks to use information to plan their day to day activities and to maximize the use of their often limited resources. Also, without an effective documentation system, gene banks cannot develop with a clear sense of direction and they cannot easily communicate or collaborate with other institutions.

Stages in construction of a documentation system:

1. ***Obtaining background information-*** In order to develop a documentation system tailored to the gene bank's needs, one need to analyze its set-up, and resources available.
2. ***Documentation objectives-*** Defining priority areas for the documentation effort is very important. These could include documentation of passport data, inventory data, and seed handling procedures, distribution data, and characterization and evaluation trials. The objectives could also include dissemination of information.
3. ***Analysis of gene bank procedures-*** Having defined the documentation objectives one can start the detailed analysis of relevant gene bank procedures. This should explore the resource requirements of each procedure, and the different types of data, which are generated or used in each procedure. The analysis will also give an idea of the information requirements of the system.
4. ***Identification of descriptors-*** Much of the data will be concerned with individual gene bank accessions. To facilitate the operation and maintenance of the documentation system, one needs to organize the descriptors into folders or forms in a manual system or separate files in a computer system.
5. ***Develop data formats entry forms and procedures-*** An important task of the documentation specialist is the design of forms for straightforward data entry, which minimizes the risk of errors creeping into the system. If these forms and procedures are designed well, the accuracy of data entry could be ensured.

CHAPTER 20

Bioinformatics

It is the combination of biology and information technology. It is the branch of science that deals with the computer-based analysis of large biological data sets. Bioinformatics incorporates the development of databases to store and search data, and of statistical tools and algorithms to analyze and determine relationships between biological data sets, such as macromolecular sequences, structures, expression profiles and biochemical pathways. It is mainly used for Map genomes and identification of genes, to determine protein structure and simulate protein interactions, to discover new therapeutic targets and design medicines aimed at the targets and to Assess the effects of virtual mutations on gene function. Computers are required in bioinformatics for their processing speed (allowing repetitive tasks to be carried out quickly and systematically) and for their problem-solving power. It requires expert human input, and the integrity and quality of the source data are also critical because bioinformatics technology uses computational tools provided by the information technology revolution, such as statistical software, graphics simulation and database management, to organize and analyze information about biological systems, which, for biotechnology, is information about cells and biological molecules. Using another product of the information revolution, the Internet, scientists broadcast this information around the world.

SEQUENCING DNA, RNA AND PROTEINS

Bioinformatics is based on the principles of **DNA sequencing**, which is performed using an automated version of the chain termination reaction, in which limiting amounts of dideoxyribonucleotides generate nested sets of DNA fragments with specific terminals bases. Four reactions are set up, one for each of the four bases in DNA, each incorporating a different fluorescent label. The DNA fragments are separated by PAGE, and the sequence is read by a scanner as each fragment moves to the bottom of the gel.

Types of DNA sequence falls under three major forms. Genomic DNA comes directly from the genome and includes extragenic material as well as genes. In eukaryotes, genomic DNA contains introns. cDNA is reverse transcribed from mRNA and corresponds only to the expressed parts of the genome. It does not contain introns. Finally, recombinant DNA comes from the laboratory and comprises artificial DNA molecules such as cloning vectors.

Genome sequencing of DNA molecules (-800 bp) can be sequences in one read. The large DNA molecules (genomes) are first broken into fragments. It is done by **Shotgun sequencing method.** It involves the generation of random DNA fragments, which are sequenced in large numbers. It provides genome of wide coverage. **Clone contigs** sequencing involves the systematic production and sequencing of subclones. High quality sequence data are generated by performing multiple reads on both DNA strands. The preliminary trace data is then base called and assessed for quality using a program such as Phred. Vector sequences and repeated DNA elements are masked off and then the sequence is assembled into contigs using a programme such as Phrap. Remaining inconsistencies must be addressed by human curators. The sequence data of lower quality are generated by single reads (single pass sequencing). Although somewhat inaccurate, single-pass sequences such as ESTs and GSSs can be generated in large amounts very quickly and inexpensively.

Sequencing of RNA are deduced from the corresponding DNA sequences. For this purpose special methods are required for the identification of modified nucleotides. These include biochemical assays, NMR spectroscopy and MS (mass spectroscopy).

Sequencing of Protein is carried out by MS. This technique is used for accurate molecular masses and are calculated from the mass / charge ratio of ions in a vacuum. **Soft ionization methods** allow MS analysis of large macromolecules such as proteins. Sequences are deduced by comparing the masses of tryptic peptide fragments to those predicted from virtual digests of proteins in databases. *De novo* sequencing is done by generation nested sets of peptide fragments in a collision cell and calculating the difference in mass between fragments differing in length by a single amino acid residue.

TECHNIQUES FOR DETERMINATION OF PROTEINS

Techniques for sequencing of DNA and proteins:

1. X-ray crystallography
2. NMR Spectroscopy
3. Electron microscopy
4. CD spectroscopy.

Techniques for gene and expression data:

1. DNA microarrays (Spotted DNA microarrays, Oligonucleotide GeneChips, printed onto the chips using solid-state chemical synthesis during chip manufacture).
2. 2D protein gels

Techniques for Protein-Protein interactions data:

1. **Genetic Methods:** Genetic analysis is used to know about protein interaction. Suppressor mutants may compensate for a deleterious original mutation by restoring a disrupted protein interaction. Similarly, synthetic lethal effects may reflect the inability of two mutant proteins to interact. Dominant negative mutations show that a proteins functions as a multimeric complex.
2. Physical Method (Affinity methods)
3. Molecular and atomic methods (Library-based methods)

Pattern Databases for Protein Families

Assigning sequences to protein families helps in predicting protein function these are stored in secondary protein family database. Consensun sequences provide condense the information from a multiple alignment into single sequence. Derivation of consensun sequences show that any protein family representation is subject to bias if the set of sequences from which it was derived is biased. They can not represent any probabilistic information apart from the most common residue at a particular position. The PROSITE database contains sequence patterns associated with protein family membership, specific protein functions and post-translational modifications. PRINTS and BLOCKS represent protein families by of multiply aligned ungapped segments (motifs) derived from the most highly conserved regions of the sequences. For this purpose WWW-based search engines for the databases are available.

Protein Domain families: Several proteins are built up from domains in a modular architecture. The protein families are best studied by protein **domain** families. Prodom is a database of protein domain sequence created by automatic means from the protein sequence databases. Sequence profiles represent complete domain sequences with scores for each amino acid at each position in a multiple alignment, and position specific measures of the likelihood of insertion and deletion. They are used as an alternative to sequence patterns in some PROSITE database entries. Hidden markov models are rigorous statistical models of protein domain family sequences. They can be viewed as sequences of match, insert and delete states that generate protein sequences according to probability distributions for each transition between states. Pfam and SMART can be used for protein domain family analysis. The integrated resource Interpro unites PROSITE, PRINTS, Pfam, Prodom and SMART.

PRINCIPLES OF GENOME ANNOTATION

Annotation is the method to obtain desirable information of genetic material specially structure and function of genes and other genetic elements from raw sequence data. Due to differences in gene structure and genome organization, the annotation of prokaryote and eukaryote genomes have different problems. To solve these problems, computers are being used to predict where the genes are located in genomic DNA. This is achieved by a combination of **signal sensing** (looking for regions with gene-like sequence context) and **homology searching** (looking for region that match the sequence of previously discovered genes). However, no gene finding method is 100% dependable.

The genomes of bacteria, genes are rarely interrupted by introns. For detection of such genes one has to carry out a six-frame translation of the genome sequence and to identify long ORFs (open reading frames). This method is helpful in detection of most of the genes, however, still more sophisticated approaches are needed to detect short genes, genes that use variations of the genetic code, and overlapping genes.

In higher eukaryote genomes, genes are widely dispersed and interrupted by large and numerous **introns. Exons** are very short and are difficult to be identified by ORF searching alone. Therefore, there is a need to combined approach involving exon detection. The preparation of whole-gene model various techniques are required like content sensing, identification of splice signals, and incorporation of auxiliary information such as cDNA sequences. Still several genes remain undetected or incorrectly delimited, and nonfunctional full or partial gene copies (pseudogenes) may be incorrectly identified as genuine genes. **RNA genes** are identified (since these do not contain a coding region) using homology searches and the prediction of secondary structure method. The annotation of human genome is very difficult. Following are the tools for gene annotation.

Gene-prediction software: Several programmes/methods have been developed for identification of genes in genomic DNA like gene prediction use ab initio methods and/or homology searches. Simple programs such as NCBI ORF finder identify ORFs by carrying out six-frame translations. For complex eukaryotic genomes, more sophisticated statistical analysis methods are required. These may incorporate simple rule-based algorithm (*e.g.* GeneFinder), neural nets (*e.g.* GRAIL) or Markov models (*e.g.* Genie, GenScan). Neural nets and Markov model algorithms must be trained on data from specific organisms.

A continuous pipeline is required to handle the large amount of data from genome projects which is to be annotate *on the fly.* Raw sequence data is fed in at one end. This is assembled and verified, annotated using one or more gene-prediction programs and the results are then fed into a database that allows sequence and annotations to be viewed using a graphical interface. Suitable formats include Artemis, Apollo, EnsEMBL, GoldenPath or AceDB.

BIOINFORMATICS: INTERNET AND WWW

The **Internet** (International computer network) is being used for a particular protocol, known as TCP/IP, to package and route data. Most academic, government and commercial institutions have access to the Internet, and individuals may also gain access by subscribing to an ISP. The Internet provides a versatile system for exchanging biological data. In case the bioinfomatic gateways do not provide the information required, general-purpose search engines can be used to locate pages containing specific key words or phrases. Otherwise, the home pages of academic institution and biotechnology companies will be useful staring points. While using a search engine it is important to refine the search and to avoid ambiguous words and phrases, as these can pull out numerous irrelevant pages. Literature databases can help by providing useful keywords and phrases to use a search terms.

The **World Wide Web** (WWW) is Internet-based system for information exchange using a protocol called **http**. Programs called browsers can access hypertext documents on the Internet. It helps in searching the relevant address, known as a URL. Hypertext documents contain text and other multimedia objects (images, audio files, etc.) but their most important property is the presence of hyperlinks that allow the browser direct access to other hypertext documents. Theses may be hosted by any computer on the Internet, and in this way, the user can rapidly jump from computer to computer around the world viewing and downloading information. Browsing, working and downloading through internet However, due to constraints in the speed of data transfer or access to the Internet, it may also be appropriate to download databases and bioinformatic software and use them on a local computer. The Internet can be daunting and intimidating. **Nine** good starting points (gateways) for bioinformatics on the Internet are listed below. Each of the sites is well maintained, simple to use and provides a wealth of resources such as links, databases and bioinormatics software.

TABLE 20.1: Important starting points for bioinformatics on the WWW

URL	Note
General Bioinfomatics 'Gatways'	
http://www.ncbi.nlm.nih.gov/	National Center for Biotechnology Infgormation homepage. It is resource for public databases, bioinformatics tools and applications. It links to various useful sites and resources for bioinformatics software.
http://www.ebi.ac.uk/	The EMBL European Bioinformatics Institute outstation. It provides resources for biological databases and software. May of them have excellent tutorial support.

URL	Note
http://www.expasy.ch/	The ExPASy (Expert protein Analysis System) Molecular Biology Server. It is maintained by the Swiss Institute of Bioinformatics (SIB). It provides links, databases and software resources for the analysis of protein sequences, structures and expression.
http://www.embl-heideberg.de/	European Molecular Biology Laboratory homepage
http://www.grnd.de/Welcome.en.html	German National Centre for Information Technology homepage
http://links.bmn.com/	The BioMedNet is the gateway to thousands of biology websites. It includes a search facility and provides descriptions of each of the web sites listed.
Genome Projects	
http://wit.integratedgenomics.com/GOLD/	Genomes On Line Database, with links to genomic database and progress reports on genome projects. Kyoto Encyclopaedia of Genes and Genomes. A very comprehensive Japanese site including metabolic maps.
Computing Notes	
http://foldoc.doc.ic.ac.uk/foldoc/index.html	FOLD (Free Online Dictionary of Computing). One can look up on the meaning of computer jargon.

Primary sequence databases are used for the storage of raw experimental data while secondary databases provide information on sequence patterns and motifs. The organism-specific databases tailored for researchers working on a particular species.

DATABASE SEARCHING AND AVAILABLE DATABASE

Database searching can be carried out to search for sequences similar to the sequence that interests you. The databases can also be utilized to find out the sequence of specific genes which are utilized for vector construction. Any database sequence and its associated documentation can be retrieved and written into a personal file or a personal data library. The National Center for Biotechnology Information (NCBI) in the NIH (National Institute of Health web (www) and others can be accessed through the Genetics Computer Group (GCG). Programs like the BLAST (Basic Local Alignment Search Tool), Entrez and NLM PubMed (MEDLINE) can be used to search databases maintained at the NCBI over the internet.), Bethesda, Maryland, USA maintain several databases. There are several programs that can be used to search these databases and some of these can be searched through the world wide ograms like FastA, TFastA and several others can be used to access the GCG database.

General NCBI information can be obtained over the internet (www) at info@ncbi.nlm.nih.gov and the NCBI home page comes up. The Entrez can be clicked on if the need is to find the sequence of a particular gene. Entrez needs information about the name of the gene etc to start the search. The sequence information can then be downloaded as needed. On the other hand if a search needs to be made to find out any similar sequences to a particular protein, the BLAST program is utilized. In this program, the choice can be made to locate protein-protein similarity by specifying blastp and swissprot or Gen EMBL. The similarity sequences in nucleotide databases can also be specified by choosing the options. The information can be downloaded and utilized. The databases of fasta or tfasta can be searched for similarity sequences through the GCG programs. The GCG program is available only on the Wisconsin package. This runs on UNIX and one can logon to the system over the network by using a terminal emulation software e. g., EM340.

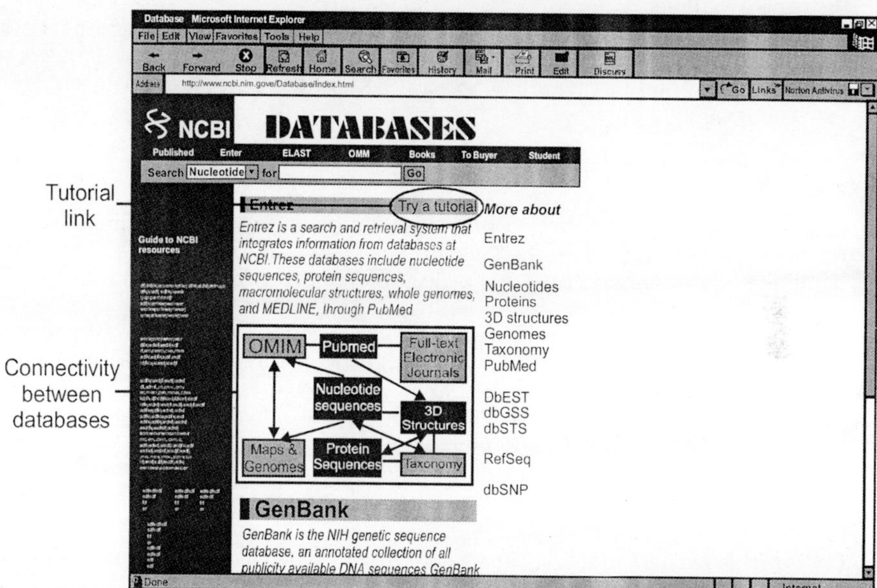

FIG. 20.1: Entrez overview page showing the tutorial link and the relationship between eight Entrez databases.

After collection of data it is essential to format data according to International standard. The collected data of nucleic acid and protein sequences are formatted in NBRF/PIR, FASTA and GDE formats. These formats have limited facilities for comments, which must include a unique identifier code and the sequence accession number. Aligned sequences are represented in NBRF/PIR, FASTA or GDE formats but there are other formats devised especially for multiple sequence alignment, including MSF, PHYLIP and ALN. The files of structural data are recorded in flat files using the PDB format. Such files contain orthogonal atomic co-ordinates together with annotations, comments and experimental details.

Annotated Sequence Database

1. **Primary raw sequence database:** The databases recorded for primary sequence are exchanged on a daily basis. These are **GenBank** (NCBI), the **Nucleotide Sequence Database** (EMBL) and the **DNA Databank of Japan** (DDBJ). Each entry is extensively annotated and has a features table to highlight the important properties of each sequence.
2. **Subsidiary sequence databases:** Specific kinds of sequence data are stored in subsidiaries of the main sequence databases. For example, ESTs are stored in dbEST, a division of GenBnak. There are also subsidiary databases for GSSs and unfinished genomic sequence data.
3. **Swiss-Prot and TrEMBL:** It is a collection of confirmed protein sequences and shows details about structure, function and protein family assignments. The

related database TrEMBL is a translation of all conding sequences in the primary nucleic acid databases. The entries in TrEMBL are less extensively annotated than those in SWISS-PROT, but are moved to SWISS-PROT when reliable annotation become available.

4. **Database interrogation** is done by using a searching tools such as **Entrez**. For cross-reference between the primary nucleotide sequence databases and SWISS-PROT is done by search of accession number,. It provides an unambiguous identifier of genes and their products.

Genome and Organism-specific Database

1. **Organism-specific resources:** It provides information and resources for those researches working on particular species. The number or organism-specific database is growing as more genome projects are initiated, and many can be accessed from general genomics gateway sites such as **GOLD**.
2. **Database formats:** Currently no universally agreed format for genome database is available, however, several viewers and browsers have been developed with graphical displays for genomic sequence analysis and annotation. The common format is **AceDB** (originally designed for the nematode *Caenorhabditis elegans*). It is an object-oriented database architecture. It is now used in many applications outside the field of genomic bionformatics.
3. **Finding organism-specific databases:** Organism-specific data are available on the Internet. For example GOLD provides gateway and illustrated data on with Ebola virus, *Escherichia coli*, *Drosophila melanogaster* and the human genome, etc.

Miscellaneous Databases

1. **Specialized sequence database:** A number of database information have been developed for the storage and analysis of particular types of sequence, for example rRNA and tRNA sequences, introns, promoters and other regulatory elements.
2. **OMIM:** It is the online Mendelian Inheritance in Man database. It is a powerful resource for the study of human genetics and human molecular biology. Each OMIM entry provides full text summary of information known about a particular gene or trait, with links to primary sequence databases and other human genetics resources.
3. **Incyte:** It is a commercial database and belongs to the LifeSeq Foundation, It provides gene sequences and transcripts with expert annotation. It is designed specifically for drug discovery research.
4. **UniGene:** It is an Experimental facility for the clustering of GenBank sequences and related to EST data. It provides little information because currently only six vertebrate and five plant species are covered by it.
5. **Structural databases:** The primary resource for protein structural data is the PDB. It contains data derived from X-ray crystallography and NMR studies. The MMDB is another structural database. It can be accessed at the NCBI web site using Entrez.
6. **Proteins and higher-order functions:** Several databases are available which store information on particular types of proteins (*e.g.*, receptors, signal transduction components and enzymes). It is the compilation of data on different types of proteins, their functions and interaction which makes possible to deduce higher-order functional networks in the cell, such as biochemical pathways, signal transduction systems and regulatory hierarchies. The important data base

is KEGG.

7. **Literature databases:** Literature databases provides information about scientific articles of various fields. One can get details about titile, authors, keywords, and abstract. MEDLINE and PubMED are widely used literature resources on the Internet and provides information of scientific literature from the 1960s up to the present day.

Data Retrieval with ENTREZ and DBGET/ LinkDB

1. **Access to distributed data:** Entrez, DBGET and SRS (sequence retrieval system) provide biological data on WWW and are the alternative to standard search engines.
2. **Entrez:** It is a WWW-based data retrieval. It is developed by the NCBI. It is used to search for information in 11 integrated NCBI databases, including GenBank and its subsidiaries, OMIM and the literature database MEDILINE, through PubMED. Entrez is accessed via the NCBI homepae. It is a simple, user-friendly system.
3. **DBGET/LinkDB:** The Kyoto University, Japan and the University of Tokyo, Japan are maintaining this data retrieval. It covers more than 20 databases and is closely associated with KEGG. A related system, LinkDB, rinds relationships between entries in the various database covered by DBGET and others. DBGET has a simpler and more limited search format than Entrez.

Data Retrieval with Sequence similarity Searches (SRS)

It is a data retrieval tool that can be used over the WWW. However, unlike these other systems, SRS is open source software and can be installed and run on a local computer network. SRS databases are grouped but using different principle to those used by Entrez and DBGET. For example, all sequences of nucleic acid and proteins are grouped together. It is used by selecting one or more of theses groupings and, within each selected group, selecting one or more of the available database. Queries can be submitted using two styles of query form, Standard or Extended. SRS is can be installed locally.

Sequence similarity searches of databases help to extract sequences which are similar to a query sequence. Such information is used to predict the structure or function of the query sequence. Prediction using similarity is a powerful and ubiquitous idea in bioinformatics. IT also helps to know facts about molecular evolution.

1. **Sequence alignment:** Any pair of DNA sequence shows certain degree of similarity. Such alignment is the first step in quantifying this *in order* to differentiate between chance similarity and real biological relationships. Alignments show the differences between sequences as changes occur during mutations [*e.g.*, insertions or deletions (indels or gaps)]. These changes are interpreted in terms of evolution.
2. **Alignment algorithms:** Dynamic programming algorithms helps in calculating the best alignment of two sequences. There are two types of **alignment** (1) Local alignments (Smith-Waterman algorithm) and (2) Global alignments (Needleman-Wunsch algorithm). Local alignments are useful when sequences are not related over their full lengths, for example proteins sharing only certain domains, or DNA sequences related only in exons.
3. **Alignment scores and gap penalties:** A simple alignment score helps in measuring the number or proportion of indentically matching residues. **Gap penalties** are subtracted from such scores to ensure that alignment algorithms

produce biologically sensible alignments without too many gaps. The gap penalties may be constant (independent of the length of the gap) or affine (containing gap opening and gap extension contributions). Gap penalties varies according to the desired application.

4. **Measurement of sequence similarity:** It is quantified using the score from the alignment algorithm, percentage sequence identities or more complex measures.
5. **Similarity and homology:** Similarity may be observed between sequences. When sequences are evolved from a common ancestor they are called homologous and often have similar biological functions (orthologs), However, the mechanism of gene duplication allows homologous sequences to evolve different functions (paralogs).

Transcription Regulatory Regions Database (TRRD)

The Transcription Regulatory Regions Database (TRRD) is a convenient tool for studying the complex process of transcription regulation. The model of structure-function organization of transcription regulatory regions of eukaryotic genes that formed the basis of the TRRD database takes into consideration a great diversity of elements involved in transcription control, their block structure, and the hierarchy essential for their functioning. At present, the TRRD database contains five interconnections: TRRDGENES (general description of genes), TRRDEXP (description of expression patterns), TRRDSITES (description of sites), TRRDFAC (description of transcription factors), and TRRDBIB. The TRRD is installed under the SRS (http://sgi.sscc.ru/), providing easy information retrieval and integration with other databases and computer systems for data processing. The TRRD database is a part of the global system GeneExpress available at Molecular Biological Server of the Institute of Cytology and Genetics, Siberian Branch of the Russian Academy of Sciences (http://wwwmgs.bionet.nsc.ru/).

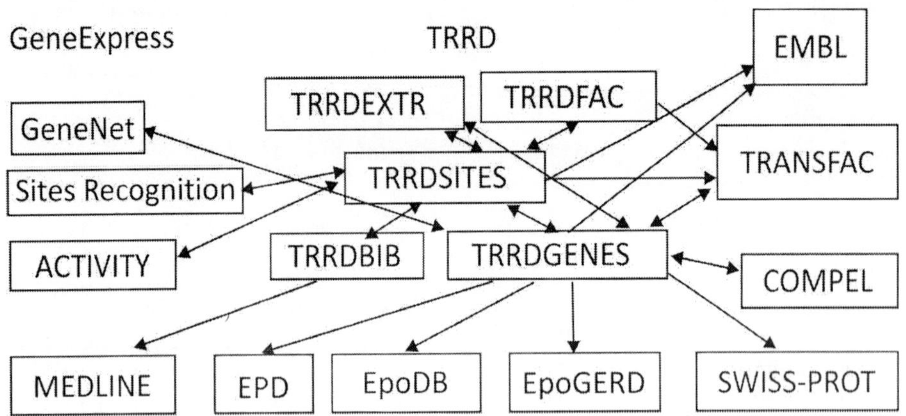

FIG. 20.2: Structure of the TRRD database and its links to other databases and computer systems.

Glucocorticoid-Controlled Gene Regulatory Regions Database (GR-TRRD)

It is used for the elucidation of the mechanisms of chemical hepatocarcinogens action. The database GR-TRRD compiling the data on regulatory regions of 39 glucocorticoid-controlled genes. Glucocorticoids influence a vast number of complex physiological and pathological processes. They help regulate protein, fat and carbohydrate metabolism, participate in immunosupression and anti-inflammation and are important for the optimum functioning of the cardiovascular, kidney, and nervous system. Glucocorticoids bring about their multiple effects by activating the intracellular glucocorticoid receptor (GR) that binds to specific glucocorticoid responsive elements (GREs) in the vicinity of regulated genes and subsequently affect their expression. For systematization, generalization, and analysis of constantly growing amount of experimental data on the glucocorticoid-controlled genes, the GR-TRRD (Glucocorticoid Regulated genes-Transcription Regulatory Regions Database) can be used: The GR-TRRD is accessible at http://wwwicg.bionet.nsc.ru/trrd/34/gluc.htm.

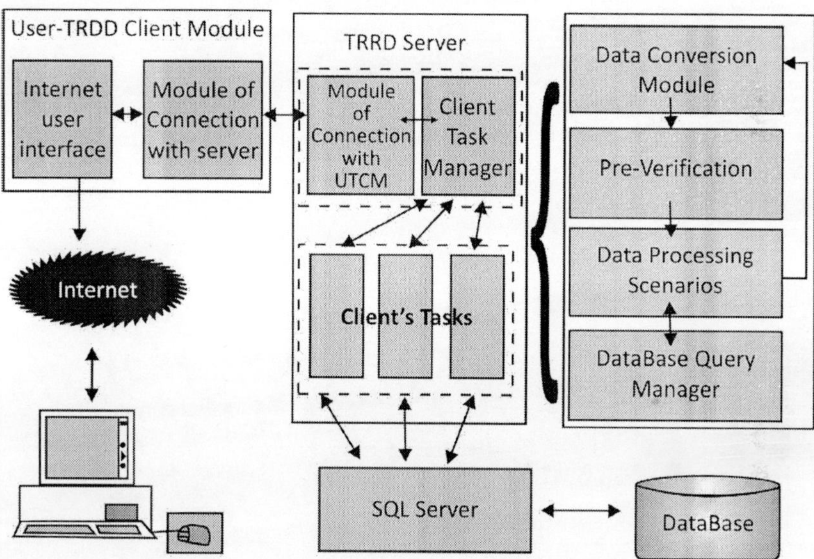

FIG. 20.3: Basic modules of the system of data input to the TRRD database.

IIG-TRRD

The volume of experimental data on regulation of functioning of both the interferon genes proper and IFN-inducible genes is growing fast. To generalize and systematize these data for description of interferon-inducible gene transcription regulation IIG-TRRD (Interferon-Inducible Genes - Transcription Regulatory Regions Database) is available. It is a constituent of the TRRD database on transcription regulatory regions of eukaryotes and consists of two major parts: (1) A formalized description of the structure of gene regulatory regions and peculiarities of their regulation. (2) A hypertext-based description of interferon system functioning containing tables, figures, and schemes.

Interferons are cytokines with a wide range of biological action. Interferons are involved in regulation of antiviral response, cell proliferation and differentiation, modulation of immune and inflammatory responses through transcription regulation of IFN-inducible genes coding for various proteins. Investigation of the mechanisms of interferon expression regulation and their effects on the target genes is of great theoretical and applied importance, since it contributes to the understanding of the causes of organism function disorders.

Analysis of the information from the IIG-TRRD database has demonstrated that the gene sets induced by type I and II interferons are partially overlapping. IFN-inducible genes code for a variety of products: enzymes, nucleotide-binding proteins, transcription factors, major histocompatibility complex (MHC) class I and II antigens, regulatory proteins, lymphocyte antigens, certain cytokines and their receptors, Fc receptors with high affinity for IgG, and a number of proteins with yet unknown functions. As a result, a complex gene network that regulates the functioning of interferon system is formed, that is, the system regulating interferon induction and action. A fragment of this gene network involving major relevant transcription factors (STAT1a, ISGF3, IRF-1, and IRF-2).

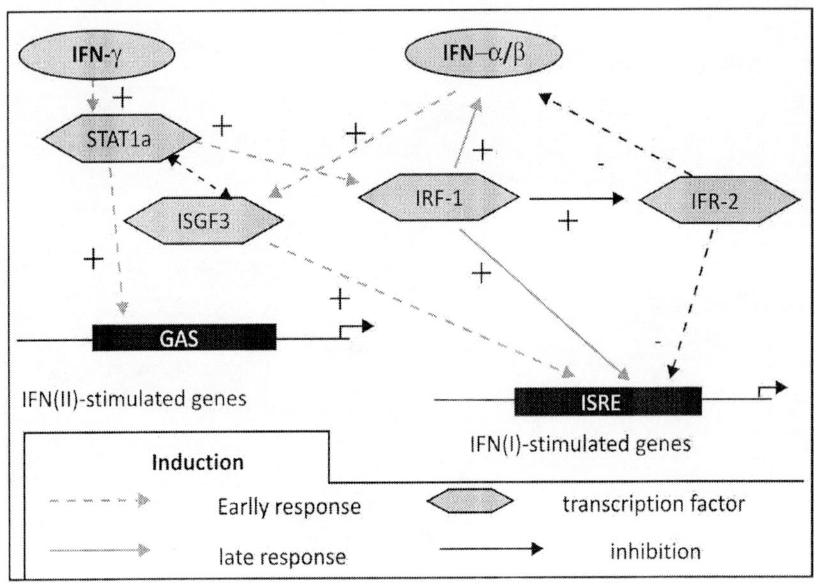

FIG. 20.4: The main components of the IFN gene network.

ESRG-TRRD

It is a database of genes with specific transcription regulation in erythroid cells. The ESRG-TRRD (Erythroid-Specific Regulated Genes) database is a constituent of the TRRD and help to accumulate the information on the genes expressed in the erythroid cells in the course of their differentiation at different stages of vertebrate ontogenesis. This information forms two blocks:

1. The data on gene expression and structure-function organization of their regulatory regions. Peculiarities of gene expression in different cell types, organs, and tissues depending on developmental stage of the organism and effect of external factors are described in the former block along with the references to the gene regulatory regions and transcription factor binding sites involved in realization of a particular expression pattern.
2. The latter block describes the gene regulatory regions, transcription factor binding sites, and the factors binding to these sites. The database also includes a system of cross-referenced hypertexts available to the Internet users at http://www.bionet.nsc.ru/trrd/.

LM-TRRD Database

The current release of the LM-TRRD (Release 4.0) compiles the data on 48 genes of the lipid metabolism system. The database contains the description of over 60 regulatory regions (promoters, enhancers, and silencers) of lipid metabolism genes. A number of regulatory regions is distanced from the transcription start by several kbp, for example the enhancer for mouse HNF-4 gene (G001173) is located 7 kbp in the 5' direction from the transcription start; the enhancer for rat hydratase-dehydrogenase gene (G000757) is located 3 kbp upstream of the transcription initiation site. Most genes are transcribed from one promoter; however, certain genes (G000684, G001191, and G001195) have alternative promoters providing the transcription either in different tissues or under the effect of different inducers. Over 300 various transcription factor binding sites occurring in the regulatory regions of LM genes are described in the LM-TRRD.

Gene networks: Based on the LM-TRRD-contained information on transcription regulation of various vertebrate genes. The lipid metabolism gene network has two fragments:

The **first fragment** describes coordination of the genes controlling the fatty acid oxidation. Involved in this process are the following genes of enzymes of fatty acid oxidation and proteins for lipid transport: ACS (G001195), HD (G000757), ACO (G000704) and apoAI (G000203), apoCIII (G000206), respectively. The PPAR transcription factors form the key component of this gene network. Glucocorticoids are known to activate the expression of the gene encoding one of the PPAR factors in liver. In addition, the interaction with fatty acids or their derivatives increases the activity of PPAR subfamily proteins. Thus, the PPAR subfamily factors provide for the expression regulation of the above-listed genes of enzymes and transport proteins depending on the presence of fatty acids in the cell and in response to glucocorticoids.

The **second gene network fragment** unites the genes controlling the cholesterol level in the cell This fragment includes the genes of enzymes involved in mevalonate pathway of cholesterol biosynthesis and fatty acid biosynthesis: HMG-CoA-S (G001198), HMG-CoA-R (G000157), FDPS (G001191), SS (G001183) and ACC (G000684), FAS (G001196), respectively, as well as LDLR (G00190). Transcription factors of SREBP subfamily are the key elements of this gene network. It has been demonstrated that SREBP transcription factors are formed in the cell from a precursor molecule (preSREBP) by sterol-regulated protease (SRP). SRP activity is, in turn, inhibited by a high cholesterol content in the cell. Thus, SREBP proteins are involved in transcription regulation of the group of genes in question and provide for either their coordinated activation or expression inhibition depending on the cholesterol level in the cell.

PLANT-TRRD Database

A large volume of information on the structure and expression regulation of various plant genes has been accumulated. The key events of expression regulation of any genes occur at the transcription level. Interaction of protein factors with DNA regulatory regions determines the time, place, and expression level of the gene under control. Ontogenesis of each organism implies the consequent ordered switching the genes on and off and modulation. Certain complexly organized transcription systems are formed in higher plants. To generalize, analyze, and systematize these data, the PLANT-TRRD database is being used as a constituent of the TRRD (Transcription Regulatory Regions Database).

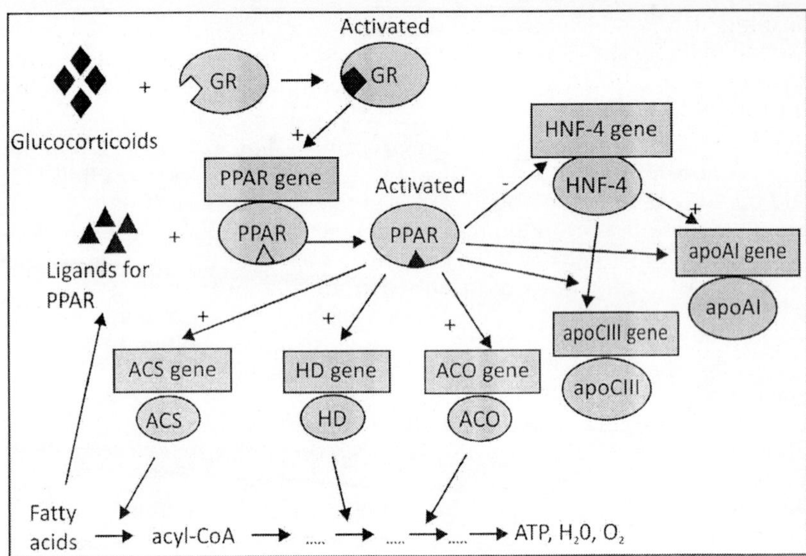

FIG. 20.5: Gene network in the system of genes regulated by PPAR factors. The circles indicate proteins; rectangles show genes coding for these proteins. ACO, acyl-CoA oxidase; ACS, acyl-coenzyme A synthetase; poAI - apolipoprotein AI; apoCIII - apolipoprotein CIII; GR - glucocorticoid receptor; HD, Hydratase-dehydrogenase; HNF4, hepatocyte nuclear factor 4; PPAR - peroxisome proliferator activated receptor.

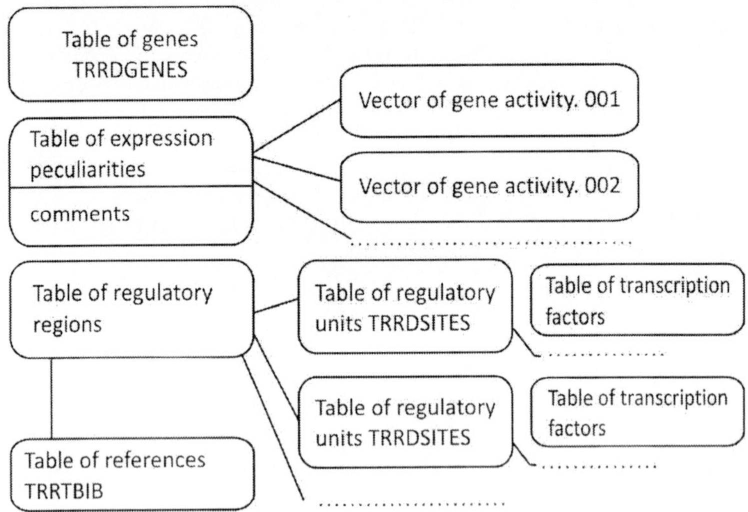

FIG. 20.6: Structure of PLANT-TRRD database.

References to GenBank, EMBL, or SWISS-PROT are given. The next block describes the peculiarities of the gene expression. The fields of this block contain the data on time and place of expression. A separate block of fields, vector of gene activity, in each card is devoted to every instance of expression described. The fields of the next block describe the entire regulatory region and particular regulatory regions that constitute it, and individual transcription factor binding sites. The PLANT-TRRD database are to be further developed depending on functional grouping of the genes into gene networks. This means that each section of the database will describe the genes providing for a certain function, such as, for example, storage proteins, heat shock genes, genes of photosynthesis system, etc.

EPOGERD

It is a database on regulation of eukaryotic gene expression. The data on structure, function, and expression of eukaryotic genes are being accumulated rapidly. A great number of molecular biological databases have been developed aiming to systematize and accumulate these data in a user-friendly format. The largest databases of general orientation - EMBL, GenBank, PIR, and SWISS-PROT - compile the information on DNA, RNA, and protein sequences as well as certain structure-function characteristics of genes and proteins. More specialized databases deal with either particular aspects or particular objects of molecular biological investigations. The data on gene structures that are necessary for specialized databases can be partially extracted from the general databases, such as EMBL and GenBank, whereas only the original publications can be the source of information on gene expression patterns and regulation.

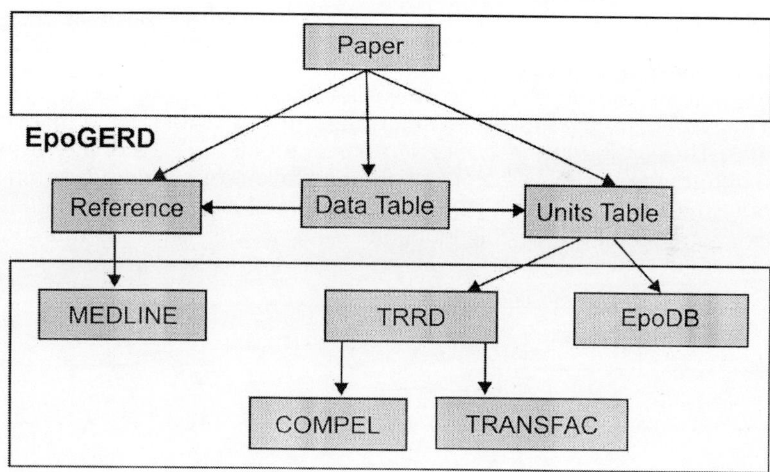

FIG. 20.7: Distribution of input information flow from an individual publication in the EpoGERD database.

HOX-PRO DB

It is used for evolution of ensembles of homeobox genes. Nowadays homeobox genes were found for species of main invertebrate and vertebrate taxa. It is established that these genes play key role in time and space orchestration of genome expression during development. Auto- and cross regulatory functional interactions join homeobox genes into genetic networks. Mutations within homeotic hox-genes in *Drosophila melanogaster* transform defined segments of the body into the character of an adjacent segments. Homologous genes there are in vertebrates, where it has been possible to mutate them, shifts in morphological character could be produced.

The discovery of the first HOX cluster outside *Drosophila* causes great hopes.

Homeobox is a *Rosetta stone* for the study of animal development. This could enable us to read the epigenetic code of other animals on the basis of our understanding of *Drosophila*. It is possible that this HOX-based epigenetic code is very ancient and was in place in the common ancestor of all modern animals. According to Slack, Holland and Graham, this character should be adopted as the defining character, or synapomorphy, of the kingdom *Animalia*.

SAMPLES and ALIGNED

These are databases for **functional site sequences**. The majority of molecular processes in the cell connected with the storage, transmission, and realization of hereditary information is controlled by relatively short regions of the DNA molecule, the so-called functional sites. A great volume of data on their location, structure, and function has been so far accumulated in such databanks as EMBL and TRANSFAC. However, these databases contain only the data on individual sites. This is a specialized database, which provide the data for recognition methods is a timely objective. It provides database SAMPLES compiling sets of 42 transcription factor binding sites using the information contained in TRANSFAC, TRRD, and EMBL databases.

Activity

It is a database for activities of functional DNA/RNA sites. Molecular genetic processes occurring in the cell, such as replication, transcription, splicing. translation, etc., are under the control of functional sites with definite specific activities. Primary sequences, location in DNA or RNA, and the values of specific activities under various conditions have been experimentally determined for thousands of actual variants of such sites. These experimental data demonstrate that the sites of the same type located in different DNA (RNA) regions of pro- and eukaryotic genomes can differ in the values of their specific activity by several orders of magnitude. This difference was observed in naturally occurring functional sites as well as in their mutational or synthetic analogues.

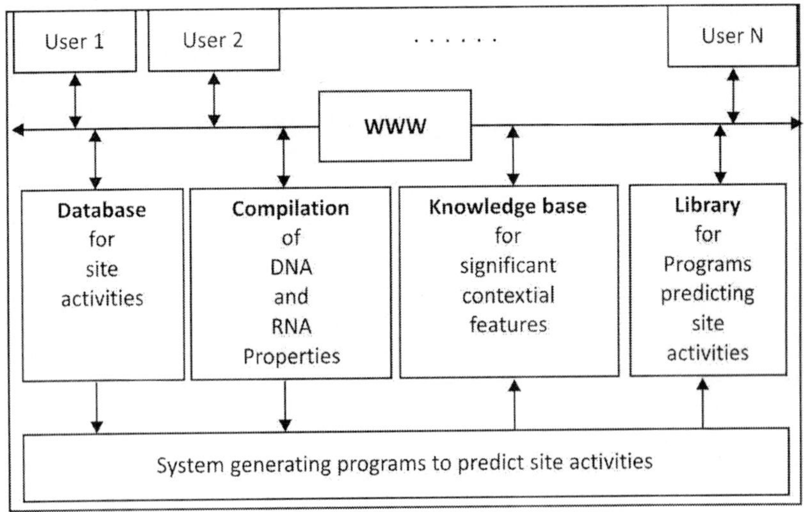

FIG. 20.8: Schematic representation of the computer system B-DNA-VIDEO.

B-DNA-Video

It is an active database for the significant b-DNA features of transcription factor binding sites. The problem of creating methods allowing precise functional site recognition in nucleotide sequences is far from being solved. The increasing volume of experimental data demonstrating that the efficient performance of functional sites is to a large extent determined by their physical and chemical and conformational properties are being accumulated. Conformational features influence the stereochemical compatibility of the sites and the interacting proteins.

Numerous data obtained so far suggest the local nonuniformity of DNA conformational and physical and chemical features and their dependence on nucleotide context. Numerous experimental and theoretical studies of B-DNA helix resulted in determination of the mean values of a considerable number of conformational and physical and chemical parameters of dinucleotides

GeneExpress System

It is database for description, analysis, and recognition of regulatory sequences in eukaryotic genomes. Currently, the number of databases on gene expression and variety of software for the analysis of these data are growing fast. That is the reason why the following problems of the modern bioinformatics of gene expression regulation are challenging:

1. Creation of the unified Internet-accessible media to provide maximal integration of information-software resources on gene expression and effective navigation of users in the integrated media.
2. Development of a technique for automated discovery of knowledge on structural and functional organization of regulatory genomic sequences and on their recognition.
3. Activation of databases on regulatory genomic sequences and their conversion from passive information carriers to the active information and software modules to provide the users with various active knowledge, for example, such as the programs for analysis and recognition of regulatory genomic sequences.

The system **GeneExpress** is developing to pursue these goals. It includes following five basic units: (1) Transcription Regulation unit contains database on transcription regulatory regions of eukaryotic genes. (2) Site Recognition unit contains programs for analysis and recognition of functional sites. (3) ACTIVITY is the module designed for site activity prediction from their nucleotide sequences. (4) mRNA Translation unit is designed for analysis of translational properties of mRNAs. (5) GeneNet is the database for accumulation of the data on eukaryotic gene networks and signal transduction pathways.

All the units are integrated into the GeneExpress system on the basis of the Network Browser of Sequence Retrieval System (SRS).

There is a possibility for user to move from TRRD to a particular knowledge base of the GeneExpress system via appropriate links. Among the currently available knowledge bases of GeneExpress are the following: knowledge bases on significant conformational, physico-chemical properties of DNA sites, knowledge base on functional sites recognition, and knowledge base on functional site activities. All this knowledge is discovered automatically by the special tools of the GeneExpress system.

GeneExpress system is also supplemented with a special tool SeqAnn for activating the TRRD database while revealing putative promoters. The core idea of this approach is to use the information from the TRRD database as a ready-to-use scenarios for promoter recognition. In a simple case, it allows a user to search a target sequence for putative promoters which are similar to actual promoters described in TRRD basing on the content and approximate location of transcription factor binding sites. In the pilot release of SeqAnn, the search for sites is performed by the homology approach. At the next stage, the entire set of the site recognition methods accumulated in GeneExpress knowledge bases will be employed. And finally, as it was noted above, the simultaneous use of a large number of recognition programs for individual sites will rise the accuracy of recognition. The information on the activity of the site to be recognized will be also used for the same purpose.

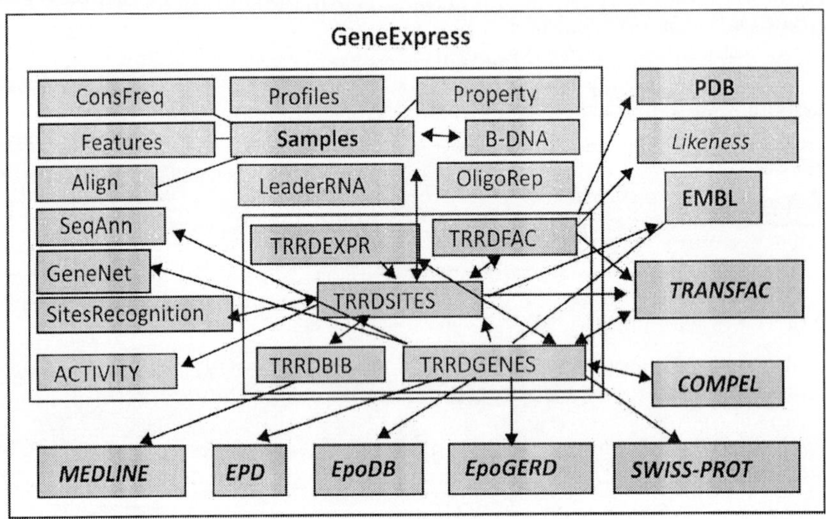

FIG. 20.9: Principle scheme of the GeneExpress system.

The **GeneNet** database, a constituent of the GeneExpress system, describes the highest level of integration responsible for the hierarchical organization of gene expression regulation. The three following hierarchical levels are considered for using the data on transcription regulation from the TRRD database: (1) organism level, (2) the single cell level and (3) the single gene level.

GENET

It is a database of **genetic networks**. Gene networks are gene ensembles functioning in a coordinated manner to provide for the vital functions, fine regulation of physiological processes, and responses to external stimuli. Any gene network is composed of following functional elements: (1) a gene ensemble interacting when certain biological functions are performed; (2) proteins encoded by these genes; (3) signal transduction pathways providing the gene activation in response to an external stimulus; (4) (4) a set of positive and negative feedbacks stabilizing the parameters of the gene network (autoregulation) or providing a transition to a new functional state; and (5) external signals, hormones, and metabolites that trigger the gene network or correct its operation in response to the changes in physiological parameters. GeNet is located at www.csa.ru/Inst/gorb_dep/inbios/genet/ genet.htm and its mirror site www.mssm.edu/molbio/genet/genet.htm.

GeNet is a hypertext database designed for the World Wide Web. The database consists of text files in HTML, image files in the GIF or JPEG formats, a suite of programs for searching and generating reports, as well as user interface implemented in JAVA and as a server-side CGI-script. The database is currently running under CONVEX OS and Solaris.

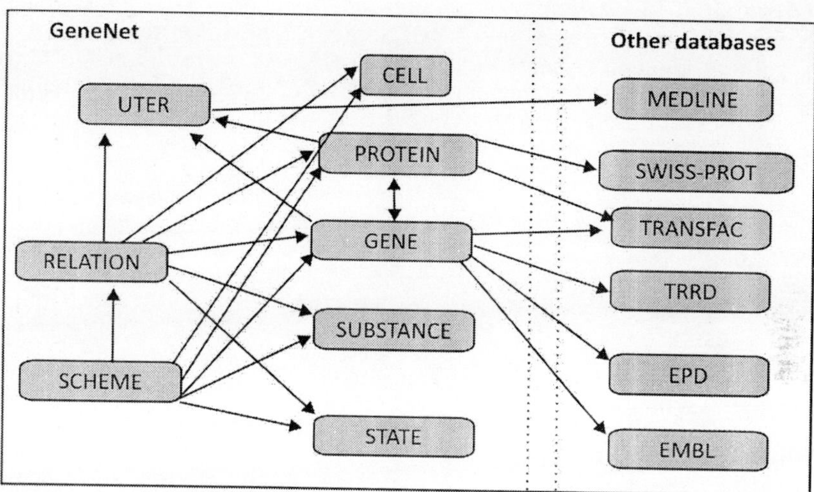

FIG. 20.10: Structure of the GeneNet database with references to other databases.

ECOCYC

It is a molecular database integration used for analysis of metabolic network control. The ECOCYC system of P. Karp [P. Karp, (1992) *A knowledge base of the chemical compounds of intermediary metabolism*. CABIOS **8** (4), 347-357] was one of the first integrative molecular database systems. Today software tools are available which allow the implementation of powerful integrative analysis tools. Moreover, for the analysis of metabolic networks models and simulation environments are available. R. Hofestadt, U. Scholz (1998) presented the architecture of the *Magdeburger Molecular Information System*" (MMIS), which allows the access and analysis of metabolic networks. One important application of our MMIS is the detection of inborn errors.

MGRS

The Molecular Genetic Regulatory Systems (MGRS) aroused in the mid-60s as an application of the ideas and approaches of cybernetics to rapidly developing molecular biology and genetics. By that time, the informational cybernetic approach had yielded the first fruits. It is based on the concept that the totality of irregular cell biopolymers (DNA, RNA, proteins, and their complexes) along with the systems carrying out various processes involving these biopolymers are considered as a **MGRS** that performs various operations with **genetic information.** Sequences of monomer's symbols are considered as **genetic texts;** the rules according to which the information is written down, as the **genetic language;** and schemes of regulatory interactions between genes, enzymes, metabolites, etc., - as **genetic networks. Genetic information** is determined as an aggregate of molecular and more complex biological properties and functions encoded in the genetic texts. The genetic texts considered from the standpoint of long-term storing of genetic information form the **genetic memory.** We are applying this approach ubiquitously.

Block-modular organization of MGRS: Actual MGRS of the cell display a hierarchical block-modular organization formed in the course of evolution. The major modules of a MGRS are: (1) the archive of genetic information (hereditary memory); (2) block of self-reproduction; (3) block for control of metabolism; (4) for control of development; and (5) for control of physiological reactions; etc. The modules (1) and (2) form the **central subsystem of MGRS;** the rest, **peripheral subsystem.** The block of self-reproduction (syser) forms the core of MGRS organization. Most known dynamic modules of lower organizational levels are codons, marks of punctuation, genes, proteins, operons, replicons, mobile genetic elements (MGE), etc.

FASTA and BLAST

BLAST and FAST are two important fast searches of sequence database. Sequences similar to a query can be found in a database by aligning it to each database sequence in turn and returning the highest scoring (most similar) sequences. For this purpose dynamic programming algorithms is used. Sensitivity and specificity quantify the success of a database search strategy.

- Sensitivity helps in measuring the proportion of real biological sequence relationship in the database that were detected as hits in the search.
- Specificity is the proportion of the hits corresponding to real biological relationships.

The types of database and query sequence of protein or nucleic acid sequences required various types and combinations. The available program is BALST server it runs at the NCBI. It is used for searching various sequence database for general-purpose. FASTA is also available at the EBI. Both these servers are used on many WWW sites to search organism-specific databases like the Sanger Centre. During the course of divergent evolution sequences of protein may be changed. For presentation of these structure and function and detection of distant evolutionary relationships between sequences BLAST and FASTA methods are sometimes helpful. This kind of detection is an important part of bioinformatic researches.

PSI-BLAST is and iterative search method. This method is found more useful than BLAST and FASTA. It has improved detection rate. PSI-BLAST has some disadvantage because it has the potential for unrelated sequences to pollute the iterative search, and has difficulties associated with the domain structure of proteins.

APPLICATIONS OF BIOINFORMATICS

Bioinformatics technology uses computational tools provided by the information technology revolution, such as statistical software, graphics simulation and database management, to organize and analyze information about biological systems, which, for biotechnology, is information about cells and biological molecules. Using another product of the information revolution, the Internet, scientists broadcast this information around the world.

Biotechnological techniques like, monoclonal antibody technologies, cell culture technology, cloning technology (molecular cloning, cellular cloning, animal cloning), genetic modification technology, protein engineering technology, hybrid technologies, biosensor technology, tissue engineering technology, DNA chip technology, etc. have provided us with massive amounts of information, in addition to useful products. Without methods for organizing and analyzing the raw data, however, we will not be able to turn it into knowledge, understanding and, ultimately, products.

Bioinformatics technology provides us with tools and methods for consistently organizing, accessing, processing and integrating data from different sources. This uniformity, in conjunction with the universal language of life at the molecular level, enables international collaboration among scientists studying any plant, animal or microbe.

Bioinformatics technology provides us with tools and methods for consistently organizing, accessing, processing and integrating data from different sources. This uniformity, in conjunction with the universal language of life at the molecular level, enables international collaboration among scientists studying any plant, animal or microbe. Bioinformatics technology helps in following ways:

1. Map genomes and identify genes.
2. Determine protein structure and simulate protein interactions.
3. Discover new therapeutic targets and design medicines aimed at the targets.
4. Assess the effects of virtual mutations on gene function.
5. Study of phylogenetics

Bioinformatics in Study of Phylogenetics

Similarities and differences can be observed among species because, if two species are very similar, they are likely to have shared a recent common ancestor. Phylogenetic analysis of various kinds of character can be used for such studies *e.g.*, nucleic acid and protein sequences. These biomolecules are the most common to all life forms and allow both closely and distantly related taxa under study and they can be compared objectively. However, caution must be exercised when inferring phylogenies from sequences because the rate of mutation may not be constant and sequences may be subject to differential selection. Evolutionary relationships is generally represented by a special type of graph called a **tree**. It has n nodes and n-1 links. A phylogenetic tree is a simple way to show evolutionary relationships, with species represented by nodes and lines of descent represented by links. Such tree may be unrooted or rooted (the position of the last common ancestor of each tree member). A phylogenetic tree that shows the evolution of species as a series of bifuncations is a **cladogram** (binary tree). The links in a cladogram may vary in length and convey a sense of evolutionary period. Classification systems of biological organisms are arbitrary in nature because there is no standard measure of difference that can define a species, genus, family or order.

Bioinformatics Helps in Building Phylogenetic Trees

Percentage of matches (similarity table) or percentage of differences (distance table) are being prepared which are used for construction of a phylogenetic tree. Distance tables are used for the analysis of macromolecular sequence data. Several methods are being used for construction of a phylogenetic tree these are as follows:

- Distance matrix methods: These methods work by selecting the two most closely related taxa in a distance matrix and clustering them. The operation is then repeated until there is only one cluster left.
- Maximum parsimony methods: In these methods, sequences are compared and clustered on the basis of the minimum number of mutations required to convert one sequence into another at any given position. The final phylogenetic tree is based on the overall number of changes required throughout the whole sequence.
- Maximum likelihood methods: These are similar to maximum parsimony methods but include a user-defined model that allows the probability of a given substitution occurring at any given position in the sequence to be built into the algorithm. The likelihood of a given sequence change is calculated for each

position in the sequence and the most reliable tree is that with the maximum overall likelihood.

Phylogenetic softwares available for phylogenetic analysis like PAUP and PHYLIP. These are suitable for distance matrix, maximum parsimony and maximum likelihood analysis.

Bioinformatics and Molecular Phylogeny

DNA accumulates mutations over evolutionary time, leading to divergence in DNA, RNA and protein sequences in different lines of descent. This principle can be used to construct phylogenetic trees. Due to differences in intrinsic mutation rate and selective constraints, different macromolecular sequences evolve at different rates, allowing the phylogenetic analysis of both closely related and distantly related organisms.

Choice of macromolecular sequences for a particular evolutionary study: The macromolecular sequence chosen for this purpose reflects the biological diversity exists in nature. Mitochondrial DNA is a rapidly evolving molecule used for such studies. For those organisms who have more diverse phylogenies, molecule of ribosomal RNA is used. Care should be taken while interpreting apparent phylogenies from poorly selected macromolecules. A few RNA viruses have rapidly evolving sequences whose repication enzymes are error-prone. Human immunodeficiency virus is an example, where up to 30 genotypes can evolve during a single infection.

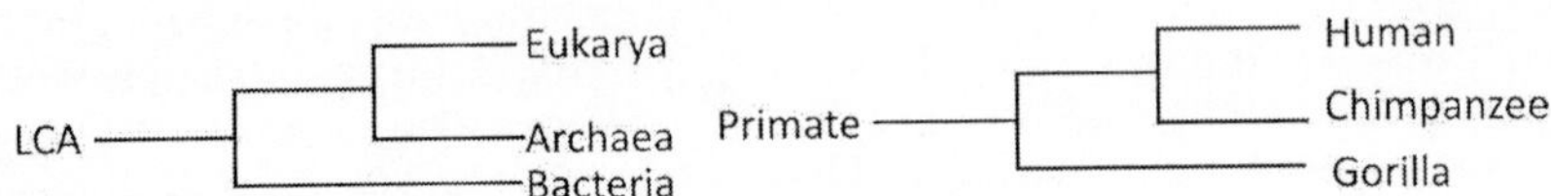

FIG. 20.11: Presentation of rooted tree. (Left) Tree of major forms of the on this planet. (Right) Three great apes with an unspecified primate ancestor.

Bioinformatics in Discovery of Drugs

Any substance, which enters human body can change either the function or structure of the human organism. It interacts with targets, usually proteins, in the body and through such interactions cause physiological responses. The pharmaceutical industry aims to discover drugs with specific beneficial effects to treat human diseases. At **first stage**, for development of a new drug the basic requirement is the identification of a suitable target, which must contribute significantly to a human disease. Ideally, altering the activity of this target should have a beneficial effect thus showing its potential for therapeutic intervention. At **second stage** the process is lead discovery. The compounds show some of the desired activity of an ideal drug. Optimization of lead compounds results in drug candidates that may be registered and submitted for clinical trials, which establish their safety and metabolic behavior in human subjects.

Bioinformatics plays important role in drug development. Pharmaceutical industry have developed massive amount of data using various techniques of Genomics, proteomics, combinatorial chemistry and HTS The role of bioinformatics is to store, track and provide tools for the analysis of these data. The large-scale functional annotation of genes is known as **functional genomics** and incorporates areas such as homology searching, structural analysis, expression analysis, large-scale mutagenesis and the analysis of protein interactions. All of these areas are important in drug development. **Genome-scale mutagenesis** is a rich source of animal disease models for target identification and validation, and large mutant collections in simple organisms can be used for the rapid high-throughout screening of potential lead compounds.

Specific applications are modeling of protein interactions with small molecules allowing rational drug design, the association of genotype and drug response patterns (pharmacogenomics), the design and assessment of chemical diversity in combinatorial libraries, and the processing and storage of data from high-throughout screens of lead compounds.

Computational Screening

Software applications such as DOCK and Autodock match potential ligands to binding sites by calculating steirc constraints and bond energies. These help in searching chemical databases of potential drugs. Some applications consider the ligand and binding site as inflexible structures, rather like pieces of a jigsaw, while others can incorporate flexibility into the molecules by calculating allowable and compatible bond torsions.

Pharmainformatics

Pharmacogenomics is the study of how variation in the human population correlates with drug response patterns. It is the study of combination of biology, chemistry, mathematics and information technology that is essential for efficient data management, processing and analysis in the pharmaceutical industry.

The analysis of genomic data and its comparison with drug response data allows patients to be clustered into drug response groups, so that appropriate drugs and dose regimens can be administered. Variation is catalogued by analyzing data on mutation (particularly SNPs) and gene expression profiles.

For advancement of pharmainformatics diverse chemical libraries are needed. Such libraries will provide insight in binding properties of the drug target. Conversely, focused libraries are required if the structure of the target is known, since this defines a particular set of ligands. Chemical diversity can be defined by comparing molecules on the basis of descriptors (functional groups) and how these fill chemical space. A number of software tools are available for the design and assessment of diverse of focused chemical libraries, and virtual screening against drug targets.

Cheminformatics

The study of cheminformatics is a combination of chemistry and information technology, is required for the processing and analysis of chemical data. Cheminformatics is relevant to biologists because chemistry data are important in many areas of molecular biology, for example in the study of protein interactions and metabolism.

Molecular Formulae

Molecules can be represented in the form of simple formulae. They represent the number and type of atoms. However, this does not show how they are connected. Structural formulae provide some information about the arrangement of atoms in a molecule and thus allow isomers to be distinguished. Molecules are represented by using simple **graphs**. They show atoms as nodes and bonds as links. For organic molecules, further simplification is achieved by assuming that carbon atoms make up the molecular backbone and that the valency of four is satisfied by hydrogen atoms unless otherwise shown. Such diagrams present all molecules as planar shapes and do not indicate the spatial distribution of atoms in three dimensions.

If four different groups are coordinated around a central carbon atom, the molecule is described as **chiral** and the concept is known as **Chirality**. Chiral molecules exist in two conformations:

1. **Enantiomers** (mirror-images of each other). Although enantiomers have the same chemical properties, many enzymes and other proteins show chiral selectivity, which is important in drug development and related fields.
2. **Diastereoisomers** (molecules may contain any number of chiral centers and a series of forms). These may have different chemical properties because of the way different groups interact within the molecule. The absolute configuration of groups around a chiral carbon atom is described by using a number of conventions. In the **DL system**, molecules are named **D** or **L** according to whether the coordinated groups are arranged in a similar fashion to those in D-glyceraldehyde or L-alanine. In the **RS system**, molecules are named **R** (*rectus*) or **S** (*sinister*) according to the size of the chemical groups surrounding the carbon atom.

Resources

- **SMILES:** This system represents chemical formulae as strings, based on a valence model in which all valencies are considered to be satisfied by hydrogen atoms unless otherwise shown. The system has conventions for representing different bond types, cyclic molecules, branches, cis/trans isomers and chirality.
- **RasMol and Chime:** There are several specialized data formats for chemical structures based on the principle of a molecular formula and associated table of connections. Viewing utilities such as RaasMol and Chime can interpret these file formats and display interactive molecular structures in a variety of user-defined schemes and colors.
- **Chemical Structure Databases:** A number of comprehensive WWW resources, including Chemical Abstracts On-Line, Chemfinder and MedChem are avilable. Each resource provides a chemical database that can be searched using a variety of query formats, for example systematic name, nonsystematic name, formula, molecular weight or CAS registry number. Search results provide physical, chemical and biomedical information with links to other databases and resources. MedChem also provides the SMILES string.
- **QSAR:** It is a statistical method used to determine how the structural features of a molecule are related to biological activity. The QSAR approach is particularly useful for categorizing the activities of related molecules with multiple functional groups. Each molecule is broken down into a series of descriptors (molecular properties) and the QSAR determines which descriptors are most likely to promote biological activity. This gives rise to a set of rules that can be used to evaluate the potential activity of new molecules.

Protein Interaction Informatics

Proteins that physically interact with each other may be involved in the same molecular pathway or network, or may form part of a multisubunit complex. Using this principle, pathways can be reconstructed based on evidence of protein interactions. However, information from other sources for example gene expression patterns and mutant phenotypes may also be useful.

Handling Y2h Data

Yeast two-hybrid (Y2H) screens produce large amounts of protein interaction data, but there is a relatively high level of spurious results (false positives and false negatives). This problem can be addressed by scoring interactions for reliability, based either on the repeatability of interactions over multiple experiments, or by the number of times a given bait will trap independent clones representing the same prey. Even so, similar large-scale screen tend to identify different (although overlapping) sets of interactions.

Protein Interaction Databases

Several databases have been set up to store the interaction data arising from large-scale Y2H screens. However, much more information on protein interactions is available in the scientific literature and a current challenge in bioinformatics is the assimilation of these interaction data from diverse sources.

The Interactome

The interactome is the sum of all protein interactions in the cell. The simplest way to represent protein interactions is a graph with proteins as nodes and interactions as links. However, when large numbers of proteins are considered, the graphs become too complex. They can be simplified by clustering functionally similar proteins, resulting in a functional interaction map that links fundamental cellular processes.

CHAPTER 21

Plant Growth Regulators for Hi-Tech Horticulture

Certain substances affect the growth quite miraculously. In early literature these were referred to as hormones. Hormone is a Greek word derived from hormao (Upu w), which means to urge on or to stimulate. Plant Bioregulators (PBRs), both natural and synthetic affect growth and developmental processes in plants. Of these, phytohormones can be classified into five groups, namely, Auxins, Gibberelins, Cytokinins, Ethylene and Abscisic acid. The application of bioregulators entails seed germination, vegetative propagation, regulations of growth, flowering, blossom/fruit thinning, fruit drop control, fruit development, enhancement of fruit quality and pre and post harvest management of fruit crops. Phytohormones are organic substances, which are naturally produced in plants, control the growth or other physiological functions at a site remote from its place of production and active in extremely minute quantities. Growth hormone as "substance which are synthesised in particular cell and which are transferred to other cells where in extremely small quantities influence developmental process." A hormone has been defined as a substance, which is produced in one part of a living organism and is transferred to another part to influence some activity. It has been called a 'chemical messenger'. However, the term hormone is quite popular and widely used. It is meant for an organic substance synthesized in one tissue and migrates to another tissue of the plant wherein very minute quantity affects the growth. Besides these five groups of plant hormones, brassinosteroides, jasmonic acid, oligosacchrides, dormin, florigen and fusicoccin etc. have been reported to possess physiological activities. But now different categories of substances affecting plant growth are known which can be broadly classified into growth promoting and growth retarding substance or into naturally occurring growth substances and synthetic growth substances. However, the term growth regulating substance is now preferred in view to include both naturally occurring and synthetic growth substances. Growth regulating substances may also be classified into following categories.

1. Auxins, e.g. Indole acetic acid (IAA)
2. Gibberellins, e.g. Gibberellic acid (GA)
3. Cytokinins, e.g. Kinetin, Zeatin etc
4. Ethylene, e.g. Ethylene
5. Dormins, e.g. Abscissic acid (ABA), Phaseic acid, Xanthoxin etc.
6. Flowering hormones, e.g. Florigen, Anthesin, Vernalin
7. Miscellaneous natural substances, e.g. Cyclitols, Vitamins, Phytochrome, Traumatic substances etc
8. Phenolic substances, e.g. Coumarin.
9. Synthetic growth retardants, e.g. CCC, AMO 1618, Phosphon D, Morphactins, Malformins, Maleic hydrazide (MH) etc.
10. Miscellaneous synthetic substances, e.g. Synthetic auxins, Synthetic cytokinins

etc.

All these are influenced by auxins, gibberellins, cytokinins, abscissic acid, ethylene etc. These have variable effects on the processes e.g. Auxins are generally used in plant cell culture at a concentration range of 0.01-10.0 mg/L. When added in appropriate concentrations they may regulate cell elongation, tissue swelling, cell division, formation of adventitious roots, inhibition of adventitious and axillary shoot formation, callus initiation and growth, and induction of embryogenesis. Cytokinins are generally used in plant cell culture at a concentration range of 0.1-10.0 mg/L. When added in appropriate concentrations they may regulate cell division, stimulate auxiliary and adventitious shoot proliferation, regulate differentiation, inhibit root formation, activate RNA synthesis, and stimulate protein and enzyme activity. Gibberellins are generally used to promote flowering, break dormancy of seeds, buds, corms, and bulbs, and causes stem elongation. There are over 120 forms of gibberellins, but GA_3 is the most commonly used form. Compounds like paclobutrazol and ancymidol inhibit the synthesis of gibberellins. Abscisic Acid (ABA) plays a role in dormancy development in embryos, buds and bulbs, and in leaf abscission. When used in tissue culture, ABA inhibits the growth of shoots and the germination of embryos. Fluridone may inhibit ABA synthesis. Polyamines are compounds that occur in high levels within plants and are used in tissue culture media at concentrations of 10-1000 mM. Polyamines may enhance regeneration of roots, shoots and embryos, delay or prevent senescence, and regulate flowering. Salicylic acid is activates genes in some plants that produce chemicals that aid in the defense against pathogenic invaders. Jasmonates are produced from fatty acids and seem to promote the production of defense proteins that are used to fend off invading organisms. They are believed to also have a role in seed germination, and affect the storage of protein in seeds, and seem to affect root growth. Plant peptide hormones are encompasses all small secreted peptides that are involved in cell-to-cell signaling. These small peptide hormones play crucial roles in plant growth and development, including defense mechanisms, the control of cell division and expansion, and pollen self-incompatibility. Polyamines are strongly basic molecules with low molecular weight that have been found in all organisms studied thus far. They are essential for plant growth and development and affect the process of mitosis and meiosis. Nitric oxide (NO) is serves as signal in hormonal and defense responses. Strigolactones are plant hormones that have been implicated in inhibition of shoot branching. Strigolactones are carotenoid derived and trigger germination of parasitic plant seeds and stimulate symbiotic mycorrhizal fungi. Strigolactones contain a labile either bond that is easily hydrolyzed in the rhizosphere meaning that there is a large concentration gradient between areas near the root and those further away. Karrikins, a group of plant growth regulators found in the smoke of burning plant material that have the ability to stimulate the germination of seeds. Currently, there are four known karrikins which are designated KAR_1, KAR_2, KAR_3, and KAR_4.

AUXINS

Auxins are defined as "An organic compound characterized by its capacity in low concentration (below 10^{-3} M or 0.001 M) to induce elongation in shoot cells and inhibition of elongation of root cells. They resemble Indole acetic acid (IAA) in physiological action. They may and generally do effect other processes besides elongation but elongation of shoot cells is considered as critical. The uses of synthetic auxins in horticulture can be traced directly to the natural roles of IAA in the plant. In general, compounds such as a-naphthalene acetic acid (NAA) are used because they resemble IAA in action but are resistant to degradation by plant enzymes.

Characteristics of Auxins: Auxins are characterized by following features-

- Polar translocation (from apex to downward movement),
- Apical bud dominance,
- Variable behaviour of root and shoot growth,
- Root initiation,
- Delay in abscission,
- Differentiation of xylem elements.

Chemical Nature of Auxins

Chemically, it was recognized as auxentriolic acid ($C_{18}H_{32}O_5$). In 1934, Kogl, Erxleben and Haagen-Smit isolated auxin b, auxenolonicacid ($C_{18}H_{30}O_4$) from corn germ oil. Later they also isolated another auxin from human urine and referred it as *heteroauxin*, which was lately called indole –3- acetic acid (IAA, $C_{10}H_9O_2N$). Auxin *a* and *b* are hardly known from plant but IAA is the principal naturally occurring auxin of all higher plants and fungi. It also occurs in human urine especially in person suffering from niacin or nicotinic acid deficiency. Indole-3-acetaldehyde, indole-3-acetonitrile, Indole-3-ethanol, 4-chloro-IAA is also found in plants.

In addition to naturally occurring auxins, certain other substance with identical properties have been reported. Some of them are- Indole butyric acid (IBA), α and β-naphthalene acetic acid (NAA), 2,4-dichlorophenoxyacetic acid (2,4-D), 2, 4, 5-trichlorophenoxy acetic acid (2,4, 5-T), indol propionic acid (IPA), naphthoxyacetic acid, 4-chloro-phenoxyacetic acid, 2,4, 6-trichlorobenzoic acid, 4-amino-3,5,6-trichloropicolinic acid etc.

Site of auxin synthesis and translocation: So for known active sites of auxin synthesis are – shoot tip region (particularly young expanding leaves of the apical bud), tip of grass or cereals seedling coleoptiles and developing embryos, developing fruits, enlarging tissues etc. The auxin synthesis occurs rapidly in green leaves in presence of light than in dark. Auxin is transported basipetally, i.e. it moves from apical to basal end. The movement is quite fast, about 1 to 1.5 cm/h (in root 0.1 to 0.2 cm/h). Recent studies have indicated that auxin transportation in plants is an overall active transport system. Some points in favour of it may be summarized as follows:

The velocity of auxin transport (1 to 1.5 cm/h) in stems and coleoptiles is about ten times faster than diffusion. Metabolic inhibitors are less effective on auxin transportation as compared to poisons. 2,3,5- triiodobenzoic acid and naphthylthamic acid are highly effective. Auxin transportation depends upon aerobic metabolism.

Auxins are able to move against a concentration gradient. It has been observed that there is good correlation between structural requirements for auxin activity and for auxin transport. The saturation effect of transport site being in between 0.1 and 1 Mm per tissue volume also supports it. Auxins are used for a variety of horticultural purposes, including:

- Promotion of rooting of cuttings (e.g., Rootone). The base of the cutting is dipped in a powder containing NAA or indolebutyric acid (IBA) prior to planting.
- Induction of flowering in pineapple (actually caused by the auxin-induced production of ethylene). NAA is generally employed as the auxin.
- Increased fruit set and induction of the pericarp in the absence of fertilization.
- Prevention of pre-harvest fruit drop.
- Auxin type herbicides (e.g., 2-4-D).
- Prevention of sprouting by inhibitions of buds and prolonged dormancy or control of flowering.
- Prevention of leaf fall or abscission and thinning of compact fruits.

GIBBERELLINS

Since the discovery of GAs in the 1920s a considerable body of information has emerged regarding the roles these hormones play in plant growth and development. As these compounds and analogs became available, either by fermentation or chemical synthesis, a number of potential practical applications were identified. Gibberellins have been primarily used for manipulating production practices and insuring the quality of high value specialty crops such as grapes, citrus, cherries, and apples. Increasing market demands for a greater variety of quality fruit may provide new opportunities for the expanded use of GAs (i.e. tropical fruits, thinning and sizing wine and new table grape varieties, and as components of integrated pest management programs for postharvest quality of fruits). Despite considerable research, GAs have not found utility in major agronomic crops. However, increased understanding of the interrelationship between GAs and other PGRs, concurrent with the development of new GAs or GA-mimics, may offer the unique activities necessary to penetrate agronomic markets.

However, despite extensive research with GAs the number of commercial uses has fallen short of their predicted impact and growth in agriculture. Unique opportunities may still result from new research and development with GAs to meet the future demands for increased food quality and production.

Characteristics of gibberellins: GA are characterized by following features.

- Prevention of genetic and physiological dwarfism,
- Breaking of dormancy,
- Induction of flowering in long day plants,
- Increase of amylase activity,
- Substitute for chilling effect.

Chemical Nature of Gibberellins

Gibberellins are cyclic diterpenes. Chemically these are represented by molecular formulae such as $C_{19}H_{24}O_6$ (GA_1), $C_{19}H_{26}O_6$ (GA_2), $C_{19}H_{22}O_6$ (GA_3 or Gibberellic acid), $C_{19}H_{24}O_5$ (GA_4) etc. They contain γ-lactones ring on cyclohexene ring. The presence of hydroxyl groups and carboxyl groups forms them esterified compound. OH group reduces their biological activity while COOH group in reduced response.

There are about 120 different gibberellins have been identified in higher plants and the *Gibberella* fungus, although only two (GA_3 and a mixture of GA_{4+7}) are available via fermentation by the fungus *Fusarium monilifome* (*Gibberella fujikuroi*) in reasonable quantity for extensive applied research. The complexity of GA chemistry has precluded the entirely synthetic routes of production. With the diversity of physiological processes in which GAs are involved, it is not surprising that their practical application spans the entire range of plant growth and development. The majority of these uses however, represent specially niche markets in high cash value horticultural commodities. Commercially, gibberellins are produced by fungal cultures, and it is the purified natural products that are applied to plants. Generally, gibberellic acid (GA_3) is used, because this is the only gibberellins obtainable in commercial quantities, although an expensive mixture of GA_4 and GA_7 is now commercially available for specific purposes.

Site of synthesis and translocation: GAs are synthesized from mevalonic acid in young tissues of the leaves (exact location uncertain), root tip and developing seeds (embryo). Gibberellins moves readily in all directions and in all tissues including phloem and xylem. GA moves in the same pattern as the carbohydrate translocation system and with a similar velocity (5 cm/h). Unlike auxin, it is passive, non-polar and diffuse type.

Following are the major effects on plants:

- Stem growth: GA1 causes hyper elongation of stems by stimulating both cell division and cell elongation. This produces tall, as opposed to dwarf, plants.
- Bolting in long day plants: GAs cause stem elongation in response to long days.
- Induction of seed germination: GAs can cause seed germination in some seeds that normally require cold (stratification) or light to induce germination.
- Enzyme production during germination: GA stimulates the production of numerous enzymes, notably a-amylase, in germinating cereal grains.

Gibberellins are used for a variety of horticultural purposes, including:

1. Enhanced production of seedless grapes. Bigger, more uniform bunches with larger fruit are produced. Among other effects, the gibberellins causes lengthening of the peduncle (stalk) attaching each grape to the cluster, thus permitting larger grapes to form. Virtually all the grapes that go to market are now treated with gibberellins.
2. Treatment of oranges to prevent rind senescence, to permit longer storage on the tree, and thus to extend the marketing period.
3. Enhancement of flower bud formation and improvement of fruit quality in cherries.
4. Improvement of fruit setting in apples and pears, particularly under weather conditions poor for setting.
5. To substitute for a chilling requirement in instances such as: (1) flower induction for seed production (radish), (2) increased elongation (celery, rhubarb), (3) earlier flower production (artichokes).
6. The production of hybrid cucumber seed. Most high-producing cucumbers are F_1 hybrids. GA sprays induce the production of male flowers on cucumber plants that normally produce only female flowers. The seed from neighboring all female plants of a different strain is then exclusively hybrid.
7. Increased malt production. The addition of GA to germinating barley during beer production enhances a-amylase production so that more malt is produced more quickly. As the malt is the raw material for fermentation, a greater production of beer is made possible by this technique.

CYTOKININS

The term cytokinin is universally used as a generic name for substance, which promotes cell division and exerts other growth regulatory functions in the same manner as kinetin. This substance was identified as 6- furfuryl aminopurine and was named kinetin. The most common cytokinin base in plants is zeatin, dihydrozeatin and triacanthene. Several substances have been found showing cytokinins like activities. The common such examples are 6-aminopurine (adenine), benzimadazole, 6-benzyladenine, 1-benzyladenine etc.

Characteristics of cytokinins: Cytokinins are characterized by following features:

- Initiation of cello division,
- Delay of senescence (Richmond –Lang effect),
- Use in tissue culture,
- Counteract apical bud dominance,
- Induces flowering in short day plants.

Chemical Nature of Cytokinins

Cytokinin has been found to contain carbon, hydrogen, nitrogen, and oxygen in the ratio $C_{10}H_9N_5O$. Cytokinin comprises a furan nucleus on extra carbon present either as a side methyl on the furan ring or as a methylene radical between the furan and purine nuclei.

Site of synthesis and translocation: Root tip is an important site of its synthesis. However, developing seed and cambial tissues are also the sites of cytokinin biosynthesis. Cytokinins have so far been extracted from coconut milk, tomato juice, flowers and fruits of *Pyrus malus*, fruits of *Pyrus communis* (pear), *Prunus cerasifera* (plum) and *Lycopersicum esculentum*, cambial tissues of *Pinus radiate*, *Eucalyptus regnans* and *Nicotiana tabacum*, immature fruits of *Zea mays*, *Juglans* sp. and *Musa* sp. Female gametophytes of *Ginkgo biloba*, fruitlets, embryo and endosperms of *Prunus persica*, seedlings of *Pisum sativum*, root exudates of *Helianthus annuus* and tumour tissues of tobacco. The cytokinins move upward (root to shoot), perhaps in the xylem stream, there are also reports of its basipetal movement in petiole and isolated stem.

The action of cytokinins is still poorly understood and insufficient evidence exists to conclusively identify any biochemical point of action. The important effects are:

1. **Cell division:** Applications of cytokinins induce cell division in tissue culture in the presence of auxin. The presence of cytokinins in tissues with actively dividing cells (e.g., fruits, shoot tips) indicates that cytokinins may naturally perform this function in the plant.
2. **Morphogenesis:** In tissue culture, cytokinins promote shoot initiation.
3. **Growth of lateral buds:** Cytokinins applications can cause the release of lateral buds from apical dominance.
4. **Leaf expansion:** Resulting solely from cell enlargement. This is probably the mechanism by which the total leaf area is adjusted to compensate for the extent of root growth, as the amount of cytokinins reaching the shoot will reflect the extent of the root system.
5. Cytokinins delay leaf senescence.
6. Cytokinins may enhance stomatal opening in some species.
7. **Chloroplast development:** The application of cytokinins leads to an accumulation of chlorophyll and promotes the conversion of leukoplasts into chloroplasts.

The major commercial uses for cytokinins derive from their ability to delay senescence and maintain greenness. The artificial, highly active cytokinin, benzyladenine, is the main compound used. The treatment of holly for festive decorations enables its harvest many weeks prior to use. Post-harvest sprays or dips are now available to prolong the storage life of green vegetables such as asparagus, broccoli, and celery.

ABSCISIC ACID

The name abscisic acid (ABA) is rather unfortunate. The first name given was "abscisin II" because it was thought to control the abscission of cotton bolls. At almost the same time another group named it "dormin" for a purported role in bud dormancy. By a compromise the name abscisic acid was coined. It now appears to have little role in either abscission or bud dormancy, but we are stuck with this name. As a result of the original association with abscission and dormancy, ABA has become thought of as an inhibitor. While exogenous applications can inhibit growth in the plant, ABA appears to act as much as a promoter (e.g., storage protein synthesis in seeds) as an inhibitor, and a more open attitude towards its overall role in plant development is warranted. It also acts as stress hormone helping the plants to cope with the adverse environmental conditions.

Characteristics of abscissic acid: Abscissic acid are characterized by following features:

- Abscissic acid interacts with IAA, cytokinin, GA in their several physiological activities.
- Control movement of guard cells and stomatal activity.
- Beginning of senescence in plants
- Decreased rate of photosynthesis and change in nucleic acid.
- Abscissic acid induce flowering during long days.

Chemical Nature of Abscissic Acid

Abscissic acid, chemically 3-methyl-5-(1'-hydroxy-4-oxo-2' 6' 6', = trimethyl-2'-cyclohexen-1'-yl) *cis,* trans-2,4-pentadienoic acid is one of the widespread and important inhibitors prevailling in plants. Abscissic acid possesses an asymmetric carbon atom and can, therefore, exist as either (+) or (-) enantiomers. Naturally occurring abscissic acid is always in (+) form (2 *cis*-ABA) and is more active than (-) form (2-*trans*-ABA). The recent work has demonstrated that 2-trans-ABA is not itself biologically active but is transformed into *cis* from in presence of ultraviolet light. A number of synthetic plant growth retardants have been developed which cause dwarfing in plant without any harmful effects. The chief compounds proven effective are the nicotinium, phosphonium, choline, hydrazine, pyrimidines and triazoles compounds.

Site of Synthesis and Translocation

Abscissic acid is synthesized from mevalonic acid in roots and mature leaves, particularly in response to water stress. Seeds are also rich in abscissic acid, which may be imported from the leaves or synthesized. Abscissic acid is exported from roots in the xylem and from leaves in the phloem. There is some evidence that abscissic acid may circulate to the roots in the phloem and then return to the shoots in the xylem. All plant parts are known to synthesise it.

The important effects of abscissic acid are:

1. Abscissic acid induce bud dormancy in plants
2. Abscissic acid accelerated leaf abscission in different horticultural plant particularly in stress condition.
3. Abscissic acid induce flowering during long days.
4. Abscissic acid stimulates release of ethylene
5. Senescence of leaf is promoted by abscissic acid.

The increase in abscissic acid concentration is an indication of senescence, characterized by loss of chlorophyll, decreased rate of photosynthesis and changes in nucleic acid and protein metabolism. In this way, abscissic acid can be considered as a chemical indicating the beginning of senescence in plants.

ETHYLENE

Ethylene (C_2H_4) is a gas at temperatures under which a plant can live. Ethylene is found in plants and act as hormone. It is now well known that ethylene is produced by flowers, leaves, stems, roots, tubers and seeds. Maximum rate of ethylene production occurs during the period of maximum respiration just before senescence in many fleshy fruits, which is followed by an increase in respiration rate called *climacteric rise*. Here it may be thought that ethylene formation may be intimately associated with the ripening process of fruits. The gas ethylene is synthesized from methionine in many tissues in response to stress. It does not seem to be essential for normal vegetative growth. It is the only hydrocarbon with a pronounced effect on plants. Most tissues in response to stress synthesize ethylene. In particular, it is synthesized in tissues undergoing senescence or ripening.

Characteristics of ethylene: Ethylene are characterized by following features:

- Ethylene is only a gaseous form plant hormone
- Ethylene causes faster abscission of leaves and flowers
- Ethylene is a ripening agent
- Ethylene increases the number of female flowers

Chemical Nature of Ethylene

Chemically, ethylene is the simplest of all plant growth substances. Other olefine molecule like acetylene and propylene can produce ethylene like action, but ethylene is 60-100 times more active than propylene. These are hydrocarbons of low molecular weight, having an unsaturated bond adjacent to terminal carbon atom. In recent years, considerable horticultural interest has been stimulated by the discovery of the ethylene releasing agent-2-chloroethyl-phosphoric acid marketed as Amchem 66-329, CEPA and ethrel. Ethephon exerts its influence by acting as a decomposition product in plant tissue and liberating ethylene close to the site of action. The new product is effective at 100-500 ppm concentrations in aqueous solution.

$H_2C = CH_2$
Ethylene

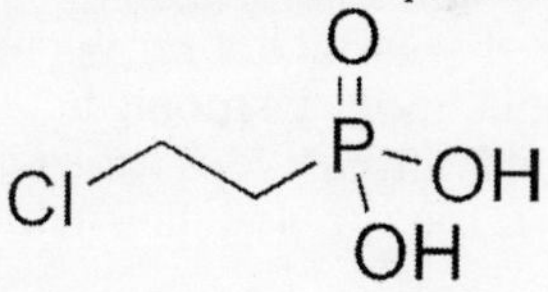

$Cl-CH_2-CH_2-P(=O)(OH)-OH$
Ethephon

Site of Synthesis and Translocation

It is now well known that ethylene is produced by flowers, leaves, stems, roots, tubers and seeds. Maximum rate of ethylene production occurs during the period of maximum respiration just before senescence in many fleshy fruits. Being a gas, ethylene moves by diffusion from its site of synthesis. Ethylene is a tiny molecule to permit ready movement through plant tissues by diffusion. Its movement through tissues may be via air spaces and thus related to tissue porosity. Its solubility in water and even its lipophilic quality should allow it to pass through cell membrane readily or perhaps stimulate the activity of permease systems.

The important effects of ethylene are:

1. Ethylene enjoys a wide variety of uses, but its gaseous nature precludes its use in non-enclosed spaces. It also influences flowering (in pine apple), apical bud dominance, inhibition of stem and root growth, petal discoloration, abscission, maturity, freeze and disease resistance. Ethylene itself can be used to enhance the ripening of fruits such as bananas in storage following their shipment in an unripe condition, his is of great benefit, since the green bananas are rugged and do not bruise or spoil easily. The tender ripe bananas can then be carried safely to market from the nearby warehouse. Recently, an ethylene-producing

liquid chemical, 2-chloroethyl-phosphonic acid (commercially called Ethrel or Ethephon) has been introduced into commerce. This compound is sprayed onto the plant at a slightly acid pH. When it enters the cells and encounters the cytoplasm at about neutral pH, it breaks down to release gaseous ethylene. Numerous commercial applications for this compound have appeared, mostly in relation to the natural effects of ethylene:

2. The most important commercial use involves enhancing latex flow in rubber trees in Southeast Asia. When a rubber tree is "tapped," the latex flows for a certain period before the cut seals and the flow stops. Ethephon delays the healing of the cut so that the latex flow continues for a longer period, thus yielding more latex with less tapping.
3. Ethylene accelerates abscission of leaves, flowers and fruits.
4. Enhancement of uniform fruit ripening and coloration. This has been shown to be of particular value in field tomatoes picked at a single time by machine.
5. Acceleration of fruit abscission for mechanical harvesting. This provides a potential area of use in a wide variety of fruits such as grapes, cherries, and citrus.
6. Promotion of female flower production in cucurbits (cucumber, squash, melon) so as to increase the number of fruits produced per plant.
7. Promotion of flower initiation and controlled ripening in pineapples.

NON-TRADITIONAL PLANT GROWTH REGULATORS

Jasmonic Acid

Jasmonic acid (JA) and its methyl ester (MeJA) were first identified as odorant compounds in the essential oil of jasmine and their chemical structures were subsequently determined (Demole *et al.*, 1962). Further studies demonstrated that jasmonates, a term that includes jasmonic acid isomers, conjugates to amino acids and sugars, and other derivatives, were widespread in Angiosperms, Gymnosperms and algae suggesting that they could fulfill an important role in plant physiology.

Jasmonic acid

Jasmonates are linolenic acid–derived, cyclopentanone–based compounds. The structural positions essential for the physiological activity of jasmonates are: a planar pentanone ring, the pentenyl chain inserted at C_7, an acetyl or longer side chain at C_3 and the keto group at C_6. However, there are exceptions to this strict structure–activity relationship. In many cell suspension cultures, certain genes are induced by an intermediate of jasmonate biosynthesis, 12–oxo–phytodienoic acids. Again in certain cell suspension cultures, 15, 16–dihydro–12–oxo–phytodienoic acids are as active as the final product, 9, 10–dihydrojasmonate. In other systems, inducing activity may be correlated with jasmonate and its amino acid conjugates.

Uses

1. Early studies showed that exogenous jasmonic acid or methyl ester can promote senescence and regulate plant growth, subsequent research has revealed that

jasmonic acid specifically alters gene expression and wounding and elicitors can cause jasmonic acid and methyl ester accumulation in plants.

2. Jasmonic acid and methyl ester inhibit the germination of non–dormant seeds and stimulate the germination of dormant seeds.
3. Jasmonic acid plays a role in the formation of flowers, fruit and seed. The relatively high levels of this compound in developing plant reproductive tissues suggest it.
4. A role for jasmonic acid in mediating the accumulation of secondary plant products has also been proposed. Exogenous jasmonic acid causes the accumulation of paclitaxel and related taxanes in *Taxus*, alkaloids in *Cantharanthus* and *Cinchona*, anthocyanins in soybean and rosmarinic acid in *Lethospermum*.
5. Jasmonates have been proposed as endogenous regulators of tuberization. They were identified as a mixture of tuberonic acid glucoside and its methyl ester, substances whose aglycone, tuberonic acid, are structurally related to jasmonic acid.
6. Jasmonic acid also plays an important role in plant insect and disease resistance. Jasmonic acid accumulates in wounded plants and in plants or cell cultures treated with elicitors of pathogen defense. Jasmonic acid activates genes encoding protease inhibitors that help protect plants from insect damage. These proteins are thiamin, osmotin and the ribosome–inactivating protein RIP60.
7. Jasmonic acid is involved in the regulation of protein storage in plants. The genes encoding vegetative storage proteins (VSPs) are regulated by jasmonic acid. VSPs accumulate in the vacuoles of paraveinal mesophyll and bundle sheath cells in soybean leaves.
8. Jasmonate–induced Gene Expression: The first observations on jasmonate–induced gene expression were made with methyl ester or JA–treated barley segments. Many of the jasmonate–induced proteins (JIPs) show a species specificity. In barley leaves, JIPs are located within the cytosol, nucleus, chloroplasts, peroxisomes and vacuoles. Tissue–specific expression of JIPs also occurs. Some putative functions have been ascribed to barley JIPs. JIP 6–is a member of the thionin class and JIP–60 when processed into a 30 KDa protein are responsible for the inactivation of certain initiation and elongation steps in ribosomal protein synthesis.
9. Wounding of plants is known to stimulate the synthesis of jasmonic acid, which in turn, elicits the expression of wound–inducible genes such as the protease inhibitor (PI–II) gene.
10. Disease resistance phenomenon: A novel induced disease resistance phenomenon was recently described by Pieterse *et al.* (1998). This phenomenon called induced systemic resistance (ISR) is induced by *P. fluorescence WCS417r*, a rhizobacterium.
11. Progress has been made in sequencing many of jasmonic acid –induced genes but understanding of the responsive domains is incomplete.

Brassinosteroids

Brassinosteroids (BRs) are a group of naturally occurring polyhydroxy steroids. Brassinolide which was originally isolated from rape pollen in 1979 was the first BR to be characterized. More than 60 BRs have been identified of these, 31 have been fully characterized including two conjugates. BRs have been identified in many plants including dicots, monocots, gymnosperms, green algae and a fern. BRs have been isolated from seeds, fruits, leaves, flower buds and galls. BR levels are relatively high in pollen. As a result of extensive investigations, BRs were found to show characteristic physiological action on the growth of plants in micro quantities. Based on a detailed analysis of their physiological action, they are regarded as a new class of plant hormones and act independently of the other endogenous plant hormones.

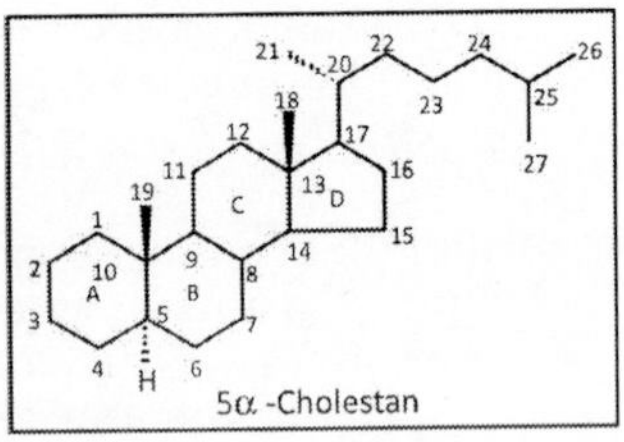

5α-Cholestan

Uses

1. The physiological effects of BRs is similar to that of other endogenous plant growth hormones such as auxins, gibberellins and ethylene but evidence indicates important differences in the mechanism of action. In many systems BRs interact synergistically with auxins. The response of BRs and GAs appears to be both independent and additive. ABA interacts strongly with BRs and prevents BR–induced effects.
2. Stimulatory effects of BRs on elongation are among the most documented physiological effects. It has been observed in many assay systems such as radish, tomato, sunflower and cucumber hypocotyls, normal and dwarf pea. Cell elongation is associated with an increase of acid secretion. Some BRs have been shown to stimulate the secretion of H^+ and to hyperpolarize the plasma membrane in different plant tissues such as maize root, azuki bean epicotyls, radish cotyledons and *Egeria densa* and *Vicia faba* leaves. These effects are completely suppressed by inhibitors of RNA and protein synthesis.
3. Inhibitory effects of BRs on adventitious root development have been observed in mung bean hypocotyls and other assays. Lateral root elongation in wheat, bung bean and maize is reduced by treatment with 24–epibrassinolide.
4. BR application to crop plants helps to overcome unfavourable cultural and weather conditions. The effects of 24–epibrassinolide on barley, bean, or lettuce plants was more prominent in soil deficient in fertilizer than in fertilized soil.
5. In cell culture of carrot, BRs induced cell enlargement without any effect on cell division.
6. BRs promote tracheary–element differentiation.
7. It has been shown that BRs stimulate ethylene production in etiolated mung bean hypocotyls segments

Oligosaccharins

Oligosaccharins are complex carbohydrates that are capable of modulating plant growth and development at low concentrations. Studies of plant–microorganism interactions yielded the first evidence that oligosaccharins could serve as biological signals. Much of this research focused on the synthesis and accumulation of antimicrobial phytoalexins in response to microbial attack. Phytoalexin synthesis and accumulation are observed not only after microbial infection but also after treatment of plant tissue with cell–free extracts of microbial origin. The active components in these extracts are commonly referred to as elicitors. The term "elicitor" is now commonly used for molecules that stimulate any plant defense mechanism. Some oligosaccharins such as oligogalacturonides act as elicitors and evoke pathogen defense responses. These defense responses include the accumulation of phytoalexins, proteinase inhibitors, lignin, peroxidase, lipoxygenase and β–1,3 glucanases. A novel class of oligosaccharin signal molecules has been discovered which plays a role in the host–specific interaction between rhizobial bacteria and leguminous plants leading to the nitrogen–fixing root nodules. These signal molecules, which were shown to be lipo–chitin oligosaccharides, are the first plant organogenesis–inducing factors discovered.

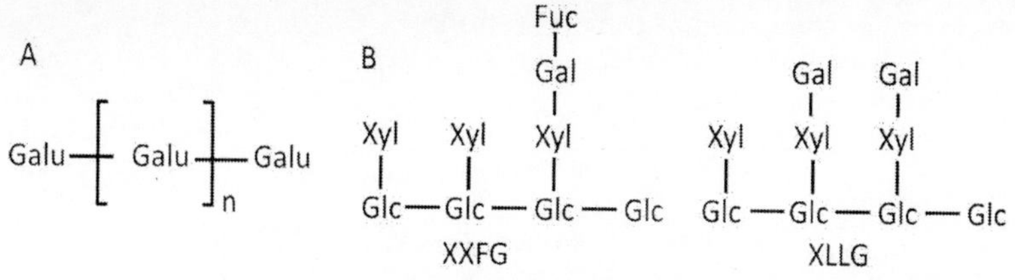

A

Galu–[Galu]$_n$–Galu

B

Fuc
Gal
Xyl Xyl Xyl
Glc—Glc—Glc—Glc
XXFG

Gal Gal
Xyl Xyl Xyl
Glc—Glc—Glc—Glc
XLLG

Oligogalacturonide

Glucose as a Hormone

In plants, sugars have conventionally been viewed as resources for respiration and metabolic intermediates as well as structural or storage components. The widely observed effects of sugar on gene expression and on plant growth and development have often been attributed to sugar metabolism and energy production. The specific effects of glucose on gene expression and development are characteristic of plant hormone actions. It has been demonstrated that hexokinase is a specific glucose sensor and recent analysis of transgenic plants and sugar response mutants have uncovered intimate cross–talk between glucose and other plant hormone signaling pathways. Recent characterization of transgenic carrot plants with antisense cell wall invertase or vacuolar invertase constructs shows malformed embryos and bushy shoots that can be corrected by hexoses. These phenotypes are apparently caused by a lack of hexose signaling which may control the balance between the plant hormones auxin and cytokinin (which are critical for embryogenesis and shoot/root development). In a recently identified gin 2 mutants, the lack of AtHXK1 causes altered sensitivity to auxin and cytokinin.

Glucose–Mediated Gene Expression

Sugar regulation of gene expression can be mediated at the transcriptional and post–transcriptional levels. Many systems including whole plants, intact tissues, tissue culture cells and freshly isolated protoplasts have been used to identify sugar responsive cis–DNA elements.

The Glucose Signal Transduction

The discovery and development of specific protein kinase (PK) and phosphatase (PP) activators and inhibitors in the past decade have provided valuable tools with which to examine the involvement of protein phosphorylation/dephosphorylation in diverse signal transduction pathways. It has been shown that PP1 and PP2A inhibitors can mimic glucose repression of photosynthesis genes in maize leaf cells and in photoautotrophic cultures of *Chenopodium rubrum.* It has been proposed that hexokinase acts as a glucose sensor. The conformational changes of hexokinase after binding to its substrate glucose may resemble the analogous ligand–induced conformational change of a typical receptor that allows modification of protein–protein interactions to trigger a signaling cascade.

Polyamines

Polyamines were accepted as plant hormones at the International Conference on Plant Growth Substances in 1982. Galston justified their classification as hormones on the following grounds.

They are widespread in all cells and can exert regulatory control over growth and development at micro molar concentrations.

In plants where the content of polyamines is genetically altered, development is affected. For example, in tissue cultures of carrot or *Vigna* when the polyamine level is low callus growth occurs. When polyamines are high embryoid formation occurs. In tobacco plants, which are overproducers of spermidine, anthers are produced in place of ovaries.

All cells contain the diamine putrescine and triamine spermidine while eukaryotic cells contain the tetramine spermine. In both prokaryotes and eukaryotes including higher plants, mutants lacking the ability to biosynthesize polyamines are unable to grow and develop normally. The addition of polyamines to these mutants generally restores normal growth and development.

Uses

At cellular pH, polyamines are polycations and thus bind to important cellular polyanions such as DNA, RNA, phospholipids and acidic protein residues and cell wall components. Through such binding, polyamines could affect the synthesis of macromolecules, the activity of macromolecules, membrane permeability and partial processes of mitosis and meiosis.

Polyamines and Flowering: Correlations between polyamines and flowering have been observed in whole plants. *Xanthium strumarium* (a short–day plant) revealed that exposure to one or two consecutive induction long nights greatly increases the titer of conjugates of the higher polyamines in leaves and later in buds. This investigation has also been extended to a long–day plant, *Sinapsis alba*. Results on leaf exudates show that free petruscine increases quite early and dramatically during the floral transition.

Polyamines and Embryogenesis: The involvement of polyamines in embryogenesis was first observed in carrot cell cultures. Studies on somatic embryogenesis in carrot reveals that embryogenic growth is suppressed by the presence of auxin in the medium, increased synthesis of polyamines is required for embryogenesis upon removal of auxin and ethylene suppresses embryogenesis. Just prior to embryoid formation, ADC activity increases rapidly accompanied by spermidine titer.

Polyamines and Senescence: Senescence in many plant organs, both *in situ* and upon their excision from the plant, is correlated with a decline in polyamine titer. Addition of exogenous polyamines delays or prevents senescence–related processes such as the decline of chlorophyll, protein and RNA content in a wide range of monocot and dicot leaves. It has been proposed that polyamine titer controls senescence at least in leaves.

Polyamines and Stress: Putrescine accumulation and ADC increase result from osmotic stress, acid feeding, high NH_4^+ feeding or exposure to the atmospheric pollutants. The response to osmotic stress in excised oat leaves within minutes. There is evidence that putrescine accumulation results not only from ADC activation but also from a decrease in spermidine syntheses activity. Plant cells can be induced to synthesize polyamines when subjected to high temperature or drought. The thermo polyamines norspermidine and norspermine have been detected in plants.

Salicylate

Salicylic or ortho–hydroxybenzoic acid belongs to a diverse group of plant phenolics. The most important mechanism for formation of benzoic acids in plants is the side chain degradation of cinnamic acids, which are important intermediates in the shikimic acid pathway. The conversion of cinnamic acid to salicylic acid is likely to proceed via benzoic or ortho–coumaric acid. A survey of salicylic acid content in the leaves and reproductive structures of 34 plant species confirmed its ubiquitous distribution in plants. The highest levels of salicylic acid were found in the inflorescences of thermogenic plants and in plants infected with necrotizing pathogens.

The interest in salicylic acid (SA) began with the discovery that this compound is a natural trigger for the metabolic explosion which raises the temperature of the thermogenic inflorescences of Arum lilies and possibly other thermogenic angiosperms by as much as 15°C (Raskin et al., 1987, 1989. 1990). This was the first demonstration of an important regulatory role for endogenous SA in plant development, which ended the long search for calorigen the elusive regulatory substance postulated more than 50 years ago (Van Herk, 1937). The indications are that exogenous application of SA stimulates flowering in the Lemnaceae and inhibits biosynthesis of ethylene, stomatal closure and ion uptake. Recent studies suggest that SA may also play a significant role in disease resistance.

Salicylic Acid and Disease Resistance

Plants react to pathogen attack by activating an elaborate defense mechanism that acts both locally and systemically. In many cases, local resistance is manifested as a hypersensitive response, which is characterized by the development of lesions that restrict pathogen growth and/or spread. Associated with the hypersensitive response is the induction of a diverse group of defense–related genes. The products of many of these genes play important roles in containing pathogen growth either directly by providing antimicrobial enzymes and secondary metabolites or indirectly by helping to reinforce the defense capabilities of host cell walls. These products include cell wall polymers such as lignin and suberin as well as phenylpropanoids and phytoalexins. Several families of pathogenesis–related (PR) proteins are also induced during the hypersensitive response. Some of these proteins are hydrolytic enzymes [β–1,3–glucanases (PR–2) and chitinases (PR–3] but the functions of other PR–proteins have yet to be determined. Most of the PR proteins have been shown to possess antimicrobial activity *in vitro* or the ability to enhance disease resistance when over–expressed in transgenic plants.

The hypersensitive response is also associated with a massive increase in the generation of reactive species (the oxidative burst), which proceeds and then accompanies lesion–associated host cell death. Over a period of hours to days after the primary infection, systemic acquired resistance develops throughout the plant. The systemic acquired resistance is manifested as an enhanced and long–lasting resistance to secondary challenge by the same or even unrelated pathogens.

The signaling pathways involved in the initiation and maintenance of the hypersensitive response and systemic acquired resistance are still poorly understood. Nonetheless, several lines of evidence suggest that salicylic acid is not the only signal involved in the induction of PR genes.

Recent advances in plant defense signaling pathway research have shown that plants are capable of differentially activating distinct defense pathways.

Salicylic Acid and Gene Expression

In plants, salicylic acid has been shown to induce the expression of many defense–related genes. The genes induced by salicylic acid can be grouped into two broad classes. The first group consists of genes whose expression is insensitive to protein synthesis inhibitors, such as the glutathione–S–transferase genes, the 35S promoter of cauliflower mosaic virus and the nopaline and octopine synthase genes of *Agrobacterium*. Promoters of this class of genes contain copies of as–1 like cis–elements that mediate salicylic acid–induced expression. Several transcription factors belonging to TGA family of bZIP proteins have been identified and shown to bind these elements.

The second class of salicylic acid–inducible genes includes the class II PR genes, whose induction by salicylic acid is sensitive to inhibitors of protein synthesis.

Nitric Oxide (NO)

Due to the diversity of its physiological functions and general ubiquity, nitric oxide has attracted a great deal of attention. In 1992, nitric oxide was recognized by science magazine as the molecule of the year and in 1998 the Noble Prize for medicine went to three pioneers in nitric oxide research. Until recently, research on the effects of nitric oxide in plants focused on atmospheric pollution by the oxides of nitrogen. It was subsequently demonstrated that plants not only respond to atmospheric nitric oxide but also produce substantial amounts of nitric oxide. Mounting evidence also suggests that nitric oxide is a novel effectors of plant growth, development and defense responses. For example, nitric oxide was shown to induce leaf expansion, root growth and phytoalexin production.

Plant Nitric Oxide Production

In plants, nitric oxide can be generated as a byproduct of denitrification, nitrogen fixation and/or respiration. In most cases, NO production in plant tissues has been linked to the accumulation of NO_2. NO is produced from NO_2 either non–enzymatically through light–mediated conversion by carotenoids or enzymatically through NADPH nitrate reductase. Nitric oxide emission from plants occurs under stress conditions, such as herbicide treatment as well as under normal growth conditions. It is believed that the ability to produce NO is common to many plant species.

Nitric Oxide and Plant Defense Response

The presence of a mammalian–type NOS (nitric oxide synthase) was first reported in 1996 (Leshem, 1996). NOS activity was found in the roots and nodules of leguminous plants, where it was thought to be induced by Rhizobium lipopolysaccharides. Thus, it was suggested that nitric oxide is involved in establishing symbiosis and in mediating nitrogen fixation by binding the home group of redox sensors such as leghemoglobin or FixL. It has also been shown that nitric oxide and possibly NOS play a prominent role in the activation of plant defenses against microbial pathogens. Infection of resistant tobacco with tobacco mosaic virus resulted in enhanced NOS activity. Similar results were obtained with soybean cells and Arabidopsis plants responding to either a bacterial pathogen (Pseudomonas syringae) or a fungal elicitor. Furthermore, tobacco-resisting infection by Ralstonia solanacearum exhibited elevated levels of NOS activity. The above findings suggest that a plant NOS is involved in defense response activation. There is sufficient evidence that nitric oxide accumulation is needed to contain pathogen growth and spread, however, it is unclear whether nitric oxide (and/or its activated derivatives) is directly toxic to microbial pathogens in plants.

Florigen, Anthesin and Vernalin

Florigen is responsible for the initiation of flowering in plants. It is synthesized in the older leaves and then transferred to the growing region where it initiates the floral bud initiation. Florigen, is therefore, also called flowering hormone. The synthesis of it depends on duration of light and specific to every plant and undergoes following steps.

High intensity light processed (Proceeds in light)
↓
Pigment conversion (Proceeds in dark)
↓
Time measuring process (Proceeds in dark)
↓
Florigen (Proceeds in dark)
↓
Florigen stabilization (Promoted by light)
↓
Post induction preparatory process (Proceeds both in dark and light)
↓
Florigen translocation (Promoted by light)
↓
Floral bud differentiation

Anthesins are a newly discovered group of hormone necessary for flower formation and florigen equals to gibberellin and anthesins (GA+A). It is believed that a substance is formed during vernalisation. This effective substance has been named as vernalin.

Morphactins

Morphactins are synthetic growth regulators, which act in variety of ways on the natural regulation mechanisms of plants. The important ones are phenoxyalkancarboxylic acid, substituted benzoic acids, maleic acid hydrazide, fluorine-9-carboxylic acid and their derivatives, chlorflurenol, chlorflurun, flurenol, methylbenzilate, dichlorflurenol etc. Morphactins have fundamental action on morphogenesis of plants and this characteristic designation (morphactins) is derived from morphologically active substances.

Use of Morphactins

1. Morphactins inhibit germination of seed particularly the emergence of the radical from the seed shell.
2. Morphactins inhibit the growth of seedling affecting the shoot and often also root.
3. They have inhibitory effect on the stem elongation. Increase concentration produces dwarfening in the plants.
4. Morphactins have also been found effective on flowering stimulation, sequence of flowering, position and number of flowers, formation of flowers, inflorescence and parthenocarpy etc.

USE OF PLANT GROWTH REGULATOR'S IN HI-TECH HORTICULTURE

In hi-tech horticulture the plant growths are used for yield enhancement, quality improvement and the facilitation of harvesting together with genetic engineering represent supplemental technologies in horticulture for increasing quality production and optimization the use of limited resources. Although plant growth regulators have been used in horticulture for as long as crop protection chemicals, their impact up to now has been relatively small and their application is limited to some specific cultures.

Most of the current uses for plant growth regulators in the high-value horticultural crops are not, however, for compounds that increase crop yield directly, either by increasing the total biological yield or the harvest index. Rather, compounds that provide economic benefit by enhancing crop quality or aid in more efficient crop management are more common. For example, gibberellic acid is used to reduce the incidence of physiological rind disorders in citrus and daminozide (N,N, dimethylaminosuccinamic acid) application to apple stimulates colour development of the fruit. Both treatments increase the value of the crop but not necessarily the yield.

Plant growth regulators that aid in crop management hi-tech horticulture fall several categories:

1. Mechanical harvesting of crops can reduce substantially the cost production. Compounds that reduce the fruit removal force have in some cases, such as sweet cherry production, allowed the development and use of mechanical harvesting equipment, which was not particularly effective, alone because the force required to remove the crop damaged trees.
2. Plant growth regulators and used for manipulation of the harvest date in a wide range of crops. Ethephon is used to accelerate and concentrate ripening of tomato prior to single mechanical harvest. Gibberellic acid is used to extend the lemon harvest season by preventing senescence of the rind.
3. Plant growth regulators can be used as direct replacements for hand labour in crop production practices other than harvesting. In apple, naphthalene acetic acid (NAA) is used to reduce excessive fruit set that would otherwise result in many small fruits.
4. Plant propagation through seed germination, cutting, layering, grafting and tissue culture.
5. Prevention of fruit drops and increases fruit set.
6. Artificial induction of seedlessness
7. Control of post-harvest development (sprouting and root formation at the time of storage).
8. Inhibition of growth of buds of nursery stock in storage or delaying growth temporarily after planting
9. Regulating of flowering and fruit size as in pineapple, pomegranate, lime etc.
10. Thinning of flowers and fruits as in mangoes, apples and peaches.

Use of auxin in hi-tech horticulture: Auxins have been used variously in hi-tech horticulture and some of the aspects are as follows:

1. **Seed germination:** Poor seed germination in some of the crop is major hindrance for sexual propagation. Know-a-days IAA, IBA, NPA, 2,4-D are most widely used in soaking seeds for germination. Auxins at a very low concentration promote germination but these effects are subjected to variation depending upon form and species of plant. Higher concentration and prolonged treatment retard and

reduce germination.

2. **Rooting:** Most of the fruit crops are vegetatively propagated except papaya, cape gooseberry, phalsa and mangosteen. The auxins stimulate root development by inducing root initials that differentiate from cells of the young secondary phloem, cambium and pith tissue. Commercial use of rooting auxin compounds is very common. They promote callus and root formation, which can improve establishment from cuttings. Dipping the cut surface into rooting compounds enhances species and cultivars difficult to root. The application of 100-500 ppm IBA for soft wood, 500-1000 ppm for semi-hard wood and 2000-5000 ppm for hard wood is common for vegetative propagation of fruit crops. Apart from auxins, a large number of physiologically and chemically unrelated compounds such as phenols and allied simple aromatic compounds, glycosides, growth inhibiting, growth retarding and ethylene producing chemicals significantly influence the rooting of cuttings.
3. **Flowering:** Induce flowering is an important feature of hi-tech production of horticultural crops. The part that growth substances play in the endogenous control of flowering has received wide attention, production of auxin by the petals in the flower bud, pollen and ovary in the developing fruit- bud and by the seed in the fruit suggests that auxin is intimately connected with flowering and also fruit setting. Exogenous application of NAA caused flowering in pineapple @ 0.05 mg/plant and higher strength i.e., 0.25 mg or 0.50 mg per plant caused 100% flowering, 2,4-D also induce flowering but NAA application induces heavier fruits. NAA and NAA based commercial formulations like planofix and celmone are widely used for induce flowering in hi-tech production. ß-hydroxyethylhydrazine also controlled flowering depending on its concentration.
4. **Fruit set:** Fruit setting has been achieved in several fruits and vegetable by the use of NAA, NOA, MCPA, CCP, 2,4-D, 4-CPA etc. in different concentrations. In mango 10 mg/1 NOA caused better fruit setting and the fruit development was parthenocarpic. For pineapple, 0.0015% NAA produced 100% fruiting. Fig, citrus, apple and pears all have been found to bear fruits by the use of auxins in different concentrations. Fruit set in citrus, grapes and phalsa is highly variable and depends largely on prevailing environmental conditions. 4-CPA, treatment serves as an alternative to caprification in fig for improving fruit set. Application of 4-CPA or B-NOA increases fruit set in strawberry, peach, plum and cherry. Auxin including 2,4-D, induced ethylene formation and fruit set in pineapple. Synthetic auxins or auxin-synergists like 2,4-D, 4-CPA (4-chlorophenoxyacetic acid), and N-metatolylphathalamic acid are used widely to increase fruit set in vegetables, for example, with tomatoes grown in glasshouses or under plastic tunnels, or with aubergines.
5. **Flower and fruit thinning:** Thinning of flower and fruits helps in optimizing the fruit size, quality and reduce the phenomenon of biennial bearing in order to maintain production levels from year to year. Application of sodium 4,6- dinitro-o-cresol (DNOC) at full bloom or NAA at post includes satisfactory thinning in apple orchards. The two auxin type compounds NAA and NAAm are used in chemical thinning of apple and pear. NAAm is effective during the same period but higher rates (17-34 ppm) are used. Biennial bearing (light and heavy crops in alternate year) is common to many tree crops. This is a great problem for horticultural crops. By corrected and timely thinning in the heavy years and timely application of NAA or other auxins can solve the problem.
6. **Prevention of fruit drop:** The mature fruit of apple, pear, mango, lemon and grape fruit abscise prior to the time of commercial harvest. This reduces the potential crop yield and may result in the tendency to begin harvesting the crop earlier than is desirable, resulting in lower quality fruit under natural conditions,

there seems to be inverse relationship between auxin content of the fruit and the tendency towards abscission. Application of auxin soon after fruit set results in an acceleration of abscission, however, when auxins such as 2,4,5-TP, NAA and 2,4-D are applied during mid-stages of fruit growth, abscission is delayed or prevented. NAA, 2,4,5-TP are used at 10-20 ppm just prior to the beginning of fruit drop in apple. In citrus application of 2,4-D at 25 ppm and in grape NAA at 50-100 ppm prevents premature fruit drop and allows an extension of the harvest season into summer.

7. **Herbicidal action:** Auxins are selective weed killers and are considerably used in the destruction of weeds in hi-tech horticultural crops. The most widespread commercial application of auxins as herbicide is well known. Auxins like 2,4-dichlorophenoxy acetic acid (2,4-D) and 2,4,5-trichlorophenoxy acetic acid (2,4,5-T), which act as herbicides when used in large doses. Dicotyledonous weeds are particularly sensitive to very high auxin level. Some of the most valuable and widely used selective herbicides in the weed control are auxins, particularly the phenoxyacetic acid analogous (e.g., 2,4-D, 2,4,5-T, and McPA). 2,4-D is one of the most widely used herbicides. It is highly selectively, non-corrosive, effective at low concentrations, safe to handle, relatively easy to formulate, and economical to use. Several benzoic acid analogs (e.g., dicamba, chloramben, and substituted picolinic acid, picloram) are also important herbicides.
8. **Parthenocarpy:** 2,4-D, 2,4,5-T, naphalene compounds, β-naphthoxy propionic acid, 4-CPA, BNOA, BNOP, BNOP, GA+ auxin have been associated with parthenocarpy. Concentration, times of application, environmental conditions, etc. are well knit. Parthenocarpy, speaking broadly, seemed to be associated with higher degree of fruit-set, advance of maturity, light coloured and astringent fruit, sugar content, vitamin C content increased at least in some cases.
9. **Auxins as inhibitors:** Higher concentrations of auxins inhibit the growth and exert toxic effect on the plants grown in hi-tech structures. In normal course in nature self-produced auxin inhibits the growth and development of lateral buds and as a result apical bud remains dominant.
10. **Breaking of dormancy:** Both 2,4-D and IAA have been found to be effective in breaking the dormancy of seeds, buds and cuttings. Growth retardants contribute to the formation of sprouts in potato and onion. Premature potato sprouting may be prevented by the application of MH, CIPC and TCNB.
11. **Abscission:** The abscission layer is a thin zone of small cells produced by newly divided cells, the walls of which becomes soft and gelatinous due to weakening of the middle lamellae of the cells so that eventually the slightest air movement or mechanical disturbance is sometime sufficient to break the stalk at this point. Spray of dilute solutions of IAA, NAA and 2,4-D neutralize or delay prevents the production of separation layer thus prolonging the life of the fruit on the trees.

Use of gibberellin in hi-tech horticulture: Gibberellins have been used variously in hi-tech horticulture for overcome genetic and physiological dwarfism, promote seed germination and break dormancy, induce flowering in lone day plants and biennial plants, affect elongation of cells and induce parthenocarpy in certain unfertilized ovaries, some of the aspects are as follows:

1. **Germination of seeds:** Gibberellin has been widely used for germination of different horticultural crops. Gibberellin increases the length of hypocotyls, cotyledonary leaf area and germination percentage of different seed propagated horticultural crops. The work on Gibberellin in relation to germination is promising and is in progress in India.
2. **Leaf expansion:** In pea, bean, tomato, pepper, sweet corn, cucumber, lettuce

and cabbage plants when treated with gibberellin, leaves become broader and elongated. This results in an increase in photosynthetic area, which subsequently heightens the plant in green houses and modified atmosper cultivation.

3. **Induce flowering:** Early flowering under the effect of gibberellins has been reported in plants irrespective of their nature of photo- sensitivity. Gibberellins may increase flowering in some species of plants. Long day plants under non-inductive photoperiods can be induced to flower by gibberellin treatments of 100-1000 mg/litre. GA_3 also inhibits flowering on several stone fruits. In young sweet and tart cherries, GA_3 is applied to prevent excessive flowering to minimize the competitive effect of early fruiting on vegetation growth. Since cherry has true flower buds, the node following flowering will not support vegetative growth. If flowering is excessive, fewer spurs develop, thereby resulting in lower yield potential during the production life of the tree. The manipulation of flowering by GA_3 has also been integrated into a management program to suppress a pollen transmitted ring spot virus infection as well as the maintenance of the desired ratio of vegetative to flower bud formation in mature tart cherries. The consequence of continuous GA_3 applications over a five years period has been a 22 per cent increase in yield. A mixture of GA_{4+7} is commercially available. It affects on varying structure in the flowering of *Lolium temulentum*, a LD-requiring plant and *Pharbitis nil*, a SD-requiring plant, as well as for flowering of Cupressaceae and Pinaceae conifers. While it is too early to predict commercial implications of such specificities in flowering response for GAs.
4. **Parthenocarpy:** Parthenocarpic fruit development in peach has been reported by the application of GA, which also affects fruit-setting and fruit development. GA induces formation of parthenocarpic fruits in case of brinjal. Some parthenocarpic fruits of guava were also produce by GA treatment. In general, GA induces parthenocarpy in pome fruits (apple and pear) and are now used to increase the fruit size and bunch length in grape.
5. **Fruit setting:** A number of deciduous fruit tree species such as apple and pear as well as some citrus species can be induces to fruit set with GA or a combination of GA and auxin. Consistently unfavourable weather during the pollination period has led the development of a hormone mixture to induce parthenocarpic fruit set in apple. In citrus, application of GA during full bloom can increase fruit set. Application of GA increase fruit set in strawberry, peach, plum and cherry.
6. **Break of seed and bud dormancy:** Seed of many fruit crops often require an extended period of after ripening at low temperature before germination. Breaking of rest is of great significance to overcome dormancy. GA are now invariably used for accelerated seed germination in citrus, cherry, grapes, ber, *Annona*, apple and peach. The treatment with GA significantly reduced/substituted the period of seed stratification. Termination of bud rest by GA sprays is a common practice in peach and apple orchards. GA treatment successfully reduces the period of chilling requirement. It is believed that principal GA effect is to enhance elongation so the radical can push through the endosperm and seed coat that restricts its growth. Similarly potato tuber can be made to germinate in winter by GA treatment.
7. **Fruit ripening:** GAs are used to delay fruit ripening in lemon in order to increase the availability of fruit during the months of May-August when demand is high but production is low. GA is applied in November- December in order to delay the harvest date and increase storage life of the fruit.
8. **Fruit russet:** The other commercially used GAs produced by *Fusarium moniliforme* cultures is a mixture of GA_4 and GA_7. An application of GA_{4+7} of increasing importance is the prevention of fruit russet, a superficial disorder in which the exocarp is interrupted by raised corky outgrowths over the surface

of green and yellow apple cultivars. Although somewhat obscure, a number of factors have been implicated for russet, such as high humidity, presence of free water, light frosts, pesticides and powdery mildew. The multiple applications of GA_{4+7} on `Golden Delicious' effectively increase the fruit finish quality and this corresponds with higher pack out gardens.

9. **Use in grape industry:** GA_3 has been used in the seedless table grape industry since 1960 to manipulate three physiological events rachis cell elongation, flower thinning, and berry enlargement. The first is a pre-bloom spray to promote elongation of the rachis to provide a larger and looser framework for the grape cluster. In wine grapes, there is a need for cluster loosening by either rachis elongation and/or flower thinning to prevent Botrytis bunch rot. To date, there is no chemical thinning agent for wine grapes. Results with GA_3 have been variable with some varieties exhibiting a dramatic reduction of return bloom the year following application others are relatively insensitive to increasing rates of GA_3. The response of wine grapes to other GAs certainly needs to be examined.
10. **Use in citrus production:** The effect of delayed fruit senescence by GA_3 has also been exploited in the citrus industry. GA_3 delays maturity of citrus rind without altering the maturation process of the internal fruit, thus preventing several rind disorders associated with fruit ripening. In lemons, this facility scheduling of harvesting to synchronize with market demands. As harvest date is delayed in grapefruit, there is added benefit in combining 2,4-D to retain the fruit on the tree. Recent findings suggest that the delay in citrus rind maturity is associated with the quantity and quality of epicuticular wax deposition and its influence on respiratory gas exchange.
11. **Seed production of conifers:** There are also minor practical uses of GAs in seed production of commercially important conifers. GA_3 can be used to promote flowering, and thereby significantly increase female cones and seed per tree in seed production orchards of Cupressaceae and Taxodiaceae species.
12. **Insect control in fruits:** Recent laboratory studies have shown that GA_3 application significantly reduces the number of probes and egg fecundity of the fruit fly due to increased exposure to toxic peel oils contained in the juvenile rind. If this effect is manifested in the field, a high level of resistance may preclude the need for post harvest treatment with an insecticide/larvicide.

Use of cytokinin in hi-tech horticulture: It prevents the breakdown of chlorophyll and causes a few effects similar to responses by light. Certain seeds that require light to germinate need cytokinin for germination. Cytokinin stimulates the formation of seedless fruits in Ficus carica and Vitis vinifera. Cytokinin also affects the abscission of leaves and transport of auxins. Cytokinin may prove to be valuable in extending the storage life of lettuce. Cytokinin can also initiate roots from African violet leaves if petioles are placed in moist medium.

Benzyladenine is used on white pine to increase lateral bud formation and subsequent growth and branching, while tetrapyramyl benzyladenine is registered for use on carnations and roses for increased lateral branching. Promalin, a mixture of benzyladenine and GA_{4+7} is used to control fruit shape in 'Delicious' apple. Promalin applied at bloom increases length to diameter ratio of the fruit. Promalin is also being used to increase lateral branching in non-bearing apple trees. Promalin stimulates branching and increases the branch angle as well increases shoot elongation, all of which aid in the development of a scaffold branching system more suitable for fruit production. Post bloom application of CPPU a derivative of cytokinins is also shown to increase the berry size more in diameter than in length.

i) **Morphogenesis:** Production of tissue culture plant is a major function of hi-tech horticulture, cytokinins play very important role in the formation of organs. In case of tobacco pith, a balanced level of IAA and kinetin (2 mg 1-1 IAA : 0.02 mg 1-1 kinetin) produces an amorphous and undifferentiated callus and if this balance is altered by raising the ratio of kinetin (either by reducing IAA or by increasing kinetin), the pith results in the formation of bud which may grow into a compete tobacco plant. Kinetin is applied to the base of cuttings of Acer rubrum, it inhibits root formation but when sprayed on the leaves, it exercises a promoting effect. Root initiation in Marsilea drummondii by cytokinins.

Some other morphogenetic responses are on the formation of plastids from proplastids, differentiation of trachieds and induction of flowering and parthenocarpy.

ii) **Breaking seed and bud dormancy:** Cytokinins have been found quite effective in breaking the dormancy of seeds and some other plant organs. For example, Lactuca sativa requires red light for breaking its seed dormancy but this effect was overpowered by the application of kinetin.

iii) **Delay of senescence:** Kinetin was able to postpone for a number of days the disappearance of chlorophyll and degradation of proteins, which normally occur, with the ageing process of leaves. Cytokinins also increasing shelf life of fruits and reduce all physiological process in fruit during storage.

iv) **Thinning of apples**: Recently, promising results were reported for thinning of apples with BA alone or in combination with carbaryl. Another cytokinin-like compound currently under development, which also thins apples, is 2-chloropyridylphenylurea (CPPU). A number of potential applications have been examined with CPPU which overlaps with several GA effects including grape berry enlargement, and increased fruit size and weight in apple. The future of this compound as a commercial product is yet to be determined and may require an identified use in a major horticultural crop to justify further development.

Use of ethylene in hi-tech horticulture: Ethylene affect root formation in many ways, hastens blanching, accelerates abscission, inhibits elongation of root and shoot, sprouting of potatoes, increases the rate of respiration of potatoes and proves effective in securing the germination of some seeds. Ethylene (100 mg/l) also induce flowering in pineapple plant. It is used for colouring lemons, loss of chlorophyll of oranges and acceleration of ripening of banana (climacteric fruit) and also citrus (non-climacteric). In banana 0.1-1000 mg/l ethylene gas applied before the preclimactric encourages ripening. Ethylene application increases the number of female flowers and fruits in cucumber plants. Ethylene in conjunction with IAA encourages rooting.

i) **Fruit ripening:** Ethylene is a ripening hormone. Sprays of ethephon during fruit development in apple, grape, strawberry and cranberry were promising for early fruit maturity and uniform ripening. This facilitates growers to make single harvest of fruits. Ethephon applied to citrus fruits prior to storing ensures post harvest degreening. In lemons, dipping in 1000 ppm ethephon resulted in attainment of marketable yellow colour. Similarly, a large variety of fruit such as banana, mango etc., can be ripened faster with ethylene treatment. The uneven ripening in coloured varieties of grape, namely, Beauty seedless, Gulabi, etc. is a major impediment in viticulture. The proportion of green berries at harvest can be reduced considerably by the application of ethephon @ 250 ppm at colour break stage of the berries. Cycocel (500 ppm) when applied twice at 15 days interval induce early maturity in Kagzi lime by about 20 days.

ii) **Harvesting of fruits:** Chemical induced harvesting can be achieved by ethrel sprays in case of apple, walnut, olive, cherry, pear and grapes. Ethrel sprays have also resulted in uniform maturation of fruits with accelerated ripening. In citrus and mango, the application of cycloheximide at pre-harvest stage facilitated an easy harvest of fruit. The majority of fruits responding to ethylene are climacteric type, as ethylene is an essential trigger of climacteric rise in respiration.

iii) Reduction in scape length: The floral axis in bulb-type flowers is often too long for their use as potted flowers. Application of ethephon to hyacinth and narcissus plants when the leaves were 10 cm long reduced the length of the floral scape by 20-30 per cent. The reduction in scape length was associated not only with a reduction in cell size, but surprisingly, also with an increase in cell number, indicating that ethephon application had actually stimulated cell division.

iv) Quality production of potato: Hollow heart and brown centre of potatoes are disorders that accompany rapid plant and tuber growth. Although application of ethephon at tuber initiation reduced yield and mean tuber size, it also substantially reduced the incidence of these disorders, leading to increased yield of marketable tubers, relative to control fields which had high disorder incidence.

v) Flowering control: For many years it has been known that ethylene stimulates flowering of Bromeliads, and treatment of pineapple plants with calcium carbide (which releases acetylene on hydration) was widely used commercially to stimulate flowering of pineapples. This response is now elicited in pineapples and flowering bromeliads by application of ethephon. A comparison of the effectiveness of ACC and ethephon in eliciting this response suggested that ACC was the preferred ethylene-releasing compound, but because of differences in application method and concentrations, this interpretation needs confirmation. Flowering is also stimulated in many other fruit crops after ethylene treatment. Smoky fires are, for example, commonly lit in mango orchards to stimulate flowering. Application of ethephon has also been shown to stimulate flowering in *Litchi*, although the results vary widely depending on application time and cultivar. Ethylene stimulates or accelerate flowering of many geophytes. In *Iris*, for example, a short treatment of harvested bulbs with ethylene will stimulate flowering in small-sized bulbs that would not normally flower. e.g., *Iris* and *Freesia*. Ethephon application has been used to extend the flowering season of plantations of *Leucospermum*, a woody proteaceous ornamental.

vi) Abscission of fruits and plant parts: The effects of ethylene in stimulating abscission are the basis of important commercial uses of ethylene-releasing compounds. Enhancement of fruit abscission in sour cherry, nut fall in macadamia, the removal of leaves from potato plants and nursery stock, and the dehiscence of the shucks of walnuts and pecans are but a few examples of the potential uses of these materials in horticulture. Defoliation of rutabaga required very high concentrations of ethephon; addition of 1 per cent ammonium peroxydisulphate greatly increased the effectiveness of the ethephon treatment, perhaps due to accelerated oxidation of IAA in the abscising organ.

Promotion of abscission as an aid in thinning or harvesting fruit crops is an important horticultural use of ethylene-releasing compounds. A recent study by Takeda and Peterson on mechanical harvesting of blackberries provides a good example of the benefits ethylene-releasing compounds provide in stimulating this response.

vii) Increase latex and gum flow: The amount of rubber latex produced in the form of coagulated latex is a function of the duration of latex flow from the taping cut that is made in the tree bark. Ethephon is applied to a resion near the tapping cut and causes latex flow to increase indurations. Resulting in an increase in the volume of latex collected. Rubber yield increases by 50-100 per cent. The mechanism for increased flow of latex by ethephon is not well understood. It is believed that lutoids, non-rubber containing bodies within the latex, are disrupted by tapping and cause coagulation or plugging of the latex vessels as a result of changes in osmotic potential or shear forces imposed by high flow rate through the narrow pores of the vessel. In *Acasia* the ethephon increase gum secretion.

viii) **Delaying flowering in fruit crops:** Application of ethephon delays bud expansion and anthesis in cherry and peach. Ethephon appears to increase the length of the dormant period of flower buds, which results in a delay in bloom, reducing the potential for spring frost damage to flowers. In some cases, ethephon also increases mid-winter flower bud cold hardiness by maintaining high levels of the cryoprotectants sorbitol and sucrose in the flower pistil.

Use of abscission in hi-tech horticulture: There are no practical uses of abscisic acid (ABA) because of the high cost of synthesis and its instability in UV light. Although ABA has been shown to have adequate anti-transpirant properties on crop such as barely and coffee. Its effect are short lived (approximately 10 days) as a result of rapid metabolism and photoisomerization of the plant growth regulator. Frequent application would, therefore, be necessary to protect plants from water stress and this would be inconvenient and expensive. Structural analogues of ABA have been synthesized but found to have similar limitations. The ABA analogs LAB 173 711 and LAB 144 143 are acetyleneacetyl –type compounds that have been found to reduce water use in crop plants, increase cold hardiness and delay flowering in peach.

i) **Stomatal Closure and Water Relations:** The drop in water potential is detected by the chloroplasts of the mesophyll cells of leaf, whose photosynthetic activities are thought to be the most sensitive of all the plant's metabolic processes to water stress. As the water potential decreases there is marked increase in permeability of chloroplast membranes to ABA. The ABA synthesized by and stored in chloroplast then diffuses into mesophyll cytoplasm and moves from cell to cell through plasmodesmata to the guard cells where it induces stomatal closure. The drop in the level of stored ABA in the mesophyll chloroplasts induce fresh biosynthesis of more ABA, which continues to be translocated to the guard cells as long as the water potential remains low. The release of ABA from the mesophyll chloroplast stops when the water potential is restored to normal and its rate of synthesis is markedly reduced. The evidences for this are as follows:
   a) The exogenous application of low concentrations (e.g. 1 mM) of ABA to leaves causes stomatal closure within 3-9 minutes.
   b) The concentration of ABA in water stressed leaves rises markedly.
   c) Mesophyll chloroplasts can synthesize ABA but guard cell chloroplasts cannot
   d) The ABA of leaves of well-watered spinach was found to be contained in the chloroplasts but after 4 h of water stress the total ABA level was found to have risen 11 fold whilst that of the chloroplast had risen only 2 fold; this suggests that the newly synthesized ABA is rapidly exported from the chloroplasts.

ii) **Guard cells:** Guard cells showed equal closure when exposed to ABA solutions over the pH range 5.0-8.0, while no significant uptake of the hormone took place at pH 8.0. This indicates that the sites of ABA action are on the outer surface of the plasmalemma. ABA prevents stomatal opening by rapidly blocking H+ extrusion and K- influx and it initiates closure by the rapid release of osmotica, in particular K+, Cl- and malate. The latter phenomenon can also be observed in the shrinking of guard cell protoplasts.

iii) **Bud dormancy**: Innate or true dormancy (i.e. temporarily suspended growth due to internal conditions) is exhibited by the buds of temperate-zone trees during winter, stem tubers (e.g. potato), root tubers (e.g., Dahlia), Corms, bulbs and rhizomes. Hemberg (1949) found that the dormancy of buds of ash (Fraxinus) and potato tubers was positively correlated with the concentration in them of growth inhibitors, identified as ABA. The external application of ABA to buds of growing seedlings caused them to become dormant. According to these findings ABA was synthesized in the plant, in the response of some environmental trigger and this ABA travelled to the buds and induced them to go dormant.

iv) **Seed dormancy:** ABA builds up in embryos of developing seeds. It is believed that increased ABA level comes primarily from de novo synthesis. However, there may be translocation of ABA or its precursors from the leaves since the radioactivity from exogenously applied 14C-ABA to leaves can be recovered in developing seeds. ABA in developing ovules inhibits the formation of germination enzymes in the embryo (Dure, 1975). Thus, ABA plays an important role in developing seeds by inhibiting vivipary. In the case of immature seeds, several evidences indicate that ABA prevents precocious germination (vivipary) of the developing embryo.

v) **Heterophylly:** Aquatic plants often produce two distinct types of leaves on the same plant. Submerged leaves are highly dissected or linear and they have undifferentiated mesophyll and an epidermis with few stomata, if present. In contrast, aerial leaves are entire and broad and they have differentiated mesophyll and a stomatous epidermis. It is interesting that the ABA treatment can induce the formation of aerial leaves on submerged shoots. Such changes in leaf morphology due to ABA have been reported for a fern, a monocot, and several dicots.

vi) **Tuberization:** ABA appears to be involved in the process of tuberization. Wareing has shown that ABA formed by the leaf of a one-node cutting taken from a short day-induced plant of Solanum andigena (a wild species of potato which requires about 20 short day cycles to fully induce tuberization) is exported to the axillary bud and causes it to develop rapidly into a small tuber. Externally applied ABA has also been shown to promote the formation of these tubers in Dahlia and Jerusalem artichoke.

vii) **Senescence:** ABA may also be involved in senescence and ripening processes. Externally applied ABA accelerates the senescence of detached leaves; however, its effect on the senescence of leaves still attached to a healthy plant is minimal. Externally applied ABA accelerates the ripening of young fruit; moreover, there is a rise in the level of endogenous ABA during the ripening of strawberries and grapes.

viii) **Role of ABA as stress hormone:** A very different role of ABA has been recorded in plants growing under stress environment, such as frost, drought, flooding, injury, mineral deficiency etc. A fairly high concentration of ABA is found in leaves of such plants when growing under stress condition, particularly with the onset stress condition, particularly with the onset of moisture stress. It is accompanied by loss of turgor and closure of stomata. ABA may also be involved in increasing the resistance of temperate-zone plants to frost damage, the external application of ABA has been shown to increase the frost hardiness of seedlings of box elder (Acer nigundo), apple and alfalfa (Medicago sativa) and to ameliorate chilling injury in cucumber seedlings.

Use of brassinosteroids in hi-tech horticulture: A considerable effort has been made in practical application of brassinosteroids and promising results have been obtained in several crops. Brassinosteroids have been evaluated foe use in improving crop yield, quality, stress tolerance and disease resistance. Application of brassinosteroids to crops helps to overcome environmental stress such as cold, salt tolerance and herbicidal injury.

Use of maleic hydrazide in hi-tech horticulture: Maleic hydrazide is used to control storage sprouting of onions, potatoes and improving self-life of different fruits and vegetables. The compound is applied as a pre-harvest foliar spray since it is rapidly translocated to the storage organs. Maleic hydrazide inhibits cell division in a wide range of plants and the ability of the effect of the compound on axillary bud growth in tobacco and the sprouting of tubers and bulbs. Maleic hydrazide is an analog of uracil and may inhibit cell division by reducing nucleic acid biosynthesis in shoot and root meristems.

SPECIAL USE OF PLANT GROWTH REGULATORS IN HI-TECH HORTICULTURE

Use in Tissue Culture Technique

Plant growth regulators are the key factors in multiplication of any plant part and genetic manipulation through tissue culture. In banana, low concentration of IAA and high level of Benzyl adenine (BA) essential for establishment of explants, while IBA helps in rapid multiplication. In mango, somatic embryogenesis from nuclear explants occurs in the presence of 2,4-D at low concentrations.

The ability of culture cells in defined media stimulated a wave of interest in somatic cell genetics and the large scale use of plant cells for production of secondary metabolites. Un fortunately, plant cell cultures are typically very heterogeneous so that there are few if any cultures of a single cell type. Haploid culture proved to be unstable reverting to higher ploidy levels at a high frequency.

Cultures enabled the development of transformation method for higher plants, most of which require regeneration of tissue from transformed cells.

(i) Auxins

Auxins cause cell enlargement and elongation but in excised, cultured systems they promote cell division. Because of their relative stability and ability to withstand autoclaving, synthetic auxins are extensively used. The most commonly used are 2,4-D, NAA and IBA. In some cases, compounds, which are not strictly auxins such as dicamba or picloram, are used as auxin substitutes. Both of the compounds are herbicides at higher concentrations. The use of auxins in tissue culture is an art and rarely can one specify a particular concentration to be used in any single case. The tissues that originally required an exogenous auxin for growth, gradually or suddenly lose this requirement. Inhibitors of auxin synthesis like 5-hydroxy-nitrobenzyl bromide (HNB) and 7- azaindole stimulate somatic embryogenesis in habituated citrus. Phenolic anti-auxins or auxin antagonists like trans-cinnamic acid, transport inhibitors like TIBA and NPA, 2,4,6-T and CMPA can affect morphogenesis and embryogenesis in addition to having effects on growth.

(ii) Cytokinines

Kinetin and zeztin which is about 10 times more potent and long considered the protype of the naturally-occurring cytokinines, are both routinely used in tissue culture. Dihydrozeatin, also naturally occurring, is not widely used. The synthetic N^6-benzylaminopurine (BAP) is much more widely used than either kinetin or zeatin N^6-Δ^2 –isopentenyl adenine (zip) is also widely used. Adenyl cytokinins are active in cell division. Dihydroconiferyl alcohol is active in various callus assays. Amino acid,

S-lathyrine from the legume lathyrus, is active callus assays. Amino acid, S-lathrine from the legume lathyrus, is active in the soybean callus growth and even potentiates kinetin, is inactive in other standard callus assays.

TABLE 21.1: Function of plant growth regulators in tissue culture

Hormone	Product Name	Function in Plant Tissue Culture
Auxins	Indole-3-Acetic Acid Indole-3-Butyric Acid Indole-3-Butyric Acid, Potassium Salt α-Naphthalene acetic acid 2,4-Dichlorophenoxyacetic Acid p-Chlorophenoxyacetic acid Picloram Dicamba	• Cell enlargement - auxin stimulates cell enlargement and stem growth. • Cell division - auxin stimulates cell division in the cambium and, in combination with cytokinin, in tissue culture. • Vascular tissue differentiation - auxin stimulates differentiation of phloem and xylem. • Root initiation - auxin stimulates root initiation on stem cuttings, and also the development of branch roots and the differentiation of roots in tissue culture. • Tropistic responses - auxin mediates the tropistic (bending) response of shoots and roots to gravity and light. • Apical dominance - the auxin supply from the apical bud represses the growth of lateral buds. • Delayed leaf senescence. • Leaf and fruit abscission - auxin may inhibit or promote (via ethylene) leaf and fruit abscission depending on the timing and position of the source. • Delayed fruit ripening.
Cytokinins	6-Benzylaminopurine 6-γ,γ-Dimethylallylaminopurine (2iP) Kinetin Thidiazuron (TDZ) N-(2-chloro-4-pyridyl)-N'Phenylurea Zeatin Zeatin Riboside	• Adventitious shoot formation • Inhibition of root formation • Promotes cell division • Modulates callus initiation and growth • Stimulation of axillary's bud breaking and growth • Inhibition of shoot elongation • Inhibition of leaf senescence
Gibberellins	Gibberellic Acid	• Stimulates shoot elongation • Release seeds, embryos, and apical buds from dormancy • Inhibits adventitious root formation • Paclobutrazol and ancymidol inhibit gibberellin synthesis thus resulting in shorter shoots, and promoting tuber, corm, and bulb formation.
Abscisic Acid	Abscisic Acid	• Stimulates bulb and tuber formation • Stimulates the maturation of embryos • Promotes the start of dormancy
Polyamines	Putrescine Spermidine	• Promotes adventitious root formation • Promotes somatic embryogenesis • Promotes shoot formation

(iii) Gibberellins

The application of GA_3 to cultural media has only been minimal. In some cases like the carrot root bioassay, they result in cell division rather than enlargement.

(iv) Ethylene

Use of ethylene inhibitors such as silver nitrate, cobaltous or nickel chloride and aminoethoxyvinylglycine (AVG) or salicylic acid increase shoot regeneration and somatic embryo production but results are contradictory. Ethylene is all pervasive and effects such as acceleration of the breakdown of cytokinins and subsequent stimulation of rooting in some systems.

(v) Abscisic Acid

ABA slows down growth and moderates the effects of cytokinins and auxins. It can bring about a more normal progression of somatic embryo development especially when hormone production has been perturbed and there is too much callus. It also prevents precocious germination in somatic embryos.

ROLE OF PLANT GROWTH REGULATORS ON SEX EXPRESSION OF CUCURBITS IN HI-TECH CULTIVATION

Maximum cucurbits are monoecious nature, they bears male and female flowers separately on the same plant. Generally, number of male flowers is more than the number of female flowers. It is desirable to have high female to male ratio. It has been observed that male flowers appear earlier then the female flowers. The lower the node number on which the female flowers appear on the main stem, the earlier is the variety. The female to male flowers ratio goes on increasing with the age of plant. Though the sex expression and sex ratio are varietal characters, they are modified by environment. Low humidity accelerates the appearance of staminate flower of cucurbits, while high humidity hastens the formation of pistillate flowers. Lower fertility higher temperature, longer light period all induce maleness. Certain gases and chemicals also affect sex in different cucurbits. Uses of some plant growth regulators at proper concentrations modify sex in cucurbits. Gibberellic acid at higher concentrations induces maleness but at the lower concentration of 10 to 25 ppm increases the number of female flowers. Two sprays one at 2 leaf stage and again at 4 leaf stage with 25 to 100 ppm MH, 100 ppm NAA or Ethrel, 3 ppm boron or 3 ppm molybdenum can suppress the number of male flowers and increase the number of female flower, fruit set and yield.

Bitter gourd (Momordica charantia): Application of growth regulator at (2-4 leaf stage) plays an important role in sex expression and sex ratio in bitter gourd. MH at 50 to150 ppm and CCC at 50 or 100 ppm increase the female: male ratio and at high concentration (200 ppm) CCC reduces it. IAA at 17.5 or 35 ppm or 3-hydroxy methyl oxindole (HMO) at 17.5 ppm increases total female flower formation in cultivator 'Kesar'of bitter gourd. Ethrel at 25 ppm also increases female flowers. GA at 60 ppm reduces the ratio of male female flowers. Seed treatment with boron at 3 or 4 ppm for 20 hours, enhancing pistillate flowers, fruits, fruit weight and yield.

1. **Sponge gourd** (*Luffa cylindrica* Roem)**:** Ethrel (250 ppm), NAA (200 ppm) when sprayed on young seeding (2-4 leaf stage) of sponge gourd stimulated the production of pistillate flowers. NAA also proved very effective in lowering down the sex ratio of male to female flowers.

2. **Round gourd** (*Citrullus vulgaris*): The sex expression and sex ratio in round gourd (squash melon) has been found to be influenced by growth substances (GA3, MH, IAA and Etheral etc.). Among them MH has been proved to be the most effective when it was applied at 50 ppm at 2 and 4 leaf stage stimulated the elongation of the main axis, induced the formation of female flower nodes and increased the number of female flowers and fruit yield. MH suppressed the male flowers and lowered the sex ratio of male to female flower.
3. **Muskmelon** (*Ccucumis melo* Linn): Weather factors play important role on sex expression and sex ratio in muskmelon. Formation of female flowers early as well as more in number can also be obtained with the application and growth regulators like 2,4,5,_ Tri-idobenzoic acid (TIBA) at 25 ppm applied at 2 leaf and 4 leaf stage of plant growth.
4. **Watermelon** (*Citrullus lanatus* Mansf.): In watermelon, application of Aminoethoxy vinyl- glycine (AVG) at 100 ppm or 200 ppm and Silver nitrate (AgNo3) at 500 ppm promote hermaphrodite flowers and reduced the number of staminate flowers. Also application of 2,4,5-Tri-iodobenzoic acid (TIBA) at 25 to 50 ppm once or twice at 2 leaf and again 4 leaf stage promoted number of female flowers, enhanced fruit set and ultimately more yield. Besides, use of micronutrients like boron molybdenum at 3 ppm proved effective in sex expression.
5. **Pumpkin** (*Cucurbita moschata*)**:** The sex expression and sex ratio in pumpkin are influenced much by the weather conditions. The cloudiness favours the production of pistillate flowers. Through application of ethephon (100-200) at early stages of plant growth of pumpkin, delayed the development of the first staminate flower and induced the appearance of pistillate flowers, the number of pistillate flowers increased to 3-fold; the lower male: female ratio and the shortest internodes, so that more plant could be grown in less space.
6. **Summer squash** (*Cucurbita pepo*)**:** Very young plants (at 2 and 4 leaf stages) of summer squash spray with growth substances in that GA_3 at 10 and 25 ppm stimulated maximum growth of vines whereas ethrel at 100 and 250 ppm inhibited the growth. However ethrel at 250 ppm induced the formation of female flowers and increased the fruit yield. IAA at 200 ppm applied at initial stage of crop gave the highest total yield. Not only growth regulators influence sex expression and sex ratio in summer squash but also micronutrients (Zn, Mn and B) applied through seed soaking had beneficial effects. The combined effect of Mn and boron treatment was the best on the number of female flower per plant and ultimately highest fruit yield.
7. **Winter squash** (*Cucurbita maxima*)**:** An application of ethephon at 300 ppm at 2 true leaf stages in winter squash enhanced earlier formation of male flowers. It was also observed that all the male main buds were already formed at the nodes at the two true leaf stage aborted and all the secondary buds developed into functional female flowers.
8. **Cucumber** (*Cucumis sativus* Linn): In cucumber two sprays, one at 2 leaf stage and again at 4 leaf stage with NAA (100 ppm), CCC (25 or 50 ppm), TIBA (25 or 50 ppm), Ethephon (50-200 ppm) resulted in an earlier appearance of the first female flowers, in respect to number of days, node number and higher percentage of female flowers and fruit yield. Besides ethrel has also been found to increase leaf area, accelerate fruit setting and increase early yields Maleic hydrazide and ascorbic acid were ineffective in sex expression in cucumber, however, GA_3 favoured maleness.

CHAPTER 22

Modern Nursery Management

INTRODUCTION

Nursery is a place where plants are cultivated and grown to usable size. Nursery techniques involve raising seedlings, saplings and grafts of economically useful and ornamental plants through scientific methods. Several new techniques are now available, which are cheap and effective. These new techniques are useful in increasing the success rate of grafts and rooting of the cutting; increasing seedling vigour; reducing transplanting shock and generally reducing the quantum of manual work. The nursery management gained a status of commercial venture where retailer nurseries sell planting materials to the general public, wholesale nurseries, which sell only to other nurseries and to commercial landscape gardeners, and private nurseries, which supply the needs of institutions or private estates.

The demand for high quality planting material is steadily increasing due to interest in vegetable gardening, fruit tree cultivation, social forestry, agro-forestry and plantation crops. The need of setting up plant nurseries to meet the demands of the people has been felt by small and marginal farmers as well as by gardeners and farmhouse owners. In order to meet this demand, there is ample scope for introduction of small nurseries, which will serve to augment the incomes of needy sections of rural society. Therefore, there is a need for low cost nursery techniques. This kind of nursery reduces manual labour and drudgery and increase vigour of seedlings. It also reduces transplanting shocks and the success rate of grafts and rooting of the cutting is high with low operating cost. There are fewer inputs like water, fertilizers etc. are low. There are following difference between conventional and hi-tech nursery with reference to tools and techniques:

Conventional nursery	Hi-tech nursery
Spade, khurpi, watering cane, fork, hoe, garden line, roller, basket, sirki, polythyne sheet, sprayer, alkathene sheet, nose-cane, duster, sticks, tags etc.	Plug trays, perforated plastic trays, strip peat pots, nursery stand, sprinklers, protected structures, water pumping motor, media mixture, rakers, temperature control devices, humidity control devices, exhausters, media pressure, seed dibbler, etc.

Though there are limitations in hi-tech nursery like the greenhouses are quite expensive and require infrastructural facilities like large space, water supply and also uninterrupted power supply. The viability of nurseries depends on market conditions. The major requisitions of nursery management are as follows:

1. Site for nursery should be selected at such places where abundant sunshine and proper ventilation is available.
2. Nursery site should be on higher location so that water stagnation is avoidable.
3. In humid and rain prone areas nursery place should be well protected from

heavy rains through protected structures.

4. The site should be nearer to irrigation facilities and easily accessible.
5. It should be protected from stray animals, snails, rats etc.
6. Soil should be sandy loam or loamy with pH range of 6 to 7 and rich in organic matter and free from pathogenic inoculums.

Necessity of a Nursery

Nursery is the necessity for every horticultural grower. The development of seedlings in nursery is not only reduces the crop span but also increases the uniformity of the crop and thus, harvesting as compared to direct sown crops. Transplanting of seedlings also eliminates the need for thinning and provides good opportunities for virus free vigorous and off-season nursery, if grown under protected conditions. Nursery is helpful and convenient to manage seedlings under small area and the grower can get timely plant protection measures are with minimal efforts. Development of a nursery provides favourable climate to emerging plants for their better growth and development. The effective utilization of unfavourable period by preparing nursery under protected conditions. Seed cost of some crops like hybrid vegetables, ornamental plants, spices and some fruits can be economized through nursery. Nursery production help in maintaining effective plant stand in shortest possible time through gap fillings. Although many fruits and vegetable seeds can be sown directly in the field, experience has shown that raising seedlings in a nursery has a number of advantages as discussed below.

1. ***Intensive care*** - Seedlings receive better care and protection (from animals, weeds and pests) in the nursery. The average garden soil is not an ideal medium for raising seedlings especially from the point of view of soil tilth. At an early stage of development most vegetable crops require special attention that is not possible in the main field.
2. ***Reduction of costs*** - Fewer seeds are used for raising seedlings in the nursery than for sowing directly in the field, because in the latter seedlings have to be thinned to one, which is wasteful. When expensive hybrid seeds are used, transplants therefore become more economically attractive. Pesticides and labour are also reduced under nursery conditions as compared to planting directly in the field.
3. ***Opportunity for selection*** - Raising seedlings in a nursery affords the grower an opportunity to select well-grown, vigorous, uniform and disease free seedlings.
4. ***Extend a short growing season for late maturing crops*** - Seedlings can be raised in a nursery under a protected environment before conditions outside become suitable for growth and transplanted into the field when conditions allow, thus reducing the amount of time spent in the field.
5. ***Forced vegetable production for an early market*** - Generally prices of horticultural produce are attractive when production or supply is low. Vegetables can be grown 'out-of-season' in a nursery when conditions are not yet favourable. Such crops will thus mature earlier after transplanting and hence stand to fetch a higher price in the market.

Type of Nurseries

In the past, nurseries were involved in almost all aspects of the production and culture of plants. They grew a wide variety of plants and sold them both wholesale and retail, as well as supplying a wide range of allied products and services. Today all but the largest nurseries tend to specialise.

Organisational structure: Nurseries can be non-private (for example, run by government, community or conservation groups) or private businesses. Private nurseries can be individually owned, partnerships, or corporate, such as a company, which has a legal identity of its own.

Nurseries can be classified in many different ways:

- According to what they grow, such as natives, exotics, seedling, cottage, garden plants, bonsai or bulbs.
- How they grow it, such as in ground production or in containers.
- The size of plants they produce, such as tube stock, small pots/containers, or advanced stock.

There are three main types of nurseries; production, growing-on and retail.

Production Nurseries (Wholesale Nurseries)

Propagate plants and grow seedlings and either sell them direct to retail outlets, landscapers and horticulture or forest departments, or wholesale them to growing –on nurseries. Important feature of successful production nurseries.

1. **Innovation**- supplyies new varieties to the market or developing new ways of growing and presenting existing varieties allow the grower to develop new markets.
2. **Specialization**- growing fewer lines in larger quantities allows the grower to improve efficiency.
3. **Meeting market demands**- knowing market demand and customers choice or are likely to want, and growing them in sufficient quantities allows the grower to meet consumer's requirements and maintain customer loyalty.

Growing-on Nurseries

Growing-on nurseries buy bulk quantities of seedlings or small plants from propagators. At the time of purchase, the plants are growing in plugs, trays or tubes the plants are then potted into larger containers and grown on for a period of time, adding value to the nursery's purchase. In addition to increasing plant size, specialized growing techniques, such as topiary, may be used to add value to the plant during the growing –on phase.

The most critical aspect of production in growing-on nurseries is developing a quality product for the retail market. At the time of resale, every plant must be at its peak, displaying healthy, vigorous and sturdy growth.

Retail Nurseries

Retail nurseries buy plants from production/propagation nurseries and resell them in small quantities as per consumer's need. In addition, the retailer also sells seedlings, bulbs, containerized and bare-rooted plants, pots, packaged potting mixes, fertilizers, sprays and bulk landscaping materials.

Nursery Management

The first step in successful production of horticultural crops is to raise healthy vigorous seedlings. Young plants whether propagated from seed or vegetative require a lot of care particularly during the early stages of growth. They have to be protected from adverse temperatures, heavy rains, drought, wind and a variety of pests and diseases. If small seeded vegetables are sown directly in the field, germination is often poor and the young plants grow very slowly and require a long time to mature. Also the season may be too short for full development in the field. To overcome these problems many vegetable crops are grown in nurseries before being transplanted in the field. A vegetables or fruits nursery is a place where plants are cared for during the early stages of growth, providing optimum conditions for germination and subsequent growth until they are strong enough to be planted out in their permanent place. A nursery can be as simple as a raised bed in an open field or sophisticated as a glass-house with micro-sprinklers and an automatic temperature control system. Often, we consider the nursery management and the plant propagation are same, though they are altogether different but inter-related. In fact, mass multiplication of quality planting materials is the central theme of nursery management but nursery management is a trade oriented dynamic process, which refers to efficient utilization of resources for economic returns. Nursery management is a team effort to reach the desired goal.

The main phases of nursery management are (i) planning-edaphoclimatic and socio-economic consideration; demand for planting materials; provision of mother block, requirement of land area, water supply, working tools, growing structures and inputs availability; accessibility; trained man power; plant protection; disposal of planting materials etc., (ii) implementation – land treatment, protection against biotic interference and soil erosion, proper layout, input supply, (iii) monitoring and evaluation-physical presence, rapid response, critical analysis, incentive to workers and (iv) feed back for further refinement. The key elements of hi-tech nursery management are (1) the place, (2) the plant and (3) the person behind.

Advantages of Nursery Management

- It is possible to provide favourable growth conditions i.e. germination as well as growth
- Better care of younger plants as it is easy to look after nursery in small area against pathogenic infection, pests and weeds.
- Crop grown by nursery raising is quite early and fetch higher price in the market, so economically more profitable.
- There is saving of land and labour as main fields will be occupied by the crops after 1 month. More intensive crop rotations can be followed.
- More time is available for the preparation of main field because nursery is grown separately.
- As vegetable seeds are very expensive particularly hybrids, so we can economize the seed by sowing them in the nursery.

Selection of Suitable Site

A number of important aspects must be considered in choosing a site for the establishment of a nursery if the outlay on seed, fertilizer, and labour is to show profitable returns. Therefore, setting up of a nursery is a long-term venture requires careful planning, because mistake committed initially cannot be rectified easily at later stage and may adversely affect the returns from the investment. These factors are broadly grouped into the following:

1. ***Environmental factors:*** This refers to natural features of the land, which may greatly influence the cost of operation and facilitate management of the nursery.
   a) Proximity to planting site (main field)- Some of the advantages of locating a nursery as close as possible to the fruit and vegetable producing area.
      i) Cost of transporting the seedlings to the field is minimized.

   ii) Less risk of loss of seedlings during transportation, and seedling failure after transplanting.
   iii) Reducing the chances of transmitting or redistributing soil-borne pathogens through seedling roots or earth balls over long distances. When, however, particular diseases occur in the nursery area it may of course be advantageous to raise the seedlings outside the affected area in order to initiate new plantings with disease-free seedling materials.

b) Land gradient (steepness of the land)- It is desirable to have the nursery on a level ground with good drainage. This will reduce the cost of establishing the nursery considerably. If the nursery is to be located on a sloping land, soil conservation measures are required, such as constructing terraces across the slope to conserve soil and moisture.

c) Nursery soils- Favourable soil conditions (good drainage, absence of toxicity, fertile, etc.) are indispensable for the success of a nursery. When nursery plants are raised in pots, polybags, seed boxes or trays, it may not be necessary for soils on the nursery site to be fertile. But in this case, a source of high quality soil must be as close to a nursery site as possible in order to lower the cost of soil transportation.

d) Water supply- a nursery should be located where a reliable, abundant and inexpensive supply of uncontaminated water is available. Water supply could be from wells, boreholes, natural streams or irrigation channels.

2. Proximity to services
   a) Labour supply- Nursery operations are labour intensive; therefore, it is very important that nurseries are sited in areas where a dependable and regular supply of experienced labour can be easily obtained.
   b) Markets- The nearness of nursery sites to potential buyers is very important to commercial nurseries intending to raise seedlings for sale to growers. Such a nursery should be located as close as possible to these growers.
   c) Supplies- it is desirable to have a nursery located close to sources of inputs (equipments or tools, and consumables such as seeds, pesticides, fertilizers, rooting media).
   d) Services- it can be advantageous to have a nursery located in an area where the services of agricultural experts (horticulturists, crop protection specialists, soil scientists, etc) can be obtained easily.

Other aspects of services are the availability of good roads necessary for the transportation of supplies, seedlings and workers.

LAYOUT OF MODEL NURSERY

Nursery is the place where all kinds of plants like trees, shrubs, climbers etc. are grown and kept for transporting or for using them as stock plants for budding, grafting and other method of propagation or for sale. The modern nurseries also serve as an area where garden tools, fertilizers are also offered for sale along with plant material. The area for models nursery production is prepared for effective utilization of inputs and to do things in proper manner. Institutions for nursery establishment as per their requirements design various location specific models. But their some important components, which should be taken into care and provision, should be made for these during planning and layout preparation for nurseries:

1. **Fence:** Prior to the establishment of a nursery, a good fence with barbed wire must be erected all around the nursery to prevent tress pass of animals and theft. The fence could be further strengthened by planting a live hedge with thorny

fruit plants (like Koranda). This also adds beauty in bearing and also provides additional income through sale of fruits and seedlings obtained from the seed.

2. **Roads and paths:** A proper planning for roads and paths inside the nursery will not only add beauty, but also make the nursery operations easy and economical. This could be achieved by dividing the nursery into different blocks and various sections. But at the same time, the land should not be wasted by unnecessarily laying out of paths and roads. Each road/ path should lead the customer to a point of interest in the nursery area.
3. **Progeny block/Mother plant block:** The nursery should have a well-maintained progeny block or mother plant block/scion bank planted with those varieties in good demand. The grafts/layers/ rooted cuttings/seedlings should be obtained preferably from the original breeder / research institute from where it is released or from a reputed nursery. One should remember that, the success of any nursery largely depends upon the initial selection of progeny plants or mother plants for further multiplication. Any mistake made in this aspect will result in loss of the reputation of the nursery. A well managed progeny block or mother plants block will not only create confidence among the customers but also reduces the cost of production and increases the success rate of grafting/ budding/layering because of availability of fresh scion material throughout the season within the nursery itself and there will not be any lag period between separations of scion and graftage. There are so many cultivars for fruit crops, grow only important cultivar in mother block as per demand and germ plasm conservation.

TABLE 22.1: Important fruit cultivars for mother block (grow only required variety)

Crop	Improved cultivars
Almond	Ne plus ultra, Non pariel, Drake, California paper shell, Marcott, Brigg's hard shell, Mission, IXL, Almond hybrid-15, Hybrid A-258, H-98, Makhdoom, Parbhat, Waris, Shalimar and Debhar
Custard apple	Arka sahan, Balangar local, Sitaphal, Red sitaphal and Mammoth
Aonla	Banarasi, Chakaiya, Fracis (Hithi Jhool), Krishna, Kanchan, NA-6, NA-7, NA- 10, Balwant, Anand -1, Anand -2, and GA-1
Apple	Ambri, Early shanburry, Golden delicious, Royal delicious, Lal ambri, Sunehari, Chaubattia princess, Chaubattia anupam, Ambrich, Ambroyal, Ambred, Ambstarking, Winesap, Rhodes island greening, Boshop, Mutsu, Branley's seedling, Winter banana, Vered, Michael, Decilious, Red gold, Gallia beauty, Red spur, Prima, Priscella, Freedom, Liberty, Co-op-12, Sir prize, Skyline supreme red delicious and Starkrimson
Apricot	New castle, Nugget, Charamgz, Early shipley, Sufeda, Chakarpara and Kaisha
Avocado	Furete, Pollock and Paradarniya
Bael	Varansi, Gorakpur, Mirzapuri, Etawah kagzi, Sewan large, Deoria large, NB-1, NB-5 and NB-9
Banana	Poovan, Dwarf cavendish, Gros michel, High gate, Low gate, CO-1, Virupakshi, Nendran, Monthan, Bodles altafort, Klue teparod, Amrit sagar, Basrai dwarf, Harichal, Robusta, Lalkela, Pisang lilin and Ney poovan
Ber	Gola, Seb, Umran, Mundia, Banarasi karaka, Kaithali, Jogia, Aligang, Early umran, Mehrun, Dodia and Tikdi
Carambola	Golden Star

Crop	Improved cultivars
Date palm	Hillawi, Barhee, Khadwary, Shamran, Zihidi, Deglet noor, Medjool, Kachchh selection, Khuneji and Khalas
Fig	Black ischia, Brown turkey, Smyrna, Common fig, Excell, Deanna, Kabul, Lucknow, Gentile, San pedro, Marseilles, Poona fig and Codadria
Grape	Pusa seedless, Banglaroe blue, Thompson seedless, Beauty seedless, Black champa, Anab-e-shahi, Perlette, Arka neelmani, Arkavati, Arka shyam, Arka kanchan, Arka hans, Arka chitra, Arka safed, Arka shweta, Arka majesty, Arka soma, Arka trishna, Arka krishna, Pusa urvashi, Pusa navrang, Loose perlette, Bokhari, Early muscat, Gold, Delight, Perlette, New perlette, Delight, Himrod, Kishmish charni, and Kishmish beli
Grapefruit	Marsh, Thompson, Ruby, Triumph, Foster, Duncan
Guava	Arka mridula, Arka amulya, Hybrid 16-1, Safed jam, Kohir safed, Allahabad safed, Lucknow-49, Habsi, Chittidar, Saharanpur seedless, Anakapalli, Harijha, Smooth green, Allahabad surekha, Behat coconut, Apple color, Red fleshed, Seedless, Banarasi red, Nasik, Dholka, Dharbar, Dharidar and Supreme
Hazelnut/Filbert	Barcelona, Daviana and Tonolo
Kiwi fruit	Abbott, Bruno, Taumuria, Monty, Crensil, Allison, Hayward and Bruno
Lemon	Libson, Meyer, Washington naval, Baramasia, Assam lemon, Gandhraj, Seville, Nepali oblong, Nepali lemon, Italian, Eureka and Villafranca
Lime	Kagzi lime, Vikram, Pramalini, PKM-1, Sai sarbti, Baramasi, Kagzi, Tahiti lime, Seedless lime.
Litchi	Swarna roopa, Dehradun, Calcuttia, Ellachi, Purabi, Late bedana, Early bedana, China, Muzzafarpur, Shahi, Dehra rose, Rose scented and Chentze
Loquat	Large agra, Improved pale yellow, Pale yellow, California advance, Large round, Fire ball, Thames pride and Tanaka
Mandarin	Kinnow, Pixie, Wilking, Page, Fortune, Sunbrust, Shamouti, Satsuma, Nagpur mandarin, Khasi mandarin, Coorg mandarin, Kamala mandarin and Darjeeling
Mango	Sindhi, Neelum, Banglora, Rumani, Amrapali, Mallika, Ratna, Au-rumani, Manjeera, Madhulika, PKM-1, Arka aruna, Arka puneet, Arka anmol, Arka neelkiran, Neelphonso, Neeleshwari, Neeleshan, Banganpalli (Banashan), Fazli, Kesar, Alphonso, Tomy atkins, Zielete, Haden, Sensation, Julie and Caribao
Olive	Frontio, Ascolano, Ascoitcrana and Coratina
Passion fruit	Keveri (A cross between purple and yellow passion fruit)
Peach	Sharbati, Saharanpur prabhat, Shimizo Hakuto, Florda sun, Florda gold, Florda red, Florda snow (Nectarine), Chinese cling, July Elberta, JH hale, Haven, Elberta, Sun red, Shan-e-Punjab, Sun red (Nectarine), Stark earliglo, Kanto-5 Partap (TA-170), Florda red, Florda prince, Earligrande, Safeda early, Cream, Halford, Spring crest and Frederica
Pear	Le-conte, Patharnakh, China pear, Gola, Smith, Baghugosha, Punjab gold, Punjab nectar, Red blush, Sensation, Conference, Fertility, Doyenne du comice, Easter beaurre, Flemish beauty, Devoe, Max red bartlett, and Manning elizabeth
Pecan nut	Mahan, Nellis, Burkett, Western schely, Moore, Money maker, Desirable, Curtis, Success and Staurt
Persimmon	Fuyu, Hachiya, Flat seedless and Hykume
Pineaple	Kew, Queen, Giant kew, Mauritius, Cayenne, Smooth cayenne, Alxanderia, Red spanish, Singapore, Spanish ruby, Sugar loaf and Abacaxy

Crop	Improved cultivars
Pistachio nut	Red allepo, Bronto
Plum	Kala amritsari, Golden lalri, Sutlej purple, Burbank, Late yellow, Kelsey, Satsuma, Victoria, Mariposa, Greengage, Yellow egg. President, Jefferson, New castle, Chakarpara, Charamgz, Moorpark, St ambroise, Shipley's early, Titron Wickson, Frontier, Zardalu, Eldarado, Methley, Kanto-5, Santa rosa, Late santa rosa, Burmosa, Stanley and Starking delicious
Pomegranate	Ganesh (GBG-1), Alandi, Muskat, Bassein seedless, Jalore seedless, G-137, P-23, P-26, GKVK-1 (Jyothi), IIHR Selection, Mridula, Ruby (Hyb. 15-9-94), Khog, Gulsha rose, Lupania, Bedana, Kabul, Spanish ruby, Muscat red, Paper shell, Dholka, Kandhari, Jodhpur local, Bedana bosec, Sindhuri
Pummelo	Walteres, Large white fleshed and Large red fleshed
Sapota	Cricket ball, Baranasi, Kallipati, Guthi, CO-1, CO-2, Oval, PKM-1, Baranasi, DSH-1, DSH-2, and Hybrid 2/4
Strawberry	Pusa early dwarf, Tioga, Torrey and Chandler
Sweet Cherry	Bing, stella, Van, Merton glory, Sue, Sam, Lambert, Black tartarian, Nepolean van lambert, Black republican, Francis and Emperor
Sweet Orange	Blood orange, Washington naval orange, Mosambi, Satgudhi, Blood red malta, Shamouti, Maltese, Hamlin, Pineapple, Valencia and Jaffa.
Tamarind	Parthisthan, Yogeshwar, PKM-1 and Urigam
Walnut	Ratna, Chakrarta No.-14, Karan, Bulbul, Roopa, Franquette, Meytte, Eureka, Hartley, Wussan-1, Placentia, Gobind, Wilson, Wonder prolific, Lake, English, Payne, waterloo, Tutle-16 and Chakarta No.-14

Source- Horticulture at a glance (by A.S.Salaria)

4. **Irrigation system:** Horticultural nursery plants require abundant supply of water for irrigation, since they are grown in polybags or pots with limited quantity of potting mixture. Hence sufficient number of wells to yield sufficient quantity of irrigation water is a must in nurseries. In areas with low water yields and frequent power failures, a sump to hold sufficient quantity of water to irrigate the nursery plants is also very much essential along with appropriate pump for lifting the irrigation water. In areas where electricity failure is a problem, which is more common, an alternate power supply (generator) is very essential for smooth running of pump set. Since water scarcity is a limiting factor in most of the areas in the country a well laid out PYC pipeline system will solve the problem to a greater extent. An experienced agricultural engineer may be consulted in this regard for layout of pipeline. This facilitates efficient and economic distribution of irrigation water to various components in the nursery.
5. **Office cum stores:** An office-cum-stores is needed for effective management of the nursery. The office building may be constructed in a place, which offers better supervision, and also to receive customers. The office building may be decorated with attractive photographs of fruit ornamental varieties propagated in the nursery with details of it. A storeroom of suitable size is needed for storing polybags, tools and implements, packaging material, labels, pesticides, fertilizers etc.
6. **Seed beds:** In a nursery, this component is essential to raise the seedlings and rootstocks. These are to be laid out near the water source, since they require frequent watering and irrigation. Beds of 1-meter width of any convenient length are to be made. A working area of 60 cm between the beds is necessary. This facilitates easy in sowing of seeds, weeding, watering, spraying and lifting of

seedlings. Irrigation channels are to be laid out conveniently. Alternatively, sprinkler irrigation system may be provided for watering the beds, which offers uniform germination and seedling growth.

7. **Nursery beds:** Rising of seedlings / rootstocks in polybags requires more space compared to nursery beds but mortality is greatly reduced along with uniformity. Nursery beds area should also have a provision to keep the grafted plants either in trenches of 30 cm deep and 1 m wide so as to accommodate 500 grafts / layers in each bed. Alternatively, the grafts/ layers can be arranged on the ground in beds of 1 meter wide with 60 cm working place in between the beds. Such beds can be irrigated either with a rose fitted to a flexible hosepipe or by overhead micro sprinklers.
8. **Potting mixture and potting yard:** For better success of nursery plants, a good potting mixture
9. is necessary. The potting mixtures for different purposes can be prepared by mixing fertile red soil, well rotten FYM, leaf mold, oil cakes etc. in different proportions. The potting mixture may be prepared well in advance by adding sufficient quantity of superphosphate for better decomposition and solubilization. The potting mixture may be kept near the potting yard, where potting/ pocketing is done. Construction of a potting yard of suitable size facilitates potting of seedlings or grafting/ budding operations even on a rainy day.
10. **Structures for model nursery**

    a) **Shade houses:** Shade houses in nurseries in tropical and sub-tropical regions offer many advantages like raising of seedlings in bags directly, protecting the grafts from hot summer months, effective irrigation through upside down overhead micro sprinklers. The shade houses made with shade nets (50% or 75%) for regulation of shade are particularly very useful in arid regions where the humidity is very low during summer months.

    b) **Green houses/polyhouses:** Grafting or budding of several fruit species under polyhouses or low cost green houses with natural ventilation will enhance the percentage of graft/ bud take besides faster growth of grafts due to favourable micro climatic conditions of polyhouse. In green house construction, a wood or metal frame work is built to which wood or metal sash bars are fixed to support panes of glass embedded in putty. In all polyhouses/ green houses means of providing air movement and air exchange is necessary to aid in controlling temperature and humidity. It is best, if possible to have in the green house heating and self-opening ventilators and evaporative cooling systems. Plastic covered green houses tend to be much lighter than glass covered ones with a build up of excessive high humidity.

    i) **Polythene film:** This is the most inexpensive covering material but it is the short lasting one. However, UV ray resisting polyethylene film of various thicknesses is usually recommended which lasts longer.

    ii) **PVC film:** This material is pliable and comes in various thickness and widths unto 6 ft. It is longer lasting than polythene and is more expensive PVC surface of film tends to collect dust and lower the light intensity in due course of time.

    iii) **Polyester film:** This is a strong material with excellent weathering properties lasting for 3-5 years and is unaffected by extremes of heat or cold though it is costly than polythene film/ PVC film.

    iv) **Fiberglass:** Rigid panels, corrugated or flat fiberglass sheets embedded in plastic are widely used for green house construction. Fibreglass is strong, long lasting, lightweight and easily applied which is coming in a variety of widths, lengths and thickness. It is costlier than poly thin film/

PVC film.

FIG. 22.1: Papaya seedling in small-scale nursery at tribal area of sirohi is covered by Agro-net

c) **Hotbeds:** The hot bed is often used for the same purpose as a green house but in a smaller scale. Amateur operations and seedlings can be started and leafy cuttings root early in the season in such structures. Heat is provided artificially below the propagating medium by electric heating cables, pot water, steam pipes or hot air blows. As in the green house, in the hot beds attention must be paid for shading and ventilation as well as temperature and humidity control.

d) **Lathhouses:** These structures are very useful in providing protection from the sun for container grown nursery stock in areas of high summer temperatures and high light intensity. Well-established plants also can require lath house protection including shade-loving plants. Lathhouses construction varies widely depending on the material used. Aluminium pre-fabricated lathhouses are available but may be more costly than wood structures. Shade is provided by appropriate structures and use of shade nets of different densities allows various intensities of light in the lathhouses.

Miscellaneous Propagating Structures

1. **Mist beds:** These are valuable propagating units both in the green house and out doors and are useful mainly in rooting of leafy cuttings.
2. **Mist chamber:** This is a structure used to propagate soft woodcuttings, difficult to root plants and shrubs. Here the principle is to spray the cuttings with a minimum quantity of water. This is achieved by providing the cuttings a series of intermittent sprayings rather than a continuous spray. The intermittent spraying can be done easily by means of a high-pressure pump and a time switch. The pump leads to a pipeline system inside the propagating structure. The mist nozzles are fitted to these pipelines and suitably spaced over the propagating material.

FIG. 22.2: locally prepared mist chamber for propogate soft wood cuttings

3. **Nursery bed:** These are raised beds or boxes made of brick and mortar, provided with drainage holes at the bottom. The dimensions of the boxes are 60 cm high, 120 cm broad and length as required preferably not exceeding 10 m. Roof structures for planting on both sides and forming ridges at the centre are constructed on the top of the nursery beds. These structures may be made permanent with angle iron or may be made of wood. Moveable bamboo mats, palm leaf mats are placed over these structures to protect the seedling from hot sun and heavy rains.
4. **Fluorescent light boxes:** Young plants of many species grow satisfactorily under artificial light from fluorescent lamp units. Although adequate growth of many plant species may be obtained under fluorescent lamps but not up to the mark compared to good green house conditions.
5. **Propagating cases:** Even in green house, humidity conditions are often not sufficiently high for rooting. The use of enclosed frames or cases covered with glass or plastic materials may be necessary for successful rooting. In using such structures, care is necessary to avoid the build up of disease organisms due to high humidity.
6. **Plastic mulch:** Mulch is a material used for covering the soil in order to prevent weed growth avoids direct evaporation of the water from the soil and to increase soil temperature. Mulching seedbeds with opaque black plastic film prevents weed growth completely and because of higher soil temperature and better conservation of soil moisture, brings about early germination and faster seedling growth.
7. **Mini-greenhouses:** Plants need carbon dioxide for making food through photosynthesis. The greenhouse helps in providing additional carbon dioxide to the plants to enhance their rate of photosynthesis. This can be achieved by enclosing the plants in a box-like structure made out of bamboo and colourless transparent plastic, with a lid at the top. The lid is closed after sunset, so that carbon dioxide produced due to respiration accumulates in the box. After sunrise, the plants start photosynthesis and since there is higher carbon dioxide content around the plant, the rate of carbon assimilation is higher. After a few hours the lid is opened to prevent overheating. Alternatively, a plot or bed of plants can be provided with plastic film skirting supported by bamboo stakes, about 12 cm. high. Since CO2 is heavier than air it tends to settle around the

plants.

8. **Light chamber:** Several plants go into winter dormancy when the day length gets short. Additional light from tube lights, given after sunset, creates long-day condition that prevents the plants from going into winter dormancy. Light, given at the end of the day (photo-period), also encourages growth of green leaves; they grow tall without developing lateral branches. On the other hand, if they are exposed to fluorescent light from tube-lights laid on the ground, they develop side branches and show a bushy habit.
9. **High humidity chamber:** This technique resolves the common problem of grafts or cuttings dying due to desiccation (drying up) when planted in the soil for rooting, by ensuring a humid atmosphere around the cuttings, thus preventing excessive evaporation. The cuttings/grafts are planted on a sand bed, enclosed on all the sides by a dome made of GI wire and covered with a transparent, colourless plastic film. The sand is watered to field capacity, and the plastic film traps evaporation inside the chamber creating a highly humid atmosphere. The dome must be shaded, since direct sunlight will heat up the internal atmosphere of the dome, killing the plants.

FIG. 22.3: High humidity chamber

Selection of Suitable Crops

Raising nursery from seeds and other planting materials is easy and convenient way for ensuring better germination and root development. The planting material of horticultural crops is multiplied under nursery conditions with proper care and management for raising healthy, vigorous and disease free seedlings. In general, fruits and vegetable crops are divided into three groups based on their relative easy for transplanting.

1. Crops like beetroot, broccoli, brussels sprouts, cabbage, cauliflower, tomato and lettuce are efficient in water absorption and rapidly from new roots after transplanting.
2. Vegetable crops that are moderately easy for transplanting are brinjal, onion, sweet pepper, chilli and celery which do not absorb water as efficiently as

crops that are easy to transplant but they form new roots relatively quickly. The vegetable crops, which are difficult to transplant, are cucurbits, sweet corn that requires special care during nursery raising and transplanting.

3. Most of fruits and tree spices are slow growing and multiplied in nursery for better seed germination and plant survival. It becomes convenient to utilize various budding and grafting tools under nursery conditions. Therefore, most of fruit crops are multiplied and propagated under nursery conditions. Besides, the shrubs and herbs of ornamental nature are multiplied under nursery conditions for their faster growth and development.

Selection of Suitable Plant Propogation Technique

The type of planting materials intended to be multiplied are based on the agro-climatic situations, type of nursery e.g. for fruit, vegetables, ornamental plants or mixed plant types and farmers demand. Normally, seed propagation is not recommended in fruit crops owing to variability in growth and fruiting because of the cross pollination, long juvenile phase, chances of transmission of seed born diseases etc. but, where vegetative propagation is unsuccessful, difficult, expensive; for raising rootstock as clonal rootstock are not standardized in most of the fruits and multiplication of clonal materials is also expensive and time consuming, for evolving new varieties etc. the seed propogation is the feasible alternative. Still, the seed propagation is commercially practiced in case of papaya, phalsa, karonda, lasoda, coconut etc. Micropropogation techniques have been standardized for various fruit crops. It is commercial exploited for banana and strawberry. It enables the production of millions of uniform and healthy plants in a short span. The important fruit crops along with methods of propogation method have been given in table 22.2.

TABLE 22.2: Methods of propagation in fruit crops.

S.N.	Name of fruit	Methods of propagation	
		Standardized method of propogation	Commercial method
1	Acid lime	Seed, budding	Seed
2	Almond	Tongue grafting and T-budding	T-budding
3	Aonla	Budding	T-budding/ Patch budding
4	Apple	Tongue grafting, stooling for rootstocks, cutting under mist	Tongue or whip grafting
5	Apricot	Budding and grafting	Grafting & T-budding
6	Avacado	Grafting (cleft, side, whip), chip budding, and cutting	Layering, T-budding
7	Bael	Root cutting, budding, layering	Patch budding
8	Banana	Suckers (sword) and micro propagation	Micropropogated plantlets
9	Ber	Budding (T, I, ring and forkert)	In-situ budding (T-budding) / Polytube budding method
10	Cape gooseberry	Seed	Seed
11	Cashew nut	Seed, grafting (soft wood, epicotyls)	Soft wood grafting
12	Cherry	Whip/tongue grafting	Layering, whip or tongue grafting

S.N.	Name of fruit	Methods of propagation	
		Standardized method of propogation	Commercial method
13	Coconut	Seed, division of growing shoots, air layering	Seed
14	Custard apple	Soft wood grafting	T-budding/ inarching
15	Date palm	Seed, offshoots	Off shoot (suckers)
16	Fig	Cutting, budding and air layering	Hard wood cutting
17	Grape	Hard wood stem cutting, leaf cutting, bud cutting, layering, chip budding and micro propagation	Hard wood cutting
18	Grape fruit	Seed and budding	T-budding
19	Guava	Cutting, air layering and budding	Stooling, Inarching
20	Jackfruit	Inarching, air layering, cutting, epicotyl grafting, budding (forkert, chip and patch)	Inarching
21	Jamun	Seed and budding	Shield & patch budding
22	Ker	Seed and suckers	Suckers/ seed
23	Karonda	Seeds	Seeds/ hard wood cutting
24	Lemon	Seed and layering	Air layering
25	Litchi	Air layering, splice grafting, inarching, budding (chip), cutting (soft wood)	Air layering
26	Loquat	Budding, grafting and seed	Inarching
27	Mandarin	Seed and budding	T/shield budding
28	Mango	Inarching, side and veneer grafting, budding, stool layering, and cutting	Inarching
29	Mangosteem	Inarching and seed	Seed and Inarching
30	Mulberry	Cutting and budding	Stem cutting/ shield budding and cutting
31	Olive	Budding	T-budding
32	Papaya	Seed and cutting with heel	Seed
33	Passion fruit	Stem cutting	Seed
34	Peach	Seed, budding, and grafting	T-budding & tongue grafting
35	Pear	Tongue grafting and double grafting	Tongue or whip grafting & T-budding
36	Pecan	Grafting and budding	Patch budding
37	Persimon	Grafting (cleft and whip) and budding (chip and T)	Crown grafting
38	Phalsa	Seed and cutting	Seed
39	Pilu	Suckers	Root suckers/layers
40	Pinne apple	Suckers (crown, root and shoot), slips	Sucker, slip
41	Plum	Budding and grafting	T-budding, tongue grafting
42	Pomegranate	Budding (chip, patch and forkert), layering (air, ground and pot),	Hard wood cutting and air layering
43	Pummelo	Seed and budding	Seed, T-budding
44	Sapota	Air layering, inarching and soft wood grafting	Inarching

S.N.	Name of fruit	Methods of propagation	
		Standardized method of propogation	Commercial method
45	Straw berry	Runners	Runner
46	Sweet orange	Seed and budding	T- budding
47	Tamarind	Seed and soft wood grafting	In-situ soft wood grafting
48	Walnut	Epicotyl grafting and patch budding	Patch budding
49	Wood apple	Root cutting, budding and seed	Seed/In-situ budding

Source- Horticulture At A Glance (By A.S.Salaria)

IMPROVISED NURSERY MANAGEMENT

Some fruits and vegetables require special cares during their early growth period. At present several improvised techniques have been developed and tested. They are suitable for small and medium level plant growers.

Nursery Facilities

There are three main facilities normally used for raising seedlings in a nursery. The choice of a particular one will depend on the available resources and prevailing environmental conditions.

Hi-tech Nursery - Environment Fully Controlled

Hi-tech nursery where the entire device, controlling the environment parameters, are supported to function automatically. To propagate plants around the year the hi-tech nursery are used. The use of hi-tech nursery is also essential to mass multiply the plants through tissue culture. The main cause of promoting optimum growth in hi-tech nursery is high relative humidity and temperature control, adequate day length and light intensity. Good light conditions are essential for the sturdy growth of the seedlings. Hi-tech nursery is well equipped with elaborate structures and has precise control on temperature, light intensity and humidity. The size and type of hi-tech nursery, primarily depends upon the need of the plant propagator. Glass covered nursery are expensive but they have long life. However, for short-term benefits plastic covered nursery can also be made. Several types of sheets are available for there construction like UV stabilized polyethylene, polycarbonate and fiberglass. These are lightweight and inexpensive as compared to glass. Plastic covered nursery are lighter than the glass covered ones but there is very high increase in humidity in such houses, especially in winters which results undesirable water drops on the plants. Major component of hi-tech nursery as:

Temperature control: The thermostat can be coupled to water circulating pumps or exhaust fan for controlling the temperature inside the nursery. However, the lowest achievable temperature in fan and pad greenhouse is not below the wet bulb temperature in any case.

Relative humidity control: The humidistat coupled to water circulating pump or exhaust fan to control the relative humidity inside the fan and pad greenhouse. The maximum achievable relative humidity is 90% only in fan regulated (FR) nursery. The RH in Non-ventilated (NV) GH can be increased by providing foggers.

Light intensity control: In certain areas where natural illumination is absent or very low, illumination for plants may be provided by artificial sources. Incandescent bulbs generate excessive heat and are unsatisfactory in most instances. Fluorescent tubes are useful as the sole source of light for fruit plant and vegetable seedlings, gloxinias and many foliage plants, which grow satisfactorily at low light intensities. Excessive light intensity destroys chlorophyll even though the synthesis of this green pigment in many plants is dependent upon light.

Quality of light: Quality of light refers to its wavelength composition. Light in the orange-red portion of the visible spectrum from either sunlight or artificial illumination is most effective in causing the long-day response in plants. Far-red radiation appears to have the opposite effect. It is probable that the wavelengths activate some hormonal mechanism within the plant, which brings about the specific effect of light on growth and development of seedlings.

FIG. 22.4: High tech nursery

Nethouse -Environment Partially Modified

Nursery users prefers to have manually or semiautomatic control arrangement owing to minimum investment. This type of nursery is constructed using galvanized iron (G.I) pipes. The canopy cover is attached with structure with the help of screws. Whole structure is firmly fixed with the ground to withstand the disturbance against wind. Exhaust fans with thermostat are provided to control the temperature. Evaporative cooling pads and misting arrangements are also made to maintain a favourable humidity inside the nursery. As these systems are semi-automatic, hence, require a lot of attention and care, and it is very difficult and cumbersome to maintain uniform environment throughout the cropping period. These nurseries are suitable for dry and composite climatic zones.

Open Field – Naturally Modified

Low cost nursery is a simple structure constructed with locally available materials such as bamboo, timber etc. Unlike conventional or hi-tech nursery, no specific control device for regulating environmental parameters inside the nursery are provided. Simple techniques are, however, adopted for increasing or decreasing the temperature and humidity. Incorporating shading materials like nets can reduce even light intensity. The temperature can be reduced during summer by opening the sidewalls and covering by green plant branchase. Such structure is used as rain shelter for growing seedling. Otherwise, inside temperature is increased when all sidewalls are covered with plastic film. This type of nursery is mainly suitable for where climatic conditions are normally favourable for the crops grown.

FIG. 22.5: NETHOUSE NURSERY

FIG. 22.6: Papaya plants in net house nursery

Soil and Soil Preparation

Soil Quality

1. Raising of fruits and vegetables seedlings requires fertile and healthy soil.
2. Preferably, the soil for nursery should be loam to sandy loam, loose and friable, rich in organic matter and well drained.
3. The soil pH should be close to the neutral i.e. about 7.0
4. The soil should be well drained, and free from water logging condition

Soil Preparation

The first step in soil preparation is to get rid of perennial weeds before turn the first spade of soil. The soil of nursery bed should be a fine tilth. It is achieved by deep cultivation of the nursery land either by soil turning plough or by spade and subsequent 2-3 hoeing with cultivator. Dead plant parts which are seem to be dwellers of pathogens and pests should be collected, removed and burnt. After that all the clots, stones and weeds from the field should be removed and land should be leveled. Mix 2 kg well rotten and fine farm yard manure/compost or leaf compost or 500 g vermi compost per square meter and mix in the top soil. If the soil is heavy mix 2-3 kg sand per square meter so that the seed emergence may not be hampered.

Soil Treatment: For Getting Healthy Seedlings

For raising the healthy seedlings, soil must be treated to make it pathogen and pest free. Treatment of nursery soil is also essential for successful management because it the base for seedlings stand, source of nutrition and pathogens. In humid condition damping-off caused by soil borne fungi like *Pythium, Rhizoctonia, Phytophthora* etc, which are in the nursery beds. The removal of pests like snails, cutworms and termite and mites which damage young seedlings, should be removed. Soil treatment may be done by methods like soil solarization, chemical treatment, biocontrol treatment etc.

1. **Soil solarization:** It is done with transparent polythene of 200 gauges during the hot and dry periods. For this soil should be moist before mulching because it increases latent heat and thermal sensitivity for resting of soil borne pathogens, harmful pests and weeds which can be reduced to a sustainable level. Suitable time period is May-June as temperature rises up to 45°C at this time. Wet the soil with water, or saturate it with water. Spread white polythene of 200 gauges on the whole nursery area for about 5-6 weeks. The margin of the polythene should be covered by wet soil (compressed mud) to check the entry of air. After 5-6 weeks remove the polythene sheet. Prepare the beds for seed sowing.
2. **Chemical treatment:** It is done by treating the nursery with by formalin solution (1:100:: Formalin: water) @ 5 lt/sqm, fungicides and insecticides
   a) **Formalin dust treatment:** This treatment should be done 15-20 days before seed sowing. Prepare formalin solution (1.5 to 2%) in one container and drench the soil @ 4-5 litre of water per square meter soil surface to saturate it up to a depth of 15-20 cm. Cover the drench area with polythene sheet of 200 gauge. Put the wet soil on the margin of the covered polythene sheet so as it does not allow the polythene film blown away by the wind and air from the covered area to out side. Removes the cover (polythene) after 15 days. Prepare the beds for seed sowing.
   b) **Application of fungicides:** Generally used fungicides: captan, thiram, ridomil, bavistin, which kill the soil borne pathogens. Use 5-6 g of any of the fungicides per square meter nursery area.
   c) **Insect Control:** Presence of certain insect pest and their egg or secondary stage insects present in the soil, which can infect the seedlings in the later stage. To save the seedlings against them, some insecticides are also used as soil treatment. Recommended insecticide is Chlorpyriphos @ 2 ml/ liter of water, depth of 15 to 20 cm in the nursery soil and then prepared the beds for seed sowing.

   The treated area should be covered with polythene sheets for 7-8 days and after that it should be remained open for next 7-10 days for facilitation of formalin emission. Fungicide treatment can be done by using Ceresan or Bavistin @ 2 g/litre before sowing. Pesticide treatment is done by Chlorpyriphos or Sewin dust powder (20-25 g/sqm) before sowing and or at the time of nursery preparation. For weed control, apply a non-selective, systemic herbicide such as Glyphosate (Round-up) to the area. Apply this material to weeds that are actively growing, generally when temperatures are consistently above 50 degrees. Spring applications are good with fall being another time when weed control is good with this material.
3. **Physical Treatment:** It is steam treatment of soil. Hot steam can be used to treat the soil against harmful insect pest. For this, cover the required area with the help of polythene sheet and stop the movement of air in the covered area. Supply the hot steam for at least 4-6 hours continuously. This way all the harmful pathogen and insect pest will be killed.
4. **Biological treatment:** Bioagents like *Trichoderma*, *Pseudomonas* and *Bacillus* formulations, NSK, Neem cake, Karanj cakes are available for soil as well as seed treatment, which performs as antagonist to harmful microorganisms. While preparing the nursery beds, neem cake /castor cake/ neem leaf/ castor leaf/ pongamia leaf/ calotropis leaf has to be incorporated @ 400 g/m^2 followed by application of spore suspension of nematophagus fungus *Paecilomyces lilacinus* or *Verticillium chlamydosporium* and then inoculation of a vesicular arbuscular mycorrhizae *Glomus fasciculatum* or *G.mossae.* This combined

treatment reduced the infestation of root-knot nematode and reniform nematode simultaneously increased the multiplication of endomycorrhizae and its colonization of tomato root which intern could protect the crop from the nematodes in the main field also. Across the beds prepared, lines are drawn at a spacing of 2.5 cm.

Seed Treatment

Seed dormancy refers to a state in which viable seeds fail to germinate when provided with conditions normally favourable to germination i.e. adequate moisture, appropriate temperature regime, a normal atmosphere and in some cases light. Dormancy has evolved as a strategy to avoid germination under conditions where seedling survival is likely to be low.

There are various degrees of dormancy varying from very slight to very strong (deep). Methods to break seed coat related dormancy:

1. **Mechanical scarification:** Manual scarification of the seed-coat by piercing, nicking, chipping, filing or burning with the aid of a knife, needle, file, hot wire burner, abrasion paper is usually considered the most effective way of overcoming physical dormancy. Since each seed is handled manually, it can be given individual treatment according to the thickness of the seed-coat. Virtually all seed can be made permeable, and the risk of over-treatment (damage) is small.
2. **Hot water:** Hot water overcomes physical dormancy in leguminosae by creating tension which consequently causes cracking of the seed coat. The method is most effective when seeds are submerged into the hot water, not heated together with the water. A quick dip is also better to avoid heat damage to the embryo. Most thick-coated *Acacia species* tolerate a brief (e.g. <1 min.) submersion into boiling water. However, for several hard coated species 2 minutes' boiling was found superior to 1 minute, and some species are pretreated by boiling for up to 5 minutes. A common procedure is to pour the seeds into boiling water and then leave them to cool and imbibe in the water for 12-24 hours. The temperature of the water decreases quickly at enough not to cause damage to most species with relatively thick seed-coats and resistant embryos.
3. **Acid pre-treatment:** Acid used for seed pretreatment is almost exclusively concentrated sulphuric acid (H_2SO_4). The acid causes some kind of wet combustion of the seed-coat and works equally well in legumes and non-legumes. However, the method is not applicable to seeds that easily become permeable because the acid then penetrates and damages the embryo.
   a) Duration of treatment varies according to the following factors:
      i) Seed-coat thickness (depending on species, maturity, age etc.)
      ii) Temperature (longer treatment is required at lower temperature)
      iii) Strength of the acid (new acid is stronger than re-used acid)
      iv) Stirring (stirring during treatment reduces duration of treatment)
      v) Relative volume of the acid (a relatively large volume of acid as related to volume of seed is likely to reduce time required for pretreatment).
   b) Acid pretreatment has several advantages in overcoming physical dormancy:
      i) It is applicable to many species, not only leguminous
      ii) The duration of treatment is short as compared to other scarification methods
      iii) It is the most effective method of bulk treatment for very hard coated seeds
      iv) It requires no special equipment

   v) Seeds can be stored for a period after treatment
  c) The method also has some major problems:
   i) It implies a serious safety hazard to workers
   ii) Seeds are in risk of being damaged by over-treatment
   iii) It can be difficult to safely dispose of waste acid
   iv) It can be expensive in some places

4. **Cold water treatment:** Soaking in stagnant or running water is less laborious method. However, germination sometimes delayed after prolonged soaking alone as compared to alternate soaking and drying. Where physical dormancy is relatively weak e.g. fresh legume seeds, soaking is often sufficient to render the seed-coat permeable. The effect of prolonged soaking on hard seed varies with species. In some species seeds become gradually permeable, in other species there is little effect of continuous soaking. Generally, however, soaking alone is a very slow procedure to overcome physical dormancy.
5. **Growth regulators** -Gibberellic acid or gibberellines (GA) is a group of naturally occurring plant hormones. The hormones play a central role in the early germination processes by activating enzyme production and mobilising storage reserves.
6. **Cold stratification** -Seeds of many other species of woody ornamentals, and pome fruit plants have embryos or endosperms that are nonfunctional, or that contain inhibitors at the time of seed maturity, and require a period of cold stratification. Cold stratification is the subjection of moist seeds to low temperatures for a specified length of time before germination.

Media for Raising Seedlings

For satisfactory germination of seed, subsequent seedling growth and anchorage or directly planted vegetative part of a plant and their root and shoot development; proper media is required. Good fertile soil is the major medium for germinating seeds and growing seedlings, although it is not the best. There are artificial media made of perlite, vermiculite, compost, well decomposed Farm Yard Manure, shredded bark, saw dust, sand, peat moss, coco-soil and sphagnum moss i.e. dehydrate residue of *Sphagnum papillosum, S. capillaceum* and *S. palustre* having better water holding with acidic pH, which are used as soil substitutes. These media are being used either alone or in combination. Some time, the amendment like gypsum and pyrite are also added in the beds while raising the nursery. For best results, a growth/rooting medium should posses the following qualities:

- The media must be sufficiently firm enough and dense to provide anchorage to seeds or cuttings.
- It should be decomposed materials with high C/N ratio.
- Its volume must be fairly constant when either wet or dry.
- Sufficiently porous to let excess water drain away
- It should have better water holding capacity.
- Free from weed seeds, nematodes and other pathogens.
- High cation exchange capacity so that it can provide nutrients
- Slightly acidic medium is preferred. The media reaction (pH of 5.0 to 7.0 and the soluble salt (EC) level of 0.4 to 1.4 dS/m is optimum for most of the plants.
- Able to withstand sterilization treatment without being altered.
- It should be readily available, reusable and cheaper.
- Not be toxic to plants

There are several materials and combinations of different materials are available for germinating seeds and rooting cuttings. Some important media for raising seedlings and rooting cuttings are follows:

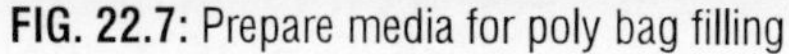

FIG. 22.7: Prepare media for poly bag filling

FIG. 22.8: Mixed media used for poly bag filling

1. **Soil mixture:** This is the most commonly employed medium for pot plants. It usually consists of red earth, well-decomposed farmyard manure, leaf mold, river sand and also charcoal in some cases. Soil mixture commonly used for pot and poly bag filling. The mixture was prepared by the use of-

   Red earth .... 2 parts
   FYM.......... 1 part
   Sand .... 1 part

2. **Sand:** It is the most satisfactory medium for rooting of cuttings. Sand is often used in media mixes to add bulk density and improve drainage. Choose particle sizes of 0.05 to 1.0 mm with at least 90 percent of the particles towards the finer range. No nutrients or buffering capacity are offered by sand.
3. **Peat:** It consists of the remains of aquatic marsh, bog or swamp vegetation, which has been preserved under water in a partially decomposed state. When such peat is derived from sphagnum, hypnum or other mosses, it is known as peat moss. It is used in mixture after breaking them and moistened. All the peats are divided into four types: 1) sphagnum peat moss, 2) hypnaceous peat moss, 3) reed and sedge peat, and 4) humus peat or muck.
   Sphagnum peat moss. This is the most desirable of the peats and most suited in container media. It is acidic, but can be easily adjusted by a number of liming materials if necessary. It is low in soluble salts, long lasting in the mix, uniform in composition, and effectively improves drainage and aeration, yet has good water and nutrient retention. Sphagnum peat has been shown to absorb seven times its weight in water.
   Although two grades of sphagnum peat moss are available, the horticultural grade should be selected, since it is finer in particle size range and grade and most appropriate for container media.
   Disadvantages to any of the peat mosses are that they are hard to wet and may shed water. Either soak the peat moss in warm water before use or add a wetting agent. Use of a wetting agent is more feasible in large-scale operations.
   Hypnaceous peat moss: This peat decomposes quicker than sphagnum peat but is still suitable for container mixes. It may be acquired from the northern states.
   The last two categories of peat, reed and sedge peats and humus peat or muck,

are not recommended in container media. They break down too quickly and interfere with proper aeration and drainage.

4. **Sphagnum moss:** Commercial sphagnum moss is the dehydrated young residue or living portion of acid-bog plants in the genus Sphagnum such as *S. papilliosum*, *S. capillacem* and *S. palustre*. It is generally collected from the tree trunks of the forest species in south Indian hills above 1500 meter above M.S.L. during rainy period. It is relatively sterile, light in with and has a very high water-holding capacity. It is the commonly used medium in air layering.
5. **Vermiculite:** It is very light in weight and able to absorb large quantities of water. This can be used as a rooting medium for air layering and also in pots for raising certain plants. It is purchased in a heat expanded form, which has a high cation exchange capacity. Unfortunately, it compresses easily and is not suitable in a long-term mix, such as that needed in container media. Like perlite, it is well suited for propagation purposes in the nursery, but not for general container nursery mixes.
6. **Bark:** Hardwood bark is a commonly used, excellent media ingredient. Aged bark must be used due to the risk of toxicity to plants when new. Bark is aged with time (at least 30 days or longer after harvest) and leaching. Also, heat and bacterial colonization help reduce toxic effects of the bark. Additional nitrogen must be added when using bark in the mix. Bacteria immobilize and use a great deal of nitrogen making the additional amount necessary for the plants' use. Softwood bark is acceptable but will lower the pH of a media more than hardwood bark. Therefore, liming may be necessary except for acid loving plants. Bark retains water, has excellent drainage, and is sterile. Some nutrients are supplied by all barks and the material is slow to decay. This is a definite advantage in a mix intended for long-term use for nursery stock. Additionally, bark is light weight which an advantage is for workers and when stock must be shipped. Bark is sometimes used alone or with sand or other media amendments.
7. **Perlite:** Perlite, heat-expanded volcanic rock is often used for increasing aeration while lowering bulk density. Because it is so light and will float when the media is watered, it should be restricted for special uses in the nursery such as for propagation media. Perlite is not appropriate in most cases for use in container nursery mixes.
8. **Coco soil:** Outer exoskeleton of coconut used for preparing coco-soil. This is the most desirable media for seed germination in portray and poly bags. It is neutral, uniform in composition, and effective improved drainage and aeration, yet has good water rotation capacity and good source of essential nutrients.
9. **Mycorrhizal fungi:** Inoculating with mycorrhizal fungi increases the effective surface area of roots. Inoculated roots take up extra nutrients and water that would not normally be available, which increases the growth rate of seedlings. Some trees used for land reclamation, like pines and cypresses, are dependent on mycorrhizal fungi and will fail to grow if they not inoculated. Using these fungi requires an organic-matter-rich soil mix. Compost for fertilization is preferred and over-fertilization should be avoided. Inoculation is used for both container planted and bed planted tree seedlings.
10. **Compost:** Compost is used in planting beds and in mixing potting soil for containerized tree stock planting. It provides a nutrient rich organic medium, as well as amending soils by retaining water, such as sandy soils, or loosening them, as with clay soils. Wastes from the nursery process can be incorporated into compost. It provides a viable alternative to expensive chemical fertilizer and can be produced on site.
11. **Livestock manure:** Livestock manure is an excellent source of nutrients for the soil, providing such nutrients as nitrogen, phosphorus, and potassium. Manure

application can also benefit the soils water holding capacity and tilth. When using livestock manure, however, one should follow good management practices in order to avoid hazards to the crop and the environment. The important sources if livestock manures are poultry waste, cattle manure, and swine manure.

12. **Slow-release fertilizers:** Advantages of slow release fertilizer are that the nutrients are available gradually over an extended period of time. This means that the gardener can fertilize less often, and the nutrients are provided gradually (at lower rate but consistently). This is how most plants prefer to be fed (like us who eat a little every day compared to just one (bigger) meal per week). Most slow release fertilizers (organic and synthetic) release at specific (warm) soil temperatures. The benefit of this is that plant roots generally are most active in warm soil and therefore the slow release fertilizer will start to make fertilizer available as soon as the plants actually needs them. As the plant roots become more active (in warmer soil) more fertilizer will automatically be released.

Since it may not be possible for one medium to have all these characteristics, different media are normally mixed together to obtain a 'near' ideal mixture. Materials found in the locality may also be used. Some media may be contaminated, therefore it may be necessary to sterilize them before they are used, to kill weed seeds, insects and numerous soil pathogens. This treatment includes the use of heat and/or chemicals. Steam sterilization and solarization are used in the heat treatment to kill all pathogens in the media. One of the advantages of this treatment is that the media can be used sooner after treatment and the treatment can be applied to the dry, wet or even cold media.

There is, however, a risk of the media structure or composition being modified by the heat. Chemical treatment, which is also referred to as fumigation, involves the use of insecticides, which may be specific to certain pathogens or broad spectrum (controls a wide range of pests). These chemicals should be handled with caution, as most of them are toxic to both humans and animals. Some of the commonly used chemicals are, chloropicrim and vapor.

Pasteurization of nursery plant growing media: Nursery growing medium may contain harmful disease causing organisms, nematodes, insects and weed seeds, so it should be decontaminated by heat treatment or by treating with volatile chemicals like methyl bromide, chloropicrin etc.

Agent	Method	Recommendation
Heat	Steam	30 min at 180° F
Methyl bromide	10 ml/cu. ft. of medium	Cover with gas proof cover for 24-48 hr. Aerate for 24-28 hr before use.
Chloropicrin	(Tear gas) 3-5 ml/ cu. ft. of medium	Cover for 1-3 days with gas proof cover after sprinkling with water. Aerate for 14 days or until no odour is detected before using.
Basamid	8.0 g/cu.ft. of medium	Cover for 7 days with gas proof cover and aerate for at least a week before use.
Formalin	20 ml/l of water (37%)	Apply 2 l/cu.ft. cover for 14 to 36 hr and aerate for at least 14 days.

Fungicides or bactericides can also achieve disinfection of the growing media

Fungicides and their effect on a few fungi

Chemical	Rate of application	Effect against
Captan	2 g/l of water	Pythium, Fusarium, Rhizoctonia and Phytophthora. Some extent to root and stem rot, white mold, black rot, crown rot and damping off.
Metalaxyl + Mancozeb (Ridomil MZ 72 WP)	1 g/l of water	Pythium, Phytophthora, Fusarium and other soil borne pathogens

Temperature necessary to kill soil pests

- 115°F for water molds (Pythium and Phytophthora)
- 120°F for nematodes
- 135°F for worms, slugs and centipedes
- 140°F for most plant pathogenic bacteria
- 160°F for soil insects
- 180°F for most of weed seeds
- 200°F for few resistant weed seeds and plant viruses

Calculating media needed- many container manufacturers provide charts for their pots indicating how many pots can be filled with a cubic yard of media. However, if containers are purchased without these guidelines or if special mix formulas, small quantities, etc. are needed, the following formula is convenient and accurate. Also, a grower can keep track of media costs invested on a per plant basis. The container's top radius (r) must be determined by measuring the diameter of the top of the pot and then dividing by two. Determine the depth of the pot (h) by measuring from the center-top to bottom.

The formula for the volume (V) is: $V = (3.1416 \times r^2 \times h)$

To calculate the volume, multiply the top radius (r) by itself (r x r = r2) and then by 3.1416. Then multiply this number by the depth at the center of the pot (h). This number is the container's approximate volume (V).

Nursery Bed Preparation

Nursery bed should be prepared according to the season and crop. In the rainy season raised beds are prepared but in the winter and summer season flat beds or sunken bed should be prepared. For the uniform and high percentage of germination the soil must be fine and moist enough. If the seedlings are to be raised in boxes during unfavourable weather condition, the flower pots, polythene bags, potting plugs, wooden treys, earthen pots etc. may be used. Prepare soil mixture in the ratio of 1:1:1 of soil, sand and well rotten FYM/leaf mould etc. and fill the mixture in these seedlings raising structure. Arrangement should be made to drain excess water from these structures by making a hole in the bottom of all types of pots.

Types of Nursery Bed

a) Flat nursery bed
b) Raised nursery bed
c) Sunken nursery bed

Flat Nursery Bed

Flat bed is commenly used in winter season for prepare rabi vegetables seedlings. Length of the bed may be kept 3 meter in sandy loam soil and 5 meter in clay soil, the width of the bed is 1 meter which easy in weeding and intercultural operations. Some time damping off disease is more prominent in flat bed.

Raised Nursery Beds

Length of the bed may be kept 3 to 5 meter; however, width is restricted to 1 meter only which facilitates intercultural operations. The beds are raised 15 to 20 cm high from the ground level. A space of 30 - 40 cm is leaft in between two beds. The space between two beds helps in weeding, nursery care against diseases and insect pest and also for draining out the excess rain water from the nursery beds. The number of beds depends on the particular crop, season and growing area of crop. The beds should be prepared in the east and west direction and line should be made from north to south direction on the beds.

Advantage of Raised Nursery Bed

1. Water movement will be uniform and drainage of excess water is possible (In the case of flat bed water moves from one end to the other and there is possibility of washing away of seeds).
2. Germination percentage of seeds is normally high. Operations like weeding and plant protection measures are easy.

Sunken nursery bed is the common method in dry and hot area. Nearly 15-20 cm deep beds of 45-50 cm width are always preferred for sunken nursery. However, its length should be made according to the requirements or size of plots but should not exceed 5-6 m. A space of 30 - 40 cm is leaft in between two beds. During bed preparation, add organic matter at a rate of about 25-30 percent by volume of soil. This translates to adding about 3-4 inches of organic matter on top of the bed and working it into about 10-12 inches of soil. Beds should be protected from heavy rains and hot sun. Major advantage of sunken nursery bed is good water retention for dry conditions and less evaporation.

FIG. 22.9: Raised nursery bed Sunken nursery beds

Poly Bag Nursery

Poly bag nursery can be adopted for producing seedlings with greater vigour. As a result of the intensive care and better maintenance of the polybag nursery, vigourous seedlings with better root system are obtained, which attain rapid reproductive development. Compared to the nursery in the field, watering, weeding and culling operations for the elimination of unwanted seedlings are easier in poly bag nursery.

The seeds can be sown in black polythene bags of 500- gauge thickness, 25 x 12 cm, 30 x 30 cm size and poly tubes as per requirement of different fruit plants. The bottom of the bags is to be provided with 8- 10 holes for draining the excess water. The commonly recommended potting media are top soil mixed with sand in 3:1 ratio or fertile top soil, sand or coir dust and well rotten and powdered cattle manure in the ratio of 3: 1: 1. red earth, well rotten and powdered cattle manure and sand in 1: 1 : 1 can also be used for bag filling. In order to produce poly bag seedlings, initially the seed are sown very closely and allowed to germinate in a pre nursery bed. The 20-35 day seedlings are then placed in the main field and other poly bag and sufficient potting mixture is added to fill the bags. Usually the poly bag seedlings are maintained for about 8 to 10 months.

FIG. 22.10: Poly bag nursery

FIG. 22.11: Filling growing media in poly bags

Protry Nursery

This technique ensures that a large number of seedlings can be raised in minimum space under optimized and controlled conditions. The seeds, saplings or cuttings are planted in plastic protary filled with a specially prepared potting medium. The protray have holes in the bottom. They are kept in shallow pits lined with a plastic film to prevent roots penetrating the soil below. The potting medium has to be porous to avoid water logging and should also have all the plant nutrients.

FIG. 22.12: Protry nursery

FIG. 22.13: Pot nursery and polybag nursery

Providing Drainage

In between beds, drains of about 30-45 cm width are prepared and connected to the main drain for removal of excess water during heavy pour. This space facilitates easy movement during intercultural operations and acts as physical barrier for inoculums spread. The drains are flooded during dry period to modify microclimate of nursery beds in favour of seedlings. In recent years various advancements have been made in nursery management for bed preparation to avoid possibilities of pathogen spread like use of soil less media, plug tray technique, perforated poly trays etc.

FIG. 22.14: Preparation of nursery beds for vegetables and ornamental flowers

FIG. 22.15: Preparation of nursery beds for fruit crops.

Sowing of Seeds in the Nursery

After the seed bed preparation seeds are sown in the nursery bed either by broadcasting or in lines depending upon the nature and season of crop.

Broad casting method: In broadcasting method seeds are broadcasted on the well prepared nursery beds and later on the seeds are covered with well rotten fine sieved and treated FYM or compost. The major disadvantages of this method are:

1. Uneven distribution of seeds in the nursery beds.
2. Growth and development of seedlings is poor.
3. Some times nursery becomes so dense to look like as patches of grasses. In such cases there is more possibility of damping off disease occurrence.

Line sowing method: Line sowing is the best method of seed sowing in nursery. Lines are made on well prepared nursery beds, 0.5 to 1.0 cm deep parallel to the width at distance of 5.0 cm from the line and seeds are sown or placed singly at a distance of about 1.0 cm apart. Cover the seeds with fine mixture of sand, soil and well rotten and sieved FYM or leaf compost etc. (1:1:1). After the seed covering a light irrigation must be given. An adequate moisture should be maintained for better germination through grass mulching, and intermittent sprinkling of water.

Seed Covering Material and its Treatments

Seed Cover

After seed sowing the seed that are sown either by broadcast method or line sowing method required cover for better emergence. Therefore, a mixture of sand: soil: FYM in the ratio of 1:1:1 is prepared, well mixed together and treated with any method of soil treatment as discussed above. It will be better to treat this mixture while treating the nursery soil. Apply 3-4 g thiram or captan per kg mixture if, it is not treated. Care should be taken that every seed is well covered by seed covering material.

Use of Mulch

To maintain the soil moisture for seed germination cover the seed bed with a thin layer of mulch of paddy straw or sugar cane trash, or sarkanda or any organic mulch during hot weather and by plastic mulch (plastic sheet) in cool weather. It has following advantages:

1. It maintains the soil moisture and temperature for better seed germination.
2. It suppresses the weeds.
3. Protects from direct sunlight and raindrops.
4. Protects against bird damage.

Removal of Mulch

Due attention is given to remove the covered mulch from the seedbed. After three days, observe the seed beds daily. As and when the white thread like structure is seen above the ground, remove the mulch carefully to avoid any damage to emerging plumules. Always remove mulch in the evening hours to avoid harmful effect of bright sun on newly emerging seedlings.

Seedling Management

Use of Shedding Net

Shading should be done to protect the young seedlings from high heat intensity in sunny areas and also from heavy rain. Polythene nets or even grass can provide shade. After seed germination during the seedling growth, if there is very high temperature (> 30°C) then beds should be covered by 50% or 60% shedding nets of green/green + black coloured, about 60 - 90 cm above ground by the use of suitable support. The shade should be removed some days before transplanting to allow the seedlings to acclimatize to field conditions.

Watering

The seedbed or seed box should be watered carefully with a fine stream of water. After the plants are well established, watering should be done thoroughly but not too often. It is advisable to irrigate seedlings in the morning and not in the afternoon as the latter leaves the soil surface moist overnight, a condition favouring damping off. The nursery beds require light irrigation with the help of rose can till the seeds get germinated.

1. Excess rainwater or irrigated water should be drained out from the field as and when it is required otherwise plants may die due to excess of water.
2. Watering in the beds depends upon the weather condition. If temperature is high, open irrigation is applied. Need not to irrigate the beds during rainy days.

Watering system: Micro irrigation system is the best for watering plants in a nursery. Micro sprinklers (fogger) or drip irrigation equipments can be used. Basically the watering system should ensure that water does not fall on the leaves as it leads to disease and scorching problems. In micro sprinkler system, water under high pressure is forced through nozzles arranged on a supporting stand at about 5 feet height. This facilitates watering at the base level of the plants.

Equipments required for drip irrigation system include

i) A pump unit to generate 2.8 kg/cm^2 pressure
ii) Water filtration system – sand/silica/screen filters

iii) PVC tubing with dripper or emitters

Nutrient Management

Seedlings planted on degraded land need well-developed root systems to survive. Proper nutrients stimulate vigorous root growth. Ideally, fertile soil should have a neutral pH level plus good soil structure and drainage, containing adequate amounts of the major plant nutrients- nitrogen, phosphorus and potassium, plus sufficient micro-nutrients (trace elements) – zinc, manganese, boron, copper, iron, magnesium, molybdenum and chlorine. Adding organic matter, humus and a wide range of soil micro-organisms will maintain healthy soil. In nursery it is important to have a good understanding of soil fertility, recognising the role of organic matter and soil structure. Usually in nursery beds normal fertilizers like urea, muraite of potash and DAP are applied. Timing of fertilization should be given in two spilt i.e. basal and top dressing (after 10 days) by broadcasting or foliar spray @ 0.5-2%. Immediate before transplanting, application of fertilizers should be avoided as it encourages diversion of plant energy toward root development in nursery and has negative impact on seedlings during exposure for transplanting. Common source of nutrients in nursery is FYM, compost, vermicompost, leaf mold, cakes etc. Besides, primary nutrients like nitrogen and phosphorus are essentially applied through straight fertilizers as these play an important role in root and shoot development.

Thinning

This is a way of regulating plant density in rows and in holes. It is an important operation to remove weak, unhealthy, diseased, insect pests damaged and dense plants from the nursery beds keeping distance of about 0.5 to 1.0 cm from plant to plant. During thinning, weak, diseased plants are pulled out to allow healthy seedlings to grow well. It is normally done when seedlings have formed a few true leaves. The thinning facilitates balance light and air to each and every plant. It also helps in watching the diseased and insect pest attacked plants while moving around the nursery.

Weed Control

Weeds are plants unwanted at a place and time. There presence in nursery increases competition with seedlings for nutrient, water, light and CO_2 results in lanky seedlings. Besides, some weeds harbour pathogens and insects and also produce allelopathic effect on crop plants. Therefore, weed control is very essential requirement for successful nursery production. Timely weeding in nursery is very important to get healthy seedling. If there are some weeds in the seed bed, remove them manually either by hand or by hand hoe (thin forked Khurpi). Pre emergence herbicides can also be sprayed soon after seed sowing to control the weeds. Stomp @ 3 ml/litre of water should be sprayed on the nursery beds after the seed sowing and seed covering with mixture of FYM. soil and sand.

The following methods control weeds in nursery;

1. Select a weed-free field or media for nursery preparation.
2. Control weeds in perimeter areas (i.e. fence rows and windbreaks).
3. To reduce weed seeds, properly store and compost manure before applying to the soil.
4. Use stallbed technique to avoid initial weed infestation.
5. Mow buffer strips to reduce seeds blown into irrigation ponds.
6. Minimize run-off from weedy fields to ponds.

7. Pump irrigation water from deep in the pond to avoid seeds on the water surface.
8. Ensure weed-free material is planted.
9. Do not move weeds between fields on equipment.
10. Cultivate fields when seedlings are small.
11. Use shallow tillage (2.5-5.0 cm) if herbicide has been applied.

Hardening

Hardening means any treatment that makes the tissues firm to withstand unfavourable environment like low temperature, high temperature and hot dry wind. Hardening is physiological process. Plants accumulate more carbohydrates reserves and produce additional quiticle on the leaves. In this process seedlings are given some artificial shocks at least 7-10 days before uprooting and transplanting. These shocks includes:

1. Gradually exposure them to higher or lower temperature and in the full sunlight.
2. Removal of all the shedding nets, polythene sheets
3. Irrigation is stopped slowly and slowly.
4. By application of 4000 ppm NaCl with irrigation water or by spraying of 2000 ppm of cycocel.
5. Place in a cold frame. Cold frame cover is enameled steel frame and has zippered openings on both sides, for easier access and is dade of woven polyethylene, like a heavy row cover. It gives about the same cold protection, but is easier to work under. An optional shade covering is available so you can use it during the growing season to protect plants from sun and heat. It provides about 27 sq. ft. of growing/pot area. Dimensions: 71-1/2" L z× 57-1/4" W.

Young, pampered seedlings that were grown either indoors or in a greenhouse will need a period to adjust and acclimate to outdoor conditions, prior to planting in the garden. This transition period is called "hardening off". Hardening off gradually exposes the tender plants to wind, sun and rain and toughens them up by thickening the cuticle on the leaves so that the leaves lose less water. This helps prevent transplant shock; seedlings that languish, become stunted or die from sudden changes in temperature. Hardening off times depend on the type of plants are growing and the temperature and temperature fluctuations.

Duration and Degrees of Hardening

1. It is very necessary that plants should be hardened according to their kind so that there is an assurance of high percentage of survival and slow growth under the condition to be expected at the time of transplanting.
2. Hardening should be gradual to prevent or check the growth.
3. Warm season crops like tomato, brinjal and chillies do not favour severe hardening. In Indian condition allowing the soil to become dry for 5-6 days does the hardening.
4. After 7 -10 days plants are ready to transplant. Try to do so on a cloudy day and be sure to water well after planting.

Effect of Hardening

The following effect may be observed by the hardening

1. Hardening improves the quality and modifies the nature of colloids in the plant cell enabling them to resist the loss of water.
2. Hardening increases the presence of dry matter and regards in the plants but decrease the percentage of freezable water and transpiration per unit area of

leaf.

3. Decreases the rate of growth in the plants.
4. Hardened plants can withstand better against unfavourable weather conditions like hot day winds or low temperature.
5. Hardening of the plants increases the waxy covering on the leaves of cabbage.

Drought Hardening

Plants that are raised under high atmospheric humidity and shade often die due to transplanting shock when shifted to the fields. To prevent this, allowing external dry air to enter the chamber gradually hardens the plants. This is achieved by lifting the plastic film on two opposite sides of the high humidity chamber to some extent to create small openings in the chamber. The openings are widened every day, in such a way that the entire film can be removed after about 8 days. In this way, the plants get adapted to dry air gradually. Such plants can be transplanted in the fields but as precautionary measure it is advisable to transplant on a rainy or cloudy day.

Nursery Plant Protection

Because of the variety of plants in the nursery, insect and disease control poses many challenges. Integrated pest management (IPM) combines chemical, cultural and biologicalcontrol techniques to address pest problems. Good sanitation and plant health reduce pest and disease problems. It is necessary to maintain vigorous, healthy plants by using proper culture and management practices to provide natural resistance to plants. Heavily insect infested or injured plants should be destroyed as earliest possible.

Insect pests, diseases are critical factors in the process of multiplication of plant material under open as well as controlled conditions. Control of insects and diseases are integral part of the nursery plant production.

Pest management- Prevention or avoidance of pests and their damage is possible by the following methods:

1. Maintain vigorous, healthy plants by using proper culture and management practices
2. Practice Integrated Pest Management (IPM) and realize that all culture and management factors can affect pests.
3. Spot spray only infested plants that exceed the threshold
4. Destroy heavily infested or injured plants
5. Conserve and promote beneficial insects by selective pesticide use

Adaptation of plant protection measures in the nursery against the incidence of insect pest and diseases is very important task to get the healthy seedlings. Damping off seedlings, leaf curl, leaf blight diseases and leaf miner and borer infect the seedling in the nursery. The care for controlling them time to time is essential.

Damping off: This is very serious disease of nursery. Pre-emergence death of seeds is seen. In first instance girdling takes place on the stem near base of the stem and seedlings bent down near the ground and die. The causal organisms are *Pythium, Phytopthora, Rhizoctonia* and *Fusarium* fungi.

Control

- Treat the nursery bed either by soil solarization, formalin solution or formalin dust or fungicides like thiram or captan.
- Treat the seeds with fungicides @ 3-5 g / kg seed. If the disease appear after the seed emergence drench the nursery beds with 0.1% solution of brassicol or 0.7%

captan or thiram after germination. It will be better to remove and burried the affected seedlings from the beds otherwise spread will be more.

The disease can be controlled to some extent by applying treated sand, soil and FYM mixture up to the level from where the seedlings are falling.

Leaf miner: It is very small sized insect enter in the leaves from margin side and move from one place to other by eating the chlorophyll. Initially the infected part of the leaves become brown and later on dry.

Control

- Spray 4% neem seed kernal extract on the plants (crush 40 g of neem seed kernal, add some water and allow them for overnight. In the morning filter the extract and makeup the volume 1000 ml.)
- Spray Endosulphan or Metasystox 1.5 ml/litre of water.

Whitefly: Whiteflies are tiny, white insects. They usually crowded in between the veins on ventral surface of leaves. They suck the sap from the plants. The affected parts become yellowish, the leaf curls downward. Whiteflies are well known a vector for transmitting the curl virus disease in nursery plants.

Control

1. Raise nursery well protected
2. Spray the plant with Malathion at 0.05 per cent or Metasystox at 0.1 per cent.

Raising of Virus Free Seedlings

Leaf curl is a white fly transmitted viral disease, infestation starts from seedling stage and continue till harvest of the crop. The disease is specially seen in the tomato and sometime in chilli too and causes great loss of the crop.

The leaves of affected plants show curling, mottling, rolling puckering etc. It can be controlled by the following ways:

- Treat the soil of the nursery by carbofuran 3-5 g/sqm.
- Seed treatment with Imidachloprid @ 2.5 g/kg seed
- Cover the seed bed after seed sowing by Agronet/anti-virus net making a tunnel like structure.
- Spray the nursery beds 15 days after seed germination at 7 days interval with Metasytox @ 1.5 ml/litre of water. Last spray is done 2 days before transplanting.
- Remove the infected plants if any in the field and burried in with soil or burn.
- In this way the raised seedlings will be healthy and free from viral diseases.

Transplanting

This refers to the operation of lifting the seedlings from the seedbeds or containers and transferring them to the field where they will grow and mature. The main aim during transplanting should be to interrupt growth as little as possible, and if the operation is not carried out properly it can severely check growth or in extreme cases cause death of transplants. Most vegetable seedlings are ready to be moved 4-8 weeks after sowing. The seedbed should be given a thorough soaking about 6-12 hours before the plants are moved to ensure that they are fully turgid and that the roots retain plenty of soil when the plants are lifted. The main field should also be irrigated at the same time so that the planting holes can be opened up easily and the plants easily 'firmed'. The best time to carry out transplanting is in the late afternoon or early evening, as this allows the plants some time to get partially reestablished before having to face heat stress during the day. A cool cloudy weather is ideal for transplanting. It is always wise to raise about 30% more seedlings than are actually required so that the weak ones can be discarded and any casualties replaced. The adaptability of vegetables to transplanting varies widely between crops. Transplanting success depends on how rapidly a plant is able to regenerate those parts of the root system that were lost/damaged during transplanting.

Selection of seedlings for transplanting: After attaining proper growth, seedlings are transplanted in main field. At the time of transplanting, seedling should be:

- Stocky and sturdy
- Should have good root system
- Should be free from any insect pests and diseases.

Summer is never the best time to move or transplant garden plants. The sun is too intense and the heat is relentless. However, sometimes we have no choice but to move plants during the hot months. Successfully transplant garden plants at any time of the year if follow a few simple guidelines:

1. Water the garden plants to be dug and/or transplanted the day before your do it. This insures that the whole plant will be hydrated, leaves and all, when it's time to transplant.
2. Dig and/or transplant when it is overcast or during the cooler evening hours.
3. Water the plant immediately before digging or removing from its pot. Soak the root ball so that the soil will adhere to the roots, when it is dug from the garden.
4. Never leave the roots exposed to sun, heat or wind. Don't remove all plants from their pots and place them in the garden. Remove just prior to planting.
5. Water the hole before place the transplant into it.
6. Place the transplant into the hole and fill it halfway with water. Allow the water to settle the soil around the roots and then finish filling the hole.
7. Lightly firm the soil around the transplant.
8. Once again, water the whole plant, leaves and all.
9. If possible, shield the new transplant from direct sunlight for 3-5 days. Use a floating row cover or lean a board in front of the transplant to block direct sun.

Check the plant daily for the first couple of weeks. Transplants will need watering every day, if not more. If it is wilting, water the plant. Depending on the weather and the plant may need to water twice a day until it becomes established. The larger the plant and/or the less roots to top growth ratio, the more water will be needed. All of this may seem extreme, but the shock of being uprooted is stressful to plants anytime of year. In the heat of summer, this extra precaution is vital to easing the transition for transplants.

Potting

Purposes of Potting

1. Preparing plants for sale such as rooted cuttings of grapes
2. Growing plants for decoration like crotons, ficus, cicus, palm etc.
3. Growing plants for experimental studies like pot -culture studies
4. For using plants as rootstocks in certain grafting methods as in inarching of mango.

Pot Mixture or Potting Compost

It is essential for potting of plants. The pot mixture is prepared by using various ingredients. The proportion of pot mixtures will vary with different kinds of plants.

1. An ideal pot mixture should have an open structure, which allows good drainage and holds sufficient moisture for plant growth and permits excess waster to drain away.
2. Should supply adequate nutrient to the plants during all stages of growth
3. Should be free from all harmful organisms and toxic minerals and
4. Should be light in weight

Potting Procedure

1. Wet the seedbed before lifting plants. Life with a ball of earth with as much of the root system intact, as possible. Do not pullout seedlings in the hot sun. Do not allow roots or the soil around the roots to dry.
2. Fill up pots by putting some crocks first, then a layer of sand (5-8 cm) and finally pot mixture (8-10 cm).
3. Place the plant with the ball of earth in the centre upon the layer of pot mixture (Place on one side of pots in the case of root stock plants used in inarching)
4. Put pot mixture around the ball of earth, press as fill up and level off, leaving one inch head space at top. Do not press over the ball of earth. It will break and damage the roots.
5. Set the stem of plant at the same height as it was in the seed bed
6. Immerse pot with plant in a tub of water gently and keep inside water till air bubbles cease to come out. Remove and place the pot under shade of trees.

Repotting

Repotting is done for changing the soil medium for pot bound plants.

Pot Bound Condition

When the potted plants are grown for more than one season or one year in pot, the root very soon become a tangled mass and exhaust all the nutrient in the limited soil, besides being circumscribed in the limited place. This stage is known as pot bound condition.

Repotting Procedure

1. It is better to west the potted plant 24 hours earlier to facilitate repotting (removal of plant from pot)
2. The technique to remove the plant with a ball of each intact is to keep the right hand palm over the soil, allowing the stem of the plant in between the first fingers and turn the pot upside down holding the pot at the bottom with the left hand and gently knocking the rim of the pot on the edge of table or any other hard surface or even on the bottom edge of another inverted pot. The ball of earth comes out of the pot. If for any reason, it fails to come out, break the pot knocking the sides with a stone or fork and free the soil from it.
3. Examine the roots, cut neatly with a secateur, the decayed, dead and dried or twisted roots. Reduce the size of the ball of earth around the roots.
4. Place the plant in the new pot at the same height at which it was in the old pot. Fill up pot with fresh pot mixture and immerse in water.

General

1. The initial reaction after potting and repotting is wilting. The transpiration loss has to be checked to help plants revive. Hence keep freshly potted plants under shade and "pot water daily".
2. After about ten days under shade, the plants should be gradually exposed to sun by keeping them for some hours under sun and then putting them under shade. The period of exposure can be increased every week until finally the plants can be kept in the open. This process is called "hardening".

Raising of rootstocks: The plants propagated by grafting or budding require raising of rootstocks, up on which scion shoot or bud of a promising cultivar is grafted or budded. This should be planned well in advance. There are several activities like, collection/ purchase of seeds to be sown for rootstock purpose, preparation of seed bed, use of poly containers, filling mixture, type of seed (orthodox or recalcitration), seed quality (boldness, viability, germination, free from pest and diseases), seed treatment, time and method of sowing, irrigation, pest, disease and weed management, thinning, removal of basal and side branches if required, after care etc.

In general, about one-year old rootstocks are used, which are sown in monsoon (July-August) of previous season and ready for grafting or budding purpose in June-September of current year. This practice is expensive and requires long time. Such rootstocks invariably lose their taproot systems as result of repeated transplanting in the nursery and plants raised in containers developed coiled roots. The plants with bigger earth ball are also not feasible for transportation at longer distance and damage of earth ball may cause of poor field survival. Therefore, the rootstocks must have root architecture with a strong tendency to penetrate into deep soil horizons.

TABLE 22.3: Important rootstocks of fruit crops

Name of fruit	Rootstocks	Distinguishing features
Almond	Peach almond hybrids GF-577 GF-667 Prunus nira Bitter almond Almond seedlings	
Apple	M-9	Chance seedling, dwarf suited to high density planting, wooly aphid resistant
	M-27	Ultra dwarf, result of M-13x M-9, suited for high density plantings
	MM-104 and MM-106	Wooly aphid resistant
	MM-111 and MM-104	Drought resistant
	Northern spy	Wooly apple aphid resistant
	Malus sikkimiensis, M. hupehensis and M. toringoides	Recurrent apomictic: propagate true to type.
	M. sergenti	Vigorous
	M. baccata, Robusta, M-16 and M-25, Merton-778	Wooly aphid resistant malling series
Apricot	Wild apricot	Commonly used

Name of fruit	Rootstocks	Distinguishing features
Avocado	West Indian avocado, Mexican avocado	
Bael	Aegle fraeglegaboensis	
Ber	Zizyphus nummularia	Dwarfing due to formation of inverted bottle neck at graft union
	Z. rotundifolia	Deep rooted suited for arid zones
Cherry	Colt and charger	Dwarf
	Wild cherry (P. puddum), Paja (P. cerasoides)	Show delayed incompatibility Produces vigorous trees
	Mahlab (P.mehalab)	Produces dwarf tree
	Mazzard (P. avium)	Recommended for cherry in H.P.
	F 12/1	
Citrus	Rangpur Lime	Susceptible to foot rot, burrowing nematode and exocortis; tolerant to tristeza, suitable for heavy and deep soils.
	Karna Khatta	Suitable for heavy soils used in U.P. Incompatible with kagzi lime and Malta Blood Red
	Rough Lemon	Commonly used in Punjab, susceptible to blight and tolerant to tristeza.
	Trifoliate orange	Cold hardy, resistant to phytophthora, tristeza and nematode; dwarfing and salt resistant suitable for shallow and high rainfall areas; incompatible with lime, lemon and mosambi.
	Flying dragon	Most dwarfing Fortunella
	Severinia bouxifolia	Salt resistant
	Soh Sarkar	Dwarfing for kinnow
	Feronia limonia	Dwarfing but not commercial
	Sour orange	Cold hardy, suitable for sandy loam soil; susceptible to tristeza
	Citranges	Susceptible to citranges, tolerant to gummosis and tristeza
	Cleopatra manadarin	Cold hardy, tolerant to tristeza and nematode
	Citrus unshiu	Freeze tolerant
Custard apple	Annona glabra (Pond apple)	Suitable for various soil conditions
Fig	Ficus glomerata	Nematode resistant
Grape	Salt creek and Dogridge	Salt and nematode resistant
	Temple	Multiple resistant/ pierce's disease
	St. George, Ripario, Gloria	Phylloxera resistant
Guava	P. friendrichsthalianum	Dwarfing and wilt resistant
	P.friendrichsthalianum var.lucidum	Wilt resistant
	P.pumilum	Most dwarfing
Jamun	Syzigium fruticosum	Termite resistant

Name of fruit	Rootstocks	Distinguishing features
Mango	Kurrukan	Salt resistant polyembryonic
	Olour	Vigourous rootstock
	Vellaikolamban	Dwarfing and allopolyploid
	Rumani	Dwarfing
	Moovandan and Nekkare	Salt tolerant
Olive	Olea huspidata	Nematode resistant
Peach	Prunus besseyi, Ruter's red leaf and Rubira	Dwarfing
	Nemaguard, Nemared Shalil and Yunnan	Nematode resistant
Pear	Quince A	Vigourous
	Quince B	Semi-vigourous
	Quince C	Dwarfing
	Pyrus pyrifolia and P.pashia	Commercially used in hills
Plum	Pixi and St. Julien K	Dwarfing for plums
	St. Julien A	Vigorous
	Brompton	Cold hardy
	Mariana	Result of cross between Prunus cerasifera x P. munsoniana, nematode resistant
	Myrobalan	Resistant to clod, collar rot, and nematodes
Sapota	Manilkara hexandra	Commercially used for most of the sapota cultivars
Walnut	Paradox (J. hindsi x J. nigra)	Shows delayed incompatibility
	Juglans nigra	Resistant to crown rot, susceptible to water logged saline or nematode infested soils.

Source- Horticulture at a glance (by A.S.Salaria)

Container and Tools for Propagation and Growing Young Plants

Earthen Pots

They are made of burnt porous clay in various sizes to provide requisite amount of soil and root space to different kinds and sizes of plants. They have straight sides and are made wider at the top than at the bottom to hold the greatest bulk of compost where the feeding roots are and also to facilitate easy removal of soil, intact with roots (ball of earth) at the time of planting or repotting. In our county, tube pots of varying sizes are used as follows.

Tube pot sizes	Height (cm)	Diameter (cm)	Cost per pot (Rs.)
Tube pot	20	13	1.50
¼ size pot	18	22	5.50
½ size pot	20	27	8.00
¾ size pot	25	32	12.00
Full size pot	35	35	15.00
Tub size pots	35	50	25.00

Seed Pan and Seed Boxes

Seeds pans are shallow earthen pots about 10 cm high and 35 cm in diameter at the top. They have one large hole for drainage in the centre or 3 holes at equidistant from each other. Seed boxes are made of wood, 40 cm wide and 60 cm long and 10 cm deep, with 6-8 properly spaced holes drilled in the bottom.

Against each of the holes is placed a crock with its concave side down. Some large pieces of crock are put over it and also the side of this crock sprinkles some coarse sand 2 or 3 handfuls sprinkled on the crock pieces forming a thin layer to prevent fine soil from clogging the drainage. Over this, required soil mixture is added. Very delicate kinds of seeds like Cineraria, Begonia, etc. are best sown in these containers.

Polythene Bags

Small polythene bags with holes punched in the bottom for drainage and filled with a porous rooting medium are used for propagation of cuttings like Jasmines, Durant, and Crotons etc. in the mist chamber. Sometimes, young seedlings, which are raised in the nursery, are subsequently transplanted in these polythene bags and kept there till they attain required growth for transplanting them to the main field (Papaya, Curry leaf etc.).

Plastic Pots

Plastic pots, round and square are used to keep mostly indoor plants. They are reusable, light weight, non-porous and they require only little storage space

Tools and implements for nursery work

Rose can/ water can	This is used for watering the nursery. Fine spray of water should be used for watering nursery of small sized seeds
Digging fork	This has prongs of 20 cm long fitted to a wooden handle. This is used for uprooting plants, rooted cuttings, harvesting of tubers etc., without damaging the root system or tubers.
Shovel	This is a curved steel plate attached to a wooden handle and used for transferring soil, manure etc.
Garden rake	This is used for leveling lands and collecting weeds. The rake consists of a number of nail like projections from a crow bar provided with long handle
Hand trowel	This is used as a small tool for making holes for planting seedlings and small plants. This is also useful for removing surface weeds in nursery beds
Secateur	This is used for cutting small shoots to regulate shoot growth in fruit trees
Budding or Grafting knife	This knife is used for budding and grafting. This has two blades in which one is with ivory edge used for lifting the bark in budding operation.

Use of Hormones

Several growth inducing hormones are now commercially available. These chemicals when applied scientifically produce remarkable results in terms of seed germination, brack seed dormancy, root formation or initiation and root growth in different nursery plants. Auxin, GA and cytokinine are known to increase seed germination and exist in roots where they participate directly in cell division and elongation. The auxins stimulate root development by inducing root initials that differentiate from cells of the young secondary pheloem, cambium and pith tissue. The plant hormones play following important role in nursery.

- Plant hormones (Auxin, GA, cytokinins) can be used to enhance seed germination.
- Growth of rootstock improved by plant hormones.
- Can be used in root initian of cuttings and layering/stooling
- Differentiation of root/shoot in micro propagation
- Enhance of root growth by plant hormones.

Uses in seed germination: Poor seed germination in some of the fruits and vegetables crop are major hindrance for propagation. Now-a-days seed socking of fruits and vegetables with IAA, IBA, NPA, 2,4-D, in range of 10 to 200 ppm are most widely used for increase germination per cent and produce healthy seedling. Gibberellic acid is widely used in the nursery, which speeds up the germination of fruits and vegetable seeds. Breaking of seed rest is of great significance to over come dormancy. The plant hormones can remove the seed dormancy caused by inhibitors. Gibberellic acid is now invariable used for accelerated seed germination in citrus (trifoliate orange, sweet orange, cleopatra mandarin and acid lime), cherry, grapes, ber, lasoda, castered apple, rayan, apple and peach.

Use in plant tissue culture: Plant growth regulators are the key factors in multiplication of any plant part and genetic manipulation through tissue culture. In banana, low concentration of IAA and high level of benzyl adenine (BA) essential for rapid growth of explants. In grape benzyl adenine and NAA are essential for establishment of explants, while IBA helps in rapid multiplication. In mango, somatic embryogenesis from nuclear explants occurs in the presence of 2, 4-D at low concentrations.

Use in rooting media: Plant growth regulators may be employed to increase the percent success by hastening root initiation and increasing the number of developed roots per cutting. It is realized that though rooting is made easy by the aid of growth regulators the plants do not have better vigour. Attempts have been made for better rooting by treatment with sugar, iron, potassium permanganate, carbon monoxide, acetylene, ethylene, etc. Rooting of majority or plant species may be stimulated by indole acetic acid, naphthalene acetic acid and indole butyric acid. The last is supposed to be better due to its non-toxic effect over wide range of concentration. The application of 100-500 ppm IBA for soft wood, 500-1500 ppm for semi hard wood and 2000-5000 ppm for hard wood is common for vegetative propogation of fruit crops. Apart from auxins, a large number of physiologically and chemically unrelated compounds such as phenols and allied simple aromatic compounds, glycosides, growth inhibiting, growth retarding and ethylene producing chemicals significantly influence the rooting of cuttings.

Application Methods of Plant Hormones in Nursery

- Dry application: the freshly cut lower end of the rooting material should be evenly sprinkled with the commercial powder (Rootex) preparation to get good results.
- Soaking method: Solution of growth regulators of the concentrations of 25-2500 ppm are used to promote rooting. More concentrated solution is used for such species which do not root easily whereas weaker concentrations are used for relatively better rooting species. The lower portion of the cutting is dipped for about 24 hours to a length of 2-3 cm of the basal end just before planting in the rooting medium.
- Dip method: this method consists in increasing the concentration of growth regulating substances for a short period of say about four to five seconds. The concentration of the freshly made solution used in such cases is kept at 500-

10,000 ppm depending on the need of the plants.

Disposal and Packing of Produce

Disposal of produce also requires skillful management, otherwise whole efforts are futile. It requires timely action so that users may get quality planting materials at proper time and at the same time, producer may gain the desired returns. Sometimes, planting materials remain unsold if they are produced delayed in the season. Further, keeping them for longer period in the nursery may not only involve additional expenditure but the quality of planting materials reduced due to coiling of the roots and over growth of plants. The nursery manager must contact the users and even he should do advance booking. The planting materials raised either in beds or in polybags should be lifted properly unwanted foliage should be removed to minimize transpiration losses. They should be graded based on size, age, vigour etc. and packed in such a way that roots remain moist. For wrapping of the planting stocks, grasses, straw, perforated polythene, gunny bags, moss grass etc.are generally used as packing materials. Now a day's wooden or perforated paper boxes and plastic try are used for long distance transit. There should be minimum time gap between lifting of the stock from the nursery and planting at desired site. To avoid exposure to sun, the planting materials should be lifted from the nursery in the evening and planted on the next day.

Cost-Efficiency of Nursery

Cost of Production

There must be a sound relationship between production cost and sales price. Both monetary figures must be constantly monitored and maintained at an acceptable level to ensure the business remains profitable. It is important to remember that a good nursery manager must not only be capable of producing and /or maintaining good quality plants and allied products, they must also be able to sell them profitable.

Cost of production + Profit = Sales price

If the cost of production becomes too high, profit will decrease. In such a situation the sales price must be increased, otherwise the profit figure can become negative amount (i.e. the nursery might be losing money rather than making it). In order to control cost effectiveness; a manager must understand and control all the factors, which influence the cost of production. They may also need to adjust the sales price figure in order to maintain an acceptable profit.

Cost of production is influenced by the following factors:

- Cost of site- lease/rent value or purchase cost
- Cost of site services- power, gas, water etc.
- Cost of materials- soil, pots, fertilizers etc.
- Cost of unsold plants- a certain proportion of stock may be lost, may die, or may just become unsaleable. Some nurseries budget as much as 20-30% of stock being thrown away.
- Labour cost- including your own time as well as employee hours, holiday pay, superannuation requirements, health cover etc.
- Advertising/promotion- printing, advertising in magazines etc.
- Selling costs- transportation, invoicing etc.
- Taxation- payroll tax, income tax etc.

Profit

This figure should be over and above money, which earn as wages. Profit should be normally being at least 15-20%. If the nursery is small, the profit margin per plant should be larger. If the nursery is large, the profit per plant can be kept lower. In that situation profit comes through quantity of sales. Profit may be increased by value addition of the plants like use decorative pots, quality media, produce bonsai, hanging baskets etc.

Sale Price

The figure for which a plant is sold can vary considerably. Retail price is normally about twice the wholesale price. This being the case, the wholesaler must sell in large numbers in order to maintain profitability. Wholesale price varies considerably. Generally speaking, the figure is largely affected by:

- The reliability of the nursery supply. If the wholesale is well established and known to be a reliable source of plants, they can demand a higher price.
- The quality of plants offered. Higher prices are paid for top-quality stock.

CHAPTER 23

Greenhouse Technology for Hi-Tech Horticulture

INTRODUCTION

The term green house refers to a structure covered with a transparent material for the purpose of admitting natural light for plant growth. The structure is, usually, heated artificially and differs from other growing structures, such as cold frames and hot beds, in that it is sufficiently high to permit a person to *work from within*.

A greenhouse is a framed or an inflated structure covered with a transparent or translucent material in which crops could be grown under the condition of at least partially controlled environment and which is large enough to permit persons to work within it to carry out cultural operations.

The European definition refers to a structure that receives little or no artificial heat. The term glass-house is used in Europe to refer to an *artificially heated structure*. Quite frequently, two or more greenhouses in one location are referred to as a *greenhouse range*. A building associated with the greenhouses that is used for storage or for operations in support of growing of plants but that is not itself used for growing plants is referred to as a head house or service building.

The future holds promise of new covering materials that will reduce the burden of heating and cooling and also of new frame designs that will be more economical.

ADVANTAGE OF GREENHOUSE

Greenhouse is useful for various advantages. Under greenhouse one can grow crops under controlled environment and throughout the year four to five crops can be grown due to the availability of required plant environmental conditions. This helps in increased crop productivity and superior quality of produce can be obtained. Gadgets for efficient use of various inputs like water, fertilizers, seeds and plant protection chemicals can be well maintained in a green house. Pests and disease can be effectively controlled in the growing enclosed area. Percentage of germination of seeds is high in greenhouses. The acclimatization of plantlets of tissue culture technique can be carried out in a greenhouse. Agricultural and horticultural crop production schedules can be planned to take advantage of the market needs. Different types of growing medium like peat mass, vermiculate, rice hull and compost that are used in intensive agriculture can be effectively utilized in the greenhouse. Export quality produce of international standard can be produce can be taken up utilizing the entrapped heat. When the crops are not grown, drying and related operations of the harvested produce can be taken up utilizing the entrapped heat. Greenhouses are suitable for automation of irrigation, application of other inputs and environmental control by using computers and artificial intelligence techniques. Self–employment for educated youth on farm can be increased.

Anthurium growing in a greenhouse

TYPES OF GREENHOUSES

Various designs of greenhouses are constructed according to the need. They may be A-shaped, Quonset, and gutter-connected types. The transparent coverings are as varied as the designs. Originally, glass was used, but now film plastics, fiberglass-reinforced plastic (FRP), acrylic panels, and polycarbonate panels are used.

Classification of Greenhouses

Greenhouses have been classified according to their applications. In general there is no single type greenhouse, which can be considered suitable for all kinds of crops and areas. Various types of greenhouse are designed to meet the specific needs.

Greenhouse Types Based on Shape

Greenhouses can be classified based on their shape or style, uniqueness of the cross section of the greenhouses, etc. The longitudinal section of the greenhouses approximately remains the same for all types of the greenhouse, therefore, this factor cannot be used for classification. The cross sections depict the width and height of the structure and the length is perpendicular to the plane of cross section. The cross section provides information on the overall shape of the structural aspects, e.g., truss or hoop, which will be repeated on very bay. Commonly followed types of greenhouse based on shape are lean-to, even span, uneven, span, ridge and furrow, saw tooth and Quonset.

Lean-to Type Green House

This type of design is used when a greenhouse is to be constructed against the side of an existing building, so that plants can make the best use of sunlight and owner can minimizes the requirement of roof supports. The roof of the building is extended with appropriate greenhouse covering material and the area is properly enclosed.

Even Span Type Greenhouse

These kinds of greenhouses are constructed in regions of India. Such small sized greenhouses are designed when owner is to use leveled ground. In this type greenhouse is constructed with the two roof slopes of equal pitch and width. These may be of several single and multiple span types. For single span types the span in general, varies from 5 to 9 m, whereas the length is around 24 m. The height varies from 2.5 to 4.3 m.

Uneven Span Type Greenhouse

This type of greenhouse is constructed on hilly terrain. The roofs of the greenhouse are of unequal width, so that owner can make use of the side slopes of hill. They are well adaptable to the hilly regions. These are not adaptable for automation.

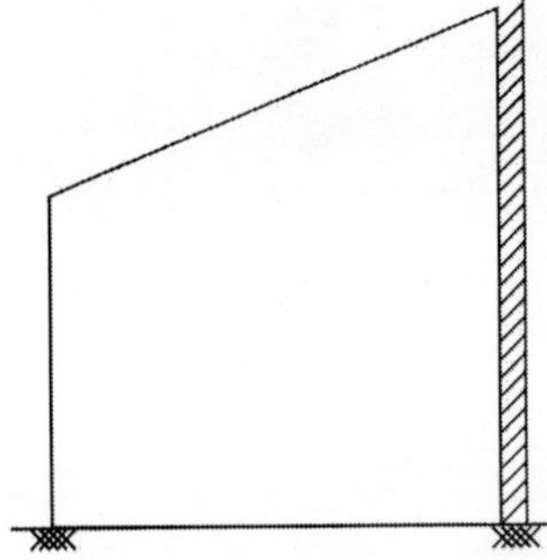

FIG. 23.1: Lean to type greenhouse.

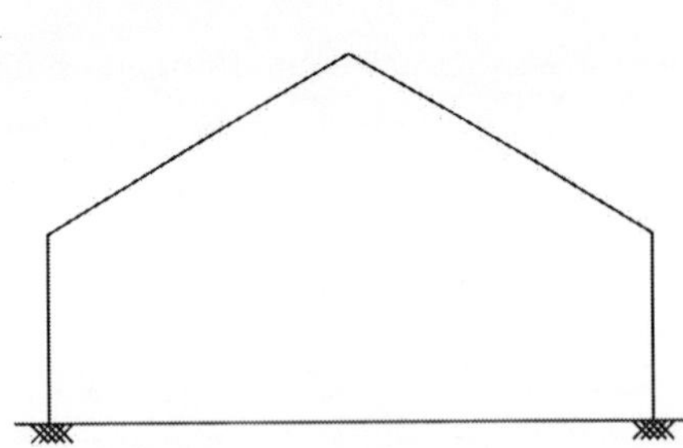

FIG. 23.2: Even span type greenhouse.

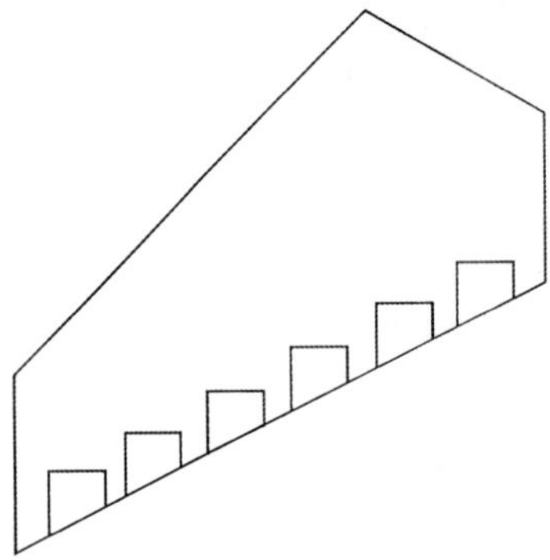

FIG. 23.3: Uneven span type greenhouse.

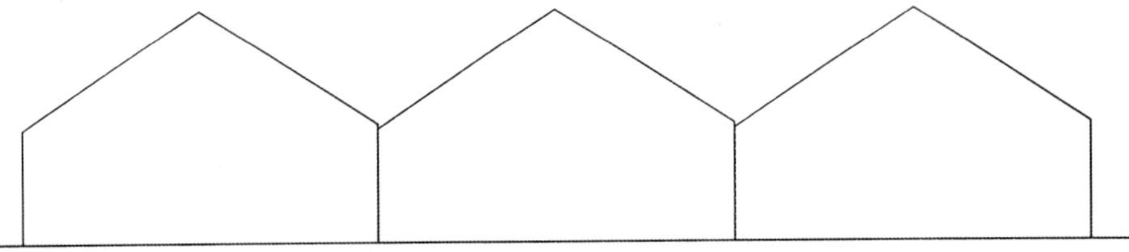

FIG. 23.4: Ridge and furrow type

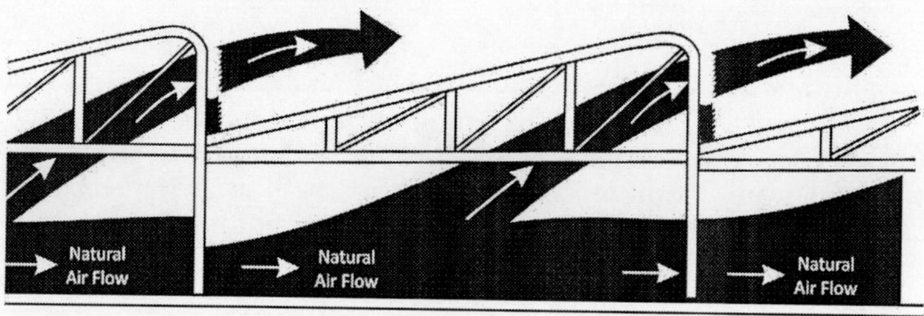

FIG. 23.5: Saw tooth type greenhouse

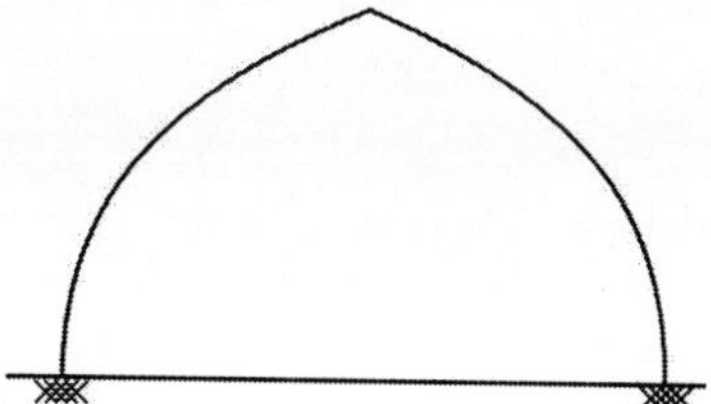

FIG. 23.6: Free standing type Quonset greenhouse

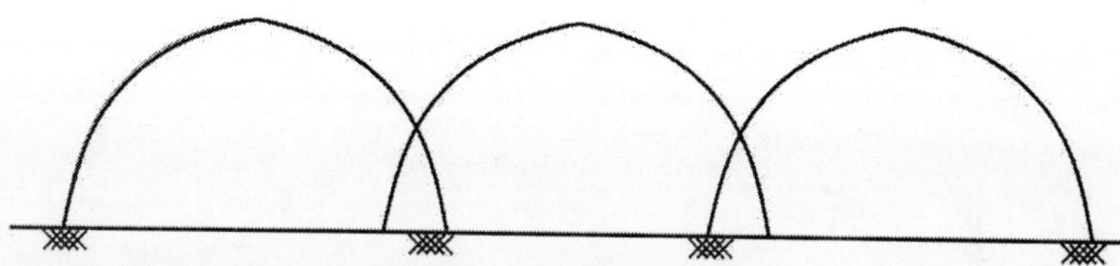

FIG. 23.7: Interlocking and furrow type Quonset greenhouse

Ridge and Furrow Type Greenhouse

These are 'A' designed greenhouses, which are connected to one another along the length of the eave. The eave serves as a furrow or gutter to carry rain and melted snow away the sidewall is eliminated between the greenhouses, which results in a structure with a single large interior. This design reduces labour cost of automation, improves personal management and reduces fuel consumption because it has less exposed wall area through which heat escapes. While construction it is necessary to know the snow loads, which will help to select the types of frames Ridge and furrow greenhouses are commonly used in northern countries of Europe and in Canada and are well suited to the Indian Conditions.

Saw Tooth Type Greenhouse

The structure is similar to ridge and furrow type greenhouses except that, there is provision for natural ventilation in this type. Specific natural ventilation flow path develops in a saw tooth type greenhouse.

Quonset Greenhouse

This kind of greenhouse is constructed by using pipe arches or trusses for support. The pipe purlins run along the length of the greenhouse. The greenhouse is covered by polyethylene sheet. It is typically less expensive than the gutter connected greenhouses and are useful when a small isolated cultural area is required. These houses are connected either in free standing style or arranged in an interlocking ridge and furrow. This kind of greenhouse provides sufficient area to plants to grow between the overlapping portions of adjacent houses. A single large cultural space thus exists for a set of houses in this type, an arrangement that is better adapted to the automation and movement of labourers.

Greenhouse Type based on Utility

Design of such greenhouses depends on the functions or utilities. The utilities include artificial cooling and heating of the greenhouse.

Greenhouse for Active Heating

In winter, air temperature inside greenhouse decreases during the night time, therefore, to avoid the cold bite to plants (freezing), some amount of artificial heat has to be supplied. To compensate the heat loss, various methods have been adopted to reduce that heat losses, viz. using double layer polyethylene, thermopane glasses (two layers of factory sealed glass with dead air space) or to use heating systems, such as unit heaters, central heat, radiant heat and solar heating system.

Greenhouses for Active Cooing

In summer for plant survival, it is necessary to reduce the temperatures of greenhouse for effective crop growth. Therefore, suitable modifications are made so that large volumes of cooled air is drawn into greenhouse. This type of greenhouse consists of either evaporative cooling pad with fan or fog cooling. This greenhouse is designed in such a way that it permits a roof opening of 40% and in some cases nearly 100%

Greenhouse Type based on Construction

The type of construction, structural and covering material used for greenhouse are important factors. Span of the house in turn dictates the selection of structural members and their construction. If the span of the greenhouse is higher we need stronger material for construction. For smaller spans, simpler design like goops can be followed. Therefore based on construction, greenhouses can be broadly classified as wooden framed, pipe framed and iruss framed structures.

Wooden Framed Structures

This kind of greenhouse is made with less than 6 m and only wooden framed structures are used. Side posts and columns are constructed of pinewood without the use of a truss. It has required strength and less expensive. Timber locally available, with goods strength, durability and machinability also can be used for the construction.

Pipe Framed Structures

This kind of greenhouse is made when the clear span is around 12 m. For this purpose pipes are used for the construction. In general the side posts, columns, cross ties and purlins are constructed using pipes. In this types also the trusses are not used. The pipe components are not interconnected but depend on the attachment to the sash bars for support.

Truss Framed Structures

This kind of greenhouse is made when the greenhouse span is greater than or equal to 15 m. For this purpose truss frames are used. Flat steel, tubular steel or angle iron are welded together to form a truss encompassing rafters, chords and struts. Struts are support members under tension. Angle iron purlins running throughout the length of greenhouse are bolted to each truss. When wide truss frame houses of 21.3 m or more are to be constructed, columns are used. Most of the glass houses are of truss frame type, as these frames are best suited for pre-fabrication.

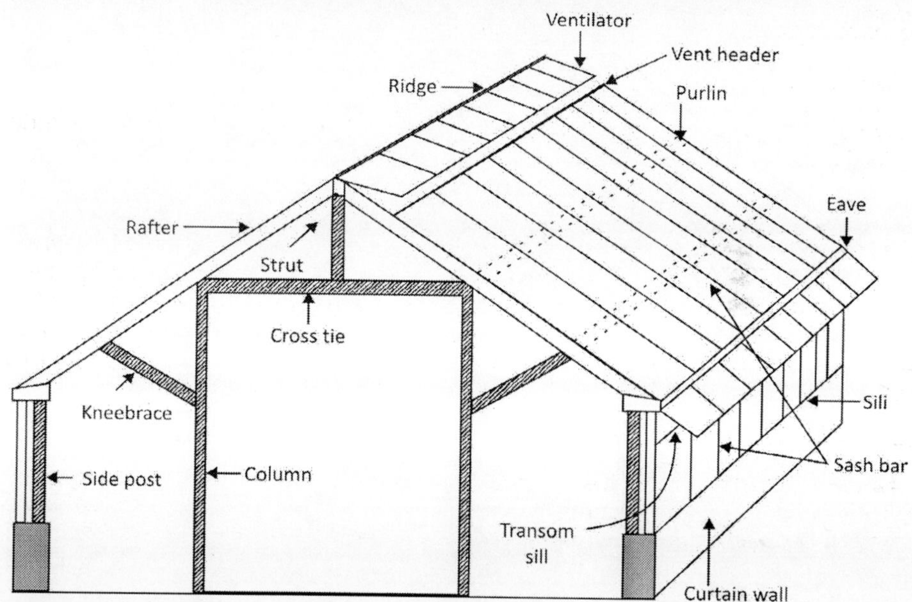

FIG. 23.8: Piped-based construction of greenhouse.

Greenhouse Type based on covering Materials

Greenhouse is to be covered with covering materials, which have direct influence on the greenhouse effect inside the structure. They alter the air temperature inside the house. The types of frames and method of fixing also varies with the covering material. Based on the type of covering materials the greenhouses are classified as glass, plastic film and rigid panel greenhouses.

Glass Greenhouses

Earlier only glass greenhouses covering with were constructed, because glass provides greater interior light intensity. These greenhouses have higher air infiltration rate which leads to lower interior humidity and better disease prevention. Lean to type, even span, ridge and furrow type of design are used for construction of glass green house.

Plastic Film Greenhouses

These greenhouses are covered with flexible plastic films (polyethylene, polyester and polyvinyl chloride). Plastic covering of greenhouse is popular because it is cheap and the cost of heating is less when compared to glass greenhouses, but such films have short life. For example, the best quality ultraviolet (UV) stabilized film can last for four years only. Quonset design as well as gutter-connected design is suitable for using covering material.

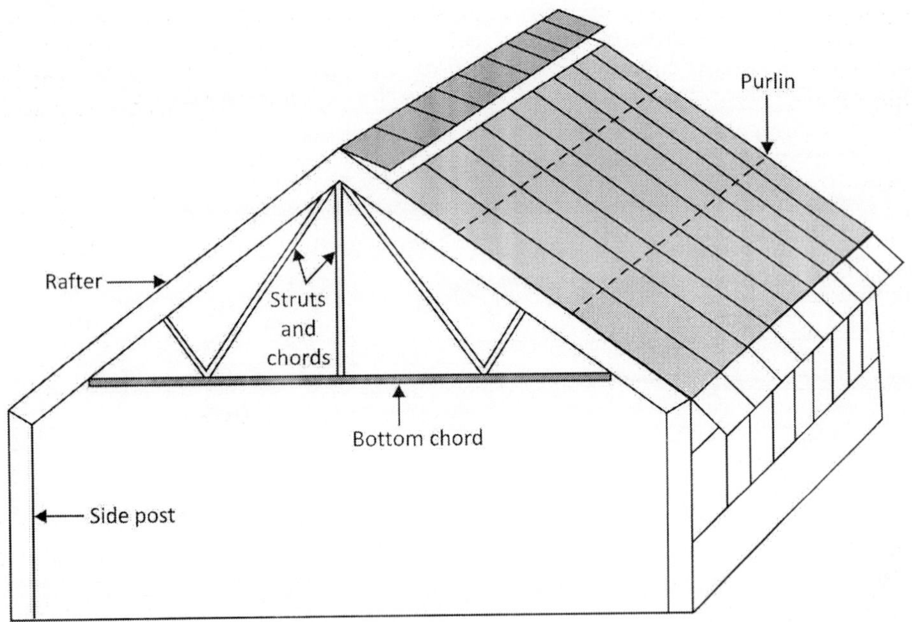

FIG. 23.9: Structure of truss framed greenhouse.

Rigid Panel Greenhouses

Polyvinyl chloride rigid panels, fiber glass-reinforced plastic, acrylic and polycarbonate rigid panels are employed as the covering material in this type of greenhouses. These panels can used in the Quonset type frames or ridge and furrow type frames. This material is more resistant to breakage and the light intensity is uniform throughout the greenhouse when compared to glass or plastic. High grade panels have long life even up to 20 years. The main disadvantage is that these panels tend to collect dust as well as to harbor algae, which results in darkening of the panels and subsequent reduction in the light transmission. There is significant danger of fire hazard.

GLASS GREENHOUSES

Glass greenhouses have been known to last 100 years. They existed prior to 1950. They are more expensive types today and need proper maintenance. Their average cost per year is still more than that of film plastic greenhouses because of fuel costs. Several styles of glass greenhouses are designed to meet specific needs. A lean-to design is used when a greenhouse is placed against the side of an existing building. It makes best use of sunlight and minimizes the requirements for roof supports. An even-span greenhouse is one in which the two roof slopes are of equal pitch and width. Glass greenhouse is seldom used today because they are not adaptable to automation. Individual greenhouses standing free of one another are well adapted to cold climates since snow easily slides from their roofs.

A ridge-and-furrow design refers to two or more. A frame greenhouse connected to one another along the length of the eave. The eave serves as a furrow or gutter to carry rain and melted snow away. The side wall is eliminated between greenhouses, which results in a structure with a single large interior. Consolidation of interior space reduces labor, lowers the cost of automation, improves personnel management, and reduces fuel consumption because there is less exposed wall area through which heat can escape. The load of the snow should be considered while preparing frame specifications of these greenhouses. Snow cannot slide off the roofs, as in the case of individual freestanding greenhouses, but must melt away. Heating pipes are usually located beneath the gutters for this purpose.

Three frame types have been used with various combinations. Wood frames were used for greenhouses under 20 feet (6 m) in width. Side posts and columns are constructed of wood without the use of a truss. Wider houses required sturdier frames. Pipe frames served well for greenhouses up to a width of about 40 feet (12 m). The side posts, columns, cross ties, and purlins were constructed from pipe. The pipe components are interconnected but the connections depend on the attachment and bars used for support.

Flat steel or tubular steel, or angle iron are welded together to form a truss encompassing the rafters, chords, and struts. Struts are support members under compression, while chords are support members under tension. Angle-iron purlins running the length of the greenhouse are bolted to each truss. The frame thus constructed can stand without support of sash bars. Columns are used only in very wide truss-frame houses of about 70 feet (21 m) and wider.

Truss-frame type of glass greenhouse is very common today. They are best suited to prefabrication and are more economical over the years. Automation has also fostered wider houses, which require the strength of truss frames.

The glass on the greenhouse is attached to sash (aluminum sash) bars. Aluminum sash bars can be made stronger than wooden ones, and thus wider panes of glass can be used with the aluminum bars. In earlier days, sash bars were made exclusively of wood, primarily cypress and redwood. Wood required periodic painting to protect it against rot. Ideally, exteriors were painted every two years and interiors every five to seven years.

All-metal greenhouses are very expensive as compare to the houses having wooden sash bars. However, all-metal greenhouses are cheaper to maintain, since they required no painting. To day, all glass greenhouse construction is of the metal type.

The structural members of the greenhouse cast shadows that reduce plant growth during the dark months of the year.

At present glass greenhouse construction are of two types: low profile or high profile.

The **low profile greenhouse** is most popular in the Netherlands. Eaves are approximately 10.5 feet (3.2 m) apart, and single panes of glass extend from eave to ridge. The lower profile slightly reduces exposed surface area, thereby reducing the heating cost. It is suggested, however, that these greenhouses are more expensive to cool in warm climates, because they need fans. Ventilator cooling during intermediate seasons is not as effective due to the lower height from ground to ventilator thus, fan cooling may be used more extensively.

High profile greenhouses are available with special sash bars that hold two or even three layers of glass which produce one or two dead-air spaces to cut heat loss.

FILM PLASTIC GREENHOUSES

Role

Flexible films of plastic including polyethylene, polyester, polyvinyl chloride, and polyvinyl fluoride have been used for greenhouse coverings. Polyethylene is principally used today for two reasons: (1) Film plastic can be used on permanent greenhouse frames at a considerable savings in the cost over a glass greenhouse. (2) The film plastic greenhouses are popular because the cost of heating them is approximately 40 per cent lower compared to single-layer glass or FRP greenhouses.

Polyethylene film is used as a greenhouse covering since 1930s and was pioneered about the middle of this century. The expanded use of polyethylene for greenhouses has been very rapid and continues to be so. Polyethylene film greenhouses constitute the largest portion of new greenhouse construction in India.

Disadvantages of Film Plastic

1. These covering materials are short-lived compared to glass and FRP.
2. Though highest quality of film plastic is ultraviolet (UV) light-resistant, 6 mil thick (6 one-thousandths of an inch, 0.15 mm) polyethylene films can last for four years. However, UV light from the sun damage, the plastic and it become brittle and dark in color and ultimately, it breaks.

Advantages of Film Plastic

1. The time required to cover a 30 ft by 100 ft (9.1 m by 30 m) Quonset-design house is minimal (about either labor hours).
2. It reduces costs of management and the use of equipment.
3. The savings in fuel as well as the lower initial purchase price of the film plastic is also an advantageous than a glass greenhouse.

Types of Film Plastic

Polyethylene is the best choice of film plastic for greenhouses. Such greenhouses have two layers. The **outer layer** is customarily 6 mil thick, while the **inner layer** may be 4 or 6 mils thick (0.1 or 0.15 mm). Polyethylene are used for covering have a UV inhibitor in it. This reduces the damage. UV inhibitor helps in longer use (life expectancy is three years) otherwise it lasts for only one heating season. Recently, the common, four-year product has been developed.

A polyethylene covering is colder in the winter than the air inside the greenhouse. When warm, moist greenhouse air contacts the cold polyethylene, it cools. As a result, water vapor condenses on the polyethylene surface. The surface is repellent to water; thus, the water forms into beads. With time, the water beads slide downward into one another and form larger beads that drop off to the plants below. The wet foliage fosters disease development, while the constantly wetted soil becomes water logged and oxygen deficient. If the plastic surface were not as repellent to water, the condensing water would form smaller droplets that would quickly flow over the surface to the ground. Spraying with a surfactant (detergent) will give this beneficial effect, but the surfactant washes off quickly. Polyethylene as well as rigid FRP, acrylic, and polycarbonate panels are now available with an anti-fog surfactant built into the surface of the film or panel.

Plants, radiate infrared (radiant) energy creates a large source of heat loss in greenhouses. Polyethylene is a poor barrier to radiant heat. Polyethylene with infrared (IR) blocking chemicals formulated into it during manufacture can stop approximately half of the radiant heat loss. On cold clear nights, as much as 25 per cent of the total heat loss of a greenhouse can be prevented in this way.

Electro magnetic radiations that are utilized in photosynthesis is termed photosynthetically active radiation (PAR). It comprises the wavelengths from 400 nm to 700 nm. Transmission of PAR through polyethylene can vary with the brand of and the chemical additives in polyethylene UV stabilized polyethylene, on average, transmits about 87 per cent of PAR. IR absorbing polyethylene, which reduces radiant heat loss, transmits about 82 per cent of PAR.

Two layers of transmit less light than one layer of glass. It is questionable whether there is less light in a polyethylene greenhouse. In polyethylene greenhouse, light transmissivity for the entire greenhouse varied from 50 per cent to 60 per cent, depending on season, orientation of the greenhouse and altitude/latitude position. The angle of incidence of light develops difference between 60 per cent and 86 per cent, and it many be due to the, sash bars, and structural members. Polyethylene greenhouses have less structural material and no sash bars.

Recently, a technology has been developed with respect to polyethylene production and their use in co-extrusion process. Three liquid resins are extruded simultaneously such that a single layer of film can have three different chemistries across it. In the present triextruded films, the inner core contains the anti-fog surfactant. This chemical is not entirely compatible with polyethylene. The repelling forces cause it to slowly bleed out of the core through the overlying zones of polyethylene that do not contain it. In this way, it can last a few years.

The IR-blocking chemical is also placed in the core. Typically, this chemical weakens polyethylene. If polyethylene containing it is stretched, it can turn cloudy. By confining the IR block to the core, a zone of unusually strong and clear polyethylene can be developed over each side of the core, which alleviates much of the clouding problem. It also gives the film an overall strength greater than that of standard three-year film of the same thickness. Some tri-extruded film is warranted for four years when it is 6 mils thick (0.15 mm) and is on either the outer or the inner covering of the greenhouse. Four mil (10 mm) can also be warranted for four years if it is used under a 6 mil, UV-inhibited film. The 4-mil tri-extruded film is sufficiently strong to equal the life of the current 6-mil, three-year film.

Average prices per square foot for 6 mil, three-year polyethylene film vary per square foot of ground covered. The lower price applies to greenhouses covered with a tube of plastic since in one step both layers of plastic can be applied. Gutter-connected greenhouse designs also frequently contribute to the lower price because less surface area per unit of ground area is exposed as compared to Quonset greenhouses. The use of channel locks for attaching the plastic as opposed to staples and batten strips can further reduce the price.

Vinyl Ultraviolet light resistant vinyl (polyvinyl chloride) films of 8 and 12 mil (0.20 and 0.30 mm) thickness are guaranteed for four and five years, respectively. This guarantee was a decided advantage years ago when polyethylene lasted for only one or two years. With the recent advent of four-year polyethylene, the advantage is nearly gone. The coast of 12 mil vinyl is 21.5 c/ft^2, which is 3 times that of 6 mil (0.15 mm) polyethylenes. Although vinyl film is produced in rolls up to 50 inches (1.27 m) wide, any width can be purchased since the supplier can seal strips of vinyl together. The vinyl films tend to hold a static electrical charge, which attracts and holds dust. This, in turn, reduces light transmittance until the dust is washed off. Vinyl films are not used to a large extent in America.

Polyester Mylar brand polyester film for a time offered the strong advantage of durability. Films of 5 mil (0.13 mm) thickness are used for roofs and can last for four years, while 3 mil (0.08 mm) films are used on vertical walls and can last for seven years. The cost of Mylar is higher than that of polyethylene. Other advantages included a level of light transmittance equal to that of glass and freedom from static electrical charge, which collects dust. Polyester is still used frequently, however, in heat retention curtains because of its high capacity to block radiant energy.

Polyvinyl Fluoride (PVF) is the most recent category of greenhouse film plastic covering and is available as Tedlar. This film acts as a protective covering on FRP panels. The anticipated life expectancy is 10 years or longer. The light transmission of 4-mil (0.10 mm) PVF is 92 per cent and is greater than that of float glass customarily used on greenhouses. PVF is available in tubes comprised of a 4-mil plus a 2 mil (0.10 mm + 0.05 mm) layer or a 3 mil plus a 2 mil (0.08 mm + 0.05 mm) layer. The 3 + 2 mil combination is used in general situation, while the 4 + 2 mil combination is used under conditions of high light intensity or wind. Four mil PVF has 4 times the tensile strength of 6 mil (0.15 mm) polyethylene. Tubes of PVF are available in widths of 8.5, 10, and 10.5 feet (2.6, 3.0, and 3.2 m) and in lengths up to 220 feet (67 m). Modest savings could be realized with PVF because of elimination of the need to re-cover 2.33 times over the 10 years and because of greater growth under higher light conditions. On the basis of a square foot of greenhouse surface area (not floor area), recovering a double-layer polyethylene greenhouse carries can average cost of about Rs. 4.50 for labor plus Rs. 6.00 for plastic. Thus, the total cost of polyethylene for the 10 years would be the initial purchase of material plus 2.33 re-coverings for a total cost of Rs. 35/ft^2. A major drawback to PVF use in its narrow widths and the need to use attachment rails to splice pieces together on the greenhouse.

PVF film can be preshrunk onto frames to form panels ideal for re-glazing glass or FRP greenhouses. The PVF film is wrapped over the frame to form two layers and is attached with tape. Exposure to a temperature of 250-275°F (121-135°C) will cause shrinkage and pull the surface taut in 5-10 minutes.

Earlier, use polyethylene, in horticultural scene, was as a temporary structure. Now it is of common use. A single layer of film plastic was generally used but it increases fuel costs. Therefore, double coverings of film plastic were used to reduce one-third of the fuel cost. The width of the dead-air space between two plastic films became a consideration. Ideally, the dead-air space should be 0.5-4 inches (1.25-10 cm) thick. When it exceeds 4 inches (10 cm), air currents can become established inside and reduce the insulating property of this space. Warm air immediately above the inner covering rises up into contact with the outer covering, and here it gives up heat. As it cools, the air becomes heavy and drops back to the inner covering to pick up more heat. This loss does not become very significant until a space of 18 inches (46 cm) is reached. Below 0.5 inch (1.25 cm), the insulating property again diminishes, and when the two layers touch, the insulation value of plastic to maneuver it around the columns. The holes were generally not sealed and thus left avenues of entry for warm air into the dead-air space, which further reduced the insulating property. A frame and scissors-truss designs necessitated dead-air spaces several feet thick. Greenhouses up to 30 feet (9.1 m) wide with a frame thickness of 4.5 inches (11 cm) could then be constructed without columns.

The greenhouses are constructed of short-life wood, which requires frequent painting to prevent rotting. White paint is used to increase interior light intensity. A mercury-base paint should be avoided. Mercury will volatilize from the paint for a considerable length of time and thus cause damage to the crop. Paints sold specifically as greenhouse paints are safe; however, other paints may be used as long as those with a mercury base are avoided. If mercury-base paint is used by mistake, injury can be avoided by painting over it with a weight mixture of 5 parts lime-sulfur fungicide and 10 parts wheat flour in 100 parts water.

Posts and other wood in contact with the ground should be treated with a wood preservative. Treated wood may be purchased for this purpose, or the wood may be treated at the time of use. Several treatments are available, but not all are safe. Pentachlorophenol and creosote should not be used. Creosote in contact with roots and foliage can burn them. Pentachlorophenol produces fumes that can last for more than a year and that are toxic to plants. Entire crops can be killed by moving them into a new house with treated posts. A single treated board can cause abnormal growth throughout the house. A very suitable wood preservative is copper naphthenate, which is sold under several trade names. Generally used as a 2 per cent solution of copper naphthenate, it can be sprayed, dipped, or applied with a brush. It is an excellent preservative for frame members as well as for wooden benches and flats Lumber pressure-treated with Wolman Salts is safe for greenhouse use.

Presently, the price of wood is quite high therefore; the price is an added burden. The film plastic greenhouse designs were mainly of following two styles:

Quonset Style Greenhouse

It is less expensive and can be purchased prefabricated or can be fabricated on the site. The trusses are constructed from water pipe that is bent to fit a 180° are modified for some-what more vertical sides. *n* designed greenhouses 20 feet (6.1 m) wide, 0.75 inch pipe is used; 1 inch pipe is used for a 30 foot (9.1 m) greenhouse width. An aluminum electrical conduit should not be used since it does not have sufficient strength to support a snow load. Slightly larger pipe is driven into the ground into which the pipe arches are inserted for support. A 2 □ 8 inch (5 cm by 20 cm) wooden plank is attached to the base of the pipe arches such that it runs along the ground partially buried. This provides a basal point of attachment for the film plastic. The pipe arches, or trusses, are supported by pipe purlins running the length of the house. Trusses are spaced 30-36 inches (75-90 cm) apart. The width of film plastic required to cover a Quonset greenhouse of given width can vary according to the height and shape of the trusses. A 20 foot (6.1 m) wide greenhouse generally requires a 32 foot (9.8 m) wide sheet of plastic. The covering width for a 30 foot (9.1 m) wide Quonset greenhouse varies greatly; the more common widths are 40 feet and 42 feet (12.2 and 12.8 m). These green houses are either constructed in a freestanding style or may be arranged in an interlocking ridge-and-furrow manner. In this latter case, the trusses overlap sufficiently to place a bed of plants between the overlapping portions of adjacent houses. A single large interior thus exists for a set of houses, an arrangement that is better adapted to the movement of labor and to automation.

Gutter Connected Greenhouse

This is the second currently popular film plastic greenhouse design. The gutters can be placed at greater heights than is possible in ridge-and-furrow Quonset ranges. This permits a roadway in either direction within the greenhouse to accommodate tractors as well as trucks. Gutters, depending upon the manufacturer, can occur at 12-40 foot (3.65-12.2 m) intervals. Columns can be placed in greenhouses with gutters spaced 12 feet (3.65 m) apart and the columns placed under each, every other, or every third gutter. A 36 foot (11 m) spacing between rows of columns, while more expensive, greatly enhances the ease with which the shading of plants with black cloth can be accomplished. Gutter-connected greenhouses greatly minimize the exposed surface area and, consequently, the heating cost. Only 20 feet (6.1 m) of film plastic is required to span a 17 foot (5.2 m) bay. Gutter-connected bays of 12,17,21,22, and 30 foot (3.7,5.2,6.4,6.7, and 9 m) widths can be covered by film plastic sheets 14, 20, 24, 35, and 36 feet (4.3, 6.1,7.3,10.7, and 11.0 m) wide. When additions are made, the film plastic can be removed from an existing side wall and the new houses connected at that point without any resulting discontinuity. In this way, a modest initial investment, inadaptable for automation, can be developed through expansions into a structure well suited to automation. The gutter-connected greenhouse brings us full circle to the category of permanent metal-frame greenhouses of which the glass greenhouse is a member. The major portion of new construction in America today, whether for temporary or permanent use, is film plastic.

Greenhouse with Double Layer Covering

All film plastic greenhouses make use of the air-inflated system. Two layers of film plastic, one applied directly on top of the other from the outside, are held apart by a cushion of air maintained at low positive pressure.

Single sheets of plastic, wide enough to span the entire truss from ground to ground, are rolled out the length of the greenhouse and are attached to the greenhouse along its length at the ground level on both sides of Quonset greenhouses and in the gutters of gutter-connected greenhouses. The ends of the sheet overlap the greenhouse ends by a few inches and is attached at that point. No attachment is made to the trusses.

Two sheets of plastic are attached to each end of the greenhouse as well. Sometimes, plastic is attached to Quonset greenhouses by placing a **batten strip** over it and nailing or stapling through it.

Some growers use thick plastic strips about 1 inch (2.5 cm) of wood strips are used. Metal or plastic channel locks are also used. The layers of film plastic are laid over the channel, and then a metal rod or plastic bead is placed over the film plastic and pushed into the channel, locking the plastic in place. The outer layer of plastic should be 6 mils (0.152 mm) thick, while the inner needs to be only 4 mils (0.102 mm) thick because of the lower UV light level. A polyethylene tube, available in widths up to 25 feet (7.6 m), is often used to cover narrower greenhouses. In this way, both layers are applied in one step, thus significantly reducing the labor cost.

The **tension** of the plastic film is dependent on temperature since film plastics contract and expands to a considerable degree with temperature shifts. When it is applied on a cold day, the film should be pulled taut. On a warm day, with temperatures near 80° F (27° C), about 2-3, about 2-3 inches (5-8 cm), of slack should be left in the covering all the way along one side of a Quonset greenhouse 20 feet (6.1 m) wide to permit contraction over the truss when cold weather comes. If this slack is not allowed, the film will tear loose from the points of attachment when it contracts during cold weather. Conversely, if it is not pulled taut when it is applied on a cold day, excess slack will occur during warm weather, resulting in an excessive air space between the two layers.

A small squirrel cage fan is used inside the greenhouse to inflate the space between the two film plastic layers. Air is maintained between 0.2 inch and 0.3 inch (5.1 mm and 7.6 mm) of water-column pressure. The high pressure should not be maintained because the plastic will stretch. The fan should have an adjustable door on the air inlet for adjusting the pressure between the two films.

During a snowstorm when snow sticks to the roof, it may be necessary to turn off the fan. This will allow the two polyethylene layers to come together, thus eliminating the insulating effect of the dead-air space. More heat will escape through the covering to melt snow and clear it from the roof.

The fan is generally mounted on the end wall of the greenhouse. A hole is cut in the end wall adjacent to the fan so that air feeding the fan is drawn in from outside.

A flexible tube, such as that used for a clothes dryer, is installed between the fan and the inner layer of plastic to be inflated. A + shaped cut is made in the inner layer of plastic, and the tube is inserted through it. The four points of plastic resulting from the cut are pulled out over the tube and taped to it to make an airtight seal. Air is conducted from the fan to the inner space through this tube. This system is sufficient to inflate the entire roof of a quonset style greenhouse.

The pressure between the layers of plastic should be sufficient to hold the layers apart under conditions of wind and yet low enough to avoid tearing the sheets. A manometer can be purchased from greenhouse supply companies for measuring this pressure. The manometer is a simple device and can be easily fabricated by the grower as follows:

1. Bend a 2 foot (61 cm) long piece of clear plastic tube into the shape of a U and attach it to a board.
2. Make a + shaped cut in the inner layer of plastic and insert one end of the plastic tube.
3. Seal the film plastic to the tube with plastic tape.
4. Put about 8 inches (20 cm) of water in the tube such that it settles at the bottom. Leave both ends of the tube open.
5. Attach a ruler to the board behind or alongside the plastic tube.

The air-inflated system offers the easiest method for covering a greenhouse with two layers of film plastic. Plastic applied by techniques other than the air-inflated system is constantly chafed against the trusses by the lifting and dropping action of the wind. This reduces its life expectancy in many cases to one heating season.

RIGID PANEL GREENHOUSES

Polyvinyl Chloride

Polyvinyl chloride (PVC) is the rigid panel sheet. It is an inexpensive covering (about 40 per cent of the cost of long-lasting FRP) and life expectancy of five years (polyethylene lasted one year). The total cost of PVC panels is 4-5 times more than that of polyethylene film and because it requires much more time to install. Rigid PVC sheets turn dark and become brittle due to UV radiations. The sheet reduces light transmission. Rigid PVC may be purchased in corrugated panels and are available in various colors however, clear panels were used for general greenhouse culture.

Fiberglass Reinforced Plastic

Role Fiberglass-reinforced plastic (FRP) can also be used for greenhouse covering. Like PVC, corrugated panels were used because of their greater strength. Flat panels are occasionally used on the end and side walls where the load is not as great. Various sizes of panels with variety of colors are available. The panels are flexible enough to conform to the shape of Quonset greenhouses, which makes FRP a very versatile covering material. FRP can be applied to the inexpensive frames of film plastic greenhouses or to the more elaborate frames of glass-type greenhouses. In the former case, the price of the FRP greenhouse lies between that of a film plastic greenhouse and that of a glass greenhouse, but the cost is offset by elimination of the need for replacement of film plastic. In the latter case, the FRP greenhouse costs about the same as the glass greenhouse.

Advantages of FRP

1. FRP is more resistant to breakage by factors such as hail or vandals.
2. Sunlight passing through FRP is scattered by the fibers in the panels, with the result that light intensity is uniform throughout the greenhouse by comparison with a glass covering.
3. Plants on the north sides of beds, and particularly in the north beds, as a whole, grow much better.

Disadvantages of FRP

1. FRP panels have the acrylic surface and are subject to etching and pitting by dust abrasion and chemical pollution.
2. Glass fibers are exposed and subject to fraying, and they begin to collect dust as well as harbor algae. The resultant effect is a darkening of the panels and a subsequent reduction in light transmission.
3. It needs regular scrubbing the FRP surface and cleaning should be done with a stiff brush or steel wool and then painting on a new surface of acrylic resin.
4. The material is inexpensive, but the labor is extensive.
5. It needs refinishing and depends on the grade of FRP.

Light Transmission

The total quantity of light transmitted through clear FRP is roughly equivalent to that transmitted through glass but diminishes in relation to its color. For greenhouse crops in general, only the clear FRP permits a satisfactory level of light transmission (88-90 per cent). Colored FRP has found a limited use in greenhouses used for growing some house plants that require low light intensity and in display greenhouses used for holding plants during the sales period.

Heat Transmission

FRP has the distinct advantage over glass of being easier to cool. The length of time that the cooling fans operated in each greenhouse in the month of June is more than in the month of December. It varies from month to month. The winter heat requirement of corrugated FRP greenhouses is about equivalent to that of structurally tight glass greenhouses.

Construction of FRP Greenhouses

It requires fewer structural members than glass greenhouses. The sash bars are not needed. The construction labor input is accordingly lower for an FRP greenhouse. Both high quality, 5 ounce FRP and double-strength glass cost about Rs. 45.00/ft^2. FRP panels are 50.5-52.6 inches (1.28-1.34 m) wide but, with overlap, have an effective covering width of only 48 inches (1.22 m). The thickness of FRP is measured in terms of weight per square foot. Trusses are spaced 8-10 feet (2.4-3.0 m) apart and purlins 4 feet (1.2 m) apart.

Construction of the greenhouse must be as airtight. Corrugated plastic closures are available for insertion between the FRP panel and frame components such as the eave and the sill to seal off outer air. Flashing is used at the ridge to cover the exposed ends of the FRP panels for the purpose of preventing water entry. The flashing can be constructed from aluminum or corrugated FRP. The FRP panels are attached to the purlins by aluminum screw nails or by aluminum wood screws. These nails and screws have a rubber washer immediately beneath the head to seal the hole made by the shaft.

When condensation flows along the inner surface of FRP, it does so along the corrugation valleys. If the FRP panels are attached directly to the purlins, the corrugation valleys are in contact with the purlins. Condensation, upon reaching this point, flows onto the purlin and drips from its lower edge, thus causing harm to plants beneath. The FRP panel must be elevated away from the purlin. Metal U-shaped supports are placed between the purlin and the corrugation ridges of the FRP panel. The nail or screw attaching the panel to the purlin passes through the support.

Fire Hazard

It is advisable to insure the greenhouse structures. Insurance rates are assessed according to the risk involved, which is greater for FRP greenhouses. Fire is the one of the causes of destruction. The glass fibers themselves do not burn, but the polyester and acrylic resins binding them together do. Fire usually, originates from a faulty electrical wire beneath a sheet or black shade cloth spread to the FRP covering of a ridge-and-furrow range on a windy night. Fire-retardant FRP panels may be used. These panels offer no support.

Acrylic and Polycarbonate

Acrylic and polycarbonate, double-layer, rigid panels are also being used since last 15 years for greenhouse construction. A number of research institutions have installed these panels. The heaviest commercial use thus far has been for glazing side and end walls on film plastic greenhouses and for retro-fitting old glass and FRP greenhouses. A modest number of new buildings have been completely glazed with these panels. While the acrylic panels are highly flammable, the polycarbonate is not flammable.

Acrylic Panels

Acrylic panels are available in thickness of 16 mm (0.63 in.) and 8 mm (0.3 in.). The thicker panels cannot be bent, but the thinner panels can be bent to fit curved-roof greenhouses. Panels are available with a coating to prevent condensation drip. The two acrylic layers of these panels are held apart by ribs spaced approximately 0.6-0.9 inch (16-24 mm) apart. The panels have a width of 47.25 inches (120 cm) and come in lengths up to 39 feet (11.9 m). The effective covering width of the panel is 48 inches (122 cm) since the metal support member takes up space between panels.

Polycarbonate Panels

They have thickness of 4, 6, 8, 10, and 16 mm (0.16, 0.24, 0.31, 0.39, and 0.63 inch). The thinner panels can be bent to fit curved-roof greenhouses, while the thicker ones cannot. Their skin thickness and rib dimensions are similar to those of acrylic panels. Panels are available with a coating to prevent condensation drip. Panels are available with an acrylic coating for extra protection from UV light. Panels are available in widths from 4 feet to 8 feet (122 cm to 244 cm). Lengths are available up to 32 feet (9.75 m).

GREENHOUSE CONSTRUCTION: RELATED ASPECTS

There are multifold considerations in selecting the site for the greenhouse. The geographical location may be chosen because of the favorable effects of that climate for plant growth or proximity to the market. Some other business considerations are availability and price of fuel, utilities, water, and labor. Sometimes considerations other than strictly business, like research ones are also important, because it influences the decision for site selection. Some of these are the suitability of living conditions, schools, places of worship, stores, cultural and athletic events, the owner may want to retain these, with business adjusted in the best possible way. When the geographical location has been selected, the items to consider in addition to the above are topography of the land price, transportation, and present and future effects of adjacent properties. A few years ago one of the requirements for location of a greenhouse site was proximity to railroad service so that a spur could be constructed and coal brought in by rail. That became of no importance when oil or gas as fuels replaced coal. However, as fuel shortages develop there may be a return to the use of coal. Regardless of the kind of fuel, it must be determined that for the present times and near future the greenhouse site is easily accessible for fuel delivery. There have been some investigations into the use of heat generated as a by-product by electricity-generating plants. When this is done the greenhouse site must necessarily be adjacent to the generating facility.

Location

The first consideration in establishing a greenhouse range is that of location. Several factors to be considered as follows:

Room for Expansion

A parcel of land larger than the immediate needs should be acquired. The ultimate size of the range should be predicted. Area should then be added to this predicted figure to accommodate service buildings, storage, and access drives. (Doubling the area covered by greenhouses would constitute a bare minimum.) Finally, extra space should be allotted to cover unforeseen needs. For instance, it may become necessary to engage in stockpiling of supplies as fostered by short. ages of materials, or the future may call for holding ponds for water effluent from the range in order to reduce the nutrient content before releasing it into streams or the groundwater table.

The floor area of service buildings required for small firms is equal to about 13 per cent of the greenhouse floor area. This requirement diminishes with increasing firm size to an area equal to 7.5 per cent of the growing area for large firms with 400,000 square feet (ft^2) (37,000 m^2) of area. On the average, service buildings are equal to 10 per cent of the growing area.

Topography of the Land

It determines the amount of work and expense that will be required to provide good surface drainage and efficient transportation. Water drainage problems are compounded when any buildings are constructed because the area covered contributes to the surface drainage water that has to be handled in adjacent areas. Before structure construction starts, the land must be worked so that there will be adequate water drainage away from the site after the present construction or any future construction.

The building site should be as level as possible to reduce the cost of grading. A level site permits the construction of large greenhouse blocks that can be easily automated. The site should be well drained. Because of the extensive use of water in greenhouse operations, providing a drainage system is always advisable. Where drainage is a problem, it is wise to install tiles below the surface before constructing the greenhouses. It is also advisable to select a site with a natural windbreak, such as a tree line or hill, on the north and northwest sides. In regions where snow is expected, trees should be 100 feet (30 m) away in order to keep drifts back from the greenhouses. To prevent shadows on the crop, trees located on the east, south, or west sides should be set back a distance of 2.5 times their height.

Land use Prediction

Laws of local government are subject to change. They affect the planning and development pressures of the greenhouse industry. Therefore, one should know them in detail of past development and local laws of the area/location in question for future direction and assessment. Some local governments classify greenhouses as agricultural businesses to protect them from prohibitive property taxation due to zoning shifts. Others, in order to change the occupants within a zone, have denied expansion permits to floral production businesses.

Climate

Worldwide geographical shifts affect climatic conditions. *The greenhouse site should be selected with specific crops in mind.* The construction depends on the local climatic conditions (intensity of sun radiations, humidity, rainfall, temperature, wind velocity, etc.) and facilities available in the area.

Areas where there is frequent fog or inclement weather or shadows from the north slope of tall mountains are poor for crops in general. The better light intensity of higher altitudes is particularly advantageous for carnation and rose crops but have little benefit for crops with a requirement for low light intensity such as African violet, begonia, gloxinia, and most green plants.

Labor Supply

Present and future labor needs should be assessed and should be in accord with the labor supply of the area. Procurement of a labor supply has been a *perennial problem* in the greenhouse/horticulture industry. While the solution has appeared to rest in locating close to an urban area, this does bring on a problem of wage level. Traditionally, greenhouse wages have been low, which has given the labor recruitment advantage to the more technologically advanced industries. The solution appears to lie in meeting the competition directly through higher wages. Higher wages can be compensated through automation, which reduces the number of employees but increases the productivity of each.

Accessibility of the Site

This is an important aspect for establishment of a greenhouse. If the proposed site is hilly, it should be anticipated that considerable land preparation will be required. Access roads to the property will need to be large enough and surfaced to handle large motor freight equipment. The type of neighboring properties may have varied effects on the greenhouse business. If it is an industrial area-or becomes one-there could be damage to crops from some kinds of air pollution. Some federal or state agencies should be able to evaluate current situations. The problems that might develop in the future are a matter of estimation. If adjacent land becomes developed for residential use, new zoning restrictions may affect operation, maintenance, or additions to the greenhouse property. Changes in land valuation affect the cost of doing business. Greenhouses that were built outside of cities become surrounded by urban development with increased pressure of urbanization may develop problems in future.

Site location is often the deciding factor in the type of fuel used. In some regions, natural gas is a cheaper source of energy than other fuels. Some greenhouse ranges are able to take advantage of this factor because their location is at a prohibitive distance from the gas line, while the competition located near the line can enjoy this advantage. In one situation, where a greenhouse range was built at a high altitude to take advantage of light conditions, the remoteness of the location necessitated the transfer of oil from large tank trucks to smaller trucks during delivery, thus raising the cost of the oil.

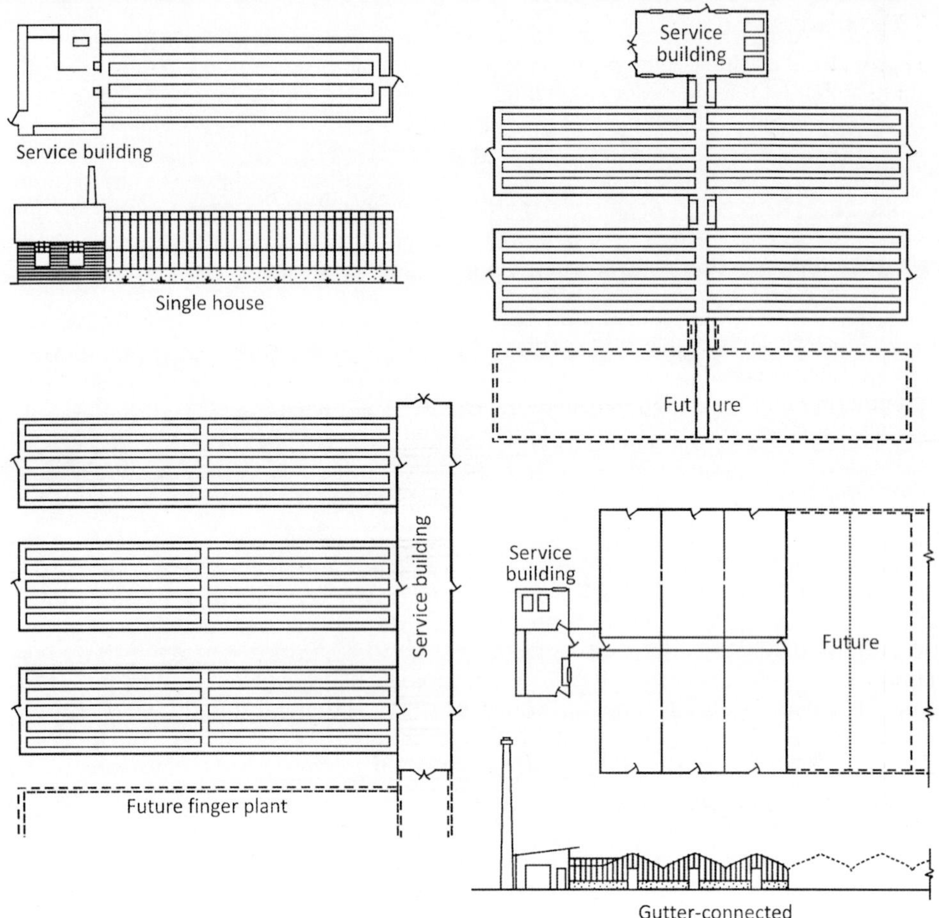

FIG. 23.10: Various arrangements of greenhouses are used depending on the crops to be grown and the land available. Separate greenhouses require more land than gutter-connected (ridge and furrow) ones, but they can provide better light, ventilation, and temperature control.

Available Water

Water is one of the most frequently overlooked commodities in the establishment of a greenhouse business. Before a site is purchased, the available water source should be tested for quality and quantity. There are several cases where businesses located in coastal and riverbed regions have been compelled to move to new locations to obtain water of suitable water quality. The cost of removing ions such as sodium, chloride, and bicarbonate can be prohibitive, but failure to do so results in plant injury. Water quantity is equally important since as much as 2 quarts of water can be applied to 1 ft^2 (20 l/m^2) of growing area in a single application. Well water is the desired source since municipal water is often too costly and may contain harmful fluoride. Pond or river water in subject to disease organisms and may require expensive chlorination.

Orientation

The greenhouse frame casts shadows. The magnitude of the shadows depends upon the angle of the sun and thus upon the season of the year. The effect can be most detrimental to growth in the winter when light is often limited. Therefore, the latitude and altitude of the area should be considered while selecting the site because it affects light shadows. In north, it should be built with the ridge running east to west so that low-angle light of the winter sun can enter along a side rather than from an end where it would be blocked by the frame trusses. The ridge of single greenhouses should be oriented from north to south since the angle of the sun is much higher. Ridge-and-furrow greenhouses (greenhouses connected to one another along their length) at all latitudes should be oriented north to south in order to compensate for a shadow that occurs from the north roof and gutter of each adjacent greenhouse. The north-south orientation permits this shadow to move across the floor during the day, whereas the east-west orientation does not.

Floor Plan for the Greenhouse

Floor plan of the greenhouse should be developed in such a way that one can able to expand it in future. This consideration best ensures that an efficient operation will always be possible. If there is a slope to the land, construction should begin at the midpoint (average elevation). In this way, the soil excavated will provide the fill needed during each expansion. In the end, there will be one final elevation for the entire greenhouse range. A plan should be prepared in such a way that it allows for the addition of service buildings and greenhouses without removal of previous buildings or the accumulation of multiple service buildings. The service building should be centrally located in a nearly square design, which minimizes distances that plants and materials need to be moved. The whole firm is one elevation and should be internally connected. This will help in internal transport system.

Doors

Doors between the service building and the greenhouse should be 10 feet (3 m) wide by 9 feet (2.7 m) high. The drive through the corridor and greenhouse should be 8 feet (2.4 m) wide to accommodate internal transport systems such as tractors. It would be well to have the greenhouse gutters 10 feet (3 m) or higher above the floor to accommodate automation and thermal blankets and still leave room for vehicles.

The greenhouse gutters need to be oriented north to south. A greenhouse block length (gutter length) of 144 feet (44 m) coincides well with the most common standard gutter unit length of 12 feet (3.7 m) and the maximum effective summer cooling distance of approximately 150 feet (46 m). The service building should have 16 foot (4.9 m) eaves to allow for doors 12 feet (3.7 m) wide by 14 feet (4.2 m) high needed to accommodate trailer trucks for both receipt of goods and shipping of plants. All plant loading should be carried out from a central point inside the service building.

Utilities

Electricity: Some sites will have better utility service than others. Greenhouse businesses cannot have interruption of electrical service for more than a few minutes without serious consequences. What has the experience been in the area in question-and what guarantees of continuous service will the utility make? Electrical shutdowns that are a nuisance for a machine shop could be a disaster for a greenhouse.

Telephone: The quality of telephone service varies from area to area. The effectiveness of this utility for the greenhouse site needs to be determined primarily because of its effect on the marketing of a perishable product.

Water

The quality and quantity of water that is available to the site must be determined. If municipal water is available, usually it is acceptable for irrigation and for use in boilers with some treatment. When water has to be supplied by well or pond on the site, there are some potential problems. Extensive treatment may be needed for boiler make-up water. It is possible, too, that sand or silt may be pumped along with the water from the well, and this may clog some equipment or cause abrasion and damage to other equipment.

Pond water can be expected to have qualities characteristic of the watershed, therefore, before the pond is constructed, the potential effects of watershed on quality of water should be analyzed. If the greenhouse drainage water contributes to the pond, it can be expected that the residues infestation of pest or pathogen may be distributed throughout the greenhouse by this means. It can be expected that some sand or silt may be pumped with the pond water with the same problems that were outlined for that occurrence in well water.

Pond Maintenance

Usually some pond maintenance is required. The most persistent problem may be the control of weeds. Weeds may be controlled with geese or some kinds of fish, or it is possible to use certain kinds of herbicides. But it must be determined that the herbicide will be completely harmless to all the greenhouse crops.

Air Bus Service

Air bus or other public transportation services are necessary at sit. In some instances this will be a consideration in obtaining workers.

Economy of Construction: Ridge and furrow type construction is usually more economical, and needs less land area. Some disadvantages are excessive shade from the gutters, lack of temperature control of individual houses, and no side ventilation. If forced-air ventilation is used, the lack of side ventilators may be no problem. The houses should be arranged in such a manner so that labor is conserved and the crop can be handled more effectively. The greenhouses should be constructed in such a way that flowers can be cut and moved to the grading area, and pot plants moved from area to area or to the shipping room very easily.

FRAMEWORK FOR VARIOUS TYPES OF GREENHOUSE

The framework of the greenhouse consists of side posts-or gutter posts in the case of connected houses-at about 3-meter intervals, to support the roof trusses. The eave or gutter extends from post to post at the point where the truss joins the side post. The ridge extends from truss to truss at the peak of the house. Midway between the ridge and eave a purlin extends from truss to truss. If the house is wide, two or more purlins may be spaced equidistant between ridge and eave. For narrow houses or houses with adequate trusses that need additional support, posts are provided at each truss from the purlin to the ground.

Usually a curtain wall of concrete or transit is used for about 1 meter at the base of the side-walls of the greenhouse. For glass houses, roof bars are installed approximately 60 centimeters apart from ridge to purlin to eave, and wall bars are placed from the eave to the sill of the side-wall. These bars provide the base for the glass. Houses that are to be covered with film plastic may have bars or rafters spaced about 120 centimeters apart with wire fabric installed beneath the plastic film to support it. Corrugated plastics should be applied to framework members that run lengthwise to the house (purlins). These should be installed about every 90 centimeters across the width of the roof depending on the weight of the corrugated plastic and the pitch of the roof.

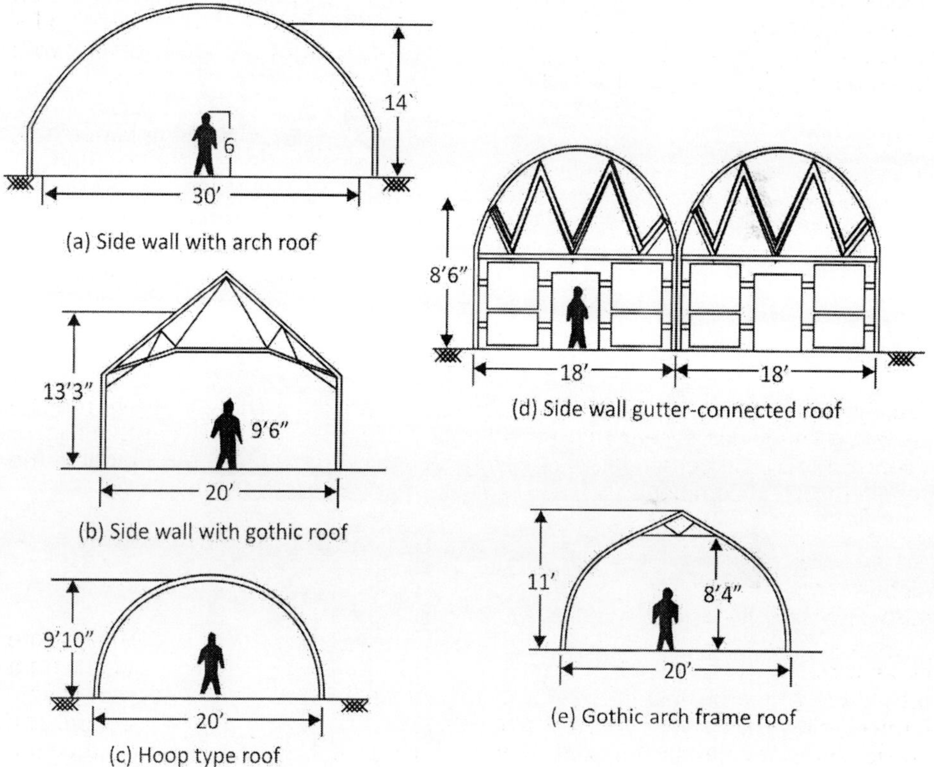

FIG. 23.11: Basic structural design of various types of greenhouses.

The post and truss designs for greenhouses may be the same whether they are to be covered with glass, film plastic, or rigid plastic; but the purlin and bar arrangement should be made specifically for the type of material that will be used to cover the structure.

Usually it is not possible to switch from one type of covering to another without making the necessary changes in bars or purlins first. However, in replacing glass with corrugated plastic in old, wooden bar houses, it is possible to remove the glass, clean and paint the bars, and then apply the corrugated plastic directly to the bars at a slight angle.

The choice of materials for the framework of a greenhouse is based on structural strength, durability, initial cost, and maintenance costs. Wooden framework is used primarily for temporary or semi permanent structures. The most durable wood with the least maintenance cost is redwood. Steel framework has been used widely.

This does require frequent painting to prevent rust, but it is very satisfactory in other respects. Steel frame houses may have redwood or aluminum roof and wall bars as well as ventilator sash.

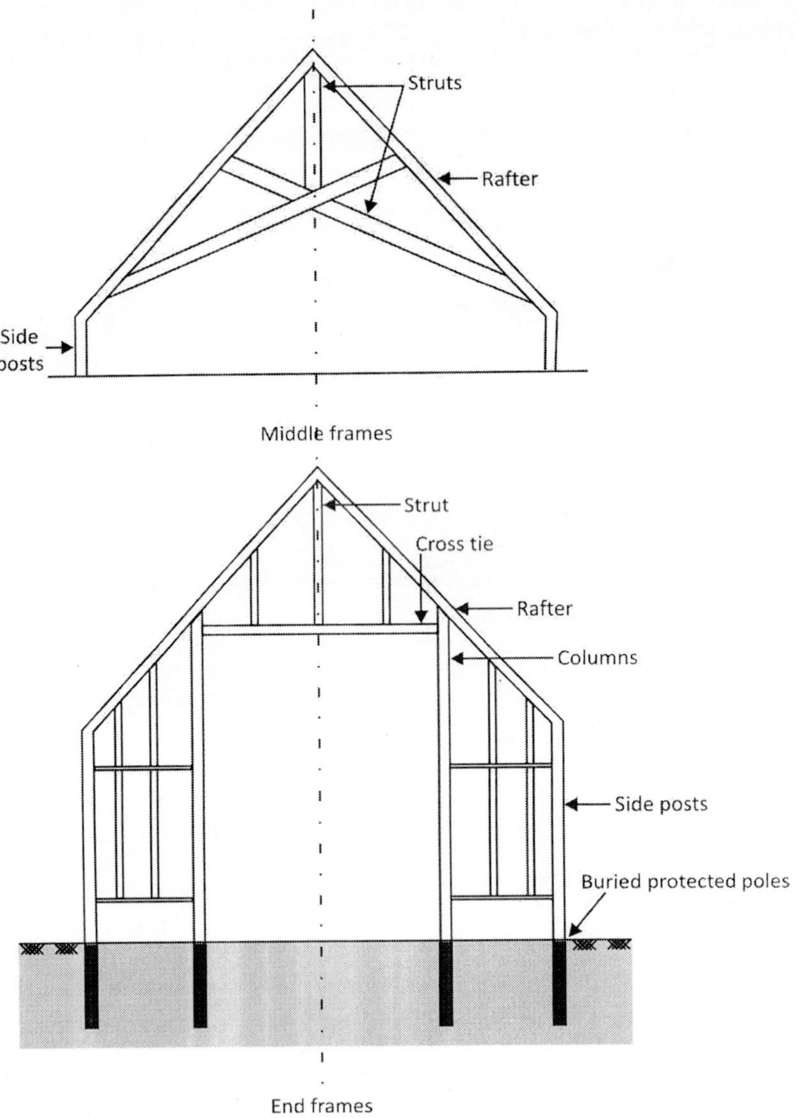

FIG. 23.12: Wooden scissors-truss type film plastic greenhouse.

Aluminum has become increasingly popular for the greenhouse framework. The roof and wall bars and ventilating sash all are made of aluminum. The corrosion problem with aluminum is slight, and the structure does not require painting. It should be kept in mind however, that the glazing procedure and maintenance are approximately the same for wood, steel, or aluminum.

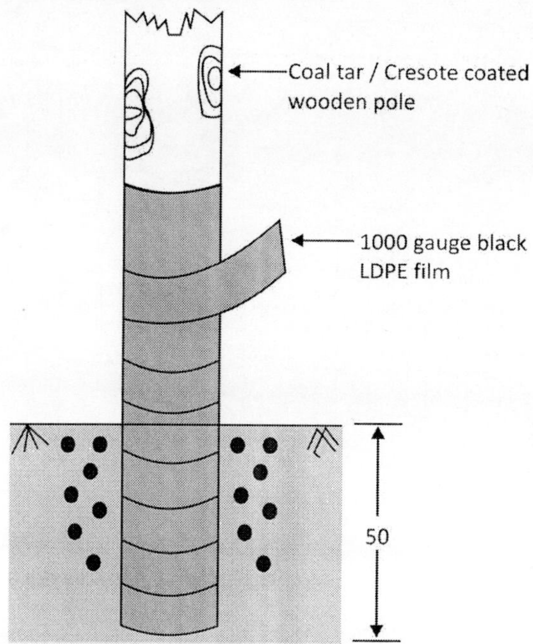

FIG. 23.13: Pole for greenhouse. Such pole is termite and moisture proof.

The greenhouse framework that has just been described has plane surfaces, and it is suitable for being covered by material that is rigid and has plane configuration. Glass panes most commonly have those characteristics. Greenhouses require coverings that transmit light, and because of the excellent light-transmission properties of glass, greenhouse structures were constructed with plane surfaces so they could be covered with glass.

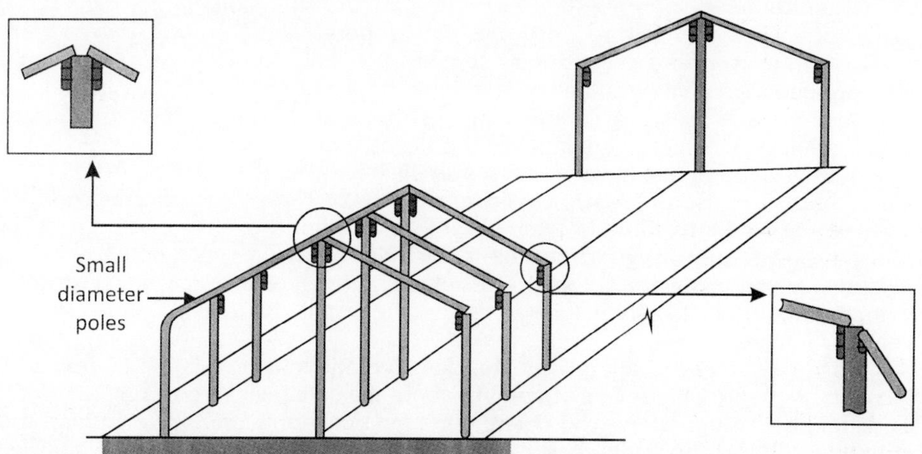

FIG. 23.14: Layout and construction of bay structure in sequence.

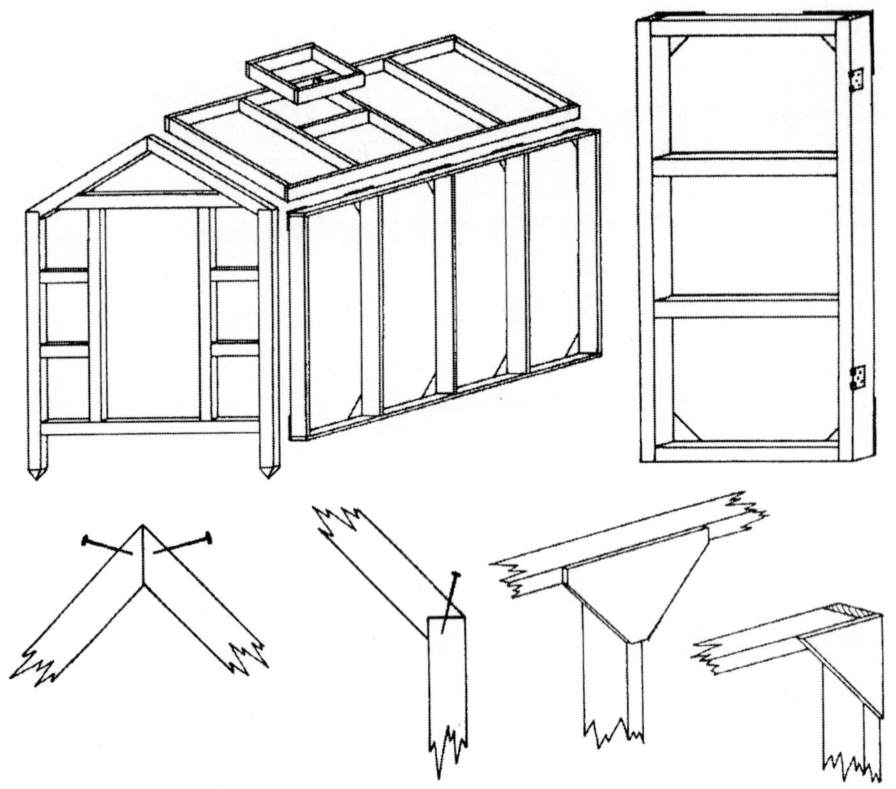

FIG. 23.15: The wood framed green house.

Greenhouse roofs must shed moisture-rain and snow-and with a structure of plane surfaces, the pitched roof provides that property.

Greenhouse requires a **venting system** of some kind. A gravity type of ventilation system needs a ventilator at the highest point in the structure, and the pitched-roof design of greenhouses provides that highest point at the ridge. Thus greenhouse structural design includes ventilators at the ridge.

Two developments have given some flexibility in the structural design of greenhouses-the use of **plastic coverings** and fans. The plastic coverings can be and are used on structures of pitched-roof design, but they also can be used on structures of arc-roof design when fans are included to handle ventilation. Regardless of kind of covering material, greenhouses of pitched-roof design may be constructed without ridge ventilators if the fan system that the installed is able to handle the total ventilation needs.

Most new greenhouse construction does include total ventilation by means of fans, but there are differences of opinion about the suitability of covering materials. The greenhouse operators who desire glass as a covering must have pitched-roof structures. Those who prefer plastic of some type as a greenhouse covering may use either pitched-or arc-roof design or both.

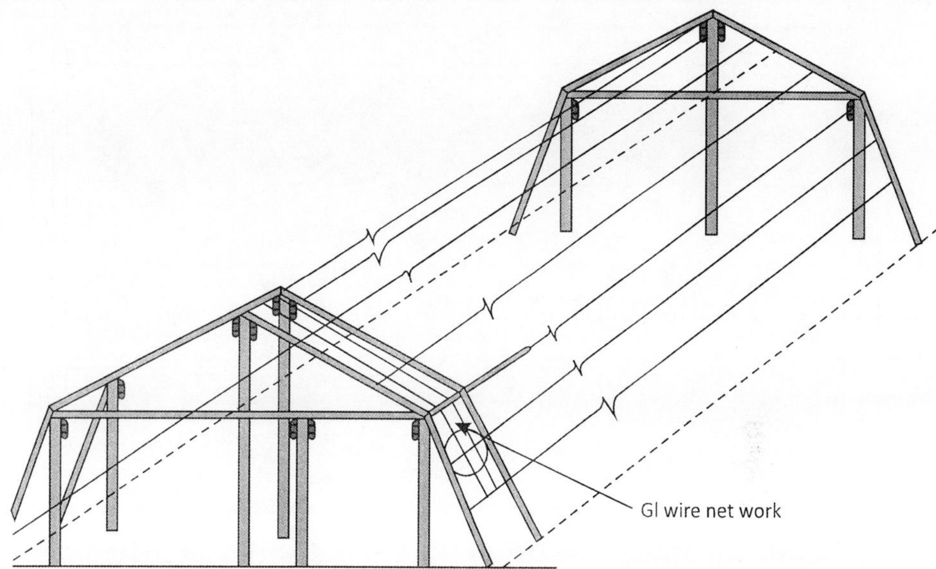

FIG. 23.16: Greenhouse can also be constructed using GI-wire network.

FIG. 23.17: Connected (ridge and furrow) greenhouses in the process of construction. These are clear-span greenhouses-the roof is supported solely by the trusses which extend from side post to opposing gutter post.

FIG. 23.18: The supporting framework for greenhouses is side or gutter posts with trusses extending across the house between opposite posts. The trusses join with the ridge at the peak, the eave or gutter at the side, and one or more purlins spaced equidistant between ridge and gutter. The roof bars, which hold the glass extends from ridge or ventilator header to purlins to gutter or eave.

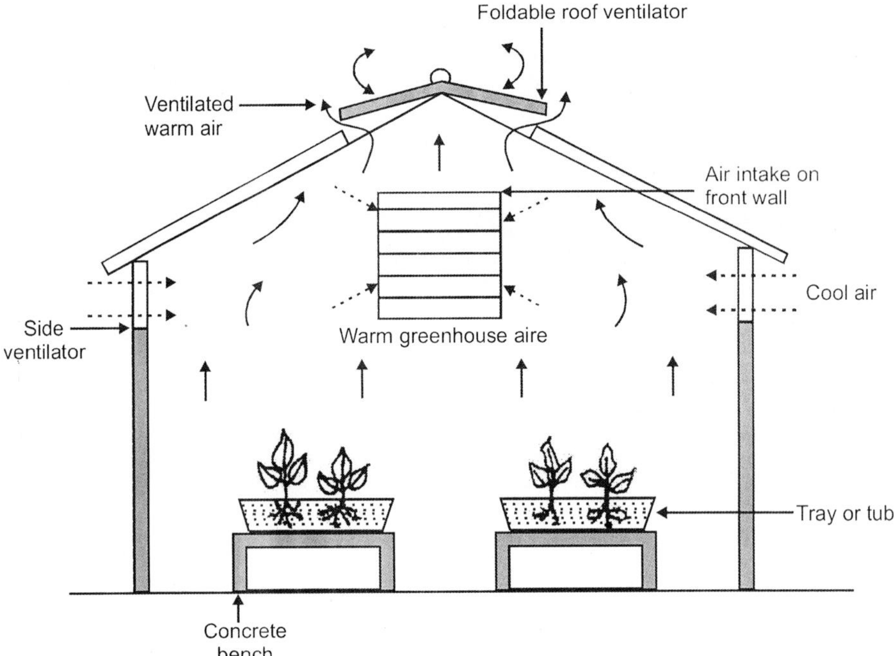

FIG. 23.19: Presentation of layout of chimney effect in general passive ventilation.

The use of plastic-films covering for greenhouses led to many innovations in greenhouse structures. Many acres of land were covered in moderate climates such as California with wooden structures in which the windward roof surface is elevated higher than the leeward surface, and the difference in elevations provides an open ventilator. In cooler climates many different types of structures have been designed for covering with plastic film. These, however, must have functional ventilation systems. Unfortunately many individuals rush into the construction of seemingly inexpensive, plastic film-covered structures without proper evaluation of the means or costs of providing ventilation, heating, irrigation, and benching.

For a structure of the same durability, the greenhouse framework to be covered by plastic must be as sound as the one that will be covered by glass. The design of the structure might be different, but the quality of materials and the strength of the framework must be the same.

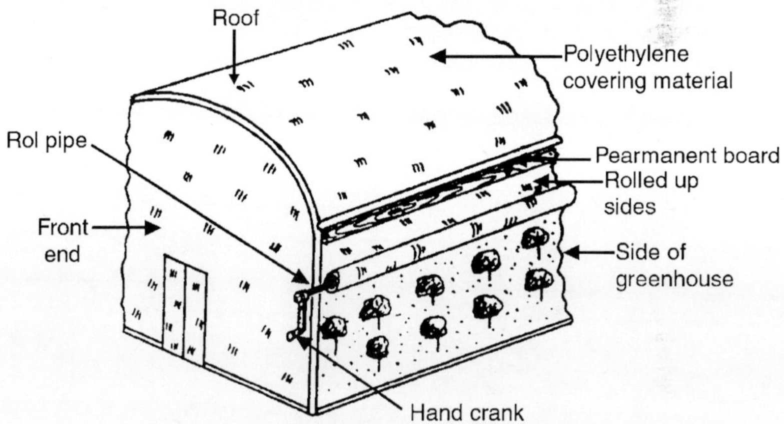

FIG. 23.20: Design of a polyhouse showing roll up side passive ventilation.

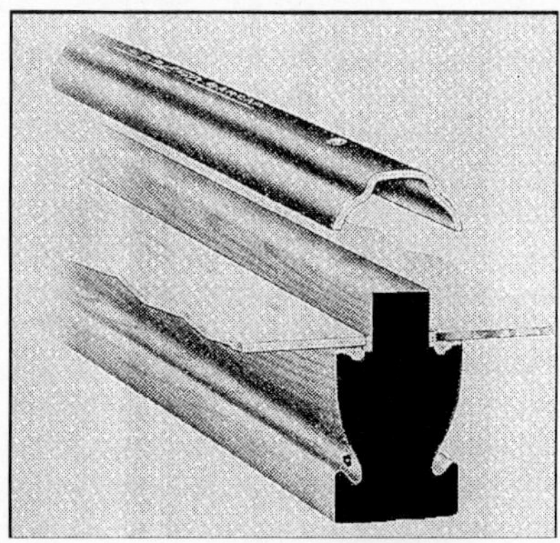

FIG. 23.21: Roof bars are designed so that the glass can be placed in bedding compound on each side, and there are drip grooves toward the bottom to conduct the condensation to the side-wall. The bar cap covers the bar to preserve the glazing.

The basic difference in framework for greenhouses to be covered with plastic rather than glass is that no roof bars are provided. The plastic is supported and/or attached to purlins that are installed from truss to truss or are to are for the length of the greenhouse. Because of this, there are a greater number of purlins in the framework of a house to be covered with plastic than one that will be covered with glass.

Types of Greenhouse Covering Materials

The greenhouse must be covered with as clear a material as possible to provide the maximum amount of sunlight for the plants. This is the primary requisite of the structure. The durability of the covering material, the initial cost, and the maintenance costs then determine the desirability of the material.

Light transmission through glass is excellent, and for that reason it has been the primary covering material used through the years. Applying a glass covering is called glazing. The initial cost of glass is reasonable, but it is difficult to install and the periodic realizing that is required is costly. It is subject to considerable breakage in areas subject to hail storms. In fact, hail damage is such a threat in certain locations that wire netting is installed above the glass to protect it from the hail.

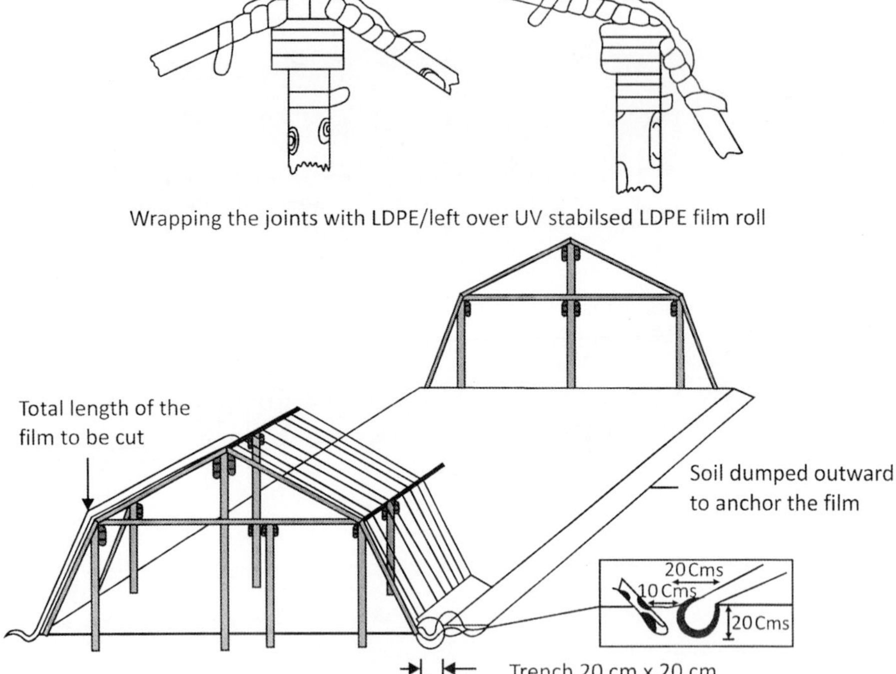

FIG. 23.22: Presentation of cladding the structure and anchoring the film using soil trenches.

Glass is placed on the greenhouse in much the same manner as shingles on a home except that the panes are not lapped as much, and they are held in place by glazing points placed at the bottom of each pane, on each side, and also at the top surface at each side. To make the house watertight the glass is set in bedding compound on the bars, and after the glass is in place glazing compound is run along each side of the bar and then painted to complete the seal. The exterior of a glass house needs repainting every 2 to 4 years to maintain a good seal between bar and glass. Occasionally, loose glazing compound will need to be replaced before repainting. If the seal between bars and glass becomes loose, it is then necessary to re-glaze the house. In re-glazing, all the glass is removed from the house, the bars are cleaned, and the glass is re-laid. This is an expensive procedure but a necessary one if the best growing conditions are to be maintained in the greenhouse. This glazing procedure is virtually the same whether the bars are wooden, steel, or aluminum, but wooden or steel bars must be painted before the bedding compound is applied.

To reduce the frequency of painting and reglazing that is necessary, aluminum bar caps are used. As the name indicates they are aluminum caps or strips that are installed over the bar. They protect the seal from the weather and reduce the amount of maintenance that is needed.

COVERING MATERIALS FOR GREENHOUSE

Through for the years there has been much interest in the use of materials other than glass for covering greenhouses. The search has been for a material that transmits light as well as glass but has improvements in other properties such as breakage, tightness of seal, and covering and recovering costs.

The hope seems to be with plastic of some type, and the various plastics have been tried as they have been developed. Some plastic materials can be used with certain success, and it is possible that other materials will be developed that are even more suitable. The **plastic greenhouse coverings** in most common usage are polyethylene film and fiberglass. Many improvements have been made in these materials since they were introduced, and it should be expected that further improvements would evolve. Degradation of both types of materials in ultraviolet (UV) light has limited their usefulness in the past, but the incorporation of UV stabilizers in the products has decreased the rate of degradation.

Fiberglass

Fiberglass is a product that has glass fibers imbedded in plastic. The glass fibers provide reinforcement, and the fiberglass panels are flexible but somewhat rigid. Fiberglass can be made in various configurations, but the most common one used for greenhouse covering is corrugated. A form such as this increases the rigidity of the product and does not require as much support as a plane (flat) farm would need. Fiberglass is manufactured in panels about 134 centimeters wide and in the length needed for the job. The panels are lapped when they are installed and the effective coverage in width is 120 centimeters per panel. With greenhouse roofs that do not have ridge ventilators, the fiberglass panels are continuous from ridge to eave or gutter. Closure with the plane surfaces of ridge and eave or gutter are made with strips that are plane on one side and corrugated on the other and are placed between the panels and ridge and eave or gutter. The ridges and furrows of the corrugations extend the length of the panel, and the attachment of the panel to the framework should be through the ridge because that minimizes leakage. For the attachment at purlins, spacers should be used between panel and purlin so that the panel does not contact the purlin. The moisture that condenses on the inner surface of the panel then can flow uninhibited to the side wall.

Fiberglass is manufactured in several colors, but it is the clear kind that is used in greenhouse construction. It is a translucent substance that transmits about 90 per cent as much light as glass does, but plant growth under fiberglass that is in good condition can be as good as it is under glass. Fiberglass does degrade in time, causing two noticeable effects. The plastic may change color gradually to tan or brown.

FIG. 23.23: Green house under construction.

FIG. 23.24: Connected Quonset structures covered with clear, corrugated fiberglass. There are no ventilators. Ventilation and cooling are by exhaust fans in one end wall and wet pads in the opposite end. Note the steam mains which service these and other houses. They are sufficiently elevated to give adequate clearance for trucks, and they are insulated to conserve heat.

The rate of this color change has been decreased by the addition of UV stabilizers in the plastic. Also the Exterior surface erodes, causing the exposure of glass fibers. This rough surface causes the collection of dirt with the resultant reduction in light transmission. When this occurs, the outer surface can be cleaned with forceful streams of water.

Fiberglass is available with Tedlar film bonded to the exterior surface, and the light transmission of this product remains good for many years with only minimum maintenance of the outer surface of the panels. The fiberglass manufacturers sell their products under limited warranty of either 10 or 20 years depending on the grade. The actual useful life of fiberglass can be expected to be something less than the full warranted time. Fiberglass has a rather low flash point as compared to other greenhouse covering materials, and the rate of combustion is fast when the flash point is reached. Special precautions need to be taken to keep fires from any source away from fiberglass-covered greenhouses.

Poly Film

Polyethylene (poly) film is by far the least expensive of the greenhouse covering materials. It also needs to be replaced most often. Ordinary grades of poly film degrade rapidly in high-intensity sunlight, and the film disintegrates. Polyethylene film used for greenhouse covering is usually in the 6-mil grade. It is a translucent substance that can transmit 80 to 90 per cent as much light as glass does. Plant growth can be as good under poly film as it is under glass. Poly film is made in widths to 12.2 meters, and the common length of a roll is 30.5 meters. This is large enough to completely cover some greenhouses. Because of the continuous covering, the poly-film covered houses have less leakage than houses covered with any other material. They are called tight houses. This has some good and some bad features. There is less heat lost by infiltration, but the relative humidity may be too high at times. Also the gas exchange with the outdoors can be so limited that there will be a shortage of oxygen or carbon dioxide or both in the plastic house. If unit heaters or carbon dioxide generators are used in plastic houses, adequate air inlet from outdoors must be provided to sustain the combustion. The manufacturers of the heating or generating equipment can supply the information on the size of air inlet that is needed. Primarily in an effort to save heat energy, some greenhouses are covered with a double layer of poly film and inflated so there is an air space between the two layers. This does cause a further reduction in light intensity in the greenhouse, and that effect needs to be evaluated together with the heat factor.

Problem of Moisture with Poly Film

Moisture that condenses on the inner surface of the poly-film covering of greenhouses can be a problem. Typically this moisture showers downward when the film ripples with the wind. The film of moisture on the polyethylene reduces the amount of light transmitted into the greenhouse, and the periodic showers can cause some plant diseases.

HEATING AND COOLING OF GREENHOUSE

Accurate temperature control in the greenhouse is of great importance because the desired, best growth of plants occurs within a limited temperature range. The degree of temperature control must be much more precise than just keeping the plants from freezing or keeping them warm enough. For each kind of plant and the type of growth desired, the best results are obtained only at the proper temperature.

Variations in this temperature may result in a complete lack of growth or extremely excessive development.

Plants grown in the greenhouse are classed by the night temperature at which they are grown. The day temperature provided is 3 to 6° C higher than night temperature depending on whether the day is cloudy or bright. Such crops as carnation and snapdragon are grown at 10° C, and some of the 16° C crops are rose, poinsettia, and chrysanthemum. Either of these night temperatures is suitable for many of the greenhouse crops, with the notable exception of most of the foliage plant crops, which require a 21° C night temperature. There are many specific effects of temperature on plant growth; but for a statement on growth in general, plants develop and grow slower in cooler temperatures and more rapidly in higher temperatures. There are specific variations, too, in the quality of growth, but in general, the plants are shorter and heavier at lower temperatures and taller and thinner at higher temperatures.

There is merit in establishing the greenhouse in an area of the country with a suitable outdoor climate so that the least possible adjustment needs to be made in temperature. Unfortunately, outdoor temperatures vary considerably from one season to the other, and a desirable outdoor temperature for one time may be a definite problem for another season.

Sensing Temperatures

Thermometers must be located in the greenhouse at plant level and in sufficient numbers so that they reflect accurately the temperature for the entire house. Placing a single thermometer in a house is not sufficient to determine temperature differentials from end to end or side to side.

Thermometers located at eye level are easy to read, but the readings are proper only if the plants are also at that level. For young plants in ground beds, the thermometer should be placed at the ground level.

Thermometers must be mounted so that the direct rays of the sun do not shine on the bulb. A suitable method of mounting thermometers is to attach them to a wooden board and orient them so that the thermometer is facing north. The proper location of thermostats which operate the ventilators or steam valves is even more important. For more precise temperature control, aspirated units are used in which a small fan draws a low volume of air through the thermometer or thermostat, providing a more representative sample of the air in the greenhouse.

Thermometers and thermostats sense air temperatures, and within limits it can be assumed that the plant temperatures are approximately the same as the air temperatures. If practical methods of sensing plant temperatures directly were available, they would be preferred. When a variable heat source is used, plant temperatures change more rapidly than air temperatures, and if plant temperatures could be sensed as rapidly as the changes occur, corrective action could be taken faster.

Sources of Heat

The primary source of heat in the greenhouse is the sun. Unfortunately it is not available for the entire 24 hours of a day, and even during the daytime the amount of heat transmitted from the sun to the plants in the greenhouse varies tremendously. During the summer days the sun causes too high temperatures in the greenhouse, and various methods are used to reduce the intensity of the rays entering the greenhouse, or cooling systems are used to reduce the temperature. On clear winter days even when the outdoor temperature is low, no supplementary heat may be required, and some ventilating may have to be used to reduce the heat.

There is much interest in the possibilities of storing solar energy for use when needed. Certainly there is a great need for such a procedure.

Some greenhouses are being heated with stored solar energy, but it appears that there will have to be further developments in methods before the irradiance from the sun can be used economically as the sole source of heat in most greenhouses.

Coal, oil, and gas are used as supplementary sources of heat in greenhouses. The choice of fuel in each greenhouse depends on the cost per Btu of heat, the availability of the fuel, transportation costs, equipment costs, possibilities of automation, and labor costs. No single fuel is best for all greenhouses. A careful study must be made at each location to determine which fuel is best suited for that situation. In many localities coal may be the most economical fuel per Btu, but because it is more difficult to handle, more labor may be required.

Methods of Distributing Heat in Greenhouse

Most commonly the greenhouse heating systems are either steam or hot water. The fuel is burned in a boiler, and the hot water or steam is piped to the greenhouses in mains and heating lines in such a way as to distribute heat evenly throughout the entire area. The cool water and condensate from the heating lines are returned to the boiler for re-heating. Steam has been used primarily in larger ranges, and hot water in the smaller greenhouses, but it is possible to use closed-system hot water at high pressure and high temperature as efficiently as steam in large ranges.

More recently, unit heaters have been used in some greenhouses. The fuel for the unit heaters primarily is gas, but there are some that use oil fuel. The most common type of unit heater used in greenhouse is the kind used in various industrial heating situations such as warehouses and garages. These units are of several different capacities ranging from 30,000 to 400,000 Btu per hour. One or more unit is installed in a greenhouse at about the bottom level of the trusses. Usually in greenhouses more uniform distribution of heat is obtained by use of several smaller units rather than one large unit. Some unit heaters have been designed for specific use with tube-fan systems in greenhouses. One type of heater installs directly in line with the tube-between the open end of the tube and the shutter. When the heater is in operation the fan directs the heat down the tube and it disperses through the holes to the greenhouse atmosphere. The other tube-fan heating system uses unit heaters on either side of the tube fan with the heat directed toward the tube fan which directs the heat down the tube.

Conservation of Heat Energy in Greenhouses

There is considerable interest in the conservation of heat energy in greenhouses primarily because of dwindling fuel supplies, but also because of greatly increased costs of fuels. Because of the need for a light-transparent covering on greenhouses, it is not possible to install conventional types of heat insulation material in greenhouse structures. The insulation either must be transparent or there needs to be a way to install and remove the insulation daily.

Insulation of Greenhouse Roofs

The greatest heat loss in greenhouses is through the roof because this is the largest exposed area and the temperature differential between indoors and outdoors is greatest at the roof. There have been mainly two approaches:

1. The installation of a second, translucent covering over the roof with inflation to maintain an air space between the two coverings, and

2. The installation of a removable covering from eave to eave or gutter to gutter.

This will help in effective conservation (about 50 per cent) of heat. Frequently the greenhouses covered with polyethylene film do have two layers of the plastic film with inflation equipment to keep the layers separated. Structures that are covered with glass may be insulated by placing a double layer of polyethylene film over the glass and keeping it inflated. The double-roof coverings are a fixed installation, but the polyethylene film will need to be replaced every 2 years The double covering reduces the light intensity in the greenhouse and during the winter this has some adverse effect on plant growth.

The other procedures for conservation of heat energy are:

- Installation of conventional insulation on north walls,
- Double covering of walls on other sides with translucent material,
- Replacement of heating lines toward the base of the structure,
- Improved maintenance programs to minimize leakage in coverings,
- Improved programmes to maximize the conversion of chemical energy of the fuel to usable heat energy in the greenhouse,
- The installation of windbreaks outside the greenhouse,
- Use of air-circulation fans in the greenhouse to return the warm, upper air toward the base.

VENTILATION AND AIR MOVEMENT EQUIPMENT FOR GREENHOUSES

A means of exchanging air with the outside is needed to regulate the temperature and humidity in the greenhouse. Recently, there has been concern with ventilation in relation to the carbon dioxide content of the air in the greenhouse. Usually the greenhouse has had ventilator sash located at the ridge. In separate houses ventilator sash may be located also just below the eave of the side wall.

The Principle of Ventilation and Air Movement

The principle involved is that the hot air, being lighter, rises and passes through the ventilator. The cold air from the exterior, being heavier, flows toward the base of the greenhouse after it enters the ventilator. This provides air movement as well as temperature adjustment in the greenhouse which is controlled by thermostats.

Effective Greenhouse-Ventilation Equipment: Plastic Ventilation Tube

Plastic ventilation tube is most effective and is used in conjunction with exhaust fans. Many greenhouse operators ventilate the houses by means of ventilation tubes from fall to spring, and from spring to fall the fan and pad system is used. They may not use traditional greenhouse ventilators at all. It was the ventilation tube that made fall to spring ventilation practical in arc roof greenhouses. And it is because of the ventilation tube plus fan and pad cooling that greenhouses of any design may be constructed without traditional greenhouse ventilators. There sizes of ventilation tube are used depending on the size of the greenhouse. The inflated diameters of these tubes are 45, 60, and 75 centimeters. The tubes are perforated along each side, the hole size and spacing are determined by the length of the tube that is used. The ventilation tubes are installed the length of the house above the bottom member of the trusses. One end of the rube is tied closed, and the other, open end is attached to a shutter in the side wall. When the shutter is open and exhaust fans are operating, air from the exterior enters the tube and then distributes into the greenhouse through the holes. One tube is used in small greenhouses and two or more tubes are installed in wider greenhouses.

The fan and rube system can be a very effective method for handling ventilation in greenhouses from fall to spring. Also shutters can be motorized and fans and shutters activated by thermostats.

Another development with the fan and tube system was the addition of a tube fan. The tube fan can be operated when the shutter is open to aid in the introduction of air into the greenhouse, or it can be operated while the shutter is closed, providing circulation of air within the greenhouse.

Air in the Greenhouse

Greenhouse needs constant air-motion. Air circulation occurred in greenhouses in the past only if there was a temperature differential. It has been good, common practice to provide some ventilation even while operating heating lines. This assures air circulation because the heated air rises and the cool air from the exterior flows toward the bottom.

FIG. 23.25: Ridge ventilators for the control of temperature, air circulation, and humidity in the greenhouse.

CARBON DIOXIDE GENERATION EQUIPMENT IN GREENHOUSE

Carbon dioxide in the greenhouse atmosphere is required for the promotion in growth and development of the plants. In the portions of the country that have cool weather from fall to spring, when venting is reduced, it is possible to increase the concentration of carbon dioxide in the air.

Various means have been used, but the most common method is by combustion of gas or oil in burners spaced throughout the greenhouse. These units are called **carbon dioxide generators**, but most of them are adaptations of heating units. The products of the combustion are heat, carbon dioxide, and water vapor.

The natural carbon dioxide concentration in the atmosphere is 300 ppm. The generators are used to increase the amount to 1,000 ppm or more. The effective time for the increased amount of carbon dioxide is during the daylight hours when food manufacturing is taking place in the plants and carbon dioxide is being used in this process.

FIG. 23.26: This is a gas-fired carbon dioxide generator. In addition to carbon dioxide, heat and water vapour also are produced.

Importance of Carbon Dioxide in Greenhouse

The general effects of the increased quantities of carbon dioxide are an increased rate of growth and development (crop production time is shorter) and larger stems, leaves, and flowers. Some of the greenhouse crop plants that benefit from increased amounts of carbon dioxide are geranium, snapdragon, chrysanthemum, and rose.

GREENHOUSE BENCHES

Greenhouse benches are used for better control of the environment around the plants, and they allow the work to proceed more efficiently. The greenhouse bench for cut flowers needs to be level or very nearly level and deep enough to hold at least 15 centimeters of soil.

Greenhouse benches provide excellent water drainage and isolation from disease or insect infestation. If the bench is constructed on the ground, the bottom and sides must be a solid piece. Concrete is the only material suitable for such a bench. To assure good drainage a V-bottom is used, with the bottom of the V at least 4 centimeters below the side. A half tile is placed over the bottom of the V, and the bottom is then filled level with coarse gravel. Lengthwise the bottom of the bench must fall 3 centimeters per 30 meters.

Materials used for Greenhouse Benches

Raised greenhouse benches are usually constructed from wood, concrete or steel. Regardless of the material, water drainage must be perfect. The width of a concrete block makes about the right elevation for a raised cut-flower bench. The bench width is usually 105 or 120 centimeters.

Greenhouse Benches for Cut-Flowers

Cut-flower benches are steamed often. For steaming it is best to have benches without posts so that they can be covered easily, and concrete benches must be constructed carefully to prevent cracking. Most cut-flower crops require some over bench structure to support cloth or wires. The method for attaching such support should be considered when the bench material is selected.

FIG. 23.27: Wooden benches for cut flowers. These are elevated enough to assure isolation from pests and pathogens as well as to provide good drainage. Benches for pot plants should be higher for ease in handling plants.

Greenhouse Benches Pot Plant

The requirements for pot-plant benches are quite different. They should be raised to about 75 centimeters for convenience in working. If they are going to be worked from both sides, the most efficient width is about 2 meters. Ideal drainage is needed, and there is interest in perfect air circulation around pot plants. Pot-plant benches are often constructed without sideboards and with their bottoms as open as possible.

FIG. 23.28: The boards in the bench bottom must be well spaced as illustrated to assure complete and immediate drainage of water from the bench.

Arrangement of the Benches

The arrangement of the benches can affect working conditions in the house considerably. In building a new structure it would really be best to plan the work area and bench arrangement, and then choose the structure to properly enclose it. Too often the greenhouse is all erected before thought is given to how the benches fit. Traditionally benches have been installed lengthwise in the greenhouse.

FIG. 23.29: Concrete, ground benches for cut flowers. Benches of this type have sides and bottoms poured in one piece for good control of pests and pathogens. The bench bottom is V shaped for adequate water drainage.

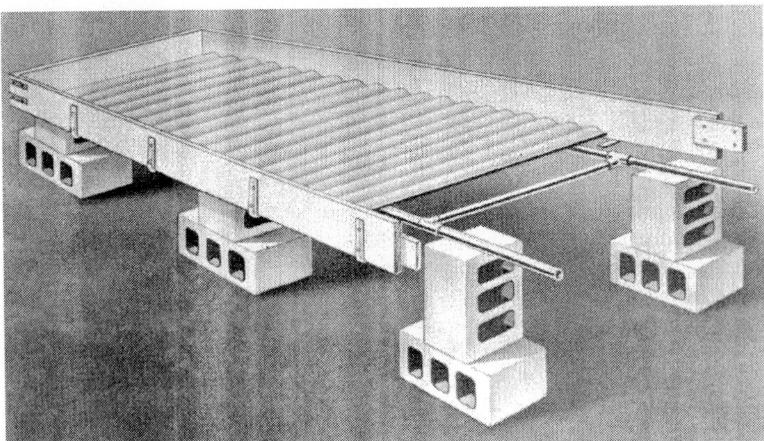

FIG. 23.30: Transite bench construction. Corrugated transite must be used on the bottom for strength. To provide good drainage, the bottom should be pitched from one side to the other. If this cannot be done, drainage holes must be drilled in each valley. There is a trend away from installation of heating lines on the bench.

FIG. 23.31: Welded-wire fabric can be used as bottoms for pot-plant benches. This provides maximum drainage and air circulation. The 14-gauge wire has 2.5 centimeter spacing.

Long benches, however, are not always the most efficient size. With some crops, cut flowers for example, long benches necessitate frequent trips through narrow walks, and long benches can be even more of a problem with pot plants. There also is an advantage in having crop size benches, which often means that the unit should be shorter than the length of the house.

FIG. 23.32: Cross-benching in a greenhouse. The principal advantages are crop-size benches and the use of wheeled carts in the wide walk to service all benches. Note that the primary source of heat is overhead unit heaters but that heating linesare provided on the side-walls as well.

FIG. 23.33: Cloth house to provide partial shade and protection from pests for crops grown outdoors during the summer in the North and the year-round in the South. This house is covered with cotton tobacco cloth, but plastic screen commonly is used.

Greenhouse Walks

Greenhouse walks are usually minimum width and for good reason. Production space in the greenhouse should be utilized to the best advantage, and it is apparent that no flowers can be grown in the walks. It is possible to use 45 to 60 centimeter walks however, at least one walk in the house should be wide enough to handle wheelbarrows, sprayers or wheeled carts. Walks should be level and should have a satisfactory surface in order to promote the best movement of workers or vehicles. In the typical greenhouse arrangement with fixed benches, approximately two-thirds of the area is benches and one-third walks.

WORKING ROOM

The working room should be located, if possible, at the north end of the greenhouse, in order to provide some protection from cold prevailing winds. Its size and design are largely dependent on the size and number of greenhouses it is to serve and the type of crops being grown. If separate from the greenhouse, the working room should be near enough to minimize walking and make transportation of plants between the two structures simple and convenient.

The working room should be heated, well lighted, and equipped with tables and storage bins for soil, sand, peat, and other similar materials. A cooler is tools, insecticides, fungicides, fertilizers, and other items necessary for the production of plants should be provided. A steam-treatment box or unit in the working room is a desirable asset.

The table space in the working room for grading, bunching, and wrapping of cut flowers and space for pot plants should be ample for all occasions such as holiday periods when a large volume of stock is handled. Substantial potting benches on wheels can be rolled to the desired location, which eliminates moving of pots in the working rooms.

The working room is often a part of the building where cut flowers, plants, pots, and other materials are received or loaded for shipping out. If such is the case, the room should be wide and long enough to accommodate one or more large trucks. Cut flowers and plants for holiday delivery may be loaded inside the warm working room, which somewhat reduces the wrapping and insulation necessary for protection from cold. Doors should be high and wide to accommodate trucks.

CONVEYORS

The use of some type of conveyor is widespread in industrial plants where mass-production methods are in vogue. The florist industry, with some exceptions, has been slow to recognize the advantages of such labor saving devices. This is partially due to the lack of knowledge of the costs of various operations, and also to the reluctance of many growers to change with the times.

Conveyor systems are usually needed where pot plants are grown. The use of conveyors to move plants for the purpose of potting, shifting from one house to another, or moving finished plants out of the houses for delivery eliminates hours of drudgery. Trolley systems are reasonably priced and save time and money in addition to increasing the efficiency of every employee.

CLOTH HOUSES

The purpose of a cloth house is to expand production area for the period when weather conditions are favorable for culture of plants without benefit of a greenhouse. Although cloth houses are no longer as popular as they once were, good-quality asters and mums can be produced under cloth.

LATH HOUSES

The precision growing of such plants as azaleas and hydrangeas may require the use of lath shade during the summer months in many localities. Cold frames with lath covering in place of glass sash are useful, although the inconvenience of frames often limits their use for this purpose. Snow fence is commonly used by supporting it 1 meter above 120-centimeter beds. In the South, slat sheds are used for growing foliage plants.

The 4-meter posts of 10- by 10-centimeter pine, pressure treated with copper naphthenate, are spaced 4 meters apart with 1 meter buried in concrete in the soil. The stringers are 60- by 180-centimeter pressure-treated pine boards, nailed on alternate sides of the posts at the top. The slats on the roof are 1- by 5- by 7.5 centimeter rough cypress spaced 2 centimeters apart. Most crops are grown under 75 per cent shade, but aglaonema is best under 90 per cent shade while peperomia should be in 50 per cent shade. Plastic screen of the proper density may be substituted for the roof slats. The south side is slatted the same as the roof, and the other sides are boarded solid. One of the best types of structure consists of a permanent frame of pipe or wood at least 2 meters high, on top of which snow fence can be unrolled or lath sash placed when it is needed.

FIG. 23.34: Lath house to provide partial shade for crops grown outdoors in the summer in the North and the year-round in the South. Southern structures usually are constructed to provide more shade and commonly are called slat sheds.

BULB STORAGE

Bulbs are very commonly stored outside in beds especially reserved for this purpose, but cold storage buildings, erected for the sole purpose of storing bulbs to condition them for forcing or holding them back for late flowering, are preferred. Storage in a naturally cool structure such as a cellar or an insulated, refrigerated structure is very satisfactory. The uniform conditions that prevail in these structures are ideal. Although they are more expensive to construct than outside beds, cool storage is more convenient, can be equipped with shelves spaced about 45 centimeters apart to accommodate a large number or pots, and in general give more satisfactory results. When shelves are used, a layer of peat on the shelf reduces the danger of drying out.

Outside beds for bulb storage consist of nothing more than shallow pits dug in the soil in which the pots or flats are placed. The variations in temperature in outside storage, the inconvenience during extremely cold weather or the muddy conditions in mild weather, and the difficulty of proper labeling are against the use of outside beds for bulbs, but because no special equipment is needed, they are common.

GENERAL STORAGE

For storage of cut flowers and potted plants that are fully developed in advance of the time they are wanted some type of structure, either naturally or artificially cooled, is necessary. Refrigerated storages are ideal, as a uniform temperature may be maintained, but a shed or heavily shaded greenhouse may be used for this purpose.

BOILERS FOR THE GREENHOUSE HEATING SYSTEM

There are many variations, but boilers are made of either steel or cast iron. The steel boilers may be fire tube, in which the heat and flue gases pass through the tubes that are surrounded by water, or water tube, in which the heat and flue gases pass around the tubes that are filled with water. The capacity of a boiler is referred to in terms of horsepower. One boiler horsepower equals a heat output of 33,475 Btu per hour. In order to know what size boiler is needed, the heat requirement of the greenhouse must be determined; it is based primarily on the area of surface of the greenhouse exposed to the outdoors, the desired temperature to be maintained within the greenhouse, and the lowest temperature to be expected outdoors.

In general, a boiler rated at 200 horsepower should be adequate to maintain a 16° C temperature in a greenhouse covering 4,650 square meters of ground when the temperature outdoors is 23°C. Instead of one boiler of the maximum capacity, two boilers are often used with about that total maximum capacity; thus either boiler can be used separately to heat the entire place in mild weather or the two boilers can be used in conjunction to provide the maximum heat in the coldest periods.

Operation of the Boiler

In order to have good combustion of fuel, the right amount of air must be supplied in the boiler. In some instances the air is furnished together with the fuel, and additional air is provided at the burner. Different methods are employed depending on the fuel and the boiler, but regardless of the method the amount of air must be carefully regulated. There must be a continuous draft that provides the new air supply to support combustion and to remove the smoke and products of combustion from the boiler.

The stack is an integral part of the heating system, and it must be designed for the specific type of boiler and operating equipment. The advice of a specialist should be sought in erecting a new stack or in connecting a different boiler to an existing stack. The requirements of the stack for a boiler that has a mechanically induced draft are different from those of the stack that produces draft by natural thermal action.

Draft System

In modern boilers the induced draft system is used which consists of a fan placed on the stack side of the boiler. The size, arrangement, and temperature of the stack are of less concern with induced draft than with natural draft, as the draft is directly related to the mechanical efficiency of the fan located between boiler and stack. Since more positive draft control is possible with the induction system, and since a tall, costly stack does not have to be provided and maintained, there is increased use of the induced draft system. It should be acknowledged that the induced draft is entirely dependent on electric power, and if there is a power failure, the boiler does not operate. Since most boilers are dependent on other electrical controls, this just emphasizes the need for an adequate, emergency, standby power generator in the greenhouse boiler room.

Operation of the Boiler

Operation of the boiler varies all the way from completely manual operation to highly sophisticated systems that are entirely automated. The choice of the method of operation of the boiler is based on the efficiency and the costs involved.

EMERGENCY HEATING EQUIPMENT FOR GREENHOUSE

The greenhouse business needs an emergency plan for providing enough heat to prevent loss of plants by freezing. The emergency plan needs to be well designed so that it can function immediately fuel would have to be on hand as well as the means for burning it. If it is an interruption in electric power, usually the best answer is the use of a standby power generator. Also it should be anticipated that there could be an interruption both in fuel and electric power. Some greenhouse operators maintain a supply of alcohol and containers that could be used to keep the house above freezing temperatures for a few hours. Others store enough temporary oil burners and oil on the place to maintain temperature during an emergency.

Emergency, Standby Power

Several essential operations in the greenhouse are dependent on electric power; and since power failures are possible, it is necessary to provide an emergency, standby power unit of sufficient size to handle these operations. The need for a continuous power supply is often associated only with the winter season as most boilers operate with electrical controls. After an emergency power unit is obtained, a routine must be established to assure its readiness for operation in any situation. In most areas power failures seldom occur, and therefore if the generator were operated only during an emergency, there would probably be a long delay in getting the unit started.

FIG. 23.35: Various types of unit heaters are available. This is a gas unit with horizontal air movement and vented to the exterior. The Quonset-type house is covered with polyethylene film.

Figuring the Heat Requirement of the Greenhouse

During the heating season the direction of heat movement or loss is from the interior of the greenhouse through the roof and walls to the cooler air outdoors. To determine the heat loss, it is necessary to know the total surface area exposed to the two different temperatures, the rate or coefficient of heat transfer through the material of which the exposed area is made, and the temperature differential between indoors and outdoors.

The rate of transmission of heat through glass varies with the temperature, but the heat transmission coefficient commonly used is 21.9, which indicates the transmission by glass of 21.9 Btu of heat per square meter per hour per 1°C temperature differential from one side of the glass to the other. The total heat loss by transmission through the exposed surfaces of the 12.2 by 61 meter greenhouse is it is going to be maintained at 16°C at 23°C outdoors would be 935,000 Btu per hour (1,094.4 times 21.9 times 39).

The traditional radiators in the greenhouse are lines of black iron pipe that are installed the length of the house and distributed across the house in a way to produce even heat throughout. The heat transfer from black iron pipe is satisfactory, and the cost of material is comparatively low. If heating lines are painted, black paint should be used. Galvanized pipe or pipe painted with aluminum or light colors does not transfer heat as readily as black pipe.

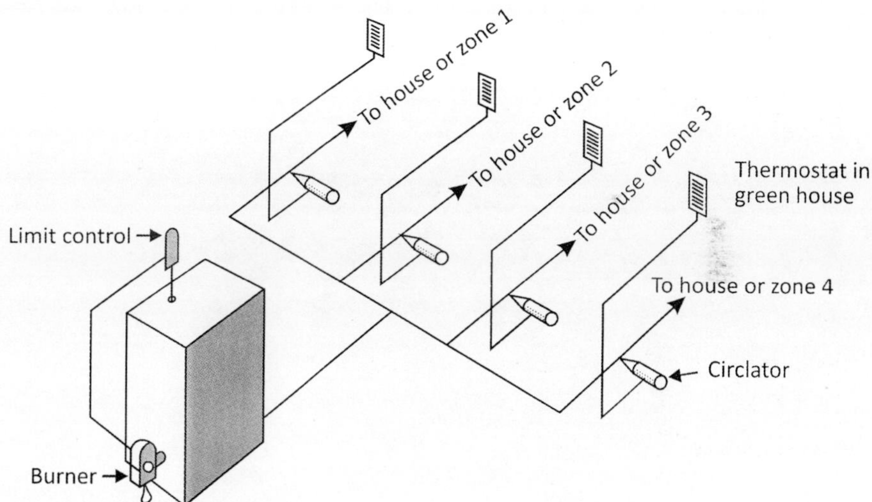

FIG. 23.36: Better control of hot-water heating is provided with circulators in the various areas to be heated.

Location of Heating Lines

The location of heating lines at the side-walls is indispensable, but the balance of the heat requirement can be located in several different ways and still produce even heating. Heating lines may be located under the benches, at the sides of the benches, or above the benches; or unit heaters may be used above the benches. If handled properly, equally good results can be obtained with any of these methods. At a steam pressure of 70 grams per square centimeter, 3 centimeter black iron pipe will give off heat at the rate of about 590 Btu per hour per linear meter of pipe in a 16°C house, and 5 centimeter hot water pipe will transfer 525 Btu per hour per linear meter of black iron pipe when the water temperature is 118°C. The heat output of the unit heaters is listed by the manufacturers.

Arrangement and Placement of Heating Lines

The heating lines must be supported sufficiently so that there are no sags, but because pipe expands and contracts with changes in heat, the pipe hangers must allow for free movement of the pipe. The side heating lines are suspended from the side walls one pipe above the other, and the bench heating lines are hung from the sides or underneath the bench. Overhead heating lines are hung from the trusses or from purlin posts.

Steam Traps

A steam trap does just what its name implies-it traps the steam in the heating lines, but it allows water to pass on to the returns. The two types of traps that are used are thermostatic and bucket. In a thermostatic trap the bimetal element is affected by temperature; it opens the outlet when water (from condensed steam) is in the trap and closes the outlet when the higher temperature steam enters the trap. The bucket trap has an inverted float bucket that opens or closes the outlet. When water enters the trap, the bucket sinks, the outlet is opened, and the water passes to the return pipe; but when steam enters, the bucket floats and the outlet stays closed. Steam traps should have screens on the inlet side to prevent the entry of dirt or scale that would interfere with the operation of the trap.

FIG. 23.37: Heating lines often are hung on the sides of benches. The heating line on the right is supplied from a header above, and the line on the left enter a steam trap connected to a return main.

Control of Heating Line

With manual control of the heating lines, hand valves are used for regulation. By checking the thermometer regularly and activating heating lines as needed by opening valves, the greenhouse can be kept within the proper temperature range. A manual system of operation does require the presence of some dedicated workers day and night-workdays and holidays-to open or close valves as needed.

Special Heat Requirements for Greenhouse

Higher temperatures are used for the propagation of plants. Propagation benches for rooting cuttings and grafting benches or cases are maintained about 6°C warmer than the temperature used for growing the crop. Seed flats also are kept in warmer temperatures for the best germination of seed. In spite of the relatively small area concerned with propagation, there is a large number of plants involved; and the temperature requirements are critical.

FIG. 23.38: Exhaust fans for a fan and pad cooling system. Note that the ridge ventilators are closed during the operation of this system.

FIG. 23.39: Whenever possible the wet pad should be located within the greenhouse so that the house can be opened and closed as the weather demands. This type of side wall ventilator operating equipment allows the wet pad to be installed inside the wall.

Providing Steam for Treatment of Soil and Equipment

Steaming soil is the best method of controlling disease organisms, pests, and weeds. In addition, it gives many soils a better structure, which promotes better soil drainage and improved root growth. For proper steaming of soil all portions of it must be heated to 118°C and maintained at that temperature for a minimum time of 30 minutes. This will require about 124,000 Btu of heat per hour (about 56 kilograms of steam) per cubic meter of soil, or, in a bench, per 6.5 square meters of soil. If steam is not available from the heating system, a steam boiler or generator should be obtained just for the purpose of providing steam for treating soil. There may be an advantage in being able to move them to the vicinity in which they are needed, but because of their small size they can steam only about at 15-meter bench at a time. Either low-pressure or high-pressure steam can be used for soil steaming because, regardless of the pressure at which the steam is piped to the bench, the pressure drops immediately when the steam is released into the soil.

Lower Down Temperatures in the Greenhouse

As noted earlier, plants produce the best growth within a fairly narrow temperature range.

Heating systems are used to increase the temperature during cool weather, and cooling systems may be used during the time of the year when the temperatures are too high. Since heat is removed from the surroundings as water changes from the liquid to the vapor state (evaporates), it is possible to cool greenhouse air and plants by introducing some water into the greenhouse atmosphere. The amount of heat removal (cooling) that is possible depends on the temperature of the air indoors and outdoors and the relative humidity of the air. Maximum cooling is possible when the relative humidity of the air is low and the temperatures are not extremely high. In most areas it is found that greenhouse air temperatures can be reduced about 6 to 12°C, which is sufficient to allow a definite improvement in plant growth. Greenhouse air temperatures often are 73°C and may reach a maximum of around 80°C during the heat of the summer.

Since the growth of many types of plants is adversely affected at temperatures above 68°C, the temperature reduction that is possible with cooling systems is sufficient to bring the greenhouse temperatures within a suitable range.

Design of the Fan and Pad Cooling System

When possible, it is best to locate the exhaust fans in one wall with the moist pad in the opposite wall, and the pad should be on the windward side. With separate houses the airflow is established lengthwise if possible, as this can be done more efficiently than across the rather short width of the house. The rate of airflow should be from 2.3 to 3.2 cubic meters per minute per square meter of ground covered by the greenhouse. The higher rate of airflow is used in situations in which the distance from pad to fan is short, and in areas of the country in which the humidity is comparatively high.

For a greenhouse 12.2 by 61 meters, the air would have to move at the rate of 1,637 (744 times 2.3) to 2,381 (744 times 3.2) cubic meters per minute, depending on the circumstances. Since approximately 1 square meter of moist pad is required per 49 cubic meters of air moved per minute, from 33 to 49 square meters of pad would be needed. The pad is moistened by water trickled through it from a trough installed above the pad.

A sufficient amount of water must be supplied to keep the pad uniformly moist, at least 4 liters of water per minute per linear meter of pad regardless of the height of the pad. There is approximately 3 per cent of water loss in the pad. In a system that re-circulates the water, the make-up water is added to the tank by a float-operated water valve.

Cool Storage

Cut flowers are stored for overnight in water in refrigerators maintained at about 7°C or lower. Some cut flowers can be held satisfactorily for several days in storages operated at 1°C, and some rooted and un-rooted cuttings can be kept in good condition for several weeks at that temperature. Some bulbs and plants require a period of cool storage in order to flower satisfactorily in the shortest time. And some holiday pot plants can be held in cool storage when they flower before the scheduled time.

Refrigerated storage may not be an absolute necessity at each greenhouse, but there are few greenhouses that could not use it to good advantage if it were available.

OTHER ESSENTIAL EQUIPMENTS FOR GREENHOUSE

Some of the equipment used in flower and plant production is standard equipment used in industrial or other types of agricultural production or adaptations of this equipment. Other items have been designed for specific uses in the greenhouse.

Equipment for the Control of Light

The sun is the primary source of light in the greenhouse, and the structure is designed so that it will admit the maximum amount of sunlight. All available sunlight is needed for plants in the greenhouse from fall to spring, but some reduction in light is usually required during the summer. This reduction in the amount of light at that time of the year can be handled by the use of screening material within the greenhouse or by the application of shading material to the outer surface of the greenhouse.

When screening materials are used within the greenhouse, they may be a fixed installation for the season or they may be removable so that they may be adjusted on short notice as needed. The density of the screening material that is needed depends on the crop or the stage of growth. Propagation areas may need considerable reduction in light intensity depending on the use of moisture in the area. About 80 per cent reduction in light intensity could be used when the area does not have a misting system and a 50 per cent reduction in amount of light should be sufficient when an effective mist system is used. Foliage plants in general should be provided with about 65 per cent light in reduction, and those that require less light should be given 80 per cent reduction. For general crop growth, 50 per cent light reduction frequently is used during the summer. For protection of standard chrysanthemum flowers, it is common practice to use from 30 to 50 per cent reduction in light intensity during the summer. Manufacturers of shading materials list the light reduction percentage of their products.

Kinds of Screening Materials

Several different kinds of screening materials are suitable. The cotton fabrics such as tobacco cloth, cheesecloth, and lightweight muslins usually cost less, but they are less durable than synthetic fabrics such as Saran, polypropylene, and polyester. Each type of fabric may have standard widths in which it is made, and these can be sewed together for continuous coverage of larger areas. A shading material may be sprayed on the exterior of the greenhouse covering instead of or in addition to the screening material that is used within the greenhouse during the summer. These commonly are called **shading compounds**, and several formulations are manufactured. In addition to the compound that is sprayed, a sprayer is required. This could be the sprayer that is used for application of pesticides, or it could be a sprayer used solely for this purpose. In some large installations, the shading spray is applied by aircraft.

Photoperiod for Flowering

The crops that need short photoperiods for flowering in naturally long days must be covered with fabric that excludes light. Black fabric is used. Cotton was the fabric used initially, and it still is used. In addition several types of plastic covers are used. There is a wide range in costs and durability of these materials. Also some materials can be handled better than others. The decision of which type of photoperiod-shading material to use needs to be based on these three factors-initial cost, durability, and handling characteristics.

Shading for the Control of Photoperiod

It needs daily handling. The covers are drawn in the evening and withdrawn in the morning. If this is a bench by bench operation, much manual labor and time are required. For this method of covering, two or three wires are installed the length of the bench about 15 centimeters above the maximum height of the plants and supported about every 3 to 4 meters. The entire length of the bench must be covered and the ends enclosed to exclude light. The length of the covers usually is about 15 meters.

Electric Light Equipment

In addition to industrial-type lighting that is used in working rooms, boiler rooms, offices, and for security purposes outdoors, there are some uses of electric lights that are specific for the greenhouse business. The main work that is done in greenhouses during the night is periodic checking of temperature and ventilation. Only minimum lighting is used for this purpose. The night worker uses a flashlight, and the only electric lights provided in the greenhouse may be by the ventilator operators and the thermometers.

Fluorescent Light

Fluorescent light has been used successfully as the sole source of light in some propagation areas. The tubes are placed together as closely as possible and within 30 centimeters of the top of the plants. The length of photoperiod usually is 16 hours. More recently **high intensity discharge** (HID) lamps have been used instead of fluorescent lamps. Either type of lamp can be used depending on the space limitations and the costs involved. The HID lamps require more space, and the cost of equipment may be greater than it is for fluorescent lamps.

Electric Lights for Photoperiod Control

Regulation of photoperiod is used for chrysanthemum, kalanchoe, poinsettia, and some other greenhouse crops. Plants vary in their sensitivity to light quantity. *Poinsettia* and kalanchoe are much more sensitive to light than chrysanthemum. Also there may be a variance in sensitivity among cultivars of a genus. The initial installations were made on mums, and it was recommended at that time that incandescent lamps be used to provide a minimum of 10 foot candles of light at the top level of the plants. For benches up to 120 centimeters wide, that light intensity could be provided by 60 watt lamps spaced 120 centimeters apart and 60 centimeters above the plant tops. The lights are controlled by a timer, and the length of the lighting period is varied with the season of the year so that the dark period is less than the minimum required for the long-day effect. This lighting may be an extension of the natural day length, or it may be an interruption of the night. It is more economical to interrupt the night than to extend the day. This lighting may be done around midnight, and it is possible to light either before or after midnight. Half the area then can be lighted at a time. Because of the reduced power demand at one time, smaller-size wire may be used, and the rate may be lower, too, during the night.

FIG. 23.40: Irrigation of pot plants by means of small diameter plastic tubes is efficient and economical.

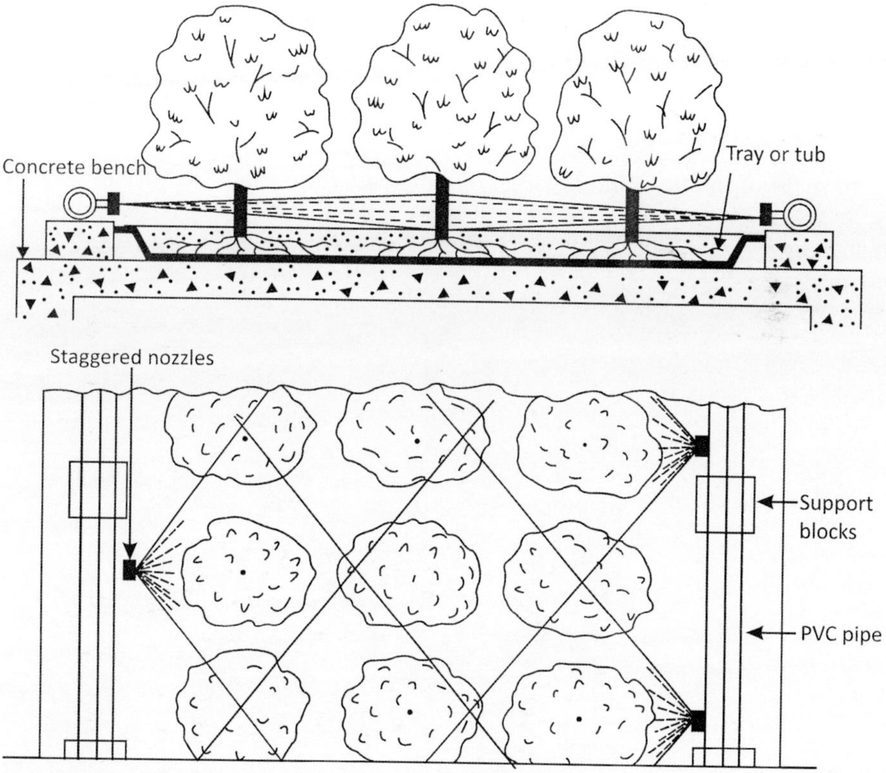

FIG. 23.41: Presentation of perimeter watering system. (Above): Sectional view. (Below) Top view.

Intermittent Light

Subsequently it was determined that photoperiodic lighting does not have to be continuous. It can be intermittent. This is known as **cyclic lighting** and a typical cycle was 6 minutes of light followed by 24 minutes of darkness with the cycle repeated for 4 hours from 10 P.M. to 2 A.M. With this arrangement five areas could be lighted in the 4 hour period; thus the power load would be reduced considerably at any one time. Cyclic lighting was not universally accepted commercially.

Problems with Fluorescent Lights

In smaller areas, the use of fluorescent lights could present some problems in shielding plants at some distance from the light. In using incandescent lamps over each bench, it is only the adjacent benches that would need to be shielded. With pink fluorescent lamps, the second and third benches distant from the tubes probably would need to be shaded from the light.

EQUIPMENT FOR MOISTURE CONTROL IN GREENHOUSE

The application of water to crop soils is known as irrigation. Irrigation equipment has been designed for distributing water to greenhouse soils and controlling the frequency and length of irrigation. When water is applied to aerial portions of plants, it is in the form of mist and the procedure is called **misting**. The objective usually is to increase relative humidity with the result that moisture loss (transpiration) from the plant is reduced. Other means for increasing humidity may be used.

How to Reduce Moisture in the Greenhouse

Irrigation Equipment: Various kinds of irrigation equipment have been used in greenhouse. A two cm hose is used, and it must be made from materials that remain pliable regardless of temperature. The irrigator uses both hands on the hose-spaced about 1 meter apart.

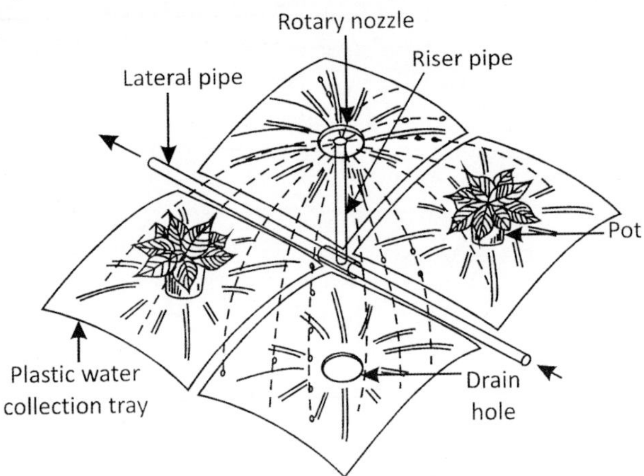

FIG. 23.42: Overhead sprinkler watering system.

1. For a right handed worker, the right hand is just in back of the delivery end of the hose and directs the stream of water to the soil.
2. For the left hand drags the hose as the irrigator walks while watering. The hose must drag freely and not kink.
3. For pot plant irrigation, each pot is watered individually with the tip end of the hose just above the soil.

4. For cut-flower irrigation, the faucet is fully opened. The tip end of the hose is just above the soil, and a water breaker must be used for this kind of watering. The water delivery from the hose in cut-flower irrigation is about 60 liters per minute.

Water Distribution Systems for Pot Plants in Greenhouse

Nozzles installed above the plants may be used for bedding plants or for small plants. This method is not practical for larger plants because the umbrella effect of the foliage causes uneven watering and the moisture on leaves and flowers may provide conditions suitable for pathogen growth.

Kinds of Nozzles: Various kinds of nozzles are available, and the system has to be designed for the characteristics of the particular nozzle that is used. A nozzle in common usage is a deflection type that delivers a horizontal spray full circle-360°-at the rate of about 2 liters per minute. These nozzles are spaced about 1 meter apart, and they may be installed on rods for individual placement throughout the area or in pipe that is placed above the plants.

Nozzles for Larger Pots: 10 centimeters and up-usually are irrigated with small diameter (about 0.1-centimeter internal diameter) plastic tubes directed to each pot. Approximately 500 pots can be supplied from one 2-centimeter poly pipe.

Capillary Irrigation System

Pot-plant soils may be **sub irrigated by capillarity** from a porous substance beneath the pots. Although capillary irrigation may be used for plants in pots of any size, it is most useful for plants in small pots-9 centimeters and smaller. The porous material on which the pots are set can be fine-textured sand or fabric that will promote capillarity.

The **requisites for a capillary irrigation system** are a level and watertight bench surface, a porous layer on the surface, and a means of distributing water evenly over the porous material. Polyethylene film usually can be used for waterproof covering on the bench. Sand or several fabrics are available for use as the porous material, and water may be distributed by various means. But whatever method is used, it must be adjustable so that the porous layer may be kept wetter or drier as needed.

Water Distribution Systems for Cut-Flower Crops

In cut-flower irrigation systems, the water is distributed either by nozzles, ooze, or by piddle. Any type of system can be used satisfactorily. The irrigation nozzles throw a horizontal spray. Some of them deliver water 360° and other either 180° or 45°. The 360° nozzles are suitable only for installation at the center of the bench, but the 180° nozzle may be placed at the center or at the side. A 360° nozzle frequently used delivers water at the rate of about 2 liters per minute and these nozzles are installed about 1 meter apart.

EQUIPMENT FOR REGULATION OF IRRIGATION

When water-distribution systems are used they may be installed with hand valves and operated manually or they may have electric valves so that they can be controlled by timers. There is a definite advantage in being able to control irrigation by means of a timer. When the grower makes the selection of the benches that are to be watered, the irrigation proceeds from one bench to the next automatically. Timers for pot plant crops should have the capability of providing irrigating times from 1 to

5 minutes. The length of irrigation time for cut-flower crops needs to be adjustable from 5 to 20 minutes.

The water flow in the misting line is controlled by an electric water valve. The misting line must be installed level, and the valve must be at that level or below. There then will be no drip from the nozzles when the valve is shut. The pipe will remain full of water and misting will be instantaneous for the entire length of the misting line when the valve opens. The timer that perates the misting system should provide misting periods of about 5 seconds with adjustable intervals between misting from about 1 minute to 1 hour.

EQUIPMENT FOR HANDLING FERTILIZER

Greenhouse growers determine the need for fertilizer applications by having analyses made of soil or leaves for the mineral nutrient content. These are laboratory procedures that are best handled by specialists in that business. Soil analysis should be made on a regular basis, and leaf analysis usually is done sonly if there is evidence of trouble that cannot be identified by other means.

Every greenhouse operator, however, should have the equipment to determine the total amount of salts (solube salts) present in the soil and the pH. The instrument of making the soluble salt determination is the Solu Bridge, and pH can be evaluated by color change with sensitive paper used for that purpose. The Soly Bridge is an electronic instrument that can measure conductivity in a soil solution. The greater the concentration of salts in the soil solution, the higher the Solu Bridge reading. Low readings indicate that a fertilizer application should be made, and high readings signify that the soil should be leached with water.

Fertilizer applications may be made either in the dry or the liquid form. Fertilizer applied in dry form may be distributed over the soil as uniformly as possible and mixed into the soil as the soil is prepared for planting, or it may be distributed evenly over the soil surface after planting. Usually these are operations by hand. Special equipment for the distribution of fertilizer in the dry form for greenhouse crops is not commonly available.

Application of Fertilizer in Liquid Forms

It has been done for years, but methods have changed as materials and equipment have been introduced. The earliest procedures were the application of manure water. Manure was added to large tanks containing water, and the manure water was pumped through pipes to the various greenhouses where it was applied to the soil. When fertilizers became available as soluble, chemical salts, they eventually replaced manure as the fertilizer source. In some instances, this method of fertilizer application still is used. The equipment required is a tank large enough for the area to be treated, a water pump large enough to supply the water volume needed, and pipe lines from tank to greenhouse area. Usually the tank method of fertilizer application is designed for periodic use rather than for use with each irrigation.

Proportions for Uniform Supply of Fertilizers

Some of the advantages in equipment of this type are that large tanks are not required and a more uniform supply of fertilizer in the soil is provided. These devices usually are known as **proportioners**. There are basically two types of proportioners, and because of the way they function they may be called positive displacement and pressure difference. The positive-displacement type of fertilizer proportioner generally is more reliable than the pressure-difference type.

The **positive displacement type** of fertilizer proportioner frequently is called an injector because it injects a uniform volume of fertilizer concentrate into a unit volume of irrigation water.

Manufacturers of this equipment use various means of sensing irrigation water volume and injection of fertilizer concentrate.

The **pressure difference type of fertilizer proportioner** commonly develops the pressure difference by means of a Venturi in which the irrigation water is routed through a reduced size pipe or orifice. An elementary kind of proportioner of this type commonly is referred to as a siphon. A variable amount of fertilizer concentrate may be proportioned into the irrigation water with this type of device depending on the rate of water flow and the water pressure.

Selection of Fertilizer Proportioner

For selection of a fertilizer proportioner, the important considerations are the suitability for the maximum and minimum water flow rates, proportioning reliability, durability, and service ability. Some fertilizer proportioners are portable, and they are moved from area to area as needed. The fertilizer proportioners that are permanently installed must be piped with a by pass so that either water or injected water can be used for irrigation as desired. The piping in the greenhouse must be arranged so that the injected water has access only to irrigation systems. The water-supply system for boiler, toilets, pads, fountains, and other uses must be separate from the irrigation water supply. Fertilizer injectors must be checked periodically for accuracy. The manufacturers have recommended checking procedures for their units, and it is important that the check be made at least quarterly.

PEST AND PATHOGEN CONTROL EQUIPMENT

In Addition to equipment for applying pest and pathogen control materials there must be a secure storage area, accurate means for measuring the materials, protective gear for personnel, and an adequate place for personnel to cleanse themselves.

Hydraulic Spraying

The most common means of applying pesticides is by hydraulic spraying because the control materials usually are in the form that can be used in that way. It is possible that some control materials are available that may be applied as a dust, granule, fog, fumigant, or aerosol. Those methods may be used as alternate means of application, but a good means of hydraulic spraying will be needed for the standard method of application.

When applying pesticides and pathocides the plant surfaces must be covered completely and uniformly without excessive run-off to the soil. This requires the dispersion of the spray liquid into very small particles, and these result when the spray liquid is forced at high pressure through the tiny orifices of the sprayer nozzles. Sprayer pumps that develop 17.6 kilograms per square centimeter pressure are satisfactory. The sprayer pumps may be powered by either electric motors or gas engines.

Stationary and Portable Sprayer

The sprayer may be stationary with the spray piped to outlets within the greenhouses, or it may be portable so that it can be wheeled to the greenhouse where it will be used. Portable power sprayers for use in the greenhouse have tanks from about 38 to 114 liters in volume. The pumps on the smaller sprayers may develop only about 7 kilograms per square centimeter pressure which makes them less efficient than the larger units. When portable sprayers are used, usually they are equipped with long enough hoses so that all areas of the benches on either side can be reached when the sprayer is positioned in the center walk of the greenhouse.

Dusting

Pesticides are not available often in dust form. If the material that is needed can be obtained as a dust, that method of application should be used because very little preparation is required and only a short time is needed for making the application. Dusting is a particularly efficient way to handle isolated infestations. It can be done promptly and should control the infestation in the initial stages. Small manual dusters may be used for this purpose.

Foggers use oil-base pesticides that were formulated for use with specific equipment. There are only a limited number of pesticides that can be used with this kind of equipment. Also the oil that is dispered along with chemical is toxic to some kinds of plants. Before investing in fogging equipment, growers should determine how suitable it is for their crops.

For Fumigators

No special equipment is required for use with fumigators, but there are very few pesticides that can be dispersed in this way. The most complete protective equipment is needed for the individual who makes the aerosol application, but the only applicating equipment needed, are the hose, tube, and nozzle assembly that attaches to the aerosol. There are very few pesticides that can be prepared for aerosol dispersion.

EQUIPMENT FOR APPLICATION OF OTHER AGRO CHEMICALS

Most of the other agricultural chemicals either are sprayed over the plants or are drenched on the soil. In either instance it is a hydraulic application. Growth regulators are used on several greenhouse crops. When they are applied as a spray, 4 liters of the spray liquid will give coverage to plants occupying 20 square meters of space. Much of the growth-regulator spraying can be handled with small compression sprayers with tank volumes from about 4 to 12 liters. However, these sprayers are more effective if they are the type that is pressurized from compressed air tanks rather than pumped by hand. The sprayers used for application of growth regulant materials should be used for that purpose only.

In the greenhouse, weed control in crop soils is one of the results of steaming the soils before planting. Non-crop areas such as under benches and walks do need some weed control measures. Most of the herbicides have the potential of damaging crop plants as well as weeds, and these materials must be used in such a way that there is no contact with the crop plants. Herbicides that volatilize cannot be used because they could affect crop plants at some distance from where they are used. Some of the herbicides that do not volatilize have been used in the greenhouse, but they must be handled very carefully. Coarse sprays are used and directed only at the weeds so there is no possibility of the spray reaching the crop plants. Small, hand pumped sprayers usually are suitable for weed control work. These sprayers, however, must be used only for weed control.

EQUIPMENT FOR PREPARATION OF SOIL FOR PLANTING

Greenhouse soils may be anything from field soils that are naturally derived to gravel. But most often they are mixtures of two or more materials. The mixing is done by hand or machine, and the prime objectives are to provide a soil that drains readily in greenhouse conditions and has a suitable mineral nutrient content. Various ingredients have been used for greenhouse soil mixtures through the years. The initial mixing procedures involved adding organic matter and possibly some aggregate to field soils to improve the water drainage characteristics of the soil. Manure was a popular ingredient in soil mixtures for years. This product varied in straw and fertilizer content, and the resulting mixtures were quite different in spite of having the same quantities of soil and manure in them each time.

Manure seldom is used in greenhouse soil mixtures anymore, but chopped straw often is used-particularly with cut-flower crops.

Peat has been used in soil mixtures for years and continues to be widely used. There are several different kinds and grades of peat. Some are powdery and others are fibrous, and pH varies considerably. It is the fibrous peats that are of most value in greenhouse soil mixtures, and the pH will need to be evaluated and adjusted as needed.

Eventually some greenhouse soil mixtures were made that contained no field soils. A single material was used in some instances. Gravel was used for some cut-flower crops, and peat was used for some container crops. They still are used for some crops. Peat has been used as one ingredient in several different soil mixtures, and the other ingredients have been various materials such as sand, perlite, vermiculite, or other materials.

Ground bark is a principle ingredient in some soil mixtures. The soil mixtures that contain no field soils usually are used only for crops grown in containers. The longer-term cut-flower crops can be grown in soil mixtures that do not contain field soils, but ingredients need to be used that provide stable drainage characteristics for the entire period of use.

Greenhouse operators who do mix soil will need a soil shredder, a soil mixer, and soil steaming equipment. The equipment required for the preparation of potting soils is different from that needed for cut-flower soils. If field soil is used in the soil mixture, a shredder will be needed. The shredder is used to break the clods into smaller pieces so that a uniform soil mixture can be made, but the shredder must not pulverize the soil.

Soil Shredders

Various sizes of soil shredders are available, from the small ones that are manually loaded by shovel to the ones that are large enough to require loading by tractor. Also the larger shredders may have conveyors or elevators to handle the soil into or out of the shredder. Shredders are used commonly in the preparation of potting soils but rarely with cut-flower soils.

Mixers of some sort are needed for potting soils. Small batches of soil can be mixed satisfactorily with shovel-in-hand. Larger soil batches are blended in mixers in which the various ingredients are placed in the ratio desired. In many instances concrete mixers have been adapted for this purpose. Some growers have found that mixers designed for blending feed mixtures are more satisfactory than concrete mixers for the ingredients that they use. Whatever mixer is used, it must provide a uniform blend of materials in the mixture. The difference in density of the ingredients may cause a mixing problem, and the moisture content of all materials needs to be correct.

With **cut-flower soils**, the mixing is done with tillers. Excellent tilling can be done with the use of spade forks by hand, but this requires more labor and time than is available sometimes. Most commonly rotary tillers powered with gasoline engines are used. The ingredients that are to be mixed with the field soil are layered on top of the soil before tilling. Some common problems with the rotary tillers are that they pulverize the soil, do not get to the bottom or corners and edges of the bench or bed, and are difficult to operate in the confined space in the greenhouse. The best answer in many situations is a combination of rotary tilling and hand spading.

MATERIALS HANDLING EQUIPMENT

The pot plant grower has much more material to handle than the cut-flower grower, and the greenhouse and benching arrangements determine the type of handling equipment to be used. Pot plant greenhouses are constructed so that wheeled vehicles can be operated in them.

Trolley Type Conveyors

Trolley type conveyors that operate from an overhead rail are rather easily installed in greenhouses and can be very efficient movers of pot plants. There has not been enough consideration given to the use of this type of system in the greenhouse.

Before building greenhouses, the pot-plant grower should spend some time viewing movement of materials in some warehouse. Some of the procedures that they use could suggest methods that are adaptable to the greenhouse.

REPAIR AND CONSTRUCTION EQUIPMENT

Any greenhouse operator needs to be a bit of an engineer. It is almost impossible to rely solely on specialist professionals to handle the maintenance of the physical facilities. Owners of smaller greenhouse businesses must be able to handle most of these jobs themselves boiler repair and operation, painting, glazing, pipe fitting, carpentry, concrete work, gas engine maintenance and motor repair, and routine maintenance of electrical and refrigeration equipment. The larger greenhouse operation will need to have one or more individuals who handle the engineering problems on the place.

CHAPTER 24

Hydroponics for Hi-Tech Horticulture

INTRODUCTION

Hydroponics by definition means *water working*. In practical use, it means growing plants in a water and nutrient solution, without soil. Hydroponics allows a gardener to grow plants in a more efficient and productive manner with less labour and time required. The science of hydroponics proves that soil is not required for plant growth but the elements, minerals and nutrients that soil contains are essential. Soil is simply the holder of the nutrients, a place where the plant roots traditionally live and a base of support for the plant structure. Today it is used to describe all ways of growing plants without soil. It is synonymous with the term soil-less culture, and both terms are restricted to the growing of plants without soil.

The term hydroponics is used to describe many different types of systems for growing plants without soil. The most common systems are:

1. Water culture or aquaculture is a system of plant growing in which the plant roots are immersed in water containing dissolved nutricul nutrients.
2. Aggregate culture, A system of plant growing in which a material such as sand, gravel, or marbles supports the plant roots
3. Aeroponics A system of plant growing in which the plant roots hang in the air and are misted regularly with a nutrient solution
4. Continuous flow systems, A system of plant growing in which the nutrient solution flows constantly over the plant roots. This system is the one most commonly used for commercial production.

Advantages of Hydroponics

Hydroponics is probably the most intensive and fascinating method of crop production in today's agricultural industry. In combination with greenhouses and protective covers (controlled environment agriculture), it is high technology and capital intensive. With the possibility of adjusting air and root temperature, light, water, plant nutrition, and adverse climate, this combination can be made highly productive, conservative of water and land and protective of the environment.

The growing of plants hydroponically has number of advantages over open field agriculture. High density maximum crop yield, crop production where no suitable land exists, crop cultivation regardless of seasonality, more efficient use of water and fertilizers and minimal use of land area are the principal advantages of hydroponics in combination with controlled environment agriculture. Another major benefit is the possibility of obtaining pesticide free products, which fetch higher prices at the increasingly ready markets, at present.

FIG. 24.1: Hydroponic culture.

The important advantages of hydroponics are given below:

1. The increased control over growing conditions makes it easier to provide the best possible environment for plants, leading to better quality produce and higher yields.
2. The production in hydroponics may be increased approximately two times as compared with soil cultivation in a comparable area with correct management practices because plants do not have to compete for moisture and nutrients
3. Hydroponic gardening can provide plants with optimum quantities of the necessary nutrients during the different seasons. This will enable maximum growth to be achieved.
4. A small hydroponic garden can be setup almost anywhere even in upstairs or balconies, open spaces and in protected structures because land is not necessary.
5. The work involved in hydroponics is generally light. Heavy manual operations associated with normal cultivation practices such as digging, making beds, watering, weeding and ploughing are eliminated.
6. Hydroponic produce generally tastes better and is higher in nutritional value than field grown crops.
7. Plants grow 50% faster than the soil grown under the same conditions because they have easy access to food and water.
8. Continuous cultivation and off-season production is possible
9. Occurrence of soil borne diseases or nematode damage is not possible, so hydroponic products are exported safely.
10. Vegetable cultivation can be done with leisure sense.
11. Many plants were found to give yield early in hydroponics system.
12. Disabled people who are unfit for open field cultivation can adopt hydroponic systems in gardening as benches may be adopted to their requirements.
13. Easy to hire labour as hydroponic system is more attractive and easier than cultivation in soil.
14. No need of electricity, pumps, etc. for the non-circulating systems of solution culture.
15. Offer possibility of growing a wide variety of vegetable and flower crops including anthurium, marigolds, etc.
16. Water wastage is reduced to minimum.

17. There is no need of crop rotations as growing medium can be reused continuously or replaced.
18. The plants are uniform in growth and maturity.
19. Erosion of top soil is not even an issue since hydroponic gardening systems use no top soil. If agricultural trends continue to erode top soil and waste water, hydroponics may soon be our only solution.
20. Hydroponic plants require very little energy to find and breakdown food. The plants then use this saved energy to grow faster and to produce more yield.
21. Hydroponics is not subject to variable weather conditions and seasonal changes due to controlled environment. For example, frost would not be a setback anymore because temperature is controlled. Strong winds and rain will no longer destroy crops due to an enclosed surrounding. These factors ensure maximum production.

Modern Hydroponics

- The recent developments in plastics have cut hydroponic system costs, which has made hydroponics financially feasible the more people.
- The development of pumps has allowed hydroponic systems to evolve to fully automatic systems.
- Presently, advances are being made in the development of highly specialized hydroponic cultures for use in atomic submarines.

Future Strategy of Hydroponics

1. Further development of the application of solar heating in hydroponic greenhouses will reduce cost and economic impact of hydroponics.
2. Currently, plans are being drawn for using the techniques of soil-less culture on space stations, and perhaps one day on surfaces of other celestial bodies (planets, moons, etc.) that do not have soil.
3. Developing countries, along with all other nations, will be able to feed many people using less land than current farming techniques.
4. Advances in lighting technology will lead to a more widespread use of hydroponics in areas with limited seasonal sunlight.
5. The application of hydroponics in providing food in areas having vast regions of non-arable land, such as deserts and mountainous terrain, will be more heavily used.

Limitations of Hydroponics

1. Hydroponic requires higher initial capital expenditure. The expenditure is increased further when the soil-less culture is combined with controlled environment agriculture.
2. Hydroponic systems require high degree of management skills especially for solution preparation, maintenance of pH and EC, nutrient deficiency judgment and correction, ensuring aeration, maintenance of favourable condition inside protected structures, etc.
3. Considering the significantly high cost, the soil-less culture is limited to high value crops of the area.
4. A large-scale cultivator may have to purchase instruments to measure pH and EC of the nutrient solution.
5. Energy inputs are necessary to run the system.
6. Yields were found to decrease when temperature of the solution rises during

warm periods.

7. The sharing of nutrient solutions can spread disease and pests a lot faster in water than soil. Since plants are grown so close together, an entire crop could be destroyed before it could be controlled.

Most of the gardeners think that hydroponics is not associated with biotechnology. Webster's dictionary defines biotechnology as biological science when applied especially in genetic engineering and recombinant DNA technology. Plants that are grown hydroponically has not been genetically altered by the additions of foreign genes to enhance a desired trait. Genetically modifying plants is an invasive process where hydroponics is not.

Hydroponics is not organic. Webster's dictionary defines organic as, relating to yielding or involving the use of food produced with the use of feed or fertilizer of plant or animal origin without employment of chemically formulated fertilizers, growth stimulants, antibiotics, or pesticides. Farmers believe organic means no pesticides or herbicides. The plant must grow naturally without using man made chemicals to control pests, weeds and diseases. Organic process feeds the soil and the soil feeds the plant. It is expected for plants to find the nutrients from the soil, assuming that the soil contains all essential nutrients. In hydroponics, the plants are fed directly without spending energy seeking nutrients with its root but rather just spends its energy growing.

HYDROPONIC TECHNIQUES AND THEIR CLASSIFICATION

Classification

Hydroponic systems are generally classified as under:

1. **Open system:** A system of hydroponics in which nutrient solution, once delivered to the plant roots, is not reused.
2. **Closed system:** A system of hydroponics in which surplus solution is recovered, replenished and recycled.
3. On the basis of nature of media used, the hydroponic systems are classified as under:
4. **Liquid hydroponics:** A system of hydroponics in which liquid medium is used.
5. **Aggregate system:** A system of hydroponics in which a material such as sand, gravel or marbles is used to support the plant roots.
6. **Aeroponic system:** A system of hydroponics in which rockwool, net pots, vine clips or other devices are used to support the plant while its roots grow completely submerged in nutrient solution.

On the basis of movement of solution, the hydroponics system are classified as under:

1. **Passive systems:** A system of hydroponics without using pump. The plant take up nutrient through a fiberous wick from the pot to the nutrient solution. The solution will generally help a bubbler to maintain proper oxygenation.
2. **Active systems:** A system of hydroponics in which nutrient solution moves actively using a pump.

Current hydroponic systems of cultivation can be classified as under:

1. Liquid hydroponics (Solution culture or aquaculture)
    a) Circulating methods (closed systems)
        i) Nutrient film technique (NFT)

   ii) Deep flow technique (DFT)
b) Non-circulating methods (open systems)
   i) Root dipping technique
   ii) Floating technique
   iii) Capillary action technique

2. **Aggregate systems (Solid media culture):** These can be open systems or closed systems.
   a) Hanging bag technique
   b) Grow bag technique
   c) Trench or trough technique
   d) Pot technique
3. Aeroponics
   a) Root mist technique
   b) Fog feed technique

HYDROPONIC TECHNIQUES

Different hydroponics systems have been described as under:

Liquid Hydroponics

Circulatory Methods (Closed Systems)

Nutrient Film Technique – NFT: The nutrient film technique (NFT) was developed by **Dr. Allen Cooper** in England in 1970, to increase the year-round production of greenhouse crops. It has since been widely adopted in Western Europe. The Nutrient Film Technique or NFT system is an active recovery type hydroponic system. Again, using submersible pumps and reusing nutrient solutions. The NFT uses a reservoir with a submersible pump that pumps the nutrient solution into a grow-tube where the roots suspended. The grow-tube is at a slight downward angle so the nutrient solution runs over the roots and back into the reservoir. The nutrient solution flows over the roots up to 24 hours per day.

In this method, no growing medium is required and the plants grow through light-proof plastic films placed over shallow, gently sloping channels. A steady flow of nutrients is maintained along the channel, and the roots grow into dense mats, with a thin film of nutrient passing over them (hence the name of the technique). A downside of the technique is that it has very little buffering against interruptions in the flow e.g. power outages, but overall, it is probably one of the more productive techniques.

Oxygen is needed in the grow-tube so capillary matting or air stones must be used. The plants are held up by a support collar or a grow-basket and no growing medium is used. The NFT system is very effective. Although, many hydroponic growers find it difficult to fine tune. It can also be very unforgiving, with no growing medium to hold any moisture, any long period of interruption in the nutrient flow can cause the roots to dry out and the plants to suffer and possibly die.

Light must be prevented from reaching of the roots of the plants as this will prevent growth of the plants. It is important to check the nutrients level regularly. This can be minimised by replacing the nutrients solution on a regular basis.

A thin film of nutrient solution is always in contact with the roots. While the nutrient solutions circulated, the root surface is exposed to air. This helps the roots to breathe. This system is very good for producing fruits and vegetables.

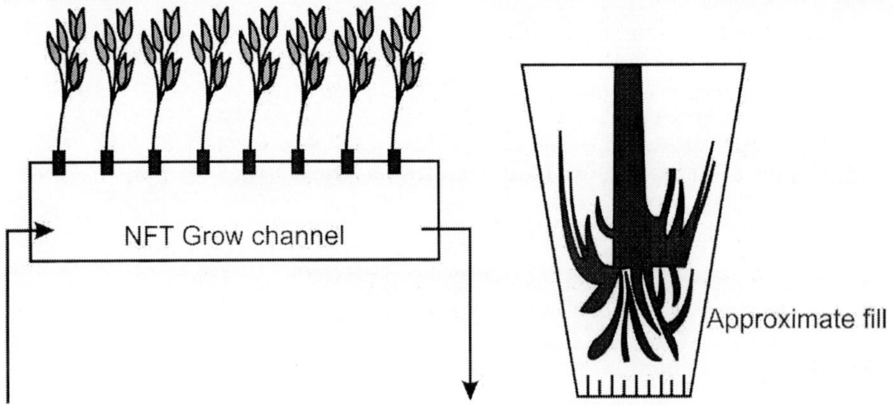

FIG. 24.2: Nutrient Film Technique

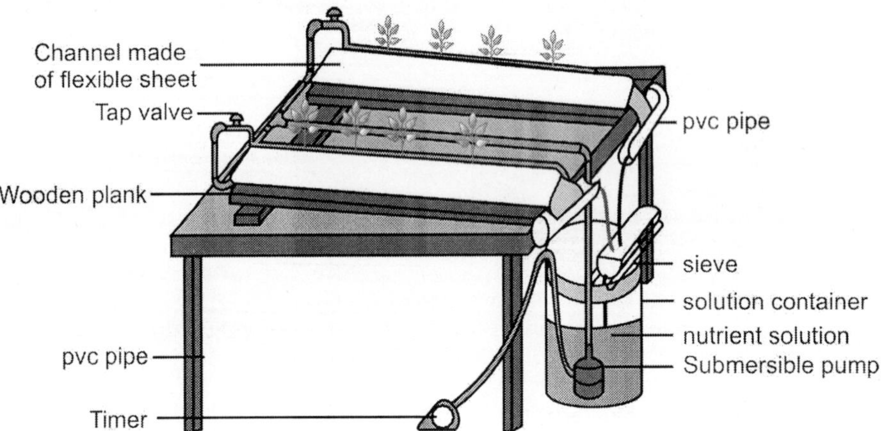

FIG. 24.3: Main features of NFT system.

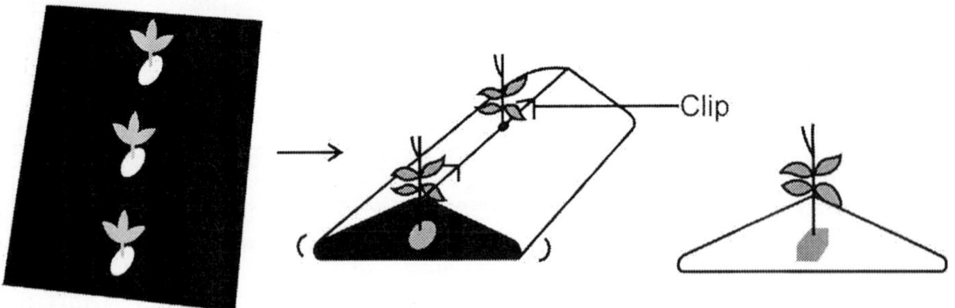

FIG. 24.4: Basic structure of NFT model.

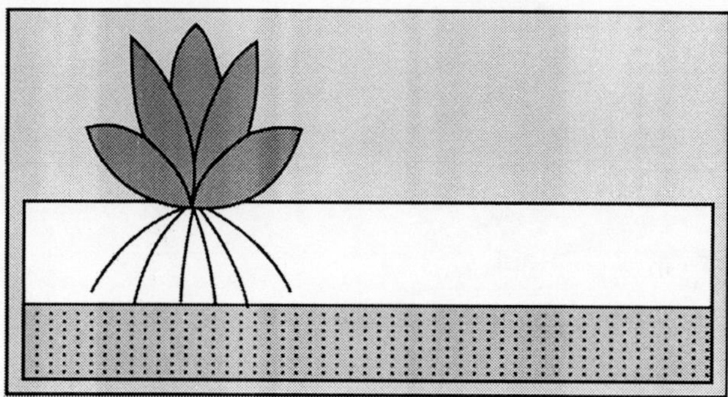

FIG. 24.5: Nutrient Film Technique – NFT

Deep Flow Technique – DFT: A depth of nutrient solution (several inches deep) is circulated around the roots by a pump and gravity drain. This method is referred to as "Dynamic Root Floatation" or "Raceway Hydroponic" technique. This system is ideal for growing leafy vegetables.

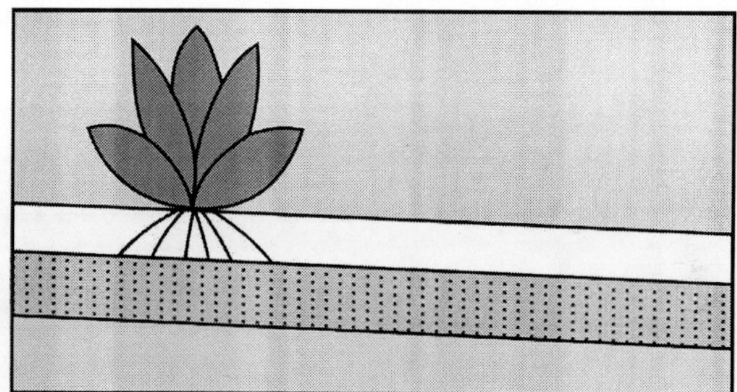

FIG. 24.6: Deep Flow Technque – DFT.

As the name implies, 2-3 cm deep nutrient solution flows through 10 cm diameter PVC pipes to which plastic net pots with plants are fitted. The plastic pots contain planting materials and their bottoms touch the nutrient solution that flows in the pipes. The PVC pipes may be arranged in one plane or in zig zag shape depending on the types of crops grown.

The zig zag system utilizes the space efficiently but suitable for low growing crops. The single plane system is suitable for both tall and short crops.

Plants are established in plastic net pots and fixed to the holes made in the PVC pipes Old coir dust or carbonized rice husk or mixture of both may be used as planting material to fill the net pots. Place a small piece of net as a lining in the net pots to prevent the planting material falling into the nutrient solution. Small plastic cups with holes on the sides and bottom may be used instead of net pots. When the recycled solution falls into the solution in the stock tank, the nutrient solution gets aerated. The PVC pipes must have a slope of drop of 1 in 30-40 to facilitate the flow of nutrient solution. Painting the PVC pipes white will help reduce the heating up of nutrient solution. This system can be established in the open space or in protected structures as part of CEA.

FIG. 24.7: Main features of Deep flow technique (Single plane pipe line system)

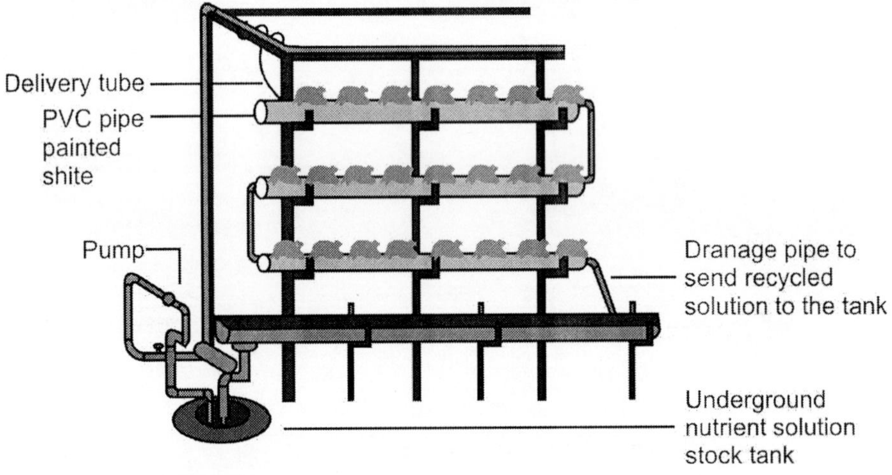

FIG. 24.8: Main features of Deep flow technique (Zig zag pipe system).

Non Circulating Methods (Open Systems)

The nutrient solution is not circulated but used only once. When its nutrient concentration decreases or pH or EC changes, it is replaced.

Root dipping technique: In this technique, plants are grown in small pots filled with little growing medium. The pots are placed in such a way that lower 2–3 cm of the pots is submerged in the nutrient solution.

Some roots are dipped in the solution while others hang in the air above the solution for nutrient and air absorption, respectively. This technique is easy and can be developed using easily available materials. This 'low tech' growing method is inexpensive to construct and needs little maintenance. Importantly, this technique does not require expensive items such as electricity, water pump, channels, etc. For root crops (beet, radish, etc.) however, an inert medium has to be used.

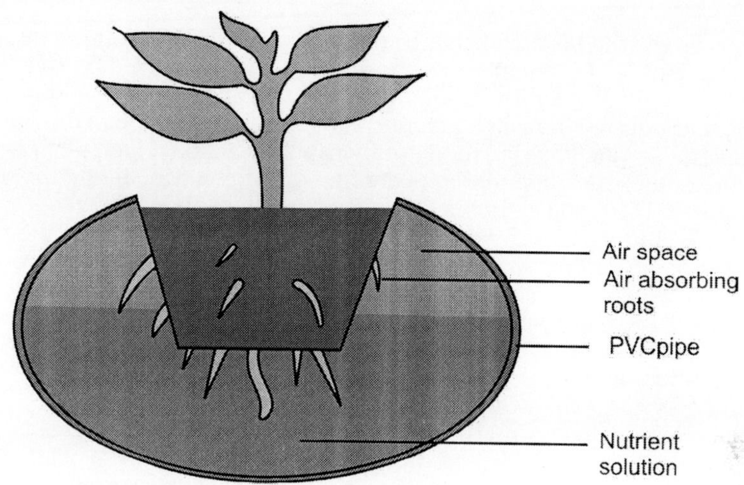

FIG. 24.9: Cross section of a PVC pipe system

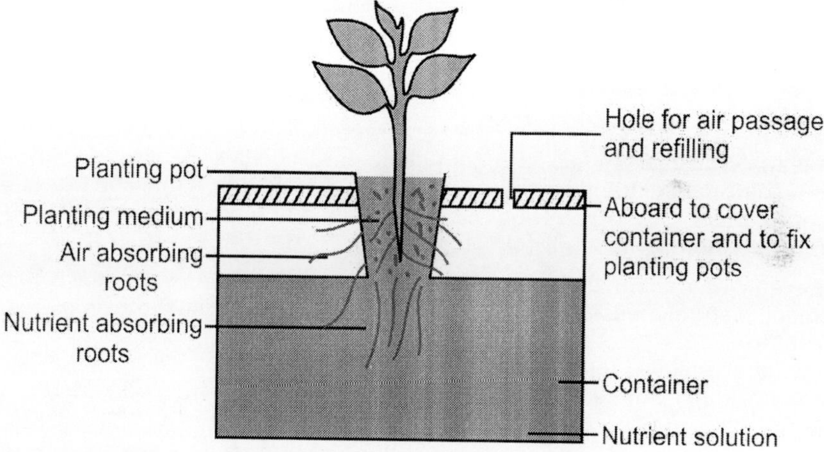

FIG. 24.10: Main features of root dipping method.

Root dipping technique for non root tuber crops: First, select a container for nutrient solution. The container can be almost any kind and shape except metal containers. Styrofoam or wooden boxes, plastic buckets or even cement troughs can be used. Styrofoam boxes are good as they can maintain the temperature of the nutrient solution. Place a black plastic sheet of at least 0.15 mm thickness as lining inside the boxes to avoid leakage and to reduce the light. The depth of the box must be about 25 – 30 cm to provide enough solution as well as enough space above the solution for oxygen absorbing roots. A board is required to place on the container to prevent light penetration. The planting pots are also fixed to this board. The number of holes in the board to fix the pots depends on the crops to be grown. An additional hole is necessary for air circulation and refilling. Seedlings are transplanted in plastic pots filled with sterilized old coir-dust or carbonised rice husk or the mixture of the two. Plastic net pots or plastic cups can be used.

Root dipping technique for root tuber crops: A 20 – 30 cm deep container can be used. It is lined with black polythene sheet and filled 1/3rd with nutrient solution. Leaving a space of about 7.5 cm above the solution level, fix a wire mesh in the box and fill with an inert medium. The seedlings / seeds are planted in the medium. At the early stage, nutrient solution will reach the planting medium through the perforated PVC pipe filled with the growing medium by capillary action. Later, plant roots will grow into the nutrient solution through wire mesh. At this stage, the perforated tube will be removed. The resulting hole will facilitate aeration and refilling.

Floating technique: Float systems take advantage the surface of the nutrient solution. Most float systems are long, rectangular reservoirs built out of cement or wood and lined with a durable, poly liner. Holes are cut in a foam board which floats on the surface of the water and plants in net pots are set in the holes. The plant roots dangle in heavily aerated nutrient solution. In areas where raw materials are limited and manufactured hydroponic systems and components are not available, the float system can be an economical means of hydroponic crop production.

Capillary action technique: Planting pots of different sizes and shapes with holes at the bottom are used. Fill these pots with an inert medium and plant seedlings / seeds in that inert medium. These pots are placed in shallow containers filled with the nutrient solution. Nutrient solution reaches inert medium by capillary action. Aeration is very important in this technique. Therefore, old coir dust mixed with sand or gravel can be used. This technique is suitable for growing ornamental flowers and indoor gardening.

Aggregate Systems

The following techniques involving inert solid media can be practiced using locally available materials. The media material selected must be flexible, friable, with water and air holding capacity and can be drained easily. In addition, it must be free of toxic substances, pests, disease causing microorganisms, nematodes, etc. The medium used must be thoroughly sterilized before use.

1. Inorganic natural media (gravel culture)
2. Organic natural media (smoked rice husk, saw dust, coconut fibre, coir dust and peat moss)
3. Inorganic artificial media (rockwool, perlite, vermiculite)
4. Organic artificial media (polyurethane, polyphenol, polyether, polyvinyl)

The most common types of media used in containerized systems of soil-less culture are peat-lite (Boodley and Sheldrake 1977), or a mixture of bark and wood chips. Tanins and acids present in the newly extracted coir-dust affect plants. Therefore, use at least 6 months old coir-dust. Dry, clean compressed coir-dust blocks are available for sale in the market. Different techniques described below, according to the method of holding the planting medium, can be practiced.

Hanging Bag Technique (Open System)

About 1 m long cylinder shaped, white (interior black) UV treated, thick polythene bags, filled with sterilized coconut fibre are used. These bags are sealed at the bottom end and tied to small PVC pipe at the top. These bags are suspended vertically from an overhead support above a nutrient solution-collecting channel. Therefore, this technique is also knows as 'verti-grow' technique. Seedlings or other planting materials established in net pots are squeezed into holes on the sides of the hanging bags. The nutrient solution is pumped to top of each hanging bag through a micro sprinkler attached inside the hanging bags at the top. This micro sprinkler evenly distributes the nutrient solution inside the hanging bag.

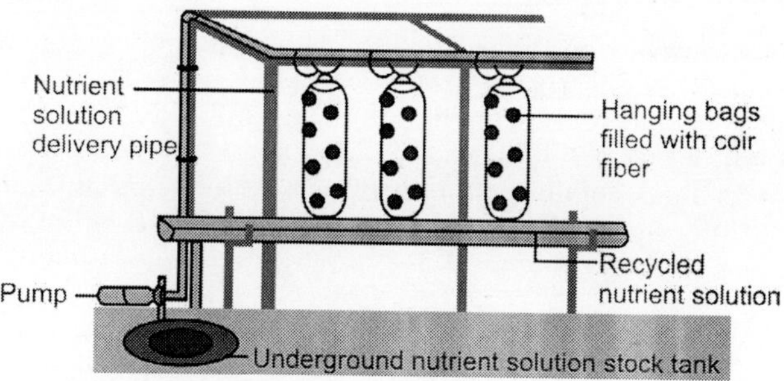

FIG. 24.11: Model of hanging bag technique.

Nutrient solution drips down wetting the coconut fibre and plant roots. Excess solution gets collected in the channel below through holes made at the bottom of the hanging bags and flows back to the nutrient solution stock tank. This system can be established in the open space or in protected structures. In protected structures, the hanging bags in the rows and amongst the rows must be spaced in such a way that adequate sunlight falls on the bags in the inner rows.

The bags are not heavy as they are filled with coconut fibre and can be used for about 2 years. The number of plants per bag varies depending on the plants. About 20 lettuce plants can be established per bag. This system is suitable for leafy vegetables, strawberry, and small flower plants. Black colour tubes will have to be used for nutrient solution delivery to prevent mould growth inside.

Grow Bag Technique

Using grow bag is a simple and cheap way to grow plants hydroponically. In this technique, 1 – 1.5 m long white (inside black), UV resistant, polythene bags are used. These bags are about 6 cm in height and 18 cm wide. The bags are filled with growing medium of choice and open ends are sealed with freezer bag ties or insulation tape.

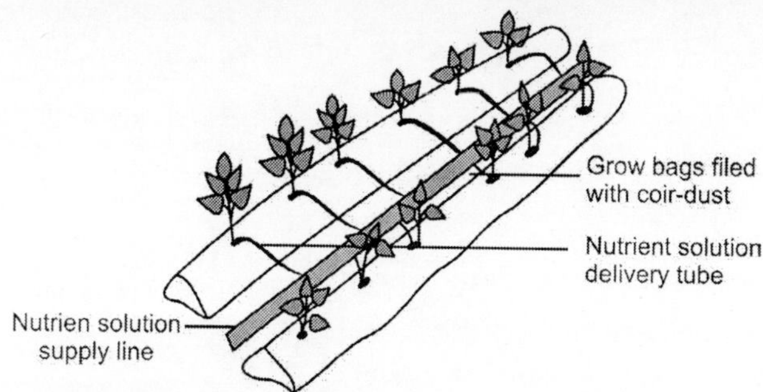

FIG. 24.12: Diagrammatic view of grow bag technique.

These bags are placed end to end horizontally in rows on the floor with walking space in between. The bags may be placed in paired rows depending on the crop to grow. Polyethylene sheet may be placed under the bags to ensure good hygiene. Make small holes on the upper surface of the bags and squeeze seedlings or other planting materials established in net pots into the coir-dust. 2-3 plants can be established per bag. Make 2 small slits around the outside as edges of the bags approximately 10 mm from the base to allow drainage of excess nutrient solution. Lay-flat bags, which have drainage slits (or overflow ports) cut along the sides an inch (2. 54 cm) or so above the base, would also benefit from a protective plastic sheet beneath them. Four to six plants per bag is usually recommended but the number will vary according to the type of plant grown or size of the bag.

Fertigation with black capillary tube leading from main supply line to each plant is practiced. The nutrient solution and water may also be added manually to these bags.

Trench or Trough Technique

In this open system, plants are grown in narrow trenches in the ground or above ground troughs constructed with bricks or concrete blocks. Both trenches and troughs are lined with waterproof material (thick UV resistant polythene sheets in two layers) to separate the growing media from rest of the ground. The width of the trench or trough can be decided depending on the ease of operation. Wider trenches or troughs will permit two rows of plants. The depth varies depending on the plants to grow and a minimum of 30 cm may be necessary. Old coir dust, sand or gravel, peat, vermiculite, perlite, old sawdust or mixture of these materials can be used as the media for this culture. The nutrient solution and water are supplied through a drip irrigation system or manual application is also possible subject to labour availability. A well-perforated pipe of 2.5 cm diameter may be placed at the bottom of the trough or trench to drain out excess nutrient solution. Tall growing vine plants (cucumber, tomato, etc.) need additional support to withstand the weight of the fruits.

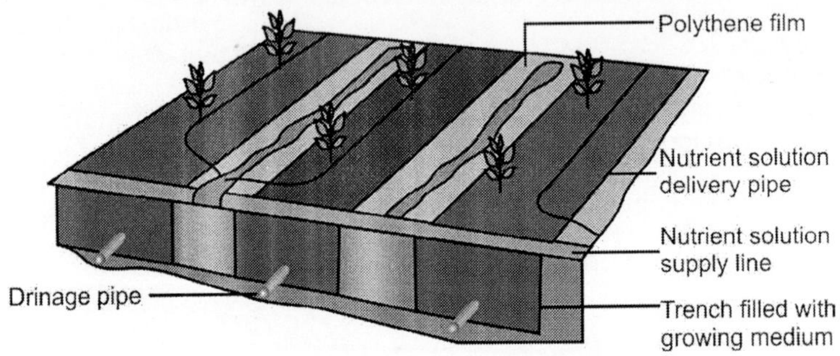

FIG. 24.13: Features of trench or trough technique

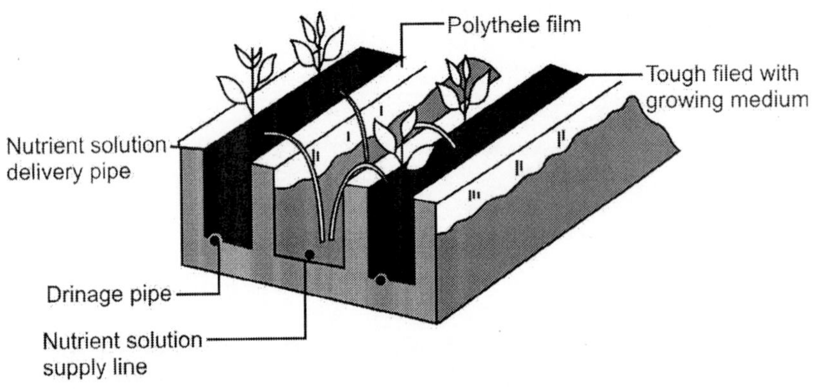

FIG. 24.14: Holistic view of hydroponic trenches.

Pot Technique

Pot technique is similar to trench or trough culture but growing media is filled in clay or plastic pots. Volume of the container and growing media depend on the crop growth requirements. The volume ranges generally from 1 to 10 litres. Growing media, nutrient solution supply, providing support to plants, etc. is similar to that of trough or trench culture.

Aeroponics

Aeroponics is the method of growing in which the plant roots are constantly misted with a nutrient solution. Designs include an A-frame with boards on each side, plant plugs set in each side and a mister between the boards spraying the roots. A round, large diameter PVC pipe set vertically with plant plugs all they way around and a mister mounted inside is another way to set up an aeroponic system. Although aeroponics is a unique way of growing, it is not a common means of commercial production.

Root Mist Technique (RMT)

A mist of nutrient solution is sprayed every 4-5 minutes onto the roots of the plants that hang from a holding frame in a root chamber. This technique is popularly known as **Aeroponics**. This system is good for initiating rooting of cuttings and also for extracting (milking) phytochemicals from the roots for medicinal purposes.

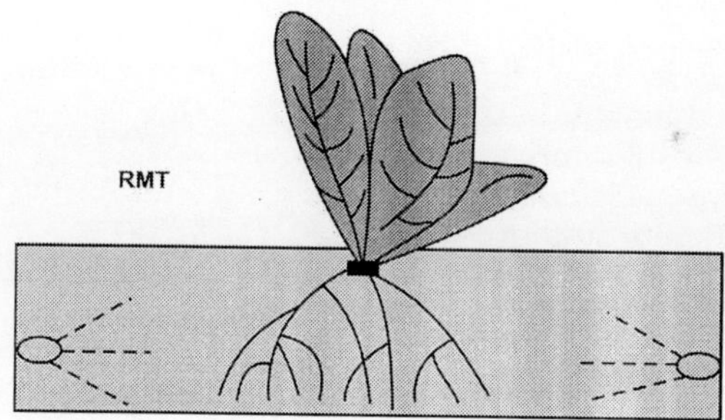

FIG. 24.15: Root Mist Technique (RMT)

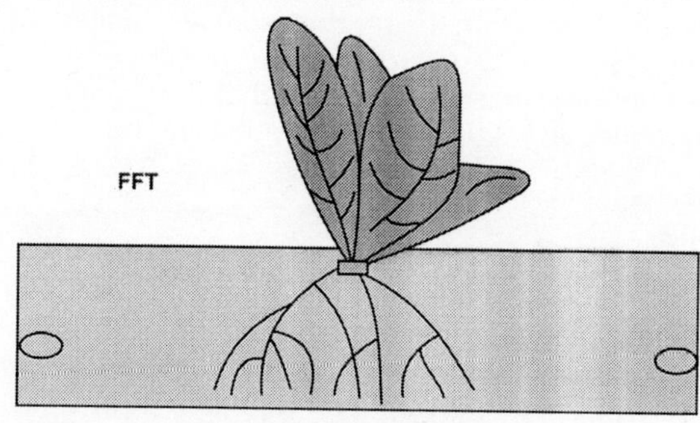

FIG. 24.16: Fog Feed Technique (FFT)

Fog Feed Technique (FFT)

This is similar to RMT but the droplet size is so very minute that the nutrient solution can hardly moisten our hand. This technique has yet to be perfected. It is good for plants having aerial roots e.g. orchids, anthurium, etc.

The other techniques are: (1) The Ebb and Flow System, Drip System, Deep Water Culture (DWC), Drip feeding, Wick feeding, Raft cultivation, Dynamic root floating system, Ridged Culture Bed, etc.

BASIC REQUIREMENTS OF HYDROPONICS

Soils naturally maintain the temperature and aeration needed for root growth. When the soil is poor, plant growth and yield decline due to unsuitable aeration and temperature. Plant cultivation is impossible under ill drained condition. Soil adjusts itself to provide suitable conditions for plant growth. It is called the buffer action of the soils. Plants also absorb nutrients released through natural mineralization.

In a solution or inert medium, maintenance of acidity or alkalinity (pH) and electrical conductivity (EC) in suitable ranges for plant root system is called buffer action. This requirement must be artificially maintained in hydroponics.

In any hydroponics system, the following basic requirements must be maintained at optimum levels:

1. Buffer action of water or the inert medium used.
2. The nutrient solution or the fertilizer mixture used must contain all micro and macro elements necessary for plant growth and development.
3. Buffer action of the nutrient solution must be in the suitable range so that plant root system or the inert medium is not affected.
4. The temperature and aeration of the inert medium or the nutrient solution is suitable for plant root system.

Growing Media

Growing medium is simply what the roots of the plants are growing in. This can be a vast variety of materials, including Rockwool, perlite, vermiculite, coconut fiber, gravel, sand or any number of other materials, even air can be a growing medium. The growing medium is an inert substance that does not supply any nutrition to the plants, all the nutrition comes from the nutrient solution (water and fertilizer mixed). You can, therefore, easily control everything the plants receive. The strength and pH of the nutrient solution is easy to adjust so that the plants receive just the right amount of nutrient and water. The watering cycles can be simply controlled by an inexpensive timer so that the plants get watered when they need to be.

Nutrient Solutions

Mineral nutrition is a broad field, which is concerned with the complex of biosynthetic events by which higher plants produce organic materials from inorganic materials absorbed from their environment. These events include the absorption process itself, the translocation of absorbed mineral nutrients from roots to stems, leaves, fruits and finally conversion of absorbed (inorganic) nutrients to organic form. The nutritional needs of plants were understood when the addition of farmyard manures, plant tissue debris, wood ashes, ground bones, composts, and other materials of plant and animal origin to soils were found to increase their productivity. In the early half of the nineteenth century it was realized that plant growth and development can proceed only when plants are supplied with certain chemical elements referred to as essential elements. These essential elements are mostly derived from mineral constituents of the soil and are absorbed principally by the roots as inorganic ions. Roots are the most important part of the plant as the nutrient uptake takes place mainly through them.

NUTRIENT SOLUTION

The standard nutrient formula used in the DRF system could be used to grow all kinds of crops, but an increase in microelement concentration was needed for some cropping seasons and vegetable species **(Kao 1990)**. The concentration of microelements was maintained at a constant level.

TABLE 24.1: Standard concentration of nutrients in nutrient solutions

Microelement	(me/liter)	Microelement	(ppm)
NO_3-N	6.0	Fe EDTA	2.5
NH_4-N	0.5	B	0.2
PO_4-P	3.0	Mn	0.2
SO_4-S	1.5	Zn	0.01
K	4.0	Cu	0.01
Mg	2.0	Mo	0.01
Ca	4.0		
EC (mmho/cm, 25°C)		1.0	
pH		6.0	

Mineral Nutrients

Through photosynthesis, green plants manufacture their own organic food, using carbon dioxide and oxygen as raw materials. The nutrients usually supplied to plants by soil are almost entirely mineral salts. Plant physiologists have discovered that plants require carbon, hydrogen, oxygen, nitrogen, phosphorus, potassium, magnesium, sulfur, calcium, iron, manganese, boron, zinc, copper, and probably molybdenum. Carbon, hydrogen, and oxygen are obtained in large quantities from water and air, but the remaining elements are ordinarily supplied as salts by the soil. The relative amount of each of these elements required for normal growth is different in each plant, but all plants require relatively large proportions of nitrogen, phosphorus, potassium, magnesium, sulfur, and calcium. Iron, manganese, boron, zinc, copper, and molybdenum are supplied in minute quantities, and are called micronutrients or trace elements. The specific salts used to supply these elements may be varied at the discretion of the grower; a typical solution of primary minerals is composed of distilled water containing potassium nitrate, KNO_3, calcium nitrate, $Ca(NO_3)_2$, potassium acid phosphate, KH_2PO_4, and magnesium sulfate, $MgSO_4$. In solution, the salts dissociate into ions; potassium nitrate, for example, is available to plants as the ions K+ and NO_3^- A solution of micronutrient salts is added to the solution of primary elements to complete the nutrient solution. A small amount of fungicide is usually added in the nutrient solution to prevent the growth of mold.

Temperature

The plants grow well only when atmospheric temperature is in optimum range. The temperatures that are too high and too low will result in abnormal development and reduced production. The ideal temperature depends on the crops to be selected for growing. Most of the common garden crops, such as tomatoes, cucumbers, lettuce, beans and peas perform well with an average daytime temperature of 78°F and an average night time temperature of 64°F. Winter vegetables, such as cabbage, brussel sprouts, spinach and broccoli should be grown in slightly cooler. A minimum / maximum thermometer will allow you to track the low and high temperatures in your growing environment.

Fresh Air

Plants require adequate air circulation (oxygen) around the plants, as well as proper aeration in the root zone. In soil, adequate oxygen is usually available but plant roots growing in water will quickly exhaust the supply of dissolved oxygen and can be damaged or killed plants unless additional air is provided. Poor ventilation in the growing environment encourages mold, mildew and plant disease. Many hydroponic gardens are located off the floor for better air circulation. Commercial hydroponic greenhouse growers use large fans and air circulation equipment to provide adequate air movement.

Shelter and Support

In open field cultivation the soil provide support to the growing plant. But hydroponically grown plants artificial support must be provided usually with string or stakes. In a commercial application, many hydroponic farmers grow their crops inside of a controlled environment greenhouse. This not only provides shelter, but also an ideal, stress-free environment for the plant. Because many hydroponic gardens are quite small and very clean, they can be set up almost anywhere indoors, on a patio or a windowsill, making it easy for the gardener to provide shelter for the plants. In a traditional garden, the soil anchors the plant and provides support. In hydroponics, the growing medium helps support the plant to some extent but most often additional support is needed. Plant stakes, strings, and clips are used for this.

Water

Providing the plants with an adequate amount of water is not difficult in the water culture system, but it can be a problem with the aggregate culture method. If the aggregate is not kept sufficiently moist, the plant roots will dry out and some will die. Even after the proper moisture level being restored, the plants will recover slowly and production will be reduced. Water quality can be a problem in hydroponic systems. Water with excessive alkalinity or salt content can result in a nutrient imbalance and poor plant growth. Softened water may contain harmful amounts of sodium. Water that tests high in total salts should not be used. Salt levels greater than 0.5 million or 320 parts per million are likely to cause an imbalance of nutrients. The amateur chemist may be able to overcome this problem by custom mixing the nutrient solutions to compensate for the salts in the water.

Light

Plants require a constant supply of energy to grow and this energy comes from light. In nature, plants receive light from the sun in a home garden, there is a need to add artificial light so the plants have an adequate amount of light to grow.

There are various types of artificial lights that provide differing light spectrums. Before learning about these artificial lights, it is important to understand how plants use light in the growth process.

Growing Media

Essential Plants Nutrients

Plants require 17 essential elements for their growth and development. Without these nutrients plants cannot complete their life cycles and their roles in plant growth cannot be replaced by any other elements.

All essential elements are classified as under:

1. **Macro nutrients:** Which required by the plants in large quantity i.e. more than 1 ppm.
   a) Carbon (C), Hydrogen (H), Oxygen (O)
      C, H, O are called structural elements
   b) Nitrogen (N), Phosphorus (P), Potassium (K)
      N, P, K are called primary elements
   c) Calcium (Ca), Magnesium (Mg), Sulphur (S)
      Ca, Mg, S are called secondary elements
2. **Micro nutrients:** Which required by the plants in small quantity i.e. less than 1 ppm. These nutrients are also trace elements: iron (Fe), chlorine (Cl), boron (B), manganese (Mn), copper (Cu), zinc (Zn), molybdenum (Mo) and nickel (Ni).

Function of Minerals

The important functions of minerals are as follows:

- Protoplasmic elements: Nitrogen, Phosphorus and Sulphur are considered as protoplasmic elements as they form part of protoplasm along with carbon, hydrogen and oxygen.
- Balancing elements: Calcium, Magnesium and Potassium counteract the toxic effect of other minerals by causing ionic balance.
- Framework elements: Carbon, Hydrogen and Oxygen are considered as framework elements because they constitute the carbohydrates, which form cell walls.
- Influence on the osmotic pressure of the cells: Plant cells contain dissolved mineral elements in the cell sap, which influences the osmotic pressure of the cell.
- Catalytic elements: Some elements function as part of the enzyme i.e., without the presence of these elements some enzyme cannot function.
- Plant nutrients are dissolved in the water used in hydroponics and are mostly in inorganic and ionic form. Primary among the dissolved cations (positively-charged ions) are Ca2+ (calcium), Mg2+ (magnesium), and K+ (potassium); the major nutrient anions in nutrient solutions are NO3− (nitrate), SO42− (sulfate), and H2PO4− (phosphate).

Numerous 'recipes' for hydroponic solutions are available. Many use different combinations of chemicals to reach similar total final compositions. Commonly used chemicals for the macronutrients include potassium nitrate, calcium nitrate, potassium phosphate, and magnesium sulfate.

TABLE 24.2: The average concentration of plant nutrient in plant solution (in ppm = parts per million)

S.No.	Nutrients	Cocnetration (PPM)
1.	Nitrogen (nitrate form)	70 –300
2.	Nitrogen (ammonium form)	0 –31
3.	Potassium	200 –400
4.	Phosphorous	30 –90
5.	Calcium	150 –400
6.	Sulfur	60 –330
7.	Magnesium	25 –75
8.	Iron	5 -5.0
9.	Boron	1 -1.0
10.	Manganese	1 -1.0
11.	Zinc	02 -.2
12.	Molybdenum	01 -.1
13.	Copper	02 -.2

TABLE 24.3: Compatibility chart for different fertilizers for hydroponics

Soluble fertilizers	AN	AS	CAN	MAP	SOP	MOP
Ammonium nitrate	-	C	C	C	C	C
Ammonium sulphate	C	-	L	C	C	C
Calcium ammonium nitrate	C	L	-	X	C	C
Mono ammonium phosphate	C	C	X	-	C	C
Potassium sulphate	C	C	L	C	-	C
Potassium chloride	C	C	X	C	C	-
Gypsum	X	X	X	X	C	C
Potassium nitrae	C	L	C	C	-	C

C = Compatible (Can be mixed in solution), L = limited compatibility (mix at the time of use or some precautions must be taken), X = Incompatible (cannot be mixed).

TABLE 24.4: Some general guidelines for EC levels

Growth stage/Condition	Fruiting Plants (such as tomatoes, cucumbers)	Leafy Plants (such as lettuce, basil)
Initial Growth (seedling stage)	1600 -1800 mMho/cm 1120 -1260 ppm	1400 -1600 mMho /cm 980 -1120 ppm
Average EC	2500 mMho/cm 1750 ppm	1800 mMho/cm 1260 ppm
Fruiting	2400 -2600 mMho/cm 1680 -1820 ppm	-
Low light conditions (winter)	2800 -3000 mMho/cm 2000 ppm	2000 mMho/cm 1320 ppm
High light conditions (summer)	2200 -2400 mMho/cm 1700 ppm	1600 mMho/cm 1120 ppm

TABLE 24.5: Chemicals needed to prepare 1000 litres of nutrient solutions proposed by Dr. Alen cooper

Nutrient chemicals	Weight in grams
Potassium dihydrogen phosphate	263.00
Potassium nitrate	583.00
Calcium nitrate	1003.00
Magnesium sulphate	513.00
EDTA iron	79.00
Manganese sulphate	6.10
Boric acid	1.70
Copper sulphate	0.39
Ammonium molybdate	0.37
Zinc sulphate	0.44

TABLE 24.6: Classification of irrigation water used for hydroponics

Quality factors	Unit	Water quality classes		
		Non-hazardous	Slight to moderate	Severe
pH			Normal range 6.5-8.4	
Salinity Ec	dSm^{-1}	0.00-0.8	0.8-3.0	>3.0
Sodium	me/lit	<3.0	>3.0	-
Chloride	me/lit	<3.0	>3.0	-
Boron	mg/lit	<0.7	0.7-3.0	>3.0
Bicorbonates	me/lit	<1.5	1.5-8.5	>8.5

Micronutrients are needed in extremely small amounts. Because it is difficult to weigh out such small quantities, stock solutions of dilute micronutrients should be made up and added to the solution tank. Use only reagent grade chemicals, which are usually available in hobby shops

Chemical	Amount for 1 gallon of stock solution	Amount of stock solution for 25 gallons of nutrient solution
Boric Acid (H_3BO_3)	2 teaspoons	1/2 pint
Manganese chloride ($MnCl_2.4H_2O$)	1 teaspoon	1/2 pint
Zinc sulfate ($ZnSO_4.7H_2O$)	2 teaspoons	1/2 teaspoon
Copper sulfate ($CuSO_4.5H_2O$)	1 teaspoon	1/2 teaspoon
Iron sulfate ($FeSO_4.7H_2O$)	4 teaspoon	1/2 pint

If tap water is used, zinc sulfate and copper sulfate may not be required. The small amounts of zinc and sulfur needed are usually present in the water or as an impurity in the other fertilizer nutrients.

An iron chelate should be substituted for the iron sulfate if your tap water is alkaline. Mix 1 1/2 ounces of Nafe (III) EDTA (13 per cent Fe_2O_3) in 5 quarts of water. Use 1/4 pint of this solution for 25 gallons of stock solution. Further other chelate products calculate the amount required to result in a final nutrient solution containing

1 part per million of elemental Iron.

As the plants take up nutrients they will release chemicals into the solution that may cause it to become more alkaline. When the pH rises above 7.0, add enough sulfuric acid (battery acid) to bring it back to a pH between 5.5 to 6.5. For large plants it may be necessary to add acid daily. If the solution becomes acidic, the pH can be raised with potassium hydroxide (KOH) or sodium hydroxide (NaOH). A variety of pH meters and testing kits are available for checking pH levels.

Note: The nutrient solution should be changed once every two weeks when the plants are small and once a week as the plants begin to grow rapidly. Add water daily to keep the amount of solution constant.

Knop's Solution

Calcium nitrate $Ca(NO_3)_2 4H_2O$ .... 0.8 gram/lt.
Potassium nitrate KNO_3 .... 0.2 gram/lt.
Magnesium sulfate $MgSO_4 7H_2O$ .... 0.2 gram/lt.
Iron Phosphate $FePO_4$ .... Traces only

Sach's Solution

Potassium nitrate (KNO_3) .... 0.5 gram/lt.
Calcium Phosphate $(PO_4)_2$ .... 0.5 gram/lt.
Magnesium sulfate $MgSO_4.7H_2O$ .... 0.5 gram/lt.
Zinc Chloride $(ZnCl_2)$ .... 0.25 gram/lt.
Iron sulfate $(FeSO_4)$ .... Traces only

Hogland Solution

$Ca(NO_3)_2 4H_2O$ .... 1.8 gram/lt.
KNO_3 .... 0.51 gram/lt.
KH_2PO_4 .... 0.14 gram/lt.
$MgSO_4.7H_2O$ .... 0.49 gram/lt.
Ferric Tartarate .... 0.005 gram/lt.

Combination developed by NBRI, lucknow for vegetables and floriculture

Magnesium Sulphate .... 0.5 gram.
Double Superphosphate .... 0.5 gram.
Potassium Nitrate .... 0.4 gram
Ammonium Sulphate .... 0.3 gram

These combination are dissolved in 4.5 liter of water.
Some non standard solution for experimental purpose can be made as follows:

Calcium Nitrate .... 1 gram
Potassium Nitrate .... 0.2 gram
Magnesium Sulphate .... 0.2 gram
Ferrous Suphate .... 0.1 gram
Urea .... 0.2 gram

Growing medium is used to lend support to the roots and plants. A variety of growing media and combinations are used for their individual qualities in various types of hydroponic systems. The purpose of growing media is to aerate and support the root system of the plant and to channel the water and nutrients. The selection of growing medium is largely depend upon availability, the type of hydroponic growing system used and plants to be grown.

The following media are generally found suitable for hydroponics

(1) Expanded Clay, (2) Rockwool, (3) Perlite, (4) Vermiculite, (5) Coco Coir, (6) Oasis Root Cubes, (7) Sand, (8) Coarse fir sawdust, (9) Crushed granite, and (10) Scoria.

EQUIPMENTS AND APPLIANCES

Following equipments are required for hydropinics:

1. Water Pumps
2. Timer and Oxygen Detection Sensor
3. PVC Pipes
4. pH Meter
5. Blowers
6. Pollinators
7. Nurtimeter
8. EC Meter
9. TDS Meter

MANAGEMENT REQUIREMENTS OF HYDROPONICS

To develop and maintain a successful hydroponics / soil-less cultivation of crops, meet the following requirements:

1. Maintain the pH of the nutrient solution in the range of 5.8 to 6.5, and electrical conductivity (EC) in the range of 1.5 to 2.5 dSm^{-1}, as these ranges are suitable for plant growth. Any pH or EC outside these ranges will reduce availability and uptake of nutrients and will also damage plant roots.
2. Plants are the best indicators of the nutrient availability. Look for nutritional disorder symptoms in plants and adjust nutrient solution accordingly.
3. Maintain adequate solution temperature. As the temperature goes up, plant respiration increases causing a higher demand for oxygen. At the same time, the solubility of oxygen decreases. This requirement is more critical in green houses and net houses where the temperature is bound to increase during mid afternoons. Steps must be taken to counter such increase.
4. Always ensure that there is plenty of dissolved oxygen in the nutrient solution as the plant roots absorb oxygen. Lack of oxygen will reduces up take of nutrients and thereby the yield and also causes root rot. In closed systems, if the recollected solution is allowed to fall into the solution tank from a height, natural aeration will take place.
5. In root dipping techniques, maintain adequate air space above the nutrient solution in the container as success depends on the rapid growth and quantity of roots that are exposed to the air. These roots absorb oxygen for the plants. Ideally, top two thirds of the young root system must be in the air and the rest must be floating in the nutrient solution.
6. Avoid any sudden changes in nutrient solution concentration as it can result in unsuitable pH and EC.
7. In root dipping techniques, during crop growth, when the solution level in the container goes down, the ion concentration may increase. Such increase

is detrimental to plant growth. If this condition is observed, siphon out the remaining solution and refill with fresh solution.

8. Ensure adequate light for the hydroponics/ soil-less culture plants. Light and all other requirements are the same as though grown in open fields.
9. Always use pest and disease free seedlings and planting materials for establishment of hydroponics crops. Remove and destroy any infected plants as soon as they are found.
10. If nematode problem is observed in solid media culture, discard the plants and sterilize the growing medium. If in doubt, discard and replace the medium. Also ensure that he water supply is also free from nematodes.
11. Algae can build up in the system and block the small tubes used for the delivery of nutrient solution. Use black colour tubes to avoid such problems. Between crops, thoroughly clean the system using a mild solution of chlorine. After cleaning, thoroughly flush the system with fresh water before replanting.
12. Adequate spacing is necessary for plant growth and when vine crops are grown, supports must be provided.
13. In open aggregate techniques, there is a possibility for nutrients to leach when water is applied. Therefore, nutrient solution may be applied continuously instead of water to supply both water and nutrients.
14. If any of these conditions are not fulfilled, one cannot obtain economical yields.

CHAPTER 25

Hi-Tech Horticulture for Postharvest Technology

INTRODUCTION

The postharvest technology to harvested fruits are mainly concerns with three main objectives:

1. to maintain quality (appearance, texture, flavour and nutritive value),
2. to protect food safety and
3. to reduce losses between harvest and consumption.

Therefore there is a need for an effective management during the postharvest period to obtain desired objectives. The process needs large scale operations, costly handling machinery and hi-tech postharvest treatments.

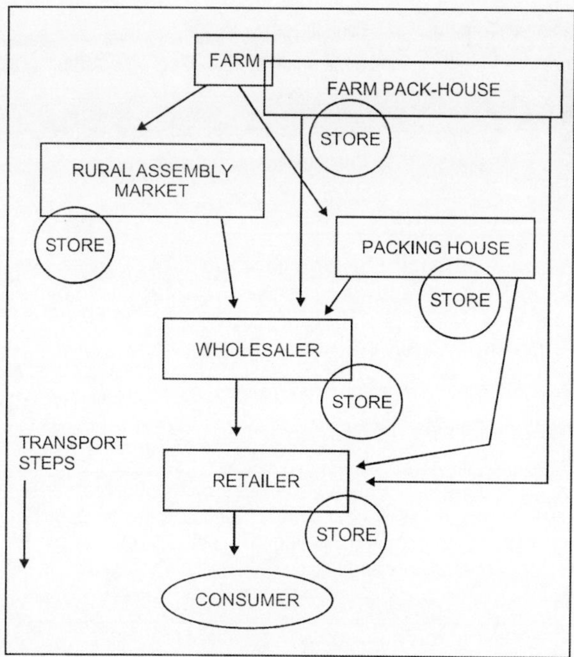

FIG. 25.1: Postharvest handling steps for a typical commodity.

This is not practical for small scale handlers. Instead, simple, low cost technologies often can be more appropriate for small volume, limited resource commercial operations, farmers involved in direct marketing, as well as for suppliers to exporters in developing countries. Following are the important steps of postharvest handling.

Fruits are the rich source of carbohydrates, minerals, vitamins and dietary fibres, these constitute an important part of our daily diet. Therefore, cultivation of fruits in diversified agro-climatic conditions, ranging from tropical to temperate, afford ample opportunities to grow a large number of fruits in different countries of the world. The production of fruits has received a greater attention even at the expense of conservation of the hard earned crops.

The postharvest losses in developing countries continue to be rough handling and inadequate cooling and temperature maintenance. The lack of sorting to eliminate defects before storage and the use of inadequate packaging materials further add to the problem. In general, minimizing rough handling, sorting to remove damaged and diseased produce and effective temperature management will help considerably towards maintaining a quality product and reducing storage losses. Storage life will be enhanced if the temperature during the postharvest period is kept as close to the optimum as feasible for a given commodity.

Group	Examples	Principal causes of postharvest losses and poor quality (in order of importance)
Immature-fruit vegetables	Cucumbers Squash Eggplant Peppers Okra Snap beans	Over-maturity at harvest Water loss (shriveling) Bruising and other mechanical injuries Chilling injury Decay
Mature-fruit vegetables and fruits	Tomato Melons Citrus Bananas Mangoes Apples Grapes Stone fruits	Bruising Over-ripeness and excessive softening at harvest Water loss Chilling injury (chilling sensitive fruits) Compositional changes Decay

Further, the relative perishability and storage life of fresh produce in air are the other reasons for losses. Following is the classification of fresh horticultural crops according to their relative perishability and potential storage life in air at near optimum temperature and relative humidity.

Relative perishability	Potential storage life (weeks)	Commodities
Very high	<2	Apricot, blackberry, blueberry, cherry, fig, raspberry, strawberry; asparagus, bean sprouts, broccoli, cauliflower, green onion, leaf lettuce, mushroom, muskmelon, pea, spinach, sweet corn, tomato (ripe); most cut flowers and foliage; minimally processed fruits and vegetables.

Relative perishability	Potential storage life (weeks)	Commodities
High	2-4	Avocado, banana, grape (without SO_2 treatment), guava, loquat, mandarin, mango, melons (honeydew, crenshaw, Persian), nectarine, papaya, peach, plum; artichoke, green beans, Brussels sprouts, cabbage, celery, eggplant, head lettuce, okra, pepper, summer squash, tomato (partially ripe).
Moderate	4- 8	Apple and pear (some cultivars), grape (SO_2-treated), orange, grapefruit, lime, kiwifruit, persimmon, pomegranate; table beet, carrot, radish, potato (immature).
Low	8-16	Apple and pear (some cultivars), lemon; potato (mature), dry onion, garlic, pumpkin, winter squash, sweet potato, taro, yam; bulbs and other propagules of ornamental plants.
Very low	>16	Tree nuts, dried fruits and vegetables.

POSTHARVEST TECHNOLOGY

It has been estimated (1994-95) that the total fruits and vegetables production in world was 388 and 486 million tonnes, respectively, while India produced nearly 39.4 million tonnes of fruits and 71.0 million tonnes of vegetables. Thus, India attained first position in fruit production and second in vegetables in the world reflecting the Indian contribution to this important segment of food industry (Kalloo, 1998). Therefore, it is necessary to prevent fruits from proper postharvest techniques.

Reasons of Postharvest Losses

Fruits are living organisms and highly perishable. They are affected by a number of factors leading to the postharvest spoilage. The reasons for postharvest losses could be many. Besides, packaging, grading and transportation also contribute to these losses. Marketing of these perishables also do not lag behind in adding to the postharvest losses. Prevention of such losses is the most appropriate method to make available more quality fruits and vegetables and is a complementary means of increased production. According to US National Research Council (Anon., 1978) the estimated postharvest loss (%) of fruits developing countries are: Apple (14), Avocado (43), Banana (20-80), Citrus (20-95), Grape (27), Papaya (40-100), etc. According to Thompson (1996) there are some of the physical causes of postharvest losses, which are as follows:

1. Loss of moisture, causing wilting/shrinkage
2. Loss of stored energy *eg.* carbohydrates
3. Loss of other food constituents (vitamins)
4. Physical losses through pest and disease attack
5. Loss in quality due to physiological disorders
6. Fibre development

The magnitude of these losses in India is comparatively higher because of lack of infrastructure required for postharvest management of the perishables and poor utilization of the fruits by processing industry. The waste generated and thrown into the environment besides causing pollution is the colossal loss of precious national resources.

To understand the other reasons of postharvest losses it is necessary to know about biological and environmental factors involved in the deterioration of the perishables; and proper use of the postharvest technological procedures. In fact, enzymes, microorganisms, pests and the environmental factors are responsible for spoilage of fruit crops. Therefore, any method which prevents enzymatic activity of plant or microbe reduces the number of microorganisms and their activities would increase the postharvest life. To understand these aspects, a sound knowledge of relevant discipline(s) is a must. Thus, to be an effective postharvest technologist, it is essential to have sound knowledge of horticulture, plant nutrition, pathology, microbiology, biochemistry and physiology, food science and technology and engineering. Even molecular biology and genetic engineering is finding increasing usefulness to find out the solution of postharvest technological problems. Biochemically and physiologically, the postharvest technologists are mainly concerned with slowing down the rate of respiration. The processing itself includes several techniques such as low temperature preservation, chemical preservation, irradiation, thermal processing, dehydration, ohmic processing and high pressure processing and preservation by fermentation.

The postharvest technology, in fact, should comprise all the in-between techniques or methods right from the field till the crops reach the hands of consumers either as fresh commodity or processed products and finally the waste is utilized in a profitable manner.

To have perfect postharvest technology, there is a need of human resource development which can provide leadership in various fields and activities of postharvest technology. Following are the fields of postharvest technology:

TABLE 25.1: Major subjects and allied fields in postharvest technology

Major subjects	Basic/ allied fields
Postharvest Physiology	Food Microbiology
Storage Systems	Food Biochemistry
Principles of Preservation and Processing	Food Chemistry
Quality Assurance of Fruits, Vegetables and their Products	Applied Physics, Food Engineering
Canning and Juice Processing	Horticulture
Fruit Fermentation Technology	Plant Physiology, Genetics and Plant Breeding
Sensory Analysis	Computer Application
Packaging Technology	Statistics
Dehydration Technology	Economics and Management
Waste Utilization Technology	Psychology
Nutritive Value of Processed Products	

Postharvest technology also needs proper knowledge to protect the perishables by proper packaging, retard their ripening or other physiological activities by low temperature storage or manipulating their physiology, eliminating the microorganisms or checking their activities are the basis of all the postharvest techniques except in case of preservation by fermentation, where the microorganisms are allowed to grow. Application of high pressure technology and the ohmic processing are the recent techniques of preservation alongwith minimal processing.

Apart from these, following are the fields by which one can have better use of produced fruits:

Importance and Loss Reduction	Postharvest loss reduction techniques Value addition Contribution to the economy Processing and farming Making availability of fruits and vegetables Tool for export earning Employment generation Adding taste, variety and providing nutrition Waste utilization Home scale preservation A vital supplier to the defence services Special canned fruits for infants and children A food supplier to the astronauts
Responsibilities of Postharvest Technologists	To provide quality, nutritious and safe foods to all customers To check the standard of cleanliness and hygiene To develop rapid and accurate tests to ensure food safety. To develop new products and technologies. To use microorganisms positively by biotechnology to modify the food. To develop machines and equipments and to determine their efficacy to run the processing unit. To interact with farmers and consumers. To transfer the technology to the users.

CONSTRAINTS FACED BY THE POSTHARVEST PROCESSING INDUSTRY

According to Joshi and Pandey, 1999, following are constrains for the postharvest processing industries:

1. Non-availability of proper infrastructure for postharvest technology (that is proper grading and packaging facilities) do not give proper prices to the farmers and quality produce to the consumers.
2. Lack of awareness for the Indian fruit varieties in the foreign countries and quarantine restrictions imposed by some countries on import of fruits from India, greatly hinders the export of perishable commodities.
3. Limited air space with high rates of freight is a significant factor, discouraging our exports.
4. Variations in quality of fresh produce lower the quality of processed products from those made in the advanced countries.
5. Low productivity of the fruits and vegetables compared to that of advanced countries and high cost of raw material, naturally increases the cost of processed products.
6. Prohibitive costs and lack of traditional taste suiting the Indian dietary habits lowers the demand of processed products resulting in sluggish growth.
7. Outmoded technologies of packaging fresh, processed and preserved products make our products very costly, of lower quality and non-competitive with international market.
8. Irregular supply and non-uniform quality of processed products do not allow committed consumers in the domestic market and discourage potential buyers in the international market.
9. Lack of trained manpower to implement the research findings in the field and processing factories.
10. Marketing limitations, financial and fiscal constraints restrict the establishment

of industry on firm footing.

11. Non-utilization of processing waste which besides causing pollution also results in loss of precious natural resources.
12. Lack of either proper research or development support to industry or non-existing problem oriented research makes our R & D efforts directionless and futile.

MATURITY INDICES AND HARVESTING OF FRUITS

Colour, flavour and texture determine the quality of fruits for processing. In contrast to these requirements, the grower is concerned with ease of cultivation, disease resistance and yield. A good quality processed product cannot be made from poor quality raw material. Requirements for processing with respect to size, shape, colour, texture and flavour are more exacting than for the fresh produce market.

The stage of harvesting determines the quality. Most of the fruits are harvested when they are mature and allowed to ripen subsequently during transit, storage or when displayed in retail shops. In such cases, the maturity standards should correlate with the quality of the fruit when ripe. On the other hand, the optimal stage of maturity of most of the fruits is transient and delay in harvesting by a day or two may render them more mature.

Maturity Indices

Maturation is the stage of development leading to the attainment of **physiological maturity** (when a plant or plant parts will continue ontogeny even if detached) or **horticultural maturity** (when a plant or plant parts possesses the pre-requisites for utilization by the consumer for a particular purpose).

Maturity indices help in deciding that when a given commodity should be harvested to provide some marketing flexibility and to ensure the attainment of acceptable eating quality to the consumer. It needs to evaluate the time of harvest for a given fruit. The grower should know the margin time of safety for marketing and to supply to the consumer with fruit of good eating quality.

Fruits picked at the wrong stage of maturity may develop physiological disorders in storage and may exhibit poor dessert quality. It has been stated that apples should be harvested at a stage of maturation and ripening which is less than ripe. For example apple picked too early may not ripen properly in storage and may develop superficial scald, bitter pit and extreme shrivelling. They are also likely to be of low dessert quality from the standpoint of colour, flavour and texture.

The ripe fruits are prone to bruising and decay. It has been suggested that if pears are permitted to ripen on the tree, the dessert and storage quality become drastically reduced through the occurrence of mealiness and core breakdown with excessive accumulation of acetaldehyde (Gerhardt and Ezell, 1934). Harvesting, storage, marketing and canning experiences have shown that this type of fruits should be removed from the tree while still in a firm hard green condition.

For selecting the harvest maturity of fruits or vegetables it should be kept in mind that harvested commodity should have its peak acceptable quality (non-toxic, size, appearance and flavour with adequate shelf life) (Thompson, 1996).

TABLE: 25.2: Maturity Indices for Selected Fruits and Vegetables (Kader et al., 1985)*

Maturity indices	Fruits
Calendar date	All fruits
DFFB	All fruits and radish
Mean heat units	Apple, pear, grape, mango, ber, litchi and many vegetables
T-Stage	Apple
Size	All fruits, beans, carrot, cucumber, cherry, asparagus and cauliflower.
Surface morphology	Grape (cuticle formation), banana and litchi.
Specific gravity	Cherries, mango and ber.
Fruit retention strength	Apple
Colour (Surface)	All fruits
(Seed)	Apple and pears
(flesh)	Mango, papaya
TSS	All fruits
Firmness	Pome and stone fruits
Juice content	Citrus
Acidity and Sugar/acid ratio	Pomegranate, citrus, and kiwi fruit
TSS/acid ratio	Grape and citrus
Sugars	Pome, stone fruits and grape
Astringency (Tannin)	Persimmon and dates.
Oil content	Avocado
Respiration and ethylene evolution rates	Apple and pears

*Modified

TABLE 25.3: Parameters to determine harvest indices

Computational methods	Calendar date, Days from full bloom to harvest (DFFB), Mean heat units, T-stage
Physical methods	Fruit retention strength, Fruit size and surface morphology, Weight, Specific gravity, Colour (Skin, flesh and seed), Flesh firmness, Total soluble solids (TSS), Juice content
Chemical methods	Titrable acidity, TSS/acid ratio, Sugars (Total and reducing), Sugar/acid ratio, Bioelectrical conductance, Starch content, Starch-Iodine test, Tannin content, Oil content
Physiological methods	Respiration rate, Ethylene evolution rate

Predicting date of first acceptable harvest using elapsed time from full bloom is the method which shows the correction in the standard figure (DFFB) on the basis of variation in the full bloom date. For every 2 days difference in average full bloom date, an adjustment of 1 day is required as given hereunder:

Suppose full bloom date of a cultivar is **May 23** (the optimum date of picking is 120 days from full bloom) but in a particular season it was recorded on May 29 then first acceptable harvesting date will be = 120 + 3 = May 26 + 123 days = **September 26**.

Another method related to the time of maturation is the date when the fruits reach the T-Stage. The T-Stage is the point at which the angle formed by the fruit receptacle and the pedicel reaches 90° and its timing is obtained by inter-polation. The angle between pedicel and receptacle is greater than 90° early in the growing season and less than 90° late in the season. The number of days between the T-Stage and the date of optimal harvesting time of apple showed a remarkable consistency for a given location. It was determined that the number of days from the T-Stage to harvest for apple varied from 120 for cv. Discovery to 130 for cvs.

The apparent effect of temperature on flowering and fruit development suggests the use of degree days or of heat summation for the more accurate prediction of harvest date.

TABLE 25.4: Heat requirement of various fruits

Fruits	Cultivar	Cut out or Base Temperature	Degree days
Apple	Red Delicious	50°F	1659 - 1705
Pear	Bartlett	45°F	2240 - 2296
Grapes	Thompson Seedless	50°F	1600 - 2000
	Bhokri		3284
	Gulabi		3508
	Banglore Blue		3562
Mango	Banganapally	18°C	1426-1445
Ber	Gola	45°F	1980 - 2236
	Kaithli		2236 - 2566
	Umran		2516 - 2920
Litchi	Shahi	15°C	813
Banana		9.8°C	1930

Physical Methods

Fruits remain attached with the main plant due retention strength. It is the force required to pull fruit from the tree (at the natural abscission site) which indicates the maturity status of the fruit. Fruit with best storage quality could be picked from the spurs using a slight rolling action. Physical methods are found as temperature dependent indices of maturity. Fruit size and weight are also the measures of fruit maturity. Fruit size and weight are dependent upon a number of variables such as soil, cultivar, fertilizer application and climatic conditions which vary from place to place and season-to-season. In banana, litchi, nuts, mango, pineapple, etc. changes in the surface morphology is the reliable maturity index and determine harvest maturity. The other features of maturity are specific gravity (by water displacement method), colour of skin, flesh and seed (by horticulture colour chart), flesh thickness (by penetrometer) and TSS (by hand refractometer) are also used frequently in different fruit crops as indices of maturity.

Using a Firmness Tester

The degree of oftness or crispiness can be estimated by squeezing produce, or by taking a bite. Objective measurements can be made with inexpensive penetrometers. The most common way to measure firmness is resistance to compression or pounds-force (lbf). The Effe-gi fruit penetrometer is a hand-held probe with a gauge for pounds-force.

To measure firmness, fruits that are uniform in temperature are used since warm fruit are usually softer than cold fruit. Fruits that are uniform in size are used, since large fruits are usually softer than smaller fruits. Two puncture tests per fruit on larger fruits, one on opposite cheeks, midway between stem and blossom ends are made. A disc of skin (larger than the tip to be used) is removed and the appropriate plunger tip is chosen. The fruit is held against a stationary, hard surface and the tip is forced into fruit at a slow, uniform speed (take 2 seconds) to the scribed line on the tip. The reading to the nearest 0.5 lb-force is taken.

Penetrometers measure the firmness or hardness of fruit. Penetrometers are ideal for determining the best time to harvest fruit or to test its progress to maturity. The firmness of fruit is one of the most widely used parameters in testing its maturity, together with measuring its sugar content using refractometers. Electronic penetrometers are available with different diameter penetration points. This allows for fruit of different size and hardness to be tested. It is important that this first part of the measurement process is done correctly if the readings obtained are to be valid in making adequate decisions as to when the fruit should be harvested. It is important the the samples that are measured are representative of the entire crop, considering a sufficient minimum quantity sample, size and quality.

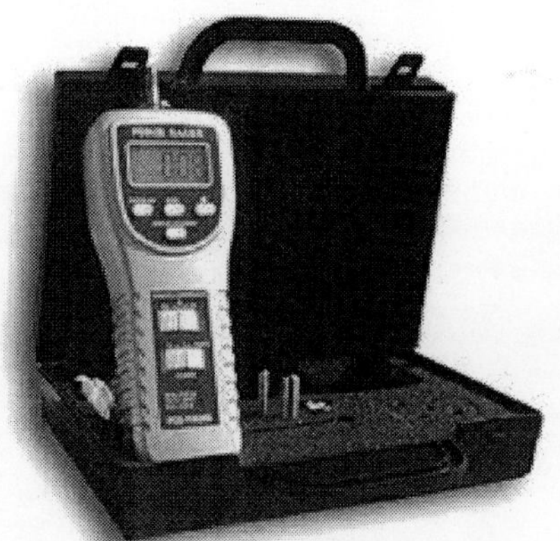

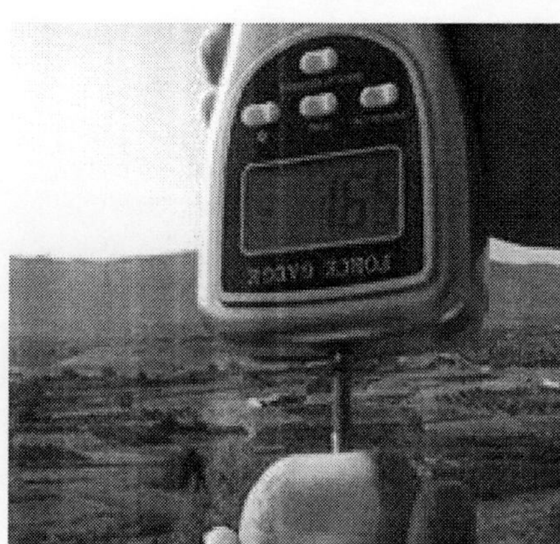

FIG. 25.2: Penetrometer measures the firmness or hardness of fruit.

Firmness testers should be calibrated by holding the tester vertically and placing the tip on the pan of a scale. Press down until the scale registers a given weight, then read the firmness tester. Repeat 3 to 5 times, if the instrument reads the same as the scale, it is ready to use. The penetrometers can be adjusted by inserting washers in the appropriate locations (follow the instructions that come with the instrument).

TABLE: 25.5: Appropriate Effi-gi Plunger tip sizes to use when measuring firmness in selected fruits

1.5 mm	(1/16 inch)	Olive
3 mm	(1/8 inch)	Cherry, grape, strawberry
8 mm	(5/16 inch)	Apricot, avocado, kiwifruit, pear, mango, nectarine, papaya, peach
11 mm	(7/16 inch)	Apple

Chemical Methods

The consumer acceptability is closely associated to the **acid content** of fruits. Attempt to assess the stage of ripeness of fruit in terms of titrable acidity and to correlate it with the time of harvest has not found general acceptance because of inconsistent results. The titrable acidity varies more with growing seasons and with the amount of nitrogen fertilizer applied to the crop, between harvest date. Therefore, the eating quality after storage of a low titrable acidity first at harvest is unacceptable, it is suggested that a minimum level of titrable acidity below which fruit should not be stored, be specified for each cultivar.

Sugar (free state or as derivatives) plays an important role in imparting attractive flavour, appearance and texture to the fruits. Sugars are the major soluble solids in fruit juices and therefore soluble solids can be used as an estimate of sweetness. The sugar concentration increases during onset of ripening of pear fruits in free space of cell. It is may be due to the enhancement of permeability of sugars across the tonoplast thus, giving the fruit a sweet and juicy taste. A hand-held refractometer is used for outdoors measurement (in percentage TSS (equivalent degrees Brix for sugar solutions) in a small sample of fruit juice. Temperature will affect the reading (increasing about 0.5% TSS for every 5°C or 10°F), so the measurement for the ambient temperature, should be adjusted.

A garlic press works well to squeeze the juice from fruit samples. For small fruits, the whole fruit should be used. For large fruits, a wedge from the stem end to the blossom end and to the centre of the fruit is to be taken. Any pulp should be removed by filtering the juice through a small piece of cheesecloth. The refractometer should be cleaned and standardized between each reading with distilled water (should read 0% TSS at 20°C or 68°F).

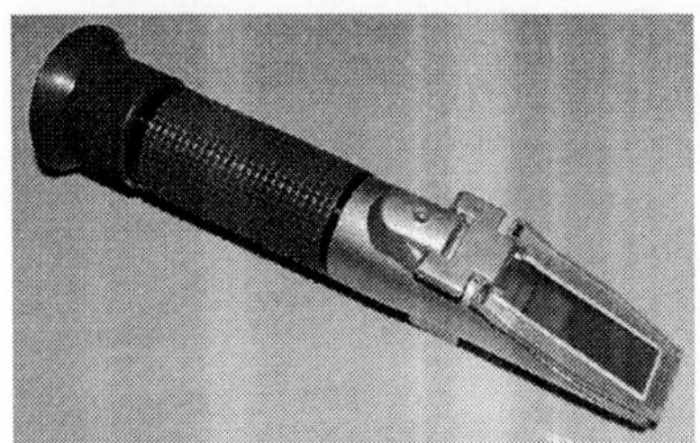

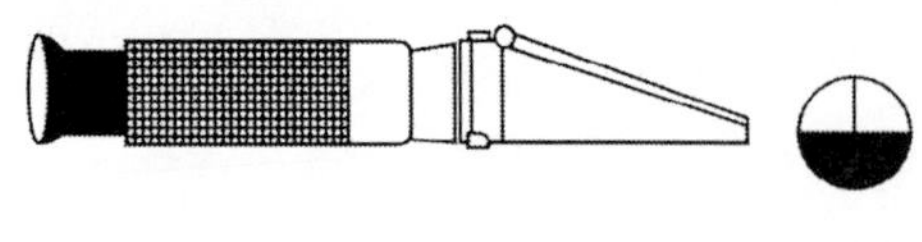

FIG. 25.3: Hand refractometer.

If reading indicates a higher % TSS, then produce is better than the minimum standard. Strawberries which are of excellent flavor, for instance, would measure 8% TSS or above. Following are the examples of proposed minimum % TSS for selected commodities: are mentioned below:

Minimum % TSS	
Apricot	10
Blueberry	10
Cherry	14-16
Grape	14-17.5
Kiwifruit	6.5

Minimum % TSS	
Mango	10-12
Muskmelon	10
Nectarine	10
Papaya	11.5
Peach	10
Pear	13
Pineapple	12
Plum	12
Pomegranate	17
Strawberry	7
Watermelon	10

Source: Kader, A. A. 1999

In summer, the sugar formed in the leaves is transferred to the fruit where it is stored as starch. As the fruit matures and ripens the starch changes to sugar. Therefore, starch content may be measured either by colorimetric method or its reaction with iodine solution. This will help to determine the optimum date of picking.

Method: For starch iodine test, 1 g of potassium iodide is dissolved in 1/8 cup of water. When it is dissolved, 2.5 g of iodine crystals are added. After dissolving, it is diluted to make 1.14 l solution and stored in a brown bottle. Fresh solution is made for each harvesting date. The prepared iodine-iodide solution is poured into a shallow pan or glass dish to a depth of 50 to 75 mm. The fruits are cut in half across the core and the surface is immersed in the iodine solution. After one minute, fruits are removed and the treated surface is rinsed with running tap water and starch iodine patterns are read and photographed immediately. The reaction of iodine with the starch on the cut surface of fruit gives dark bluish black colour and used as an indication of starch content thus, maturity. The changes in intensity and pattern of the dark colour as the season progresses has been charted and found reliable index of maturity for several cultivars of apple and pear. ISO quantification is provided by allocating a value to each distinct stages of starch disappearance. This produces a starch index chart. Most of the cultivars have a characteristic pattern of disappearance, so starch charts are often cultivar specific, hence the starch iodine value at any optimum harvest date can vary between cultivars.

The level of **oil content** increases with the onset of maturity. It can be used as a reliable index for the maturity determination of avocado. It should not be less than 8% by weight excluding the skin and seed. During maturity in citrus fruits there is an increase in the juice content of the fruit. The minimum juice content has been specified for Navel oranges (30%), grapefruit and other oranges (35%), lemons (25%), mandarins (33%) and clementines (40%) below which they should not be harvested.

Physiological Methods

The physiological method is based on the evolution of CO_2 and ethylene. This evaluation rates have been successfully used as the most reliable methods to determine the optimum physiological state for the maturity of fruits.

In climacteric fruits at physiological maturity and ripening processes there is an increase in their respiratory activity. The processs result in the increased evolution of CO_2. The preclimacteric minimum (PCM), when CO_2 production is at its lowest level, is, therefore, a useful indicator of physiological state of the fruit. It was initially thought to provide a good method of determining the suitability of fruit for storage, the optimum time of harvest being at the PCM, before ripening commenced.

The variability among fruit samples and patterns of evolution CO_2 means it may take several days to ascertain whether the climacteric has been initiated, by what time fruit should be harvested at the optimum date. Thus, measurement of respiratory activity is a useful test 'in retrospect' to determine, for example, what was physiological state in fruit with the best storage when harvested, but not as a practical maturity index.

During the climacteric, the temporary but marked increase in CO_2 production is not the only phenomenon occurring. Production of ethylene, a ripening hormone in climacteric fruit (Burg and Burg, 1965), follows a similar pattern concurrently and hence, can be a useful indicator of the physiological maturity of the fruit. The ethylene production was found to be the most useful of all the indices examined for Royal Delicious apple.

It has been suggested that fruit should be harvested just before the commencement of the climacteric errespective of internal ethylene concentration (low or higher) is still low, for best storage. The consentration of ethylene (low or higher) production does not mean that fruit will have good eating quality after storage. Ethylene measurements are related with with other maturity indices and help to determine the storage period, and which would be more suited for immediate consumption or other end uses. The method for assessing maturity based on ethylene production is not necessarily superior to other methods (soluble solids, acidity, starch index, firmness and colour) in determining the optimal picking date for a good storage.

The other indices are available water into free water in tissue, changes in optical density, skin colour, etc.

HARVESTING

The best part about growing fruit trees and shrubs is harvesting the fruit. The exact harvesting time depends on the variety you choose and your climate. Harvesting is an important operation in fruit crop production and any insufficiency during this time may lead to the loss of whole year's work. Use of improper harvesting method may result in the damage of crop. Bruising which is caused by compression (due to overfilling of boxes or in bulky stores), impact (due to dropping of crop or from something hitting the crop) or vibration (due to loose packing during transportation). So during harvesting, factors like delicacy of crop, maturity criteria, time and method of harvesting, mode of packaging and transportation, the importance of speed during and directly after harvesting, economy of the operations and the need for the harvesting method to fulfill the market requirement should be taken into considerations (Thompson, 1996).

Bruising in Fruit

Fresh-market fruit growers have long been concerned about bruising. Processing-fruit growers also have grown concerned, because unbruised fruit commands the best prices. The vast majority of bruising in the harvest process falls into two categories:

1. Picking bruises associated with rough handling and detrimental impacts.
2. Compression bruises associated with significant vibrations during transport.

Bruising is an ever-present problem. One study showed that bruising of fruit after harvest ranged from 0.6 to 13 per cent, with an average of 7.1 per cent. A study conducted of packing sheds indicated that bruising caused 8.1 per cent of the culls, while another study found bruising to cause only 2.7 per cent of the culls. At the retail level in supermarkets, bruising was found to range from 29 to 78 per cent, averaging 61 per cent. While bruising is a concern, it must be regarded as a defect that can be controlled through basic management principles. The damage inflicted on fruit is related to the energy available for bruising and the characteristics of the product. The energy available for bruising is in turn related to:

1. The suspension characteristics of the vehicle transporting the fruit.
2. The energy input to the system (a function of roughness of the road and vehicle speed).
3. Engineering factor involving both the properties and the packaging of fruit.

The damage suffered by fruit is dependent on the number of individual shocks and their severity, and is directly related to the energy absorbed by the fruit. For example the packers of Golden Delicious have learned that packing apples directly on removal from storage may produce more bruised fruit than if fruit is packed after being held at a relatively low humidity for a few days to create an outer layer of bruise-resistant cells. Reducing the amount of bruising in fruit appears to be attainable by reducing the amount of energy that fruit receives in handling.

In practical terms, bruising can occur during any of six operations in which fruit is removed from the tree and moved into storage. Listed below are seven locations of fruit and the six steps involved in moving the fruit from tree to storage:

1. Fruit on the tree Step 1 Harvesting.
2. Fruit in the bins Step 2 Moving bins out of the orchard.
3. Bins at the edge of the orchard Step 3 Moving bins to loading area.
4. Bins in loading area Step 4 Loading bins on truck.
5. Bins on truck in loading area Step 5 Trucking bins to storage.
6. Bins on truck at storage Step 6 Forklift hauling bins to storage.

Bins in storage need to educate employees in the proper method of performing assigned tasks. The proper harvesting involves the following:

1. Wearing proper clothing and a hat.
2. Adjusting the bucket. (Picking buckets with rigid sides and of a reasonable size are recommended)
3. Checking all ladders before using.
4. Carefully setting ladders and setting them at the proper angle.
5. Keeping your body centered on the ladder.
6. Handling fruit like eggs.
7. For apples, using stem-on picking methods.
8. Getting your hands in the bucket to prevent bruising.
9. Picking a tree from the bottom up.
10. Releasing fruit carefully and slowly into the bulk bin.
11. Reporting all accidents to the grower.

Excessive Heat and Worker Safety

A combination of very high temperatures (significantly above normal) and a higher than normal humidity can severely reduce the body's ability to maintain a proper internal temperature. Prolonged exposure to these conditions can lead to heat cramps, heat exhaustion, and heat stroke.

The **Heat Index** (HI), also referred to as the "apparent temperature," is a measure of how humidity acts along with high temperatures to reduce the body's ability to cool itself. The HI is the temperature (in degrees) the body senses, based on normal humidity levels. For example, if the actual temperature is 100°F with 40 per cent relative humidity, the effect of these conditions on the body is the same as 110°F with normal humidity (about 20 per cent). The basic assumption in computing HI is that the person is in the shade, at sea level, with a wind speed of 6 mph. Exposure to full sunshine can increase the HI about 5° to 15°F. Various wind speeds can also alter the HI but usually have small effect.

Harvesting Methods

The harvesting methods and the time of harvest play a crucial role in maintaining the best quality of crop during the course of postharvest handling and storage. For the determination of harvest maturity, number of methods has been evolved, which may vary from crop to crop.

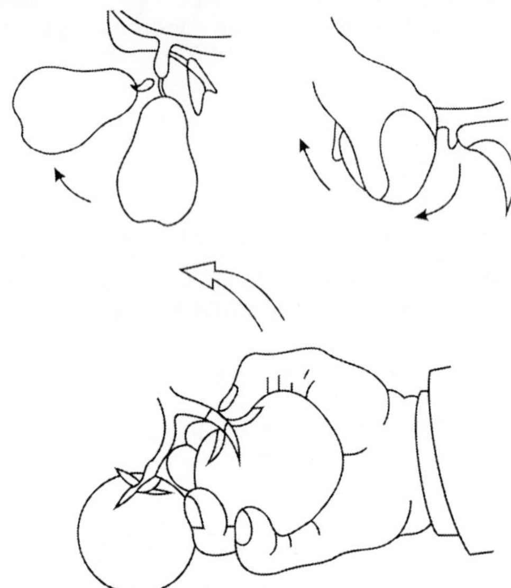

FIG. 25.4: Careful picking to avoid damage (Source: FAO, 1989).

Harvesting practices cause as little mechanical damage to produce as possible. Gentle digging, picking and handling will help reduce crop losses. In some fruit crops, a natural break point forms at the junction of the stem and the stalk when produce is mature. Harvesters should grasp the product firmly but gently and pull upward. Wearing cotton gloves, trimming fingernails and removing jewelry such as rings and bracelets can help reduce mechanical damage during harvest.

Harvesting Containers

Picking baskets, bags and buckets come in many sizes and shapes. Buckets are better at baskets in protecting produce, since they do not collapse and squeeze produce.

These harvesting containers can be made by sewing bags with openings on both ends, fitting fabric over the open bottom of ready-made baskets, fitting bags with adjustable harnesses, or by simply adding some carrying straps to a small basket.

Plastic crates are relatively expensive but are durable, reusable and easy to clean. When empty, they can be nested to save space in storage or transport. When filled they can be stacked if every other crate is turned in the direction opposite to the one below. If plastic crates are well vented along the sides and/or bottom, they can also be used to wash and/or cool produce after harvest

FIG. 25.5: Plastic picking baskets with bottom and sides slots for ventilation and drainage.

Harvesting Tools

Some fruits need to be clipped or cut from the parent plant. Clippers or knives should be kept well sharpened. Penduncles, woody stems or spurs should be trimmed as close as possible to prevent fruit from damaging neighboring fruits during transport.

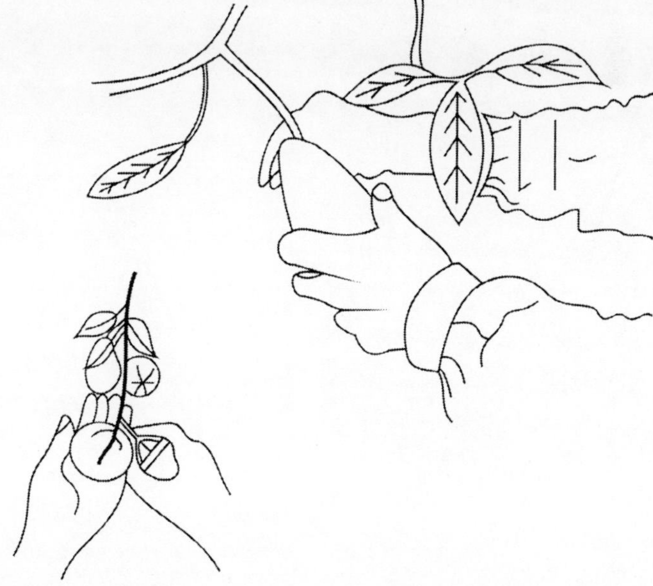

FIG. 25.6: Harvesting using clipper

Pruning shears are often used for harvesting fruits. A variety of styles are available as hand held or pole models, including shears that cut and hold onto the stem of the cut product. This feature allows the picker to harvest without a catching bag and without dropping fruits. Using a cutting tool attached to a long pole can aid picking of crops such as mangoes and avocados when the fruit is difficult to reach. Cutting edges should be kept sharpened and the catching bag should be relatively small The angle of the cutting edge and the shape of the catching bag can affect the quality of the fruit harvested, so it is important to check performance carefully before using any new tools. Picking poles and catching sacks can be made by hand or purchased from horticultural supply companies. The collection bags illustrated below were hand woven from strong cord or sewn from canvas. The hoop used as the basket rim and cutting edges can be fashioned from sheet metal, steel tubing or recycled scrap metal.

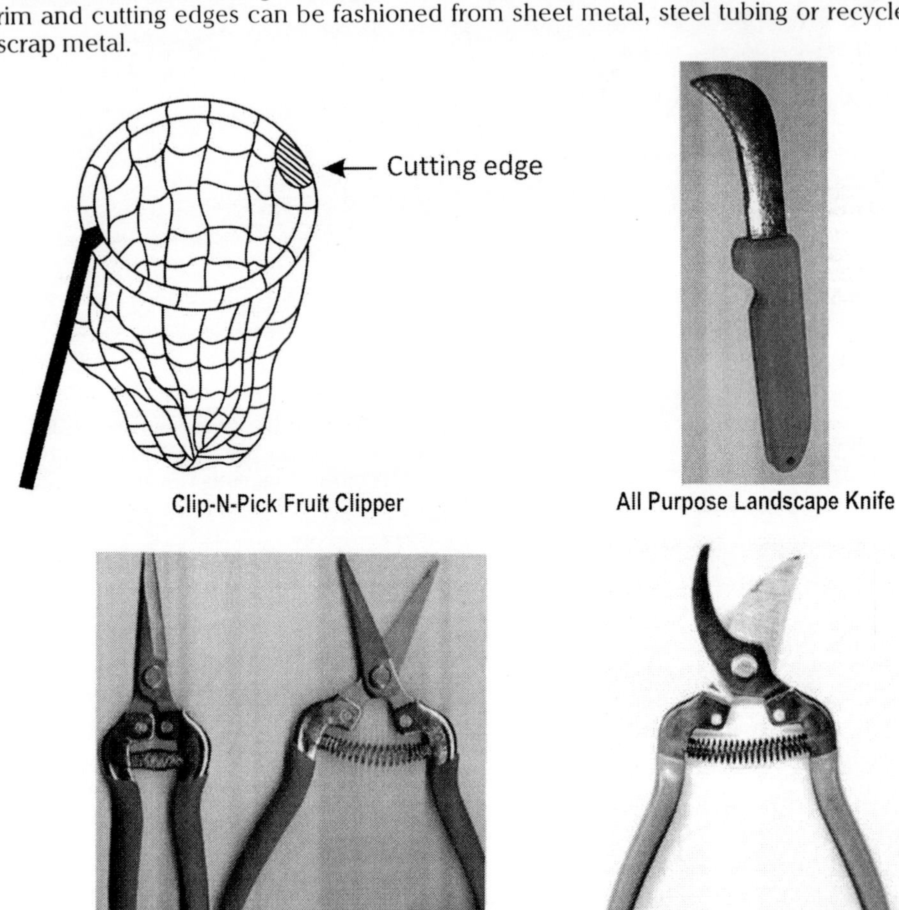

FIG. 25.7: Harvesting tools.

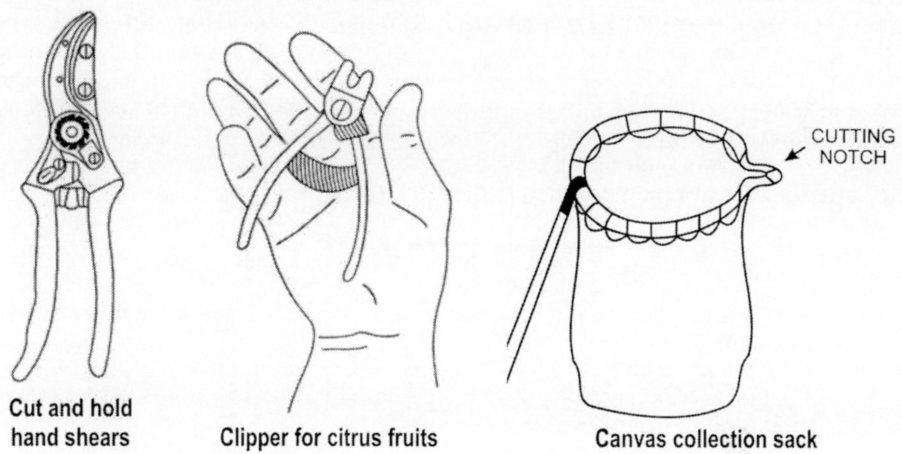

FIG. 25.8: Various harvesting tools.

Fruit trees are sometimes quite tall and letting fruit fall to the ground when it is cut from the tree will cause severe bruising. If two pickers work together, one can clip or cut the fruit from the tree and the other can use a sack to break its fall. The catcher supports the bag with his hands and one foot, catches the falling fruit, then lowers the far end of the bag to allow the fruit to roll safely to the ground.

FIG. 25.9: Harvesting from fall trees using a sack (Source: FAO, 1989).

The nut crops (pistachio nuts) should not be knocked to the ground during harvest because of their open shells and relatively high moisture content. They should be harvested putting plastic sheeting or canvas tarpaulins, which is spreaded below the tree during harvesting.

Various methods are in use to gravest fruits. These are following:

Hand harvesting: It is economic with low damage. For small sizeed fruits or picking from thorny plants are the major obstacles.

Mechanical harvesting: It needs mechanical assistance to hand pickers by ladders and worker positioners (tree towers and platforms) and self-propelled carts (to reduce non harvest time by assisting with materials) handling and movement of the workers and to increase productivity during harvesting time by providing optimum working conditions and by making detachment of the fruits easier. Mechanical harvesting devices employ direct contact methods such as combing, cutting, pulling snapping, twisting, stripping and compacting.

PRE-HARVEST FACTORS AFFECTING POSTHARVEST LIFE

The overall quality and condition of fresh produce remain intact, however, it decline after harvest and cannot be improved. Harvesting also depends on the market demand and value of the produce in the market. Growers, in general, rely on their own experience and local traditions in selecting crops and in cultivation practices, but if they want or need assistance they may need to be referred to agricultural extension officers or possibly to research and development specialists of their national department of agriculture or its equivalent. Market factors for the produce depends on many factors like neighbours, town people, retailers, jobbers or middlemen, commission agents, size, shape, maturity, appearance, perishability of the produce and pricing limitations of the buyer. It also depends on the "too good" or "too bad" quality of the produce, acceptability of the produced varieties, etc. The markability also depends on international trade, this specification of variety. If a grower have grown new variety of any fruit crop, it may or may not be easily acceptable to the market because of traditional human conservatism unless there are compelling economic incentives.

Pre-Harvest Production Practices

Pre-harvest production practices also affect postharvest returnsced crop. The reasone are following:

Water supply (Irrigation): Lack of rain or irrigation may lead to low juice content and thick skin in citrus fruit. For example in Bartlett pear, size and TSS are related to the level of water stress experienced by the tree. Increased tree water stress increases TSS, firmness and yellow fruit colour, decreased fruit size and vegetative growth. A higher TSS may result in premiums paid to producers, because of the link between applied water and yield, irrigation practices typically aim at the best overall economic balance between productivity and quality. Management of water frequently poses a dilemma between yield and postharvest quality. A deficiency or excess of water may influence postharvest quality of strawberry crops.

Soil fertility and use of fertilizers: Lack of plant foods in the soil can seriously affect the quality of fresh produce at harvest. On the other hand, too much of fertilizers can harm the development and postharvest condition of produce. For example lack of potash can bring about poor fruit development and abnormal ripening; calcium-moisture imbalance can cause blossom-end rot in tomatoes and bitter pit in apple; boron deficiency can lead to lumpiness in papaya, hollow stem in cabbage and cauliflower; the cracking of outer skin in beets. Therefore, nutritional status is an important factor in quality at harvest and postharvest life of various fruits crops. The nutrient with the single greatest effect on fruit quality is nitrogen. High nitrogen levels stimulate vigorous vegetative growth, causing shading and death of lower fruiting wood.

Foliar nutrient sprays: Calcium is often considered to be the most important mineral element in determining fruit quality, especially in apples and pears, where it has been demonstrated to reduce metabolic disorders, maintain firmness and reduce decay. In apples, to significantly affect fruit firmness or decrease fruit decay caused by postharvest wound pathogens, it is necessary to raise the level of flesh calcium (dry weight basis) to 800 to 1,000 $\mu g\ g^{-1}$.

Cultivation practices: Good crop husbandry is important in achieving good yields and quality of fresh produce. Certain aspects are particularly important, such as, weed control and crop hygiene.

Agricultural chemicals: The use of proper pesticides and herbicides and growth regulating chemicals may improve the marketability of fruit in order to control the time of fruit set and to promote uniform ripening. They are of little importance to small-scale production.

Canopy manipulations: It is done by proper pruning and fruitlet thinning. They increase fruit size while also reducing total yield; a balance between yield and fruit size must be achieved. The fruit count : leaf count ratio (F : L) influences high bush blueberry fruit quality more than mineral nutrition. A high F : L results in later ripening, lower TSS and smaller berries. Fruit canopy position plays important olin fruit production, for example large differences in SSC, acidity and fruit size have been observed in fruits obtained from the outside versus inside canopy positions of open-vase trained peach, nectarine and plum trees. Peaches grown under a high light environment (outside canopy) have a longer storage and market life than peaches grown under a low light environment (inside canopy). The use of more efficient training systems that allow sunlight penetration into the center and lower canopy areas is recommended to reduce the number of shaded fruit. Summer pruning and leaf pulling around the fruit increases fruit light exposure and, when performed properly, can increase fruit colour without affecting fruit size and SS. Girdling (a commercial practice in which the phloem of the tree or vine is removed) 4 to 6 weeks before harvest can increase peach and nectarine fruit size and advance and synchronize maturity. In some cases, girdling increases fruit SSC but also increases fruit acidity and phenolics so that the taste resulting from the additional sugars may be masked. Girdling can also cause the pits of peach and nectarine fruits to split, especially if it is done too early during pit hardening. In grapes, the balance between vegetative and fruit growth can be altered by girdling vines that have excessive vigor and a history of poor berry and bunch size. In these cases, girdling may improve bunch shape and berry size.

Crop rotations: Crop rotation may be an effective management practice for minimizing postharvest losses by reducing decay inoculum in a production field.

TIME OF FRESH PRODUCE HARVEST

Any type of fresh produce is ready for harvest when it has developed to the ideal condition for consumption and is called harvest maturity. Therefore, it is the time when the fruit is ready to harvest. Harvest maturity is based on grower's judgments which are based on:

1. Sight colour, size and shape
2. Touch texture, hardness or softness
3. Smell odour or aroma
4. Taste sweetness, sourness, bitterness
5. Resonance sound when tapped.
6. The current market value of the expected yield.

PERISHABILITY AND PRODUCE LOSSES

Perishable fruits are subjected to decay, destruction or death because all fruits are living plant parts and contain 65 to 95 per cent water and they continue their living processes after harvest. Their postharvest life depends on the rate at which they use up their stored food reserves and their rate of water loss. When food and water reserves are exhausted, the produce dies and decays. The causes of loss are as follows:

1. Physiological deterioration and postharvest physiology of fresh produce
2. Mechanical damage (physical injury)
3. Diseases and pests
4. Types of fresh produce.

POSTHARVEST MANAGEMENT OF MUSHROOMS

According to Savreet Sandhu and K. S. Khehra (*Agrobios Newsletter, June 2010*) mushrooms are most important microbes used as a source of human food. They are rich source of nutrition. Mushroom production in India had a beginning in late sixties in Himachal Pradesh and Jammu Kashmir. Commercial mushroom cultivation and production, however, is of comparatively recent origin.

In India only three mushroom types-white button mushroom (*Agaricus bisporus* and *Agaricus bitoquis*), oyster mushroom (Pleurotus spp.) And paddy straw mushroom (*Volvariella* spp.) are being cultivated commercially. Button mushroom comprises more than 90% of India's total mushroom production. Commercially, button mushroom is most viable mushroom the world over, but its method of cultivation is not as simple as those of oyster and straw mushrooms. Its growing requirements are also much more exacting than the others. In fact, its commercial cultivation has already attained a hi-tech farming status. Oyster mushroom cultivation is confined to small and marginal mushroom growers. Majority of the commercial units and all export oriented units grow white button mushroom which has maximum acceptability both in domestic and export market both in fresh and canned form.

Mushroom production in India has increased from 100 tonnes in 1970 to more than 40,000 tonnes in 1998, but this production is very low compared to other mushroom growing countries. Since mushrooms are living entities, even after harvesting, they are affected by a number of factors leading to their post harvest spoilage and losses. Once the fruiting body matures and harvested, degradation process starts and it becomes inconsumable after sometime if not handled properly.

Postharvest Technologies

During the peak period of harvesting, the gluts in the market can be checked by adopting appropriate post harvest handling practices.

Handling Fresh Mushrooms

Harvesting: Harvesting of mushrooms at optimum stage of maturity is of great significance. White button mushrooms should be harvested when their cap size is 30-45 mm in diameter or when their cap diameter is twice the length of stem. Oyster mushrooms are harvested, when their fruit bodies have curled under edges with formed gills. Paddy straw mushrooms are harvested at button or egg sized stage.

Pre-cooling: Although, mushrooms are grown at lower temperature, even they should be pre-cooled immediately after harvesting, to their optimum storage temperature (5 °C) to check the rate of high respiration and changes taking place at ambient room temperatures.

Sorting: Sorting is done to remove undesirable bodies in the produce. This helps in getting premium price to the growers.

Dipping treatments: Fresh whit button mushrooms are white or light buff. To maintain whiteness, mushrooms are dipped in dilute solution of hydrogen peroxide (1:3) for half an hour and then stepping in 0.25% citric acid solution containing 500 ppm sulphur dioxide gives significant effect.

Packaging: Properly washed mushrooms are packed in suitable containers. Since mushrooms are very sensitive to desiccation and drought, selection of suitable package is very important. Mushrooms are usually packed in polythene bags of 200-500 gm capacities. For distant markets, polystyrene or fibre board punnets over-wrapped with partially permeable polyvinyl chloride packing material is recommended.

Transportation: Transportation in refrigerated vans is recommended for distant markets.

Storage: A temperature of 5°C along with 85-90 % relative humidity is generally recommended for their storage. Pre-storage treatments with salt, potassium metabisulphite, citric acid and tartaric acid individually or in combination reduces discoloration during storage at room or cool temperatures. Irradiation with gamma rays (1-2 KGy) also increase their storage life for 9-10 days at 15 °C.

Processing: Pickling and sun-drying are economically viable methods of preserving mushrooms. However, freezing and sun-drying give them an excellent quality. They can be used commercially for export market. Different technologies which have a potential for processing mushrooms are described below.

- Low temperature: The basic principle of freezing is the speedy removal of heat from mushrooms. Freezing stops microbial activity. In quick freezing mushrooms passes through zone of 0-3°C in 30 minutes or less. Mushrooms are blanched prior to freezing. The storage temperature is maintained at -35°C to -40°C for best performance.
- High temperature: This is well practiced technique all over the world to destroy micro-organisms and to prevent the recontamination by hermetic sealing. Generally for canning purpose, small sized buttons are required without stem attached. Ascorbic acid and citric acid are recommended as useful adjunct for improving color in canning of mushrooms. Steeping buttons prior to canning results in increasing the yield. Food additives - agar (0.125%) and pectin (0.5%) - improve the flavour of canned products.
- Drying: Drying is done to reduce the water activity in the product to check the microbial activity. Freeze drying yields a product of excellent quality but the cost is very much. sun drying is appropriate technology in our country because of free source of energy. blanching and sulphuring are important treatments before drying. Drying at 60 ± 2°C for 8.5 hours yields best product with longer shelf life.
- Chemicals: Several chemical additives which are non-nutritive and are added in small quantities in foods to improve their appearance, flavour, texture, storage and preserving properties. potassium metabisulphite is commonly used preservative in mushroom processing. salt, sugar, acetic acid, vinegar and essential oils from spices are other main additives.
- Pickling and lactic acid fermentation: Fermentation is process of anaerobic or partial anaerobic oxidation of carbohydrates. during fermentation, sufficient quality of lactic acid is produced to prevent the product from further spoilage during storage. Pickles, chutney and ketchup can be stored for 6 months at ambient temperature.
- Irradiation: Application of gamma radiation retards the deteriorative processes

and helps extend their shelf-life. Doses of 1-2 KGy delays cap opening and stem elongation in button mushrooms, extending their marketability.

Mushrooms can also be utilized in the preparation of weaning foods, biscuits and soup powders. The recipies for preparation of dishes like mushroom kidney fry, mushroom curry, mushroom salads, cheese sandwiches, mushroom stuffed capsicum, stuffed morels, mushroom burgers and pizzas are also available.

Future Thrust and Export

At present, most of the mushrooms in India are consumed fresh. An overview of post-harvest scenario indicates that handling and processing technologies have been standarized to some extent. Introduction of refrigerated vans to collect the produce from the growers directly and delivering it to the terminal markets boost the production and make available quality produce to the consumers. For export of processed mushrooms, sophisticated technologies-individual quick freezing, freeze-drying and vaccum drying are required. Recently, tremendous interest has been generated amongst the entrepreneurs and many export oriented units with modern methods of preservation and packing are coming up. Development and introduction of new products with wider acceptability and comparatively at low price further increase the demand and consumption of mushroom products. Therefore, without employing proper handling and processing techniques such a highly perishable commodity cannot be made available to common masses at a reasonable cost.

CHAPTER 26

Organic Horticulture

Organic farming is a farming in harmony with nature respecting natural cycle, natural land fertility, animal welfare, landscape features and rural development.

Organic production involves a system that uses a combination of management techniques to maintain soil quality and fertility, and control weeds, pests and diseases. Crop rotation plays a big role in achieving these goals. Conventional chemical fertilizers, herbicides and pesticides are eliminated, although "organic" products are generally allowed, subject to compliance with the organic standard.

Organic farming is a holistic production management system which promotes and enhances health of agro-ecosystem related to biodiversity, nutrient biocycle and soil biological and microbial activities. *United States Department of Agriculture (USDA)* has defined this holistic production system as "a production system which avoids or largely excludes the use of synthetically compounded fertilisers, pesticides, growth regulators, and live stock feed additives. To the maximum extent feasible organic farming system rely upon crop rotations, crop residues, animal manures, legumes, green manures, off-farm, organic wastes, mechanical cultivation, mineral-bearing rocks, and aspects of biological pest control to maintain soil productivity and tilth, to supply plant nuteints, and to control insects, weeds, and other pests."

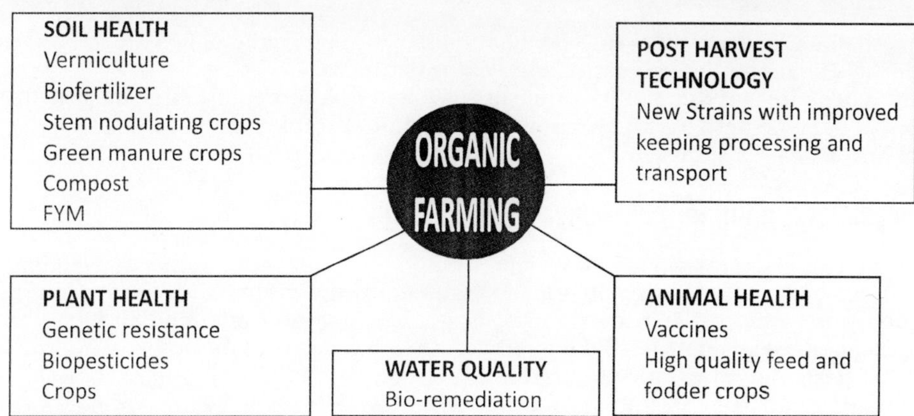

FIG. 26.1: Inputs used in organic farming. (Swaminathan, 2001) (modified).

Organic fruit production is a system devoid of the use of any chemical or genetically modified inputs, in which the biological potential of the soil, organic sources and underground water resources are conserved and protected by adopting suitable cropping pattern including agro-forestry and methods of organic replenishment. The US Dept. of Agriculture defines organic farming as a system that is designed to produce agricultural products by the use of methods and substances that maintain the integrity of organic agricultural products until they reach the consumer. Organic farming is a method of crop and livestock production that involves much more than choosing not to use certain pesticides, fertilizers, genetically modified organisms, antibiotics, and growth hormones that are not permitted by organic standards.

The organic fuit prodiction is a comprehensive management approach to improve the health of underlying productivity of the soil. Lampkin *et al.* (1999) mentioned that organic agriculture is a production system which avoids or largely excludes the use of synthetic compounded fertilizers, pesticides, growth regulators and livestock feed additives.

Organic fruit production relies on crop rotation, crop residues, animal manure, legumes, green manure, off farming organic waste and aspects of biological pest control. It includes economics, orchard design, and cultural considerations, including crop varieties, site selection, site preparation, soil fertility, weed control, and pest management (insects, diseases, and vertebrates).

The term 'organic' is best thought of as referring not to the type of inputs used but to the concept of the farm as an organization in which all the components: the soil minerals, organic matter, microorganisms, insects, plants, animals and humans, interact to create coherent, self-regulating and stable whole. Reliance on external inputs, whether chemical or organic, is reduced as far as possible. Organic farming is a holistic production system (Lampkin *et al.* 1999).

WHY ORGANIC FARMING?

There are several reasons for people to adopt organic farming practices, however, there are some reasons that people farm organically like:

Economics: Demand is growing for organic produce. Most organic farms need fewer inputs from outside the farm, which saves money for the farmer. Prices are often higher for organic produce, so profits may be satisfactory even if yields are lower.

Beliefs: Some people believe that organic farming is the 'right' way to farm or that it is better for the long term future of the earth.

Lifestyle: Not using any chemicals may lead to a more enjoyable lifestyle and better health. There may be more work to do, for example weeding with a scarifier or hoe instead of spraying.

Organic Farming *Verses* Conventional Farming

Organic and conventional farms can coexist quite happily. Some conventional farmers believe that having an organic farm nearby means that they will have more problems with weeds, insects and diseases. There is no evidence that this is the case-good organic farmers, like good conventional farmers, do not let weeds and other pests run rampant. Some farmers fear that the practice of organic farming is spreading and that conventional farmers will be forced to change the way they farm. Again, there is no indication that this is the case.

Conventional farmers use techniques and products, such as artificial fertilisers, insecticides, herbicides and fumigants. These are not available to organic farmers, who therefore have to be cunning in handling potential problems, for example heading off pests before they become too great a problem. Organic farmers need to know the habits and life cycles of pests and apply this knowledge to pest control. They have to replace chemicals with labour and management.

Good organic farming, like any other good farming, requires a whole-farm approach. This means a crop or animal is managed as part of the whole farm rather than in isolation. Organic farmers must use long-term strategies and preventive measures, and balance their approaches carefully. For example, if cultivation to control weeds is carried out too frequently, the soil structure can be damaged.

ORGANIC FARMING TACTICS

Organic farmers use many tactics to deal with potential problems and to improve soil, plant and animal management. Some of the important organic farming tactics are as follows:

1. Rotations help break disease cycles (for example, planting sunflowers alternately with cereal crops) and manage weeds (canola can shade out weeds so that fewer survive to cause problems for a following wheat crop). Rotations with legumes can increase soil nitrogen, particularly if the legume is ploughed in as a green manure.
2. Green manure crops are ploughed in to increase organic matter and soil fertility.
3. Cover crops protect young plants, reduce weeds and encourage predators of pests.
4. Plants with long taproots (for example, canola, safflower and sunflower) can help break up compacted soil and bring nutrients up from lower layers.
5. Weeds can be controlled by hand weeding, cultivation, rotations, grazing management, and by using specialised tools such as brush weeders and flame weeders.
6. Parasites of animals can be managed by selecting resistant animals, varying the stocking density, adopting suitable grazing strategies, and appropriately timing shearing and crutching.
7. Trees act as windbreaks and provide shelter for birds and other animals which can help in pest management.

HOW TO CONVERT TO ORGANIC FARMING?

If organic farming is to be followed, then it is necessary to make some decisions and take some steps to ensure that the transition is as successful as possible. If you are already farming without the addition of chemicals, you may want to have your farm certified. If you have previously used chemicals and wish to convert to organic farming, the conversion will usually take a few years. Your certifying agency will regard your produce as 'converting to organic' or 'in-conversion', rather than 'organic', during this time. There is less demand for 'in-conversion' produce than for 'fully organic' produce. The wholesaler may have purchased enough organic product and not require the same line of in-conversion product, or may offer a lower price for it. Otherwise, produce may have to be sold on the conventional market, any premium being lost.

To step into organic farming practices following steps are to be taken:

Review of Farm Situation

Have a good look at your farm and your present farming methods, work out what organic farming can offer you, and realise what is involved. Moving to organic farming cannot happen overnight - it takes time. You need to understand your present position as well as where you want your farm to be (that is, organic, biodynamic, certified etc.) before you decide on the best way to get there. You will probably need to make big changes to the way you farm.

Getting Certified

Decide how you will have your farm certified and then commit yourself to meeting the standards of your certifying group and resisting the temptation to reach for a chemical when things get tough.

Start Organic Farming at Small Scale

Start with a small area to discover your limitations and identify possible problems. What are your constraints in terms of farming organically? Here are some of the aspects to look at:

Knowledge

Do you know enough about farming organically? Do you know where to find out more? Who can help?

Markets Status

Have you identified markets for your produce, and decided how and when you are going to market your produce? For example, will you supply organic wholesalers or a processor, or sell at farmers markets or at the farm gate? What are the requirements of these markets (*e.g.* quality and packaging)? Will you add value to your product by processing on-farm? How will you promote your product? What price do you expect to receive?

Tools to be Used

Do you have the equipment you need? Some farmers need extra implements for weed management; others have all they need. The tools you need will depend on the size of your organic enterprise and on the crops you are planning to grow.

Labour and Time

Can you get staff who will work with and understand organic methods? Most people find that organic farming is more time-consuming than conventional farming.

Inputs

Do you have access to suitable seed, livestock and other inputs such as manure? Note that organic farming is not a 'no-input' system.

Financial Aspects

Do you have the financial reserves to cover the drop in income which usually occurs during conversion to organic farming? Do you know the costs and benefits of farming organically? If you are converting for other than financial reasons, how much income can you forgo to satisfy that need? For how long can you forgo that income?

Your Property

What are the likely problems given the history and surrounds of your farm? There may be patches of soil with high levels of chemical residues, or you may have poor fences which allow neighbouring stock to pass diseases and parasites on to your animals.

PRINCIPLES OF ORGANIC FARMING

Organic farming is based on enhancing the natural biological cycles in soil (*e.g.* nutrient cycling in the soil), crop (*e.g.* encouraging natural predators of crop pests) and **livestock** (*e.g.* development of natural immunity in young animals); on building up soil fertility through the use of nitrogen (N) fixation by legumes and enhancing soil organic matter; and on avoiding pollution.

Thus the aim is to work with natural processes rather than seek to dominate them, and to minimise the use of non-renewable natural resources such as the fossil fuel used for the manufacture of fertilisers and pesticides. Organic farming principles also encompass high animal welfare standards and the improvement of the environmental infrastructure of the farm.

In a process of several decades, the international organic community, organized in the IFOAM Movement, agreed on a common understanding on what the principles of organic agriculture are? They formulated clearly the minimum requirement as written down in the IFOAM Basic Standards. These standards are based on a number of principles. They address both the general level as well as the field level. The principles show that organic farming is much more than the renunciation of agro-chemicals.

TERMINOLOGY USED IN ORGANIC FARMING

- Accreditation: Procedure by which an authoritative body gives a formal recognition that a body or person is competent to carry out specific tasks
- Accreditation Programme: A certification programme, which has been accredited by the National Accreditation Programme and has agreed to comply with the accreditation contract.
- Appeal: Requested from a certified operator to reconsider the decision taken by the certification programme
- Application Programme: A certification programme that has applied for National Accreditation
- Biodiversity: The variety of life forms and ecosystem types on Earth. Includes genetic diversity (i.e. diversity within species), species diversity (i.e. the number and variety of species) and ecosystem diversity (total number of ecosystem types).
- Buffer zone: A clearly defined and identifiable boundary area bordering an organic production site that is established to limit application of, or contact with, prohibited substances from an adjacent area.
- Certification: The procedure by which a third party gives written assurance that a clearly identified process has been methodically assessed, such that adequate confidence is provided that specified products conform to specified requirements.
- Certification body: The body that conducts certification, as distinct from standard-setting and inspection.
- Certification mark: A certification body's sign, symbol or logo that identifies product (s) as being certified according to the rules of a program operated by that certification body.
- Certification program: System operated by a certification body with its own rules, procedures and management for carrying out certification of conformity.
- Certificate of Registration: Document issued by certification programme, declaring an operator that he /she is licensed to use the certification mark on specified products.

- Certificate Transference: The formula recognition by a certification programme of another certification programme or of individual projects or products, certified by that programme, for the purpose of permitting its own certified operators to trade or process under the programme's own certification marks, the prodcts certified by the other programme.
- Contamination: Pollution of organic product or land; or contact with any material that would render the product unsuitable for organic certification.
- Conventional: Conventional means any material, production or processing practice that is not certified organic or organic "in-conversion".
- Conversion: The process of changing from conventional agricultural system to organic farming. This also called transition.
- Conversion period: The time between the start of the organic management and the certification of crops and animal husbandry as organic.
- Crop rotation: The practice of alternating the species or families of annual and/or biennial crops grown on a specific field in a planned pattern or sequence so as to break weed, pest and disease cycles and to maintain or improve soil fertility and organic matter content.
- Culture: A micro-organism, tissue, or organ, growing on or in a medium.
- Declaration of Interest: A declaration of interest in the organic industry made by those involved in the certification process to enable determination of the individual's objectivity.
- Direct source organism: The specific plant, animal, or microbe that produces a given input or ingredient, or that gives rise to a secondary or indirect organism that produces an input or ingredient.
- Disinfect: To reduce, by physical or chemical means, the number of potentially harmful micro-organisms in the environment, to a level that does not compromise food safety or suitability.
- Exception: Permission granted to an operator by a certification body to be excluded from the need to comply with normal requirements of the standards. Exceptions are granted on the basis of clear criteria, with clear justification and for a limited time period only.
- Farm unit: The total area of land under control of one farmer or collective of farmers, and including all the farming activities or enterprises.
- Food additive: An enrichment, supplement or other substance which can be added to a foodstuff to affect its keeping quality, consistency, color, taste, smell or other technical property.
- Genetic diversity: Genetic diversity means the variability among living organisms from agricultural, forest and aquatic ecosystems; this includes diversity within species and between species.
- Genetic engineering: Genetic engineering is a set of techniques from molecular biology (such as recombinant DNA) by which the genetic material of plants, animals, micro-organisms, cells and other biological units are altered in ways or with results that could not be obtained by methods of natural mating and reproduction or natural recombination. Techniques of genetic modification include, but are not limited to: recombinant DNA, cell fusion, micro and macro injection, encapsulation, gene deletion and doubling. Genetically engineered organisms do not include organisms resulting from techniques such as conjugation, transduction and natural hybridization.
- GMO (Genetically Modified Organism): A plant, animal, or microbe that is transformed by genetic engineering.
- Genetic resources: Genetic resources means genetic material of actual or potential value.
- Green manure: A crop that is incorporated into the soil for the purpose of soil

improvement. May include spontaneous crops, plants or weeds.

- Indirect Certification: Process where by the license with the operator includes the services sub-contracted by the operator or the production of produces organized under the operator.
- Inspection: Visit on site/farm to verify that the programme of an operator is in accordance with specific standards of a certification programme.
- Inspection Body: Body that performs inspection services on behalf of a certification programme.
- Inspector: Person appointed by a certification programme or by an inspection body to undertake the inspection of an operator.
- Labeling: Any written, printed or graphic representation that is present on the label of a product, accompanies the product, or is displayed near the product.
- License: An agreement or contract that grants a certified operator that right to use the certificates or the certification marks in accordance with the rules of that programme.
- Operator: An individual or business enterprise, responsible for ensuring that products meet the certification requirements.
- Organic: "Organic" refers to the farming system and products described in the IFOAM Basic Standards and not to "organic chemistry".
- Organic Management: A management system, which incorporates a farm / processing design, and practices that seek to create an ecosystem, which achieve environmentally sourced and sustainable productivity.
- Organic product: A product which has been produced, processed, and/or handled in compliance with organic standards.
- Organic seed and plant material: Seed and planting material that is produced under certified organic management
- Parallel production: Any production where the same unit is growing, breeding, handling or processing the same products in both a certified organic system and a non-certified or non-organic system. A situation with "organic" and "in conversion" production of the same product is also parallel production. Parallel production is a special instance of split production.
- Part Conversion: When conventional, in-conversion and/or organic production or processing occurs in the same unit.
- Processing aid: Any substance or material, not including apparatus or utensils, and not consumed as a food ingredient by itself, intentionally used in the processing of raw materials, foods or its ingredients, to fulfill a certain technical purpose during treatment or processing and which may result in the non-intentional, but unavoidable presence of residues or derivatives in the final product.
- Split production: Where only part of the farm or processing unit is certified as organic. The remainder of the property can be: (a) non-organic, (b) in conversion or (c) organic but not certified.
- Standards: A set of rules of regulations under which operator is producing organic products.
- Synthetic: Manufactured by chemical and industrial processes. May include products not found in nature, or simulation of products from natural sources (but not extracted from natural raw materials). (Source: IFOAM, IIRD)

ORGANIC FRUIT PRODUCTION

The general information on cultural practices for fruit production (choosing varieties, spacing, pruning, training, irrigating, harvesting, postharvest handling, etc.) is relevant to both organic and conventionally managed operations. The organic producers should verify with the appropriate certification bodies that their practices and any materials they intend to use are compliant with applicable standards for their intended markets. This is especially true if those markets are international, where there may be additional production and labeling requirements.

Organic production is defined by USDA's National Organic Program (NOP) as "*A production system that is managed... to respond to site-specific conditions by integrating cultural, biological, and mechanical practices that foster cycling of resources, promote ecological balance, and conserve biodiversity.*"

Organic fruit production involves more than simply excluding synthetic pesticides and fertilizers. Benign neglect does not meet NOP standards for production. Organic agriculture is an integrated approach to active and observant management of a farming system. It begins with good soil management for nutrient cycling, productivity, and tilth. It involves an integrated, preventative approach to pest management to protect the health and productivity of the orchard.

Orchard Management

There has been a shift in recent years from the traditional 'top-down' approach to a more open management system where all levels have a say in the allocation and use of the **resource**. If properly done, this system ensures that the needs and concerns of those most affected by the use of the resource are addressed, *without loosing sight* of the wider issues touching the society as a whole. The effective communication is the key to managing a resource shared between various users and managed by different levels. Only once the needs of each user are understood can the resource be allocated and managed in a sustainable manner.

Land Management

It is defined as the process of managing the use and development (in both urban and suburban settings) of land resources in a sustainable way. Land resources are used for a variety of purposes which interact and may compete with one another; therefore, it is desirable to plan and manage all uses in an integrated manner.

Erosion is a major cause of poor soil quality. It affects not only the eroded areas which lose their topsoil, but lowland soils and rivers downstream. It is not easy to protect lowland areas from non-point source pollution. The best approach is to intercept soil erosion, by improved land management and land use. The use of soil conservation practices can reduce soil erosion, water loss and nutrient loss.

Some successful practices found in Asia today include the use of grass strips on the lower parts of cultivated slopes, and the use of forest buffer strips as vegetative barriers around lakes and reservoirs. These buffer strips intercept nutrients and pesticides dissolved in run-off from cultivated fields on higher slopes. There is also widespread use in Asia today of mulches and cover crops to protect the soil.

Principles of Sustainable Soil Management

1. Soil livestock cycle nutrients and provide many other benefits.
2. Organic matter is the food for the soil livestock herd.
3. The soil should be covered to protect it from erosion and temperature extremes.
4. Tillage speeds the decomposition of organic matter.
5. Excess nitrogen speeds the decomposition of organic matter; insufficient nitrogen slows down organic matter decomposition and starves plants.
6. Moldboard plowing speeds the decomposition of organic matter, destroys earthworm habitat, and increases erosion.

7. To build soil organic matter, the production or addition of organic matter must exceed the decomposition of organic matter.
8. Soil fertility levels need to be within acceptable ranges before a soil-building program is begun.

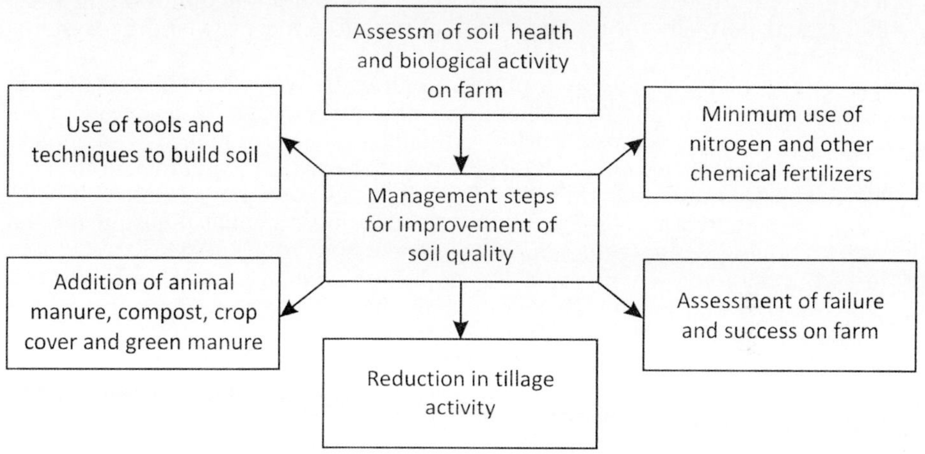

FIG. 26.2: Management Steps for improvement of soil quality.

Physical Land Management

The physical means of land management can be done by following means:

Buffer Strips: A buffer strip is an area of land maintained in permanent vegetation that helps to control air, soil, and water quality and other environmental problems primarily on land that is used for agriculture.

Cover Crops: Cover crops are crops grown to protect the soil when a crop isn't normally growing. Cover crops help to maintain soil structure, add organic matter to the soil, tie up excess nutrients and control pests. Several plant species are used as cover crops. Cover crop is any annual, biennial, or perennial plant grown as a monoculture (one crop type grown together) or polyculture (multiple crop types grown together), to improve any number of conditions associated with sustainable agriculture. Cover crops are fundamental, sustainable tools used to manage soil fertility, soil quality, water, weeds (unwanted plants that limit crop production potential), pests (unwanted animals, usually insects, that limit crop production potential), diseases, and diversity and wildlife, in agroecosystems.

Advantages for Soil and Water Management

1. Cover crops can improve soil quality by increasing soil organic matter levels through the input of cover crop biomass over time. Increased soil organic matter enhances soil structure, as well as the water and nutrient holding and buffering capacity of soil.
2. Crop cover increases soil carbon sequestration, which has been promoted as a mitigation strategy to help offset the rise in atmospheric carbon dioxide levels. Despite the limitations in both annual rates and attainable soil carbon levels, sequestration of carbon in soil organic matter can contribute to climate change mitigation. In addition to removing CO_2 from the atmosphere soil carbon

accumulation has positive influences on soil quality and wider environmental benefits. In the event of severe weather changes, better soil quality can counteract the risk of losing soil fertility and stability – an important factor in safeguarding the productivity of agricultural land.

3. Crop cover reduces soil erosion.
4. It often also reduces both the rate and quantity of water that drains off the field, that would normally pose environmental risks to waterways and ecosystems downstream.
5. Cover crop biomass acts as a physical barrier between rainfall and the soil surface, allowing raindrops to steadily trickle down through the soil profile.
6. Cover crop root growth results in the formation of soil pores, which in addition to enhancing soil macrofauna habitat provides pathways for water to filter through the soil profile rather than draining off the field as surface flow.
7. Crop cover increases water infiltration, and is the potential measure for soil water storage and the recharging of aquifers.
8. In agroecosystems where water for crop production is in short supply, cover crops can be used as a mulch to conserve water by shading and cooling the soil surface.
9. Crop cover reduces evaporation of soil moisture. In other situations farmers try to dry the soil out as quickly as possible going into the planting season.
10. Cover crops helps to conserve water, in temperate regions (particularly in years with below average precipitation) they can draw down soil water supply in the spring, particularly if climatic growing conditions are good.
11. Use of drip irrigation in organic farming is also beneficial. It rnsures accurate irrigation and nutrigation, reduces weeds, rctivates compost as the main source of nutrients, saves water, reduces runoffs, maintains the correct water and air balance in the soil and prevents diseases; drip irrigation keeps leaves and fruit dry- Economizes on manpower costs and minimizes hands-on involvement.

Drainage: Although excessive soil water is often a problem on poorly drained soils, weather conditions are extremely variable, and crops sometimes suffer from drought, which can also reduce yield. Intensive drainage systems are necessary to ensure access to many fields during wet periods. But past drainage practices have not always encouraged water conservation. As a result, these systems have tended to over-drain many areas and increase drought damage during dry growing seasons.

These problems are result in a transition from conventional drainage methods to water table management systems. The latter provide drainage during wet periods but also include control structures to manage the water level at the drainage outlet, making it possible to reduce over-drainage. In some cases, the system can be used to provide sub-irrigation during dry periods. Collectively, these management practices are referred to as ***water table management*** and include any combination of management practices such as surface drainage, subsurface drainage, controlled drainage, or sub-irrigation- that influence the level of the shallow water table.

1. Excess rainfall leaves a field either as surface drainage (surface runoff) or as subsurface drainage (shallow groundwater flow). This difference is important from a water quality stand point because the characteristics of the drainage waters differ. In practice, it is usually difficult to differentiate between surface and subsurface drainage because the outflow in drainage ditches or canals is a combination of the two.
2. Drainage waters from surface drainage systems contain higher concentrations of organic nitrogen, phosphorus, and sediment than those from subsurface drainage systems. Outflow from subsurface drainage systems contains higher

concentrations of nitrates than that from surface drainage systems. Both systems result in increased nutrient concentrations in drainage outflows compared to undeveloped sites, a result of fertilizer use on the developed sites.

3. Drainage control has little net effect on total nitrogen and phosphorus concentrations in drainage outflow. Controlled drainage may reduce nitrate-nitrogen (NO_3-N) concentrations in drainage outflow by up to 20 per cent, but total Kjeldahl nitrogen (TKN) concentrations are somewhat increased.
4. Controlled drainage tends to decrease phosphorus concentrations on predominately surface systems but has the opposite effect on predominately subsurface systems.
5. Seasonal variations may also occur, depending on rainfall, soil type, and the relative contribution of surface or subsurface drainage to total outflow.
6. Controlled drainage, when managed all year, reduces total outflow by approximately 30 per cent compared to uncontrolled systems, although outflows vary widely depending on soil type, rainfall, type of drainage system and management intensity.
7. Drainage reduces the time that excess rainfall remains on site before it reaches the receiving system, resulting in higher outflow rates (more water over a shorter time) than from undeveloped sites.
8. Any drainage management strategy can help to keep nutrients on site and prevent them from reaching sensitive receiving surface waters, is a positive step.
9. Controlled drainage management should be practiced with a crop rotation of corn, wheat, and soybeans, although increasing acreages of potatoes, peanuts, and vegetable crops are being included.

For lands which have naturally low lying or have high water tables need drainage. Drainage benefits your crops and adds value to agricultural land. Land can be drained in many ways. Talk to an experienced, licensed drainage contractor for cost-effective drainage options for your fields.

Erosion Control Structures: Erosion control structures are measures taken to help control surface runoff to reduce soil erosion. Such control can be done making water and sediment control basins, terraces, grassed waterways, stabilization of stream banks and livestock and machinery crossings. These structures are usually used in combination with cultural techniques and conservation tillage systems to reduce water erosion.

Green Manure Crops: Green manure crops are short-term cover crops used to cover and protect the soil between crops, particularly after short-season crops such as peas. These crops are grown for the plant material produced, which can then be returned to the soil to maintain soil organic matter levels - an excellent source of foodstuff for soil life.

Irrigation: Irrigation is the practice of adding water to moisture-deficient soils to improve production. Adequate moisture reduces crop stress and prevents disease. Over irrigation can lead to nutrient leaching and increased disease. You need to strike the right balance when planning an irrigation program. Irrigation must be applied properly to be cost-effective and prevent harm to the environment. Before irrigation management it is necessary to known the soil type and water-holding capacity. The irrigation should be done when critically needed to the crop.

There are many types of irrigation systems. For this purpose sprinkler, trickle, surface, or sub-irrigation systems can be used. The most common ones used in India are trickle (drip) or sprinkler (overhead). The type you choose will depend on the method of application, land slope, and crop to irrigate.

Manure Management

The proper application of livestock manure can benefit soil by returning nutrients removed by crops, supplying organic matter to feed the soil life, which in turn will help to improve soil structure. However, livestock manure must be handled properly to prevent pollution and loss and to ensure the greatest economic benefit.

Manure is a valuable by-product of a livestock operation. It provides a source of crop nutrients and organic matter, valuable in sustaining soil productivity. The physical characteristics of manure may be described according to its solids content. Manure may be handled as a liquid, slurry or semi-solid and as a solid.

Liquid manure: It is produced when little or no bedding is used. It contains less than 5 per cent solids. The manure, as excreted, typically has a solids content from 9 to 12 per cent. The additional liquid comes from washing floors and equipment and spillage from water.

Semi-solid (Slurry): It contains from 5 to 20 per cent solids.

Solid manure: It has sufficient bedding added so it is stackable. It contains a solids content greater than 20 per cent. To produce a solid, the liquid is drained off or absorbed into the bedding material.

Manure acts as a fertilizer. Several major benefits are associated with the proper management of livestock manure. Manure is a valuable organic source of crop nutrients, particularly nitrogen, phosphorous and potassium, and several secondary micro-nutrients. When recycled to the land, manure reduces the need to purchase commercial fertilizers. Soil quality improves with added organic matter. When considering manure as a fertilizer the important characteristics are: (1) nitrogen content, (2) phosphorus content, (3) potassium and sulphur content, (4) organic matter content, and (5) microorganisms.

Nitrogen: The nitrogen in manure exists in two forms, organic nitrogen which forms the largest portion, and ammonium nitrogen. Organic nitrogen is not available for use by plants until it mineralizes to the nitrate form, the same form as the nitrogen in some commercial fertilizers. The rate of mineralization will depend upon the composition of the manure, soil type, temperature and moisture content. Generally, 30 to 50 per cent of the nitrogen will be mineralized during the first cropping season with the balance mineralized during the next three years.

As the soil warms in spring, the nitrogen undergoes a conversion from the ammonia form to the nitrate form. Nitrate-nitrogen is dissolved in water and can move away from the root zone in the soil. This process is known as leaching. When manure is handled and spread on land, some of the nitrogen in the manure will be released to the atmosphere as ammonia gas.

To maximize the fertilizer value, manure must be stored and handled to reduce nitrogen losses. Nitrogen can be conserved by using the following practices:

1. Transfer the manure from the barn to storage as often as possible, especially during the summer;
2. Use transfer systems such as bottom loading storages that minimize contact between the manure and the air;
3. Use storage facilities with a minimum of exposed surface area, or cover the manure storage to reduce contact with air; and
4. When spreading, minimize exposure time by incorporating the manure into the soil as quickly as possible. Most nitrogen losses occur within the first 24 hours after the manure is spread.

Phosphorus: The phosphorus in manure exists in both the organic and inorganic forms, but unlike nitrogen, phosphorus is not released to the air. As with nitrogen, the inorganic form is readily available to be used by plants, but the organic form is not available until the phosphorus is mineralized. Phosphorus is not leached from the root zone unless very high levels build up in the soil. Surface waters can be polluted if runoff occurs from fields where the manure has not been incorporated.

Microorganisms: Manure contains large numbers of micro-organisms. The micro-organisms that carry or cause diseases are known as pathogens. The easiest transfer of pathogens from animals to people is through direct contact. Consumption of contaminated products which are uncooked, such as vegetables irrigated with manure is one example. The potential for disease transmission is reduced when pathogens are rapidly destroyed by drying and exposure to sunlight. Disease transmission risk is also largely reduced when manure is incorporated into soil.

Organic Matter: The application of manure to crop land can help maintain soil organic material, improve soil tilth and control soil erosion. When fields are cropped, a portion of the organic matter is removed from the field. The use of commercial inorganic fertilizers does not replenish the lost organic material. Depending on the soil type, a steady depletion of soil organic matter can result in a soil structure that is more susceptible to erosion. Repeated long-term application of manure at reasonable rates will improve the soil condition and water-holding characteristics as a result of the regular addition of organic matter.

Other Organic Materials: Materials such as compost, cannery waste, sewage sludge, and other organic wastes can help to build and maintain soil structure. Take confirmation that the materials should not contain contaminants that are harmful to plant or soil life.

Storage of Manure

Due to the nature of Indian climate as well as our seasonal cropping systems, it is necessary to store manure for various periods of time. The length of manure storage time depends on many factors. The storage capacity should allow for 7 months of liquid manure accumulation to efficiently utilize the manure for crop production. A solid manure storage should allow for a minimum of 60 days accumulation of manure in combination with field storage.

Existing livestock operations that have adequate manure storage facilities and are not causing environmental concerns are deemed to be in compliance with these guidelines. However, existing livestock operations that do not have adequate manure storage facilities, and are causing pollution of surface water or groundwater or other environmental concerns, are deemed to be in non-compliance with these guidelines. Therefore, a well planned manure storage system is needed to realize the potential value of manure and to avoid pollution problems. A manure storage system should be able to provide: (1) protection of groundwater and surface water, (2) minimum odour levels, (3) capacity that ensures manure can be stored until it can be efficiently used on the land, and (4) minimum nutrient losses during storage and handling. The appropriate system should be adopted which can keep the moisture content of the manure. It should have safety and security of persons and livestock, labour efficiencies and costs of complete systems. The location of the storage in relation to streams and watercourses, wells, neighbours, roads, etc. should be available.,

The storage should be located close enough to the barns to allow for convenient filling and still permit future expansion of the facilities. Leachate, manure and manure contaminated liquid from a manure storage must not enter a watercourse. The soil, bedrock and groundwater conditions should be evaluated by a qualified professional to ensure that the site is suitable for the type of storage planned.

Manure can be handled as a solid, semi-solid or a liquid, depending upon the chosen system of animal management and individual operator preferences.

Reduced Tillage Systems

Reduced tillage covers a wide range in tillage systems, including no-till, ridge-till, and reduced tillage forms like chisel plowing or *soil saving*. These systems leave residue cover on the soil surface and help to reduce soil erosion, by both water and wind, reduce tillage erosion and improve soil structure (over time with good management).

Residue Management: Increasing the levels of crop residues left on the soil surface will protect the soil from erosion, improve soil structure and add organic matter. Residue protects the soil by intercepts the raindrops and wind impact, preventing detachment of soil particles and creates thousands of tiny dams and windbreaks on the soil surface, slowing the movement of water, wind, and soil across the field. Residue cover moderates soil temperature and encourages higher earthworm populations, which benefit the soil structure.

Strip Cropping: Strip cropping is the practice of alternating strip widths of row crops with forages or cereal crops. There are four kinds of strip cropping: (1) contour, (2) field, (3) contour buffer and (4) wind. The selection of method depends on the crops that can be grown, kind of erosion you are eliminating and topography of the field and soil type.

Timely Tillage: Tillage is the agricultural preparation of the soil by ploughing, ripping, or turning it. Tillage can also mean the land that is tilled. There are two types of tillage: primary and secondary tillage. Tillage may be classified as follows:

- **Intensive tillage:** Intensive tillage systems leave less than 15% crop residue cover or less than 560 kg/ha of small grain residue. These types of tillage systems are often referred to as conventional tillage systems but as reduced and conservation tillage systems have been more widely adopted, it is often not appropriate to refer to this type of system as conventional. These systems involve often multiple operations with implements such as a mold board plow, disk, and/or chisel plow. Then a finisher with a harrow, rolling basket, and cutter can be used to prepare the seed bed. There are many variations.
- **Reduced tillage:** Reduced tillage systems leave between 15 and 30% residue cover on the soil or 560 to 1100 kg/ha of small grain residue during the critical erosion period. This may involve the use of a chisel plow, field cultivators, or other implements. See the general comments below to see how they can affect the amount of residue.
- **Conservation tillage:** Conservation tillage is a term that covers a broad range of soil tillage systems that leave residue cover on the soil surface, substantially reducing the effects of soil erosion from wind and water.

Windbreaks and Shelter Belts: Windbreaks are such structures which break the wind-flow and reduce wind speed while shelterbelts are rows of trees or shrubs planted for protection of crop against wind. They provide a protective shelter against wind and suitable for birds and honeybees as well as produce cattle feed and fuel wood. In the wind erosion areas, windbreaks and shelterbelts are to be planted. They prevent soil erosion, arrest the march of desert and protect the agricultural and residential lands from dust storms. A row of trees and shrubs planted across the wind-direction is the most effective. It reduces the wind speed up to 60-80% on leeward side.

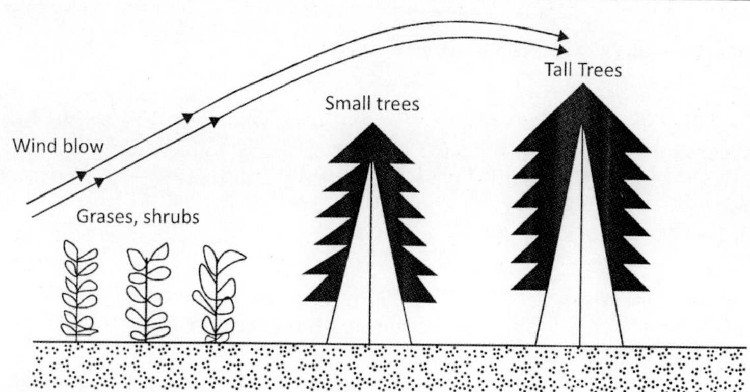

FIG. 26.3: Planting of trees and shrubs as sheltrbelt.

Wind Abatement Systems: Wind abatement systems are used with horticultural crops to protect tender plants. The practice includes using strips of cereal crops between beds or rows of crop. These strips slow wind and soil movement. Cereal strips can also reduce pest damage, and improve soil and air temperatures in the early spring.

Windbreaks: Planting trees in strategic areas on the farm will benefit crop production, because trees act as barriers to the wind, thereby reducing wind erosion, providing protection to the crops and preserving moisture in the soil, which is beneficial to drought-prone soils.

Hedges: A **hedge** or **hedgerow** is a line of closely spaced shrubs and tree species, planted and trained in such a way as to form a barrier or to mark the boundary of an area. Hedges used to separate a road from adjoining fields or one field from another, and of sufficient age to incorporate larger trees, are known as hedgerows. It is also a simple form of topiary. Hedges are usually divided into two classes: Outward fences, planted either with hawthorn or black-thorn.

Agro-forestry

Agro-forestry may be defined as an integrated self-sustained land management system, which involves deliberate introduction/ retention of woody components with agricultural crops including pasture/livestock, simultaneously or sequentially on the same unit of land, meeting the ecological and socio-economic needs of people.

It is also defined as a collective name of land use systems and technologies where woody perennials are deliberately used from the same land management units as agricultural crops and/or animals in some form of temporal sequence. In agro forestry systems, there is both ecological and economic interaction between different components.

Based on the kind of associated agricultural products, major function of the tree component, spatial arrangement of trees and duration of combination, several systems have been identified. In India, agroforestry systems are classified as:

Agri-silviculture	:	Crops + trees
Silvi-pasture	:	Trees + pasture
Agri-silivi-pasture	:	Crops + trees + pasture/animals
Horti-pasture	:	Fruit trees + Multipurpose tree species + pasture/animals
Agri-horti-silviculture	:	Crops + fruit trees + MPTS

Agro forestry is a system where woody and non-woody interacting plant combinations are carried out for achieving certain objectives like:

1. Conservation of soil and water.
2. Recycling of nutrients.
3. Maintenance of organic matter and some important physical properties of soil.
4. Sustained crop productivity, In addition to the production of timber and green leaf forage for cattle.

Fuel trees like Acacia, Subabool and Eucalyptus etc. can be planted to meet the fuel and forage requirement of farm family and cattle respectively. Intercrops with pulses and oil seeds can be taken up initially, till the canopy of the trees obstructs the penetration of sunlight.

Sustainable Water Management (SWM)

It is simply to manage our water resources while taking into account the needs of present and future users. SWM involves a whole new way of looking at how we use our precious water resources. Water problems cannot be solved by quick technical solutions. It requires the consideration of cultural, educational, communication and scientific aspects. Given the increasing political recognition of the importance of water, it is in the area of sustainable freshwater management. Therefore, SWM attempts to deal with water in a *holistic fashion*, taking into account the various sectors affecting water use, including political, economic, social, technological and environmental considerations. The following management measures should be done:

1. Freshwater is a finite and valuable resource that is essential to sustain life, the environment and development.
2. The development and management of our water resources should be based on a participatory approach, involving users, planners and policy makers at all levels.
3. Women play a central role in the provision, management and safeguarding of water resources.
4. Water has an economic value and should therefore be seen as an economic good.

Watershed Management

Watershed means an area of land whose runoff drains to a common point *i.e.* in the river or waterpool. Watershed may be defined as a natural unit of land whose runoff collects & flows out of the area through a single common outlet into another water body. It is a drainage basin which is demarcated by ridges or gullies. Watershed, drainage area or catchment area are generally interchangeably used meaning for the same. Ordinarily the cachment area (water collecting area) is larger but the area of watershed is comparatively smaller. Watershed area is the unit from the both angles *i.e.* land unit and water body unit; but there are separate units for both land and water in the catchment area. Therefore catchment area comprises of all the areas from which water flows out into a common river or water pool. The different river's tributaries having different water cachment areas flow out into a common lager river. The larger drainage area is called river basin. The water catchment area of a tributary is called sub-catchment area and the different cachment areas may be identified in it for it's certain parts. The catchment area of this certain part is called watershed area. There may be variations in the watershed area. When these variations are more in number it means the area is hilly, undulating or rolling and we get smaller watershed area where as in the plain we get larger.

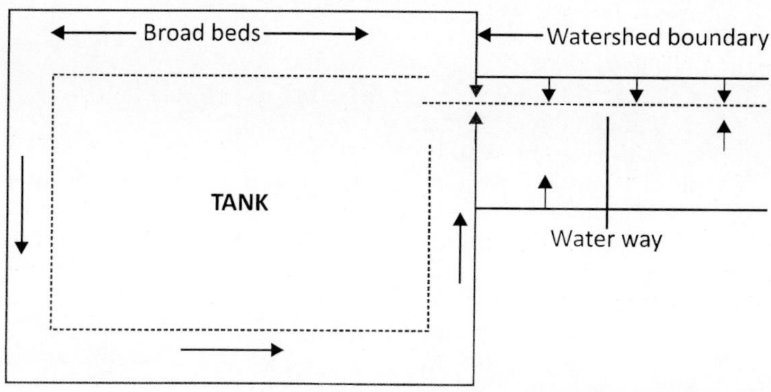

FIG. 26.4: Water shed.

The watershed for the agricultural purpose may be several fields of farmers that contribute runoff water to the flow at the lower point. The runoff movement depends on the land topography. Runoff from the upper fields enters into the lower fields in watershed causes soil and water erosion

Soil conservation measures in an isolated manner are not sufficient to check the erosion. Therefore, it is necessary that soil conservation measures should be adopted on the watershed basis instead of personal individual field. In the dry land areas, soil and water erosion, flood etc. are the major problems. Considering these point, the Govt. of India launched a Watershed Management Programme in the seventh five year plan to develop dry land areas on the basis of watershed. For soil and water conservation measures, watershed is demarcated into sub-watersheds and micro-watersheds. Micro-watersheds are the basis for planning and execution. In the beginning, there was a plan to develop 4000 watersheds in different agro-climatic regions. The Department of Agriculture provided the funds for this programme and the responsibility of evaluating 47 watersheds management programmes initiated in different states was given to ICAR. This programme was accelerated in 1983 when World Bank gave the assistance in the development of four watersheds. Watershed management programme has following features:

1. **Soil and Water Conservation Measures:** Soil and water conservation measures on the watershed basis include all those measures which are effective in preventing or delaying the movement of soil and rock particles. In it the emphasis is given on soil surveying, contour and graded bunding, repairing of old bunds and dams, land reclamation, digging of farm ponds etc.
2. **Scientific Dry Farming:** The agricultural activities like contour ploughing, pre-monsoon ploughing, deep ploughing, mulching, development of new varieties according to agro-climatic zones, weed control, integrated nutrient management and integrated pest management improve the production in dry land areas.
3. Forestry and development of pasture; and
4. **Animal husbandry and development of dairy:** Proper animal husbandry, artificial insemination, economic assistance to the milk producing cooperative societies etc. are to be adopted.

Principles of Watershed Management

Main principles of watershed management based on resource conservation, resource generation and resource utilisation are:

1. Utilising the land based on its capability,
2. protecting fertile top soil,
3. minimising silting up of tanks, reservoirs and lower fertile lands,
4. protecting vegetative cover throughout the year,
5. *in situ* conservation of rain water,
6. safe diversion of surface runoff to storage structures through grassed water ways,
7. stabilisation of gullies and construction of check dams for increasing ground water recharge,
8. alternate land use systems for efficient use of marginal lands,
9. water harvesting for supplemental irrigation,
10. ensuring sustainability of the ecosystem,
11. maximising farm income through agricultural related activities such as dairy, poultry, sheep and goat farming,
12. improving infrastructural facilities for storage, transport and agricultural marketing,
13. setting up to small scale agro industries, and
14. improving socio-economic status of farmers.

Aims and Objectives of Watershed Management

1. To protect, conserve & improve the land resources for efficient and sustained production;
2. To protect and enhance water resource, moderate floods & reduce silting up of tanks, increase irrigation and conserving rainwater for crops and thus mitigate droughts.
3. To utilise the natural local resources for improving agriculture and allied occupation or industries so as to improve socio-economic conditions of the local residents.

Thus, watershed management is the planned use of watershed lands in accordance with pre-determined objectives such as the control of erosion, stream flow, sedimentation, and the improvement of vegetative cover and other related resources. Such watershed management integrate plans for soil conservation, soil improvement, irrigation, drainage, flood prevention and development of water supplies for all purposes. In case of agricultural micro-watersheds, important aspects are soil conservation measures, soil improvement, storage and utilisation of runoff water. In dry farming areas, watershed approach provides an ideal means for integrated development. There is a need for multipronged approach to maximise crop production & also to ensure stability in rural income for integrated watershed management having following steps:

1. Water harvesting
2. Location specific technology for crop production.
3. Adoption of intercropping and crop substitution according to soil suitability.
4. Afforestation on cultivable wastes and marginal lands.
5. Practice of dry land horticulture to minimise risk. Mango, guava, ber, pomegranate, sapota, and tamarind have good potential on marginal land.
6. More efficient exploitation and use of ground water and irrigation.
7. Development of dairying and pasture management.
8. Stall-fed goat and sheep farming needs to be popularised.
9. Popularisation of sericulture and mulberry cultivation.

10. Processing of farm and horticultural produce.
11. Sound extension and training programmes for the farmers.

Water Harvesting/Runoff Concentration

Water Harvesting/Runoff Concentration/Rainfall Precipitation: In humid climates, runoff usually occurs only when rain falls on a saturated soil. Under semi-arid conditions, the total annual precipitation may or may not be sufficient, but most of the rain falls in a short period causing humid climate conditions and resulting into runoff. In both conditions, the excess rain water should be safely guided and collected in nearby ponds for recycling in drier period for life saving irrigation.

Water harvesting is defined as the collecting and storing water on the surface of the soil for subsequent use. It is a method to induce, collect, store and conserve local surface runoff for agriculture in arid and semiarid regions. Water harvesting was developed to a fine art about 2000 years ago by the Nabateans. The term water harvesting was first time used by Myors.

Normally water harvesting is more effective in areas situated near hill side or in areas where large portion is bare soil and cultivation is done on small portions. The technology of water harvesting differs according to soil types, topography, annual rainfall and the crops to be grown. In India, three types of water-harvesting techniques are generally practised:

1. **Inter-row water harvesting:** Practised in arid areas with light soil where annual rainfall does not exceed 400-450 mm. The crop is sown in narrow strips between wide intervals that are ridged as artificial miniature watersheds. Later on these are compacted to increase runoff to the crop rows. It is more practical since no land is sacrificed for harvesting water.
2. **Inter-plot or Micro plot water harvesting:** In this case water is harvested in the passages or furrow between the plots when rainfall is comparatively more. Runoff from the sloping area supplements rainfall for raising crop on level land,
3. **Water harvesting in farm ponds and reservoirs:** Surface runoff from small watersheds is stored in farm ponds & reservoirs for utilisation as supplemental or life saving irrigation. This harvested water can either be used as life saving irrigation to *kharif* crops at the time of drought due to irregular monsoon or as supplemental irrigation to winter crops. Since the stored water is limited, steps should be taken to reduce losses from seepage and evaporation. Suitable lining materials for pond beds and anti-evaporantes should be found out and used.

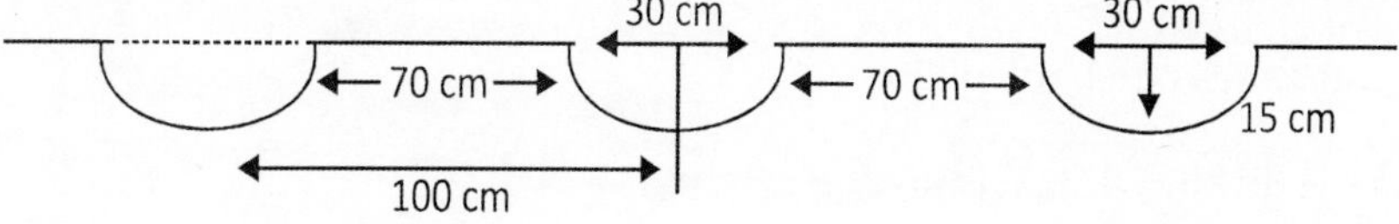

FIG. 26.5: Inter-row water harvesting system.

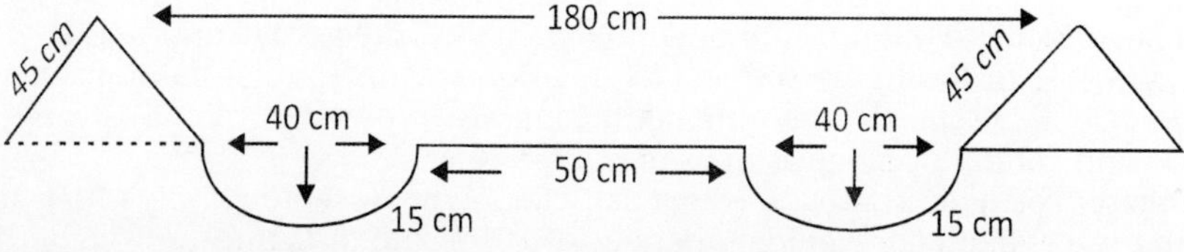

FIG. 26.6: Modified inter-row system.

Runoff can be increased either by land alteration or by chemical treatment. Impermeable catchments are prepared by spraying asphalt compounds. Thin plastic films or metal sheets can also be used to cover ground surface for water collection. Waterborne asphalt emulsions are much effective in controlling seepage loss. Bentonite has excellent sealing properties if kept continuously wet.

In the case of Broad Bed and Furrow system, excess water during rainy season is collected through the furrows and passed through grassed waterways into farm pond or reservoirs.

The stored water should be recycled most judiciously so as to get maximum return out of it. Only one or two irrigations are possible with the limited water stored. Therefore supplemental irrigation can be scheduled by two approaches viz. soil moisture depletion approach and critical stage approach. Irrespective of the crop stage, irrigation is scheduled when soil moisture approaches permanent wilting point to save the crop. This supplemental irrigation is known as life-saving irrigation.

Jalshakti: A chemical which when applied (mixed) in soil, improves the aeration, infiltration and water holding capacity of the soil.

Evapo-traspiration: Soil moisture is the most limiting factor in dry land agriculture. About 60-75% of the rainfall is lost though evaporation. The evapo-transpiration is the evaporation from the soil surface and transpiration from the plant surface. Such losses can be reduced by (i) mulches (ii) antitranspirants (iii) wind breaks and (iv) weed control.

Antitranspirants: Such materials are applied to transpiring plant surfaces to reduce water loss from the plant. Antitranspirants are of four types:

1. **Stomata Closing Type:** Such antitranspirants reduce water loss through stomatal closing. *e.g.* phenyl mercuric acetate (PMA), herbicides like **Atrazine** in low concentrations, **ABA** and CO_2.
2. **Film Forming Type:** Retard moisture loss due to formation of thin film as physical barrier *e.g.* Mobileaf, hexadecanol, silicone, oils, waxes.
3. **Reflecting Type:** Such materials reflect the radiation and thus reduce leaf temp and vapour pressure gradient from leaf to atmosphere and ultimately transpiration is reduced *e.g.* celite (a diatomaceous earth product) and Kaolin. Application of 5% Kaolin spray reduces transpiration loss.
4. **Growth retardants:** Such chemicals reduce shoot growth and increase root growth and thus enable the plants to resist drought. It also induces stomatal closure *e.g.* cycocel (CCC).

Antitranspirants generally reduce photosynthesis. Therefore, their use is limited to save the crop from death under severe moisture stress. They have some practical use in nurseries and horticultural crops.

Organic Farming and Water Use Efficiency

Sustainable agricultural development requires technologies and practices that make more efficient and productive use of resources and an enabling environment that encourages the adoption of these technologies.

A successful solution to bio-organic agriculture means ensuring optimal plant nutrition while restoring the soil and its microorganisms to their natural state. With synthetic fertilizer, pesticides and herbicides are prohibited, compost and other organic and natural nutrient sources play a key role.

Organic farming improves water use efficiency. Research shows that organic systems use water more efficiently due to better soil structure and higher levels of humus and other organic matter compounds. Soil water held in the crop root zone remains consistently higher by a statistically significant margin in the organic plots than the conventional plots, due to the higher organic matter content in the organic treated soils. The open structure allows rainwater to quickly penetrate the soil, resulting in less water loss from run-off and higher levels of water capture. The organic systems capture about twice as much water as the conventional treatment. Organic farming increases soil fertility, soil moisture retention, aeration, nitrogen fixation, mineral availability, disease suppression, soil tilth and general structure.

Nevertheless, using natural nitrogen sources for fertilization in the irrigation system requires special attention, such as proper filtration and maintenance as well as periodical flushing of the drip system to avoid potential clogging.

Globally, in both irrigated and rain fed agriculture only about 10–30% of the available water (as rainfall, surface or groundwater) is used by plants as transpiration. In arid and semi-arid areas, where water is scarce and population growth is high, this figure is nearer 5% in rain fed crops. There is, therefore, great potential for improving water use efficiency in agriculture, particularly, in those areas where the need is greatest.

The environmental and social benefits of more efficient on-farm water and energy use are a necessity to addressing the increasing demands for water and reduction in energy inputs.

Need for Organic Package

1. There is a need for hydraulic planning and solutions to suit individual farm requirements followed by the proper assessment of environmental conditions, topography, varying soil types, and nature and availability of water sources.
2. A flexible systems should be designed to produce maximum yields at a minimum costs.
3. To manage and develop an accurate irrigation system to control weeds and can carry different water qualities.
4. Comprehensive support for irrigation and nutrient management of all crop types in open field and greenhouse cultivation, compost application, equipment maintenance and disinfestations should be maintain.
5. Application of an irrigation system that adapts itself to the crop type. These can be either above-surface or subsurface drip irrigation (SDI) systems.
6. There should be an accurate spread of compost and installation of dripper lines above the compost to provide the crop with all its required nutrients through the drip irrigation system.
7. For fruit plantations, feeding channels should be designed to provide trees with a three-year compost supply. This technology uses open ditches with above-surface drippers and, when necessary, micro-sprinklers.
8. Development of water efficient system that accurately places the dripper near the target to control weeds and prevent leaves from getting wet for open field cultivation.
9. There is a need for field experience in organic crops from supporting farmers.

Water Savings

Until now, supportive water polices have allowed farmers access to plentiful water, but increasing urban and environmental demands are putting greater pressure on limited supplies. In addition, other policies have essentially ignored agriculture's potential to address climate change through soil carbon sequestration and reduction of nitrous oxide emissions from more efficient fertilizer and water use practices. The confluence of these two shifts in farming considerations provides a perfect economic and environmental opportunity for farmers to adopt more efficient water management practices. In addition, building soil organic matter through carbon sequestration enhances soil quality and has a long-term positive effect on crop production.

Mulching, cover cropping and reduced tillage can all serve to minimize the need for applied water. Building soil organic matter and improving soil structure are key drought-management strategies.

According to the Encyclopedia of Soil Science, each kilogram of soil organic matter can absorb 20 times its weight in water. In effect, the soil can be viewed as a *sponge or a water reservoir* in its own right. Furthermore, improving soil structure, particularly in the surface layer, can improve infiltration and percolation.

According to Rodale Institute studies report that the open soil surface structure resulting from building soil organic matter increases percolation 25 to 50% over unimproved soils.

One of the challenges in implementing reduced tillage practices is finding compatible irrigation practices. Both subsurface drip and overhead sprinkler are compatible with reduced tillage since crop residues would not interfere with water application as with furrow irrigation.

A comparative studies of drip systems versus furrow irrigation have shown that drip systems can be more efficient in terms of N use (10-20% less) and water use (20-50% less) with comparable yields. Similar water savings can be realized with overhead sprinkler technology.

ORGANIC FARMING AND SOIL FERTILITY

The foundation of organic farming lies in the health of the soil. A fertile soil provides essential nutrients to a growing crop plant and helps support a diverse and active biotic community. Strategies the transitional farmer will employ to build the soil are crop rotations, animal and green manures, and cover cropping.

Organic farming is the process by which crops are raised using only natural methods to maintain soil fertility and to control pests. The amount of crops produced by conventional farming methods is often larger than that of organic farming. But conventional farming, with its heavy use of manufactured fertilizers and pesticides (agrochemicals), has a greater negative effect on the environment. In comparison, organic farming produces healthy crops while maintaining the quality of the soil and surrounding environment.

Soil fertility is a measure of the soil's ability to grow crops and plants. Fertility is affected by a soil's tilth and the amount of nutrients it contains. Tilth refers to the physical structure of soil. Good tilth means that soil is loose and not compacted. It holds a great amount of water without becoming soggy and permits air to penetrate to plant roots and soil organisms. It also allows plant roots to grow and penetrate deeper.

The nutrients in soil are directly related to the soil's concentration of organic matter (living or dead plants and animals). Plants require more than 20 nutrients for proper growth. Some of these nutrients are obtained primarily from the soil, especially inorganic compounds of nitrogen, phosphorus, potassium, calcium, magnesium, and sulfur. In natural ecosystems, microorganisms (bacteria and fungi) in the soil break down organic matter, releasing the inorganic nutrients necessary for plant growth.

In conventional farming, soil tilth is destroyed by the use of heavy machinery, which compacts the soil. Very little organic material is added to the soil in conventional farming, decreasing the amount of nutrients that are naturally produced. Instead, inorganic nutrients are added directly to the soil in the form of synthetic fertilizers, which are manufactured from raw materials. These fertilizers are often applied at an excessive rate. As a result, they pass through the soil to contaminate groundwater and flow along the surface of soil to pollute surrounding bodies of water, threatening native species.

In contrast, organic farmers try to increase soil fertility by increasing the organic matter in the soil. They do so by adding the dung and urine of animals (which contains both organic matter and large concentrations of nutrients), by plowing under growing or recently harvested plants (such as alfalfa or clover), or by adding compost or other partially decomposed plants. These methods rely more heavily on renewable sources of energy and materials rather than on nonrenewable materials and fossil fuels.

Today, the list of alternative strategies for improving soil fertility includes the use of compost, crop residues, animal manure, biomass, chipped wood, hedgerow intercropping (alley farming), and cover crops — such as legumes.

Organic Farming Practices for Soil Fertility

Animal Manures

In India, livestock manure was traditionally used to maintain fertility of the farm soil. It was earlier applied to farm in either raw or composted form. Animal manure is the economical source of crop nutrients. There are three key steps to utilizing manure in an environmentally and economically sound manner:

1. Know the nutrient content of the manure,
2. Apply a uniform rate based on crop nutrient needs, and
3. Adjust the rate of supplemental fertilizer to compensate for the nutrients applied in the manure.

Raw manure is useful in supplying nutrients and adding organic matter to the soil, as well as encouraging healthy biological processes in the soil. Livestock manure may contain contaminants, and the level of the nutrient content also varies with the animal, bedding, and storage. Therefore, it is essential to compost the manure, because the heat created during composting may kill weed seeds and break down contaminants such as antibiotics. Nutrient content of animal manure is variable, depending on the type and age of animal, feed source, housing type, handling method, temperature, and moisture content. Because of this variability in nutrient content, individual land application decisions should be based on the nutrient content of the manure to be applied.

The crop nutritional requirements are the same for organic and conventional farms, organic producers need to be more creative due to the limitations on allowable inputs. The fresh manure has higher available nitrogen, but overapplications can lead to salt buildup and leaching. Therefore, it is essential that before application of livestock manure, soil should be tested to monitor soil fertility so that farmer can add the right amount of raw or composted manure to plots and avoid nutrient imbalances. Soil pH is also important because it influences nutrient solubility, microbial activity, and root growth.

Average nutrient contents are useful in determining overall waste management plans, but should not be used to determine actual land application rates. Application rates are usually based on nitrogen (N), phosphorus (P expressed as P_2O_5) and/or potassium (K expressed as K_2O) contents. Quantities of secondary and micronutrients applied in manure are generally sufficient to prevent deficiency.

Improper storage of manure can result in water pollution. The nutrient content of the manure, as well as the ability to spread it uniformly, are reduced with rainfall.

Cover Cropping

Cover crops also called green manure are used to manage a range of soil macronutrients and micronutrients. One of the biggest challenges of cover cropping is to fit cover crops into your current rotations, or to develop new rotations that take full advantage of their benefits.

This crop provides soil cover and can help loosen compacted soil through the growth of roots and improved water filtration. Cover crops can maintain or increase soil organic matter if they are allowed to grow long enough to produce high herbage. Cover crops also help prevent soil erosion caused by both water and wind, suppress weeds by keeping the sun from reaching weed seeds, and reduce insect pests and diseases. In addition, a legume used as a cover crop can provide nitrogen to the soil. Nonlegumes can take up excess nitrogen, phosphorus, and potassium from previous crops and recycle them to the following crop. Common cover crops and green manures include rye, buckwheat, hairy vetch, crimson clover, subterranean clover, red clover, sweet clover, cowpeas, millet, and forage sorghums.

Nitrogen is often the most limiting nutrient in crop production. Cover crops are grown and incorporated (by tillage) into the soil before reaching full maturity which improve soil fertility and quality. They are commonly leguminous plants (fabaceae or pea family). They have following features:

1. Leguminous cover crops are typically high in nitrogen and can often, to varying degrees, provide the required quantity of nitrogen for crop production that might normally be applied in chemical fertilizer form (called fertilizer replacement value).
2. The leguminous cover crops form symbiotic relationships with rhizobial bacteria that reside in legume root nodules. Lupins are nodulated by the soil microorganism *Bradyrhizobium* sp. (Lupinus). Bradyrhizobia are encountered as microsymbionts in other leguminous crops (*Argyrolobium, Lotus, Ornithopus, Acacia, Lupinus*). These bacteria convert biologically unavailable atmospheric nitrogen gas (N_2) to biologically available mineral nitrogen (NH_4^+) through the process of biological nitrogen fixation.

Catch crops are typically fast-growing annual cereal species adapted to scavenge available nitrogen efficiently from the soil. The nitrogen tied up in catch crop biomass is released back into the soil once the catch crop is incorporated as a green manure or otherwise begins to decompose

Green Manures

When a cover crop is tilled into the soil while it is still lush and green, it is referred to as a green manure. Green manures are important under an organic farming system because they help to add organic matter and nutrition to the soil. When a green plant is incorporated into the soil, it has high amounts of nitrogen and moisture and becomes a food source for soil microorganisms and earthworms. During the process of decomposition by the organisms in the soil, organic matter and nutrients become available to the crop plants. An additional benefit from using green manures is the suppression of weeds and soilborne diseases.

In situ green manure: Green manures are either grown and used *in situ*, or used as green leaf manure. In the *in situ* method, green manure crops are grown in a field prior to crop cultivation and then cut and buried when approximately 50 per cent of all plants are flowering. The use of sunn hemp (*Crotalaria juncea*) is popular because of its ability to grow fast and its efficient nitrogen fixing capacity. These plants are grown and sacrificed to improve the living condition of the main crop. Sunn hemp is grown in tomato, tobacco and onion fields. After ploughing, the sunn hemp seeds are broadcast. Farmers then cut and incorporate the sunn hemp into the soil. After churning, the soil is left for one to two months to allow for the decomposition process.

Green gram and black gram are also used as green manure *in situ* in paddy fields. Grasses such as *Mimosa pudica*, *Tephrosia pumila* and other non-legumes are allowed to grow as a cover crop on the soil. At the end of the hot period the soil is ploughed once, and this vegetation is turned into the soil as green manure.

Green leaf manure: The practice of incorporating green leaf manure is different from green manure grown *in situ*. *Thespesia* and *Gliricidia* are grown in

Crop rotations: Many farmers find that rotating crops improves the tilth or aggregation of the soil. Planning a crop rotation requires a farmer to plant crops at different times and in different locations in the same field. When they are used in tandem, crop rotations using cover crops, (1) Decrease pest pressure from insects, weeds and diseases. (2) Enhance biological activity in the root zone to improve nutrient transfer from soil to plants and expand root growth. (3) Improve soil physical condition through increased organic matter that absorbs water better, holds more moisture into dry times, drains better in wet times and creates better earth worm habitat in all times. (4) Create biologically and economically durable crop systems.

Required Elements for Plant Nutrition

Sixteen elements are considered necessary for plant growth. Carbon, hydrogen, and oxygen make up most of the elements in plants and are used in the formation of organic compounds. The other 13 elements are classified as macro or micronutrients based on the amount found in plants. Macronutrients of greatest concern are nitrogen, phosphorus, and potassium. Other macronutrients include sulfur, calcium, and magnesium. Micronutrients include iron, zinc, manganese, copper, boron, chlorine, and molybdenum. In general, fertilizers or amendments for organic production must come from natural carbon containing nonsynthetic materials or nonsynthetic inorganic materials. Organic fertilizers usually are needed in larger quantities than conventional fertilizers since they are lower in nutrients. It is good to know the nutrient analysis of the organic fertilizer before its application so that the right amount is applied.

Nitrogen	It is the most frequently limiting nutrient for crop production. Organic farms need to supply N through sources such as legumes, animal wastes or by-products, plant-processing by-products, or limited additions of mined mineral deposits.
Phosphorus	The sources include manure, bone meal, fish and poultry meal, and rock phosphate.
Potassium	The sources include manure, alfalfa meal, kelp meal, greensand, wood ash, potassium sulfate, and granite dust.
Sulfur	This element is found in rainwater, especially in acid rain. When a farmer uses organic sources of nitrogen, the sulfur requirements are usually met.
Calcium	Generally supplied through lime applications used to adjust acidic soils. Sources include calcitic lime, and dolomitic lime that also contains magnesium. Other sources include colloidal phosphate, bone meal, gypsum, and wood ashes.
Magnesium	The sources are dolomitic lime and langeinite.
Micronutrients	They are mostly satisfied through manure, compost, and liming amendments.

Soil pH

It is important to manage the pH of the soil since it can affect the plant's ability to take up nutrients and the microbial activity in the soil that affects the processes needed for plant nutrition. To correct acidic soil, farmers apply lime based on a soil test. An alkaline soil can be more difficult to correct since adding sulfur to make the soil more acidic is only a temporary remedy and can be more expensive than liming. When adjusting the pH, it is important to know the crop's pH requirement since different crops grow best at different pH levels. For most crops, optimum pH levels are between 6.0 and 7.0.

Organic Farming: Nutrient Recycling

Farm nutrient balances have become a central tool for assessing and reducing the environmental impact of intensive farming, they also serve as indicators of sustainable land management. For organic farms in particular an analysis of internal nutrient cycle is essential since purchased fertilizers and fodder have limited nutrient inputs limited. Nutrient management strategies need to be developed to use nutrients efficiently and avoid nutrient losses, ensuring high system productivity. Nutrient balances are necessary to assess nutrient use efficiency, nutrient turnover and utilization, nutrient enrichment in one part of the farm at the expence of another, and to estimate nutrient losses at farm level. For organic farms, common fram-gate balances yield only global figure because the main nutrient flows take place within the farm.

In organic agriculture, the internal farm nutrient cycle must be quantified to ensure high system productivity accompanied by environmentally sound production processes. In contrast to common farm-gate and field balances, budgeting at the stall level is seldom undertaken. When budgeting mixed farming systems, a substantial lack of nutrients can be detected in the forage and straw input – stall – manure output nutrient flow chain. Therefore, stall balances focus on a central component of whole-farm nutrient budgets for developing efficient nutrient management strategies. Mutrient balances are very sensitive to variations in mass flow and nutrient content for components with high nutrientcontents and/or a large contribution to total mass flow (e.g. manure, silage). In developing strategies to minimise N losses, by reducing N surplus in the ration, one must consider, that, in contrast to dairy farms, a suckler herd forbeef production integrated in an organic farm has to adapt to crop production demands.

Nutrients cycle among pools within an ecosystem, and losses of nutrients to the environment accompany each transfer from pool to pool. Efficient recapture of nutrients by plants is critical in extensively managed grasslands if these swards are to persist. In intensively managed systems, the greatest contribution of efficient recapture of nutrients may be minimizing loss of nutrients to the environment and associated negative impacts.

Soil Organic Matter Dynamics and Management

In a balanced ecosystem, the litter deposition is equal to the litter degradation and soil organic matter is also in equilibrium. Changes in native vegetation, soil-tillage system decrease soil organic matter. The rate of soil organic matter decomposition is associated with greater microbial activity and soil perturbation and lesser residue deposition. After an initial reduction in the soil organic matter levels (disturbance phase), a new equilibrium is established between litter production and decomposition rates. Soil organic matter stabilizes at lower, the same, or higher levels than the original ones. A considerable amount of N from fertilizer can be incorporated into soil organic matter provided that sufficient residual herbage is left following each grazing. In addition to fertilization, forage use and management may also alter C accumulation.

Recalcitrant compounds are hard to **decompose** substances and originate either from compounds found in plants (e.g., tannins, lignin, polyphenols), or are formed during the decomposition process. **Lignin**, C, N, P, and polyphenols and their ratios are often used as indicators of litter quality. As a general rule, legumes have better-quality residue than grasses, and aboveground residues have better quality than roots and rhizomes. However, large variability exists among species. Thus, the use of plants with low quality residues and higher allocation of biomass to the root system could be proposed as an alternative to increasing soil organic matter. **Tropical grasses** (e.g., gambagrass, ***Brachiaria*** spp.) can increase C storage in the soil not only due to their large root system but also due to the low quality residue originating from this root system.

The organic matter deposited beyond this limit may still undergo biochemical protection by forming recalcitrant compounds; however, this quantity is not well established for particular soils. Unprotected organic matter with higher turnover rates (light fraction) also increases with greater primary productivity, providing more nutrients to the grassland ecosystem.

Improved vegetative cover, land configuration management and alley farming will form the base for sustained production achieving better soil and **moisture conservation and improved nutrient recycling**.

Plants require 13 essential nutrient elements (6 major, 7 micronutrients) and can profit from some beneficial nutrients. The amount of nutrients required by crops depends largely on the target yield level. Only under rare circumstances the nutrients required for high yields are "automatically" supplied from the soil nutrient reserves.

The response of crop growth and yield to the nutrient supply can be expressed in growth or yield curves and described by "yield laws". In practical cropping, nutritive minimum factors are often responsible for unsatisfactory yields and thus present growth problems. Therefore, a proper nutrient management (of the nutrients added to the soil as well as in order to keep many nutrients, especially microelements, on a sufficient natural supply level) is of fundamental importance for effective, successful and sustainable agriculture. Exploitation of soil nutrients is frequently applied which decreases the level of available nutrients quickly or slowly and is therefore non-sustainable.

Incorporation of Plant Residues in Agricultural Systems

Incorporation of plant residues in agricultural systems is an important factor in the control of soil fertility and maintenance of soil organic matter. If such measures are not taken into account at the right time, then ultimately soil is depleted with deterioration of fertility and productivity side by side. Plant residues are known to affect soil physical properties, availability of nutrients and soil biological activity. However, effects of plant residues on soil and crop differ and depend on their decomposition and nutrient release-rates.

Ground or finely chopped residues material, for example, is likely more susceptible to microbial attack than intact plant parts due to a better soil-residue contact and lack of intact lignified barrier tissues. In contrast to this, however, fine particles are also more likely to be protected against decomposition through physical protection by clay and other particles. The rate of CO_2 efflux, under conditions where moisture and temperature are not limiting, can provide an indication of organic matter quality and whether the soil environment is conducive to the decomposition process.

Carbon Nitrogen Ratio

Plant residues with high C:N ratio and high lignin and polyphenol contents decompose and release nutrients slowly. Such residues have a low direct nutrient effect and a high indirect mulching effect on crop. In turn, residues with low C:N ratio, and lignin and polyphenol contents decompose rapidly, and have a high direct nutrient effect and a low indirect mulching effect. Therefore, the decompositions of plant residue is related to their C:N ratio and lignin and phenol contents.

Geographical Nutrient Transfer

In smaller or greater region there are natural nutrient flows and nutrient transfers by trade, e.g., into cities or even export into other countries. These mainly unilateral transfers are of environmental concern since they are not only the cause of nutrient losses from a farming area (and partly completely lost for production), but also unwanted from water eutrophication point of view.

Natural Transfers (Nutrient Flows in Landscapes): A steady flow occurs naturally with surface or ground water movement in hilly or mountainous areas as part of the slow natural erosion process even under natural vegetation cover. The annual losses from the hilly pat of the landscape are relatively small, and so re the gains for the low-lying land. In geological periods, however, this transfer has produced many fertile alluvial sols in river basins that now belong to the best agricultural lands. If this process is accelerated by recent man-made strong soil erosion, it can lead to serious soil fertility decrease in the hilly area and to an excessive input on nutrient (especially N and P) into water. The extent of nutrient transfers in small watershed areas or whole river basins is difficult to determine because of many factors involved (sources not only from agriculture, but also from towns, industry, etc.).

From nutrient flows to cycles: A main objective is converting unwanted flowers into useful cycles. Recycling of nutrients within the farm area tends to be not uniform. None of the cycles is closed, most of them are leaking. Some losses are unavoidable (and may be considered as natural load, a minimum tribute for primary production), whereas, others are avoidable by adequate nutrient management. There is a widespread opinion that decreasing N-fertilization is the best preventive measure for reducing leaching losses of nitrate. This holds only true, however, for fertilization systems where N is over-emphasized (unfortunately not a rare practice in intensive fertilization cropping), whereas, even high application of N together with a sufficient supply of other nutrients (i.e., with no yield limiting minimum factor) may result only in still tolerable losses. Cycling is a very complex system. Although it is not so difficult to design a complex model of nutrient flows, their quantitative assessment will partly have to be based on some more or less uncertain estimates. Therefore, the result can only be approximate. For practical use, cycling models should be restricted to the important nutrient fluxes.

Organic farming and waste recycling: Organic farming is the backbone of sustainable agriculture. Organic farming mainly depends on organic recycling. Industrial agricultural chemicals like fertilizers, pesticides, herbicides, etc. are not used or used to the minimum extent necessary in this kind of farming.

Harnessing the nutrient energy from biological and industrial wastes is of prime importance for maximizing production. When these wastes are recycled as manure for crop production and are subjected to the degradation and assimilative capacity of soil, pollution of streams and/or rivers receiving these wastes is reduced to a large extent as compared to their direct disposal in water bodies.

Organic recyclable wastes include crop residues, animal wastes, farm/ industrial wastes, municipal and sewage wastes. They are valuable sources of plant nutrients and humus. In tropical and sub-tropical soils found in India, there is a general deficiency of organic carbon and plant nutrients due to rapid loss of these components by biodegradation. To make up for these losses, extensive utilization of organic residues in agriculture is essential. In addition, they also protect the soil from erosion. The manurial value and quality of these wastes could be improved by composting and enriching these organic sources along with inexpensive materials such as rock phosphate.

In India, there is a great potential for utilization of crop residues/straw of some of the major cereals and pulses. Approximate availability of straw is to the tune of 141.2 mt, which contribute about 0.7, 0.84 and 2.1 mt. of N, P_2O_5 and K_2O respectively. Considering that about 50 % of these residues are utilized as animal feed, the rest could be very well utilized for recycling of nutrients. Crop residues have wide C:N ratio and when applied directly, nutrients may get immobilized leading to initial adverse effect on crop growth. Hence, adequate care is required to use the residues after proper composting with efficient microbial inoculants. Inoculation of these wastes with efficient cellulotic microorganisms produces enough simple sugars for growth and multiplication of beneficial microflora like free living N fixers and phosphate solubilizers in the soil which ultimately increases crop yields.

While the incorporation of crop residues e.g. wheat and rice straw, as such or inoculated with fungal species had beneficial effects on crop yields and improvement in physico-chemical properties of soils. Incorporation of legume residues viz., green gram or soybean stalks in rice, cowpea stalks in pearl millet or wheat saves energy input besides improving the soil physical and chemical properties.

The Cycle system for nutrient management: With a system approach i.e. synergistic integration of more than one component, the management of nutrient is to be done in such away so that there is minimum loss along with addition (in organic form) of the nutrient in the soil, taken out in terms of yield. This makes a cyclic system of nutrient management, which has major role in long term sustainable production.

Minimizing Nutrient Losses: Important loses of nutrients occur through removal of the harvested products, soil erosion, gaseous losses and leaching.

Factors affecting soil erosion are rainfall intensity and quantity of rainfall, erodability of the soil, length and steepness of slopes and soil cover. Traditional soil conservation projects have been emphasizing the use of physical structures (terraces, etc.) which mainly reduce soil erosion by changing the steepness and length of slopes. Current approaches to soil conservation, however, give more attention to reducing soil erosion by improving soil cover rather than the construction of physical structures.

Changes in soil slope: Reducing soil slope by terracing has been the traditional approach to control soil erosion in many soil conservation projects. Terracing is a very effective way of reducing soil erosion if properly constructed. The major disadvantage is the high labour requirement. The current approach is to give less emphasis to complete terracing and instead promote the use of progressive terracing. Progressive terracing implies that terraces are formed gradually as sediments are deposited in front of a contour barrier. This will often also tend to increase the fertility of the soil in front of the barriers because the deposited sediments are often rich in nutrients.

Soil cover: Improving soil cover can be just as effective in controlling soil erosion as changing the slope. The initial process in water erosion is detachment of soil particles by the force of raindrops. Improving soil cover will reduce the velocity of the raindrops, thereby halting soil erosion in its initial phase. Soil cover and be altered by changing the crop, increasing the yield by use of soil fertility enhancing methods and use of surface mulch.

Harnessing Untapped Nutrient Sources: If used appropriately, the recycling of organic waste from urban to rural areas is a potential, largely untapped, source of nutrients for farm and crop needs, especially on agricultural lands near urban centers. For example, environmentally undersirable wastewater has been used to irrigate fields and return nutrients and organic matter to the soil. Like organic manure, urban waste sludge is a source of primary nutrients, albeit a relatively poor source in comparison with commercial fertilizers. Stabilized municipal waste sludge typically contains about 3.3 per cent nitrogen, 2.3 per cent phosphorus, and 0.3 per cent potassium, although some concentrations can reach as high as 10 per cent nitrogen and 8 per cent phosphorus on a dry weight basis (EPA 1984). Urban waste also contains organic compounds such as dyes, inks, pesticides, and solvents that are often found in commercial and industrial sludge. These pathogens have been shown to cause genetic damage, while others, such as bacteria, protozoa, and viruses can cause salmonellosis, amoebic dysentery, and infectious hepatitis.

Exploring internal nutrient sources: Although new sources of nutrients can be developed, genetic engineering offers the potential for plants themselves to generate some of the nutrients they require through nitrogen fixation. In this process, rhizobium bacteria infect, invade, and draw energy from leguminous plants, and in return the bacteria convert and store atmospheric nitrogen in a form that the plant can use for growth.

Besides helping the plants themselves, cereals grown in rotation with leguminous plants can absorb the nitrates released from the decaying roots and nodules of the leguminous plants. Experiments have shown that rice-legume rotations can result in a 30 per cent reduction in chemical fertilizer use.

Better use of Nutrient Sources

Soil Nutrient and Soil Fertility Aspects: Soil nutrients are the basic source of the farm nutrient supply. Part of them is utilized by crops, i.e., the easily available portion (water-soluble, exchangeable) as well as the easily mobilizable fraction. The mobilization of nutrients from (mineral and organic) slowly available sources can be enhanced to a certain extent by activating soil life (in general by organic matter of by special biofertilizers), by crop varieties with strongly mobilizing capacity, by better accessibility of nutrients after structure improvement, by deepening of the plow layer or by fallow periods. The best use of nutrients for crop growth can be obtained on the basis of a high soil fertility level.

Optimum Soil Reaction: Many soils, due to a continuous acidification caused by several factors, are a little or much too acidic for high productivity. Therefore, liming is required and the soil should have its optimum soil reaction which varies with soil texture, cropping intensity, etc. The usual (local) recommendations for optimum pH are generally higher with higher clay content and yield level.

Organic Materials: There is a large variety of materials produced on the farm as well as "imported" from outside, especially communal and food processing waste materials. The following are the organic materials which can be used:

1. Crop residues and farm manures
2. Commercial and industrial waste products
3. Other organic manures

Organic residue means the remaining after the harvesting and processing of a crop. It is hardly a good product for human consumption. Agricultural crop residues are abundant both in kind and quantity. These residues form the major part of agricultural production and are susceptible to a common method of bioconversion.

The bioconversion of residues is important for organic farming. Residues are not used as foods because they are inedible without some form of bioconversion.

Cellulose-rich Substrates: They form a total of more than 1,800 million tons annually of renewable resources. They are to a great extent found in Asia, and it is therefore not surprising that they consist primarily of rice straw. The present use is often none; in some areas it is used for fuel. Straw could form an extensive base for reeding ruminants. There is no doubt that bioconversion would greatly improve the use of these materials, particularly in rural areas. To what extent in vitro SCP (single-cell protein) production can play a major role here depends greatly on local circumstances and on the results of research efforts in this field. In practice, cellulosic residues are mainly lignocellulosic and sometimes include starchy and/or sugary components. The following are major waste materials under each of these categories with high potential for bioconversion. They are rice straw, rice husks, corn stalks, corn cobs, bagasse, cane filter cake (mud), kenaf (***Hibiscus cannabinus***) stalks, animal manure, pineapple peels.

Starchy and Sugary Wastes: The carbohydrates portion of these residues is more easily accessible. They require a somewhat less difficult form of SCP bioconversion. Starchy residues are obtained from cassava meal (residue), cassava effluent while sugary residues are obtained from molasses, distillery slops, pulp waste liquor, coconut water (juice). Cassava and sugar beets provide the greatest amount of residue. High productivity in relatively poor soil has made cassava a popular staple food, especially in countries most in need of food. Non-cellulosic carbohydrate residues, such as starch in cassava and sugar-containing residues (molasses), are generally used as feeds for non-ruminants, i.e., pigs and poultry. Broken rice rejected in the milling process, barley, maize, and cassava meal are commonly used as feeds in Thailand. The use of these residues may be considered a form of bioconversion in common practice throughout the world.

TABLE 26.1: Chemical Composition (% dry matter) of various residues

	Grain* (Straw)	Leaf* (Grass)	Citrus** (Pulp)	Manure*** (Poultry)
Organic matter	95	91	93	77
Ash	5	9	7	23
Crude protein	3	17	7	32
Crude fibre	48	27	14	-
Nitrogen-free-extract	43	44	69	27

* O. Kellner and M. Becker, Grundüge der Futterungslehre (1959).
** CVB Cattle Feed Table.
*** F. de Boer and A. Steg (Hoorn), report, "Megista mestdag," part II.

The sources of organic residues are rice, maize, cassava, soybeans, groundnuts, coconuts, oil palm, sugar cane, rubber, coffee, forestry products, livestock, etc.

How to use crop residues: Several research and development programmes and projects are being carried out by universities and research institutions in India. Those specifically involving carbohydrate raw materials are as follows:

1. Anaerobic digestion (and biogas production) of distillery slops,
2. microbial preservation of bagasse for pulp production,
3. single-cell protein (SCP) production from pineapple waste,
4. protein enrichment of cassava residue,
5. microbial fertilizer (*Rhizobium*) from cane filter cake.

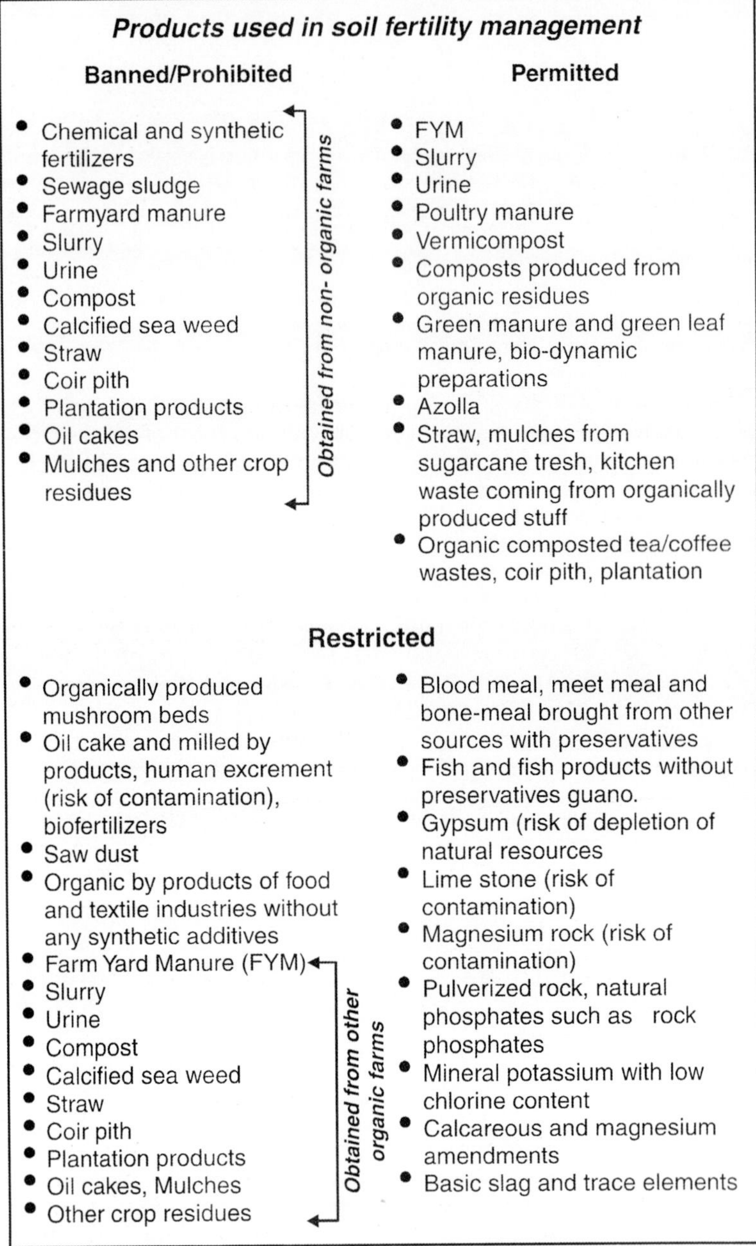

FIG. 26.7: Summary of standards for organic farming (Bhattacharrya, 2004).

The residues are converted into useful by-products. For successful implementation several factors should be taken into consideration in order to ensure better results. There are four major factors which are interrelated to one another are: technical, economic, socio-cultural, and, finally, political - all, in most cases,

Organic Manures

Organic manures are natural products used by farmers to provide food (plant nutrients) for the crop plants. There are a number of organic manures like farmyard manure, green manures, compost prepared from crop residues and other farm wastes, vermicompost, oil cakes, and biological wastes - animal bones, slaughter house refuse.

Organic manures increase the organic matter in the soil. Organic matter in turn releases the plant food in available from for the use of crops. However, organic manures should not be seen only as carriers of plant food. These manures also enable a soil to hold more water and also help to improve the drainage in clay soils. They provide organic acids that help to dissolve soil nutrients and make them available for the plants.

Organic manures have low nutrient content and therefore need to be applied in larger quantities. For example, to get 25 kg of NPK, one will need 600 to 2000 kg of organic manure where as the same amount of NPK can be given by 50 kg of an NPK complex fertilizer.

The nutrient content of organic manures is highly variable from place to place, lot to lot, and method of preparation. The composition of fertilizers is almost constant. For example, urea contain 46% N regardless of which factory makes it any where in the world.

Different fertilizers contain different amounts of plant nutrients, organic manures are also not alike. Average quality of farmyard manure provides 12 kg nutrients per ton and compost provides 40 kg per ton. Most of the legume green manures provide 20 kg of nitrogen per ton. Each ton of sorghum/rice/maize straw can be expected to add 26 kg of nutrients.

Green manuring is the practice of growing a short duration, succulent and leafy legume crop and ploughing the plants in the same field before they form seeds. Sesbania, Crotalaria, 'Pillipesara', Cowpea etc are good for green manuring. **Green leaf manuring** refers to adding the loppings from legume plants or trees to a field and then incorporating them into the soil by ploughing. Glyricidia, Pongamia, Leucina are common green leaf manuring plants.

Manuring is of two kinds:

1. **In–situ Manuring:** This is the method of manuring to the soil *at the site.* On the basis of origin of manure there are two types of in-situ manuring: (1) manuring by animals at site, (2) Manuring with plant material grown at site, popularly known as **green manuring**.
   a) **In-situ Mnuring by animal:** This is a widespread practice in areas where herders traditionally migrate their flocks each year for the search of better pastures. There is a mutual advantage to both herders – who need a place to settle their sheep each night – and farmers, who get urine and manure right in their fields for a small fee The manuring of fields by tethering the animals directly in the fields, is also an important strategy that farmers developed to suit their conditions. The droppings are later ploughed directly into the soil. This practice is undertaken when there are no crops in the field. It has many advantages: (1) Useful to small and marginal farmers who do not own cattle and have limited land. (2) Useful to farmers having limited farm yard manure. (3) Transportation and labour charges are avoided. (4) Pastorals can generate income by charging a fee for this service. (5) Manure decomposes easily. (6) Release nutrient into the soil through decomposition. (7) Promotes easy infiltration in vertisols. (8) Utilises urine which is normally

wasted. The limitations are:(1) It is difficult for those areas where sheep population is low. (2) Farmers have to depend on migratory practices of the herders.

b) **Insitu Manuring with Plants (Green Manures):** Green manuring can be defined as a practice of ploughing or turning into the soil undecomposed green plant tissues for improving physical structure as well as soil fertility. From the time immemorial, the turning in a green crop for improvement of the conditions of the soil has been a popular farming practice. Greenmanuring, wherever feasible, is the principal supplementary means of adding organic matter to the soil. It consists in the growing of a quick-growing crop and ploughing it under to incorporate it into the soil. The green-manure crop supplies organic matter as well as additional nitrogen, particularly if it is a legume crop, which has the ability to fix nitrogen from the air with the help of its root nodule bacteria. The green-manure crops also exercise a protective action against erosion and leaching. Green manuring offers an inexpensive way of improving crop yields and it takes little extra effort. Green manures are especially important on farms where there is not enough animal manure available, and when it is not possible to bring in natural fertilizers from elsewhere. Although the use of green manures may seem to create extra work, they enhance soil fertility, supplement for nutrients improved soil structure, prevention of soil erosion and weed control.

Green manuring *in situ* is followed in Northern India while green leaf manuring is common in eastern and central India. Green manuring is a practice of ploughing or turning into the soil undecomposed green plant material for improving the physical conditions of the soil or for adding nutrients. Any crop or plant (generally leguminous) grown and ploughed *in situ* is called a green manure, for example, dhaincha, sesbania, sunnhemp, wild indigo and pillipesara. A leguminous crop producing 10-25 tonnes of green matter per ha will add about 60 to 90 kg N. *green leaf manure* refers to turning under of green leaves and tender green twigs collected from shrubs and trees grown on the bunds, wastelands and nearly forest areas. Common shrubs and trees useful for this purpose are *karanj*, *glyricidia*, *neem*, *subabul*, etc.

2. **Green Leaf Manuring:** Green-leaf manuring refers to turning into the soil green leaves and tender green twigs collected from shrubs and trees grown on bunds, waste lands and nearby forest areas. The common shrubs and trees used are Glyricidia (*Glyricidia maculata*), *Sesbania speciosa*, Karanj (*Pongamia pinnata*), etc.

TABLE 26.2: Nutrient potential of green manures

Green manure	Biomass (tonnes)	N accumulobase (Kg/ha)
Sesbania aculeata	22.50	145.00
S. rostrata, Crotalaria	20.06	146.00
Juncea, Tephrosia	18.40	113.00
Perpurea	6.80	6.00
Green gram	6.50	60.20
Black gram	5.12	51.20
Cow pea	7.17	63.30

TABLE 26.3: Nutrient content of important green manure and green leaf manure crops

Crop	Nutrient content (% on dry weight basis)		
	N	P	K
Green manure			
Sesbania aculeate	3.3	0.7	1.3
Crotalaria juncea	2.6	0.6	2.0
Sesbania speciosa	2.7	0.5	2.2
Tephrosia purpurea	2.4	0.3	0.8
Phaseolus trilobus	2.1	0.5	-
Green leaf manure			
Pongamia glabra	3.2	0.3	1.3
Glyricidia maculeata	2.9	0.5	2.8
Azadirachta indica	2.8	0.3	0.4
Calatropis gigantea	2.1	0.7	3.6

Ex-situ Manuring: Ex-situ is literally *out of place* and the opposite of in situ (*in place*). Many a times in-situ manuring not able to fulfill the nutrient requirement of the soil on the other side plenty of organic waste from agriculture, animals and human being is available which is rich in nutrient and needs safe disposal and crop field are the best place for this. In the organic farming the nutrient cycle has to be maintain in which the nutrient taken out in terms of economic yield has to be returned to soil and use of agriculture waste is the appropriate solution for maintaining the cycle.

Organic manures from agriculture/animal/human waste, contain essential plant nutrients and other growth promoting agents like enzymes and harmones, while no synthetic chemical fertilizer can supply all together thus they are indispensable from the manural schedule for any crop production. The organic manure through the process of decomposition and humification gives humus which helps to improve the physical, chemical and biological properties of soil. The organic acid released during decomposition of organic manure controls certain fungal pathogens and nematode infestations. Commonly available organic manures like FYM, crop residues, poultry and sheep manure, oil cakes and other farm wastes will effectively be utilized for nutrient supplement.

Agricultural wastes (animal manures and crop residues or mixtures of the two) are now considered quite an important component for horticultural crop production especially that based on organic systems. Agricultural wastes can now be considered to include crop after harvest and primary processing, tree residues organic/plant residues from social forestry, animal excreta and processing left-overs from the slaughter houses and agro-industrial wastes. Thus, agricultural wastes comprise all organic wastes produced and disposed of or used in primary agricultural production. Of course, these materials are essentially by-products of production rather than wastes in the strict sense of the word. The important sources are, livestock and human wastes, crop residues, tree wastes and aquatic weeds, urban and rural wastes, agro-industries byproducts and marine wastes.

The important agricultural wastes are crop residues, agro-industrial wastes, rice husk, bagasse, pressmud, tea wastes, coir waste, etc. Agricultural wastes may contain solids, volatile and fixed and suspended and dissolved, settleable solids nitrogen (organic nitrogen, ammonia nitrogen, nitrite nitrogen, nitrate nitrogen, phosphorus and bio plant growth promoters, etc.

Concentrated Organic Manures

Concentrated organic manures are those materials that are organic in nature and contain higher per centages of essential plant nutrients such as nitrogen, phosphorus and potash, as compared to bulky organic manures. These concentrated manures are made from raw materials of animal or plant origin. The concentrated organic manures commonly used are aquatic weeds, oilcakes, blood-meal, fish meal, meatmeal and horn and hoof meal.

Economic Value of Organic Waste

The economic value of an organic waste or residue to a farmer is the value of the increase in crop yield and/or crop quality that is derived from its use. Since crop yield response to an organic amendment follows the law of diminishing returns, the average yield increase is always larger than the yield increase attributable to the incremental unit of organic material. If the price of the farmer's product is the same regardless of the quantity produced, which is usually the case, then the revenue derived from the average unit of organic material will be greater than the revenue from the incremental unit. Thus, the value of the average unit of organic material will be higher than the value of the incremental unit of organic material.

Farmers can be expected to utilize organic materials to the point where the revenue from the incremental unit is equal to its price, assuming application is included in the price. They would certainly use no more than that quantity since additional application of the organic material would yield a loss. Likewise, they would not wish to use less since total profits would decline. Thus, an estimate of the value of the incremental waste is necessary by management agencies when analyzing the marketability of a recycled product. However, public decision-makers should also be interested in the estimates of the total economic benefits from resource recycling alternatives, such as composting, compared with ultimate of the value of the average unit of organic material would be the appropriate figure to use in public project analysis.

Composting

Composting is a biological decomposition process that converts organic matter to a stable, humus-like product under controlled conditions. During the composting process, microorganisms utilize the decomposable microbial substrates present in the organic compost both as an energy source and for conversion to microbial substances.

The composting process of simply a means of converting raw organic compost, a potential source of odor and public health problems, into a safe product called humus. By definition, composting is an alternative system for solving some of the compost handling problems existing on today's farms.

It is a natural process, which occurs in nature in which organic matter is decomposed by microorganisms forming a humus-like substance. The process itself is not new. It has been in practice for many centuries by farmers who have stacked animal manure into piles or gardeners who have placed garbage, leaves, grass cuttings, etc., into pits.

Despite its wide usage, no advances were made in the treatment process until 1925 when Sir Albert Howard, a British agronomist stationed in India, developed a systemized process for composting. The improved process was labeled the Indore Method after the stage in central India where it was conceived and tested. The method described by Howard (1935) involves the formation of a layered pile about

1.5 m high using garbage, animal composts, sewage sludge, straw, and leaves. Initially the process was anaerobic and required 6 months for completion to occur. The process was later modified by turning the pile over twice, which reduced the composting time to 3 months.

Classification of Composting

It is classified into following general categories:

1. **Oxygen Availability:** Composting, categorized by the amount of oxygen available, is either **aerobic** or **anaerobic** in nature.
2. **Aerobic composting:** The composting of municipal refuse and large-scale agricultural composts should be carried out under aerobic (with oxygen) conditions. Aerobic composting is governed by the activity of aerobic microbes and hence required the availability of atmospheric oxygen during the period of decomposition. If the environmental factors are optimal, aerobic composting is characterized by high temperature, the absence of foul odors, and a short stabilization period. The high temperatures have a sterilizing effect by destroying weed seeds and pathogenic organisms.
3. **The Aerobic Heap:** A heap should contain enough bulk of food for the heat-loving bacteria for theit growth and activity. If the pile is too small, there will be insufficient heat build-up. To obtain a suitable C:N mix on a garden scale, the materials may first need to be assembled over a period of time. This can best be achieved by creating a preliminary stockpile. Garden debris, mowings, weeds and so forth can be loosely stacked and covered in one pile or placed in plastic bags; likewise kitchen compost.
4. **Anaerobic composting:** It is governed by anaerobic bacteria that operate in the absence of atmospheric oxygen. The process is characterized by lower temperatures, the production of odorous gases, and longer stabilization times. Since anaerobic composting does not require atmospheric oxygen, the pile or bed can be sealed to prevent the escape of foul smelling gases and left alone. The major advantage of anaerobic composting is that the process can be carried on with a minimum of attention and as such requires little or no energy once the compost bed is established.

Use of Biofertilizer

Biofertilizer is a large population of a specific or a group of beneficial microorganisms for enhancing the productivity of soil either by fixing atmospheric nitrogen or by solubilising soil phosphorus or by stimulating plant growth through synthesis of growth promoting substances. Bio-fertilizers based on renewable energy sources are cost effective supplement to chemical fertilizers, eco-friendly and can help to economise on the high investment needed for chemical fertilizer use as far as nitrogen and phosphorus are concerned.

Biofertilizer is a 100% natural and organic fertilizer that helps to provide and keep in the soil all the nutrients and microorganisms required for the benefit of the plants.

Biofertilizers do not come under the purview of the definition of the term fertilizers, as they do not contain substantial quantity of plant nutrients as other fertilizers like urea, diammonium phosphate, muriate of potash etc. contain. The term biofertilizer is a misnomer but is in wider use. They should be termed as inoculants, after the name of the microorganisms they contain, viz., *Rhizobium* inoculant, *Azospirillum* inoculant or blue green algae inoculant.

Bioinoculants or biofertilizers are different from chemical fertilizers. Biofertilizers on application remain in soils, multiply and keep benefiting the growing crops. They do not get depleted as in the case of fertilizers and, therefore, if the optimum soil conditions prevail, population of added microorganisms builds up and frequent application of biofertilizers can be avoided.

There are millions of microscopic organisms near the plants that not only provides nutrients to the plants but also helps to keep the water and retain the nutrients in the soil, for its easy availability to the plants.

Biofertilizer can be defined as a substance which contains living microorganisms which, when applied to seed, plant surfaces, or soil, colonize the rhizosphere or the interior of the plant and promotes growth by increasing the supply or availability of primary nutrients to the host plant. This definition is based on the logic that the term biofertilizer is a contraction of the term biological fertilizer.

As biology is the study of living organisms, biofertilizer should contain living organisms which increase the nutrient status of host plant through their on-going existence in association with the plant. This definition separates biofertilizers from organic fertilizer (fertilizer containing organic compounds which directly, or by their decay, increase soil fertility). Like wise, the term biofertilizer should not be used interchangeably with the terms, green manure, intercrop, or organic supplemented chemical fertilizer. Oken and Landera Gonzalez (1994) argue that rhizosphere organisms which improve utilization of soil nutrient but do not replace soil nutrients (like chemical fertilizers) should not be called biofertilizers.

Some plant growth promoting rhizobacteria (PGPR), that represent a wide variety of soil bacteria when grown in association with a host plant, result in stimulation of growth of their host. However, not all PGPR can be considered as biofertilizers. Bacteria that promote plant growth by control of deleterious organisms are biopesticides, but not biofertilizers. Interestingly some PGPR appear to promote growth by acting as both biofertilizer and biopesticide. For example, strains of *Burkholderia cepacia* have been shown to have biocontrol characteristics to *Fusarium* spp., but also can stimulate growth of maize under iron-poor conditions via siderophore production.

The biofertilizer has a relative high nutrient concentration, and even so, it can be used directly over soil before planting. Once diluted, it constitutes a high quality foliar fertilizer, and in this form, it is known as diluted biofertilizer. Diluted biofertilizer also has all the needed conditions to be used as a complete nutrient solution in organic hydroponics. The advantages of using the biofertilizer are enormous. Not only it is very economical, but also for the high agricultural yields which it produces.

Most fertilizers add nitrogen to the soil. This can be done via chemical fertilizers, or through a process called **biological nitrogen fixation** (BNF). On a worldwide basis it is estimated that about 175 million tons of nitrogen per year is added to soil through biological nitrogen fixation (BNF). The term *bio* means living; so biofertilizers refer to living, microbial inoculants that are added to the soil. These biofertilizers are products consisting of selected and beneficial microorganisms, which are known to improve plant growth through supply of plant nutrients.

The soil microorganisims commonly used in biofertilizers are: Phosphate Solubilizing microbes, Mycorrhizae, *Azospirillum, Azotobacter, Rhizobium,* blue green algae, and *Azolla*.

Biofertilizers are ready to use live formulates of such beneficial microorganisms which on application to seed, root or soil, mobilize the availability of nutrients by their biological activity in particular, and help build up the micro-flora and in turn the soil health in general.

Need for Biofertilizers

Biofertilizers have definite advantage over chemical fertilizers. Chemical fertilizers supply nitrogen whereas biofertilizers provide, in addition to nitrogen certain growth promoting substances like hormones, vitamins, amino acids, etc. Crops have to be provided with chemical fertilizers repeatedly to replenish the loss of nitrogen utilised for crop growth. On the other hand biofertilizers supply the nitrogen continuously throughout the entire period of crop growth in the field under favourable conditions. Continuous use of chemical fertilizers adversely affects the soil structure whereas biofertilizers when applied to soil improve the soil structure. The deleterous effects of chemical fertilizers are that they are toxic at higher doses. Biofertilizers, however, have no toxic effects.

It may be borne in mind that biofertilizers are no substitute for chemical fertilizers. At present, the use of chemical fertilizers is far below the recommended level. Therefore, the aim and object of spread of biofertilizer technology as a Industry, is to build up efficiency in use of chemical fertilizers, supplemented by low cost inoculants to the extent possible. Biofertilizers are used for integrated farming-systems mainly in the agricultural sector or areas with little or no access to chemical fertilizer.

Role of Biofertilizers

Biofertilizers are inputs containing microorganisms, which are capable of mobilizing nutritive elements from non-usable form to usable form through biological processes. They are less expensive, eco-friendly and sustainable. The beneficial microbes in the soil, which are of greater significance to horticultural crops are, biological nitrogen fixers, phosphate solubilizers and mycorrhizal fungi which are the phosphate scavengers.

1. Depleting soil fertility due to widening gap between nutrient removal and supplies.
2. Depleting feedstock/fossil fuels and increasing cost of fertilizers.
3. Growing concern about environmental hazards.
4. Increasing threat to sustainable agriculture.

A number of products are now available that are generally referred to as soil and plant additives, of non-traditional nature. These products include:

1. Microbial fertilizers and soil inoculants contain unique and beneficial strains of soil microorganisms.
2. Microbial activators that supposedly contain special chemical formulations for increasing the numbers and activity of beneficial microorganisms in soil.
3. Soil conditions that claim to create favourable soil physical and chemical conditions which result in increased growth and yield of crops.
4. Vermicompost helps in improving soil health and fertility.

Nitrogen fixing organisms can be provided to the farmers in the name of microbial inoculants otherwise termed as biofertilizers. The biofertilizers containing biological nitrogen fixing organisms are of utmost important in agriculture in view of the following advantage:

1. They help in the establishment and growth of crop plants and trees.
2. They enhance biomass production and grain yields by 10-20%.
3. They are useful in sustainable agriculture.
4. They are suitable in organic farming.
5. They play an important role in agroforestry / silvipastoral systems.

After the advent of the acetylene test for assessing nitrogen fixing potential, rapid progress has been made in identifying nitrogen fixing organisms. Out of a large number of microorganisms possessing the property of nitrogen fixation, only a few such as *Rhizobium, Azotobacter, Azospirillum,* BGA, *Azolla*, etc. have been commercially exploited. The reaction in biological nitrogen fixation is essential the same as in production of chemical fertilizers (Haber-Bosch process) i.e. the catalytic reduction of dinitrogen (N_2) to ammonia (NH_3).

Biological Weed Control

Biological control of weed is a long-term solution which is most effective as part of an integrated weed management approach. Biological control is rarely a 'silver bullet', as not all weeds have biological control agents that would be considered safe for introduction in India. There are three main methods used for control of weeds: biological, mechanical, and chemical. Mechanical control includes mowing, hoeing, cultivation, and hand pulling. Chemical control is the use of herbicides. You can also use a combination of these three methods in an integrated weed management approach which some authorities would consider as the fourth method.

Tactics of Integrated Weed Management Systems

Prevention: The backbone of any successful weed management strategy is prevention. It is important to prevent the introduction of seeds into the field through sources like irrigation water or manure.

Crop rotation: A practical and effective method of weed management (discussed in previous sections).

Flame weeding: Good for control of small weeds.

Delayed planting: Early-germinating weeds can be destroyed by tillage. And with warmer weather, the subsequently planted crop (depending on the crop, of course) will grow more quickly, thus competing better with weeds.

Staggered planting schedule: This will allow more time for mechanical weed control, if needed. This also lessens the weather risks and spaces out the work load at harvest time.

Surface residue management: As mentioned earlier, a thick mulch may shade the soil enough to keep weed seeds from germinating. In addition, some plant residues are allelopathic, releasing compounds that naturally suppress seed germination.

Altered plant spacing or row width: An example is narrow-row (7–18" between rows compared to conventional 36–39" between rows) soybean plantings. The faster the leaves shade the ground, the less weeds will be a problem.

Herbivore: Cattle, geese, goats, and insects can be used to reduce populations of specific weeds in special situations. Cattle, for example, relish Johnson grass. Weeder geese were commonly used in cotton fields before the advent of herbicides. Musk thistle populations can be satisfactorily reduced by crown- and seed-eating weevils. Goats may be used for large stands of various noxious weeds.

Adjusting herbicide use to situation: Herbicide selection and rate can be adjusted depending upon weed size, weed species, and soil moisture. Young weeds are more susceptible to chemicals than older weeds.

Integrated Weed Management Systems

Weeds as competitors in crops present a number of unique challenges that need to be recognized when developing management strategies. The intensity of weed problems during a growing season will be influenced by weed population levels in previous years. The axiom "one year's seeding equals seven years' weeding" is apt.

Mechanical and Physical Biocontrols

Mechanical and Physical Biocontrols included in this category utilize some physical component of the environment, such as temperature, humidity, or light, to the detriment of the pest. Common examples are tillage, flaming, flooding, soil solarization, and plastic mulches to kill weeds or to prevent weed seed germination.

Biological Control

Biological Control of weed is necessary because herbicide application has been the most widely adopted management tool against many rangeland and pasture weeds. Unfortunately, the marginal economic rewards and inaccessibility of many of the infested rangelands has limited the value of chemical application. Simultaneously, the elimination of desirable broad leaved species and the potential contamination of the water table by herbicidal application are deleterious side effects. To be successful, an introduced biological control agent need not kill its weedy host outright. Biological control using insects is slower than other weed control methods. Biological Control of weeds provides specific insects that attack only certain weeds. Not all weed problems can be addressed by the insects.

The biological control of weeds involves the use of living organisms, such as insects, herbivorous fish, other animals, disease organisms, and competitive plants to limit their infestations. Currently, there are two approaches to this method of weed control, viz., (i) the classical biological control philosophy approach, and (ii) the bioherbicide philosophy approach. Any biological control approach shall yield best results when it is integrated with a comprehensive weed management programme.

The classical biological weed control aims at restoring the balance between the target alien weed and its natural enemies in the ecosystem by purposeful introduction of suitable, exotic bioagents. The immediate response to successful introduction of a native bioagent is a decrease in the weed population, followed after some period by a decline in the bioagent population due to its food shortage. then often the weed population tends to recover. The process continues in a cyclic fashion till the bioagent-weed populations get stabilized at a low level. At that time, man enters the cycle and he puts the cleared area to some economic use. The biological weed control should not be expected to eliminate the target weed from an area; in fact, success of biological control of weeds depends upon continued presence of the weed existing in small numbers and shifting with time. The method being a slow opreating one, is persently used mainly in non-crop situations. On crop lands, even if successful, the bioagent will not get opportunity to work on the host weed because of frequent use of insecticides and fungicides in modern agriculture. Otherwise, *Cyperus rotundus* can possibly be controlled in crop fields with a moth borer, *Bactra verutana*. Another promising example of selective biocontrol in crop fields is complete denudation of *Ludwigia parviflora* 'waterpurslane' by *Haltica cyanea* Web. 'steelblue beetle', in rice fields.

An important aspect of biological weed control is that at a time, it is applicable to the control of only one major weed species that has spread widely. Also, one should be prepared to accept cyclic resurgence of the weed under attack by the classical biocontrol approach. Further, while with perennial weeds the main objective of biocontrol is the destruction of the existing vegetation, in the case of annual weeds prevention of their seed production is generally more important.

Successful Bioagent

Host Specificity: Biocontrol agents recommended for weed control should be host-specific, *i.e.* they should not attack other plant species, particularly the economic ones. The bioagents released so far (chiefly insects) had successfully passed the 'starvation tests' before they left the laboratories for field use. They were found to prefer to starve to death rather than feed upon plant species other than the target weed. Starvation test is a good safeguard against an introduced bioagent becoming a pest on some economic crop plant.

Bioagent hardiness: It is important that a classical bioagent withstands elements of its new environment in which it is transplanted. This includes freedom from its own parasites and predators. Further, the bioagent should be hardy enough to survive short and long periods of food shortage when the target weed population is brought to a low level. This has a significant implication in the case of carp which cannot survive even a short starvation period. To continue their survival, the grasscarp must be shifted to a new weedy waterbody soon after the weed biomass in the previous one is lowered sufficiently.

Feeding habit: Great stress has been placed on the selection of insects as bioagents that attack either the flowers and seeds of the weed or bore into their stems. Insect bioagents with such feeding habits have been found more host specific and efficient in controlling weeds than the leaf or root feeders. But lately certain leaf eating insects have been found equally valuable in the control of weeds, particularly of the perennials. Likewise, insects that attack roots are likely to be very effective on perennial weeds but insufficient attention has been paid to them. In any case, the destructive capacity of a bioagent on the weed should be very high so that its small population could destroy large weed infestations.

Ease of multiplication: High rate and ease of natural reproduction of bioagent is important in insects, pathogens, snails, and competitive plants. But it is very much unwanted in the case of carp employed for aquatic weed control because their excessive population may compete with cultivated fish. Fortunately, all present researches indicate that a bioagent carp like the Chinesegrasscarp, is a extremely slow, natural breeder in the tropics and sub-tropics, although under artificially controlled conditions it can be multiplied rapidly (Gupta, 2001).

Types of Classical Bioagents

Insects: Insects have been more frequently used than other animals for the biological control of weeds. The first successful example of weed control with deliberately introduced insects was reported from Hawaii in 1902 when a project for the control of *Lantana camara* L. was initiated with a moth, *Crocidosema lantana* Busck. Since then insects belonging to widely different families and genera have been employed to control some other noxious weed species. Alien weeds have been the main subject of biocontrol with insects imported from the evolutionary centres of the weed concerned. But there is at least one example when a native weed has been brought under control with an alien insect species. The weed is manuka, *Leptospermum scoparium* Forester, and it was controlled on its native land, Newzealand, with mealybug *Eriococcus orariensis* Hoy, introduced from Australia. Insects selected as biocontrol agents of weeds are largely host specific *ie* one insect species is employed to destroy only one weed species.

Carp Fish: Certain fresh water carp fish consume large quantities of aquatic weeds and they are very suitable for relieving waterbodies of their massive growth of noxious, aquatic vegetation. Whiteamur (Chinesegrasscarp) *Ctenopharyngodon idella* Val is the most promising species of carp for aquatic weed control.

Plant Pathogens: Many fungi are observed to attack specific weed species, but examples of successful, practical weed control with fungi are few. Of late, it has been discovered that *Acacia glauca* could be controlled successfully by injecting a suspension of spores of *Cephalosporium zonatum* Sawada.

Competitive Plants: Certain plants may be highly competitive in nature; capable of suppressing some specific weed/s wthout causing any meaningful hinderance to our interests.

Bioherbicide

Bioherbicides, in variance with the *Bioagents*, are herbicidal chemicals released by some living organisms. At the present, these organisms comprise largely certain microbes (fungi/ bacteria) because of which the bioherbicides are also termed as *Mycoherbicides*. The possibility of contolling weeds with exudates of certain higher plants, called allelochemicals.

The mycoherbicides were first promoted by certain corporates in the USA as 'made to order' products to control specific noxious weeds. Sometimes, the mycoherbicides are also offered as fermentation products of the relevant microbes. In either case, the farmer is advised to spray the formulation, after making recommended dilutions, on the target weed just like a herbicide, soon after procuring it, to obtain best results. It may be brought home here that mycoherbicides act like synthetic, contact herbicides, requiring repeat applications to tackle the new flush of the weed, if any.

The *bioherbicide philosophy* differs from the classical biocontrol philosophy referred to earlier, in certain ways as follows:

1. Bioherbicide remains active only on the current weed populations, without any chance of cyclic perpetuation of the weed (or of the bioagent); each new flush of the weed thus requiring retreatment with it.
2. Bioherbicide can be developed for selective control of weeds in a crop just like any other selective herbicide, which is not the case with the classical philosophy bioagents.
3. The development of bioherbicides is of great interest to industrialists since it involves every season requirement of the product for field use. In variance with it, the classical biological control approach has no incentive to the private, profit-oriented organisations; it must depend solely upon public sector support.

Parasitic weeds, allergenic weeds, crop mimics, and narcotic plants are excellent targets for activating the bioherbicidal approach. Besides, it has always been the goal of scientists to develop products also for broad-spectrum weed control in agriculture, competitive to chemical herbicides. But thus far it has been elusive, save a few individual weeds that have been successfully curbed with bioherbicides in agriculture and horticulture.

Biological Control of Diseases and Insects

Biological control is defined as the reduction in attack of a crop species by a pathogen achieves using another living organism, or organisms. Biological control covers the use of any organism to control a pathogen. This definition includes host plant resistance as a natural, and highly effective, form of biological control. This concept itself includes both direct and indirect effects, due either to introduced antagonists, or manipulation of existing populations to reduce disease. Agronomic practices such as soil amendment with organic matter to boost microbial populations are therefore covered by this definition.

Introduction is not theoretically limited to insects and vertebrate predators; the use of the myxomatosis virus to control rabbits in Australia suggests that microbes may be able to play similar roles. However, introductions have two fundamental limitations:

1. They have an extra dimension of unpredictability because an organisms, hitherto unknown, is being brought to a new area which it might find futile in unanticipated ways and/or may not acclimatized in changed conditions;
2. There is no clear way to make a profit from one-of introductions.

Substantial contributions can also come from fundamental biotechnological research.

Increases in potency by modification of the plasmid complement of the bacterium, which controls the synthesis of the protein both reduce the cost of production and provide the starting place for new, more easily used formulations.

Changes in the spectrum of activity of the toxic protein by genetic engineering can open wholly new markets.

Production of asporogenic strains may reduce production costs by avoiding the waste of metabolic energy on spore production and may make the product more acceptable in certain countries.

Table 26.4: Important pesticidal and plant growth regulatory compounds of microbial origin

Class	Type	Example
Insecticides	Miticidal antibiotics	Tetranactin
	Broad spectrum insecticide	Avermectin, Milbemycin,
	nematicide, miticide	Bacillus thuringiensis
	Bacterial toxins	(B.t) endotoxin
	Fungal toxins	Beauverolides and related depsipeptides
Fungicides	Antifungal antibiotics	Griesofulvin, Blasticidin S, Kasugamycin, Polyoxins, Validamycin
	Antibacterial antibiotics	Streptomycin
Herbicides	Herbicidal antibiotics	Bialaphos etc.
Growth regulators	Fungal products	Gibberellins, abscisic acid

Types of Biological Control

There are two main types of biological control, **inundation** and **augmentation**, both of which may be practiced with microbes or other organisms. They are essentially self-descriptive.

1. **Inundation**: Involves the use of very large numbers of organisms over a short time to suppress or destroy a population.
2. **Augmentation:** It is being the supplement of an existing natural population, which has the power to control a pest or disease.

Following are the approaches used for the biological control of plant pathogen:

Direct	Introduction of antagonists able to suppress activity of the pathogen
Indirect	Manipulation of the microbial balance by means of crop practices. Cultural measures, soil supplements, etc.
Host-mediated	Induction of localized and systemic resistance, Virus cross-protection

The **classical biological** control is based on potential natural control agents and needs good understanding of the population biology and ecology of both the target pest or pathogen, and the biocontrol agent. A number of microbial products have been released onto the market for the control of insects and weeds. Microbial pesticides are based on the toxin-producing bacterium, *Bacillus thuringiensis*, and insect-pathogenic fungi and viruses. All the products for weed control are based on plant-pathogenic fungi and are therefore described as **mycoherbicides**.

Microbial Antagonism

Soil-borne plant pathogens interact with other microorganisms found in the soil. The living organisms present in soils are capable of reducing the viability or infectivity of plant pathogens. These pathogens have a large influence on the incidence and severity of disease in the fields. It has further indirect evidence for such microbial antagonism comes from surveys of disease in continuous monoculture. been known for many years that certain soil can reduce disease caused by vascular wilt fungi, such as *Fusarium*, and that the suppressive properties of such soils can be removed by sterilization. It is also possible to convert a *conductive* soil into a *suppressive* one by mixing in a proportion of the suppressive type.

The simplest explanation for this phenomenon is that as the pathogen population increases, there is a corresponding increase in the amount or activity of antagonistic microorganisms. The exact mechanism of such antagonism is still not clear.

Selection of Biocontrol Agents for Plant Disease Control

There is little difference, in principle, between screening chemicals and testing biological agents for disease control. Both selection processes require an appropriate assay system to detect activity against the target pathogen, or pathogens.

Screening for biocontrol agents should, therefore, focus on specific natural sources chosen to optimize the changes of isolating strains with the correct biological properties. **Random screening** of various microorganisms for suppressive effects on plant disease has often been used to select the most promising strains for further development. The success of screening should be aimed on rational approach based on ecology. For effective results, a biocontrol agent (BCA) should able to colonize a particular habitat, or to occupy a specific niche, in sufficient numbers to interfere with the growth or survival of the target pathogen. Rather than introducing a randomly selected microbial antagonist, it would be better to introduce one known to be adapted to the habitat concerned. Hence the best place to look for potential BCAs is in the specific environment in which they are to be used. For example, if the target is a pathogen, which infects plant roots, the logical place to look for an antagonist is in the rhizosphere. Any microorganism isolated from this habitat is likely to have biological and physiological attributes enabling successful multiplication in the root zone, and is therefore, *rhizosphere competent*. Similar concepts apply to leaf surfaces or other substrates such as straw where a particular stage in the pathogen life cycle might be disrupted.

Organic Control of Diseases

Plant Diseases

Pear Fireblight: Fireblight, caused by the bacterium *Erwinia amylovora*, is one of the most serious and economically damaging diseases of pears. It can be transmitted by bees, aphids, psylla, or other insects, and can also be spread by blowing wind and rain. Pruning can be another source of infection. The bacteria gain entry to the tree through blossoms or lush new growth and, once inside, begin to work toward the roots. If the disease spreads unchecked to the trunk and roots, it can kill the tree;

however, in resistant varieties the bacteria rarely invade beyond young wood. Under the bark, the bacteria form a canker where they will survive the winter, only to infect more trees the next year. If on a warm day you see pustules on a tree, oozing an orange-brown liquid, you are looking at fireblight.

Fireblight development is greatly favored by the presence of young, succulent tissues. Cultural practices that favor *moderate* growth of trees are recommended. These include using only half as much compost as for apples, never using manure, and avoiding heavy populations of clovers and other legumes around the base of the tree. Pruning should also be limited, and only minimum fertilization applied. Choosing fireblight-resistant pear cultivars is a good start to managing fireblight, but it is by no means a panacea.

Asian-pear cultivars also vary in their resistance to fireblight. Preventing infection is the next key to fireblight control. Sprays of agricultural-grade streptomycin, applied at early bloom to prevent infection. Both streptomycin and terramycin are listed as approved substances for fireblight control in the National Standards on Organic Agricultural Production and Handling. Using these two substances together can reduce the amount of streptomycin sprayed each year, which may help to protect the antibiotic's effectiveness. By itself, BlightBan can provide up to 50% suppression, but not control.

Pear Blossom blast: Another bacterial disease, blossom blast (causal organism: *Pseudomonas syringae*) may afflict pears, usually as a blossom blight resulting in reduced fruit set. It can also cause twig dieback and bark cankers, and may lead to severe wood damage of Asian-pear cultivars in particular. Controlling this disease is difficult because its occurrence is widespread on many plant species and not easily predicted; once symptoms appear, control efforts are too late. Protecting orchards from frost damage can limit injury. An early application of BlightBan® A506 can help reduce frost damage potential by excluding the ice-nucleating bacteria.

Pear scab: Pear scab (*Venturia pirina*), a fungal disease closely related to apple scab, is neither as common nor as economically important on pears as apple scab is on apples. Nevertheless, it can cause economic damage by marring the appearance of the fruit. Pear scab causes lesions on leaves, shoots, and fruit, and -unlike apple scab- infects twigs, where it can overwinter. Pear cultivars differ in their susceptibility to scab. Organic control is identical to that for apple scab, usually consisting of treatment with sulfur, lime-sulfur.

Other Pear diseases: Two foliar diseases, fabraea leaf spot (*Fabraea maculata*) and mycosphaerella leaf spot (*Mycosphaerella pyri*), are usually no problem in sprayed orchards but can reach damaging levels in unsprayed ones. Susceptible cultivars, which include nearly all European pears, can be defoliated, resulting in reduced buds and dwarfed or- if the fruit itself is infected- unmarketable fruit. Sprays for pear scab, such as Bordeaux mixture, will generally control these two maladies as well. Also, Surround™, a kaolin-clay-based insect repellant, is registered and labeled for suppression of fabraea leaf spot.

Pear Insect and Mite Pests

Many of the same pests that affect apples also prey on pears, although often to a lesser degree.

Pear psylla: The pear psylla (*Cacopsylla pyricola*), an aphid-like insect whose only host is the pear, is the crop's most significant insect pest. In conjunction with fireblight, pear psylla is largely responsible for the decline in Eastern pear production. The honeydew left by the psylla damages the fruit by supporting growth of sooty mold and causing a black russeting; these two effects account for most of the economic damage caused by the psylla. It is also an important vector of fireblight and "pear decline disease," and can weaken trees in areas of heavy infection. Although the psylla develops resistance to insecticides, it can usually be controlled either with dormant oil sprays or with sulfur sprays (but not a combination of the two, which damages plants.) Insecticidal soap can reduce active populations during the growing season. Fortunately, the recently-introduced Surround™. In fact, use of Surround for pear psylla control has made organic pear production viable for the first time.

Pear True bugs: Stink bugs, tarnished plant bug, and other true bugs (insects in the order Hemiptera) will readily feed on pears throughout the growing season. Early feeding damage may result in a pucker or dimple in the fruit. Mid- and late-season feeding often results in the development of so-called "stone cells" immediately beneath the feeding site. These stone cells are very hard and can seriously compromise the marketability of affected fruit. Surround™ is labeled for suppression of stinkbugs and lygus bugs in pears. Also, some of the relatively new biological insecticides derived from the seeds of the neem tree are effective against true bugs on fruit crops, for example, Aza-Direct™.

Pear Mites: In high enough numbers, pearleaf blister mites (*Phytoptus pyri*) and pear rust mites (*Epitrimerus pyri*) can reduce the photosynthetic efficiency of leaves and cause russeting on fruit. Both species will likely be more troublesome in dusty conditions, as the dust interferes with their natural predators. These predators, including green lacewings, predatory mites, and various species of ladybird beetles, as well as bigeyed bugs, minute pirate bugs, and predatory thrips—will usually control mite outbreaks if they are not disrupted by dust or pesticides. Dormant horticultural oils and insecticidal soaps can both help control spider mite outbreaks. Neem-based biological insecticides are also labeled for controlling mites on pears. European-pear cultivars are more sensitive to mite feeding than Asian pears, and any pear that is drought-stressed will be more susceptible to mite damage.

Other Pear insects: Other insect pests that may affect pears include codling moth, apple maggot, scale, and plum curculio. The pear slug (actually a sawfly larva) causes leaf damage; it can be controlled with a forceful stream of water, soap, or wood ashes. The leafroller complex, including tufted apple bud moth, variegated leafroller, and redbanded leafrollers, represents another potential, but comparatively minor, pear pest problem. Mating disruption tools that are effective for some leafrollers are being developed. In addition, there are biological insecticides labeled for control of leafhoppers, aphids, and leafrollers, among other pests.

Examples of Successful Biocontrol Agents

The examples which have shown promise in experimental trials, but have yet to make a significant impact in the market place have been developed from *Bacillus thuringiensis*

Bacillus thuringiensis: (Bt) based insecticidal formulations are available in USA, France, Japan, and many other developed and developing countries. The toxicity of Bt strains has been attributed to a glycoprotein, protoxin, often referred to as delta endotoxin, which is produced during sporulation/fermentation. It is a crystalline product. When the larvae ingest Bt spores, they are activated within the gut to produce toxins. The histological studies conducted on the mid-gut epithelial cells of *Aedes aegypti* revealed that the toxins disrupted the cells, which became distorted and eventually burst.

Agrobacterium radiobacter (Agrocin): Crown gall is plant tumour caused by the soil-borne bacterium *Agrobacterium tumefaciens*; galls occur on the roots or often at the crown of the plant between roots and shoot. The pathogen has a wide host range but the disease is especially important in some fruits such as peaches, on grapevine, and in woody ornamentals such as roses. A highly effective biocontrol method has been developed using a related non-pathogenic bacterium, *Agrobacterium radiobacter*, to treat roots during transplanting. The BCA is commercially available and gives cheaper, more effective control than antibacterial chemicals.

Pseudomonas fluorescens (Phenazine): Damping-off disease caused by *Pythium* species and *Rhizoctonia solani*, and *Gaeumannomyces graminis* (in cereals) soil-borne fungal pathogens. These pathogens infect seeds and roots are a serious constraint to agricultural production. They affect crop establishment, leading to patchy growth and delayed development. One feature common to the infection cycle of these diverse pathogens is the need to colonize the zone surrounding seeds or roots prior to penetration of host tissue. Interfering this step by introducing aggressive microbial competitors, or to manipulate resident microbial populations can reduce infection.

Bacillus (Bacteria): *Bacillus* is an alternative to *P. fluorescens* and has been used for seed treatment. *Bacillus* produce resistant endospores, which can be kept as a dry formulation for long periods without losing viability. This property might also aid use in situations exposed to drying, such as on leaves and other aerial plant surfaces.

Protozoa: The phylum Protozoa consists of about 15,000 free-living and parasitic species; about 300 members of the sub-phyla Ciliophora (ciliates), Cnidispora (microsporidians) and Sporozoa (sporozoa) and the super classes Mastigophora (flagellates) and Sarcodina (amoebae) have been isolated from insects. These members produce entomopathogens.

Fungi: Over 400 species of fungi are known to attack insects and mites. Fungi usually infect their hosts by direct invasion and are, therefore, able (unlike most bacteria and viruses) to attack pests without first being ingested. In addition by their ability to sporulate on the dead bodies of their houses, also control the insect. The timely application of sufficient *fungal propagules* at an early stage can control pests to sub-economic levels for the duration of the crop. However, to be effective, fungi have fairly stringent requirements for humidity and temperature. If the humidity falls, or the temperature varies too far from the fungus, optimum, control of pests may be sluggish or non-existent and the desired epidemic (termed **epizootic**) may fail to develop.

- Trichoderma: Nectria galligena, which causes silver-leaf disease of fruit trees. It grains entry to the host via pruning wounds. These sites can be treated with a fast-growing antagonist, such as formulations of the saprophytic fungus Trichoderma, to prevent infection by the pathogen.
- Peniophora (Phlebia) gigantea (Basidiomycete): Heterobasidion annosum is a soil-borne fungus and causes serious problem in forestry plantations and causes root and butt rot of conifers. The pathogen is not aggressive to intact trees but instead exploits cut tree stumps to establish a food base from which it can spread along the roots and infect adjacent hosts. As forestry practice involves frequent thinning and falling of trees, this provides the pathogen with ample opportunity to gain access. The use of chemical fungicides declines the pathogen problem to an ineffective level. Biocontrol agent as microbial antagonists has been tried. The most effective agent proved to be another wood-rotting basidiomycete, Peniophora (Phlebia) gigantea. Spores of this fungus are painted or sprayed onto cut stumps. Around 106 spores are used per square metre of stump surface, which ensures rapid colonization of the timber by the antagonist, thereby occupying the infection court and denying access to the pathogen.
- Metarhizium anisopliae: Nevertheless, fungi have great promise as economical pest control agents. Metarhizium anisopliae is the best known of all entomopathogenic fungi and was the first fungus to be produced on a large scale. It infects a wide range of species and has been successfully commercialized by a

number of small companies in Brazil for the control of spittlebug in pasture and in sugar cane; it has been combined effectively with a virus to obtain control of the rhinoceros beetle on palms in the Pacific.

- Beaveria bassiana: Another extensively researched fungus is Beaveria bassiana; its product has been formulated in the USSR to control Colorado potato beetle. Experiments in the United States have shown good control of this beetle to be possible. Verticillium lecanii has been produced commercially in the U.K. for several years. Strains of this fungus can control aphids and whitefly, sometimes for months at a time. Hirsutella thompsonii was produced briefly in the United States by Abbott Laboratories for the control of mites in citrus; environmental conditions did not favor the growth of the fungus and it has not yet found a place in citrus pest control programmes. Nomuraea rileyi was investigated for many years by researchers in the U.S. primarily in collaboration with Abbott. This fungus can be very effective against caterpillars particularly in soybeans.

Arthropod Toxins from Fungi

Tetranactin: A macrotetrolide antibiotic was isolated from the fermented broth of *Streptomyces aureus* S-3466. It exerts a potent miticidal action (LC_{50} = 4.8 ppm) against the carmine spider mite (*Tetranychus cinnabarinus*) and concentrations of < 10 ppm were found lethal to related four mites such as two-spotted spider, European red mites etc. It is also known as an ionophore antibiotic because of its facile formation of complexes with alkaline cations.

Avermectins and Milbemycins - Multipurpose Biocides from Fungi: These are members of the class of sixteen- ring macrolides, having a significant role in crop protection, treatment of parasitic diseases of animals and of humans. Natural avermectins fall into two groups, designated as A and B, the former possessing a methoxy-group at C-5 and the latter possessing a hydroxy group instead. These groups are further sub-divided into the 1-series with a 22 double bond and a 2-series with a 22 double bond and a 2-series, which has an axial C-23-OH. A final sub-division into *a* and *b* series designates the presence of a sec-butyl or an isopropyl group respectively at C-25.

- Avermectins were isolated from an actinomycete found in soils of Kawana, Ito city, Japan. This novel actinomycete was named Streptomyces avermitilis MA-4680. Extensive work on this species, such as causing mutation with various agents and modification of the growing medium gave more than a fifty-fold improvement in yield from the mutated strain, S. avermitilis M-4848, which could produce 53 mg ml-1 of avermectins.
- Mibemycins have relatively simpler structures but with a greater diversity of functionalisation. Over 20 naturally occurring milbemycins known can be sub-divided into a and b series, on the basis of respective presence or absence of tetrahydrofuranyl ring as a part of the structural skeleton. A comparison of structures of avermectins and milbemycins would reveal the only major difference as the lack of L-oleandrosyl - L- oleaandrosyl group at C-13 in the latter.

Viruses: Viruses are strong regulators of insect populations and have often been identified as key determinants of the periodicity of upsurges of pests of such relatively undisturbed habitats as forests and pastures. Habitat disruption by normal agricultural practices generally diminishes the **epizootic capacity** of insect viruses but nevertheless economically telling epizootics do occur as for instance in *Trichoplusia ni* on cabbage and *Phthorimaea opercullela* on potatoes.

- **Baculoviruses (BVs):** Baculovirus infections are common among Lepidoptera and Hymenoptera sp., when larvae eat food contaminated with virus containing inclusions, they get infected. Once infected, these viruses have the ability to multiply in their host, sometimes keeping the target pest at a low level for several years making further control measures unnecessary. The bioactivity of baculovirus is attributed to the matrix protein hedrin, which dissolve in the dissolves in the insect gut releasing virus particles, which infect and multiply in gut epithelial cells. In lepidopteran insects the infection quickly spreads to other tissues.

Biological Control of Pests

In agriculture, pests are any living thing that causes injury or disease to crops. This can occur when insects eat foliage or stored produce, when bacteria or fungi cause plant diseases, or when weeds interfere excessively with the growth of crop plants. In conventional farming, pests are usually managed using various types of pesticides, such as insecticides, herbicides, and fungicides. In the short term, these methods can be effective in reducing the influence of pests on crops. However, the long-term use of these chemicals has been shown to have a severe effect on the environment. Pest management is an ecological matter. The size of a pest population and the damage it inflicts is, to a great extent, a reflection of the design and management of a particular agricultural ecosystem.

Types of Biocontrol

Natural biological control results when naturally occurring enemies maintain pests at a lower level than would occur without them, and is generally characteristic of biodiverse systems. Mammals, birds, bats, insects, fungi, bacteria, and viruses all have a role to play as predators and parasites in an agricultural system. By their very nature, pesticides decrease the biodiversity of a system, creating the potential for instability and future problems. Pesticides, whether synthetically or botanically derived, are powerful tools and should be used with caution.

Applied biological control, also known as augmentative biocontrol, involves supplementation of beneficial organism populations, for example through periodic releases of parasites, predators, or pathogens. This can be effective in many situations—well-timed inundative releases of *Trichogramma* egg wasps for codling moth control, for instance.

1. Substances added to agricultural lands to encourage plant growth and higher crop production.
2. Any biomass of plants or animals, living or dead.
3. Substances used to reduce the abundance of pests, any living thing that causes injury or disease to crops.
4. The physical structure of soil.

Organic farmers do not use synthetic, manufactured pesticides to manage their pest problems. Rather, they rely on other methods. These include using crop varieties that are resistant to pests and diseases, introducing natural predators of the pests, changing the habitat of the crop area to make it less suitable for the pest, and (when necessary) using a pesticide derived from a natural product

Slugs

Trapping, Natural Enemies and Biological Control: Traps containing either milk or beer have been used for decades to catch slugs. These traps are labour-intensive and costly. **Controlled refuges** may be used, as an alternative or in addition. These are simply pieces of wood, tile or carpet placed on the soil among crops; they act in two ways. They offer within-plot refuges for predatory beetles, which help to control slugs and many other pests. They also attract slugs, which use them as daytime refuges and sites for egg laying. Regular weekly inspection, and collection of any slugs or their eggs found, offers a quick and effective means of control.

Marsh flies (*Tentanocera elata* and *Euthycera cribrata*, Schiomizidae) are parasites of slugs and snails, and are attracted by the wildlife garden, with its ponds and flowers. These flies can lay up to 600 eggs in a season, while each fly larva may kill from nine to twenty-five slugs. The newly hatched larva waits for a slug to pass by, then grabs hold of it with special mouth hooks, and burrows straight inside. At this stage it is a parasite, consuming the living slug tissue. Eventually the slug will die, and the larva then leads more of a predatory existence, waiting to ambush passing slugs, or following them along their slime trails.

Carabid beetles are effective biological control agents. Many carabids eat slugs, however, certain species, such as *Pterostichus niger* and *Abax parallelepipedus*, have been shown to be particularly important slug predators. **Microorganisms**, like nematode worms, a microsporidium, and certain strains of *Bacillus thuringiensis* (BT) all show promise, and may one day make organic slug control much simpler. **Birds**, like thrushes, blackbirds, robins, starlings, crows, jackdaws, gulls and even owls, feed on slugs. Encourage these in the garden with shrubberies to nest in and provide winterberries, plus nest-boxes where appropriate. If you have space and opportunity, you can use chickens or ducks as biological control agents.

Integrated Pest Management (IPM)

IPM, as it was originally conceived, proposed to manage pests though an understanding of their interactions with other organisms and the environment. However, IPM has strayed from its ecological roots.

Pest does not necessarily constitute a problem. Before a potentially disruptive control method is employed, appropriate decision-making criteria are used to determine whether or not pest management actions are needed.

Biointensive IPM incorporates ecological and economic factors into agricultural system design and decision making, and addresses public concerns about environmental quality and food safety. The benefits of implementing biointensive IPM can include reduced chemical input costs, reduced on-farm and off-farm environmental impacts, and more effective and sustainable pest management. An ecology-based IPM has the potential of decreasing inputs of fuel, machinery, and synthetic chemicals- all of which are energy intensive and increasingly costly in terms of financial and environmental impact. Such reductions will benefit the grower and society. Over-reliance on the use of synthetic pesticides in crop protection programs around the world has resulted in disturbances to the environment, pest resurgence, pest resistance to pesticides, and lethal and sub-lethal effects on non-target organisms, including humans.

A systems approach to pest management based on an understanding of pest ecology. It begins with steps to accurately diagnose the nature and source of pest problems, and then relies on a range of preventive tactics and biological controls to keep pest populations within acceptable limits. Reduced-risk pesticides are used if other tactics have not been adequately effective, as a last resort, and with care to minimize risks.

Components of Biointensive IPM

An important difference between conventional and biointensive IPM is that the emphasis of the latter is on proactive measures to redesign the agricultural ecosystem to the disadvantage of a pest and to the advantage of its parasite and predator complex. At the same time, biointensive IPM shares many of the same components as conventional IPM, including monitoring, use of economic thresholds, record keeping, and planning.

Pest Identification: A crucial step in any IPM program is to identify the pest. The effectiveness of both proactive and reactive pest management measures depend on correct identification. After a pest is identified, appropriate and effective management depends on knowing answers to a number of questions. These may include:

1. What plants are hosts and non-hosts of this pest?
2. When does the pest emerge or first appear?
3. Where does it lay its eggs? In the case of weeds, where is the seed source? For plant pathogens, where is the source(s) of inoculum?
4. Where, how, and in what form does the pest overwinter?
5. How might the cropping system be altered to make life more difficult for the pest and easier for its natural controls?

Monitoring: Monitoring involves systematically checking crop fields for pests and beneficials, at regular intervals and at critical times, to gather information about the crop, pests, and natural enemies. Sweep nets, sticky traps, and pheromone traps can be used to collect insects for both identification and population density information. Leaf counts are one method for recording plant growth stages. Square-foot or larger grids laid out in a field can provide a basis for comparative weed counts.

Economic Injury and Action Levels: The economic injury level (EIL) is the pest population that inflicts crop damage greater than the cost of control measures.

Time and Resources: A successful biointensive IPM program takes time, money, patience, short- and long-term planning, flexibility, and commitment. The pest manager must spend time on self-education and on making contacts with Extension and research personnel. Be aware that some IPM strategies, such as increasing beneficial insect habitat, may take more than a year to show results.

Botanical Controls

Botanical pesticides are prepared in various ways. They can be as simple as pureed plant leaves, extracts of plant parts, or chemicals purified from plants. Pyrethrum, neem formulations, and rotenone are examples of botanicals. Some botanicals are broad-spectrum pesticides. Others, like ryania, are very specific. Botanicals are generally less harmful in the environment than synthetic pesticides because they degrade quickly, but they can be just as deadly to beneficials as synthetic pesticides. However, they are less hazardous to transport and in some cases can be formulated on-farm. The manufacture of botanicals generally results in fewer toxic by-products.

Chemical Controls

Included in this category are both synthetic pesticides and botanical pesticides.

Synthetic pesticides comprise a wide range of man-made chemicals used to control insects, mites, weeds, nematodes, plant diseases, and vertebrate and invertebrate pests. These powerful chemicals are fast acting and relatively inexpensive to purchase. Pesticides are the option of last resort in IPM programs because of their potential negative impacts on the environment, which result from the manufacturing process as well as from their application on the farm. Pesticides should be used only when other measures, such as biological or cultural controls, have failed to keep pest populations from approaching economically damaging levels.

Others

Compost teas are most commonly used for foliar disease control and applied as foliar nutrient sprays. The idea underlying the use of compost teas is that a solution of beneficial microbes and some nutrients is created, then applied to plants to increase the diversity of organisms on leaf surfaces. This diversity competes with pathogenic organisms, making it more difficult for them to become established and infect the plant.

ORGANIC VEGETABLE CROP PRODUCTION

While home gardeners can grow the majority of vegetables "organically," the list of crops suitable for organic production is more limited for commercial production. Markets for organic produce have increased in the last decade, but some vegetables are considered too high risk for commercial organic production, particularly cabbage and related vegetables. Pest control for these crops is very difficult and is compounded if the crops are located in regions where canola and mustard are being grown. If they are to be attempted, these crops need to be well isolated from crops which share the same pests.

Carrots

Carrots are considered to be one of the easiest crops to grow organically due to the limited number of pests. However, they should not be planted near alfalfa or other hay and winter cereals, due to the incidence of leafhoppers in these crops which can spread aster yellows disease to carrots. The disease causes stunting or discoloration of carrot foliage, and extremely hairy roots which are unmarketable. As carrots are slow to germinate and are poor weed competitors when young, weed control is generally the greatest challenge in growing carrots.

Pumpkins and Squash

Pumpkins and the various types of summer and winter squash, including zucchini, are also potential crops for organic production. They are not prone to many insect pests or diseases in Saskatchewan. Selecting early maturing winter squash is important.

Potatoes

While not highly recommended for organic production in Saskatchewan, potatoes have been successfully grown organically by a few producers. Disease control, particularly for blight, is of major concern. Insects such as Colorado potato beetles, flea beetles and wireworms can also be major problems. This crop is best grown well isolated from other potato fields and with long rotations. Careful cultivar selection may assist in disease control.

Sweet Corn

Sweet corn is a popular item at farmers' markets, roadside stands, supermarkets and market gardens. One of the greatest challenges in growing sweet corn is providing it with sufficient nitrogen to meet its very high requirements. Growers also need to look for untreated seed. Corn seed maggot, corn earworm and European corn borer are potential pests.

Onions

Organic production of onions is very risky, due to the high likelihood the crop will be infested with onion maggots. Without the use of conventional pesticides, it may be very difficult to produce a profitable, marketable crop. If onions are to be planted, it is suggested to plant the crop about 1.6 km (1 mile) from previous onion cropland. Although they prefer onions, onion maggots can also infest garlic, leeks, chives and shallots. Onion thrips are another common problem. They will migrate from alfalfa and cereal crops at harvest time. Heavy irrigation is known to reduce thrips populations.

Cole Crops (Cabbage Family)

Vegetables in this group are not recommended for organic production in Saskatchewan, due to the high incidence of pests. This family includes: cabbage, cauliflower, broccoli, Brussels sprouts, kohlrabi, kale, radish, rutabaga and turnip. They are very prone to imported cabbageworm, diamondback moth larvae, root maggots, red turnip beetle, flea beetles, cabbage thrips and wireworms. The incidence will be higher in regions where canola and mustard crops are being grown. Of the group, rutabagas and turnips are particularly difficult to protect organically from root maggots.

Other Vegetables

Many other vegetables are somewhat intermediate in their ease of cultivation by organic methods. Some of them are listed here, along with potential problems.

Asparagus: This perennial crop is not widely grown in Saskatchewan. The early spears can be a target for overwintering onion thrips. Asparagus beetles, exclusive to this crop, can also feed on spears and make them unmarketable. Crown and root rot may be managed to some extent by cultural practices.

Beans and Peas: Untreated seed must be used, making them more prone to seed and/or root rot. Powdery mildew is a common problem in peas, particularly late in the season.

Beets: If marketed with tops attached, beet leafminer control will be necessary. Row covers may be considered. Adjacent weeds such as lambsquarters are also hosts for leafminer. Rotations are helpful in minimizing problems with leaf and root diseases. Swiss chard and spinach share some of the same pests and diseases.

Celery: This crop needs an ample water supply at all times to keep growth succulent. Water stress will cause tough, strong flavoured stems. Tarnished plant bug can cause serious damage to this crop; thus it should not be grown near alfalfa fields.

Cucumbers: They are generally easy to grow, but need a consistent water supply, especially when fruit is developing. Cucumber beetles can spread bacterial wilt disease. Transplanting and the use of row covers early in the season will provide earlier harvests.

Lettuce: Leaf lettuce is the easiest to produce but highly perishable when harvested. Romaine and head lettuce are later maturing. Lettuce should be marketed as soon as possible after harvest. Tipburn resistant varieties should be grown. Lettuce is prone to various diseases including Sclerotinia (white mold). Using raised beds and rotating fields with grain crops may be helpful. Aster yellows disease, spread by leafhoppers, may also be a potential problem, but the incidence is quite variable from year to year.

Melons: Both muskmelon (often called cantaloupe) and watermelon are long-season crops that need warm weather to produce well. Production in tunnels, on plastic mulch or season-extending greenhouse-type structures has given earlier yields. These crops are best grown on sandy land that warms up easily.

Peppers: They are grown by both organic and conventional methods, unless grown in a protected structure, peppers are somewhat of a gamble in Saskatchewan. They prefer a warm, long growing season, although excessively hot summer temperatures can cause flower bud drop.

Spinach: Beet leafminers are a major problem in this crop which may not be profitable unless it can be grown in isolation or under row covers.

Tomatoes: They are susceptible to early and late blight, cutworms, and sometimes to Colorado potato beetles. In Saskatchewan, early maturing and cold-tolerant cultivars are required unless the plants are to be grown in greenhouses.

HERB PRODUCTION

Culinary Herbs

A wide range of herbs can be grown outdoors or in greenhouses. Some herbs such as sweet basil and rosemary are very adaptable to greenhouse production. Organic garlic is in demand in some areas, but production may not be practical on a large scale because garlic must be kept weed free and can be affected by some of the same pests as onions. Growers may find the garlic seed market to be more lucrative than the culinary market.

Most culinary herb production needs to take place near urban centres where farmers' markets or restaurants can be supplied, unless the grower plans to dry and package the herbs for later use. Other value-added products are also possible, including herbal tea mixtures, flavoured vinegars and oils, as well as packaged seeds.

Medicinal Herbs

The vast majority of medicinal herbs grown in India are produced organically. However, markets for medicinal herbs are very volatile and growers need to use caution before planting any medicinal herbs on a large scale. As with vegetables, medicinal herbs are mostly grown as row crops. Pest and disease problems are known for some and the incidence could increase if acreages become significant.

Herb crops for cosmetic uses are also gaining prominence. Willow herb (fireweed) is one example. Much of it is gathered from the wild, rather than being cultivated.

CHAPTER 27

Nutrient Management in Horticultural Crops

Adequate supply of plant nutrients is important to ensure efficient crop production. Adding animal and vegetable manure to the soil to restore fertility is the practice from time immemorial. As the use of the chemical fertilizers increased due to their large availability and easy to adopt technique, organic manure application slowly declined. Since the nutrient turn over in soil plant system is considerably high under intensive farming, neither the chemical fertilizers nor the organic or biological sources alone can achieve a sustainable production. Choice of manures and fertilizers and their application at right quantity and time are important to get higher production.

Good use of nutrient management will improve both your production and the environment. Cost-effective application will produce the best yield potential while minimizing costs. In turn, this helps reduce nutrients lost to the soil through leaching and water erosion.

Proper nutrient management provides nutrients at the appropriate rate, timing, and with the appropriate method to produce an economically optimal crop in terms of both yield and quality. It minimizes the risk of causing pollution by loss of nutrients via runoff, leaching, emissions to the air or other mechanisms.

For horticultural crops the would be equivalent to the amount of a nutrient that would be supplied to a crop to meet its needs for that nutrient while maintaining soil nutrient levels at pre-application levels. This is a difficult value to determine particularly for perennial crops (tree fruits, berries, etc) and may require consultation with local crop specialists.

Soil testing is an important first step. Soil test results will give you a base on which to analyze soil needs. Also do tissue tests, if available for your crop. This shows what plant nutrient levels are at that point in time, as opposed to what is available in the soil. This is particularly important for perennial crops.

Record keeping of soil and tissue tests will help track trends. Include observations on crop growth, yield, quality and weather conditions during the growing season. A few general best management practices include:

1. If possible, use split applications of nitrogen to reduce the possibility of loss by leaching. This will also increase management and application costs and may not be suited to all crops.
2. Keep soil healthy so that root systems will be most effective in using nutrients.
3. If nutrients are left after a crop harvest, use cover crops to hold them for the next crop.

To be competitive, horticultural crop growers have to produce high quality products efficiently. To be sustainable, soil and water resources must be protected and conserved. Best management practices need to be chosen and adapted to suit each farm operation.

It is necessary to evaluate and refine the nutrient application schedule for horticultural crops. Once an initial nutrient application schedule has been established for all fields and harvests, the next step is to assess the feasibility of that schedule. The following questions should be considered in this assessment:

1. Is the level of nutrients being supplied similar to the crop nutrient requirement?
2. Does the schedule utilize all the manure volume that is generated in a year and is it applied in such a manner that the manure storage facilities never exceed their capacity or alternatively fall below empty?
3. Are some nutrients being supplied adequately while others are applied in excess? If so, what kind of strategy will be adopted to avoid the long-term accumulation of certain nutrients, particularly phosphorus or potassium?
4. Is the nutrient application equipment chosen to place nutrients where their utilization by the crop will be maximized?
5. Considering time of year, rate of application, and method of application, will nutrient applications elevate the risk of causing pollution by runoff, leaching or other mechanisms?

ESSENTIAL ELEMENTS IN PLANT NUTRITION

Plants contain small amounts of 90 or more elements, but 16 elements are known to be essential for their growth and completion of life cycle. They are: Carbon, hydrogen, oxygen, nitrogen, phosphorus, potassium, calcium, magnesium, sulphur, iron, manganese, zinc, copper, boron, molybdenum and chlorine. In addition, four more elements viz., sodium, cobalt, vanadium and silicon are absorbed by some plants for special purposes. All these elements are not required for all plants, but all have been found essential for one plant or the other. Among these, all carbon atoms and most of the oxygen atoms are derived from carbon dioxide which is assimilated principally in photosynthesis. More specifically, approximately one-third of oxygen atoms in organic material in higher plants are derived fro soil water and two-thirds from carbon dioxide of the atmosphere. The chemical basis for this conclusion is the reaction occurring during photosynthesis, wherein one molecule each of carbon dioxide (CO_2) and water (H_2O) are combined in the presence of enzymes. The elements C, H and O are not minerals. The rest of the elements are absorbed from the soil and these are called mineral elements since they are derived from minerals. These mineral elements are mainly absorbed in ionic form and to some extent in non-ionic form as shown in Table 27.1.

TABLE 27.1: Forms of mineral elements absorbed by plants

Mineral element	Ionic form	Non-ionic form
Nitrogen	NH_4^+, NO_3^-	$CO (NH_2)^2$
Phosphorus (P)	$H_2PO_4^-$, HPO_4^{2-}	Nucleic acid, phytin
Potassium (Kalium-K)	K^+	
Calcium (Ca)	Ca^{2+}	
Magnesium (Mg)	Mg^{2+}	
Sulphur (S)	SO_4^{2-}	SO_2
Iron (Fe)	Fe^{2-}, Fe^{3-}	$FeSO_4$ with EDTA
Manganese (Mn)	Mn^{2+}	$MnSO_4$ with EDTA
Zinc (Zn)	Zn^{2+}	$ZnSO_4$ with EDTA

Mineral element	Ionic form	Non-ionic form
Copper (Cu)	Cu^{2+}	$CuSO_4$ with EDTA
Boron (B)	$B_4O_7^{2-}$, $H_2BO_3^-$, HBO_3^{2-}	
Molybdenum (Mo)	MoO_4^{2-}	
Chlorine (Cl)	Cl^-	

EDTA = Ethylenediaminetetracetic acid.

CLASSIFICATION OF ESSENTIAL NUTRIENTS

Essential nutrients are divided into two groups on the basis of their quantity required by the plants. Those required in large quantities are classified as **macronutrients** and those in small amounts as **micronutrients**. The Big Three: nitrogen, phosphorus and potassium are the primary fertilizer elements used in agriculture. Air, water and soil are the three sources from which the growing plants can obtain their requirements.

Sodium, silicon, cobalt and vanadium are **beneficial** for some plants but have not been established as essential elements for all higher plants. Macronutrients are needed in concentrations of 1000 ug/g of dry matter or more, where as micronutrients (trace elements) are needed in tissue concentrations equal to or less than 100 ug/g of dry matter. The low requirement of plants for trace elements can be accounted for by the participation of these elements in enzymatic reactions and as constituents of growth hormones rather than as components of major plant products such as structural and proptoplasmic tissue.

Mengel and Kirkby (1978) classified plant nutrients based on their biochemical behaviour and physiological functions into **four** groups.

1. **Group 1:** C, H, O, N and S. These are major constituents of organic material involved in enzymatic processes and oxidation-reduction reactions.
2. **Group 2:** P and B. These are involved in energy transfer reactions and esterification with native alcohol groups in plants.
3. **Group 3:** K, Ca, Mg, Mn and Cl. This group plays osmotic and ion balance roles, plus more specific function in enzyme conformation and catalysis.
4. **Group 4:** Fe, Cu, Zn and Mo. Present **as** structural chelates of metalloproteins. These elements enable electron transport by valency change.

Nutrients are also classified, based on ionic forms absorbed by plants into cations and anions and metals and non-metals.

Cations	NH_4^+, Ca^{2+}, Mg^{2+}, K^+, Cu^{2+}, Fe^{2+}, Mn^{2+}, Zn^{2+}.
Anions	NO_3^-, HPO_4^{2-}, $H_2PO_4^-$, SO_4^{2-}, BO_3^{3-}, MnO_4^{2-}, Cl^-.
Metals	K, Ca, Mg, Fe, Mn, Zn, Cu.
Non-metals	N, P, S, B. Mo. Cl.

CRITERIA OF ESSENTIALITY

Even though, some of these elements have been known since long, their essentiality has been established in the last century. Arnon and stout (1939) and Arnon (1952) proposed the following criteria of essentiality of mineral elements.

1. A deficiency of the element in question results in failure to complete the life cycle,

2. Deficiency of element in question can be corrected only by supplying that particular element, and
3. The element must extend its effect directly on growth or metabolism and not by indirect effect such as antagonism of another element present at a toxic level.

Arnon's criteria of essentiality of elements for plant growth may be more correct. However, it appears to be too rigid from practical point of view. Nicholas (1961) proposed the term **functional** or **metabolic nutrient** to include any mineral element which functions in plant metabolism, whether or not its action is specific. Elements such as sodium, cobalt, silicon and vanadium are classified as essential when the less restrictive definition of essentiality is used.

ESSENTIAL NUTRIENT ELEMENTS AND THEIR SOURCES

Essential elements required in relatively large amounts: **Macronutrients**

Mostly from air and water:	**Carbon (C)**
	Hydrogen (H)
	Oxygen (O)
From soil solids:	Nitrogen (N)
	Phosphorus (P)
	Potassium (K)
	Calcium (Ca)
	Magnesium (Mg)
	Sulphur (S)

Essential elements required in relatively small amounts: **Micronutrients**

From soil solids	**Iron (Fe),**	**Manganese (Mn),**
	Zinc (Zn),	Copper (Cu),
	Boron (B),	Molybdenum (Mo),
	Cobaly (Co),	Chlorine (Cl).

Essential Elements from the Soil

1. **Macronutrients:** Of the fourteen essential elements obtained from the soil by the plants, six are used in relatively large quantities and are thus referred to as **Macronutrients**. They are nitrogen, phosphorus, potassium, calcium, magnesium and sulphur.
2. **Primary or fertilizer elements:** N, P and K are commonly supplied to the soil as farm yard manure and compost or as commercial (Chemical) fertilizers and are therefore, called primary or fertilizer elements.
3. **Secondary elements:** Ca, Mg and S are referred to as secondary elements. Ca and Mg are added to acid soils in limestone and are called lime elements.
4. **Micronutrients:** The other nutrient elements, viz., iron, manganese, copper, zinc, boron, molybdenum, chlorine, and cobalt are used by higher plants in very small amounts, thereby justifying the name: **Micronutrient** or **Trace Elements**.

General Conditions Conducive to Micronutrient Deficiency

- Highly leached acid sandy soils.
- Muck soils.
- Soils high in pH.
- Soils intensively cropped and heavily fertilized with micronutrients only.

Antagonistic Effects of Micronutrients

- Excess copper or sulphate may adversely affect the utilization of molybdenum.
- Iron deficiency is encouraged by an excess of zinc, manganese and copper.
- Excess phosphate may encourage deficiency of zinc, iron and copper.
- Heavy N fertilization intensifies copper deficiency.
- Excess sodium or potassium may adversely affect manganese uptake.
- Excess lime reduces boron uptake.
- Iron, copper and zinc may reduce the absorption of manganese.

CHELATE

A chelate is an organic compound which combines with and protects certain metallic cations including Fe, Mn, Zn and Cu.

Synthetic Chelating Agents

- Fe, Zn, Mn, Cu – EDTA – Ethylenediaminetetracetic acid.
- Fe = EDDHA- Ethylenediamine di-O-hydroxyphenyl acetic acid.
- Fe = DTPA – Diethylenetriaminepenta acetic acid.
- Zn = NTA – Nitrilotriacetic acid.
- Fe, Zn = Hydroxyethylenediaminetetracetic acid.

FUNCTIONS OF ESSENTIAL NUTRIENTS

Nitrogen (N)

- Nitrogen is an essential constituent of proteins and chlorophyll and is present in many other compounds of great physiological importance in plant metabolism such as nucleotides, phosphatides, alkaloids, enzymes, hormones, vitamins etc.
- Impart dark green colour to plants.
- Promotes leaf, stem and other vegetative growth but retains small root system.
- Produces rapid early growth.
- Improves quality, succulence of leafy vegetables and fodder crops.
- Increases protein content of food and fodder crops.
- Governs to a considerable degree, the utilization of potassium, phosphorus and other elements.

Phosphorus (P)

- Phosphorus is a constituent of nucleic acid, phytin and phospholipids. It is important in laying down the primordial for the reproductive part of plant.
- It is also an essential constituent of majority of enzymes which are of grat importance in the transformation of energy, in carbohydrate, metabolism, in fat metabolism and also respiration in plants.

- Stimulates early root development and growth.
- Gives rapid and vigorous start to plants, strengthens straw and decreases lodging tendency.
- Brings about early maturity of crops, particularly the cereals and counteracts the effects of excess nitrogen.
- Stimulates flowering and aids in seed formation.
- Increases ratio of grain to stalk.
- Improves quality of food grains.
- In legumes it increases the formation of root nodules.
- In excessive amount it may cause trace element deficiencies (mainly iron and zinc) and may at times alleviate the detrimental effects of overliming.

Potassium (K)

- Imparts increased vigour and disease resistance to plants.
- Produces strong stiff straw in cereals specially rice and wheat, thus reduces lodging in cereals.
- Imparts winter hardiness to legumes.
- Regulates water conditions within the plant cell and water loss from the plant by maintaining the balance between anabolism, respiration and transpiration.
- Essential in formation and transfer of starches and sugars. Thus potassium is required in large quantities for potato, sweet potato, turnip, banana and tapioca.
- Helps in formation of protein and chlorophyll.
- Increases plumpness of grains and seeds.
- Acts as an accelerator of enzyme action.
- Improves quality of final products such as improvement in quality of tobacco leaf, quality of fibre in fibre crops, taste, size and keeping quality of fruits.

Calcium (Ca)

- It is a constituent of cell wall and as such it increases stiffness of straw.
- Promotes early root development and growth.
- Provides a base for neutralization of organic acids.
- Essential to activate growing points especially root tips.
- Influences the water economy of the plant, protein-carbohydrate ratio in fat metabolism as well as many other physiological processes.
- Improves intake of other plant nutrients especially N and trace elements such as iron, boron, zinc, copper and manganese by correcting soil pH.
- Encourages seed production.

Magnesium (Mg)

- It is a constituent of chlorophyll and is essential for all green plants.
- Plays a part in the production of carbohydrates, proteins, fats, vitamins and certain catalytic reactions in the enzyme system.
- Acts as a part of carrier of phosphorus in plants particularly in connection with the formation of seeds of high oil content.
- Helps translocation of starches.
- Regulates the uptake of other nutrients and the base economy of plants.

Sulphur (S)

- It helps in chlorophyll formation and encourages vegetative growth.

- It is an essential constituent of many proteins, enzymes and certain volatile compounds such as, mustard oil and helps in the reduction-oxidation system in respiration.
- Increases root growth.
- Stimulates seed formation.
- Promotes nodule formation on roots of legumes.

Boron (B)

- The primary role of boron appears to be in calcium metabplism, both with its uptake by roots and its efficient use in plants.
- Tends to keep calcium soluble and increases its mobility in plants.
- Acts as a regulator of potassium/calcium ratio to the plant.
- Helps in absorption of nitrogen.
- It is a constituent of cell membrane and is essential for cell division.
- Boron is concerned with precipitating excess cations, buffer action, mainetance of conducting tissues and with regulatory effect on other nutrients.
- It is necessary for translocation of sugars in plants.
- Helps the vascular system in roots to give out branches to supply nodule bacteria with carbohydrate food so that bacteria may not become parasitic.
- It plays a vital role in fertilization of certain crop plants.

Manganese (Mn)

- The function of manganese is being closely associated with that of iron. It also supports the movement of iron in plants.
- Helps in chlorophyll formation.
- Acts as catalyst in oxidation and reduction reactions within the plant tissues.
- A good manganese supply sometimes helps in counteracting the bad effect of poor aeration.

Iron (Fe)

- Though it is not a constituent of chlorophyll, it helps in its formation.
- Helps in absorption of other nutrients.
- As a constituent of enzyme systems which bring about oxidation-reduction reactions in plant, it regulates respiration, photosynthesis and reduction of nitrates and sulphates.
- It is essential for the synthesis of proteins present in the chloroplasts.

Zinc (Zn)

- Zinc is a constituent of several enzyme systems which regulate various metabolic reactions in the plant.
- Influences the formation of some growth hormones like IAA in plants.
- Helpful in reproduction of certain plants.
- Associated with water uptake and water relations in the plant.

Molybdenum (Mo)

- Acts in enzyme systems which bring about oxidation-reduction reactions especially the reduction of nutrients of nitrates to ammonia prior to amino acid and protein synthesis in the plant cells.

- It is essential for atmospheric nitrogen fixation - symbiotic as well as non-symbiotic.
- Two components of Nitrogenase (1) Fe and (2) Fe + Mo. When 1000 atoms of Fe is present then 4 atoms of Mo is required. In bacteria, Mo can be replaced by vanadium, obvious role in N fixation.

Copper (Cu)

- Copper forms many compounds with amino acids and proteins in plant.
- It acts as 'electron carrier' in enzymes which bring about oxidation-reduction reactions in plants.
- Helps in utilization of iron in chlorophyll synthesis.
- When applied in relatively large quantity. It neutralizes the inactivating certain toxins present in the soil.

Chlorine (Cl)

- The role of chlorine is still somewhat obscure; however, both root and top growth seem to suffer, if it is absent.
- Essential for photosynthesis and as an activator of enzymes involved in splitting water. Associated with osmo-regulation of plants growing on saline soils.
- Plants like tobacco, tomato, sugarbeet, wheat, peas etc. generally require about 1 kg Cl- for each 4000 kg of dry matter (weight). In tobacco, it increases the water content of tssues and affects the carbohydrate metabolism.

Cobalt (Co)

- Cobalt is essential for the symbiotic fixation of nitrogen. This element is a component of vitamin B12, which is thought to be nodule tissue known as leg hemoglobin.
- Structural component of vitamin B12, which in turn essential to the formation of leg hemoglobin needed for N fixation; essential for growth and metabolism of micro organisms; required by Rhizobia, activator of certain enzymes like orginase, Lecithinase etc.

Sodium (Na)

- Essential for sugarbeets and such crops; influences water relations in sugarbeet and increases their resistance to drought.
- In higher plants, Na is essential and to some extent it can supplement for K.

Vanadeum (V)

- In symbiotic N fixation, reported in BGA e.g. nostoc, anaeena etc. required 0.05 to 0.1 ppm and 0.1-1 ppm
- Vanadium can replace Mo, obvious role in N fixation.

Silicon (Si)

- Essential to certain diatoms. Silicious cell wall (made of Si) protects food material which is oils in diatoms, gives resistance to rice and perhaps other crops like cucumber, barley etc.
- Si is reported to increase the top length, number of stems, fresh and dry weight

of rice plant grown in nutrient cultures.

Selenium (Se)

- Can be replaced with 'S' in the production of Methionine, cysten because it is analogous to 'S' (i.e., same function) but origin different.(Homologous means same origin).

Aluminium (Al)

- Activator of glandular system but more amount has toxic effect.
- Functional Nutrients = 16 essential elements + Cobalt (Co^{++}), Sodium (Na^{+}),
- Vanadium (V) and Silicon (Si).
- Thus Functional nutrients are 20 in number for the plants.
- Role of Essential Plant Nutrients

ROLE OF ESSENTIAL PLANT NUTRIENT

Followings are the four major roles:

1. Structural components of cell constituents and its metabolically active compounds
2. In maintenance of cellular organization *e.g.* turgidity etc.
3. In energy transformation; and
4. In enzyme action.

Hidden Hunger

The term **Hidden Hunger** is used to describe a plant that shows no obvious symptoms yet the nutrient content is sufficient to give the top profitable yield. Fertilization with the sure rate rather than bare economic optimum for on average yield helps to detain the top profitable yield. Sure rate is a rate a little above critical limit to be sure that there is no reduction on yield.

The Hidden Hunger is an expression in soil science to designate a condition in plants brought about through the lack of certain essential elements in the soil and is not easily detected by out ward appearance.

DEFICIENCY SYMPTOMS

The deficiency symptoms can be distinguished on the basis of the following:

- Region of occurrence.
- Presence or absence of dead spots and
- Chlorosis of entire leaf or interveinal chlorosis.

Deficiency Symptoms on Old Leaves

a = Without dead spots:

N = Uniform chlorosis of leaves including veins. The leaves become stiff and erect especially in cereals. Cereals show characteristic 'V' shaped yellowing at the tip of the lower leaves.

P = Small, erect, usually dark green leaves with a greenish red, greenish brown or purplish tinge. The rear side develops bronzy appearance premature ripening of crop.

Mg = Yellowing (Chlorosis) in between the veins and veins remain green. Leaf is not erect, Necrosis in extreme case only n the margins; upward curling of leaves along the leaf margin, most prevalent towards the end of growing crop season. Cereals fairly resistant but oat is more susceptible.

With Dead Spots

K = Yellowing starts from tips or margins of leaves extending to the centre of leaf base; Necrosis very soon, sharp difference between green and yellow and yellow and necrotic portions, curling of leaves towards under surfaces. **Tip burn, margin scorching.** Brown spots near margin, rolling of laminae, Barley most susceptible, Potato shows abnormal dark green colour of foliage followed by browning, cotton bolls either small, knotty or poorly opened.

Mo = Translucent spots of irregular shape in between the veins of leaves; these spots are light green, yellow or brown, chlorotic mottling between veins of middle leaves. The affected spots are impregnated with resinous gum which exudes from rear side of the leaf from the reddish brown spots, Brassica sp. most susceptible, **yellow spot of citrus, whiptail of cauliflower** (rat disease).

Symptoms on New Leaves

Veins Remaining Green: Interveinal Chlorosis

1. **Iron (Fe):** The principal veins remain conspicuously green and other portions of the leaf area turn yellow tending towards whiteness, chlorosis starts from base of leaf.
2. **Manganese (Mn):** The principal veins as well as the smaller veins are green. Interveinal portion is yellowish not tending towards whiteness; **Complete Interveinal Chlorosis;** chequered; appearance to leaf; oat and soybean most affected. The minimum level in healthy oat at flowering is 14 ppm, leaf top remains green and base dies, **Grey speek of oat, Pahala blight of sugarcane.**

Veins not Remaining Green

1. **Sulphur (S):** Leaves small and veins are paler than intervainal portion. No dead sports; plants not loose the lower leaves as in N-deficiency. Vegetable leaves develop yellowish green colour and become thick and firm.
2. **Copper (Cu):** Leaf yellowish tending towards whiteness; In extreme deficiency chlorosis of veins occurs and leaf loses luster. Chlorosis spreads alongside the margins towards the base of leaves; development of several auxillary buds when shoot apex dies, wheat very susceptible; multiple bud formation in leaf axils, **resetting and excess gumming.**

Symptoms on Terminal Buds

1. **Calcium (Ca):** Bud leaf becomes chlorotic white with the base remaining green. About one-third chlorotic portion of the tip hooks drown ward and becomes brittle, root systems poorly developed or lack fibre.
2. **Boron (B):** Yellowish of chlorosis which starts from the base to tip. Tip becomes very much elongated into a whip like structure and becomes brownish or blackish brown; leaves become thick and margin upwardly rolled; Hollow stem

of cauliflower, Browning of cauliflower curd, Top sickness of tobacco and Hard fruit of citrus.

Deficiency Symptoms on both Old and New Leaves

Zinc (Zn): Lamina becomes chlorotic and veins remain green; dead spots over leaf including veins, tips and margins; Strucrural aberrances in root tips; dwarfing of vegetative growth: New leaves emerge white in colour (White bud of maize): Khaira disease of rice; In cereals deficiency appears in 2-4 leaves from top during vegetative stage. Plant bushy due to reduced internodal elongation (malformation).

TOXICITY SYMPTOMS

1. **Nitrogen (N):** Lodging and abortion of flowers, susceptible to pests and diseases.
2. **Phosphorus (P):** Deficiency of Fe and Zn.
3. **Iron (Fe):** Tiny brown spots on lower leaves of rice starting from tips.
4. **Manganese (Mn):** Stunted plant and tillering is often limited in lowland rice.
5. **Boron (B):** Chlorosis of tips of older leaves especially along the margins.

MOBILITY OF NUTRIENTS

Mobility in Soil

1. Mobile nutrients: Such elements are highly soluble and are not adsorbed on clay complex *e.g.* NO^{-3}, SO_4^{2-}, BO_3^{3-}, Mn^{2+}, Cl^-
2. Less mobile: Such elements are also soluble but adsorbed on clay complex and thus their mobility is reduced *e.g.* NH_4^+, K^+, Ca^{2+}, Mg^{2+}, Cu^{2+}.
3. Immobile: Highly reactive and get fixed in the soil *e.g.* $H_2PO_4^-$, HPO_4^{2-}, Zn^{2+}.

Mobility in Plants

1. Highly mobile: N, P and K.
2. Moderately mobile: Zn.
3. Less mobile: S, Fe, Cu, Mn, Mo, and Cl.
4. Immobile: Ca and Boron.

CLASSIFICATION OF DIAGNOSTIC METHODS

Various methods used to diagnose deficiency of nutrients are listed below:

Field Experiments with Fertilizers

- Simple fertilizer trials on cultivators' fields.
- Complex experiments.

Pot Culture Experiments Methods Using Plants

- Visual method or hunger signs of plants.
- Use of indicator plants.
- Application of nutrients to leaf surfaces.
- Nutrient injections.
- Foliar diagnosis: In this comparison between the nutrient status of comparable

leaves of high and low yielding crop plants is done which is termed as crop logging.

- Rapid plant tissue tests.
- Using plants as nutrient extractants from soils.
- Methods of Mitscherlich.
- Method of Neubauer.

Soil Analysis Methods

- Determination of available nutrients.
- Determination of exchangeable cations.
- Rapid soil tests with the help of soil testing kits.

Methods Using Micro-Organisms an Extractants

TABLE 27.2: List of indicator plants for nutrient deficiencies

Deficient nutrient	Indicator plants
N	Cauliflower, Cabbage
P	Rape seed
K	Potato
Ca	Cauliflower, Cabbage.
Mg	Potato
Fe	Sugarbeet
Mn	Sugarbeet, Oat, Potato
B	Sugarbeet

Plant nutrients (being mobile) showing deficiency symptoms on younger leaves: N, P, K, Mg, Zn.

Plant nutrients (being immobile) showing deficiency symptoms on youner leaves: Ca, S, Fe, Cu, B, Mn.

Nutrient Uptake by Plants

Inorganic nutrients alone can be absorbed by plants; organic forms have to undergo mineralization for uptake by plants. Nutrient transport to roots, absorption by roots and translocated for assimilation in the shoot occurs simultaneously,

Nutrient Supply to Plant Roots

Nutrient ions must come in contact with root surface for absorption by the roots. Nutrient supply to roots depends on nutrient concentration in soil solution, soil moisture status and capacity of plants for absorption. There are three ways in which nutrient ions in soil solution may reach root surface: **mass flow, diffusion and root interception.**

Mass Flow

Mass flow occurs when nutrient ions and other dissolved substances are transported in the flow of water to the roots due to transpiration water uptake by the plant. It may also take place in response to evaporation and percolation of soil water. Mass flow is passive (without using energy) transport in soil water. The contribution of mass flow in the process of nutrient supply to roots can be calculated from the product of the soil solution concentration and the volume of water transpired by the plant.

MF = C x WU

Where, MF is mass flow, C is the solution concentration of any given ion and WU is total water uptake, which is the water content in plant plus the water transpired. Mass flow can meet the nutrient needs of the crop for all the nutrients except N, P, K, Fe and Mn, provided sufficient nutrient concentrations are present in soil solution.

Diffusion

Diffusion is the movement of ions from a region of high concentration to a region to low concentration by random thermal motion. As the plant roots absorb nutrients from surroundings soil solution, a concentration gradient is set up and nutrients move by diffusion along the concentration gradient. Movement of nutrients by diffusion depends on diffusion coefficient, concentration of nutrient in soil solution and buffering capacity of soil solid phase for the nutrient in the soil solution phase.

Diffusion coefficient describes the diffusivity of a homogeneous medium. For a given spacing of roots in the soil, it determines the fraction of nutrients in the soil that can reach the root during a specified period of plant growth. It is influenced by volumetric soil water percentage, impedance factor (tortuosity factor) which expressers the irregular path of diffusion in soil pores and the buffering capacity. Diffusion can be increased by increasing soil moisture content for reducing tortuosity. Buffering capacity usually decreases as nutrient levels in soil are raised. Reduction in buffering capacity is associated with a rise in rate of diffusion. The distance for diffusive nutrient through soil to the roots is usually in the range of 0.1 to 1.5 mm. Phosphorus has the lowest diffusion rate. Plant requirements for P and K are mostly met the process of diffusion.

Root Interception

As roots grow in the soil, they push the soil aside and the root surfaces come in direct contact with soil particles and plant nutrients. Nutrient interception by roots depends on the soil column occupied by roots, root morphology and concentration of nutrients in the root occupied soil volume. On an average, soil volume occupied by roots of important food crops is around 0.7 to 0.9 per cent. The only nutrient which might be supplied completely by interception is calcium, although the process may provide a significant part of the requirement for Mg, Zn and Mn.

Nutrient Absorption by Plants

Nutrient ion uptake by intact plants is a catenary process. After reaching the plant roots, ions have to enter the root cells, transported from cell to cell towards xylem, be exerted into xylem vessels and finally to the growing organs. Ion movements in plant cells are active and passive. In **Passive process,** ions move from a higher to lower concentration or down a chemical gradient of potential energy. In the case of **active process,** ions move against a concentration gradient and the movement depends on electrochemical potential gradient. In the active process, cations are attracted to a negative electro-potential, whereas anions to a positive electropotential.

Carrier Theory

The term **carrier** is usually used to refer an agent responsible for transporting ions from one side of the membrane to the other. Carriers have similar properties to those of enzymes but unlike enzymes, carriers have not been isolated and characterized. According to carrier theory, a carrier meets the particular ion for which it has affinity, forms a carrier ion complex and moves across the membrane. The enzyme phosphatase, located at the inner membrane boundary, splits off the phosphate group from the carrier complex and the ion is released. The energy required in this process of transport is due to involvement of adenosine triphosphate (ATP). The ATP is regenerated from ADP plus inorganic phosphate (Pi) via respiration (oxidative phosphorylation). The entire process may be described as indicated below (Mengel and Kirkby, 1978).

$$\text{Carrier} + \text{ATP} \xrightarrow{\text{Kinase}} \text{Carroer (P)} + \text{ADP}$$

$$\text{Carrier} + \text{ATP} \quad \text{Carrier (P)} - \text{ion}$$

$$\text{Carrier (P)} - \text{ion} \xrightarrow{\text{Phosphatase}} \text{ion} + \text{ADP} + \text{Pi}$$

$$\text{Net: Ion} + \text{ATP} \xrightarrow{\text{Transport}} \text{ion} + \text{ADP} + \text{Pi}$$

SOURCES OF PLANT NUTRIENTS

Plant nutrients are essentially supplied through manures and fertilizers. Manures are organic in nature and applied in large quantities. They are also called as organics or organic manures. They are of animal or plant origin and contain more than one nutrient element. Nutrient content in the organic manures is low. Fertilizers are inorganic or synthetic and the nutrient content is higher than in manures. Fertilizers are available for a particular nutrient or combination of nutrients.

Manures, Fertilizers and Biofertilizers

Manures

The term **'Manure'** was used originally for denoting materials like cattle manure and other bulky natural substances that were applied to land, with the object of increasing the production of crops. Later chemical substances like $(NH_4)_2\ SO_4$ and super phosphate containing plant nutrients in abundance were used for increasing crop production and were also called manures. So in a general way, manure is a substance containing plant nutrients. Later, the organics were specifically called 'manures' and inorganics were called fertilizers. Therefore, manures are defined as the plant and animal wastes which are used as sources of plant nutrients.

Advantages of Manuring

1. Manures supply plant nutrients including micro-nutrients.
2. They improve soil physical properties like structure, water holding capacity etc.
3. Increase nutrient availability.
4. CO_2 released during decomposition acts as a CO_2^{-} fertilizer.
5. Provide food for soil micro organisms.
6. Plant parasitic nematodes and fungi are controlled to some extent by altering the balance of micro organisms in soil.
7. Provide buffering action in soil reaction.
8. Prevent loss of nutrients by leaching or erosion.

On the basis of concentration of nutrients, **Manures can be grouped into two:**

1. Bulky organic manures and

2. Concentrated organic manures.

Bulky Organic Manures

Contain small percentage of nutrients and they are applied in large quantities. Farm Yard Manure, Compost and Green Manure are the most important and widely used bulky organic manures.

Farm Yard Manure (FYM)

Decomposed mixture of dung and urine of farm animals along with litter and left over material from roughages or fodder fed to the cattle. FYM contains 0.5 % N, 0.2 % P_2O_5 and 0.5 % K_2O. Urine contains 1 % N and 1.35 % K_2O. N present in urine is mostly in the form of urea which is subjected to volatilization losses. Size of trench: 6-7.5 m long, 1.5 to 2.0 m wide and 1 m deep, heaped up to the height of 45 -60 cm above the ground level, dome shaped and plastered with cow dung earth slurry. The manure becomes ready for use in about 4=5 months after plastering. Chemical preservatives are used to reduce losses and enrich FYM *e.g.* gypsum, Kainite, and superphosphate. Gypsum absorbs urine and prevents volatilization loss of urea and also adds Ca and S. Superphosphate also acts similarly and increases P content. Bacteria and actinomycetes play active role in decomposition. Generally, 60-70 % moisture in the initial stage and 30-40 % moisture in decomposed manure (ready to use) as well as 50-60 °C temperature under the heap are favourable for the activities of these micro-organisms. Generally, 10-20 tonnes well decomposed FYM per ha is applied but > 20 tonne FYM/ha in case of fodder grasses and vegetables, **at least 15 days** before the sowing to avoid immobilization of N. Entire amount of nutrient is not available immediately. About 30 % N, 60-70 % P_2O_5 and 70 % K_2O are available to the first crop.

FYM is the most commonly used organic manure in India. One tonne of cattle dung can supply only 2.95 kg N, 1.59 kg P_2O_5 and 2.95 kg of K_2O.

- Hot and Cold Manure: The manure obtained from the excreta of horses and sheep is called hot manure in temperate countries where as Pig and cattle manure is called cold manure due to comparatively less vigorous decomposition and less rise in temperature in manure.
- Short and Long Manure: Decomposed manure that has lost the structure of the original materials is called short manure while the fresh manure having pieces of straw and other materials is called long manure.
- Fire Fanging: There is profuse fungal growth on the surface of moist manure, giving it an ashy grey appearance. It is referred to as fire fanging.

Compost

Composting is the process of reducing vegetable and animal refuse (rural or urban) except dung to a quickly utilizable condition for improving and maintaining soil fertility and decomposed material is called **compost**. Compost is like well decomposed cattle manure in general appearance, more powdery and lighter in colour. Ordinary compost can be enriched with N and P through Azotbacter and superphosphate respectively. When superphosphate is used during compost making it is called **super compost** and compost prepared by using N-fixing bacteria is called **AZBO-Compost**. AZO-Compost is the cheapest source of N among all organic manures (N – 1.5 %).

Superphosphate or rock-phosphate at the rate of 10-15 kg/tonne of raw material is applied at the initial stage of filling the compost pit. Size of pit is about the same of FYM.

The compost made from farm waste like sugarcane trash, rice straw, weeds and other plants and other waste is called **Farm Compost** (0.5 % N,),15 % P_2O_5 and 0,5 % K_2O). Farm compost prepared by Bangalore methodcontains 0.80-1.24 % N, 0.40-0.59 % P_2O_5 and 2.0-3.3 % K_2O.

The compost made from town refuses like night soil, street sweepings and dustbin refuse is called **Town compost** (1.4 % N, 1 % P_2O_5 and 4 % K_2O).

Night Soil (or Poudrette)

Night soil is human excreta, both solid and liquid. It contains 5.5 % N, 4.0 % P_2O_5 and 2.0 % K_2O). The dehydration of night soil, as such or after admixture with absorbing materials *e.g.* soil, ash, charcoal and saw dust produces a poudrette that can be used easily as manure proudrette contains 1.32 % N, 2.8 % P_2O_5 and 4.1 % K_2O.

Sewage and Sludge

The solid portion in the sewage is called **sludge** and liquid portion is **sewage water.** Both the components are given a preliminary fermentation and oxidation treatments to reduce bacterial contamination and offensive smell, otherwise, soil quickly becomes **'sewage sick'** owing to the mechanical clogging by colloidal matter in the sewage and the development of anaerobic organisms which not only reduce the nitrates already present in the soil but also produce alkalinity. These defects can be removed by thoroughly aerating the sewage in the settling tank by blowing air through it. The sludge that settles at the bottom in this process is called **'activated sludge' (3-6 % N, 2 % P_2O_5 and 1 % K_2O).**

Sheep and Goat Manure

It contains 3 % N, 1 % P_2O_5 and 2 % K_2O. It is applied to the field in two ways:

1. Sweeping of sheep and goad sheds are placed in pits for decomposition and it is applied later to the field.
2. Sheep penning wherein sheep and goats are allowed to stay overnight in the field and urine and feacal matter is added to soil.

Poultry Manure

It contains 3.03 % N, 2.63 % P_2O_5 and 1.4 % K_2O. Litter is the straw, peat, sawdust; dry leaves etc. are used as bedding material for farm animals and birds. It absorbs urine and feces voided by animals and birds.

Green Manuring

Green undecomposed plant material used as manure is called **green manure.** It is obtained in two ways:

1. By growing green manure crops in the field and incorporating it in its green stage in the same field. It is called green manuring.
2. **Green Leaf Manuring** is the application of green leaves and twigs of trees, shrubs and herbs collected from elsewhere especially wasteland fields, bunds and forests to the field. Forst tree leaves are the main sources for green leaf manure *e.g.* neem, mahua, Glyricidia, Karanji (pongamia), sesbania, subabul, indigo etc.

Lentil is used in Kashmir for green manuring in rice. Sunnhemp is most outstanding green manure, well suited to almost all parts of the country. Dhaincha does well in alkaline and waterlogged soils. Berseem for orchards and irrigated crops of cotton and sugarcane as inter cropping.

Plants at flowering stage contain the greatest bulk of succulent organic mattere with low C:N ratio. The incorporation of the green manure crop at this stage allows a quick liberation of N in available form. The fertilizer value of the legume crop can be increased by manuring it with superphosphate (at the rate of 100 kg/ha). It increases of green manure crops and thus converting an inorganic fertilizer into organic manures. It has residual effects also.

Green manuring gives 60-80 kg N/ha on an average. Green manuring crops having **stem nodulation are:**

1. **Sesbania rostrata:** It is fastest N-fixing plant at the rate of 100-285 kg N/ha in 45-55 days.
2. *Aeschynomene afraspera*.

Green Manuring is a practice of ploughing or turning into the soil undecomposed green plant material for improving the physical conditions of the soil or for adding nutrients. Any crop or plant (generally leguminous) grown and ploughed *in situ* is called a **'green manure'. *e.g.*** dhaincha, sesbania, sunnhemp, wild indigo and pillipesara. A legumous crop producing 10-25 tonnes of green matter per ha will add about 60 to 90 kg N.

Green leaf manure refers to turning under of green leaves and tender green twigs collected from shrubs and trees grown on the bunds, wastelands and nearby forest areas. Common shrubs are trees useful for this purpose are karanj, glyricidia, neem, subabul, etc. The nutrient contains of green manure and green leaf manure crops are given in Table 27.3.

TABLE 27.3: Nutrient content of important green manure and green leaf manure crops

Crop	Nutrient content (% on dry weight basis)		
	N	P	K
Green Manure			
Sesbania aculeate	3.3	0.7	1.3
Crotalaria juncea	2.6	0.6	2.0
Sesbania speciosa	2.7	0.5	2.2
Tephrosia purpuria	2.4	0.3	0.8
Phaseolus trilobus	2.1	0.5	-
Green Leaf Manure			
Pongamia glabra	3.2	0.3	1.3
Glyricidia maculeata	2.9	0.5	2.8
Azadirachta indica	2.8	0.3	0.4
Catatropis gigantean	2.1	0.7	3.6
Weeds			
Parthenium hystorophorus	2.7	0.7	1.5
Eichhornia crassipes	3.0	0.9	0.2
Trianthema portulacastrum	2.6	0.4	1.3
Ipomoea	2.0	0.3	0.4
Calotrophis gigantean	2.1	0.5	0.3
Cassia fistula	1.6	0.2	1.2

Promising Green Manure Crops

Dhaincha (*Sesbania aculeate*)

It is suitable for loamy and clayey soils. Dhaincha is an ideal green manure crop for rice based cropping system. It is fairly resistant to drought as well as stagnation of water. The green manure crop grows well even in alkaline soils and corrects alkalinity if grown repeatedly for 4-5 years. The roots have plenty of nodules. It yields about 10-15 tonnes of green manure per ha and requires a seed rate of 30-40 kg/ha.

Sesbania speciosa

It is a valuable green manure for waste lands and can be grown in a wide range of soils. Seed production is prolific, however, pods are frequently attacked by insects. The green manure can be raised on the field borders along the bunds. Sesbania seedling (21 days) can be planted in a single line 5-10 cm apart in the boarders of the fields close to the bunds. In about 90 days it produces about 2-4 tonnes of green manure per ha. It does not affect the rice yield by shading or root effect. If second rice crop is planted immediately after the first crop, the manure can be incorporated into the field. About 300-400 g of seeds are sufficient to raise nursery and plant the seedlings along the bunds of one hectare.

Sesbania rostrata

One of the important features of this green manure is that in addition to the root nodules, it produces in the stem. The stem nodulation is an adaptation for water logged situation since flooding limits growth of green manures and may reduce root nodulation. Under normal condition, both root and stem nodules are effective in N fixation. It can produce about 40-50 tonnes of fresh biomass from one ha in about 50-60 days. It has higher N amount of 3.56 % on dry weight basis. Biomass production is higher during summer (April – June) than in winter (Dec-Jan) season. This green manure can also be produced by raising seedlings (30 days old) and planted in the rice field along the bunds or as intercrop with rice.

Sunnhemp (*Crotalaria juncea*)

It is a quick growing green manure crop and gets ready for incorporation in about 45 days after sowing. It does not withstand heavy irrigation leading to flooding. The crop is at times subject to complete damage by leaf eating caterpillars. The crop can produce about 8-12 tonnes of green biomass per ha. The seed requirement is 30 kg/ha.

Wild Indigo (*Tephrosia purpurea*)

This is a slow growing green manure crop and cattle do not prefer to graze it. The green manure is suitable for light textured soils, particularly in single crop wetlands. It established itself as a self sown crop and the seeds remain viable till the harvest of rice. On an average about 3-4 tonnes of green manure is obtained in one hectare. The seed rate is 30 kg/ha. The seeds have a waxy impermeable seed coat and hence scarification is required to induce germination. Soaking seeds in boiling water for 2-3 minutes is also equally effective in promoting germination.

Indigo (*Indigofera tinctoria*)

It resembles wild indigo and is a long duration crop with more leafy growth. It comes up well in clayey soils with one or two irrigations.

Pillipesara (*Pheseolus trilobus*)

This is a duel purpose crop yielding good fodder for the cattle and green manure. Pillipesara comes up well in hot season with sufficient soil moisture. Loamy or clayey soils are best suited. After taking one or two cuttings for fodder or light grazing by animals, the crop can be incorporated into the soil. About 5-8 tonnes of manure can be obtained from one hectare.

Green Leaf Manure Crops

A number of leguminous and non-leguminous plants are grown on bunds or wastelands with the prime objective of utilizing their foliage as green manure.

Glyricidia (*Glyricidia maculeata*): This is a shrubby plant that comes up well in moist situations. Under waste lands, farm road sides, field bunds, etc. The crop can be established by stem cuttings or seedlings planted in the field borders. It can be pruned for its tender loppings and compound leaves for green leaf manuring at the time of puddling rice. On an average, a well established plant yields 12-15 kg green matter. About 400 plants on the peripheral bunds yield 5-6 tonnes green manure/ha.

Karanj (*Pongamia glabra*): It is a leguminous tree grown in wastelands. On an average, a tree can yield 100-120 kg of green matter. The leaves contain about 3.7 % N (on dry weight basis).

Neem (*Azadirachta indica*): It is an indigenous tree with wide range of adaptability. Almost all parts of the tree have been used for several millennia for preparation of medicine and its leaves and leaf-extracts to repeal insect pests of crops and stored grains. The leaves are pruned at the time of planting rice and incorporated as green leaf manure.

Calatropis (*Calotropis gigantean*): On road sides and fallow lands, the plant grows wild under different soil and climate conditions. Leaves are more succulent and a plant can produce about 4-5 kg of green matter.

Management of Green Manuring

Time of Sowing and Seed Rate: March is the best suited period for raising green manure.

Seed Treatment: Germination is a problem for the green manure seeds like *Tephrosia purpurea* of their hard seed coat. Scarification of the seeds with sand and pre-soaking in hot water (50 °C) for 5-minutes are found to improve seed germination up to 98%. Acid seed treatment is also found effective.

Stem Cutting: *Glyricidia* is one of the green leaf manure tree crops propagated through seedlings and stem cuttings. A large percentage of the cuttings fail to root readily. The cut ends have been kept immersed in 0.005 % IAA and 0.1 % seradix A for one day and planted in small beds and irrigated for six days. The hormonal treatment aided root formation to a remarkable degree (90%). The green manure crops can be established using cutting of soft wood (from freshly formed branches) and hard wood (leaving top and bottom 30 cm in the stem and older branches). Soft wood cutting of S. *aculeate* and hard wood cuttings of S. *rostrata* are superior. The establishment of the cuttings can be improved by dipping in fresh cowdung (10 % suspension).

Mixed Cropping: It has been found feasible to cultivate green manure mixed with rice fallow pulses. Green manure crops like indigo and pulses (black gram) can be mixed with the ratio of 1:3. Pillipesara can also be sown in the same proportion. By adopting such practice in rice fallow pulses, no yield reduction has been noticed. Growth of green manure crop is also not spoiled. By the time pulse is harvested, the green manure put forth growth only to a few cm. After the harvest of pulses, the green manure put on growth.

Intercropping: Field investigations at Coimbatore, reveal the possibility of intercropping green manure in transplanted rice. Seedlings of S. *rostrata* were planted at an interval of 1.5 m. On 309 days after planting, the green manure is cut at a height of 1.5 cm from the ground level and incorporated as manure. The ratoon is allowed and the refresh incorporated while planting second crop of rice.

Border Planting: Border planting of green manure, particularly *Sesbania speciosa* in the rice field bunds was advocated even during 1950s in Thanjavur, Tamil Nadu. The seedlings of the green manure were planted at 60-90 cm apart on the bunds at the time of planting *Kuruvi* crop during June. About 70 % of the crop was used as green manure for the *thaladi* crop planted in October. About 1.5-2.0 tonnes of green leaf was obtained from the bund crops and incorporated. The remaining left standing for producing seeds. It was also possible to produce in one ha, the seeds sufficient for raising green manure for 2.5 ha.

Phosphorus Response: Incorporation of green manure with single superphosphate or rock phosphate facilitates higher response to added phosphorus. Recommended dose of phosphorus for rice can be very well applied to green manure crop and incorporated for greater efficiency.

Vermicompost

Compost that is prepared with the help of earthworms is called **Vermicompost.** The earthworms may be of local species or more vigorous exotic ones. Earthworms consume large quantities of organic matter and excrete soil as casts. The weight of the material passing through the body each day is almost equal to the weight of the earthworm. The casts of the earthworm have several enzymes and are rich in plant nutrients, beneficial bacteria and mycorrhizhae.

Vermicompost is made in small pits of suitable size, say 2 m x 1 m x 0.5 m in a shady area in the farm. The different waste materials available in the farm like cowdung, plant residues, weeds, etc., are placed in the pit layer wise and soil is added for each layer. Earthworms are released for each layer and water is applied. Compost is ready within 2-3 months.

The advantages of vermicomposts are:

1. Vermicompost is a rich mixture of major and minor plant nutrients. On an average, vermicomost contains 3 % nitrogen, 1 % phosphorus and 1.5 % potash.
2. Vermicompost is an excellent base for the establishment of beneficial free living and symbiotic microbes. Application of vermicompost increases the total microbial population of nitrogen fixing bacteria and actinomycetes. The symbiotic association of microrrhiza on plant root system increases. Earthworm casts harbour large number **vesicular arbuscular microrrhizal** (VAM) propagules. The VAM propagules survive up to 11 months on earthworm casts. The increased microbial activity improves the availability of soil phosphorus and nitrogen.
3. A number of plant growth promoters are observed in earthworm casts.
4. Quick response can be obtained with vermicompost comparewd to ordinary compost or farm yard manure.
5. Presence of earthworms helps in aerating the soil.

Concentrated Organic Manures

Concentrated organic manures have higher nutrients than bulky organic manures.

1. **Bird guano:** The excreta and dead remains of the birds. 7-8 % N, 11-14 % P_2O_5 and 2-3 % K_2O.
2. **Fish guano:** The refuse left over after the extraction of oil from the fish in factories, is dried in paved yards and used as manure 7 % N and 8 % P_2O_5.
3. **Fish Manure:** 4-10 % N, 3-9 % P_2O_5 and 0.3-1.5 % K_2O.
4. **Raw Bone meal:** 3-4 % N, 20-25 % P_2O_5.
5. **Steamed Bone meal:** 1-2 % N, 25-30 % P_2O_5.
6. **Blood Meal:** 13-20 % N, rich in iron and its application gives a deep rich colour to the foliage, much appreciated by ornamental gardeners.
7. **Meat meal/Meat guano/tankage:** 8-9 % N and 7 % P_2O_5.
8. **Calcined bone: 37 % P_2O_5.**
9. **Oil cakes:** Safflower cake – 7.9%N, Sunflower cake – 7.8% N
10. **Groundnut cake** – 7.2% N, Cotton cake – 6.5% N.

Oil Cakes

After oil is extracted from oil seeds, the remaining solid portion is dried as cake which can be used as a manure. The oil cakes are of *two types:*

1. Edible oil cakes which can be safely fed to livestock; e.g.: groundnut cake, coconut cake, etc., and
2. Non-edeble oil cakes which are not fit for feeding livestock; e.g., Castor cake, Neem cake. Mahua cake, etc.

Both edible and non-edible oil cakes can be used as manures. However, edible oil cakes are fed to cattle and non-edible oil cakes are used as manures especially for horticultural crops. Nutrients present in oil cakes, after mineralization, are made available to crops 7 to 10 days after application. Oil cakes need to be well powdered before application for even distribution and quicker decomposition. The average nutrient content of different oil cakes are presented in Table 27.4

Amendments

Soil amendments are substances that influence plant growth favourably by increasing nutrient availability either by altering soil reaction or by changing non-available to available form and by improving the physical condition of the soil. Liming materials such as limestone, burnt lime, slaked lime, etc., are used to reduce acidity. Lime is broadcasted on soil surface and ploughed in. If large quantities of lime are to be applied for strong acidic soils, half is applied at first ploughing and the remaining half is applied at second or third ploughing. Judicious liming increases the crop yields and also improves fertilizer efficiency. Amendments like gypsum, sulphur, iron sulphate, etc., are used for reclaiming alkaline soils. These materials reduce pH resulting in higher nutrient availability and better soil physical conditions. Materials like sand, tank silt, pressmud, crop residues are added to the soil either for improving soil physical condition or increasing nutrient availability or both.

Biofertilizers

The term **'biofertilizers'** includes selective microorganisms like bacteria, fungi, and algae which are capable of fixing atmospheric N or convert insoluble phosphate in the soil into forms available to plants. Biofertilizers are cost effective, eco-friendly and renewable sources of plant nutrients to supplement chemical fertilizers. Biofertilizers also play a vital role in maintaining long term soil fertility and sustainability.

Groups of Biofertilizers

Nitrogen fixing organisms are either free lining or having symbiotic association with plants. They directly or indirectly contribute N nutrition to crop plants.

Free Living: This group includes numerous species of blue green algae and certain bacteria such as *Azotobactor.*

Symbiotic Association: The bacteria, *Rhizobium*, live symbiotically on the roots of the legumes. Azolla, a water fern, fixes atmospheric N in symbiotic association with blue green algae (BGA), *Anabaena azollae* in rice ecosystem.

Associative Symbiotic

One of the species Azospirillum, A. lipoferum has associative symbiosis with higher plant systems.

Based on the type of microorganisms, the biofertilizers can also be classified as follows:

1. **Bacterial biofertilizers**: *e.g.* Rhizobium, Azospirilium, Azotobacter, Phosphobacteria.
2. **Fungal biofertilizers:** *e.g.* Mycorrhiza
3. **Algal biofertilizers**: *e.g.* Blue Green Algae (BGA) and Azolla.
4. **Actinimycetes biofertilizers:** *e.g.* Frankia.

Biofertilizers are mostly cultured and multiplied in the laboratory. However, blue green algae and azolla can be mass-multiplied in the field.

Characteristics of Features of Biofertilizers

Rhizobium

Rhizobium is relatively more effective and widely used. Rhizobium, in association with legumes, fixes atmospheric N. The legumes and their symbiotic association with the rhizobium bacterium result in the formation of root nodules that fix atmospheric N. Successful nodulation of neguminous crop by rhizobium largely depends on the availability of a compatible strain for a particular legume. Rhizobium population in the soil depends on the presence of legume crops in the field.

In the absence of legumes, the population of rhizobium in the soil diminishes. Most strains of rhizobia native to the soil are effective in N fixation. Some of them may be ineffective and produce ineffective nodules. These nodules are smaller in size, soft and whitish in colour. Their N fixing capacity is much low. The effective strains produce large, pink and firm nodules. Their N fixing capacity is far greater. Artificial seed inoculation is often needed to restore population of effective strains of rhizobium near the root zone to ensure high yields. Each legume requires a specific species of rhizobium to form effective nodules.

Azospirillum

Azospirillum is known to have a close associative symbiosis with the higher plant system. These bacteria have association with cereals like sorghum, maize, pearl millet, finger millet, foxtail millet and other minor millets and also fodder grasses. Bacteria colonizing the roots not only remain on the root surface but also a sizable proportion of them penetrates into the root tissues and live in hormoney with the plant. They do not, however, produce any visible nodule or out-growth on the root tissue.

TABLE 27.4: Cross inoculation group of Rhizobium

Rhizobium species	Cross inoculation group	Legume types
R. leguminosarum	Peas	Pisum, Vicia
R. phaseoli	Beans	Phaseolus
R. trifolii	Clover	Trifolium
R. meliloti	Alfalfa	Melilotus, Medicago, Trigonella
R. lupine	Lupini	Lupinus, Orinthopus
R. japonicum	Soybean	Glycine
R. sp.	Cowpea	Vigna, Arachis

Azotobacter

It is a common soil bacterium and the species A. *chrococcum* is present widely in Indian soils. Soil organic matter is the important factor that decides the growth of this bacteria. Poor organic matter content in the Indian soil is a limiting factor that not only affects the proliferation of Azotobacter in soil but also limits its N fixing capacity.

Blue Green Algae (BGA)

Blue green algae are referred to as rice organisms because of their abundance in the rice fields. Many species belonging to the genera, *Tolypothrix*, *Nostoc*, *Schizothrix*, *Calothrix*, *Anoboenopsis* and *Plectonema* are abundant in tropical conditions. Most of the N fixing BGA are filamenters, consisting of chain of vegetative cells including specialized cells called heterocyst which function as a micronodule for synthesis and N fixing machinery. BGA, *Anabaena azollae,* also occurs in symbiotic relation with the aquatic fern, Azolla.

In addition to N fixation, BGA also synthesis and liberate growth promoting substances such as auxin and amino compounds which stimulate the growth of rice.

Multiplication of BGA

Algae may be multiplied in troughs or cement tanks for small quantity and in the field for large scale production.

Trough Method

Shallow trays of 2 x 0.2 m of galvanized iron sheet or permanent tanks are constructed. They are kept in the open air and exposed to the sun. The size may vary according to the quantity of algae material to be produced. About 10 kg of soil is taken and mixed well with 200 g of superphosphate. Water is stagnated to a depth of 5-15 cm in the trays depending on the local conditions to meet the evaporative demand. The soil should be neutral in pH and if it is acidic, lime should be added. To prevent predators, carbofuram 3 G is to be applied at the rate of 25 g/tray. About 250 g of algal culture is sprinkled on the surface of the standing water.

Growth of algae is rapid during hot summer moths (April-June). In about 7-10 days, a thick mat of algae is formed. If the daily evaporation loss is high, water is added intermittently. When the algal growth becomes sufficiently thick, watering is to be stopped. Water is allowed to evaporate completely. The dry algae crack into flakes which are collected and stored for further use. About 1.5 to 2.0 kg of algae material can be collected from each trough of 0.4 m^2 area.

The troughs are again filled with water and a small amount of dry algal flakes are added as inoculum. Multiplication can be continued in this manner. Once the soil in the trough is exhausted (usually after three or four harvests) it can be replaced with fresh soil mixed with super phosphate and continued as above.

Mass Multiplication of BGA in the Field

Field production of algae is adopted for multiplication of the material on a commercial scale. In the wetland, about 40 m^2 areas are demarked and puddle to facilitate water stagnation. The area is encircled with strong earthen bunds of about 15 cm height. Super phosphate is applied at 2 kg for 40 m^2 area. About 5 kg of composite algal culture is required for 40 m^2 area. To control predators like daphnids, snails, and mosquitoes, carbofuran 3 G is applied at 200 g. In about 7-10 days period, a thick mat of algae develops in clay soils. In loamy soils, about 3-4 weeks are required for full growth of algae. Once the algae have attained full growth with complete coverage, algal flakes are collected and stored in sacks for further use.

During summer months, the average yield of algae ranges from 20 to 30 kg/ 40 m^2. Continuous harvest of algae may be made from the same area by reflooding the plot and applying super phosphate and pesticides. Addition of further inoculum is not necessary for subsequent production.

Azolla

Azolla is a free floating aquatic fern. Azolla in symbiotic association with blue green algae (*Anabaena azolla*) assimilates atmospheric N and enrich rice field. Azolla is widely distributed in the rice growing tracts of tropics and temperate zones and grows on the irrigated rice fields. There are six species of Azolla: A. *caroliana*, A. *nilotica*. A. *filiculoides*, A. *mexicana*, A. *microphylla* and A. *pinnata*. The common species occurring in India is A. *pinnata*. A heat tolerant species, A. *microphylla* is introduced to India from the Latin America. A. *filiculoides* is a temperate species got from Europe. New crosses between A. *microphylla* and A. *filiculoides* are also available. Several of hybrids have high nutritive value (4.0 – 4.5 % N on dry weight basis) and grow fast. The hybrids can also withstand both heat and cold.

Azolla mutants with a high degree of tolerance to salinity have been developed at International Energy Agencies, Vienna, Scientists in China and U>S> are now working on introducing a *Bacillus thuringiensis* (Bt) toxin in Azolla to fight against insect pests.

The water fern is mainly propagated by vegetative means. It may yield on an average about 1-5 kg per square meter in a week. Under normal conditions, the fern doubles itself in three years. Multiplying Azolla through sporocarp technique have also been developed in many research institutions in the world.

Multiplication of Azolla

Azolla can be multiplied in small cement cisterns or galvanized iron trays of convenient size. Azolla can also be multiplied in the field nursery.

Cement Cisterns

Azolla can be multiplied in cement cisterns of 2 × 1 × 0.3 m. A layer of 7.5-10.0 cm soil is provided at the bottom of the cisterns. About 5.0-7.5 cm depth of water is maintained in the cisterns. Super phosphate at the rate of 8-12 g and carbofuran at 2.0 g are applied. About 200 g of fresh azolla is inoculated to each cistern. In about 12-15 days, about 2.0-2.2 kg of azolla can be harvested. After each harvest, super phosphate is applied to facilitate fast multiplication. Every time, 200 g of azolla is inoculated for further multiplication.

Mass Multiolication in Nursery Plots

The field selected for azolla nursery should be nearer to shade, thoroughly ploughed and inoculated. The field is divided into plots of 40 m^2 area) 20 × 2 m) by providing suitable bunds and irrigation channels. Water is maintained to a depth of 10 cm. About 10 kg of fresh cattle dung is mixed with 20 litres of water and sprinkled uniformly throughout the plots. Fresh azolla is inoculated at 8 kg /plot and super phosphate (100 g) is applied twice at an interval of 4 days. Carbofuran is added at 100 g per plot, 7 gays after inoculation to control insect pests. About 50=75 kg of fresh azolla can be harvested 15 days after inoculation. Azolla can be again inoculated with the addition of cattle dung, super phosphate and carbofuran granules.

P - Solubilising Microorganisms

A group of heterotropic microorganisms are known to have the ability to solubilize P from insoluble sources. The organisms include bacteria: *Bacillus megatherium*, B. *circulans*, and B. *subtilis*, and fungi: *Aspergillus awamori*, *Penicillium digitum* and *Trichoderma spp*. These organisms secrete various organic and inorganic acids. They act on insoluble phosphates and convert them into soluble phosphates in the rhizosphere. Among the microorganism, bacteria are found to be more efficient in the secretion of organic acids. They are generally called **phosphobacteria.** Adding of organic manures helps in increasing the solubilising power of the microorganisms.

Vesicular Arbuscular Mycorrhiza (VAM)

Many of the graminaceous and leguminous plants harbour VAM. These plants contain special structures which help in the transfer of nutrients from soil into the root system. The VAM fungi are intercellular and obligate endosymbionts. Most of the economically important crops, particularly perennial such as apple, rubber, coffee, tea and oil palm are infected by VAM. The fungi are mass produced only in the presence of living roots. The production of inoculum needs a host plant and a growth medium, usually soil, which provides congenial conditions for growth and reproduction of fungi.

Frankia

Some of the actinorrhizal plants like Casuarina and Albus are found to have nodules infected by Frankia. It has been reported that N fixed by Frankia in Alnus ranges from 50 to 150 kg/ha/year. The potentiality of these microorganisms has been demonstrated in increasing the biomass yield of Casuarina and Alnus.

Biofertilizers and their Usages in Field Crops

Niofertilizers are used in different ways in the field crop production:

1. Seed treatment or seed inoculation.
2. Seedling root dip.
3. Soil application.

Seed treatment or seed inoculation: Each packet (200 g) of inoculant is mixed with 200 ml of rice gruel or jaggery solution. The seeds required for one hectare are mixed in the slurry so as to have uniform coating of the inoculant over the seeds and then shade dried for 30 minutes. The treated seeds should be used within 24 hours. One packet of inoculant is sufficient to treat 10 kg seeds. Rhizobium, Azospirillum, Azotobacter and Phosphobacteria are applied as seed treatment.

Seedling root dip: This method is used for transplanted crops. Five packets (1.0 kg) of the inoculant are required for one ha and mixed with 40 litres of water. The root portion of the seedlings is dipped in the solution for 5 to 10 minutes and then transplanted. Azospirillum is used for seedling root dip particularly for rice.

Soil application: Ten packets (2.0 kg) of the inoculant are required for one ha of main field. The inoculant is mixed with 20 kg of well dried, powdered FYM and broadcast in the field just before transplanting/sowing.

Use of VAM Biofertilizer

The inoculum should be applied 2-3 cm below the soil at the time of sowing. The seeds are sown or cuttings planted just above the VAM inoculum so that the roots may come in contact with the inoculum and cause infection. Bulk inoculum of 100 g is sufficient for one meter square area.

Seedlings raised in the polythene bags need 5-10 g of bulk inoculum for each bag. For large scale usage, 10 kg of the inoculum can be mixed with one tonne of potting mixture. This can be packed in polythene bags before sowing. At the time of planting of saplings, VAM inoculum is to be applied at the rate of 20 g/seedling in each spot. In the existing tree, inoculum of 200 g is required for each tree. The inoculum is to be applied near the root surface.

Use of Blue Green Algae (BGA)

Algal culture is applied as dried flakes at 10 kg/ha over the standing water in the rice fields. This is done two days after transplanting in loamy soils and six days after planting in clayey soils. Addition of excess quantity of algae is not harmful; instead it may accelerate the build up of population. The field is kept water logged for atleast a couple of days immediately after algal application. The biofertilizer is to be applied for 3-4 consecutive seasons in the same field. Further application may not be required once soil build up the inoculum is achieved.

Use of Azolla

Green manure: Azolla is applied at the rate of 0.6-1.0 kg/m^2 (6.25-10.0 t/ha) and incorporated before transplanting of rice.

Dual crop: Azolla is applied at the rate of 100 g/m^2 (1.25 t/ha), one to three days after transplanting of rice and allowed to multiply for 25-30 days. Azolla fronds can be incorporated into the soil at the time of first weeding.

Fertilizers

Liebig in Germany and Lawes in England developed independently the idea of treating phosphate with H_2SO_4 and producing a water soluble phosphate by about 1840. It is known as super phosphate or simply super. Lawes set up a factory at Deptford in 1843, for the production of super phosphate. Possibly this was the first factory to be set up for the production of artificial fertilizers on a commercial scale.

Fertilizers are the **organic** or **inorganic** materials of natural or synthetic origin which are added to the soil to supply certain elements essential to the plant growth of plants. The term **'Fertilizer'** is now commonly restricted to commercial products.

Amendments are the substance other than manures and fertilizers which are added to soils for the improvement of their conditions. Amendments are also termed as **ameliorants**, improves but the main objective of applying them is for correcting the soil condition.

Classification of Fertilizers

1. **Straight Fertilizers:** Such fertilizers have declarable content of only one major nutrient, e.g., urea, ammonium sulphate.
2. **Binary Fertilizer:** contains two major nutrients e.g., Potassium Nitrate.
3. **Ternary Fertilizer:** contains three major nutrients e.g., Ammonium potassium phosphate.
4. **Compound/Complex Fertilizer:** Such fertilizer has a declarable content of at least two of the major nutrients obtained **chemically** and generally granular in form, e.g., Nitro-phosphate, Amm-phosphate and Diammonium phosphate (DAP).
5. **Mixed Fertilizer:** Individual or straight fertilizer materials are blended together **physically** to permit application in the field in one operation. Such fertilizers supply two or three major nutrients in a definite proportion of grade, *e.g.* Ntrophosphate with potash 15:15:15 of NPK
6. **Complete Fertilizer:** having all the three primary major nutrients viz. N, P and K.
7. **Incomplete Fertilizer:** Containing any two primary nutrients.
8. **Low**-analysis Fertilizer: having less than 25 % of the primary nutrients e.g., SSP (16 % P_2O_5), sodium nitrate (16 % N).
9. **High-analysis Fertilizer**: having more than 25 % of the total primary nutrient content e.g., urea, anhydrous ammonia (82.2 % N), Ammonium phosphate (20 % N + 20 % P_2O_5) and DAP (18 %N + 46 %P_2O_5).
10. **Fertilizer grade:** refers to the guaranteed analysis of its plant nutrients. It is the minimum guarantee of the plant nutrient contents in the term of N, available P_2O_5 and K_2O *e.g.* 6:24:24
11. **Fertilizer Ratio:** refers to the relative percentage of N, P_2O_5 and K_2O i.e., 1:4:4 if fertilizer grade is 6:24:24.

Materials Used in Manufacturing of Fertilizers (Mixed)

1. **Supplier of plant nutrient:** Straight fertilizers are used for this puposes.
2. **Conditioners:** To check absorbing moisture and making one, conditioners like straw, groundnut husk, peat soil etc. are used they just reduce caking and applied in drilling conditions and these conditioners are of low organic materials.
3. **Neutraliser of acidity or basicity:** Dolomitic limestone is used to reduce residual acidity. Most fertilizers leave residual acidity or basically.
4. **Filler materials or make weight materials:** Sand, soil, earth, coalash, charcoal such waste materials are added to make up the difference between the weight of the added fertilizers required to supply the plant nutrients and the desired quantity of the fertilizer mixture.

Precautions in Mixing Fertilizers

1. Hygroscopic fertilizers should not be mixed because they form cakes after mixing. Hygroscopic fertilizers are in descending order:
   a) Ammonium Nitrate: NH_4NO_3
   b) Urea: Co $(NH_2)_2$
   c) Ammonium sulphate: $(NH_4)_2$ SO_4
   d) Ammonium sulphate nitrate: $(NH_4)_2SO_4$ NH_4NO_3
   e) Calcium ammonium nitrate: (CAN)
   f) The most hygroscopic fertilizer is Ammonium nitrate and urea stands second. All these fertilizers have NH_4^+
2. Fertilizer containing NH_4^+ should not be mixed basically reactive materials like

lime, basic slag and rock phosphate because mixing results in the loss of N through escape of NH_4 gas.

3. All water soluble phosphate fertilizers like super phosphate should not be mixed with those fertilizers that contain free lime because it will convert the soluble phosphate into insoluble form.
4. Slightly acidic fertilizer containing chloride may damage the gunny bags and drilling equipments.

Acidity and Basicity of Fertilizer

Equivalent acidity: The amount of $CaCO_3$ required to neutralize the acid residue caused by application of acidic fertilizers in the soil e.g., 100 kg $(NH_4)_2$ SO_4 produces acidity which needs 110 kg Ca CO_3 to neutralize it. Therefore, equivalent acidity of $(NH_4)_2SO_4$ is 110. Anhydrus NH_3 has more equivalent acidity than other fertilizers.

Equivalent basicity: Residual basicity caused by application of basic fertilizers, expressed in terms of $CaCO_3$ equivalent of basic residue left by a fertilizer material (in kg/100 of fertilizer salt).

Equivalent acidity and equivalent basicity of fertilizer

Sl No.	Fertilizer	Equivalent acidity	Equivalent basicity
1.	CaCN		63
2	$NaNO_3$		29
3	KNO_3		29
4.	$Ca(NO_3)_2$		21
5.	Anhydrous Ammonia	148	
6.	Nitrare of soda potash		26
7.	NH_4Cl	128	
8.	$(NH_4)_2SO_4$	110	
9.	$(NH_4)_2$ SO_4. NH_4NO_3	93	
10	Urea	80-84	
11	DAP	77	
12	Ammonium nitrate	60	
13.	Mono-ammonium phosphate	55	

Groups of Fertilizers

Fertilizers are inorganic/synthetic substances containing one or more plant nutrients in easily soluble and quickly available form. Because of their availability in concentrated form they have the advantage of being small for bulk storage and transport and handling are easier. In addition, crops may be supplied with exact quantity of nutrient required for their growth and development in soils of varying fertility levels. Fertilizers are grouped according to the nutrients they contain. They are:

1. Nitrogenous Fertilizers.
2. Phosphatic Fertilizers.
3. Potassic Fertilizers.
4. Complex or Mixed Fertilizers.
5. Micronutrients.

Commercial Nitrogenous Fertilizers

Except rice and potato, the ammoniacal form is not directly absorbed by crops. Ammonia always produces acidity. Except NO_3, CaCN and CAN; all other nitrogenous fertilizers are acidic in nature, NO_3^- and CaCN are basic in reaction, whereas CAN is neutral.

Nitrogenous Fertilizers

Nitrogenous fertilizers may contain N in NH_4^+ or NO_3^- forms or both. Some of the fertilizers contain N in amide form. These fertilizers, when applied to the soil, are rapidly converted into ammoniacal form. Commonly used N fertilizers are grouped as follows:

1. **Ammoniacal fertilizers:** Ammonium sulphate, ammonium chloride, ammonium phosphate, and anhydrous ammonia.
2. **Nitrate fertilizers:** Calcium nitrate, sodium nitrate, and nitrophosphate.
3. Fertilizers containing ammonium and nitrate: Ammonium nitrate, calcium ammonium nitrate and ammonium sulphate nitrate.
4. **Amide:** Urea, Urea phosphate.
5. **Nitrogen solutions:** Anhydrous ammonia, aqueous ammonia, and solutions containing urea, nitrate, ammonia of their combinations.
6. Slowly available N fertilizers:
7. Chemicals having inherent capacity of slow release: Ureaform, IBDU, CDU, Oxamide, etc.
8. Coated urea fertilizers: Sulphur coated urea.
9. Modified urea form: Urea super granules, nitrification inhibitor (N serve, AM, thiourea) treated urea.

Inorganic Nitrogen Carriers

Fertilizer	Chemical form	Nutrient content (%)		
		N	P_2O_5	S
Nitrate fertilizers				
Sodium nitrate	$NaNO_3$	16	-	-
Calcium nitrate	$Ca(NO_3)_2$	15.5	-	-
Ammonium fertilizers				
Ammonium sulphate	$(NH_4)_2SO_4$	20	-	24
Ammonium phosphate	$NH_4\,H_2PO_4$	20 Or 16	20	-
Ammonium chloride	NH_4Cl	24-26	-	-
Anhydrous ammonia	NH_3	82	-	-
Ammonia solution	-	20-25	-	-
Nitrate and ammonium fertilizers				
Ammonium nitrate	NH_4NO_3	33-34	-	-
Calcium ammonium nitrate		20	-	-
Ammonium sulphate nitrate		26	-	-
Amide fertilizers				
Urea	$CO(NH_2)_2$	46	-	-
Calcium cyanamide	$CaCN_2$	21	-	-

Characteristics of N Fertilizers

1. **Nitrate Fertilizer:** highly mobile in soil, suitable for top dressing, highly soluble and subjected to leaching and denitrification in water logged soils, increase alkalinity, more suitable for dry soils.
2. **Ammoniacal Fertilizer:** Readily soluble in water, less leaching losses as NH_4^+ are absorbed on clay complex, suitable for water logging fields, acidic in nature, rice in early stage is capable of utilizing NH_4^+ form.
3. **Amide Fertilizers:** Organic fertilizers, first converted into ammoniacal and then to nitrate form. Urea is acidic but CaCN is basic. Highly soluble in water. Amide $-(NH_4)_2CO_3$ ---- NO_3^-.
4. **Nitrate-ammoniacal group:** NO_3^- nitrogen readily available to plants for rapid growtyh and NH_4^+ nitrogen at the later stage.

Properties of Important N Fertilizers

1. **Ammonium sulphate:** It is one of the most important and oldest N fertilizers. It is a white crystalline salt containing 20-21 % ammoniacal N and 24.2 % S. It is readily soluble in water. The main advantage of this fertilizer is low hygroscopicity and chemical stability. This fertilizer is particularly suitable for wet land crops like ricesince the ammonium ion is retained by the clay particles. Since the fertilizer is of acidify in nature, it is advantageous to apply it to soils having alkaline pH and for acid requiring crops. Ammonium sulphate is recommended for tobaccoand potato to improve the quality of the produce. Ammonium sulphate can be used as a part of the basal dressing and also for top dressing.
2. **Ammonium chloride:** It is a white crystalline compound containing 26 % ammoniacal N and is acid forming. Like ammonium sulphate it can be applied to rice crop. It is also suitable for other crops viz., barley, wheat, corn, sorghum, fibre crops and sugarcane. Since the fertilizer contains both N and Cl it can be recommended for crops that love chloride e.g., coconut, oil palm and kiwi fruit. It is generally not recommended for tobacco and potato as the added Cl may affect the burning quality of tobacco and keeping quality of potato.
3. **Ammonium nitrate:** It is a while crystalline salt, containing 33-35 % N with equal proportion of nitrate and ammoniacal N. This is a quick acting fertilizer. It is highly hygroscopic and hence adequate care is required to prevent caking and physical deterioration while storing and handling. It is prone to leaching and denitrification losses than other ammonia products.
4. **Calcium ammonium nitrate (CAN):** It is also called nitro-lime or nitrochalk (locally called Kisan Khad) and consists of a mixture of ammonium nitrate and powdered chalk made into granules coated with soap stone. Nutrient content is 25.0 % N. It is hygroscopic and hence adequate care is required while handing. As half of N is present in nitrate form, it is a quick acting fertilizer.
5. **Ammonium sulphate nitrate (ASN):** It is a double salt made from a combination of ammonium nitrate and ammonium sulphate containing 26 % N and 12 % S. Nitrogen is both in ammoniacal (19.5 %) and nitrate form (6.5 %). This fertilizer is recommended for basal application prior to sowing or as top dressing. However, it should not be applied along with the seeds.
6. **Urea:** Urea was the first organic compound to be synthesized from inorganic substances (ammonium cyanate) by the German Chemist, Wohler in 1828. It is white, crystalline in appearance containing 46 % N, all in amide form. In India, it is the most popular nitrogenous fertilizer. Urea is completely soluble in water. In the wetland condition it is subjected to loss through leaching and in dry condition by volatilization. It has been reported that a loss of 16-25 % could occur from the

surface application of prilled urea in wet land. Urea should, therefore, be worked into the soil under moderately moist conditions. Because of the incorporation, it is quickly converted into ammonia. It is retained by the soil colloids and does not get leached out or volatilized.

Urea is applied basally or as top dressing and is suitable for most of the crops and soil types. Application of urea in combination with the seed resulted in a poor germination of wheat and barley. Hence, it is advisable to apply the fertilizer as side-dressing. Biuret (NH_2-CO-NH-CO-NH_2) content urea is an important factor to be considered because of its phytotoxicity. When urea is applied as foliar nutrition the biuret content should not exceed 1 %, preferably lower than 0.5 %. Otherwise, scorching may occur on the leaves. It has been reported that injury to citrus occurred even when the biuret content was as low as 0.25 %. Citrus and other crops including pine apple are sensitive to biuret in urea when applied as foliar spray. Soil applied urea does not cause any deleterious effect to citrus even when the biuret content is 2.5 per cent.

7. **Anhydrous ammonia:** The highest N containing fertilizer is the anhydrous ammonia (82 % N). It is a colourless gas at normal pressure and temperature and easily liquidified under pressure. To avoid ammonia losses, it should be applied at some depth (10-20 cm) depending on soil conditions. It has been observed that losses are much less when ammonia is applied below 7.5 cm. The moisture content of the soil is also important. Losses are minimum when the soil is worked slightly below the field capacity. The fertilizer can also be applied through irrigation water. Application of ammonia in sprinkler irrigation may result in greater losses of ammonia. Considerable care is required while handling this material as it gives irrigation.
8. **Liquid nitrogenous fertilizers:** Principal nitrogenous solutions are produced from urea and NH_4NO_3 in water and they known as UAN solution. A mixture of ammonium nitrate and urea is much more soluble in water than either of them alone. The liquid nitrogenous fertilizers have certain advantages over the conventional solid fertilizers. Solution form can be easily handled and applied more uniformly and accurately. They are suitable for various types of irrigation systems. Many pesticides are compatible with N solutions and can be applied together. Unit cost of N is usually less and handling is much easier. But they require special application equipment and storage containers since they are corrosive. The equipment made of aluminium or certain stainless metals is recommended for solutions containing dissolved nitrogen.
9. **Slow release N fertilizers:** Fertilizers which slowly release N have been develoed to supply N continuously over an extended period. An ideal source of slow release fertilizer is the one that will release nutrients, at the rate approximately equal to the crop removal.

Nitrogenous Fertilizers Having Slow Availability

Slow rate of release of N is achieved by the use of substances having low solubility or due to physical barriers formed by coating an insoluble material or modifying the size of urea.

1. **Low solubility N fertilizers:** Low solubility materials are developed by condensation products of urea with formaldehyde, crotonylidene diurea and isobutylidene diurea and urea-Z. (Table 27.5). Among them, urea formaldehydes (ureaform) are most widely used in USA.
2. **Coated urea materials:** Nitrogenous fertilizers are coated with substances that reduce their rate of solubility. Sulphur coating of prilled urea reduces the

dissolution rate. It also minimizes the toxicity arising due to heavy application of urea. Sulphur coated urea (SCU) is a controlled release N fertilizer (36-38 % N) containing of S shell around each urea granule. The advantage of this fertilizer is that in addition to N, it also contains S.

3. **Urea super granules (USG):** Urea in bigger size granular form is called urea super granule. The size of the urea granule varies from 1-3 g. Because of the bigger size, it is slowly soluble. Deep placement of these granules in puddle soil enables higher use efficiency. Machines have also been developed for placement of granules in the reduced zone of puddle soils for rice cultivation.

Table 27.5: N content of slow release fertilizers

Fertilizers	N content (%)
Ureaform (Urea-formaldehyde)	35.0
Crotonylidene diurea (CDU)	30.0
Isobutylidene diurea (IBDU)	30.0
Urea-Z (UZ)	35.0
Sulphur coated urea (USU)	36-37

Nitrification Inhibitors

Nitrogen loss from the applied fertilizers takes place mainly after conversion to nitrate form by the activity of the bacteria, *Nitrosomas* and *Nitrobactor*, together known as nitrobacteria. Losses can be reduced by inhibiting the growth or activity of these bacteria and thereby reducing conversion of ammonium to nitrate.

Several chemicals have been found to be toxic to the nitrifying bacteria. When these chemicals are added to the soil, they temporarily inhibit nitrification. E.g., N serve (2-chloro-6(trichloromethyl) pyridine) referred as nitrapyrin and AM (2-amino-4-chloro-6 methylpyrimidine). These materials are, however, expensive and not cost-effective. Hence, some of the indigenous materials like neem (*Azadirachta indica*) and karanj (*Pongamia glabra*) oil cakes have been tested and found to reduce the bacterial activity. Coating or mixing of urea with powdered neem cake has been found to give better response.

1. **Neem coated urea (NCU):** To prepare neem coated urea, the required materials are urea 100 kg, powdered neem cake 20 kg, coal tar 1 kg and kerosene 1.5 litres. First coal tar is melted in an iron pan. After a slight cooling, kerosene is added. This solution is poured in the heaped urea and powdered neem cake mixture. It is then thoroughly mixed well. The coated urea is shade-dried and used for basal application in rice.
2. **Neem blended urea (NBU):** For the preparation of neem blended urea, 100 kg urea and 20 kg neem cake (5:1) are required. The finely powdered neem cake is thoroughly mixed with urea and kept overnight. This can be used for top dressing rice at tillering stage.

Relative Efficiency of N Fertilizers

A number of experiments have been conducted on different crops under different situations. General conclusions from these experiments are:

1. Under most conditions, urea, calcium ammonium nitrate and ammonium sulphate, on equal nitrogen basis, are equally effective,

2. Ammoniacal fertilizers are more effective than nitrate fertilizers for low land rice,
3. Nitrate nitrogenous fertilizers are better suited for top dressing, and
4. For crops such as tea, ammonium sulphate is more effective than others.

Phosphatic Fertilizers

Phosphatic fertilizers are classified into three groups depending upon the form in which phosphoric acid is combined with calcium and their solubility.

1. **Water soluble phosphoric acid or monocalcium phosphate [$Ca(H_2PO_4)_2$]:** These fertilizers contain P in the available form of phosphoric acid. E.g., Super phosphate and ammonium phosphate.
2. **Citric acid soluble phosphoric acid or dicalcium phosphate ($CaHPO_4$):** Fertilizers of this group are particularly soluble in citric acid. e.g., basic slag and diacalcium phosphate.
3. **Insoluble phosphoric acid or tricalcium phosphate [$Ca_3(PO_4)_2$]:** Rock phosphate, raw bone meal and steamed bone meal.

P-Fertilizers

Fertilizer	Nutrient content (%)		
	P_2O_5	S	N
P-fertilizers containing water soluble phosphoric acid or monocalcium phosphate			
Single super phosphate	16-18	12	-
Double super phosphate	32	-	-
Triple super phosphate	46-48	4	-
Ammonium phosphate	20	-	-
P-fertilizers containing citric acid soluble P_2O_5 or dicalcium phosphate			
Basic slag	14-18		
Diacalcium phosphate	34-39		
Rhenania phosphate	23-26		
Bone meal	23-30		
P-fertilizers containing tricalcium phosphate or P_2O_5 which is not soluble in water or citric acid			
Rock phosphate	20-40		
Raw bonemeal	20-25		
Steamed bonemeal	22		

Properties of Important P Fertilizers

1. **Super phosphate:** Super phosphate is the oldest commercially available fertilizer and it is manufactured in three grades: Single super phosphate (16-20 % P_2O_5)., dicalcium phosphate 35-38 % P_2O_5) and triple super phosphate (44-49 % P_2O_5). Single super phosphate (SSP) is the most commonly used phosphatic fertilizer in India. It also contains sulphur (12 %). It is available in the form of grey to brownish powder as well as in granulated form. Granulated SSP is the most suitable for application by machine, but is somewhat costlier than powdered SSP. The fertilizer is suitable for a wide range of soils and crops. It should be applied before or at the time of sowing/planting as basal aoolication. To increase

the efficiency of applied fertilizer in acid soils, it should be used with organic manures.

2. **Ammonium phosphates:** The available fertilizers in this group are monoammonium phosphate (MAP) containing 11 % N and 52 % P_2O_5 and diammonium phosphate (DAP) containing 18 % N and 46 % P_2O_5. These fertilizers can be used for direct application to the soil. DAP is recommended for rice nursery and foliar spray for pulses.
3. **Rock phosphate RP):** In India, rock phosphate deposits are found in different parts of the country. Mussorie deposit, found in Uttar Pradesh, is exclusive of sedimentary origin and has relatively high reactivity. The calcium phosphate present in phosphate rocks is less stable and gets easily acidulated by the soil acidity and organic acids present in the soil organic matter. Thus, its direct application to acid soils is a cheaper and efficient alternative to application of water soluble forms.
4. The particle size plays a vital role in the effectiveness of phosphate rock for direct application. Finer the particle, more effective is the rock phosphate for direct application. Agronomic evaluation of indigenous rock phosphate has been done for Mussorie rock phosphate (MRP) containing 20-21 % P_2O_5 for different crops all over the country. Available experimental evidences indicate the economic importance of MRP in the intensive agriculture.
5. **Basic slag:** It is obtained as a bye-product from steel industry. The high grade European slag (15-18 % P_2O_5) is widely used as a phosphatic fertilizer in Central Europe. But the product obtained from Indian steel plants has a low content of phosphorus (3-8 %) and is not used as a fertilizer. It contains considerable quantity of CaO and hence it is suitable for acidic soils.

Relative Efficiency of P Fertilizers

From the results of the experiments conducted under different situations, general conclusion are:

1. For short duration crops and those with restricted root system, fertilizers with high proportion of water soluble phosphorous are advantageous,
2. Higher water soluble phosphorous is less important for long duration crops,
3. Higher water solubility is desirable for crops requiring quick start,
4. Localized placement of water soluble phosphate fertilizers is more effective when the rate of application is limited,
5. On acid soils, granular fertilizers with high degree of water solubility are more effective than powdered fertilizers when the fertilizer is to be mixed with soil,
6. On acid to neutral soil, band placement of powdered fertilizers with high degree of water solubility will give better results than mixing the fertilizer with soil,
7. On calcareous soils, granular water soluble phosphates are more effective,
8. Fertilizers with low water solubility will give good results when applied in powdered form mixed with soil,
9. Monoammonium phosphate is better than diammonium phosphate on calcareous soils, and
10. Rock phosphate and bone meal are ideal for strong acid soils and for long duration crops.

Potassic Fertilizers

Potassium is absorbed by plants in larger amounts than any other element except nitrogen. Unlike phosphorus, potassium is present in relatively large quantities in most soils. On an average, potassium of the earth's crust is about 1.9 per cent. Its concentration in soils normally varies from 0.5 to 2.5 per cent with a typical value of 1.2 per cent. More than 95 per cent of the total potassium in the soil is within the crystal lattices of silicate materials. The micas (muscovite and biotite) and the feldspars (orthoclase microcline) constitute the major K bearing materials. Potash is also found in soils in the form of secondary or clay minerals such as illites, vermiculites, chlorites and interstratified minerals.

Commercially available potassic fertilizers are salts of potassium. They are completely soluble in water and contain nutrients in readily available form. Commonly used potassic fertilizers are: (1) Muriate of potash (potassium chloride) and (2) Sulphate of potash (potassium sulphate). The major potassic fertilizers and their nutrient contents are as follows:

Fertilizer	Chemical form	Nutrient content (%)	
		K_2O	Others
Potassium chloride	KCL	60	-
Potassium sulphate	K_2SO_4	48-52	-
Potassium Nitrate	KNO_3	44	13 N
Sulphate of potash-magnesia	K_2SO_4 and $MgSO_4$	25-30	10-12 MgO

1. **Potassium chloride or muriate of potash (MOP):** The principal fertilizer source of potassium is potassium chloride or muriate of potash. It is a low cost and high analysis (60-63 % K_2O) K fertilizer. It also contains 45-47 % chloride, depending on the original rock composition. It is white to slight red in colour depend on the impurities present in the potash minerals and the method of refining. The salt is neutral in reaction and can be applied to all soils and crops. However, for tobacco and potato this fertilizer should not be used since the quality is affected due to the presence of chloride. Muriate of potash is generally applied basally. But in light textured soils, where the loss due to leaching is more, it is recommended for top dressing also.
2. **Potassium sulphate or sulphate of potash (SOP):** It is a white crystalline salt containing 50-53 % K_2O and 17 % S. It can be used for all soils and crops as basal as well as top dressing. This fertilizer is costlier than muriate of potash. It is generally used in crops like potato and tobacco which are sensitive to high amount of chloride.
3. **Potassium nitrate (KNO_3) NOP):** 13 % N and 44 % K_2O, excellent source of K and N mainly used for fruit trees and crops such as tobacco and vegetables.

In addition to above two fertilizers, soluble potassium nitrate (13-0-46) is markwted by some companies. This is specially suited for foliar application and fertigation (application through drip irrigation system), as the fertilizer dissolves completely in water without leaving any residues. As it is very costly, it is recommended only for high value crops.

Complex or Mixed Fertilizers

Complex Fertilizers

Fertilizers containing atleast two or more of the primary essential nutrients (N, P_2O_5 and K_2O) are called complex fertilizers. These fertilizers contain two or more plant nutrients (of which two are major nutrients) in chemical combination. They are generally granular in form and can be uniformly applied to the field. Unit nutrient cost of complex fertilizer is lower than that of the conventional fertilizers. Complex fertilizers produced in India can be broadly grouped as (i) ammonium phosphates, (ii) nitrophosphate and (iii) NPK complex fertilizers.

1. **Ammonium phosphates:** These fertilizer materials contain ammonium and phosphate radicals in varying proportion: (a) Ammonium phosphate sulphate (18-9-0, 16-20-0, 19.5-19.5-0, 20-20-0); (b) urea ammonium phosphate (24-24-0, 28-28-0); (c) Ammonium phosphate sulphate nitrate (20-20-0); and (d) Diammonium phosphate (18-46-0).
2. **Nitrophosphates:** These fertilizers are prepared by acidulation of rock phosphate and ammoniation of the resultant slurry. If needed, slurry is combined with potassium salt. The final product is made up of ammonium nitrate, dicalcium phosphate, mono-ammonium phosphate, etc.
3. **NPK Complex fertilizers:** Several grades of these fertilizers are being produced in India. NPK composition varies with the fertilizers: e.g., 10-26-26, 12-32-16, 14-35-14, 14-28-14, 17-17-17, 19-19-19, and 22-22-11. These fertilizers possess good physical properties.

In addition to these complex fertilizers, new specially fertilizers with varying compositions of NPK and micronutrients, specially designed for specific crops, are produced and marketed by many companies in the last few years. These are completely soluble in water and are suitable for foliar application and fertigation. As the cost of these fertilizers is very high, they are recpmmended only for high value crops.

The fertilizers which contain all the three primary elements are called **complete complex fertilizers** and those which contain only two nutrients are referred to as **incomplete complex fertilizers.**

Methods of Preparation of Complex Fertilizers

1. By chemical reaction or change.
2. By purely mechanical mixing of straight fertilizers known as **fertilizer mixtures.**

Chief Characteristics of Complex Fertilizers Manufactured by Chemical Reaction

1. They usually have high content of plant nutrients. As such they are also called also **high analysis fertilizers.**
2. They usually have a uniform grain size and good physical condition.
3. They supply to the soil nitrogen and phosphorus in an available form in one operation.
4. They are cheaper.
5. They are non-caking and non-hygroscopic, thus, safer storage.

Complex Fertilizers Manufactured in India

1. Ammonium phosphate sulphate
2. Diammonium phosphate.
3. Nitrophosphate.
4. Urea ammonium phosphate.
5. NP/NPK fertilizers.
6. Urea and NP/NPK fertilizers.

Mixed Fertilizers

Fertilizer materials containing two or more plant nutrients are physically mixed manually or mechanically. There is no chemical combination of nutrients. Special mixtures are prepared for specific crops. The mixture can be prepared in the farm itself. Preparation of such mixtures requires great care and adequate knowledge about the chemistry of the ingredients used. This will avoid uneven mixing of incompatible fertilizers causing nutrient loss in the form of gases, conversion of soluble nutrients into insoluble forms and caking of materials. The following materials should not be mixed and used for making mixed fertilizers.

1. Ammonium sulphate, ammonium chloride, other ammonium fertilizer and nitrogenous organic manures with lime.
2. Sodium nitrate or potassium nitrate with super phosphate.
3. Ammonium sulphate nitrate with lime.
4. Urea with super phosphate.
5. Super phosphate with lime or calcium carbonate or wood ashes.

Bone meal, sulphate of potash and muriate of potash can be mixed with all the fertilizers.

Secondary Nutrients: Calcium, Magnesium and Sulphur

Calcium, magnesium and sulphur are called **secondary nutrients** as these are of secondary importance in the manufacture of commercial fertilizers. These are added to the soil through some of the commercial fertilizers. Calcium, magnesium and sulphur are supplied to the plants incidentally by the application of N P K fertilizers and as such, fertilizers are not manufactured to supply these nutrients. Fertilizer materials that supply these nutrients are presented in Table 27.6.

TABLE 27.6: Ca, Mg and S contents of different fertilizer materials

Fertilizer	Nutrient content (%)			
	Ca	Mg	S	Others
Cacium nitrate	19.4	1.5	-	-
Gypsum	29.2	-	18.6	-
Rock phosphate	33.1	-	-	25.2 (P_2O_5)
Single super phosphate	19.5	-	12.5	16.0 (P_2O_5)
Triple super phosphate	14.0	-	1.0	43.5 (P_2O_5)
Epsom salt	-	9.6	13.0	-
Poatassium magnesium sulphate	-	11.1	22.3	31 (K_2O)
Potassium sulphate	-	-	17.7	48 (K_2O)
Ammonium sulphate	-	-	24.2	21 (N)
Ammonium sulphate nitrate	-	-	12.1	26 (N)
Basic slag	-	-	3.0	15.6 (P_2O_5)
Copper sulphate	-	-	11.4	21 (Cu)
Ferrous ammonium sulphate	-	-	16.0	6(N) 16 (Fe)
Ferrous sulphate	-	-	18.8	32.8 (Fe)
Elemental sulphur	-	-	100.0	-
Urea-gypsum	4.6	-	0.6	36.8 (N)
Urea-sulphur	-	-	10.0	40 (N)
Zimc sulphate	-	-	17.8	36.4 (Zn)

MICRONUTRIENTS

There are four main classes of micronutrient sours: inorganic, synthetic chelates, natural organic complexes and fritted glass products (frits). Inorganic sources are usually the least expensive but are not always the most effective sources for crops. The organic sources include oxides and metallic salts such as sulphates, chlorides and nitrates. Sulphates are the most common of the metallic salts. Oxides of Zn and Mn are also widely used. Synthetic chelates are formed by combining a chelating agent with a metal through coordinate binding. They are generally the most effective micronutrient sources, but are also the most expensive. E.g., Zn EDTA (Na_2 Zn-ethylenediamine tetra-acetic acid) and Fe EDDHA (Fe-ethylenediamine di-orthrohydrorxy phenyl acetate).

Natural organic complexes are made by reacting metallic salts with some organic by-products of the wood-pulp industry. Several classes of these complexes are the lignosulphonates, phenols and polyflavonoids. These complexes cost less per unit of micronutrient but they are usually less effective than synthetic chelates.

Fritted micronutrients are glassy products and are generally used only on sandy soils in regions of high rainfall where the leaching loss of nutrient is more. Frits have only a small share of the micronutrients market.

The common micronutrient fertilizers and their nutrient content are given in Table 27.7.

TABLE 27.7: Micronutrient fertilizers and their nutrient content

Fertilizers	Element/form	Content (%)
Zinc sulphate (heptahydrate)	Zn	21.0
Manganese sulphate	Mn	30.5
Ammonium molybdate	Mo	52.0
Borax	B	10.5
Solubor	B	19.0
Copper sulphate	Cu	24.0
Ferrous sulphate	Total Fe	19.5
	Ferrous	19.0
	Ferric	0.5
Chelated Zn (EDTA form)	Zn	12.0
Chelated Fe (EDTA form)	Fe	12.0
Zinc sulphate (monohydrate)	Zn	33.0

Deficiencies of Micronutrients

Since mid 1960s, deficiencies of micronutrients in both soils and crops have been mounting of alarming rate. Analysis pf soil samples from all the states in India showed that zinc deficiency is very wide spread followed by iron, boron, manganese, copper and molybdenum. Zinc deficiency is the most serious in Andhra Pradesh, Bihar, Haryana, Madhya Pradesh, Punjab, Tamil Nadu and Uttar Pradesh. Iron deficiency is observed in all these states except Andhra Pradesh and Madhya Pradesh. Deficiencies of Fe, Mn and Cu are of lesser intensity than that of Zn. However, Fe deficiency is wide spread in Entisols, Inceptisols, Aridsols, especially calcareous soil with high pH and sandy soils of Punjab and Haryana, Vertisols and Alfisols of Tamil Nadu and calcareous soils of Bihar.

Manganese and copper deficiencies are not as widespread as that of Zn and Fe. Soils of Karnataka (19 %) are deficient in Mn and that of Kerala (13%) and Tamil Nadu (14 %) in Cu. Boron deficiency ranges from 49 to 84 per cent in Tarai alluvial, red and lateritic and Alfisols of West Bengal, Orissa, Meghalaya and Madhya Pradesh. Boron deficiency is more critical in highly calcareous soils, leached sandy soils, limed acid soils and reclaimed lateritic soils.

Conditions Conducive to Micronutrient Deficiency

There are several reasons for the widespread concern for micronutrients of which the important are:

1. Intensive crop production using improved crop varieties increased crop production and hence micronutrient depletion from soil,
2. The trend towards high analysis fertilizers reduced the use of low analysis fertilizers, which used to supply some micronutrients, and
3. Increased knowledge of plant nutrition has helped in diagnosis of micronutrients that formerly might have gone unnoticed.

The following soil conditions are more conducive to micronutrient deficiency:

1. Highly leached acidic sandy soils,
2. Soils with high organic matter (Organic soils),
3. Soils very high in pH,
4. Intensively cropped soils with heavy rates of commercial fertilizers, and
5. Soils with high water table or poor drainage.

The micronutrients can be applied to the soil as that of NPK fertilizers, foliar spray, mixed with fertilizers, seed soaking and seed coating. However, the first three methods are commonly used because of simplicity.

FERTILIZER APPLICATION

Fertilizers are used to make certain that soil fertility is not a limiting factor in crop production. The objective of fertilizer use can be obtained only when the fertilizer is efficiently used by the crop.

Principles of Fertilizer Application

Kind of Fertilizer

Fertilizers containing nitrate N are not ideal for sandy soils as the applied nitrogen is lost by leaching. Ammonia or ammonia farming fertilizers are ideal for low land rice. Acid forming fertilizers can be used for neutral and alkaline soils. Water soluble phosphatic fertilizers like super phosphate is best suited to neutral soils. Citrate soluble phosphates give best results on acid soils. Fertilizers containing chloride should be used for crops such as potato and tobacco.

Quantity of Fertilizers

Quantity of fertilizer depends on uptake of nutrients by crops. It should be based on soil analysis for different nutrients. Traditional cultivars do not respond to high rate of application.ds and improved cultivars give high yield at higher rates of application. Irrigated crops require higher quantities of fertilizers than dry land crops. Legumes require more phosphorus than nitrogen. Four types of recommendations are:

1. **Threshold dose:** It refers to minimum fertilizer for a positive crop response. There may not be any economic response with a lower dose than this.
2. **Minimum recommended dose:** This dose provides a positive response with maximum profit per rupee invested in fertilizer application. It is for farmers with limited resources for investment on fertilizer application.
3. **Optimum dose:** It is the quantity of nutrient applied for realizing maximum benefit from its application, irrespective of returns per rupee invested. It is usually the economic optimum dose preferred by farmers with resource base.
4. **Yield maximizing dose:** It refers to the quantity of nutrients applied for obtaining the highest yield. Financial implications are not considered in obtaining the higher yield. In general, yield increment is progressively less per unit increase in fertilizer level. Therefore, the law of diminishing returns operate in such a situation.

Frequency of Application

Number of applications depends on the nature of soil and duration of the crop. Higher quantities of fertilizers can be applied to heavy soils at longer intervals because of their ability to retain the nutrients for longer use. Sandy soils require frequent applications to minimize leaching losses of nutrients. In general, long duration crops require more number of applications, especially nitrogen, than short duration crops.

Time of Application

A close relationship exists between available soil moisture and response of crops to fertilizers. In arid and semiarid regions, soil moisture is the limiting factor for fertilizer use. All the fertilizers are recommended as basal application for rainfed crops grown on stored soil moisture. Early application is more effective for short duration crops.

Rate of crop growth is proportional to the rate of assimilation of nutrients, especially nitrogen. Demand for nitrogen increases from early vegetative to flowering stage. Hence, nitrogen fertilizers should be applied atleast twice, one at sowing or planting and again before flowering. Phosphorus and potash requirements, in general, are higher during early phases of crop growth. As such entire dose may be applied at sowing.

Placement of Fertilizers

Placement of fertilizers in the root zone improves their efficiency. There should be minimum contact of fertilizer with the soil to avoid fixation of nutrients. Phosphorus fixation in acid soils can be minimized by placement of water soluble phosphate fertilizers. Ammoniacal fertilizers should be placed in reduced zone for low land rice to avoid losses of applied nitrogen. Response of dry land crops to fertilizers increase with their placement at about 5-10 cm below the surface, where soil moisture will be available for over a long period of time.

Methods of Application

An important item in efficient use of fertilizer is that of placement in relation to plant.

Solid Fertilizers

Broadcasting is the method of application of fertilizer uniformly over the entire field. It may be at planting or in standing crop as top dressing. **Broadcasting at planting** is adopted under certain conditions.

1. Soils highly deficient, especially in nitrogen,
2. Where fertilizers like basic slag, dicalcium phosphate, bone meal and rock phosphate are to be applied to acid soils, and
3. When potassic fertilizers are to be applied to potash deficient soils.

Top dressing is application of fertilizer to the standing crop. Usually, nitrate nitrogen fertilizers are top dressed. Depending on the duration of the crop and soil type, top dressing may be more than one to meet the crop needs at times of greatest need of the crop.

Placement: Fertilizers are placed in the soileither before sowing or after sowing the crop.

Plough-sole placement consists of placing the fertilizer in a continuous band at the bottom of thje furrow during the process of ploughing, which is usually covered by the next furrow adjacent to it. Since the fertilizer is placed in subsoil where soil moisture is available for over a longer period, it will be available to the plant for longer time, especially during dry season.

Deep placement is application of fertilizers, especially nitrogen, in the reduced zone to avoid nitrogen losses in low land rice. Sub soil placement refers to placement of fertilizer in the subsoil with the help of machinery. It is usually adopted where the subsoil is strongly acidic. Usually, phosphatic and potassic fertilizers are applied with this method.

Localised placement: In this method fertilizers are applied close to the seed or plant. It is usually adopted when relatively small quantities of fertilizers are to be applied.

Contact placement or drill placement refers to drilling seeds and fertilizer simultaneously at sowing. Care must be taken to place the seed and fertilizer at different depths to avoid salt injury to the germinating seed. Ferti-seed drills are popular in dry land agriculture for drill placement.

Band placement consists of applying the fertilizer in continuous bands, close to the seed or plant. When the plants are widely spaced, fertilizer is placed on one or both sides of the plant. It is called **hill placement.** Crops like cotton, castor, etc, are well suited for this method. **Row placement** consists of placing the fertilizer in a row in continuous bands on one or both sides of row by hand or seed drill. This method is ideal for sugarcane, tobacco, maize and vegetable crops.

Pellet placement is application of fertilizer, especially nitrogen in pellet form in the low land rice to avoid nitrogen loss from applied fertilizer. Soil and fertilizer are mixed, usually, in the ratio of 1:10 or 15 and made into a dough. Smmall pellets are then made and placed in the reduced zone. **Side dressing consists** of spreading the fertilizer between the rows or around the plants. As such it is a broad term covering methods such as row placement and hill placement.

Liquid Fertilizers

Starter solution: These are solutions of fertilizers prepared in low concentrations used for soaking seed, dipping roots or spraying on seedlings for early establishment and growth.

Foliar application: This method, nutrients are applied to the standing crops in the form of spray for quick recovery from the deficiency. It avoids fixation of nutrients in the soil.

Soil application: Liquid fertilizers such as anhydrous ammonia are applied directly to the soil with special injecting equipment. Liquid manures such as urine, sewage water and shed washing are directly let into the field.

Fertigation: This is the application of fertilizer in irrigation water in either open or closed system. The open system includes lined and unlined open ditches and gated pipes that are used for furrow and flood irrigation. Sprinkler and trickle systems are the main closed systems. Nitrogen and sulphur are the principal nutrients applied by fertigation.

BALANCED FERTILIZER APPLICATION

Crop yields are correlated with fertilizer application. The role of chemical fertilizers in increasing crop yield has become obvious with the introduction of HYVs responding to fertilizers. Most of the soils are only able to supply a fraction of nutrients required by the arable crops. Even available nutrients are rarely present in appropriate proportion to meet the crop needs.

Balanced nutrition is provision of adequate amount of nutrients in required proportion for realizing maximum yield. Studies on uptake of nutrients by crops indicated huge nutrient depletion from soil.

Crop	Yield (t /ha)	Nutrient Uptake (kg)		
		N	P_2O_5	K_2O
Wheat	5	125	45	165
Rice	5	100	55	150
Maize	5	130	70	180
Sorghum	5	90	53	136

The results of long term fertilizer trials indicate that Indian soils could only supply about 78 kg N /ha, 14 kg P_2O_5/ha and 76 kg K_2O/ha when continuously cropped without fertilizer application. The highest response of 14.8 kg/ha nutrient at 50 kg N/ha has been with 25 kg/ha each of P_2O_5 and K_2O. For obtaining a corresponding response, at higher levels of 95 kg N/ha, the requirement of P_2O_5 and K_2O are 75 and 50 kg/ha respectively. Balanced fertilization should take care of all nutrient deficiencies. Calculation of nutrient balance shows that the contribution of different sources to phosphorus removed by crops in India is 78.0 per cent from soil reserves, 14 per cent from fertilizers and 8 per cent from organic manures. Results of long tems fertilizer experiments indicate that nitrogen alone can increase the phosphorus uptake upto 130 per cent. Balanced fertilization can be practiced through soil testing.

$$\text{Nutrient requirement (NR) in kg/q} = \frac{\text{Nutrient uptake (kg/ha)}}{\text{Yield (q/ha)}}$$

$$\text{Soil efficiency (SE) per cent} = \frac{\text{Nutrient uptake}}{\text{Soil test value (kg/ha)}} \times 100$$

$$\text{Fertilizer efficiency (FE) per cent} = \frac{\text{Nutrient uptake - soil test value x SE}}{\text{Rate of fertilizer applied}} \times 100$$

$$\text{Fertilizer dose (kg/ha)} = \frac{\text{NR}}{\text{E} \; / 100} \times \text{target} - \frac{\text{SE}}{\text{E}} \times \text{Soil test value}$$

INTEGRATED NUTRIENT MANAGEMENT (INM)

Soil fertility is a prerequisite to its productivity. From the stand point of crop yields and quality, nutrient supply both from organic and inorganic sources are important. As such it is essential to substitute the inorganic fertilizers with nutrients from organic sources. The **integrated nutrient management (INM)** system envisages use of organic manures, green manures, biofertilizers (Organics) along with chemical fertilizers (inorganics).

A number of diverse organic sources are available for use in agriculture. These sources can reduce the mining of soil nutrients and improve overall soil productivity. Recycling of wastes and their conversion into easily transportable and usable forms should be an integral part of soil fertility management programme. Use of legume inoculants and biofertilizers along with chemical fertilizers offers an opportunity to achieve long term sustainability in crop production systems without land degradation.

Integrated nutrient management system improves the soil physical conditions in terms of soil structure, aggregate stability, soil moisture retentivity and hydraulic conductivity. Land degradation and environmental pollution can be minimized for sustainable agriculture.

However, integrated **nutrient management (INM)** aims at efficient and judicious use of all the major sources of plant nutrients in an integrated manner so as to get maximum economic yield without any deleterious effect on physico-chemical and biological properties of the soil. The basic concept underlying INM is maintenance and possible improvement in soil fertility for sustained productivity on a long term basis. Major sources of plant nutrients are fertilizers, FYM/compost, green manures, biofertilizers and crop residues/ recyclable waste. Proper integration of one or more sources will ensure optimum nutrient supply.

Fertilizer is the most important component in integrated nutrient supply system under intensive cropping. Use of balanced fertilization is required to boost the crop productivity. Chemical fertilizers are to be properly integrated with other sources.

Integration of fertilizers with bulky organics: Farm yard manure and compost are the traditional organic resources that have been used in crop production. About 50 % of the manorial potential of livestock population is only utilized for crop production. Sizable amount of plant nutrients are lost because of improper storage, handling and incorporation techniques. Results of various field experiments have clearly indicated the importance of integrated use of FYM and fertilizers for overall improvement in soil fertility. Results also indicate that a part of fertilizer requirement of rainy season (*kharif*) crop can be met with the use of FYM in different cropping systems without any adverse effect on the total productivity. Based on a number of experiments on rice, it is concluded that 1 tonne of FYM supplying 12 kg nutrients ($N+P_2O_5+K_2O$) produces as much grain as 3.6 kg fertilizer nutrients (Singh and Dwivedi, 1996). In addition to supplying plant nutrients to the current crop, FYM leaves substantial residual effect on succeeding crops in the system. Meelu and Morris (1984) reported that 12 t FYM and 80 kg N/ha gave residual effect on the succeeding wheat equivalent to 30 kg/ha each of N and P_2O_5. There was also a build up in soil fertility with FYM application. Kulkarni et al.,(1983) on the basis of 240 trials concluded in cultivators' fields in rice growing regions of the country concluded that application of 12 tonnes of FYM along with 90 kg N/ha to the *kharif* crop followed by 60 kg N/ha to *rabi* crop could produce an yield comparable to 120 kg N, 60 kg P_2O_5 and 60 kg K_2O in *kharif* and 60 kg N/ha to *rabi* crop in rice-rice cropping system.

Besides supplying plant nutrients, FYM application also improves the physico-chemical and biological properties of the soil. Application of FYM along with recommended level of fertilizers enables to maintain the organic matter in the soil compared to fertilizer application alone.

Green manuring with inorganic fertilizers: Before the introduction of inorganic fertilizers, green manuring was invariably practiced for cropslike rice, wheat, sugarcane, etc. With the easy availability of fertilizers and adoption of highly intensive cropping, the practice was almost given up. In the recent years, there is a renewed interest on green manuring. Sunnhemp and dhaincha are the most common green manure crops in India. Green manures produce large amount of green biomass of narrow C:N ratio. When incorporated into the soil, green manures contribute sizable amount of plant nutrients, particularly N and also improve the soil physical conditions. Experiments carried out at several locations have distinctly established the advantage of introducing green manure crop as a supplement to fertilizers in cereal crop rotations. Green manuring along with recommended level of fertilizers benefits the crops in the rotation and increase the total productivity of the system to the tune of 10-20 per cent. Incorporation of 10 tonnes/ha of green manure through Sesbania or Crotolaria supplement fertilizer N equivalent to 40-45 kg/ha. Integrating P with green manure in the first crop of rice produces higher yields of second season rice.

Green manuring in drylands is rare, but is practiced in small pockets in India. On vertisols receiving more than 1000 mm of rainfall in Khammam district of Andhra Pradesh, green manure is grown during the South-West monsoon season in June followed by its incorporation before sowing sorghum in Septemeber. Depending on sowing of green gram, it is either incorporated as green fodder or harvested as pods. This practice has been observed to supplement 50 % N (Venkateswaralu, 1987). Inclusion of a grain legume has been found to be advantageous for sustained yields under rainfed situations' e.g., cowpea-finger millet (Bangalore), green gram-finger millet (Bhubaneswar), sunnhemp-whear (Hoshiarpur) and black gram-wheat (Rewa). The benefit of green manuring is felt only after continued applications for 3-5 years (Singh and Venkateswaralu, 1985).

Wherever it is not possible to include green manure in the multiple cropping system, there is a possibility of inclusion of legume in the system. Because of their shorter duration, small stature and erect growth habit, legumes could be easily accommodated as intercrops. Legumes of short (Vigna group) or long (Pigeonpea) duration groups offer greater advantage as intercrops in cereals by providing additional profit to the farmers. The Vigna group is harvested earlier and provides some nitrogen to the longer duration companion crop. Pigeonpea is of longer duration. It is least competitive during the early growth phase and starts using the space effectively by putting up additional growth after harvest of the early duration companion crop. These pulses provide N up to 50 kg/ha to the associated crops in vertisols. Higher benefits have been reported for crop combinations of cowpea (fodder) + sorghum, green gram/black gram + sunflower and pigeon pea + groundnut.

Green manure crops can be very well accommodated as inter crops in wide spread crops like sugarcane and maize. Since initial crop growth upto 100 days after planting of sugarcane is very slow, this period can be utilized for growing legumes in inter-row spaces as intercrops for grain or fodder purposes. Summer legume crops like black gram and green gram intercropped with spring planted sugarcane increased the cane yield markedly besides producing 0.4-0.5 tonne of pulse grain/ha (Yadav, 1980).

In addition to their capacity to fix atmospheric N, the legumes are capable of extracting less soluble form of P from the soil. They also extract the higher velency nutrient cations better.

Biofertilizers with inorganics: Various biofertilizers can be used according to needs of the crops. Biofertilizers act as complimentary and supplementary sources of plant nutrients. Biofertilizers help in increasing the biologically fixed atmospheric

N and enhancing native P availability to the crop. Among the biofertilizers for increasing N supply, N-fixing bacteria (Rhizobium, Azotobactor and Azospirillum), blue green algae (BGA) and Azolla are important. The availability of P is improved by P solubilising microorganisms (PSM) and mycorrhizae.

Rhizobium is of vital importance to various pulses and some of the oil seeds and fodder legumes as 80 to 90 % of their N requirement is met through biological nitrogen fixation (BNF). Inoculation of legumes with specific species of Rhizobium is widely used and is a successful practice throughout the world. Responses to Rhizobium under field conditions have been quite variable and depend on soil conditions, quality of inoculum and effectiveness of native Rizobium. It fixes atmospheric N ranging from 50 to 60 kg/ha in groundnut and 100-300 kg/ha in alfalfa. Pulse crops fix 50-110 kg N/ha under optimum conditions. It is generally accepted that the fixation is about 50 kg/ha in a season. Field studies have indicated an increase in pulse yields by 14-53 % due to Rhizobium inoculation.

Azospirillum and Azotobactor are the other important groups of bacteria. *Azospirillum* fixes N in associative symbiosis with the plant roots and *Azotobactor* is a free-living N fixer. On an average, they contribute 15-20 kg N/ha to different crops when inoculated with efficient strains. Application of organic manure along with Azospirillum enhances N fixation to the tune of 20-40 kg/ha. Promising results have been obtained with the use of Azospirillum in crops like wheat, rice, maize, barley, sorghum and millets. Azotobactor has given promising results in rice, sugarcane, wheat and pearl millet. Response to the bacterial inoculant appears to be enhanced when applied along with N and P in rice and cotton and with P in sugarcane.

Blue green algae and Azolla are very well suited for rice ecosystem. In lowland rice, the average N contribution by these N fixers is equivalent to 20-40 kg/ha. Various field studies have shown positive influence of BGA on rice plants under different agro-climatic conditions. If this technology is popularized, nitrogen equivalent of 1.5-1.7 m.t. urea could be supplied to rice through BGA (Venkataraman and Tilak, 1990). Besides supplying N, BGA also synthesizes and liberates growth promoting substances such as auxin and amino compounds which stimulate the growth of rice plant. It also supplies additional humus to the soil. Results of the studies conducted by the International Net work on Soil Fertility and Fertilizer Evaluation for Rice (INSFFR) have shown that incorporation of 10 tonnes of fresh azolla before or after transplanting is equivalent to split application of 30 kg fertilizer N/ha. It can be applied along with fertilizer N to step up rice yields. Based on field trials, an increase in rice yield by 13-54 % has been reported, besides building up organic carbon, total N and availability of P in the soil (Satapathy and Singh, 1980).

In recent years, several strains of P solubilizing bacteria and fungi have been isolated. Phosphobacteria solubilize the residual and insoluble phosphates available in the soil-water solution very near to root surfaces. The phosphate solubilizing power is further enhanced by addition of organic matter. Studies reveal that integrated use of P solubilizing cultures with low grade rock phosphate can add 30-35 kg P_2O_5/ha in neutral to mildly alkaline soils. Vasicular-arbuscular mycorrhizae (VAM) exploit the soil phosphate better, leading to more efficient use of added fertilizers, particularly of slowly mobile ions like NH_4^+, Cu^{2+} and Zn^{2+}. VAM infected sorghum had significantly high P in roots and shoots and increased the dry matter production. Because of problems in isolation and multiplication of pure strains of VAM fungi, large-scale use on commercial basis is yet to be established.

SOIL SAMPLING FOR NUTRIENT MANAGEMENT

In Nutrient Management Planning, soil sampling is done to collect a soil sample for lab analysis that represents the variability in the soil of the field being sampled. To do this many small samples are collected and mixed together to make one composite sample for each field. The results are used to determine what level of additional nutrients, if any, are required for the crop(s) being grown. There are a number of steps to follow as outlined below to ensure that a representative sample is collected.

When to Collect Soil Samples

Recommendations for fertilizing fruit and vegetable crops are based in part on soil test results. Soil testing provides information on lime and fertilizer needs prior to planting and is particularly well calibrated for nutrients such as phosphorus, potassium, magnesium, calcium, sulfur, zinc, and boron. Soil testing prior to planting takes the guesswork out of making fertilizer recommendations and leads to more efficient nutrient management. Fertilizer recommendations are intended for field-grown fruit and vegetable crops. For container-grown crops, such as transplants or vegetables grown in the greenhouse in pots, different soil tests should be used. For fertilizer requirements of established perennial crops and for fine-tuning fertilizer needs of annual crops, a combination of soil testing and tissue analysis should be used.

Proper interpretation of soil test results for making fertilizer recommendations is dependent on collecting a representative sample. Soil samples can be collected any time of the year, although spring and fall sampling are usually the most convenient. If soil test results from a given field are to be compared over the years, it is best that samples be collected at the same time of year. Each field to be sampled should be divided into uniform areas. Each area should have the same soil texture and color, cropping history, and fertilizer, manure, and lime treatments. One sample should not represent more than 20 acres on a level uniform field, or 5 acres on hilly or rolling land.

Samples are most easily collected using a soil tube, soil auger, or a garden spade. To take the soil sample, scrape off all surface residue and litter and take the sample to a depth of 6-8 inches for annual crops and 10-12 inches for perennial crops. Usually 15 to 20 subsamples (one core per subsample) should be collected from randomly selected areas in the field. The soil should be thoroughly mixed in a clean plastic pail and about 1 pint of this mixture should be placed in a sample bag or box. Samples can be sent directly to the Soil Testing laboratory nearer to your place with complete information of your area.

Check with the chosen laboratory as to their turnaround time for soil analysis. Samples may need to be collected 2-3 weeks before the intended time of manure application to give the lab enough time to complete the required analyses. Annual sampling should occur at approximately the same time each year.

Sampling Frequency

Annually cropped fields should be sampled once per year, in late spring just prior to manure application. Fields in perennial forage should be sampled before they are reseeded and thereafter once every 3 years until they are reseeded.

Sampling Strategy

Before beginning soil sampling, consider the variability within each field on the farm. Fields up to 10 ha (25 acres) in size that are fairly level, have a similar soil type throughout and have been farmed for several years as a single unit can be sampled as one unit. Fields that have significant variability in them should be subdivided into similar units. Variability may be due to topography, eroded areas, sandy vs. clay-rich areas, or sections of a field that have previously been farmed separately and may have received different amounts of manure or fertilizer over the years. Divide such fields into similar units, and sketch the boundaries on a permanent map that is kept with the soil analysis records. Assign a number or name to each field unit. Fields that are otherwise similar but that are greater than 10 ha in size should be subdivided into sampling units of 10 ha in size. Once fields have been subdivided, always sample them in this configuration.

Sampling Pattern

For most horticultural fields that are relatively uniform, samples should be collected from throughout the field using a zig-zag or Z pattern. Avoid areas that are not cropped such as wet spots, or areas that might skew the analysis such as old manure stockpile sites or fertilizer spills.

Sampling Equipment

Samples can be collected using a soil sampling tube, an auger or a trowel. A soil sampling tube or soil probe works best in well cultivated soils without rocks but is difficult to use in rocky, very dry or very wet soil. An auger or trowel is better for less well cultivated or rocky soils.

When sampling with a tube or auger, follow manufacturers' directions. When sampling with a trowel, make a V-shaped hole where the sample is to be taken. Take a 2-3 cm (1 inch) thick slice down one side of the hole to 15-20 cm (6-8 inches), and trim the slice to form a 2-3 cm (1 inch) wide core.

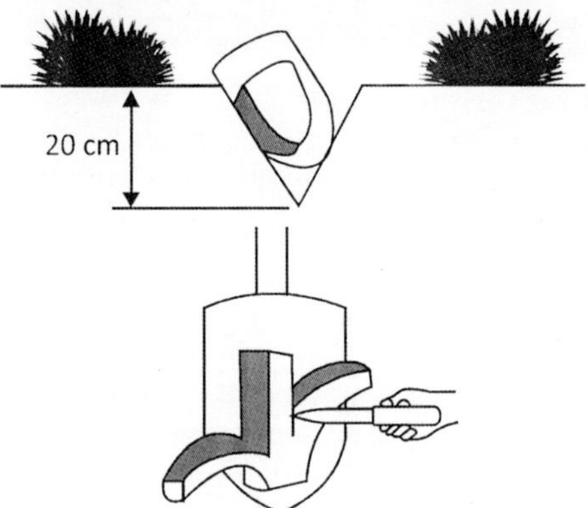

FIG. 27.1: Shovel method of soil sampling

Lift out the soil slice and place it into the sample bucket. If sampling for micronutrients or metals, ensure that the sampling equipment is clean, and has no rust on it as metals from the equipment itself can contaminate the sample. Wear rubber gloves for micronutrient and metal sampling

Sampling Depth

For routine nutrient analysis, and metal or micronutrient determination, collect samples to 15 cm deep (6 inches). Most manure and fertilizer nutrients will remain in the top 15 cm (6 inches) of soil. If sampling soil in the fall to monitor soil nitrate-N level, sample at two deeper increments, 15-30 cm (6-8 inches) and 30-60 cm (1-2 feet) if soil conditions permit as excess soil nitrate will move below the top 15 cm (6 inches) with fall rain. When sampling below 15 cm (6 inches), collect separate samples from each additional depth, and composite these separately.

Making a Composite Sample

Collect 5 soil samples per ha for small fields or 15 to 20 samples per 10 ha (25 acres) field to make up the composite sample for each field. Each soil sample should be approximately the same size or weight. Place all samples in a clean plastic pail and mix well. From the mixed samples, take out a small amount of soil (1/2 kg or 1 lb) and place into a heavy plastic bag or clean plastic container. This is the composite sample that will be used for lab analysis. It is very important that it be well mixed as the lab will use only a few grams of soil, and that small amount must represent equally all the cores that were initially placed in the pail.

If the soil is very wet at the time of sampling, it may mix more easily if it is air dried for a day. Use a trowel to break up lumps.

Sample Handling and Containers

Composite samples should be placed in plastic freezer bags or ½ to 1 litre clean plastic containers with screw top lids. Close samples tightly in bag or container. If samples are not to be taken to the lab immediately, place them on ice or freeze until transported to the laboratory. If shipping samples, ensure they reach the laboratory as quickly as possible.

Labeling

Clearly label samples with date, field or site identification, sampling depth (0-15 cm or other) and your name or the farm name. Bagged samples should be double bagged and a paper label placed between the two bags. Do not put paper labels in with soil as moisture may destroy label. Do not rely solely on information written on the outside of a plastic bag as it can easily rub off.

What Analyses are required?

To complete the nutrient management planning process, the following soil test information is required:

1. Available phosphorus (P)
2. Available potassium (K)
3. Nitrate-nitrogen (NO_3-N)
4. For spring soil tests in the Interior, ammonianitrogen (NH_4-N)

Other information may be included with the soil test report, such as soil concentrations of secondary and micro-nutrients, and metals, bulk density, pH and % organic matter. This is important information, and should be kept on record.

Fertilizer Analyses and Calculating Fertilizer Rates

By convention, fertilizer phosphorus and potassium are expressed on the oxide basis, P_2O_5 and K_2O, respectively. In contrast, fertilizer nitrogen is expressed on the elemental basis, N. The percentages of each nutrient on the fertilizer label are referred to as the grade of fertilizer and are guaranteed by the manufacturer. All fertilizer recommendations are based on the amount of N, P_2O_5, and K_2O to apply per given area (usually per acre). Fertilizers can be sold as complete fertilizers, i.e., they contain all three primary nutrients- nitrogen, phosphorus, and potassium- such as 10-10-10, or they can be sold as single nutrient fertilizers such as 46-0-0 (only N), 0-0-60 (only K_2O), etc. To determine the actual amount of nutrient in a given weight of fertilizer, multiply the percentage of nutrient by the weight of fertilizer and then divide by 100.

Soil pH Modification

Soil pH is an important chemical property that affects nutrient availability and microbial activity. In general, the optimum pH for most fruit and vegetable crops is between 5.8 and 7.0 for mineral soils and 5.4 and 6.2 for organic soils (peats and mucks). Two exceptions are blueberries and potatoes. Blueberries are adapted to acid soil conditions and grow best at a soil pH between 4.5 and 5.2. Potatoes can tolerate a wide range in soil pH; however, potato scab can become more of a problem in scab-susceptible varieties as soil pH increases above 5.3.

Interpreting Lab Results for Nutrient Management Planning

To develop a Nutrient Management Plan, the following information is required from the soil test reports for each field:

1. Available phosphorus in ppm, ug/mL or mg/L (all are equivalent)
2. available potassium in ppm, μg/mL, or mg/L
3. Nitrate-nitrogen in kg/ha (multiply value in ppm by 1.5)

On the soil lab report, the section where this information is found may be called soil test results, available nutrients, nutrient analysis or another similar term. The fertilizer recommendations are not required other than as a comparison with the recommendation generated by the nutrient management plan. Many labs rate the level of each nutrient in the soil as low, medium, high or very high, or deficient, marginal, optimum or excess to indicate how well the soil in each field is supplied with each nutrient. The soil nutrient rating scales for phosphorus and potassium used in the nutrient management planning process are close to those used by most commercial labs.

For further information, contact District Agricultural Officer/Krishi Vigynan Kendra of our area.

Liming

Liming materials are used to increase the pH of soils. Not all soils in India need to be limed. In general, soils which receive lower amounts of rainfall need liming. Soils formed under these conditions have a high native pH. The soil of higher rainfall areas tend to be acidic. Intensive cropping and continuous use of manure and/or ammonium based fertilizers will acidify soils over time. In contrast, if high pH irrigation water is used on acid soils, soil pH may actually increase over time. The need for lime is determined from a routine soil test.

The pH value reported in a soil test indicates whether lime is needed, but cannot be used to determine how much lime is needed. The amount of lime to apply is dependent on reserve soil acidity, which is measured in the laboratory by the "buffer index" or SMP buffer test. The lower the buffer index, the more lime is required. In most cases, buffer index is related to soil texture and organic matter content. At the same soil pH value, soils with a high clay and organic matter content will require more lime than soils with a high sand and low organic matter content. The difference in lime application rates is determined by the SMP buffer index. The SMP buffer test is not used if the soil pH is higher than 5.9 because of the high relative error above this level. The amount of lime recommended for mineral soils is that needed to raise the pH to 6.0 or 6.5. For peat and muck soils with a soil-water pH of 5.4 or less, lime is recommended to raise the pH to 5.5.

The amount of lime to apply is dependent on the type or quality of the liming material. The quality of a limestone is based on its **calcium carbonate equivalent** (CCE) and particle size. Higher the CCE, greater will be liming potential. The smaller the particle size, the faster it will react with the soil to raise pH. The **effective neutralizing power** (ENP) takes into account the fineness factor, CCE, and per cent dry matter. The ENP can be thought of as a way to compare one lime source to another for effectiveness, similar to the way a comparison between fertilizers is made.

CHAPTER 28

Hi-Tech Horticulture: Micro-Irrigation (Drip and Sprinkler Irrigation)

INTRODUCTION

Water is one of the most important factors limiting production of horticultural crops. Lack of water can lead to declining tree health, lower yield and poor quality fruits. In fact, every aspect of plant growth is affected by water stress. The long-term viability of orchard high production of vegetable and flowers, therefore, depends on availability of water, particularly, during critical stages of plants production cycle. Water is used in photosynthesis to form sugars, as a solvent for biochemical reactions and for the translocation of cell constituents and the support of the plant by turgor, but only about one per cent of the water absorbed by the plants is utilized for these metabolic activities. However, frequent water stress may upset the nutrient status of the plant, resulting in various nutrient deficiencies and metabolic disturbances.

ECONOMIC CHARACTERISTICS OF WATER

Water provides goods (e.g. drinking-water, irrigation water) and services (e.g. hydroelectricity generation, recreation and amenity) that are utilized by agriculture, industry and households. Provision of many of these goods and services is interrelated, determined by the quantity and quality of available water. Management and allocation of water entails consideration of its unique characteristics as a resource. Water used for irrigation can be pumped from reserves of groundwater, or abstracted from rivers or bodies of stored surface water. It is applied to crops by flooding, via channels, as a spray or drips from nozzles. Crops also obtain water from precipitation. Water infiltrates into the soil, evaporates, or runs off as surface water. Of the water that infiltrates the soil, some is taken up by plants (and later lost through transpiration) and some percolates more deeply, recharging groundwater. This water can be polluted with agrochemicals (fertilizers, herbicides and pesticides), with salts leached from the soil and with effluent from animal waste.

Characteristics of demand for water for irrigation relate to quantity, location, timing and quality. Irrigation generally requires large volumes of water, which can be low in quality. This is in contrast to household use of water, for example, which requires low quantities of water of high quality. The large volumes of water required for irrigation usually have to be transported over some distance to the field. For surface water, canals and pipes can enable conveyance; in the case of groundwater, extraction is provided via tubewells. In terms of timing, demand for irrigation water can extend through the growing season and, where adequate supplies are available, extend into the dry season for multiple cropping. Peak demand for irrigation water does not usually coincide with peak flows of surface water.

THE DOMINANCE OF AGRICULTURAL IN WATER USE

Irrigation is a vital component of agricultural production in many developing countries. In 1997-99, irrigated land provided two-fifths of crop production in developing countries, and accounted for about one-fifth of the cultivated area. The divergence in these statistics reflects the high crop yields and multiple cropping that are achieved through irrigation (FAO, 2002a). Developing countries are particularly dependent on irrigation. Food production in developing countries is increasing in response to the demands of an expanding population and rising prosperity. Some of this demand will be met by increased productivity of rainfed agriculture, some by increased imports, but irrigated agriculture will be a major contributor.

- Agriculture is the largest user of water in all regions of the world except Europe and North America (FAO, 2002b). In 2000, agriculture accounted for 70 per cent of water withdrawals and 93 per cent of water consumption worldwide, where consumption refers to withdrawals net of returns flows and evaporation. This is in contrast to industry, which accounted for 20 per cent of withdrawals and 4 per cent of consumption worldwide in 2000, and household use, which accounted for 10 per cent of withdrawals and 3 per cent of consumption (FAO 2004 (AQUASTAT-database) FAO, 2002b). The water requirements of agriculture are large relative to water requirements for other human needs. The human body needs about 3 litres of water per day;
- For domestic uses people use approximately 30 - 300 litres of water per person per day;
- To grow their daily food needs people require 3000 litres of water per person per day. (FAO 2003)

Water Withdrawals and Consumption

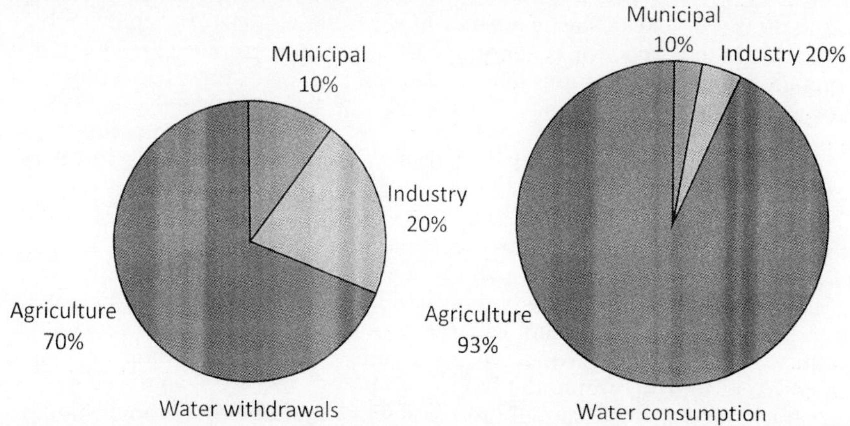

FIG. 28.1: Water withdrawals and consumption

However, the agriculture sector is often criticized for high wastage and inefficient use of water at the point of consumption (i.e. at farm level) encouraged by subsidized low charges for water use or low energy tariffs for pumping. Hence, irrigation plays an important role in fruit cultivation. Micro-irrigation (MI), aims at increasing the area under efficient methods of irrigation viz. drip and sprinkler irrigation.

Micro Irrigation System (MIS)

Drip/Micro irrigation one of the major sectors of application of plastics in horticulture is in the area of water management and conservation. Drip/Micro irrigation is a technology for providing irrigation to plants through network of pipes. The term drip/ micro irrigation includes emitting water by drippers, micro-sprinklers, mini-sprinklers, micro-jets, misters, fan jets, micro-sprayers foggers, emitting pipes, micro-tubes and similar other emitting pipes. The use of different components depends upon the requirement, which may vary due to crop type, water requirement, plant spacing, soil type etc. Drip/Micro irrigation technology can be used both on the surface as well as sub-surface. Technology helps in increasing productivity by 30 to 100 per cent with 50 to 70 per cent saving of water. There is enhanced water use efficiency (WUE). Micro irrigation saves the energy as compared to sprinkler and conventional methods. Sprinkler irrigation enables water to be sprinkled with the help of a network of pipes and sprinklers (nozzles) on the plant foliage. Sprinkler irrigation system has become popular as a method of irrigation especially under undulating topography particularly for light textured soils for a variety of horticultural crops. Sprinkler system is ideally suitable for close spaced crops like vegetables, spices, medicinal & aromatic plants and flowers. It is now popularized in some fruit and plantation crops. Systematic research on micro irrigation of horticultural crops was conducted at Tamil Nadu Agriculture University, Coimbatore in seventies and thereafter a large number of trails have been conducted for wide varieties of crops under varying agro-climatic conditions by Water Technology Centre, IARI, New Delhi and MPKVV, Rahuri. All the trials have proved that micro irrigation improves productivity and quality of produce, reduces labour cost, salt concentration in the root zone, disease incidence, and economizes on fertilizer if used through drip system. Since, drip irrigation permits frequent application of water without soil moisture stress, enhanced growth is achieved. However, high investment at initial stage and frequent clogging of emitters, have become impediment in wider uses of this system. Fertilizer use efficiency is also enhanced when it is applied through drip irrigation. Increased productivity with saving of 30 to 40 per cent of fertilizer is achieved through fertigation. Thus, drip irrigation is becoming a major component in precision farming. Realizing the usefulness of drip irrigation, several incentives have been provided by Government of India to popularize this system with farmers. Initially, GOI launched a Centrally Sponsored Scheme on Use of Plastics in Agriculture during the VIII Plan, duly earmarking 200 crore (out of Rs.250 crores) for drip irrigation. The Scheme was continued during IX Plan under the title Development of Horticulture through Plasticulture Applications with an outlay of Rs. 375 crore out of which Rs. 300 crore was for drip irrigation. The scheme was thereafter subsumed under the Centrally Sponsored Scheme on Macro Management in Agriculture during 2000-01. In July 2003, Government of India constituted a Task Force on Micro Irrigation to assess the scope and potential of the technology and to suggest strategies for it's promoted in the country. The Task Force, which was headed by the then Chief Minister of Andhra Pradesh, gave several recommendations for promoting micro- irrigation in the country. It is recommended that additional area of 6.65 m ha could be brought under the micro-irrigation. The cost for that purpose would be about 1,828 crores. Expert group recommends that banana, grape, be brought under the purview of micro irrigation. Sprinkler irrigation system does not categorized under use of plastics in agriculture. Some of the major recommendations of the Task Force on Micro irrigation are as under.

- A total of 17 million ha need to be covered under Micro Irrigation in the country by the end of XI Plan including three million ha during X Plan
- Financial Assistance should be available on all crops. The rate of financial assistance to the farmers should be 50% of unit cost, 40% share should be from GOI and the balance 10 % by the respective state governments. The states may raise their share through RIDF loan from NABARD.
- The funds need to be released through NABARD/leading nationalized banks rather than state implementing agencies or macro management scheme; there should be tax concessions to make the system affordable by farmers.
- An Apex Body in the form of National Council for Precision Farming needs to be constituted as a registered society under Societies Registration Act 1860.

In line with the recommendations of the Task Force on micro-irrigation, the Government of India launched a Centrally Sponsored Scheme on micro-irrigation during the X Plan, with effect form 2005-06 with outlay of Rs 850.00 crore for the plan period. The scheme envisages covering an area of 6.2 lakh ha under MI (3.8 under drip and 2.4 under sprinkler) during the X Plan. The horticultural crops would share 50% drip irrigation with allotment of 60% share of budget due to closely spaced cash crops. Through the GOI support, good awareness has been created about the usefulness of micro-irrigation in different parts of the country. Due to governmental support, presently more then 3 lakh ha land is covered under micro-irrigation system. India is emerging as a leading country in the use of micro-irrigation system. In view of paucity of water availability, drip irrigation has to be given priority to make the technology more effective it will be necessary to address the methods for predicting the temporal and spatial variation of soil moisture, the minimal and optimal fraction of the soil volume to be wetted, and management system in different horticultural crops under varying weather and soil conditions, application of nutrients, critical stages of growth and development etc.

TYPES OF MICRO-IRRIGATION SYSTEM (MIS)

Drip Irrigation: It involves technology for irrigating plants at the root zone through emitters fitted on a network of pipes (mains, sub-mains and laterals). Drip irrigation is sometimes called trickle irrigation and involves dripping water onto the soil at very low rates (2-20 litres/hour) from a system of small diameter plastic pipes fitted with outlets called emitters or drippers. Water is applied close to plants so that only part of the soil in which the roots grow is wetted, unlike surface and sprinkler irrigation, which involves wetting the whole soil profile. With drip irrigation water, applications are more frequent (usually every 1-3 days) than with other methods and this provides a very favourable high moisture level in the soil in which plants can flourish. Drip irrigation system is used for cultivation of apple, grapes, banana, sugarcane, tea, coffee, cotton, mango, teakwood, vegetables, flowers, etc. The emitting devices could be drippers, which are designed to discharge water at prescribed rates. The use of different emitters will depend upon specific requirements, which may vary from crop to crop. Water requirement, age of plant spacing, soil type, water quality and availability are some of the factors which would decide the choice of the emitting system. Sometimes micro-tubes are also used as an emitter, though it is inefficient. Micro-irrigation technologies, commonly in use in water scarce areas of developed countries, constitute one such intervention with the ability to use water more efficiently in irrigated agriculture. These technologies can improve productivity; raise incomes through crop yields and outputs; and enhance food security of households. Numerous studies have established the gains from micro-irrigation adoption and several government and non-government organizations are engaged in actively promoting the technologies. In India, micro-irrigation technologies have been marketed for more than three decades. The main vehicle of government policies to promote micro-irrigation systems is product subsidies in certain cases up to 90 per cent. However, there has been a lukewarm response to such initiatives from farmers, especially smallholders. This can be attributed to several causes: lack of access to groundwater, lack of cash, crop specificity of the available micro-irrigation technologies, lack of know-how, poor product quality and absence of adequate credit facilities. Studies show that despite active promotion, the appeal of these technologies has remained confined to "*gentlemen farmers*" - wealthier farmers who produce commercial crops. Despite these constraints, in certain pockets of India, these technologies have become a popular choice among farmers. It is notable that, in some of these cases, the technologies have been adopted in the absence of government subsidies. However, in general special efforts are required to market cost appropriate technologies to the poor and smallholder farmers. Drip irrigation is often promoted for reasons that do not match with the farmers' main concerns. While the government promotes drips as long-term investments for water saving and sustainable agriculture, the farmers look for more immediate and assured benefits, such as lower costs and increased incomes.

Components of Drip Irrigation System (Listed in Order from Water Source)

1. Pump or pressurized water source
2. Water Filter(s) - Filtration Systems: Sand Separator like Hydro-Cyclone, Screen filters, Media Filters
3. Fertigation Systems (Venturi injector) and Chemigation Equipment (optional)
4. Backwash Controller (Backflow Preventer)
5. Pressure Control Valve (Pressure Regulator)
6. Main Line (larger diameter Pipe and Pipe Fittings)
7. Hand-operated, electronic, or hydraulic Control Valves and Safety Valves
8. Smaller diameter poly tube (often referred to as *laterals*)
9. Poly fittings and Accessories (to make connections)
10. Emitting Devices at plants (ex. Emitter or Drippers, micro spray heads, inline drippers, trickle rings)

Note that in Drip irrigation systems pump and valves may be manually or automatically operated by a controller.

Distribution line and fittings: Distribution lines consist of a network of graduated pipe size starting with a single, large main line followed by smaller submain and lateral lines. The buried main and submain lines are normally permanent and area made up of PVC and HDPE. Main and submain line range from 40 mm OD to 110 OD sizes depending upon required water flow to the laterals. Both main and submain line should have valved outlets for periodic flushing. Lateral lines are usually made of LDPE or LLDPE and ranges from 10 mm to 20 mm diameter. Appropriate fitting are available for connecting PVC, HDPE and polyethylene distribution lines. Using heat butt-welding can join PVC pipes. Polyethylene tubing, etc. which are not cementable, can be connected with barbed or compression. These common fitting are elbows, tees, crosses, unions and plugs or end caps. Most large drip irrigation systems employ some type of filter to prevent clogging of the small emitter flow path by small waterborne particles. New technologies are now being offered that minimize clogging. Some residential systems are installed without additional filters since potable water is already filtered at the water treatment plant. Virtually all drip irrigation equipment manufacturers recommend that filters be employed and generally will not honor warranties unless this is done. Last line filters just before the final delivery pipe are strongly recommended in addition to any other filtration system due to fine particle settlement and accidental insertion of particles in the intermediate lines.

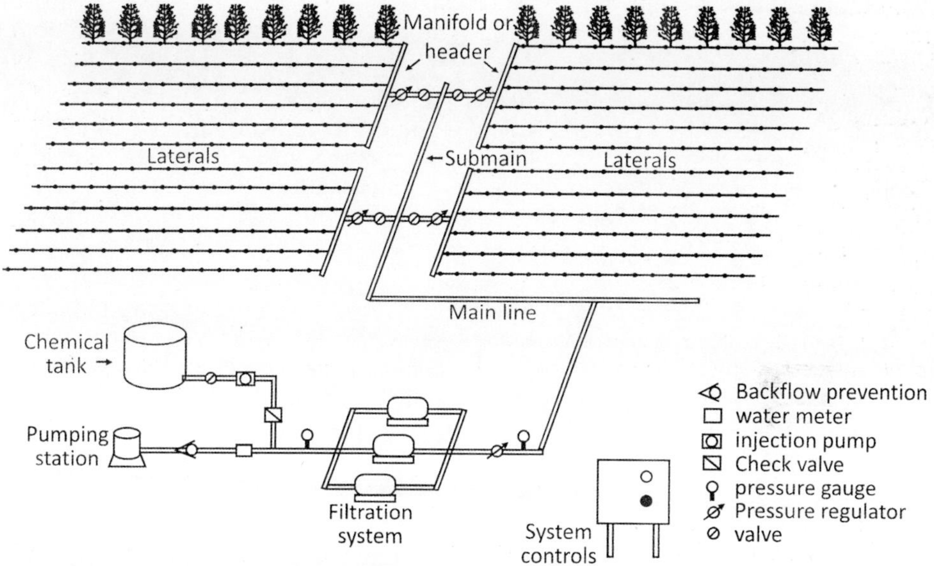

FIG. 28.2: A typical micro-irrigation system layout for vegetable production

Pumps and Power Unit

Micro-irrigation systems are typically designed to make the best use of the amount of water available. The type and size of pump selected will depend on the amount of water required, the desired pressure and the location of the pump relative to the distribution network. Electric power units or internal combustion engine driven pumps are equally adaptable. However, the electric power unit is preferred because it is easier to automate.

Filters

Filters remove sand and larger suspended particles before they enter the distribution network. However, the filters cannot remove dissolved minerals, bacteria and some algae. The three types generally used are screen filters, disk, sand filters and centrifugal filters.

Distribution Lines

The water distribution system is a network of pipes and tubes that can range in size from 1/2 inch to 6 inches in diameter. Water from the pump may be carried to the edge of the field by a single large main. Smaller sub-mains may then carry the water to laterals and ultimately to the emitters.

Control Components

The control portion may include a combination of the following devices: pressure regulator, valve, vacuum relief valve and timing clock or controller. A flow meter should be used to measure the amount of water. Pressure gauges monitor the water pressure at the pump and other locations. Equipment to inject fertilizers into the water line is also frequently used. Backflow prevention devices are used to prevent contamination of the water source.

Emitter/ Dripper

An emitter is also called a dripper and is used to transfer water from a pipe or tube to the area that is to be irrigated. Typical emitter flow rates are from 2-20 litre/hr. In many emitters, flow will vary with pressure, while some emitters are *pressure compensating*. These emitters employ silicone diaphragms or other means to allow them to maintain a near-constant flow over a range of pressures, for example from 10 to 50 psi (70 to 350 kPa).

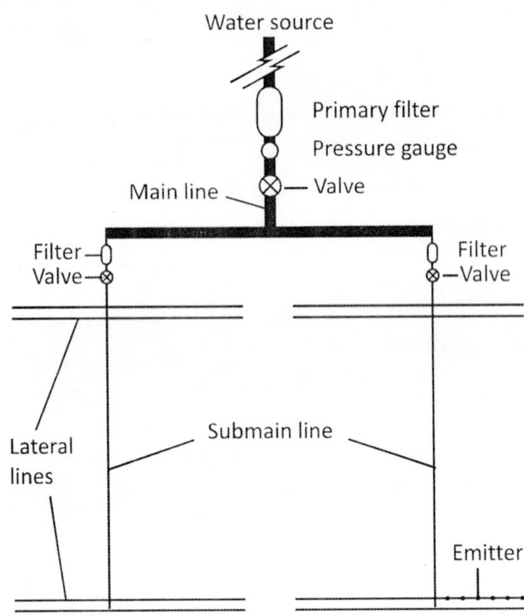

FIG. 28.3: Typical water distribution line of a micro-irrigation system

Advantages of Drip Irrigation

Drip irrigation is most suitable for row crops (vegetables, soft fruit), tree and vine crops where one or more emitters can be provided for each plant. Generally only high value crops are considered because of the high capital costs of installing a drip system. Drip irrigation is adaptable to any farmable slope. Normally the crop would be planted along contour lines and the water supply pipes (laterals) would be laid along the contour also. This is done to minimize changes in emitter discharge as a result of land elevation changes. The major advantages of drip irrigation systems are:

1. It gives recorded increase in yield up to 230%.
2. It saves water up to 70% compare to flood irrigation. More land can be irrigated with the water thus saved. High water application efficiency.
3. Crop grows consistently, healthier and matures fast.
4. Early maturity results in higher and faster returns on investment.
5. Fertilizer use efficiency increases by 30%. Minimized fertilizer/nutrient loss due to localized application and reduced leaching.

6. Cost of fertilizers, inter-culturing and labour use gets reduced.
7. Fertilizer and chemical treatment can be given through micro-irrigation system itself.
8. Undulating terrains, saline, water logged, sandy and hilly lands can also be brought under productive cultivation.
9. Leveling of the field not necessary, ability to irrigate irregular shaped fields, Allows safe use of recycled water.
10. Moisture within the root zone can be maintained at field capacity.
11. Soil type plays less important role in frequency of irrigation and minimized soil erosion
12. Highly uniform distribution of water i.e., controlled by output of each nozzle.
13. Foliage remains dry thus reducing the risk of disease.
14. Regulating the valves and drippers can regulate variation in supply.

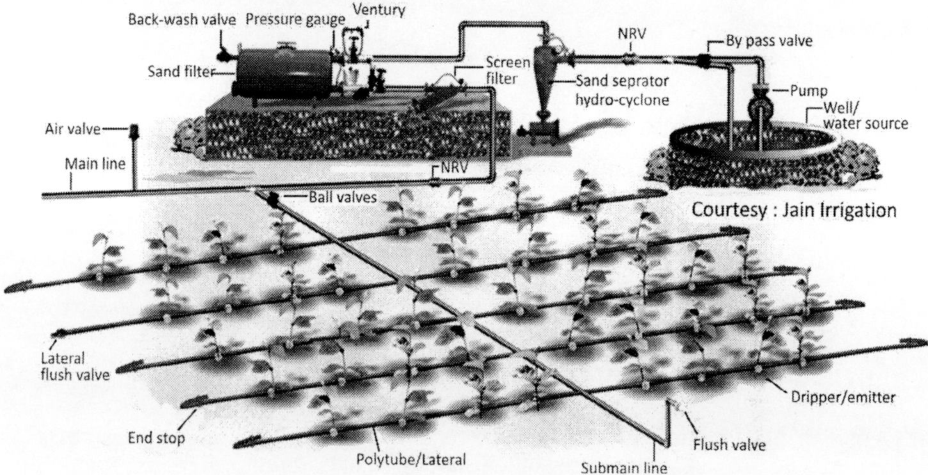

FIG. 28.4: Drip irrigation (Adopted from Jain Irrigation systems Ltd., Jalgaon)

Disadvantages of Drip Irrigation System

1. **Management:** Micro-irrigation systems normally have greater maintenance requirements. Soil particles, algae, or mineral precipitates can clog the emission devices.
2. **Potential for damage:** Animals, rodents and insects may cause damage to some components. The drip and bubbler irrigation systems need additional equipment for frost protection.
3. **High initial cost:** Micro-irrigation systems are ideal for high value installations such as orchards, vineyards, greenhouses, and nurseries where traditional irrigation methods may not be practical. However, the investment cost can be high.
4. **Germination Problems**: In lighter soils subsurface drip may be unable to wet the soil surface for germination. Requires careful consideration of the installation depth.
5. **Salinity:** Most drip systems are designed for high efficiency, meaning little or no leaching fraction. Without sufficient leaching, salts applied with the irrigation water may build up in the root zone, usually at the edge of the wetting pattern.
6. **Clogging:** If the water is not properly filtered and the equipment not properly

maintained, it can result in clogging.

7. Waste of water, time & harvest, if not installed properly. These systems require careful study of all the relevant factors like land topography, soil, water, crop and agro-climatic conditions, and suitability of drip irrigation system and its components.

FIG. 28.5: Drip irrigation system in a small village.

PLASTIC MULCHING WITH DRIP IRRIGATION SYSTEM

The use of plasticulture as plastic covered raised beds in drip irrigation is now common. The use of plasticulture in the production of horticultural crops (vegetables, small fruits, flowers, tree fruits, and ornamentals) presents a lot of benefits. The beds typically are from 6- to 10- inches in height and 24- to 36-inches in width. The plastic mulch serves to retain injected fumigants and nutrients, minimize weed growth, maintain bed shape and keeps the lower fruit away from the soil. This type of bedding practice allows easy adaptation to micro-irrigation systems that utilize line-source or drip type irrigation laterals. Drip irrigation laterals can be installed at the same time as the plastic mulch by modifying the mulch-laying implement to hold a reel of tubing. On sandy soils, where capillary raise of water is small, laterals are placed directly under the mulch. Some tubing manufacturers recommend burying the tube 1- to 2-inches below the soil surface. On heavier soils, the irrigation tubing can be buried much deeper and sometimes left in place for several years. This type of drip irrigation is called Subsurface Drip Irrigation (SDI).

Farmers can truly maximize efficiencies and growth. Plasticulture add benefit of evaporative cooling and improve plant respiration (means plants do not overheat and shut down), fruit quality is better and shelf life is extended. Using daily drip irrigation for 2-6 hours farmers can maintain soil water capacity at 90 –100%. At renovation no re-establishment of the planting occurs. Plants are never hurt or water stressed.

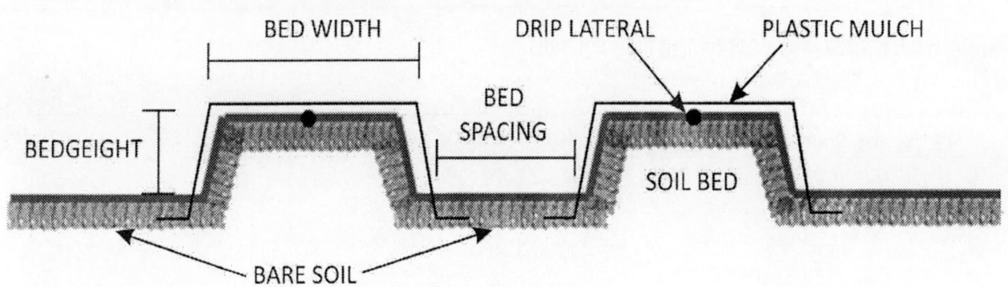

FIG. 28.6: Illustration of drip irrigation and soil bed.

TABLE 28.1: Result of studies on micro irrigation under PFDCs funded by NCPAH

S.No	Crops	Yield (q/ha)		Irrigation (cm)		WUE (q/ha/cm)		Advantage of MI (%)	
		Surface	Drip	Surface	Drip	Surface	Drip	Surface	Drip
1	Beet	5.70	8.90	86.00	18.00	0.07	0.50	79.10	36.00
2	Bitter gourd	32.00	43.00	76.00	33.00	0.42	1.30	56.60	25.60
3	Bottle gourd	380.0	558.0	84.00	74.00	4.50	7.50	12.00	31.90
4	Brinjal	91.00	148.0	168.0	64.00	0.55	2.30	62.00	38.50
5	Broccoli	140.0	195.0	70.00	60.00	2.00	3.25	14.30	28.20
6	Cauliflower	83.00	116.0	39.00	26.00	2.10	4.50	33.30	28.40
7	Chilly	42.30	60.90	109.0	41.70	0.39	1.50	61.70	30.50
8	Cucumber	155.0	225.0	54.00	24.00	2.90	9.40	55.60	31.10
9	Okra	100.0	113.1	53.50	8.60	1.87	13.20	84.00	11.60
10	Onion	284.0	342.0	52.00	26.00	5.50	13.20	50.00	17.00
11	Potato	172.0	291.0	60.00	27.50	2.90	10.60	54.20	41.00
12	Radish	10.50	11.90	46.00	11.00	0.23	1.10	76.10	11.80
13	Sweet potato	42.40	58.90	63.00	25.00	0.67	2.40	60.30	28.00
14	Tomato	45.00	58.00	102.0	77.00	0.44	0.75	24.50	22.40
15	Banana	277.0	329.0	186.0	172.0	1.50	1.90	7.50	15.80
16	Grape	264.0	325.0	53.00	28.00	5.00	11.60	47.20	18.80
17	Papaya	130.0	230.0	228.0	73.0	0.60	3.20	685.0	43.50
18	Pomegranate	34.00	67.00	21.00	16.00	1.62	4.20	23.80	49.30
17	Water melon	82.10	504.0	72.00	25.00	5.90	20.20	65.30	16.30
18	Ber	13.7/ plant	18.0/ plant	15.4 m^3/ plant	12.5 m^3/ pl.	0.90	1.40	18.80	23.90
20	Guava	0.16/ plant	0.22/ plant	6.4 m^3/ plant	5.2 m^3/ plant	0.03	0.04	18.60	27.30
21	Kinnow	68.00	98.00	22.10	17.30	3.10	5.70	21.70	30.60
22	Citrus	15.00	27.00	23.00	17.50	0.65	1.54	23.90	44.40

Source: Report of Task Force on Micro-irrigation, GOI (2004)

TABLE 28.2: Area under micro irrigation in India

(Area as on March, 2006 in 000 ha)

S.No.	Name of state	Drip system	Sprinkler system	Total area
1	Andhra Pradesh	155.44	124.51	279.951
2	Arunachal Pradesh	0.613	0	0.613
3	Assam	0.116	0.129	0.245
4	Chhattisgarh	1.979	3.765	5.744
5	Goa	0.746	0.306	1.052
6	Gujarat	53.707	96.374	150.081
7	Haryana	4.258	503.877	508.135
8	Himachal Pradesh	0.116	0.581	0.697
9	Karnataka	114.433	157.028	271.461
10	Kerala	10.562	1.548	12.11
11	Madhya Pradesh	6.483	100.00	106.483
12	Maharastra	341.848	153.507	495.355
13	Manipur	0.03	0	0.03
14	Mizoram	0.072	0.106	0.178
15	Nagaland	0	3.962	3.962
16	Orissa	2.036	20.22	22.256
17	Punjab	5.101	10.0	15.101
18	Rajasthan	10.025	554.708	564.733
19	Sikkim	0.80	10.03	10.83
20	Tamil Nadu	116.665	26.332	142.997
21	Uttar Pradesh	4.609	10.0	14.609
22	Uttaranchal	0.038	0.006	0.044
23	West Bengal	0.110	150.02	150.13
Grant Total		**829.788**	**1927.009**	**2,756.797**

Low-cost micro-irrigation technologies: Micro-irrigation technologies can be broadly categorized into two types based on their technical and socioeconomic attributes: low-cost micro-irrigation technologies and the commercialized, state-of-the-art micro-irrigation systems. Low-cost systems include the *pepsee* easy drip technology, bucket and drum kits, micro sprinklers, micro tube drip systems and others that have been designed by organizations such as the International Development Enterprises (IDE), along with innovative farmers. The more sophisticated, capital-intensive systems are conventional drip and sprinkler systems. Low-cost micro-irrigation technologies are largely promoted to poor farmers, hence their competitive pricing and compatibility with smallholder farming systems. Farmers can generally recover their initial investment capital between one and three years, although the extent of economic gains from investment depends on the type of crop. Subsidies and options for financing from organizations and government schemes.

Indigenous drip method of irrigation: This method is cheap, consists of mud pots of 20 litres capacity with pencil size hole made a little above the bottom. These pots are kept buried neck deep in the basins of fruit trees. The pot is filled with water and covered with lid. Water slowly trickles from the pot and replenishes the root zone. This method is suitable for young fruit trees when water is in shortage. By this method, considerable amount of water is saved. Evaporation and deep percolation losses are less. Weed growth in the basins of fruit trees is also less.

Sprinkler Irrigation: Under sprinkler irrigation water is sprinkled under pressure into the air and plant foliage through a set of nozzles attached to network of aluminum or High Density Poly Ethylene (HDPE) pipes in the form of rainfall. These systems are suitable for irrigating crops where the plant density is very high where adoption of drip irrigation systems may not be economical. Sprinkler irrigation is suitable for horticultural crops like vegetables and seed spices. Sprinkler irrigation is a method of applying irrigation water, which is similar to natural rainfall. Water is distributed through a system of pipes usually by pumping. It is then sprayed into the air through sprinklers so that it breaks up into small water drops, which fall to the ground. The pump supply system, sprinklers and operating conditions must be designed to enable a uniform application of water. Water is distributed through a system of pipes usually by pumping. It is then sprayed into the air and irrigated entire soil surface through spray heads so that it breaks up into small water drops, which fall to the ground. Sprinklers provide efficient coverage for small to large areas and are suitable for use on all types of properties. It is also adaptable to nearly all-irrigable soils since sprinklers are available in a wide range of discharge capacity. Sprinkler system is designed to ensure maximum water saving, combining high quality, affordability and ease of installation. The systems are made out of high strength and chemical resistance engineering plastics to achieve functional satisfaction and to maintain cost economics.

There are many different type of sprinklers, viz., over-tree or under-tree and fixed or moveable. The over-tree sprinkler heads (nozzles) are mounted on high standards and spray over the vegetable crown. The purpose is to achieve a greater uniformity of distribution of water without interference from the trunks and branches. The height of the risers can be increased progressively in a new plantation. For the mature tree, they are provided with lateral support. For mango, the sprinklers are spaced 20 m apart both ways. For a typical tree spacing of 10 m x10 m each sprinkler covers two rows of trees in either direction. Similarly, for lime and lemon spaced at 4 m x 6 m, spacing of the sprinklers can be kept at 12 m x 18 m to cover three rows of trees in either direction. Where water is expensive and wind conditions are severe, it is preferable to use under-tree type of sprinkler. This has a vertical trajectory angle from 40 to 70 and a typical radius of 15 m. Typical spacing for these low sprinklers are from 6 to 12 m depending upon the crop. These sprinklers are usually of low intensity ranging from 3 to 8 mm/hr. A humid microclimate is formed under the crowns which also reduces evaporation from the ground surface.

Where to use Sprinkler Irrigation

Sprinkler irrigation is adaptable to any farmable slope, whether uniform or undulating. The lateral pipes supplying water to the sprinklers should always be laid out along the land contour whenever possible. This will minimize the pressure changes at the sprinklers and provide a uniform irrigation. Sprinklers are best suited to sandy soils with high infiltration rates although they are adaptable to most soils. The average application rate from the sprinklers (in mm/hour) is always chosen to be less than the basic infiltration rate of the soil so that surface ponding and runoff can be avoided. A good clean supply of water, free of suspended sediments, is required to avoid problems of sprinkler nozzle blockage and spoiling the crop by coating it with sediment.

Sprinklers are not suitable for soils, which easily form a crust. If sprinkler irrigation is the only method available, then light fine sprays should be used. The larger sprinklers producing larger water droplets are to be avoided.

Components ofsprinkler irrigation system- A typical sprinkler irrigation system consists of the following components:

1. Pump unit
2. Mainline and sometimes sub-mainlines
3. Laterals
4. Sprinklers

The pump unit is usually a centrifugal pump, which takes water from the source and provides adequate pressure for delivery into the pipe system.

FIG. 28.7: Micro-sprinklers of different sizes.

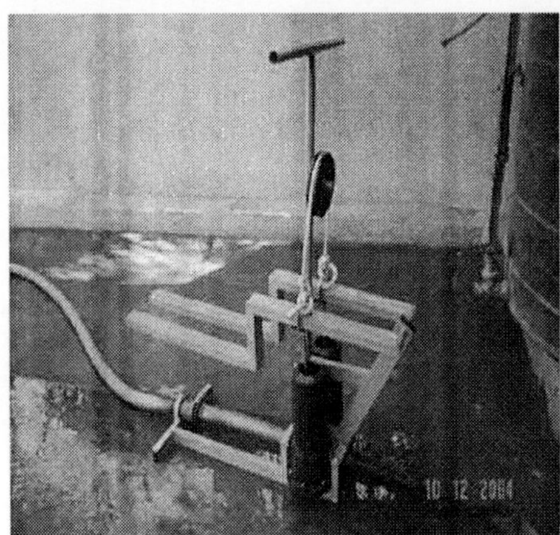

FIG. 28.8: Fig. Sprinkler pump (Left), Sprinkler in lawn field (Right)

The mainline - and sub-mainlines are pipes meant for delivery of water from the pump to the laterals. In some cases these pipelines are permanent and are laid on the soil surface or buried below ground. In other cases they are temporary, and can be moved from field to field. The main pipe materials used include asbestos cement, plastic or aluminium alloy. The laterals deliver water from the mainlines or sub-mainlines to the sprinklers. They can be permanent but more often they are portable and made of aluminium alloy or plastic so that they can be moved easily.

The most common type of sprinkler system consists of a system of lightweight aluminium or plastic pipes, which are moved by hand. The rotary sprinklers are usually spaced 9-24 m apart along the lateral, which is normally 5-12.5 cm in diameter. This is so it can be carried easily. The lateral pipe is located in the field until the irrigation is complete. The pump is then switched off and the lateral is disconnected from the mainline and moved to the next location. It is re-assembled and connected to the mainline and the irrigation begins again. The lateral can be moved one to four times a day. It is gradually moved around the field until the whole field is irrigated. This is the simplest of all systems. Some use more than one lateral to irrigate larger areas.

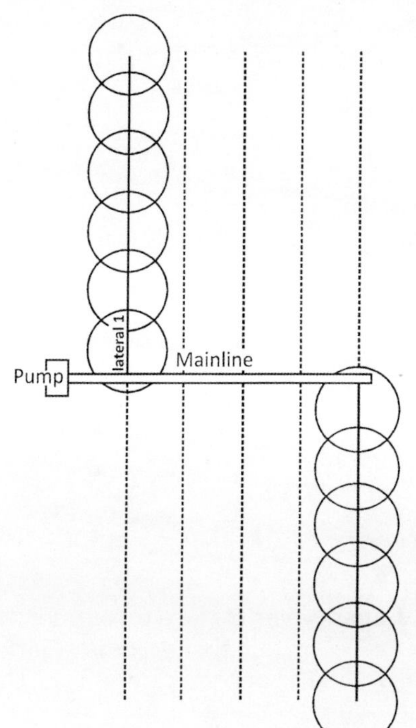

FIG. 28.9: Sprinkler System Layout using

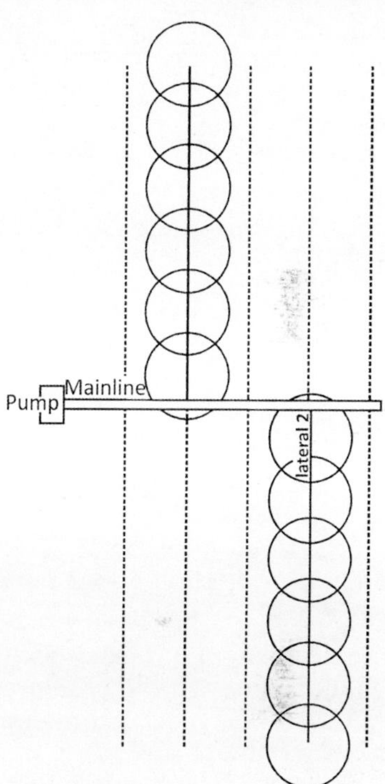

FIG. 28.10: Hand-moved sprinkler system two laterals

OPERATING OF SPRINKLER SYSTEMS

The main objective of a sprinkler system is to apply water as uniformly as possible to fill the root zone of the crop with water.

Wetting patterns: The wetting pattern from a single rotary sprinkler is not very uniform. Normally the area wetted is circular (see topview). The heaviest wetting is close to the sprinkler (see sideview). For good uniformity several sprinklers must be operated close together so that their patterns overlap. For good uniformity the overlap should be at least 65% of the wetted diameter. This determines the maximum spacing between sprinklers.

The uniformity of sprinkler applications can be affected by wind and water pressure.

Spray from sprinklers is easily blown about by even a gentle breeze and this can seriously reduce uniformity. To reduce the effects of wind the sprinklers can be positioned more closely together.

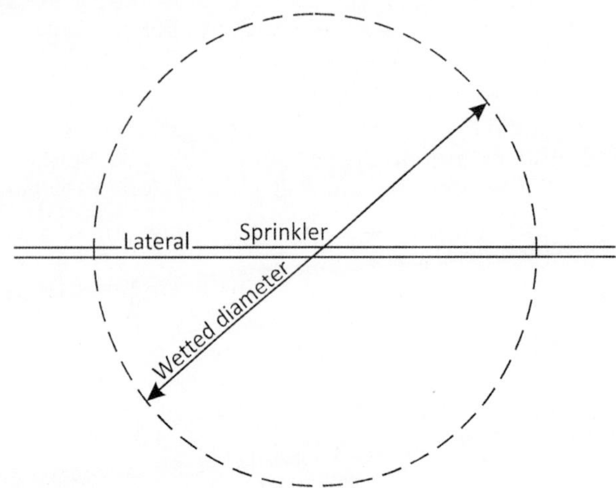

FIG. 28.11: Wetting pattern for a single sprinkler (TOP VIEW)

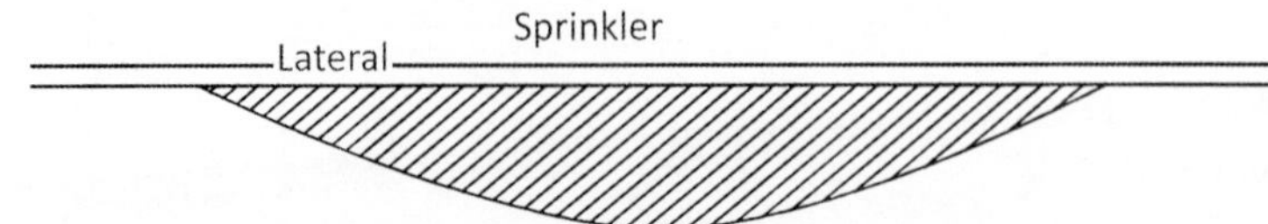

FIG. 28.12: Wetting pattern for a single sprinkler (SIDE VIEW)

Sprinklers will only work well at the right operating pressure recommended by the manufacturer. If the pressure is above or below this then the distribution will be affected. The most common problem is when the pressure is too low. This happens when pumps and pipes wear. Friction increases and so pressure at the sprinkler reduces. The result is that the water jet does not break up and all the water tends to fall in one area towards the outside of the wetted circle. If the pressure is too high then the distribution will also be poor. A fine spray develops which falls close to the sprinkler.

Application Rate

This is the average rate at which water is sprayed onto the crops and is measured in mm/hour. The application rate depends on the size of sprinkler nozzles, the operating pressure and the distance between sprinklers. When selecting a sprinkler system it is important to make sure that the average application rate is less than the basic infiltration rate of the soil. In this way all the water applied will be readily absorbed by the soil and there should be no runoff.

Sprinkler Drop Sizes

As water sprays from a sprinkler it breaks up into small drops between 0.5 and 4.0 mm in size. The small drops fall close to the sprinkler whereas the larger ones fall close to the edge of the wetted circle. Large drops can damage delicate crops and soils and so in such conditions it is best to use the smaller sprinklers.

Drop size is also controlled by pressure and nozzle size. When the pressure is low, drops tend to be much larger as the water jet does not break up easily. So to avoid crop and soil damage use small diameter nozzles operating at or above the normal recommended operating pressure.

Control of Sprinklers Cycle

If you're irrigating using sprinklers, the water will probably start to run off into the gutter, or into a low spot, before the soil is wet to a 6 inch depth. This is because the sprinklers put out more water in a given amount of time than the soil can absorb. In technical terms the precipitation rate of the sprinklers is greater than the infiltration rate of the soil. Continue this run-stop-wait-run cycle until the soil is saturated to a 6 inch depth. This process is referred to as cycling the sprinklers. Almost all sprinkler systems need to be cycled for proper irrigation.

Advantages of Sprinkler System

Water savings: Conveyance loss is minimal. Evaporation, runoff and deep percolation are reduced as compared to other traditional irrigation systems. A water supply source with limited flow rates such as small water wells or city/rural water can be used.

Energy savings: A smaller power unit is required compared to sprinkler irrigation systems.

Weed and disease reduction: Because of limited wetted area from non-spray type of micro-irrigation, weed growth is inhibited and disease incidences reduced.

Can be automated: Fertilizers and chemicals can be applied with water through the irrigation system. Micro-irrigation systems can be automated which reduces labor requirements.

Improved production on marginal land: On hilly terrain, micro-irrigation systems can operate with no runoff and without interference from the wind. The fields need not be leveled.

Disadvantages of Sprinkler System

A common problem with sprinkler irrigation is the large labour force needed to move the pipes and sprinklers around the field. In some places such labour may not be available and may also be costly. To overcome this problem many mobile systems have been developed such as the hose reel raingun and the centre pivot.

Another system, which does not need a large labour force, is the drag-hose sprinkler system. Main and laterals are buried PVC pipes, one lateral covers three positions. Sprinklers on risers carried by skids are attached to the laterals through hoses (similar to garden sprinklers). Only the skid with the sprinkler has to be moved from one position to another, which is an easy task.

The wetting pattern from a single rotary sprinkler is not very uniform. Normally the area wetted is circular. The heaviest wetting is close to the sprinkler. For good uniformity several sprinklers must be operated close together so that their patterns overlap. For good uniformity the overlap should be at least 65 per cent of the wetted diameter. This determines the maximum spacing between sprinklers.

Bubbler irrigation: Bubbler irrigation, a relatively new system, is designed to reduce energy requirements through inexpensive, thin walled, corrugated plastic pipe with a diameter that even the low pressure head from a surface ditch might suffice. The disadvantages in drip irrigation *i.e.* high energy consumption to pump water with pressure for distribution, clogging of drip emitters and damages to delicate screen filters, are all avoided. Bubblers typically apply water on a *per plant* basis.

Bubblers are very similar to the point source external emitters in shape but differ in performance. Water from the bubbler head either runs down from the emission device or spreads a few inches in an umbrella pattern. The bubbler emitters dissipate water pressure through a variety of diaphragm materials and deflect water through small orifices. Most bubbler emitters are marketed as pressure compensating. The bubbler emission devices are equipped with single or multiple port outlets. Most bubbler heads are used in planter boxes, tree wells, or specialized landscape applications where deep localized watering is preferable. The typical flow rate from bubbler emitters is between 2 and 20 gph. Bubbles system can cater to irrigation of widely spaced crops such as mangoes, sapotas, oranges, coconuts, grapes etc.

FIG. 28.13: Bubbler emitters of different shapes with single and multiple port exits

Microjet irrigation: In microjet irrigation, water leaves the jets at a pressure of nearly one bar. This gives throw distances of 1 to 4 m with a correspondingly large wetted area of ground. The water discharge of the jets is 5 to 160 litres per hour *i.e.* much higher than that of the drippers in a drip irrigation system. Depending on the design and operating pressure, the microjects produce coarse of fine spray *i.e.* varying degrees of atomization of the jet and varying droplet sizes.

Microsprikler irrigation: The majority of the devices for the distribution of irrigation water which are marketed and designed for mini or microsprinklers by the manufacturers differ from microjets in that the water is distributed by moving, mostly rotating, parts which usually produce a rotating jet of water. The microsprinklers differ clearly from the customary type of sprinklers used for irrigation which deliver more than 1,000 litres/hour, operate at pressures in excess of 2.5 bar, and has a throw distance of more than 10 m. The discharge in microsprinklers currently available varies from 28 to 223 litres/hour, operating pressure 0.8 to 4.0 bar and throw distance 0.9 to 4 m. The individual microjet is able to wet a much larger area of ground than a dripper. The desire to retain the advantages of drip irrigation and microjet irrigation while overcoming the disadvantage with regard to distribution of water led to the development of microsprinklers. These are suitable for the irrigation of forest trees.

Importance of Micro-Irrigation System

Changes in Cropping Patterns

An interesting outcome of micro-irrigation adoption in the study area is its impact on cropping patterns. Farmers who adopted alternative technologies in the study locations changed their crops and also the extent of cultivation. More specifically, micro-irrigation adoption proved to encourage farmers to increase their overall cropping intensity or to shift their cropping patterns to high-value, water intensive crops.

For example, in Maharashtra, the main change in cropping pattern observed was a shift from groundnut and oil seeds to high-value, water intensive crops, such as banana. In Gujarat, an increase in vegetable production was observed. If the result is an increase in total water use, there could be conflict between the positive impact on poverty and food security and the sustainability of water resource use, especially groundwater, when micro-irrigation is adopted. Studies have shown that crop yields improve when drip irrigation is used. Yields are higher because the systems allow for multiple crops to be grown; for crops to be grown under circumstances where it was not possible when there is an early withdrawal of the monsoon; and for cropping to be intensified in the same field. Farmers who adopt the system also have the possibility of extending irrigated or cultivated area on their land. This has a significant effect on the incomes of farmers, as higher yields bring higher earnings. Drip systems have the potential to improve the quality of the harvest because it is a form of precision irrigation, and applies water directly to the root zone. Each plant is able to receive the right amount of water at regular intervals, and less water is lost due to conveyance. When farmers use traditional irrigation methods, such as flood irrigation, they often tend to over irrigate or under irrigate their crops depending on the amount of water they are able to pump and not necessarily on the amount that their crop requires.

Increasing incomes and Reducing Poverty

Low-cost drip systems increase income for poor farmers by enabling more efficient use of water resources, improving yield, improving quality, and reducing labour costs. Drip systems have a particular niche in monsoonal climates. They allow farmers to plant earlier so that the crop is already established at the onset of rains and can make efficient use of rainwater. This helps to avert a crop loss, or a decline in yield that could arise from a dry spell or the early withdrawal of rain. One of the key advantages of using such technologies is that it helps to extend the use of water during times of drought or water scarcity, and mitigates the risk of losing a crop. Micro-irrigation can thus improve livelihood security to poor farmers vulnerable to rainfall variability.

In addition to quality and yield, farmers may choose to use drip systems because it reduces labour. Micro-irrigation allows for early harvesting of the crop, which reduces the labour costs of farmers. Weeds, insects and other plant diseases have also proven to occur less frequently with the adoption of this technology, cutting down the efforts that farmers have to make to protect their crop. The lower energy expended when micro-irrigation is used in cultivation has an effect on the overall cost of production, making this one of the main reasons why farmers switch to the technology. Interviews with mulberry farmers in Kolar, India recorded that the labor requirement reduced drastically from using drip irrigation over flood irrigation.

A drip system is usually permanent. When remaining In place during more than one season, a system is considered permanent. Thus it can easily be automated. This is very useful when labour is scarce or expensive to hire. However, automation requires specialist skills and so this approach is unsuitable if such skills are not available.

Water can be applied frequently (every day if required) with drip irrigation and this provides very favourable conditions for crop growth. However, if crops are used to being watered each day they may only develop shallow roots and if the system breaks down, the crop may begin to suffer very quickly.

IRRIGATION CRITERIA AND IRRIGATION SCHEDULING

Irrigation criteria are the indicators used to determine the need for irrigation. The most common irrigation criteria are soil moisture content and soil moisture tension. Less common types are irrigation scheduling to maximize yield and irrigation scheduling to maximize net return. The final decision depends on the irrigation criterion, strategy and goal. Irrigators need to define a goal and establish an irrigation criterion and strategy. To illustrate irrigation scheduling, consider a farmer whose goal is to maximize yield. Soil moisture content is the irrigation criterion. Different levels of soil moisture trigger irrigation. For example, when soil water content drops below 70 per cent of the total available soil moisture, irrigation should start. Soil moisture content to trigger irrigation depends on the irrigator's goal and strategy. In this case, the goal is to maximize yield. Therefore, the irrigator will try to keep the soil moisture content above a critical level. If soil moisture falls below this level, the yield may be lower than the maximum potential yield. Thus, irrigation is applied whenever the soil water content level reaches the critical level.

How much water to apply depends on the irrigator's strategy. For example, the irrigator can replenish the soil moisture to field capacity or apply less. If no rain is expected and the irrigator wishes to stretch the time between irrigations, it is advantageous to refill the soil profile to field capacity. If rain is expected, it may be wise not to fill the soil profile to field capacity, but leave some room for rain. If the irrigator's goal is to maximize net return, an economic irrigation criterion is needed, such as net return. This is the income from the crop less the expenses associated with irrigation. The importance of irrigation scheduling is that it enables the irrigator to apply the exact amount of water to achieve the goal. This increases irrigation efficiency. A critical element is accurate measurement of the volume of water applied or the depth of application. A farmer cannot manage water to maximum efficiency without knowing how much was applied.

Also, uniform water distribution across the field is important to derive the maximum benefits from irrigation scheduling and management. Accurate water application prevents over- or under irrigation.

Over irrigation wastes water, energy and labor; leaches expensive nutrients below the root zone, out of reach of plants; and reduces soil aeration, and thus crop yields. Under irrigation stresses the plant and causes yield reduction.

Advantages of Irrigation Scheduling

Irrigation scheduling offers several advantages

1. It enables the farmer to schedule water rotation among the various fields to minimize crop water stress and maximize yields.
2. It reduces the farmer's cost of water and labor through fewer irrigations, thereby making maximum use of soil moisture storage.
3. It lowers fertilizer costs by holding surface runoff and deep percolation (leaching) to a minimum.
4. It increases net returns by increasing crop yields and crop quality.
5. It minimizes waterlogging problems by reducing the drainage requirements.
6. It assists in controlling root zone salinity problems through controlled leaching.
7. It results in additional returns by using the "saved" water to irrigate non-cash crops that otherwise would not be irrigated during water-short periods.

Irrigation scheduling provides an average 35 per cent savings in water and energy. In fuel costs alone, this is a per-season savings of about 550 kwh per acre for a center pivot sprinkler or about 200 kwh per acre for a gated pipe.

Irrigation Scheduling Methods

All irrigation-scheduling methods consist of an irrigation criterion that triggers irrigation and an irrigation strategy that determines how much water to apply. Irrigation scheduling methods differ by the irrigation criterion or by the method used to estimate or measure this criterion. A common and widely used irrigation criterion is soil moisture status.

TABLE 28.3: Compares different methods of irrigation scheduling by monitoring soil moisture content or tension. The methods described in the table measure or estimate the irrigation criterion.

Method	Measured parameter	Equipment needed	Irrigation criterion	Advantages	Disadvantages
Hand feel and appearance of soil.	Soil moisture content by feel.	Hand probe.	Soil moisture content.	Easy to use; simple; can improve accuracy with experience.	Low accuracies; field work involved to take samples.
Gravimetric soil moisture sample.	Soil moisture content by taking samples.	Auger, caps, oven.	Soil moisture content.	High accuracy.	Labor intensive including field work; time gap between sampling and results.
Tensio-meters.	Soil moisture tension.	Tensiometers including vacuum gauge.	Soil moisture tension.	Good accuracy; instantaneous reading of soil moisture tension.	Labor to read; needs maintenance; breaks at tensions above 0.7 atm.
Electrical resistance blocks.	Electric resistance of soil moisture.	Resistance blocks AC bridge (meter).	Soil moisture tension.	Instantaneous reading; works over larger range of tensions; can be used for remote reading.	Affected by soil salinity; not sensitive at low tensions; needs some maintenance and field reading.
Water budget approach.	Climatic parameters: temperature, radiation, wind, humidity and expected rainfall, depending on model used to predict ET.	Weather station or available weather information.	Estimation of moisture content.	No field work required; flexible; can forecast irrigation needs in the future; with same equipment can schedule many fields.	Needs calibration and periodic adjustments, since it is only an estimate; calculations cumbersome without computer.
Modified atmometer.	Reference ET.	Atmometer gauge.	Estimate of moisture content.	Easy to use, directreading of reference ET.	Needs calibration; it is only an estimation.

TABLE 28.4: Ranges in available water capacity and intake rate for various soil textures

Soil Textures	Available water capacity (mm of water/cm of soil)		Intake Rate (mm/hr)	
	Range	**Average**	**Range**	**Average**
Sands	0.5 - 0.8	0.65	12 - 20	16.0
Loamy Sand	0.7 - 1.0	0.85	7 - 12	9.5
Sandy loam	0.9 - 1.2	1.05	7 - 12	9.5
Loam	1.3 - 1.7	1.50	7 - 12	9.5
Silt loam	1.4 - 1.7	1.55	4 - 7	5.5
Silty clay loam	1.5 - 2.0	1.75	4 - 7	5.5
Clay loam	1.5 - 1.8	1.65	4 - 7	5.5
Clay	1.5 - 1.7	1.60	2 - 5	3.5

TABLE 28.5: Irrigation depth for various fruit crops

Crops	Depth to Irrigate (cm)
Apples	90
Cherries	60
Grapes	90
Peaches	60
Pears	60
Raspberries	60
Strawberries	30

FERTIGATION

Fertigation is the application of fertilizers, soil amendments, or other water soluble products through an irrigation system. For intensive and economical crop production, the best solution for higher productivity is fertigation, where both water and fertilizers are delivered to growing crops through microirrigation system. Fertigation provides Nitrogen, Phosphorus, Potash as well as the essential trace elements (Mg. Fe, Zn, Cu, MO, Mn) directly to the active root zone, thus minimizing losses of expensive nutrients, which ultimately helps in improving productivity and quality of farm produce. Fertigation ensures higher and quality yield along with savings in the time and labour, which makes Fertigation economically profitable. Experiments have proved that the system economises use of fertilizer and water ranging from 40-60 per cent. Few progressive farmers of grapes, pomegranate, and banana are experiencing this and still it is beyond the reach of resource poor farmers. Fertigation is ideally suited for hi-tech horticultural production systems since it involves not only the efficient use of the two most precious inputs i.e., water and nutrients but also ensures their simultaneous availability to plants. Though micro irrigation has found widespread use in plantation and horticultural crop production in India, fertigation is confined to few high value crops. Significant yield response coupled with enhanced quality of produce is possible through hi-tech productivity using fertigation. One of the main reasons for slow adoption of fertigation is the non- availability of totally water-soluble fertilizers at affordable cost. Due to the non-availability of good fertigation compatible soluble NPK fertilizers, imported fertilizers are being, used which escalates the cost to Rs.50/- to Rs.60/- per kg cost of nutrient whereas, the average per kg cost of nutrients in India is around Rs.4/- to Rs.6/- per kg. Another important reason for lack of enthusiasm in fertigation is the high investment on machinery tools for injecting the nutrients into the system, and the lack of precise information about fertigation in different crops

- The grape, pomegranate and banana growers in Maharashtra have adopted fertigation to some extent. Based on the studies on fertigation carried out on different horticultural crops using several formulations of water-soluble fertilizers
- By and large a saving of at least 20-40 per cent in fertilizers could be made if fertigation is adopted using water-soluble fertilizers (WSF) due to better fertilizer use efficiency.
- The water soluble fertilizers being completely soluble in water are ideally suited for fertigation as they do not cause any clogging of the system due to the high acidic urea phosphate used in formulating these fertilizers.
- Frequent and split application of fertilizers through fertigation near the root zone of the crops help in reduced leaching and consequently better absorption of nutrients resulting in increased yields by 25 to 35 per cent besides improvement in the quality of the produce in almost all the crops.
- Keeping in view the NPK requirement of various horticultural crops and several formulations that are available for evaluating their efficacy, fertilizers in the NPK ratio of 1:1:1; 2:1:3 and 1:2:0 are found to be more desirable as these could be used for majority of the crops by supplementing either nitrogen (N) or potash (K) nutrients wherever necessary through fertigation. The studies indicate that fertigation holds ample scope for adoption especially in high value horticultural crops for getting high productivity and quality produce. It would also be cost effective if type, level, split applications and cost of water soluble fertilizers are optimized for various crops/regions.

Chemigation, a related and sometimes interchangeable term, is the application of chemicals through an irrigation system. Chemigation is considered to be a more restrictive and controlled process due to the potential nature of the products being delivered - pesticides, herbicides, fungicides - to cause harm to humans, animals, and the environment. Therefore chemigation is generally more regulated than fertigation.

Fertigation is used extensively in commercial agriculture and horticulture and is starting to be used in general vegetable and ftuit production, applications as dispenser units become more reliable and easy to use.

1. Fertigation is used to spoon feed additional nutrients or correct nutrient deficiencies detected in plant tissue analysis.
2. Usually practiced with high value crops such as vegetables, turf, fruit trees, and ornamentals.
3. Injection during middle one-third or the middle one-half of the irrigation recommended for fertigation using micro propagation.
4. Water supply for fertigation kept separate from domestic water supply to avoid contamination.
5. Change of fertilizer program during the growing season in order to adjust for fruit, flower, and root development.

Fertigation allows the landscape to absorb up to 90% of the applied nutrients, while granular or dry fertilizer applications typically result in absorption rates of 10% to 40%. Daily feeding in small, scientifically consistent doses lead to healthier, more beautiful growth. Fertigation will even fertilize your trees and shrubs in your deserts cape as well as those yards that have lawn.

FIG. 28.14: Large-scale fertication plant (Left), fertication injector (Right).

Once a drip irrigation system has been installed, it makes economic and environmental sense to fertilize the crop via irrigation water. If done properly, this results in more efficient use of fertilizers, and probably reduces fertilizer contamination of groundwater. More nutrients are taken up by the crop, and fewer leach down below the plant root zone.

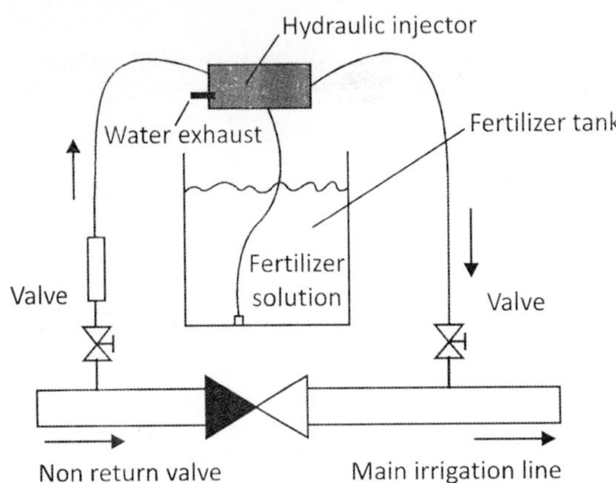

FIG. 28.15: Illustration of fertigation system

In its broadest sense, fertigation means feeding a crop by injecting soluble fertilizers into water in the irrigation system. There are a variety of ways to introduce chemicals into a drip system, based on different kinds of pump - small electric powered pumps or those powered by irrigation water, venturies, pressure differential tanks, bladder tanks, and gravity. Each drip system may utilize a different method or a combination of methods. If fertigation is to be successful, irrigation scheduling must be coordinated closely with the nutrient needs of the crop. To be a good fertigator, a grower first needs to be a good irrigator.

Methods used in Fertigation

1. Drip irrigation which reduces per water and nutrient application rates relative to sprinklers.

2. Sprinkler systems increase leaf and fruit quality.
3. Other methods of application include lateral move, traveler gun, and solid set systems.
4. All systems should be placed on a raised and/or sealed platform, not in direct contact with earth.
5. Fitted with chemical spill trays.
6. Continuous application- fertilizer is supplied at a constant rate • Three stage application-irrigation starts without fertilizers and then the later in process fertilizers are applied.
7. Proportional application-injection rate is proportional to water discharge rate.
8. Quantitative application-nutrient solution is applied in a calculated amount to each irrigation block.
9. To determine the injection rate for the particular fertilizer being used use the formula: Maximum injection rate = $(5 \times Q \times L) / (f \times 60)$. Q = irrigation pump discharge in liters per second - L = fertilizer tank volume in liters - F = amount of fertilizer in grams.

Preparation and Uses of Fertilizers

Stock Solution Preparation

Farmers mix solid fertilizers as ammonium sulfate, urea, potassium chloride and nitrate, and liquid phosphoric acid to prepare a *tailor made* stock solution. The stock solution is then injected into the irrigation system, at rates of 2-10 L/m^3, depending on the desired concentrations of N, P and K Clear NK, PK and NPK fertilizer solutions with at least 9-10% nutrients (N, P_2O_5, K_2O) based on cheap solid fertilizers (urea, phosphoric acid and KCl) can be easily prepared on the farm site with limited facilities under "grass roots" field conditions, with minimal mixing.

Compound Solid Fertilizer Mixtures

Manufactured for use in fertigation, with different ratios between the three major elements. The first mixture used in fertigation was 20-20-20 and was produced in the mid-sixties. Some compositions contain microelements in the form of chelates.

Compound Liquid Fertilizer Solutions

Due to solubility, the total nutrient concentration is much lower (5-3-8; 6-6-6; 9-2-8, etc.). Specified for use in greenhouses. Some compositions contain microelements in the form of chelates.

Generally two fertilizer tanks that contain the concentrated fertilizer solutions are used to separate those fertilizers that can interact. A possible combination is: a tank "A" containing calcium nitrate, potassium nitrate magnesium nitrate and microelements, whereas tank "B" contains ammonium sulfate, phosphoric acid and nitric acid; in this way P and Ca/Mg are in different tanks to avoid their precipitation. A third tank "C" contains an acid solution to control the pH of the fertilizer solution and to wash the irrigation system to avoid drippers clogging.

Final fertilizer mixing water should have pH of 5.5 to 6.5, EC $\leq$ 0.1, low carbonate ($\leq$ 2.25), borone (1 ppm), sodium (SAR $\leq$ 15), chloride ($\leq$ 5 ppm), nitrate (15 ppm) and free from heavy metals.

Fertilizers Suitable for Fertigation

Name	Chemical form	N-P_2O_5-K_2O Content	Solubility (g/l) at 20°C	Remark
Ammonium Nitrate	NH_4NO_3	34-0-0	1830	Incompatible with acids
Ammonium Sulfate	$(NH_4)_2SO_4$	21-0-0	760	Clogging with hard water
Urea	$CO(NH_2)_2$	46-0-0	1100	Contains nitrogen at high solubility
Diammonium Phosphate	$(NH_4)_2HP_2O_5$	18-46-0	575	Contains phosphorous at high solubility
Potassium Chloride	KCl	0-0-60	347	Chloride toxic for some crops, Cheapest K source
Potassium Nitrate	KNO_3	13-0-44	316	Expensive, high Nitrate
Potassium Sulfate	K_2SO_4	0-0-50	110	Excellent source of sulfur, clogging with hard water
Phosphoric acid	H_3PO_4	0-52-0	457	Incompatible with Calcium

Interaction between the Fertilizers and Irrigation Water

Water Quality

When the water sources contain high contents of calcium, magnesium and bicarbonates (hard waters), the reaction of the water is alkaline with pH values between 7.2 and 8.5. The interaction of these waters with fertilizers can cause diverse problems, such as formation of precipitates in the fertilization tank and clogging of the drippers and filters. In waters with high calcium content andbicarbonates, use of sulphate fertilizers causes the precipitation of $CaSO_4$ obtruding drippers and filters. The use of urea induces the precipitation of $CaCO_3$ because the urea increases pH.

Problem with Phosphorus Application

The presence of high concentrations of calcium and magnesium and high pH values lead to the precipitation of calcium and magnesium phosphates. Recycled waters are particularly susceptible to precipitation due to its high bicarbonate and organic matter content. The resultant precipitates are deposited on pipe walls and in orifices of drippers and can completely plug the irrigation system. At the same time, P supply to the roots is impaired. When choosing P fertilizers for fertigation with high calcium and magnesium concentrations, acid P fertilizers (phosphoric acid or monoammonium phosphate) are recommended.

Clogging

This is specially critical for drip systems that must be kept free from suspended solids and microorganisms that plug the small orifices in the emitters. In the case of clogging of the drip system by bicarbonate precipitation the use of fertilizers with acid reaction partially corrects this problem. However, acid fertilizers cause corrosion of the metallic components of the irrigation system and damage the cement and asbest pipes. Therefore, the periodic injection of acid in the fertigation system is recommended in order to dissolve the precipitates and to unclog the drippers. The following acids can be used phosphoric, nitric, sulfuric and chlorhydric. The irrigation and injection system should be carefully washed after the injection of acid.

Fertigation under Saline Conditions

Crops vary widely in their tolerance to plants, reference tables are available defining individual crop sensitivity to total soluble salts and individual toxic ions. When brackish waters are used for irrigation, we must bear in mind that fertilizers are salts and therefore they contribute to the increase of the EC of the irrigation water. Nonetheless, calculation of the contribution of chloride from KCl to the overall load of chloride from irrigation water shows its relative by low share.

When irrigation water has an EC > 2 dS/m (with high salinization hazard), and crop is sensitive to salinity, we must decrease the amount of accompanying ions added with the N or K. For example, in avocado - a very sensitive crop to chloride - KNO_3 is preferred on KCl to avoid Cl accumulation in the soil solution. This practice diminishes leaf burning caused by Cl excess. Also in greenhouse crops grown in containers with a very restricted root volume we must choose fertilizers with low salt index. Sodium fertilizers as $NaNO_3$ or NaH_2PO_4 are unsuitable due to the adverse effect of sodium on the hydraulic conductivity and the performance of the plant.

A correct irrigation management under saline conditions includes water application over the evaporation needs of the crop, so that there is excess water to pass through and beyond the root zone and to carry away salts with it. This leaching prevents excessive salt accumulation in the root zone and is referred to as leaching requirement.

Fertilizers compatibility

When preparing fertilizer solutions for fertigation, some fertilizers must not be mixed together. For example, the mixture of $(NH_4)_2SO_4$ and KCl in the tank considerably reduce the solubility of the mixture due to the K_2SO_4 formation. Other forbidden mixtures are:

1. Calcium nitrate with any phosphates or sulfates
2. Magnesium sulfate with di- or mono-ammonium phosphate
3. Phosphoric acid with iron, zinc, copper and manganese sulfates

The use of two fertilization tanks allows separating the fertilizers that interact and cause precipitation, placing in one tank the calcium magnesium and microelements, and in the other tank the phosphorus and the sulfate

Advantages of Fertigation

The fertigation allows applying the nutrients exactly and uniformly only to the wetted root volume, where the active roots are concentrated. This remarkably increases the efficiency in the application of the fertilizer, which allows reducing the amount of applied fertilizer. This not only reduces the production costs but also lessens the potential of groundwater pollution caused by the fertilizer leaching. Fertigation allows adapting the amount and concentration of the applied nutrients in order to meet the actual nutritional requirement of the crop throughout the growing season. In order to make a correct planning of the nutrients supply to the crop according to its physiological stage, we must know the optimal daily nutrient consumption rate during the growing cycle that results in maximum yield and production quality These functions are specific for each crop and climate, and were determined in different experiments for the main crops in Israel like tomatoes, cucumbers, melons maize, etc. The optimal curve of consumption of nutrients defines the minimal application rate of a certain nutrient that is required to maintain a constant nutrient concentration in the soil solution. These data constitute the base of the recommendations given by the Israeli Soil Extension Service for the farmers regarding the fertigation regime for the different crops. Other advantages of the fertigation are:

1. The saving of energy and labor.
2. The flexibility of the moment of the application (nutrients can be applied to the soil when crop or soil conditions would otherwise prohibit entry into the field with conventional equipment).
3. Convenient use of compound and ready-mix nutrient solutions containing also small concentrations of micronutrients, which are otherwise very difficult to apply accurately to the soil.
4. The supply of nutrients can be more carefully regulated and monitored. When fertigation is applied through the drip irrigation system, crop foliage can be kept dry thus avoiding leaf burn and delaying the development of plant pathogens.

Drip and micro-irrigation have a characteristic not shared by other irrigation methods: fertigation is not optional, but is actually necessary. Fertigation provides the only good way to apply fertilizers physically to the crop root zone. On high value drip irrigated crops, such as lettuce, tomatoes, and peppers, the level of fertigation management for achieving high yields and crop qualities exceeds to what is found with other irrigation methods and crops.

Disadvantages of Fertigation

1. Concentration of solution decreases as fertilizer dissolves, leading to poor nutrient placement
2. Results in pressure loss in main irrigation line
3. Limited capacity
4. Use of chemical fertilizers of low-sustainability, instead of organic fertilizers.
5. Dependent on water supply not being restricted by drought rationing.

CHAPTER 29

Plasticulture for Hi-Tech Horticulture

Plasticulture is the use of plastics in agriculture, horticulture, water-management, food grains storage and related areas. *It is defined as the use of plastics in plant and animal agriculture.*

The plastic materials themselves are often and broadly referred to as *ag plastics*. Plasticulture *ag plastics* include soil fumigation film, irrigation drip tape/tubing, nursery pots and silage bags, but the term is most often used to describe all kinds of plastic plant/soil coverings. Such coverings range from plastic mulch film, row coverings, high and low tunnels, to plastic greenhouses

A variety of plastics materials and end products are deployed in plasticulture applications - for water conservation, irrigation efficiency, crop and environment protection, as well as end product storage and transportation. Plastics can play a major role in energy conservation. They require minimum energy in production and conversion to finished products. Plastics have definite advantages over conventional materials like higher strength / weight ratio, superior electrical properties, superior thermal insulation properties, excellent corrosion resistance, superior flexibility, impermeability to water, gas, etc., resistance to chemicals, and less friction due to smoother surface. Therefore, plasticulture applications are one of the most useful indirect agricultural inputs which hold the promise to transform Indian agriculture and bring in the *Second Green Revolution*. It is used in water management, drip irrigation, controlled environment agriculture and -LEPD film mulching.

Plastics make agriculture more productive and efficient by cutting costs and saving time. They also conserve precious natural resources, such as water, nutrients, fossil fuels, and many forms of energy, especially sunlight. In some cases, plastic can be used to limit sunlight and retain moisture, especially in arid regions. Plasticulture also reduces competition from invasive weeds and insects.

The early development of plasticulture strawberry farming in the eastern United States was led by Dr. Barclay Poling in North Carolina. Dr. Poling developed growing techniques that were effective for both reducing crop losses and increasing farmers' ability to meet market place demand for fruit. Early plasticulture methods developed around planting "plugs" (small green plants growing in plastic trays) under black plastic mulch. Farmers saw notable success and many growers throughout the mid-Atlantic and Southeastern states began experimenting with plasticulture farming. Today thousands of farmers throughout the country grow small-fruit and other produce using plasticulture methods.

In recent years, plasticulture farming has evolved dramatically through experimentation, field trials, and working relationships between commercial growers, nurseries and experimental stations. The term *plasticulture* refers to planting and growing techniques that may, or may not involve the use of any plastic material. The old idea that plasticulture farming must mean plastic mulch has changed. For example, for a progressive strawberry grower, plasticulture farming begins with raised beds of dual-row strawberries planted to stand with drip tape irrigation. Plastic mulch is an option. There is no single best method for all growers.

The key benefits of Plasticulture are: superior weed control, reduced loss of soil moisture and nutrients, longer potential picking season, efficient and convenient harvesting, increased fruit size and yield and more marketable fruit.

Typical products used in plasticulture include:

1. Bunk silo covers
2. Hay and silage bags and wraps
3. Baling twine and ties
4. Low tunnel, high tunnel, nursery house and greenhouse covers
5. Mulch films
6. Drip irrigation tape and tubing
7. Row and floating row covers
8. Sun shades and screens
9. Nursery pots, trays, tray inserts, and flats
10. Baskets and clamshells
11. Fumigation films
12. Pails and barrels
13. Degradable and biodegradable plastics
14. Edible films
15. Plastic fittings and parts associated with many of these systems.

The National Committee on use of Plastics in Agriculture (NCPA) was set up in 1981 for promoting use of plastic in agriculture. Its mandate is to provide research and training support to the State Governments through its Plasticulture Development Centers (PDCs), overseeing the implementation of the schemes on plasticulture and helping in transfer of technology. The NCPA has presently 11 PDCs mostly located in State Agricultural Universities. Six more new PDCs are being set up to cover the major States.

Total outlay of Rs. 250 crore has been made during VIII Five Year Plan for taking up various programmes for popularizing use of plastics including drip irrigation, greenhouse, mulching etc. in horticultural crops.

USE OF PLASTICULTURE

Plastic Mulch

Plastic mulches have been used commercially on vegetables since the early 1960s. Much of the early university research before 1960 was conducted on the impact of color (black or clear plastic film) on soil and air temperature, moisture retention and vegetable yields (Emmert 1957). Most plastic mulches used in the United States are made of linear low- or high-density polyethylene. They are 0.012 to 0.031 mm thick, 122 to 152 cm wide, and on rolls 607 to 1463 m long, depending on the thickness of the mulch. Linear high-density polyethylene is used to reduce weight and cost, and is stronger than the same thickness of low-density polyethylene. The plastic mulch is either slick (smooth), or embossed with a diamond-shaped pattern. This pattern helps reduce expansion and contraction, which results in the loosening of the mulch from the raised bed. The raised bed is generally 10 - 15 cm high and 75 cm wide, and has a slope of 3 cm from the center to the edge. Soil under a raised bed will warm up more quickly in the spring, and will also shed excess water from the middle of the bed, thus keeping the crop plants drier and preventing deterioration in the quality of the product

Some of the other benefits of plastic mulch include earliness, reduced weed pressure in the crop bed, reduced evaporation from the soil surface, reduced splashing of soil onto crops, etc. Plastic mulch is now a component of plasticulture crop production. Plasticulture crop production often uses several different tools to increase production. The most common products used are plastic mulch and drip irrigation. Although, other products like floating row-cover, low tunnels, high tunnels, etc. are all used.

FIG. 29.1: Plastic mulch.

Melons are one of the crops that often show a big response to the use of plastic mulch. Black plastic mulch is most commonly used but, clear, white and other colors are common. Plastic mulch helps increase soil temperatures early in the growing season which benefits warm season crops like melons. Different mulches warm the soil in different ways. Black plastic warms the soil through direct contact. The sun heats up the black plastic and where the plastic touches the soil that heat is transferred. Areas where there is a gap between the soil and the plastic heat much slower. The air gap acts like insulation. The air must be heated before the heat is transferred to the soil.

Clear plastic mulch warms soils the most. The clear plastic allows the suns rays to shine through directly warming the soil. The plastic helps to hold the heat. Soils under clear plastic can be 5 degrees, or more, warmer than bare soil in the same field. The disadvantage of clear plastic is that the sunlight passing through the mulch also allows weeds to grow. A crop that will quickly cover the plastic can reduce weed growth. Or, some herbicides can be used underneath the plastic to reduce weed growth.

Wavelength Selective Mulches

Another option is the use of wavelength selective mulches. The wavelength selective mulches try to combine the benefits of both black and clear plastic mulch. The idea is that the wavelengths that warm the soil pass through the mulch but the wavelengths that allow plant growth are blocked by the mulch. These products fall in-between clear and black plastic in their soil warming properties. They will stop most weeds from growing but there a few weed species that will grow under them. These products are often green or brown in color.

Other Colors of Mulch

There have been several studies looking at different colors, red, yellow, green, blue, etc., for different crops. Some studies have found that different crops do best with certain colors. The results have been variable. One of the problems is that there haven't been standard colors established. Red is not always red! There is red, pink, rose and of course.... scarlet! There are also white mulches to help cool soils during hot periods. To avoid weed growth under white mulches grower should try white on black mulch. There are also reflective mulches which may be used to help insect and disease control. These products work to confuse virus vectors like aphids. These products can be effective, but, only as long as the crop canopy does not cover them.

Other Commercial Mulch

Other commercial mulch products include photodegradable mulch that is broken down by sunlight. These types of mulch have met with mixed results because of erratic breakdown. Sometimes they break at the desired time. Other times they may break down too early or too late. Also, the buried edges will not break down until they are exposed to sunlight. There is also work being done to develop paper mulch. This would be a big benefit because it could be worked into the ground instead of having to be picked up after the growing season. So far paper mulch has not been perfected. The product we tested was difficult to lay, often tearing. Also, the roll is much larger in diameter and heavier than plastic. Hopefully, a paper mulch product will be on the market soon.

As with all things there are some down sides to plastic mulch. The biggest problem is removal from the field and disposal. Plastic can be difficult to pick up and is quite dirty which limits is ability to be recycled. Along with potential contamination by pesticides. The most common means of disposal is the landfill.

While plastic mulch can inhibit weed growth it also can create some difficult to control weed problems. Weeds tend to grow through holes in the plastic. Yellow nutsedge will grow right through plastic mulch. Weeds also tend to grow right along the edge of the plastic where they escape mechanical cultivation. Some herbicides can be used under plastic mulch while other cannot. Growers also need to be careful about apply herbicides to row middles and getting herbicides on the mulch. Roundup will stay on the mulch until it is washed off. If young melon vines reach it first they pick the roundup up. Or, if curbit is applied to the plastic mulch rain may wash it into the plant hole and kill the melon transplant.

Weed Control

One advantage of the plasticulture system is the excellent weed control from the dark plastic mulch and fumigants (where used). However, weeds can still establish in the hole by each plant, in the row middles and along the field edges. Weeds often establish in bare soil at the ends of the rows or field edges. Farmers encourage germinating and killing weeds before planting. It is recommend that laying of plastic 30-40 days before planting. However, shorter intervals can also be successful. Herbicides are often used between rows, but care must be taken that these chemicals do not run off the plastic and concentrate in holes.

Weeds reduce melon yield, decrease the number of melons, and reduce melon size. Weeds can delay maturity and may decrease quality. Severe weed infestations interfere with harvest. With an effective weed management program it is possible to avoid these negative effects of weeds.

An effective weed management program for melons includes:

1. Identification of the major weed species in the field.
2. Use of cultural practices to minimize growth and seed production of weeds.
3. Use of weed control practices to reduce weed emergence and kill emerged weeds.

A good time to identify weed species in a field is the year prior to growing melons. Periodically during the season, note what weeds are present, and whether they are scattered across the field or concentrated in a few patches. Notes about weed species will help in choosing herbicides and other weed control practices for the melon crop.

Cultural practices such as fertilization, irrigation and crop spacing affect weed growth. If these practices can be chosen to minimize weed growth, weeds may be easier to kill. Weeds that escape control will compete less with the crop and produce less seed. For instance, applying fertilizer and irrigation on beds only rather than broadcasting it across the entire field can reduce weed growth between the beds. Crop spacing is important because the closer together rows and plants are, the sooner the canopy will shade the ground. A complete canopy severely reduces the amount of light small weeds receive, limiting their growth and seed production. Row spacing is important also because it influences what tractors and cultivation equipment will fit between the rows.

Weed control practices include use of herbicides, cultivation, plastic mulch, and preparation of a stale seedbed.

The stale seedbed method may provide some control of weed species that emerge early in the season. The stale seedbed method should be combined with other techniques to achieve adequate weed control. It is more effective with larger plantings, when there is time to allow weeds to emerge.

Opaque plastic or paper mulches effectively control weeds in and near the crop row.

Low and High Tunnels

Low tunnels are ideal for the early production of many vegetable crops. They are constructed by installing 6-foot wide sheets of either slitted or perforated plastic over wire hoops that are spaced about 4 feet apart. To prevent wind damage, the cover is pulled taut over the hoops (but not stretched), and the edges are secured with soil. Generally, the cover is left over the plants for 3-4 weeks. The slits and perforations provide daytime ventilation, eliminating the need for manual venting. Low tunnelsare generally used in conjunction with black plastic mulch for weed control and enhanced soil warming.

Frost protection of 2 to 4°F is afforded by the covers, but the biggest benefit is growth enhancement provided by the daily increments of solar heating. Daytime temperatures might be excessive under the cover if ambient temperatures are above 90 degrees F. Under those conditions, partial removal of the cover might be necessary, especially for tomatoes and peppers that have open blossoms.

Slitted and perforated row covers at 0.8 mil are available in rolls 6' × 1000', 500',& 250'. Number 10 smooth galvanized wire for hoops is available in bundles of 100. It is 76" long and may be used with machines designed for installing hoops and covers.

Row covers, high tunnels and low tunnels can make possible earlier crops of vegetables by creating a mini-greenhouse effect. The first row covers used were solid polyethylene sheeting that needed support and required venting during the day.

To eliminate the need for venting by hand, a variety of materials have been developed, including slitted polyethylene covers that require wire hoops; floating nonwoven sheets; a white point-bonded, polypropylene material and a polyethylene sheet with tiny pores. Row covers also help keep out insect pests.

FIG. 29.2: Use of plastic for row cover and frost protection.

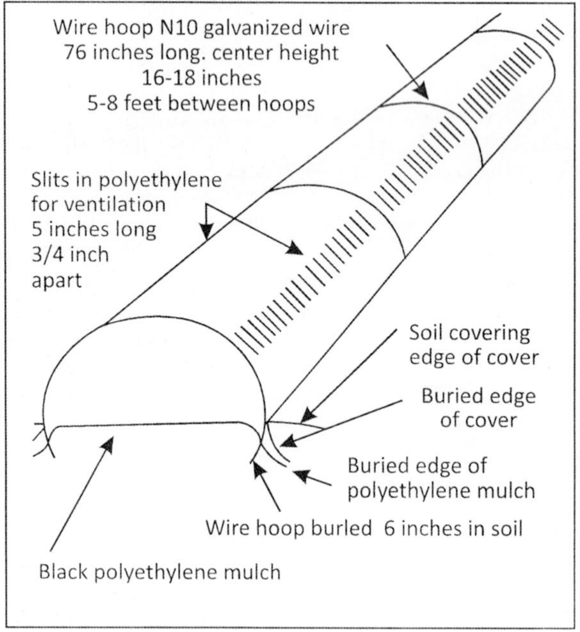

FIG. 29.3: Wire for Hoops: Number 10 gauge smooth galvanized steel wire is packaged in bundles of 100. Wires are 76" long and can be inserted 6" into the soil at each end to form a support hoop that is 16" - 18" high at the center. Wires should be spaced 5' to 8' apart.

Method of Installation

1. Space hoops about 4 feet apart to prevent sagging of the covers.
2. Apply covers on a calm day.
3. Apply the cover taut and secure the edges with soil.
4. Machines are available for hoop and cover installation.

5. To prevent weed growth under the cover, install the cover right along the edges of the black plastic mulch.

Advantages of Slitted/Perforated Row Covers

1. Outdoor crops may be planted about 10 days earlier than normal
2. Highest heat retention with clear perforated covers and slitted covers
3. Earlier maturity and higher yields
4. No manual ventilation necessary
5. Excellent for tomatoes, pepper, cucumbers, melons, and many other crops

Crops that are Suited for Slitted/Perforated Row Covers

1. Vegetables - warm season and cool season crops that are transplanted
2. Small fruits - strawberries in matted row
3. Cut flowers

The results from the use of slitted or perforated row covers on various crops will vary, but in general all crops will benefit. The primary precaution is to provide extra ventilation when there are daytime spikes in temperature. Drip irrigation and mulch are recommended as part of the growing system.

High tunnel system may also be constructed for crop protection with the use of plastic. High tunnels are another option for producing vegetables in a plasticulture system. They can be used to extend the spring and fall growing seasons. High tunnels are covered with a single layer of polyethylene film. The use of high tunnels is widespread in many parts of the world, especially in Asia and the countries of Spain and Italy.

The Haygrove Tunnel System

Haygrove is imported multi-bay structures from Southern Europe. The concept was a good one but the tunnels needed radical improvement to meet our needs. Over several years new structures were tested and improved at the farm until we had a low cost, field scale tunnel system within which tractors can be used and that stand up to strong winds.

FIG. 29.4: View of a high tunnel used for tomato production.

The impact on British horticulture soft fruit has been spectacular and this has spread to other crops with grower over 4,000 acres of Haygrove Tunnels built in the last 5 years covering small fruit, cherries, plums, flowers, herbs, ornamentals, tomatoes, peppers and some high value vegetable crops. Haygrove tunnels are now used across the world, helping growers to produce reliable supplies of high quality flowers, fruits and vegetables. North American growers are now seeing the benefits with success reported by strawberry, raspberry, cherry, cut flower and tomato growers.

TABLE 29.1: Use of Haygrove tunnels and crop production

Crop	Benefits
Tomato	Increased yield Dramatic reduction in early and late blight infection (Phytophthora and Alternaria). Improved skin finish
Cherry	Increased yield Fruit allowed to size by safely delaying harvest. Losses due to cracking and rots virtually eliminated. Better quality Larger, sweeter fruits
Raspberry	High quality late season production from fall bearing raspberries
Strawberries and Raspberries	Reliable supply to end customers and more efficient use of labour. Clear economic benefits: 14% yield increase, 34% increase in grade one yield, 2-3 weeks earlier start to picking, 31% more income

Runner Control

Plants cannot root on plastic. Runners will only root where the grower allows. The use of growth regulators should be considered to inhibit runner production. Research and strawberry physiology indicates runner initiation begins after flowers are produced. Experience dictates that monthly applications are better than one large dosage.

Pest Management

Plasticulture system must have a good integrated program for insect, disease, and weed control. To obtain good insect and disease control, it is important to use a sprayer that generates sufficient pressure so that pesticide sprays penetrate to, and cover, the whole crop. This means using sprayers with pumps capable of generating over 200 psi, with appropriately sized nozzles. It is important to use chemical sprays efficiently, to control the targeted pest without damaging the environment. The use of an integrated pest management (IPM) approach, which combines the use of disease-resistant varieties with chemical and biological control practices, crop rotation, and effective monitoring, is recommended. Only approved herbicides for the vegetable being grown should be used between rows of mulched vegetable beds, because this should not be considered a fallow area. The use of low-pressure sprayers, coupled with shielded application of herbicides, is recommended for spaces between vegetable beds. This approach will protect the mulched crop from herbicide, so it will not be damaged by any concentration of herbicide in the planting hole.

Reflective plastic mulches such as silver mulch have been shown to interfere with the movement of aphids. Aphids are common vectors of virus diseases on various vegetable crops (e.g., watermelon mosaic virus II, which causes green streaks in summer squash and mottling and green streaks in yellow squash, melons and pumpkins

Strip Fumigation and Soil Solarization

It is necessary to sanitize the soil in which greenhouse or mulched crops are grown. Plastic mulches are used with chemical fumigants, or as covers during soil solarization. In row or strip application of a fumigant, the amount of material actually applied per hectare will depend on the row width, and will be a percentage of the broadcast rate. The temperature of the soil should be at least 10° C, and soil should be well worked, free from undecomposed plant debris, and have adequate moisture for seed germination. If the weather and soil are warm, the fumigant should escape through the plastic mulch in 12-14 days. Fumigation is used primarily for nematode control, but a multipurpose fumigant can also provide good control of soil-borne diseases. Soil solarization is another way to control soil pests. Solarization describes a hydrothermal method of cleansing soil that occurs in moist soil which is covered by mulch film (usually clear) and exposed to sunlight during the hot summer months.

Windbreaks

The use of windbreaks, whether permanent (trees) or annual (grain crops), is an important part of the plasticulture system of production, but is often overlooked. Windbreaks consisting of strips of winter wheat, rye, or barley are often planted to protect young vegetable seedlings from prevailing winds. A combination of permanent and annual windbreaks can modify wind profiles and influence temperatures and other microclimate features. Windbreaks may also serve as a habitat for both beneficial and pest insects. For maximum effectiveness, the grain strips should be planted in the fall. Each strip of grain crop should be 3 to 3.5 m wide (the width of a sowing drill). Enough room should be left between the strips for five or six mulched beds of vegetables, each around 2 m wide. Topdressing the strips in the spring helps to ensure a dense stand of grain.

Another option is to plant a solid grain cover crop in the fall. It is important to till the crop area early in spring, so that residues from the cover crop will not interfere with application of the plastic mulch, drip irrigation and fumigation (if required). Once wind protection is no longer required, the grain strips are mowed, and used as drive rows for spraying for insect and disease control, and later for harvesting.

Management of Ripening Times and Season Extension

Plasticulture allows growers the advantages of more control and more flexibility for managing fruiting schedules. If straw mulch is left on top of beds, covering the plastic mulch, plants will closely follow the growth and fruiting schedule of plants being grown with traditional methods. If straw is removed from the beds covered with plastic mulch in early spring, the black mulch will absorb and retain the sun's warmth. Plants will grow more quickly and fruit ripening will be around 7 days earlier than traditionally grown plants of the same variety. Leaving straw on the plastic mulch a few days longer will delay plant growth and fruiting. Your harvest season can be extended. Caution! Never leave straw mulch on plants as they begin to grow. This will reduce potential yields! To summarize, we can say that with plasticulture the same variety of fruits can be harvested at three different times.

1. **Normal Fruiting Schedule:** Leave straw through early spring to have same production schedule as bare ground (recommended for all growers that don't have frost protection.)
2. **Early Fruiting Schedule:** Remove straw in the early spring to allow early growth and fruiting. Adding row cover can add and additional 5-7 days to earlier fruiting. Damage from frost is also possible!

3. **Late Fruiting Schedule:** Leave straw over plastic until late spring to promote slower growth and late fruiting.

Increased yield: Average harvest efficiency can increase by 20% or more. Larger fruit, more fruit per plant, increased picking rate, and controlled timing of fruiting contribute to this increase.

Higher quality fruit: Larger, sweeter, fruit, with less damage, is more attractive to buyers and PYO customers.

Increased sales and profitability: There is a consistent increase in sales of fruit every year. Combining all of the appropriate elements of plasticulture will lead to higher yield potentials, higher quality fruit, and increased revenue.

Clean Picking Surface

Removing runners and dead leaves reduces incidence of fruit rot, Fungi cannot survive on plastic. It is recommended that in the spring when the mulch or row cover is removed, prior to applying a row cover. This is not as important if straw mulch is maintained for harvest. When the plastic is clean it also allows water to move away faster, so fruit is damaged less by downpours and excessively wet periods.

Labor Issues Concern Small and Large Scale Plasticulture Growers

1. Small scale producers can usually manage planting and weeding on plasticulture with limited to no addition labor.
2. Harvesting the increased yields can be problematic and some small growers have noted significant crop losses due to the lack of harvest labor.

FIG. 29.5: Various of types of plastic flower pots.

Plastic Flower Pots and Field Harvesting Containers

Plastic flower pots are now commonly used for flower production because plastic is lighter and more robust, therefore transport cost are lower, and so are damage losses. Clay is a lot heavier and requires much more volume for the same numbers of items. Plastics raw materials are going to increase in cost to a marked extent, and even with recycling. Plastic pots are going to rise in price. There will be a price increase for clay pots, but mainly as a result of greater transportation costs. Manufacture of clay pots is a very old technology, although the firing process will cause problems on both cost and pollution considerations. Plastic pots are only economic when produced in large numbers, requiring investment in very expensive machinery. Regarding the aesthetics and function, without touching the items, it is difficult to tell the difference between a terracotta clay and plastic pot, at least in the larger sizes and when new, and there is not a whole lot to choose between the two from a horticultural pov, so it's down to your own preferences.

FIG. 29.6: Agricultural field harvesting containers.

DISPOSAL OF THE USED PLASTICS FROM PLASTICULTURE

Disposal of the used plastics from plasticulture is certainly a worldwide problem. There have been many attempts to solve if, including the development and use of photodegradable or biodegradable materials (which just disappear), using the plastics several times (delaying the problem), reducing the weight of films (less material to deal with), recycling (making the material into other products) and incineration. Plastic mulches and drip irrigation tapes are the hardest to deal with. After a season in the field, these materials are dirty and often wet, which makes them difficult to recycle. One option is to incinerate them to recover their very substantial fuel value. A pound of plastic has as many BTU's (thermal units) as an equivalent amount of fuel oil.

Used plastics have been burned in waste-to-energy plants, but one problem is that they create hot spots in the waste stream. Another option is to create a "fuel nugget" that can be burned to supplement coal or other waste products, or used alone to heat various kinds of structures. More research needs to be done on the proper collection, preparation and transportation of the plastics from the site of use to the point where it is processed, and then to the point of incineration. Incineration may be the answer to the disposal problem, but much more work is needed.

Current and Future Status of Biodegradable Plastics

Recently, biodegradable polymers have been developed for agricultural uses. Biodegradable films are often thinner than traditional polyethylene, but otherwise are quite similar. Biodegradable films are available in clear, black and a variety of colors. They may be made from renewable resources such as starch, cellulose, or degradable polymers. There are two types of biodegradable polymer films presently being investigated. One is a polyethylene which has a proprietary additive which allows oxidative degradation of the polyethylene film. The breakdown is a two step process. The first step is the oxidative degradation breaking the long carbon chains down into smaller pieces. The oxidative degradation is triggered by sunlight, heat and mechanical stress. The film becomes brittle, and the molecular fragments are then further broken down by microbial action to CO_2, water and natural substances.

The second step is the microbial action. Depending on how the film is formulated will affect the length of time required for it to break down (from several months to several years). These are not the same as the photodegradable mulches that were previously available that left plastic residues in the fields.

The other types of films under test are starch based polyester. Starch is added to the formulation to break up the long polymer (carbon) chains to shorter lengths which are then subject to further breakdown through microbial activity to CO_2 and water. Polyester is also biodegradable. There are also polyethylene films with starch added to enhance breakdown, but the problem is that molecular fragments of polyethylene may remain in the soil.

There are compostable standards in the US (ASTM) and Europe (EN) which state that in order for a product to be compostable the following criteria must be met:

1. **Disintegration**, the ability to fragment into non-distinguishable pieces and safely support bioassimilation and microbial growth.
2. **Biodegradation**, conversion of carbon to CO_2 to the level of 60% over a period of 180 days (ASTM standard) and 90% in 180 days (EN standard).
3. **Safety**, that there is no evidence of eco-toxicity in the finished compost and soils can support plant growth.
4. **Toxicity**, that heavy metals concentrations are less than 50% recommended values. Few products presently meet the compostable standards. However, products may be biodegradable, but in a slower time frame.

Biodegradable films generally are thinner and therefore, extra care must be exercised when laying the films to prevent tears and punctures. Weed control must be good under the mulch as weeds will grow through the biodegradable films rather than being smothered as often occurs under the stronger polyethylene films.

CHAPTER 30

Value Added Products of Horticultural Crops

The average per capita availability of fruits and vegetables in India is inadequate to meet requirements for nutritional and protective diet of the population. Of the total horticulture production, only 67 % is used for local consumption while 30 % is subjected to post harvest losses, 2 % for export and 0.8 % for processing. There is immense scope for developing horticultural produce market. Changing food habits, life styles and health consciousness and purchasing power have created an unprecedented opportunity for farmers and horticultural entrepreneurs.

As farmers struggle to find ways to increase farm income, interest in *adding value* to raw horticultural products has grown tremendously. The value of farm products can be increased in endless ways by cleaning and cooling, packaging, processing, distributing, cooking, combining, churning, culturing, grinding, hulling, extracting, drying, smoking, handcrafting, spinning, weaving, labeling, or packaging. Today, more than ever, adding value means *selling the sizzle, not the steak*. The "sizzle" comes from information, education, entertainment, image, and other intangible attributes.

There are several regulations involved with food processing, some people may choose to add value in other ways. On a larger scale, producer-controlled processing for energy, fiber, and other non-food uses are options. On a smaller scale, items such as flower arrangements, garlic braids, grapevine wreaths, etc. are a few examples. In addition, ideas for providing entertainment, information, and other services associated with direct marketing are abundant.

Besides offering a higher return, value-added products can open new markets, create recognition for a farm, expand the market season, and make a positive contribution to the community. However, adding value is not a panacea for all the problems rural India is facing. It is a long-term approach. It requires the willingness and ability to take on risk, as well as adequate capital, management skills, and personal skills, such as the ability to interact with the public to succeed.

Value-added agriculture generates several billion rupees in economic impact for India each year. In fact, the economic impact of adding value beyond the farm gate is usually several times the value of agricultural production at the farm gate alone.

Horticultural crop producers receive a much smaller portion of the consumer's income than do food processors, especially those who produce brand name items. Capturing those additional income or adding value to farm or ranch products is a goal of many producers. Even for "commodity" products, such as juices, the farm-to-retail price spread indicates that those further up the supply chain earn much more than producers at the ranch or farm level.

WHAT IS VALUE ADDED?

Value added means adding value to a raw product by taking it to, at least, the next stage of production. This can be as simple as retaining ownership of your farm. Value can be added through membership in a cooperative that processes your products. Or, adding value may be as elaborate as going all the way to the consumer with a case-ready food product.

If you are considering a value-added enterprise, there are two key questions to answer:

1. What is customer value?
2. What creates a value-added product?

In our efforts to diversify our products and to meet the market requirements, there is an increased need to add value to our products in order to fetch better prices. These opportunities are through pre-packaged and high care products in addition to processing into juices, concentrates, chutneys, canning, and other products. Indeed the bulk of our exports to the EU supermarkets are now pre-packaged. This has been due to the setting up of high care facilities by our exporters. These facilities meet food processing and handling standards. The potential of the US and the Japanese market are high on the priority for future exploitation of the value addition opportunities. The phytosanitary requirements for access to these markets are currently being addressed.

What is Customer Value?

Customer value reflects the relationship between the benefits customers receive from and the price they pay for a product. The more benefits relative to the price, the higher customer value. This does not necessarily mean that greater value results from a low price. The price of a particular product may be high, but if the associated benefits are high as well, the customers perceive value in the product. This interaction creates customer value and, thus, the opportunity to add value to your product.

Creating customer value is critical in building a profitable and substantial business. However, one must bear in mind that it is the customers' perception of value, not the producers', that is critical. It is also important to recognize that different customers have different perceptions of added value. These perceptions correspond to their expectations of quality, service, convenience and selection.

What Creates Added Value?

Value is usually created by focusing on the benefits associated with the agribusiness product or service that arise from:

- Quality: Does the product or service meet or exceed customer expectations?
- Functionality: Does the product or service provide the function needed of it?
- Form: Is the product in a useful form?
- Place: Is the product in the right place?
- Time: Is the product in the right place at the right time?
- Ease of possession: Is the product easy for the customer to obtain?

A product must have one or more of these qualities to generate additional value. Remember that a product is simply a bundle of benefits, and that the more benefits there are the more customers will perceive the product as having value.

What's Driving Value Added?

Agribusiness, particularly the food sector (including horticultural produce) is rapidly consolidating and increasingly responding to the changing tastes and preferences of consumers. Consumers have higher incomes than ever before. They are focusing more on convenience, quality, variety, service, health, and social consciousness. They are also faced with the increasing value of (and demands on) their time. In a nutshell, consumers are more value conscious than ever.

I) CREATING A VALUE ADDED PRODUCT

To take advantage of value-added opportunities, one must know and understand customers as follows

1. What consumer segments might want your product?
2. What are the benefits desired by these potential customers?
3. What criteria are they looking for when purchasing?
4. Writing these down as part of a business plan is an important first step in beginning a value-added business.

Post-Harvest and Value Addition

Horticulture contributes important role in total GDP of India. It substantially contributes to generation of rural income, creation of employment opportunities, foreign exchange earnings in addition to providing raw materials to the agro-processing industry. Fruits and vegetables also provide food security and nutritive diets to households. These benefits have made the government to focus on policies that enhance horticulture development.

Post-Harvest Losses

Post-harvest losses have many different forms. The most obvious loss is a quantitative loss, such as reduced weight and partial or total waste of a product due to decay or senescence. A less discernible loss is a qualitative one, such as deterioration in texture, flavor or nutritional value. Other forms include loss of viability of propagules and reduction in monetary value due to reduced prices. Post-harvest losses can occur in the field, in packing areas, in storage, during transportation, and in wholesale or retail markets. Severe losses occur because of poor facilities, lack of know-how, poor management, market dysfunction, or simply the carelessness of farmers or workers. Postharvest losses may also occur at consumers' homes, in the kitchen or on dining tables. However, losses after produce has left the retail market are generally difficult to control by agricultural means. The causes of post-harvest losses can be divided into four following categories:

Mechanical Injury: Owing to their tender texture and high moisture content, fresh fruits and vegetables are very susceptible to mechanical injury. Poor handling, unsuitable containers, improper packaging and transportation can easily cause bruising, cutting, breaking, impact wounding, and other forms of injury.

Physiological Deterioration: Fruit and vegetable cells are still alive after harvest, and continue their physiological activity. Physiological disorders may occur due to mineral deficiency, low or high temperature injury, or undesirable atmospheric conditions, such as high humidity. Physiological deterioration can also occur spontaneously by enzymatic action leading to over ripeness and senescence, a simple aging phenomenon.

Parasitic Diseases: High post-harvest losses are caused by the invasion of fungi, bacteria, insects and other organisms. Microorganisms attack fresh produce easily and spread quickly, because the produce does not have much of a natural defense mechanism, and has plenty of nutrients and moisture to support microbial growth. Post-harvest decay control is becoming a more difficult task, because the number of pesticides available is falling rapidly as consumer concern for food safety increases.

Lack of Market Demand: Poor planning or inaccurate production and market information may lead to overproduction of certain fruits or vegetables which cannot be sold in time. This situation occurs most frequently in areas where transportation and storage facilities are inadequate. Produce may lie rotting in production areas, if farmers are unable to transport it to people who need it in distant locations. Nobody knows exactly the extent to which harvested fruits and vegetables are lost rather than consumed. Many people believe that postharvest losses are higher in less industrialized countries, this generalization may not be always true. Higher losses may occur in developing countries for lack of good facilities and technologies. However, post-harvest losses may be lower in less urbanized developing countries, where the products need to be transported a shorter distance to market, and there is a shorter time lag period between harvesting and consumption. Consumers in such countries are also more likely to buy lower-grade products. In industrialized countries, on the other hand, produce is shipped for longer distances, the marketing system is more complicated, and consumers demand good quality when they buy food. All these factors contribute to greater postharvest losses.

It is safe to say that post-harvest losses occur in every country, but the magnitude and major causes of losses and the effective remedial methods differ greatly from one country to another. In order to solve specific problems in specific areas effectively and economically, a comprehensive knowledge of the nature of post-harvest losses, a grasp of various kinds of technologies, and a tactical selection of strategies is necessary.

The post-harvest and value addition is important for perishable crops because after harvest it is essential to reduce losses, enhance financial or nutritional crop-value, and assure food safety. There should be proper storage and preservation of quality throughout the marketing chain.

TABLE 30.1: Annual Value addition in Horticulture Crops (Rs. in crors)

Year	Fruit Crops		Vegetable Crops		Spices Crops		Total	
	Value Production	% Growth	Value Production	% Growth	Value Production	% Growth	Value Production	% Growth
2000-01	1474	-	737	-	828	-	3039	-
2001-02	1664	12.9	792	7.5	911	10.0	3367	10.8
2002-03	1992	19.7	844	6.6	809	-11.2	3645	8.3
2003-04	2303	15.6	1099	30.2	1433	77.1	4835	32.6
2004-05	2611	13.4	1122	2.1	1359	-5.10	5092	5.3
2005-06	2869	9.9	1332	18.7	1844	35.7	6045	18.7
2006-07	3402	18.6	1521	14.2	1864	1.1	6787	12.3
2007-08	4063	19.4	2035	33.8	3771	102.0	9869	45.4
2008-09	4048	-0.4	1872	-9.9	3565	-5.4	9485	-3.9

Grading

Essentially all fruits and vegetables sold in modern markets are graded and sized.

Sophisticated marketing systems require precise grading standards for each kind of product. More primitive markets may not use written grade standards, but the products are sorted and sized to some extent. Typical grading facilities in large packinghouses include dumpers and conveyors. Produce is graded by human eyes and hands, while moving along conveyor belts or rollers. "Electric eyes" are sometimes used to sort produce by color. In small-scale packing operations, one or a few grading tables may be enough. Dumping, conveying and grasping can cause mechanical injury to some products. Equipment should have a smooth, soft surface, and dumping and grading operations should be gentle, to minimize injuries.

Many products are sized according to their weight. Automated weight sizers of various capacity are used in packinghouses. Round or nearly round fruits are often sized according to their diameter, using automated chain or roller sizers or hand carried ring sizers. An inefficient sizing operation can also cause significant injuries.

Packaging

There are two very different types of packaging:

1. The first is when produce is packed in containers for transportation and wholesale.
2. The second is when produce is packed into small retail units.

Ideal containers for packing fruits and vegetables should have the following attributes:

1. They are easy to handle.
2. They provide good protection from mechanical damage.
3. They have adequate ventilation.
4. They are convenient for merchandising (i.e. they can easily carry printed information and advertising about the product etc.).
5. They should also be inexpensive, and easily degradable or recyclable. Many kinds of containers have been used, but the ideal is yet to be found.

Methods of packing can affect the stability of products in the container during shipping, and influence how much the container protects their quality. In fiberboard boxes, for example, delicate and high-priced products are often packed in trays, while other products are simply all put in the box together.

Precooling

Good temperature management is the most effective way to reduce post-harvest losses and preserve the quality of fruits and vegetables. Desirable storage and transportation temperatures for major fruits and vegetables have been identified and published. Temperatives which are low, but not low enough to cause chilling injury, slow down physiological activity and hence the rates of senescence of the products. Low temperatures also reduce microbial growth rates and the rate of decay.

Storage

Many horticultural crops have a relatively short harvesting season. Storage is needed to extend the marketing period. Various storage methods have been used on a commercial scale.

Air-cooled common storage houses are often used, or underground or cave storage using natural cold air. Storage humidity is sometimes regulated by controlled ventilation and dehumidifiers. Refrigerated storage (cold storage) controls temperature and humidity precisely by mechanical means. Controlled atmosphere

(CA) storage controls the concentrations of oxygen and carbon dioxide, in addition to temperature and humidity. Modified atmosphere (MA) storage also controls oxygen and carbon dioxide concentrations, although not as precisely as in CA, by using semipermeable polymeric films. A good control of temperature, humidity and atmospheric composition maximizes the storage life span of a product.

Capturing Value and Creating Value

There is a distinction between a strategy to capture value and a strategy to create value. This distinction is important to understand, because each strategy offers specific opportunities and risks that influence the success or failure of the value-added venture.

For horticultural producers, capturing value usually means capturing some of the value added by processing and marketing. For example, the producer's share of the food rupee value has seen a steady decline. The average farmer's share of the income in past years has declined due to high price rise in the market and marketing chains (processing, distribution, and marketing). Farmer is trying direct marketing to consumers, turning farm products into food products, and joining producer alliances and cooperatives that invest in facilities to process their farm products on a larger scale. The horticulture produce based industry may be set up, observing the following criteria:

1. Importance of fruits and vegetables in human diet.
2. Understanding principles of food preservation and processing.
3. Low cost methods of preservation and processing such as chemical preservation; preparation and preservation of fruit pulps, juices and beverages; pickles, chutney and sauces; preservation of vegetables by lactic fermentation; drying and dehydration; jams, jellies, marmalades and preserves.
4. Creating other value added products.
5. Know about safe food production.
6. The product should be based on nutrients, dietary recommendations and deficiency symptoms.

Marketing directly to the consumer can be done on a small or large scale and in a variety of ways. Options for the producer who enjoys direct contact with consumers include selling at farmers' markets and through community supported agriculture systems. Other options include sales directly to restaurants and local institutions, as well as mail order and Internet sales.

Large scale processing through producer alliances, such as agricultural cooperatives or limited liability companies, has seen growing interest among producers. These alliances can offer a way to pool resources and manage risks. In some cases, producers lose marketing or processing facilities when corporate agribusinesses close local facilities.

Learning New Production and Marketing Skills

It usually requires learning new production and marketing skills, dealing with food safety, labeling, and other regulations, and coping with liability issues and insurance. Demand for the innovative product or service must usually be created through advertising, promotion, and consumer education, and this is a lengthy, expensive process. Marketing risks may be lower with a created-value strategy, for if this demand can be established, there is potential for higher, stable prices and little direct competition. Contract agreements for identity-preserved products such as Juices, jams, etc. However, producers will need to learn new marketing skills, carefully assess feasibility, and develop marketing plans for created-value products or services without established marketing channels.

Many producers combine aspects of both capturing and creating value in their ventures. A producer may create an unusual food product, such as local specialty tomato sauses, juices, etc and sell them at the farmers' market, capturing more of the food marketing money. Food processors may use organic or unusual ingredients, and so on.

Launching of Food Business

The food business is extremely competitive and dominated by a few large companies that are driven by cost and price considerations. Food is considered a mature industry, which means very little growth in demand. Food manufacturing continues to face narrowing margins and decreasing profits. Retailers are wielding more and more power over food wholesalers and manufacturers, and requiring manufacturers to pay more for shelf space ("slotting fees"), bear more of the product development risk, and provide product uniformity and quality. Smaller processors could have trouble finding outlets for their products, if they cannot meet scale requirements by the large retail operations that dominate the industry. In this mature industry, the only real growth is in niche food markets, where producers create value by adding special services for consumers, offering quality attributes like organic certification, specialized health products, and so on. They are able to charge prices 30 per cent or more over mainline markets. Industry analysts agree that these markets are likely to continue to grow.

Consumers are more conscious of the need for vegetables in their diet to maintain good health. However, today's time-pressed shopper may face some barriers when preparing vegetables for home consumption--among them, their lack of convenience. The incorporation by food manufacturers of "value-added" vegetables into easy-to-prepare products, such as frozen dinners, instant soups and other prepared foods, provides consumers with a convenient way to work toward getting their five-a-day.

Foods, beverages, or confections are used by common men, therefore, high grade, style, and/or quality may be prepared from fruits. They may be prepared using particular processing and design. Consumers with special dietary needs - both for specific health conditions and the general health concerns of our aging population offer more opportunities for new products.

CHAPTER 31

Hi-Tech Horticulture: Harvest Technology and Value Addition of Medicinal Plants

HARVESTING

Harvesting from the Wild

Harvesting medical herbs requires careful planning to ensure the parts are processed in peak condition and fast enough to retain their active ingredients. Wild plants offer a free and natural source of herbal remedies and give the satisfaction of colleting herbs in the traditional way. Furthermore, active constituents are often more highly concentrated in wild plants as the herb is generally likely to be growing in its preferred environment.

Identification

Proper identification of wild plants is essential. Use a field or wild flower guide to help you, as many plants look similar. If in double, do not pick the plant, as poisoning can result from misidentification.

Collect material from healthy plants that are free from disease, insect damage and, in the case of wild plants, pollution. It is important to discard any damaged plants as they can lead to disease or decay in dried plant material. Do not mix cut plant material to avoid mistakes in identification.

Time of Harvest

Harvesting should be done in dry weather, preferably on a sunny morning after the dew has evaporated. Picking in the right weather conditions, when the plant is at its peak of maturity, ensures that it will have a high concentration of active constituents. Unless otherwise stated in the individual plant entries on leaves are best collected as they open during the spring or summer months, flowers as they start to bloom, fruit and berries just as they become ripe, and roots in the autumn once the plant has drawn its vitality back beneath ground. Bark must be gathered with great are if the shrub or tee is to survive – in most cases harvest it in spring or autumn.

Harvesting Correct Useful Parts

In many cases, different parts of the same plant for example the leaves and seeds, can have quite different actions and uses. Make sure that you harvest the correct medicinal part of the plant for your purposes.

PROCESSING

In order to make use of its latent medicinal qualities, a plant must be treated and modified in such a way that its specific curative substances can be extracted. Such transformations form part of what is known as the 'galenic' branch of pharmacy. The active principles contained in every medicinal plant consist of a number of compounds, which, individually or in groups, can have a specific action on the organs or our bodies. Different methods have, therefore, been devised to extracts these substances, either singly or collectively, depending on the result required. Some forms of extraction, being particularly complex, are beyond the skills of laymen and are exclusively the province of the pharmaceutical chemist or similar specialists. Others, of a more simple nature, can be carried out by anyone. There are four basic, simple preparation techniques: decoction, maceration, infusion and juice extraction. Although these are very straightforward methods, they are just as effective as the more elaborate systems involving considerably more complicated processes. These four methods will be discussed separately, with instructions as to the exact procedures for extracting the active principles and the circumstances in which each should be adopted.

Decoction

Decoction is the method normally used for those medicinal herbs whose active principles are difficult to extract because they are contained in woody parts of the plant, or which require prolonged heating in order to pass into solution. The procedures dealt with here only involve water extraction, as more complex forms requiring various solvents are too technical for a book of this nature.

Extraction by decoction involves boiling the whole plant or a part of it in water for a given time and then allowing it to macerate for a further period before filtering. A slight variation of this method is to soak the plant beforehand in cold water for a specific time, depending on the drugs required, in order to soften the leathery part and thus make extraction easier at the decoction stage. This method is reserved for certain plants such as those with a high mucilaginous content.

Decoction is, understandably, quite a drastic method of extraction and one, which is capable of modifying some of the plant compounds. It is important to ensure that the main active principle to be extracted is not affected by other harmful compounds induced by the heat. The boiling process during, which water is added, nearly always destroys the volatile active principles, that is to say, those with a low boiling point. This means, therefore, that the resulting decoction will never possess all the original principles of the drug.

Filtering Operation

The filtering operation is equally important, as it serves to strain off not only the leathery residues of the plant but also the mucilage or deposits of amphorous substances that are carried in solution during the boiling. The liquid obtained is, therefore, always left to clear before the next stage when it is filtered through paper of finely woven linen.

Extraction

Extraction is much more suitable method for obtaining the active principles when the parts of the plant being used are soft and fragile, such as leaves bud or flowers. It is preferable, except in certain cases, to use herbs that have been slightly dried, as the reduction of water content will concentrate the principal constituents. Unlike decoction, this method allows retention of many of the volatile principles, which would otherwise be lost, but does not permit extraction of those requiring a continuously high temperature. An expert, with extensive knowledge of medicinal plants and their medical compounds, can decide, which of these two methods is more suitable for obtaining the active principles required.

Infusion

Infusion entails pouring boiling water over that part of the plant known to contain the required drug. The container should be made of pottery of glass to avoid the formation of tannate of iron, which due to the presence of tannin in almost all plants, would occur it metal were used. As soon a s the water ahs been poured over the plants, the container should be covered, thereby allowing the steam, rich in volatile compounds, to condense. After a given time, which varies for each plant used, the liquid is filtered before being used.

Maceration

The third simple method, known as maceration, is used for medicinal plants whose active principles are soluble in cold water for several hours during, which time all the principles that do not need heat to release them (i.e. that are not thermolabile) will be released into the solution. The yield of mucilage from certain plants, such as the mallow (*Malva*) and marsh mallow (*Althaea officinalis*), is in fact greater when extraction is carried out by means of cold maceration rather than by infusion.

The method of maceration described above is hardly ever used nowadays in the pharmaceutical industry. Instead, tinctures are prepared by diluting a certain amount of fluidextract of the requisite drug with alcohol of a predetermined strength. When, as is sometimes done, a sugar solution is added to spirit solutions, the resulting preparation is known as an *elixir* or *ratafia*. These preparations are, in fact, genuine liqueurs and have been almost abolished in pharmaceutical practice.

If the vehicle of maceration were to be a mixture of alcohol and ether, andethereal tincture' or, more scientifically, an ethereal spirit solution would be obtained. Many varieties of such preparations could once be found in all the pharmacopoeias, but are now little used. The best known is ethereal tincture of valerian.

Extraction

It is a straightforward method used for the extraction of juices. In this case the plants, which contain the required medicinal ingredients must be fresh and have high water content. The Juice, which can be extracted by various methods, contains the mineral salts and vitamins of the plant, together with all the other substances that can be obtained through pressure. Again, the juice will not contain all the active principles of the plant but, on the other hand, cold extraction does not modify the structure of those compounds, which as we have seen, can be altered by heat. In the home, juices can be produced by first parboiling the fresh plants very briefly and then applying pressure with an appropriate utensil such as a small press. Because of the additional heat used for parboiling, however, this method will modify some of the constituents.

Better still, of course, is the modern electric blender, which extracts almost the entire juice content of the plant.

All the methods described above have a time limit; hence they are considered short-term procedures. In fact, the decoction, the infusion, the maceration and the juice all contain biological compounds that immediately undergo processes of oxidization and fermentation that denaturalize the medicinal substance, often making the liquid taste repugnant and, therefore, unusable. To avoid this, it is possible to add preservatives but this, of course, would be quite contrary to the whole concept of non-adulteration, which is the aim of the methods described here. Additions of sugar and alcohol admittedly also enable juices to be kept for a slightly longer period and, if vacuum-bottled with sugars, their properties and the freshness of their components will remain unaltered; unfortunately, this kind of preservation is almost impossible in the average household because the apparatus required is often expensive.

The Problem

The problem of conserving medicinal principles, once extracted, is of the utmost importance. The pharmaceutical industry has, therefore, done much work in this field, producing further preservation techniques so that active principles can be used at any time and in any dosage. The difficulty of tracking down exotic or rare drugs, combined with the relatively short harvesting period of healing herbs, has necessitated the development and regulation of highly advanced methods of extraction and preservation.

Forms of Preparations

Such preparations are known as extracts. They are divided into semi-solid, dry and fluid types; another means of classifying them is by their vehicle of evaporation: water, water and alcohol, or ether. Strictly speaking, an extract is the concentration of all the substances and salts diluted in the juice obtained from it. This concentration is achieved by the use of two fundamental processes: the first is based on freezing and the other on evaporation. The latter, more generally used, consists of putting the juice into an evaporating dish from, which the water content is evaporated at a low heat. Obviously, depending on the length of the evaporation process, different concentrations of the juice will be obtained. Concentrations will thus range from those still possessing a high water content to concentrates in which water is no longer present. The extract obtained is termed soft extracts have a semi-solid honey-like consistency, while dry ones are often composed of solid masses that can be reduced to powder. A classic example of a dry extract is that of aloes, which has the appearance of black, solid lumps, almost vitrified, but which are friable and easily pulverized; an even better known instance is liquorice, which can be reduced to a powder or to liquorice sticks.

Water Content of Juice

The water content of juice is not, however, always the vehicle of evaporation used in obtaining an extract. Instead, to facilitate concentration through evaporation, a solution of water and alcohol is often used. Alternatively, ether may be used and then recovered through distillation. The addition of these substances to the water enables compounds to dissolve, which are hardly soluble at all in water alone, but which become very soluble in a combination of the two liquids.

The Solvents

Extracts in which ether is used are described as ethereal and those in which alcohol are used as hydroalcoholic. Certainly the most important type of extract is the fluidextract, which has only relatively recently been formulated and accepted in pharmaceutical practice. In comparison with the type of extract previously described it has one property, which makes it extremely useful: one gramme of fluidextract is equivalent to one gramme of the original plant material. In other words, all the active principles contained in a precise quantity of plant matter to be treated are present in the same quantity of fluidextract obtained. The fluid extract makes it possible to have any drug available at any given time (it does not deteriorate quickly and fluid extract obtained. The fluid extract makes it possible to have any drug available at any given time (it does not deteriorate quickly and is not modified, except very slowly, by external agents); it is of a purify and accuracy of titration that would be impossible to achieve in drugs fresh from the plant.

Alcohol-soluble Substances

When, for particular purposes, it is necessary to extract only the alcohol-soluble substances such as alkaloids, bitter principles, and organic acids and colouring substances, another specific method of preparation is used. This is generally known as an alcoholic tincture or, more correctly, spirit solution. Apart from the reason already given, these tinctures are produced because their high alcohol content makes them easy to preserve. A spirit solution can be prepared either by maceration, in which a measured amount of the medicinal plant is immersed in alcohol of a suitable strength for a predetermined length of time and subsequently filtered, or by digestion, passing it through a still. This apparatus is equipped with a cooling flask, which causes the alcohol that ahs been condensed in it to fall back into the receiver flask in which the drugs to be digested have been placed. Because tinctures are very sensitive to light they are always kept in bottles made of dark glass. The alcohol in which the plants have been macerated develops varying degrees of strength, depending on the drug being used and the ease with which it releases its essential constituents, but it is usually not less than 60% vol. Although these are alcoholic preparations they cannot.

The alcohol content is sufficient to obtain an extraction of the soluble compounds if the medicinal plants are allowed to macerate in the wine for the length of time appropriate to each type. Unfortunately, medicated wines are subject to acetic fermentation (that is, they go vinegary) and form sediment just like any other wine. It is, therefore, important to prepare only small quantities and to keep the wine completely airtight and away form the light.

Fixed Solvents

Essential oils are soluble in solvents that are described as fixed if they do not evaporate at a normal temperature, and as volatile if the reverse is the case. Fixed solvents consist of solid or liquid odourless fats such as lard, tallow or refined olive oil. These solvents enable essence to be extracted by means of a method known as *hot enfleurage*. The essence-producing plants are placed in copper containers and left to macerate in fat, which is brought up to a temperature of about 104°F. (40°C.), with more plants being added until the fat is saturated. The fat is then separated from the essence to give a good yield.

Volatile Solvents

Extraction with volatile solvents, while presenting much greater difficulties in obtaining the essential oil, has the advantage of making it possible to extract essences from plants that have only very low percentage of the fragrant principle. The volatile solvents normally used are benzene and petroleum ether, which carry both the essence in solution and other products, making it necessary to carry our further, rather laborious tasks in order to obtain a pure products. The first phase of extraction by this method is the soaking of the flowers in an apparatus known as a *digester*. The solution obtained from this is distilled, during, which process the solvent is recovered. The essence' because of the traces of extraneous material contained in it. In order to purify the concrete essence, it is treated with pure alcohol, which takes up the essence, leaving many un-dissolved impurities. The alcohol is then brought to a low temperature, which enables the remaining impurities to become concentrate. As this stage, the alcohol can be called an extract. To complete the process, the extract is distilled to give the resulting essence a high degree of purity.

Uses of Extracts

These extracts can be used either by themselves or as the basic ingredient of various medicinal compounds, both of which are guaranteed as a homogenous product. Although the method of preparation varies according to the drug, it can be summarized as follows: 1,000 parts of the finely chopped dried plant are left to macerate for several hours in alcohol of the appropriate strength. This alcohol is then strained off and discarded. A similar amount of fresh alcohol is then poured over the same plant matter and a part of it strained off a day later. This alcohol, the weight of which is in direct relation to the type of herb used, is held in reserve. New alcohol is added until the dried plant material is exhausted of all the soluble extracts. The next stage is to distil this second liquid, recovering the alcohol and allowing the rest to evaporate until an extract is obtained, which, when mixed with the part held in reserve, will be of exactly the same weight as the original herb material. Some experts maintain that the solvent action of the alcohol can be improved by the addition of certain substances such as inorganic or organic acids, salts and ammonia. The various pharmacopoeias certainly use one or other these additives. The fluidextract will remain unaltered for a long time If kept in dark glass bottles, which are firmly stoppered.

The above explanation demonstrates the superiority of this method of extraction over the other methods described, and accounts for are almost unreserved use throughout the world.

Unlike fluidextracts, be preserved indefinitely. They react quickly to light and air, which cause flocculation and acetic fermentation.

Looking at the field of galenical preparations (those deriving from plant of natural origins) as a whole, there is one particular preparation reassessed by modern phytotherapy, both because of its successful results and its ease of preparation. It can even be made at home. This is medicinal wine. As long ago as the thirteenth century, Arnold of Villanova wrote his *Incipit* Tractatus de Vinis, the earliest published book on wines, which dealt particularly with 'methods of making preparations with wine basis'. Such wines are made by simply macerating plant drugs in wine, preferably white, with a high alcohol content, such as a fortified wine.

Examination of Galenic Products

These are examined are the *alcoholates* or *spirits*, produced by the distillation of spirit solutions. The difference between spirits and spirit solutions is that the former, through distillation, contain only the volatile principles of the drug and not the fixed principle, as do the latter. For the distillation, 70-80% vol. alcohols are generally used, after appropriate maceration. Only a certain part of the distillate is retained, so as not to change its characteristics with substances of low volatility, which remain in the flask, following the concentration of the watery part. Some of the best known spirit solutions still used today are aqua vitae of lemon balm and Hungary water, or spirit of rosemary, the former being used as a digestive antispasmodic and the latter as a perfume component. Spirits have excellent preservation qualities due to their high alcohol content.

Syrups Preparation

Instead of using alcohol as a preservative for garden herbs, sugar can be sued in a concentrated form. By adopting this alternative, sugary solutions can be made, which contain medicinal substances; they are known as *syrups*. Syrups also serve to disguise the unpleasant taste of many medicines.

Ordinary syrup is made by dissolving a certain quantity of sugar in water, at a low heat; the mixture needs to be stirred frequently and allowed to evaporate sufficiently to obtain the concentration necessary to achieve the characteristic density f syrup. The curative substances are then dissolved in it and it is these that determine the particular qualities of each type of syrup. If the resulting liquid is cloudy, it can be clarified. Syrups should be stored in hermetically sealed bottles and kept in a cool dry place, preferably away from the light. Despite such precautions, syrup is prone to alteration, especially when the additives are organic. In such cases, the syrup will have to be re-boiled in order to increase its density and, therefore, its preservation potential. Instead of sugar, honey can be substituted using the same procedure. Such preparations are called *mellitus* or honeyed syrups. For instance, 'melrosette' or 'honey of roses', which is still made for children, belongs to this category.

Electuaries of Conserves

Nowadays there is a growing demand for plants and vegetables that have not been contaminated with insecticides or fungicides, to be used in a natural form as natural form as possible. The interest in macrobiotics ash even brought back a number of preparations, which, due to their short shelf life, had become neglected. These confections are known as *electuaries* of *conserves* and are in the form of paste made by mixing the pulp of fruits, which have special nutritive of medicinal properties, with certain powders, honey of syrups. Some of these are sold commercially but, because of the high percentage of organic substances in them, preservatives have to be added. The macrobiotic diet makes use of several types of electuaries in this double role of supplementary food and medicine.

Use of Oil as Solvent

When medicinal substances are dissolved in oil rather than the usual solvents, the preparations can be made either by dissolving one part of an oily fluid extract of a particular plant in a given quantity of refined olive oil or, alternatively, by boiling the plant itself in the oil and then filtering the as rheumatism, in which an oily preparation is required to facilitate localized massages.

Closely related to the medicated oils are the ointments, also known as *salves* or *unguents*, the base of which is usually a semi-solid, fatty or greasy substance such as lard or wool-fat (lanolin). They may be prepared simply by mixing powdered vegetal drugs into the fat, or the fat these ointments are all intended for external use.

At one time simple ointments such as these were very well known and widely used, but nowadays they have been largely replaced by commercial products containing chemical additives. The creams and liniments found on the market today tend to have quite different bases. Sometimes, however, we are reminded of remedies, which must have been familiar to our ancestors, when we see an old apothecary jar bearing a name such as unguent of polar or unguent of lily root. In Britain there is, increasingly, a tendency towards these simple creams containing no chemical additives.

Distillation Process

Distillation process is of fundamental importance for obtaining those ubiquitous ingredients of so many medicines and cosmetics known as essential oils or essences. These are not only used for their aromatic qualities, and s the main ingredients in perfumery, but also as the main medicinal ingredients in modern aromatherapy. The rest of this section, therefore, is devoted to a fairly detailed description of various processes necessary for the production of these oils.

Essences

Essences are complex substances, which are often fragrant and, although not oily, when placed on paper will produce a translucent effect. It is now possible to reproduce a great many essential oils chemically, and although this might suggest that distillation is an outdated method, that is not the case.

The essences derived from plants do not merely consist of open or more compounds, which it is possible to reproduce, but of a whole combination of substances, which are often undefined. It is not difficult to differentiate between the natural and the synthetic essences, because the natural perfume is altogether more delicate. There are also cases were chemistry has to make use of natural raw materials in order to reproduce some of the compounds synthetically produced.

The reasons why plants manufacture essential oils are still being investigated. A Few of the many theories formulated have some credibility, but what is certain, in any event, is that essences are beneficial to the economy of a plant, enabling it to manufacture other products. Some authorities maintain they are either waste materials, which have not yet been eliminated or that they are a means of protection against excessive sunshine. It is, in fact, true to say that nearly all-alpine plants are scented and the perfume that surrounds them serves as a filter through, which the solar rays must penetrate.

Essences are quite easily altered, and the least change in temperature or humidity is enough to allow the enzymes present to become activated. Another factor to be borne in mind when gathering herbs is the effect of light on the amount of essential oil contained in plants at different times of day. Experienced collectors, knowing just when the healing properties of each type of plant are at their peak, can decide exactly when to harvest it in order to derive the most benefit. Thus in some plants, the amount of essence will be reduced by light, whereas in others it will be concentrated. The composition of the soil may also affect the size of the yield.

Having made these points, it has to be said that the most satisfactory extraction is obtained when there is the least possible alteration to the characteristics of the essence itself. This type of extraction, which is described as *cold*, entails no change in the composition of the oil because no heat is involved. Unfortunately this method can only be applied to certain plants, mainly citrus, that possess a great many oil-producing glands. With the cold method the components present in the oils undergo only the slightest alteration, thus making it possible to produce essences of great purity, which can even be used internally. The two main methods of cold extraction are:

Expression

The former is carried out by putting the skins of the fruit into special containers where they are pressed, thus causing their glands to rupture and allowing the essence to be collected together with water and other substances. Essences extracted in this way are, inevitably, cloudy, although their fragrance and substance remain unaltered and they clarify when subsequently decanted. This method is mainly used now for homemade essences. Industry has adopted the

Ecuelle Process

This involves the use of machines in which the fruit is revolved in metal cups or drums lined with spikes, which delicately rupture the oil-bearing glands to release their essences. As this is a fast process and the fruit are not identical in size or shape, there is considerable wastage of essence; this is, however, compensated for by the speed of the operation.

Cold Enfleurage

This method alters neither the perfume nor the characteristics of the oils. This procedure is based on the special is based on the special quality possessed by some substances to condense essential oils on their surface area. These substances are purified waxes, which are spread over flat glass surfaces; the plants are placed on top and enclosed in special containers. After a day or two, the plants are removed and replaced by fresh ones; the process is then repeated over several months or until the wax has reached saturation point. The wax is then removed from the glass and the essence recovered by following the same procedure as for volatile solvents.

The problems involved in obtaining the products of essences producing plants are such that the methods described are only adopted for particular purposes.

Distillation

Normally, even though the process may change the composition of the essences due to the heat employed, *distillation* is used. This can be divided into three categories: simple distillation over direct heat, steam distillation and vacuum distillation. In distillation, heat both ruptures the oleiferous glands and, unfortunately, slightly modifies the essence. We shall only deal with the type of distillation in which water is the transporting vehicle. The distilled liquid will be a mixture of condensed water and essential oil, which will separate naturally because of their different specific gravities. The plants to be distilled must be used while still fresh and should have been gathered when most fragrant. In order to avoid fermentation, they should never be piled on top of one another.

The apparatus used for distilling is known as an alembic, from the Arab word *al-anbiq*, a vase. This is divided into three parts: boiler, distilling column and condenser. Vaporization of the water and volatization (rapid evaporation) of the essential oils are achieved in the distilling flask. The vapours then rise up the column and come down through the condenser, whence, having recondensed rapidly, the liquid falls into a receiver.

The boiler is usually made of copper or stainless steel, but may be of glass if laboratory apparatus is being used. It is generally cylindrical in shape and its base, to which the heat is applied, is never in direct contact with the plants, as these are kept slightly raised on wire mesh racks. Part of the space between the base and the first rack contains water, which produces the necessary steam. The column, which is firmly set in the neck of the boiler, leads to the condenser and allows for the expansion of the steam and for its initial cooling. If the boiler is metal, the column is clamped to it to form an airtight join, but if it is glass, either the foot of the column is shaped to fit the neck of the heater flask, or it is passed through a close-fitting stopper or cork. The condenser consists of a spiral tube, which is completely immersed in a vessel containing cold running water.

FIG. 31.1: An old-fashioned copper distilling apparatus at the Fragonard Perfumeries Museum in Grasse (Cote D'Azur). Its parts are easily identifiable as the boiler (right), connecting tube and condenser (left).

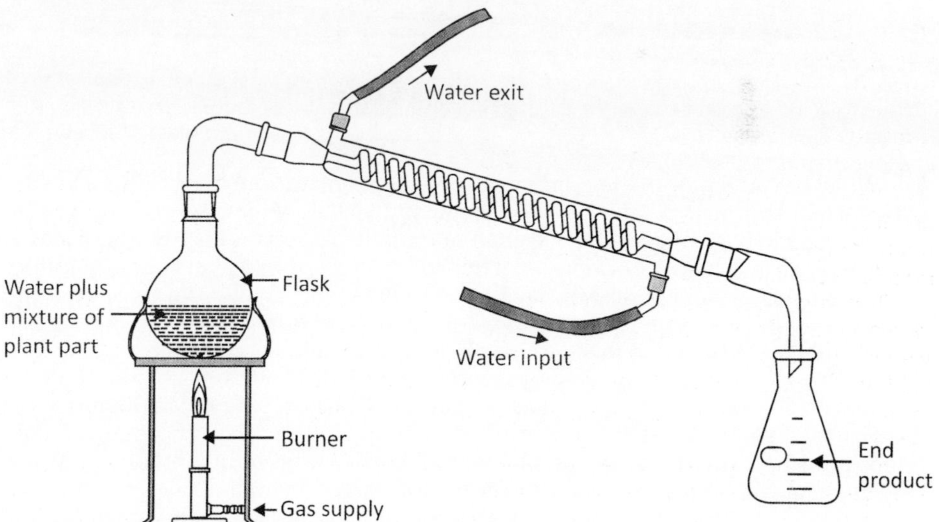

FIG. 31.2: Distillation over direct heat with a serpentine condense to prolong the time taken for the vapour to pass through, thus aiding condensation.

FIG. 31.3: Perfume distilling apparatus in operation at the Fragonard Perfumeries in Grass. In the foreground can be send the liebing condensers and the distillation flasks containing the substances to the distilled.

When the steam from the column comes into contact with the cold walls of the condenser, it condenses and, by force of gravity, runs down into a receiver, known as a 'Florentine flask', where the essence automatically separates from the water. If this type of flask is not used, separation has to be effected by means of a separatory funnel. Under normal conditions, boiling point is reached when the tension of the water equals the atmospheric pressure and to achieve this, the heat source is placed directly under the boiler; in this case distillation is brought about by direct heat.

In the laboratory the boiler is replaced by a glass distillation flask, and the heat is also applied from below to avoid breakages an asbestos, wire mesh or gauze mat is interposed so that the heat is not applied directly to the glass. In order to obtain uniform heating, the flask may be placed in a bed of sand or in water-bath. In this way the heat will be diffused evenly over the lower surface of the flask.

The two main types of condenser used in a laboratory for liquids with a low boiling point are the Liebig and the serpentine. If the steam, rather than being produced in the same container as the herbs, as described above, is produced in a separate boiler, this prevents the essence-containing plants form being 'cooked', which might adversely affect the result.

The method, known as *steam distillation*, makes it possible to distil those liquids, which have higher boiling point than water at temperatures lower than the boiling point of water. This method also makes it possible to obtain essences that are not excessively altered and is, therefore, preferable to the previously described system because of the greater purity of the resulting product.

Distillation can also be carried out at a pressure lower than that of the atmosphere. This is done by vacuum pumps and is known as 'distillation under reduced pressure' or 'vacuum distillation'. By decreasing the atmospheric pressure, the boiling point is also lowered, so enabling distillation to take place at a lower temperature to obtain a product that is even less modified than that produced by steam distillation.

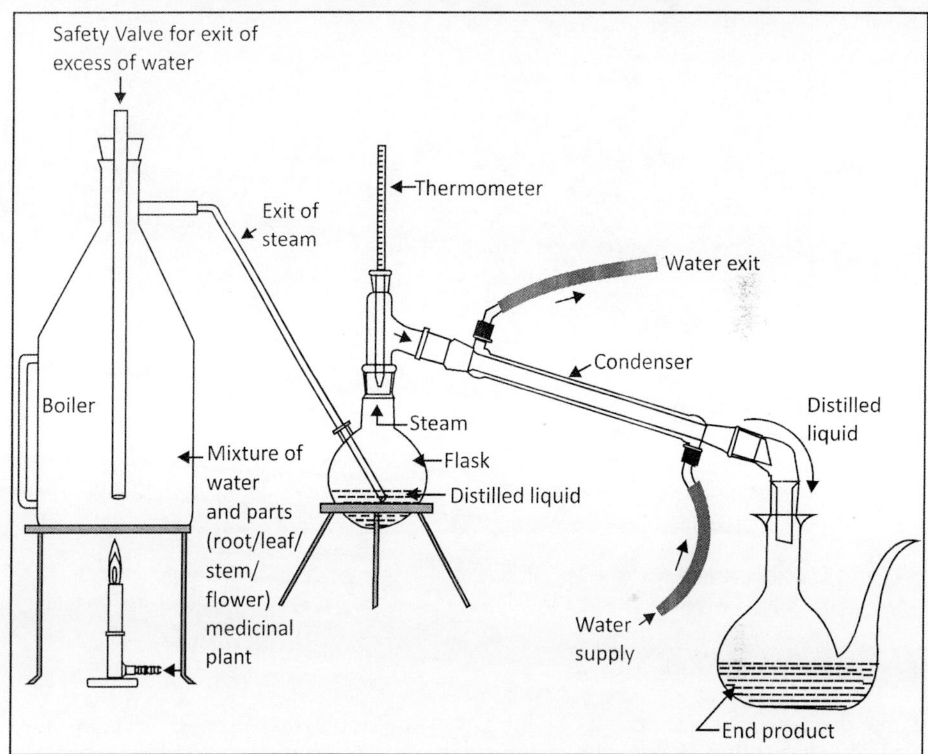

FIG. 31.4: The water contained in the boiler is brought to the boil by means of a Bunsen flame. The steam passes through the tube into the flask. If the pressure inside the boiler becomes excessive, the water is pushed up the central tube until the pressure is equal to the external pressure; the tube thus acts as a safety-valve. The distillation flask must be pre-heated inorder not to create unwanted internal condensation and differences in temperature: inside it are the substances to be distilled. As the steam breaks down the oleiferous glands in these, the mixture of water and essence in the form of vapour moves towards the connecting piece between the distillation flask and condenser. The thermometer (b) allows the operator to judge the temperatures of distillation of the various compounds. The vapour is finally condensed in the condenser and the resulting distillate falls through, via the connecting piece, into the Florentine flask where the substances decant to form two layers, due to the difference in their specific weights. Separation of distilled water and essence will occur automatically when the level reaches the top of the lateral overflow.

FIG. 31.5: An example of a deterpentor used to obtain essences that are known as 'absolute' that is, completely without terpenes; these are hydrocarbon substances used in various industrial sectors.

Fractional Distillation

Once the essences have been obtained, they are not always used in their complete state but are often broken down into their various components. This is done by fractional distillation, based on the already discussed principle of boiling, which occurs as a result of the equality between the atmospheric pressure and the sum of the vapour tensions of the components. The distillation temperature will initially be lower than that of the most volatile component and will then gradually rise. By carefully controlling the temperatures, the fractions of the essences can thus be collected. As the various components separate, they are retained individually by changing the receiving vessel when the appropriate temperatures are reached. To carry out such work in a laboratory, a special container in used, which is known as a Pauly apparatus or, more commonly a 'spider'. However, on an industrial scale such a method would be totally unsuitable because it is lengthy, laborious and, what is worse, discontinuous. For practical reasons, therefore, the *dephlegmator* or *fractionating column* is used. The most usual type is the Egrot dephlegmator, in which the essence vapours are cooled by a flow of water. This causes only the less volatile constituents to condense, thus effectively separating them from the more highly volatile ones.

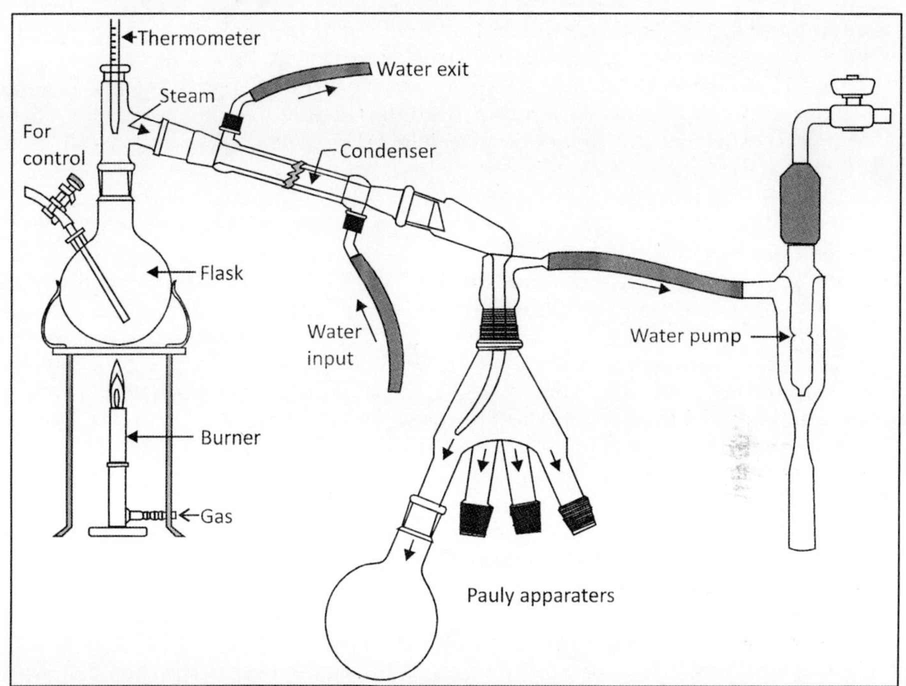

FIG. 31.6: Fractional distillation. The substance to be distilled in the distillation flask is heated by means of a Bunsen flame. Rich vapours of essential oil are formed which begin to cool down in the ground joint and then to condense in the condenser. The pauly apparatus anables the various fractions of distillation to be directed to the collecting flasks. In this type of distillation, all the apparatus is completely watertight. Boiling is facilitated by reducing the temperature at which this takes place; this is done by lowering he internal pressure by means of water pump that incorporates the principle of the Venturi tube – a device used to measure the flow rate of water in conduits that are closed and under pressure. As the reduced internal pressure could cause the apparatus to implode, air is constantly fed into the distillation flask, controlled by a spring clip (6)., the thermometer enables the operator to known when distillation of the various fractions of the essence has begun.

Distilled water is the secondary product of distillation and is used a great deal in perfumery. It is, of course, merely water in which a very small amount of essence has remained and which gives it a fragrance and character of its own. Distilled waters do not keep well and are prone to alteration as they are subject of fungi and moulds. The only precaution possible is to store them away from the light in dark glass bottles. Their fragrance can be made more subtle by diluting a little of the essence itself in the water but this will deprive the water of its active volatile and other unknown principles that are typical of distillation. Distilled waters are widely used today in phytotherapeutic technology because they are readily available in large quantities. Although they contain only a low level of active principles, their medicinal effect is consistent with the concept of homoeopathic medicine.

PARTS OF MEDICINAL PLANT TO BE USED

1. **Aerial Parts:** These include all the parts of the plant growing above ground – stems, leaves, flowers, berries and seeds. The stems are normally cut 5-10 cm (2-4 in) above ground shortly after the plant has begun to flower, when it is

putting most effort into growth. Perennials may be cut higher above ground to encourage further crops. Removes and dry large flowers crops. Remove and dry large flowers and leaves separately; smaller cones can be dried on the stem.

2. **Large Flowers:** In most cases, flowers are picked just after they have opened, usually during the spring or summer. Sometimes only specific parts of the flower are used, such as the petals of marigold (*Calendula officinalis*), while other flowers are used whole.
3. **Fruit and Berries:** Harvest fruit and berries in early autumn when rope but still firm. If allowed to become over-ripe, they may not dry properly. They can be picked individually or in bunches.
4. **Roots, Rhizomes, Tubers and Bulbs:** The underground parts of the plant are usually gathered in autumn after the aerial part have withered or because inactive and before the soil is waterlogged or frozen. Many roots may also be collected in early spring before the root loses its vitality to aerial growth. Dig deeply around the root, prising it carefully out of the ground. Some tap roots are very difficult to uproot completely. Remove the required amount and replant the remaining underground part.
5. **Seeds:** Collect ripe seed pods, capsules or flowering stems in late summer before the seed have been scattered.
6. **Sap and Gel:** Only harvest sap from your own garden. Collect sap in the spring as it rises, or as it falls in the autumn. Trees such as silver birch (*Betula pendula*) produce huge quantities of sap if tapped, although this reduces the tree's vitality. Bore a deep hole into the trunk of the tree – no more than a quarter of the diameter of the trunk – and place a collecting cup under the mouth of the hole. In spring, litres of sap may be produced and it is essential to stop the hole with resin or wood filler once about a little of fluid has been removed. Collect milky juices or latex from plants such as dandelion (*Taraxacum officinale*), by squeezing the stems over a bowl. Wear gloves as latex or sap can be corrosive. The gel from aloe vera (*Aloe vera*) is scarped out after slicing the leaf lengthwise and peeling back the edges.
7. **Bark:** Only harvest bark from your own shrubs or trees as it carries the risk of losing the whole plat through over-stripping or "*ringing*" (removing a whole band of bark). It is far better to collect bark from outlying branches, which can then be pruned back. If stripping bark from a plant, gather it in autumn when the sap is falling. Remove insects, lichen and moss from the bark, cut it into small pieces and place on a tray to dry.

Other Method of Processing

Apart from simply air-drying herbs, there are a number of other ways to preserve their medicinal benefits.

Dehumidifying

An effective but expensive way to dry herbs is to use a dehumidifer, which literally sucks water out of the plant. The dehumidifer should be placed in a more or less sealed small room in which the herbs are hung in loose bunches or placed on mesh trays. Herbs will dry quickly with this method and, as no heat is used, there is little deterioration or decay.

Freeze-Drying

Freeze-drying retains colour and flavour but is more suited to culinary than to medicinal herbs. Whole sprigs of herbs such as parsley (*Petroselinum crispum*) or sage (*Salvia officinalis*), can be frozen in plastic freezer bags. There is no need to defrost before use as the leaves crumble easily when still frozen. Chickweed (*Stellaria media*) can also be frozen and used topically for itchy and weeping skin conditions. Many plant may be juiced, frozen as ice cubes and thawed as required.

Storing Herbs

Storage method is very important for maintaining the quality of dry medicinal plants Leaves, flowers, roots and other parts should be stored in sterilized, dark glass containers with airtight lids. They may also be stored in new brown paper bags, which must be kept dry and away from light. Metal and plastic containers are inadvisable as they may contaminate the herb. If stored in a cool dark place, herbs can be kept for about 12 months after harvesting. Herbs frozen in plastic freezer bags can be used for up to 6 months. Make sure the container you use is labeled with the herb. Source, strength of preparation if appropriate, and date of harvesting. Watch out for insect infestation. If this occurs, put all affected material in a sealed plastic bag and discard. Sterilize the containers and check all other herbs for contamination.

PRESERVATION

Medicinal plants can be preserved in a number of ways, the most common and simple, being air or oven drying. A warm, dry place, for example an airing cupboard, is ideal. Use plain paper for drying herbs, never printed newspaper. Dried herbs can be stored for man months in a dark glass jar or a brown paper bag.

STANDARDIZATION OF MEDICINAL PLANTS

When a medicine is administered to a patient, the doctor who prescribes the medicine needs to be assured that the medicine contains the correct amount of the right substances as only then would the medicine induce its therapeutic effect. This is assured both to the doctors and the patient by the pharmaceutical company producing the medicine. In the laboratories of the pharmaceutical house, the medicine – or batches of the same – have been tested for quality and quantity and the drug is released for use only if the quality control laboratories of the pharmaceutical house approve the batch in question. Overseeing the situation nationally are national drug control laboratories run by government scientists who also monitor the equality of the drug being released by carrying out random checks to ensure that the results conform to the standards laid down by both the pharmaceutical house and the national drug regulatory authority at a time when the pharmaceutical house requests for the release of a particular drug in the market. This is known as standardization – a system to ensure that every packet of medicine that is being sold has the correct substance in the correct amount and will induce its therapeutic effect.

This system has worked well in developed countries and reasonably well in countries in the third world, particularly if the country has a strong drug regulatory agency with its own laboratories. However, the system has been developed for synthetic medicines and its has not been difficult to "*fingerprint*" the synthetic medicines and lay down standards. It is much more difficult to ensure that this system works in regard to herbal medicine.

It is very important that a system of standardization is established for every plant medicine in the market because the scope for variation in different batches of medicine is enormous. Plant material, as has already been mentioned earlier, may vary in its alkaloid content and therefore in its therapeutic effect according to different places of collection, with different times in a year for collection, with collection at the same time and place but in different years, and with different environmental factors surrounding the cultivation of a particular medicinal plant. Adding to this variability is the fact that in herbal medicine several plants may be used together in the same preparation. This means that there should be a quality control test for the entire preparation to ensure quality of the preparation.

As long as herbal medicines were grown locally, perhaps even in a kitchen garden and used fresh, for example, as a tea, whenever needed, it may not have been necessary to carry out extensive procedures of standardization. During the author's first visit to China in 1981, the Chinese scientists could not always fully comprehend the scope of standardization when he raised this topic since they were largely using freshly-prepared herbal medicines. However, the position changed and by the time of this next visit in 1985, most scientists involved with the development of herbal medicine were very concerned about standardization. Herbal medicines were now being prepared in factories and pharmaceutical houses and packed for use all over the country and also for export. Quality control had therefore to be carried out with their products and this could only be carried out if standards had been laid down to ensure quality. These standards had to be developed for different plants and for the final products and the whole issue of standardization of herbal remedies and medicinal plants came to the fore.

Before considering this issue in more detail, two things should be kept in mind. If plants are used fresh, after brewing as a tea or as a juice, it is probably not necessary to carry out tests for quality control. However, the moment plant substances are prepared for storage and subsequent use, it is essential to carry out tests to check that the standards are being maintained.

The second point to remember is that when plants are used fresh – and perhaps cultivated in the kitchen garden or village garden – these lead to cheap remedies. However, if tests for standardization of plant preparations has to be carried out and if standards are maintained, this requires the use of some sophisticated equipment and the time of suitably-trained technicians and supervisors. This inevitably means that this cost will be added on to the price of the herbal medicine which does not remain a cheap medicine any more. That is why it has often been stated that herbal medicines are, in fact, not cheap. This depends on the way herbal medicine is used. With elaborate quality control tests needed for four to five ingredients and the product, a herbal medicine may indeed be more expensive than a synthetic drug for the same purpose.

It is generally felt that standards can only be established for an isolated compound. Interest in developing extracts for use – both at the primary health care level and more widely – is therefore tempered by a feeling that these extracts cannot be standardized and therefore the extracts are not a marketable commodity. Development of therapeutic substances concentrates on isolated compounds and those plants in which an isolated compound cannot be obtained are not followed up. It has been stated earlier that biological activity present in some extract will be present when the plant is in an extract form only.

The full pharmacological effect will be decreased or even disappear if further extraction and fractionation is carried out. If specialists in the modern system of medicine want to use some of the medicinal plants in use today e.g. an extract of *Momordica charantia* for diabetes mellitus, they will need to consider working with extracts. It should be clearly stated that extracts can be standardized and the lack of standardization methodology today should not be a reason for not working with extracts. There may be other reasons for not working with extracts but the lack of standardization is not one of them. If the method of production of the extract is properly standardized and if there is an appropriate chemical assay method developed, there is no reason why standards cannot be developed for an extract or a fraction. Further, this will be augmented by a bioassay method, which would certainly be available. Once this reservation in thinking is overcome, a wide range of medicinal plants becomes available for investigation, for isolating maximum activity in an extract and for using the extract.

Another apparent consistent in standardization appears to be a preoccupation of chemists and pharmacologists with a medicinal plant having only one active principle which needs to be standardized. Again, if modern medicine is to take advantage of medicinal plants used in the traditional systems of medicine, then one has to accept that, in many instances, the pharmacological and therapeutic activity will not be residing in one active substance but may, in fact, be present in a combination of a few active substances. As far as standardization is concerned, this again need not be an insurmountable barrier because all the active substances could be identified and standards set for each of these.

Once scientists in the field of medicinal plants research accept these two "*facts of life*" – that standards have to be set for extract and that standards may have to be set for several active substances in a plant – and once they work towards these goals, there will certainly be advances in the field. Such advances have been lacking in the last two decades. Participants at two recent international meetings organized by Who have clearly pointed out the problems relating to standardization of herbal medicines and the tremendous need for urgent immediate work in this field. This first was an Interregional Meeting on Standardization and Use of Medicinal Plants, held in Tianjin, China, in November 1980, while the second was a Conference on Research in Traditional Medicine, held in Tokyo in March 1986. This meeting identified three areas of priority for research, and research on standardization was one of these. Standardization of a plant produce entails standardization of the raw material, standardization of the method of production and quality control of the final product.

Identification of the actual plant is not easy. Sometimes two plants are known by the same local name. When the plant is being collected, it is very possible that another plant is being collected at the same time and this will be mixed up with the plant which actually has the activity. One example of this is manifest when one collects the plant, *Boerhaavia diffusa*, which is used widely as a "*quality of life enhancer*" in the traditional systems of medicine. However, both *Boerhaavia diffusa* and the plant, *Trianthema portulacastrum*, are known as "*Punarva*", and so both plants may be collected at the same time.

A second area of concern is that plant collectors may substitute cheaper plant material that looks the same, for the actual plant material which has the therapeutic effect. In this case, it is not a case of inadvertent mistaking a non-medicinal plant for a medicinal plant, as described in the last paragraph, but one of adulteration and cheating. One example which is very well known concerns the bark of *Caesalpina sappen*. When this substance is collected, bark from other plants such as *Pterocarpus, Gluta Travancorea* and *Toona cilita* are substituted for the bark of *Caesalpina sappen*. Other examples with commonly-used medicinal plant sin India are substitution of the bark of *Saraca indica* by the bark of Trema orientalis and substituon of the plant, *Halarrhena antidysentarica*, by the plant *Wrightia tinctoria*. It is very important, therefore, when laying down standards for a herbal product – either for experimental research, clinical evaluation or widespread commercial use – to lay down clear-cut standard nomenclatures and clear-cut criteria to actually "*finger print*" that the substance being used in actually the substance which should be used. Once it has become clear that the substance being used is the substance being used is the substance meant to be sued, the first hurdle in standardization of the preparation is overcome. Every crude ingredient going into the herbal preparation must be clearly identified.

The next step is to standardize the method of production. This could be the method, as it usually is today, of isolating the compound which is to be used or the method for preparing the extract which is to be used. The steps leading up to this should be clearly described and the end points defined.

Finally, procedures have to be developed for laying down standards for laying down standards for the final product – if it is a single substance or for the final formulation – if it is a mixture of substance. These standards would attest to the quality of the final product and would comprise both physiochemical methods and biological tests of efficiency. If mixtures or extracts are to be used as the final product, then methods for standardization of the mixtures or extracts should be developed. These standards should be strictly adhered to.

It has been stated very often that the lack of standardization of herbal remedies and plant medicines is holding back the use of medicinal plants in the modern system of medicine. It has also been said that practitioners of the modern system of medicine cannot have confidence in the use of herbal the medicines and do not want to use such medicines because they are not certain that the substance being used is what it is supposed to contain. This sentiment is not confined to practitioners of the modern systems of medicine. At the III International Congress of Traditional Asian Medicine, held in Bombay from 4 to 9 January 1990, similar views were expressed by a large number of participants who die not belong to the allopathic medical profession. Botanists brought out of the difficulty in the identification of Artemesias as the herbs and shrubs belong to at least ten species of Artemesia. A group of Ayurvedic research workers stressed the importance of developing good methods for standardizing crude extracts, decoctions and compound formulations which consists of different ingredients. It was clearly stated that the technology currently available is not able to develop suitable quality control criteria for mixtures and compound preparations. Research, therefore, should be carried out to develop the methodology needed for the standardization of such preparations as otherwise these could never be used either for primary health care or more widely.

Another investigator suggested that extensive work be put into the methodology of standardization keeping in view the vary intricate and complicated system being dealt with. He pointed out the biological activity of a plant depended not only on the use of the proper plant and its contents but also on the presence of the required quality and nature of secondary metabolities in the raw drug. The availability of these secondary metabolities depended on environmental and other factors such as time and season for collection, stage of plant growth, storage, drying procedure and geographical variation. Unless clear-cut standards are laid down for the evaluation of the final product, these factors could certainly modify and affect the efficiency of the product and lead to failures in therapy. Still another investigator stressed the need to develop methods for rapidly determining whether the plants being collected were authentic and whether the quality of the material used was up to the standard.

There appears, therefore, to be a general consensus that the lack of standardization constitutes a major constraint to the development and use of medicinal plants; that this need is felt by practitioners of both the modern and the traditional systems of medicine and that the area which needs attention is that for developing a methodology. It is very interesting to note that researchers and practitioners of the traditional systems of medicine do not have any reservation about the use of modern technology for developing methods for standardization. In fact, these workers have often stressed not only that all modern method should be used and that these methods are not always adequate and that more sophisticated and more advanced test should be devised.

ACTIVE CONSTITUENTS OF MEDICINAL PLANTS

Active ingredients are the main effective compounds of medicinal plants. Quantification of Active ingredients is the first step of standardization. Various chemical tests are available for quantification of active ingredients. Some of the main groups of active ingredients are described here.

1. **Mucilage:** Found in many plants, mucilage is made up of polysaccharides (large sugar molecules) that soak up water, producing a sticky jelly-like mass. Mucilage lines the mucous membranes of the digestive tract, protecting against irritation, acidity and inflammation. This soothing and protective action appears to extend to other areas, including the mucous membranes of the throat, lungs, kidneys and urinary tubules. Slippery elm (*Ulmus rubra*), is a typical mucilaginous herb.
2. **Phenols:** This group of compounds includes salicylic acid – the natural forerunner of aspirin. Salicylic acid is found in many plants, for example wintergreen (*Gaultheria procumbens*), and white willow (*Salix alba*). Another phenol is thymol – a constituent of thyme (*Thymus vulgaris*). Phenols are antiseptic and reduce inflammation when taken internally yet they have an irritant effect when applied to the skin.
3. **Tannins:** Tannins are produced to a greater or lesser degree by all plants. The harsh, astringent taste to tannin-laden bark and leaves makes them unpalatable to insects and grazing animals. Tannins contract the tissues of the body – hence their use to "*tan*" leather. They draw the tissues closer together and improve their resistance to infection. Oak bark (*Quercus robur*), and black catechu (*Acacia catechu*) are both high in tannins.
4. **Coumarins:** Coumarins of different kinds are found in many plant species and have widely divergent actions. The coumarins in melilot (*Melilotus officinalis*), thin the blood, while bergapten, found in celery (*Apium graveolens*), is used as a sunscreen, and khellin, found in visnaga (*Ammi visnaga*), is a powerful smooth muscle relaxant.
5. **Anthraquinones:** Anthraquinones are the main active constituents in herbs such as senna (*Cassia senna*), and Chinese rhubarb (*Rheum palmatum*), both of which are taken to relieve constipation. Anthraquinones have an irritant laxative effect on the large intestine, causing contractions of the intestinal walls and stimulating a bowel movement approximately 10 hours after being taken. They also make the stool more liquid, easing bowel movements.
6. **Flavonoids:** Found in many plants, flavonoids have a wide range of actions. They are anti-inflammatory and are especially useful in maintaining healthy circulation. Rutin, a flavonoid found in plants including buckwheat (*Fagopyrum esculentum*), and lemon (*Citrus limon*), strengthens capillary walls.
7. **Anthocyanins:** These pigments, which give flowers and fruits a blue, purple or red hue, help to keep the blood vessels healthy. Blackberry (*Rubus fruticosus*), and grapes (*Vitis vinifera*) contain appreciable quantities anthocyanins.
8. **Glucosilinates:** Found exclusively in species of the mustard family, glucosilinates have an irritant effect on the skin, causing inflammation and blistering. Applied as poultices of painful or aching joints, they increase blood flow to the affected area, helping to remove the build-up of waste products (a contributory factor in joint problems). Glucosilinates also help to reduce thyroid function. Both radish (*Raphanus sativus*), and mustard (*Sinapis alba*) contain significant quantities of glucosilinates.
9. **Volatile Oils:** Volatile oils – which are extracted from plants to produce essential oils – are some of the most important plant constituents of all. Tea tree (*Melaleuca alternifolia*), for example, is known to contain over 60 different

volatile compounds within its volatile oil, many of them being strongly antiseptic. Some volatile oils contain sesquiterpenes, such as azulence, found in German chamomile (*Chamomilla recutita*). These constituents have an anti-inflammatory effect.

10. **Saponins:** There are two types of saponins – triterpenoid and steroidal saponins. The latter get their name from their similarity to the human body's own naturally occurring steroid hormones. Many plants containing steroidal saponins have a marked hormonal activity, liquorice (*Glycyrrhiza glabra*), being one of the best known. Triterpenoid saponins, for example those in cowslip root (*Primula veris*), are often strong expectorants, and may also aid in the absorption of nutrients.
11. **Cardiac Glycosides:** Found in various medicinal plants, most famously in common foxglove (*Digitalis purpurea*), yellow foxglove (*D. lutea*), and woolly foxglove (*D. lanata*), cardiac glycosides such as digitoxin, digoxin, and gitoxin have a strong, direct action on the heart, helping to support its strength and rate of contraction when it is failing. Cardiac glycosides are also significantly diuretic. They help to transfer fluids from the tissues and circulatory system to the urinary tract, thereby lowering blood pressure.
12. **Cyanogenic Glycosides:** Though these glycosides are based on cyanide, a very potent poison, they have a helpful sedative and relaxant effect on the heart and muscles in small doses. Wild cherry bark (*Prunus serotina*), and elder (*Sambucus nigra*), both contain cyanogenic glycosides, which contribute to both plants' ability to suppress and soothe irritant dry coughs.
13. **Vitamins:** Some plants contain significant levels of vitamins. Watercress (*Nasturtium officinale*), for example, contains an appreciable quantity of vitamin E, and the hips of dog rose (*Rosa canina*), have particularly high levels of vitamin C. Most other medicinal plants contain at least some vitamins. While the content may be small it contributes to overall daily intake. For other plants that are rich in vitamins.
14. **Bitters:** Bitters are a varied group of constituents linked only by their pronounced bitter taste. The bitterness itself stimulates secretions by the salivary glands and digestive organs. Such secretions can dramatically improve the appetite and strengthen the overall function of the digestive system. With the improved digestion system. With the improved digestion and absorption of nutrients that follow, the bodyis nourished and strengthened. Many herbs have bitter constituents, notably wormwood (*Artemisia absinthium*), and chiretta (*Swertia chirata*).
15. **Alkaloids:** A very mixed group, alkaloid mostly contains a nitrogen-bearing molecule (-NH_2) that makes them particularly pharmacologically active. Some are well-known drugs and have a recognized medical use. Vincristine, for example, derived from Medagascar periwinkle (*Vinca rosea*), is used to treat some types of cancer. Other alkaloids, such as atropine, found in deadly nightshade (*Atropa belladonna*), have a direct effect on the body, reducing spasms, relieving pain and drying up bodily secretions.
16. **Minerals:** Some herbs are particularly rich in minerals. Horsetail (*Equisetum arvense*), for example, has high levels of silica. Dandelion (*Taraxacum officinale*), has large quantities of potassium, and unlike other diuretics which flush this mineral out of the body, it helps to maintain high levels of potassium. These plants act as mineral supplements in their own right, while other herbs with a small concentration contribute to overall intake. For other plants with a high mineral content.

Tests for Standardization

Generally, the tests for standardization of plant medicines in use today are macroscopic, microscopic, physiochemical and biological. It is not the place to deal critically and exhaustively with these test nor is the author competent to do so. The list of tests generally used is given below.

- Macroscopic examination
- Microscopic examination, e.g.
  - Anatomical structure
  - Detailed study of fragments
- Physiological testing, e.g.
  - Boling point
  - Freezing point
  - Absorption coefficient
  - Loss on drying
  - Ash value
  - Refractory index
  - Optical ratio
  - Water/alcohol soluble solid
  - Spectroscopic analysis
  - Thin layer chromatography – R.F. value
  - Gas liquid chromatography
- Biological testing where appropriate

Bioassay

If a herbal remedy or a plant medicine is sought to be prepared for clinical trial or for wide use, it is essential that it passes the tests appropriate for the particular plant described above. Safety and efficiency need to be evaluated and documented. Only then will practitioners have confidence in the herbal medicine they use. Even these test methods are not adequate. More tests have to be developed to cope with enormous variability that may be, this has to be undertaken if we are interested in utilizing the heritage of medicinal plants available for use. If we are not prepared to expend the time, resources and superior calibre manpower needed, then we will never be able to develop more medicines from plants. The great effort and crusade for saving the rain forests – one reason for which is that these forests contain plants which would be the source of medicines in the future – would have been in vain and would remain a hallow sentiment. The best brains and sizable resources need to be put, not only in the areas of standardization but also in allied areas such as pharmacokinetic studies with plants, clinical evaluation methodology, development of toxicology models and development of clinical pharmacology in the third world where these plant are being used. Only then could we hope to use properly the herbal remedies we have or discover new medicines from plants in the coming century.

Preparation of Herbal Remedies

Identification

Before using medicinal plants that have been collected from the wild, it is essential that they are correctly identified. If inn doubt, do not use the herb. The wrong identification of herbs has led to many cases of poisoning. Foxglove leaves (*Digitalis purpurea*), for example, are often mistaken for comfrey (*Symphytum officinale*).

Utensils

Use glass, enamel or stainless steel pots and pans, wooden or steel knives and spatulas and plastic or nylon siees. A wine press is useful for making tinctures. Do not use aluminium utensils, as this potentially toxic element is easily absorbed by herbs.

Sterilization

All utensils used to make herbal remedies should be sterilized for at least 30 minutes in a well-diluted sterilizing solution, such as the type used for a baby's bottle. After soaking rinse thoroughly with boiled water and dry in a hot ovenor wash in a dishwasher. Proper sterilization maintain hygiene and prevents remedies, especially creams and syrups, from turning mouldy.

Weights and Measures

For most purposes, ordinary kitchen scales are suitable, although electronic scales are more accurate. Metric measurements of grams and litres are generally much easier to use than imperial measures when making remedies. If it is difficult to weight a small quantity, such as 10 g, on your scales, measure double the weight, i.e. 20 g, then halve the quantity. Liquids can be measured in a kitchen measuring jug, although conical or straight-sided glass measures are more accurate. Very small volumes of liquid can be measure in drops (*Measuring Remedies*, right).

Storage

Different preparation may be kept for varying periods of time before they begin to lose their medicinal properties. Infusions should be made fresh each day and decoctions must be consumed within 48 hours. Store both in a refrigerator or cool place. Tinctures and other liquid preparations, such as syrups and essential oils, need to be stored in dark glass bottles in a cool environment away from sunlight, but can be kept for a number of months or years. Ointments, creams and capsules are best kept in dark glass jars, although plastic containers are also acceptable.

Infusions

The medicinal value of many herbs lies chiefly in their volatile oils, which will disperse into the air if a lid is not used. This is especially important in the case of German chamomile (*Chamomilla rectita*). Use a teapot, or place lid or saucer over a cup if making a small quantity. Use water that has just boiled.

Popular herbal teas, such as German chamomile, are often taken as much for their refreshing taste as for their medicinal value and can be safely consumed in quantities of up to 5 or 6 cups a day. Some herbs, however, such as yarrow (*Achillea millefolium*), are significantly stronger and must be taken in less frequent doses. Other herbs, such as feverfew (*Tanacetum parthenium*), are so strong that they are not suitable for use in infusions. Always check the recommended dosage and quantity of herb to use, as infusions have medicinal actions and can produce unwanted effects at the wrong dosage.

Decoctions

Decoctions are generally made using roots, bark and barriers, but something, leaves and flowers may be included. Add these more delicate parts of a plant once the heat is turned off and the decoction has finished simmering and is beginning to cool. Then strain and use as required.

Chinese Decoctions

In traditional Chinese medicine, decoctions are the main way in which herbal medicines are the main way in which herbs medicines are prepared. Large quantities of herb are often used to produce a highly concentrated liquid, or the decoction is further reduced so that there is only 200 ml of liquid remaining. This increases the preparation's concentration. This process is useful for astringent barks such as babul (*Acacia arabica*), and common oak (*Quercus robur*), which may be used externally to tighten gums or wash weeping skin rashes. (Do not take internally.)

Infused Oils

- Got infused oils: Although hot infused oils can last up to a year, they are most potent when used fresh. If only using infused oils very occasionally, make a smaller quantity than the standard amount with the same proportion of herb to oil. The wine press may be replaced with a jug – when cool enough to touch, squeeze the oil through the jelly bag as illustrated in Cold Infused Oils below.
  Many herbs make effective hot infused oils, especially spicy herbs such as ginger (*Zingiber officinale*), cayenne (*Capsicum frutescens*), and pepper (*Piper nigrum*). These oils can be rubbed into the skin to relieve rheumatic and arthritic pain, improve local blood flow and relax muscles. Other hot infused oil from leafy herbs, such as comfrey (*Symphytum officinale*), speed wound healing. Oil infused with mullein (*Verbascum thapsus*), is used for earache and ear infections, and chickweed (*Stellaria media*), ointment may be produced from a hot infused oil.
  Cold infused oils: Making a cold infused oil is a slow process and involves leaving a jar packed with herbs and oil to stand for several weeks. Sunlight encourages the plant to release its active constituents into the oil. It is the most suitable method of oil infusion for fresh plant material, especially the more delicate parts such as flowers. St John's wort (Hypercum perforatum), marigold (Calendula officinalis), and melilot (Melilotus officinalis), are three of the most commonly produced cold infused oils. St John's wort oil is anti-inflammatory and analgesic and may be applied topically or taken internally (after consulting a herbalist) for peptic ulceration.
- **Ointments:** Ointments can be made with dozens of bases and they vary in consistency, depending on the constituents and proportions used. The simplest way to make a soft, all-purpose ointment is to use petroleum jelly or soft paraffin wax (other methods are explained below). Petroleum jelly is impermeable to water and provides a protective barrier for the skin. Single herbs or mixtures of herbs may be used as required, provided they are finely cut and essential oil can be stirred into the ointment just before straining.
- Consistencies: A solid and relatively grease-free ointment will spread easily and is useful for preparations such as lip balms. This may be made by using alternatives to mineral oils. Melt 140 g of coconut oil with 120 g of beeswax and 100 g of powered herb. Simmer gently for 90 minutes in a glass bowl set in a pan of boiling water, or a double boiler, then strain and pour into jars.
- **A less solid** ointment, for conditions such as skin rashes, may be made by combining olive oil and beeswax. Melt 60 g of beeswax with 500 ml of olive oil and 120 g of dried or 300 g of fresh herb in a glass bowl. Cover and place in a warm oven for 3 hours, then remove strain and pour into jars. This ointment

can also be made by combining 500 ml of hot infused oil with 60 g of melted beeswax.

Creams: Small quantities of additional ingredients such as tinctures, powders and essential oils can be added to a cream before or after it is put in jars. Adding an essential oil, such as 1 ml of tea tree (Melaleuca alternifolia) to 100 ml of cream, counters mould growth and lengthens shelf life, as does 5 ml of borax. Other recipes for making cream use infusions, tinctures or infused oils.

Compresses and lotions: After an accident or sports injury, bruising and swelling can often be reduced or prevented if a hot compress is swiftly applied, provided the skin is unbroken. Cold compresses are particularly useful in soothing inflammation, cooling fever and easing headaches. Both hot and cold compresses should be frequently soaked and re-applied for maximum benefit.

Other Preparations

Stem Inhalations

Steam inhalations are an effective way to clear catarrh, and relieve sinusitis, hay fever and bronchial asthma. The combination of steam and antiseptic ingredients clears the airways throughout the respiratory system.

Baths and Skin Washes

Herbal baths and skin washes can relieve many conditions, including aching limbs and stuffy sinuses. They are made from diluted essential oils or infusions. Eyebaths soothe sore, inflamed or irritated eyes.

- To make a herbal bath Add 500 ml of strained infusion or 5-10 drops of essential oil to a running bath.
- To make a skin wash Make an infusion, strain it and bathe the affected area.
- To make an eyebath Make a small quantity of an infusion or use a herbal teabag. Strain the liquid carefully into a sterilized eyebath. Alternatively, add 2-3 drops of a tincture to an eyebath filled with water that has just boiled. Allow to cool and place the eyebath firmly over the eye. Tip the head back and eye by continuously blinking. Repeat up to 3 times a day.

General caution Eyebaths should be very weak, so as not to sting the eyes. Always use boiled, cooled water in a sterilized container. Do not bathe eyes over a period of more than 2-3 weeks at a time. If bathing eyes frequently, add a tiny pinch of salt to each eyebath to counter leaching of salts and minerals from the eye.

Juices

The juices extracted from many herbs can be taken internally or applied externally. **To make** Pulp the plant preferably using a mechanical juicer. Otherwise use a food processor. Squeeze the pulp through a jelly bag to collect the juice. Some herbs need to be cooked in order to extract their juice.

STANDARDIZATION OF MEDICINAL PLANTS

When a medicine is administered to a patient, the doctor who prescribes the medicine needs to be assured that the medicine contains the correct amount of the right substances as only then would the medicine induce its therapeutic effect. This is assured both to the doctors and the patient by the pharmaceutical company producing the medicine. In the laboratories of the pharmaceutical house, the medicine – or batches of the same – have been tested for quality and quantity and the drug is released for use only if the quality control laboratories of the pharmaceutical house approve the batch in question. Overseeing the situation nationally are national drug control laboratories run by government scientists who also monitor the equality of the drug being released by carrying out random checks to ensure that the results conform to the standards laid down by both the pharmaceutical house and the national drug regulatory authority at a time when the pharmaceutical house requests for the release of a particular drug in the market. This is known as standardization – a system to ensure that every packet of medicine that is being sold has the correct substance in the correct amount and will induce its therapeutic effect.

This system has worked well in developed countries and reasonably well in countries in the third world, particularly if the country has a strong drug regulatory agency with its own laboratories. However, the system has been developed for synthetic medicines and its has not been difficult to "*fingerprint*" the synthetic medicines and lay down standards. It is much more difficult to ensure that this system works in regard to herbal medicine.

The system of standardization should be established for every plant medicine in the market because the scope for variation in different batches of medicine is enormous. Plant material, as has already been mentioned earlier, may vary in its alkaloid content and therefore in its therapeutic effect according to different places of collection, with different times in a year for collection, with collection at the same time and place but in different years, and with different environmental factors surrounding the cultivation of a particular medicinal plant. Adding to this variability is the fact that in herbal medicine several plants may be used together in the same preparation. This means that there should be a quality control test for the entire preparation to ensure quality of the preparation.

Following factors should be given due consideration while using plants:

1. If plants are used fresh, after brewing as a tea or as a juice, it is probably not necessary to carry out tests for quality control. When plant substances are prepared for storage and subsequent use, it is essential to carry out tests to check that the standards are being maintained.
2. When plants are used fresh – and perhaps cultivated in the kitchen garden or village garden – these lead to cheap remedies.

It is necessary that tests for standardization of plant preparations should be carried out and if standards are maintained, this requires the use of some sophisticated equipment and the time of suitably-trained technicians and supervisors. This inevitably means that this cost will be added on to the price of the herbal medicine, which does not remain a cheap medicine any more.

Herbal medicines are not cheap, it is more expensive than a synthetic drug for the same purpose. It also depends on the way herbal medicine is used. Quality control tests should be conducted for four to five ingredients and the products.

Standards can only be established for an isolated compound. Interest in developing extracts for use – both at the primary health care level and more widely – is therefore tempered by a feeling that these extracts cannot be standardized and therefore the extracts are not a marketable commodity. Development of therapeutic substances concentrates on isolated compounds and those plants in which an isolated compound cannot be obtained are not followed up. It has been stated earlier that biological activity present in some extract will be present when the plant is in an extract form only.

Standardization is the basic need of a chemists and pharmacologists to know the contents of active principle of medicinal plant. Pharmacological and therapeutic activity will not be residing in one active substance but may, in fact, be present in a combination of a few active substances. As far as standardization is concerned, this again need not be an insurmountable barrier because all the active substances could be identified and standards set for each of these.

Identification of the actual plant is not easy. Sometimes two plants are known by the same local name. When the plant is being collected, it is very possible that another plant is being collected at the same time and this will be mixed up with the plant which actually has the activity. One example of this is manifest when one collects the plant, *Boerhaavia diffusa*, which is used widely as a "*quality of life enhancer*" in the traditional systems of medicine. However, both *Boerhaavia diffusa* and the plant, *Trianthema portulacastrum*, are known as "Punarva", and so both plants may be collected at the same time.

It is observed that plant collectors may think for substitute for cheaper plant material. In this case, it is not a case of inadvertent mistaking a non-medicinal plant for a medicinal plant, as described in the last paragraph, but one of adulteration and cheating. One example which is very well known concerns the bark of *Caesalpina sappen*. When this substance is collected, bark from other plants such as *Pterocarpus, Gluta Travancorea* and *Toona cilita* are substituted for the bark of *Caesalpina sappen*. Other examples with commonly-used medicinal plant sin India are substitution of the bark of *Saraca indica* by the bark of Trema orientalis and substituon of the plant, *Halarrhena antidysentarica*, by the plant *Wrightia tinctoria*. It is very important, therefore, when laying down standards for a herbal product – either for experimental research, clinical evaluation or widespread commercial use – to lay down clear-cut standard nomenclatures and clear-cut criteria to actually "*finger print*" that the substance being used in actually the substance which should be used. Once it has become clear that the substance being used is the substance being used is the substance meant to be sued, the first hurdle in standardization of the preparation is overcome. Every crude ingredient going into the herbal preparation must be clearly identified.

Standardize for the method of production is also necessary for isolation of the compound, which is to be used or the method for preparing the extract which is to be used.

Standard procedures should be developed for getting the final product – if it is a single substance or for the final formulation – if it is a mixture of substance. These standards would attest to the quality of the final product and would comprise both physiochemical methods and biological tests of efficiency. If mixtures or extracts are to be used as the final product, then methods for standardization of the mixtures or extracts should be developed. These standards should be strictly adhered to.

For standardization of plant medicines macroscopic, microscopic, physiochemical and biological methods are used. It is not the place to deal critically and exhaustively with these test nor is the author competent to do so.

Decoction

Decoction is the method normally used for those medicinal herbs whose active principles are difficult to extract because they are contained in woody parts of the plant, or which require prolonged heating in order to pass into solution. The procedures dealt with here only involve water extraction, as more complex forms requiring various solvents are too technical for a book of this nature.

Extraction by decoction involves boiling the whole plant or a part of it in water for a given time and then allowing it to macerate for a further period before filtering. A slight variation of this method is to soak the plant beforehand in cold water for a specific time, depending on the drugs required, in order to soften the leathery part and thus make extraction easier at the decoction stage. This method is reserved for certain plants such as those with a high mucilaginous content.

Decoction is, understandably, quite a drastic method of extraction and one, which is capable of modifying some of the plant compounds. It is important to ensure that the main active principle to be extracted is not affected by other harmful compounds induced by the heat. The boiling process during, which water is added, nearly always destroys the volatile active principles, that is to say, those with a low boiling point. This means, therefore, that the resulting decoction will never possess all the original principles of the drug.

Filtering Operation

The filtering operation is equally important, as it serves to strain off not only the leathery residues of the plant but also the mucilage or deposits of amphorous substances that are carried in solution during the boiling. The liquid obtained is, therefore, always left to clear before the next stage when it is filtered through paper of finely woven linen.

Extraction

Extraction is much more suitable method for obtaining the active principles when the parts of the plant being used are soft and fragile, such as leaves bud or flowers. It is preferable, except in certain cases, to use herbs that have been slightly dried, as the reduction of water content will concentrate the principal constituents. Unlike decoction, this method allows retention of many of the volatile principles, which would otherwise be lost, but does not permit extraction of those requiring a continuously high temperature. An expert, with extensive knowledge of medicinal plants and their medical compounds, can decide, which of these two methods is more suitable for obtaining the active principles required.

Infusion

Infusion entails pouring boiling water over that part of the plant known to contain the required drug. The container should be made of pottery of glass to avoid the formation of tannate of iron, which due to the presence of tannin in almost all plants, would occur it metal were used. As soon a s the water ahs been poured over the plants, the container should be covered, thereby allowing the steam, rich in volatile compounds, to condense. After a given time, which varies for each plant used, the liquid is filtered before being used.

Maceration

The third simple method, known as maceration, is used for medicinal plants whose active principles are soluble in cold water for several hours during, which time all the principles that do not need heat to release them (i.e. that are not thermolabile) will be released into the solution. The yield of mucilage from certain plants, such as the mallow (*Malva*) and marsh mallow (*Althaea officinalis*), is in fact greater when extraction is carried out by means of cold maceration rather than by infusion.

The method of maceration described above is hardly ever used nowadays in the pharmaceutical industry. Instead, tinctures are prepared by diluting a certain amount of fluidextract of the requisite drug with alcohol of a predetermined strength. When, as is sometimes done, a sugar solution is added to spirit solutions, the resulting preparation is known as an *elixir* or *ratafia*. These preparations are, in fact, genuine liqueurs and have been almost abolished in pharmaceutical practice.

If the vehicle of maceration were to be a mixture of alcohol and ether, andethereal tincture' or, more scientifically, an ethereal spirit solution would be obtained. Many varieties of such preparations could once be found in all the pharmacopoeias, but are now little used. The best known is ethereal tincture of valerian.

Extraction

It is a straightforward method used for the extraction of juices. In this case the plants, which contain the required medicinal ingredients must be fresh and have high water content. The Juice, which can be extracted by various methods, contains the mineral salts and vitamins of the plant, together with all the other substances that can be obtained through pressure. Again, the juice will not contain all the active principles of the plant but, on the other hand, cold extraction does not modify the structure of those compounds, which as we have seen, can be altered by heat. In the home, juices can be produced by first parboiling the fresh plants very briefly and then applying pressure with an appropriate utensil such as a small press. Because of the additional heat used for parboiling, however, this method will modify some of the constituents.

Better still, of course, is the modern electric blender, which extracts almost the entire juice content of the plant.

All the methods described above have a time limit; hence they are considered short-term procedures. In fact, the decoction, the influsion, the maceration and the juice all contain biological compounds that immediately undergo processes of oxidization and fermentation that denaturalize the medicinal substance, often making the liquid taste repugnant and, therefore, unusable. To avoid this, it is possible to add preservatives but this, of course, would be quite contrary to the whole concept of non-adulteration, which is the aim of the methods described here. Additions of sugar and alcohol admittedly also enable juices to be kept for a slightly longer period and, if vacuum-bottled with sugars, their properties and the freshness of their components will remain unaltered; unfortunately, this kind of preservation is almost impossible in the average household because the apparatus required is often expensive.

The Problem

The problem of conserving medicinal principles, once extracted, is of the utmost importance. The pharmaceutical industry has, therefore, done much work in this field, producing further preservation techniques so that active principles can be used at any time and in any dosage. The difficulty of tracking down exotic or rare drugs, combined with the relatively short harvesting period of healing herbs, has necessitated the development and regulation of highly advanced methods of extraction and preservation.

Forms of Preparations

Such preparations are known as extracts. They are divided into semi-solid, dry and fluid types; another means of classifying them is by their vehicle of evaporation: water, water and alcohol, or ether. Strictly speaking, an extract is the concentration of all the substances and salts diluted in the juice obtained from it. This concentration is achieved by the use of two fundamental processes: the first is based on freezing and the other on evaporation. The latter, more generally used, consists of putting the juice into an evaporating dish from, which the water content is evaporated at a low heat. Obviously, depending on the length of the evaporation process, different concentrations of the juice will be obtained. Concentrations will thus range from those still possessing a high water content to concentrates in which water is no longer present. The extract obtained is termed soft extracts have a semi-solid honey-like consistency, while dry ones are often composed of solid masses that can be reduced to powder. A classic example of a dry extract is that of aloes, which has the appearance of black, solid lumps, almost vitrified, but which are friable and easily pulverized; an even better known instance is liquorice, which can be reduced to a powder or to liquorice sticks.

Water Content of Juice

The water content of juice is not, however, always the vehicle of evaporation used in obtaining an extract. Instead, to facilitate concentration through evaporation, a solution of water and alcohol is often used. Alternatively, ether may be used and then recovered through distillation. The addition of these substances to the water enables compounds to dissolve, which are hardly soluble at all in water alone, but which become very soluble in a combination of the two liquids.

The Solvants

Extracts in which ether is used are described as ethereal and those in which alcohol are used as hydroalcoholic. Certainly the most important type of extract is the fluidextract, which has only relatively recently been formulated and accepted in pharmaceutical practice. In comparison with the type of extract previously described it has one property, which makes it extremely useful: one gramme of fluidextract is equivalent to one gramme of the original plant material. In other words, all the active principles contained in a precise quantity of plant matter to be treated are present in the same quantity of fluidextract obtained. The fluid extract makes it possible to have any drug available at any given time (it does not deteriorate quickly and fluid extract obtained. The fluid extract makes it possible to have any drug available at any given time (it does not deteriorate quickly and is not modified, except very slowly, by external agents); it is of a purify and accuracy of tiration that would be impossible to achieve in drugs fresh from the plant.

Alcohol-soluble Substances

When, for particular purposes, it is necessary to extract only the alcohol-soluble substances such as alkaloids, bitter principles, and organic acids and colouring substances, another specific method of preparation is used. This is generally known as an alcoholic tincture or, more correctly, spirit solution. Apart from the reason already given, these tinctures are produced because their high alcohol content makes them easy to preserve. A spirit solution can be prepared either by maceration, in which a measured amount of the medicinal plant is immersed in alcohol of a suitable strength for a predetermined length of time and subsequently filtered, or by digestion, passing it through a still. This apparatus is equipped with a cooling flask, which causes the alcohol that ahs been condensed in it to fall back into the receiver flask in which the drugs to be digested have been placed. Because tinctures are very sensitive to light they are always kept in bottles made of dark glass. The alcohol in which the plants have been macerated develops varying degrees of strength, depending on the drug being used and the ease with which it releases its essential constituents, but it is usually not less than 60% vol. Although these are alcoholic preparations they cannot.

The alcohol content is sufficient to obtain an extraction of the soluble compounds if the medicinal plants are allowed to macerate in the wine for the length of time appropriate to each type. Unfortunately, medicated wines are subject to acetic fermentation (that is, they go vinegary) and form sediment just like any other wine. It is, therefore, important to prepare only small quantities and to keep the wine completely airtight and away form the light.

Fixed Solvents

Essential oils are soluble in solvents that are described as fixed if they do not evaporate at a normal temperature, and as volatile if the reverse is the case. Fixed solvents consist of solid or liquid odourless fats such as lard, tallow or refined olive oil. These solvents enable essence to be extracted by means of a method known as *hot enfleurage*. The essence-producing plants are placed in copper containers and left to macerate in fat, which is brought up to a temperature of about 104°F. (40°C.), with more plants being added until the fat is saturated. The fat is then separated from the essence to give a good yield.

Volatile Solvents

Extraction with volatile solvents, while presenting much greater difficulties in obtaining the essential oil, has the advantage of making it possible to extract essences from plants that have only very low percentage of the fragrant principle. The volatile solvents normally used are benzene and petroleum ether, which carry both the essence in solution and other products, making it necessary to carry our further, rather laborious tasks in order to obtain a pure products. The first phase of extraction by this method is the soaking of the flowers in an apparatus known as a *digester*. The solution obtained from this is distilled, during, which process the solvent is recovered. The essence' because of the traces of extraneous material contained in it. In order to purify the concrete essence, it is treated with pure alcohol, which takes up the essence, leaving many un-dissolved impurities. The alcohol is then brought to a low temperature, which enables the remaining impurities to become concentrate. As this stage, the alcohol can be called an extract. To complete the process, the extract is distilled to give the resulting essence a high degree of purity.

Uses of Extracts

These extracts can be used either by themselves or as the basic ingredient of various medicinal compounds, both of which are guaranteed as a homogenous product. Although the method of preparation varies according to the drug, it can be summarized as follows: 1,000 parts of the finely chopped dried plant are left to macerate for several hours in alcohol of the appropriate strength. This alcohol is then strained off and discarded. A similar amount of fresh alcohol is then poured over the same plant matter and a part of it strained off a day later. This alcohol, the weight of which is in direct relation to the type of herb used, is held in reserve. New alcohol is added until the dried plant material is exhausted of all the soluble extracts. The next stage is to distil this second liquid, recovering the alcohol and allowing the rest to evaporate until an extract is obtained, which, when mixed with the part held in reserve, will be of exactly the same weight as the original herb material. Some experts maintain that the solvent action of the alcohol can be improved by the addition of certain substances such as inorganic or organic acids, salts and ammonia. The various pharmacopoeias certainly use one or other these additives. The fluidextract will remain unaltered for a long time If kept in dark glass bottles, which are firmly stoppered.

The above explanation demonstrates the superiority of this method of extraction over the other methods described, and accounts for are almost unreserved use throughout the world.

Unlike fluidextracts, be preserved indefinitely. They react quickly to light and air, which cause flocculation and acetic fermentation.

Looking at the field of galenical preparations (those deriving from plant of natural origins) as a whole, there is one particular preparation reassessed by modern phytotherapy, both because of its successful results and its ease of preparation. It can even be made at home. This is medicinal wine. As long ago as the thirteenth century, Arnold of Villanova wrote his *Incipit* Tractatus de Vinis, the earliest published book on wines, which dealt particularly with 'methods of making preparations with wine basis'. Such wines are made by simply macerating plant drugs in wine, preferably white, with a high alcohol content, such as a fortified wine.

Examination of Galenic Products

These are examined are the *alcoholates* or *spirits*, produced by the distillation of spirit solutions. The difference between spirits and spirit solutions is that the former, through distillation, contain only the volatile principles of the drug and not the fixed principle, as do the latter. For the distillation, 70-80% vol. alcohols are generally used, after appropriate maceration. Only a certain part of the distillate is retained, so as not to change its characteristics with substances of low volatility, which remain in the flask, following the concentration of the watery part. Some of the best known spirit solutions still used today are aqua vitae of lemon balm and Hungary water, or spirit of rosemary, the former being used as a digestive antispasmodic and the latter as a perfume component. Spirits have excellent preservation qualities due to their high alcohol content.

Syrups Preparation

Instead of using alcohol as a preservative for garden herbs, sugar can be sued in a concentrated form. By adopting this alternative, sugary solutions can be made, which contain medicinal substances; they are known as *syrups*. Syrups also serve to disguise the unpleasant taste of many medicines.

Ordinary syrup is made by dissolving a certain quantity of sugar in water, at a low heat; the mixture needs to be stirred frequently and allowed to evaporate sufficiently to obtain the concentration necessary to achieve the characteristic density f syrup. The curative substances are then dissolved in it and it is these that determine the particular qualities of each type of syrup. If the resulting liquid is cloudy, it can be clarified. Syrups should be stored in hermetically sealed bottles and kept in a cool dry place, preferably away from the light. Despite such precautions, syrup is prone to alteration, especially when the additives are organic. In such cases, the syrup will have to be re-boiled in order to increase its density and, therefore, its preservation potential. Instead of sugar, honey can be substituted using the same procedure. Such preparations are called *mellitus* or honeyed syrups. For instance, 'melrosette' or 'honey of roses', which is still made for children, belongs to this category.

Electuaries of Conserves

Nowadays there is a growing demand for plants and vegetables that have not been contaminated with insecticides or fungicides, to be used in a natural form as natural form as possible. The interest in macrobiotics ash even brought back a number of preparations, which, due to their short shelf life, had become neglected. These confections are known as *electuaries* of *conserves* and are in the form of paste made by mixing the pulp of fruits, which have special nutritive of medicinal properties, with certain powders, honey of syrups. Some of these are sold commercially but, because of the high percentage of organic substances in them, preservatives have to be added. The macrobiotic diet makes use of several types of electuaries in this double role of supplementary food and medicine.

Use of Oil as Solvent

When medicinal substances are dissolved in oil rather than the usual solvents, the preparations can be made either by dissolving one part of an oily fluid extract of a particular plant in a given quantity of refined olive oil or, alternatively, by boiling the plant itself in the oil and then filtering the as rheumatism, in which an oily preparation is required to facilitate localized massages.

Closely related to the medicated oils are the ointments, also known as *salves* or *unguents*, the base of which is usually a semi-solid, fatty or greasy substance such as lard or wool-fat (lanolin). They may be prepared simply by mixing powdered vegetal drugs into the fat, or the fat these ointments are all intended for external use.

At one time simple ointments such as these were very well known and widely used, but nowadays they have been largely replaced by commercial products containing chemical additives. The creams and liniments found on the market today tend to have quite different bases. Sometimes, however, we are reminded of remedies, which must have been familiar to our ancestors, when we see an old apothecary jar bearing a name such as unguent of polar or unguent of lily root. In Britain there is, increasingly, a tendency towards these simple creams containing no chemical additives.

CHAPTER 32

Marketing of Horticultural Produce

In India, the organised marketing of agricultural commodities has been promoted through a network of regulated markets. Most state governments and UT administrations have enacted legislations to provide for the regulation of agricultural produce markets. While by the end of 1950, there were 286 regulated markets in the country, their number as on 31 March 2006 stood at 7566. In addition, India has 21780 rural periodical markets, about 15 per cent of which function under the ambit of regulation. The advent of regulated markets has helped in mitigating the market handicaps of producers/sellers at the wholesale assembling level. But the rural periodic markets in general, and the tribal markets in particular, remained out of its developmental ambit. markets in particular, remained out of its developmental ambit.

MARKET

The term market is derived from the Latin word *Marcatus,* meaning merchandise. The market provides a mechanism for the sale of goods, but the actual delivery of goods may not take place in all the cases. It also refers to any region in which buyers and sellers are brought in contact with one another, and by means of which the prices of goods and services are finalized easily and quickly. Markets can be classified in several ways from different approaches.

On Geographic or Area Basis: From the stand point of geographical area, markets are divided into (a) Local Markets, (b) National Market and (c) International Market.

1. **Local Market**: These markets relate to a particular locality. In the case of these markets, commodities are sold within geographical limits. Such commodities are difficult to be sold outside local limits. Generally, commodities which are heavy and perishable have local markets. For example bricks, vegetables, fruits, milk etc have local markets.
2. **National Market**: The growth of industries has widened the scope of market on national level. With the growth of transporation and communication, most of the goods are marketed at national level.
3. **International Market**: These are known as foreign markets where goods are sold beyond national boundaries. With the growth of transportation and communication systems, a number of products have acquired an international level.

On the Basis of Importance: On the basis of importance markets may be divided into (d) Primary Market. (e) Secondary Market. (f) Terminal Market.

1. **Primary Markets:** In primary markets, primary producers of agricultural products or manufactured goods sell to wholesalers, who assemble the goods

from different sources of production. These markets are generally found in villages.

2. **Secondary Markets:** In the secondary markets, wholesalers sell the goods to retailers for further selling. Semi-processed and Semi-manufactured goods are generally sold and purchased in secondary markets E.g. Yarn market.
3. **Terminal Market:** It is the market where final products are sold to final consumers i.e. consumers purchase goods in the terminal markets from the retailers.

On The Basis Of Business: On the basis of volume of business, the market may be divided into (a) Wholesale Market. (b) Retail Market.

1. **Wholesale Market:** In wholesale market goods are bought and sold in huge quantities. In these markets sellers are wholesalers and the buyers are retailers. Wholesalers purchase goods in bulk quantities and sell the same to retailers in small quantities.
2. **Retail Market:** In this market retailers who purchase goods from wholesalers, sell to ultimate consumers in individual units i.e. Very small quantities.

On Economic, Basis: In economics, markets are classified into (a) Perfect Market (b) Imperfect Market.

1. **Perfect Market:** In perfect market there will be perfect competition between buyers and sellers who have full knowledge of other buyers and sellers. Due to this only one price will prevail in the market for the commodity. The following are the essential features of perfect market.
2. Group of buyers and sellers.
3. Effective competition between buyers and sellers.
4. One price for the commodity throughout the market.
5. **Imperfect Market:** Imperfect market is a market which is not a perfect market. In this market we find some kind of maladjustment in demand and supply; buyers and sellers have no knowledge of other buyers and sellers.
6. **On Time Basis:** On the basis of time markets may be classified into (a) Very Short Period Markets.(b) Short Period Markets and (c) Long Period Markets.
7. **Very Short Period Markets:** It refers to markets which exist for a very short period, normally a day. Such markets generally sell fruits, flowers, vegetables, milk etc.
8. **Short Period Markets:** These markets include weekly markets held in villages. Fairs are also included in this category.
9. **Long Period Markets:** Durable goods are purchased and sold in long period markets. In these markets goods may be held for a long period without any deterioration in quality.

On The Basis Of Nature of Goods: On the basis of the nature of goods that are purchased and sold, markets may be divided into (a) Commodity Markets. (b) Capital Markets. (c) Foreign Exchange Markets.

MARKETING

According to William Stanton Marketing is a total system of interacting business activities designed to plan, price, promote and distribute want satisfying products and services to present and potential customers. Thus the main idea of modern marketing concept is customer satisfaction.

Selling

Selling is concerned with the transfer of goods and services to the consumers. It is mainly concerned with the plans to get the customers to exchange his money to goods and services. It is primarily concerned with the seller's interest.

Functions of Marketing

A marketing function is an act or operation or service by which the original producer and final consumer are linked together. If marketing functions are not properly carried out, the business unit may not be in a position to dispose off its products and all the efforts made for production may not bear fruits. The marketing functions can be divided into two types viz (i) Concentration and (ii) Dispersion. The process of concentration is concerned with gathering raw materials, manufactured goods at a central place namely market. Dispersion means distribution of goods to final consumers. Concentration involves a number of marketing functions like (a) Buying (b) Trading (c) Storing (d) Grading (e) Financing etc.

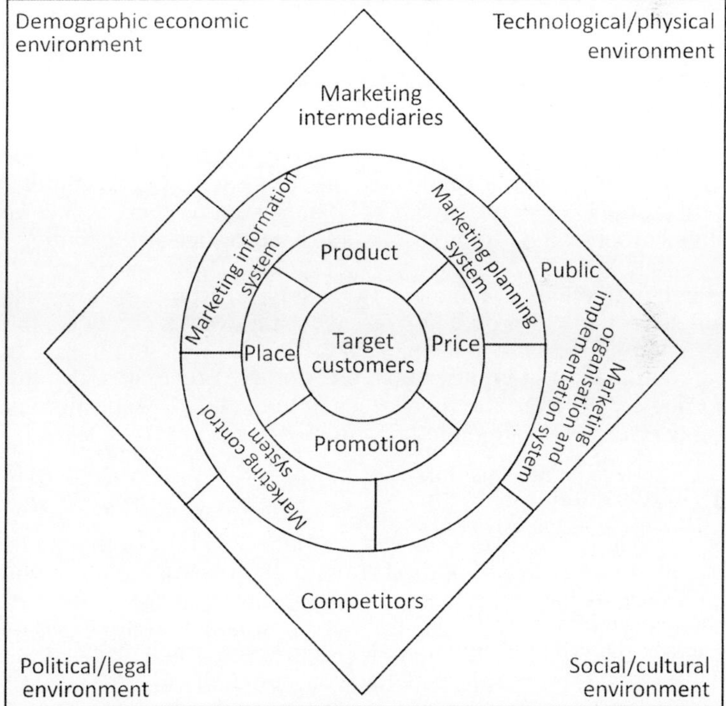

FIG. 32.1: Factors influencing a firm's marketing strategy.

The functions of marketing also include. There are two important functions of physical supply (a) Transportation and (b) Storage and ware housing. The storage is done in warehouse which creates time utility by storing the goods through out the year and releasing them as and when they are needed. The warehouses may be private, public or bonded warehouses.

AGRICULTURAL MARKETING

India is the second largest producer of fruits and vegetables in the world. However, this is not matched with the availability of post-harvest infrastructure (PHI) leading to enormous losses valued at more than 30 billions of rupees annually. The extent of these problems faced by farmers is serious.

Agricultural marketing may be viewed as a process encompassing all the steps involved from the producer to the consumer including pre and post harvest operations such as assembling, grading, storage, transportation and distribution. According to the National Commission on Agriculture (XII report) agricultural marketing is a process which starts with the decision to produce a saleable farm commodity and it involves all the aspects relating to pre and post harvest operation including assembling, grading, storage, transportation and distribution. By performing these operations, it adds value to the produce in terms of time, place and farm utilities. Storage facilities are also essential in marketing infrastructure.

Agricultural marketing has assumed increased importance after launching of the new economic policy and consequent opening up of India's markets to foreign suppliers and buyers and access by Indians to world markets. To enable Indian farmers to derive the full benefits from the new liberalized world trade regime, it is necessary to remove various constraints and deficiencies in the existing domestic markets and marketing practices.

Marketing involves finding out what your customers want and supplying it to them at a profit. The importance of good marketing can be conveniently considered from the four different perspectives of the national economy, the farmer, the product and the consumer. The whole marketing process has to be customer oriented. Production must supply customers with what they want or need. In fact, marketing is a commercial process and is only sustainable if it provides all the participants with a profit.

Agricultural marketing is achieved by a series of processes which include planning and scheduling of production, good post harvest techniques, the grading and sorting of crops and the packing, evaluation and dissemination of market information, coordination of inputs, transport, storage, processing, distribution and selling of products. These are the mechanics of marketing. For this purpose proper market executive should be appointed or can be organized at own level.

Marketing Environment

The marketing environment is the sum total of internal and external factors which the organization operates. Some of the factors of the environment are controllable and some of them are uncontrollable. The marketing manager must obtain deep and up-do-date knowledge of all these forces as his marketing strategy is influenced by these at every step. The four P'S namely product, price, place and promotion are forced to change as per the changes in the environmental forces.

Environment in general can be defined as *surroundings which influence a particular activity*. Marketing environment is defined as *the forces and actors that affect the ability of a company to develop and maintain successful relationships with its target customers*. The producer organizations produces a number of products and the services connected with it are influenced by a group of factors which are operating within and outside the organization. The environment within which the organizations are operating is dynamic and uncertain. The different forces of environment provide a number of opportunities and threats to the organizations. Hence, a marketer must develop marketing mix decisions as per the changes in the marketing environment. Therefore, analysis of the environment should be attempted to give an extensive insight as the current market conditions as well as of impact of external factors that are uncontrolled by marketers. In analysing the environment five important stages are identified for the purpose of analysis.

1. Audit of Environment
2. Assess Nature of Environment
3. Key Environmental Factors
4. Identify opportunities and threats
5. Strategic Decision making

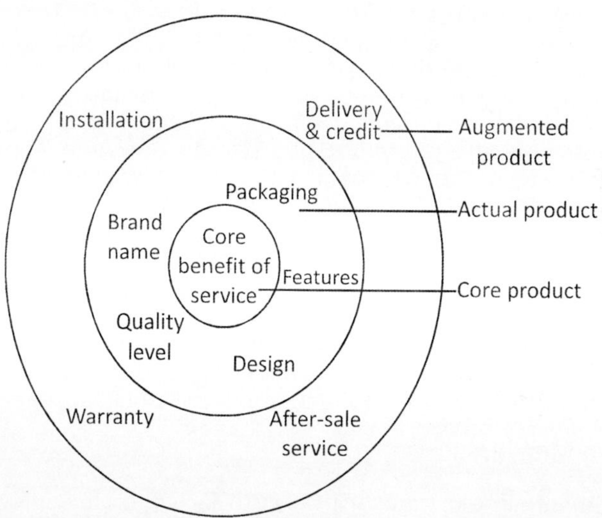

FIG. 32.2: Marketing Environment. (Source: Varshney and Gupta, Marketing Management – An Indian Perspective, p: 24).

The marketing environment may be internal or external. It has also been classified as micro or macro environment. Some has classified as controllable or uncontrollable forces.

Soon after harvesting, farmer intends to shift the produce to market, as early as possible. Farmer expects an environment to market the produce, where his decision should not be effected by any other variable factor or forces and is assured of a competitive price. This objective is enshrined in the creation of regulated market. In the management of these markets, a farmer representative serves as a member and this gives him a sense of belonging in the market, which makes the market environment more conducive for completing the transactions. Marketing environment also includes, required infrastructure for efficient handling of the commodity, such as covered auction platform, tested and trusted weighing machines, cleaning, grading and storage / cold storage facility, loading and unloading facilities, internal roads to decongest the vehicular movement, pledge finance availability, banking, postal services, cattle-shade, farmer's rest house and link road from farm to the market.

EMERGING ALTERNATIVE AGRICULTURAL MARKETING INSTITUTIONS IN INDIA

According to Vilas Jadhav, Mahantesha Naik of GKVK University of Agricultural Sciences Bangalore (*Agrobios Newsletter, June 2010*), alternative marketing is the traditional marketing system of farm produce has paved way for other alternative marketing system, which is more need based and sometimes suits the individual partners in the trade. The need of the alternative marketing of farm produce is to reduce the distributive system. Another reason is to explore the possibility of bringing the buyer to the production place that is, bringing the market at farm. Even though, the marketing committee is represented by majority of the farmers' representative they lack of accountability to the users of the market. In actual operation of the market, farmers lack of effective voice and hence bargaining power. Majority of small farmers have no voice in marketing there produce. Intermediary and system inefficiency consume a disproportion share of the consumer's price, there by eliminating incentive for improved productivity and quality farmers have no freedom to choose the best buyer and are denied of proper price information and technical advice. There is ample scope for the public-private partnership to enhance the efficiency, especially for the creation of marketing infrastructure and management by the private sector. This pre-supposes a change in the government policy in its role from financier having control over the management of markets to that of facilitator in this respect, of the late the Government of India has come out with a model APMC Act which redefines the role of present agriculture produce market committee to promote the alternative marketing institutions to benefit of the farmers.

Various Reform Measures

Suitable amendments in the state APMC Act to facilitate efficient and alternative marketing channels for farm produce. The progressive dismantling of controls and regulations under act, law for implementing negation warehouse receipts system passed by the parliament. Enactment of integrated food law to promote quality and single window regulatory system. Allowing future trading in agriculture commodities for a better price risk management and price discovery. Legal persons, grower and local authorities permitted to establish new markets in any area. No compulsion on grower to sale their produce through existing regulated markets.

Reforms through Model APMC Act

1. Setting up of special commodity market.
2. PPP in management and extension activities promotion of trading.
3. Encouraging the professional management in APMCs
4. Single point levy of market fee or single point registration of functionaries.
5. Establishment of private market or farmer consumer markets direct marketing.
6. Powers to state government for exemption of market fee setting up of state marketing extension cell state agricultural produce standards bureaus.

Market liberalization, globalization and marketing reforms are causing a transformation in agriculture and agri-food markets in India. Food markets are changing towards the high value food product like fruits, vegetables and animal products and in response, the agriculture production portfolio is diversifying towards high value of agriculture commodities. The question arises, whether small holders participate in high value.

The different forms of alternate model that could influence efficiency in the institution are:

1. Direct marketing by producers.
2. Marketing through farmers' interest group.
3. Setting up of terminal markets by private sector.
4. Forward market and futures exchanges.
5. Electronic trading or E-commerce.
6. Vertical integration and contract farming.
7. Organized retailing.

Direct Marketing

Farmers come into direct contact with the consumers and receive the payment directly from the consumers. This system is prevailing in many parts of the country viz:

- Apni Mandi: In these mandies, farmer-producers bring the produce for sale directly to the consumers, as in Punjab and Rajasthan
- Rythu Bazars: In Andhra Pradesh, such markets have been established to provide direct link between farmers and consumers in the marketing of fruit and vegetables and other essential food items.
- Uzhavar Santhaigal: Government of Tamil Nadu had established farmers markets called Uzhavar Sandies in selected municipal and panchayat areas. In these markets, better marketing infrastructure is provided to farmers free of cost. Farmer get other inputs and quality seeds in these markets. (However it has been discontinued by the present State Government)
- Raithara Santhegalu: In Karnataka, marketing Board has established farmes market or Raithara Santhe without any middle men and provides a place where farmers and consumers directly interact.

Group Marketing

Group marketing includes, joint planning, funding, implementation, pricing, sharing risk equally in the marketing and reaping the benefit of collective bargaining. This indirectly results into more returns and economizes the marketing cost.

Cooperative Marketing

This marketing system is pursued through cooperative societies registered under Cooperative Societies Act. This system is pursued on the principle of "self help by mutual help". It reduces the marketing cost, enhances the bargaining power and there is equitable distribution of the proceeds.

Mega Food Parks

A new scheme of Mega Food Parks is proposed which is envisaged to be a well defined agri/horticulture proceesing zone. The mega food park is designed to link the farmer with minimizing the intermediaries, they will act as sourcing hubs for the retail outlets. It is tentatively envisaged to support 30 Mega Food Parks, spread across the country.

Concept of Modern Terminal Market

Terminal markets (the hub) to be linked to number of collection centers (the spokes) which would be conveniently located at key production centers to allow easy access to farmers and provide service to them. To provide state of art facilities for grading, transportation, storage, domestic marketing and export at both hub and spoke centers.

Some Emerging New Agricultural Markets in India

1. Mumbai flower Auction market
2. Thane Terminal Market
3. Noida fruit and vegetable Market
4. Solan Wholesale Market
5. Parwano wholesale market
6. Bhopal Terminal market
7. Sambalpur Terminal Market

Commodity Futures

Commodity exchange is any exchange where various commodities and derivatives products are trader. Exchange traded quality derivatives are 4-6 yrs old. Organized commodity derivatives in India started as early as 1875. Bombay Cotton Traders Association.

Forward Contract

A bilateral agreement in which the buyer and seller agree the delivery of a specified quantity and quality of an asset at a specified date and agreed price. Price of contract is called a delivery price: usually is equally to spot price plus cost of carry. Contracts are custom designed. Actual delivery has to take place. No margin systems exist here. No exchange guarantee.

Futures Contract

Future contracts are highly uniform and well specified commitments for a carefully described good (quantity and quality of the goods) to be delivered at a certain time and the permissible price fluctuations are specified (minimum and maximum daily price changes). They are exchange traded version of forward contract.

Strategy Proposed during the State's 10th Plan in Agriculture Marketing and Agri-business

1. Enhancing marketability of agriculture commodities by providing infrastructure facilities, revamping the regulated markets.
2. Provision of post harvest handling facilities for value addition and prevention of wastage (like cold storage).
3. Provision of backward and forward linkages through marketing, agro-processing and export.
4. Ensuring better realization for agriculture produce through alternative markets like product-wise Terminal Markets.
5. Stepping up export of agri / horti produce-with setting up of AEZ and establishment of Food Laboratories.

Policy to attract private sector in storage and agro-processing industries Integrated approach from planting to marketing, which includes choice of crops, grading, packaging, storage and marketing for domestic and international, is required.

Food Chain Revolution

A new revolution of the *Food Chain Revolution* is needed in India. It is the important part of this revolution and will reduce the wastage of food grains, fruits, and vegetables. It is essential to modernize the existing post harvest operations at farm level to minimise the post harvest losses. New and improved value addition techniques should be adopted by the farmers at farm level. This will increase their income. Rural markets (weekly, monthly) need to be strengthened. An efficient agriculture marketing system requires a healthy environment, smooth channels for the transfer of produce, physical infrastructure to support marketing activities, easy cash support to the widely scattered community of producers and also a sense of market orientation among the farmers. The rural road connectivity and other infrastructure should be improved so that farmers can able to sale the products in cities. An extensive education programme should be organized to educate farmers for development of their skill to market intelligence. An adequate credit support should be given to the farmers during unfavourable seasons. The small and marginal farmers are more affected in adverse climatic situation, therefore, proper financial support should be given to them.

ROLE OF DEPARTMENT OF AGRICULTURAL MARKETING AND AGRI-BUSINESS

The department helps the farmers in marketing their agricultural produce in a fair manner and to ensure remunerative returns to them. The agricultural produce trade is usually organized under the Tamil Nadu Agricultural Produce Marketing (Regulation) Act 1987 and Rules 1991 (which replaced the earlier Act of 1959).

The Department also focuses on the other related activities. The department is also responsible for creation of modern marketing infrastructure, post harvest management, food processing, agri exports, which indirectly help the farmers to realize better returns. The important roles are as follows:

1. Enhancing marketability of Agri commodities, by providing necessary infrastructural facilities.
2. Preventing the wastage through post harvest facilities.
3. Value addition through post harvest handling facilities like cold storage, grading and packing centres.
4. Provision of backward and forward linkages through marketing, agro processing, and contract farming.
5. Educating the farmers on marketing practices - with extension network.
6. Better realization to agri produce through alternative markets like product -wise Terminal Markets/Mega Markets.
7. Revamping Regulated Markets with structural changes in the existing system.
8. Stepping up export of agri / horti produce - with setting up of AEZ and establishment of Food Parks.
9. Providing marketing extension services at the cutting edge level by integrating extension network of Agriculture and Horticulture departments.
10. Commercialization of agriculture - market driven approach.
11. Policy interventions to attract private sector into storage and agro processing industries
12. Integrated approach - from planting to marketing - which includes choice of crops, grading, packaging, storage and marketing in domestic and international markets.

Development of Market Committees and Regulated Markets

There is an urgent need to frame Market Committees which can regulate the markets of horticultural produce. For example, at present 20 Market committees are functioning in Tamil Nadu covering various districts and functioning with nominated Members and Chairperson elected by the Members from January 2003. There are 273 Regulated Markets, 15 Sub-Markets, 16 Check posts, 108 Rural Godowns and 108 Grading centres functioning under 20 Market Committees. The committee gives remunerative prices and ensured the farmer to sale the produce through closed tender system in the Regulated Markets. The Market Committees are collecting 1% of the value of produce transacted as market fee from the traders. In Tamil Nadu 96 Commercial Grading Centres, 11 Kapas Grading Centres and 1 Tobacco Grading Centre are functioning in the Regulated Markets.

Creation of Infrastructure for Marketing

For proper marketing infrastructure like transaction sheds, input shops, godowns, office, cum godowns, payment counters, rest sheds, drinking water facility, sanitary facilities, drying yards etc. are required These facilities will encourage farmers and traders for ensure better marketability of horticultural produce. During the year 2002–2003 Rs.566.65 lakhs was spent towards creation of above such infrastructures in the regulated markets. Works costing to the tune of Rs.1692.00 lakhs were taken up during 2003 – 2004.

POST-HARVEST MANAGEMENT FORMARKETING OF HORTICULTURAL CROPS

The management of post harvest losses of horticultural crops is of paramount importance. These practices will minimize post harvest losses. Therefore, adequate post harvest, cold chain and cold storage facilities are required. For example in Tamil Nadu there are about 133 cold storage units with a combined capacity of 1 lakh MT. In these cold storages mostly dry commodities like tamarind, pulses, eggs, chillies and seafood are stored. Participation of private sectors should be encouraged to establish the cold storage units in the State, however, the visible response is very poor till date. Though National Horticulture Board, Government of India is extending a subsidy of 25% of capital cost or Rs.50 lakhs whichever is less, but the available funds are inadequate to attract the entrepreneurs, because cold storage does not need just cold storage units but it also requires the cold chain, which comprises pre cooling chambers, reefer vans and cold storage units with latest technology, which again hikes the capital cost.

Need for Drying Yards in the Villages

The efforts have been made to minimize the post harvest losses of horticultural produce by constructing drying yards. About 580 drying yards at village level have been constructed in the year 2003-2004. Locating suitable site in villages for putting up drying yards takes longer time and hence it is considered that mechanical or solar driers can be installed in villages for drying commodities like chillies. Solar driers are also helpful for this purpose.

Development of Product Specific Markets

Product specific markets help to procure one specific kind of horticultural produce at one place. The area specific crop can be marketed in that yard. This will facilitate higher sale of a specific produce.

Agricultural Production and Marketing Information Centre

Market awareness plays a vital role in marketing of agricultural produce. Proper information on commodity prices and demand in various markets etc. should be made available to the farmers so that they can plan well in advance which crop is to be cultivated for better returns. Each centre will have Internet facility and electronic display board. The daily price and arrivals that prevail in different regulated markets will be transmitted to all information centres which will be displayed in the electronic display boards for the benefit of farmers and traders.

Agro Food Processing Units

Food Processing Industries are the vital link between farm and industry. It helps to enhance overall agricultural growth by adding value to the produce, generating employment opportunities and ensuring assured income to the farmers. It has been estimated that in India, only 2% of the total horticultural produce is processed whereas Philippines (50%) and Brazil (70%) are ranging higher. In India, there is an increasing demand for processed products. Currently the value addition is estimated at 7% of the total production and it is hoped that in next 5 years it will increase to 20%. By establishing agro food processing units India can reduce the post harvest wastage both in domestic and international markets. More new project proposals should be encouraged for setting up new food industries and expansion and modernization by Ministry of Food Processing Industries for subsidy.

Contract Farming

Contract farming is defined as a system for the production and supply of agricultural/ horticultural produce under forward contracts between producers/suppliers and buyers. The essence of such an arrangement is the commitment of the producer/ seller to provide an agricultural commodity of a certain type, at a time and a price, and in the quantity required by a known and committed buyer. Contract farming usually involves the following basic elements - pre-agreed price, quality, quantity or acreage (minimum/maximum) and time.

Contract farming involving business agreements for purchase of output and often provision of inputs and extension advice is increasing in the State, particularly for sugarcane, cotton and horticultural crops. Contract farming needs to be encouraged by provision of incentives, if necessary. A valid risk sharing mechanism should be formulated to avoid legal wrangle between the producers and contracting firms, which might crop up due to price fluctuations.

Key principles of the contract farming may be:

1. One village, one group (SHG).
2. One village, one variety product.
3. Crop loan at 12% per annum on Group's guarantee.
4. Door delivery of quality inputs at discounted rates.
5. Crop insurance
6. Synchronized sowing
7. Integrated crop management through competent Farm Service Centres.
8. Contamination control measures from farm to factory
9. Assured buyback of final produce from farmers' doorsteps.
10. The sponsor plays the role of a perfect coordinator/facilitator between the producer and the consumer.
11. Core R&D team.
12. Unique partnership with local agencies including a public sector enterprise.

13. Execution of technology transfer through well-trained extension personnel.
14. Supply of all kinds of agricultural implements free of cost to contracted farmers.
15. Supply of timely and quality farm inputs on credit.
16. Prompt dispatch/delivery/procurement of the mature produce from every individual contracted farmer through the system of 'Quota Slips.'
17. Effective adoption/use of modern communication technology like pagers for communication with field executives
18. Regular and timely payment to contracted farmers through computerised receipts and transparent system.
19. Maintenance of perfect logistics system and global marketing standards.

According to Pawar, Surve and Ingle (*Agrobios Newsletter*, July, 2010) the Indian agri-food system is undergoing rapid transformation and there is growing evidence that contract farming will have an important role in this transformation. Contract farming is defined as system for the production and supply of agricultural /horticultural produce under forward contracts between producers /suppliers and buyers. The essence of such an agreement is the commitment of the producers/ sellers to provide an agricultural commodity of a certain type at a time and price, and in the quantity required by a known and committed buyers. According to the contract, the farmer is required to plant the contractor's crop on his land and to harvest and deliver to contractor a quantum of produce, based upon anticipated yield and contracted acreage. This could be at a pre-agreed price. Towards these ends, the contracture supplies the farmer with selected input, including the required technical advice. Thus the contracture supplies all the inputs required for cultivation, while the farmer supplies land and labors. However, the term and nature of the contract differ according to variation in the nature of crop to be grown, agencies, farmers, and technologies and the content in which they are practiced.

Types of Contract

The contract can be classified into three, not mutually exclusive categories, viz., market specification, resource providing and production management. The market specification contracts are pre-harvest agreement that binds the firm and grower to a particular set of condition governing the sale of the crop. The condition specify price, quality and pricing. Resource providing contracts oblige the processor to supply crop inputs, extension or credit, in exchange for a marketing agreement. Production management contracts bind the farmer to follow a particular production method or input management, usually in exchange for a marketing agreement or resource provision.

Why Contract Framing in India

In our country the farmers face the problems of traditional technology and management practices, little bargaining power with input supplies and produce market, in adequate infrastructure and market information, lack of post-harvest management expertise, poor package of produce an inadequate capital to grow a quality crop. They are waiting for change for better living standard. Contract farming helps small framers to participate in the production of high value crop like vegetable, flowers, fruits etc and benefit from market led growth. Contract farming will maximize the profit to the farmers and minimize risk in farming like production related risk, transfer price risk & production risk. There is a tendency amongst the uses to go in far environmental friendly, value added quality agro products in there daily life. The farmers find it easy to get under are root inputs, technological and extensive services, post harvest processing facilities and more importance, the marketing of there produce with assured cash returns. Contract farming facilities more and more private companies to develop backward linkages with the farmers.

Present Stage of Contract Farming in India

The farm ministry details an agenda for expansion of agricultural credit to the tune of Rs. 7, 36,570 crores during 10th plan and the official note to he finance ministry gave financial of contract farming by bank priority. Agricultural and processed food product development authority is developing policy guideline on contract farming for forwarding to state government for implementation. The guidelines will focus on regularizing the relation between producer and processor of food material. Contract framing is already undertaken in tea estates by major companies including Pepsi food, ITC, Hindustan lever and for crop diversification by mahindra subhalabh services with Punjab agro for basmati rice and durum wheat besides drawing a plan to setup grain handling and storage facilities. In Karnataka wide varieties of vegetable, gherkins, lime, pomegranate, grapes for resins, pearl onion, asparagus and mangoes for pulp are already covered under contract farming.

Development of Food Parks

The Food Park have common facilities of cold storage, pack house, grading centre, quality control laboratory, drying yard, warehouse, effluent treatment facilities, post harvest management unit, technical information centre, and special common facilities (e.g., fruit processing and aseptic packaging, printing and labeling, dehydration of fruits and vegetables, value add centre, spray drying units etc.). These facilities help a small entrepreneur, small-scale industries and farmers to take up food processing without much capital investment. In India, there is an urgent need of such parks and about 39 Food Park project have been sanctioned. In Virdhunagar (TN) such Food Park in private sector at has been established with assistance from MFPI, GOI.

Development of Agri Export Zones

To compete with Global Trade Scenario efforts should be are made to provide improved access to the produce / products of the Agriculture, Horticulture and allied sectors in the global market by setting up of Agri Export Zones (AEZ). It helps to have a comprehensive look at a particular Agriculture / Horticulture produce located in contiguous area for the purpose of developing and sourcing the raw materials, their processing, packing, for exports.

Concept

Promotion of agricultural exports is important for creating conditions for providing remunerative prices to farm products. Sporadic efforts have been made in the past for promoting exports of agricultural produce/products from the country. Thus on one hand Research and Development has taken place with little bearing on the development of a particular agricultural produce for the purpose of export, on the other hand financial and fiscal incentives are being provided for exporting a particular produce without actually addressing pre-harvesting and post-harvesting practices.

Though India ranks as the world's second largest producer of horticultural crops, the country fares less impressively on processing fronts. Lack of adequate storage facilities is one of the main problems faced by food processing industry leading to almost 30-40 per cent wastage of the production. Simultaneously, most of the varieties grown in India are simply not suitable for processing. Therefore the concept of Agri Export Zones developed by APEDA aims to improve the levels of food processing to reduce waste, increase marketability and help farmers to enjoy higher value of realization. It takes a comprehensive look at a particular produce/product located in a contiguous area for the purpose of developing and sourcing the raw materials, their processing/packaging, leading to final exports. Thus, the entire effort is centered on the cluster approach of identifying the potential products, the geographical region in which these products are grown and adopting an end-to-end approach of integrating the entire process right from the stage of production till it reaches the market.

However, the entire effort begins with market analyses to assess the market potential for that produce. Subsequently, steps are taken to ascertain the quality parameters that are required for making this produce/product acceptable in the international market.

The attempt then is to see what sorts of interventions are required at each stage of the value chain to come up with such a produce/product.

These interventions commence right from availability of inputs like seeds, fertilizers, pesticides, credit etc. This would then entail research into the pre-harvest protocols, if they do not already exist and application and extension of these protocols and practices in the fields of identified farmers.

Development of harvesting protocol, its implementation and the post-harvest management are all critical links in the value chain. These would include activities like storage (both intermediate and final), grading, sorting, processing, packaging and transportation. Each of these activities would involve a specialized intervention, which could be technological (hence would involve research) or financial.

It will also identify/enlist difficulties/ problems encountered at each stage. These difficulties could be procedural in nature or may relate to a particular quality standard. A package needs to be developed to suggest solutions to these problems. Corrective time bound measures will then be delegated to specific government agencies. The APEDA will ensure that the measures are completed on schedule. Export quality products will have specific niche markets identified and will be promoted extensively. Under the programme, international Agricultural experts may also be invited to demonstrate modern harvesting technologies.

Finally, it is the marketing, which is the key to the success. The capability to position the product in an appropriate market and getting an in an appropriate market and getting an appropriate price would ultimately determine whether the AEZ could get going.

Advantages of Agri Export Zones

1. Strengthening of backward linkages with a market oriented approach.
2. Product acceptability and its competitiveness abroad as well as in the domestic market.
3. Value addition to basic agricultural produce.
4. Bring down cost of production through economy of scale.
5. Better price for agricultural produce.
6. Improvement in product quality and packaging.
7. Promote trade related research and development.
8. Increase employment opportunities.

TABLE 32.1: Various proposed/working Agri export zones

Place	Type of Produce	Proposed Project Cost
Hosur in Dharmapuri	Cut flowers	Rs. 24.85 crores
The Nilgiris	Flowers production	Rs. 15.89 crores
Theni district (T.N.) (Covering Theni, Dindigul, Madurai, Virudhunagar, Tirunelveli and Kanniyakumari)	Mango	Rs. 24.60 crores
Cuddalore district (T.N.)	Cashew	Rs. 10.36 crores
Tiruchirappalli District (T.N.)	banana	Rs. 10.00 crores
Theni,	Grapes	
Erode	Turmeric	
Tuticorin	Medicinal Plants	
Coimbatore and Dindigul Districts	Onion	

The Agri export zone should have common processing unit, common marketing and leasing the lands for floriculture units, cold storage, reefer vans, common marketing facility, collection centres, processing units, cold storage, etc.

Role of Seed Certification Department

Seed is the basic input for increasing agricultural productivity. To define quality seed, four components viz., physical purity, germ inability and vigour, genetic purity and seed health should be taken in to account.

The function of seed certification is to encourage the production of adequate quantities of genetically pure and good quality seeds, particularly high potential, improved and hybrid varieties of different crops. In order to achieve this requirement the Seed Certification Department carries out the certification work in various crop varieties notified under Indian Seeds Act 1966 and in accordance with the provisions of the Seed Rules 1968 to maintain the quality of the seeds produced in the state. There exists a wide gap between the requirement and availability. By increasing the production of certified seeds through private sector, the total certified seed production could be increased. Hence the private sector seed production will be concentrated in order to fill up the gap between availability and requirement of quality seeds. Out of the total certified seeds produced, 62.51 per cent is contributed through private sector, 34.32 per cent is contributed by Government and 3.17 per cent is contributed through Quasi Government sector. During the year 2007-08 as against the target of 60,500 metric tones of seeds to be certified, a quantity of 52,891 metric tones of seeds were certified upto February 2008. During 2008-09 it is proposed to certify and distribute 62,500 metric tones of seeds under various crops.

Parameters of Seed Certification

Physical purity: The physical purity means that farmers should sow only pure seed and not any kind of mixture of other crop seeds, weed seed and inert matter. The physical purity of seed is improved by seed processing during which all the undesirable contaminants and immature shrivelled seeds are removed.

Germination and vigour: Germination potential of seeds determine seedling emergence, crop stand and ultimately the yield. The seed lots ability to withstand inclement germination and seedling establishment conditions is provided by seed vigour. The seed germination deteriorates during handling and storage of seeds. Therefore, seed should be stored in areas which have low relative humidity and temperatures.

Genetic purity: An improved variety has certain unique characters of economic importance such as high productivity, disease resistance, adaptability to unfavourable weather conditions, high grain or fruit quality, etc. It is also expected to possess certain diagnostic characters which distinguish it from other varieties. Therefore the genetically pure seed of a variety is expected to have all the unique economic and diagnostic characters. Genetic purity of seed crop is maintained by isolation of seed crop by a specified distance from crops of other varieties / contaminants. This help in avoiding undesirable cross pollination during flowering or physical admixture at the time of harvesting and threshing. The isolation distance for seed multiplication of self pollinated crops like rice is lower i.e. 3 m and is higher for cross pollinated crops like maize i.e. 200 m for certified seed and 400 m for foundation seed.

Seed health: The seeds which are free from seed borne diseases not only help in raising disease free crop but also help in checking further spread of diseases. Seed infested with pests will loose viability very fast during storage and will become unfit for planting within a very short time. The health of the seeds can be maintained by proper seed storage and seed treatment with fungicides. The insect infestation should be controlled by proper fumigation and / or seed treatment.

Seed Certification Agencies in India

State Agency	Address
Andhra Pradesh	Andhra Pradesh State Seed, Certification Agency, House No. 5-10-193, 1st Floor, HACA Bhawan, Opposite Public Gardens Hyderabad-500004
Assam	Assam State Seed Certification Agency, Dr. B.K. Kskoti Road Ulubari, Guwahati-781007
Bihar	Bihar State Seed Certification Agency, Tranquality Riding Road, Sheikhpura, Patna-800014 4
Chhattisgarh	Chhattisgarh State Seed Certification Agency, Indira Gandhi Agri. University Campus, Krishak Nagar, Raipur-492006 (C.G.)
Delhi	Seed Certification Unit Delhi Administration, Room No. 5 & 6 Old Civil Supply Building, New Courts Tis Hazari, Delhi, 110004
Gujarat	Gujarat State Seed Certification Agency, Outside Shahpur Gate, Opposite Chisti Chambers Narayan Kunj, Ahmedabad-380001
Haryana	Haryana State Seed Certification Agency, SCO 1038-39, Sector 22-B Chandigarh-160020
Himachal Pradesh	Himachal Pradesh State Seed Certification Agency, Nalagarh House, Simla-171005
Jammu & Kashmir	(a) Seed Certification Wing Department of Agriculture Jammu Division, Talab Tilloo Jammu (b) Seed Certification Wing Department of Agriculture Kashmir Division, Lalmandi Post Office Jawahar Nagar Srinagar-190008
Karnataka	Karnataka State Seed Certification Agency, (K.A.I.C. Premises) Opp. Baptist Hospital Bellary Road, Hebbal Bangalore-560 024
Kerala	Department of Seed Certification Directorate of Agriculture, Vikas Bhavan Trivandrum-647041

State Agency	Address
Madhya Pradesh	Madhya Pradesh State Seed Certification Agency, HIG/A-7, BDA Colony, Near 1250 Hospital, Tulsinagar, Blopal-462003
Maharasthra	Maharashtra State Seed Certification Agency, In front of NilKanth Cotton Mill, NH-6 Muzhzapur Road, Shivani Akola – 444 005
Orissa	Orissa State Seed Certification Agency, Cold Storage Campus, Samantrapur Bhubaneshwar-751002
Puducherry	Puducherry State Seed Certification Agency, New Light House Road, Vamba Keerapalayam, Puducherry - 605001
Punjab	Punjab State Seed Certification Authority, SCO-837-838, Sector 22-A Chandigarh-160022
Rajasthan	Rajasthan State Seed & Organic Production Certification Agency, 3rd Floor, Pant Krishi Bhawan, Jan Path Jaipur-302004 - Rajasthan
Sikkim	Seed Certification Wing Department of Agriculture Government of Sikkim Gangtok-737101
Tamilnadu	Department of Seed Certification 1424 A THADAGAM Road, GCT Post Coimbatore-641013
Uttar Pradesh	Uttar Pradesh State Seed Certification Agency, A-284, Sector 5, Indira Nagar Lucknow
West Bengal	West Bengal State Seed Certification Agency, 238, Netaji Subhash Chandra Bose Road, Calcutta- 700040

Global Seed Business

Formal and informal exchange of seeds and planting materials provided opportunities for the rate of seeds at global level. According to International Seed trade Federation (FIS), it is now valued at US$ 30 billion. While the global domestic markets have grown slowly, international seed exchanges have quadrupled in the last 25 years from a modest US$ 880 million to over US$ 30 billion at present. Thus the trade between the countries now accounts for over 11 per cent of the total seed trade.

The Indian Seed Industry

Tracking the revolutionary changes in the Indian agricultural sector, it is quite clear that technology, institutions, and markets have had a very important role to play. The Indian seed industry is currently valued around Rs 2500 crores ($ 500 million) and is proposed to be around 3750 crores ($ 750 million) by 2002. There are about 150 organized seed companies in India today. Several companies have Government of India (DSIR) recognized research and development departments and have produced and released a large number of varieties and hybrids in several crops. The contribution of private research in terms of value is steadily increasing.

The National Seed Corporation was established in 1963. The Government of India enacted the Seeds Act in 1966 to regulate the growing seed industry. The sixties were the most eventful times for Indian agriculture, not only because of introduction of high-yielding cereals, particularly wheat and rice but also for many other positive developments related to seed such as, constitution of Seed Review Team, enactment of Seeds Act, 1966 and formation of National Commission on Agriculture. This was the period, during which the private sector significantly stepped into the seed business. The Seeds Act stipulated that seeds should conform to a minimum stipulated level of physical and genetic purity and assured percentage germination either by compulsory labelling or voluntary certification. Further, the Act provided a system for seed quality control through independent State Seed Certification Agencies which were placed under the control of state departments of agriculture. Presently, both private and public sector companies/corporations are involved with the production of seed.

Today, the Indian seed programme boasts one of the biggest seed markets being eighth largest in the world with an estimated value of INR 49 billion (USD 1.06 Billion) and with an annual growth rate of 12% to 13 %. The development of private seed industry is no more confined to just production and marketing of seed. It has well acquired technological strength to cater to the varietal needs of tomorrow. Along with industries Indian farmers have in recent years adopted intensive cultivation practices in order to meet the growing demand for agricultural produce.

Indian seed export figures are showing healthy trend. Our share in International seed trade is very meager but there has been a phenomenal increase in exports from 13.9 to 83.9 crores in a span of only five years. It is interesting to note that our exports are shifting to high value quality seeds. We started with export of open pollinated wheat seeds and now the major component of export is F_1 hybrid seeds. Many of the multinational seed companies have now identified the Indian potential in seed trade.

The Indian seed industry at present consists of two National Originations namely National Seed Corporation (NSC) and State Farms Corporation of India (SFCI). Besides these 13 State Seeds Corporations, 147 private seed companies, 19 State Seed Certification Agencies and 63 Notified Seed Testing Laboratories are present. Public sector produces bulk of self pollinated crops of which wheat and rice account for about 60 per cent while private sector deals mostly on low volume and rice account for about 60 per cent while private sector deals mostly on low volume and high return crops like hybrids, vegetables and flowers accounting for about 50 per cent of total production of certified seed in India.

The department of Seed Certification started functioning with effect from 24th October 1979 and is implementing the following activities in order to make available quality seeds to the farming community:

Seed Certification: Seed certification is done to encourage the production of adequate quantities of genetically pure and good quality seeds, particularly of high yielding improved varieties and hybrids of different crops.

Seed Inspection: It ensures the quality of the seeds distributed to the farmers, seed selling points are inspected periodically and seed samples are drawn and sent for analysis in the notified Seed Testing Laboratory and based on the results legal actions are being initiated against the offenders.

Seed Testing: The seed testing is being carried out to analyse the quality of the seed lots. Factors like germination, physical purity, moisture, seed health and admixture of other distinguishable varieties are being analyzed in the notified seed testing laboratories as per the hand book on seed testing released by Government of India.

Training: The Training wing organizes various kinds of trainings like, orientation training, refresher training, training to seed producers, quality control training to seed dealers, training is given to the seed dealers on quality maintenance, storage of seeds and also on the regulatory aspects of seed legislation.

CROP INSURANCE SCHEME IN INDIA

According to Shipra (*Agrobios Newsletter, June, 2010*) the Government of India started offering widespread crop in insurance in 1985 with the Comprehensive Crop Insurance Scheme. The CCIS has been replaced by the National Agriculture Insurance

Scheme (NAIS). The NAIS is considered to be an improvement over the CCIS, but it has simply replaced one flawed scheme with another slightly less flawed one. Government crop insurance has proved to be a failure worldwide, but India seems to have ignored both its own failure and the failure of other countries. The main flaws of the NAIS are the goal of financial viability, its mandatory nature, its failure to address adverse selection, arbitrary premiums, and the area approach. Internationally, private crop insurance is not highly developed but varied successful private programs do exist. Even if India withdrew from crop insurance schemes, it could still support farmers through an income guarantee or investment in infrastructure.

Farmers face floods, drought, pests, disease, and a plethora of other natural disasters. The weather is their greatest adversary, something that can never be controlled by man. Yet, farming has been in existence since the caveman turned his spear in for a hoe. Farming has come a long way since then; nevertheless; farmers are still at the mercy of the heavens. Crop insurance is a risk management tool that farmers can use in today's agricultural world. For a premium, farmers can pass their weather-related risk onto a third party. Farmers in India have been subjected to publicly administered insurance schemes since 1972. Every scheme has been flawed, yet the government of India is still attempting to strengthen agriculture by protecting its farmers from the weather. India's failure at providing public crop insurance does not stand alone. In both the developing and developed world, governments' crop insurance schemes have run at huge losses while not delivering an effective product. The inadequacy of such schemes is a well-established fact. On the other hand, private insurance does exist in situations where it is feasible and no subsidized insurance is offered. The farmers stand to benefit even more from private insurance when there are several The government's most current crop insurance scheme, the National Agriculture Insurance Scheme has only been implemented since the Rabi season of 1999-2000.

Within five year the NAIS is supposed to become financially sustainable, charging Farmers premiums based on actuarial rates and administrative costs. A data based Analysis of the NAIS is not possible, as data for only two seasons exists. However, shortcomings of previous crop insurance schemes, general trends of agricultural insurance in other countries, and inherent theoretical flaws in the NAIS all point towards disaster.

The main flaws of the NAIS can be summarized as follows:

1. Lofty goal of financial viability
2. Mandatory for loanee farmers
3. Adverse selection, in the case of non-loanee farmers
4. Premiums do not equal Risk level
5. The area approach

The CCIS was mandatory for loanee farmers growing covered crops and insured 100% of the crop loan. The NAIS is also mandatory for loanee farmers growing covered crops in implementing states. The indemnity is based upon the value of the threshold level for each crop grown in a set area. The threshold yield is based upon a moving average of the yield over past five years. Anyone remotely familiar with agriculture would understand that five years yield data does not accurately represent complex weather patterns. Additionally, non-loanee farmers are allowed to participate in the NAIS, but up to this point very few, except in Maharashtra, have chosen to do so.

The CCIS was often criticized as being "bank insurance," and the NAIS is no better on this count. Producers taking out loans have no choice--if the state authorities decide to implement the NAIS the producer must purchase insurance.

Insurance is not the only risk-management tool available to farmers in India. Diversification, fragmented land holding, off-farm employment, and savings are just of few of the options. Requiring loanee farmers to prove their ability to manage risk is good business--forcing one option upon them is not. Farmers with adequate risk management capabilities should not be forced to purchase crop insurance in order to receive a loan. Twenty-two states/union territories participated in the CCIS, while only 16 are participating in the NAIS. Punjab and Haryana are two of the leading agricultural states, but have yet to participate in the CCIS or NAIS. Agriculture in Punjab and Haryana is less risky than that of other states. However, one cannot assume that no demand exists for crop insurance in these states, as no cropping system is without risk. Three levels of indemnity do exist within the NAIS, but three risk levels can in no way address the diverse climate and agriculture in India. If premiums were based upon true risk levels, farmers in any agricultural system could avail of crop insurance if it fit their risk-management needs. If a farmer faces such high risks that he cannot survive without subsidized insurance the cropping system is not sustainable adverse selection can be observed when a group of farmers is offered crop insurance at the same premium, as is often the case in government administered/subsidized crop insurance. The worst farmers will avail of the insurance, making claims unreasonably high. Publicly administered crop insurance schemes worldwide have been largely unsuccessful partially due to adverse selection. In India adverse selection can clearly be seen at the state level, as Punjab and Haryana do not participate in the NAIS. If adverse selection were unavoidable, the entire insurance sector would not exist. History tells us that private insurance has been better able to address the problem of adverse selection. The NAIS makes insurance mandatory for loanee farmers, so adverse selection is more apparent at the state level. Non-loanee farmers have no obligation to purchase crop insurance and will most likely follow the pattern of adverse selection. If the NAIS can become financially viable, premiums would have to continually rise due to adverse selection. The NAIS has several shortcomings and its success is at best questionable.

However, the government could easily argue that farmers would otherwise remain uninsured and that private insurance would not enter the market. Options to the NAIS do exist that would help farmers, especially in the long run. The government has admitted that it lacks the resources to administer a proper insurance scheme at the individual level. For various reasons a second-rate scheme is deemed as necessary. A better option would be an income guarantee not based upon yield, crop grown, or farm size. Considering the various subsidies that are given to farmers through various means--fertilizers, seed, price supports, etc.—an income guarantee should not be an unfeasible option.

The value of crop insurance, private or subsidized, is much debated by academics and policy makers. The concept of index-based contracts for natural disasters in place of crop insurance has been recently introduced. Farmers would purchase a contract and be compensated when a certain event or natural disaster occurs. Rainfall contracts are one example. Rain is relatively simple to monitor and the history of rainfall in most areas is well known. Farmers would be compensated if the rainfall in an area would go below a set level, with varying levels of payment depending upon the level of rainfall. The faults of this approach lie in its similarity to the area approach. However, the benefits are significant, including reduction of moral hazard, adverse selection, and transaction costs. This alternate model could be adopted as an improvement over the NAIS but would still deter the private sector from entry into crop insurance Private crop insurance can be observed worldwide, even though it is not highly developed. Private crop insurance has tended to cover more specific risks and not cover management related risks. These insurance policies offered must fit needs of farmers and be beneficial- otherwise they would not exist. This is not necessarily the case with government sponsored crop insurance. Private insurance works in a wide range of countries for a wide range of agricultural activities.

Insurance programs vary from tropical plantation crops in Latin America to tree crops in the USA. The NAIS will not fix the ills of Indian agriculture, nor will any other grand insurance scheme planned by the authorities. Private crop insurance may or may not develop if all government crop insurance is abolished. Abandoning insurance schemes does not mean abandoning farmers. Farmers could be given an income guarantee not based on yield, price, or area planted. Even now an income insurance scheme is being considered in India. Investment in agricultural infrastructure/research would be more equitable as opposed to subsidies to crop insurance and may yield more long-term benefits. Farmers deserve the chance to farm on their own. They know the weather better than anyone- it is their greatest foe and their greatest friend. The government should stop trying to play God and help farmers help themselves.

INDIA/STATE GOVERNMENT ORGANIZATIONS FOR AGRICULTURAL MARKETING

Directorate of Marketing and Inspection (DMI)

The Directorate of Marketing and Inspection (DMI) is an attached Office of the Ministry of Agriculture. It was set up in the year 1935 to implement the agricultural marketing policies and programmes of the Central Government. Since its very inception, the Directorate continues to be responsible for bringing about an integrated development of marketing of agricultural and allied produce in the country with a view to safeguard the interests of producer-sellers as well as the consumers. It maintains a close liaison between the Central and the State Governments in the implementation of agricultural marketing policies in the country. The Head Office is at Faridabad and the Branch Head office is at Nagpur, while regional offices are at Chennai, Delhi, Guntur, and Mumbai. Following are the main activities of DMI:

1. Promotion of Standardization and Grading of agricultural and allied produce under the Agricultural Produce (Grading & Marking) Act, 1937 as amended in 1986;
2. Market Research, Surveys and Planning;
3. Agricultural Marketing Reforms
4. Agricultural Marketing Information Network;
5. Promotion of Cold Storage
6. Construction of Rural Godowns
7. Development of Marketing Infrastructure, Grading and Standardization
8. Training in agricultural marketing;
9. Marketing Extension;

The standardization and grading of agricultural and allied produce for agricultural commodities are framed based on their intrinsic quality. Food safety factors are being incorporated in the standards to compete in World trade. Standards are being harmonized with international standards keeping in view the WTO requirements. Certification of agricultural commodities is carried out for the benefit of producer/ manufacturer and consumer. Certification of adulteration prone commodities viz. Butter, Ghee, Vegetable Oils, Ground-Spices, Honey, Wheat Atta etc. is very popular. Blended Edible Vegetable Oils and Fat Spread are compulsorily required to be certified under Agmark. The Certification mark under the Agricultural Produce (Grading & Marking) Act is popularly known as "AGMARK". Grading is carried out in accordance with the standards notified, following meticulous procedure of sampling, testing, packaging, marking and sealing as per the instructions issued under the Act and Rules.

National Institute of Agricultural Marketing (NIAM)

The Institute started functioning at Jaipur (Rajasthan) with effect from 8th August, 1988. It is an autonomous body under the administrative control of the Ministry of Rural Areas and Employment. NIAM focuses on following areas:

1. Agricultural Marketing system in States
2. Post Harvest Loss Reduction Aspects
3. Information Technology Application in Agricultural Marketing
4. Future and Forward Markets and Commodity Exchanges
5. Food safety, quality certification & standardization
6. Legal reforms
7. Market infrastructure
8. Modern terminal market and their operations and management
9. Public-private partnership and channel partners.
10. Ministry's programmes and their capacity building
11. Organic, medicinal and aromatic plants
12. Coop marketing
13. Market led extension
14. Sanitary and phyto-sanitary measures
15. Grading, standardization and certification
16. WTO
17. Warehousing and storage

Various Department of Government of India have taken several initiatives for the development of this sector. For example, the state government is setting up a hi-tech floriculture park at Mungpoo in North Bengal. This park would provide common infrastructure to the entrepreneurs, who lack resources to build the required infrastructure. Similarly, the state government through the West Bengal State Food Processing and Horticulture Development Corporation Ltd is planning to set up a market complex for flowers with cold chain infrastructure for preserving flowers at Mullickghat, Kolkata. A comprehensive multistoried flower market with facilities for cold storage has recently come up at Panskura under the state government initiative. The Department of Food Processing Industries and Horticulture, Govt. of West Bengal is planning to set up a market complex, exclusively for flowers at Ranaghat.

Karnataka is a progressive state in the field of modern horticulture in the country. The diverse agro-ecological conditions prevailing in Karnataka has made it possible to grow different types of horticultural crops such as fruits, vegetables, flowers, spices, plantation crops, root and tuber crops, medicinal and aromatic crops etc. Karnataka is the 3rd largest producer of fruits and stands 5th position in area and production of vegetable crops. State stands 1st in area and 3rd in production of flower crops. It occupies 2nd and 3rd position with respect to area and production of plantation crops. Karnataka is the largest producer of Spices, Aromatic and Medicinal crops.

Government of Karnataka is considering fruit and vegetable processing sector as a thrust area for development. With growing urbanization and increasing quality consciousness, the market for processed fruit and vegetable is expected to grow rapidly. Therefore, there is enormous potential for developing agro products.

MARKETING OF HORTICULTURAL CROPS

Marketing involves finding out what your customers want and supplying it to them at a profit. Marketing is governed by whole marketing process and should be customer oriented. Production must supply customers with what they want or need. Therefore, people spend their money. Marketing is a commercial process and is only sustainable if it provides all the participants with a profit. Agricultural marketing is achieved by a series of processes which includes harvesting techniques, the grading and sorting of crops and the packing, transport, storage, processing, distribution and selling of products. These are the mechanics of marketing. Marketing also involves a series of activities in making available services and information which influence the desired level of production relative to market requirements, and the movement of the product (or commodity) from the point of production to the point of consumption.

TABLE 32.2: Important horticultural products

Cut flowers	Rose, chrysanthemum, carnation, gladiolus, tulip, narcissus, orchid.
Cut follage	Asparagus tern, leather feat fern, soft ruscus.
Fruit that are normally considered as vegetables	Breadfruit, avocado, tomato, egg plant (brinjal or aubergines) hot pepper, sweet pepper, karella (or bitter gourd), squash, marrow, gourd, cucumber, luffa, pumpkin, plantain, christophine or choyote.
Herbs	Parsley, mint, coriander, dill, basil, rosemary, thyme, sage.
Leaf and stem vegetables	Lettuce, cabbage, spinach, chard, brussels sprout, endive, watercress, celery, asparagus, celeriac, green onion, leek, amaranthus, bean sprout, bokchoy, Chinese cabbage, Chinese celery, spinach, chicory, kohlrabi, fennel.
Other fruit	Strawberry, pineapple, Cape gooseberry, watermelon, sweet melon, raspberry, blackberry, blackcurrant, gooseberry, cranberry, blueberry, rhubarb, loganberry.
Root vegetables	Arrowhead, onion, potato, sweet potato, cassava, yam, taro, garlic, radish, carrot, turnip, parsnip, beetroot, Jerusalem artichoke, dasheen, eddoe.
Spices	Black pepper, chili pepper, cardamon, ginger, clove, cinnamon, bay leaf, turmeric.
Temperate plants for garden	Roses, ornamental shrubs, herbaceous flowers, bedding plants, conifers, flowering bulbs.
Tree fruit	Orange, lemon, lime, mandarin, grapefruit, apple, mango, banana, guava, soursop, lichee, peach, apricot, pear, plum, rambutan, fig, quince, persimmon, durian, chiku, pawpaw, pomegranate, mangosteen, loquat, carambola, cherimoya, cherry, date, mulberry.
Tree nuts	Cashew, walnut, hazelnut, macadamia, pistachio, pecan, coconut, almond, Brazil.
Tropical plants for house and garden	Dieffenbachia, coleus, yucca, cordyline, dracaena, monstera, fatshedera, ficus, maranta.
Vegetables derived from seeds and flowers	Broccoli, cauliflower, artichoke, pea, bean, lentil, chickpea, broad bean, okra, mangetout pea, asparagus pea, yardlong bean, sweetcorn.
Vine fruit	Grape, passion fruit, kiwi fruit (or Chinese gooseberry).

Horticultural produce is heterogeneous in nature. Horticultural crops not only provide us with nutritional and healthy foods, but also generate a cash income to growers. Appropriate production practices, careful harvesting and proper packaging, storage and transport all contribute to the good produce quality. Once a crop is harvested it is impossible to improve its quality. The horticultural crops, because of their high moisture content are inherently more liable to deteriorate especially under tropical conditions. Moreover, they are biologically active and carry out transpiration, respiration, ripening and other biochemical activities, which deteriorate the quality of the produce.

Characteristics of Horticultural

1. Horticultural crop products are eaten for their contribution to the flavour and interest of food because they are rich in minor but essential nutrients, especially vitamins.
2. They are not basic treated as food commodities but for the health purpose people buy them even paying high price.
3. Their consumption vary from person to person depending on the selling price and the income of the buyer.
4. Not all crops are in liking of every person and many of the crops are not traded in large volumes because of the limited market.
5. The horticultural products are perishable. There is always a reduction in quality if not sold immediately. They usually lead to a fall in price.
6. Wide range and variety of horticultural products are available, therefore, consumer will try to go for low priced products.
7. The products are usually freely available in the market and the prices are determined by supply and demand.

Above factors have impact on the prices of horticultural crops which directly influence farmers profit. For example a farmer who has high quality tomatoes to sell may easily get good price equivalent to many times of his growing cost but a farmer gets less price when the market is oversupplied or the fruits are two days old.

The market wholesale prices of horticultural crop fluctuate. It may be double or halve in the same day, depending on the skill of the salespeople and on consumer demand. The prices depends on the season to season, from year to year; from the start of the season to the main supply period; from day to day; and from market to market.

Their prices are based on the quality assessment and whole process of marketing. Therefore, horticultural produce needs sorting, cleaning and grading to impart a uniform appearance to the produce. Inspection, sampling, analyses, packing, marking, labeling etc., are the primary elements of any quality control system. In order to enforce systematic and scientific grading of farm produce and quality control, matching with the present day requirement either at national or international level, agricultural marketing personnel should have a clear understanding and knowledge of the basic concepts and modus operandi of quality system It needs final inspection/ testing. A common Indian farmer is not well aware of all these steps, therefore, to make horticulture produce hi-tech, it necessary to educate the farmer from grass-root level so that he may able to get better economic return of horticultural produce. Marketing price also include harvesting, packaging, grading, transport and market commissions.

Therefore, for hi-tech horticulture proper marketing strategies should be prepared. Since horticultural products are mainly sold fresh; some are eaten raw while others are cooked. It is essential to give main emphasis here is on the commercial implications of different harvest and post-harvest practices, like:

1. **Harvesting:** The timing, technique and conditions at harvesting can significantly affect prices.
2. **Harvesting prices:** With some crops, harvesting can be undertaken early to

take advantage of high-priced opportunities. Exploiting these short-term market opportunities requires a close link with the market.

3. **Harvesting and crop maturity:** Shelf life and long-term storage is affected by the maturity of the crop at harvest. The storage characteristics of root vegetables are generally improved by only harvesting fully mature crops.

The fruits which are suitable for long-term storage (apples, pears, citrus and grapes), there are significant differences between cultivars, growing region and sometimes season in the optimum harvesting time. The expert assistance is necessary to know about proper term crop storage which can significantly improve farmer incomes. The important are:

1. **Harvesting and quality:** The harvesting and handling effect the quality of the produce in the market. Once a fruit is plucked from a plant or a root or leaf vegetable is harvested, it is cut off from its source of food and, particularly, water.
2. **Proper Treatment:** The effects of poor treatment normally show themselves some days later, when the produce is being presented for sale or is in storage. Poor treatment has two effects:
   a) Reduction in price.
   b) In the long term, the reputation of the production area is diminished leading to lower prices.
3. **Timing of harvesting during the day:** Harvesting should be done in cool climate and the plants are most turgid (high moisture content). The suitable time is the early morning. The other factors should also be considered as: the dew should be dry, availability of labour, minimize the time produce to left standing in the field.
4. **Harvesting techniques:** On high trees fruit can be harvested with a hook and a catching bag on a pole or similar harvesting aid. This prevents fruit falling to the ground.

The techniques of harvesting vary from crop to crop. For other crops knives and clippers can improve harvesting practices because they can cut through fibrous tissue, stems and leaves can be trimmed and clean cuts reduce the likelihood of infection. Tools like this are used for harvesting lettuce, cabbage, sweet pepper, egg-plant, honeydew melons and banana. In the case of bananas a slight cut is made in the upper pseudostem to allow the bunch to ease down gently. Individual hands are cut off from the bottom hand upwards. The leafy vegetables are harvested by cutting the plant with a sharp knife as close to the root as possible. Uprooting results in soil coming into contact with the produce. The bulb crops such as garlic and onions are harvested by pulling the leaves at the neck and then cutting the leaves about 3 cm from the bulb.

Other Aspects

There are other several aspects of good marketing of horticultural products:

1. **Field containers:** The crop damage associated with moving sacks of produce through the field is reduced. With picking bags it is preferable to be able to release the bottom so that the produce can be let out gently, rather than upending the bag. Baskets or boxes with sharp or rough edges should not be used. Proper care should be taken during transferring produce from one container to another and transportation. The produce may also be harvested into the container for storage or transportation.
2. **Harvesting system:** The perishable produce should be covered with damp cloths for protection against the sun's heat. The containers should be kept in

a shaded area. The leafy vegetables should be sprinkled with water at regular intervals to maintain leaf turgidity. Field assembly points or shade house should be prepared with natural materials or a canvas tent with ventilation. Containers should be periodically cleaned.

3. **Curing and drying:** Bulb crops such as onions and garlic can be dried in the field by being spread one layer thick over about six dry days. The root crops respond to warm moist conditions after harvest by thickening and hardening their skins. This provides protection against dehydration and infection.
4. **Trimming and sorting:** Cabbages, cauliflower, chinese cabbage and lettuce will have their outer leaves trimmed, except for three or four wrapper leaves, to give some protection to the head. Long stalks attached to the fruit, as in citrus, should be cut as close to the fruit as possible to prevent damage to other fruit.
5. **Grading and packing:** Prepare specialized grading areas or sheds for this purpose, preferably thatched roofs. Grading tables should be covered with polythene sheeting with good light system. Tin roofs can be painted white to reflect heat while water trickled down the outside of a shed helps reduce the heat inside the building.
6. Packaging and presentation: The two main functions of packaging are to help prevent mechanical damage and to sort the produce into an acceptable size for the market and for handling. Good packaging can also enhance the attractiveness of the produce.

The horticultural products are traditionally processed due to non availability of storage facilities e.g. dried fruit and jams. Marketing of these products may be done through society. The efforts should be made to enrich the society for market development of the processed products. Te flower market depends on aesthetic aspects of the buyers. Therefore, promotion and marketing are essential in the commercializing horticultural products. They are expensive and require specific organization of growers and supermarket/retail groups. They address to following basic needs:

1. Deliver produce regularly and reliably to the market;
2. Provide a feed-back of information between the market and the growers so that production can be constantly oriented towards consumer demand
3. Establishing an ‘image’ (brand names and trademarks, labels, etc.) and building confidence in the consumer through education.
4. Providing a guarantee of produce to the consumer through education.
5. Providing a guarantee of the production process as related to food safety.

MARKETING CHANNELS

Marketing distribution channel consists of different levels. Each level of marketing intermediaries is representing a channel level. Channel level is a layer of intermediaries that performs some work in bringing the product and its ownership closer to the final buyer. The number of levels determines the length of the channel; the more the levels, the longer is the channel. A marketing channel that has no intermediary levels is known as direct marketing channel. A channel containing one or more intermediary levels is known as indirect marketing channel.

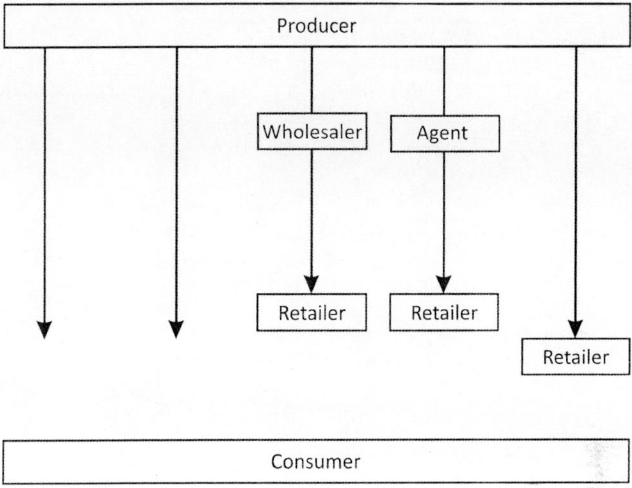

FIG. 32.3: Consumer Marketing Channels.

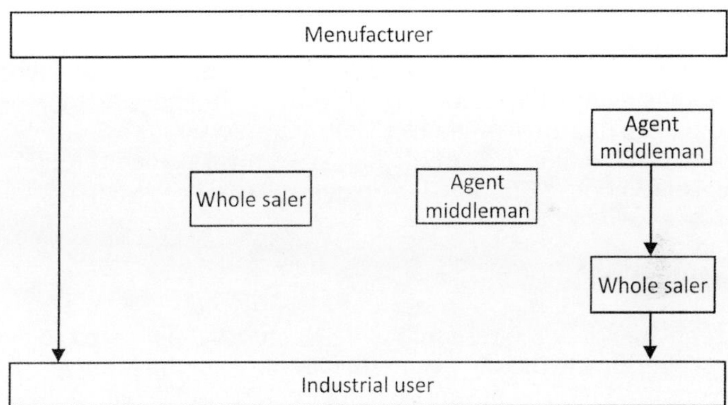

FIG. 32.4: Industrial Marketing Channels.

The horticultural product's distribution channel should be objective and efficient. The selected channels must have lowest cost with maximum overall profit. It should also be remembered that there is no single channel of distribution that will always result in optimum profit.

The integrated marketing concept has prompted many manufacturers to employ several kinds of channels. There are various constraints that are to be considered before deciding channel objectives. These are:

Nature of Market: The selection of channels depends firstly on the requirements of the market i.e. what the consumer wants and how much is wanted? The producer must also make it sure how best he can reach the market in order to sale the product. Besides these one should also assess various factors such as buying habits of consumers, the size of average sale, the concentration of purchases, repeat sales, the seasonality of sales, scope of distribution and competition.

Type of Products: Product features influence on the decision for suitable channels, like perishability, size, style of presentation to the consumer, unit value, and newness of the products, etc.

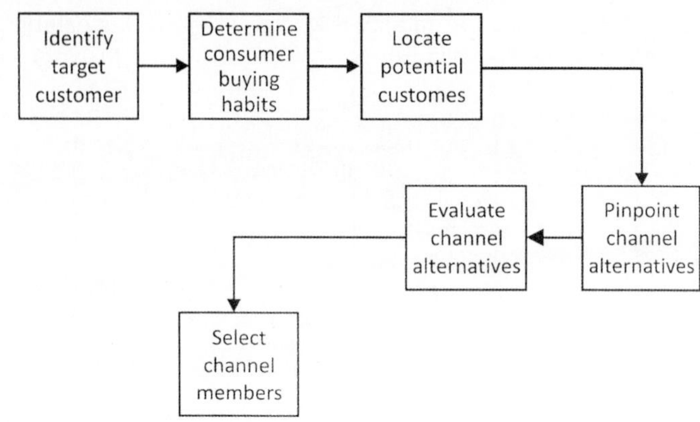

FIG. 32.5: Channel Selection Process.

Consumer's Buying Habits: The sale of the horticultural products depends on buying habits of the consumer.

The sale of horticultural produce (fruits, vegetables, bedding plants, fresh cut flowers or culinary herbs) can be sold using following marketing types:

Alternative Marketing: It is a traditional marketing system of farm produce.

Group Marketing: Group marketing includes, joint planning, funding, implementation, pricing, sharing risk equally in the marketing and reaping the benefit of collective bargaining. This indirectly results into more returns.

Cooperative Marketing: This marketing system is pursued through cooperative societies registered under Cooperative Societies Act. This system is pursued on the principle of *"self help by mutual help"*.

Direct Marketing

Direct marketing needs some research which includes the research on your product compares to others in the market places, the demand for the product and the unique benefits of your product. The advantages of direct marketing are:

1. It lower down marketing costs.
2. It has large customer base and potential for increased income.
3. The farmer can earn more profit on products.
4. Farmer can judge the test of the market think for launching new products.
5. Farmer can have direct contact with customers

The disadvantages of direct marketing are:

1. Farmer needs additional time to sell products.
2. Farmer needs additional capital for buildings and to pay extra costs for transportation of the products to customers.
3. Farmer needs to comply with legal restrictions and licensing requirements.

Direct marketing is done using various sources:

1. **Apni Mandi:** In these *mandies*, farmer-producers bring the produce for sale directly to the consumers, as in Punjab and Rajasthan
2. **Rythu Bazrs:** In Andhra Pradesh, such markets have been established to provide direct link between farmers and consumers in the marketing of fruit and vegetables and other essential food items.

3. **Raithara Santhegalu:** In Karnataka, marketing Board has established farmers market or *Raithara Santhe* without any middle men and provide a place where farmers and consumers directly interact.
4. **Farmers' market:** A farmers' market is a focal point where producers and processors gather on a regular basis to offer home-grown produce, agricultural and home-crafted products to sell directly to consumers. The advantages of farmers' markets are direct contact with a larger, regular and loyal customer base, lower marketing and overhead costs, increased income, a great way to sell excess produce and a market to test new products. The disadvantages of farmers' markets are extra time away from your horticulture business, possible low sales volume and the costs of transporting product to the farmers' market.
5. **On-farm sales (farm store or U-pick):** If your farm has plenty of parking space and is easy to get to, it may be ideal to encourage customers to come to the farm. To offer on-farm sales you need a cash register, weigh scale and likely a cooler, depending on your produce. On-farm sales offer lots of flexibility on length of season, size, hours of operation and product mix.
6. **Mail order sales:** Growers of perennials and small shrubs can use a mail order system to expand your customer base without spending extra capital on storefront buildings.
7. **Food service or door to door sales:** When you sell and deliver your horticulture product to homes, retail businesses, institutions and restaurants, you are in the door to door sales business.

Indirect Marketing

The indirect marketing is done by wholesaler, retailer, institutional buyer and broker. Indirect marketing is not profitable for small or marginal farmer because it adds an extra cost to the product and farmer looses contact with the customer in indirect marketing channels. Indirect marketing is profitable also because some indirect sellers are cost effective and efficient due to economies of scale; and buyers and sellers are linked together very efficiently. The farmer has to search for indirect sellers to take responsibility for marketing the produce.

On Line Commodity Exchange of India Limited (OCEIL)

It is India's first demutualized, on line multi-commodity exchange started at Ahmedabad on 26th Nov 2002 and has an in-house Clearing House. In is an electronic trading and demutualized system and makes the exchange unique, ensures flawless trading. It enhances the confidence of the trade participants. In demutualized setup, the ownership, management and trading are in the hands of three different sets of people. OCEIL has connectivity with all members. It has provision of adequate margin to ensure that the chances of default are reduced.

e-Commerce

According to WTO, E. commerce is production, distribution, marketing, sale or delivery of goods and services by electronic means. Electronic commerce, commonly known as (e-shopping) **e-commerce** or **eCommerce**, consists of the buying and selling of products or services over electronic systems such as the Internet and other computer networks. The amount of trade conducted electronically has grown extraordinarily with widespread Internet usage. The use of commerce is conducted in this way, spurring and drawing on innovations in electronic funds transfer, supply chain management, Internet marketing, online transaction processing, electronic data interchange (EDI), inventory management systems, and automated data collection systems. Modern electronic commerce typically uses the World Wide Web at least at some point in the transaction's lifecycle, although it can encompass a wider range of technologies such as e-mail as well.

A large percentage of electronic commerce is conducted entirely electronically for virtual items such as access to premium content on a website, but most electronic commerce involves the transportation of physical items in some way. Online retailers are sometimes known as e-tailers and online retail is sometimes known as e-tail. Almost all big retailers have electronic commerce presence on the World Wide Web.

Electronic commerce that is conducted between businesses is referred to as business-to-business or B2B. B2B can be open to all interested parties (e.g. commodity exchange) or limited to specific, pre-qualified participants (private electronic market). Electronic commerce that is conducted between businesses and consumers, on the other hand, is referred to as business-to-consumer or B2C. This is the type of electronic commerce conducted by companies such as Amazon.com. Online shopping is a form of electronic commerce where the buyer is directly online to the seller's computer usually via the internet. There is no intermediary service. The sale and purchase transaction is completed electronically and interactively in real-time such as Amazon.com for new books. If an intermediary is present, then the sale and purchase transaction is called electronic commerce such as eBay.com.

Electronic commerce is generally considered to be the sales aspect of e-business. It also consists of the exchange of data to facilitate the financing and payment aspects of the business transactions. It is an advance state of electronic technology. It is just an extension of business conducted on net. It is an interactive TV with Internet taking over computing sphere. A complete reinvention of how one does business through E.Com makes communication, information gathering and trade between companies and consumer easier and faster.

International Marketing Options

International market provides good price for the horticulture products. One can get expanded market with increase profits. International markets offer a larger and more diverse market base and can give you more security through diversifying into different global economies.

Fixing of the Price Range of Horticultural Products

1. Analyze the main competitors.
2. Assess whether new competition will enter your market.
3. Observe trade show activity.
4. Prepare a main competitor list.

Above steps you will know the prices competitors charge and why they charge what they do? Use this analysis to develop the upper limit of your price range. Be sure you compare your products to competitors.

Evaluation of Customer Demand

Demand is a term used to show how badly the customer wants your bedding plants or organic vegetables. Looking into customer's demand you can set the price of the product that is price according to demand. The proper assessment should be done as follows:

1. Does my customer believe price indicates the quality of a product?
2. How does my customer like to spend money?

3. How many buyers are there and where do they live?
4. How much do they earn and what is their education level?
5. How old are my customers.
6. What does my customer do in their spare time?
7. What size and type of family does my customer have?
8. What type of service does my customer value?
9. Where could I improve on competitor's products?

To get answer of these questions it needs proper evaluation of the customer.

QUALITY OF FARM PRODUCE

Quality encompasses both inherited and acquired characteristics of farm produce. The inherited characteristics are commodity and variety specific and remains unchanged or unaffected from the impact of environment and these are shape, size, colour, odour, weight, and taste. Inherited characteristics very often form the basis of the marketing decision. Since such quality factors reflect the consumer's preference, they play a dominating role in identifying the demand potential of that particular commodity

On the other hand, acquired characteristics are the result of the impact of the external environment, which varies from place to place and area to area. Such characteristics are by and large, responsible for heterogeneity and alter the quality of the produce. This includes both animate sources e.g., insects, microbes, fungi and their metabolites such as aflatoxin, uric acid, excreta etc. and autolysis of the produce itself.

Cultural practices such as harvesting technique and post harvest practices like storage packaging and handling also contribute to the quality of the produce, and play a major role in quality maintenance and rendering the produce marketable at acceptable price.

Quality Factors / Refractions

Inherited Characteristics: These include shape, size, colour, odour, taste, texture, weight, strength, luster, etc.

Aquired Characteristics: These include foreign matter, admixture, slightly damaged, damaged or discoloured, immature and shriveled, green grains, chalky, red kernels, broken, fragments, splits and wholes, weevilled, etc.

In case of fruits and vegetables, blemish, damaged, diseased, mechanical injury, cuts, spotted, wrinkled, foul smell, oozing, punctured etc. are factors considered to be responsible for down grading the quality of the produce

STANDARDISATION

Standard is a measure that is generally accepted as having fixed value. The measure is in units of intrinsic qualities or characteristics of a product or a service. According to ISO, a standard is a result of a particular standardization effort approved by a recognised authority and may take the form of a document containing a set of conditions to be fulfilled. Standards are of three types:

1. Object standard, such as mass, length and time.
2. Documentary standards, include specifications, methods, nomenclature and administrative aspects and.
3. Conceptual standards, which is neither in the form of object nor document but

is passed from mouth to mouth and generation to generation such as behaviour norms.

According to ISO, standardization is the process of formulating and applying rules for an orderly approach to a specific activity for the benefit and cooperation of all concerned, and in particular for the promotion of the over all economy, considering due safety requirements. The important objectives are:

1. To achieve overall economy in terms of cost, human resources and conservation of essentials. It involves judicious choice of raw material, production and handling techniques more economically.
2. Simplification –convenience in use, rationalization, interchangeability of parts, increased productivity and cutting unnecessary waste.
3. To evolve best possible solution to recurring problems, involving standardization of basic terminology, codes, practices, models, contracts.
4. To define requisite levels of quality for practical evaluation and attainment of quality.

Standard System

The standard system is system of systems, which pervades all aspects of marketing and provides infrastructure for their operation and regulation of marketing activities. These standards are framed under Agricultural Produce (Grading and Marking) Act, 1937, by the central Govt. They are popularly known as AGMARK- STANDARD. These standards are implemented in accordance with grading and marking rules and instructions stipulated for the commodity. These specifications are revised from time to time and amended suitably, keeping in view qualitative changes, consumer preferences etc.

TABLE 32.3: Names and addresses of the National Authorities for inspection of the commercial quality of fresh fruit and vegetables

Country	Designated authority/contact
Austria	Federal Ministry of Agriculture, Forestry, Environment and Water Management, Stubenring 12 1012 Vienna Vienna
Belgium	Federal Agency for Safety of the Food Chain WTC III, Simon Bolivarlaan 30 1000 Brussels Belgium
Bulgaria	Ministry of Agriculture and Food Chief Expert Quality Control FFV 55, Hristo Botev Bulv 1040 Sofia Bulgaria
Canada	Canadian Food Inspection Agency 2nd Floor 159 Cleopatra Drive, Nepean, Ontario (Canada)
Cyprus	Ministry of Agriculture, Department of Agriculture, Louki Alrita, 1412 Nicosia, Cyprus
Denmark	Danish Plant Directorate, Skovbrynet 20, 2800 Lyngby, Denmark
Czech Republic	Ministry of Agriculture Food Production and Legislation Department Těšnov 17 11705 Prague Czech Republic
Finland	Finnish Customs/Customs Laboratory E-mail: kristiina.ala-fossi-aalto@tulli.fi
France	DGCCRF, 59 boulevard Vincent Auriol – télédoc 251, 75703 Paris Cedex 13, Bureau D4, France
Germany	Bundesanstalt fuer Landwirtschaft und Ernaehrung, Deichmanns Aue 29 53179 Bonn, Postal address: 53168 Bonn Germany

Country	Designated authority/contact
Israel	Plant Protection and Inspection Service (P.P.I.S), P.O. Box 78, Bet Dagan 50250, Israel
Italy	Bentivoglio, n. 41, 00165 ROMA
Kenya	Kenya Plant Health Inspectorate Service (KEPHIS), P.O. Box 49592 – 00100, GPO, Nairobi, Kenya
Luxembourg	Administration des Services techniques de l'Agriculture, Service de l'horticulture, 16, route d'Esch, L-1470 Luxembourg
Morocco	Food Exports Control and Coordination Organization 72, Angle Boulevard Mohammed Smiha et Rue Moulay Mohamed El Baâmrani Casablanca Morocco
Netherlands	Kwaliteits–Controle–Bureau P.O. Box 43133 2504 AC Den Haag Netherlands
New Zealand	New Zealand Food Safety Authority, E-mail: Kay.shapland@nzfsa.govt.nz_
Poland	Agricultural and Food Quality Inspection, 30 Wspolna St., 00-930 Warsaw, Poland
Slovak Republic	State Veterinary and Food Administration of the Slovak Republic, Botanicka 17842 13 Bratislava
South Africa	National Department of Agriculture, Directorate Food Safety and Quality Assurance, Private Bag X 343, Pretoria, 0001, Website: www.nda.agric.za
Spain	General Sub-Directorate of Inspection, Certification and Technical Assistance for Foreign Trade (Inspection Service SOIVRE) E-mail: tzapatero@mcx.es
Switzerland	Qualiservice GmbH, E-mail: Petra.sieghart@qualiservice.ch_
United Kingdom	Horticultural Marketing Inspectorate, Rural Payments Agency, Room SCF3, South Core, Produce Hall, Western International Market, Hayes Road, Southall, UB2 5XJ, United Kingdom, E-mail: hmi-tdc@rpa.gsi.gov.uk

INSPECTION AND SAMPLING

Inspection

Inspection of commodity means "the process of determining whether a commodity in a particular lot complies with the standard specifications applicable to it or any other specification stipulated in the trade contract".

Sample

Sample is a specimen or small quantity of something. Selected sample represents the whole lot. It may be inspected before drawing a representative sample in order to observe whether the lots offered for grading are in sound and in merchandisable condition. The horticultural product sample should be free from moulds, weevils, obnoxious smell and excessive pesticidal residues. In case any part of the consignment is found sub-standard, it should be segregated and sampled separately.

Grade and Grading of Samples

Grade means the lots having approximately the same quality characteristics and grading is the process of dividing a quantity of the same kind of commodity into different homogenous groups according to prescribed standards. The commercial grading is subjective by nature. Usually, it is done by inspection method. No technical expertise is needed however, adequate experience to judge the commercial quality and to allot the grade on the spot. For grading standard samples should be prepared in respect of the important commodities of that market area, which will help in comparing the quality of the produce offered for grading. All the details of grading should be recorded for examining next lot. Then the grades may be announced / allotted for sampling. The important norms of grading are:

1) Grading is done on the basis of standards, which are largely used and easily understood with low cost of grading operations.
2) Grading is based on measurable quality factors in which inter grade variations should be less.
3) Grading standards covers maximum percentage of the produce.
4) Grading norms should be based on maximum percentage of the market users.
5) Unless the standards are economically meaningful to the trade they are not accepted.
6) Grading is sorting of unlike lots o products into uniform categories according to quality standards of the same product.

Grading helps to get a higher price for their produce and it facilitates marketing. A farmer widens the market for the product. He can get orders even on the telephone without any inspection of the quality of the product from distant places. Grading also reduces the cost of marketing by minimizing the expenses on the physical inspection of the produce, minimizing storage lose, reducing its bulk, minimizing advertisement expenses and eliminating the cost of handling and weighing at every state. If his produce is graded, the farmer can get easy finance when commodities are stored; and can get the claims settled by the railways and insurance companies. The farmer can improve the quality of the stored products by removing the inferior goods from the good lot.

Grading helps to evaluate the standard and quality of products and to get fair prices. The farmer can assess market competition and pricing. After grading the product becomes homogenous, stand with perfect competition, encourages price competition among sellers, and reduces extraordinary profits. Thus, the grading of product is beneficial to all the sections of society, i.e. the producers, traders and consumers of the product.

Grading needs accuracy because the sample truly represents the entire lot. Though correct grading can not be done in short time. The general characteristics of the commodity should be preliminarily inspected before drawing samples from the lot. In the case of packed material or the packing material, stitching of the bags or metallic linings, the gross and net weight of the contents etc., should be invariably noted and verified.

Grading and Storage

1. Storage of graded produce is essential to assess the value and provide pledge loan and loan against the warehousing receipt.
2. It resolves quality disputes.
3. Storage is also done according to grade types and as per their requirements.
4. It helps in planning the plant protection measures.
5. Low-grade produce needs different care.
6. Grading helps to plan the blending and supply to different consuming segments and consumers can able to select their choice and farmers can able to plan the production matching with the demand.
7. E. commerce is possible in well defined quality of the produce at Global level.

Techniques of Sampling

Sampling may be done using Tube Sampler, Deep Bin Probe, Parkhi or Scoop,

Thermo-sampler and Sample Divider. Samples are of various kinds. They are as follows:

1. **Primary sample:** When sample is drawn from the heap or bag from a single position of the lot.
2. **Composite sample:** Primary samples drawn from the same lot shall be thoroughly mixed and blended to form homogeneous composite sample in a sample divider.
3. **Test sample**: One portion of composite sample weighting 500 gms and packed in cloth bag.
4. **Sample for Moisture**: Part of the composite sample weighing about 250 gms packed in polythene bag and heat-sealed kept in airtight container.
5. **Labelling of sample**: Gives all the details of the commodity like lot number, quantity, name etc. Label is affixed to the sample pack and sealed.
6. **Sampling from bags:** from bulk or heap depending upon lot size r = N/n, where N= No. Containers and n = Number of Containers to be sampled. Every bag should be sampled.

Grading Machines

In grading, sorting is done in unlike lots according to the quality specifications. Small scale farmers do grading by hands but the large producer done grading by electronic sorter for grading of fruits like.

For example, apples are graded into three categories or rejected, according to European standards, on the basis of their external quality. Besides colour and shape specifications, this involves the presence of defects, their origin and their number or size. The category *extra* should have no defects, category 'I' allows small skin defects while category 'II' permits fruits with more serious defects. Fruits, which do not reach the minimum requirements, are rejected and are traditionally used in processing. The automation of the grading could be achieved by colour machine vision. This would require at least three steps: image acquisition, its segmentation (to locate the fruit, the defects, the background colour vs the blush) and the classification of the fruit.

An automatic fruit sorting and grading machine has been developed by Partha Bhattacharjee, Coordinator - CSIO, Chandigarh and CEERI, Chennai. This automatic machine for fruit grading and sorting has been developed as per international standards. This development is unique in the sense that similar machine is working in the national market. Automatic fruit grading and sorting has not been implemented widely for all types of fruits. In the present scenario manual sorting is more popular. India being an agricultural country, exports a huge quantity of fruits abroad. But still in our country @40% produce is wasted for want of facilities for preservation and processing. The processing industry processes a very small portion of produce. Fruits have to be graded and sorted before being packed and sent to the international market for pricing. If done manually the process is slow and grading is done by visual inspection that could be error prone. Grading is done on the basis of various criteria like weight, shape, color, size etc. In this machine these factors are taken care by image processing and weight measurement through load cell. In this development the sorting and grading is done principally by photographing the entire surface of the fruit and subsequently by processing the image.

The main advantage of this machine is that it will facilitate automatic grading and sorting of apple in a non destructive method. The whole machine is cheap compared to other existing machines where robotic arms are used. It will also increase export of apple as it can process large volume in short time and does the quality assessment as per the requirement of export. In our country the entire produce of apple can not be taken to the storage place because of the lack of processing. But with this machine this can be achieved. It also facilitate the machine sorting of apple where human error is not introduced.

In Japan (Naoshi Kondo, Kyoto University, Kitashirakawa-Oiwakecho, Sakyo-ku, Kyoto 606-8502, Japan) automatic grading system has been developed using robot system, which automatically provides fruits from containers and inspects all sides of the fruits. The robot system consists of two fruit providing robots and a grading robot. Both robots had 3 DOF Cartesian coordinate manipulators, and 12 suction pads as end-effectors for transporting fruits. The grading robot had 12 color TV cameras, and 28 lighting devices as a machine vision system. The grading robot sucked 12 fruits up at a time and 12 bottom images of all the fruits were acquired during the manipulator moving from a halfway stage to carriers on a conveyor line. Before releasing the fruits to the carriers, 4 side images of each fruit were acquired by rotating the suction pads for 270°. The stroke of the manipulators was about 1.2 m and it took about 4.3 s to move back to the initial position that meant that this robot could grade three fruits per second. This grading robot system made following effects:

1. Labor substitution
2. Objective grading operation without human subjective judgment.
3. Data accumulation on fruit grading for traceability and farming guidance to producer.

Analysis

Every sample should be analyzed. The correct sample size should be taken for analysis. To analyse the sample for its various refraction and percentage by weight of these refractions is calculated to know the tolerance limit prescribed for various grades and finally the grade is assigned for that commodity. Sample for moisture assessment is taken separately and moisture percentage is determined by any of the available moisture meter, before assigning the grade.

Certification

The purchaser usually mention his specification of the horticultural produce, therefore the producer should obtain the certificate to show the purchaser. Certification involves the issue of, a certificate or a mark (or both) to demonstrate that a specific product meets a defined set of requirements for that product, usually a standard. The producer should obtain MARK from the certification authority. The mark carries a reference to the number or name of the relevant product standard against which the product has been certified. Before issuing the certificate, Certification Agency uses different evaluation techniques. These include testing of the product, in process quality control, surveillance visit to factory and testing the samples drawn from the open market.

Certificate for Export

Taking into consideration that European Commission (EC) has instructed its member countries to introduce checking operations with regard to import of fresh fruits and vegetables at their respective points of entry and, whereas the European Commission shall allow in the normal course entry of such commodities if checks have been carried out in the exporting country and, all relevant information is provided to them in this regard, Directorate General of Foreign Trade, Ministry of Commerce and Industry. All the fresh fruits and vegetables exported to European Unions may be inspected and certified by Directorate of Marketing & Inspection (DMI), Department of Agriculture and Cooperation, Ministry of Agriculture.

The inspection and certification is voluntary in nature. Ministry of Commerce and Industry set up a Standing Committee on Fresh Fruits and Vegetables under the Chairmanship of the Chairman, Agricultural and Processed Food Products Export Development Authority (APEDA) to look into formulation of export standards for fresh fruits and vegetables. The Standards could be notified under Agricultural Produce (Grading & Marking) Act, 1937.

Accordingly, standards of 18 important fruits and vegetables viz. Grapes, Litchi, Mangoes, Pomegranate, Guava, Pineapple, Banana, Papaya, Plums, Shelling peas, Sugar snap peas, Ribbed Celery, Headed cabbage, Spinach, Brussels Sprouts, Tomatoes, Garlic and Onion were drafted by DMI and were approved by the Standing Committee. These Standards are harmonized with EC and Codex Standards and are in the process of formal Notification in Gazette of India. Standards of other 7 fruits and vegetables viz. Pears, Cherry, Strawberry, Melons, Water melons, Beans and Cauliflower have been approved by the Standing Committee.

European Commission has designated Directorate of Marketing and Inspection (DMI) as an official authority and inspection body in India for checking conformity to marketing standards applicable to fresh fruits and vegetables prior to import into the community vide Commission Regulation (EC) No. 761/2003 of 30th April, 2003. Taking above facts into consideration, DMI is implementing Certification Scheme for export of fresh fruits and vegetables to EC countries. Interested persons can contact nearest office of the DMI for getting details of the Certification Scheme.

A product bearing a mark carries a third party guarantee that:

- The product has been produced according to specified standard
- Production has been supervised and controlled
- Product has been tested and inspected

Customers can approach certification agency, if the product does not meet the declared standards. In India, for agricultural commodities, Directorate of marketing & Inspection, Govt of India, is the designated certification body, which is empowered to issue the certificate to the quality of notified commodities graded under AGMARK. Generally, AGMARK standards are accepted internationally, however, with the advent of WTO, the international requirements for quality have become more specific (Codex Standard) and need to be incorporated in the domestic or national standards strictly, to match them with the standards prevailing in the international parlance.

For products in the regulated sector, one can contact the enquiry points (In India, it is Director (Quality system), Bureau of Indian Standards, Manak Bhavan, 9- Bahadur Shah Zafar Marg, New Delhi – 110002) setup in the country under WTO agreement on TBT and from these points, the needed information, technical regulation and conformity procedures may be obtained from its counterparts in the target market.

A supplier's declaration of conformity is a procedure by which a supplier gives written assurance that a product, process or service conforms to specified requirements. The details include:

- Name and address of supplier
- The identification of the product, process or service
- Conformity statement
- The reference normative document (standards, technical regulation etc)
- Date and place of issue of declaration
- The authorized signatory on behalf of the supplier.

PRICING HORTICULTURE PRODUCTS

Marketing of horticultural products has generally been free of direct government interventions. The Government has not been directly involved in the pricing or performance of physical functions of horticultural marketing. Its role has been minimal, mainly confined to regulatory and facilitative functions. The remarkable performance of the industry has been ascribed to this policy, which engendered autonomy in production and marketing decisions thus fostering significant local private initiatives and dynamism within the industry.

The local horticultural market is very open; hence prices are determined by supply and demand factors. The domestic market has concentrated on vegetables and fruits while cut flowers are mainly for the export market.

Pricing need not be an onerous task. How you price your horticulture products is worth some thought and effort since it directly affects your ability to make a profit. Horticulture products include bedding plants, fresh fruits and vegetables, fresh-cut flowers and culinary herbs. Take some time to research the following six management questions.

1. How much does it cost to grow my product?
2. What is my break even point?
3. What are my profit goals?
4. How will I market my product?
5. What price range do my competitors charge?
6. What is the customer demand for my product or service?

The Cost to Grow Product

It is important to keep good records for your business. By reworking the records you use to keep track of costs you can calculate complete product costs. A product cost is a subtotal of the money you spend to grow, package, distribute and promote your horticulture products. It's important that you include all costs in the product cost subtotal. Even the cost for unsold plants should be listed.

Product costs include both variable and fixed costs. Most of your expenses will fall into these three categories.

Materials costs: All material or supply costs used to make a product should be listed in the materials cost group. Materials expenses are variable costs. Each time you spend money to make a product you increase your variable costs. To identify variable costs, look to see which costs keep pace with the amount you produce. When the number of tomatoes grown in your greenhouse drops by half, your variable costs will also drop by half.

Example: A grower of specialty geranium bedding plants would include the costs for containers, planting medium, fertilizer, water, plant cuttings, boxes and the material shipping costs in the materials cost group.

Labor costs: Each time you pay people (including yourself) to seed, fertilize and harvest the crop, you should add the amount to labor costs. Include labor costs for machine operators and for the people who package or prepare bedding plants for shipping. Most labor costs are variable costs since they change with the number of units grown. The Agriculture Statistics Yearbook can provide you with average labor costs for some industries.

Overhead: This is a catch-all for the money you spend to operate the business. Common overhead expenses are land and equipment expenses, insurance, utility costs, depreciation and salaries for management staff. If you can't directly trace cost to labor or materials, it likely belongs in overhead. All overhead costs are fixed costs.

Fixed costs are a part of operating a business. They can't be tied to the number of units produced since they are the same whether 10 or 10,000 are produced. You have to pay the fixed costs whether or not you sell anything.

Product cost is an important first step in setting product price. It tells you the minimum your product can be priced at. You need to know your product price before you can calculate how many units you must sell to break-even. The break-even point is where there is neither a loss nor a profit from the sale of your products.

Profit Goals

Profit goals will help you make two other business decisions. Based on your goals you can set an accurate price and decide the amount of product you will grow. A goal is a clear statement about your future business plans. You need to consider personal goals, cash flow needs and product position to develop an accurate profit goal.

Profit goals are an important type of business goal. They state how much profit the business should earn. You can set your profit goals as a percentage above the product cost subtotal or set a total profit figure for the entire business. Use industry price ranges to help set the profit level most suitable for your crops or horticulture products. A profit goal will guide decisions on the amount of produce you will grow and the price you will charge.

IMPORTANT PRODUCTS OF HORTICULTURE INDUSTRY

1. Canned Fruits and Vegetables, Canned Fruits Pulps
2. Dehydrated Fruits and Vegetables
3. Dry Apricot
4. Fruit Bars, Sundried Mango Papad.
5. Fruit Jams and Jellies
6. Fruit Toffees
7. Fruit vinegar
8. Ginger and Garlic paste
9. Grape raisins/Kismis
10. Osmo-air dried fruits
11. Pappain Powder
12. Pickles and Chutneys
13. Potato/Banana/Topioca Chips & Wafers
14. Pre-packing of Fruits and Vegetables
15. Preserves/ Murabha, Petha
16. Processed Anardhana
17. RTS Beverages, fruit juices, etc.
18. Squashes and Lime juice cordials
19. Tamarind Products
20. Tomato Products
21. Tuti-Fruity
22. Vegetable sauces

CHAPTER 33

e-Marketing of Horticultural Produce

The "e" word has become increasingly evident on the lives of students in ways many could not have imagined less than ten years ago. With relative ease, the "e" is attached to activities like real estate, retailing, banking, entertainment and now education. The "e" stands for electronic and it relates to the use of the Internet to undertake the wide range of activities. As we become more familiar with the language of the Internet we find just how much it pervades our daily lives in the dot. com age. We readily recognise http://www......... as an Internet web site and see it plastered on vehicles, billboards, hot air balloons, merchandise and in the screen and print media. Businessman are now beginning to hear terms like e-marketing as it subtly becomes part of our regular vocabulary.

E-LEARNING

e-Learning is learning which takes place as a result of experiences and interaction in an Internet environment. It is not restricted to a regular school day and can take place in a variety of locations including home, school and community locations e.g. libraries, cafes etc. e - Learning refers to anything delivered, enabled, or mediated by electronic technology for the explicit purpose of learning. Basically, e-learning refers to use of information and communication Technology in education management. The letter 'e' in e-Learning stands for "electronic" version of learning. The use of e-learning in academic activities is of two kinds namely online learning and offline learning, which are based on individual student and group of student based learning.

The term includes:

1. Online learning
2. Virtual learning
3. Distributed learning
4. Web-based learning
5. Computer – based learning

Rosenberg (2001) described how e-learning *"refers to the use of Internet technologies to deliver a broad array of solutions that enhance knowledge and performance"*. To suggest that e-learning is about having something "delivered" to the farmer infers a less interactive environment and one where the farmer may well spend considerable time teaching themselves! *"e-learning is a form of distance learning" he also made it clear that what we know as "distance learning is not e-learning"*

Learner control, access, cost savings, flexibility in scheduling, speed in delivery, and no travel requirements are among the many benefits of e-learning. The mantra of many trainers throughout the past 20 years has been learner- centered instruction. Learner – centered instruction is vital to the success of adult training.

E-EDUCATION IN AGRICULTURE

As part of an e-education that is delivered via e-education to farming community and agricultural professionals in India, these approaches can help farmers to better manage agriculture in the face of current challenges and leveraging the agricultural production towards the upcoming food and economic need of the country.

In India, agriculture is one of the most important economic sector as well as the principle employer. Currently, Indian agriculture is under pressure from growing populations, increasing affluence, a dwindling natural resource base, and globalization. Indian farmers need to improve their economic performance, increase profits and production, and move towards sustainable agriculture. Previous advances in agriculture have come through improvements in traditional inputs, but future improvements will relay on intelligently used information to better manage agricultural production.

Knowledge-intensive management (KIM) can improve profits, production, and sustainability in the agricultural sector. Until human resource development and Information and Communication Technology (ICT) infrastructure is sufficient to deliver knowledge-intensive management skills to farmers, the information and skills can be delivered to agricultural professionals, who in turn can help farmers access the information and skills they need for better management. ICTs can be used to deliver knowledge-intensive management skills to agricultural professionals through e-education.

India's food production and productivity may be increased by an effective use of ICT for agricultural purposes. The developed nations are using laser technology instead of tractors to plough lands. This helps in optimising the use of various inputs such as water, seeds, fertilisers, etc. The problem is that Indian farmers cannot afford this technology and unless government comes in support for agricultural infrastructure, the same remains a dream only.

Further, power and electricity also remains a major problem for Indian farmers and alternative means of power like solar energy panels, regulated and optimised by ICT, can be a blessing for them. Thus, e-agriculture in India can put India on the higher pedestal of Green Resolution making India self-sufficient in the matters of food grains.

e-Education is a flexible form of education that allows participants with different schedules in different locations to interact and learn in a collaborative, online environment.

There is sufficient empirical evidence that investments in e-education yield positive results for agricultural growth throughout the world. Studies have demonstrated improved agricultural developments in:

1. Application and production of knowledge for the agricultural development
2. Ability of participants to manage learning and understanding of information
3. Access to agricultural information that increases knowledge, inquiry and depth of investigation.
4. Improvement of inventive thinking skills in several agricultural sector
5. Improvement of effective communication between farmers and agricultural professionals
6. Improvement of interpersonal skills between farmers (teamwork and collaboration)
7. Improvement of productivity skills for high yield agricultural crops
8. Creation of a pool of globally competitive agricultural professionals

There are several approaches developed for distributing the agricultural information to farming community and agricultural professionals.

1. Setting up of Agricultural Internet Booth in all villages
2. Supply of printed reports based on the queries of farmers
3. Setting up of Kiosk machines in rural centres
4. Online training with multimedia support (e-learning)
5. Internet/Intranet communication services
6. Compact Disc based training materials
7. e-Content development of agricultural information
8. Regional Website based information dissemination
9. Simulation based agricultural practices for farmers demonstration
10. Management Information System
11. Decision Support System
12. Video Conferencing
13. Integrating agri market through internet
14. Conducting awareness programmes about e-education
15. Encouraging NGOs to work together with Government agencies to bridge the gap between farmers literacy and ICT advancement.

The following are the focused areas of agriculture where in which e-education can be effectively implemented:

1. Soil and crop fertility
2. Hydro-ecology
3. Crop – weather dynamics
4. Energy budgeting
5. Bio-pest and disease management
6. Crop diversification
7. Climate change and its impact
8. Post harvesting
9. Agribusiness markets
10. Panchayat Raj – A catalyst for sustainable agriculture development.

E-CONTENT DEVELOPMENT THROUGH WEB PAGES

Knowledge-intensive management through web pages can improve profits, production, and sustainability in the agricultural sector. ICTs through web pages can be used to market agricultural products.

e-Content (Electronic Content) is a digital content that can be transmitted over a computer network such as the Internet. This is a continuing need to develop relevant, useful and authentic content. As more Indians use the Internet and as worldwide more people use the Internet to know about India, There is a growing need for the development of a critical mass of localized content – both in English and Indian Languages.

The definition of 'content' is very wide and include all categories of popular content – not just software and news, but also games, entertainment, sports, education, agriculture, programming, directories, classified advertising, science and technology and online communities denoted to major interests. Content development and management lies at the base of web, Intranet and CD ROM publishing; it also underlies e-commerce and e -learning – two growing trends, worldwide. e-Content that is facilitated by the use of digital tools and content. Typically, it involves some form of interactivity. e-Content provides the core of the business and competitive advantage lies in the immediacy of access, exchange and providing a meaningful flow of real-time to the people who make decisions and the people who execute them.

The primary sources of e-content development are, CD ROM, Audio, Video, Animations, Web Sites, Blogs, On-line Forums, e-Mails, Chat Rooms, News Groups, SMS (Short Message Service), MMS (Multimedia Message Service), Internal Documents, Public Opinions, Television, Radio, Mobile Service, Questionnaires and Interviews, Books, Journals and Magazines.

e-Content is cost effective and time saving approach. The content can be obtained by the individual according to their convenience and digested slowly. Infinite number of users can be targeted by an individual. In e-Content, photographs and video clips of horticultural produce are inserted wherever necessary. The individuals are able to retrieve all the information necessary from one place.

A typical e-content development web site may consists of the following sequence of processes right from web page design to host the page in a Web server. The processes are as follows:

1. Web Development and Design
2. Graphic Design
3. Web Site Optimization
4. Content Resources
5. Domain Names
6. Web Hosting
7. System Maintenance

Web Development and Design Tools - To Create Web Site

1. Hyper Text Markup Language (HTML)
2. Cascading Style Sheet (CSS)
3. Dynamic Hyper Text Markup Language (DHTML)
4. Adobe Dream Weaver
5. Microsoft Front Page
6. Web Browsers e.g. Internet Explorers and Mozilla Firefox (for reviewing)

Anyone who can create a well-formatted document in a word processor can create a website. Web development and design is a little more complicated, but the essentials steps are the same. With a little patience and perseverance, web skills can be nurtured and matured.

Graphic Design Tools – Digital Image Resources for Web Design

- Adobe Photoshop
- Adobe Flash
- Adobe Page Maker
- Coral Draw
- Sound Forge

The tools listed in this category will help beginners create and manage a wide variety of imaging tasks: editing photos, creating web images, finding the right colors for a website, perform screen captures, and even a recommended audio/video tool.

Web Site Optimization

Web Page Analyzer

Website optimization as the process by which a website is streamlined to become more efficient for three audiences.

- The website developer,
- The website reader,
- The search engine or to put it in question form.
- Does the website work?
- Is the website usable?
- Can the website be found?

Content Ressources

- Microsoft Word
- Printable Document Format
- Images Files (jpeg, tiff, bmp, gif, psd, cdr)
- Audio Files (wav, mp3, mp4, avi)

Any website that disseminates information is an e-content website. Running e-content websites requires a set of tools and resources in order to make it successful. While the principles of basic website development and design still pertain, e-content websites needs to concern themselves with moving visitors from landing page to indepth knowledge page.

Domain Names

Domain Hosting Providers (Free and Paid Services): The domain name resources in this section will help users to manage all their domain name assets. It is important to consider domain names as assets. They have value. They can be considered an investment of sorts. They are virtual real estate. Everyone from beginners to professionals need to take on this understanding.

Web Hosting – Uploading the Web Pages in Web Server: There are essentially six items to look at when deciding on a cheap web hosting plan in a Web server:

- Reliability,
- Price,
- Bandwidth,
- Disk Space,
- Services,
- Hosting Location.

System Maintenance

System maintenance in this context emphasis the maintenance of both Web site and PC maintenance to make the web information live. Life on the web requires a certain software toolset. While most of these tools are not web development related, they are necessary for anyone who wants to maintain a safe and efficient web experience. This utilities section provides recommendations for software that fits in one of two categories.

- Web Utilities
- PC Utilities
- Anti-virus tools
- Anti-spyware

E-COMMERCE (MARKETING)

ICT helps the farmers with the latest information on the rates, arrivals etc. in the neighbouring state / national Mandis; ensuring prompt calculations of arrivals and actual sales; providing traders with information on various taxes payable by them on the actual purchases made by them; reconciliation of bank accounts; and issue and reconciliation of transit permits issued etc. This has not only improved the effectiveness of trading in the Mandis but has also benefited the farmers, traders and the government in collection, processing, analyzing and dissemination of information, policy making and revenue collections as part of good Governance.

Most people think of e-commerce as buying goods over the Internet. Certainly, any search engine, if asked to look for e-commerce sites, will find thousands of relevant websites. Even back in May 2000, one search engine found more than a million web pages containing the word e-commerce. The use of WAP (Wireless Application Protocol) standard, a global specification for providing information content to wireless devices, including mobile phones, TV sets, etc., will soon be commonplace. It is expected that by 2005, mobile Internet access will communicate at speeds up to 200 times faster than today. Emerging technologies have redefined business by erasing traditional boundaries of time and geography. They have created new virtual communities of customers, distributors and suppliers, linked by a demand for products and services.

India is one of the top information technology markets in the world. The Education Ministry has integrated IT into the teaching process to promote IT skills among the population. In response to the surge in Internet use, TV cable, which offers high-speed and high-quality data transfer.

Developed countries all over the world have drawn up measures to meet the rapid developments in information technology and e-commerce in recent years, as they jockey for leadership in world markets. According to the analysis in the OECD's book "*Electronic Commerce-Opportunities and Challenge for the Government*", the government should play a role in the promotion of e-commerce. The major roles of the government can be described as follows:

1. Provide an environment conducive to the development of e-commerce
2. Governments must quickly open up e-services, including information, telecommunications, and electronic financial exchanges, and should encourage the use of these services.
3. Promote the standardization of solutions dealing with electronic trade, to prevent copyright from becoming a hindrance to new businesses or an impediment to e-commerce.
4. E-commerce is being widely used for the sale of consumer products. Governments need to set up a comprehensive transportation infrastructure to support this. This will prevent inefficient transportation or distribution systems from delaying the rate of e-commerce development.
5. Governments need to announce what action they intend to take in regard to e-commerce, so that industry and government can have a common understanding and consensus in regards to the objectives of this development.
6. Increase the visibility of e-commerce, and promote electronic interaction between industry and government agencies
7. Government agencies responsible for activities related to e-commerce should be used to stimulate all potential industries to use e-commerce.

Therefore, it is essential to take the following measures:

1. To construct a demonstrative e-business system from selected farm marts, outlets, associations or organizations.
2. To assist farm-product organizations in realizing e-business operations.
3. To propagate e-business concepts and cultivate needed personnel.
4. To help the industry to frame the product specifications and data standards required in the electronic trading.
5. To improve the foundation-constructing environment for e-agriculture.
6. Provide consultation and guidance by means of technical service teams and the Internet.
7. Enhance the operations of electronic supply chains, to expand business by effectively integrating market demands.
8. Help farmers and farmers' organizations to apply e-commerce or information technology.

In general, the development of e-agriculture will take advantage of existing technical resources to establish mutually beneficial strategic alliances. The goal is to create an e-agriculture environment for modern farmers, so they can integrate and become highly competitive suppliers of farm products on the market.

The farmers should contact Internet companies actively engaged in business-to-business e-commerce through a network of partner companies and farmers' organizations in India. The primary goal is to increase efficiency and reduce costs.

Types of Electronic trading or e-Commerce

MARKNET

The Maharashtra state agricultural marketing board held initiated use of IT in agricultural marketing. MARKNET (agricultural marketing intelligence network in Maharashtra) is one of the leading projects under which APMC have been computerized and connected through internet for information exchange. MSAMB has launched www.MSAMB.com website on this website comprehensive information of MSAMB is incorporated. All information is available in Marathi and English version.

AGMARKENET

Under this programmed, computer sets been made available free of cost to 164 APMCs similarly; projection TV system of MARKMET information display is being implemented. Presently 47 APMCs have installed projection TV system.

Digital Mandi

It is a unique e-trading platform having special features that enable the farmers to reap the most out of the opportunity price of major mandies are transparently provided on computer screen, giving the farmer the choice to sell his stock in more profitable market.

e-Choupal

Indian tobacco company (ITC) pioneering agribusiness initiative e-choupal is unique in many ways. The e-choupal is a complete end solution as it delivers real time information and customized knowledge to improve farmers; decision making ability to align farm output with a market demands, and to improve productivity; aggregates demand like a virtual producer' cooperative and provides access to high quality farm inputs at lower cost: acts as direct marketing channels with more efficient price discovery and lower transaction costs in output marketing: is scalable as it is built on market principles is replicable across different crops and geography in India because of its conceptual strength and integrates farmers to the market place and simultaneously, meets the needs and challenges at the bottom of economic pyramid.

Contract Farming

Contract farming, is a type of farming wherein the industry or perspective buyer enters into a contract with the farmer and promises to buy the farmer's produce at a pre-negotiated price under pre-negotiated conditions. Besides, the buyer agrees to supply the required farm inputs at the required time. In this process farmers are assured of an established market and a fixed price for their produce. The buyers would be able to procure the produce of a specified quality at much cheaper rate.

The contract could be of three types:

1. **Procurement contracts** under which only sale and purchase conditions are specified,
2. **Partial contract** where in only some of the fruits are supplied by the contracting firm and produce is brought at pre determined price.
3. **Total contract** under which the contracting firm supplies and manages all the inputs and the farmers becomes just a supplier of basic production facility and labour.

The farmers have realized higher profits in a short period of two months period under contract. Majority of the farmers involved were small and marginal farmers. The income derived from this short duration crop has facilitated the farmers to take up one or more crops after harvesting of baby corn.

OPPORTUNITIES AND CHALLENGES

Globalization is making the market very competitive. Farmers have to face this challenge by reorienting themselves, so that what they grow becomes more marketable. We believe that marketing and IT are very important components for the modernization of agriculture.

The government's role is to make sure that producers and buyers meet in a market-place. The government needs to facilitate and create market encounters and investment missions where the different producers and buyers can meet. However, this is easier said than done. With e-commerce, we can explore the possibility of creating a virtual market place.

The future is bright for e-commerce in agricultural marketing. However, there are three main issues that need to be addressed.

Telecommunications in the countryside are not as developed as in the cities. Since farmers live in the countryside, giving them access to the Internet is a major concern. Improving the infrastructure so that they have fast, reliable Internet access will be a big advance in the use of e-commerce.

Compared to the service workers or industrial workers, farmers have a low literacy rate. Relatively few farmers have computer skills. Making farmers computer literate will be a challenge.

Ensuring that the information on a marketing website is timely and relevant. Maintaining a marketing website is a lot of work. It requires timely information, so that farmers can take immediate advantage of opportunities as they arise.

India/s farming is small-scale. If India does not apply IT to agricultural marketing, to make it more efficient in today's global environment, many thousands of farmers in India will fall behind and eventually lose to the competition.

When the government is drawing up its promotion program for IT in agriculture, it should study global opportunities, as well as relevant IT and telecommunications applications and processes. It should promote India's resources, and help growers identify niche markets.

Wholesaling is a very important step in the process of distributing agricultural products. In a developing country, farmers and suppliers are numerous but operate on a small scale, as do jobbers and retailer. Food processing is not yet well established. Hence, major food items, whether for domestic consumption or for export, remain mostly in fresh form or have only very basic processing.

However, fresh produce such as vegetables and fruits decay within days unless they are refrigerated. They have to be sold as soon as possible, or else farmers suffer great losses. In this regard, a dual system, including both the traditional channels (producers → assembly points → wholesale markets → jobbers → retailers) and an e-commerce marketing system may complement each other, to form together a more effective marketing system.

CHAPTER 34

Export of Horticultural Crops

India exported horticultural produce and product worth Rs. 90315.20 million during 1998-99. Tea, fetched the highest exchange of Rs. 21918 million followed by spices, coffee and cashew. During this period, fresh fruits and vegetables earned foreign exchange of Rs. 5360.20 and 7056.80 million, respectively. With the opening of global market and removal of quantitative restrictions under the WTO agreement, export-import scenario is likely to change at much faster pace. Horticulture stands as the major beneficiary in such a scenario to contribute to the much needed for ex-reserves. So far, Indian horticulture has been insulated from forces of outside world. Time bound removal of quantitative curbs on imports and other barriers to access to domestic market under WTO, of which India is a signatory, will require Indian horticultural produce/product to be competitive both in the domestic and export market. This would call for use of hi-tech horticultural technologies. Such technologies can be defined as '*those, which are modern, less-environment dependent, capital intensive and have the capacity to improve productivity and quality*'. These will include, use of genetically modified crop varieties, micro-propagation, integrated nutrient and water management, integrated pest management, protected cultivation, organic farming, use of modern immuno-diagnostic techniques for quick detection of viral diseases and hi-tech post-harvest technologies, including cold chain.

Horticultural crops provide a better alternative for diversification of Indian agriculture in view of higher returns available from them. Horticulture sector helps in improving productivity of land, generating employment, improving economic conditions of the farmers and entrepreneurs, enhancing exports and foreign exchange earnings and above all providing nutritional security to the people.

Horticultural products have a high income elasticity of demand. As income goes up, demand rises rapidly, especially in the middle-and high-income groups in developing and developed countries. In developed countries, the growing concern for health and nutrition has caused consumer preferences to shift from high-fat, high-cholesterol foods, such as meat and livestock products, to low-fat, low-cholesterol foods, such as fish, fruits and vegetables. Also, there is an increasing tendency in developed countries to diversify the diet by consuming a wide variety of fruits and vegetables, a change partly stimulated by the increase in international travel and communications. This, in turn, has led to and is facilitated by increasing Imports of new and non-traditional horticultural products, especially from the tropical developing countries.

Conditions for increasing production of horticultural crops are very favourable in the country. This is partly because production of horticultural crops in general is labour-intensive. India endowed with abundant labour in relation to capital has competitive advantage in production and exports. Horticultural products not only have good potential for generating employment in cultivation but also in processing, marketing, and distribution. And they are frequently produced on small farms, thus providing an important source of additional income for poor farmers in developing countries. There are many horticultural products, especially fruits, that fetch high prices in world trade.

WORLD EXPORTS FROM INDIA

Fruits

World exports of fresh fruits are on the increase. There are complementarities in the supply and marketing of fresh and processed fruits. With modern packaging and fruit treatment technologies there is considerable value added attached to the exportation of fresh fruits.

In grapes, India attained the highest yield level in the world at 20,548 Kg. per hectare. India ranked seventeenth in production and her share in global production was only 1.33% in 1994.

Tropical or semi-tropical fruits play a substantial role in international trade. Bananas and oranges are very important from export point of view. The only other tropical fruit of considerable significance from export point of view is pineapple in fresh form. Exports of apples and grapes are also increasing but on a lower growth rate.

Tropical fruit other than bananas and citrus that enter international trade whose value is significant is the pineapple. The volume of fresh pineapple imports is over 4 million tonnes, a sizeable portion of which is used for processing purposes.

There is a wide range of tropical fruits of a more exotic nature which has attracted increasing attention in the past few years, the market for exotic fruits, in fresh or processed form, is as yet limited but the potential is there. Among the most promising ones are mangoes, guavas, papayas, and pasion fruits. But there are scores of non exotic fruits which could qualify as well, although with more limited applications and among them, mention may be made of tamarinds, sour sops, cashew apples, acerola cherries, jack fruits, anonas, as well as the lychees, longans, and mangosteens of Far Eastern reputation.

Developing countries are expected to have a comparative advantage in tropical fruits and, therefore, to perform well in export markets. But tropical fruits constituted a small per cent of total world trade in fresh fruits. This indicates that the demand for tropical fruits in industrialized countries has been limited. However, familiarity with tropical fruits is growing in developed countries as the result of, first, an increase in international travel; second, an increase in immigrants from tropical developing countries, who have created a demand for tropical foods; and third, rising affluence in developed countries, which has fostered a desire for diversification of food consumption patterns and thus a trend towards the more exotic fruits.

Vegetables

Of the various vegetables that enter into world exports in fresh form, three are important namely tomatoes, potatoes and onions.

India's share in world exports of potatoes and tomatoes is insignificant. In the case of onions, however, India has a significant share of over six per cent.

Among fresh vegetables the miscellaneous vegetable categories had the largest share in the value of world exports - 75 per cent -followed by roots and tubers at 20 per cent. More than 90 per cent of roots and tubers were potatoes. At 28 per cent, tomatoes were the most important single item in the miscellaneous vegetables category. Another 31 per cent was distributed as follows:

Lettuce (8 per cent), cucumbers (8 per cent), chillies and pepper (6 per cent), cabbages (3 per cent), cauliflowers (3 per cent) and carrots (3 per cent).

In potatoes, yield level's in India at 15,000 kg. per hectare is good and equals that of world average. In U.S.A., yield rate was the highest at 37,394 Kg. per hectare in 1994. India ranked sixth in world production and her share in world production was only 5.7% in 1994.

In tomatoes, yield per hectare is very low compared to any major producing country. Yield in U.S.A. was 63661 kg. per hectare.

India is the largest producer of cauliflower in the world. Yield at 17778 Kg per hectare compares well with the world average of 17956 Kg. per hectare. In China yield was 25709 kg. per hectare in 1994.

In India, yield rate of onion was low at 11026 kg. per hectare. In China it was 17407 Kg., in U.S.A. 44142 Kg., in Turkey 20000 kg. and in Iran 36607 kg. per hectare. Yields in peas and cabbages are very low in India compared with any major producing country.

Thus the efficient production of high-value horticultural products on small farms can help alleviate rural poverty. The growing interest in horticultural exports in the country also reflects the search for diversified non-traditional agricultural exports in order to expand foreign exchange earnings so that the country can meet rising import requirements for accelerated economic growth and mounting debt service payments.

TABLE 34.1: Three Year Export Statement of APEDA Products (floriculture, fruits and vegetables)

Qty in MT, Value in Lakhs

Product	2006-07		2007-08		2008-09	
	Qty	Value	Qty	Value	Qty	Value
Floriculture	42545.28	65269.73	36240.71	34014.42	30798.34	36881.41
Fruit and vegetable seeds	8104.09	12158.95	10157.13	14212.29	8535.53	11999.09
Total	**50649.37**	**77428.68**	**46397.84**	**48226.71**	**39333.87**	**48880.5**

TABLE 34.2: Three Year Export Statement of APEDA Products (Fresh fruits & vegetables)

Qty in MT, Value in Lakhs

Products	2008-2007		2007-2008		2008-2009	
	Qty	Value	Qty	Value	Qty	Value
Fresh onions	1378373.17	116330.57	1008606.48	103577.89	1670186.29	182752.21
Other fresh vegetables	276824.60	43314.38	350235.47	48949.01	505285.47	68020.32
Dried nuts (walnuts)	5062.86	11803.79	6716.48	16207.80	5696.34	14123.63
Fresh mangoes	79060.88	14193.95	54350.80	12741.76	83703.18	17071.25
Fresh grapes	85897.79	30192.45	96963.57	31782.51	124627.97	40861.28
Other fresh fruits	177638.30	30997.39	207700.78	30452.60	256768.53	43086.84
Total	2002857.6	246832.53	1724573.58	243711.57	2646267.78	365915.53

TABLE 34.3: Three Year Export Statement of APEDA Products (Processed fruits and vegetables)

Qty in MT, Value in Lakhs

Products	2008-2007		2007-2008		2008-2009	
	Qty	Value	Qty	Value	Qty	Value
Dried and preserved vegetables	119270.43	42754.17	125726.28	42993.81	147861.21	49641.51
Mango pulp	156835.51	50582.79	166752.17	50968.51	173013.60	75298.90
Other processed fruits and vegetables	318067.57	95550.82	311756.29	96281.65	387126.42	137179.00
Pulses	255084.47	78999.40	170614.39	54900.85	136880.08	54232.50
Total	849257.98	267887.18	774849.13	245144.82	844881.31	316351.91

GOVERNMENT OF INDIA'S EFFORTS

There is a vast potential for growing a larger variety of horticultural crops like fruits, vegetables, flowers, spices, cashewnut, coconut, areca nut, root and tuber crops, medicinal and aromatic plants etc. The diversity of physiographic, climatic and soil characteristics make it possible to grow horticultural crops almost throughout the year.

Major Development Programmes in Central Sector

Fruits

The Central Sector Scheme on Integrated Development of Tropical, Arid and Temperate Zone Fruits was launched during VI Five Year Plan with an outlay of Rs. 74 crore. The Programmes envisaged under this scheme include:

1. Establishment of big nurseries under public sector.
2. Establishment of small district level nurseries under private sector.
3. Setting up of tissue culture units under public and private sectors.
4. Training of farmers, (e) establishment of oil extraction plants.
5. Area expansion; (g) improving productivity; and (g) demonstration plots.

Between 1992 and 1995, 13 big nurseries, 188 small nurseries, 21 tissue culture units were established. 758 demonstration plots have been established and an area of 11,747 hectare under area expansion and 330,321 hectare under rejuvenation respectively were covered.

Vegetables

The production of vegetables, including potato, during 1993-94 was 650.90 lakh tonnes, covering an area of about 48.20 lakh ha. The production in the country is expected to increase significantly due to popularisation of hybrid seeds of various vegetables.

The Central Sector Scheme on Production and Supply of Vegetable Seeds is being implemented with VIII Plan outlay of Rs. 15 crore. The Scheme envisages distribution of vegetable seed mini-kits including seeds of hybrid varieties, production of foundation seed and setting up of hybrid seed production units in private sector.

Root and Tuber Crops

Root and tuber crops including potato have been identified as most important group of staple food in the tropical countries of the world. The production of potato alone during 1993-94 was 180.03 lakh tonnes. For promoting cultivation of root and tuber crops, a Central Sector Scheme is being implemented with VIII Plan outlay of Rs. 2.5 crore. The Scheme envisages setting up of Main/Sub Bio-centres, distribution of tuber seed mini-kits and laying out of demonstration plots in the States.

Mushroom

The Central Sector Scheme on Mushroom cultivation standard in VIII Five Year Plan with an outlay of Rs. 10 crore, aims primarily at creating the infrastructure and for spawn and pasteurised compost production and providing training to farmers for mushroom cultivation. Work is under progress for establishment of a total of 20 units each of spawn production and pastuerised compost and training of 800 farmers in 21 States/UTs.

CHAPTER 35

Hi-tech Horticulture and Intellectual Property Rights

The simple way to explain intellectual property is to understand that all great expressions and products have an origin. A work of music, art, writing, film, or computer software belongs to its creators. Your creative expressions and inventions have a lot of value and must be protected however; your ideas are not considered intellectual property. Anyone who tries to copy or "borrow" your work without permission is profiting from your hard work – your intellectual property – without any credit to you.

Intellectual property is divided into two groupings: Industrial property (patents, trademarks, industrial designs, and geographic indications of source) and Copyright (literary and artistic works) / Rights related to copyright (performing artists, producers, and broadcasters).

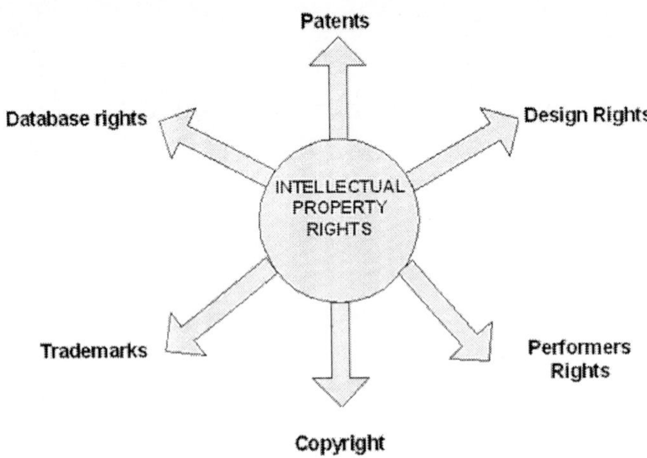

FIG. 35.1. Various parameters of Intellectual property rights

The adoption of IPRs (Intellectual property rights) in agriculture has a recent origin, even in the developed world. Industrial property shall be included within the broadest sense and shall apply not only to industry and commerce proper, but likewise to agricultural and extractive industries and to all manufactured or natural products, for example, wines, grain, tobacco leaf, fruit, cattle, minerals, mineral waters, beer, flowers and flour.

The first significant application of intellectual property to agriculture occurred with the evolution on the initiative of associations of horticulturalists and plant breeders of the UPOV convention for the protection for plant varieties. The traditional practice of farmers in replanting, exchange and sale of seed from the previous harvests impacted upon the capacity of breeders in recouping investments through repeat sales of improved varieties. Consequently, the UPOV convention was amended to permit limitations to the extent of seed saving. These limitations have had an impact on food security in circumscribing the availability of saved seed to farming communities.

In any event, the development of the modern biotechnology, which has led to the patenting of the genetic composition of new varieties, provides on the threat of a patent infringement action an effective bar to the capacity of farmers to save seed.

Intellectual Property Rights

Intellectual property rights are a legal concept that confers rights to owners and creators of the work, for their intellectual creativity. Such rights can be granted for areas related to literature, music, invention, variety, methods, plants etc, which are used in the business practices. In general, the intellectual property law offers exclusionary rights to the creator or inventor against any misappropriation or use of work without his/her prior knowledge. Intellectual property law establishes equilibrium by granting rights for limited duration of time. Every nation has framed their own intellectual property laws. But on international level the World Intellectual Property Organization (WIPO) governs it. The Paris Convention for the Protection of Industrial Property in 1883 and the 'Berne Convention for the Protection of Literary and Artistic Works' in 1886 were first conventions, which have recognized the importance of safeguarding intellectual property. Both the treaties are under the direct administration of the WIPO. The WIPO convention lays down following list of the activities or work which are covered by the intellectual property rights -

- Industrial designs
- Scientific discoveries
- Protection against unfair competition
- Literary, artistic and scientific works
- Inventions in all fields of human endeavour
- Performances of performing artists, phonograms and broadcasts
- Trademarks, service marks and commercial names and designations
- All other rights resulting from intellectual activity in the industrial, scientific, literary or artistic fields.

HISTORY OF IP RIGHTS RELATED TO FOOD AND AGRICULTURE

Intellectual property rights impact food security in three principal areas: plant variety rights protection, patenting and geographical indications.

Plant Variety Rights Protection

Intellectual property rights play key role in the inclusion of agricultural innovations.

The Congress pomologique de France, held in 1911, had called for special protection for plant varieties. The International Union of the Horticultural Profession also considered the matter at its Congresses in Luxemburg (1911), London (1912) and Ghent (1913). The International Institute of Agriculture in its 1927 congress had stated that the protection of a denomination was insufficient and that a way had to be found to require "any grower who engaged in reproduction of those breeds for the purposes of sale to pay a royalty to the producer". A meeting was held in Amsterdam on 17 November 1938, of the International Association of Plant Breeders for the Protection of Plant Varieties (ASSINSEL). The first ASSINSEL congress, held in Paris on July 8 and 9, 1939 adopted a three-point resolution:

1. To accept internationally the filing of trademarks and appellations as a means of protection (pending introduction of a patent);
2. To adopt the principle of a licence, to be drawn up by ASSINSEL for the purposes of multiplication and sale; and
3. To accept internationally the definition of the word 'original' (as) seed produced, offered or sold by the breeder of the variety or under his control by his licensees or successors in title.

The Second World War interrupted these developments. At its semmering congress in June 1956 a resolution of ASSINSEL called for an international conference to promulgate an international system for the protection of plant varieties. The French government had been approached by ASSINSEL, because it had indicated a favourable attitude. On 22 February 1957, the French government issued invitations to 12 Western European countries to attend a diplomatic conference in Paris conference from May 7 to 11, 1957 to consider establishing an international regime for the protection of plant varieties.

Participation was limited by the French to those states who were known to have similar concerns to it on this subject. The conclusions of the 1957 conference were set down in its Final Act, adopted on May 11, 1957. This recognised the legitimacy of breeders' rights and established as the preconditions for protection that a variety had to be distinct from preexisting varieties and sufficiently homogenous and stable in its essential characteristics. It defined the rights of the breeder and acknowledged the principle of the independence of protection. Following three meetings of the Drafting Committee and two meetings of Committees of Experts, the second session of the Conference was held in Paris from 21, November to 2, December 1961. An International Convention for the Protection of New Varieties of Plants (UPOV) was presented for the Consideration of the Conference.

Definitions

Therefore, most plant variety protection (PVP) laws are modeled on the UPOV convention. Generally, under this legislation the plant breeder is conferred an exclusive right to do or to license the following acts in relation to propagating material of the variety:

1. Produce or reproduce the material
2. Condition the material for the purpose of propagation
3. Offer the material for sale
4. Sell the material
5. Import the material
6. Export the material
7. Stock the material for the purposes described above.

Under the 1978 and 1991 versions of UPOV, seed saved by a farmer from harvested material and treated for the purpose of sowing a crop on that farmer's own land may be permitted under legislation as a non-infringing use. In some countries plant variety rights might be exempted from infringement when propagating material is used as a food, food ingredient or fuel, or for any other purpose not leading to or involving the production or reproduction of propagating material. The general duration of PBR under the UPOV Convention is 25 years in the case of trees and vines and 20 years for any other variety.

A plant variety is considered to be registrable, if it has a breeder, is distinct, uniform, stable and has not been or has only recently been exploited. A plant variety is considered distinct if it is clearly distinguishable from any other variety whose existence is a matter of common knowledge.

It is uniform if, subject to the variation, which may be expected from the particular features of its propagation, it is uniform in its relevant characteristics on propagation. A plant variety is stable if its relevant characteristics remain unchanged after repeated propagation. A plant variety is taken not to have been exploited if it or propagating material has not been sold to another person by or with the consent of the breeder.

SALIENT FEATURES OF PROTECTION OF PLANT VARIETIES AND FARMERS' RIGHT ACT

Protection of Plant Varieties and Farmers' Rights Authority

This authority will have a chairperson and 15 members including eight ex-officio representatives from various departments and institutions of the government of India and seven others representing farmer's associations, tribal organizations, industry, agricultural universities, national-level women's organizations and state governments. The registrar-general of the Plant Varieties Registry will be its member's secretary. The authority will be assisted by committees and officers such as the registrars.

Function of the Authority

- The authority will promote the development of new varieties and protect the rights of farmers and breeders.
- In particular it will provide for
    - Registration of extent and new plant varieties and essentially derived varieties subject to the relevant terms and conditions.
    - Characterization and documentation of varieties registered under the Act.
    - Documentation, indexing and cataloguing of farmer's varieties.
    - Compulsory cataloguing facilities for all plant varieties.
    - Ensuring that seed of varieties registered under this Act are available to farmers, and providing for compulsory licensing of such varieties if the breeder dose not arrange for production and sale of seeds in the manner prescribed.
    - Collecting data and information on plant varieties, including the contribution of any person at any time to their evolution or development in India or abroad for compilation and publication.
    - Ensuring the maintenance of the National Register of Plant Varieties.

Power of the Authority

- The Plant Varieties and Farmer's Rights Protection Authority will have all the powers of civil court of receive, administer oaths, enforce the attendance of witnesses, compel the discovery and production of documents and issue commissions for the examination of witnesses.
- Its orders would be executable as the decrees of a civil court.

Registration of Varieties

A National Register of Plant Varieties will be kept at the head of office and branch offices of the plant varieties registry in which the particulars of all varieties will be registered. The register will maintain under the direction of central government. It will be kept under the control and management of the Plant Varieties and Farmers Rights Protection Authority.

Criteria for Registration

A variety will be registered if it fulfils the criteria of novelty, distinctiveness, uniformity and stability.

Persons Who can Apply

An application for registration can made by:

1. The breeder of a variety;
2. A successor of the breeder
3. An assignee of the breeder
4. A farmer or a group or community of farmers claiming to be breeders of a variety;
5. A person authorized by any of those specified above to make an application on their behalf;
6. A university or publicly funded agricultural institution claiming to be the breeder of a variety.

Duration of Registration

Registration of a variety will be valid for a period of

1. Nine years for trees and vines
2. Six years for other crops.

This can be renewed further for the remaining period of

1. 18 years in respect of trees and vines;
2. In the case of extent varieties, 15 years from the date of notification of the variety under section 5 of the seeds Act, 1966;
3. In other cases, 15 years from the date of registration. It would be mandatory for the breeder to deposit the seeds or propagating material in the National Seed Bank.

Rights of Breeder or his Successor

Registration of a variety confers an exclusive right on the breeder or his successor, agent or licensee to

- Produce, sell, market, distribute, import or export the variety
- A breeder may also authorize any person to produce, sell, market, or otherwise deal with the registered variety subject to the relevant conditions.

Exclusion of some Varieties

Registration will not be granted to some varieties if it necessary to prevent their commercial exploitation in order to protect order to protect public order or public morality or human, animal and plant life and health or to avoid serious prejudice to the environment.

Researchers' Rights

For the purpose of research, any person can use a registered variety as an initial source in an experiment to create other varieties. For further repeated use of the variety as a parental line, the breeder's permission would be required.

Farmer's Rights

1. A farmer who has bred or developed a new variety would be entitled to registration and other protection extended to a breeder
2. A farmer's variety shall be entitled for registration if the application contains a declaration as specified in clause (h) or subsection (l) of section 18 of the Act
3. Farmers who conserve landraces and wild relatives of plants of economic importance and work on their improvement through selection and preservation would be entitled to recognition and reward from the National Gene Fund if their material is used as gene donors to varieties registered under this Act
4. Farmers would continue to be entitled to save, sow, resow, exchange and share or sell their farm produce including the seed of a protected variety. However, they would not be entitled to sell branded seed of a protected variety.

IMPLICATIONS OF PPV AND FR ACT

Agricultural Research

Implementation of the PPV and FR Act promises to bring monetary returns to institutions and individuals on their investment in research. By granting plant to institutions and individuals on their investment in research. By granting plant breeders rights (PBRs), the legislation is likely to encourage plant-breeding activity and stimulate research. Henceforth, researchers in the public and private sector would not only complete with each other but also with farmers in developing location-specific varietals and sustainable crop management technologies and innovations. There would be increased interest in realizing the commercial potential of research.

However, there is concern that projects with commercial potential may dominate the research agenda at the cost of basic research with little direct income potential.

DUS and Multi-location Testing

With the enactment of the PPV & FR Act, India is gearing up to give effect to plant breeder's right (PBRs) to recognize and protect the rights of farmers and breeders. These rights can granted for a variety that is new, distinct, uniform and stable in other words, one that measures up to a test of distinctness, uniformity and stability (DUS). Farmer's varieties need not comply with the novelty provision but all other types of varieties claiming PBR have to do so. The DAC of the ministry of agriculture and the Indian Council of Agricultural Research (ICAR) have initiated the process for DUS testing. The department has been allocating funds since 2000 to various ICAR institutes to strengthen their DUS test facilities. The council has identified 35 crops for finalizing the DUS tests guidelines and 43 centers for undertaking the tests.

Plant Breeder's Rights

In order to claim PBR, a variety must be clearly distinguishable for at least one essential character from the varieties commonly known in India and abroad. Such distinctiveness has to be bred into the variety while incorporating other economically important traits. Establishing such distinctive features is likely to be increasing difficult in future. Initially there may be a rush of varieties for DUS testing, but this is expected to slow down eventually for want of distinctiveness. Further, distinctiveness does not guarantee commercial viability; breeders have to combine both qualities. Unless this is achieved through planned breeding activity, the genotypes developed cannot be commercially exploited fully through PBR.

Sharing Germplasm

The most important ingredient for a crop improvement program is the availability of genetic diversity. Until recently, germplasm worldwide was considered the heritage of mankind to be shared and used by all. With the new world order that is emerging in the post-WTO era, the availability of germplasm may be restricted. Any breeder trying to develop new varieties should obtain explicit permission to use germplasm or landraces from the region from where they were originally collected. Although Prior Informed Consent (PIC) and Material Transfer Agreements (MTAs) do assure benefit sharing, the willingness to share germplasm rests with the breeder/farmer and /or the community. Unless the material is available in the National Gene Bank (NGB), from where it can be accessed by signing an MTA, there is no getting material if the party is not willing to share it.

Further, the second step in the development of new varieties is test elite lines in the All India Coordinated Research Projects (AICRPs). Breeding nominates the materials in good faith, and further selection would depend on the performance of cultures in the program. Simultaneously, other breeders also nominate the material and are free to use available material in the trial if they wish to do so. Now, in the PPV and FR regime, if a particular line is doing extraordinarily well in station-or state level testing, the breeder may not be as readily agreeable as in earlier days to part with the line for multi-location trials because his peers may want to identify it for their use or select and modify as an essentially derived variety (EDV). Unless divulged, benefit sharing becomes a bone of contention between breeders and organizations. Thus, the free exchange of germplasm and breeding line may become the casualty of the PPV and FR Act unless some innovative practices are devised.

Transgenics and EDVs

The increasing use of biotechnology in producing transgenic crop varieties and genetically modified organisms (GMOs) also requires the development of bio-safety norms to regulate trade in such crops, animals and products. The trade in GMOs will have to be strictly regulated, and that capacity needs to be created urgently. A responsible regulatory system will ensure that we attract investment in this sector. The prior information consent of farmers must be taken while pursuing on-farm trials of transgenic. Public notice must be given of all such trials and an informed debate generated on the issues involved rather than allow populist propaganda to be disseminated. It must, however, be remembered that much greater environmental damage takes place due to chemical pesticides currently in use than might possibly be caused by a transgenic pest tolerant crop.

The concept of EDV was developed to prevent biotechnologically produced varieties from taking away the benefits of conventionally bred varieties by transferring one or a few genes into or from them. However, EDV does not deal with the incorporation of genes from a protected variety into a non-protected variety. It has to be recognized that it was conventional breeding by farmers or plant breeders that made the expression of a particular critical gene possible. Therefore, the claimant for plant variety protection for a biotechnologically produced variety should disclose the source parents and must agree to contribute part of the gain to the breeders of the source variety.

Farmer's Rights

Provisions of the PPV and FR Act would encourage conservers, farmers and farmer-breeders. Farmers are the custodians of local germplasm. Through pure line selection many of them have developed very important landraces with unique biotic/abiotic stress tolerance or quality attributes or medicinal value. As a result of the PPV and FR Act, such farmers as individuals or as members of a community or through an NGO will be able to claim PBRs or benefits accruable as per clause 26(2) of the Act in respect of any new variety in the development of which local germplasm or a landrace was used.

Other explicit provision in the Act safeguard the farmer's right to grow seed and preserve produce to grow the next crop and also to share, exchange and sell it without the risk in infringement of PPV and FR.

Criteria for Deciding Protection of Plant Varieties

Distinctness

The variety shall be deemed to be distinct if it is clearly distinct from any other variety whose existence is a matter of common knowledge at the time of filling of the application. The object of this criterion is to ensure that the candidate variety can be identified amongst all other varieties whose existence is known, whether or not they are protected. An application for protection or for the entry of a variety in an official register in any country causes the variety to be recorded as a matter of common knowledge. In other words, the application for the protection should be filed with UPOV before disclosing it to any other agency.

Uniformity

The variety shall be deemed to be uniform if, subject to the variation may be accepted from the particular features of its propagation it is sufficient uniform in its relevant characteristics. The objective of this criterion is to ensure that the individuals representing the variety which is a candidate of protection from a group which is identifiable on the basis of the description of its characteristics. In other words the variation between individuals within a variety must be less than that within a species. In the absence of this condition it would become impossible to identify distinct. The degree of uniformity is determined taking into account the mode of reproduction of the species and all the genetic structure of varieties. The same levels of uniformity cannot be required for a strictly self-pollinating species or for a species, which is vegetatively propagated. An acceptable level of uniformity would ensure that it could be used for agricultural production. In this regard the difference between the protection, given by UPOV and patent system can be noted.

Stability

The variety shall be deemed to be stable if its relevant characteristics remain unchanged after repeated propagation or, in the case of a particular cycle of propagation at the end of each such cycle. The idea is to ensure that the variety will be identical to the description established at the moment of granting protection after repeated propagation. Stability, as well as uniformity may be lost if the rights holder fails to maintain the variety true to the description established when the rights were granted.

Novelty

The variety shall be deemed to be new if, at the date of filling of the application for breeder's right, propagation or harvesting material of the variety has not been sold or otherwise disposed of to others, by or with the consent of the breeder for the purpose of exploitation of the variety. It is also understood that a variety to which people have had free access in the in the past cannot be protected because then the interest of those who have relied on the free access, will suffer.

As it is some time necessary to see the response of the market to new varieties before deciding whether or not to apply for protectional, grace period has been included. The period is one year prior to the date of application in the country where the application is filed and in countries other than that in which the application has been filed and six years in case of trees and vines and four years for all other species.

Appropriate Denomination

The variety shall be designated by a denomination, which will be its generic designation. The premise that the variety denomination must be its generic designation class for a requirement that denomination must enable the variety to be identified. Users and consumers need to have some method of knowing that a sample is a sample of a particular identified plant variety, because it is often non-possible to identify it from its appearance. This is facilitated by requiring that a specific denomination and only that denomination be used to identify a variety in trade.

Governing Acts of Intellectual Property Rights

- Patents: The Patents Act. 1970. Amended in March 1999, early 2002 and 2004 by 2005 by Act.
- Design: Design Act. 2000
- Trade Marks: A new Trademarks Act, 1999
- Copyright: The Copyright Act, 1957 As Amended in 1983, 1984, 1992, 1994 and 1999 and the Copyright Rules, 1958.
- Layout Design of integrated circuits: The semiconductor Integrated Circuit Layout Design Act 2000.
- Protection of undisclosed information: No exclusive legislation exists but the matter would be generally covered under the Contract Law (Contract Act. 1872)
- Geographical indication: The Geographical Indication of Goods (Registration and Protection) Act. 1999.
- Plant variety Protection and Farmers Rights Act 2001:

Types of Intellectual Property Rights

Intellectual Property Rights signifies to the bundle of exclusionary rights, which can be further categorized into the following heads-

Copyright

India copyright law, laid down in the Indian Copyright Act, 1957 as amended by copyright (Amendment) Act, 1999, fully reflects the Berne Convention on Copyrights, to which India is party. Additionally, India is party to the Geneva Convention for the Protection of rights of producers of Phonograms and the Universal Copyright Convention. Copyright, one of the forms of intellectual property law offers exclusive rights for protecting the authorship of original & creative work like dramatic, musical and literary in nature. Symbolized as "©", here the term 'exclusive rights' mean that the holder has the right to determine who will be credited with the work, who will perform the work and who will be benefited financially from it. However, copyright does not extend any protection to the facts, methods of operation, system, ideas except to the ways in which they can be expressed. The Indian copyright act facilitates the owner for reproducing or reusing their copyrighted items, to prepare its derivate, to public their work and to distribute copies of their creative items. Copyright aims to protect the work of creator, transformed in a tangible form of expression. It includes art work, plays, movies, shows, various types of music, sound and songs, books, manuscripts, written work and all types of images, photos, pictures, drawings, graphics, verities, breeding method etc.

Patent

The basic obligation in the area of patents is that, invention in all branches of technology whether products or processes shall be patentable if they meet the three tests of being new involving an inventive step and being capable of industrial application. A patent is termed as the exclusionary rights given by the government or the authorized authority to its inventor for a particular duration of time, in respect of his invention. It is the part of the intellectual property right, which connotes with all those rights which are granted to any person for protecting its invention, process, discovery, composition or new useful development etc. from its further usage without any authentication. If more than two persons have jointly applied for patent license, both will own the patent separately. Generally either natural person or a whole entity can apply for a patent. However, it is somehow necessitated in almost all nations, to give the name of inventor(s) in the public record and what is the procedure they have followed in acquiring the exclusive rights for inventions. To successfully enacting the patent laws within its national territorial, every country has their own patent offices.

Trademark

Trademarks have been defined as any sign, or any combination of signs capable of distinguishing the goods or services of one undertaking from those of other undertakings. Such distinguishing marks constitute protectable subject matter under the provision of the TRIPS Agreement. The trademark or trade mark, symbolized as the â, ¢ and Â ®, is the distinctive sign or indication which is used for signifying some kind of goods or/and services and is distinctively used across the business organization or by an individual for identifying and uniquely classifying the source or their products and/or services among consumers and making a distinction of its products or services from the other entities. One of the part of the intellectual property law, trademark signifies to the name, word, phrase, logo, image, design, symbol or combination of any or all of these elements. The trademark grants rights to the owner, which in turns may take or can commence legal proceedings in case of infringement of trademark. In India, only the trademark proprietor whose trademark has been registered can put the symbol Â ® into use. If any use the symbol without the registration of mark he/she will be held under illegal use of the trademark. If anyone is engage in selling or providing services by using the false trademark he/she will be entitle to the penalty of minimum 6 months which may extend to maximum of 3 years and with the fine of not less than Rs.50, 000 and which can extend to Rs.2, 00,000.

Trade Secrets

Trade secret points towards a formula, pattern, any instrument, design which is kept confidential and through which any business or trade can edge over its rival and can enjoy economic gain. Trade secrets can be anything from a chemical compound, manufacturing process, design or preserving materials or even a list of consumers or clients. It is also known as "confidential information" or "classified information". To be safeguarded under trade secrets, the matter should be 'secret'. Though the definition of trade secret is variable as per the jurisdiction but there are following elements that are found to be same -

- Is not known by the public.
- Provides some financial sort of gain to its holder.
- Involves reasonable efforts from the holder side for maintaining secrecy.
- Importance of data or information to him or for his rivals.
- The ease by which information could be learned or duplicated by others.

Utility Model

The utility model is the intellectual property right for protecting the inventions. It is somehow described as the statutory monopoly which is bestow upon for the fixed duration of time in exchange to the inventor for the offering of the sufficient teaching of the invention and permitting the other person, possessing the ordinary skills of the relevant art, of performing the invention. The rights granted under the utility model are somewhat identical to those conferred upon by the patent but are more considerable for using the term 'incremental inventions'. Sometimes words like 'petty patent', 'innovation patent', 'minor patent' and 'small patent' are used in reference of the utility model. Such models are considered to be more suitable particularly for the small-scale enterprises, which in turns make the 'minor' improvements with the adoption of the existing products. Utility models are more commonly used for the mechanical innovations.

Geographical Indication

The agreement contains a general obligation that parties shall provide the legal means for interested parties shall provide the legal means for interested parties to prevent the use of any means in the designation or presentation of a good that indicates or suggests that the good in question originates in a geographical area other than the true place of origin in a manner which misleads the public as to the geographical origin of the goods. Geographical Indication (GI) signifies to the name or sign, used in reference to the products, which are corresponding to the particular geographical area or somewhat related to the origin like town, region or nation. Thus GI grants the rights to its holder, which acts as the certification mark and shows that the specified product consists of the some qualities and is enjoying good reputation due to its origin from the specified geographical location. The Trade Related Aspects of Intellectual Property Rights (TRIPs) agreement has defined the geographical indications rights as the exclusionary rights for the indicator which identify the goods originated within the member nations territories, or area or region of that territory, where the reputation or other attributes of the goods is essentially related to the geographic origin of the place. Geographical indications are the part of the intellectual property law therefore like any other law the regulation and govern conditions of GI also varies from one country to another as high differences have been found out in the use of generic terms across the world. Such case is prominent for food and beverage, which more commonly use the geographic terms. Geographical Indications are aimed towards identifying the source of the product and is considered as the valuable business tool.

Industrial Design Rights

Obligations envisaged in respect of industrial designs are that independently created designs that are new or original shall be protected. Industrial design rights are defined as the part of the intellectual property rights which confers the rights of exclusivity to the visual designs of objects, which are generally not popular utilitarian. It safeguards the appearance, style, design of the industrial object such as spare parts, textiles, and furniture. According to the Industrial Design Society of America (IDSA), "Industrial Design (ID) is the professional service of creating and developing concepts and specifications that optimize the function, value and appearance of products and systems for the mutual benefit of both user and manufacturer." As these designs consist of the aesthetic features therefore they do not provide any protection to the technical features of the article. The right accruing to the right holder is the right to prevent third parties not having his consent from making, selling or importing articles being or embodying a design, which is a copy or substantially a copy of the protected design when such acts are undertaken for commercial purposes. The origin of design rights can be traced back in the United Kingdom as 'Designing and Printing of Linen Act' (1787).

A new designs law repealing and replacing the Designs Act, 1911 has been passed by Parliament in the Budget Session, 2000. This Act has been brought into force from 11, May 2001.

ADVANTAGES OF INTELLECTUAL PROPERTY RIGHTS

Intellectual property rights help in providing exclusive rights to creator or inventor, thereby induces them to distribute and share information and data instead of keeping it confidential. It provides legal protection and offers them incentive of their work. Rights granted under the intellectual property act helps in socio and economic development.

Intellectual Property Rights and Agriculture

- Gene and gene sequences- Patents, Copyrights
- Chemical components such as peptides nucleoproteins anti-bodies, etc. - Patents, TM and copyrights in case of peptide sequences and similar products.
- Extraction products for health care such as drugs, medicines, vaccines, growth promoters, etc. – Patents, TM.
- Farm and nursery produce- Patents, TM.
- Plant and animal breeding- Trade secrets, TDK, Sui Generis system

Intellectual Property Rights in India

India has defined the establishment of statutory, administrative and judicial framework for protecting the intellectual property rights in the Indian territory, whether they connotes with the copyright, patent, trademark, industrial designs or with other parts.

Tuning with the changing industrial world, the intellectual property rights have continued to strengthen its position in the India. In 1999, the government has passed the important legislation in relation to the protection of intellectual property rights on the terms of the worldwide practices and in accordance to the India's obligations under the Trade Related Aspects of Intellectual Property Rights. It consists of:

- The Patents (Amendment) Act, 1999 which was passed on 10th March, 1999 in the Indian Parliament for amending the Patents Act of 1970 which in turns facilitate to establish the mail box system for filing patents and accords with the exclusive marketing rights for the time period of 5 years.
- The Trade Marks Bill, 1999 was passed in the India parliament during the winter

session for replacing the Trade and Merchandise Marks Act, 1958. It was passed on 23rd December 1999.

- The Copyright (Amendment) Act, 1999 was passed by both upper house and lower house of the Indian parliament and was later on signed by the Indian president on 30th December 1999.
- The sui generis legislation was approved by both houses of the Indian parliament on 23rd December 1999 and was named as the Geographical Indications of Goods (Registration & Protection) Bill, 1999.
- The Industrial Designs Bill, 1999 was passed in the Upper House of the Indian parliament for replacing the Designs Act, 1911.
- The Patents (Second Amendment) Bill 1999 was introduced in the upper house of the parliament for further amending the Patents Act 1970 and making it compliance with the TRIPS.

The relationship Between Intellectual Property Rights (TRIPS) Agreement

Probably the most notorious requirement of the TRIPS agreement is that in Article 27.3(b) which requires that members shall provide for the protection of plant varieties either by patents or by an effective *sui generis* system or by any combination.

The TRIPS agreement provides an indication of a possible meaning of the term "effective". TRIPS employs the term "effective" in particular in the context of the national enforcement of rights and procedures for the multilateral prevention and settlement of disputes, in which the rights to be conferred by an IPR are either defined in detail, or as "equitable remuneration". This formulation argues that a *sui generis* system needs to allow effective action against any act of infringement, as required by the relevant articles of the TRIPS agreement. The major limitation of this approach is that the effectiveness of a *sui generis* system thus assessed does not depend on the requirements for, or on the level of, protection.

The agreement on TRIPs requires WTO members to introduce an "effective *sui generis* system" for the protection of plant varieties. This commitment by WTO members implies that most developing countries, which hitherto had not extended intellectual property rights (IPRs) to their agricultural sector would have to do so. They would thus have to adopt intellectual property protection (IPP) regimes more like those prevailing in the industrialised countries.

The response of some developing countries to these developments in the Union International pour la Protection des Obtentions Vegetables (UPOV) convention has been the adoption of alternative *sui generis* options for the protection of plant varieties as is discussed in section 3. Initiatives have been taken both by governments as well as civil society organisations. These *sui generis* options take into consideration the contribution made both by traditional farmers and commercial plant breeders in shaping present day agriculture. Recognition of farmer's contribution to agriculture through on-farm innovation has only happened recently, partly stimulated by global initiatives aimed at protecting biodiversity and recognizing indigenous knowledge. Farmers Rights, and how to protect them; have become an issue, as has how to make the formal system of agricultural research benefit resource poor farmers.

Intellectual Property, Plant Breeders and Farmers

The framework for agricultural development in developing countries is changing rapidly as a result of major changes in the international regime governing genetic resources. A key element in this change apart from the Convention on Biological Diversity (CBD) and the International Treaty on Plant Genetic Resources for Food and Agriculture (ITPGR) is the requirement for WTO members to extend intellectual property protection (IPP) in agriculture by protecting improved varieties of plants either through an effective *sui generis* system of protection, or by patents or both. This is just one of many requirements of the Agreement on Trade-Related Aspects of Intellectual Property Rights (TRIPS), which involves far reaching changes in the protection of intellectual property in many developing countries. Most have to amend their existing IPP regimes and adopt ones more like those in the industrialised countries. TRIPS represents a significant step towards the adoption of a more harmonised system of IPP with a wider scope of protection and stronger rights for the owners of intellectual property than previously existed in developing countries.

The evolution of Plant Breeders' Rights (PBRs) has a long and often controversial history. At least two critical questions lay behind the introduction of PBRs as a form of intellectual property rights (IPRs). The first concerned the rationale for introducing IPRs to cover improvements in plant varieties and the second, the form of protection to be adopted. Here, the key consideration was whether or not a patent-like protection was to be extended to new plant varieties. Although the first multilateral system for the protection of PBRs was created through the convention establishing the International Union for the Protection of New Varieties of Plants (UPOV Convention) in 1961 its origins go back much further.

Today's major considerations for developing countries in excluding agriculture from IPP, with countries like India, for instance, explicitly excluding food from patenting. The Plant Patents Act gave impetus to the process of acceptance of IPP in agriculture by a larger set of countries with two factors moving the process forwards, eventually to UPOV. The first was the successful enactment of laws protecting the rights of breeders in the Netherlands and Germany.

Farmers as Innovators

On-farm innovation by farmers has happened continuously since settled agriculture began. The farmer-innovation process, in which farmers adopt clearly defined criteria to identify the improved varieties they developed, has a certain resemblance to that followed by formal plant breeders. The latter rely on the three-fold criteria of distinctiveness, uniformity and stability (DUS) of the plant varieties they develop to lay claim to PBRs. But similar farmer innovation receives no recognition or reward. The recognition that formal plant breeders receive is facilitated by the extensive codification of their knowledge. The lack of such codification lies at the heart of the relative neglect of the contribution that farmers have made.

One impediment to codification of farmers' knowledge is the nature of the farmer innovation process. Unlike the formal breeder, who conducts his research under controlled conditions, the farmer uses the available environmental conditions to conduct his experiments. Furthermore, farmers' use of environmental conditions is based on a detailed knowledge of the environment in which they practice their farming. Natural selection occurs by the action of environmental stress on inherent variation caused by gene recombination and mutation, but this process occurs not only in the plants that are directly useful to humankind but also in other organisms, which directly affect the process of agricultural production. These include weeds, pests and micro-organisms. Farmer's reliance on natural selection has resulted in a continuous and evolving process of change in agriculture, and has brought about adaptation within a specific or given environment.

Recognition of farmer's contribution to the advancement of agriculture has come at both national and international levels. Developing countries are trying to create the administrative structures that would be necessary to mainstream the farmer's contribution and are taking necessary legislative action. One of the first such efforts has been made by India in enacting the Plant Varieties and Farmers' Right Act. The

Organisation of African Unity (OAU) has also developed model legislation for the protection of the rights of farmers and breeders, as well as other local communities, which Namibia has translated into a draft law.

Therefore, the rights to be protected should be those set out in the obligations members have assumed for instance under the agreement on TRIPS and equally the CBD. The rights should be protected in accordance with national objectives referred to in article 7 and the principles in article 8 (Agreement on TRIPS) as apply in the country should be protected within the overall framework of the CBD and the first recital of the preamble to the WTO Agreement on sustainable development, and the protection should be consistent with international obligations that members have assumed, for instance under the CBD.

Features of UPOV Systems of Protection

UPOV 1961

There are three alternative systems of protection of plant varieties under the UPOV convention the 1961 Act (with the 1972 amendments), the 1978 Act and the 1991 Act. The 1961/72 and the 1978 Acts are essentially similar, while the 1991 Act makes fundamental departures from the earlier ones. Over the past decade, the UPOV membership has changed character. From 18 developed countries in 1991, UPOV had a membership of 50 countries at different stages of development early in 2002.

Forms of protection: Each member state could recognise the right of the breeder by the grant of a special title or of a patent. But countries whose national law allowed protection under both these forms were allowed to use only one of them for the same botanical genera or species.

Scope of protection: Prior authorisation from breeders had to be sought for production and commercial marketing of the reproductive or vegetative material, as such, of the new variety, and for the offering for sale or marketing of such material. The breeder's rights extended to ornamental plants or parts of plants marketed for purposes other than propagation when they were used as propagating material in the production of ornamental plants or cut flowers. Breeder's authorisation was required only if the new varieties were used as an initial source of variation in creating new varieties or for marketing such varieties. In other words, the use of the varieties for research purposes was allowed. The framework developed in UPOV provided the minimum standards of protection, and any member state could provide a higher level of protection to the new varieties, even extending the breeders' rights to the marketed products.

Duration of protection: A minimum of 18 years protection for vines, fruit trees and their root-stocks, and 15 years for all other plants were provided. Member states could adopt longer periods of protection and could fix different periods for some classes of plants to take into account the requirements of regulations concerning the production and marketing of seeds and propagating materials.

Conditions for protection: The 1961 Act allowed protection of varieties that were (a) new, (b) distinct, (c) homogenous and (d) stable. These attributes of plant varieties that could be granted protection were elaborated in the model law that UPOV developed after the 1978 Act was finalized.

UPOV 1978

The UPOV began in 1974, ostensibly to make the convention more attractive to non-members. The needs of two countries in particular, the USA and Canada, were addressed in the revision of the convention. Two sets of conditions for PVP had to be met under UPOV '78. The first was the minimum number of genera or species to which any UPOV member state had to extend protection in their national legislation. The second related to the characteristics of the plant varieties that could qualify for protection.

Number of genera/species to be protected: Initially, the provisions had to apply to at least five genera or species. Within three years, this number had to increase to 10 and after a further three years to 18. Within eight years, at least 24 genera or species had to be covered.

Conditions for protection: UPOV '78 allowed protection of plant varieties (Article 6) that were: (a) new, (b) distinct from any other variety that was in common knowledge (c) sufficiently homogenous and (d) stable in their essential character. Any plant variety that met these criteria could qualify for protection, irrespective of the origin, artificial or natural, of the initial variety from which it had resulted (Box 4). This implies that unlike patents, which are normally not granted to discoveries, plant varieties could be protected even when they were "discovered".

Nature of protection: UPOV '61 allowed countries to protect the interests of plant breeders, either through the grant of a special title or a patent but not both. The USA, however, had extended protection to sexually propagating plants through the Plant Variety Protection Act of 1970 besides providing for plant patents. An amendment in Article 37 paved the way for the accession of the USA to the treaty by including an exception to Article 2(1), which allowed a state already providing dual protection to continue to do so provided "it notifies the Secretary General (of the UPOV) of that fact". Furthermore, countries using the patent laws to protect plant varieties were allowed to use the patentability criteria and the period of protection as was provided for under their patent laws.

Scope of breeder's rights: The rights provided, as spelt out in Article 5(1), allowed breeders control over the following activities associated with reproductive or vegetative propagating material: (a) production for the purposes of commercial marketing, (b) offering for sale, and (c) marketing. Article 5(4) states that any member of UPOV "may grant to breeders a more extensive right" than set out in Article 5(1), "extending in particular to the marketed product".

Safeguarding public interests: Article 9 of UPOV '78 allows the exclusive rights of breeders to be restricted in the public interest. The model law of UPOV '78 offers three interpretations of Article 9: through the grant of a voluntary licence by the right holder for the exploitation of the variety; licences of right; and, compulsory licences.

UPOV 1991

The decision to revise the provisions of the 1978 Act was taken in 1986. The 17 members took this decision at a time when biotechnology was increasingly being used for plant breeding activities, which prompted demands for the adoption of the patent system in agriculture. These demands were strengthened after the first industrial patent was granted for an improved crop variety in the USA. The main goal of the revision was to strengthen the breeder's right.

Coverage of varieties: Until 1991, members had flexibility in the coverage of genera and species subject to PBRs in part due to concerns about the impact of PBRs on genetic diversity. Till then, varietal protection could be adopted by member countries in a phased manner and did not require comprehensive coverage of all varieties. UPOV '91 requires a comprehensive coverage of plant varieties by the member states, but not immediately.

Nature of rights enjoyed by the breeder: UPOV '91 marks a major departure from UPOV '78 in the nature of rights provided to the breeder. Article 14 defines these in four areas: (a) the propagating material, (b) the harvested material, (c) certain other products, which are discussed below, and (d) EDVs, which are discussed separately. Breeder's rights on propagating material include: (a) production or reproduction (multiplication), (b) conditioning for the purposes of propagation, (c) offering for sale, (d) selling or other marketing, (e) exporting, (f) importing, and (g) stocking for any of the purposes referred to above. Propagating material, as understood in UPOV '91, included "parts of the plant intended for the production of new plants, for example seeds, and certain parts of plants that may be used either for consumption or sowing. The rights of the breeder over the propagating material have been extended to all acts involving commercialisation.

Essentially derived varieties: The inclusion of essentially derived varieties (EDVs) in UPOV '91 is generally regarded as the single most important change to UPOV. Under this provision, the so called research exemption available under UPOV '78, which allowed breeders to freely use protected varieties for research purposes and for breeding new varieties, was excluded. This has major ramifications for developing countries where farmer innovators have been an integral part of the innovation systems (Section 1.2). Article 14(5), which provides for the inclusion of EDVs of protected varieties within PBRs, seeks to strengthen the rights of the breeder by bringing within protection "essentially derived and certain other varieties" of the protected varieties.

Is UPOV Appropriate for Developing Countries?

According to Suman Sahai, Convener of Gene Campaign (*Current Science*, Vol. 84, NO. 3, 10 February 2003) the interests of developing countries are not served by UPOV, which is completely insensitive to their needs. In all fairness, UPOV was not created for developing countries and therefore does not address itself to their concerns. In understanding the UPOV system, it is crucial to understand that right from 1961, even when it was more flexible than it is today, UPOV granted only one right, the right to the plant breeder. There was never any concept of Farmers' Rights. What was granted at best was an exemption to farmers and researchers, from the otherwise exclusive rights granted to the breeder.

In light of all this and the path breaking Indian legislation, the recommendation made by the Agriculture Ministry to join UPOV, has stunned the international community and those who have fought hard to get a balanced legislation in India. The Indian legislation is incompatible with UPOV in the farmer's rights it provides and the strong public interest clauses it contains, which UPOV does not support.

An Alternative to UPOV

The Gene Campaign (Gene Campaign, J-235/A, Sainik Farms, Khanpur, New Delhi 110 062, India), along with other civil society organizations, believes that UPOV does not have conditions favourable to developing countries. Gene Campaign (since 1993), a leading research and advocacy organisation, has been working to empower local communities to retain control over their genetic resources in order to ensure food and livelihood security. Closely involved in policymaking and legislation with respect to biological resources, Gene Campaign has enabled rural and adivasi communities to participate in policies relating to these resources.

According to Suman Sahai (*Current Science*, Vol. 84, NO. 3, 10 February 2003) it is absolutely necessary that developing countries need to craft their own platform, to address their special needs, which are very different to the needs and requirements of, industrialized countries for whom UPOV was developed. Some efforts have already been made to provide an alternative to UPOV. The model law drafted by the Organization of African Unity (OAU) and the Gene Campaign's Convention of

Farmers and Breeders (CoFaB) is being discussed in various forums. CoFaB has been analysed along with UPOV and described by the UNDP as 'a strong and coordinated international proposal which offers developing countries a far better alternative to European legislation, by focusing on the need to protect farmers' rights and food and nutritional security goals of their people.

INDIAN LAWS AND PROTECTION OF PLANT VARIETIES

Indian legislation to fulfill its commitments under TRIPS Article 27.3(b) is the Protection of Plant Varieties and Farmers' Rights (PPVFR) Act. This legislation is an attempt by the Indian government to recognise the contribution of both commercial plant breeders and farmers in plant breeding activity. It is the outcome of twin pressures on the government one set from the world system to introduce IPP to recognise commercial plant breeder's contribution in the development of new varieties, which was reinforced by the emerging private seed industry in India, and the other from farming communities opposed to the introduction of any form of IPRs in the agricultural sector.

The Indian Parliament has finally passed the Plant Variety Protection and Farmers' Rights Bill. India has now, for the very first time, put in place a law to grant Plant Breeders' Rights on new varieties of seeds. The law also grants Farmers' Rights. These have been included in the legislation as a result of the determined and sustained campaign by NGOs, spearheaded by the Gene Campaign. Gene Campaign's position right from the start was that if the *status quo* had to be changed and India had to grant Plant Breeders' Rights, our legislation would have to grant a strong Farmers' Rights at the same time.

The pivotal importance of the farmer having the right to sell seed has to be seen in the context of seed production in India, where the farming community is the largest seed producer, providing about 87 percent of the country's annual requirement. Denying the farmer the right to sell seed would displace the farming community as the country's major seed provider.

Farmers' Rights

In section 39 (iv) of the chapter on Farmers' Rights, the right to sell seed even protected seed has finally been provided The farmer shall be deemed to be entitled to save, use, sow, resow, exchange, share or sell his farm produce including seed of a variety protected under this Act in the same manner as he was entitled before the coming into force of this Act. However, the farmer is not entitled to sell *branded* seed of a variety protected under this Act' (editor's italics).

Other Kinds of Farmers' Rights

1. The Act acknowledges the role of rural communities as contributors of landraces and farmer varieties in the breeding of new plant varieties. Breeders wanting to use farmers' varieties for creating Essentially Derived Varieties (EDVs) cannot do so without the express permission of the farmers. Anyone can register a community's claim and have it duly recorded at a notified center. If the claim is found to be genuine, a share of profits made from the new variety has to go into a National Gene Fund.
2. Breeders must to submit an affidavit that their variety does not contain a Gene Use Restricting Technology (GURT) or terminator technology.
3. Rightly assuming that farmers may unknowingly infringe Breeders' Rights since they will not be used to the new situation, the law provides for protection from

prosecution for innocent infringement.

4. The clause protecting the farmer from spurious seed leaves too much to the discretion of the Authority. There should be specific guidelines, such as that compensation should amount to at least twice the projected harvest value of the crop. In addition, a jail term should be provided for repeated offence.
5. Breeders' Rights are fully protected by the legislation. On registration, the breeder has complete rights of commercialization for the registered variety. These include the right to produce, sell,

Indian legislation to fulfill its commitments under TRIPS Article 27.3(b) is the Protection of Plant Varieties and Farmers' Rights (PPVFR) Act. This legislation is an attempt by the Indian government to recognise the contribution of both commercial plant breeders and farmers in plant breeding activity. It is the outcome of twin pressures on the government one set from the world system to introduce IPP to recognise commercial plant breeders' contribution in the development of new varieties, which was reinforced by the emerging private seed industry in India, and the other from farming communities opposed to the introduction of any form of IPRs in the agricultural sector. The private seed industry in India argued for the introduction of IPP to cover the agricultural sector ever since its entry on a major scale was facilitated following the amendment of the Seed Act in 1988, which provided greater space to the private sector to operate in the industry. The main argument of the seed companies was that incentives for supply of improved varieties of seeds could only be provided by appropriate IPRs. The farmers, however, were beneficiaries of the breeding activity undertaken by publicly funded institutions, which from the mid 1960s had provided the improved varieties of seeds that made the Green Revolution in India a reality. These publicly funded institutions did not depend on IPRs; their activities were determined by government policy making. The balance, however, shifted in favour of extending IPRs in agriculture after India assumed membership of the WTO.

Environment-Related Trade Implications

Although WTO membership will provide opportunities to increase exports and many sectors are likely to benefit from it, these same sectors will face challenges from increasingly stringent environmental standards and various voluntary measures. The imports of forest and food products are growing fast in India. In order to protect human, plant and animal health and the environment, India needs to strengthen its sanitary and phytosanitary regulatory regime. A high degree of policy preparedness will be necessary to enhance policy and administrative capacity to address the issues above. WTO membership may also bring excellent opportunities for India to gradually upgrade its environmental management systems and standards.

IPRs and Biotechnology

Agriculture has been an important engine of economic development, and the mainspring of economic progress in agriculture has been productivity improvements driven by technological change that is fueled by research and development (R&D).

Technological change has driven economic progress in agriculture and will continue to play a crucial role in the 21st century. The latest wave of technological change in agriculture is based in molecular biology. Will horticulture participate? Genetically engineered crop varieties have been adopted on a wide scale in some agronomic crops, but horticultural crops face larger hurdles. High costs for research, development and regulatory approval combined with the small acreages planted and the diversity of varieties, will limit the potential for profitable applications of biotechnology to many fruits and vegetables, tree fruits and nuts, and nursery crops. In addition, there are market barriers. Like most important changes in agriculture, modern biotechnology has met with spirited political opposition from some quarters. Threats of political action may discourage food manufacturers and retailers from adopting biotech products that are wanted by some consumers and may be profitable for growers.

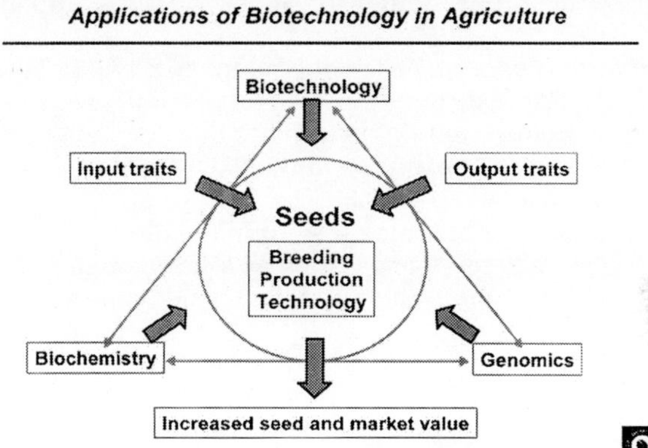

Applications of Biotechnology in Agriculture

Biotechnology
Input traits
Output traits
Seeds
Breeding
Production
Technology
Biochemistry
Genomics
Increased seed and market value

FIG. 35.2: The applications of biotechnology in agriculture will flow from various areas of research as input and output traits through seed as a delivery system to the farmer.

Biotechnology projects in horticulture being shelved because of intellectual property (IP) conflicts. In a typical situation, a plant scientist at a university agricultural experiment station or a smaller seed firm has developed a remarkable new variety using the cutting edge scientific tools of plant biotechnology. Then, as they or the nursery or the growers association with whom they work take the next steps to develop and release the new variety to commercial growers, their efforts are quickly and quietly shut down by a letter from an attorney. The letter alleges that the new variety contains a piece of technology that infringes upon a client's IP claims.

Biotech R&D Trends

Recent U.S. Department of Agriculture (USDA) registrations for field trials of transgenic crops show that R&D in horticultural crops is lagging when compared with the major row crops. Even leading transgenic horticultural crops such as melon, lettuce, strawberry, grape, apple and sunflower are hardly represented in field trials. Horticultural crops are completely dwarfed by corn, the single most commonly tested transgenic crop, which by itself is the subject of almost half of all transgenic field trials. Of course, U.S. production of any single horticultural crop is far less valuable than U.S. production of corn. Less field testing is to be expected for less valuable crops. But, even when applying a rough calculation to account for the differences in size and value of individual crops dividing by the annual value of each crop's U.S. production horticultural crops tend to show a greater farm-gate value per field trial. In other words, horticultural crops are subject to fewer genetic field trials, and presumably receive less biotech R&D, for every dollar of crop production. There is a less investment in biotechnological techniques for horticultural crops than for major row crops, both in absolute terms and relative to overall crop values, while a greater proportion of that smaller R&D investment in horticultural crops comes from the public sector. Involvement by commercial firms in horticultural crops seems to be missing.

Current practices in patenting and intellectual property (IP) protection have created barriers to the use of biotechnology and advanced agricultural technologies for the creation and commercialization of new crop varieties. The complex and cumulative nature of biological innovation requires access to multiple technologies that are often exclusively owned or licensed. For example, commercializing a single variety of transgenic tomato could involve obtaining the rights to use a variety of technologies and genes from numerous life-sciences companies, government agencies and universities.

Obtaining *freedom to operate* (**FTO:** the ability to clear all IP barriers and bring a product to market) for transgenic crop varieties is difficult. There is considerable uncertainty as to who holds what rights to particular technologies, and negotiating access to those rights is time consuming and costly. This is a problem for the major international agricultural companies that focus primarily on high volume crops such as corn, soybeans and cotton for research institutions that work on specialty crops grown on much smaller acreages, such as tomatoes, strawberries, apples and cabbage and for public institutions that work on staple crops for humanitarian use in developing countries.

The international agricultural companies have taken steps to solve their FTO problems through mergers and cross-licensing agreements that bring together essential IP components within one company. However, public sector institutions such as universities, government agencies, international agricultural research centers and others working on specialty and staple crops are still struggling to find ways to gain FTO. In addition, donor agencies such as the Rockefeller and McKnight Foundations, which have a long history of investing in agricultural research that benefits subsistence farmers in developing countries, have also found that IP constraints are reducing the flow of technology.